Lecture Notes in Civil Engineering

Volume 144

Lecture Notes in Civil Engineering (LNCE) publishes the latest developments in Civil Engineering - quickly, informally and in top quality. Though original research reported in proceedings and post-proceedings represents the core of LNCE, edited volumes of exceptionally high quality and interest may also be considered for publication. Volumes published in LNCE embrace all aspects and subfields of, as well as new challenges in, Civil Engineering. Topics in the series include:

- Construction and Structural Mechanics
- Building Materials
- Concrete, Steel and Timber Structures
- Geotechnical Engineering
- Earthquake Engineering
- Coastal Engineering
- Ocean and Offshore Engineering; Ships and Floating Structures
- Hydraulics, Hydrology and Water Resources Engineering
- Environmental Engineering and Sustainability
- Structural Health and Monitoring
- Surveying and Geographical Information Systems
- Indoor Environments
- Transportation and Traffic
- Risk Analysis
- Safety and Security

To submit a proposal or request further information, please contact the appropriate Springer Editor:

- Pierpaolo Riva at pierpaolo.riva@springer.com (Europe and Americas);
- Swati Meherishi at swati.meherishi@springer.com (Asia - except China, and Australia, New Zealand);
- Wayne Hu at wayne.hu@springer.com (China).

All books in the series now indexed by Scopus and EI Compendex database!

More information about this series at http://www.springer.com/series/15087

Organization

Executive Committee

Chairman

Prof. Hemanta Hazarika, Kyushu University, Japan

Co-chairmen

Prof. Gopal Santana Phani Madabhushi, University of Cambridge, UK
Prof. Kazuya Yasuhara, Prof. Emeritus, Ibaraki University, Japan
Prof. Dennes T. Bergado, Prof. Emeritus, Asian Institute of Technology, Thailand

Secretaries

Dr. Yoshifumi Kochi, K's Lab Inc., Japan
Dr. Babloo Chaudhary, National Institute of Technology Karnataka Surathkal, India
Mr. Takashi Fujishiro, Geo-disaster Prevention Institute, Japan

Co-secretaries

Dr. Masanori Murai, Shimizu Corporation, Japan
Mr. Tsuyoshi Tanaka, Tokyo City University, Japan

Members

Mr. Tadashi Akagawa, Japan Foundation Engineering Co., Ltd., Japan
Prof. Yasuhide Fukumoto, Kyushu University, Japan
Mr. Hideo Furuichi, Giken Corporation, Japan
Mr. Shinichiro Ishibashi, Nihon Chiken Co., Ltd., Japan
Dr. Tomohiro Ishizawa, National Research Institute for Earth Science and Disaster Resilience, Japan
Dr. Yusaku Isobe, IMAGEi Consultant Corporation, Japan
Prof. Haruichi Kanaya, Kyushu University, Japan
Mr. Hiraoki Kaneko, Japan Foundation Engineering Co., Ltd., Japan
Mr. Daisuke Matsumoto, Japan Foundation Engineering Co., Ltd., Japan
Mr. Yoshikazu Ochi, Kawasaki Geological Engineering Co., Ltd., Japan

Mr. Seiichiro Oiyama, Giken Corporation, Japan
Mr. Taisuke Sasaki, Nihon Chiken Co., Ltd., Japan
Dr. Tadaomi Setoguchi, Former Vice Mayor, Sasebo City, Japan
Prof. Yoshihisa Sugimura, Kyushu University, Japan
Dr. Sugeng Wahyudi, NITTOC Construction Co., Ltd., Japan
Dr. Naoto Watanabe, KFC Ltd., Japan
Mr. Shigeo Yamamoto, Chuo Kaihatsu Corporation, Japan

Treasurers

Ms. Mika Murayama, Kyushu University, Japan
Mr. Takashi Fujishiro, Geo-disaster Prevention Institute, Japan

Technical Committee

Chairman

Prof. Haruichi Kanaya, Kyushu University, Japan

Co-chairman

Prof. Yasuhide Fukumoto, Kyushu University, Japan

Secretary

Dr. Babloo Chaudhary, National Institute of Technology Karnataka Surathkal, India

Co-secretaries

Dr. Sugeng Wahyudi, NITTOC Construction Co., Ltd., Japan
Mr. Divyesh Rohit, Kyushu University, Japan

Members

Prof. Fauziah Ahmad, Universiti Sains Malaysia, Malaysia
Mr. Tadashi Akagawa, Japan Foundation Engineering Co., Ltd., Japan
Prof. Guangqi Chen, Kyushu University, Japan
Dr. Takuro Fujikawa, Fukuoka University, Japan
Mr. Takashi Fujishiro, Geo-disaster Prevention Institute, Japan
Prof. Yasuhide Fukumoto, Kyushu University, Japan
Mr. Hideo Furuichi, Giken Corporation, Japan
Dr. Zentaro Furukawa, Kyushu University, Japan
Dr. Stuart Haigh, University of Cambridge, UK
Mr. Shinichiro Ishibashi, Nihon Chiken Co., Ltd., Japan
Dr. Tomohiro Ishizawa, National Research Institute for Earth Science and Disaster Resilience, Japan
Dr. Yusaku Isobe, IMAGEi Consultant Corporation, Japan
Prof. Haruichi Kanaya, Kyushu University, Japan

Dr. Kazuhiro Kaneda, Takenaka Corporation, Japan
Mr. Hiroaki Kaneko, Japan Foundation Engineering Co., Ltd., Japan
Dr. Yoshifumi Kochi, K's Lab Inc., Japan
Mr. Daisuke Matsumoto, Japan Foundation Engineering Co., Ltd., Japan
Dr. Masanori Murai, Shimizu Corporation, Japan
Prof. Juichi Nakazawa, Prof. Emeritus, National Institute of Technology, Maizuru College, Japan
Dr. Shunsaku Nishie, Chuo Kaihatsu Corporation, Japan
Prof. Masashi Sakai, Prof. Emeritus, Kyushu Sangyo University, Japan
Mr. Taisuke Sasaki, Nihon Chiken Co., Ltd., Japan
Dr. Satoquo Seino, Kyushu University, Japan
Prof. Takayuki Shimaoka, Kyushu University, Japan
Prof. Naoaki Suemasa, Tokyo City University, Japan
Dr. Makoto Takada, Chuo Kaihatsu Corporation, Japan
Prof. Takeshi Tsuji, Kyushu University, Japan
Dr. Naoto Watanabe, KFC Ltd., Japan
Dr. Sugeng Wahyudi, NITTOC Construction Co., Ltd., Japan
Mr. Shigeo Yamamoto, Chuo Kaihatsu Corporation, Japan
Dr. Norimasa Yoshimoto, Yamaguchi University, Japan

International Advisory Committee

Advisory Board

Prof. Juichiro Akiyama, Prof. Emeritus, Kyushu Institute of Technology, Japan
Prof. Akira Asaoka, Prof. Emeritus, Nagoya University, Japan
Prof. Ramanathan Ayothiraman, Indian Institute of Technology Delhi, India
Prof. G. L. Sivakumar Babu, Indian Institute of Science Bangalore, India
Prof. Richard J. Bathurst, Royal Military College, Canada
Prof. Adimoolam Boominathan, Indian Institute of Technology Madras, India
Prof. Ross W. Boulanger, University of California Davis, USA
Prof. Deepankar Choudhury, Indian Institute of Technology Bombay, India
Prof. Misko Cubrinovski, University of Canterbury, New Zealand
Mr. Shunta Dozono, Ministry of Land, Infrastructure, Transport and Tourism, Japan
Prof. Tuncer B. Edil, University of Wisconsin-Madison, USA
Prof. Ahmed W. Elgamal, University of California San Diego, USA
Prof. Shailesh. R. Gandhi, Sardar Vallabhbhai National Institute of Technology Surat, India
Prof. Noriaki Hashimoto, Kyushu University, Japan
Prof. Yasuhiro Hayashi, Kyushu Sangyo University, Japan
Prof. Takenori Hino, Saga University, Japan
Prof. Akihiko Hirooka, Kyushu Institute of Technology, Japan

Dr. Sanjay Kumar Shukla, Edith Cowan University, Australia
Prof. Francesco Silvestri, University of Naples Federico II, Italy
Prof. Thallak G. Sitharam, Indian Institute of Technology Guwahati, India
Prof. Motoyuki Suzuki, Yamaguchi University, Japan
Mr. Yukihisa Takahashi, Taisei Corporation, Japan
Mr. Hideki Tanigawa, JAFEC USA Inc., USA
Prof. Vikas Thakur, Norwegian University of Science and Technology, Norway
Prof. Ikuo Towhata, Specially Appointed Professor, Kanto Gakuin University, Japan
Prof. Lanmin Wang, Lanzhou Institute of Seismology, China
Prof. Koichiro Watanabe, Kyushu University, Japan
Prof. Kentaro Yamamoto, Nishinippon Institute of Technology, Japan
Prof. Atsushi Yashima, Gifu University, Japan
Prof. Susumu Yasuda, Tokyo Denki University, Japan
Prof. Noriyuki Yasufuku, Kyushu University, Japan
Prof. Koki Zen, Prof. Emeritus, Kyushu University, Japan
Prof. Askar Zhussupbekov, Eurasian National University, Kazakhstan

Members

Prof. Ioannis Anastasopoulos, Swiss Federal Institute of Technology in Zurich, Switzerland
Dr. Panjamani Anbazhagan, Indian Institute of Science Bangalore, India
Prof. Netra Prakash Bhandary, Ehime University, Japan
Dr. Gabriele Chiaro, University of Canterbury, New Zealand
Dr. Rajan Kumar Dahal, Tribhuvan University, Nepal
Dr. Kiyoshi Fukutake, Shimizu Corporation, Japan
Prof. Chandan Ghosh, Indian Institute of Technology Jammu, India
Prof. Seyed Mohsen Haeri, Sharif University of Technology, Iran
Dr. Padavala Hari Krishna, National Institute of Technology Warangal, India
Prof. Manzul Kumar Hazarika, Asian Institute of Technology, Thailand
Prof. Shinya Inazumi, Shibaura Institute of Technology, Japan
Dr. Kiyonobu Kasama, Kyushu University, Japan
Dr. Suman Manandhar, Tribhuvan University, Nepal
Dr. John McDougall, Edinburgh Napier University, UK
Dr. Anil Kumar Mishra, Indian Institute of Technology Guwahati, India
Dr. Adapa Murali Krishna, Indian Institute of Technology Tirupati, India
Prof. Mitsu Okamura, Ehime University, Japan
Prof. Kiyoshi Omine, Nagasaki University, Japan
Dr. Rolando P. Orense, University of Auckland, New Zealand
Prof. Devendra Narain Singh, Indian Institute of Technology Bombay, India
Dr. Doni Prakasa Eka Putra, Gadjah Mada University, Indonesia
Dr. Hing-Ho Tsang, Swinburne University of Technology, Australia
Dr. Bui Trong Vinh, Ho Chi Minh City University of Technology, Vietnam
Dr. Jayan S. Vinod, University of Wollongong, Australia
Prof. Chungsik Yoo, Sungkyunkwan University, Korea

Preface

Advances in Sustainable Construction and Resource Management is a compilation of peer-reviewed papers to be presented in the 1st International Symposium on Construction Resources for Environmentally Sustainable Technologies (CREST 2020) at Kyushu University, Fukuoka, Japan. The organizer of the symposium is Kyushu University, Japan. It is co-organized by University of Cambridge, UK, and International Society for Soil Mechanics and Geotechnical Engineering (ISSMGE). It is supported by Ministry of Land, Infrastructure, Transport and Tourism (MLIT), Japan, Fukuoka Prefecture, Fukuoka City, Embassy of India in Japan, UN-Habitat in Fukuoka, International Press-in Association (IPA), Kyushu Branch of The Japanese Geotechnical Society (JGS), Global and Local Environment Co-creation Institute (GLEC), Ibaraki University and Japan Federation of Construction Contractors.

The main purpose of this symposium is to disseminate information and exchange ideas on issues related to natural and man-made disasters, and to arrive at solutions through the use of alternative resources, towards building a sustainable and resilient society from the geotechnical engineering perspectives. The symposium aims to bring together scientists, researchers, engineers and policy-makers throughout the world under one umbrella for debate and discussion on those issues. The symposium focuses on the sustainability, promotion of new ideas and innovations in design, construction and maintenance of geotechnical structures with aim of contributing towards climate change adaptation and disaster resiliency to meet the Sustainable Development Goals (SDGs) of the UN. It is hoped that through active participation of all the stakeholders (industries, government and academia), the symposium will be able to inculcate some changes to the existing perception of dealing with problems faced by humankind in this ever-challenging world.

The book focuses mainly on the issues related to (i) cascaded and material recycling in geo-engineering, (ii) natural disaster and resiliency and (iii) climate change adaptation and innovation. It contains the latest information, innovation and technology development in these broad areas from all over the world. The volume comprises 79 contributions which will be delivered during the symposium by the scholars and experts in their fields. All the manuscripts were thoroughly reviewed by at least two reviewers selected from an international panel of experts to check for the relevance to the themes as well as the quality of technical contents and presentation.

The publication of *Advances in Sustainable Construction and Resource Management* has been possible through the sustained efforts of the staff of the Research Group of Adaptation to Global Geo-disaster and Environment (Geo-disaster Prevention Engineering Laboratory), Kyushu University, Japan. The editors express their sincere thanks to all the members of the research group. Special thanks go to Mr. Divyesh Rohit and Dr. Sugeng Wahyudi, Co-secretaries of CREST 2020 Technical Committee, for their tireless efforts and dedication, which were instrumental in the timely publication of the book. The efforts and suggestions by Dr. Babloo Chaudhary and Dr. Masanori Murai, Secretary and Co-secretary, respectively of CREST 2020 Executive Committee, towards speedy publication of this book are also gratefully acknowledged. Last but not least, the editors also would like to express their sincere gratitude to all the reviewers for their time and effort to review the manuscripts and improve their contents.

Advances in Sustainable Construction and Resource Management provides a wealth of knowledge and information, which will be instrumental in disseminating the latest information and technology towards building a sustainable and resilient society from the geotechnical, geological and hydrological perspectives. The book is aimed at researchers, designers, consultants, government officials, academicians and students involved in those fields. The editors hope that in the years to come, the knowledge, findings and the development of technologies that have been compiled in this book will contribute towards achieving the SDGs set forth by the UN. We further hope that, with such achievements, the realization of sustainable and resilient infrastructures across the globe will not be a distant dream.

Fukuoka, Japan — Hemanta Hazarika
Cambridge, UK — Gopal Santana Phani Madabhushi
Ibaraki, Japan — Kazuya Yasuhara
Khlong Luang, Thailand — Dennes T. Bergado

Acknowledgements

Financial Supports

The chairman of the 1st International Symposium on Construction Resources for Environmentally Sustainable Technologies (CREST 2020) and the editors of this book gratefully acknowledge the financial support provided by Kyushu University under two special projects (Progress 100 and SHARE-Q) and JAFEC USA Inc., California, USA.. The editors also would like to acknowledge all our sponsors (Diamond, Gold, Silver and Bronze), without which holding of the symposium and this publication would not have been possible.

Panel of Reviewers

The manuscript for each chapter included in this book was carefully reviewed for the quality and clarity of technical contents by at least two members of a review panel consisting of the following international experts. The editors wish to express their sincere gratitude to all the reviewers for their valuable time and contributions.

Dr. Christelle Abadie, University of Cambridge, UK
Dr. Amizatulhani Abdullah, Universiti Malaysia Pahang, Malaysia
Dr. Ardy Arsyad, Hasanuddin University, Indonesia
Dr. B. Munwar Basha, Indian Institute of Technology Hyderabad, India
Prof. Gokhan Baykal, Bogazici University, Turkey
Prof. K. S. Beena, Cochin University of Science and Technology, India
Prof. Netra Prakash Bhandary, Ehime University, Japan
Dr. Arup Bhattacharjee, Jorhat Engineering College, India

Dr. Babloo Chaudhary, National Institute of Technology Karnataka Surathkal, India
Dr. Jitesh Chavda, Sardar Vallabhbhai National Institute of Technology Surat, India
Prof. Guangqi Chen, Kyushu University, Japan
Dr. Gabriele Chiaro, University of Canterbury, New Zealand
Dr. Do Minh Duc, Hanoi University of Science, Vietnam
Dr. Mahmoud Fawzy, National Research Centre, Egypt
Dr. Takuro Fujikawa, Fukuoka University, Japan
Mr. Takashi Fujishiro, Geo-disaster Prevention Institute, Japan
Dr. Zentaro Furukawa, Kyushu University, Japan
Prof. Chandan Ghosh, Indian Institute of Technology Jammu, India
Dr. Stuart Haigh, University of Cambridge, UK
Dr. Tri Harianto, Hasanuddin University, Indonesia
Prof. Hemanta Hazarika, Kyushu University, Japan
Prof. Manzul Kumar Hazarika, Asian Institute of Technology, Thailand
Dr. Yi He, Southwest Jiaotong University, China
Dr. Amarnath Hegde, Indian Institute of Technology Patna, India
Dr. Prasad Hiremath, Fugro Survey (India) Pvt. Ltd., India
Prof. Masayuki Hyodo, Yamaguchi University, Japan
Dr. Taichi Hyodo, Toyama Prefectural University, Japan
Prof. Isao Ishibashi, Old Dominion University, USA
Dr. Yusaku Isobe, IMAGEi Consultant Corporation, Japan
Dr. Kazuya Itoh, Tokyo City University, Japan
Prof. Sangseom Jeong, Yonsei University, South Korea
Dr. Arvind Kumar Jha, Indian Institute of Technology Patna, India
Dr. Kazuhiro Kaneda, Takenaka Corporation, Japan
Dr. Kiyonobu Kasama, Tokyo Institute of Technology, Japan
Prof. Takeshi Katsumi, Kyoto University, Japan
Prof. Seiki Kawagoe, Fukushima University, Japan
Dr. Vilayvong Khonesavanh, Kyoto University, Japan
Dr. Takashi Kiyota, The University of Tokyo, Japan
Dr. Yoshifumi Kochi, K's Lab Inc., Japan
Dr. Eiji Kohama, Port and Airport Research Institute, Japan
Prof. Takaji Kokusho, Prof. Emeritus, Chuo University, Japan
Dr. Abhishek Kumar, Indian Institute of Technology Guwahati, India
Dr. Shiv Shankar Kumar, National Institute of Technology Patna, India
Dr. Wei Feng Lee, Ground Master Construction Co., Ltd., Taiwan
Prof. San-Shyan Lin, National Taiwan Ocean University, Taiwan
Dr. Guojun Liu, Changshu Institute of Technology, China
Prof. Santana Phani Gopal Madabhushi, University of Cambridge, UK
Dr. Suman Manandhar, Tribhuvan University, Nepal
Dr. Anil Kumar Mishra, Indian Institute of Technology Guwahati, India
Dr. Masanori Murai, Shimizu Corporation, Japan
Prof. Kasinathan Muthukkumaran, National Institute of Technology Trichy, India
Dr. Kentaro Nakai, Nagoya University, Japan

Prof. Juichi Nakazawa, Prof. Emeritus, National Institute of Technology, Maizuru College, Japan
Prof. Toshihiro Noda, Nagoya University, Japan
Dr. Atsushi Nozu, Port and Airport Research Institute, Japan
Prof. Kiyoshi Omine, Nagasaki University, Japan
Dr. Rolando P. Orense, University of Auckland, New Zealand
Prof. Rathish Kumar Pancharathi, National Institute of Technology Warangal, India
Dr. Jignesh Patel, Sardar Vallabhbhai National Institute of Technology, Surat, India
Dr. Satyajit Patel, Sardar Vallabhbhai National Institute of Technology, Surat, India
Dr. Suchit Kumar Patel, Central University of Jharkhand, India
Prof. Kyriazis Pitilakis, Aristotle University Thessaloniki, Greece
Dr. Jalilinasrabady Saeid, Kyushu University, Japan
Dr. Mizuki Sakai, National Institute of Technology, Nagano College, Japan
Prof. Kazunari Sako, Kagoshima University, Japan
Dr. Srikrishnan Siva Subramanian, Chengdu University of Technology, China
Dr. Bali Reddy Sodom, Madanapalle Institute of Technology and Science, India
Prof. Chandresh H. Solanki, Sardar Vallabhbhai National Institute of Technology, Surat, India
Prof. Naoaki Suemasa, Tokyo City University, Japan
Dr. Yuichi Sugai, Kyushu University, Japan
Dr. Mohd Khaidir Abu Talib, Universiti Tun Hussein Onn Malaysia
Mr. Tsuyoshi Tanaka, Tokyo City University, Japan
Dr. Jidong Teng, Central South University, China
Dr. Thuy Tran Thi Thanh, People's Committee of District 12, Ho Chi Minh City, Vietnam
Mr. Vamsi Nagaraju Thotakura, Sagi Ramakrishnam Raju Engineering College, India
Prof. Tetsuo Tobita, Kansai University, Japan
Prof. Ikuo Towhata, Specially Appointed Professor, Kanto Gakuin University, Japan
Dr. Balunaini Umashankar, Indian Institute of Technology Hyderabad, India
Dr. Sugeng Wahyudi, NITTOC Construction Co., Ltd., Japan
Prof. Lanmin Wang, Lanzhou Institute of Seismology, China
Dr. Lin Wang, Chuo Kaihatsu Corporation, Japan
Dr. Naoto Watanabe, KFC Ltd., Japan
Prof. Noriyuki Yasufuku, Kyushu University, Japan
Prof. Kazuya Yasuhara, Prof. Emeritus, Ibaraki University, Japan
Dr. Norimasa Yoshimoto, Yamaguchi University, Japan
Prof. Feng Zhang, Nagoya Institute of Technology, Japan

Contents

Recycled Geomaterials in Constructions

A Review of Data on Biodegradable Resin Concrete and Future Tasks 3
Mariko Suzuki, Kazuya Inoue, and Toshinori Kawabata

California Bearing Ratio of a Cohesive Soil Reinforced with Waste Tyre Fibres 13
Mohit K. Mistry, Ankit Vasoya, Shruti J. Shukla, Chandresh H. Solanki, and Sanjay Kumar Shukla

Effect of Installed Geotextile/Polyester and Biodegradable Materials for Dewatering Soft Clay 23
Flemmy Samuel Oye, Kiyoshi Omine, and Zicheng Zhang

Effect of Moisture Content on Particle Breakage of Recycled Concrete Aggregates During Compaction 33
Syed Kamran H. Shah, Taro Uchimura, and Ken Kawamoto

Environmental Implications of the Recycling of End-of-Life Tires in Seismic Isolation Foundation Systems 43
Laura Banasiak, Gabriele Chiaro, Alessandro Palermo, and Gabriele Granello

Estimating Landfill Height Considering the Heterogeneous Physical Property of Dredging Soil as Reclamation Material 53
Yoshihisa Sugimura, Yasuhiro Sega, Masaaki Katagiri, and Katsuhide Nishizono

Evaluation for Effective Use of Dredged Soil as Containment Dikes of Disposal Pond 63
Katsuhide Nishizono, Yasuhiro Sega, Masaaki Katagiri, and Yoichi Watabe

Experimental and Numerical Evaluation of Clay Soils Stabilized with Electric Arc Furnace (EAF) Slag 73
Muhammed Mahmudi, Selim Altun, and Tugba Eskisar

Experimental Study on Engineering Properties of Soft Soil Stabilized by Overboulder Asbuton 83
Noor Dhani, Tri Harianto, A. R. Djamaluddin, and Ahmad Gasruddin

Fundamental Study on the Effect of CO_2 Curing on the Strength Development of Alkaline Construction Sludge 95
Nguyen Duc Trung, Kimitoshi Hayano, and Hiromoto Yamauchi

Fundamental Study on the Mechanical Characteristics of Sand Treated by a PS Ash-Based Improving Material 107
Maliki Otieboame Djandjieme, Kimitoshi Hayano, Yoshitoshi Mochizuki, and Hiromoto Yamauchi

Performance Assessment of Recycled Tire Materials in Marine Landfill Application 117
Chunrui Hao, Hemanta Hazarika, and Yusaku Isobe

Performance of Rubble Brick Drains in Earthquake-Induced Liquefaction Mitigation Under Existing Buildings 127
Samy Garcia-Torres and Gopal Santana Phani Madabhushi

Performance of Sand–Rubber Mixture Infill Trench for Ground Vibration Screening 139
J. S. Dhanya, A. Boominathan, and Subhadeep Banerjee

Potential Application of Bauxite and Iron Mining Residues in Civil Construction in Brazil 147
Julian Buritica Garcia, Virginie Queiroz Rezende Pinto, and Juan Félix Rodríguez Rebolledo

Reuse of Waste Tires to Develop Eco-rubber Seismic-Isolation Foundation Systems: Preliminary Results 159
Gabriele Chiaro, Ali Tasalloti, Alessandro Palermo, Gabriele Granello, and Laura Banasiak

Shear Characteristics of Geomaterials Mixed with Fibrous Wood Chip and Converter Steelmaking Slag 171
Tomotaka Yoshikawa, Yoshiaki Kikuchi, Shohei Noda, and Akihiro Oshino

Shear Strength Enhancement of Soil Mixed with Palm Oil Clinker 183
Amizatulhani Abdullah, Cut Ainul Mardziah Amir, and Mohd Yuhyi Mohd Tadza

Study on Bio-cementation of Ex-coal Mining Soil as a Road Construction Material 193
A. M. Indriani, Tri Harianto, A. R. Djamaluddin, and A. Arsyad

Utilization of Coal Gangue for Earthworks: Sustainability Perspective 203
Mohammed Ashfaq, M. Heera Lal, and Arif Ali Baig Moghal

Physical and Numerical Modeling of Disaster Mitigation Techniques

A Consideration on Numerical Model for the Relationship Between Evaporation Efficiency and Volumetric Water Content 221
Yuta Jikuya, Kazunari Sako, and Shinichi Ito

A Finite Difference Scheme for the Richards Equation Under Variable-Flux Boundary 231
Yasuhide Fukumoto, Fengnan Liu, and Xiaopeng Zhao

Centrifuge Modelling Influence of Various Integration Schemes of Retaining Walls on Seismic Behaviour Using Tilting Table Test 247
Kazuya Sano, Kazuya Itoh, Tsuyoshi Tanaka, Naoaki Suemasa, and Takeharu Konami

Effect of Bearing Pressure on Liquefaction-Induced Settlement in Layered Soils 261
Yutao Hu, Hemanta Hazarika, Siavash Manafi Khajeh Pasha, Stuart Kenneth Haigh, and Gopal Santana Phani Madabhushi

Evaluation of Ground Anchor Residual Tension by Vibration Method 271
Kazuki Nawa, Atsushi Yashima, Yoshinobu Murata, Keizo Kariya, Hideki Saito, Kunio Aoike, Yoshinori Sone, and Mitsuru Yamazaki

Jet Grouted Columns with Horizontal Slab as Ground Improvement Towards Liquefaction Mitigation 283
Myat Myat Phyo Phyo, Hemanta Hazarika, Hiroaki Kaneko, and Tadashi Akagawa

Mechanical Behaviors of Assembled Multi-step Cantilever-Retaining Walls for High Embankments by Numerical Simulation Method 295
Zhaoying Li and Shiguo Xiao

Microscopic Analysis of the Influence of Pore Size Distribution on SWCC Using Extended DDA 305
Longxiao Guo and Guangqi Chen

Physical Model Tests for Newly Developed Breakwater Foundation Subjected to Earthquake and Tsunami 315
Babloo Chaudhary, Hemanta Hazarika, Akira Murakami, and Kazunori Fujisawa

Simulation of Crack Initiation and Propagation Using the Improved DDA 327
Mingyao Xia and Guangqi Chen

Stability Assessment on Expressway Embankment by Automatic Survey System 337
Kosuke Nakashima, Atsushi Yashima, Yoshinobu Murata, and Keizo Kariya

Stability of Reinforced Soil Quay Wall Subjected to Combined Action of Earthquake and Tsunami 349
Manu K. Sajan and Babloo Chaudhary

The Influence of Coupled Horizontal and Vertical Components of Strong Ground Motions on the Ground Response Analysis: Experimental and Numerical Approaches 361
Tsubasa Maeda, Hazarika Hemanta, and Yoshinori Kato

Verification of Seismic Deformation of Road Embankment Laying on the Coast Using Numerical Analysis 375
Kentaro Kuribayashi, Tadashi Hara, and Shuichi Kuroda

Characterization of Construction Materials

Behavior of Unsaturated Silty Soil Due to Change in Water Content and Suction 397
Ibrar Ahmed, Jiro Kuwano, Adnan Anwar Malik, and Dipak Raj Shrestha

Cyclic Strength and Deformation Characteristics of Sand Containing Fines with Plasticity 411
Hideo Nagase, Akihiko Hirooka, Keigo Fukumoto, Keiichiro Miyaji, and Naoya Kawano

Effect of Mineral Additives on the Strength Characteristics of a Laterite Soil 423
Tri Harianto and Widya Dwi Utami

Effect of Spatial–Temporal Discretization Order on the Selection of Lattice Boltzmann Forcing Strategy in Convective Flow Simulation Within Internal Geometrical Arrangement of Concrete Structure 433
Aditya Dewanto Hartono, Kyuro Sasaki, and Ronald Nguele

Effects of Water Absorption and Retention Performance of Paper Sludge Ash in Combination with Cement to Stabilize Dredged Soil 455
Phan Nguyen Binh, Kimitoshi Hayano, Mochizuki Yoshitoshi, and Hiromoto Yamauchi

Fundamental Study of Adsorption Thin Layers for Safe Storage of Heavy Metal Contaminated Soil 467
Osamu Otsuka, Ryo Nishizato, Minoru Okuno, Naoto Watanabe, Tsutomu Matsuo, Tomomichi Obuchi, Takashi Mizobuchi, Yoichi Kitamura, Eriko Kikawa, Ken Sato, and Hidekazu Shimohara

Influence of Sand on the Behavior of Buffer Material Used for the Nuclear Waste Disposal 477
Padavala Hari Krishna and Adla Nandini

Swell-Compressibility Behavior of Geopolymer Blended Expansive Clays 493
T. V. Nagaraju, B. M. Sunil, and Babloo Chaudhary

Swelling Behavior and Permeability of Polymerized Bentonite with Due Monomers 501
Nie Jixiang and Chai Jinchun

Suitability of Mechanically Biologically Treated Waste for Landfill Covers 511
T. G. Parameswaran, N. Anusree, P. Sughosh, G. L. SivakumarBabu, and T. K. K. Chamindu Deepagoda

Synergy Between Poly Vinyl Alcohol and Bentonite in Drilling Fluid Formulation: An Experimental Study 521
Danielle Poungui, Kyuro Sasaki, Yuichi Sugai, and Ronald Nguele

Advancement in Sustainable Construction Techniques

Bearing Capacity Determination of Jointed-Timber Piles in the Saga Lowland 533
Sailesh Shrestha, Yuki Matsumoto, Shunsuke Moriyama, and Norihiko Miura

Experiment of a Hard Rock Excavator by Using an Edge Excavation ... 547
Yudai Ihara, Takahisa Shigematsu, Shinichi Kawamura, Yoshihiro Ohnishi, Sho Miyanaga, and Muhamad Afif Bin Osman

Full-Scale Testing of a Structure on Improved Soil Replaced with Rubber–Gravel Mixtures 561
Athanasios Vratsikidis, Angelos Tsinaris, Anastasios Kapouniaris, Anastasios Anastasiadis, Dimitris Pitilakis, and Kyriazis Pitilakis

Interdependence and Rationality Between Sustainable Indicators and Criteria—A Fuzzy AHP Approach 575
Suchith Reddy Arukala and Rathish Kumar Pancharathi

Introduction of the Recycling and Reusing Method of Screw Steel Pile EAZET 591
Shuo Teng and Tadashi Maejima

Mechanical Behavior of Cement-Treated Soils with Nanosilica—A Green Binder 609
Gizem Aksu and Tugba Eskisar

Structural Stability Against Earthquake and Tsunami Using Environmentally Sustainable Materials ... 619
Kiran Hari Pradhan, Hemanta Hazarika, and Yasuhide Fukumoto

Tip Shape Effect on Screw Pile Installation and Ultimate Resistance ... 627
Adnan Anwar Malik, Yahya Ndoye, and Jiro Kuwano

Climate Change Independent Natural Disasters

Application of Deep Mixing Method for Mitigation of Potential Seismic and Hurricane-Induced Hazards ... 641
Rakshya Shrestha, Nozomu Kotake, and David Yang

Characteristics of Surficial Mass Movements on Cut Slopes of the Prime National Highway 1 (PNH-1) in Bhutan ... 653
Kiyoharu Hirota, Tomohiro Nishimura, Takeshi Kuwano, and Tomoharu Iwasaki

Estimation of Liquefaction Area Using the Screw Driving Sounding Data in Kumamoto ... 663
Kohei Tsuji, Naoaki Suemasa, and Tsuyoshi Tanaka

Deterministic Seismic Hazard Analysis of Ankleshwar City, Gujarat ... 673
Manali Patel, Chandresh Solanki, and Tejas Thaker

Large Strain Torsional Shear Tests on the Mechanism of Long-Distance Flow Slide in Palu, Central Sulawesi, Indonesia ... 681
Risqi Faris Hidayat, Takashi Kiyota, Muhammad Umar, and Hasbullah Nawir

One-Dimensional Soil Column Simulation on Water Film Formation During Earthquake ... 693
Divyesh Rohit, Hemanta Hazarika, Chengjiong Qin, Tsubasa Maeda, Takaji Kokusho, and Yuichi Yahiro

Risk Assessment of Landslide Disaster for Detecting Isolated Communities in Wakayama Prefecture After Nankai Trough Earthquake ... 705
Katsuki Hozumi, Hisakazu Sakai, and Yoshio Kajitani

Innovative and Sustainable Techniques in Geoengineering

DCM Columns and PVDs Hybrid Ground Improvement for Embankment Construction ... 719
Junfeng Ni and Jinchun Chai

Evaluation of Hybrid Pile Supported System for Protecting Road Embankment Under Seismic Loading ... 733
Chengjiong Qin, Hemanta Hazarika, Siavash Manafi Khajeh Pasha, Hideo Furuichi, Yoshifumi Kochi, Daisuke Matsumoto, Takashi Fujishiro, Shinichiro Ishibashi, Naoto Watanabe, and Shigeo Yamamoto

Evaluation and Short-Term Test on Potential Utilization of Ground Source Heat Pump for Space Cooling in Southeast Asia ... 745
Arif Widiatmojo, Yutaro Shimada, Isao Takashima, Youhei Uchida, Srilert Chotpantarat, Punya Charusiri, Juraluk Navephap, and Trong Thang Tran

Experimental Study on Strengthening Near-Surface of Slopes Using Bio-grouting Technique ... 771
Sivakumar Gowthaman, Kazunori Nakashima, Hiromi Nakamura, and Satoru Kawasaki

Sustainable Improvement Technology for Soft Grounds—The Geo Drain SPD Method ... 787
Hironobu Iitsuka and Yoshiki Nara

Climate Change Induced Disasters

Development of a Junior High School ESD Program to Raise the Disaster Prevention Awareness Based on the Experience of Heavy Rainfall in Lowland ... 801
Takashi Shimoyamada and Satoquo Seino

River Education on Sustainability and Disaster Prevention at Elementary Schools in Island Regions ... 815
Hidefumi Hatashima and Satoquo Seino

Instrumentation for Disaster-Affected Reinforced Slopes in Uttarakhand, India ... 829
Minimol Korulla, K. S. Beena, and Pankaj Kr. Mourya

Proactive Conservation of Main Tower of Bayon, Angkor, Against Global Warming ... 839
Yoshinori Iwasaki, Mitsuharu Fukuda, Mitsumasa Ishizuka, and Takeshi Nakagawa

Geological and Hydrological Phenomena on Hazard Estimation

Groundwater Level Changes in the Coastal Construction Site of Coal-Fired Power Plant, Cilacap, Indonesia; Natural or Construction Effect? ... 855
Doni Prakasa Eka Putra, Rilo Restu Surya Atmaja, Fania An Nisaa, Kurnianto Dwi Setyawan, and Pramono Hadi

Landslide Susceptibility Assessment in Trenggalek, East Java, Indonesia: A Geological Overview 873
Wahyu Wilopo, Egy Erzagian, Diyaning Ratri, and Teuku Faisal Fathani

Riverbed Fluctuation Analysis of Small Rivers Equipped with Stream Barb Groins 885
Mizuki Sakai, Yugo Hashimoto, and Eiji Matsushita

The Mechanism of Riverbank Erosion Caused by Ship-Generated Waves Along Hau River's Entrance Navigation Channel, Southern Vietnam 897
Son Huu Nguyen, Tin Trung Huynh, Vinh Trong Bui, and Ngo Van Dau

Application of Information Technology Towards Natural Disaster Mitigation

Battery-Less Infrastructure Monitoring Sensor Platform 907
Haruichi Kanaya, Mohamed M. Mansour, Shunsuke Hatanaka, Ataru Nakashima, and Osamu Takiguchi

Developing a Sustainable System for Early Warning Against Landslides During Rainfall 917
Yan Liu, Hemanta Hazarika, Osamu Takiguchi, and Haruichi Kanaya

Evaluation of Slope Failure and Driftwood Disaster Susceptibility Zone Using Geo-spatial Information 927
Lu Tian, Yasuhiro Mitani, Yuki Okajima, Hisatoshi Taniguchi, and Taiga Tabuchi

Innovative Case Studies in Earthquake Geotechnical Engineering

Change in Shear Wave Velocity During Consolidation and Undrained Cyclic Loading on Cemented Sand 941
Masataka Shiga and Takashi Kiyota

Contribution of Geotechnical Engineers in Restoration After Large-Scale Earthquake Disaster—Example of Aso Ohashi Bridge Replacement Project 953
Koji Yamashita, Shingo Tanaka, and Takefumi Yamaguchi

Liquefaction Evaluation of Residential Land Using Geophysical Survey 965
Shotaro Higuchi, Atsushi Yashima, Yoshinobu Murata, Keizo Kariya, and Hiroshi Yokawa

About the Editors

Hemanta Hazarika is currently a Professor at the Graduate School of Engineering and Department of Interdisciplinary Science and Innovation, Kyushu University, Fukuoka, Japan. His research activities include disaster prevention and mitigation, stability of soil-structures during earthquakes and tsunami, ground improvement, application of recycled waste and lightweight materials in constructions, and landslides and their mitigations. He has more than 350 technical papers in various international journals, International conferences, workshops and symposia to date. He has also authored two textbooks on Soil Mechanics and their Japanese versions. In addition, he also served as the editor of three technical books on disaster mitigation and waste tire recycling. Prof. Hazarika pioneered the use of waste tire derived recycled materials for earthquake protection of infrastructures and holds two patents on those. Prof. Hazarika has several years of experience in teaching, research as well as geotechnical practice and consulting both within and outside Japan. He has also successfully led a World Bank project on "Improving the Resilience of Nepal's Strategic Roads Network".

Gopal Santana Phani Madabhushi is a Professor of Civil Engineering at the University of Cambridge, UK and the Director of the Schofield Centre. His expertise extends from dynamic centrifuge modelling to the time domain finite element analyses of earthquake engineering problems. He has an active interest in the areas of soil liquefaction, soil-structure interaction and liquefaction resistant measures and their performances. His research interests also include the biomechanics of hip replacement surgeries. He has acted as an expert consultant to the industry on many geotechnical and earthquake engineering problems. He has 150+ Journal Publications and 270+ papers in International conferences and workshops to date. He has authored a remarkable book on the *Design of Pile Foundations in Liquefiable Soils,* Geotechnical Chapters in the book on *Designing to Eurocode 8* and a recent book on *Centrifuge Modelling for Civil Engineers.*

Kazuya Yasuhara is an Emeritus Professor at Ibaraki University, Japan. After retiring from Ibaraki University in 2010, Prof Yasuhara was the International Project Coordinator at the Institute of Global Change Adaptation Science (ICAS) from 2010 to 2019. Since 2019, he has been working as a Specially appointed researcher at Global and Local Environment Co-creation Institute (GLEC) of Ibaraki University, as well as Professor Emeritus. He has also been teaching Climate Change Risk Management in the Master Course of Climate Change and Development in Vietnam-Japan University (VJU), Hanoi, Vietnam, since retirement. His research interests are: climate change-induced compound geo-disasters in Asia-Pacific regions and their adaptation, earthquake-induced deformation, and instability of infrastructures on fine-grained soils including the countermeasures against damage. He also served as a Review Editor (Chapter 29 and Technical Summary) for IPCC AR 5 from 2010 to 2014.

Dennes T. Bergado is an Emeritus Professor at Asian Institute of Technology, Thailand, where he was hired as a faculty since 1982 until retirement. His research areas include probabilistic and reliability analyses of geotechnical properties and structures as well as ground Improvement and geosynthetics. He established the Asian Center for Soil Improvement and Geosynthetics (ACSIG) Research Center including the first geosynthetics research laboratory in Thailand. He has pioneered the use of prefabricated vertical drain (PVD) for ground improvement in soft Bangkok clay with subsequent combinations of vacuum and heat preloading. Prof. Bergado published 2 Books, 7 Chapters in Books, 22 Edited Books, 5 Guest Editorships of Journals, 156 Journal Articles, 310 Conference/Invited Papers, 73 Invited/Keynote Lectures. Currently, he has been involved in multiple infrastructure projects in Bangkok, Thailand and Laos PDR funded by ADB, EIB and World Bank as a Geotechnical Specialist.

Recycled Geomaterials in Constructions

A Review of Data on Biodegradable Resin Concrete and Future Tasks

Mariko Suzuki, Kazuya Inoue, and Toshinori Kawabata

Abstract Biodegradable resin concrete is made of biodegradable resin and aggregate. The experiments on this new concrete were started in 2012, and six papers were published. The strength degradation of biodegradable resin concrete was evaluated using the results of the three-point bending and compression tests, and the surface degradation was evaluated using the binarized photo data. The results showed that the decomposition of the biodegradable resin by microorganisms and the bond strength reduction between the biodegradable resin and aggregate by hydrolysis caused the degradation mechanism. Moreover, to estimate the fatigue life and clarify the strength retention period of biodegradable resin concrete, the statistical deterioration prediction formulas were introduced using the Weibull distribution. This provided the appropriate estimates associated with the strength retention period of biodegradable resin concrete. This paper summarizes the experimental results on biodegradable resin concrete and describes future tasks.

Keywords Biodegradable resin concrete · Strength degradation · Surface degradation · Statistical deterioration prediction

1 Introduction

Biodegradable resin has a short history and was discovered in the 1980s [1]. It has been used in many fields, from medical science to agriculture. In the field of medical science, biodegradable resin is used as a needle and suture thread [2]. In the field of agriculture, it is used as mulching materials, seedling pots, and nets for voluble stems [3, 4]. The previous studies are elaborated on the mechanical properties of biodegradable resin [5, 6]. In the fields of engineering, the biodegradable resin is used for products such as sandbags and ropes [7, 8] that are flexible materials. On the other hand, biodegradable resin concrete that is the target in this study is a rigid material. Therefore, it is critical to clear its mechanical characteristics.

M. Suzuki (✉) · K. Inoue · T. Kawabata
Graduate School of Agriculture Science, Kobe University, Kobe, Japan
e-mail: msuzuki@peridot.kobe-u.ac.jp

H. Hazarika et al. (eds.), *Advances in Sustainable Construction and Resource Management*, Lecture Notes in Civil Engineering 144,
https://doi.org/10.1007/978-981-16-0077-7_1

Temporary materials such as sheet piles made of steel or concrete are abandoned on-site after construction if they adversely affect the surrounding soil when they are extracted during construction work. Commonly used temporary materials abandoned on-site can cause various problems for future redevelopments such as a decline in land worth and industrial waste generation. To address these problems, biodegradable resin concrete prepared from biodegradable resin and aggregate was developed. Experiments on this new concrete began in 2012, and six papers were published. The strength degradation of biodegradable resin concrete was evaluated by three-point bending and compression tests [9, 10]. The surface degradation was evaluated from binarized photo data [11]. Furthermore, to estimate the fatigue life and clarify, the strength retention period of biodegradable resin concrete and statistical deterioration prediction formulas were introduced using the Weibull distribution [12]. This study summarizes the experimental results of biodegradable resin concrete and describes future tasks.

2 Outline of Experiments

2.1 Materials and Mixture Proportions of Specimens

In this study, two types of biodegradable resin used as a bonding material were used for the biodegradable resin concrete, namely polybutylene succinate adipate (PBSA) and polylactic acid (PLA). PBSA was used because the biodegradable resin has a fast biodegradability [13]. The melting point of PBSA is 95 °C, and its melt flow rate (MFR) is 1–3 [14]. PLA, generated from corn, is a raw ingredient harvested in large quantities every year and is an aliphatic polyester resin. The melting point of PLA is 170 °C, and its MFR is 17. The PBSA and PLA used in this study are shown in Figs. 1 and 2.

Fig. 1 PBSA

Fig. 2 PLA

Table 1 Mixture proportions of specimens

	Ratio of mass (%)	
Biodegradable resin	12	10
Calcium carbonate	20	20
Fine sand (fineness modulus: 1.09)	20	20
Coarse sand (fineness modulus: 4.91)	48	50

The specimens were prepared from biodegradable resin, fine sand, coarse sand, and calcium carbonate. The composition of the biodegradable resin concrete is listed in Table 1. Calcium carbonate was used to compensate for fine-grain fraction, and it does not chemically react with the biodegradable resin.

2.2 Exposure Conditions and Tests Method

Table 2 shows the adopted exposure conditions. The ligneous compost used Case A is composed of bark compost and peat moss, and full of microbes. Case C, aridity conditions, meant that specimens were placed in a box filled with silica gel. The

Table 2 Exposure conditions

	Ratio of mass				
	A	B	C	D	E
Washed sand	85	90	0	0	0
Ligneous compost	5	0	0	0	0
Silica sand	0	0	0	95, 90, 80	0
Water	10	10	0	5, 10, 20	100

specimens were placed in plastic boxes in each of the five exposure conditions. The insides of the boxes were maintained at 20 °C. The exposure period was set to a maximum of two years.

First, before the mechanical tests, the surface of the specimens exposed under Cases A, C, and E was observed with a digital microscope. The photos were taken every two weeks, and the shooting points were marked on the specimens so that photos could be taken at the same position. Pixels of 1024 × 1280 were captured, and the central portions were cropped to 512 × 640. The pixels of 512 × 640 were binarized at a threshold of 128 after 256 shades of gray images were made [15]. The threshold of 128 was determined by trial and error to correctly discriminate between the aggregates and resin. This binarization was conducted for both PBSA specimens and PLA specimens at the same conditions. For further details of binarization, refer to the reference [11].

Three-point bending and compression tests were conducted to clarify the deterioration of the biodegradable resin concrete specimens. The sizes of the specimens were 40 mm × 40 mm × 160 mm for the bending test and 40 mm × 40 mm × 40 mm for the compression test.

3 Statistical Estimation of Degradation

It is critical to estimate the fatigue life and reveal the period for which the biodegradable resin concrete retains its strength. There are some prediction methods of degradation, namely physical and chemical formulas, statistical formulas, and turning numbers into an easily existing formula. This study used the Weibull distribution by considering the complexity of the degradation of the biodegradable resin concrete. It is one of the most widely used distributions in dependability analysis and flexibly applicable to fault curves. W. Weibull suggested the Weibull distribution in 1939 during his experiments of metallic fatigue life [16]. At that time, he concluded that the breakdown of brittle materials occurred due to independent flaws in materials. The two-parameter Weibull probability distribution function is given by

$$F(t) = 1 - R(t) = 1 - \exp\left\{-\left(\frac{t}{\beta}\right)^{\alpha}\right\} \quad (1)$$

where $R(t)$ is the degree of confidence, t denotes time, α is the shape parameter, and β is the scale parameter.

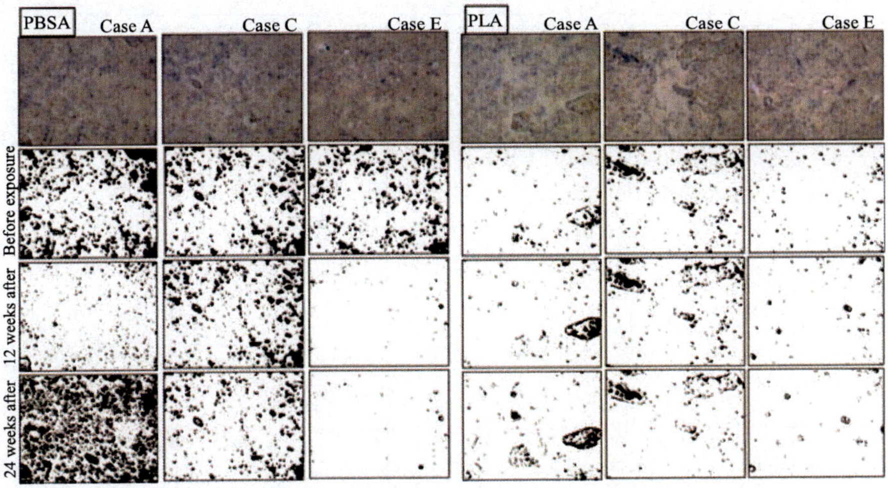

Fig. 3 Micrograph of the specimens' surface (PBSA and PLA)

4 Results and Discussion

4.1 Binarization

The micrographs of the specimen surface are shown in Fig. 3, and the proportions of white pixel count to total pixels after binarization are shown in Fig. 4. In Case C, the aridity condition and both the PBSA and PLA specimens remained almost the same as the initial state. In other words, the deterioration might not progress in an air-dried state. The specimen showing the most change is the PBSA specimen exposed in Case A. This specimen turned white after two weeks. After that, its surface slowly became black. This means that aggregates outcrop owing to the degradation of the resin. However, no PLA specimens in Case A became black, indicating that no surface deterioration occurred. The PLA specimen surface is not deteriorated because PLA is hydrolyzed without the direct involvement of microorganisms [17]. The aqueous exposure of PBSA and PLA specimens (Case E) also did not become black. It is inferred from the observation of the specimen surface that microorganisms in soil decompose PBSA specimens.

4.2 Three-Point Bending Test

The bending test results are shown in Fig. 5. A plot in the figure is the mean of five specimens. PBSA specimens were exposed for two years, and PLA specimens were exposed for one year. The bending tests were conducted every two months.

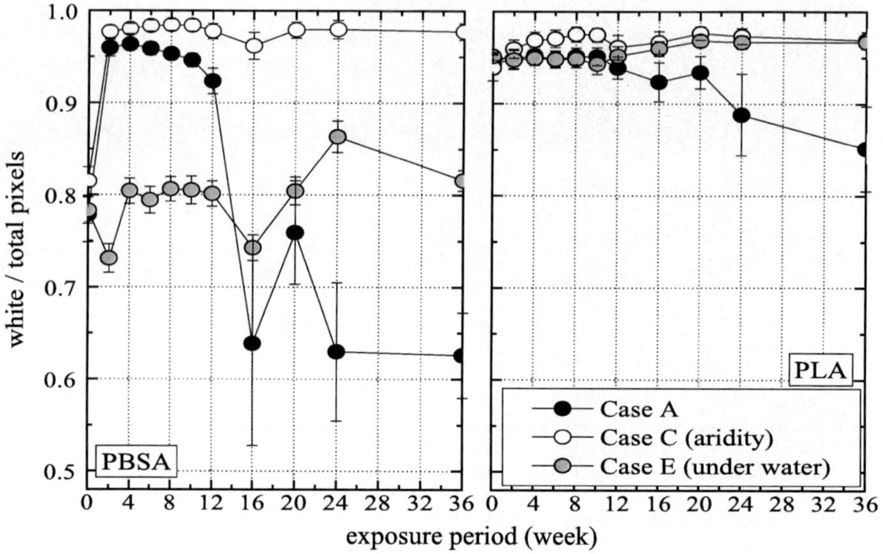

Fig. 4 Results of binarization (PBSA and PLA)

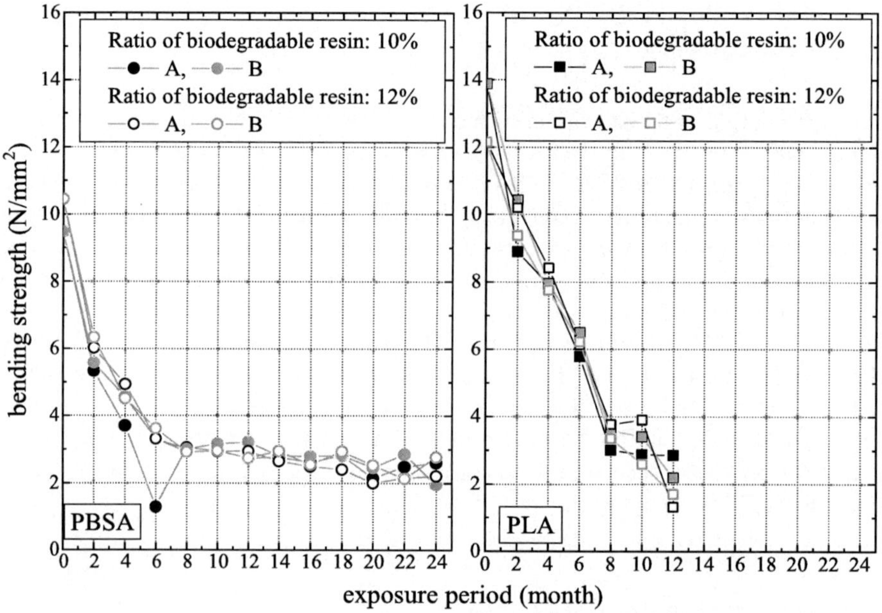

Fig. 5 Bending test results

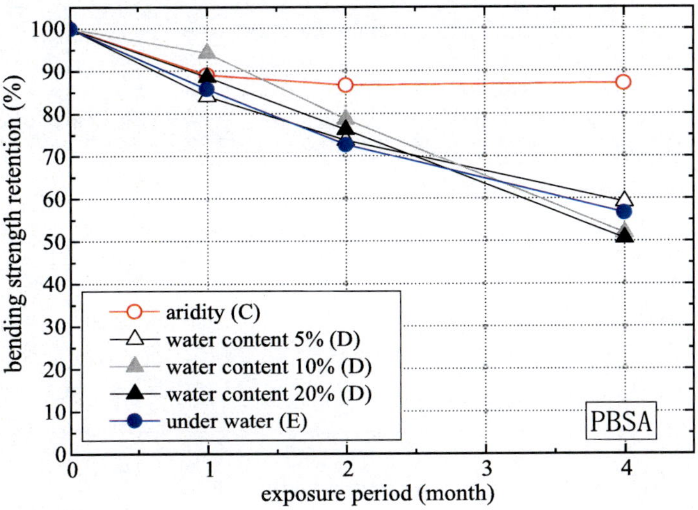

Fig. 6 Strength retention by a variation in water content

The bending strength was reduced in all specimens with different resin ratios and soil patterns. The bending strength significantly decreases for six months after exposure. This strength reduction is not caused by microorganisms because the microbial degradation speed is slow. This significant strength reduction, therefore, could be because of the water affecting the bond strength between the biodegradable resin and aggregates. This is because the exfoliation of aggregates from the resin was confirmed at the broken-out section after the bending test.

Figure 6 shows the strength retention by a variation in water content. The remarkable strength reduction occurred for about six months. Therefore, the effect of water on the biodegradable resin concrete will occur a few months after exposure. Therefore, the experiment was terminated to clarify the bending strength reduction by varying the water content for up to four months. The specimens of atmospheric curing (Case C) had little deterioration. The tendency of strength reduction from the Cases D and E was similar. From this result, this biodegradable resin concrete could be degraded if it is exposed to soil with a water content of more than 5%.

4.3 Compression Test

Figure 7 shows the compression test results. The compressive strength reduction was observed in all the specimens with different resin types and soil patterns. Compressive strength stems from the strength of the aggregates. However, bending strength is the ascribable tensile strength of the resin. This is because the residual strength was

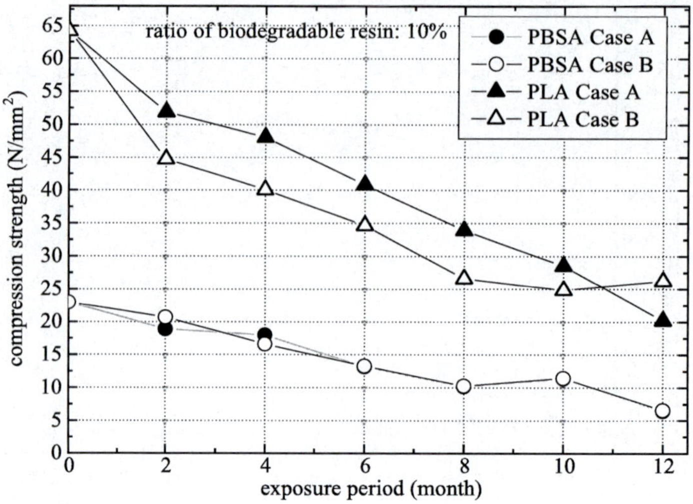

Fig. 7 Compression test results

kept higher than the bending residual strength. In other words, the inner part of the specimen retained its strength.

4.4 Estimation of Degradation

The mean times to failure of PBSA and PLA are listed in Figs. 8 and 9. It shows how much time will be required to drop below each strength. It will be quite a long time before the strength of biodegradable resin concrete becomes zero from the present estimation model. This is not a surprising result because the biodegradation rate is slow after strength reduction using water. The biodegradable manufacture in the previous studies by other researchers retained the shapes and strength even though the ropes were exposed for three years [18]. However, biodegradable resin concrete will not be harmful to land development because it is expected to reduce strength.

5 Future Tasks and Perspectives

The final goal of this study on biodegradable resin concrete is to achieve the practical use as the temporary construction materials such as sheet piles. The biodegradable resin concrete is the new construction material, and the biodegradable resin concrete was made through trial and error in this study. Biodegradable resin concrete requires further investigation. The biodegradable resin concrete is known to deteriorate owing to many factors such as microbial action, water, temperature, light, and ultraviolet

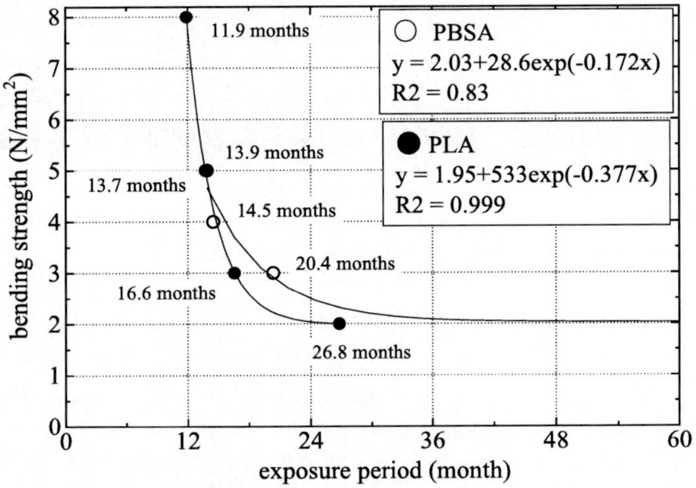

Fig. 8 Mean time to failure (MTTF) computed from the bending strength

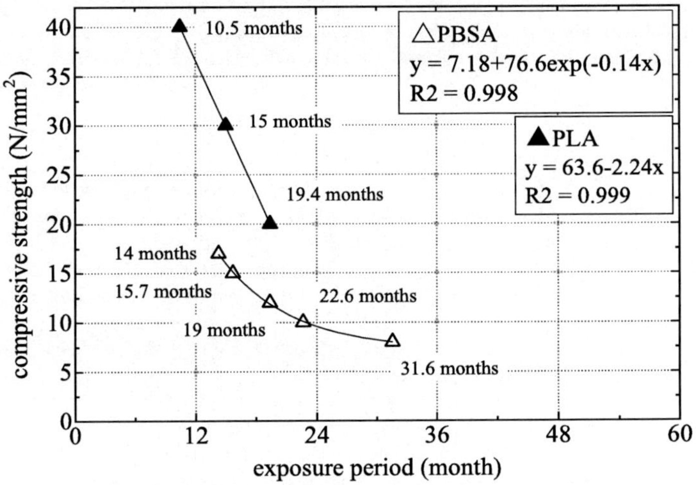

Fig. 9 Mean time to failure (MTTF) computed from the compressive strength

rays. This study is reported on microbial degradation and the action of water; therefore, the effects of other factors on the biodegradable resin concrete must be clarified. By establishing a deterioration control method, biodegradable resin concretes would be easy to use.

References

1. Japan BioPlastics Association: The book of Bio Plastics, p. 56. Japan BioPlastics Association (2009) (in Japanese)
2. Aoyagi, S., Izumi, H., Aoki, T., Fukuda, M.: Development of micro Lancet needle made of biodegradable polymer for medical treatment, the transactions of the institute of electrical engineers of Japan. Publ. Sens. Micromach. Soc. **127**(2), 53–62 (2007)
3. Japan BioPlastics Association: The book of Bio Plastics, Japan BioPlastics Association, pp. 106–110 (2009) (in Japanese)
4. Bernardo, D.T., Nagata, M., Mitarai, M., Ishino, F., Hiyoshi, K.: Study on mechanization of mulching cultivation using biodegradable film for early season culture rice. Part 2. Development of mulch slit cutting mechanism for biodegradable film using nylon cord: development of mulch slit cutting mechanism for biodegradable film using nylon cord. J. Jpn. Soc. Agric. Mach. **62**(2), 154–166 (2000)
5. Kitamoto, H.: Practical research and current status of environment-friendly plastic. Agric. Hortic. **84**(1), 67–71 (2009) (in Japanese)
6. Satho, N., Akama, T., Yaguchi, H.: Research on the degradablilty control of biodegradable plastics, the annual research report of Industrial Technology Institute. Miyagi Prefect. Gov. **1**, 6–9 (2003). (in Japanese)
7. Japan BioPlastics Association: The Book of Bio Plastics, p. 114. Japan BioPlastics Association (2009) (in Japanese)
8. Nishimura, Y., Hiraishi, N., Ishida, T., Ishida, K.: Biodegradation and photodegradation of biodegradable ropes. Aichi Ind. Technol. Inst. Res. Rep. **6**, 196–199 (2007). (in Japanese)
9. Suzuki, M., Kubo, K., Suzuki, T. and Kawabata, T.: Mechanical approach to eco-friendly biodegradable resin concrete. Concr. Res. Technol. **25**, 119–124 (2014) (English Abstract)
10. Suzuki, M., Kubo, K., Suzuki, T., Sawada, Y., Kawabata, T.: Effect of different exposure conditions on mechanical properties of biodegradable resin concrete. Irrigation Drainage Rural Eng. J. **296**, I_19–I_24 (2015)
11. Suzuki, M., Kubo, K., Kawabata, T.: Comparison of degradation of biodegradable resin concrete made from PBSA or PLA. Concr. Res. Technol. **27**, 7–12 (2016) (English Abstract)
12. Suzuki, M., Kubo, K., Kawabata, T.: Statistical estimation approach of degradation of biodegradable resin concrete. Concr. Res. Technol. **27**, 1–6 (2016) (English Abstract)
13. Okino, Y.: PBS type biodegradable aliphatic polyester. Environ. Conserv. Eng. **34**(6), 406–410 (2005) (in Japanese)
14. Ikada, Y.: Basics and Applications of Biodegradable Polymer, pp. 248–249. Industrial Publishing & Consulting, Inc. (1999) (in Japanese)
15. Okuno, R., Mori, M., Tokashiki, M., Asano I.: Evaluation of degradation degree of organic surface coating materials by image processing. In: Japan Concrete Institute Proceedings of Annual Conference, vol. 33, no. 1, pp.791–796 (2011) (in Japanese)
16. Weibull, W.: A statistical distribution function of wide applicability. J. Appl. Mech. **18**, 293–297 (1951)
17. Kimura, T., Ihara, N., Ishida, Y., Saito, Y., Shimizu, N.: Hydrolysis characteristics of biodegradable plastic (poly latic acid). Nippon Shokuhin Kagaku Kogaku Kaishi **49**(9), 598–604 (2002) (in Japanese with English summary)
18. Nishimura, Y., Hiraishi, N., Ishida, T., Ishida, K.: Biodegradation and Photodegradtion of Biodegradable Ropes, pp. 196–199. Aichi Industrial Technology Institute (AITEC) (2007) (in Japanese)

California Bearing Ratio of a Cohesive Soil Reinforced with Waste Tyre Fibres

Mohit K. Mistry, Ankit Vasoya, Shruti J. Shukla, Chandresh H. Solanki, and Sanjay Kumar Shukla

Abstract This paper investigates the influence of waste tyre fibres on the geotechnical properties of clayey soil used as a subgrade material. Laboratory California bearing ratio (CBR) tests were conducted for the unsoaked condition as a measurement of pavement performance under the normal rainfall. Three types of tyre fibres as *TFA*, *TFB* and *TFC* used in the present study were mixed in a proportion of 0.25, 0.5, 0.75 and 1% by dry mass of clayey soil. In each test, the specimen was prepared in accordance with the compaction characteristics achieved from the modified compaction energy. The test results reported that the compaction characteristics of cohesive soil were getting increased with an increase in fibre content. In general, regardless of the fibre size, the CBR values get increased with an increase in fibre content. The experimental results showed that clayey soil achieves five times higher CBR value for the addition of 0.75% *TFC*. From the test results, it can be concluded that the use of waste tyre fibres considerably reduces the pavement thickness and hence reduces the construction cost.

Keywords Cohesive soil · Waste tyre fibres · Compaction characteristics · California bearing ratio

1 Introduction

The effective application of soil mechanics in geotechnical projects purely depends upon the laboratory soil results and their interpretation. In many site-specific problems, laboratory tests mainly assist in predicting probable soil behaviour by means of

M. K. Mistry (✉) · A. Vasoya · S. J. Shukla · C. H. Solanki
S V National Institute of Technology, Surat, Gujarat, India
e-mail: mohit.aryan1@gmail.com

S. K. Shukla
School of Engineering, Edith Cowan University, Joondalup, Perth, WA 6027, Australia

V.R. Siddhartha Engineering College, Vijayawada, India

Fiji National University, Suva, Fiji

H. Hazarika et al. (eds.), *Advances in Sustainable Construction and Resource Management*, Lecture Notes in Civil Engineering 144,
https://doi.org/10.1007/978-981-16-0077-7_2

Table 1 Engineering properties of soil

Property	Values
Specific gravity (G_s)	2.63
Sand (%)	12
Silt (%)	54
Clay (%)	34
Plasticity index (I_P)	34
IS classification	CH
Unconfined compressive strength	114 Kpa

associated basic principles. In the last few decades, the use of waste tyre products as a reinforcing material in subbase has been increased in pavement construction [1–7]. Application of waste tyre products in subbase material as a reinforcing material is an economical and eco-friendly solution [8]. When road networks are constructed on expansive soil, it is very important to control swell–shrink nature. The studies conducted by [9–11] reported that the use of waste tyre fibres significantly controls the swelling and shrinkage nature of expansive soils. This paper investigates the influence of waste tyre fibres on the geotechnical properties of clayey soil used as a subgrade material. Laboratory California bearing ratio (*CBR*) tests were conducted for the unsoaked condition as a measurement of pavement performance under the normal rainfall. Three types of tyre fibres as *TFA*, *TFB* and *TFC* used in the present study were mixed in a proportion of 0.25, 0.5, 0.75 and 1% by dry mass of clayey soil. In the present study, the attempt has been made to check the feasibility of waste tyre fibres in pavement application.

2 Material

2.1 Soil

Soil used in the present study was collected from the Surat city of Gujarat state, India. The characterization of soil was done in accordance with IS 2720-part IV (1985). The engineering properties of soil are mentioned in Table 1.

2.2 Fibres

In the present study, the waste tyre fibres were used as a reinforcing material. The collected fibres were having length varying from 425 μm to 20 mm. Further, with conventional sieve analysis, fibres were divided into three different sizes (Table 2). The particle gradation curve of fibres is illustrated in Fig. 1. In accordance with ASTM

Table 2 Geometrical properties of fibres

Property	Values		
Length (L) (mm)	9	12.8	18.7
Diameter (D) (mm)	1	2.0	4.2
Aspect ratio (L/D)	9	6.4	4.45
Categorization of fibres	*TFA*	*TFB*	*TFC*

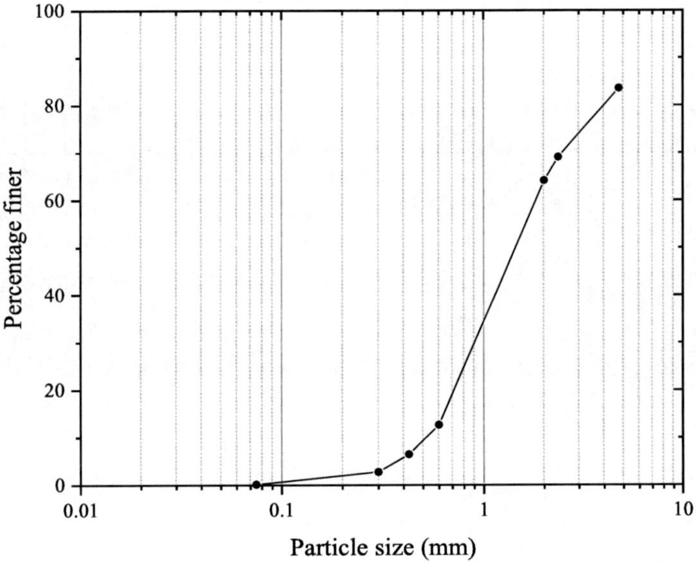

Fig. 1 Grain size distribution of fibres

D6270-08, the fibres were classified as buffing rubbers. In general, the specific gravity (G_f) of waste tyre fibre is ranging from 1 to 1.5. The waste tyre fibre used in the present study is free from wire mesh, and hence it has specific gravity (G_f) as 1.13. The water absorption capacity of this fibre is 3%. Figure 2 represents the waste tyre fibres.

3 Experimental Work

For the requirements of flexible pavement design, one of the simple strength tests would be a plate bearing test on the subgrade using the plate area and the loading approximately equal to the anticipated contact area and the wheel load. The California bearing ratio (C.B.R.) test is generally considered as a small-scale bearing test in which the ratio of penetration to the size of bearing area is much greater than in the

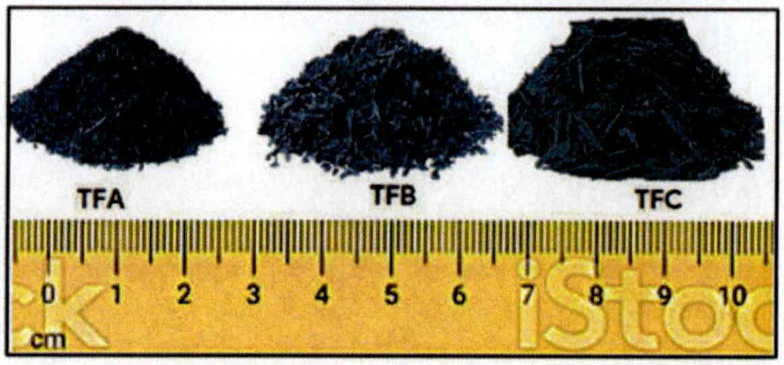

Fig. 2 Waste tyre fibres

plate bearing test [12]. The *CBR* method is a good substitutive for a heavy field test. It effectively determines the strength properties at a point in the soil mass. To mimic the on-field condition such as subgrade or pavement performance during normal rainfall, the unsoaked *CBR* tests were conducted on high plasticity clayey soil reinforced with three types of fibres, namely as *TFA*, *TFB* and *TFC,* and the resulted reinforced soils were denoted as *FRC* (fibre-reinforced clay), *FRCA*, *FRCB* and *FRC,* respectively. Each type of fibres was added in a dosage of 0.25%, 0.50%, 0.75% and 1.00% by dry weight of soil. Each *CBR* test was conducted as per the procedure described in IS 2720-part 16_(1987). Prior to the *CBR* test, the maximum dry unit weight for each type of *FRC* was determined using modified compaction test in accordance with IS 2720-part 8_(1980). To achieve the homogeneous mixture, soil mass for each layer was separated, and then fibres were spread in each separated soil mass. Later, water was added in each separated mass and then mixed thoroughly. The process of mixing soil sample was adopted from Mistry et al. [13].

4 Results and Discussion

4.1 Influence of Waste Tyre Fibres on Compaction Characteristics

Figure 3 shows the variation in maximum dry unit weight with an increase in fibre content for each type of fibres. It can be observed that with an increase in fibre size, *MDU* gets increased. However, for the different size of the fibres, the fluctuation with increasing fibre percentage can be observed. But in general, it can be stated that in comparison with unreinforced soil, the fibre-reinforced soil achieves the higher density. The applied high compaction energy may be the probable reason for the increment in *MDU*.

The differing results were observed by Mistry et al. [13] that with the application of standard compaction energy tyre, fibre-reinforced soil achieves the lower density than unreinforced soil. Hidalgo et al. [14] reported that irrespective of applied energy,

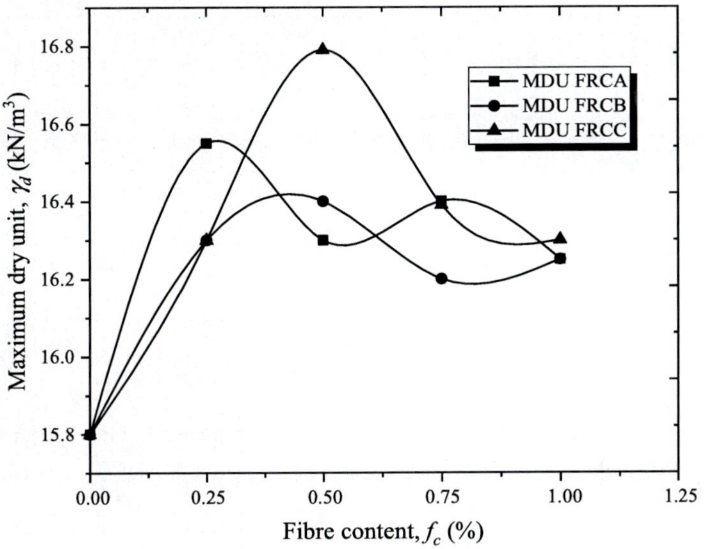

Fig. 3 Variation in maximum dry unit weight with respect to fibre content

the waste tyre fibre-reinforced clayey soil shows the linear trend in compaction behaviour. Ajmera et al. [15] reported that *MDU* of reinforced soil increases initially when mixed with 2–4% crumb rubber, and with further addition of crumb rubber, it decreases.

The variation in optimum mixing moisture content is illustrated in Fig. 4. It is observed that due to the high water absorption capacity and rough surface of tyre fibres, the reinforced soil requires more water for the proper mixing. Hence, the optimum mixing moisture content increases with an increase in fibre size as well as content. It is observed that in comparison with *FRCA* and *FRCB*, in *FRCC* required *OMMC* is higher. As reported by Mistry et al. [16], the surface morphology of the tyre fibres plays an important role in fibre-reinforced soils. In addition to high water absorption capacity, *FRCC* possesses more surficial roughness which absorbs more water in its surficial gaps. Few researchers [3, 13, 17–19] reported the similar observations. Seda et al. [20] and Özkul and Baykal [21] reported that irrespective of compaction energy, both unreinforced soil and tyre buffing reinforced soil demanded roughly same water content.

4.2 *Influence of Waste Tyre Fibres on Load–Penetration Curve in Unsoaked Condition*

The observations of *CBR* tests are illustrated in terms of the load–penetration curve in Fig. 5. It can be seen that in comparison with unreinforced soil, with an increase in

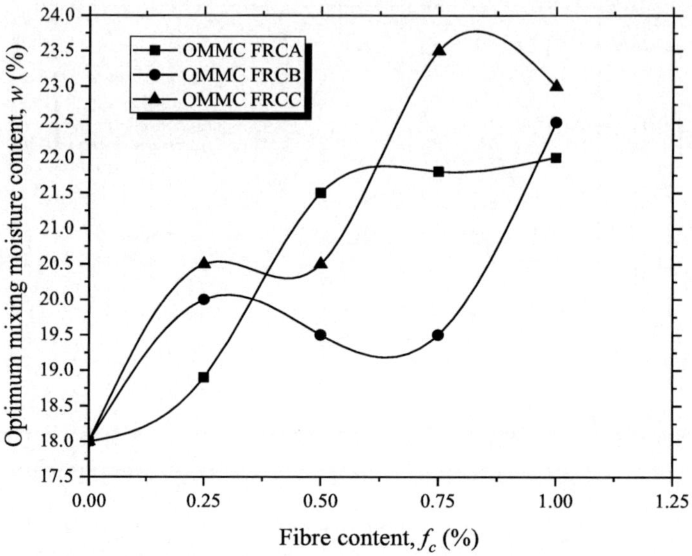

Fig. 4 Variation in optimum mixing moisture content with respect to fibre content

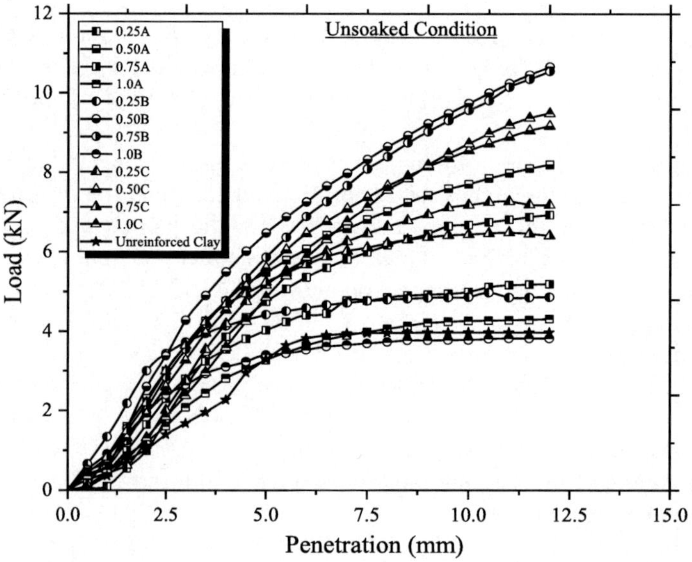

Fig. 5 Load versus penetration curve for varying percentage of tyre fibres in unsoaked condition

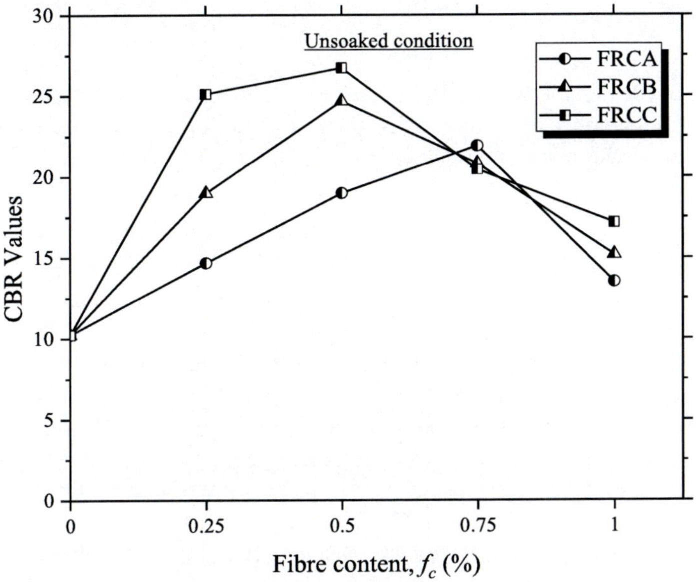

Fig. 6 Variation in *CBR* values with respect to fibre content

fibre size and content, the load absorption capacity of fibre-reinforced soil increases. The increase in penetration resistance capacity is attributed to the high energy absorption capacity of tyre fibres. The load absorption capacity of tyre fibres depends upon their thickness. In comparison with *TFA* and *TFB*, *TFC* is thicker and hence possesses more energy absorption capacity which further leads to higher resistance against penetration. Similar load–penetration behaviour was reported by Jha et al. [22] in their study of plastic waste fibre and by Mistry et al. [23] in their waste tyre fibre-reinforced subgrade materials. The influence in *CBR* values with varying fibre size and content is illustrated in Fig. 6.

The test results reveal that maximum *CBR* value was achieved with the 0.50% *TFC*. It can be seen that in case of *TFB* and *TFC*, *CBR* values get increased up to 0.50% fibre addition, and with further increment in fibres, it drops the *CBR* values, whereas in case of *TFA*, it drops after 0.75% fibre addition. The reduction in CBR values at higher fibre content may be attributed to the increased flexibility of fibre-clay matrix. It can be observed that for *TFB* and *TFC*, the behaviour of reinforced soil with respect to increase in fibre content was similar, whereas the improvement with *TFA* was very less in comparison with *TFB* and *TFC*. Similar trend in *CBR* value was reported by [4, 18, 24]. Ravichandran et al. [6] and Razali et al. [24] stated that reduced dry density of fibre-clay matrix was responsible for reduction in *CBR* value at higher tyre fibre content.

5 Conclusions

In the present study, the feasibility of waste tyre fibres was examined by conducting modified compaction tests and *CBR* tests. Based on the experimental results, following conclusions can be made:

- Irrespective of fibre content, the addition of waste tyre fibres in high plasticity clay soil shows the improvement in maximum dry unit weight.
- The increase in fibre size and content increases the optimum mixing moisture content significantly. The larger size of fibres increases the number of voids in fibre-clay matrix and hence water absorption.
- The high energy absorption capacity of tyre fibres increases the penetration resistance of fibre-clay matrix.
- The waste tyre fibres (except the smaller size of the fibre *TFA*) improve the *CBR* values up to 0.50% and there onwards it decreases with an increasing content.
- However, the maximum improvement in *CBR* value of clayey soil was achieved as 2.5 times with the addition of *TFC*.
- The considerable reduction in overall cost and thickness of flexible pavement can be achieved if the subgrade material is reinforced with waste tyre fibre and can be a one of the best tyre disposal options.

References

1. Eldin, N.N., Senouci, A.B.: Use of scrap tires in road construction. J. Constr. Eng. Manag. **118**, 561–576 (1992). https://doi.org/10.1061/(ASCE)0733-9364(1992)118:3(561)
2. Khan, R.A., Shalaby, A.: Performance of a road base constructed with shredded rubber tires, Proceedings. Annu. Conf. - Can. Soc. Civ. Eng. **2002**, 2513–2522 (2002)
3. Subramaniam, R.M., Jeyapriya, S.P.: Study on effect of waste tyres in flexible pavement system. Indian Geotech. Soc. Chennai Chapter, 20–24 (2009)
4. Cabalar, A.F., Karabash, Z., Mustafa, W.S.: Stabilising a clay using tyre buffings and lime. Road Mater. Pavement Des. (2014). https://doi.org/10.1080/14680629.2014.939697
5. Teja, S.S., Siddhartha, P.: Stabilization of subgrade soil of highway pavement using waste tyre pieces. IJIRSET, 3265–3273. https://doi.org/10.15680/IJIRSET.2015.0405045
6. Ravichandran, P.T., Prasad, A.S., Krishnan, K.D., Rajkumar, P.R.K.: Effect of addition of waste tyre crumb rubber on weak soil stabilisation. Indian J. Sci. Technol. **9** (2016). https://doi.org/10.17485/ijst/2016/v9i5/87259
7. Tajdini, M., Nabizadeh, A., Taherkhani, H., Zartaj, H.: Effect of added waste rubber on the properties and failure mode of kaolinite clay. Int. J. Civ. Eng. **15**, 949–958 (2017). https://doi.org/10.1007/s40999-016-0057-7
8. Shukla, S.K.: Fundamentals of fibre reinforced soil engineering, 1st edn. Springer Nature, Singapore (2017). https://doi.org/10.1007/978-981-10-3063-5
9. Mukherjee, K., Mishra, A.K.: The impact of scrapped tyre chips on the mechanical properties of liner materials. Environ. Process. **4**, 219–233 (2017). https://doi.org/10.1007/s40710-017-0210-6
10. Shukla, T., Mistry, M., Solanki, C., Shukla, S.K., Shukla, S.: Influence of randomly distributed waste tire fibres on swelling behaviour of expansive soils. In: Gali, M.L., Raghuveer Rao, P. (eds.), Problematic Soils Geoenvironmental Concerns. Springer Nature, Singapore (2020)

11. Mistry, M., Shukla, S., Solanki, C., Shukla, S.: Consolidation and hydraulic conductivity of high plastic clay reinforced with waste tyre fibres. J. Test. Eval. **49** (2020). https://doi.org/10.1520/JTE20200079
12. Bowles, J.E.: Physical and Geotechnical Properties of Soils. McGRAW-HILL International Book Company (1979)
13. Mistry, M., Shukla, T., Venkateswalu, P., Shukla, S., Solanki, C., Shukla, S.K.: A new mixing technique for randomly distributed fibre-reinforced expansive soil (2019). https://doi.org/10.1007/978-981-13-7010-6_15
14. Hidalgo, C., Garzón-roca, J., Martínez, P., Elvira, M., De, G., Insa, R.: Applied Clay Science Swelling potential reduction of Spanish argillaceous marlstone Facies Tap soil through the addition of crumb rubber particles from scrap tyres. Appl. Clay Sci. **132–133**, 768–773 (2016). https://doi.org/10.1016/j.clay.2016.07.027
15. Ajmera, et al.: Geotechnical Properties of Clays Modified with Recycled Crumb Rubber Beena. Geotech. Struct. Eng. Congr. **2016**(483), 483–489 (2016)
16. Mistry, M., Shukla, S., Solanki, C., Shukla, S.: Study of fibre-clay interface behaviour and reinforcing mechanism. Geotech. Geol. Eng. **38**, 1899–1917 (2020). https://doi.org/10.1007/s10706-019-01138-y
17. Vakili, R., Abdollahi, A.S., Jafarian, Y., Baziar, M.H.: Evaluating conservatism of SPT-based simplified liquefaction procedures using laboratory and field evidences. Electron. J. Geotech. Eng. **14 L**, 1–11 (2009)
18. Priyadarshee, A., Gupta, D., Kumar, V., Sharma, V.: Comparative study on performance of tire crumbles with fly ash and kaolin clay. Int. J. Geosynth. Gr. Eng. **1** (2015). https://doi.org/10.1007/s40891-015-0033-3
19. Mistry, M., Venkateshwarlu, P., Shukla, S., Solanki, C., Shukla, S.K.: Effect of placement of waste tyre fibres on unconfined compressive strength of clayey soil (2020). https://doi.org/10.1007/978-981-13-7480-7_11
20. Seda, J.H., Lee, J.C., Antonio, J., Carraro, H.: Beneficial use of waste tire rubber for swelling potential mitigation in expansive soils. Soil Improv. (2007)
21. Özkul, Z.H., Baykal, G.: Shear strength of clay with rubber fiber inclusions. Geosynth. Int. **13**, 173–180 (2006). https://doi.org/10.1680/gein.2006.13.5.173
22. Jha, J.N., Choudhary, A.K., Gill, K.S., Shukla, S.K.: Behavior of plastic waste fiber-reinforced industrial wastes in pavement applications. Int. J. Geotech. Eng. (2014). https://doi.org/10.1179/1939787914Y.0000000044
23. Mistry, M., Vasoya, A., Shukla, S., Solanki, C., Shukla, S.K.: Sustainable utilization of waste tyre products for pavement applications. In: Innov. Civ. Eng. Mater. Ed. B. Vol. as a Part Spec. Issue J. Second ASCE India Conference on 'CRSIDE 2020.' Cambridge Scholar Publishing (2020)
24. binti Razali, N., binti M. Sa'don, N., bin Karim, A.R.A.: Strength and durability effects on stabilized subgrade soil J. Civ. Eng. Sci. Technol. **7**, 9–19 (2016)

Effect of Installed Geotextile/Polyester and Biodegradable Materials for Dewatering Soft Clay

Flemmy Samuel Oye, Kiyoshi Omine, and Zicheng Zhang

Abstract Geotextile, polyester, and biodegradable (paper) materials are prominent in dewatering slurry sludge/sediment and soft clay. This is because of their high permissibility properties. The purpose of this study is to increase the efficiency of the dewatering setup by inserting geotextiles materials and paper into a dewatering setup with woven geotextile on the base before filling with soft clay, respectively. In this study, simple dewatering system by pouring soft clay into a 0.0135 m^3 dewatering box and tube with the installation of geotextile material and paper at 100 mm spacing to promote the dewatering setup efficiency is investigated. A soft clay of 200% water content was poured into the boxes and thereafter, geotextiles and paper were installed separately in the 0.0135 m^3 dewatering setup. The test results show that string polyester material in a tube dewatering setup can dewater the slurry clay below its liquid limit (142%) within a week, while box setup reduces the slurry clay water content its liquid limit after 14 days. Tube dewatering setup is more effective because it contains small volume of slurry clay. The suction test results show that the dewatering performance of a dewatering setup is subject to the dewatering days and the amount of void that can be reduce by the dewatering setup.

Keywords Geotextile container · Dewatering · Slurry clay consolidation · Suction pressure

1 Introduction

1.1 Research Background

Suitable lands and sustainable construction materials for civil engineering works are scarce. The scarcity of land and construction material is common in developed

F. S. Oye (✉) · K. Omine · Z. Zhang
Nagasaki University, Nagasaki 852-8521, Japan
e-mail: oyesamuel82@gmail.com

H. Hazarika et al. (eds.), *Advances in Sustainable Construction and Resource Management*, Lecture Notes in Civil Engineering 144,
https://doi.org/10.1007/978-981-16-0077-7_3

countries. The shortage of construction lands and high procurement cost of construction materials is due to rapid human growth, industrial growth, disposal law over the disposal of dredged sediment, and the excessive demands of infrastructural facilities.

Generally, dredged sediment is difficult to handle because of high amount of water in it. Disposing dredged sediment without treatment can need to percolation of unwanted substances into the soil and water. Therefore, dredged sediment most be treated before disposing or used as recycled landfilling material.

Several conventional dewatering methods like electroosmosis, centrifugal, mechanical and chemical dewatering methods have been used successfully to treat dredged sediment. These methods are energy intensive, costly, and environmentally unfriendly.

For keeping the dewatering environment safe and accelerate the dewatering time, vertical drains are used for treating soft clay soil. However, dewatering by installing vertical drain without preloading is not effective.

Naturally, dredged sediment is dewatered by sun drying. This method is totally environmentally friendly except that it takes longer time say 2–3 months to dewater dredged sediment satisfactorily and it requires large space of land for drying. In reducing land space requirements using sun drying dewatering method, geotextile bag or tube dewatering method is used. However, geotextile tube or bag can only dewater small volume of slurry and it is slow.

To accelerate the dewatering time and keeping the immediate environment free from percolation of unwanted chemical or substance into the soil or water, this study is proposed. In this study, geotextile, paper, and composite polyester drainage materials were installed separately in woven geotextile bag filled with slurry clay. This method is cheap, eco-friendly, and fast in removing water from slurry clay.

1.2 *Description of Soil Material*

The soil sample used in this study was taken from Kumamoto Prefecture, Japan. The soil samples were taken at the different spots of the site. This was done to know the slurry clay properties randomly. From the result of the specific density test conducted for sample A, B, and C, the test results show that the soil varies in specific density regardless that they were extracted from same location. Sample A has a specific density of 2.27 g/cm^3, sample B is 2.52 g/cm^3 while sample C is 2.51 g/cm^3. In this study, mixed sample C was used as the test sample and it is a composition of sample A and C because both constituents have vast differences in organic content and consistency as shown in Tables 1 and 2 of this paper.

Table 1 Clay sample ignition loss

Sample	Site A	Site B	Site C
Ignition loss	23.4	19.2	17.33

Table 2 Soil consistency test result

Sample	Liquid limit (%)	Plastic limit (%)	Plastic index
Sample A	121.98	88.12	39.86
Site C	158.0	96.92	61.08
Mixed sample (A and C)	142.0	97.01	44.99

2 Methodology

In this study, series of test was conducted as thus

1. Drainage material water absorption test
2. Tube dewatering test
3. 0.0135 m^3 dewatering box test
4. Field assessment of the drainage materials dewatering performances.

2.1 *Drainage Material Water Absorption Test*

In this study, numbers of drainage materials water absorption rate were investigated. This was done to know the drainage material with good absorption feature. The slurry soil of 200% water content was poured into a plastic tube of 750 mm length and 400 mm diameter. Then, each drainage material was installed in the tube filled with slurry. After this was done, the surface of the tube was sealed with plastic tape. The plastic seal prevents evaporation of water from the tube and allow water to drain out only through the installed drainage material 200 mm outlet provided. From the test result shown in Fig. 1, composite polyester string material also known as chemical polymer mixture was more effective in draining out water in the slurry clay. Therefore, composite polyester material was used in this study for dewatering slurry soil.

2.2 *Tube Dewatering Test*

This test was conducted to know the influence of clay density on the dewatering capacity of the installed drainage material. In this section of the test, slurry clay of 200% water content was poured into 76 mm diameter plastic tube. The tube contains 1200 g slurry clay. Then, paper drain (newspaper) and composite polyester string were installed into the tubes, respectively. The surface of the tube was sealed with plastic tape to prevent evaporation of water from the surface of the setup. The dewatering rate of the setup was measured in a period of 7 days using a digital weighing balance machine (Fig. 2).

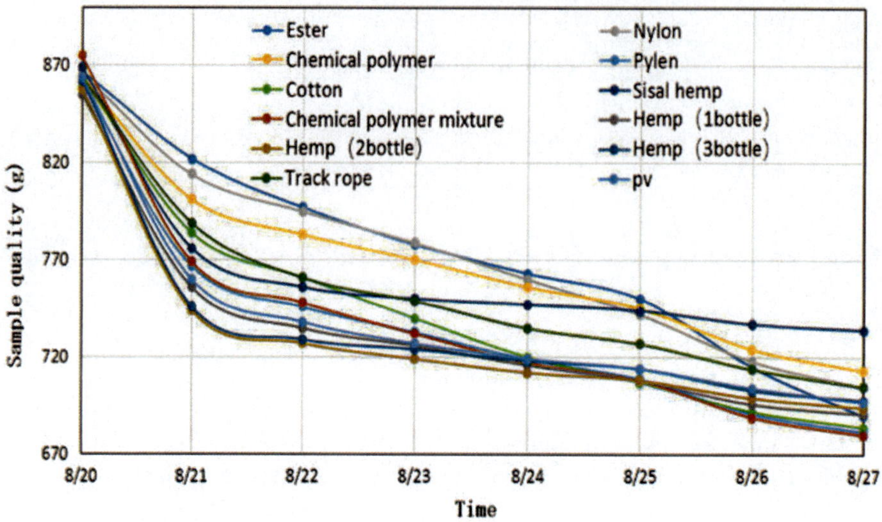

Fig. 1 Dewatering/drainage material water absorption rate

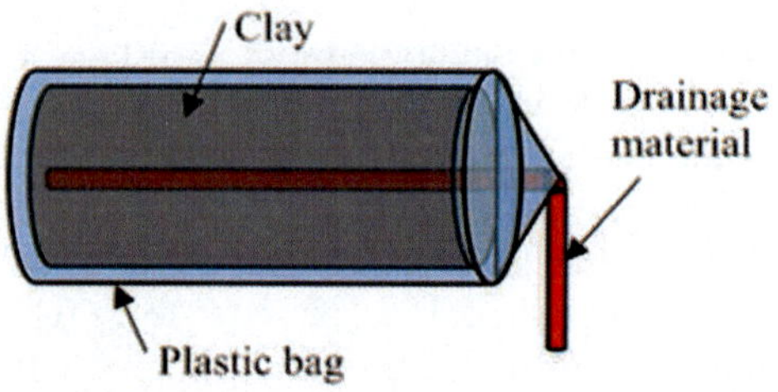

Fig. 2 Schematic tube test diagram

2.3 0.0135 m³ Dewatering Box Test

This is conducted to compare the effectiveness of the drainage material in small tube setup and 0.0135 m^3 dewatering box.

In this text, woven geotextile material was laid on the bottom of a dewatering box of 300 mm height, 450 mm long, and 150 mm wide. Then, slurry clay of 200% water content was poured into the 0.0315 m^3 dewatering box with woven geotextile material spread on it base. The soft soil in the box weighed 15,000 g. Thereafter, nonwoven geotextile material, paper, and composite polyester material were installed separately in the dewatering setups filled with slurry clay 100 mm apart. 200 mm outlet drainage material was provided. The provided outlet drainage material promotes secondary consolidation stage of the dewatering setup. Suction sensor was inserted in the dewatering setup for measuring strength reclamation as water deplete from the clay (Fig. 3).

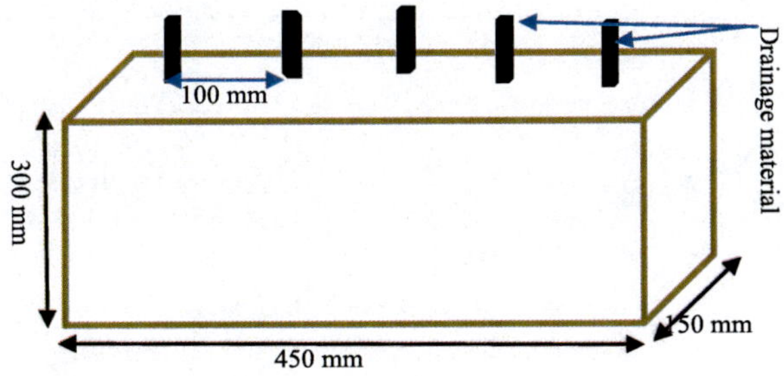

Fig. 3 0.0135 m^3 schematic dewatering box diagram

2.4 Field Assessment of the Drainage Materials Dewatering Performances

The dewatering effectiveness of vertical drains (nonwoven geotextile and composite drainage material) used in the study was investigated in the field. It was difficult to install paper in the field test, so it was not used.

During this, 145% natural water slurry clay was poured into 1 m by 1 m box made with woven geotextile material with the aid of excavating machine. Thereafter, nonwoven geotextile and composite drainage materials were installed separately in the 0.0153 m^3 dewatering setup filled with slurry clay at 100 mm apart. Suction sensors were installed in the dewatering setups filled with slurry clay and their suction readings were taken. The field test lasted for 31 days. Soil samples were taken in 7 days and 31 days for water content tests.

3 Results and Discussions

3.1 Influence of Slurry Volume on Vertical Drain Performance

As shown in Fig. 4, the installed vertical drains in tube setup dewatered the slurry clay below its liquid limit within 7 days, while 0.0135 m^3 dewatering setup could not dewatered the same slurry clay within 7 days but in 14 days. This different dewatering performance of the drainage material in both dewatering setups is caused by the different amount of slurry clay in the respective setup. According to Henbo et al. [4] and Radhakrishan et al. [6], the slurry clay closer to the drainage material dewater faster compared to those far from drainage material. In other words, if the water pressure length is short, water will move out from the sediment faster. Furthermore, the water absorption capacity of drainage material also influences dewatering rate. In tube and 0.0135 m^3 dewatering test, composite drainage material was more effective

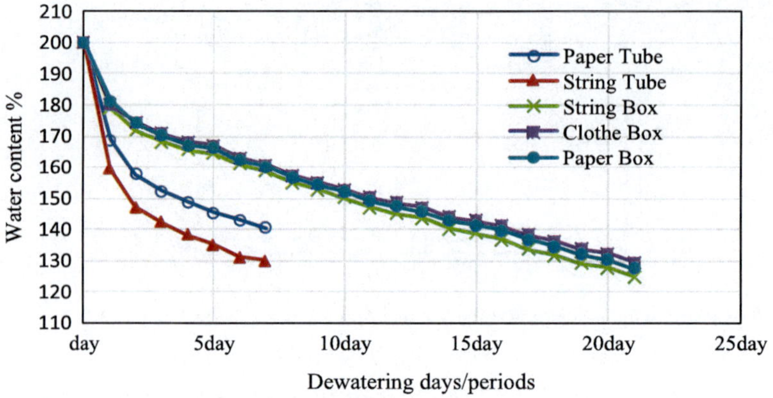

Fig. 4 Relationship between clay volume, dewatering rate, and time

than other dewatering materials used. This is so, because composite drainage material is made up of cotton and polymer material. According to Kevin [5], Flemmy et al. [3], and Bhatia et al. [2] work, every dewatering material has their different ability to absorb and drag out water from slurry. The choice of drainage material to use for treating slurry sediment depends on the soil properties of the slurry and the objective of the project.

3.2 Influence of Drainage Materials in Field (in Situ) Dewatering Condition

As shown in Figs. 5 and 6, the geotextile materials in the dewatering setups were unable to reduce the natural water content of the sample below it liquid limit within 7 days but in 31 days. This explains that slurry clay dewatering rate is influenced by dewatering time and depth. In addition, the soil sample closer to the drainage material dewaters quickly. This confirms that, reduction in the water pressure length of the soil by inserting vertical drains accelerates sediment dewatering time [1]. This result is good compared to past dewatering activities using geotextile tube dewatering system without vertical drain that takes 2 or 3 months to consolidate dredged sediment. The suction pF increases as the clay strength increases. During the first and second day of the test, the suction pF was zero, it increases to 35 kPa after 31 days.

3.3 Relationship Between Void Ratio and Suction pF

The suction of the treated slurry clay increases as the water in the sample decreases. In the first and second day of the dewatering test, the pF did not increase both in ex

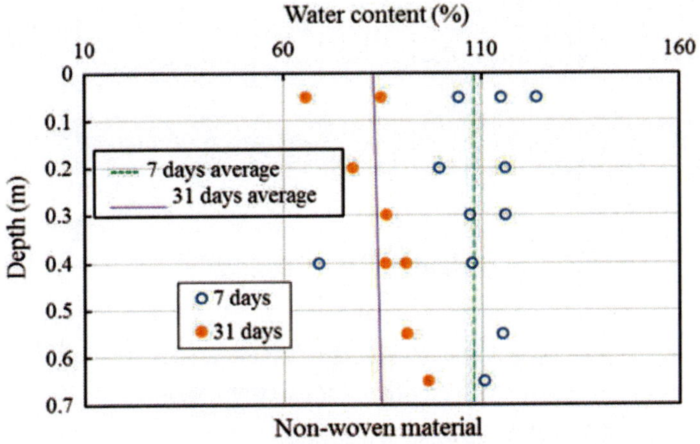

Fig. 5 Relationship between clay depth and nonwoven geotextile material dewatering rate

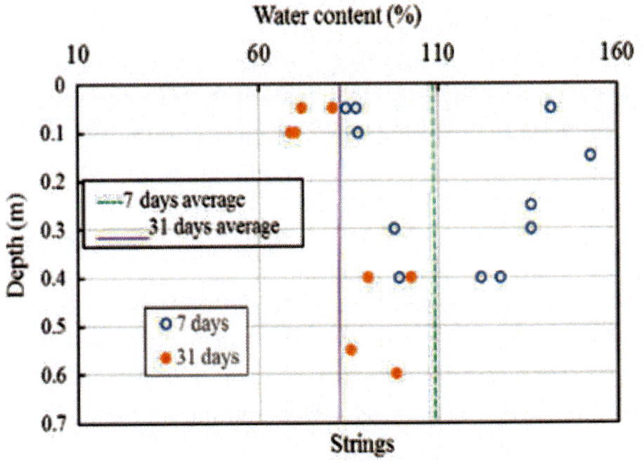

Fig. 6 Relationship between clay depth and composite polyester material on dewatering rate

situ and in situ dewatering conditions. However, the pF value increases to 25 kPa after 14 days in the ex situ dewatering test and 35 kPa after 31 days in in situ test condition.

The suction pF increasing rate is in relationship with the volume of void in the slurry clay. In the early stage of the test, the slurry is fully saturated. At this stage, the soil pore space is fully occupied by water. The water in the soil pore space reduced the soil's strength. However, as the water depletes from the soil pore replaced by fine particles, the pF increases. Therefore, the increasing rate of soil suction from the pF test is directly proportional to the extends at which the soil void (water) is reduced by the dewatering setup. The test result is shown in Fig. 7.

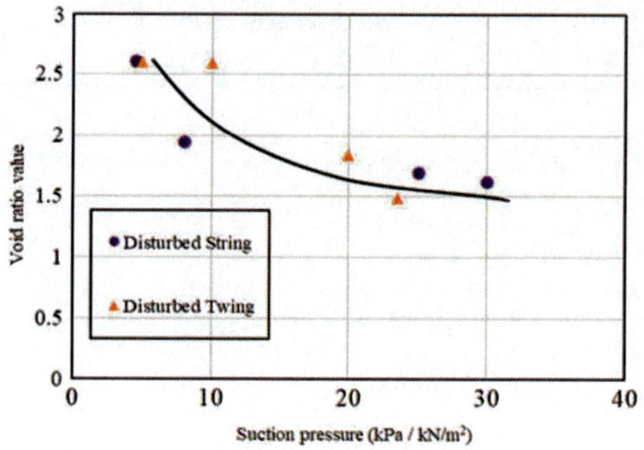

Fig. 7 Relationship between pF and void ratio

4 Conclusion

In this study, the important of installing vertical drain (geotextile, polyester, and paper) in dredged slurry is investigated. Installing drainage material in geotextile bag filled with slurry clay accelerates the dewatering rate. The drainage length and volume of clay in a dewatering setup yield low or fast dewatering rate. This is because soil water flow quickly to shorter drainage length. In addition, there is a quick removal of water from the dewatering setup with a small volume of slurry clay than those with large volume of clay.

The relationship between the pF and soil void describes the dewatering efficiency per day of the dewatering setups. In ex situ and in situ dewatering test, 25 and 35 kPa vacuum pressure can be obtained by installing vertical drains in the slurry clay without applying external force or energy on the treated slurry clay.

References

1. Ali, A., Mohammed, S., Ameer A.: Evaluation of vertical drain with different materials. In: International Conference on Civil Engineering, vol. 584, p. 1012011 (2019). https://doi.org/10.1088/1757-899X/584/1/012011
2. Bhatia, S.K., Smith, J.L., Christopher B.R.: Geotextile characterization and pore-size distribution: Part 3. Comparison of methods and Application to design. Geosynthetics Int. **3**(3), 301–328 (1996)
3. Flemmy, S.O., Omine, K., Zicheng, Z.: Soft clay improvement technique by dewatering and mixing sand soil. Int. J. Geomate **17**(63), 9–16 (2019)
4. Henbo, S.O.: Consolidation of clay with reference to influence of vertical drainage. Swedish Geotech. Inst. Proc. **18**, 160 (2004)

5. Kevin: Geotextile tubes and their application. Georgia Southern University (2014). https://digitalcommons.georgiasouthern.edu/etd
6. Radhakrishan, G., Kumar, M., Anjan, R., Prasada, G.V.R., Prasad, D.S.V., Venkateswarlu, D.: Study of consolidation accelerated by sand drains. Indian Geotechnical Conference, (2010)

Effect of Moisture Content on Particle Breakage of Recycled Concrete Aggregates During Compaction

Syed Kamran H. Shah, Taro Uchimura, and Ken Kawamoto

Abstract Construction and rehabilitation of roads demand a large volume of crushed rock which is used in granular layers of the road pavement. If recycled concrete aggregates which are extracted from construction and demolition waste satisfy all the engineering aspects of the material, then their use not only reduces the depletion of natural resources but also effectively helps to utilize the waste. However, the use of recycled concrete aggregate (RCA) as a granular material highly depends upon its performance under field loading conditions. The particle breakage is one of the important factors that affects the performance of recycled concrete aggregate, especially in long-term pavement performance. However, prior to long-term performance evaluation of RCA, the particle breakage during compaction process shall be estimated. In this regard, the present study focuses on the particle breakage of recycled concrete aggregate during compaction. Energy equivalent to modified proctor is used to investigate the particle breakage of recycled concrete aggregates. Moreover, influence of moisture content on the particle breakage during compaction is also studied. Results indicated that particle breakage is affected by variation in moisture content. With increase in moisture content, the particle breakage decreases.

Keywords Particle breakage · Recycled aggregates · Moisture content

1 Introduction

The large amount of construction waste is generated after the demolition of existing infrastructure or occurrence of natural disasters in recent times. This construction waste material mostly containing demolished concrete from roads and buildings is often dumped as waste. This large amount of construction waste generated causes a disposal problem. It is difficult to dispose of all waste materials in designated landfill sites due to the shortage of land with increasing world population and an increase in public resistance. To solve the problem of utilization of construction waste materials

S. K. H. Shah (✉) · T. Uchimura · K. Kawamoto
Graduate School of Science & Engineering, Saitama University, Saitama, Japan
e-mail: skamranhs@gmail.com

H. Hazarika et al. (eds.), *Advances in Sustainable Construction and Resource Management*, Lecture Notes in Civil Engineering 144,
https://doi.org/10.1007/978-981-16-0077-7_4

mixed with industrial by-products such as wastewater sludge and incineration ash in construction industry is also gaining attention [1].

Utilization of recycled aggregates in construction activities leads back to the end of World War II, when the extensive demolition of infrastructure happened and a high need to both get rid of the waste material and rebuild devastated countries especially Europe. In the 1970s, the USA began to reintroduce the use of RCA in non-structural uses, such as fill material, foundations, and base course material [2]. Since then, researchers became interesting regarding how viable RCA is as an option to replace unused natural aggregate (NA) in construction activities [3].

Since last few decades an increase in utilization of recycled aggregates has been observed as alternative to natural aggregate in pavement construction. The increase in demand and use of natural aggregates are due to depletion of natural resources, as for large construction projects a big amount of material is required. Therefore, more use of construction and demolition waste, especially recycled concrete aggregates (RCA) in geotechnical applications, is increasing [4].

Certainly, the use of different recycled materials and industrial by-products in construction has many benefits. Major benefits are minimizing the use of natural resources, reduction in carbon dioxide emissions to avoid damage to environment, and sustainable infrastructure. However, some barriers to adopt such materials especially in pavement construction still exist worldwide. Such barriers are that lack of clients' confidence in such materials, lack of specifications and legalization, etc. [5].

Long-term durability of the material is important. That is why past researches focused on the long-term durability of aggregates through element testing (triaxial test). Several studies have been done on particle breakage. Most of these previous studies mainly focused on particle breakage during triaxial or any other associated testing [6, 7], therefore giving less attention to the compaction induced breakage. It is obvious that the particle degradation starts from the very initial stage, i.e., compaction. Moreover, in case of recycle concrete aggregates (RCA), the importance to know the particle breakage is high in both phases, i.e., construction phase and service phase as its strength is generally less than the natural aggregates. Therefore, it is important to estimate the particle breakage of RCA at construction phase and service life phase (long term) separately. Hence, the objective of this study is to investigate the particle breakage during compaction phase before proceeding to access the long-term durability of recycled concrete aggregates (RCA) as pavement base material.

2 Material Properties

Recycled concrete aggregate (RCA) was collected from a recycling plant in Japan (Fig. 1). This material was sieved in the laboratory before use, and the material was adjusted from 19.0 to 0.075 mm in accordance with RM-30 gradation [8] suggested by Japan Road Association (2010) (refer Fig. 2). The properties of the RCA used are described in Table 1.

Fig. 1 Recycled concrete aggregate (RCA) graded in accordance with RM-30 (2010)

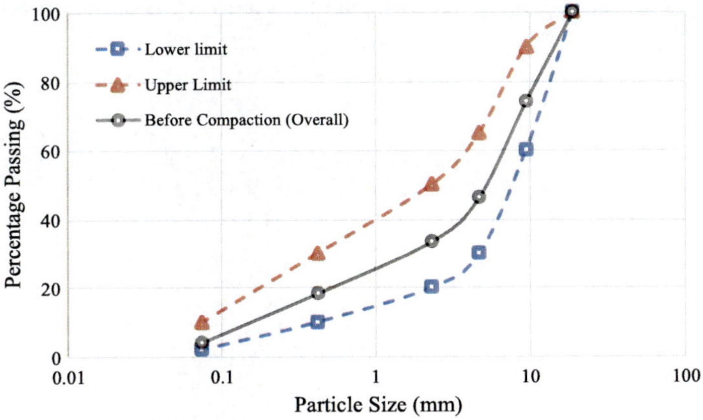

Fig. 2 Gradation curves of the RCA samples with upper and lower limit curves of RM-30 suggested by Japan Road Association (2010)

Table 1 Properties of RCA used

Material properties	
Specific gravity	2.59
Absorption of RCA (particles > 2.36 mm) [%]	5.9
Absorption of RCA (particles < 2.36 mm) [%]	15.7
Maximum dry density, MDD [g/cm^3]	1.84
Optimum moisture content (OMC) [%]	10

3 Experimental Setup and Testing Conditions

Compaction tests were conducted at modified proctor energy of 2700 kJ/m^3. The moisture content range varies from 5 to 15% in order to cover the dry and wet side of

Table 2 Experimental setup specifications and testing conditions

Experimental setup specifications			
Sample mold		*Rammer*	
Height (cm)	15	Drop height (cm)	45
Diameter (cm)	7.5	Weight (kg)	4.5
Test conditions			
Moisture content range (%)		5–15	
Compaction energy (kJ/m^3)		2700	

the optimum value. It consists of a two-way split sample mold and a rammer, details described in Table 2 and experimental setup shown in Fig. 3.

Fig. 3 Experimental setup, i.e., rammer and split mold

4 Testing Procedure

To assess the breakage of particles during compaction, coarser fraction of the material ranging from 19 to 4.75 mm was colored using fluorescent colors. The color codes used for three fractions are illustrated (refer Fig. 4a).

Series of compaction tests were conducted under different moisture contents ranging from 5 to 15% with compaction energy (2700 kJ/m^3) equal to modified proctor. At each target moisture content, material was compacted in three equal layers with designated compaction energy (Fig. 4b).

The purpose of using color scheme is to identify the particle size which are more prone to breakage due to compaction effort. Due to smaller size, fine particles ranging from 2.36 to 0.075 mm were not considered in color scheme.

5 Results and Discussions

A series of compaction tests was conducted at different moisture contents. After compaction, the sample was retrieved in layers and then each layer was sieved again. After sieving in accordance with layers, particles which shifted from one color scheme to another color scheme were picked up by color identification and weighed (refer Fig. 5.). With this methodology, it can be clearly observed that the fraction size which has moved to the next size range and actual breakage or mass reduction can be calculated.

The particle size distribution curve shifted toward left from the initial position after compaction test (refer Fig. 6). It is observed that the shift of curve is due to particle breakage. It is also observed that the particle breakage decreases with the increase in moisture content. It is presumed that with the increase in moisture content water provides a cushion and/or slippage surface around the particles which result in less breakage of particles.

There are certain reasons which may affect this behavior. One of the facts for more breakage in dryer state might be that particles are in direct contact and there is more surface friction. Due to this close contact, impact from particle to particle causes more breakage. However, in case of higher moisture content, there is a chance that water provides a cushion around the particles which also act as lubrication.

In order to identify the particle size that contribute more in breakage, coloring technique was used.

For easy understanding, material is divided into two portions, coarse fraction ranging from 19.0 to 2.36 mm and finer fraction 2.36–0.075 mm. As shown in (Fig. 7), the right axis shows the mass accumulation in fine fraction whereas the left axis shows the mass reduction in coarse fraction. The results indicated coarser fraction such as 19.0, 9.5, and 4.75 mm broke and this breakage decrease with increase in moisture content. The mass reduction percentage is described as;

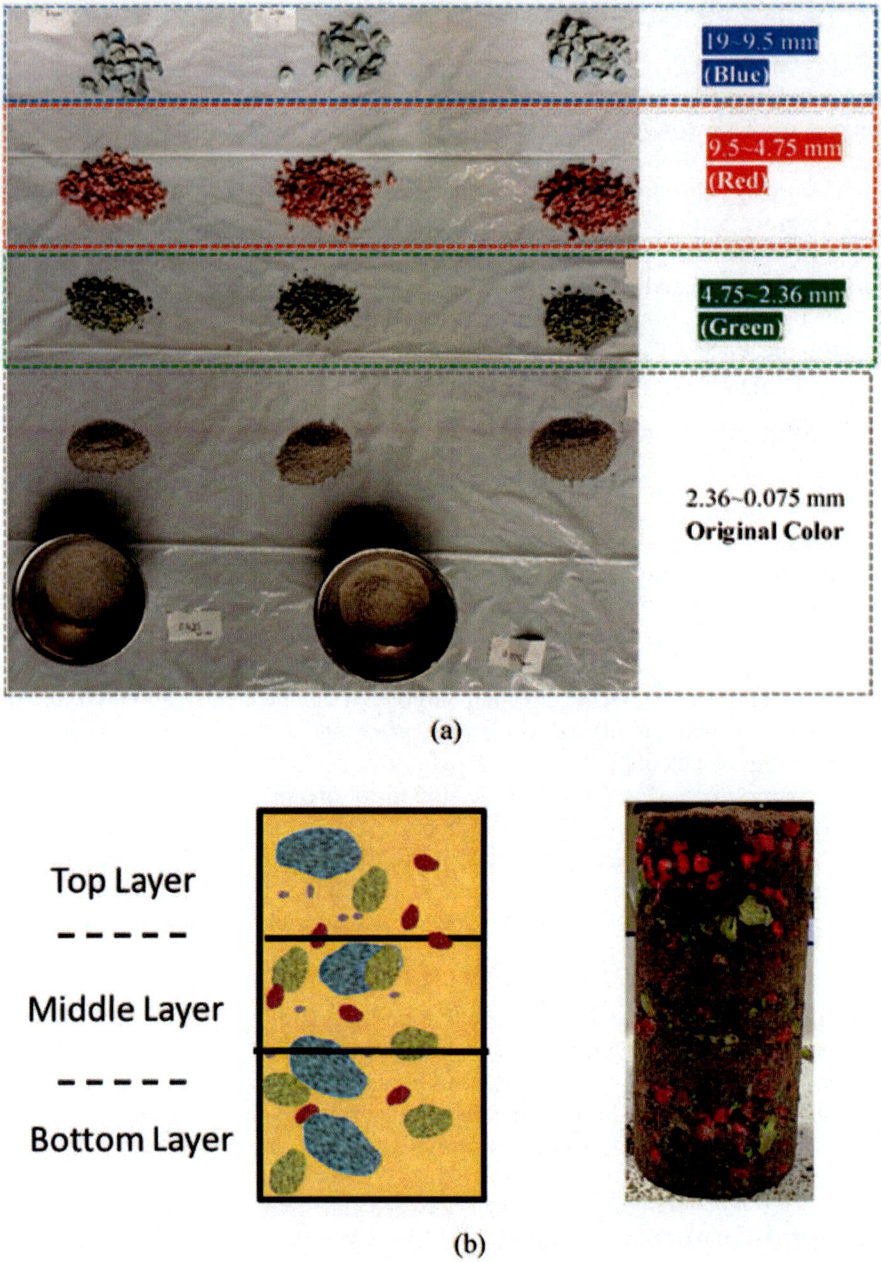

Fig. 4 **a** Color scheme for each particle size range and **b** layerwise illustration and prepared specimen

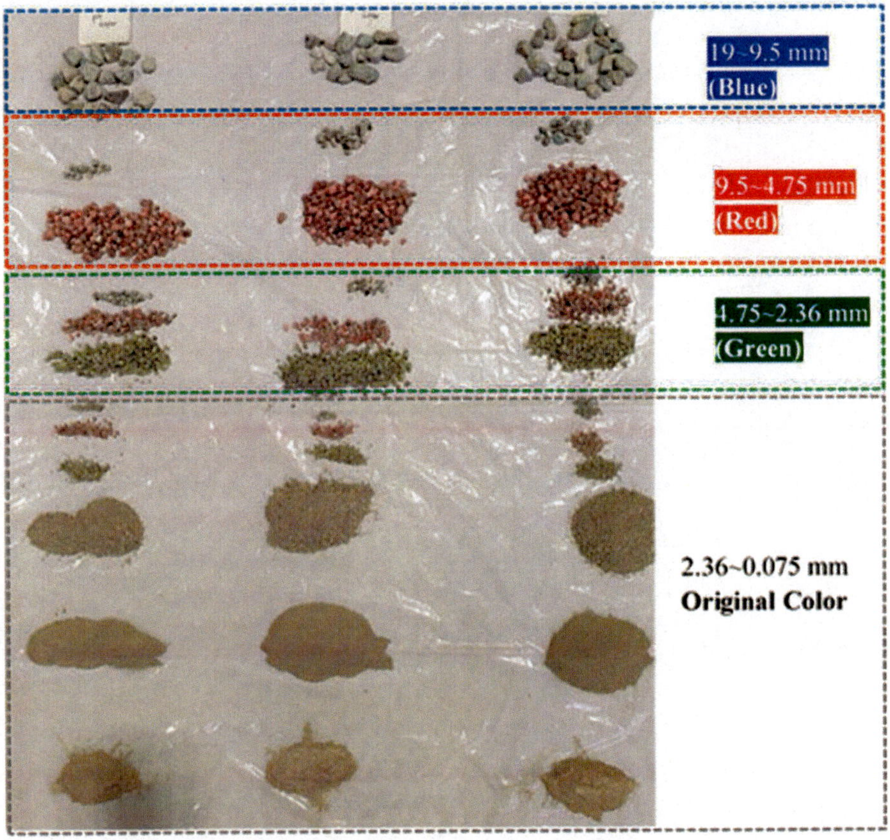

Fig. 5 Each particle size fraction range after compaction test based on color scheme

$$\text{Mass reduction (\%)} = \left[(\text{Initial mass} - \text{final mass})/\text{initial mass} \times 100\right] \quad (1)$$

The finer particles (2.36–0.075 mm) did not show considerable breakage but their final mass increased due to the accumulation of broken particles from top sizes of coarse fraction (19.0, 9.5, and 4.75 mm) [9]. Particularly at optimum moisture content (OMC) particle size range 4.75–2.36 mm of the coarse fraction shows highest mass reduction as compared to 19.0–9.5 mm and 9.5–4.75 mm size ranges of coarser fraction.

In order to observe the effect of moisture content on overall breakage of particles, Marsal's breakage index is adopted [10]. The Marsal's breakage, B_g, is a quantitative breakage index and is defined as the percentage by weight of the solid phase that has broken.

$$B_g = \sum (\Delta W_k > 0) \quad (2)$$

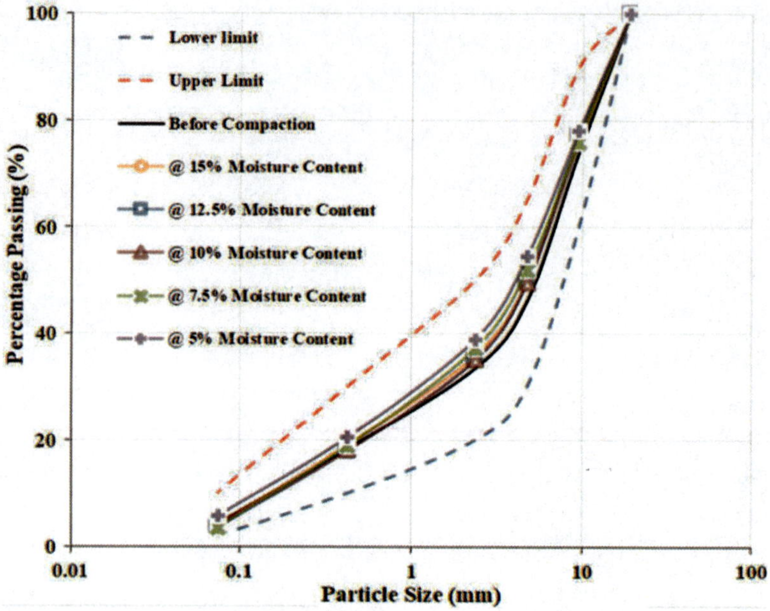

Fig. 6 Particle distribution after compaction test

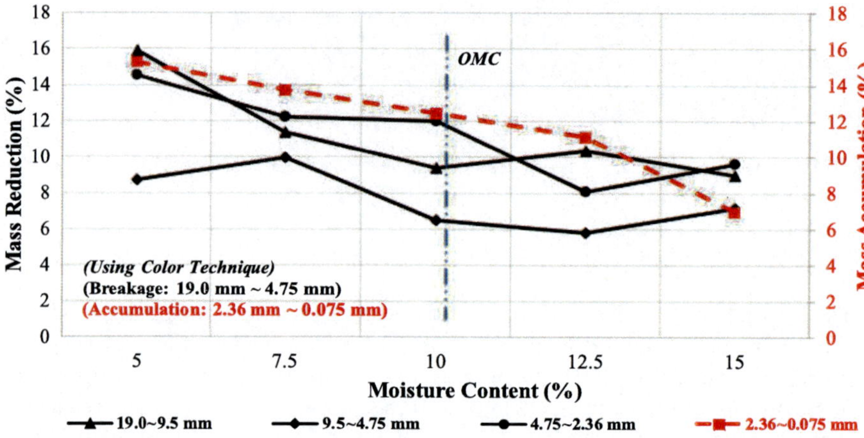

Fig. 7 Mass reduction and accumulation using color technique

and

$$\Delta W_k = W_{ki} - W_{kf} \tag{3}$$

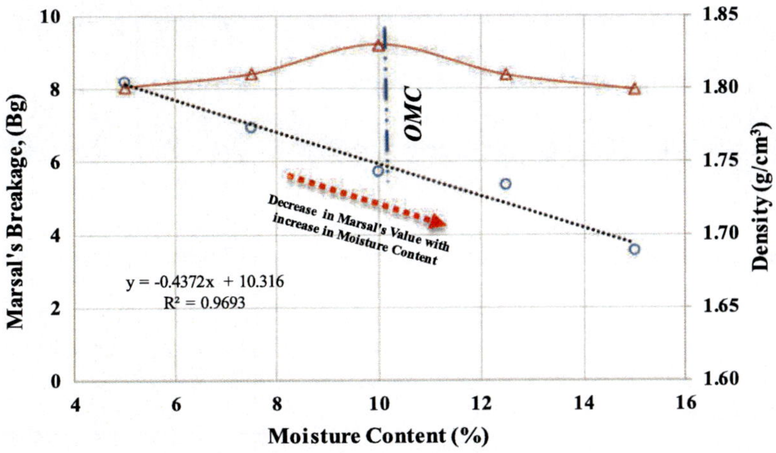

Fig. 8 Marsal's breakage index as a function of moisture

wherein W_{ki} is the percentage of sample weight retained before test in each sieve size and W_{kf} is the percentage of sample weight retained after test in each sieve size, respectively.

The Marsal's index showed a clear relationship between breakage index and moisture contents as shown in Fig. 8.

6 Conclusions

The current study focused on the particle breakage of recycled concrete aggregates (RCA) during compaction process. The results of the compaction tests performed indicate that:

(1) Breakage of particles decreases with increase in moisture content as evident from the particle size distribution curves drawn before and after the test. Therefore, it is important to consider the particle breakage of recycled concrete aggregates (RCA) at compaction stage prior to studying the long-term durability as a base material.

(2) The coloring technique adapted in this study is capable to indicate the particle size more prone to breakage and particle tracing is possible to track the particle movement after testing from one size range to the other especially in coarse fractions.

(3) At optimum moisture content (OMC), smaller size range of coarse fraction shows more breakage as compared to other coarse fraction size ranges.

(4) Mass accumulation happens in finer content due breakage of large particles as indicated by the coloring technique used.

(5) Marsal's breakage index, B_g, also showed a clear relationship between overall breakage and moisture content.
(6) The coloring technique and Marsal's breakage index, B_g, shows a good agreement in terms of effect of moisture content on particle breakage.
(7) Along with the traditional breakage indices, the coloring technique can be used to study the breakage of particles in long-term durability tests.

Acknowledgements This work is partially supported by the SATREPS project of Japan Science and Technology Agency (JST) and Japan International Cooperation Agency (JICA).

References

1. Iqbal, M.R., Hashimoto, K., Tachibana, S., Kawamoto, K.: Geotechnical properties of sludge blended with crushed concrete and incineration ash. Int. J. Geomate **16**(57), 116–123 (2019)
2. Buck, A.D.: Recycled concrete as a source of aggregate. ACI J. **74**, 212–219 (1977)
3. McNeil, K., Kang, T.H.K.: Recycled concrete aggregates: a review. Int. J. Concr. Struct. Mater. **7**(1), 61–69 (2013)
4. Abukhettala, M.: Use of recycled materials in road construction. In: Proceedings of the 2nd International Conference on Civil, Structural and Transportation Engineering, ICCSTE'16. Paper No. 138. Ottawa, Canada (2016)
5. El-Badawy, S., Gabr, Ala., El-Hakim, R.A.: Handbook of Eco-materials. Springer (2017)
6. Zhang, Z., et al.: Experimental evaluation on aggregate particle breakage in unbound granular mixture compaction. In: Proceedings of 1st International Conference on Transportation Infrastructure and Materials, ICTIM 2016. DPI-Proceedings. Xi'an, China (2016)
7. Indraratna, B.: Effect of confining pressure on the degradation of ballast under cyclic loading. Geotechnique **55**(4), 325–328 (2005)
8. RM-30, Gradation Standard for Recycled Materials: Japan Road Association (JRA) (2010) (in Japanese)
9. Sun, Y., Nimbalkar, S., Chen, C.: Particle breakage of granular materials during sample preparation. J. Rock Mech. Geotech. Eng. **11** (2019)
10. Tawk, M., Indraratna, B., Rujikiatkamjorn, C., Heitor, A.: Review on compaction and shearing-induced breakage of granular material. In: Geotechnics for Transportation Infrastructure, vol. 2, Part 2, pp. 259–270. Springer Publication (2019)

Environmental Implications of the Recycling of End-of-Life Tires in Seismic Isolation Foundation Systems

Laura Banasiak, Gabriele Chiaro, Alessandro Palermo, and Gabriele Granello

Abstract The extensive generation of end-of-life (ELT) tires worldwide has resulted in adverse environmental effects and threats to public health and safety including tire fires and leaching of contaminants into the soil environment, groundwater, and surface water. It is becoming imperative to investigate more sustainable and large-scale opportunities for the reuse of ELTs. One novel engineering solution is their reuse in structures with enhanced seismic resilience. This is particularly important in countries such as New Zealand and Japan where past earthquakes have caused widespread damage and socioeconomic loss. Research is being carried out to investigate a seismic-resilient engineered foundation-soil system for low-to-medium-density low-rise residential housing composed of a layer of granulated tire rubber (GTR) and gravel, and a flexible rubber–concrete raft foundation. It is essential to ensure that such innovations do not result in long-term negative impacts on the environment as tires contain hazardous compounds used in their manufacturing and the steel fiber within the tires can leach heavy metals (e.g., zinc, manganese, lead, cadmium). Preliminary batch leaching tests undertaken on (1) GTR and (2) GTR: gravel mixtures (20–40% rubber content by volume) indicated leaching of calcium (Ca), sodium (Na), magnesium (Mg), and potassium (K) which was contributed to the gravel and low levels of zinc (Zn), which were attributed to the tires.

Keywords Waste tires · End-of-life tires · Tire leachate · Foundation systems

1 Introduction

The generation of end-of-life (ELT) tires worldwide has given rise to large piles of ELTs that do not readily degrade. The current rate of ELT production in New

---

L. Banasiak (✉)
Institute of Environmental Science and Research (ESR), Christchurch, New Zealand
e-mail: laura.banasiak@esr.cri.nz

G. Chiaro · A. Palermo · G. Granello
University of Canterbury, Christchurch, New Zealand

H. Hazarika et al. (eds.), *Advances in Sustainable Construction and Resource Management*, Lecture Notes in Civil Engineering 144,
https://doi.org/10.1007/978-981-16-0077-7_5

Zealand is over 5 million/year and of this an estimated 70% (3.5 million) are destined for landfills, stockpiles, illegal disposal or are otherwise unaccounted for [1]. The accumulation of ELTs has resulted in adverse environmental effects and threats to public health and safety. Problems associated with ELTs include tires as breeding grounds for pests (e.g., mosquitoes, rodents) [2, 3], tire fires and air emissions from tire fires [4], and the leaching of contaminants into groundwater sources or nearby waterways [5, 6]. Due to these issues, it is necessary to investigate sustainable and large-scale opportunities for the recycling and reuse of ELTs.

While there are several recycling methods available for ELTs, i.e., retaining walls [7], reinforcing layers in earthfill [8], artificial reefs [9, 10], additives in road pavement [11], playground surfaces [12], rubber roofs [13], draining systems in landfills [14], and fuel for cement kilns [15] only 1.5 million tires/year are exported or reused/recycled in New Zealand and there are no national regulations currently in place to efficiently manage the recycling of ELTs. One novel engineering solution is their reuse in structures with enhanced seismic resilience [16]. This is particularly important in countries such as New Zealand and Japan where past earthquakes have caused widespread damage and socioeconomic loss.

The Institute of Environmental Science and Research Ltd (ESR) and the University of Canterbury are undertaking a multidisciplinary joint research project to recycle ELTs in an eco-rubber geotechnical seismic isolation (ERGSI) foundation system for low-to-medium-density low-rise residential housing [17]. The foundation system will be composed of a layer of granulated tire rubber (GTR) and gravel, and a flexible rubber–concrete raft foundation. Five methodologies will be implemented during this project, namely: (1) geotechnical laboratory investigations to understand the macro-mechanic properties of the GTR:gravel mixtures; (2) structural laboratory tests to identify the mechanical characteristics of rubber–concrete; (3) environmental laboratory tests to identify/quantify the degradation profile of the GRT:gravel mixtures; (4) numerical modeling to optimize the proposed foundation system; and (5) proof-of-concept testing of the physical model of the ideal foundation system. Further details about the project can be found in [17, 18]. While the introduction of recycled waste materials in building foundations may have benefits in terms of cost reductions and increased seismic resilience [19, 20], it is essential to ensure that such innovations do not result in long-term negative impacts on the environment resulting from leaching of contaminants.

The composition of a tire will determine the contaminants in tire leachate. Tires are predominantly composed of carbon black, vulcanized rubber, rubberized fabric containing reinforcing textile cords, antioxidants, silica, pigments, process and extender oils, accelerators, and steel wire [21] and may contain petroleum residues acquired through use [5]. While tire rubber itself can be considered inert under ambient foundation conditions [1], tires contain approximately 1.5% by weight of hazardous compounds. This includes additives used in the tire manufacturing process (e.g., organohalogen compounds and acidic solutions) and heavy metals (e.g., Zn, Mn, Pb, Cd) from the steel fibers [21].

A comprehensive literature review has been undertaken to determine if the GTR-gravel mixtures are likely to leach hazardous contaminants. Zn has been widely identified as an element of environmental concern in tire leachate studies [6]. When the steel components of tires are exposed, there may be elevated Mn and Fe levels within tire leachate [22]. Tire leachate may be toxic to some fish species, bacteria, invertebrates, and green algae [23], and studies have shown that tires immersed in fresh water can leach chemicals that cause endocrine disrupting activity in aquatic organisms [24]. In contrast, tire rubber also has a significant capacity for sorption of organic and inorganic contaminants [25], thus making them a potential material for environmental protection and remediation when in contact with contaminated waters.

The main objectives of this study are to examine the leaching of selected inorganic constituents and organics from mixtures of granulated tire rubber and gravel under representative water chemistry conditions. Specifically, this research will investigate: the leaching of total dissolved organic carbon from the tire rubber and gravel mixtures; the leaching of organic and inorganic constituents from the tire rubber and gravel mixtures as a function of time during one month period; the effect of rubber particle size on leaching; and the effect of gravel particle shape on leaching. Environmental Laboratory Testing.

1.1 Test Materials

For the batch leaching tests, two different solid test materials were used:

Test material 1: Two sizes of GTR were investigated, namely: R1 6–8 mm and R3 1–2 mm. A rubber content of 40% (by volume) was used in all the batch leaching tests (except for the blank tests).

Test material 1: Two shapes of gravel were investigated, rounded, and angular.

Prior to undertaking the batch leaching tests, the tire rubber and gravel were rinsed with deionized (DI) water in order to remove any dirt, sand, grit, and other foreign particles.

1.2 Batch Test Procedure

Batch leaching tests of selected GTR:gravel mixtures were conducted to determine if they contain contaminants that may cause environmental harm if leachable. The factors that were tested are the rubber content (20–40% content by volume), size of the tire chips, and the contact time with the aqueous solution. See Table 1 for the list of tests to be undertaken in this project (only results for Tests #1–3 shown). For each test variation, i.e., gravel size and shape (apart from the blanks), three replicates will be undertaken. Blank tests with GTR and DI water (Test #13), and gravel and DI water (Test #14–17) will also be undertaken to be able to ascertain the contribution of leaching from the GRT and/or gravel. Minimal leaching is expected from the inert gravel, which is predominantly greywacke.

Table 1 GTR:gravel batch leaching tests to be undertaken (Bolded text—results presented in this paper)

Test #	Tire rubber (volume, %)	Gravel (volume, %)	Gravel size D_{50}	Gravel shape
1	**40**	**60**	**R1, 6–8 mm**	**Rounded**
2	**40**	**60**	**R1, 6–8 mm**	**Rounded**
3	**40**	**60**	**R1, 6–8 mm**	**Rounded**
4	40	60	R3, 1–2 mm	Rounded
5	40	60	R3, 1–2 mm	Rounded
6	40	60	R3, 1–2 mm	Rounded
7	40	60	R1, 6–8 mm	Angular
8	40	60	R1, 6–8 mm	Angular
9	40	60	R1, 6–8 mm	Angular
10	40	60	R3, 1–2 mm	Angular
11	40	60	R3, 1–2 mm	Angular
12	40	60	R3, 1–2 mm	Angular
13–Blank	100	0	–	–
14–Blank	0	100	R1, 6–8 mm	Rounded
15–Blank	0	100	R3, 1–2 mm	Rounded
16–Blank	0	100	R1, 6–8 mm	Angular
17–Blank	0	100	R3, 1–2 mm	Angular

To determine the batch contact time to be used in the leaching tests, Tests #1–3 were firstly undertaken and leachate samples analyzed. The appropriate contact time for the remaining tests corresponds to the point at which leaching of constituents from the GTR:gravel mixtures in Test #1 reaches a plateau.

For Tests #1–3, 40% tire rubber (by volume) and 60% R1 rounded gravel (by volume) were soaked in DI water at room temperature for a period of 1 month. Three GTR:gravel mixes were placed in 5 L wide mouth glass bottles and filled with 2 L of leaching solution (DI water). The mass volume conversion for the GTR used in Tests #1–3 is shown in Table 2. The appropriate volume of (1) GTR and (2) GTR–gravel mixtures (according to Table 1) was placed in glass containers with deionized water for 30 days. For Test #103, 100 g GTR and 355 g gravel were used.

The 5 L glass test bottles were sampled periodically (regular sampling days: 0, 0.5, 1, 3, 5, 7, 9, 12, 15, 18, 20, 25, and 28; total of 34 samples). The pH and electrical conductivity of the solution in the glass test bottles were measured and recorded at the beginning of the test and before a sample was taken using a YSI Pro Plus Quatro meter with pH and conductivity sensors. Each sample was filtered with syringes (50 mL Terumo® syringe) through a sterile 0.45 μm membrane filter (Millex® HA filter unit, MF-Millipore™ PVDF membrane, Merck Millipore) into two separate bottles: 100 mL acid washed glass bottles) for total organic carbon (TOC) and ultraviolet absorbance at 254 nm (UV254) and 30 mL Thermo Scientific Nalgene™ Wide-Mouth Lab Quality HDPE bottles for the analysis of inorganics/metals. The samples

Table 2 Mass volume conversion table for GTR (R1, $D_{50} = 6–8$ mm) used in batch leaching tests. Bolded text-GTR used Test #1–3

GTR content (volume, %)	GTR content (weight, %)	Min. density (g/cm^3)	Max. density, proctor (g/cm^3)	Mass of GTR in test (g)	Mass of gravel in test (g)
0	0	1.57	1.75	–	–
10	5	1.44	1.62	100	1900
25	12	1.28	1.49	100	733
40	**22**	**1.09**	**1.31**	**100**	**355**
100	100	0.51	0.65	100	0

in the glass bottles were acidified using sulfuric acid (H_2SO_4) to pH < 2. The sample in the HDPE bottles were acidified with nitric acid (HNO_3) to pH < 2. Following acidification, the samples were refrigerated until analysis.

1.3 Leachate Analysis

The components of interest to be analyzed in the tire leachate included: DOC, UV254, zinc (Zn), cadmium (Cd), chromium (Cr), arsenic (As), potassium (K), phosphorus (P), sodium (Na), manganese (Mn), iron (Fe), calcium (Ca), magnesium (Mg), aluminum (Al), copper (Cu), and lead (Pb). Inorganic constituents/metals were measured using inductively coupled plasma-mass spectroscopy (ICP-MS). TOC was measured using a Shimadzu TOC high temp combustion analyzer equipped with an autosampler.

2 Preliminary Results

2.1 PH and Electrical Conductivity

Figure 1 shows the pH and EC of the leachate from Test #1–3. The pH of the leachate remained relatively stable throughout the batch leaching tests (Test #1: pH 5.99 ± 0.20, Test #2: pH 6.09 ± 0.18, Test #3: pH 6.16 ± 0.15), whereas the EC of the leachate significantly increased from 13.7–19.8 to 29.2–40.8 μS/cm between Day 0 and Day 7, whereby it continued to gradually increase. EC in the leachate did not seem to reach a plateau, with the EC of leachate in Test #1 continuing to increase significantly to 68.4 μS/cm at Day 28. This is indicative of leaching from the tires.

Figure 2 shows that the TOC concentration of the leachate increased significantly in the first day of the tests from 0.93 to 7.19–9.14 mg C/L. However, between Day 1 and Day 28, the TOC concentration, although increased, not significantly.

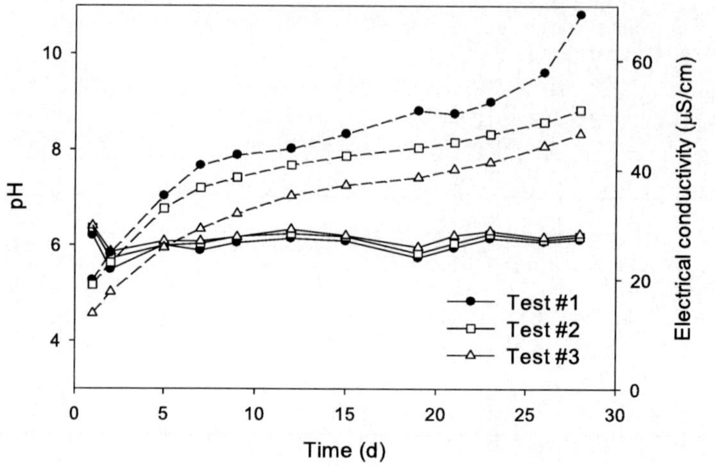

Fig. 1 pH and EC (μS/cm) of leachate in batch leaching tests (Test #1–3; GTR R1, $D_{50} = 6–8$ mm; gravel rounded)

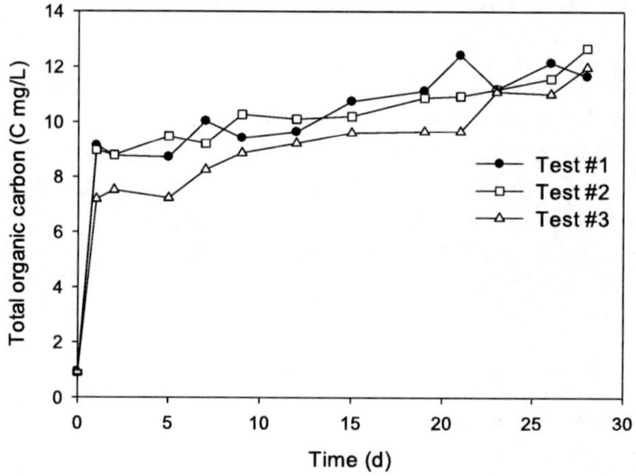

Fig. 2 Total organic carbon (TOC, mg C/L) in the leachate in batch leaching tests (Test #1–3; GTR R1, $D_{50} = 6–8$ mm; gravel rounded)

2.2 Metals

Figure 3 shows the average concentration of elements in the leachate during Tests #1–3. The predominant element leached was Ca (maximum concentration C_{max} = 5772 μg/L, followed by Na (C_{max} = 2091 μg/L), K (C_{max} = 1644 μg/L), Zn (C_{max} = 1581 μg/L), and Mg (C_{max} = 540 μg/L). All of these concentrations were observed in Test #1. The Ca, Na, and K were due to the presence of the greywacke gravel, which is composed of sandstone, quartz, and feldspar. The presence of Zn was attributed

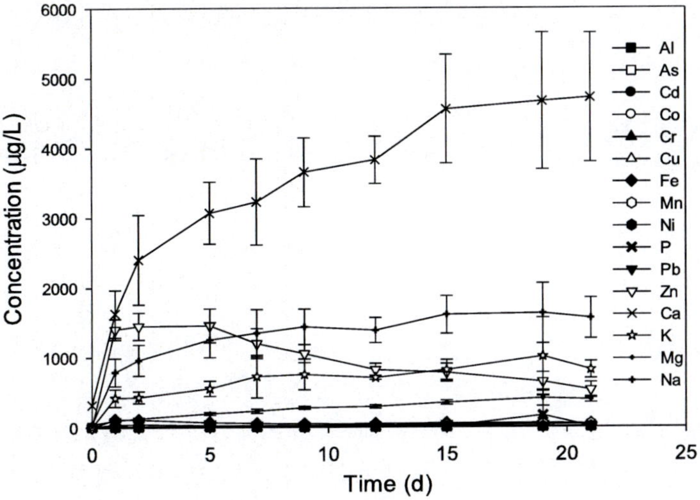

Fig. 3 Average element concentration (μg/L) in leachate during batch leaching tests (Test #1–3; GTR R1, $D_{50} = 6–8$ mm; gravel rounded)

to the tires, as zinc oxide is used as a vulcanization aid in the tire rubber production process [25] and the concentration was consistent with previous studies [5, 6, 27].

The Zn C_{max} observed during these batch tests (1581 μg/L) was slightly above the NZ drinking water aesthetic guideline value of 1500 μg/L (taste threshold value) but below the value at which the presence of Zn may affect the appearance of the water (3000 μg/L) [28]. The test results were also assessed by comparison to NZ landfill waste acceptance criteria (WAC) for Class A and B total concentration and leachability limits criteria. Class A landfills are sited in areas that reduce the potential for adverse environmental effects, have engineered systems designed to provide a degree of redundancy for leachate containment, and collect landfill leachate and landfill gas. While Class B landfills have limited or no engineered systems designed to collect landfill leachate or gases and may be in areas that pose a risk to the environment (i.e., over highly permeable sands and/or gravels, active faults, or floodplains) [29]. The concentration of Zn in the leachate in Tests #1–3 was below the leachability limit for Class A landfills (10^5 mg/L) but above the leachability limit for Class B landfills (1 mg/L). Therefore, the GTR:gravel mixtures used in Tests #1–3 could only be disposed of in Class A landfills or in Class B landfills following pretreatment.

Since the concentration of leachate from tires is highly dependent on the exposed surface area, further tests undertaken on the GTR:gravel mixtures (Tests #4–6, Tests #10–12 with smaller particle size and greater surface area) could potentially leach more contaminants.

The total mass of metals leached during the experiments (results available only up until Day 21) is shown in Fig. 4. Again, the predominant element leached was Ca with an average mass of 143 ± 14.4 mg (315 ± 31.6 mg/kg GTR:gravel material used in test), followed by Na (54.1 ± 8.11 mg, 119 ± 17.8 mg/kg), Zn (42.9 ±

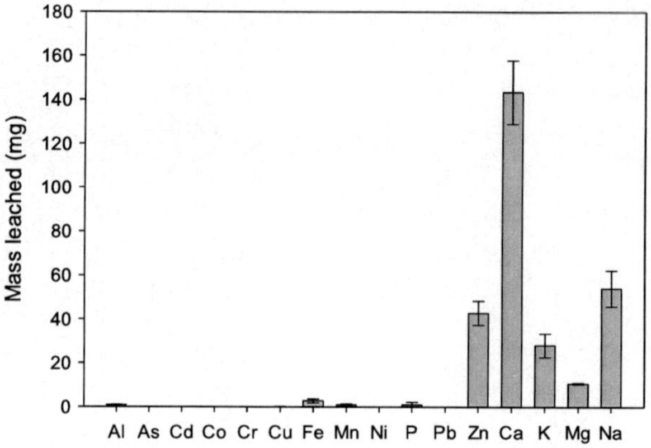

Fig. 4 Mass of metals (mg) leached during batch leaching tests up until Day 21 (Test #1–3; GTR R1, $D_{50} = 6$–8 mm; gravel rounded)

5.42 mg, 94.3 ± 11.9 mg/kg), K (28.0 ± 5.41 mg, 61.6 ± 11.9 mg/kg), and Mg (10.7 ± 0.37 mg, 23.4 ± 0.8 mg/kg). Only a small amount of Fe was found in the leachate (2.76 ± 0.89 mg), with the other elements of concern less than 1 mg.

3 Conclusions

The disposal of approximately 5 million ELTs yearly in New Zealand, either through landfills or through illegal disposal, is a serious environmental concern. The project on the "eco-rubber geotechnical seismic isolation (ERGSI) foundation system" for low-to-medium-density low-rise residential housing aims to develop a geostructural solution that will be cost-effective. Preliminary batch tests undertaken on granulated tire rubber and gravel (GTR:gravel) mixtures indicated minimal leaching of zinc (Zn) from the tires with concentrations observed slightly above the NZ drinking water aesthetic guideline value of 1500 μg/L. The batch testing setup is not representative of field conditions since in the field the GTR:gravel mixture would be compacted, not in continuous contact with groundwater and there is potential for adsorption and dispersion of contaminants. Therefore, following completion of the batch tests, laboratory column tests will be undertaken to more accurately simulate the field conditions. This will be done by using data from less conservative (more realistic) leaching tests which include site parameters (e.g., geometry and hydrology) to model the release, transport and fate of the contaminants over time from the source to the potential receiving environments/receptors. Following the environmental assessment, a suitable engineering countermeasure (e.g., geomembrane) to remediate or to contain the leaching material will be carefully selected and tested.

Acknowledgements The authors are grateful for the research support provided by the Ministry of Business, Innovation and Employment of NZ (MBIE Smart Ideas Research Grant No. 56289). The authors would also like to thank Mona Hosseinpour Moghaddam for the literature review on tire leachate, Angela Robinson and Hasara Ranchagoda (University of Canterbury summer students) for conducting the batch leaching tests, Robert Stainthorpe (University of Canterbury) for undertaking the ICP-MS analysis and Aude Thierry and Peter McGuigan (University of Canterbury) for undertaking the TOC analysis.

References

1. Ministry for the Environment: Waste tyres economic research. Report 3, pp. 87 (2015)
2. Lampman, R., Hanson, S., Novak, R.: Seasonal abundance and distribution of mosquitoes at a rural waste tire site in Illinois. J. Am. Mosquito Control Assoc. **13**(2), 193–200 (1997)
3. Torretta, V., Rada, E.C., Ragazzi, M., Trulli, E., Istrate, I.A., Cioca, L.I.: Treatment and disposal of tyres: two EU approaches. A review. Waste Manage. **45**, 152–160 (2015)
4. Steer, P.J., Tashiro, C.H.M., Mcillveen, W.D., Clement, R.E.: PCDD and PCDF in air, soil, vegetation and oily runoff from a tire fire. Water Air Soil Pollut. **82**(3–4), 659–674 (1995)
5. Edil, T.B.: A review of environmental impacts and environmental applications of shredded tire scraps. In: Hazarika, H., Yasuhara, K. (eds.) Scrap Tire Derived Geomaterials—Opportunities and Challenges: Proceedings of the International Workshop IW-TDGM 2007 (Yokosuka, Japan, 23–24 March 2007). Taylor & Francis Group, London, UK (2008)
6. Selbes, M., Yilmaz, O., Khan, A.A., Karanfil, T.: Leaching of DOC, DN, and inorganic constituents from scrap tires. Chemosphere **139**, 617–623 (2015)
7. Li, L., Xiao, H., Ferreira, P., Cui, X.: Study of a small scale tyre-reinforced embankment. Geotext. Geomembr. **44**(2), 201–208 (2016)
8. Garga, V.K., O'Shaughnessy, V.: Tire-reinforced earthfill. Part 1: Construction of a test fill, performance, and retaining wall design. Canad. Geotech. J. **37**(1), 75–96 (2000)
9. Collins, K.J., Jensen, A.C., Mallinson, J.J., Roenelle, V., Smith, I.P.: Environmental impact assessment of a scrap tyre artificial reef. ICES J. Mar. Sci. **59**, S243–S249 (2002)
10. Fabi, G., Spagnolo, A., Bellan-Santini, D., Charbonnel, E., Çiçek, B.A., García, J.J.G., Jensen, A.C., Kallianiotis, A., Santos, M.N.D.: Overview on artificial reefs in Europe. Braz. J. Oceanogr. **59**, 155–166 (2011)
11. Agudelo, G., Cifuentes, S., Colorado, H.A.: Ground tire rubber and bitumen with wax and its application in a real highway. J. Clean. Prod. **228**, 1048–1061 (2019)
12. Bocca, B., Forte, G., Petrucci, F., Costantini, S., Izzo, P.: Metals contained and leached from rubber granulates used in synthetic turf areas. Sci. Total Environ. **407**(7), 2183–2190 (2009)
13. Romero-Flores, M., Becerra-Lucatero, L.M., Salmón-Folgueras, R., Lopez-Salinas, J.L., Bremer-Bremer, M.H., Montesinos-Castellanos, A.: Thermal performance of scrap tire blocks as roof insulator. Energy Build. **149**, 384–390 (2017)
14. Hudson, A.P., Beaven, R.P., Powrie, W., Parkes, D.: Hydraulic conductivity of whole and shredded tyres for use in landfill drainage systems. Proc. ICE Waste Resour. Manage. **160**(2), 63–70 (2007)
15. Georgiopoulou, M., Lyberatos, G.: Life cycle assessment of the use of alternative fuels in cement kilns: a case study. J. Environ. Manage. **216**, 224–234 (2018)
16. Tsang, H.-H., Lo, S.H., Xu, H., Sheikh, M.N.: Seismic isolation for low-to-medium-rise buildings using granulated rubber–soil mixtures: Numerical study. Earthquake Eng. Struct. Dynam. **41**(14), 2009–2024 (2012)
17. Hernandez, E., Palermo, A., Granello, G., Chiaro, G., Banasiak, L.: Eco-rubber seismic isolation foundation systems, a sustainable solution for the New Zealand context. Struct. Eng. Int. (2020). https://doi.org/10.1080/10168664.2019.1702487

18. Chiaro, G., Palermo, A., Banasiak, L.J., Granello, G.: Direct shear behaviour of gravel-granulated tyre rubber mixtures. In: Proceedings of the 13th ANZ Geomechanics Conference, pp. 221–226, 1–3 April, Perth, Australia (2019)
19. Tasalloti, A., Chiaro, G., Palermo, A., Banasiak, L.J.: Effect of rubber crumbs volumetric content on the shear strength of gravelly soil in direct shear apparatus. In: Proceedings of Geo-Congress 2020, 22–25 Feb, Minneapolis, Minnesota, USA (2020)
20. Tasalloti, A., Chiaro, G., Palermo, A., Banasiak, L.J.: Dynamic properties of gravel mixed with granulated tire rubber. In: Proceedings of the 7th International Conference on Recent Advances in Geotechnical Earthquake Engineering and Soil Dynamics, July 13–16, Bangalore, India (2020)
21. Basel Convention Working Group: Basel Convention Technical Guidelines on the Identification and Management of Used Tyres. Basel Convention on the control of transboundary movements of hazardous wastes and their disposal. Document No. 10 (1999).
22. Humphrey, D.N., Katz, L.E.: Field study of water quality effects of tire shreds placed above the water table. In: Beneficial Use of Recycled Materials in Transportation Applications, 13–15 Nov 2001, Arlington, Virginia (2001)
23. Birkholz, D., Belton, K.L., Guidotti, T.: Toxicological evaluation for the hazard assessment of tire crumb for use in public playgrounds. J. Air Waste Manag. Assoc. **53**(7), 903–907 (2003)
24. Li, W., Seifert, M., Xu, Y., Hock, B.: Assessment of estrogenic activity of leachate from automobile tires with two in vitro bioassays. Fresenius Environ. Bull. **15**(1), 74–79 (2006)
25. Gunasekara, A.S., Donovan, J.A., Xing, B.: Ground discarded tires remove naphthalene, toluene, and mercury from water. Chemosphere **41**(8), 1155–1160 (2000)
26. U.S. Tire Manufacturers Association: What's in a tire. Accessed July 2019. https://www.ustires.org/system/files/USTMA-TireGraphic-v2.pdf (2019)
27. Edeskär, T.: Technical and environmental properties of tyre shreds focusing on ground engineering applications, Technical report, Lulea University of Technology (2004)
28. Ministry of Health: Drinking-water Standards for New Zealand 2005 (Revised 2008). Wellington, NZ (2008)
29. Ministry for the Environment: Module 2—Hazardous waste guidelines: landfill waste acceptance criteria and landfill classification, Wellington, NZ (2004)

Estimating Landfill Height Considering the Heterogeneous Physical Property of Dredging Soil as Reclamation Material

Yoshihisa Sugimura, Yasuhiro Sega, Masaaki Katagiri, and Katsuhide Nishizono

Abstract When pump-dredged clays are discharged into a disposal pond in the sea, soil particles are initially suspending in the water and gradually settles to the sea bottom. According to subsequent discharging, a new sediment is piled on the top surface of the sediment already formed, and the sediment already formed consolidate due to the own weight of new one over a long period of time. To predict the capacity of such disposal pond, it is important to estimate the time-dependent volume change of dredged clays during the reclamation process. In order to estimate the high accuracy the time when the current disposal pond will be full, it is necessary to use the reclamation analysis developed based on a general consolidation theory. In this paper, targeting the S-3 area of Shinmoji-Oki disposal pond project currently underway, we assume that the model of dredged clay deposited in the future on the ground model can be identified by back analysis. Predictive analysis is performed using consolidation parameters assumed from the physical properties of the collected clay, and different filling periods are confirmed due to differences in the physical properties. Furthermore, we also propose a method to improve the prediction accuracy.

Keywords Dredged clay · Reclamation analysis · Liquid limit of clay

Y. Sugimura (✉)
Kyushu University, Fukuoka-city, Fukuoka, Japan
e-mail: y.sugimura@civil.kyushu-u.ac.jp

Y. Sega
Ministry of Land, Infrastructure, Transport and Tourism, Kitakyushu-city, Fukuoka, Japan

M. Katagiri
Nikken Sekkei Civil Engineering, Koraku, Bunkyo-ku, Tokyo, Japan

K. Nishizono
Coastal Development Institute of Technology, Minato-ku, Tokyo, Japan

H. Hazarika et al. (eds.), *Advances in Sustainable Construction and Resource Management*, Lecture Notes in Civil Engineering 144,
https://doi.org/10.1007/978-981-16-0077-7_6

1 Introduction

Ports are the basic infrastructure that supports the Japanese economy, but vessels are becoming increasing larger, and for ports to maintain their functions, deepening and widening navigation channels and basins are required. To realize a sustainable society, the dredged material generated by these port development projects should be effectively used as reclamation material, including consolidated clay, and not dumped into the ocean. On the other hand, there is also a problem that disposal pond capacity is limited. Therefore, promoting dredging works while accurately discerning the remaining capacity of the operating disposal pond makes sense from an environmental perspective in terms of the effective use of the total amount of dredged soil and from a social perspective in terms of the smooth and reliable implementation of the project. However, the effective use of dredged soil, especially dredged clay with many fine particles, entails geotechnical issues.

A common method for dredging soft seabed soil is to suck up the seabed and seawater using a drag suction dredger. When this material is put into a disposal pond constructed on the surface of the sea, coarse-grained soil is sedimented near the outlet of the sand pipe, and the dredged soil is classified and is not deposited uniformly. Egashira et al. [1] conducted a reclamation prediction analysis using the generalized one-dimensional consolidation theory to predict the mud surface level rise in a disposal pond when developing the new Kitakyushu Airport. In the analysis, the dredged soil layer was assumed to be a uniform dredged clay layer, and the consolidation constant was set to be the average value of the varied measured values and used for subsequent prediction analysis. In addition, Sugimura et al. [2] reported the consolidation constants under low confining pressure of several dredged clays collected from a disposal pond to predict its capacity with high accuracy at Nanao Port. Even in this report, the liquid limit of the collected samples was 90–145%, and there were variations that could not be considered homogeneous.

It is important for the subsequent disposal plan to know with high accuracy the time when the current disposal pond will be full. However, since the dredged clay that accumulates in the disposal pond is not homogeneous, how to set up the ground model required for the analysis becomes an issue. In this paper, targeting the S-3 area of Shinmoji-Oki disposal pond project currently underway, we assume that the model of dredged clay deposited in the future on the ground model can be identified by back analysis. Predictive analysis is performed using consolidation parameters assumed from the physical properties of the collected clay, and different filling periods are confirmed due to differences in the physical properties. Furthermore, we also propose a method to improve the prediction accuracy.

2 Outline of the S-3 Area of the Shinmoji-Oki Disposal Pond

Figure 1 [3] shows the location of the S-3 area of the Shinmoji-Oki disposal pond and the positional relationship between the Kanmon Waterway and the Shinmoji Waterway, in which dredging operations are underway. The area is adjacent to the Kanda disposal pond, which was established in 1979, and the S-1 and S-2 areas of the Shinmoji-Oki disposal pond, which were operational from 1996 to 2002.

The previous ground level of the S-3 area is approximately DL-7 m, and the thickness of the alluvial clay layer under the seabed is 5–7 m. Beneath this layer, there is a diluvial clay layer. The crest height of the constructed revetment was DL + 8.0 m, and the receiving height of the dredged soil was DL + 7.5 m. The seabed soil was dredged in the S-3 area from 1998 to 2001, and there was a deep excavation area of approximately DL-22 m because the dredged soil was used as reclamation material for Kitakyushu Airport.

After that, from 2011 to 2013, an extra fill dike of 4 m was constructed to increase the capacity of the disposal pond. Furthermore, from 2013 to 2015, accelerated consolidation work was carried out to quickly drain pore water from the dredged soil layers accumulated in the S-3 area [4]. The current plan is to put 330,000 m^3 of dredged soil into the S-3 area per year.

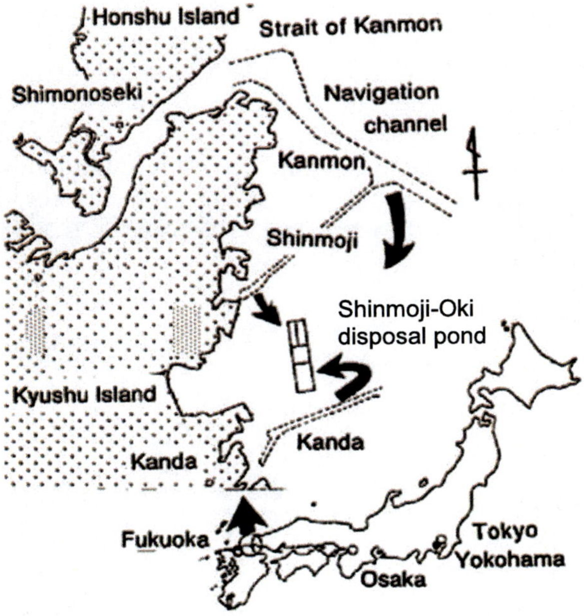

Fig. 1 Location of the Shinmoji-Oki disposal pond [3]

3 Analytical Calculation of Reclamation

3.1 Reclamation Analysis Code

For the reclamation analysis, we used the analysis code [1] that was confirmed to be effective in the construction of Kitakyushu Airport. The code is based on generalized one-dimensional consolidation theory and can express the self-weight consolidation behavior of ultra-soft clay.

The general one-dimensional consolidation theory can be expressed as following three equations; mass conservation, Darcy's law and balance of momentum neglecting accelerations and constitutive equation of soil skeleton. To solve these equations, the coupling method proposed by Imai [5] was used. In the coupling method, the three equations were converted into partial differential equations on the first floor, and each partial differential equation was coupled the other at any time. The merit of this method is that the reclamation process where accumulated layer thickness is increasing with time can be expressed easily.

To analyze the consolidation of new fresh sediment piled on the top of existing sediment, the determination of the boundary condition of the fresh sediment becomes important. In this paper, the boundary condition at the top of the fresh sedimentation is fixed at 9.8 Pa as proposed by Yamauchi et al. [6].

3.2 Assumed Consolidation Parameters

Negi et al. [7] investigated the stress dependence of the specific volume, f (=$1 + e$) and coefficient of consolidation, c_v for dredged clays in various parts of Japan, as shown in Fig. 2, and showed the relationship with the liquid limit, w_L, as shown in Fig. 3. Here, the α and c_{v1} were determined by Negi et al. [7], shown in Fig. 2.

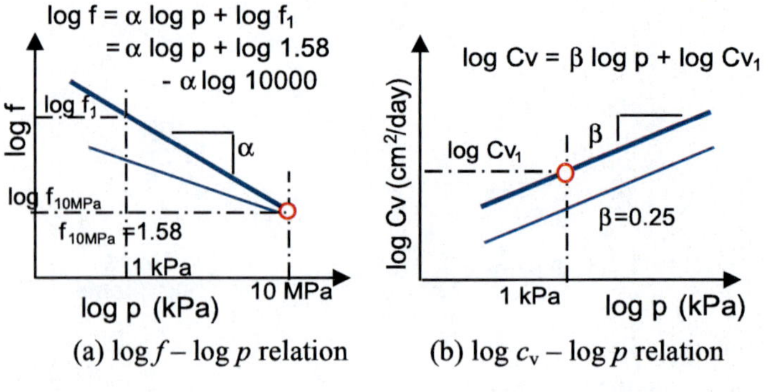

Fig. 2 Assumed consolidation constitutive model [7]. **a** log f—log p relation, **b** log c_v—log p relation

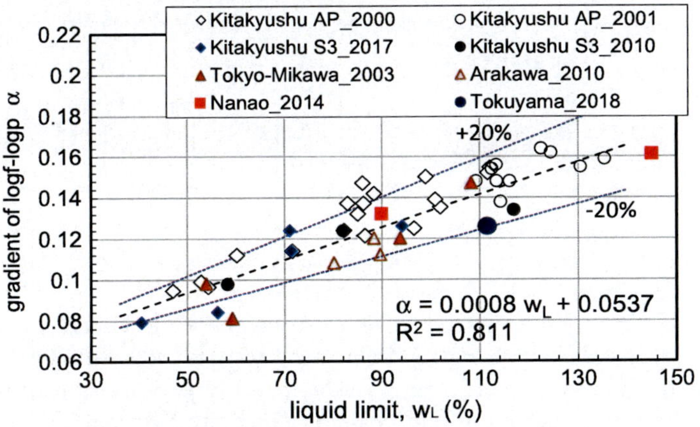

(a) Gradient of the $\log f - \log p$ relation

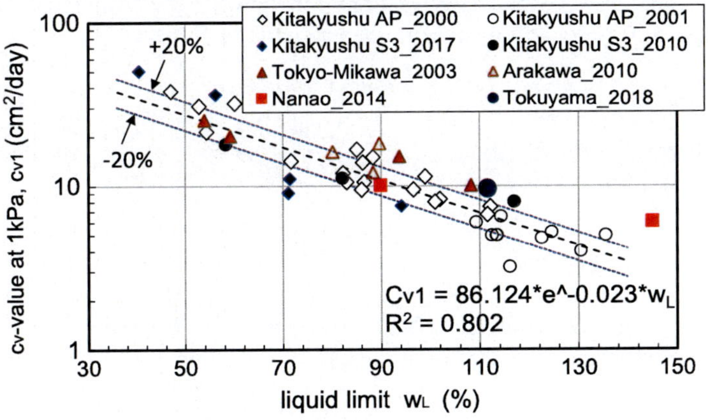

(b) c_v – value at 1 kPa

Fig. 3 Consolidation parameters and the liquid limit relation [7]. **a** Gradient of the log f—log p relation, **b** c_v—value at 1 kPa

In this paper, the consolidation parameters used for the analysis were set based on the liquid limit of the dredged clay. The ground model and the consolidation constant of the dredged clay that had been used by 2015 were the results of the back analysis performed by Nakamichi et al. [4].

The coefficient of consolidation, c_v in the clay layer installed Plastic Board Drains (PBD), was identified so that the settlement to 4 years from the installation of PBD predicted by the one-dimensional reclamation analysis was consistent with that obtained by the Barron's equation. This method could not express strictly the actual behaviors, but the settlement of ground at 4 years later was equivalent to 80% of consolidation degree, and the stress ratio at the drain and undrain parts became less than 20%. It was thought that this approximation could be expressed the actual

Table 1 Assumed consolidation parameters of dredged clay disposed from 2015

	Liquid limit (%)	α-value	c_{v1} (cm^2/day)
M-material	65	0.104	20
H-material	75	0.112	15
L-material	55	0.096	30

behaviors from engineering point of view. Here, for the 4 years applied at the Barron's equation the function of PBD was set at effectiveness.

For the consolidation parameters of dredged clay to be discharged after 2015, the liquid limit of the planned dredging site was set from the liquid limit of the sample taken from the borrow area. The range of the liquid limit of the collected clay was 60–70%. Therefore, the liquid limit of the representative M material is 65%, and the liquidity limits of the high-plasticity H material and the low-plasticity L material are set to 75 and 55%, respectively, assuming variations. The assumed consolidation parameters for the prediction analysis are collected in Table 1.

4 Reclamation Analysis

Figure 4 shows the analysis results. The figure is also drown the average mud height measured in any year up to 2019 as each mark and the planned reclamation height as a red line. In the cases of M, H and L materials, the predicted ground surface will be reached in April 2027, October 2025 and April 2027, respectively. Furthermore, it is found that these differences in the liquid limit will be only 10% in just 2 years.

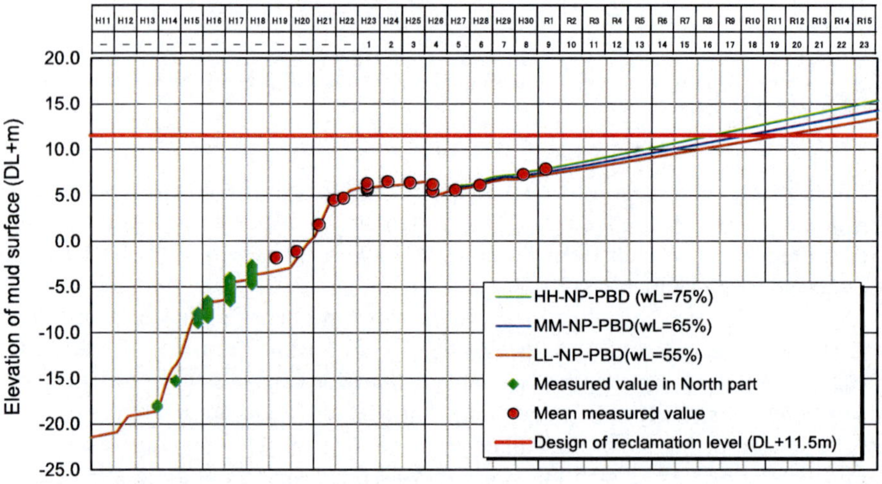

Fig. 4 Monitored data and prediction of the time elevation curve of the mud surface

Figure 5a, b are the comparison between water content distributions of accumulated clay layer at No.1 and No.2 points in 2016 August and the results of back-analysis at the same point and time. The numbers shown in these figures are the α-values indicated the slope of $\log f$—$\log p$ relations, and the blue number in the parenthesis is the assumed α-values at the first prediction. The black numbers are the identified α-values obtained from the back-analysis. On the other hand, the blue line is indicated as the predicted water content distribution with the assumed α-values at the first prediction in the accumulated clay layer before the install of PBD's, and the black bold line shows the calculated water content distribution of accumulated clay layer at the same time of ground survey. Here, the identified α-value in the range from DL-14 m to -9 m of the No.1 point shown in Fig. 5a is modified to 0.098 and those in other parts are used the assumed values at the first prediction. And the identified α-value of the No.2 point in Fig. 5b are the same as those of the first prediction.

Figure 5c shows the distributions of liquid limit at the No.1 and 2 points. The magnitudes of liquid limit in the whole layer at the No.1 are lower than those at the No.2. Specially the liquid limits of clays in the range from DL-14 m to DL-9 m at the No.1 point are around 60% and are lower than those in another layer. As the characteristics of clay between physical properties and consolidation behavior, the modification of α-value to change from 0.105 to 0.080 is reasonable.

Figure 6 is the relationship between liquid limit and consolidation parameters of clays sampled from the concerned and neighboring areas by the Multi-Sedimentation Test (MST) proposed by Yamauchi et al. [6]. Here, the consolidation parameters of clay were the α and c_{v1} shown in Fig. 2, and each mark shows the test result on

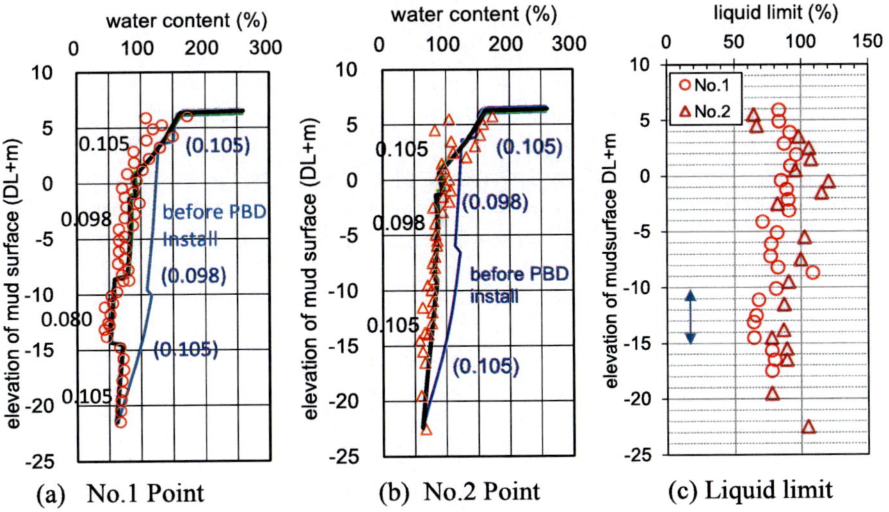

Fig. 5 Measured and back-analyzed water content distributions of accumulated clay layer at No.1 and No.2 points in 2016(H28), and liquid limit distributions of the same points. **a** No.1 point, **b** No.2 point, **c** liquid limit

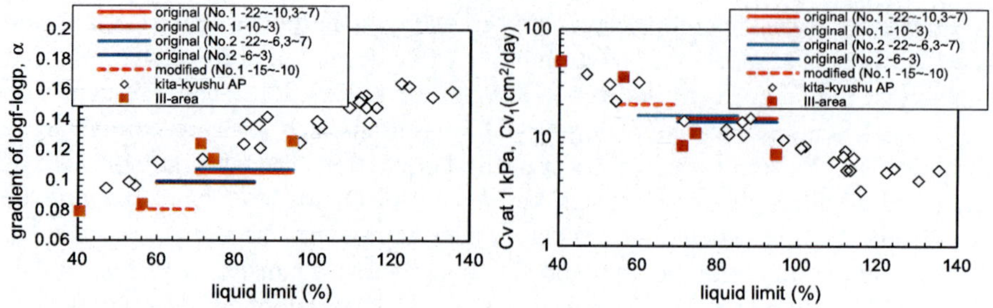

(a) Gradient of $\log f - \log p$ relations (b) c_v-value at 1 kPa

Fig. 6 Relationship between liquid limit and α, c_{v1}. **a** Gradient of $\log f$—$\log p$ relations, **b** c_v—value at 1 kPa

the dredged clay sampled from disposal pond. The α-values identified by the back-analysis are drown with lines against the range of liquid limit of clays corresponding to the sampling depths. The solid line means the assumed value at the first prediction, and the broken lines are the identified values at the modified prediction.

As shown in Fig. 6a, the assumed α-values at the first prediction are located in lower part of the range of examined values. The identified α-value from the DL-14 m to -9 m at the No.1 point becomes lower than the first value, as the liquid limit of clays in this range becomes lower than other depth. The assumed α-values including with identified values are located in the lower part of existing range.

This study predicts when dredged soils with different physical properties are discharged and do not directly represent the dispersion of physical properties in the disposal pond reported by Sugimura et al. [2], in which the physical properties of the scattered and deposited clays are set uniformly. Therefore, it cannot be said that the prediction accuracy is sufficient, but at the present time, it can be judged sufficient for predicting the receiving capacity when formulating future disposal pond plans and subsequent site utilization processes. To make more accurate predictions, it is useful to use the consolidation parameters identified using the observation results made by Katagiri et al. [8]. To do so, it is important for us to observe results such as changes in the reclamation height over time and the water content distribution in the accumulated clay ground.

5 Conclusions

In this paper, the c_v value in the layer installed the PBD's was modified so that the settlement of clay ground for a certain period predicted by one-dimensional consolidation analysis was consistent with that obtained by the Barron's equation, and was identified by the back analysis using the monitored results. Based on these analyzed results, the time elevation of mud surface in the future was predicted.

The consolidation characteristics of new accumulated dredged clay were selected from the relationships between liquid limit and consolidation parameters. The behavior of the ground level in the future disposal pond was predicted using the code developed on generalized one-dimensional consolidation theory. To predict the time when the ground face of accumulated dredged clay reaches at the planned elevation, the ground model contributed by the identified consolidation parameters was used, and the consolidation parameters of dredged clay in future were determined by the physical properties of dredged clays, using the relationship estimated from the results of dredged clay around Japan. In this simulation, when the liquid limit of dredged clay was changed by 10%, it was confirmed that the time to complete reclamation differed by approximately 2 years.

It is effective that the one-dimensional consolidation analysis using the ordinary one-dimensional consolidation parameters so to represent the behavior estimated by the Barron's equation. And it was argued that the accuracy of prediction improved by the identification using the actual data.

This similar result will not only contribute to the effective use of dredged soil, including cohesive soil, but it can also be used to examine the volume balance when formulating a port plan. In the future, we plan to improve the method of setting the consolidation parameter by investigating the physical properties of the discharged soil, by conducting a predictive analysis that incorporates the consolidation parameter from them, and by comparing it with the actual measurement results.

References

1. Egashira, K., Iwataki, K., Sato, T., Katagiri, M., Terashi, M., Yoshifuku, T.: Prediction and evaluation of landfill by dredged sediment. J. JSCE 715 / III-60, 147–164 (2002).
2. Sugimura, Y., Shirai, H., Shibuya, K., Katagiri, M., Kodama, D.: Multi-layer sedimentation experiment of Nanao Port dredging clay. In: The 48th Geotechnical Research Conference, pp. 1023–1024 (2013).
3. Egashira, K., Yamagata, N., Takada, T., Katagiri, M., Terashi, M., Toshifuku, T., Murakawa, S.: Settlement analysis and observational method for the reclamation with dredged clay—a case record at New-Kitakyushu Airport—, Soft Ground Engineering in Coastal Areas (Proc. of the Nakase Memorial Symposium), pp. 119–126 (2002).
4. Nakamichi, M., Yamaguchi, M., Umeyama, T., Yamamoto, S., Kawahara, O., Katagiri, M., Ohishi, K., Yoshifuku, T.: Capacity increase project for Shinmoji-oki disiposal pond. In: Proc. of 15th ARC on SMGE, Paper No. ESD-43 (2015).
5. Imai, G.: A unified theory of one-dimensional consolidation with creep. In: Proceedings of 12th ICSMFE, vol. 1, pp. 57–60 (1989).
6. Yamauchi, H., Imai, G., Yano, K.: Effect of the coefficient of consolidation on the sedimentation consolidation analysis for a very soft clayey soil. In: Proceedings of 25th Annual Meeting of JSSMFE, pp. 359–362 (1990) (in Japanese).
7. Negi, T., Higuchi, A., Kawashima, Y., Mori, H., Furusawa, T., Nanno, Y., Ohishi, K., Katagiri, M.: Determination method of consolidation parameters used for landfill analysis of dredged clay. In: Ocean Development Symposium, Paper No. 33 (2019).
8. Katagiri, M., Kunita, A., Yamamoto, S.: Modification method of ground model for reclamation analysis of dredged clays with high accuracy. In: Proceedings of 16th ARC on SMGE, Paper No. SA11-02-006 (2019).

Evaluation for Effective Use of Dredged Soil as Containment Dikes of Disposal Pond

Katsuhide Nishizono, Yasuhiro Sega, Masaaki Katagiri, and Yoichi Watabe

Abstract An extra fill on revetment was constructed by using mechanically dewatered clay (MDC) lumps made from dredged material in the third area of the Shinmoji-Oki disposal pond (S-3) 9 years ago. Since this sample is an artificial material, its long-term stability has not been confirmed. In order to understand the evolution of the properties of the constructed revetment over time, mechanical tests, such as direct shear tests, have been conducted on block samples taken from two designated areas of the extra fill on revetment over the nine years since its construction. As a result, it was confirmed that the shear strength under certain overburden effective stress was sufficiently satisfied with the value assumed in the original design and the strength properties did not change much after 9 years, although the strength constants varied.

Keywords Dredged clay · Disposal pond · Effective use · Mechanical dewatering

1 Introduction

The Kanmon Waterway, which is an important route for ship traffic, requires dredging in order to adapt to a water depth of −14 m. Since the dredged seabed soil must be disposed of without causing marine pollution, the Shinmoji-Oki disposal pond was constructed in three stages. The soil is disposed of in the third area of the pond (S-3) (see Fig. 1). The S-3 area is 482 m by 1485 m and is surrounded by revetments. Initially, the revetment height was DL + 8.0 m, and the pond capacity

K. Nishizono (✉)
Coastal Development Institute of Technology, Minato-ku, Tokyo, Japan
e-mail: nishizono@cdit.or.jp

Y. Sega
Ministry of Land, Infrastructure, Transport and Tourism, Kitakyushu-city, Fukuoka, Japan

M. Katagiri
Nikken Sekkei Civil Engineering, Bunkyo-ku, Tokyo, Japan

Y. Watabe
Hokkaido University, Sapporo-city, Hokkaido, Japan

H. Hazarika et al. (eds.), *Advances in Sustainable Construction and Resource Management*, Lecture Notes in Civil Engineering 144,
https://doi.org/10.1007/978-981-16-0077-7_7

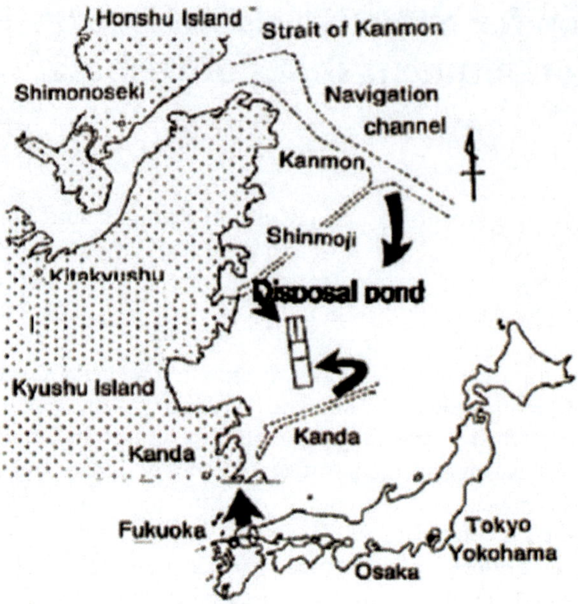

Fig. 1 Location of Shinmoji-Oki disposal pond

was 12,000,000 m^3 (see Fig. 2). After that, to increase its capacity, the dike was raised using mechanically dewatered clay (MDC) lumps made by forcing out pore water from the dredged clay [1]. As a result, the height of the revetment with extra fill is DL + 12.0 m, and the capacity has been expanded by approximately 3,100,000 m^3 (see Fig. 3).

Typically, mountain soil is used as the material for dikes; MDC was used for the first time in Japan in the S-3 area. However, since MDC is an artificial material, its long-term characteristics are unknown. Therefore, in this paper, the authors will evaluate the stability of MDC by assessing the changes over time in its mechanical properties from 2011, when the extra fill was added to the revetment, to the present [2].

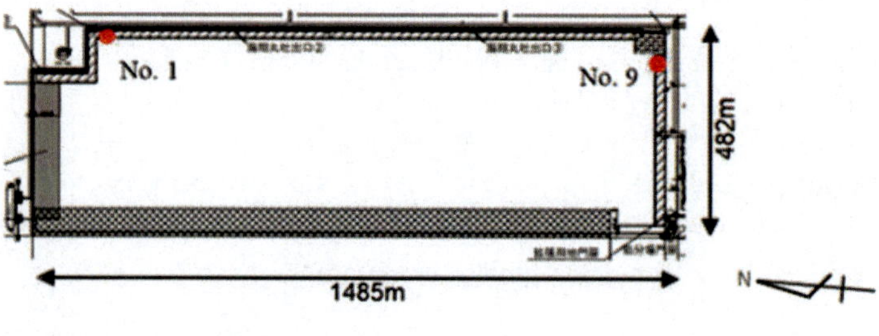

Fig. 2 S-3 area of Shinmoji-Oki disposal pond

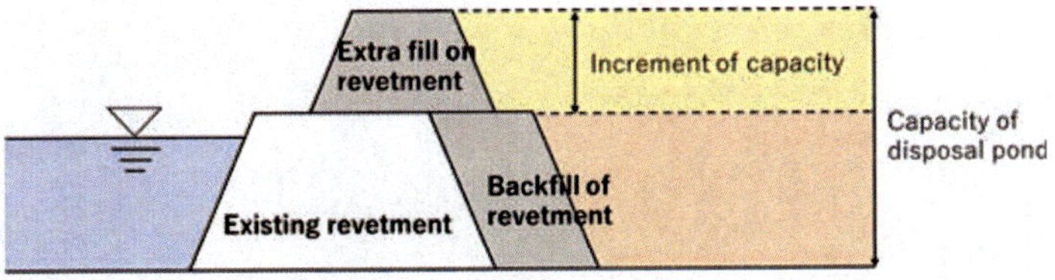

Fig. 3 Concept of capacity increment method using MDC

2 Outlines of the Use of Mechanically Dewatered Clay Lumps for Dikes

2.1 Mechanically Dewatered Clay Lumps Production Process

The dewatered soil was prepared by a high-pressure filter press method, which is classified as a pressurized dewatering and filtration method. Specifically, the clay slurry is fed at high pressure into a filter chamber surrounded by a filter cloth, where it is filtered and dewatered. This principle is the same as that of osmotic consolidation, where the pore water flow imposes an effective stress on the clay particles, which in turn consolidates the clay sample in the filter chamber. Therefore, the quality of the soil mass produced depends on the size of the feed pressure and the degree of consolidation of it when the clay slurry feed is stopped. Moriki et al. [3], using dredged sediment with the physical properties shown in Table 1, showed from various experiments and investigations that the predetermined failure criterion can be maintained if the water content of the individual clods is less than 53% and the needle penetration gradient of the soil mass is greater than 0.92 N/mm and proposed that the point at which the clay slurry feed stop is when the drainage rate is less than 45 L/min under a feed pressure of 4 MPa.

Moriki et al. [3] described the manufacturing process of MDC lumps used as extra fill for the dikes as follows (see Fig. 4):

(i) Gravel was removed from the raw mud to obtain a clay slurry with a water content of about 300% (three times of w_L).
(ii) To this clay slurry, polyaluminum chloride (PAC) (5%) and slaked lime (2%) were added to the weight of clay particles as an auxiliary material in the dehydration process.

Table 1 Physical properties of dredged soil

Water content (%)	113–214
Density of soil particles (g/cm^3)	2.638–2.681
Liquid limit (%)	82–131
Soil content (%)	1.6–11.3

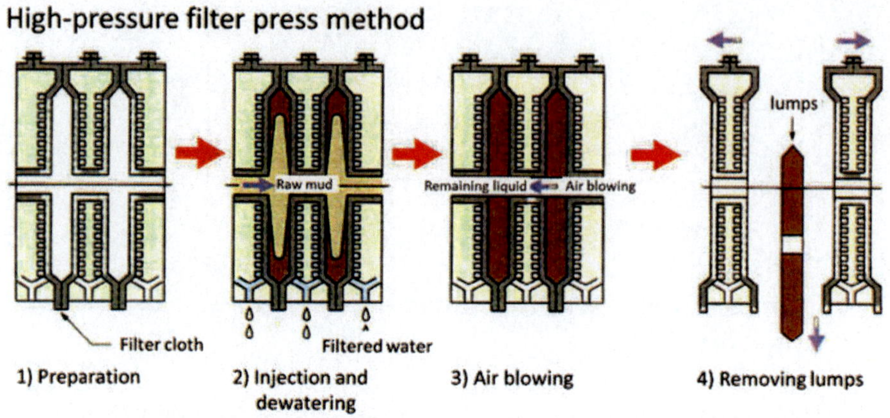

Fig. 4 Mechanically dewatered clay lumps production process

(iii) During the dewatering process, the clay slurry was sent into a filter chamber for approximately 20 min with low-pressure pumping.

(iv) Switching to high-pressure pumping at 4 MPa, the clay slurry was fed into the chamber until the drainage rate reached 45 L/min (total filtration chamber volume 6.8 m^3; volume strain rate = 0.7%/min) or less.

(v) After that, the frame was opened, and the lumps were removed.

2.2 *Characteristics of Mechanical Dewatered Clay Lumps*

Figure 5 shows the state of the dewatered soil immediately after removal. Depending on the width of the filter chamber, the thickness of the dewatered soil was about

Fig. 5 Dewatered soil immediately after completion

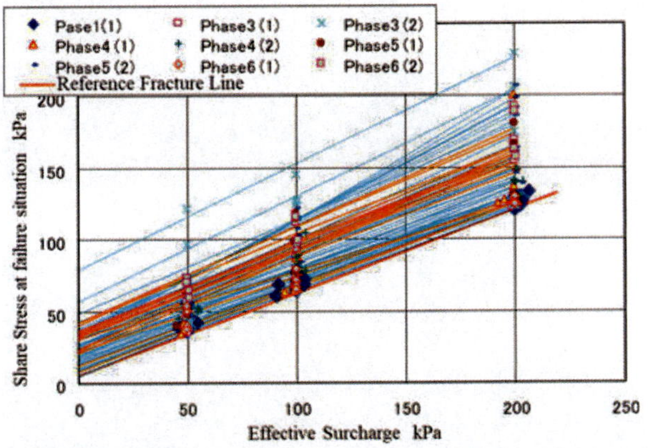

Fig. 6 Fracture point and fracture line of dewatered soil immediately after completion

25 mm, and its size was up to 70 mm, as it was crushed during the process of opening and transportation. The average particle size of MDC that was transported/placed temporarily after production was 31 mm, and the uniformity coefficient was 18.5. The water content of the soil mass as a quality control value when producing MDC was lower than the standard value (53%) shown by Moriki et al. [3]; the average value was 45.3%, and the maximum value was 48.4%. On the other hand, the needle penetration gradient of the soil mass was higher than the standard value (0.92 N/mm), with an average value of 1.17 N/mm and a minimum value of 1.09 N/mm.

Figure 6 shows the results of the constant pressure direct shear tests performed immediately after the fabrication process at each production period. It shows failure points and their linearly approximated failure lines for dewatered soils packed to a unit weight of 14.5 kN/m^3 with internal diameters of 200 mm and depths of 160 mm in a direct shear box under vertical stresses of 50, 100, and 200 kPa. In the same figure, the design failure criterion[1)] for the revetment is shown as a red line. Both the failure point and the failure line obtained from the experiment were located above this reference line.

2.3 Extra Fill on Existing Revetments

The extra fill on revetments with MDC was applied in accordance with the method by Moriki et al. [3]. The main features are as follows:

Spreading thickness: 30 cm, Unit volume weight: 16.5 kN/m^3, Slope: 1.5;

Failure criteria: $c_d = 5$ kPa, $\phi_d = 30°$.

The construction of the revetment over DL + 4.5 m began in 2010, and the embankment of the revetment, including No.1 and No.9 points, was completed in 2011. Figure 7 shows the construction status of the extra fill on revetments near No.9 point in Fig. 2.

Fig. 7 Construction status of the extra fill on revetments near No.9

3 Ground Survey Results of Extra Fill

3.1 Chronological Change of Share Strength

Figure 8 shows the water content of the block sample taken every year after construction. Figure 9 shows the results of the direct shear test of the block sample and the design strength ($c_d = 5$ kPa, $\phi_d = 30°$) when constructing the dikes using MDC. The sampling locations were No. 1 near the northeast corner and No. 9 near the southeast corner, and the sampling depth was 50 cm from the surface. The water content of the collected samples tended to decrease with time. On the other hand, the envelope line of failure points was above the design value at 50 kPa upper loading pressure, although there was some variation.

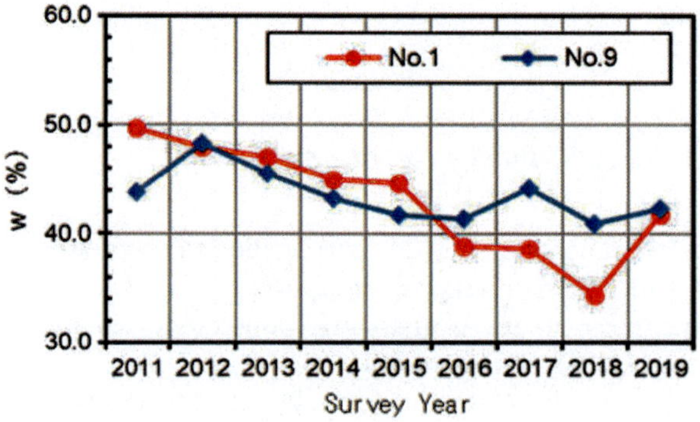

Fig. 8 Change in water content of block sample over time

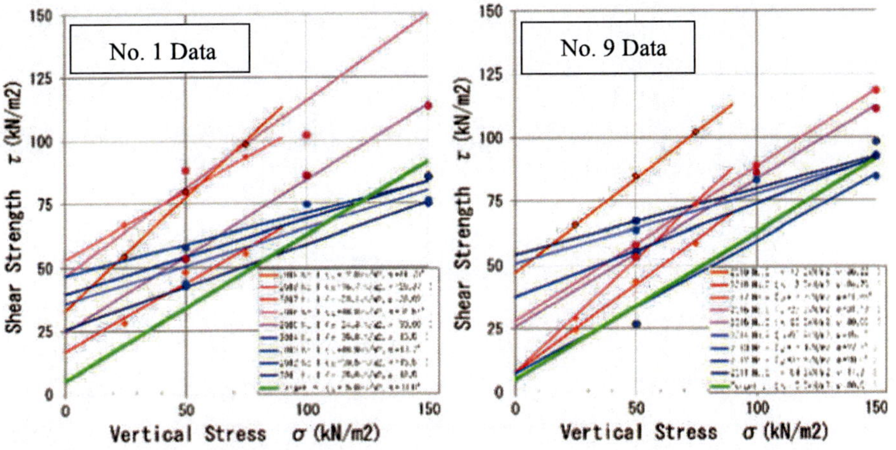

Fig. 9 Change in the failure strength with time and the failure criterion of the block sample. **a** Results from No.1, **b** results from No.9

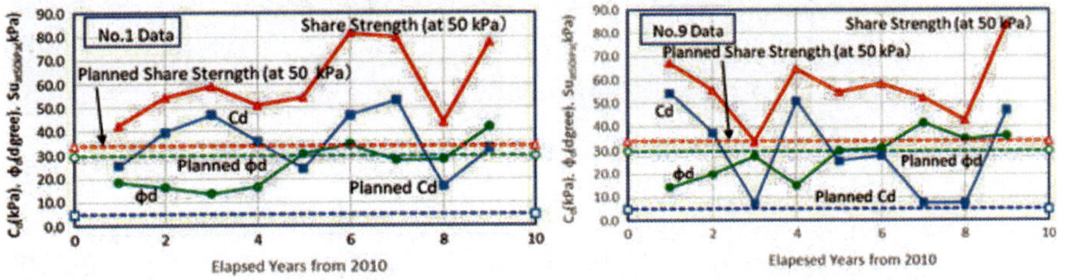

Fig. 10 Changes in the strength constant and shear strength at 50 kPa with a direct shear test. **a** Results from No.1, **b** results from No.9

Figure 10 shows the changes over time of the fracture criterion line, c_d and ϕ_d in Fig. 9, and the shear resistance under a loading of 50 kPa. Although the strength constants varied, the shear resistances, which contribute to stability, were all higher than the designed values.

3.2 Condition of the Site

Figure 11 shows the changes in monthly rainfall and average temperature around the site from January 2010 to December 2019. In most years, the monthly precipitation varies by around 200 mm between summer and winter, indicating that the area is repeatedly dry and wet.

Figure 12 shows the situation of the dikes and block sample near No. 9 in 2017. It is clear that the white areas are the clods themselves and contain uncrushed material.

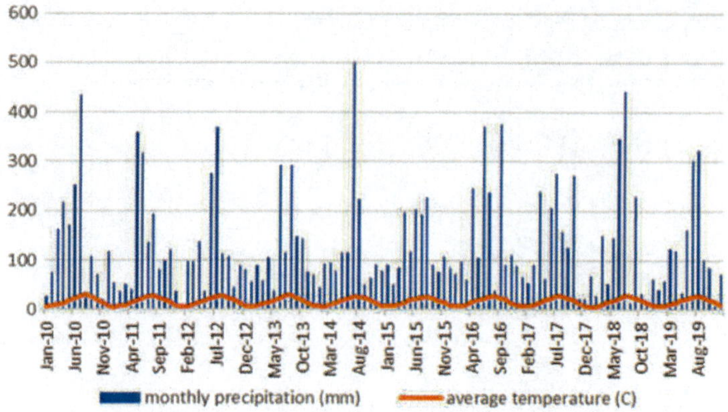

Fig. 11 Changes in monthly rainfall and average temperature at Kitakyushu Airport near the site

Fig. 12 Situation of the dike and a block sample taken near No. 9 in 2017

Figure 13a shows the photograph taken immediately after the construction of the dike (2011), and Fig. 13b shows a block sample near No.9 in 2015. From these photographs, it can be considered that there is no significant change in the soil mass situation.

Monthly visual inspections and yearly inspections using simple measurements have been carried out to check for any deformation of the extra fill on revetments. As a result, no cracks, sagging, or deflections of the slope have been observed so far.

4 Conclusions

It has been nine years since the extra fill on revetments was completed, and the site has gone through dry and wet conditions every year, but there has been no change in the slope, such as cracks, sagging, or deflections. As a result of annual observation

Fig. 13 Situation of the dike in 2011 and a block sample taken near No. 9 in 2015. **a** Dike in 2011, **b** block sample in 2015

of the ground block samples at a depth of 50 cm from the surface layer at two points in the dike, the shear resistance of MDC was higher than the design value in all cases and the extra fill on revetments was confirmed to be stable.

In the future, we would like to assess the physical properties of MDC by measuring the surface wave velocity (v_s) over a wide range and the tip resistance (q_t) and evaluate the stability of the bank.

References

1. Murayama, I. et al.: Mechanical behavior of a filter-pressed clay lump and their interactions. In: 49th JGS-Symposium, pp. 91–98 (2004) (in Japanese)
2. Nakamichi, M. et al.: Capacity increment project of Shin-Mojioki disposal pond. In: 59th JGS-Symposium, pp. 667–674 (2014) (in Japanese)
3. Moriki, A., et al.: Design method of additional embankment using dredged marine clay lumps. Tsuchi-To-Kiso **59**(7), 22–25 (2011). (In Japanese)

Experimental and Numerical Evaluation of Clay Soils Stabilized with Electric Arc Furnace (EAF) Slag

Muhammed Mahmudi, Selim Altun, and Tugba Eskisar

Abstract In this paper, engineering characteristics and settlements of clay soils stabilized with EAF slag were evaluated by laboratory experiments and numerical analyses. The strength parameters obtained from laboratory tests constituted the input parameters of the numerical analyses that displayed the settlements of clay soil layers under shallow foundations. In the experimental part, clay soil was replaced with EAF slag at 30 and 50% ratios, and triaxial compression tests and one-dimensional consolidation tests were conducted at various initial confining stresses to obtain the strength parameters of the clay-EAF slag mixtures. The test results showed that partial replacement of EAF slag with clay soil provided significant improvements on the strength and compressibility parameters of the clay soils. In the numerical analyses, the settlements of clay layers under the shallow foundations were evaluated by replacing the mechanical properties of clay-EAF slag mixed soil instead of clay soil. The findings of the numerical analyses showed that the maximum excess pore pressure development in clay-EAF slag mixtures was 77% (70C-30EAF) and 90% (50C-50EAF) less than the clay soil, and the development area was shifted to the untreated part of the soil. The consolidation duration was two times shorter in 50C-50EAF specimens and the level of settlement decreased by 20% (70C-30EAF) and 36% (50C-50EAF) compared to the untreated soil.

Keywords EAF slag · Consolidation test · Triaxial test · Numerical analysis

1 Introduction

Construction in saturated clay soils is one of the important geotechnical problems due to their low compressive strength and poor engineering characteristics. Soil stabilization can be defined as the modification or improvement of the soil properties in order to increase the engineering performance under static or cyclic loadings. Adding cementing agents such as cement [1, 2], lime [3, 4], industrial byproducts

M. Mahmudi · S. Altun · T. Eskisar (✉)
Civil Engineering Department, Ege University, Izmir, Turkey
e-mail: tugba.eskisar@ege.edu.tr

H. Hazarika et al. (eds.), *Advances in Sustainable Construction and Resource Management*, Lecture Notes in Civil Engineering 144,
https://doi.org/10.1007/978-981-16-0077-7_8

like fly ash [5–8], slag [9–11] or adding reinforcement elements such as fibers [12, 13] into soil results in improved geotechnical properties.

There are several kinds of slag materials, obtained after the different processes of steel making processes, namely as blast furnace, basic oxygen furnace, and electric arc furnace, etc. Electric arc furnace slag is produced during the manufacture of crude steel by the electric arc furnace (EAF) process. As the mechanical strength and specific density of slag is high, thus this material can be refined into a wide range of aggregate sizes. Therefore, EAF slag could be used as an ideal aggregate for replacement purposes.

Literature on previous research about steel slags have concentrated mainly on their use as concrete and asphalt aggregates [14–17]. In the geotechnical field, slag was occasionally studied to evaluate the improvements in the engineering properties of soft soils or as a geo-fill materials [9–11].

The experimental test results showed that partial replacement of EAF slag with clay soil provided significant improvements on the strength parameters of the clay soils. In the numerical part, the settlements and the bearing capacity of clay layers under the shallow foundations were evaluated by replacing the mechanical properties of clay soil with clay-EAF slag mixed soil.

It was found that the consolidation settlements and consolidation duration decreased when the clay soil was stabilized with EAF slag. The result of the present investigation indicated that EAF slag additive improved the engineering characteristics of clay soils, and therefore, it could be recommended as a low-cost stabilizer of cohesive and clay like soils.

2 Material Properties and Experimental Program

The natural clay soil and EAF slag for this study were obtained from Salihli to Manisa and Izmir cities in Turkey, respectively. EAF slag was grinded by using the Los Angeles device, then the particles were sieved by No. 4 sieve, and the particles that were less than 4.75 mm were used as an additive material. The natural unit weight of EAF slag was 2.85 t/m^3, the specific gravity was 3.41, and the porosity of EAF slag was 0.36.

The scanning electron microscope (SEM) micrograph of clay and the light microscope (LM) image of EAF slag are shown in Fig. 1. The light microscope of EAF slag–clay mixtures implied that the slag grains varied from angular to sub-angular particles. The LM image of EAF slag implied that the slag grains varied from angular to sub-rounded particles. Many macro-pores were also visible on the surface of steel slag particles which specifies that steel slag is a porous material compared to clay particles. The SEM image showed that a clay particles have a well-crystalized structure with plates in a framework form.

The effective size (D_{10}), the diameter corresponding to 30% finer (D_{30}), mean grain size (D_{50}) of EAF slag are 0.44 mm, 1.20 mm, and 1.80 mm, respectively. The grain size distribution curves and properties of the materials used in this study are

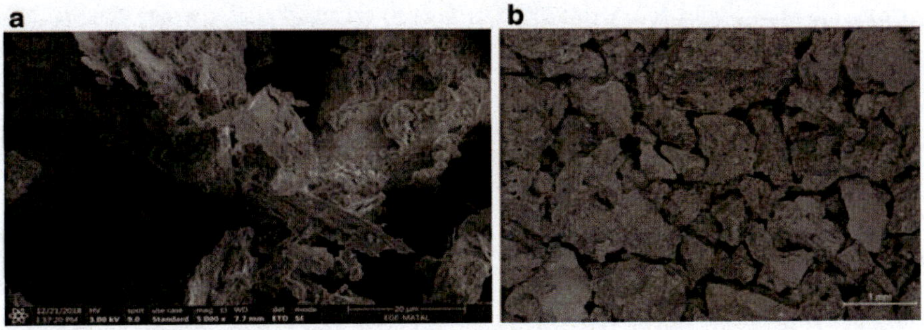

Fig. 1 Material images, **a** SEM micrograph of clay soil, **b** LM image of EAF slag

presented in Fig. 2 and Table 1. As seen in Table 1, the host soil and EAF slag were classified as a low plasticity clay (CL) and poorly graded clean sand (SP) according to the Unified Soil Classification System (USCS). Additionally, the classification

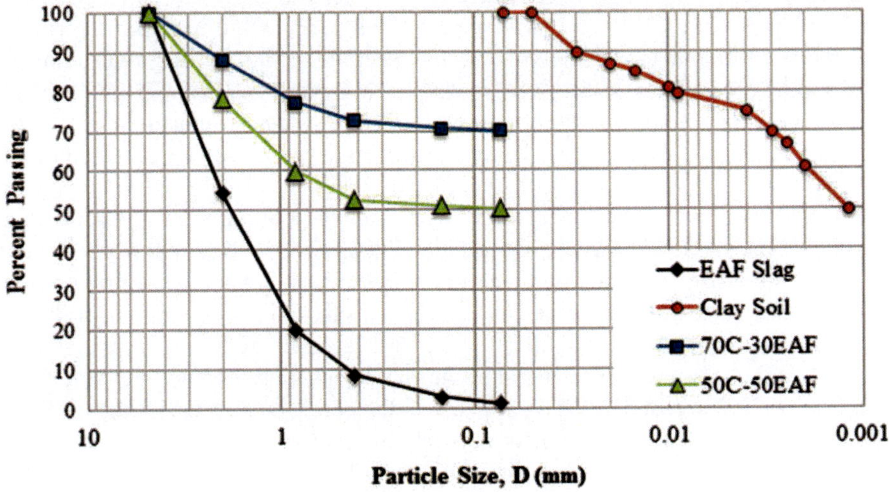

Fig. 2 Grain size distributions of EAF slag, clay, and clay-EAF slag mixtures

Table 1 Properties of the materials used in this study

Specimens	PL	LL	PI	OMC	USCS symbol
	(%)	(%)	(%)	(%)	
EAF slag	–	–	–	–	SP
Clay	43.5	23.6	19.9	15.9	CL
70C-30EAF	25.6	14.1	11.5	10	CL
50C-50EAF	21.1	11.3	9.8	9	CL

of mixtures with 30% (70C-30EAF) and 50% EAF slag (50C-50EAF) was graded as low plasticity sandy clay (CL). According to the results obtained from Atterberg limits tests and standard Proctor tests, it should be noted that with the addition of EAF slag, index properties and optimum moisture content (OMC) of clay soil have changed. EAF slag additive improved the engineering properties of the clay soil.

This study can be divided into two components. The first component is the determination of the strength parameters, engineering characteristics, and consolidation parameters of the clay-EAF mixtures with the static compression triaxial test and consolidation tests. The results of the aforementioned experiments were used as the input parameters in the second component in order to perform numerical analyses and evaluate the consolidation settlements of clay soil under different circumstances.

3 Experimental Test Result

Static compression triaxial testing was conducted on the isotropically consolidated-undrained specimens, and each specimen was sheared under 50, 100, and 200 kPa of confining pressures. CU triaxial tests were conducted on cylindrical specimens with a diameter of 50 mm and a height of 100 mm, keeping a constant height-to-diameter ratio of (H/D) = 2. Triaxial tests were performed according to ASTM D 4767, and all specimens were prepared at their optimum moisture contents by using the standard compactive effort. The specimen was saturated with flowing CO_2. After this step, de-aired water was allowed to flow through the specimen from the bottom to the top of the specimen. Skempton's coefficient, B, value was assured to be at least 0.95 at the end of the saturation step. The pressure inside the specimen was kept constant by keeping the effective confining pressure at 20 kPa throughout the saturation step. Static triaxial tests were performed under strain-controlled conditions at a strain rate of 1 mm/min. The tests were finalized when an axial strain of 20% was achieved. As seen from Table 2, the friction angle of the specimens increased and the amount of cohesion tended to reduce with an increase in the percentage of EAF slag.

Consolidation tests were performed on clay specimens and clay-EAF slag mixtures. The tests were held regarding the standard of ASTM D2435. The specimens had a height of 20 mm, and they were placed in a consolidometer ring with a diameter of 64 mm (2.5 in.). LIR was kept as unity, and the duration of each loading was 24 h.

Table 2 Strength parameter of the specimens

Specimen name	Friction angle, φ_{cu} (°)	Cohesion, c_u (kPa)
EAF slag	37	–
Clay	13	39
70C-30EAF	18	35
50C-50EAF	19	30

Table 3 Consolidation parameters of the specimens

Specimen name	Compression index (C_c)	Recompression index (C_r)	Coefficient of consolidation (c_v)
			cm^2/s
EAF slag	0.113	0.009	1.141
Clay	0.189	0.035	0.019
70C-30EAF	0.136	0.024	0.025
50C-50EAF	0.125	0.021	0.039

In order to examine the effect of EAF slag additive on the consolidation parameters of the soil, consolidation tests were conducted on the specimens containing 30 or 50% EAF slag additive, and all specimens were prepared at their optimum water contents. In order to saturate the test specimens with water, the tests on clay soil and EAF slag added soil specimens were carried out after 1 day of curing. For each consolidation test, the loading pressure levels applied during the test in general were 50, 100, 200, 400, 800 kPa and unloading pressures were 400 and 50 kPa. The coefficient of consolidation was determined from time deformation curves using the square root of time method for each applicable load increment. Table 3 presents the coefficient of consolidation (c_v) which was used in numerical analyses, compression index (C_c), and recompression index (C_r) of the specimens decreased with an increase in the percentage of EAF slag.

4 Numerical Analysis

In this study, an improved zone was formed by soil replacement from ground surface to a depth of 3.0 m. Because, under suitable ground conditions, it may be possible to remove the soil by excavation and mix it with EAF slag. Afterward, laying it in smaller layers (ca. 50 cm) and then compacting it to a desired density will result in a replacement in the original location. This type of improvement was considered in the numerical analyses.

The finite element analyses were carried out by using an effective stress based on PLAXIS 2D [18] software program. In order to generate the mesh, fine and very fine mesh options were chosen for investigation area, to achieve accurate results with higher sensitivity. After defining the geometry model, standard boundary fixities were applied; material properties were attained to all clusters from the data sets. The fixed boundaries were closed from 3B distances from the right, left, and bottom of the model. The model boundary conditions were assumed such that the vertical boundaries were free in vertical direction and fixed in horizontal direction, while the bottom horizontal boundary was fixed in both the horizontal and vertical directions. In this study, 60 kPa uniformly distributed static loading was applied to a raft foundation with a width of 10.0 m and with a thickness of 1.0 m. This structure represented a

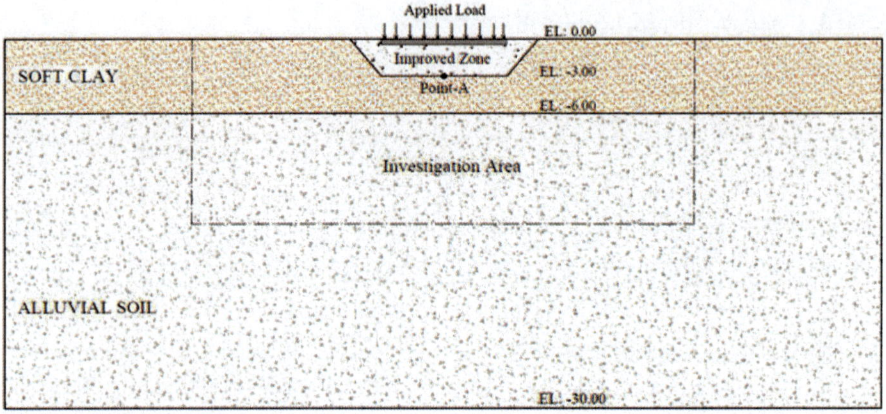

Fig. 3 Model used in this study

six-story building that was constructed on soft saturated soils. The overview of mesh set up and the simplified finite element model are illustrated in Fig. 3.

The ground water table level was at the surface, so that the fully saturated conditions could be applied to the soil layers. Therefore, all analyses were carried out in undrained conditions in the clay layer. The geological profile that represented the area under investigation consisted of two distinct layers. On the surface, there was a 6.0 m clay layer which properties were introduced in Table 1. The next layer was a dense alluvial soil layer up to −30.0 m depth.

The Mohr–Coulomb (MC) and soft soil (SS) models were preferred in the analyses. The results of the aforementioned experiments were used in MC and SS soil models. For SS soil model, $\lambda = 1.e{-}9$, $\kappa = 1.e{-}10$, and $e_{\text{init.}}$ varied from 0.45 to 0.55 depending on the soil mixture. As the soil is normally consolidated, OCR was 1. For MC soil model, Young's modulus increased with the addition of EAF slag from 16,500 (clay) kPa to 15,300 kPa for 70C-30EAF and to 14,500 kPa for 50C-50EAF.

In the initial condition, closed consolidation boundaries were applied at the right and the left sides of the soil model. The pore water pressure and initial stresses were calculated. The bottom of the model was open, because the excess pore pressures were considered flow freely into the deep and permeable alluvial layer. In the initial situation, the upper structure and its related stress increment were neglected. After the application K_0 conditions in the initial phase and displacements were reset to zero (phase-0), the consolidation analysis was performed in two phases. In order to perform a consolidation analysis, a proper time step must be selected. In phase-1 of consolidation analysis, the structural load was applied under the undrained conditions for 180 days of construction. In phase-2, excess pore pressure was allowed to dissipate with time to the minimum excess pore pressure level which was 1 kPa. In both phase-1 and phase-2, the consolidation settlements and the excess pore water pressure changes have been checked in the middle of the clay layer (Point A, Elevation Level (EL): −3.0 m).

4.1 Consolidation Settlements

In order to evaluate the response of a foundation constructed on an improved clay soil layer, it was important to analyze the consolidation settlement of the foundation without any improvements. The vertical settlement increments of untreated and treated soils during the construction stage (180 days) and consolidation phase are illustrated in Fig. 4. As seen from Fig. 4, the untreated and treated soils had different levels of settlements (u_y). The decrement of settlement level was also followed by a reduction of consolidation duration. This condition indicated that the settlements decreased and, possibly, the bearing capacity increased when the clay soil was stabilized with EAF slag. After the end of construction stage (after 180 days), the settlement amounts of treated soils increased linearly with time and the settlements of the clay layer, 70C-30EAF soil layer, and 50C-50EAF soil layer were 8.5 cm, 6.8 cm, and 5.4 cm, respectively. The corresponding consolidation rates were 85, 93, and 95%. The u_y-time profile progressed reaching to constant values of the settlement after the construction stage was completed. The constant settlement values were achieved after 400, 300, and 180 days in clay, 70C-30EAF slag mixture, and 50C-50EAF slag mixture layers. Therefore, the 50C-50EAF slag mixture had the shortest time to reach to a constant settlement value of 5 cm. Untreated clay required 490 days to subside 10 cm at the end of the primary consolidation stage. The consolidation was almost fully completed in the treated soil layers much more earlier; therefore, the settlement problem of the structure was successfully accomplished.

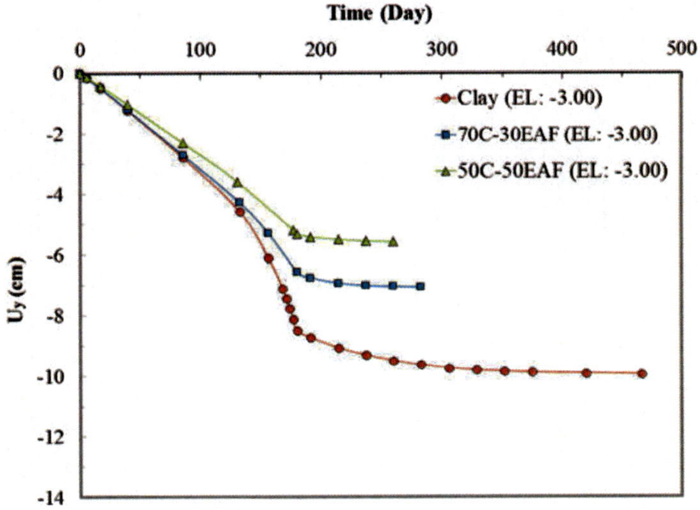

Fig. 4 Settlement-time relations calculated for the midpoint of the soft soil. (EL: −3.0 m)

4.2 Development of Excess Pore Pressure

The generation and distribution of excess pore pressure in clay and clay-EAF slag mixture layers during the construction stage are illustrated in Fig. 5a–c.

With the loads acting on the improved soil, the excess pore pressure development in clay-EAF slag mixtures was less than the clay soil. It is also noteworthy to mention that the focal point of excess pore pressure was shifted down to the untreated zone between −3.0 m and −6.0 m in treated soils (Blue to red color ranges denote Fig. 5 (a) 0–50 kPa, Fig. 5 (b) 0–17 kPa, Fig. 5 (c) 0–11 kPa.

The development of the excess pore pressure of untreated and treated soils during the construction stage (180 days) and consolidation phases are illustrated in Fig. 6.

During the consolidation period, the excess pore pressure increased with in time. After the end of construction the maximum excess pore pressure of −47.46 kPa was generated in the middle of the clay layer, but with the improvement of EAF slag,

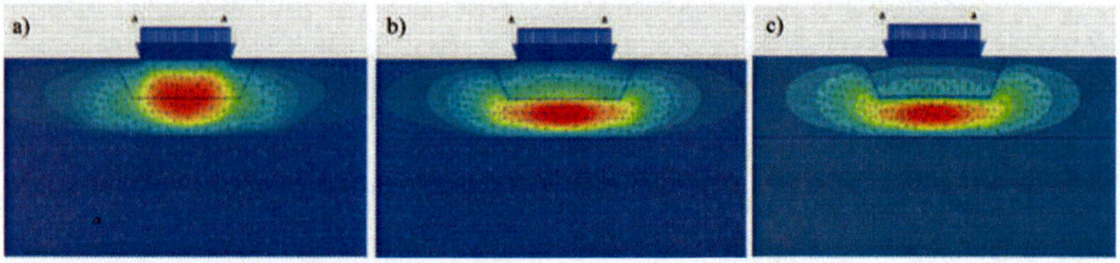

Fig. 5 Generation and distribution of excess pore pressure in the investigation area, **a** clay, **b** 70C-30EAF slag mixture, **c** 50C-50EAF slag mixture

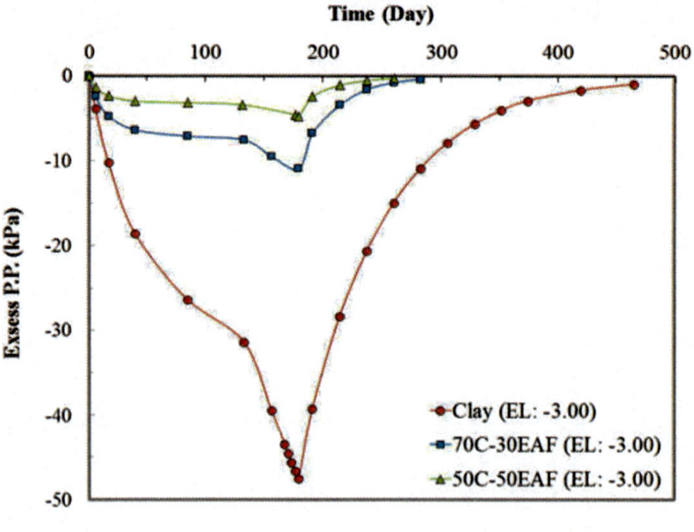

Fig. 6 Excess pore pressure development in the middle of clay layer

the excess pore pressure was reduced to −10.97 kPa (70C-30EAF) and −4.77 kPa (50C-50EAF) depending on the increasing rate of EAF slag. The acquisition of lower pore pressures helped in the completion of consolidation settlements.

5 Conclusions

The effects of slag content on the engineering properties and settlements of clay-EAF slag mixtures were investigated both experimentally and numerically in the current study. It should be noted that with the addition of EAF slag, engineering properties, compressibility, and shear strength parameters of the clay soils have been improved. The numerical analyses were evaluated based on the effect of EAF slag on the settlement of clay-EAF slag mixture layers under static loading conditions. The finite element analysis showed that using clay-EAF slag mixtures as a soil replacement under a foundation could reduce the vertical and total settlements of the soil layers as well as decreasing the time for the completion of the consolidation settlements.

Partial or complete replacement of soils with electric arc furnace (EAF) slag is increasingly drawing attention due to its benefits, including improvements in engineering properties of soils. The result of the present investigation indicated that EAF slag additive was recommended as a low-cost modifier or stabilizer of cohesive soils and clay like soils with poor engineering properties.

References

1. Al-Rawas, A.A., Hago, A.W., Al-Sarmi, H.: Effect of lime, cement and Sarooj (artificial pozzolan) on the swelling potential of an expansive soil from Oman. Build. Environ. **40**(5), 681–687 (2005)
2. Ronoh, V., Too, J.K., Kaluli, J.W., Victor, M.R.: Cement efects on the physical properties of expansive clay soil and the compressive strength of compressed interlocking clay blocks. Eur. Int. J. Sci. Technol. **3**(8), 74–82 (2014)
3. Khemissa, M., Mahamedi, A.: Cement and lime mixture stabilization of an expansive over consolidated clay. Appl. Clay Sci. **95**, 104–110 (2014)
4. Saeeda, K.A.H., Kassima, K.A., Yunusa, N.Z.M., Nurb, H.: Physico-chemical characterization of lime stabilized tropical kaolin clay. J. Teknol. **72**(3), 83–90 (2015)
5. Bose, B.: Geo engineering properties of expansive soil stabilized with fly ash. Electron J. Geotech. Eng. **17**, 1339–1353 (2012)
6. Hossain, M.T., Hoq, A., Akhter, M., Hossain, A.F.: Investigation on different properties of organic soil by adding fly ash. Int. J. Eng. Sci. Technol. **7**(1), 1 (2015)
7. Pei, X., Zhang, F., Wu, W., Liang, S.: Physicochemical and index properties of loess stabilized with lime and fly ash piles. Appl. Clay Sci. **114**, 77–84 (2015)
8. Solanki, P., Zaman, M., Dean, J.: Resilient modulus of clay subgrades stabilized with lime, class C fly ash, and cement kiln dust for pavement design. Transp. Res. Rec. J. Transp. Res. Board **2186**, 101–110 (2010)
9. Shalabi, F.I., Asi, I.M., Qasrawi, H.Y.: Effect of by-product steel slag on the engineering properties of clay soils. J. King Saud Univ.—Eng. **29**, 394–399 (2017)

10. Mozejko, C.A., Francisca, F.M.: Enhanced mechanical behavior of compacted clayey silts stabilized by reusing steel slag. Constr. Build. Mater. **239**, 117901 (2020)
11. Rajakumaran, K.: An experimental analysis on stabilization of expansive soil with steel slag and fly ash. Int. J. Adv. Eng. Technol. (2015).
12. Anggraini, V., Huat, B.B.K., Asadi, A., Nahazanan, H.: Efect of coir fibre and lime on geotechnical properties of marine clay soil. In: 7th International Congress on Environmental Geotechnics: ICEG2014, Engineers, Australia. p. 1430 (2014)
13. Estabragh, A.R., Bordbar, A.T., Javadi, A.A.: A study on the mechanical behavior of a fiber-clay composite with natural fiber. Geotech. Geol. Eng. **31**(2), 501–510 (2013)
14. Anastasiou, E.K., Papayianni, I., Papachristoforou, M.: Behavior of self-compacting concrete containing ladle furnace slag and steel fiber reinforcement. Mater. Des. **59**, 454–460 (2014)
15. Kavussi, A., Jalili Qazizadeh, M., Hassani, A.: Fatigue behavior analysis of asphalt mixes containing electric arc furnace (EAF) steel slag. J. Rehab. Civil Eng. **3**, 74–86 (2016)
16. Rondi, L., Bregoli, G., Sorlini, S., Cominoli, L., Collivignarelli, C., Plizzari, G.: Concrete with EAF steel slag as aggregate: a comprehensive technical and environmental characterisation. Compos. Part B **90**, 195–202 (2016)
17. Roy, S., Miura, T., Nakamura, H., Yamamoto, Y.: Investigation on material stability of spherical shaped EAF slag fine aggregate concrete for pavement during thermal change. Constr. Build. Mater. **2019**(215), 862–874 (2019)
18. Brinkgreve, R.B.J., Broere W.: PLAXIS Delft University of Technology & PLAXIS b.v., The Netherlands, Plaxis 2D (2015).

Experimental Study on Engineering Properties of Soft Soil Stabilized by Overboulder Asbuton

Noor Dhani, Tri Harianto, A. R. Djamaluddin, and Ahmad Gasruddin

Abstract Buton Island in Southeast Sulawesi, Indonesia, has a large amount of over 660 million metric tons of asphalt deposit in many forms. One of the types is overboulder which about 30% of the total deposit of asbuton. Overboulder occurred naturally when limestone reacts with bitumen in the topsoil layer of asbuton itself, making it a waste material which could be used as stabilizer material, due to its lime content with low bitumen content at about below 2%. The objective of this research was to analyze the effect of overboulder addition to soft soil as a pozzolanic agent and certain curing time to its unconfined compressive strength and CBR value. The soft soil was compacted according to ASTM D-698 standard method, as well as soft soil with overboulder addition. The UCT and CBR specimen was remolded based on its optimum moisture content and maximum dry density obtained from the compaction test. The unconfined compressive test was conducted according to ASTM D-2166 methods, and CBR test was conducted according to ASTM D-1883 methods. All specimen was treated with 0, 3, 7, 14, and 21 days of curing time in order to maximize its pozzolanic reaction between soil and overboulder asbuton. The overall result of the test shows that overboulder asbuton addition to soft soil increases its mechanical/engineering properties. Overboulder asbuton has potential as a local alternative to cement and/or lime in soil–cement and/or soil–lime stabilization due to its pozzolanic properties.

Keywords California bearing ratio (CBR) · Overboulder asbuton · Soft soil · Stabilization · Unconfined compressive strength (UCS)

N. Dhani (✉) · A. Gasruddin
Dayanu Ikhsanuddin University, Baubau City, Southeast Sulawesi, Indonesia
e-mail: noordhani_iptek@yahoo.co.id

T. Harianto · A. R. Djamaluddin
Hasanuddin University, Makassar, South Sulawesi, Indonesia

H. Hazarika et al. (eds.), *Advances in Sustainable Construction and Resource Management*, Lecture Notes in Civil Engineering 144,
https://doi.org/10.1007/978-981-16-0077-7_9

1 Introduction

Expansive soil is a common problem faced in developing in Indonesia. Soft soil may be existed in the form of peat soil, or soft clay distributed in east Sumatra, Kalimantan and West, West Coast Sulawesi, North Sumatra Beach, and Southern Papua covering an area of +20,000,000 ha.

Expansive soil thickness varies with low bearing capacity and compressive strength, the consequences cause many problems for construction built on peat soil due to high compression (consolidation) properties and long secondary compression.

Buton Island in Southeast Sulawesi, Indonesia, has a large amount of over 660 million metric tons of asphalt deposit in many forms. In Indonesia, natural asphalt derived from Buton Island was called asbuton (Asphalt Buton). The asbuton deposit is divided into several types. One of the types is overboulder which about 30% of the total deposit of asbuton. Overboulder occurred naturally when limestone reacts with bitumen in the topsoil layer of asbuton itself, making it a waste material which could be used as stabilizer material, due to its lime content with low bitumen content at about below 2%. Recently in Indonesia, asbuton, in general, started to be studied for either stabilization or asphalt content extraction. The objective of this research was to analyze the effect of overboulder addition to soft soil as a pozzolanic agent and certain curing time to its unconfined compressive strength and CBR value.

2 Literature Study

Soil cement has been used as a massive fill to provide foundation strength and uniform support under large structures. An extensive laboratory testing program was conducted to determine static and dynamic design characteristics, liquefaction potential, and durability of soil cement. Results showed that with only 5% cement content by dry weight, cohesion increased significantly, and it was possible to obtain a material with enough strength to prevent liquefaction [1]. Stabilized base or sub-base materials may be used to provide support for either flexible or rigid pavements, but are more frequently used with flexible pavements. The key to strength development in stabilized base or sub-base mixtures is in the matrix that binds the aggregate particles together. The strength of the matrix is affected by the cement material used in the mixture [2].

The pozzolanic reaction process, which can either be modest or quite substantial depending on the mineralogy of the soil, is a long-term process. This is because the process can continue as long as a sufficiently high pH is maintained to solubilize silicates and aluminates from the clay matrix, and in some cases from the fine silt soil. These solubilized silicates and aluminates then react with calcium from the free lime and water to form calcium silicate hydrates and calcium aluminate hydrates, which are the same type of compounds that produce strength development in the hydration of Portland cement [3]. However, for the pavement of high traffic loads, stabilization based on sand with 10% cement is recommended. It is recommended that future research can mix more than one types of sand for the stabilization and find out the compressive strength for that mixed sand sample. On the other hand, sand with higher F. M. value and lower F. M. value can mix with different percentages [4].

Soil stabilization is the alteration of the properties of soil to improve its engineering performance through its strength, stiffness, compressibility, permeability, and workability. In the last few decades, soil–cement stabilization has been more widely used in pavement construction compared to other methods, because of the significant improvements it can make to soil properties. It is found that the engineering behavior of a cement-stabilized base and that of a granular base when under traffic loading is not similar [5]. The strength of road subgrades, soil bases and sub-bases are commonly assessed in terms of the California bearing ratio (CBR) which is dependent on the type of soil used and its density and moisture content. Thus, it is important to use correct test procedures to assess the properties of a cement-stabilized road base [6].

In most applications, Type I or Type II Portland cement is normally used, and the cement content can vary from 4 to 16% of the dry weight of soil. Generally, as the clay portion of the soil increases, the quantity of cement required also increases. The particle size distribution of ordinary Portland cement is quite well defined falling within a range of 0.5–100 μm with a mean of 20 μm [7].

After it is stabilized, a soil layer will act as a soft and low strength concrete with linear elastic properties [8]. The tensile stress at the bottom of cement treated layers can cause fatigue cracking. The elastic modulus and the tensile strain at the bottom of cement treated layers are considered for the detailed analysis of a stabilized layer. The strength of the stabilized base and the sub-base are commonly assessed using the unconfined compressive strength (UCS) [9].

Overboulder asbuton dominated by calcium content of 79.64% and silicon content of 9.63%, so that in the composition contained by these minerals can be used as pozzolanic material that can function as a binder and filler from soft soil material that can increase the bearing capacity of the material [10]. However, compressive strength and durability were the two major approaches made, with the Americans more inclined toward durability technique. Other techniques employed were California bearing ratio, tensile and flexural tests [11].

3 Materials and Methods

In order to provide detail of materials used in the experimental study, laboratory investigation program was carried out to evaluate the basic properties and mechanical properties of the untreated soil and stabilized soil, and in this case, soft soil stabilized by overboulder. Each material used are then classified using unified soil classification system (USCS). In this case, the material used was soft soil with overboulder asbuton as stabilizer. The initial engineering properties was then compared to the stabilized soil.

3.1 Location

The asbuton overboulder material was brought from Buton Island and sampled at Lawele with coordinate 5° 13′ 53.56″ S and 122° 58′ 0.40″ E. Soft soil used was taken from around Gowa, South Sulawesi, Indonesia.

3.2 Methods

Unconfined compression test and California bearing ratio specimen was remolded according to standard method ASTM D 1883 and ASTM D 2166 with 6 in. in diameter and 7 in. high for CBR and 2 in. diameter with 4 in. high for UCT. Specimen is tested at curing time 0, 3, 7, 14, and 21 days (Figs. 1 and 2).

Fig. 1 Overboulder quarry at Lawelle

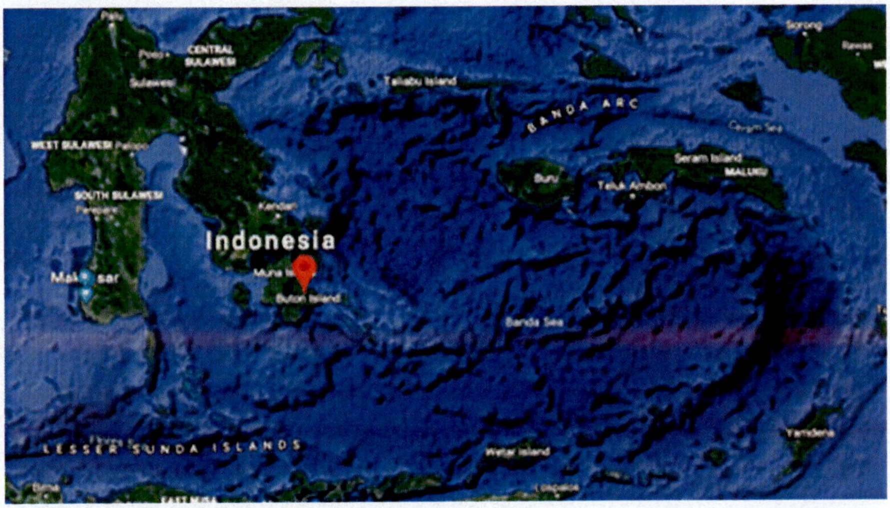

Fig. 2 Overboulder quarry location at southeast Sulawesi, Indonesia

4 Result and Discussion

The results of the basic properties and mechanical properties of soft soil are shown in Table 1.

According to laboratory analyzes, the plastic limit value is 46.35%, and the plasticity index is 14.40%. By linking the plastic limit values to the plasticity index in the plasticity diagram, the soil type is classified as MH (high plasticity silt), which corresponds to silty soils with high plasticity. Soil grains are dominated by a silt fraction of 34.55% and clay of 30.25%. By overall properties result, soil used in this study is a fine-grained soil, while the overboulder properties are shown in Table 2.

The results of the tests on the physical and mechanical characteristics of the overboulder indicate that it is classified as SP/sand with poor quality. However, the CBR value shows a higher value than soft soil (Fig. 3).

Microstructure characteristics was tested using XRD, SEM, and EDS. Based on the test results of energy dispersive X-ray spectroscopy of overboulder, it was described that overboulder asbuton dominated by calcium content of 79.64% and silicon content of 9.63%. From this point, overboulder asbuton assumed to have a pozzolanic properties (Table 3).

The addition of overboulder shows behavior that tends to show a cross-trend to changes in moisture content and dry weight. As the overboulder increases, the dry density value decreases while the optimum moisture content increases. Figure 4 illustrated that the dry density increases with the addition of overboulder, the dry density has increased by 0.6%, compared to the maximum dry density of untreated soil which is 13.21–13.29 kN/m^3.

Table 1 Basic properties and engineering properties of soft soil

Test		Test results	
		Result value	Unit
Basic properties of soft soil			
Initial water content (w)		35.72	%
Specific gravity (Gs)		2.65	–
Sieve analysis and hydrometer			
a	Sand	35.20	%
b	Silt	34.55	%
c	Clay	30.25	%
Atterberg limits			
a	Liquid limit (LL)	60.76	%
b	Plastic limit (PL)	46.35	%
c	Plasticity index (PI)	14.40	%
d	Shrinkage limit (SL)	26.51	%
Standard Proctor			
a	Maximum dry density, (γd)	1.35	gr/cm^3
b	Optimum moisture content (OMC)	29.84	%
Classification according USCS: MH, AASHTO: A-7–5			
Engineering properties of soft soil			
Unconfined compressive strength (qu)		47.35	kN/m^2
California bearing ratio–Unsoaked (CBR)		7.79	%
Elasticity modulus (E)		2064.50	kN/m^2

Figure 5 shows that by adding overboulder would increase the compressive strength of soft soil. With 0 days of curing time, the compressive strength reaches 854.73 kN/m^2 which is 18 times of magnitude compared to untreated soil. With 3 days of curing time, reaches 929 kN/m^2, 7 days of curing time significantly reaches 2139.52 kN/m^2, 14 days of curing time increasing to 2264.76 kN/m^2 and in 21 days increasing to 2313.2 kN/m^2. The highest compressive strength achieved in 21 days of curing time with the compressive strength increased 48 times compared to the untreated soil.

Figure 6 shows that adding overboulder would increase the CBR value of soft soil. With 0 days of curing time, the CBR value reaches 8.77% which is 112.5% higher compared to untreated soil. With 3 days of curing time, reaches 10.34%, 7 days of curing time significantly reaches 16.19%, 14 days of curing time increasing to 20.68% and in 21 days increasing to 21.13%. The highest CBR value achieved in 21 days of curing time with the CBR value increased 271% compared to the untreated soil.

Figure 7 illustrated the relation of UCS and CBR value with 5% overboulder added to soft soil at certain curing time. It shows that the engineering properties of

Table 2 Basic properties and engineering properties of overboulder asbuton

Test		Test results	
		Result value	Unit
Basic properties of overboulder asbuton			
Specific gravity (Gs)		2.65	–
Sieve analysis and hydrometer			
a	Sand	81.04	%
b	Silt	13.70	%
c	Clay	5.16	%
Standard Proctor			
a	Maximum dry density, (γd)	1.37	gr/cm^3
b	Optimum moisture content (OMC)	19.13	%
Classification according USCS: SP, AASHTO: A-1-B			
Engineering properties of overboulder asbuton			
Unconfined compressive strength (qu)		60.79	kN/m^2
California bearing ratio–Unsoaked (CBR)		12.06	%

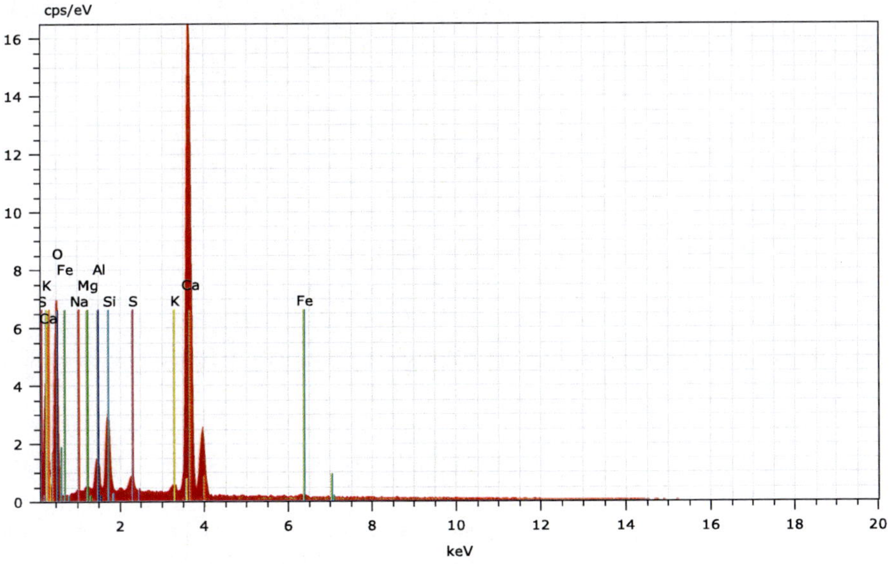

Fig. 3 Energy dispersive X-ray spectroscopy of overboulder asbuton

Table 3 Chemical composition of overboulder asbuton

Spectrum: test

Element	unn. C [wt.%]	norm. C [wt.%]	Atom. C [at.%]	Compound	norm. Comp. C [wt.%]	Error (3 Sigma) [wt.%]
Sodium	0.24	0.40	0.46	Na2O	0.54	0.17
Magnesium	0.21	0.35	0.38	MgO	0.57	0.14
Aluminium	1.43	2.34	2.29	Al2O3	4.42	0.32
Silicon	2.76	4.50	4.23	SiO2	9.63	0.46
Sulfur	0.52	0.85	0.70	SO3	2.13	0.16
Potassium	0.92	1.51	1.02	K2O	1.82	0.19
Calcium	34.90	56.92	37.46	CaO	79.64	3.18
Iron	0.60	0.97	0.46	FeO	1.25	0.19
Oxygen	19.72	32.16	53.02		0.00	8.18
Total:	61.30	100.00	100.00			

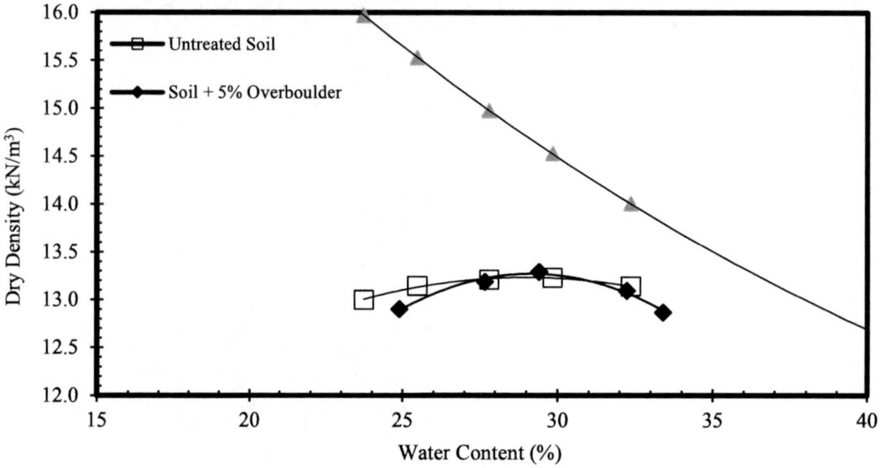

Fig. 4 Maximum dry density and optimum water content of untreated soft soil compared to soil + 5% overboulder

overboulder-stabilized soft soil accelerates in 3–7 days of curing time. The CBR and UCS value increased significantly with addition of overboulder. This showed that overboulder indeed has pozzolanic properties and is able to be used as cement or lime replacement as soil-stabilizing material. With the abundant amount of overboulder, it could possibly become useful local content to be used widely in construction.

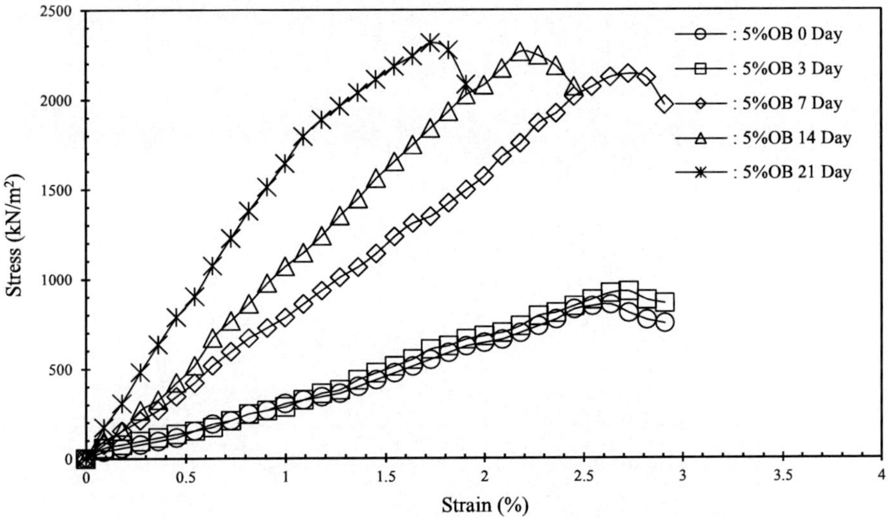

Fig. 5 Stress–strain behavior of overboulder-stabilized soft soil at certain curing time

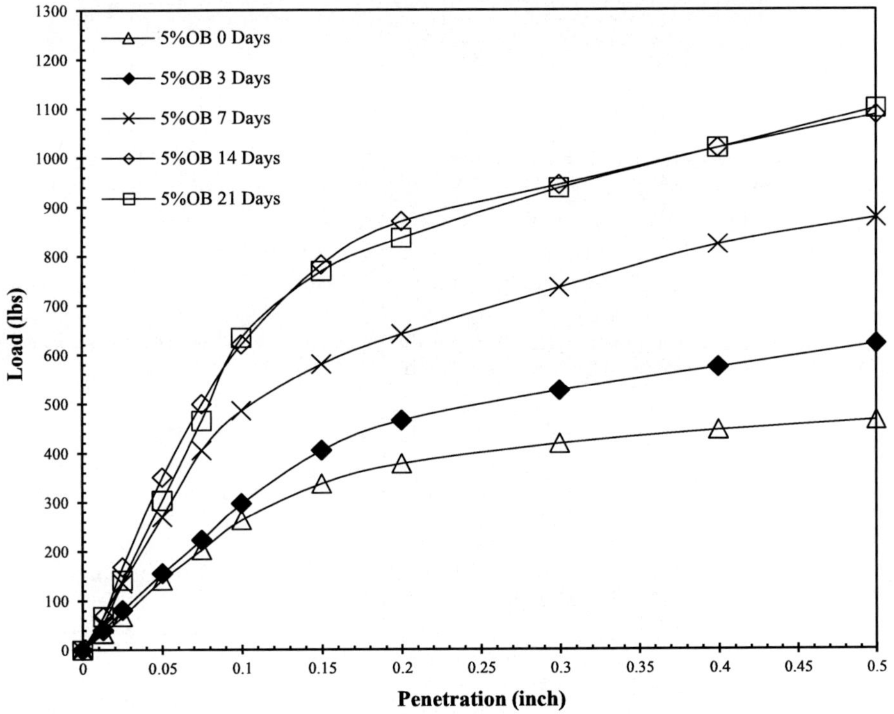

Fig. 6 Load–penetration behavior of overboulder-stabilized soft soil at certain curing time

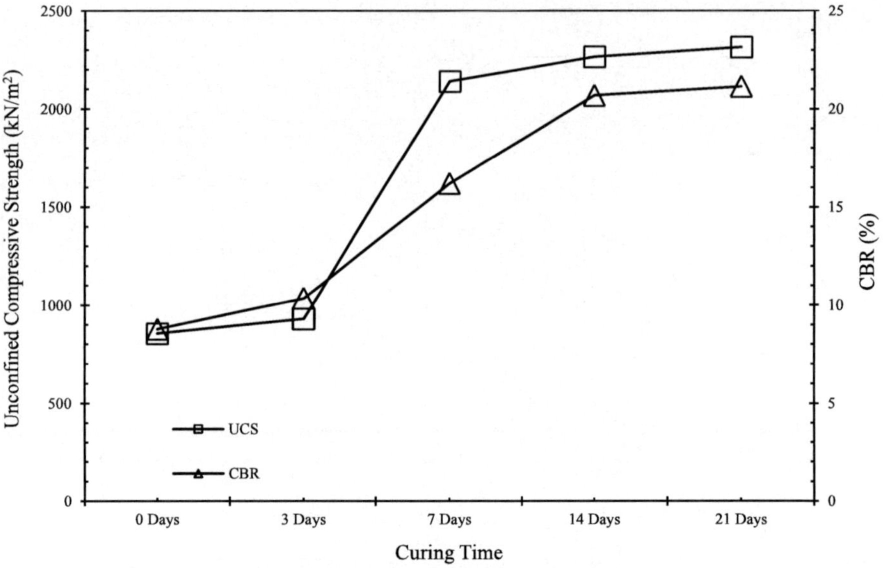

Fig. 7 Load–penetration behavior of overboulder-stabilized soft soil at certain curing time

5 Conclusion

Overboulder asbuton is considered as a pozzolanic agent due to its chemical composition. It is also proofed by the increasing value of UCS and CBR after overboulder addition to soft soil. This showed that overboulder is able to be used as cement or lime replacement as soil-stabilizing material. With the abundant amount of overboulder, it could possibly become useful local content to be used widely in construction.

Acknowledgements The authors gratefully acknowledge our colleagues and laboratory crews for their technical support during the experimental works.

References

1. Dupas, J.M., Pecker, A.: Static and dynamic properties of sand-cement. In: Proceedings, ASCE, vol. 105, No. GT3, Mar., pp. 419–436 (1979).
2. Soil Stabilization in Pavement Structures: A User's Manual. Volumes 1 and 2. Federal Highway Administration, Report No. FHWA-IP-80-2, Washington, DC (1980)
3. Little, D.N., Nair, S.: Recommended Practice for Stabilization of Subgrade Soils and Base Materials. National Cooperative Highway Research Program (NCHRP), Web only document—144, Texas A&M University, Texas (2009)
4. Mahmud Hasan Mamun M.: Improvement of sub base soil using sand-cement stabilization. Am. J. Civ. Eng., Bangladesh (2016). ISSN: 2330-8737

5. Bandara W.W.: Cement stabilized soil as a road base material for use in Sri Lankan Roads. J. Inst. Eng., Sri Lanka (2017)
6. Standard Specification for Construction and Maintenance of Road and Bridges (SSCM): Road Development Authority (RDA) (1989).
7. Ingles, O.G., Metcalf, J.B.: Soil Stabilization: Principles and Practice. Butterworth, Sydney (1972)
8. Croney, D., Croney, P.: The Design and Performance of Road Pavements, 3rd edn. Published by McGraw-Hill, UK (1998)
9. Technical Basis of Austroads Pavement Design Guide: TP-T33, 2011 United Kingdom Road Note 31, Transport & Road Research Laboratory (1977).
10. Dhani, N. et al.: Characteristics study of overboulder asbuton as pozzolanic material for soft soil stabilization. In: International Symposium of Infrastructure Development, Indonesia (2018).
11. Noor, M.J.M.M.: Durability and strength characteristics of cement stabilized modified Malaka series. J. Islamic Acad. Sci., Malaysia (1994).

Fundamental Study on the Effect of CO_2 Curing on the Strength Development of Alkaline Construction Sludge

Nguyen Duc Trung, Kimitoshi Hayano, and Hiromoto Yamauchi

Abstract Construction sludge is known as waste material generated from construction industry. Since the construction sludge is usually very soft and at a high water content, it is often necessary to treat it with cement or lime to be recycled as construction materials. However, after the treatment, the increased pH sometimes cannot satisfy environmental regulation. By injecting high-pressure CO_2 gas into the specimens, neutralization of alkaline construction sludge was accelerated. This research aims to study the effectiveness of using cement, CO_2 gas, and paper sludge (PS) ash for not only lowering the pH level but also strengthening the alkaline construction sludge to be re-used in long-term durability. Alkaline simulated sludge having similar pH levels and state with the actual sludge were produced in the laboratory. The observed results showed that pH was decreased from more than 11.5 to around 8.6 after 7 h curing by the injection of CO_2, whereas the addition of PS ash shortened the CO_2 curing time. The strength of simulated samples was decreased during CO_2 curing but then rapidly increased under air curing conditions. Strength of the samples with PS ash was seen significantly improved compared to that of samples without PS ash. After CO_2 curing, under air curing conditions, the pH of treated samples tended to rebound to a certain level less than the original value, but then gradually decrease. It can be seen that the maximum value of pH rebound depends on the pH value obtained at the end of the CO_2 curing.

Keywords Alkaline construction sludge · CO_2 curing · Cement-treated sludge · Paper sludge ash · Strength development

N. D. Trung (✉)
Graduate School of Urban Innovation, Yokohama National University, Yokohama, Japan
e-mail: nguyen-trung-hy@ynu.jp

K. Hayano
Faculty of Urban Innovation, Yokohama National University, Yokohama, Japan

H. Yamauchi
DOMI Environmental Solutions, Sodegaura, Chiba, Japan

H. Hazarika et al. (eds.), *Advances in Sustainable Construction and Resource Management*, Lecture Notes in Civil Engineering 144,
https://doi.org/10.1007/978-981-16-0077-7_10

1 Introduction

The development of infrastructure system as well as civil construction generates a huge industrial waste which is widely known as construction sludge. Due to its existence in liquid state, the construction sludge should be immediately treated with cement or lime to improve its strength before piling on the tipper truck. This process increases the pH level of construction sludge. However, in developed countries such as Japan, the law has regulated the pH of sludge to range between 5.8 and 8.6 for an effective dispose into the environment.

Based on the statistical data collected from in situ specimens, the on-site treated construction sludge was simulated by mixing Ao clay and blast furnace cement type B following the appropriate proportion that produces the pH of 11. Under air conditions (atmospheric conditions), the natural alkaline neutralization of construction sludge spends such a long time decreasing the pH value to reach 8.6. The previous research pointed out that it took at least 30 days to decrease the pH of alkaline construction sludge from 10 to 8.6 under air curing conditions [1]. The curing time was seen to be much higher for specimens having higher pH value. Indeed, specimens having pH > 10 cannot be naturally lowered to 8.6 under air curing conditions. By applying CO_2 curing, the pH of alkaline sludge was dramatically lowered. Some key factors affecting the acceleration of alkaline neutralization were investigated to be the degree of CO_2 concentration, CO_2 curing time, the layer thickness of specimens, and the initial water content. Especially, a decrease in the initial water content strongly accelerated the CO_2 curing time.

From that point of view, Ogasawara et al. [2]. attempted to reduce the initial water content of alkaline construction sludge by adding paper sludge ash of various proportions into simulated sludge. Paper sludge ash has high water absorption capacity and hence showed the effectiveness in setting a low initial water content of the mixture. On the other hand, PS ash also granulates the original construction sludge. This increases the surface area that accelerates the alkaline neutralization [3].

Moreover, the addition of PS ash into alkaline construction sludge could strengthen this mud to be used as a recycling material. It was investigated that the effectiveness of adding PS ash into cement-treated sludge for reducing the free water content consequently enhanced the strength. However, the strength development and pH of alkaline sludge treated by cement, PS ash, and CO_2 curing have not been discovered yet.

Carbon dioxide (CO_2) generated from the operation of factories all over the world cause the greenhouse effect, global warming, and other extreme weather events. Paper sludge ash is also a waste product from the paper industry. Thus, the combination of two kinds of waste material is expected to help mitigate the negative environmental impacts and contribute to a sustainable development.

In this research, treated construction sludge was simulated by a mixture called cement-treated sludge which included Ao clay at liquid limit state along with a weighted 3% of blast furnace cement type B. Then, a weighted 20% PS ash is also added to Ao clay in cement-treated sludge to form hybrid-treated sludge. The hybrid-treated sludge and cement-treated sludge were cured under sealed conditions before CO_2 curing. After the treatment using CO_2 incubator with CO_2 degree concentration of 10%, the pH of both, the cement-treated sludge and hybrid-treated sludge were measured to assess the acceleration of alkaline neutralization. Cone index tests were also conducted before and after the CO_2 treatment to understand the effect of CO_2 curing on the strength development of the mixtures. An effort to retain the authentic in situ state is made by curing the specimens under air conditions to study the strength development, pH stabilization, and durability in long term. At regular intervals during air curing time, the pH levels of CO_2-treated sludge were repeatedly measured, followed by the cone index test to investigate the strength improvement, pH variation, and durability in the long term. A comparison of alkaline sludges treated with and without PS ash was analyzed in terms of strength development and pH stabilization to assess the effectiveness in treating and recycling alkaline construction sludge.

2 Experiment Procedure

2.1 Material and Preparation

As mentioned previously, based on the statistical data of in situ treated construction sludge collected from the site, the simulated sludge called cement-treated sludge was produced in the laboratory corresponding to the pH and moisture content of site samples. This sludge consists of Ao clay at the liquid limit state ($w_L = 40.7\%$) and blast furnace cement type B. In order to strengthen the cement-treated sludge, PS ash is added to form a new composite mixture called hybrid-treated sludge.

Materials used for simulated sludge

The main material used for simulating alkaline sludge is Ao clay. It is categorized into CL (clay (low liquid limit)). The liquid limit (w_L) and plastic limit (w_P) of Ao clay are 40.7% and 23.7%, respectively. In order to form a clay slurry, distilled water is mixed with Ao Clay.

Cement used in this research was blast furnace cement type B. This is a mixture of ground granulated blast furnace slag (BFS) and Portland cement with the amount of BFS ranging between 30 and 60% as specified in JIS R 5211 [4]. This kind of cement is considered an eco-friendly cement since slag is used to produce cement instead of limestone and coal. Thus, the CO_2 emissions generated due to the decarboxylation of limestone or the incineration of coal are reduced.

Paper sludge ash is the cinder generated from the incineration of paper sludge. The main advantage of using PS ash is its high water absorption performance as mentioned previously. Moreover, adding PS ash granulates the new composite mixture, thereby increasing the surface area, and thus accelerates the alkaline neutralization. The proportion of PS ash in the hybrid-treated sludge that affects the acceleration of alkaline neutralization was assessed by Ogasawara et al. [2], according to which, the shortest CO_2 curing time was obtained at a corresponding PS ash to Ao clay ratio of 20% in weight. Thus, this research initially focused on this proportion of PS ash, and the others will be covered in future research.

The particle density of Ao clay, PS ash, and blast furnace cement type B are 2.716 g/cm^3, 2.603 g/cm^3, and 3.04 g/cm^3, respectively. The particle size distributions of Ao clay, PS ash, and hybrid-treated sludge are shown in Fig. 1, whereas that of cement-treated sludge was not measured since the mixture was not granulated. The typical chemical compositions of Ao clay, blast furnace cement type B, and PS ash are shown in Table 1.

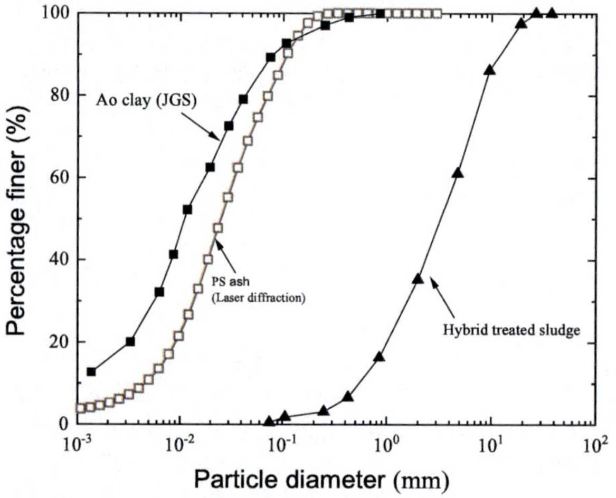

Fig. 1 Particle size distribution of Ao clay, PS ash, and hybrid-treated sludge [2]

Table 1 Chemical compositions of materials (%mass ratio)

(a) PS ash material

CaO	SiO_2	Al_2O_3	SO_3	FeO	MgO	P_2O_5	Others
40.6	20.7	17.2	6.2	4.6	2.7	1.4	6.6

(b) Blast furnace cement type B [5]

CaO	SiO_2	Al_2O_3	SO_3	Fe_2O_3	MgO	P_2O_5	Others
54.1	25.6	9.2	2.1	1.7	3.6	0.2	3.5

(c) Ao clay [6]

CaO	SiO_2	Al_2O_3	FeO	MgO	K_2O	Na_2O
1.9	57.1	20.1	11.1	4.1	4.1	1.7

Table 2 Cases of mixed condition

Types of specimen	Cement-treated sludge (%)	Hybrid-treated sludge (%)
Ratio of blast furnace cement type B to Ao clay in weight	3	3
Ratio of PS ash to Ao clay in weight	0	20

The pH of Ao clay, PS ash, and blast furnace cement type B are 7, 12.3, and 11.3, respectively [1, 2]. It can be seen that Ao clay is neutral, while that of PS ash and blast furnace cement type B are alkaline.

Preparation of simulated sludge

To produce cement-treated sludge, Ao clay at the liquid limit (w_L) is mixed with an amount of 3% in weight of blast furnace cement type B. Firstly, dry Ao clay was stirred with an appropriate amount of distilled water for 3 min to achieve the liquid limit state ($w_L = 40.7\%$) by using a bench-top mixer. Then, blast furnace cement type B was added and homogeneously stirred for 7 more minutes. Sequentially, the hybrid-treated sludge was simulated by adding PS ash to the cement-treated sludge with a ratio of 20% in weight of dry Ao clay, and the new composite mixture was then stirred for 5 more minutes. The mixing portion of cement-treated sludge and hybrid-treated sludge is presented in Table 2. Simulated sludge was then stored in plastic bags for curing under sealed conditions. This keeps the pH steady, and it helps make comparison of pH development under various conditions such as hermetic, sealed, and CO_2 curing conditions. After such periods of sealed curing time, the resultant simulated sludge was taken out to conduct the experiments.

2.2 *Methodology and Test Procedures*

Methodology

The objective of this research is to treat the construction sludge, so it can be used as a recycling material. In order to do so, the pH of construction sludge must be treated to be maintained at a stable value below 8.6, whereas the strength needs to be improved. To do so, the idea of using CO_2 gas to neutralize the alkalinity and PS ash to enhance the strength of the construction sludge was initiated with trying to find a solution to manage the huge amount of industrial wastes primarily generated in developed countries.

As mentioned, the cement-treated sludge was used to simulate the treated construction sludge on the field site. To enhance the strength of the cement-treated sludge further, PS ash was added due to two reasons. First, the strength development of sludge is obviously induced by cement. However, a previous study showed that the strength development of cement-treated soil is influenced by the water/cement (W/C) ratio [7]. Before adding cement for the immediate treatment at field site, construction sludge is present mostly in the liquid state. The high initial water content of construction sludge obstructs the hydration of cement that affects the strength development. To overcome this, the addition of an absorption material such as PS ash to the cement-treated sludge can solve this problem by setting the lower initial water content of the sludge. In other words, the *W/C* ratio is decreased which supports the hydration of cement. On the other hand, PS ash has a similar chemical composition as cement that also influences the hydration and contributes to the strength development.

Not only does PS ash improve the strength of cement-treated soil, it also granulates the treated mixture that may catalyze the carbonation. The hydration of cement takes place between the components of cement and water which generates calcium hydroxide $Ca(OH)_2$ and calcium silicate hydrates (C–S–H) [8]. The calcium hydroxide crystals formed in the process dissolve in water providing hydroxyl (OH^-) ions which may increase the pH level of treated sludge. By injecting CO_2 gas, the alkalinity is expected to be neutralized due to the carbonation of calcium hydroxide and calcium-silicate-hydrate (C–S–H) gel in the alkaline construction sludge. The mechanism of carbonate acceleration can be explained by chemical reactions as follows:

$$Ca(OH)_2 + CO_2 \rightarrow CaCO_3 + H_2O \tag{1}$$

$$(CaO)_x(SiO_2)(H_2O)_y + xCO_2 \rightarrow xCaCO_3 + (SiO_2)(H_2O)_t + (y - t)H_2O \tag{2}$$

Naturally, the spontaneous carbonation of calcium hydroxide suspension could gradually occur since the CO_2 gas exists in the air conditions at a concentration degree of 0.03% [9]. Thus, the pH of alkaline sludge could be decreased slowly to a certain value under the air conditions. In order to accelerate the alkaline neutralization, specimens are placed in the CO_2 incubator where the degree of CO_2 concentration is controlled to catalyze the carbonate reaction. Obviously, the pH of treated sludges are decreased during CO_2 curing, but the following pH levels after CO_2 treatment are important and must be monitored. The effect of PS ash on the pH maintenance of treated alkaline sludge after CO_2 treatment was then investigated by a comparison of pH levels of simulated sludges treated with and without PS ash. Also, the CO_2 curing time was adjusted to evaluate the effectiveness of pH reduction as well as the influence of the following pH after neutralization.

The strength development of cement-treated sludge and that of hybrid-treated sludge was measured by the cone index test followed by the Japanese Geotechnical Society Standard (JGS 0716-2009). And hence, the contribution of PS ash to the strength development in the short and long term of durability was also evaluated.

In this research, pH level and strength of cement-treated sludge and hybrid-treated sludge were measured simultaneously for comprehensively evaluating the effectiveness of using CO_2 gas together with PS ash in treating alkaline sludge.

Test procedures

After the preparation, simulated sludge was cured under sealed conditions up to 14 days. Each specimen was poured into a plastic bag with a weight of 3 kg and then stored in room at the temperature of 25 °C. After 3, 7, and 14 days of the sealed curing, simulated sludge was sequentially poured into trays, placed in a CO_2 incubator for curing under the CO_2 curing. Before placing in the CO_2 incubator, the height of mixtures in trays was adjusted around 25 mm. The CO_2 concentration degree in the incubator was set at 10%, and CO_2 gas could penetrate to the sludge from the surface of specimens.

The selection of CO_2 concentration of 10% vol is based on the results of the previous research [1]. It concluded that the acceleration of alkaline neutralization is induced with the increase of CO_2 concentration but in the range of degree less than 10%. The difference in neutralization rate in cases of CO_2 concentration degree between 10 and 20% was negligible.

In this study, the cement-treated sludge was deliberately crushed before the CO_2 curing to increase the neutralization speed so that it can be similar to that of the hybrid-treated sludge. Temperature and humidity in the CO_2 incubator were set at 25 °C and 90%, respectively. The curing time was set to be 7 h because the pH level of most specimens lowered to 8.6. Specimens in trays were then taken from the top to the bottom to measure the pH level. The pH measurements were implemented during the sealed curing and the CO_2 curing following the Japanese Geotechnical Society's Standard. Cone index tests were also implemented on the sludge during the sealed curing and the CO_2 curing.

The CO_2-treated sludge was then cured under air conditions for certain periods corresponding to 3, 7, and 14 days by storing in opened plastic bags in room at temperature of 25 °C. Again, pH measurement and cone index test were conducted to understand the behavior of cement-treated and hybrid-treated sludge after being treated with CO_2 and curing under air conditions. Moreover, the other 21-day sealed treated sludges were cured under CO_2 curing conditions with different periods of curing time to investigate the effect of initial treated pH value after CO_2 curing to the following pH thereafter. The procedure of conducting the experiment is shown in Fig. 2.

3 Results and Discussions

During 28 days under sealed condition, the pH level of cement-treated sludge and hybrid-treated sludge was almost equal and maintained to be stable between 11.5 and 11.9. It can be explained that the alkaline neutralization cannot be occurred in the absence of CO_2, thereby maintaining the stability of pH at a high value. In

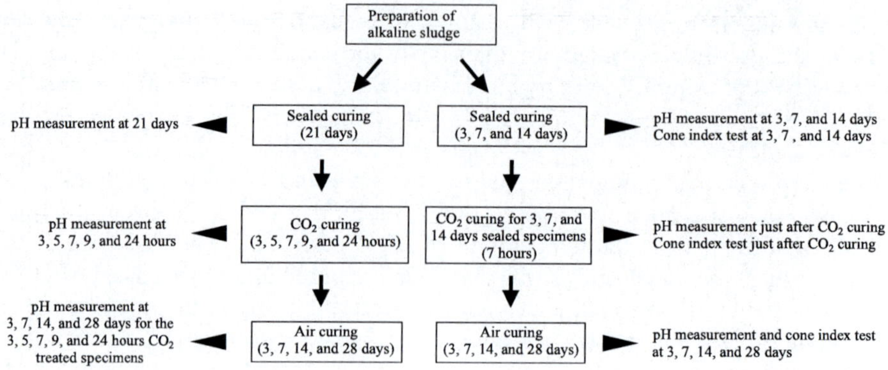

Fig. 2 Experiment procedure flowchart

general, the cone index of hybrid-treated sludge was much higher than that of cement-treated samples because a certain amount of free water was absorbed by the PS ash. Moreover, by absorbing water, the initial *W*/*C* ratio was reduced that influenced the hydration of cement as well as the strength development. However, the results showed that the strength of simulated sludges developed much slowly in sealed conditions compared to that in the air conditions as shown in Fig. 5.

Figure 3 shows the pH level of treated sludge under CO_2 curing and air curing conditions. The pH level was dramatically decreased from more than 11.5 to around 8.6 during CO_2 curing. It could be explained that the OH^--ions causing high alkalinity were dissolved by CO_2 gas. With the same CO_2 curing conditions and curing time, the pH of hybrid-treated sludge was seen to be decreased more significantly than that of the cement-treated sludge. This can be explained by the porous structure of hybrid-treated sludge arising from its size particles that allows the penetration of

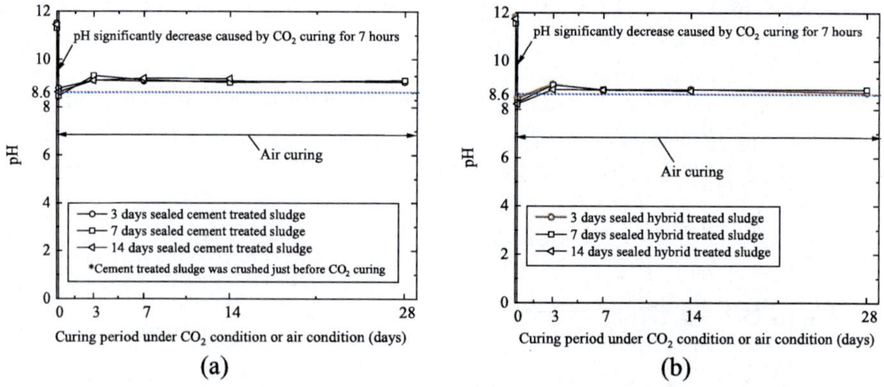

Fig. 3 pH rebound of simulated sludge after CO_2 curing: **a** cement-treated sludge, **b** hybrid-treated sludge

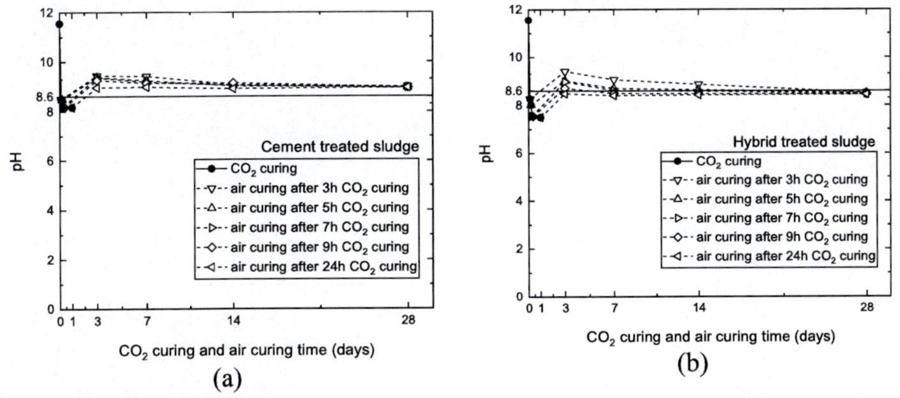

Fig. 4 pH rebound of simulated sludges after such periods of CO_2 curing time: **a** cement-treated sludge, **b** hybrid-treated sludge

CO_2 gas. In contrast, the c cement-treated sludge which was not granulated prevents the approach of CO_2 deeper into the particles. However, the results showed that the pH level of all samples rebounded but did not return to the original value. It reached a maximum value after being cured for almost 3 days under air conditions and then gradually decreased. This phenomenon probably arose from the residual hydration after CO_2 curing.

Figure 4 expresses the pH rebound of the 21-day sealed treated sludges cured under different periods of CO_2 curing time. The pH fell dramatically (from 11.56 to 7.55 and 11.55 to 8.21 in the case of hybrid-treated sludge and cement-treated sludge, respectively) after 7 h of CO_2 curing but then continued slightly decrease (from 7.55 to 7.5 and 8.21 to 8.17 in the case of hybrid-treated sludge and cement-treated sludge, respectively) during the rest of curing time up to 24 h.

At the end of 24 h CO_2 curing, pH of cement-treated sludge could not be decreased to less than about 8, whereas that of hybrid-treated sludge fell to 7.5. It can be seen that the pH at the end of the CO_2 curing is the crucial factor influencing the following pH in long-term durability. Although affected by the pH rebound phenomenon, the upcoming pH of specimens having the pH less than about 8 at the end of the CO_2 curing was maintained at the values less than 8.6 several days after air curing to satisfy the requirement of Japanese Law.

Although the effectiveness in reducing pH after 7–24 h of CO_2 curing was negligible, the results showed that a longer curing time shows a better effect in suppressing the pH rebound and hence contributes to a low maintenance of pH thereafter.

Figure 5 shows the strength development of simulated sludges under the sealed, CO_2 curing, and air conditions. It can be seen that the strength decreased during CO_2 curing in case of cement-treated sludge. This strength reduction was induced by the dissolution of hydroxyl ions (OH^-). Furthermore, the reduction of the strength of hybrid-treated sludge was negligible. This is likely due to the absorption of water produced from CO_2 curing process by PS ash. This mechanism may continuously

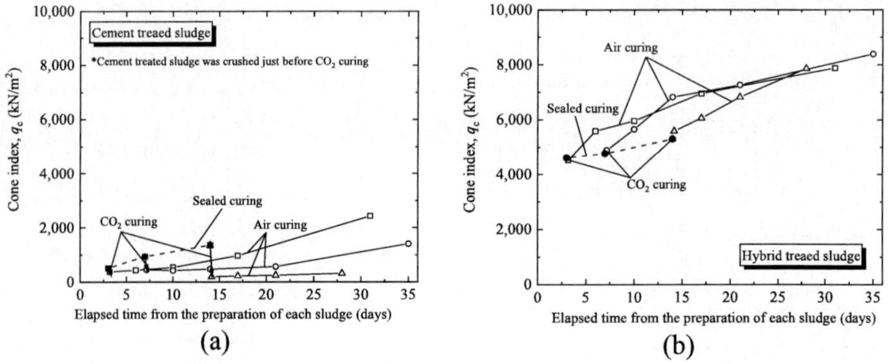

Fig. 5 Strength development of simulated sludge: **a** cement-treated sludge, **b** hybrid-treated sludge

take place after CO_2 curing which caused the strength of hybrid-treated sludge to increase abruptly, whereas the strength of cement-treated sludge developed gradually. Indeed, by reducing the free water content of hybrid-treated sludge, the water/cement ratio reached an appropriate value that accelerates the hydration of cement as well as increases the strength effectively.

4 Conclusions

The effects of CO_2 curing and PS ash on the strength development of alkaline construction sludge were investigated in this research based on experimental research. Depending on the characteristics of actual treated construction sludge, simulated sludge named cement-treated sludge was prepared. An attempt to enhance the strength of cement-treated sludge was conducted by adding PS ash to form hybrid-treated sludge.

Under the sealed conditions, the pH of simulated sludges was maintained to be stable between 11.5 and 11.9 after 14 days of curing. There was no difference between the pH values of cement-treated sludge and that of hybrid-treated sludge.

The alkaline neutralization was successfully accelerated by using the CO_2 incubator set-up with the CO_2 concentration degree of 10%. pH was dramatically decreased from 11.56 to 7.55 and 11.55 to 8.21 in the case of hybrid-treated sludge and cement-treated sludge, respectively, after 7 h of CO_2 curing. However, the slowdown in neutralization was seen thereafter.

After CO_2 treatment, the pH was rebounded to a certain value less than the original value after curing for 3 days under sealed conditions. The following pH depends on the pH value at the end of the CO_2 curing. The hybrid-treated sludge could be neutralized to reach a pH less than about 8, and then, the following pH was seen maintained below 8.6 several days after air curing.

A reduction in strength of cement-treated sludges was observed during the CO_2 curing. In addition, strength reduction of hybrid-treated sludge was negligible because of the absorption capacity of PS ash.

Strength developed significantly after the CO_2 treatment, especially for hybrid-treated sludge. This can be attributed to the absorption of free water present in the sludge that influenced the appropriate *W*/*C* ratio accelerating the strength development.

References

1. Imai, K., Hayano, K., Yamauchi, H.: Fundamental study on the acceleration of the neutralization of alkaline construction sludge using a CO_2 incubator. Soils Found. **60**(4), 800–810 (2020)
2. Ogasawara, T., Hayano, K., Yamauchi, H.: Proposal of neutralization acceleration method for alkaline construction sludge using Carbon dioxide and PS ash based stabilizing material. In: Proceeding of the 55th Annual Meeting of Japanese Geotechnical Society, 2201-07-01, 2020 (in Japanese)
3. O'Connor, W.K., Dahlin, D.C., Rush, G.E., Dahlin, C.L., Collins, W.K.: Carbon dioxide sequestration by direct mineral carbonation: process mineralogy of feed and products. Mining, Metall. Explor. **19**(2), 95–101 (2002)
4. Maeda, Y.: Slag Cement-related products which utilized a property of the ground granulated blast furnace slag. In: Nippon steel & Sumitomo Metal Technical Report No. 109 July 2015, pp. 114–118 (2015)
5. Shinmi, T., Chabayashi, T., Hiroyoshi, K.: Properties of blast furnace slag cement type B using the low burning-temperature type clinker. Cem. Sci. Concr. Technol. **71**(1), 302–307 (2017)
6. Kawai, S., Hayano, K., Yamauchi, H.: Fundamental study on curing effect and its mechanism on the strength characteristics of PS ash-based improved soil. J. Jpn. Soc. Civ., Eng. Ser. C (Geosphere Engineering) **74**, 306–317 (2018)
7. Miura, N., Horpibulsuk, S., Nagaraj, T.S.: Engineering behavior of cement stabilized clay at high water content. Soils Found. **41**(5), 33–45 (2001)
8. Duggal, S.K.: Building Materials. New Age International (P) Limited (2009)
9. Nakarai, K., Yoshida, T.: Effect of carbonation on strength development of cement-treated Toyoura silica sand. Soils Found. **55**(4), 857–865 (2015)

Fundamental Study on the Mechanical Characteristics of Sand Treated by a PS Ash-Based Improving Material

Maliki Otieboame Djandjieme, Kimitoshi Hayano, Yoshitoshi Mochizuki, and Hiromoto Yamauchi

Abstract This paper presents the potential of using paper sludge ash (PS ash) to improve the properties of sand for the usage as a backfill material. So far cement and lime have been used to increase the liquefaction strength of the backfilled sand around underground pipes and manholes. However, the cement- or lime-treated backfilled sand sometimes solidifies too much resulting in the problem of re-excavation ability. In this study, Toyoura sand improved by adding PS ash was investigated to evaluate the ability to be re-excavated by analyzing the unconfined compression strength. Moreover, a comparison of the results with the cement improvement case was conducted. Experimental results show that the q_u of treated sand increases with the increase of PS ash addition ratio. However, the curing effect on strength depends on the addition ratio. It is revealed that the PS ash addition ratio 2.5–5.0% is recommended for PS ash-treated specimen since their unconfined compressive strength is moderate over time.

Keywords Paper sludge ash · Sand · Density · Re-excavation · Unconfined compressive strength

1 Introduction

Construction of public water supplies, waste collection, pipelines, or manholes, etc., required backfilled sand. Historical behavior of backfilled sand during a significant

M. O. Djandjieme (✉)
Graduate School of Urban Innovation, Yokohama National University, Yokohama, Japan
e-mail: maliki-djandjieme-mv@ynu.jp

K. Hayano
Faculty of Urban Innovation, Yokohama National University, Yokohama, Japan

Y. Mochizuki
Sustainable Eco Corp, Yokohama, Japan

H. Yamauchi
DOMI Environmental Solutions, Sodegaura, Japan

H. Hazarika et al. (eds.), *Advances in Sustainable Construction and Resource Management*, Lecture Notes in Civil Engineering 144,
https://doi.org/10.1007/978-981-16-0077-7_11

natural event such as earthquake has shown its weakness. These weaknesses were described as a liquefaction phenomenon occurring around the underground pipe due to the buildup of pore water pressures during dynamic loading inside the backfilled sand. The consequences of the phenomenon have been investigated in Japan. According to the survey, the nature of damages linking to sewage pipes systems, including manholes, varied from region to region. In Kanto region, especially along the coast and in the lower reaches of the Tone River Basin far from the epicenter, road surface subsidence, pipe sag, and sand plugging occurred by massive liquefaction [1]. The analysis of the results of the surveys shows two categories of liquefaction pattern: partial liquefaction in backfilling soil and massive liquefaction in the surrounding. Partial liquefaction occurred in backfilling soil. It is described as the boreholes that were excavated for placing the sewage pipes and manholes. This led to pipe sag, manhole uplift, and road surface subsidence, affecting the closure of drainage function and traffic disruption. Statistically, 90% of the underground structures were damaged by liquefaction with an average of 65% of partial and 25% of massive liquefaction. Both types of liquefaction need to be limited by improving backfilled sand with additional material susceptible to prevent liquefaction. Therefore, the application of paper sludge (PS) ash material as a ground improvement material is promoted in this research for its capacity to improve the mechanical behavior of soil.

From the previous studies, PS ash can increase the reactive surfaces of the soil grains and can increase the production of hydration and pozzolanic reactions among soil particles [2]. These characteristics show that PS ash could be a suitable potential replacement for Portland cement [2]. Furthermore, it was found that a small amount of treatment PS ash-based improving material can absorb the excess water in the fine-grained gravel and avoid segregation between coarse and fine particles during the compaction of filled material. Therefore, the strength of the treated fine-grained gravel increases. Moreover, curing of the specimen showed a better bonding of PS ash and clayey soil by absorbing the water content of clay soil by air-dried powdered paper sludge ash, and recently, PS ash is adopted as an innovative material that can be used as material for ground improvement [3].

However, this research project aims to apply PS ash as a binder material in the treatment of backfill sand. As a preliminary step, mechanical properties of sand treated by PS ash are investigated from a series of unconfined compression tests to evaluate the required additional ratio of PS ash and to analyse its re-excavation ability after treatment. A comparative study with cement will also be evaluated to approve its usage as cement replacement in backfilled sand around underground pipes and manholes.

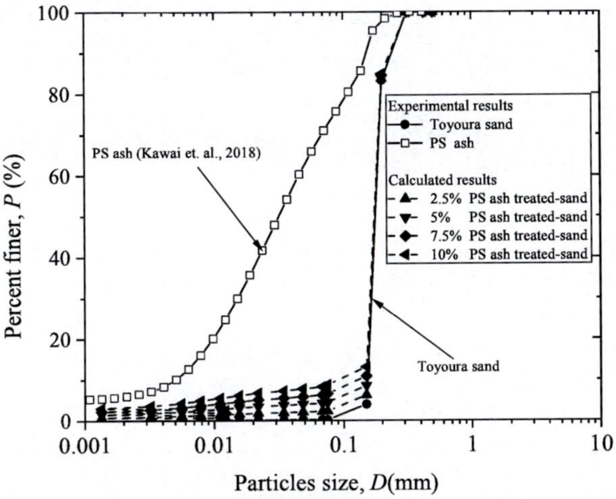

Fig. 1 Particle size distribution of the materials used

2 Experimental investigation

2.1 *Materials*

Toyoura sand is well-known sand for use in research in Japan. According to the unified soil classification system, the sand is classified as SP (poorly graded sands), with the following properties: mean diameter, $D_{50} = 0.16$ mm, a coefficient of curvature, $C_c = 0.83$, and a coefficient of uniformity $C_u = 1.47$. A commercial type of PS ash-based improving material [4], described as PS ash in this paper was used with ordinary Portland cement to improve the properties of Toyoura sand. The particle size distribution of Toyoura sand, PS ash, and calculated results of PS ash-treated sand is shown in Fig. 1.

2.2 *Methodology of Testing*

The minimum and maximum density tests were first conducted to estimate the minimum and maximum void ratio. Based on the results, a series of unconfined compression tests (UCTs) was then performed on both specimens of sand treated by PS ash and treated by cement following Japanese Geotechnical Standards [6] and using a wet tamping to limit the increasing volume inside the specimen. The water content of sand was adjusted to 10% before mixing with stabilizing agents under various ratios. The mixture then was transferred into a cylindrical plastic mold with the dimension of 50 mm of diameter and 100 mm of height for each treated sand. The mixture in the mold was filled and smoothly compacted in three layers by hitting side

Table 1 Conditions of treated specimens for UCTs

Addition ratio of stabilizing agent, A_{PS} (%)	Relative density, D_r (%)	Underwater curing days, t (day)
2.5, 5.0, 7.5, 10.0	50, 90	7, 14, 28

by side on the surface of the mold using a hammer to achieve the required relative density. The specimens in the molds were placed inside a rubber container for three days before being filled with water and stored at a controlled room temperature of 20 °C. Then, the container was filled with water, and the specimens in the molds were cured underwater for 7, 14, and 28 days. Details of the treatment condition are shown in Table 1. Three specimens were prepared for each treatment condition.

3 Experimental Results and Discussion

3.1 Influence of PS Ash to Void Ratio Development of Treated Sand

Figure 2a shows the minimum dry density ρ_{dmin} and the maximum dry density ρ_{dmax} plotted against the addition ratio of PS ash, A_{PS}. Here, A_{PS} is defined as the ratio of the PS ash to the Toyoura sand in dry weight. Figure 2b shows the ρ_{dmin} and the ρ_{dmax} plotted against the addition ratio of cement, A_{cement}. Here, A_{cement} is defined as the ratio of the cement to the Toyoura sand in dry weight. Comparing Fig. 2a and b, it can be seen that the sand-treated PS ash has a lighter weight characteristic than the sand treated by cement as the addition rate increases.

Figure 3a shows the maximum void ratio e_{max} and the minimum void ratio e_{min} plotted against the addition ratio of PS ash, A_{PS}. As seen in Fig. 3a, e_{min} changes little

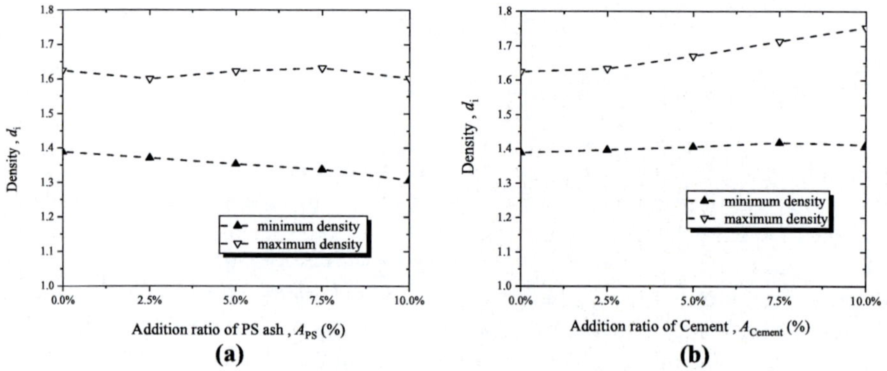

Fig. 2 ρ_{dmin} and ρ_{dmax} of sand treated by **a** PS ash and **b** cement

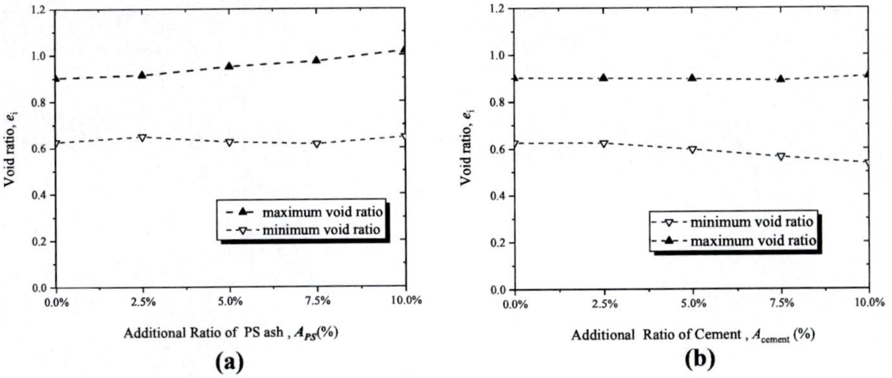

Fig. 3 e_{min} and e_{max} of sand treated by **a** PS ash and **b** cement

with the increase of A_{PS}, while the e_{max} increases gradually with high A_{PS}. Therefore, the difference, $e_{max}- e_{min}$ is increasing with the increase of A_{PS}. The increase of $e_{max} - e_{min}$ with treated material is important so that the compaction effort of the PS ash-treated sand as backfilled material around underground pipes could be rewarded.

The e_{max} and e_{min} void ratio showed in Fig. 3b confirms an evolution of the difference $e_{max} - e_{min}$ with cement ratio in the mixture. For cement treatment, e_{max} changes little with the increase of the addition ratio, while the e_{min} decreases. The tendency is different between PS ash and cement which is not clear, but the difference in the specific gravity of both particles may affect. That is, the specific gravity of ordinary Portland cement is $G_s = 3.150$, which is larger than the specific gravity of Toyoura sand, $G_s = 2.641$, while the specific gravity of PS ash is $G_s = 2.603$, which is smaller than the specific gravity of Toyoura sand, $G_s = 2.641$.

3.2 *Mechanical Properties of PS Ash and Cement-Treated Sand from Unconfined Compression Tests*

For different mixing ratios, Figs. 4 and 5 show the stress–strain behavior after 7, 14, and 28 days of curing period, respectively. Figure 4 represents the strain–stress relationship of PS ash-treated sand, and Fig. 5 serves as a comparative graph of the cement-treated sand. The analysis of Fig. 4 shows that for a small mixing ratio, low peak stress about 11.9 kPa is obtained. In fact, after the addition of PS ash, the peak strength increases with an increase in mixing ratio content. The unconfined compressive strength, q_u, of the specimen with the relative density $D_r = 90\%$ is higher than that of the specimen with a relative density, $D_r = 50\%$. As a comparative point of view, Fig. 5 shows that for a small mixing, a ratio high peak stress about 82.3 kPa is obtained. Comparing Fig. 4 and Fig. 5, the cement-treated specimens showed higher strength and rigidity than the PS ash-treated specimens.

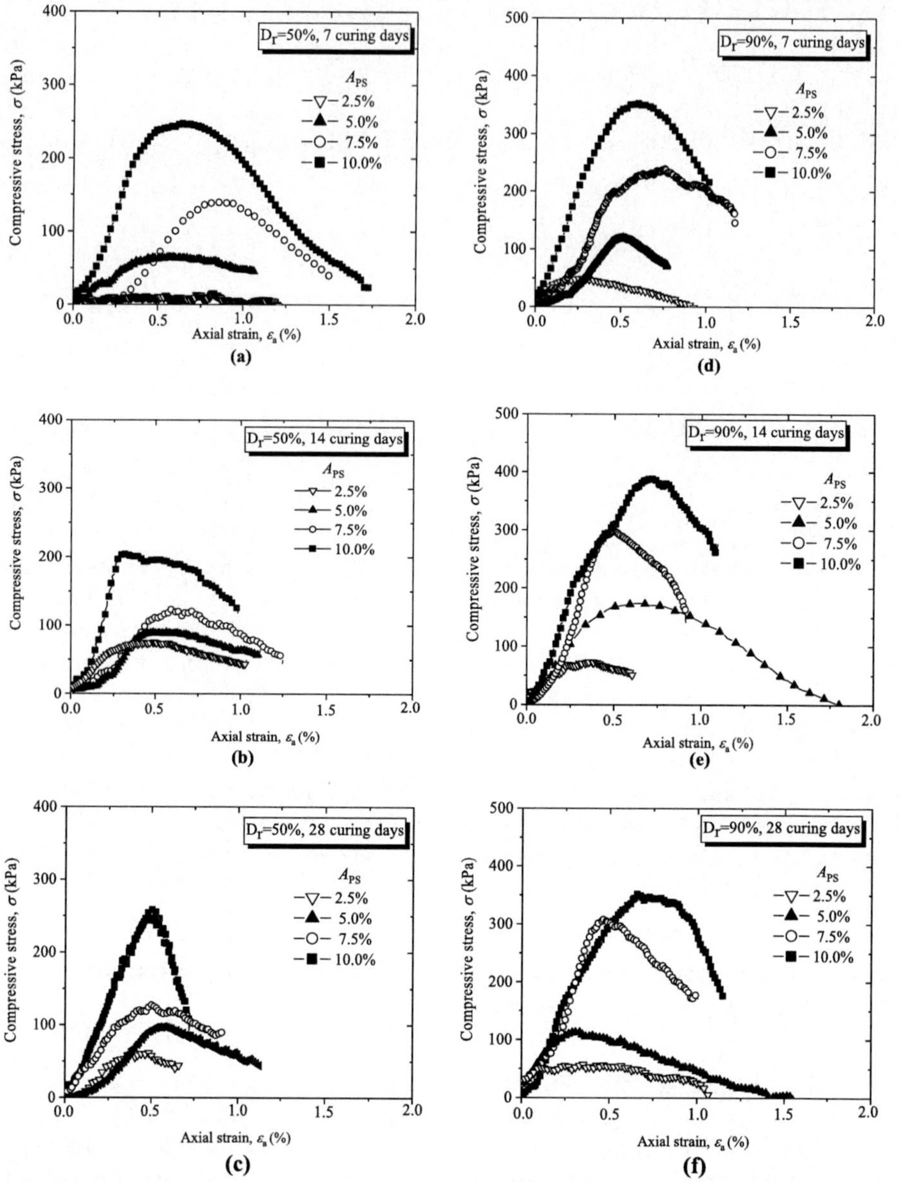

Fig. 4 Stress–strain relationships of PS ash-treated sand obtained from UCTs (**a**, **b**, **c** for D_r = 50% – **d**, **e**, **f** for $D_r = 90\%$)

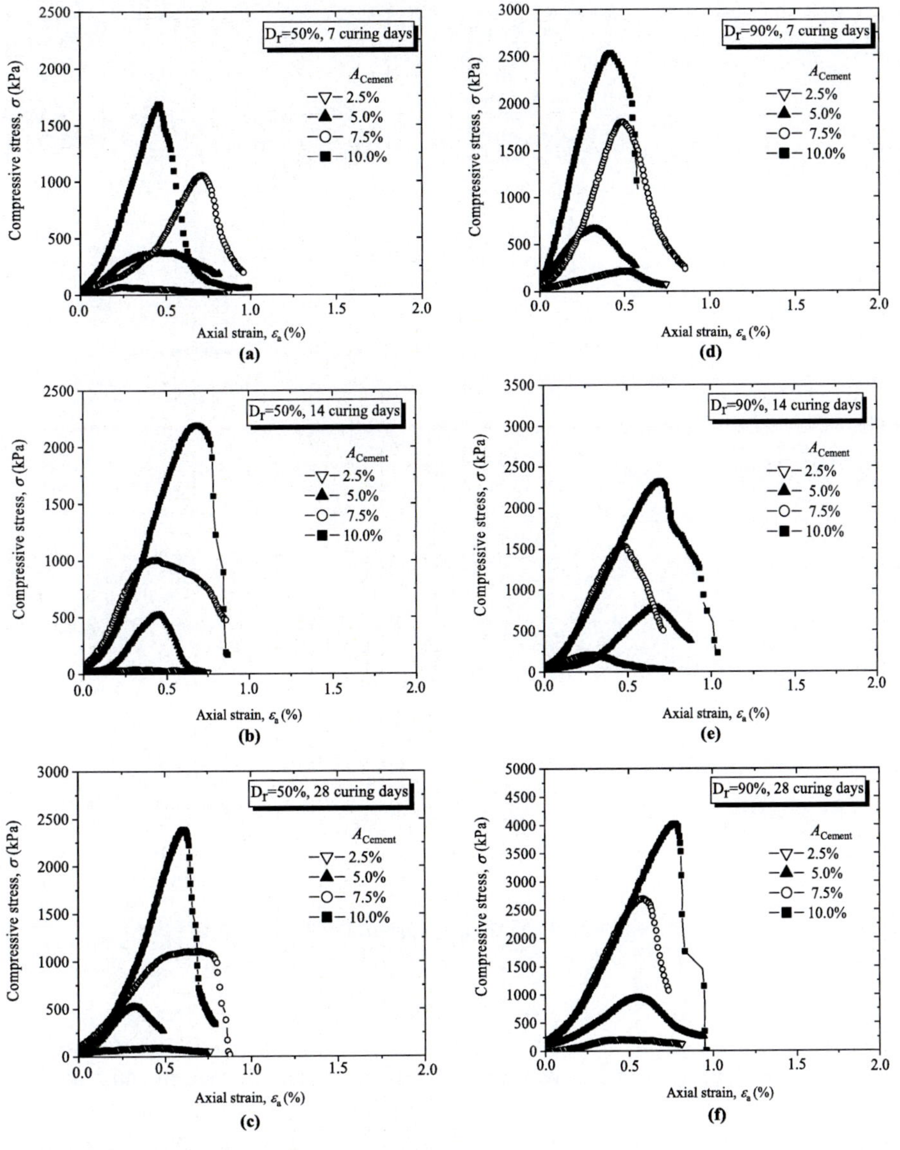

Fig. 5 Stress–strain relationships of cement-treated sand obtained from UCTs (**a**, **b**, **c** for D_r = 50% – **dd**, **e**, **f** for $D_r = 90\%$)

Figure 6 shows the average unconfined compressive strength q_u versus curing time for PS ash-treated sand with 2.5, 5, 7.5, and 10% PS ash mixing ratio, respectively. Figure 7 shows the average unconfined compressive strength, q_u, versus curing time for cement treated sand with 2.5, 5, 7.5, and 10% cement mixing ratio, respectively. Here, the average unconfined strength, q_u, means the average value of q_u of the

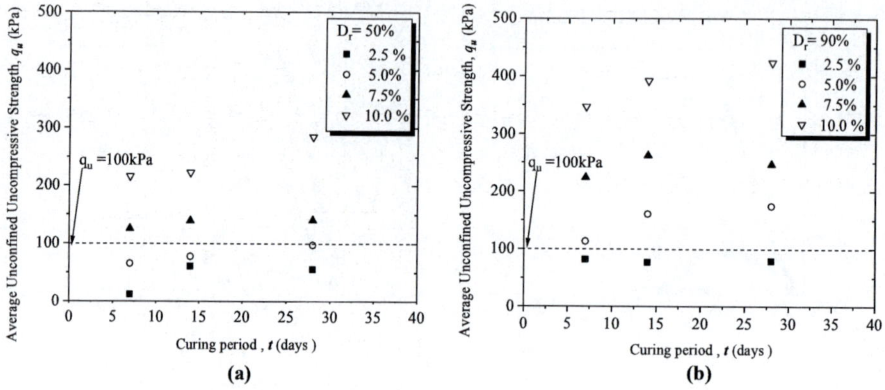

Fig. 6 Average q_u—curing time t relationships of PS ash-treated sand

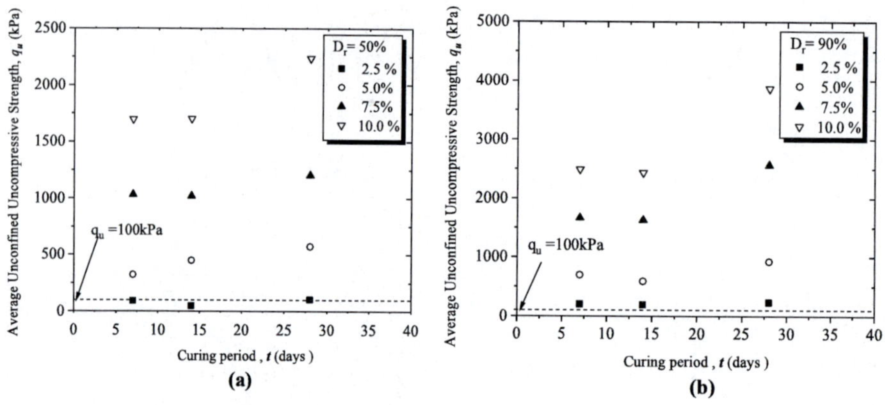

Fig. 7 Average q_u—curing time t relationships of cement-treated sand

three specimens for each treatment condition. The treated sand strength gradually increases with the curing time except for the specimens with a PS ash mixing ratio of 2.5%. It is found that the average, q_u does not increase significantly with the curing time if the A_{PS} is 2.5%. Comparing Fig. 6 and Fig. 7, when the mixing ratio increases, the difference between the q_u of cement-treated sand and the q_u of PS-treated sand is large. This behavior of cement-treated sand can be a great concern when the backfilling ground needs to be re-excavated.

From the viewpoints of liquefaction resistance and re-excavation ability, the unconfined compressive strength of the treated backfilled sand is better too kept around 100 kPa over time. Thereby the, test results suggest that the appropriate range of PS ash might be around 2.5–5.0%.

Figure 8a and b show the average unconfined compressive strength, q_u versus the average Young's modulus E_{50} for PS ash-treated sand and cement-treated sand,

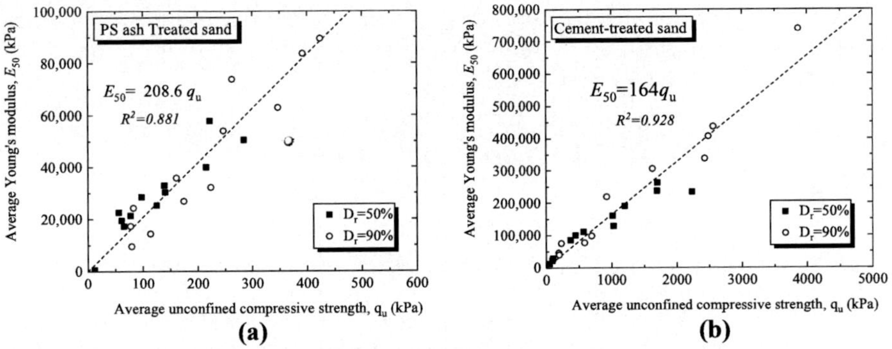

Fig. 8 Average q_u—average E_{50} relationships of sand treated by **a** PS ash and **b** cement

respectively. In order to analyze the deformation over the range of unconfined compressive strength, treated sand indicates that the E_{50} increases with the increase of the q_u. The analyses showed a well-defined correlation between E_{50} and q_u of the compiled treated sand. As expected, the coefficient of the relation $q_u \sim E_{50}$ of the PS ash-treated sand is slightly higher than that of the cement-treated sand.

4 Conclusions

The following conclusions can be drawn from the present study:

1. With the increase of PS ash addition, e_{min} of the treated sand changes little, while the e_{max} increases. Therefore, the difference, $e_{max} - e_{min}$, of the treated sand is increased with an increase of PS ash content. This point is essential so that the compaction effort of the PS ash-treated sand as backfilled material around underground pipes could be rewarded.
2. Experimental results show that the q_u of PS ash-treated sand increases with an increase in PS ash mixing ratio. However, the curing effect on strength depends on the addition ratio. It is revealed that the PS addition ratio of between 2.5 and 5% is recommended since the unconfined compressive strength is moderate and does not exceed 100 kPa over time. On the other hand, cemented-treated sand has a disadvantage because the strength obtained for cement treatment is much higher than that of PS ash-treated sand even for a small mixing ratio between 2.5 and 5%. Thereby, it may be difficult to manipulate it during the maintenance of an underground structure.
3. E_{50} increases with the increase of unconfined compressive strength, q_u; moreover, the E_{50} is slightly higher in the case of PS ash-treated sand compared with the cement-treated sand.

Focusing on the improvement of backfilled sand, PS ash is more suitable as a backfilled material not only due to its ability to keep its compression strength almost constant but also to simplify the excavation work for maintenance purposes. However, it is important to make a further study on its effect over a long curing period such as 90, 180 days, etc., and on its resistance to liquefaction.

References

1. Matsuhashi, M., Tsushima, I., Fukatani, W., Yokota, T.: Damage to sewage systems caused by the Great East Japan Earthquake, and governmental policy. Soil. Found. **54**(4), 902–909 (2014)
2. Singh, T.S., Pant, K.K.: Solidification/stabilization of arsenic-containing solid wastes using Portland cement, fly ash and polymeric materials. J. Hazard. Mater. **131**(1–3), 29–36 (2006)
3. Zuber, S.Z.S., Kamarudin, H., Al Bakri Abdullah, M.M., Binhussain, M.: Review of soil stabilization techniques. Australian J. Basic Appl. Sci. **7**(5), 258–265 (2013)
4. Yuki, W.: Effects of paper sludge ash-based improving material on the compaction and mechanical characteristics of fine-grained gravel. Japan. Soc. Civ. Eng. JSCE (2018)
5. Al-Jabban, W. (2017). Soil modification by adding small amounts of soil stabilizers: impact of Portland cement and the industrial by-product Petrit T (Doctoral dissertation, Luleå University of Technology).
6. Japanese Geotechnical society standards: Laboratory testing standards of geomaterials, 2015th ed., vol. 1. Maruzen Publishing Co., Ltd.

Performance Assessment of Recycled Tire Materials in Marine Landfill Application

Chunrui Hao, Hemanta Hazarika, and Yusaku Isobe

Abstract In the marine landfill sites, the alluvial clay layer is mainly used as the impermeable layer at the bottom of the site. This paper provides initial findings regarding the utilization of tire-derived geomaterials (TDGM) in the form of a horizontal layer placed between waste and alluvium clay layers in marine landfill sites. The purpose of installing the reinforcement layer is to protect the alluvium clay layer (Impermeable layer) from waste input. On the other hand, the reinforcement layer is also placed underneath the waste to ensure the collection and drainage of leachate during the degradation and the consolidation of waste. The effectiveness of such configuration is investigated numerically with the PLAXIS 2D software. The Sekiguchi-Ohta model (Viscid model) and hardening soil model are used as the material model, so that the settlement behavior was captured. The purpose of this research is to evaluate the effectiveness of horizontal reinforcing inclusion made from TDGM in mitigating settlement and damage of the alluvium clay layer.

Keywords Marine landfill · Tire materials · Leachate · Consolidation · Settlement

1 Introduction

As a result of economic development and population growth, the contemporary world is faced with a global increase in the amount of waste. One of the biggest challenges for a megacity is securing space for final disposal sites. Marine landfill sites in Japan create the potential for new city development in addition to the appropriate disposal of waste [1]. In recent years, the number of marine landfill sites has increased in Japan due to non-availability and a shortage of suitable land onshore. As compared to the land disposal sites, marine landfill sites also have a huge waste capacity. According

C. Hao (✉) · H. Hazarika
Kyushu University, Fukuoka, Japan
e-mail: haochunrui1992@gmail.com

Y. Isobe
IMAGEi Consultant Corporation, Tokyo, Japan

H. Hazarika et al. (eds.), *Advances in Sustainable Construction and Resource Management*, Lecture Notes in Civil Engineering 144,
https://doi.org/10.1007/978-981-16-0077-7_12

to the investigation of general waste marine landfills in 2013, there are 34 locations in Japan, the area is 653 ha, and capacity is 128 million m^3 [2]. It is expected to be a viable replacement for the land disposal site in the future.

In the marine landfill sites, the alluvial clay layer is primarily used as the impermeable layer at the bottom of the site. However, there is always a risk of damage to impermeability during the waste deposition or due to the consolidation induced settlements. Once the fluid permeation has increased, it will lead to the leakage of harmful contaminants from waste to the surrounding ecosystem. Therefore, the protection of the alluvial clay layer is a significant issue in the feasibility of a marine landfill site. Furthermore, to improve drainage performance during waste deposition and consolidation, it is also needed to adopt appropriate countermeasures. However, there is no economical construction method to solve these problems.

In the past few years, research, regarding the utilization of scrap tire products in construction projects, has gained popularity. Tires are considered as highly durable and non-degradable materials. The thermal recycling of scrap tires, however, is harmful to the environment because it releases more CO_2 and further leads to global warming compared to the other efforts of recycling. Therefore, reusing scrap tires is a preferred option to prevent environmental issues created by thermal recycling. Tire-derived geomaterial (TDGM) from recycled tire materials has been chosen as the preferred material due to its basic properties and beneficial performance such as lightweight, high incompressibility and permeability, and excellent vibration absorption capacity. In addition, these materials are non-dilatant in nature unlike other granular geomaterials. Because of the advantageous physical and mechanical characteristics of scrap tires, it has been utilized in the area of construction in multiple forms. Construction using waste tire chips instead of conventional aggregates maybe both environmentally and economically beneficial [3–7]. It also has high drainage capacity which means that more water will be able to flow freely through these materials compared with fine-grained soils. In order to exploit the permeability of the tire-derived materials, it has been served as a drainage layer for landfill [8, 9]. Other than that, research conducted by Pasha et al. [10] and Hazarika et al. [11] also highlighted the potential of scrap tire products in enhancing the stability of the foundation, preventing liquefaction in soil, and reducing settlement.

In this study, rubber deterioration caused by continuous permeation of waste leachate was considered. And a marine landfill site with the TDGM reinforcement layer placed between waste and the alluvial clay layer was modeled in 2D. The purpose of this study is to evaluate the effectiveness of TDGM in reducing settlement of alluvial clay during the long duration. Figure 1 shows the marine landfill site with a reinforcement layer.

2 The Physical Properties of TDGM

In this study, the grain size distribution of TDGM was determined by a sieving analysis conducted according to JGS-0131- 2009. Particle size distributions of TDGM

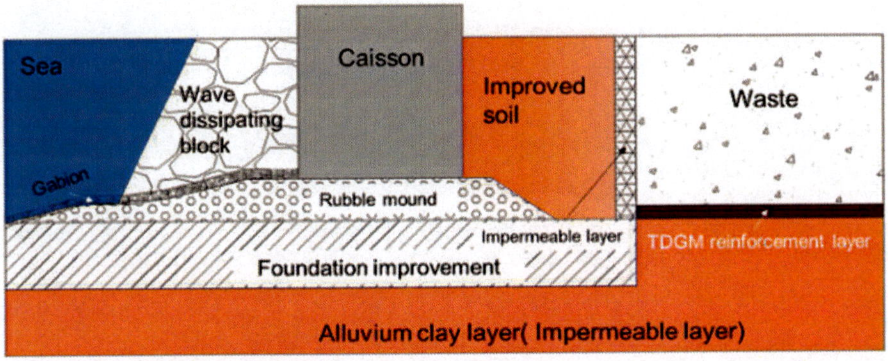

Fig. 1 Marine landfill site with a reinforcement layer

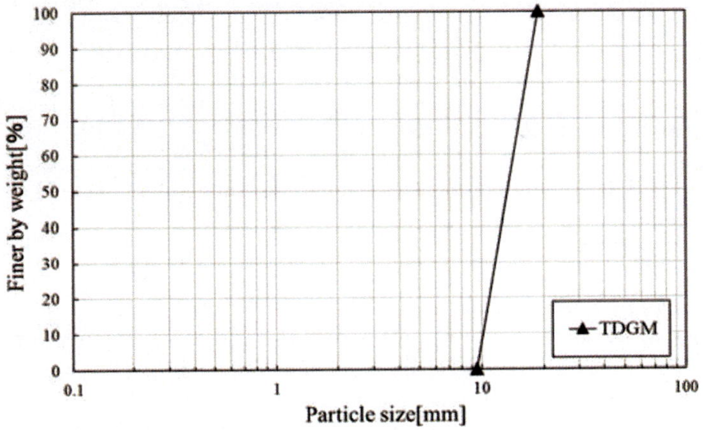

Fig. 2 Particle size distribution of TDGM

is shown in Fig. 2. The specific gravities (Gs) of TDGM were determined according to JGS-0111–2009. TDGM has specific gravities of 1.65. A series of minimum and maximum densities tests were conducted according to JGS-0162–2009. The ρ_{dmin} and ρ_{dmax} are 0.73 and 0.84 g/cm^3.

3 Leachate and Deterioration

Table 1 indicates the main leachate ion components of waste in a marine landfill site in Fukui Prefecture, Japan [12]. The rubber will undergo significant changes over time when exposed to heat, light, oxygen, or ozone. These changes can make a dramatic effect on the service life and properties of the rubber. The deterioration that

Table 1 Main leachate ion components of waste [12]

Ion components	Concentration (mg/L)
SO_4^{2-}	110.40
NO_3^-	0.00
CL^-	131.35
NH_4^+	6.48
Na^+	80.50
K^+	12.09
Ca^{2+}	38.00
Mg^{2+}	1.46

occurs in leachate is less affected by oxygen and light, various chemical ions [13], which means that the properties of rubber are less influenced by deterioration in the long term.

4 Finite Element Modelling and Numerical Simulation Cases

In this study, numerical models were created using the dynamic module of a finite element software, PLAXIS 2D. The alluvium clay layer was modeled using the Sekiguchi-Ohta model (Viscid model). The TDGM reinforcement layer and sand layer (comparative case study) were modeled using the hardening soil model. The selection of these models and parameters was made according to previous research conducted by Murakami et al. [14] and Abdullah et al. [15].

The hardening soil model is superior to the Mohr–Coulomb model or linear elastic model because the hardening soil model is capable of modeling modulus reduction with increasing strain. When subjected to primary deviatoric loading, soil shows a decreasing stiffness and plastic strain development. The Sekiguchi-Ohta model is a cam-clay-type effective stress model for the behavior of clay-type soil. It is utilized to time-dependent behavior, utilizes effective stresses, and has anisotropic hardening features. There are two versions of the model, the inviscid model for time-independent behavior (elasto-plastic), and the viscid model for time-dependent behavior (creep, elastic-viscoplastic) (PLAXIS Material Models Manual 2019) [16]. According to the Material Models Manual, the hardening soil model and the Sekiguchi-Ohta model are briefly explained in the following part.

The basic idea of the hardening soil model is that there is a hyperbolic relationship between the axial strain ε_1 and the deviatoric stress q under triaxial loading. Herein, standard drained triaxial tests usually yield the following curves:

$$-\varepsilon_1 = \frac{1}{2E_{50}} \frac{q}{1 - q/q_a} \quad \text{for} : q < q_f \tag{1}$$

where q_a is the asymptotic value of the shear strength, q_f is ultimate deviatoric stress, and E_{50} is the confining stress-dependent stiffness modulus for primary loading.

$$E_{50} = E_{50}^{\text{ref}}\left(\frac{c\cos\varphi - \sigma_3^{'}\sin\varphi}{c\cos\varphi + p^{ref}\sin\varphi}\right)^m \quad 0.5 < m < 1 \tag{2}$$

where p^{ref} is reference pressure, E_{50}^{ref} is a reference stiffness modulus which corresponds to p^{ref},$\sigma_3^{'}$ is the minor principal stress.

$$q_{\text{f}} = \left(c\cot\varphi - \sigma_3^{'}\right)\frac{2\sin\varphi}{1-\sin\varphi} \text{ and } q_{\text{a}} = \frac{q_{\text{f}}}{R_{\text{f}}} \tag{3}$$

where R_f is the failure ratio and $R_f = 0.9$ is chosen as a standard default setting in PLAXIS 2D.

The stress-dependent stiffness modulus for unloading and reloading stress paths is used:

$$E_{\text{ur}} = E_{\text{ur}}^{\text{ref}}\left(\frac{c\cos\varphi - \sigma_3^{'}\sin\varphi}{c\cos\varphi + p^{\text{ref}}\sin\varphi}\right)^m \quad 0.5 < m < 1 \tag{4}$$

where $E_{\text{ur}}^{\text{ref}}$ is the reference Young's modulus for unloading and reloading which corresponds to p^{ref}. In many practical cases, PLAXIS 2D gives default setting of $E_{\text{ur}}^{\text{ref}} = 3E_{50}^{\text{ref}}$.

The Sekiguchi-Ohta model can be considered as a generalized cam-clay model type for the behavior of clay soils. In the Sekiguchi-Ohta model, it is assumed that there is a relation between volumetric strain ε_v and mean effective stress p', and is given as:

$$\varepsilon_v - \varepsilon_v^0 = -\lambda^*\ln\left(p'/p^0\right) \tag{5}$$

where λ^* is modified compression index, which determines the compressibility of the material in primary loading.

During isotropic unloading and reloading, a different path is calculated as:

$$\varepsilon_v^e - \varepsilon_v^{e0} = -\kappa^*\ln\left(p'/p^0\right) \tag{6}$$

where κ^* is the modified swelling index, which determines the compressibility of the material in unloading and reloading.

Based on Eq. (6), linear stress dependency on the tangent bulk modulus is given as:

$$K_{ur} = \frac{E_{ur}}{3(1 - 2\upsilon_{ur})} = \frac{p'}{\kappa^*} \quad (7)$$

where ur means unloading/reloading, K_{ur} is the elastic bulk modulus, E_{ur} is the elastic Young's modulus, and υ_{ur} is Poisson's ratio.

Numerical simulation cases in this study are given in Table 2 and only case 5-1 placed sand reinforcement layer. The layout of the numerical model is presented in Fig. 3. The reinforcement layer was constructed on the alluvial clay layer. A

Table 2 Numerical simulation cases

Case	Saturated unit weight of waste (kN/m^3)	The thickness of reinforcement (m)	The thickness of alluvial clay (m)	The length of alluvial clay (m)
Case 1-1 (TDGM)	14	0	5	50
	17	0	5	50
	20	0	5	50
Case 1-2 (TDGM)	14	0.5	5	50
	17	0.5	5	50
	20	0.5	5	50
Case 2-1 (TDGM)	14	0	10	50
	17	0	10	50
	20	0	10	50
Case 2-2 (TDGM)	14	0.5	10	50
	17	0.5	10	50
	20	0.5	10	50
Case 3-1 (TDGM)	14	0	15	50
	17	0	15	50
	20	0	15	50
Case 3-2 (TDGM)	14	0.5	15	50
	17	0.5	15	50
	20	0.5	15	50
Case 4-1 (TDGM)	14	0	10	200
	17	0	10	200
	20	0	10	200
Case 4-2 (TDGM)	14	0.5	10	200
	17	0.5	10	200
	20	0.5	10	200
Case 5-1 (Sand)	14	0.5	10	200
	17	0.5	10	200
	20	0.5	10	200

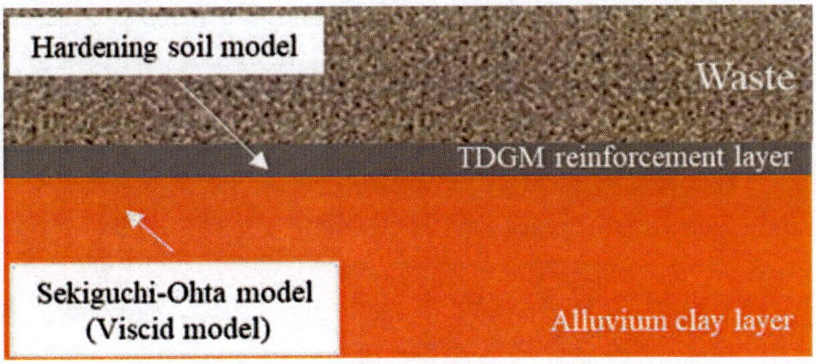

Fig. 3 Layout of the numerical model

parametric study was performed by varying the length and thickness of alluvial clay layers and saturated unit weight of the waste using the hardening soil model and Sekiguchi-Ohta model (Viscid).

Numerical simulations of alluvium clay layer settlement due to the consolidation of solid wastes were carried out in the following stages:

1. Stage 1: Settlement of alluvium clay layer considering the vertical accumulation of solid wastes at the rate of 1 m/year during its lifetime (15 years).
2. Stage 2: Settlement due to the consolidation over a period of 15.

5 Results and Conclusions

A total of nine cases (reinforced cases and non-reinforced cases) were simulated under three different saturated unit weight of waste, which are given in Table 2. In the table, Case 1-1, Case 1-2, Case 2-1, Case 2-2, Case 3-1, and Case 3–2 represent the cases in which the length of alluvial clay layers was 50 m and the thickness of alluvial clay layers was different (5, 10 and 15 m). Case 4-1, Case 4-2, and Case 5-1 (sand comparative case) represent the conditions in which the length of alluvial clay layers was increased to 200 m and the thickness of alluvial clay layers was 10 m.

The models were simulated for consolidation analysis to evaluate the effectiveness of the reinforcement layer in reducing the settlement. The maximum vertical settlement of the alluvium clay layer is illustrated in Figs. 4, 5, and 6. As shown in Figs. 4 and 5, the maximum settlement with TDGM reinforcement in 30 years is smaller than the unreinforced case. The maximum settlement increases with an increase in a saturated unit weight of waste from 14–20 kN/m^3. In the unreinforced case, the maximum settlement decreases as the clay thickness increases. On the other hand, the maximum settlement increases as the clay length increases. It can be obviously seen that the maximum settlement is similar in the TDGM reinforcement case. The computed maximum settlement of alluvium clay layer is less than 0.2 m for the TDGM reinforcement case, and 0.05 m for the unreinforced case in 30 years.

Fig. 4 Maximum settlement of alluvial clay with different clay thicknesses

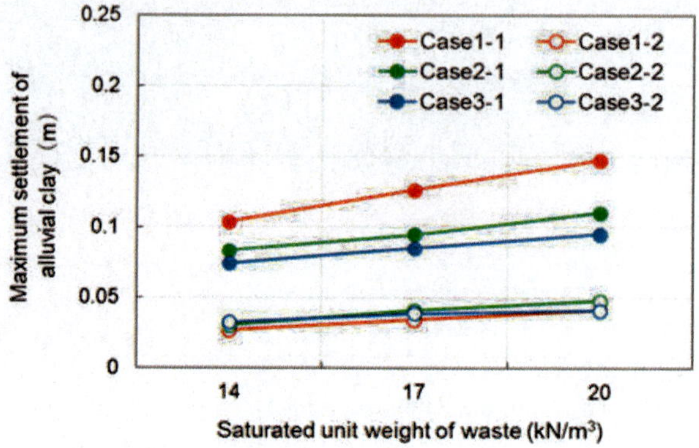

Fig. 5 Maximum settlement of alluvial clay with different clay lengths

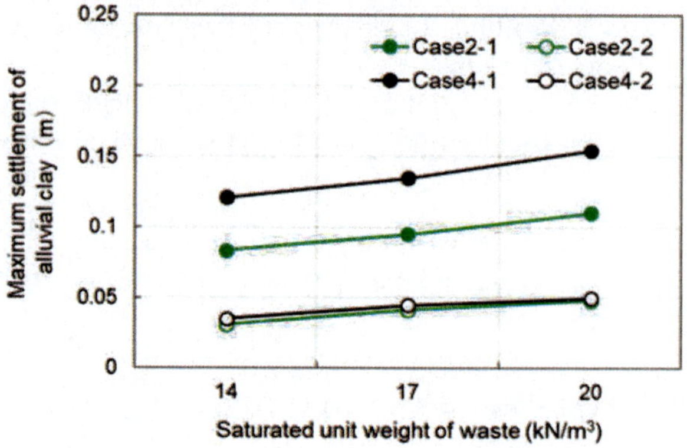

Fig. 6 Maximum settlement of alluvial clay with different reinforcement materials

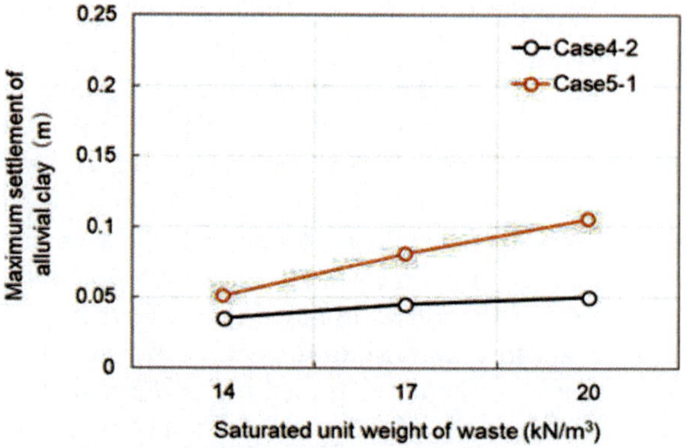

Figure 6 shows the maximum settlement of TDGM case is less than the sand case which means TDGM works better than the traditional reinforcing material.

This paper provides the initial findings regarding the improvement of the marine landfill site using TDGM. The main conclusions from this study are as follows:

1. The properties of rubber (TDGM reinforcement) are less influenced by deterioration in marine landfill sites.
2. Compared with the reinforced cases, the maximum settlement of alluvium clay layers in unreinforced cases is larger. And TDGM is preferable than the traditional reinforcing material. However, according to the maximum settlement of alluvium clay layer in 30 years, the TDGM reinforcement shows the limited effect on the alluvium clay settlement under waste deposition and consolidation.

Thermal recycling is detrimental to the environment because it releases more CO_2. The application of recycled tire materials is expected to contribute toward significantly reducing CO_2 emission. The technique using tire-derived geomaterial (TDGM) in marine landfill sites also possesses tremendous potentials for leachate collection systems. The TDGM reinforcement layer is placed underneath the waste also to ensure the collection and drainage of leachate during the degradation and the consolidation of waste. However, not enough study has been conducted to investigate the permeability of TDGM as a new geomaterial. In future studies, permeability tests were performed to measure the hydraulic conducting of TDGM. These experimental findings could be very helpful for the long-term performance of TDGM reinforcement in marine landfill sites.

Acknowledgements The authors deeply appreciate Ms. Chu Chu, former graduate student of Geo-disaster Prevention Engineering Laboratory (Research Group of Adaptation to Global Geo-disaster and Environment) of Kyushu University for her assistance in performing the analysis.

References

1. Tokyo Metropolitan Government: Tokyo Bay-side Landfill-Structure, Maintenance, Land-use. https://www.tokyokankyo.jp/tokyoprogram_contents/image/jp/pdf/4-4Bayside-1.pdf
2. Endo, K.: Structure, management and role of the marine landfill sites. In: The 27th Annual Conference of Japan Society of Material Cycles and Waste Management, pp. 17–20. Japan Society of Material Cycles and Waste Management, Wakayama (2016) (in Japanese)
3. Humphrey, D.N., Manion, W.P.: Properties of tire chips for lightweight fill. In: Grouting, Soil Improvement and Geosynthetics, pp. 1344–1355. American Society of Civil Engineers, New Orleans (1992)
4. Ahmed, I.: Laboratory study on properties of rubber-Soils. Joint highway research project, Indiana Department of Transportation and Purdue University, West Lafayette, Indiana (1993)
5. Edil, T., Bosscher, P.: Engineering properties of tire chips and soil mixtures. Geotech. Test. J. **17**(4), 453–464 (1994)
6. Kaushik, M.K., Kumar, A., Bansal, A.: Performance assessment of gravel–tire chips mixes as drainage layer materials using real active MSW landfill leachate. Geotech. Geol. Eng. **33**(4), 1081–1098 (2015)
7. Hazarika, H., Yokota, H., Endo, S., Kinoshita, T.: Cascaded recycle of waste tires—Some novel approaches towards sustainable geo construction and climate change adaptation. Geotechnics for Natural and Engineered Sustainable Technologies, pp. 63–81. Springer, Singapore (2018)
8. Thakur, D., Kaushik, M.K.: Assessment of drainage properties of tires derived aggregates to be used as drainage layer material in leachate collection system of MSW Landfill. Int. J. Innovative Res. Sci. Eng. Technol. **5**(2), 2219–2230 (2016)

9. Chu, C., Hazarika, H., Isobe, Y.: Application of the waste tire cascade recycling to the seafloor protection at seashore landfill for final disposal. In: 53rd Proceedings of the Japan National Conference on Geotechnical Engineering, 0285, D-04. The Japanese Geotechnical Society, Takamatsu (2018) (in Japanese)
10. Pasha, S.M.K., Hazarika, H., Yoshimoto, N.: Dynamic properties and liquefaction potential of Gravel-Tire Chips Mixture (GTCM). J. Jap. Soc. Civ. Eng. Ser. A1 Struct. Eng. Earthquake Eng. (SE/EE) **74**(4), 649–655 (2018)
11. Hazarika, H., Pasha, S.M.K., Ishibashi, I., Yoshimoto, N., Kinoshita, T., Endo, S., Karmokar, A.K., Hitosugi, T.: Tire-chip reinforced foundation as liquefaction countermeasure for residential buildings. Soils Found. **60**(2), 315–326 (2020)
12. Tanaka, H., Kamura, K.: The dissolution of inorganic ions in the stabilization of controlled landfills: impact of the amount of infiltration water on the leachate ion concentration. J. Jap. Soc. Mater. Cycles Waste Manage. **28**, 13–25 (2017). (in Japanese)
13. Mitariai, Y., Yasuhara, K., Kikuchi, Y., Karmokar. A. K.: Development of cement-treated clay mixed with scrap tire chips as novel geomaterial and study of its mechanical properties. J. Jap. Soc. Civ. Eng. C, **63**(3), 881–900 (2007) (in Japanese)
14. Murakami, T., Niihara, Y., Yamada, T., Ohno, S., Noguchi, T., Miyata, M.: Prediction of horizontal deformation of large-scale sea wall by elasto-viscoplastic finite element analysis. J. Jap. Soc. Civ. Eng. Ser. C (Geosphere Eng.), **68**(2), 224–238 (2012) (in Japanese)
15. Abdullah, A., Hazarika, H.: Improvement of shallow foundation using non-liquefiable recycle materials. In: The 15th Asian Regional Conference on Soil Mechanics and Geotechnical Engineering, 2(54), pp. 1863–1867. Japanese Geotechnical Society Special Publication, Fukuoka (2016)
16. PLAXIS 2D: Material Models Manual 2019. https://www.plaxis.com/support/manuals/plaxis-2d-manuals

Performance of Rubble Brick Drains in Earthquake-Induced Liquefaction Mitigation Under Existing Buildings

Samy Garcia-Torres and Gopal Santana Phani Madabhushi

Abstract Countermeasure techniques against earthquake-induced liquefaction aim to mitigate damage on infrastructures. Vertical drains performance has been evaluated during the last decades showing effectiveness in the reduction of structural damage. The performance of alternative drain arrangements considering the influence of existing buildings have been recently analyzed showing positive results on foundations settlement reduction. Although geotechnical mitigation measures main objective is to reduce detrimental effects in buildings, it is also important to incorporate economic and environmentally sustainable solutions. In this paper, the use of recycled coarse material inside drains such as high permeable rubble brick obtained from post-earthquake debris or buildings demolition is examined. Dynamic centrifuge test series was carried out to evaluate the performance of rubble brick vertical drains placed around existing foundations with different loads. The study shows the influence of the foundation load variation on the excess pore pressure generation due to earthquakes, the following post-seismic dissipation, and the foundation response.

Keywords Liquefaction · Earthquakes · Soil–structure interaction

1 Introduction

The use of mitigation techniques against earthquake-induced liquefaction is crucial for buildings located in liquefiable areas. Perimeter drains are an optimum alternative against liquefaction damage as they are highly effective in controlling the generation of high excess pore pressures below buildings and allowing their rapid dissipation. Previous work focused on the performance of perimeter drains suggests the effectiveness of the technique in terms of the reduction of complete liquefaction potential and structural settlement [1, 2].

S. Garcia-Torres (✉) · G. S. P. Madabhushi
University of Cambridge, Cambridge CB3 0EL, UK
e-mail: sssg2@cam.ac.uk

H. Hazarika et al. (eds.), *Advances in Sustainable Construction and Resource Management*, Lecture Notes in Civil Engineering 144,
https://doi.org/10.1007/978-981-16-0077-7_13

Current solutions developed in the geotechnical field required to follow a sustainable course. Concerning this, the reuse of construction materials from debris or building's demolition is an ideal option for coarse material inside drains. The use of gravel drains containing crushed concrete as high permeable material was previously analyzed by Orense et al. [3]. An improved response of the structure was observed due to the effective performance of the columns. This paper evaluates the performance of rubble brick inclined drains as a mitigation technique for existing buildings against earthquake-induced liquefaction. The analysis considers the generation of excess pore pressures due to an earthquake and their following dissipation below foundations of two different bearing pressures. In addition, the foundation settlement and dynamic response was evaluated.

2 Experimental Setup

Dynamic centrifuge tests were conducted using the University of Cambridge Turner Beam Centrifuge, considering a model/prototype scale of 1/50. Two models containing an arrangement of eight inclined drains placed in loose Hostun sand around a foundation were evaluated. Foundations of 150 kPa (SG6) and 50 kPa (SG7) were placed over the soil surface in each model, and instruments such as pore pressure transducers and accelerometers were situated at different depths of the stratum (Fig. 1). In addition, linear variable displacement transducers (LVDTs) and microelectromechanical system accelerometers (MEMs) were located in the foundation to measure displacement and rotation. The cam-sat system was used for the model saturation [4], and hydro propyl methylcellulose was employed to increase the fluid viscosity in the model.

Input motions given in Table 1 were carried out using the servo-hydraulic actuator [5]. Results of earthquake 2 are considered in this analysis as it represents the strongest motion. For the model preparation, Hostun sand was poured in a laminar box [6] using the sand pouring machine at Schofield Centre [7]. Inclined tubes were placed at a depth of 96 mm (4.8 m in prototype scale) and filled with high permeable rubble brick as coarse material. The tubes were sealed at the top to avoid fine material inside the drains during the sand pouring of upper layers. After reaching the required level of sand, the tubes were pulled up from the model. Soil properties are presented in Table 2.

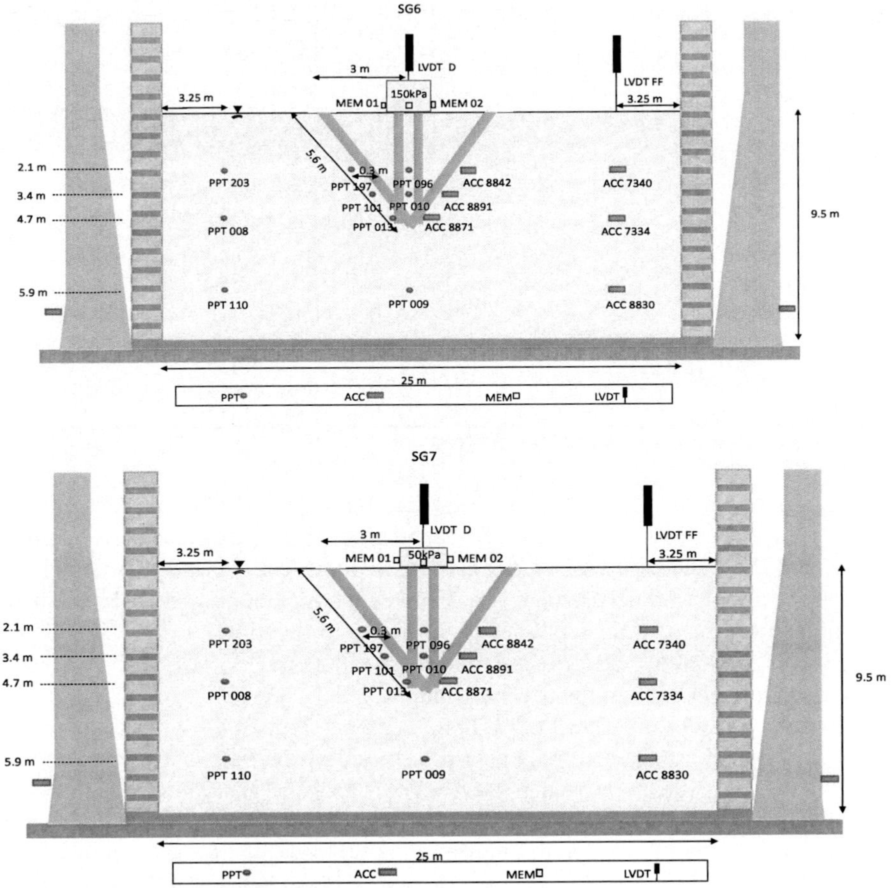

Fig. 1 Test layout for SG6 and SG7

Table 1 Test input motions

Property	Hostun sand	Rubble brick
Density (kg/m^3)	1420	1100
Hydraulic conductivity, k (m/s)	0.5×10^{-3}	5×10^{-3}
Relative density, Dr (%)	44	46
Void Ratio, e	0.87	1.27
Average particle size, D_{50} (mm)	0.48[a]	2.00
Specific gravity, Gs	2.65[a]	2.50

[a]Haigh et al. [8]

Table 2 Soil properties

Earthquake	Duration (s)		Peak acceleration (g)	
	Model	Prototype	Model	Prototype
1	0.2	10	2.0	0.04
2	0.2	10	9.5	0.19
Imperial valley	1.2	60	4.5	0.09

3 Results

3.1 Soil Response

3.1.1 Excess Pore Pressure Generation

Excess pore pressure ratios (r_u) time-histories for SG6 and SG7 are presented in Fig. 2, considering the soil below the foundation central axis, near the inclined drains and the free field. Dashed lines and dots represent the end of shaking and dissipation initiation times, respectively. Complete liquefaction was observed in the free field along the stratum depth due to the lack of mitigation in both cases. At the top layer, larger excess pore pressure ratios (r_u) of 0.3 and 0.7 were observed below the foundation and close to the perimeter drains in SG7 (Fig. 2b.), compared to SG6 that presented values of 0.2 and 0.5 at both locations (Fig. 2a.).

The greater additional confining pressure exerted by the foundation of 150 kPa in SG6, allowed significant resistance to excess pore pressure generation. Similar trend of lower excess pore pressure generation was observed below the heavy foundation at a depth of 3.4 m. The r_u values generated at this level were lower compared to the top layer due to the less confinement pressure. At a depth of 4.7 m, the excess pore pressure ratio near the inclined drains reached a value of 0.65 in SG7 (Fig. 2f), while the soil in SG6 presented a lower ratio of 0.5 (Fig. 2e). Although the soil was influenced by the inclined drains in both arrangements; the generation is mainly managed by the foundation bearing pressure. At the lower layer (depth of 5.7 m), greater ratios were reached below the foundation central axis in SG7 (Fig. 2h.) compared to SG6, suggesting a minimal influence of the foundation of 50 kPa at this level.

3.1.2 Bearing Pressure Influence in the Dissipation of Excess Pore Pressures

Dissipation of excess pore pressures at a depth of 2.1 m is presented in Fig. 3 for SG6 and SG7. Contours of excess pore pressures ratios were plotted considering dissipation initiation time for the soil below the foundation, outside the perimeter drains, and the free field in SG6 (150 kPa). The analysis reflects the behavior of excess

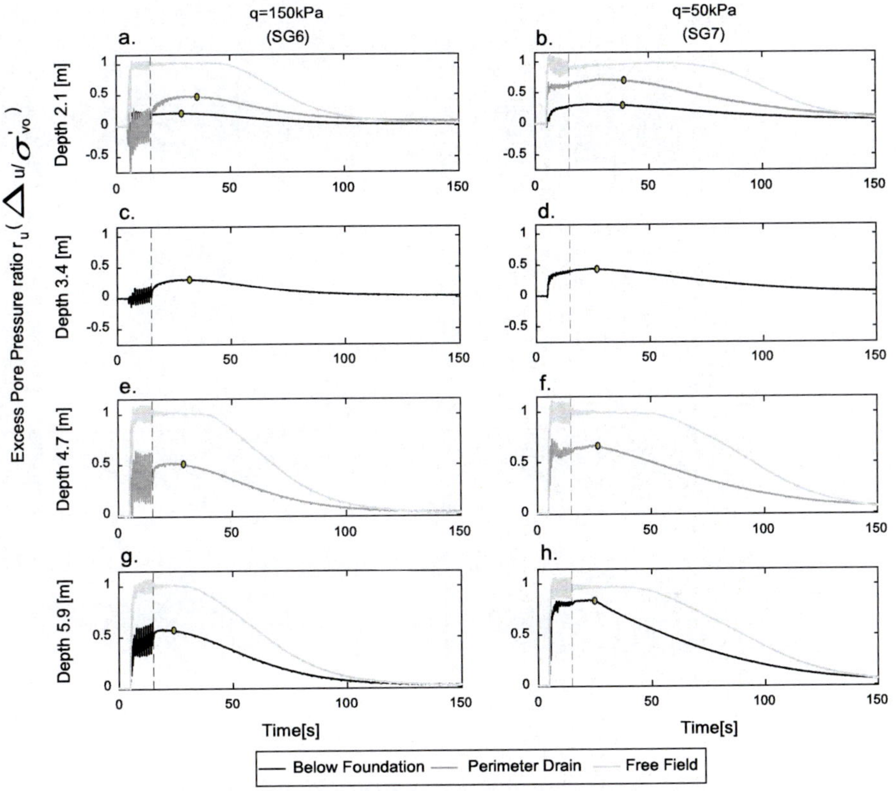

Fig. 2 Excess pore pressures ratios (r_u) time-histories for SG6 and SG7

pore pressures in the soil containing inclined drains and influenced by a foundation bearing pressure of 50 kPa and 150 kPa. Distance from the drain to the foundation central axis was 1.9 m at this depth.

At the end the shaking, a uniform behavior was observed below the foundation and around the inclined drains in SG6, reaching a r_u value of 0.2 (Fig. 3a). At this time, larger ratios of 0.3 and 0.6 were observed at both locations in SG7 (Fig. 3b). Dissipation started at $t = 28.5$ s in SG6, from the area enclosed by the drains at the top layer (Fig. 3c), while excess pore pressures were still generating below the foundation of 50 kPa, showing a ratio of 0.4 (Fig. 3d). Rapid dissipation and faster regained of stiffness in the soil was presented under the foundation of 150 kPa due to the high confining pressure exerted by the foundation. A rapid reconsolidation of the soil particles is expected to occur in presence of significant additional bearing stress. Dissipation near the inclined drains in SG6 started at $t = 35.4$ s, presenting significant dissipation below the foundation and a ratio of 0.5 near the inclined drains (Fig. 3e). Meanwhile, a moderate reduction of excess pore pressures was observed below the foundation in SG7 and near the inclined drains excess pore pressures continued to generate (Fig. 3f). High excess pore pressures below the light foundation were

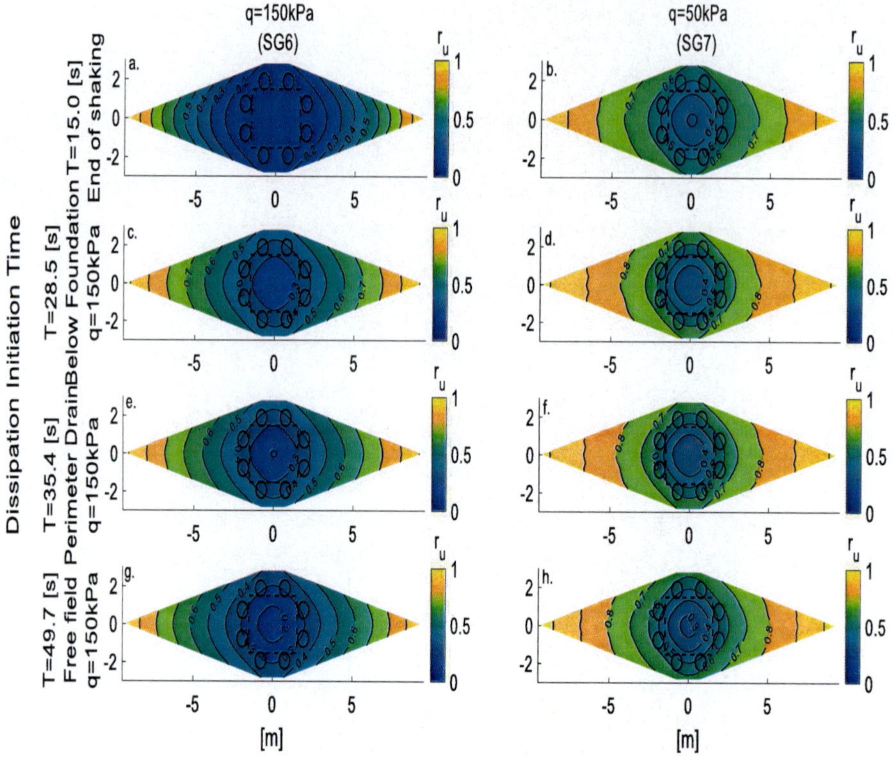

Fig. 3 Horizontal contours of excess pore pressure ratios (r_u) in SG6 and SG7 at different dissipation times

maintained for a longer time compared to the foundation of 150 kPa. At t = 49.7 s, the free field started to dissipate in SG6, showing a significant reduction of excess pore pressure near the drains and below the foundation (Fig. 3g). On the other hand, greater r_u values were observed in SG7 at the entire layer, with no dissipation of excess pore pressures in the free field (Fig. 3h).

3.1.3 Vertical Dissipation Path

Figure 4 shows the vertical dissipation of excess pore pressures along the stratum depth. Contours of excess pore pressures ratios were plotted for SG6 and SG7, considering dissipation times of t = 15 s, t = 30 s, t = 45 s, and t = 75 s. The area enclosed by the inclined drains below the foundation of 150 kPa (SG6) presents an optimal control of pore pressures at shallow layers (Fig. 4a). Greater r_u values were reached in the soil close to the perimeter drains due to the significant radial and upward fluid flow from lower layers. Moreover, complete liquefaction (r_u= 1) was observed in the free field at this time, due to the absence of drainage enhancement

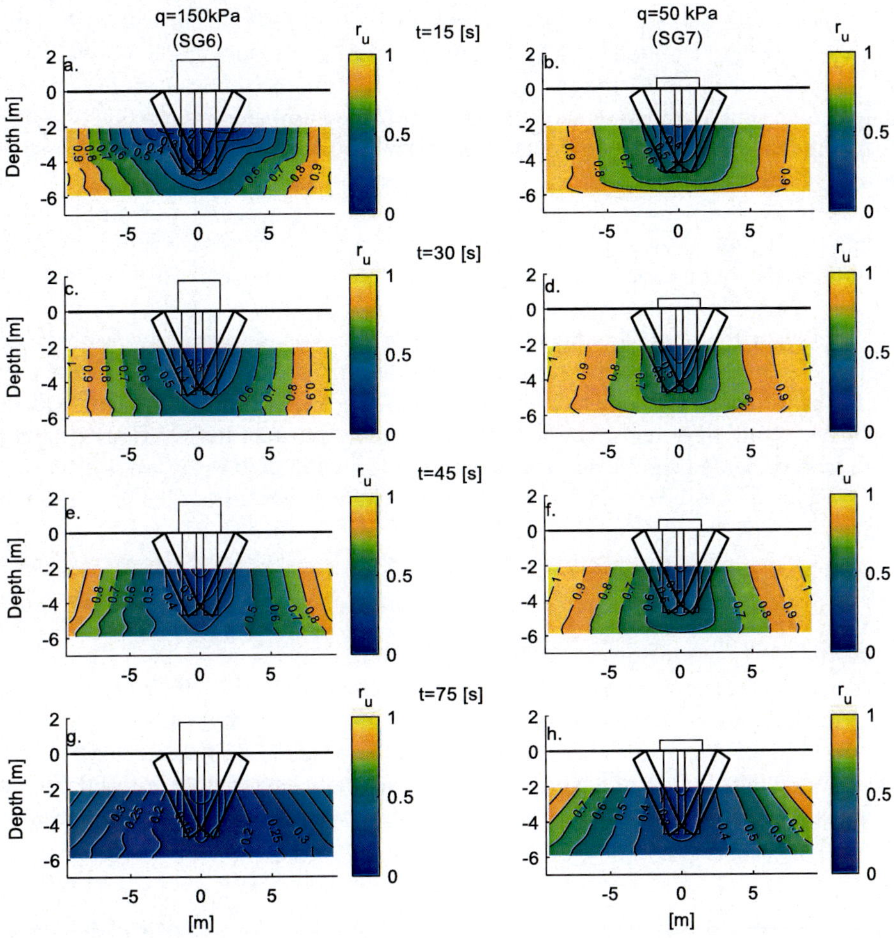

Fig. 4 Vertical contours of excess pore pressure ratios (r_u) in SG6 and SG7 at different dissipation times

and confining pressure exerted by the foundation. Similar trend of higher excess pore pressure close to the perimeter drains compared to the zone under the foundation was observed in SG7 (Fig. 4b); however, greater generation was registered at both locations in SG7 compared to SG6, due to the lower confining stress exerted by the foundation of 50 kPa.

In addition, a bulb of low pressures was generated below the foundation in the area enclosed by the inclined drains in SG6, after the shaking (Fig. 4c). This behavior suggested the importance of the drains radial proximity to the foundation in limiting high levels of r_u. On the other hand, at $t = 30$ s, the soil under the foundation in SG7 was still presenting pore pressure generation showing ratios of 0.4 and 0.7 below and near the perimeter drains, respectively. Similar to SG6, a bulb of low pressures was generated in the soil enclosed by the drains in SG7 (Fig. 4d); nevertheless, the bulb reached a lower magnitude due to the lower confining pressure of the foundation.

The bulb of low pressures started to dissipate from the top in SG6. Dissipation began under the foundation and continued near the perimeter drains (Fig. 4e). The inclined drains generated a protected area of low pressures below the foundation that was easily dissipated by the inclined drains highly influenced by the great confining stress. In SG7, a passive dissipation of the bulb was registered compared to SG6 (Fig. 4f). The low bearing pressure and the weak action of the overlapped area at the bottom of the inclined drains allowed a slower dissipation in SG7. The free field was still showing greater r_u values at the top layer in both cases.

Finally, at $t = 75$ s, significant reduction of excess pore pressure was observed below the foundation and near the perimeter drains in SG6 due to the effective action of the inclined drains in accelerating dissipation (Fig. 4g). Meanwhile, greater values near the drains were registered at the bottom and top layers in SG7 (Fig. 4h). The free field presented significant dissipation at this time in SG6 at the top layer, while no dissipation was observed for the same zone in SG7 because of the upward fluid flow.

3.1.4 Soil Acceleration

Acceleration of the soil below the foundation, close to the perimeter drain, and in the free field are presented together with the input motions at depths of 2.1 m and 4.7 m in Fig. 5 for SG6 and SG7. The free field presented a complete acceleration decoupling in response of the complete liquefaction behavior at the top and middle layers in SG6 and SG7. The acceleration response was managed by the softening in the soil as similar shear reinforcement was provided by the drains arrangement in

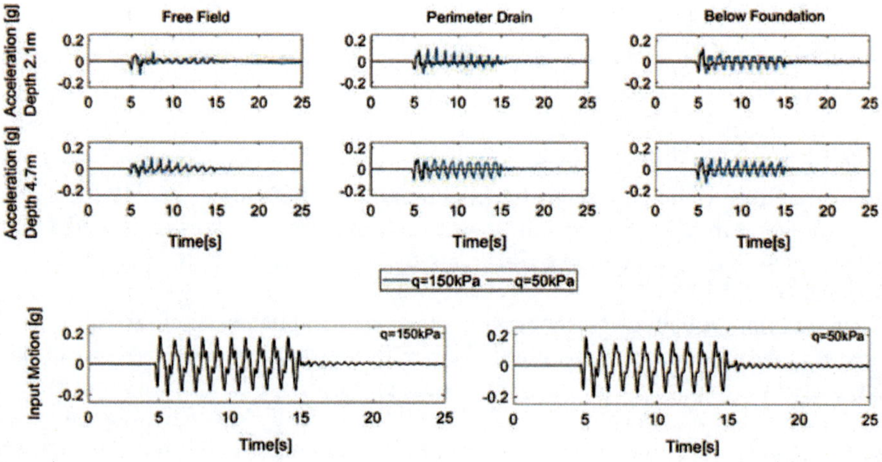

Fig. 5 Soil acceleration for SG6 and SG7

both cases. Significant acceleration reduction near the perimeter drains and below the foundation was presented in SG7, while a greater acceleration response was observed in SG6 for both locations due to the lower soil softening.

In addition, greater soil acceleration was observed below the foundation compared to the area near the perimeter drains at the top layer particularly in SG6, as a response of the lower level of softening below the foundation central axis along the stratum depth.

3.2 Foundation Response

3.2.1 Settlement of the Foundation

Figure 6 shows the free field and foundation settlement together with input motions for SG6 and SG7. During the shaking, the foundation of 50 kPa reached a settlement of 271 mm (87% of the total settlement), due to volumetric deformations and deviatoric strains (Fig. 6b). The greater settlement obtained during the shaking compared to the post-shaking stage verified the effective work of the technique during dissipation. On the other hand, settlement of the foundation reached a value of 257 mm during the shaking in SG6, as a result of the lower soil softening reached below the foundation, enabling less reconsolidation volumetric strains in the soil (Fig. 6a).

During the dissipation stage, the foundation of 50 kPa presented a slightly greater settlement response compared to the heavy foundation. The rapid dissipation of excess pore pressures below the foundation of 150 kPa let to a faster reduction of volumetric and deviatoric deformations and rapid regained of stiffness in the area enclosed by the perimeter drains. In SG7, high excess pore pressures were maintained for a longer time enabling a greater settlement of the foundation. The similar settlement response obtained for a foundation of 50 kPa and another one three times heavier suggests the relevance of the bearing pressure in the effective

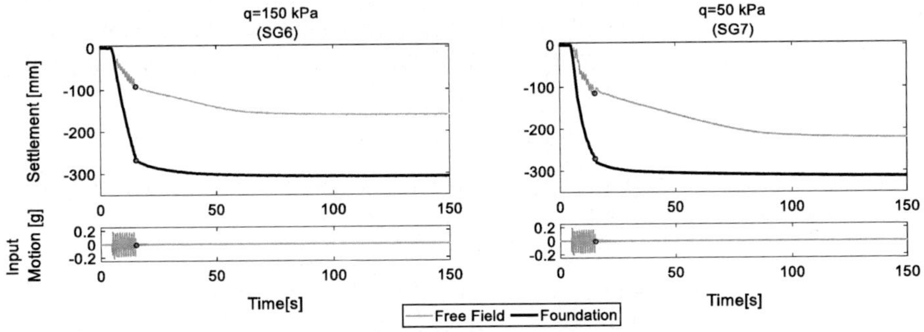

Fig. 6 Settlement time-histories for free field and foundations in SG6 and SG7

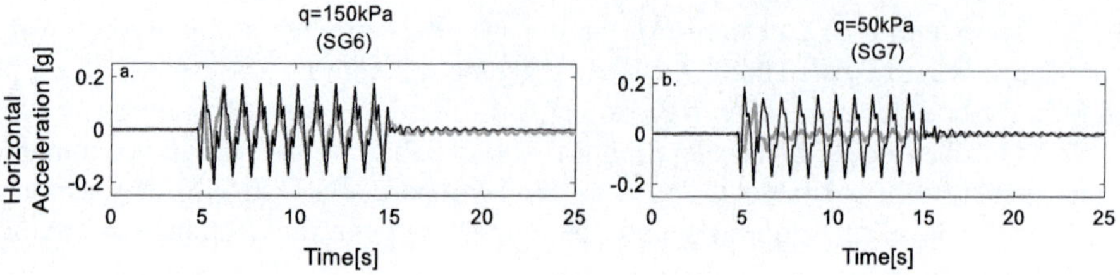

Fig. 7 Foundation horizontal acceleration for **a** SG6 and **b** SG7

performance of the technique and the importance of considering this parameter in the arrangement design.

3.2.2 Dynamic Response of the Foundation

Horizontal acceleration of the foundation together with the input motion is presented in Fig. 7 for SG6 and SG7. Lower transference of the input motion to the foundation was observed in SG7 due to the large soil softening generated below the foundation, reaching a mean value of 0.03 g after the second cycle of the shaking. Great horizontal acceleration response was observed for the foundation of 150 kPa (Fig. 7a). The limited soil softening in the area enclosed by the drains allowed a large acceleration response of the foundation, showing a peak value of 0.15 g (Fig. 7b). The inclined drains were unable to provide sufficient shear reinforcement in the soil as no amplification of the seismic demand was observed.

The rotational response of the foundation required to be evaluated as significant horizontal acceleration was observed for the foundation in SG6. Figure 8a presents rotation time-histories for the foundations of 50 and 150 kPa. A peak rotation of $\mathbf{5 \times 10^{-3}}$ rad was registered for the foundation of 150 kPa, as a response of the great performance of the inclined drains in limiting significant soil softening and allowing a great transference of the input motion to the foundation. The light foundation presented a lower peak rotation of $\mathbf{2.5 \times 10^{-3}}$ rad as a result of the minor horizontal acceleration in the foundation. In addition, the heavy foundation reached a permanent rotation of $\mathbf{1 \times 10^{-3}}$ rad, while the foundation of 50 kPa presented a negligible permanent rotation of $\mathbf{5 \times 10^{-4}}$ (Fig. 8b).

4 Conclusions

Inclined drains containing rubble brick as coarse material revealed an efficient performance in presence of a high bearing pressure compared to light foundations. The effective control of high excess pore pressures and their rapid dissipation in the area enclosed by the inclined drains allowed an improvement in the foundation settlement response. Nevertheless, the inclined drains permitted a great transference of the input

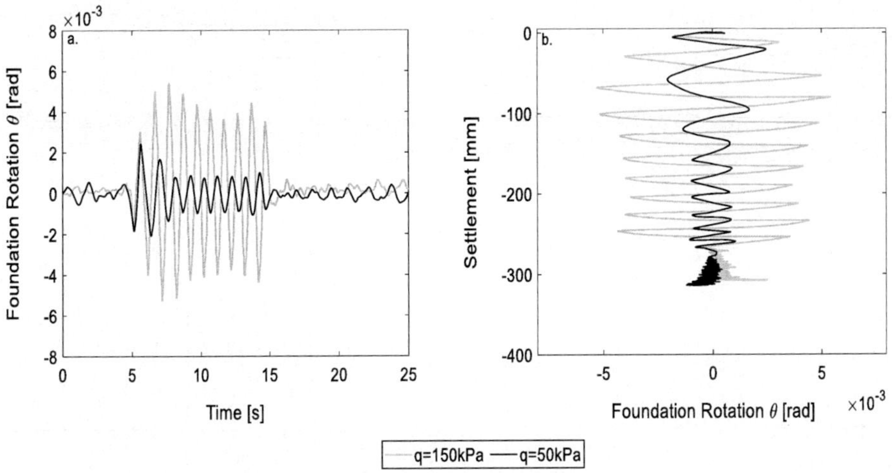

Fig. 8 **a** Foundation rotational response and **b** foundation rotation vs settlement for SG6 and SG7

motion to the heavy foundation enabling a larger rotational response and increasing the risk of damage in the superstructure. The bearing pressure plays a relevant role in the performance of the technique and requires to be considered in the design of the drains configuration.

References

1. Olarte, J., Paramasivam, B., Dashti, S., Liel, A., Zannin, J.: Centrifuge modeling of mitigation-soil-foundation-structure interaction on liquefiable ground. Soil Dynam. Earthq. Eng. **97**, 304–323 (2017)
2. Paramasivam, B., Dashti, S., Liel, A.: Influence of prefabricated vertical drains on the seismic performance of structures founded on liquefiable soils. J. Geotechn. Geoenviron. Eng. **144**(10) (2018)
3. Orense, R.P., Morimoto, I., Yamamoto, Y., Yumiyama, T., Yamamoto, H., Sugawara, K.: Study on wall-type gravel drains as liquefaction countermeasure for underground structures. Soil Dynam. Earthq. Eng. **23**(1), 19–39 (2003)
4. Stringer, M.E., Madabhushi, S.P.G.: Novel computer-controlled saturation of dynamic centrifuge models using high viscosity fluids. Geotech. Test. J. **32**(6), 559–564 (2009)
5. Madabhushi, G.S.P.: Centrifuge Modelling for Civil Engineers. Taylor & Francis, London (2014)
6. Brennan, A.J., Madabhushi, G.S., Houghton, N.E.: Comparing laminar and equivalent shear beam (ESB) containers for dynamic centrifuge modelling. In: Proceedings of the 6th International Conference ICPMG, 6, pp. 171–176 (2006)
7. Madabhushi, S.P.G., Houghton, N.E., Haigh, S.K.: A new automatic sand pourer for model preparation at University of Cambridge. In: Proceedings of the 6th International Conference on Physical Modelling in Geotechnics, pp. 217–222. Taylor & Francis Group, London, UK (2006)
8. Haigh, S.K., Eadington, J., Madabhushi, G.S.P.: Permeability and stiffness of sands at very low effective stresses. Géotechnique **62**(1), 69–75 (2012)

Performance of Sand–Rubber Mixture Infill Trench for Ground Vibration Screening

J. S. Dhanya, A. Boominathan, and Subhadeep Banerjee

Abstract Ground vibrations arising from construction and industrial activities and road/rail traffic can induce settlement issues, cracks, and severe damage to adjacent and remote structures. One of the well-established methods to eliminate such unwanted ground-borne vibrations is to incorporate trench barriers between the source of vibration and the structure to be protected. Recently, the use of shredded rubber from recycled tires has gained prominence in various geotechnical applications. The high energy absorption capacity of rubber is well established in the past, making it an ideal material in vibration mitigation studies. In the present study, 2D finite element analysis was carried out to investigate the use of sand–rubber tire mixture (SRM) infill trench barriers for the screening of ground-borne vibration due to vertical ground vibrations. In the present study, the typical soil profile from the Indo-Gangetic plain region is considered. 1 m width open and SRM infill trenches with a depth of 1–3 m are considered. The rubber content in the SRM fill trenches was chosen as 30% and 50%. The hyper elastic material model was adopted for the modeling of the SRM infill trench, while the soil medium was modeled using the hypoelastic constitutive model. The ground excitation was created by applying sinusoidal vertical motion with 2 m/s amplitude and a frequency of 50 Hz at the ground surface away from the trench. During the excitation, the vibration levels were computed at different locations in front of and away from the trenches. It was found that SRM infill trench with 50% rubber content performs similar to the open trenches to reduce the vertical vibration amplitude.

Keywords Vibration mitigation · Infill trench · Sand–rubber mixture · FEM

J. S. Dhanya (✉) · A. Boominathan · S. Banerjee
Department of Civil Engineering, IIT Madras, Chennai 600036, India
e-mail: dhanyacivil@gmail.com

H. Hazarika et al. (eds.), *Advances in Sustainable Construction and Resource Management*, Lecture Notes in Civil Engineering 144,
https://doi.org/10.1007/978-981-16-0077-7_14

1 Introduction

Ground-borne vibrations arising from construction activities such as blasting, demolition of buildings, dynamic compaction, excavation, and driving of piles very closer to civil engineering structures often get transmitted to nearby buildings. Besides, road/rail traffic, machine induced vibrations, and other industrial activities can damage the neighboring buildings, streets, underground pipelines, and as well as on sensitive equipment and cause disturbance to the occupants of the buildings. In severe cases, ground-borne vibrations can cause soil settlement and soil densification leading to damage to the surrounding structures. The adverse effects of dynamic impact loading, such as pile driving, even lead to shaking of structures, settlement due to liquefaction, and the formation of localized heave. The structural damage can be chiefly attributed to vibratory cracking from ground vibrations, resonant structure vibrations, and vibratory settlement of foundation [1]. It is essential to estimate the transmitted vibration intensity to avoid structural damage, which depends on the attenuation characteristics of the soil that act as the transmitting medium. To achieve the levels of ground vibration specified in the standard and to minimize the unwanted ground-borne vibrations for reducing structural damage and improving building functionality, vibration control through cost-effective vibration screening methods is essential.

Wave barriers such as trenches can be used as a successful technique to minimize the problems due to ground vibrations. The trench barriers are installed between the source of vibration and the structure to be protected. The problem of screening of vibrations by the use of trench barriers can be collectively classified into active isolation and passive isolation. Active isolation involves the installation of wave barriers close or near to the vibration source to reduce the propagation of waves away from the source. In contrast, passive isolation involves providing wave barriers near the structure where the impact caused by the vibration should be reduced [1, 2]. Such trenches on the path of wave propagation typically attenuate the surface waves, thereby reducing the intensity of the ground vibration.

The most efficient wave barriers to screen ground vibration are open trenches. The introduction of open trenches poses several limitations for practical applications. Due to the localized collapse of the trench walls, safety issues, and unexpected filling owing to rain or construction activities, it is often difficult to install and maintain open trenches to the desired depth and width. Baker [3] has conducted a series of field model tests to investigate the effectiveness of barriers infilled with bentonite (i.e., soft barrier) and concrete (i.e., stiff barrier) installed near and far from the source of the disturbance. Later, infilled trenches with geofoam, sawdust, and bentonite were widely investigated [4, 5].

Several experimental studies were conducted to examine the key parameters influencing the efficiency of wave barriers in vibration isolation, which suggested that the depth of the trench and the wavelength of Rayleigh waves have a significant impact on vibration screening [2]. Woods [1] reported that a minimum depth of 0.6 times

Rayleigh wavelength should be provided for effective isolation. It was reported that the width of the trench has little effect on vibration screening.

Lately, rubber was used to a large degree as an anti-vibration product to limit noise, shock, and vibration isolation in a wide variety of industries. One of the main applications of scrap tire with rubber as its principal component is its utility in vibration isolation due to high damping and energy absorption capacity [6, 7]. The scrap tire products such as tire shreds, chips, and aggregates have found its way into the civil engineering field since the 1990s. By and large, the unrecycled scarp tire known as 'black pollutant' [8] posing global pollution due to its sheer volume was found to be an excellent additive to soil mainly due to its non-biodegradability. In the geotechnical field, the tire-derived geomaterials have witnessed rapid growth in applications such as lightweight landfills, backfilling of retaining walls and buried pipeline, and ground improvement material for highway embankments [9, 10]. The use of tire-derived geomaterials mixed with sand for earthquake protection of buildings has been the topic of interest in recent years [11–13]. The liquefaction mitigation potential of tire chips mixed with soil for foundation soil and as backfill material was explored by recent studies [14, 15]. The high damping characteristics of scrap tires mixed with sand point out its promising potential in vibration mitigation.

Recently, a few studies have investigated the use of sand–rubber mixture (SRM) as a wave barrier. Mahdavisefat et al. [16] conducted a series of full-scale field experiments to investigate the effect of open and infilled trenches on vibration screening. The authors proposed that sand–rubber mixture can be used to fill the trench as it is a lightweight, high energy absorbing and environmentally friendly material for a wide range of vibration frequencies (10–600 Hz). It was also reported that the SRM with 30% rubber content performed better than other mixtures. Chew and Leong [17] conducted a full-scale field experimental study to investigate the performance of sand–rubber mixture as a vibration barrier, and it was found that the infilled trench with high rubber content significantly mitigates the vibration and suggested that optimum depth of SRM infill trench was similar to the findings of Woods [1].

Limited studies reported on SRM in fill trench barriers arrived at different optimum proportions of SRM for vibration isolation. In view of this, the present study aims to investigate the effectiveness of sand–rubber tire mixture (SRM) as the fill material for the infill trench barrier for the screening of ground-borne vibration by carrying out numerical analysis using FE Code ABAQUS.

2 SRM for Vibration Mitigation

2.1 *Problem Statement*

The present study focuses on the performance of an innovative screening technique using SRM infill trench barrier for ground-borne vibration mitigation. Numerical analysis was carried out on a FE model of the trench barrier underlain by a layered

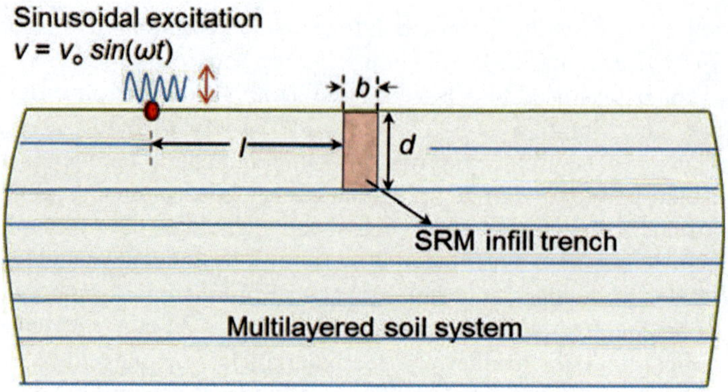

Fig. 1 SRM-isolation system used for vibration mitigation

soil medium. Figure 1 shows the schematic representation of the trench barrier with width w and depth d.

The vertical sinusoidal loading in terms of velocity, $v(t) = v0\sin(\omega t)$, was applied at the ground surface at a distance of l from the trench. Ground vibrations generated due to the vertical excitation were obtained at a fixed distance on either side of the wave barrier.

The active vibration screening efficiency of the SRM infilled trench barrier was quantified in terms of amplitude reduction factor (ARF) i.e., ratio of velocity amplitude after and before the installation of the trench at different locations caused by the harmonic excitation.

2.2 *Finite Element Modeling*

A 2D numerical study was carried out on the trench barrier system with SRM fill material for vibration mitigation using ABAQUS (Fig. 2.). The width (b) of the trench

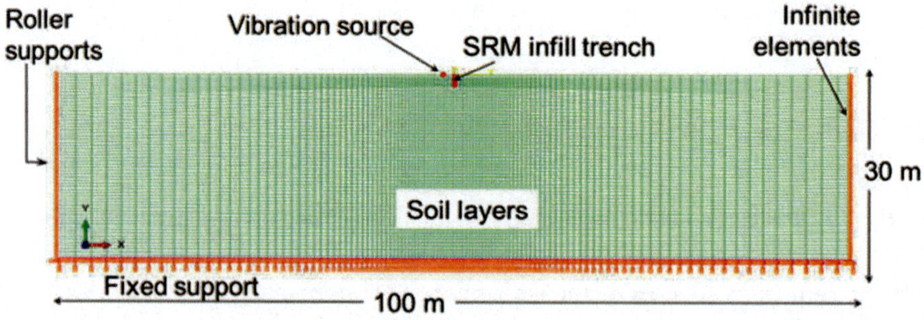

Fig. 2 Finite element model of the SRM isolation system

was considered as 1 m, while the depth of the trench (d) was varied from 1 to 3 m. The source to barrier distance was kept constant at 6 m. The entire trench system was underlain by a 30 m deep layered soil medium corresponding to a typical soil profile from the Indo-Gangetic plain region reported in Dhanya et al. [13]. The shear wave velocity of the top layer of soil was 200 m/s, and the Rayleigh wave velocity was estimated to be 187 m/s. The length of the soil medium is considered as 100 m to ensure free-field conditions. In the numerical analysis, the rubber content of SRM infill trench was considered as 30 and 50%.

The maximum mesh size for the problem was determined using CFL criteria [18], considering the wavelength of the vibrations and the shear wave velocity of the soil medium to ensure accurate wave propagation. The soil medium and the trench were discretized using four-node plane strain continuum elements. The mesh size of the soil medium was varied from 1 m × 1 m at the center to 5 m × 1 m toward the edges of the model. The default boundary conditions were used to represent the soil matrix, while infinite elements were provided at the far-field to ensure the absorption of outgoing waves, thereby preventing wave reflection.

The hyperelastic material model was adopted for the modeling of the SRM infill trench layer. The soil medium was assumed as homogeneous and the hypoelastic constitutive model was adopted to model the soil. The details of the material properties for the soil medium, sand, and SRM in terms of shear modulus degradation curves, stress–strain curves for SRM, Poisson's ratio, and density, were adopted from Dhanya et al. [13]. The Rayleigh damping coefficients method was adopted to account for the damping of vibrations in the material.

In the present study, vertical sinusoidal ground vibration of amplitude 2 m/s with frequency of 50 Hz was applied at the ground surface (Fig. 1). During the excitation, the time history of velocity before and after the trench and at different points of interest was obtained.

2.3 *Results and Discussion*

The velocity response measured just after the trench barrier with a depth of 1.5 m (SRM with 30% rubber content) for the sinusoidal vertical excitation on the ground surface is presented in Fig. 3. It can be easily noticed from the figure that the reduction in velocity amplitude is more significant due to the introduction of the open trench which is well established in past studies [1, 19, 20].

The efficiency of vibration screening using trenches was analyzed in terms of amplitude reduction factor (ARF), i.e., the ratio of the amplitude of vibration after and before the installation of trench barriers. Figure 4 presents the variation of ARF with the depth of trench normalized with the Rayleigh wavelength (d/λ). It can be noted that as the depth of trench increases, there is a general trend of reduction in ARF for open and SRM infill trench cases. The ARF of 0.3 was achieved between d/λ of 0.6 to 0.8, similar to the findings of Woods [1] in all three cases. The SRM50 infill trench was able to achieve 0.85 times ARF as that of the open trench, while SRM

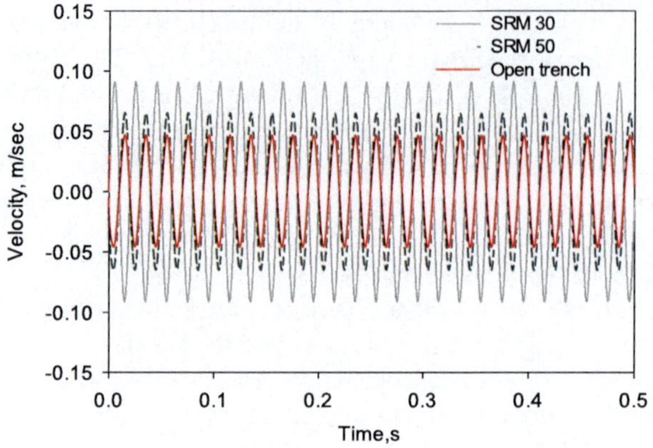

Fig. 3 Velocity–time histories after the trench barrier at $l = 1$ m and excitation frequency = 50 Hz

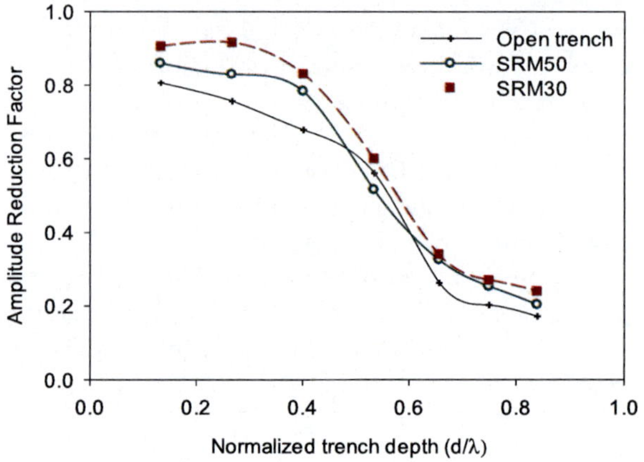

Fig. 4 Variation of amplitude reduction factor with depth of the trench ($l = 2$ m)

30 infill trench was able to reach around 0.7 times ARF as that of the open trench. Therefore, SRM infill trench with the rubber content of 50% screen the vibration amplitude by only about 15% less than that for open trenches.

Figure 5 presents the variation of ARF at varying distance from the source of vibration excitation. The sudden reduction in ARF with introduction of trench at a distance of 2 m from the source of excitation is evident for all the three cases followed by a gradual reduction of ARF. ARF of 0.3 was achieved at a distance of 1 m away from the open trench, while for SRM50 and SRM 30 trenches, it was achieved at a distance of 1.2 and 2 m, respectively, away from the trench location. At a distance of 2 m away from the trench the vibration reduction due to SRM 50 is 20% less than that of open trench while vibration reduction due to SRM 50 is 30% less than that of open trench. The SRM50–filled trench was found to provide best barrier effects than

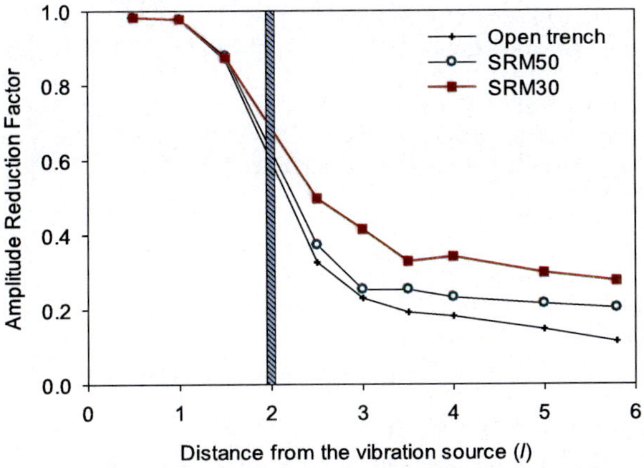

Fig. 5 Variation of amplitude reduction factor with distance from the trench (Depth = 2 m)

SRM 30 and exhibits comparable efficiency to that of open trench. Similar trends were also observed by Chew and Leong [17].

3 Conclusions

In the present paper, finite element studies were carried out to evaluate the effectiveness of SRM infill trenches to mitigate the ground-borne vibrations. Overall, it was found that the introduction of SRM infill trench barriers can significantly reduce the intensity of vertical ground vibrations. SRM infill trench with the rubber content of 50% screen the vibration amplitude by only about 15% less than that for open trenches, and hence, it can be used for screening of high frequency vibrations. However, further studies are required to arrive optimum dimensions of SRM infill trenches for effective screening of ground vibrations at wide range of frequency of excitation under different ground conditions.

References

1. Woods, R.D.: Screening of surface waves in soil. J. Soil Mech. Found. Eng. **94**(SM4), 951–979 (1968)
2. Richart, F.E., Hall, J.R., Woods, R.D.: Vibrations of Soils and Foundations, Englewood Cliffs. Prentice-Hall, NJ (1970)
3. Baker, J.M.: An experimental study on vibration screening by in-filled trench barriers, M.Sc. Thesis, State University of New York at Buffalo, USA (1994)
4. Murillo, C., Thorel, L., Caicedo, B.: Ground vibration isolation with geofoam barriers: centrifuge modeling. Geotext. Geomembr. **27**, 423–434 (2009)

5. Celebi, E., Firat, S., Beyhan, G., Cankaya, I., Vural, I., Kirtel, O.: Field experiment on wave propagation and vibration isolation by using wave barriers. Soil Dyn Earthq. Eng. **29**(5), 824–833 (2009)
6. Hazarika, H., Yasuhara, K., Karmokar, A. K., Mitarai, Y.: Shaking table test on liquefaction prevention using tire chips and sand mixture. In: Proceedings of the International Workshop on Scrap Tire Derived Geomaterials—Opportunities and Challenges, IW-TDGM 2007; pp. 215–222 (2008)
7. Kaneko, T., Orense, R.P., Hyodo, M., Yoshimoto, N.: Seismic response characteristics of saturated sand deposits mixed with tire chips. J. Geotech. Geoenviron Eng. **139**(4), 633–643 (2013)
8. Xiong, W., Li, Y.: Seismic isolation using granulated tire–soil mixtures for less-developed regions: experimental validation. Earthq. Eng. Struct. Dynam. **42**, 2187–2193 (2013)
9. Edil, T., Bosscher, P.: Engineering properties of tire chips and soil mixtures. Geotech. Test. J. **17**(4), 453 (1994)
10. Hazarika, H., Yasuhara, K., Kikuchi, Y., Karmokar, A.K., Mitarai, Y.: Multifaceted potentials of tire-derived three-dimensional geosynthetics in geotechnical applications and their evaluation. Geotext. Geomembr. **28**(3), 303–315 (2010)
11. Tsang, H., Lo, S.H., Xu, X., Sheikh, M.N.: Seismic isolation for low-to-medium-rise buildings using granulated rubber–soil mixtures: numerical study. Earthq. Eng. Struct. Dynam. **41**, 2009–2024 (2012)
12. Dhanya, J.S., Boominathan, A., Banerjee, S.: Performance of geo-base isolation system with geogrid reinforcement. Int. J. Geomech. **19**(7), 1–13 (2019)
13. Dhanya, J.S., Boominathan, A., Banerjee, S.: Response of low-rise building with geotechnical seismic isolation system. Soil Dyn. Earthq. Eng. **136**, 106187 (2020)
14. Hazarika, H., Kohama, E., Sugano, T.: Underwater shaking table tests on waterfront structures protected with tire chips cushion. J. Geotech. GeoEnviron. Engg., ASCE, **134**(12), 1706–1719 (2008)
15. Hazarika, H., Pasha, S.M.K., Ishibashi, I., Yoshimoto, N., Kinoshita, T., Endo, S., Karmokar, A.K., Hitosugi, T.: Tire chips reinforced foundation as liquefaction countermeasure for residential buildings. Soils and Founds **60**(2), 315–326 (2020)
16. Mahdavisefat, E., Salehzadeh, H., Heshmati, A.A.: Full-scale experimental study on screening effectiveness of SRM infilled trench barriers. Geotechnique **68**(10), 869–882 (2018)
17. Chew, J.H., Leong, E.C.: Field and numerical modelling of sand-rubber mixtures vibration barrier. Soil Dynam. Earthq. Eng. **125**, 105740 (2019)
18. Courant, R., Friedrichs, K., Lewy, H.: On the partial difference equations of mathematical physics. IBM J Res Dev **11**(2), 215–234 (1967)
19. Dasgupta, B., Beskos, D.E., Vardoulakis, I.G.: Vibration isolation using open or filled trenches Part 2: 3-D homogeneous soil. Comput. Mech. **6**, 129–142 (1990)
20. Saikia, A., Das, U.K.: Analysis and design of open trench barriers in screening steady-state surface vibrations. Earthq. Eng. Eng. Vib. **13** (2014)

Potential Application of Bauxite and Iron Mining Residues in Civil Construction in Brazil

Julian Buritica Garcia, Virginie Queiroz Rezende Pinto, and Juan Félix Rodríguez Rebolledo

Abstract Alumina is generated from the processing of an ore known as bauxite; the process generates a highly alkaline red-colored waste containing iron, titanium, and silica that can change any material's properties that comes into contact with the waste. According to Abal (Aluminio Associação Brasileira de Aluminio: Perfil da indústria Brasileira do. São Paulo 2019 [1]), Brazil is the third-largest producer of alumina in the world and the fourth-largest bauxite producer. On the other hand, iron is one of the most abundant natural elements being obtained from minerals such as hematite, magnetite, limonite, and siderite. Brazil is the second-largest producer of iron ore globally, mainly in the region known as the Quadrilatero Ferrífero QF in the Minas Gerais State. The storage and disposal of waste from iron and aluminum production has become a major environmental problem. South America is a region rich in mineral resources, so the number of tailings dams is very high and, in most cases, built without respecting international standards and good engineering practice. Exploring mineral resources and destroying the native forest, contaminating water resources, and throwing gases and particulate matter into the atmosphere also consume much energy and contribute to global warming. The main methods of disposal of mineral products in Brazil are the cheapest wet path and have a much higher potential for environmental impact than the dry path. In recent years in Brazil, there have been significant tragedies involving tailings dams of iron, such as the failure of the Fundão dam on November 5, 2015, and the Brumadinho dam's failure on January 25, 2019. Both cases caused the deaths of approximately 270 people and 25 missings. Thus, better mechanisms should be evaluated for the disposal of mining tailings such as dry processing and to reduce the use of dams in Brazil and South America. This work aimed to investigate the potential uses of alumina and iron tailings in civil construction. The waste has reasonable characteristics to be used as embankment fill material in infrastructure works, foundation material of some engineering works, ground improvement, or calcined additives in cement manufacturing and flexible pavements.

J. B. Garcia (✉) · V. Q. R. Pinto · J. F. R. Rebolledo
University of Brasilia, Brasilia, Brazil
e-mail: julianburitica@gmail.com

H. Hazarika et al. (eds.), *Advances in Sustainable Construction and Resource Management*, Lecture Notes in Civil Engineering 144,
https://doi.org/10.1007/978-981-16-0077-7_15

Keywords Mining · Bauxite · Iron · Environmental impact

1 Introduction

South America is rich in natural resources. Due to this fact, mining represents an essential source of income for all countries and is the engine of the region's economic development at the expense of immeasurable environmental impacts. There are vast reserves of minerals such as iron, manganese, aluminum, gold, phosphate, niobium, among others in Brazil. Brazil is the third-largest alumina producer in the world behind China and Australia [1] and the sixth-largest aluminum producer. Alumina is generated mainly from the processing of an ore known as bauxite. The essential condition for ore formation is the existence of a tropical climate with high temperatures and drying and wetting cycles that favor the leaching process Sampaio et al. [2].

Bauxite is a reddish rock with high percentages of aluminum. Bauxite mining in the country is carried out in Minas Gerais, Goiás, and Pará that represents 94.2% of Brazilian production (ANM [3]). The Amazon region is located in the northern part of South America in the equatorial region and has high temperatures and two defined periods of rain and drought. On the other hand, iron is one of the most abundant elements in nature being obtained from minerals such as hematite, magnetite, limonite, siderite, among others. Brazil is the second-largest producer of iron ore in the world, mainly in the region known as the "quadrilateral Ferrífero" in the Minas Gerais State.

Mineral exploration inevitably generates waste, storage, and mineral waste disposal which has become a major environmental problem for Brazil. Most of the waste is disposed of in dams. Thus, this paper presents a bibliographic review of the potential application of bauxite and iron residues, besides the classification, physical, and mechanical characterization of waste piles and compacted residue of bauxite to be used as embankment foundation or as a foundation bearing for operational buildings in large mining areas. Following the breakdown of tailings dams in Brazil in recent years, the study of alternatives for use and transformation into co-products of mining waste is of paramount importance to reduce the amount of waste stored and the risks associated with dams, as well as alternatives to dry production and storage of filtered waste in piles. The collapse of the iron ore tailings dam on the "córrego do feijão" in Brumadinho occurred on January 25, 2019. According to a national mining dam register survey, the dam failure released around 12 million cubic meters of tailings, and the Brumadinho disaster caused more than 300 fatalities (counting the victims already identified and those still disappeared) and immeasurable impacts on the environment and the local economy. On November 5, 2015, the Fundão dam failure occurred at Mariana MG, releasing an estimated 34 million cubic meters of tailings, killing 17 people and leaving approximately 10,482 affected people (Minas Gerais State Government [4].

2 The Potential Application of Iron Mining Residues in Civil Construction

In the region known as the "Quadrilátero Ferrífero" alone, over 700 iron tailings dams constitute real safety challenges for dams and the environment for mining managers and public agencies [5].

When tailings from a dam are brought to high temperatures in the kiln, and the granular materials from pozzolan and pigment are separated, the co-products generated can be used to manufacture concrete, the ceramic industry, and the manufacture of bricks and blocks. Infrastructure such as embankment material, reinforcement of subgrade in road construction, production of asphalt mixtures, construction of gravity barriers to contain mud or rocks, among others.

Some investigations have been carried out, replacing the sand with ore waste to make high strength interlocking blocks for paving, as shown in Fig. 1.

Filho [5] used iron ore waste collected from a dam in the municipality of Igarapé/MG (Fig. 2) to construct concrete blocks for paving with sand substitution contents of 10%, 50%, and 80% tailings, and also for use as a pavement layer through CBR and expansion testing. The tailings were favorable for its use in the two applications mentioned, requiring only a screening process to remove roots and organic material.

The manufacture and resistance tests carried out on the blocks were in accordance with standard NBR 9781 [6].

Fig. 1 Process of manufacturing of blocks with mining waste [7]

Fig. 2 Collection site at the sterile dam [5]

The interlocking blocks produced with the ore have regular physical appearance without cracks and lamination, and average compressive strength close to 50 MPa, well above the minimum established level of 35 MPa.

The pure material for use in pavement layers obtained CBR results between 9 and 53% and expansion between 0.11% and 0.34% values that qualify the material for use as reinforcement of the subgrade.

Salgado [7] conducted research replacing the sand from the manufacture of concrete blocks with a mining tail composed mainly of iron and silica, and concludes that the use of the tail did not influence significant changes in the studied properties. Therefore, the sand may be replaced entirely in the manufacture of paving blocks for paving.

3 The Potential Application of Bauxite Mining Residues in Civil Construction

Several studies show that bauxite residues can improve the manufacture of sulfate-resistant cement and various special cement. The addition of 2–10% bauxite contributes to lowering concrete costs, lowering the temperature, and lowering carbon dioxide emissions [8].

The waste can be used as an embankment foundation, material for impact gravity walls for rocks or debris flows, or a foundation for operational or conventional buildings in the extensive mining areas where the stability is guaranteed.

4 Physical and Mechanical Characterization of a Pile and Compacted Waste of Bauxite

This paper presents the analysis of the geotechnical behavior of bauxite samples collected in northern Brazil. Pinto [9] presented the physical characterization (grain size, consistency limits, humidity), resistance, and deformability (triaxial and consolidation test) and field tests (SPT, CPTu, and vane test). The material tested samples correspond to loose-form bauxite sterile (waste piles) and compact form in embankments. An example of the samples is shown in Fig. 3. The undisturbed block samples were obtained following the recommendations of the ASTM D7015-04 [10].

Figure 4 shows a typical section of the plowed area with the location of some tests in profile on the waste pile and on the landfill carried out with the pile material itself to conform to the pile tops and regularize the surface. It is possible to observe where the loose and compacted samples of Fig. 3 were taken.

Fig. 3 Undisturbed sample in loose and compacted state [9]

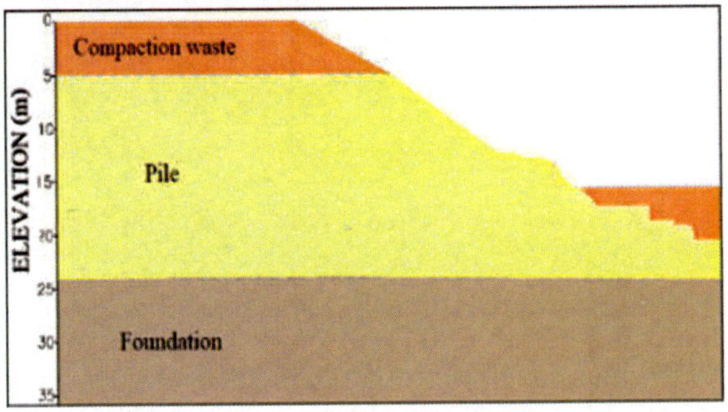

Fig. 4 Typical section of plowed area (Modified -Pinto [9])

Twenty-nine undisturbed samples were collected, 65 SPT tests, 45 Piezocone CPTu tests, and 45 vane tests, 22 compaction tests, 12 consolidation tests, and 25 CIU triaxial tests were performed. The sterile bauxite material has been classified as a silty clay of high plasticity. The sterile samples were divided into two groups, samples with a low percentage of gravel less than 30 (% G > 30) that present typical sandy silts, high density, and low moisture content. In contrast, samples with a percentage of gravel below 30 (% G < 30) present typical compost of silty clays. In the studies carried out, it was observed that 80% of the samples taken have a percentage of gravel less than 30%, concluding that the sterile bauxite has a more clayey behavior. Physical characterization, resistance, and deformability tests were carried out to obtain geotechnical parameters in both states, compacted and loose, and to be able to evaluate the mechanical behavior of the material. To define the optimal compaction parameters, the compaction test was performed on the deformed samples, and the result of the compaction test on the material is shown below in Fig. 5.

With the definition of the optimum moisture content for the two materials studied, an average compaction degree equal to 104% was calculated. In contrast, the loose material showed an average degree of compaction of 91%.

The processing of data from laboratory and field tests performed presents the results of the parameters obtained for the waste pile and compacted residue with their respective coefficient of variation that shows the heterogeneity of the waste (Table 1).

Although the waste has enormous variability, the geotechnical parameters are reasonable in strength, especially the compacted waste.

When analyzing the results of the reed tests in the region of the pile of stevia, considerable variability of results was observed, according to Pinto [9] some tests were carried out in regions partially saturated by partial drainage of the land. The

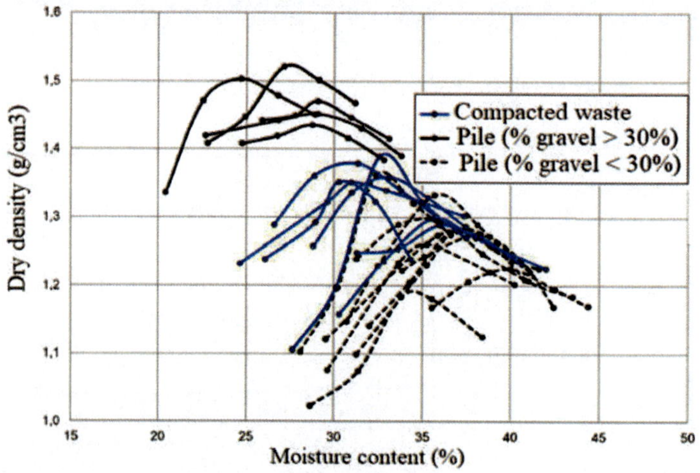

Fig. 5 Compacted curve of the material (Modified -Pinto [9])

Table 1 Geotechnical parameters for the bauxite [9]

Geotechnical parameters	Compacted residue		Waste pile	
	Average	Coefficient of variation CV (%)	Average	Coefficient of variation CV (%)
ϒ (kN/m³)	18	4	17	4
G	2,7	4	2,7	3
w (%)	0,3	9	0,4	14
Compaction degree (%)	103,9	6	91,3	5
c (kPa)	24,5	35	15,2	64
φ(°)	32,6	7	27,2	15
Void ratio	1	4	1,2	11

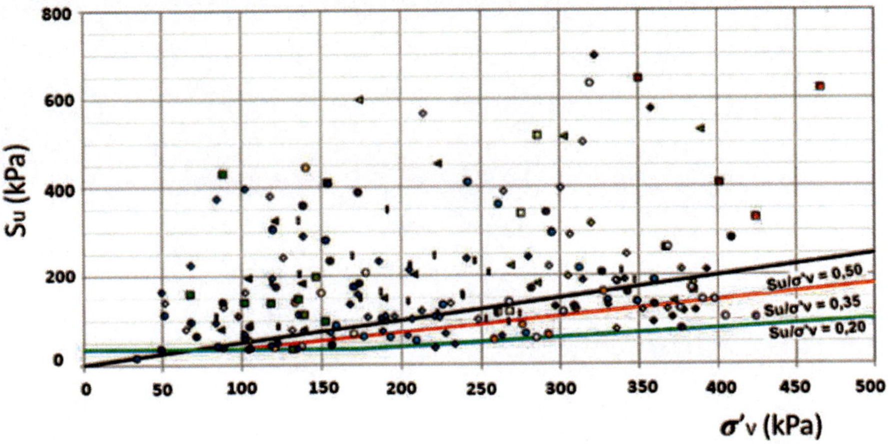

Fig. 6 Variation of the sterile pile resistance ratio [9]

tests carried out in saturated regions in the waste pile showed undrained resistances between 100 and 200 kPa. Figure 6 shows the waste pile's resistance ratio, which shows excellent dispersion and minimum resistance ratio values of 0.22 and values well above 0.5. (S_u = undrained shear strength and σ_v = total vertical stress)

5 Geotechnical Behavior of the Bauxite Waste as a Foundation Layer

Pinto [9] performed numerical simulations of finite elements in the Plaxis 3D software to evaluate the bauxite sterile's performance as a foundation layer for the execution of a shallow slab foundation and a deep pile raft foundation; the two modeled cases were hypothetical. The author calibrated the modified cam-clay constitutive model, as shown in Figs. 7 and 8.

Pinto [9] performed parametric analyzes for two types of foundation (raft system and piled raft system) changing characteristics, pile length (L), and pile diameter (D) as shown in Fig. 9.

When a shallow foundation with a rigid and flexible slab was considered, the maximum total settlements were high (between 50 and 80 mm) for the drained conditions. On the other hand, in the undrained analyzes, the settlements were much lower than the established maximums (30 mm), showing that it is not a critical condition for the problem studied (Fig. 10).

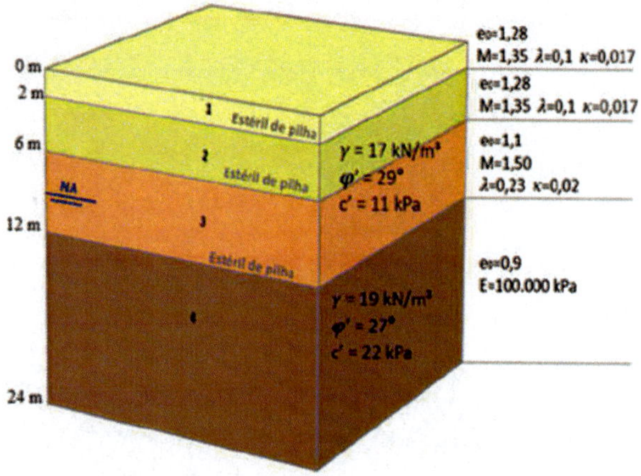

Fig. 7 Modified cam-clay model parameters [9]

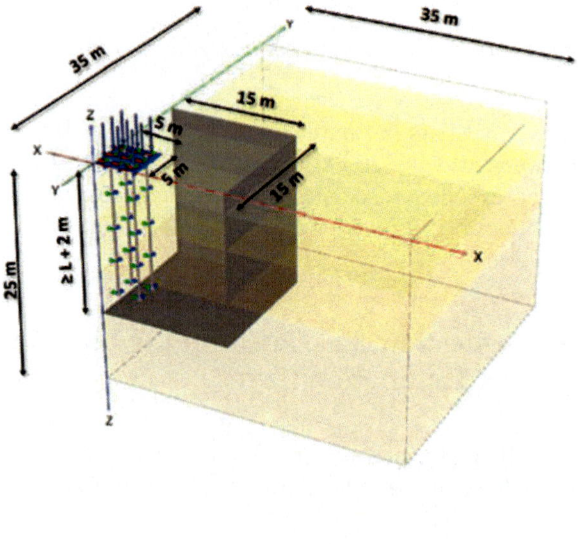

Fig. 8 Geometric configuration Pinto [9]

Fig. 9 Foundations cases analyzed Pinto [9]

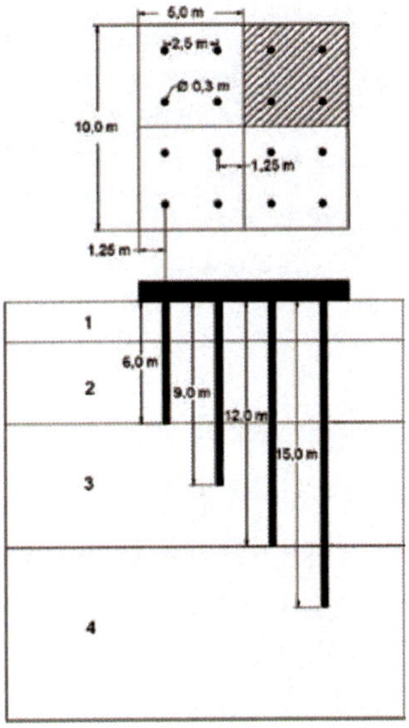

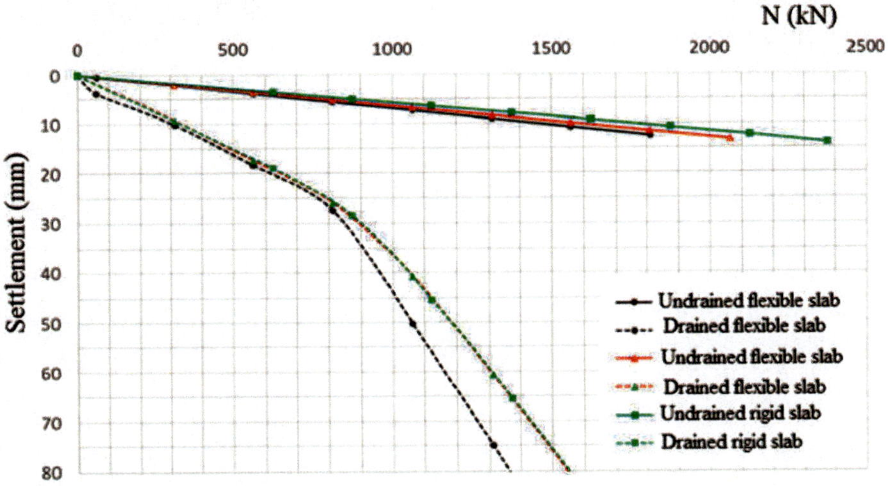

Fig. 10 Load-settlement curves for shallow foundation [9]

When a deep foundation with a rigid slab was considered, the maximum total settlements were high. On the other hand, if a flexible slab is considered (L/D = 40 and L/D = 50), the settlements obtained are within the tolerable limits which are within the criterion adopted as a maximum settlement of 30 mm (Fig. 11).

For the criterion of maximum allowable total settlement, considering the load established design, the flexible raft configuration, with L/D = 40, was the one that

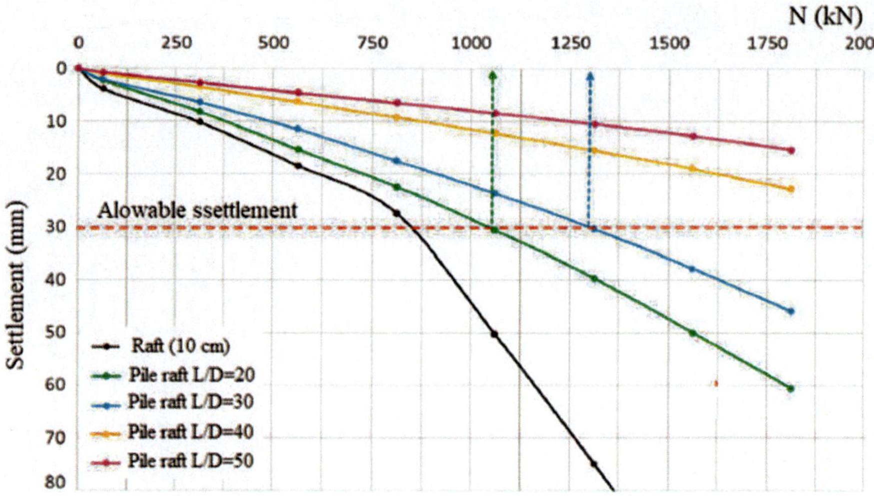

Fig. 11 Load-settlement curves for deep foundations with flexible slab [9]

presented better results, since the use of piles with L/D = 30 resulted in a settlement of 0.5 mm above the allowable value.

By performing numerical simulations with various types of foundations based on the material studied, Pinto [9] concluded that the material of the sterile bauxite pile is competent as foundation material for deep foundations that support loads from a mine operational structure, and it may be possible to build some large concrete structure in a mining area.

6 Conclusions

Iron-mining dam residues is a quality material for the production of high-strength interlocking blocks following NBR 9781 recommendations and as a constituent material in the reinforcement of subgrade in the pavement's structural layers.

Bauxite waste in the compacted state may be competent to build significant engineering works in restore areas. Waste in the loose sterile pile state may be competent as the foundation layer of intermediate engineering works as operational structures whenever and when had been stability or as embankment material for the construction of impact gravity barriers for rocks, mud, and debris flows.

This work and all data mentioned herein are strictly for educational use.

Acknowledgements The support of the foundation for research support of the Federal District (FAP-DF) is acknowledged and appreciated.

References

1. Abal: Aluminio Associação Brasileira de Aluminio: Perfil da indústria Brasileira do. São Paulo (2019)
2. Sampaio, J.A., Andrade, M.A.C., Dutra, A.J.B.: Rochas e minerais industriais. CETEM/2008, 2 Ed, 311–314 (2008)
3. ANM: Agência Nacional de Mineração: Sumário Mineral Brasileiro 2017. Brasília, DF, 2p (2017)
4. Governo do estado de Minas Gerais. Relatório: Avaliação dos efeitos e desdobramentos do rompimento da Barragem do Fundão em Mariana-MG. Grupo da Força Tarefa, Decreto 46.892/2015. Belo Horizonte, Brasil (2016)
5. Filho, J.N.S.: Estudos de reaproveitamento dos resíduos das barragens de minério de ferro para uso na pavimentação de rodovias e fabricação de blocos intertravados. Dissertação de mestrado, Centro Federal de Educação Tecnológica de Minas Gerais. CEFET – MG (2013)
6. Associação Brasileira De Normas Técnicas (ABNT): NBR 9781. Peças de Concreto para Pavimentação—Especificação (1987)
7. Salgado, L.M.: Blocos de concreto para pavimentação produzidos com rejeito de mineração e cinzas de bagaço de cana de açúcar. Dissertação de mestrado, Universidade Federal de Lavras. Lavras—MG (2018)
8. International Aluminium Institute: Opportunities for use of bauxite residue in special cements. World Aluminium (2018)
9. Pinto, V.Q.R.: Comportamento Geotécnico de estéril de bauxita e seu potencial uso como estrato de fundação. Dissertação de mestrado em Geotecnia, Universidade de Brasília. Brasília. Brazil (2019)
10. ASTM: D7015-04 standard practices for obtaining undisturbed block (cubical and cylindrical) samples of soils. West Conshohocken, PA; ASTM International (2004). https://doi.org/10.1520/D7015-04

Reuse of Waste Tires to Develop Eco-rubber Seismic-Isolation Foundation Systems: Preliminary Results

Gabriele Chiaro, Ali Tasalloti, Alessandro Palermo, Gabriele Granello, and Laura Banasiak

Abstract This paper presents preliminary experimental investigations carried out to evaluate the feasibility of using gravel–rubber mixtures as low-cost geotechnical seismic-isolation foundation systems for low-rise buildings in New Zealand. One-dimensional compression and direct shear tests are performed to evaluate the compressibility and strength properties of three types of rigid-soft granular mixtures prepared using a combination of uniformly-graded rounded and angular gravels as well as large and small recycled rubber particles. The mixtures are prepared at volumetric rubber contents (*VRC*) of 10%, 25%, and 40%. The concept of skeleton material is adopted to establish whenever the gravel or rubber matrix would govern the overall mechanical behavior of the mixtures. It is shown that the compressibility of the mixtures highly depends of the skeleton material. Mixtures with *VRC* < 10% (gravel skeleton) perform better than those with *VRC* ≥ 25% (rubber skeleton), i.e., smaller vertical strains are developed under the same applied vertical stress level. Moreover, within the range of *VRC* tested, the mixtures made of large rubber particles are less compressible than those consisting of small rubber particles. On the other hand, under the testing conditions investigated, all mixtures exhibited high strength (i.e., friction angle > 35°) irrespective of the *VRC* and rubber particle size.

G. Chiaro (✉) · A. Tasalloti · A. Palermo · G. Granello
Department of Civil and Natural Resources Engineering, University of Canterbury, Private Bag 4800, Christchurch 8041, New Zealand
e-mail: gabriele.chiaro@canterbury.ac.nz

A. Tasalloti
e-mail: ali.tasalloti@canterbury.ac.nz

A. Palermo
e-mail: alessandro.palermo@canterbury.ac.nz

G. Granello
e-mail: gabriele.granello@canterbury.ac.nz

L. Banasiak
Environmental Science and Research Institute Ltd, Christchurch 8041, New Zealand
e-mail: laura.banasiak@esr.cri.nz

H. Hazarika et al. (eds.), *Advances in Sustainable Construction and Resource Management*, Lecture Notes in Civil Engineering 144,
https://doi.org/10.1007/978-981-16-0077-7_16

Keywords Waste tires · Seismic-isolation · Foundation systems · Gravel–rubber mixtures · Compressibility · Strength

1 Introduction

The current annual rate of waste tires production in New Zealand is over 5 million and is expected to grow over time with increased population and number of vehicles on roads. It is estimated that 70% of such waste tires are destined for landfills, stockpiles, illegal disposal or are otherwise unaccounted for, giving rise to piles of tires that do not readily degrade or disintegrate [1]. With the ever-growing volume of waste tires, environmental, socio-economic, and health concerns have urged the reuse of waste tires through sustainable recycling projects.

Waste tires are a great source of environmentally friendly and sustainable building materials that may provide novel and effective engineering solutions to attain structures with enhanced seismic resilience [2]. This makes them ideal materials for developing affordable, medium-density, low-rise buildings that are in high demand in New Zealand.

Aimed at addressing this challenging problem, a joint geo-environmental-structural experimental research program has been carried out by researchers of the University of Canterbury and the Institute of Environmental Science and Research Ltd. to investigate sustainable options for the reuse/recycling of waste tire in civil engineering applications. One such option is to employ waste tire (in the form of granulated rubber) mixed with gravelly soils to develop eco-rubber geotechnical seismic-isolation (ERGSI) foundation systems [3–6] for low-rise buildings composed of: (i) a shallow and resilient layer of mixed shredded tires and gravel, and (ii) a flexible rubber-concrete raft foundation (Fig. 1).

The feasibility of using recycled rubber to develop ERGSI foundation systems is being assessed by:

- Geotechnical and environmental engineering investigations to identify optimum gravel-recycled rubber mixtures, having excellent mechanical properties and minimal leaching attributes;
- Structural engineering tests to design flexible, fiber-reinforced, rubberized-concrete raft foundations with satisfactory structural performance;
- Numerical and physical models to prove the concept, evaluate the seismic performance of the entire ERGSI foundation system, and quantify the level of reduction in the seismic response of prototype buildings.

In this paper, results of the geotechnical investigation that include the physical, compressibility, and strength characteristics of selected gravel-rubber mixtures will be presented and discussed in details. Preliminary findings from the environmental and structural studies have been reported in Banasiak et al. [6] and Hernandez et al. [3], respectively. Results of the physical and numerical studies will be presented elsewhere in due course.

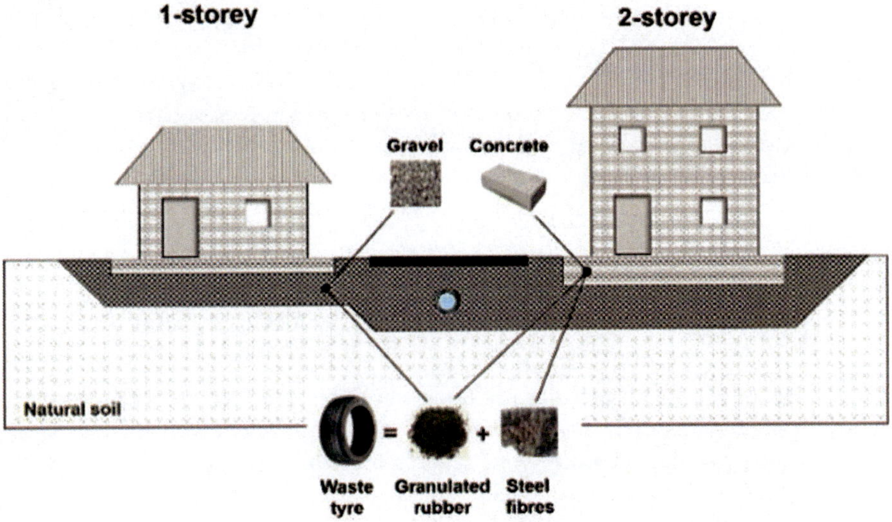

Fig. 1 Proposed ERGSI foundation system for residential buildings in New Zealand

2 Use of Soil–rubber Mixtures in GSI Foundation Systems

Scrap tire derived materials (in the form of chips, crumbs, granules, and shreds) mixed with granular soil (primarily sand) have been traditionally used in civil/geotechnical applications such as light backfill material, drainage layer, slope stabilization, and landfill construction [6, 7]. However, the attractive strength and dynamic properties along with environmental benefits, low cost, and material availability of soil–rubber mixtures support their use for geotechnical seismic-isolation (GSI) foundation systems of residential structures in developing [2] and developed countries [3]. Tsang et al. [2] carried out a numerical investigation on the feasibility of using soil–rubber mixtures for GSI systems of middle-rise residential buildings, concluding that GSI systems can reduce considerably both the horizontal and vertical ground motion acceleration response of structures subjected to earthquake shakings. However, to date, no specific studies have included experimental verification and/or exact specifications of the actual grain size of soil and rubber, their aspect ratio ($D_{50,\mathrm{Soil}}/D_{50,\mathrm{Rubber}}$), and volumetric fractions that should be used to design GSI systems [9].

Thus far, much research has focused on the mechanical characterization of sand–rubber mixtures. Hazarika and Adbullah [10], however, pointed out that in the selection of the soil and recycled rubber to form soil–rubber mixtures for use in geotechnical applications, the availability and the cost efficiency of both materials should be carefully considered. Actually, to avoid segregation, the recycled rubber should be cut into smaller pieces when mixed with sandy soils, which will inevitably increase the implementation costs. Hence, it was proposed to use gravel–rubber mixtures instead.

To date, however, very limited studies have focused on the geotechnical characterization of gravel–rubber mixtures [6, 10–12], making their physical properties and mechanical behavior largely unknown. Thus, given the above background, this paper will contribute in filling a major gap in the knowledge by providing new data on and insights into the physical and mechanical engineering properties of gravel–rubber mixtures that would facilitate their adoption to develop GSI systems for residential building, including the ERGSI foundation system proposed in this study.

3 Experimental Study: Materials and Procedure

Three types of rigid-soft granular mixtures were prepared using a combination of uniformly graded rounded (G_{RND}) and angular (G_{ANG}) gravels as well as large (R_{LRG}) and small (R_{SML}) recycled rubber particles. Figure 2 shows the particle size distribution and photographic images of each material. The index properties of the gravels and recycled rubbers are summarized in Table 1.

Specifically, to evaluate the effects of gravel particle shape on compressibility and strength shape of gravel–rubber mixtures, both the rounded gravel and angular gravel

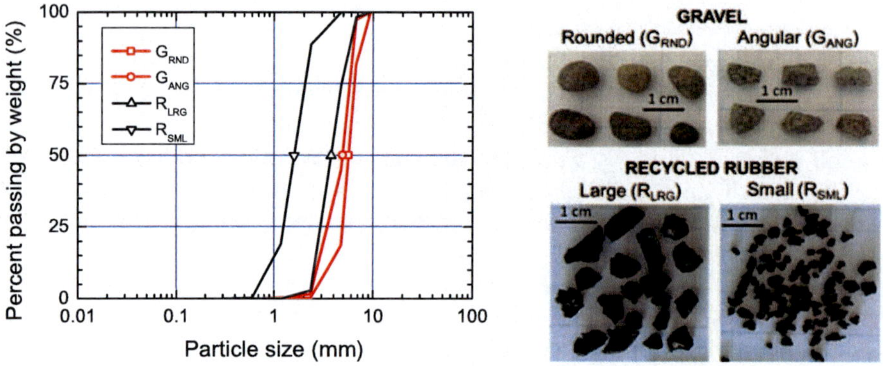

Fig. 2 Particle size distribution and photographic images of tested materials

Table 1 Properties of gravels and recycled rubber used in this study

Material	Gravel		Recycled rubber	
	Rounded	Angular	Large	Small
Symbol	G_{RND}	G_{ANG}	R_{LRG}	R_{SML}
Mean diameter D_{50} (mm)	5.6	4.2	3.7	1.6
Specific gravity G_s	2.72	2.76	1.15	1.14
Maximum dry density ρ_{dry} (kg/m^3)[a]	1753	1598	649	602

[a]by Proctor compaction tests

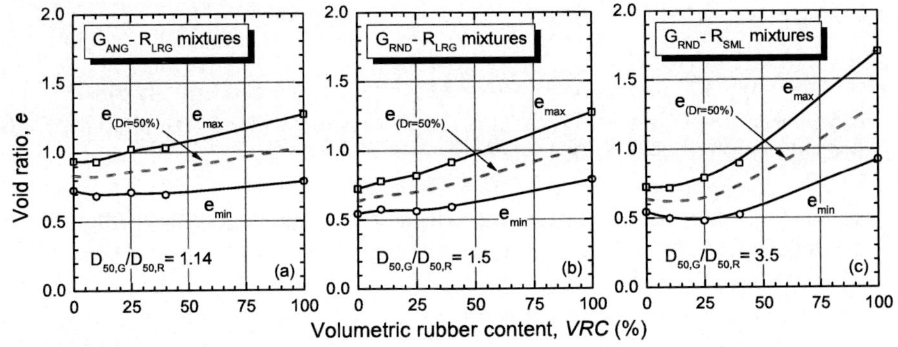

Fig. 3 Void ratio variation with *VRC* for the gravel–rubber mixtures tested in this study

were mixed with the large rubber particles to form $\mathbf{G_{RND}}$-$\mathbf{R_{LRG}}$ ($D_{50,G}/D_{50,R} = 1.5$) and $\mathbf{R_{ANG}}$-$\mathbf{R_{LRG}}$ ($D_{50,G}/D_{50,R} = 1.14$) assortments. Moreover, to assess the effects of the particle size ratio ($D_{50,G}/D_{50,R}$) on compressibility and strength of gravel–rubber mixtures, the rounded gravel was also mixed with small rubber particles to create $\mathbf{R_{RND}}$-$\mathbf{R_{SML}}$ ($D_{50,G}/D_{50,R} = 3.5$) compounds. The mixtures were prepared at volumetric rubber contents (*VRC*) of 10%, 25%, and 40%, where *VRC* is defined as the ratio between the volume of rubber and the total volume of solids.

All specimens were prepared at a relative density (D_r) of 50% by dry tamping method and tested in dry conditions. As reported in Fig. 3, the relative density was calculated based on the values of minimum void ratio (e_{min}—obtained by standard Proctor compaction tests) and maximum void ratio (e_{max}—obtained by carefully pouring the materials in the compaction mold with zero depositional height). Note that, for the tested gravel-mixture assortments, vibratory table tests were found not effective to obtain e_{min}. This is in contrast to the response of hard-grained stiff granular materials [13] for which vibratory compaction is most effective; however, it can be attributed to the elastic nature of the recycled rubber particles in the mixtures.

Segregation is an inherent difficulty in granular mixtures that may be triggered by size, density, stiffness, and shape features [14]. Segregation in the specimens was prevented by minimizing any vibration and avoiding granular flow.

The compressibility of the investigated materials was evaluated by using an appositely-built medium-size one-dimensional compression cell (inner diameter 250 mm; specimen height from 150 mm) up to a vertical stress of 500 kPa. On the other hand, the strength properties were estimated by means of a medium-size direct shear box (100 mm × 100 mm in cross section and 53 mm in height) under three normal stress levels of 30, 60, and 100 kPa. The horizontal displacement rate was 1 mm/min.

4 Experimental Results

The gravel–rubber compounds investigated in this study are essentially binary granular mixtures consisting of rigid and soft particles of different size and shape. As described hereafter, their packing density (i.e., void ratio) and mechanical behavior (i.e., compressibility and strength) are influenced by the inherent properties of the materials (i.e., size and shape), the proportion of large/small and rigid/soft particles in the total volume of solids, as well as the size ratio of large and small particles ($D_{50,\text{large}}/D_{50,\text{small}}$).

4.1 *Packing of Gravel–Rubber Mixtures with Different Particle Size Ratios*

Experimental evidences from previous studies have indicated that three packing states exist for two-size particle mixtures: floating state (large-particle dominated), non-floating state (small-particle dominated), and transitional state. In the floating state, small particles inhibits contact between large particles. In the non-floating state, the large particles are in contact, and the small particles mainly fill the voids between large particles. In the transitional state, small particles partially fill the gaps between large particle, and large particle are only partially disconnected.

For the examined mixtures, the variation of void ratio with increasing *VRC* is shown in Fig. 3. The values of the e_{max} and e_{min} depend on the inherent properties of the materials. For the R_{ANG}-R_{LRG} and R_{RND}-R_{LRG} mixtures that have similar particle size ratio (i.e., 1.14 and 1.5, respectively), e_{max} and e_{min} increase with increasing *VRC*. Because the large particles (gravel) are surrounded by small particles of almost the same size (large rubber), no packing phenomenon occurs, and a clear distinction between floating and non-floating state, as described above, cannot be made. Similar trends have been reported by Pasha et al. [11].

For the R_{RND}-R_{SML} mixtures ($D_{50,\text{G}}/D_{50,\text{R}} = 3.5$), the minimum of the e_{max} and e_{min} is achieved at *VRC* of 10% and 25%, respectively. These values of *VRC* indicate that a transition from non-floating to floating packing state occurs in the mixtures with increasing *VRC*.

4.2 *Compressibility Characteristics of Gravel–Rubber Mixtures*

The packing properties of the mixture described earlier may provide useful information regarding the expected mechanical behavior of the mixtures. However, the way the load is transferred with in the mixtures depends primarily on the skeleton material [14, 15]. According to the concept of skeleton material, a skeleton is formed when particles of the same material are in contact with each other and are able to transfer loads. The material forming the skeleton becomes the matrix material that governs the overall mechanical behavior of a mixture. In the case of gravel–rubber mixtures, two matrix material can be expected: i) a gravel matrix, which will lead to a stiffer gravel-like behavior of the mixtures; and ii) a rubber matrix that will produce a softer rubber-like response of the mixtures.

The stress–strain response of gravel–rubber mixtures under zero lateral strain loading—k_0 conditions—is shown in Fig. 4. While side friction is expected to lessen deformation (being currently evaluated in the laboratory), trends clearly show increase in compressibility with increasing *VRC* for any given vertical shear stress level. However, the extent of compressibility highly depends of the skeleton material. It can be observed that mixtures with *VRC* < 10% (gravel skeleton) perform better than those with *VRC* ≥ 25% (rubber skeleton), i.e., much smaller vertical strains (ε_v) are developed under the same applied vertical stress (σ_v') level. Moreover, within the range of *VRC* tested, it is also observed that the mixtures made of large rubber particles are less compressible than those consisting of small rubber particles. For example, if a reference $\sigma_v' = 100$ kPa is considered, the $\varepsilon_v < 2.5\%$ for all mixtures with *VRC* < 10%, while for those with *VRC* ≥ 25% ε_v is between 5 and 14%, being $\varepsilon_v(R_{RND}\text{-}R_{SML}) > \varepsilon_v(R_{RND}\text{-}R_{LRG}) > \varepsilon_v(R_{ANG}\text{-}R_{LRG})$.

Figure 5 shows the variation with *VRC* of the constrained modulus ($M = \Delta\sigma_v'/\Delta\varepsilon_v * 100$) computed between two successive loading stages during loading for all mixtures. The data show that *M* drastically decreases with increasing *VRC*, indicating that when *VRC* ≤ 10%, the mixture supports the load mainly though the gravel skeleton. The increased *VRC*, however, gradually reduces the mixture stiffness until eventually *VRC* is high enough, so that the load is transferred primarily through the rubber-rubber-contact (rubber skeleton).

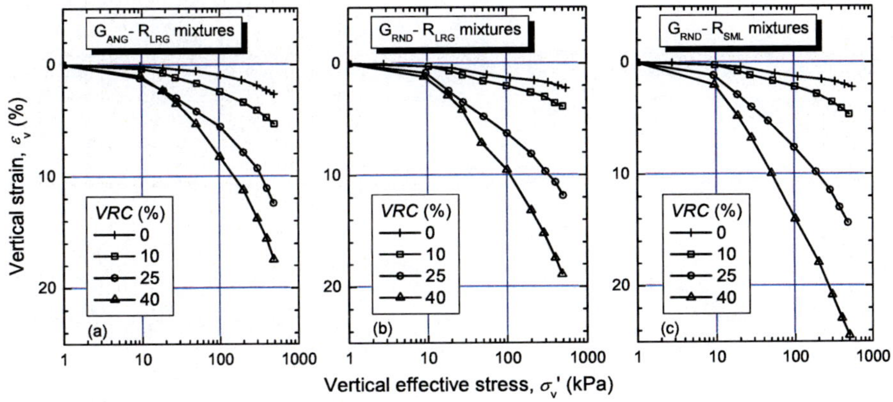

Fig. 4 Compressibility characteristic of the gravel–rubber mixtures tested in this study

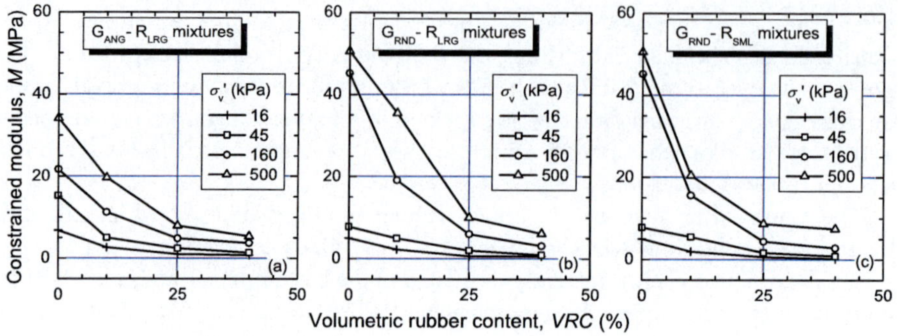

Fig. 5 Variation of constrained modulus with *VRC* for the gravel–rubber mixtures tested in this study

4.3 Shear Strength Characteristics of Gravel–Rubber Mixtures

Typical stress–strain–volume change behavior obtained by direct shear tests at a vertical stress of 60 kPa for all tested mixtures is reported in Fig. 6. It can be observed

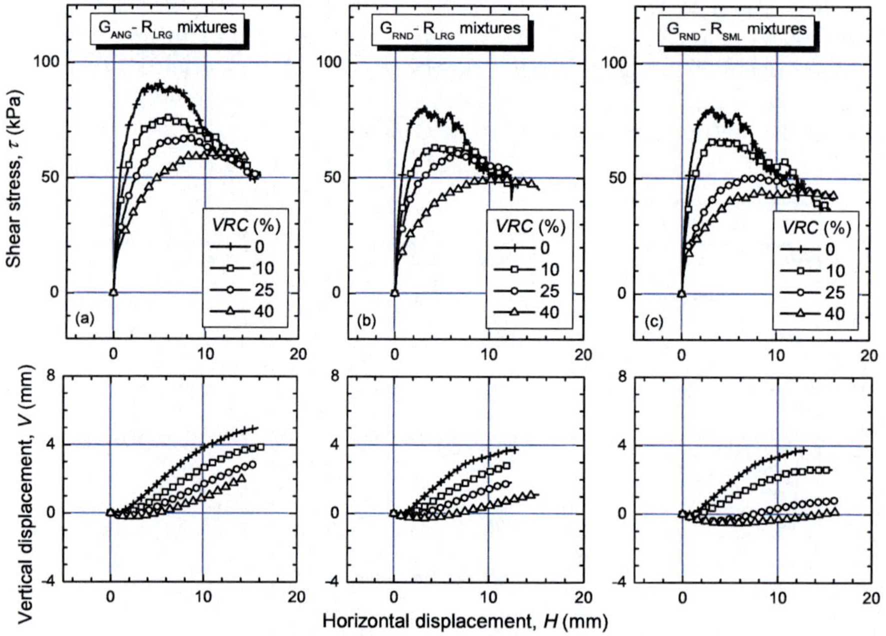

Fig. 6 Stress–strain–volume behavior of the gravel–rubber mixtures evaluated in this study in the direct shear apparatus

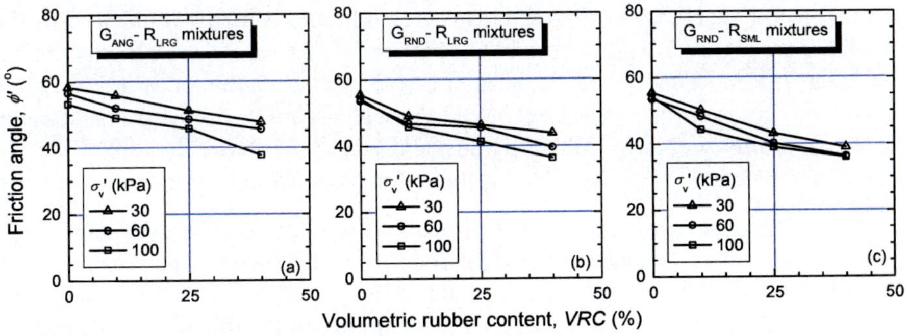

Fig. 7 Variation of friction angle with *VRC* for the materials tested in this study

that by increasing *VRC*, the material response gradually changes from dilative with clear peak shear state to contractive without peak shear state. It is observed that not only the stiffness of the materials drastically decreases, but also the peak shear stress. Moreover, the lowest peak shear stress is measured for the R_{RND}-R_{SML} mixtures, implying that the large rubber particles are able to provide higher strength compared to the small rubber particles.

Figure 7 reports the variation with *VRC* of the friction angle of all mixtures for three different levels of vertical stress. The friction angle drastically decreases with increasing *VRC* and decreases with increasing vertical stress level. However, under the testing conditions used in this study, gravel–rubber mixtures have high strength (i.e., friction angle > 30–35°) irrespective of the *VRC* and rubber particle size making them acceptable for many geotechnical engineering application [16], including ERGSI foundation systems.

5 Conclusions and Recommendations

The experimental investigations presented in this paper are part of a broader study aimed at evaluating the feasibility of using gravel–rubber mixtures as low-cost geotechnical seismic-isolation foundation systems for low-rise buildings in New Zealand. Although not exhaustive, the data reported in this paper provides important insights on key factors such as gravel shape, rubber size, and gravel–rubber particles size ratio that affects the mechanical behavior of gravel–rubber mixture, which should be taken into consideration in the design of such synthetic materials for use in ERGSI foundation systems.

In conclusion, the results of the investigation have shown that:

- All the gravel–rubber mixtures exhibited friction angle value > 35° acceptable for use in ERGSI foundation systems, irrespective of the rubber size and content, gravel particle shape, and size ratio between gravel and rubber particles;

- On the other hand, the compressibility of the mixtures is highly influenced by the rubber content and the rubber size. The higher the *VRC* and the smaller the rubber size are, the more compressible the mixtures are. However, this study indicates that under a static vertical stress level of 20 kPa (typical for 1–2 timber frame house in NZ), mixtures with large rubber particle and $VRC \leq 25\%$ would undergo vertical strain of 3% or less, making them suitable for use in ERGSI foundation systems.

Despite this encouraging results obtained in this study, the ultimate adoption of these gravel–rubber mixtures in ERGSI foundation engineering applications will greatly depend on the elastic and dynamic properties of the materials. Therefore, the authors are currently evaluating the small-strain shear stiffness, shear modulus degradation, and damping properties of these materials by bender elements and triaxial tests. Moreover, impact tests and small-scale shake table tests are conducted to evaluate the dynamic properties of a prototype ERGSI foundation system.

Acknowledgements The authors are grateful for the research fund provided by the Ministry of Business, Innovation and Employment of New Zealand (MBIE Smart Ideas Research Grant No. 56289). Laboratory assistance provided by Dr Sean Rees, Mr. Siale Faitotonu, Mr. Maysen Bloemen, and Mr. Ryan Thom is also greatly appreciated.

References

1. Ministry for the Environment: Waste Tyres Economic Research. Report **3**, 87 (2015)
2. Tsang, H.H.: Seismic isolation by rubber-soil mixtures for developing countries. Earth Eng. Struct. Dyn. **37**, 283–303 (2008)
3. Hernandez, E., Palermo, A., Granello, G., Chiaro, G., Banasiak, L.: Eco-rubber seismic isolation foundation systems, a sustainable solution for New Zealand context. Struct Eng Int (in press) (2020)
4. Chiaro, G., Palermo, A., Granello, G., Tasalloti, A., Stratford, C., Banasiak, L.J.: Eco-rubber seismic-isolation foundation systems: a cost-effective way to build resilience. In: Proc. 2019 Pacific Conf Earth Eng, Auckland, New Zealand, pp. 8 (2019)
5. Tasalloti, A., Chiaro, G., Palermo, A., Banasiak, L.J.: Effect of rubber crumbs volumetric content on the shear strength of gravelly soil in direct shear apparatus. In: Proc. Geo-Congress 2020, Minneapolis, Minnesota, USA, ASCE GSP, vol. 319, pp. 259–266 (2020)
6. Banasiak, L.J., Chiaro, G., Palermo, A., Granello, G.: Recycling of end-of-life tyres in civil engineering applications: environmental implications. In: Proc. 2019 WasteMINZ Conf, Hamilton, New Zealand, pp. 1–15 (2019)
7. Senetakis, K., Anastasiadis, A., Pitilakis, K.: Dynamic properties of dry sand/rubber (SRM) and gravel/rubber (GRM) mixtures in a wide range of shearing strain amplitudes. Soil Dyn. Earth Eng. **33**(1), 38–53 (2012)
8. Mashiri, M.S., Vinod, J.S., Sheich, M.N., Tsang, H.H.: Shear strength and dilatancy behaviour of sand–tyre chip mixtures. Soils Found **55**(3), 517–528 (2015)
9. Triavos, A., Alexander, N.A., Diambra, A., Ibrahim, E., Vardanega, P.J., Gonzales-Buelga, A., Sextos, A.: A sand-rubber deformable granular layer as low-cost seismic isolation strategy in developing countries: experimental investigation. Soil Dyn. Earth Eng. **125**, 1–12 (2019)
10. Hazarika, H., Abdullah, A.: Improvement effects of two and three dimensional geosynthetics used in liquefaction countermeasures. Jpn. Geotech. Soc. Special Publ. **2**(68), 2336–2341 (2016)

11. Pasha, S.M.K., Hazarika, K., Yoshimoto, N.: Physical and mechanical properties of gravel-tire chips mixture (GTCM). Geosynthetics Int. **26**(1), 92–110 (2019)
12. Chiaro, G., Tasalloti, A., Banasiak, L., Palermo, A., Granello, G., Rees, S.: Sustainable recycling of end-of-life tyres in civil (geotechnical) engineering applications: turning issues into opportunities in the New Zealand context. N.Z. Geomech. News **99**, 38–47 (2020)
13. Iolli, S., Modoni, G., Chiaro, G., Salvatore, E.: Predictive correlations for the compaction of clean sands. Transp. Geotechnics **4**, 38–49 (2015)
14. Kim, H.K., Santamarina, J.C.: Sand-rubber mixtures (large rubber chips). Can. Geotech. J. **45**, 1457–1466 (2008)
15. Lee, J.S., Doods, J., Santamarina, J.C.: Behavior of rigid-soft particle mixtures. J. Mater. Civ. Eng. **19**(2), 179–184 (2007)
16. Chiaro, G., Indraratna, B., Tasalloti, S.M.A., Rujikiatkamjorn, C.: Optimisation of coal wash–slag blend as a structural fill. Ground Improv. **168**(1), 33–44 (2015)

Shear Characteristics of Geomaterials Mixed with Fibrous Wood Chip and Converter Steelmaking Slag

Tomotaka Yoshikawa, Yoshiaki Kikuchi, Shohei Noda, and Akihiro Oshino

Abstract Under-sieve residue is a sediment containing large amounts of wood chips generated by natural disaster. The mechanical properties of this residue have not been well understood, and thus, it has usually been landfilled. The development of a method for effective utilization of under-sieve residue is believed to facilitate quick recovery and reconstruction during disasters. Accordingly, this study aimed to devise an effective approach to convert under-sieve residue into a geological material. Consolidated-drained triaxial compression (CD) tests were performed using mixtures of coconut fiber, which closely resembles under-sieve residue, with steelmaking slag and blast furnace slag fine powder to understand the effect of the mixing ratio of coconut fiber and curing periods. The results indicated that the maximum shear strength of the slag mixed with coconut fiber at 5% (coco5) by volume was higher than that of a "slag only" specimen regardless of curing periods. The maximum shear strength of the slag mixed with coconut fiber at 10% (coco10) or 33% (coco33) by volume was higher than that of "slag only" without curing. However, the maximum shear strength of coco10 or coco33 did not increase with curing.

Keywords Wood chips · Steelmaking slag · Solidified geomaterials

1 Introduction

The 2011 Tohoku earthquake off the Pacific coast of Tohoku generated a large amount of disaster waste, including tsunami deposits [1]. In Japan, the government, the construction industry, and related industries actively worked on the disposal of disaster waste. The process, which usually takes 10 years or more, was completed in about 4 years, and the recycling rate was nearly 90% [2]. However, most of the disaster waste containing a large amount of wood chips, called under-sieve residue, was not effectively used because its mechanical properties as well as the decay process and effects of the wood chips were unknown. According to the Ministry

T. Yoshikawa (✉) · Y. Kikuchi · S. Noda · A. Oshino
Tokyo University of Science, 2641, Yamazaki, Noda 278-8510, Chiba, Japan
e-mail: 7619703@ed.tus.ac.jp

H. Hazarika et al. (eds.), *Advances in Sustainable Construction and Resource Management*, Lecture Notes in Civil Engineering 144,
https://doi.org/10.1007/978-981-16-0077-7_17

of the Environment [3], the Great Nankai Trough Earthquake and an earthquake with the epicenter immediately below that of the Tokyo Metropolitan Earthquake are expected to occur in the future, and they will generate more disaster waste than the Tohoku-Pacific Ocean Earthquake did. Developing an effective method of using under-sieve residue will reduce the pressure on the function of the final disposal site at the time of disaster, and also it will accelerate disaster recovery and reconstruction processes. Since 2011, research on disaster waste has been actively conducted. For example, Takahashi et al. [4] showed the potential of recycling tsunami deposits containing large amounts of wood chips by applying a fibrous solidification method. Nakano et al. [5] reported that wood-rotting fungi did not progress into a wood chip-mixed material and that the unconfined compression strength of the specimens did not decrease when the wood chip-mixed material was sufficiently compacted.

Steelmaking slag is an industrial by-product of crude steel production, and its amount of production is 14 million tons per year. At present, it is effectively used as a roadbed material and a ground improvement material. However, the expansion of sales channels has become an issue with the reduction of public works expenses. With characteristics such as hydraulic properties [6] and excellent wear resistance [7], steelmaking slag can be used as a high-quality geomaterial.

The authors believe that the mechanical properties of under-sieve residue can be improved by the hydraulic solidification property of steelmaking slag, and the mixture can be effectively used as geomaterials. There are no cases to study the mechanical characteristics of the under-sieve residue solidified with other material focusing on the mixing ratio and shape of wood chips in detail, although there are some studies aimed at effective utilization of under-sieve residue. The authors have focused on wood chips in under-sieve residue and performed mechanical experiments on steelmaking slag mixed with wood chips, which were used as imitated under-sieve residue, under various conditions such as the shape and mixing ratio of the wood chips [8]. Although the initial stiffness of the mixture decreased by increasing with wood chip mixing ratio after 7-day curing, the uniaxial compressive strengths of the mixture were higher than that of a "slag only" specimen. In addition, steelmaking slag has the property of solidifying in the mid to long term [9]. In this study, CD tests were conducted to further understand longer-term mechanical properties of materials mixed with fibrous wood chips (coconut fiber) and slag in more detail.

2 Materials and Test Methods

Four kinds of materials were used in this research: coconut fiber ($\rho_s = 0.53$ g/cm^3, 0.27 mm average diameter and 24.4 mm average length in 100 samples), steelmaking slag without aging treatment ($\rho_s = 3.35$ g/cm^3, $D_{50} = 0.75$ mm), blast furnace slag fine powder (BFSFP) ($\rho_s = 2.89$ g/cm^3, Blaine value 4000 cm^2/g), and ion-exchanged water. Figure 1 shows the grain size accumulation curve of the steelmaking slag used in this research. The "slag only" specimen is a mixture of steelmaking slag and BFSFP at a mass ratio of 96:4 based on previous research results [10].

Fig. 1 Grain size distribution curve of steelmaking slag

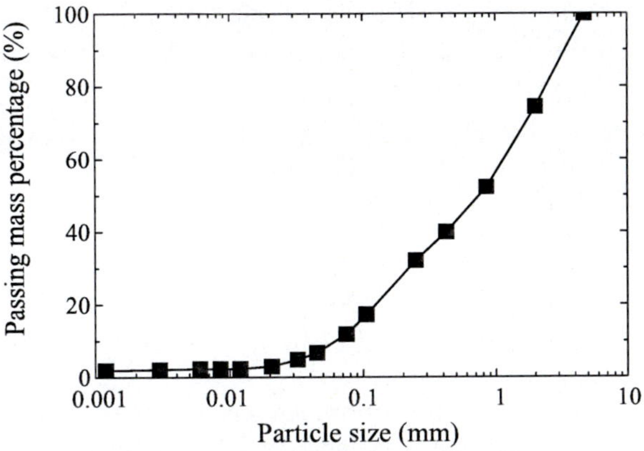

The appearance of each sample is shown in the previous paper [8]. Coconut fiber is a plant fiber obtained from coconut peel [11]. Steelmaking slag has a ridged shape, and its main components are CaO, SiO_2, FeO + Fe_2O_3, MgO, and Al_2O_3. BFSFP is produced by finely pulverizing granulated blast furnace slag from an ironworks blast furnace. It is a hydraulic admixture, widely used as a raw material for blast furnace cement and ready-mix concrete admixture [12].

Figure 2 shows the results of compaction tests (JIS A 1210: A-b method) conducted prior to CD tests. To determine the maximum dry density of each mixing condition, only one or two consolidation conditions were selected for each mixing condition, because the material used in this study showed no clear dry density peaks under the compaction curves from previous studies [13] and the amount of steelmaking slag was limited. The compaction tests were performed under the water content immediately lower than the water content to be over compaction. To determine this water content, simple compaction experiments were conducted using a hollow cylindrical mold with a diameter and a height of 50 and 100 mm with changing

Fig. 2 Compaction curves of slag with different mixing ratio of coconut fiber

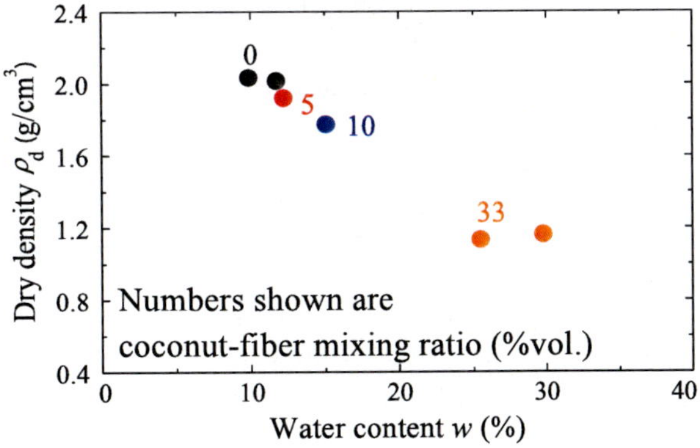

Table 1 Specimen preparation conditions

Coconut fiber mixing ratio (%vol.)	Maximum dry density ρ_d (g/cm^3)	Water content w (%)	Curing periods (day)
0 (slag only)	2.02	12.0	0, 7, 28, 84, 334
5 (coco5)	1.92	13.0	0, 7, 28, 84, 335
10 (coco10)	1.78	14.7	0, 7, 28, 84
33 (coco33)	1.13	25.0	0, 7, 28, 84, 337

the water content by 5% before the compaction test determining a maximum dry density.

Table 1 shows the specimen preparation conditions. Specimens (50 mm in diameter and 100 mm in height) for CD tests mixed the materials with predetermined mixing conditions, and they were compacted by wet tamping to a thickness of 20 mm per layer in a hollow cylindrical mold. The target values of ρ_d and w are used in the experiment in Table 1. The target dry density ρ_d in Table 1 was calculated by multiplying the maximum dry density ρ_{dmax} obtained from Fig. 2 by the degree of compaction $D_c = 95\%$. CD tests were conducted on cured specimens for a predetermined cured period. In CD test, after saturation, each specimen was isotopically consolidated at a back pressure of 200 kPa and an effective confining pressure of 50 kPa. The axial strain rate was 0.01%/min.

3 Effect of Curing Periods of Specimen

Figure 3a–e shows the relationship between the axial strain ε_a and the deviator stress q for each curing period obtained from the CD tests. The difference in the colors of the solid lines in the figure indicates the difference in the mixing ratio of coconut fiber. The mark plotted with numbers in the figures shows the point of the maximum shear stress in each test, and the numbers show coconut fiber mixing ratio. From Fig. 3, "slag only", coco5, and coco10 exhibited the softening behavior after showing the maximum shear strength in all curing periods. On the other hand, the deviator stress of coco33 did not show clear peak stress until the end of the test. The stress–strain relationship varied significantly for the coconut fiber mixing ratio between 10 and 33% by volume. In addition, in the case of coco33, q_{max} at 84-day curing and 334-day curing was smaller than that without curing. The volume of specimen did not expand and the dry density did not change due to the curing. Because of this, the cause of the decrease in q_{max} was unclear.

Figure 4 shows the relationship between the axial strain ε_a and the volumetric strain ε_{vol} in the case of without curing. The volume of "slag only", coco5, and coco10 exhibited variations from shrinkage to expansion, while the volume of coco33 continued to shrink until the end of the test. No peak strength was obtained, and the volume continued to shrink until the end of CD test using only wood chips in 2016

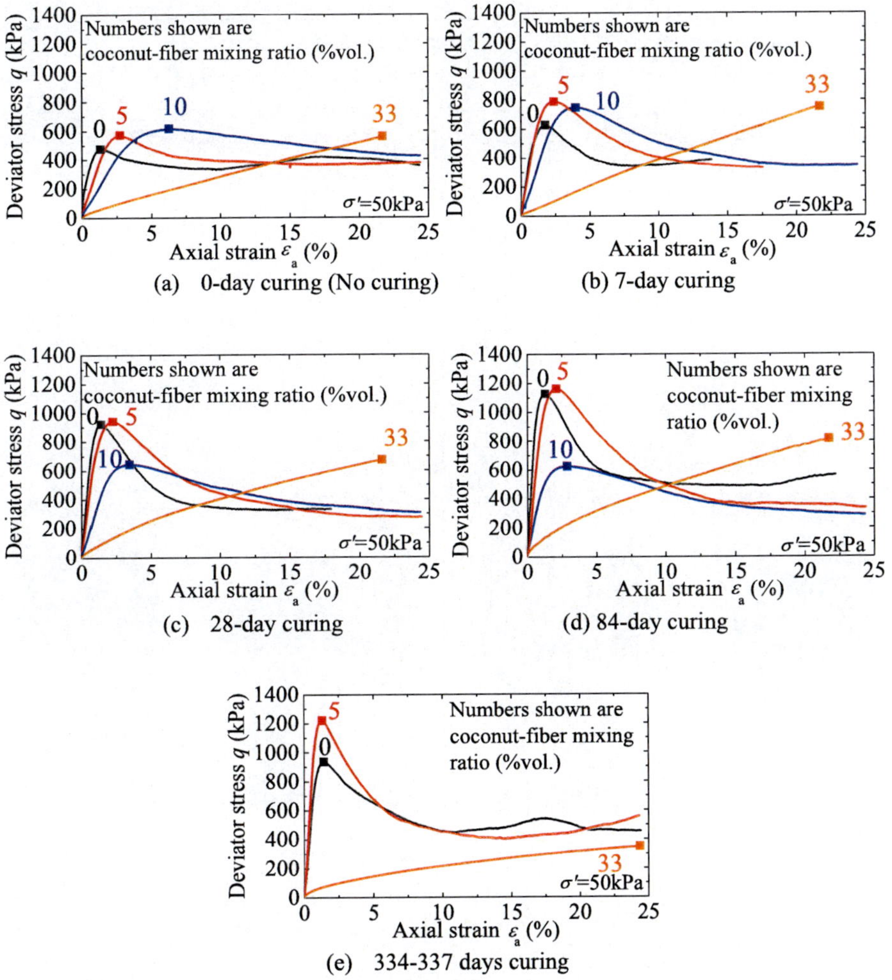

Fig. 3 Relationship between axial strain and deviator stress

and 2017 [13]. These results suggested that the effect of coconut fiber was prominent in the stress–strain relationship in the case of coco33.

Figure 5 shows images of the specimens cured for 7 days at the end of the CD tests. Reason of the stress–strain relationship difference in Fig. 4 is evident from the specimen shown in Fig. 5. When the coconut fiber mixing ratio was 10% by volume or less, shear failure was occurred. However, in the case of coco33, no apparent shear failure was occurred.

Comparing the maximum shear stresses and the failure strains of solidified and non-solidified geomaterials, the maximum shear stress was larger and the failure strains were smaller than those of non-solidified geomaterials. As for the maximum deviator stresses, the maximum shear stresses of "slag only" and coco5 were

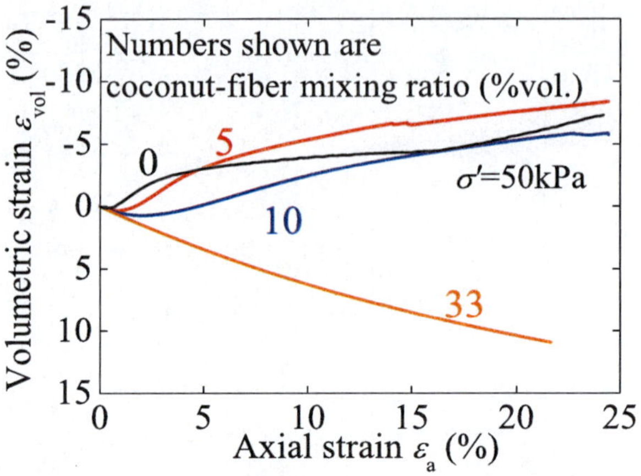

Fig. 4 Relationship between axial strain and volumetric strain without curing

Fig. 5 Specimens with different coconut fiber mixing ratio at 7-day curing after CD tests

increased as increasing in curing periods. On the other hand, those of coco10 and coco33 were almost constant regardless of curing periods. This difference would be because the effect of slag solidification was large when coconut fiber mixing ratio was 5% by volume or less. As for the failure strains, the failure strains of coco5 and coco10 were larger than that of "slag only" in all curing periods. However, the difference of them was getting smaller as increasing in curing periods. The specimens were considered to be granular material with small amount of coconut fiber in the early stage of curing. The slag contained in the specimen would be solidified, and the effects of solidification were clear under the coconut fiber mixing ratio less than 5% by volume with curing periods. In the case of coco10, although the slag solidification effect was clear in the failure strain, it is not so clear in the maximum shear stress. Moreover, in the case of coco33, there were small solidification effects on both the maximum shear stress and failure strain, because of too much amount of coconut fiber.

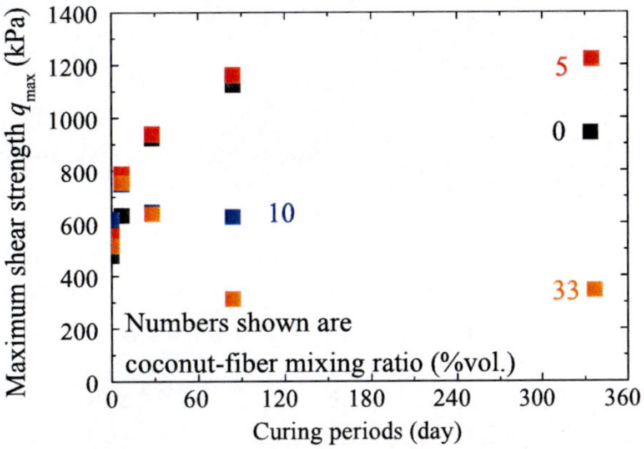

Fig. 6 Relationship between curing periods and maximum shear strength

Based on the test results in Fig. 3, Fig. 6 shows the relationship between the curing periods and the maximum shear stress q_{max}. In the case of coco33, q_{max} did not appear until the end of the test in the case of coco33; thus, the axial stress at $\varepsilon_a = 21.69\%$ is plotted as q_{max}. From Fig. 6, in the case of "slag only" and coco5, q_{max} increased up to 84-day curing, and q_{max} hardly changed after 84-day curing. In the case of coco10 and coco33, q_{max} hardly changed or decreased as increasing in curing periods. In this way, the maximum shear strength increase by curing was the same as "slag only" when coconut fiber mixing ratio was less than 5% by volume, and it was smaller than that of "slag only" when coconut fiber mixing ratio was more than 10% by volume. It means the effect of mixing coconut fiber was getting prominent when the fiber mixing ratio was more than 10% by volume.

Figure 7 shows the relationship between the curing periods and the deformation modulus, E_{50} (MPa), which is calculated as follows:

$$E_{50} = (q_{max}/2)/\varepsilon_{50} \quad (1)$$

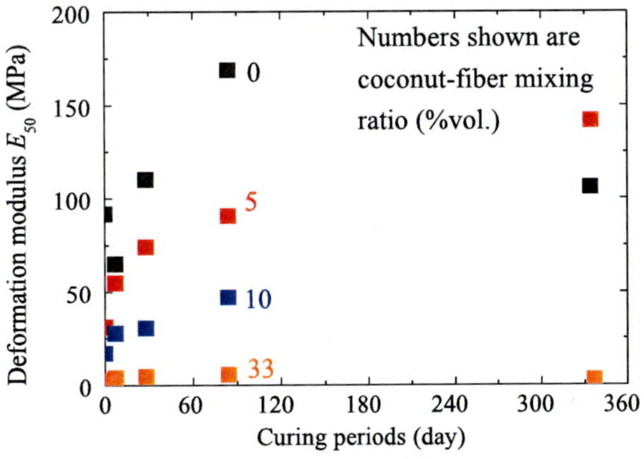

Fig. 7 Relationship between curing periods and deformation modulus

where ε_{50} (%) represents the axial strain when the axial stress has a value of $q_{max}/2$.

In this calculation, the axial deviator stress of $\varepsilon_a = 20\%$ was used as q_{max} in the case of coco33. Focusing on up to 84-day curing, the lower the coconut fiber mixing ratio was, the higher the value of E_{50} was. Focusing on after curing for 84 days or more, E_{50} increased with curing periods when the coconut fiber mixing ratio was 5% or 10% by volume, on the other hand, E_{50} of "slag only" was decreased. The reason of variance is unclear. In the case of coco5, although E_{50} at 335-day curing was larger than that of at 84-day curing, q_{max} at 335-day curing was the almost the same as that of at 84-day curing (Fig. 6). It means a kind of delayed hardening occurred in coco5.

In the case of coco33, E_{50} was too small (2.5 to 4.5 MPa) compared to other geomaterials even it increased with the curing period up to 28-day curing. As is shown up to here, the mixture with 33% by volume of coconut fiber showed different mechanical behavior from "slag only".

Figures 8a–d show the relationship between q_{max} and E_{50}. The approximate straight line passing through the coordinate origin and the coefficient of determination, R^2, are shown in the figures. Other than the case of coco10, there were linear relationships between qmax and E50. In the case of coco10, range of qmax was too

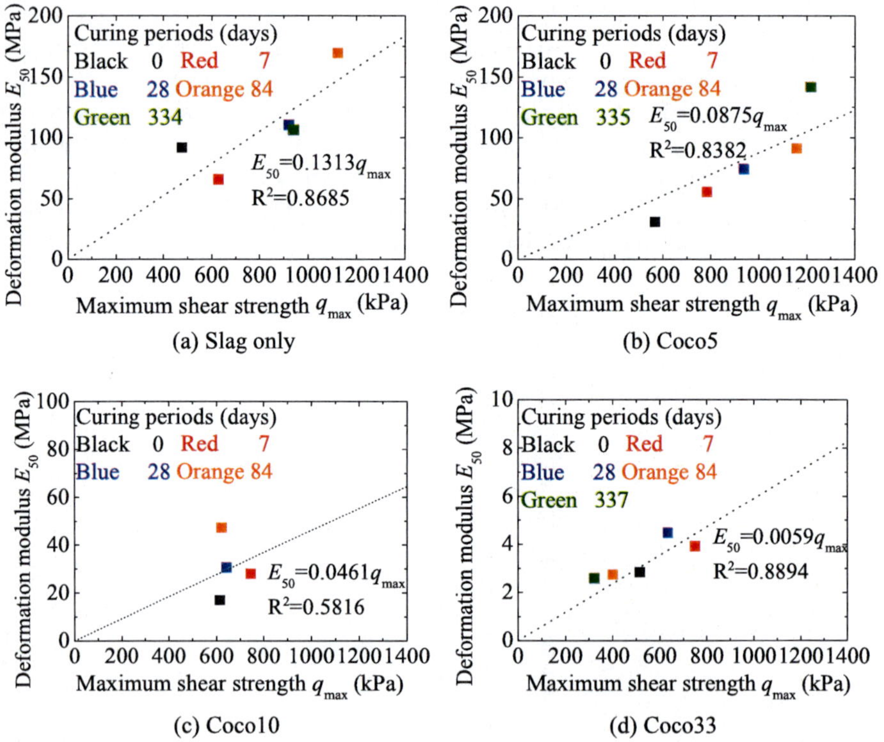

Fig. 8 Relationship between maximum shear strength and deformation modulus

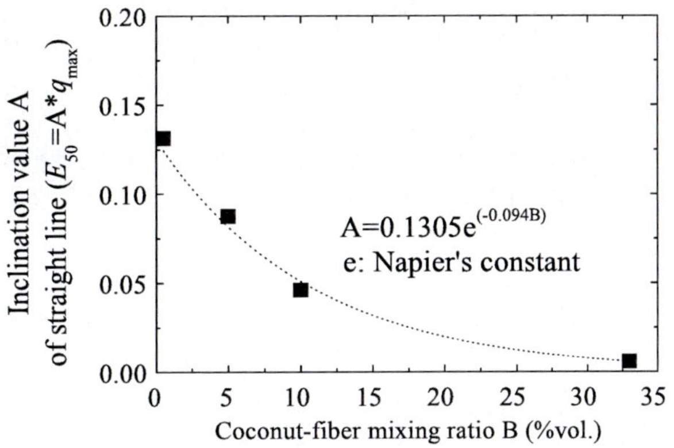

Fig. 9 Relationship between coconut fiber mixing ratio and inclination value

small, and the authors should conduct more experiments before getting a conclusion. E_{50} at the same q_{max} tended to decrease as the mixing ratio of coconut fiber increased. In particular, in the case of coco33, the stress–strain relationship is significantly different from the other cases, as can be seen from the stress–strain curve in Fig. 3. Above results show that more mixing ratio of coconut fiber shows more ductility.

Figure 9 shows the relationship between the inclination, *A*, of the straight line ($E_{50} = A^*q_{max}$) obtained based on Fig. 8 and the coconut fiber mixing ratio, *B*, of fibrous wood chips. From Fig. 9, the inclination, *A*, decreases exponentially as the coconut fiber mixing ratio, *B*, increases. This means that E_{50}/q_{max} decreases exponentially as the coconut fiber mixing ratio increases. In addition, this result shows that the q_{max}–E_{50} relationship can be inferred from the mixing ratio of coconut fiber. However, it is necessary to verify whether this relational expression of *A-B* ($A = 0.1305^*e^{-0.094^*B}$) has the physical significance by obtaining data under conditions in which the shape of wood chips and the degree of slag solidification are changed.

4 Conclusions

This study conducted CD tests using coconut fiber mixed solidified materials to understand the effect of the mixing ratio of coconut fiber and curing period on the mechanical properties of the mixed material.

The key findings of this study are as follows:

(1) q_{max} barely changed after curing for 84 days with or without coconut fiber. In other words, q_{max} of the mixed materials can be estimated in a relatively short period of time.

(2) q_{max} of coco5 was higher than that of "slag only", regardless of the curing period. However, when the mixing ratio of coconut fiber was increased, q_{max} was lower than that of "slag only".

(3) When the coconut fiber mixing ratio was 10% by volume or less, the effects of mixing ratio and curing period on $q_{\max}$ and E_{50} were much complex. To get clear conclusion, more experiments were required.
(4) $q_{\max}$–E_{50} relationship of the mixed material with slag and coconut fiber could be estimated from the wood chips mixing ratio. However, the physical significance of the relational expression ($A = 0.130{*}e^{-0.093{*}B}$ in Fig. 9) between the inclination, *A*, and the coconut fiber mixing ratio, *B*, obtained in this experiment needs to be examined separately.

These results suggested that the solidificated geomaterials mixed with slag and fibrous wood chips could be used as embankment materials on the soft ground and backfill materials because it has deformation followability. However, the use of this material in practice requires extensive testing to understand the properties of the material. The authors will carry out experiments under various conditions to theoretically understand the strength characteristics of wood chips slag mixed materials. In the future, it will be possible to estimate the strength characteristics of the materials mixed with under-sieve residue and slag from the strength characteristics of under-sieve residue and the strength characteristics of slag.

Acknowledgements This research was supported by the Steel & Slag Association and the Obayashi Foundation. The authors express their appreciation for their support.

References

1. Kazama, M., Noda, T.: Damage statistics (Summary of the 2011 off the Pacific Coast of Tohoku Earthquake damage). Soil Foundations **52**(5), 780–792 (2012)
2. Ide, K.: Treatment of disaster waste generated by the Great East Japan Earthquake -Treatment of disaster waste by member corporations of the Japan Federation of Construction Contractors. In: The 15th Asian Regional Conference on Soil Mechanics and Geotechnical Engineering, vol. 2, no. 2, pp. 135–142 (2015)
3. Ministry of the Environment in Japan: Movement over the disaster waste management. https://kouikishori.env.go.jp/en/movement/. referred at 2020/08/31
4. Takahashi, H., Kuribara, H., Satomi, T.: Development of ground materials and cover soils by recycling waste woods and tsunami sludge. J. JSEM **13**, s108–113 (2013)
5. Nakano, M., Sakai, T., Hamashima, K., Jin-no, T., Nonoyama, H.: Long-term mechanical behavior of disaster debris soil with wood chips deterioration. In: 51st Japan National Conference on Geotechnical Engineering, pp. 575–576 (2016) (in Japanese)
6. Karamacharya, M.R., Uchida, I., Idemitsu, T.: On the studies of utilisation of converter furnace slag in road bases. J. Jpn. Soc. Civ. Eng. **282**, 101–113 (1979). (in Japanese)
7. Nishi, M., Kawabata, K.: Some basic and mechanical properties of iron and steel slags as base-course materials. J. Jpn. Soc. Civ. Eng. Issue **414**, 89–98 (1990). in Japanese
8. Yoshikawa, T., Kikuchi, Y., Noda, S.: Effects of wood chips types and mixing ratio to the compression characteristics of steelmaking slag mixed with wood chips. In: Geotechnics for Sustainable Infrastructure Development - Geotec Hanoi 2019, Phung (edt), pp. 737–744 (2019). ISBN 978-604-82-0013-8
9. Yoshikawa, T., Kikuchi, Y., Hyodo, T., Oobayashi, S., Kakihara, Y., Iwai, D., Kaneda, T.: The effect of aging treatment on the mechanical properties of steelmaking slag mixed with crushed slag. In: Proceedings of the 7th China-Japan Geotechnical Symposium, pp. 410–415 (2018a)

10. Yoshikawa, T., Kikuchi, Y., Noda, S., Kakihara, Y., Oshikawa, R., Oshino, A.: Unconfined compression strength of converter steelmaking slag mixed with subsidiary solidifying materials. In: Research Presentation of the 15th Kanto Branch of Japanese Geotechnical Society, pp. 298–301 (2018b) (in Japanese)
11. Irwin, M., Hutten: Handbook of nonwoven filter media, 2nd edn. In: Advancing Chemical Engineering Worldwide, pp. 158–275 (2016)
12. Nobata, K., Ueki, Y.: Basic property and the method of effective use on portland blast-furnace slag cement and ground granulated blast furnace slag. Shinnittetsu Giho **376**, 41–44 (2012). in Japanese
13. Yoshikawa, T., Kikuchi, Y., Hyodo, T., Kuroiwa, Y., Tanaka, T.: Geomechanical properties as geomaterial of wood chips mixed with steelmaking slag. In: Proceeding of the 12th JGS Symposium on Environmental Geotechnics, pp. 199–204 (2017) (in Japanese)

Shear Strength Enhancement of Soil Mixed with Palm Oil Clinker

Amizatulhani Abdullah, Cut Ainul Mardziah Amir, and Mohd Yuhyi Mohd Tadza

Abstract Palm oil has been a major industry that contributed to Malaysia's economic growth since quite a long time. Approximately, 23% of the 5.8 million hectares of Malaysia oil palm plantations are certified with Malaysian Sustainable Palm Oil (MSPO) certificate. However, there are still a huge number of plant oil plantations that have not been shifting towards sustainable palm oil consumption and production. Palm oil industry is normally being associated with negative impacts such as deforestation which resulted to the loss of wildlife as well as haze issue that imposed health risks to the communities. Beside that, the uncontrolled dumping of palm oil waste (palm oil clinker) is also causing a big problem. Palm oil clinker is a waste by-product from the processing stages of palm oil industry. Recycling the material into construction industry may help to promote sustainability and solving waste disposal issues. In this research, palm oil clinker was mixed randomly in soft soil. Raw palm oil clinker was taken from the nearby palm oil factory. The palm oil clinker as well as the soft soil was tested for its basic properties (particle size, specific gravity and soil consistencies). Furthermore, several number of soil sample were prepared, and each of the sample was mixed with fine grain size of palm oil clinker in several proportions. All the sample mixtures were tested by using unconfined compression test (BS 1377-7:1990). Enhancement of the soft soil's shear strength after been added with palm oil clinker was investigated and compared. It is found that the addition on palm oil clinker into soft soil is able to improve the shear strength of the soil. This study hopes to give some alternatives to the utilization of palm oil clinker in geotechnical engineering field. Previous researches by other researchers were conducted in utilizing the lightweight properties of the palm oil clinker (as aggregate replacement) to produce a lightweight structure. Recycling such waste in several civil engineering fields may help to preserve environment and contribute to sustainable industry.

A. Abdullah (✉) · C. A. M. Amir · M. Y. M. Tadza
Faculty of Civil Engineering Technology, Universiti Malaysia Pahang, Kuantan, Pahang, Malaysia
e-mail: amizatulhani@ump.edu.my

M. Y. M. Tadza
e-mail: dryuhyi@ump.edu.my

H. Hazarika et al. (eds.), *Advances in Sustainable Construction and Resource Management*, Lecture Notes in Civil Engineering 144,
https://doi.org/10.1007/978-981-16-0077-7_18

Keywords Palm oil clinker · Shear strength · Soft soil improvement · Sustainable construction · Material recycling

1 Background of Study

Palm oil (as shown in Fig. 1) is the world largest industry exceeding the rest of the oils and fat production recorded in year 2017 [4]. Major world producers of palm oil are Indonesia, Malaysia, Thailand, Colombia and Nigeria, while some major world importers of palm oil include India, European Union (EU), China, Pakistan, United States of America (USA), Bangladesh, Nigeria and Philippines [2].

Oil palm has been commercially cultivated in Malaysia since 1917 and after 100 years of cultivation, it brought tremendous income growth to the country. In Malaysia, exporting of palm oil and palm oil-based products is worth approximately RM65 billion in year 2018 [2].

Oil palm planted area was reported to increase from 5.74 ha in year 2016 to 5.81 hectares in year 2017 (as tabulated in Table 1). From the number, majority of the area was owned by private estates (approximately 61%) while approximately 16% was owned by government schemes such as Federal Land Development Authority (FELDA), Federal Land Consolidation and Rehabilitation Authority (FELCRA) and Rubber Industry Smallholders Development Authority (RISDA). Meanwhile, another 17% was owned by independent smallholders and state schemes/government agencies (6%).

Parallel to the increment of oil palm plantation area, the number of palm oil mill also has shown significant development. As reported by Kushairi et al. [1], in the year 1990, there were only 261 palm oil mills operated with a total processing capacity of

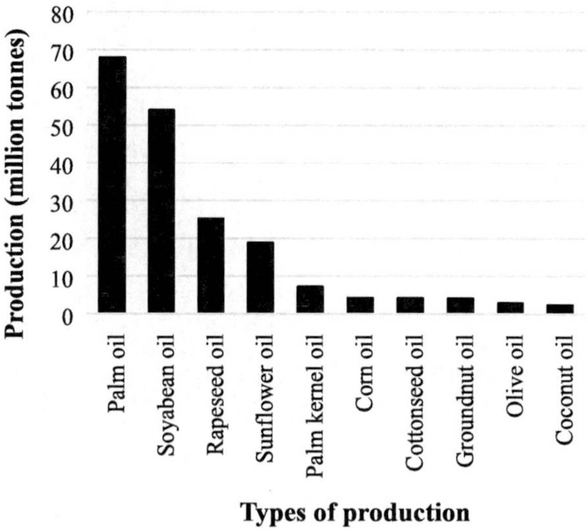

Fig. 1 Production of various types of oils and fats

Table 1 Distribution of oil palm planted area by category in Malaysia from year 2016 to year 2017 [2]

Category	2016		2017	
	Area (ha)	%	Area (ha)	%
Private estates	3,508,554	61.2	3,543,429	61.0
Government schemes	951,169	16.5	940,326	16.1
State schemes/government agencies	344,314	6.0	347,632	6.0
Independent smallholders	933,948	16.3	979,758	16.9
Total	5,737,985	100	5,811,145	100

Table 2 Number of palm oil mills, refineries, kernel crushers and oleochemical plants and their capacities in Malaysia in year 2017 [2]

Facility	Number	Processing capacity (tonnes per year)
Palm oil mill	454	112,187,800
Palm oil refinery	53	27,328,200
Kernel crusher	45	7,269,700
Oleochemical plant	19	2,668,929

42.87 million tonnes of fresh fruit bunches per year. After two decades, this number had increased to 421 palm oil mills with a total processing capacity of 97.38 million tonnes per year. Recently, in year 2017, there were 454 palm oil mills reported to operate with processing capacity of 112.19 million tonnes per year as shown in Table 2.

The Third Industrial Master Plan (IMP3) was launched in 2006 for the period of 2006 until 2020 to emphasize on the downstream manufacturing activities into wider range of highly value-added products through commercialization, development and collaboration between government research agencies and the palm oil industry [3]. During this period, Economic Transformation Programme (ETP) was also launched to energize Malaysia to become a high-income nation by the year 2020. Therefore, to ensure the success of the program, oil palm industry was classified under the 12 National Key Economic Areas (NKEA). Under NKEA, the palm oil sector is aimed not only to improve productivity but also to focus on the sustainable development of the industry. The government has taken proactive steps by introducing Malaysian Sustainable Palm Oil (MSPO) certification scheme in year 2015. Approximately, 23% of the 5.8 million hectares of Malaysia oil palm plantations are certified with Malaysian Sustainable Palm Oil (MSPO) certificate. However, there are still a huge number of plant oil plantations that have not been shifting towards sustainable palm oil consumption and production.

Therefore, sustainability issue should be one of the main concerns that needs to be addressed fully and in an effective manner in order to ensure that the industry will remain resilient in the future. Palm oil industry is normally being associated with negative impacts such as deforestation which resulted to the loss of wildlife as well as haze issue that imposed health risks to the communities. Besides that, the uncontrolled dumping of palm oil waste (palm oil clinker) is also causing a big problem. One aspect which can support the sustainability issue is by transforming the abundantly generated oil palm waste by-products through versatile approaches. It is quite challenging due to the increase in number of oil palm mills resulted to significant increase of waste generated by the mills. In this research, this oil palm waste by-product is used to enhance soil properties.

2 Methodology

This section describes on the material preparation and mixture as well as the laboratory test that conducted upon the samples.

2.1 Material Preparation and Mixture

Palm oil clinker (POC) was obtained from a palm oil factory located in Lepar Hilir, Pahang. Raw POC before being crushed is shown in Fig. 2. Large size of POC was crushed to smaller size (coarse grain size and fine grain size) manually by using a hammer. The POC was mixed into soft soil (which was purchased directly from the manufacturer). Soft soil was mixed randomly with POC in the following ratio of 100:0, 90:10, 80:20, 70:30, 60:40 and 50:50 (by volume).

Fig. 2 Large size of POC

Table 3 Properties of the clays

Properties	MX80 bentonite	Yellow bentonite	Speswhite kaolinite
Specific gravity	2.80	2.84	2.61
Liquid limit	437	135	51.4
Plastic limit	63	58	32
Shrinkage limit	12.2	13.6	26.5
Grain size distribution < 75 μm < 2 μm	100 96.4	100 97.18	100 97.92

2.2 *Laboratory Test*

The materials used in this study were tested for its basic properties and strength characteristics according to the standard, particle size distribution (ASTM D422-63), soil consistency (ASTM D4318-00), specific gravity (ASTM D854-00, ASTM C127-15), compaction (ASTM D698-07) and unconfined compression test (UCT).

3 Results and Discussion

3.1 *Basic Properties of the Material*

Soft soil used in the study is known as kaolin FMC soil. This kind of soft soil can be easily found and purchased in Malaysia. According to the Atterberg limit test that had been conducted upon the soil sample, it is found that the liquid limit, plastic limit and plastic index of the soil were 63.32%, 50.4% and 9.92%, respectively. The specific gravity of the soil is found to be approximately 2.41.

The comparison of the soft soil with other type of soft clay (MX80 bentonite, yellow bentonite and Speswhite kaolinite) is shown in Table 3 [5].

Figure 3 shows the particle size distribution of POC. Median size of POC is found to be approximately 15 mm. Specific gravity of POC is 1.96 while the maximum dry density and optimum moisture content are 1.35 g/cm^3 and 20%, respectively. The compaction curve of POC is shown in Fig. 4.

3.2 *Shear Strength of Soft Clay Mixed with Palm Oil Clinker*

Unconfined compression test was conducted upon samples of soft soil mixed with POC of fine grain size at several volume proportions (100:0, 90:10, 80:20, 70:30, 60:40 and 50:50).

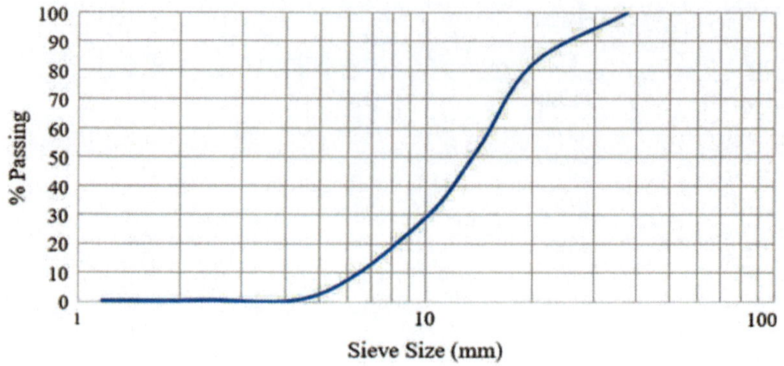

Fig. 3 Particle size distribution of POC

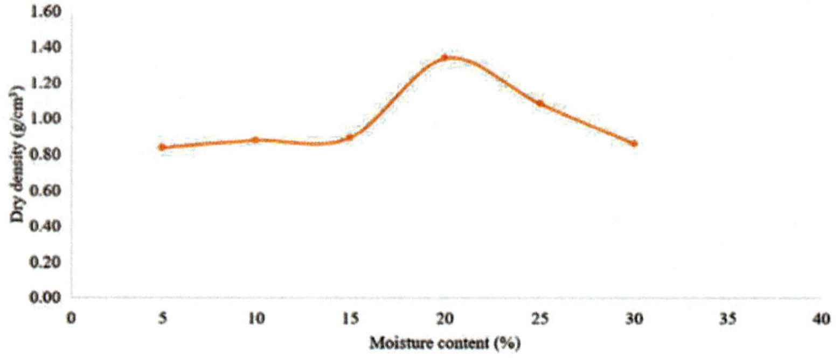

Fig. 4 Compaction result of POC

Figure 5 shows the condition of the 100% soft soil sample at the beginning and at the end of the shear strength test, respectively. The stress–strain curve of the untreated soft soil sample is shown in Fig. 6.

Figure 7 shows the comparisons of the sample for 50:50 mixtures at the beginning and at the end of the shear strength test. The shear failure plane can be seen clearly from the figure. In addition, the condition of 80:20 mixture sample at the end of the test is shown in Fig. 8.

Figure 9 shows the variations of the stress–strain curve of soft soil added with various proportions of POC. Based on the graph, it can be observed that adding fine grain size of POC in the amount of 20% gives the best improvement towards the strength of the soft soil. Other value of proportion did not make any significance improvement to the soft soil.

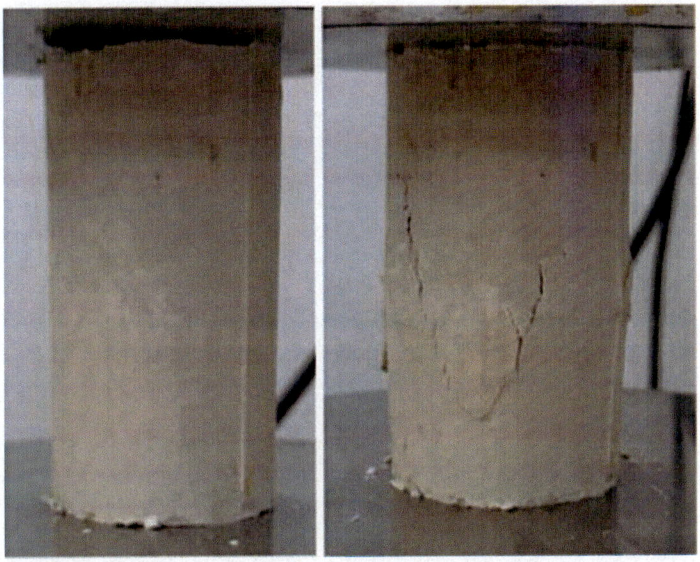

Fig. 5 UCT of 100:0 mixtures (before and after the test)

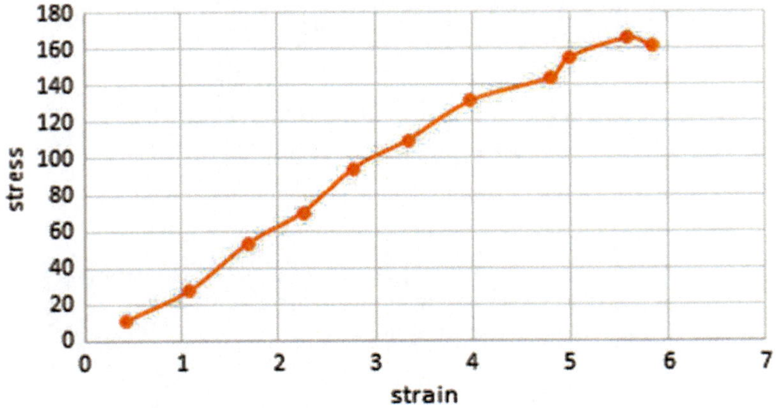

Fig. 6 Stress–strain curve of 100:0 mixtures

4 Conclusion

Several tests to determine basic properties of soft soil and POC have been conducted which are particle size distribution, specific gravity, soil consistencies, compaction and unconfined compression test. From this study, it is found that by adding 20% of fine grain POC is able to improve the shear strength of the soft soil. It shows that the addition on palm oil clinker into soft soil is able to improve the shear strength

Fig. 7 UCT of 50:50 mixtures (before and after the test)

Fig. 8 UCT of 80:20 mixtures (before and after the test)

of the soil. This study hopes to give some alternatives to the utilization of palm oil clinker in geotechnical engineering field. Recycling such waste in several civil engineering fields may help to preserve environment and contribute to sustainable

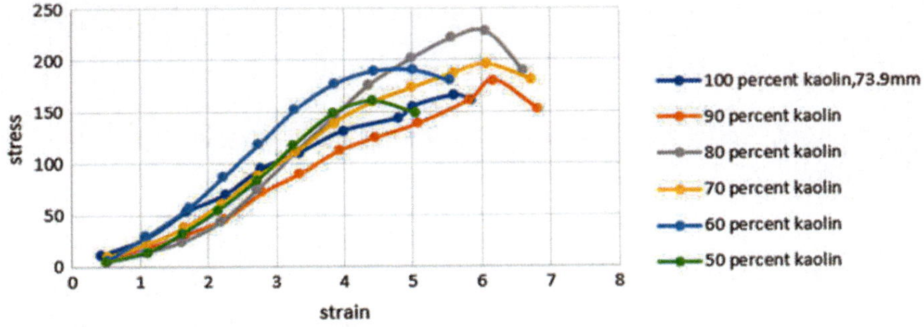

Fig. 9 Variation of stress–strain curve of soft soil added with palm oil clinker (fine grain size)

industry. Further study needs to be conducted to assess the enhancement made by different size of POC when added with soft soil.

Acknowledgements This work was supported by Universiti Malaysia Pahang under Grant of RDU1703305. Special thanks also go to Lim Li Yang, Asyraf Hazwan and Aizuddin for their effort in conducting laboratory works.

References

1. Kushairi, A., Loh, S.K., Azman, I., Hishamuddin, E., Abdullah, M.O., Izuddin, Z.B.M.N., Razmah, G., Sundram, S., Parveez, G.K.A.: Oil palm economic performance in Malaysia and R&D progress in 2017. J. Oil Palm Res. **30**(2), 163–195, June 2018
2. MPOB: Malaysian Oil Palm Statistics 201, 37th edn., p. 205. MPOB, Bangi (2018)
3. Nambiappan, B., Ismail, A., Hashim, N., Ismail, N., Shahari, D.N., Nik Idris, N.A., Omar, N., Salleh, K.M., Mohd Hassan, N.A., Kushairi, A.: Malaysia: 100 years of resilient palm oil economic performance. J. Oil Palm Res. **30**(2), 13–25, June 2018
4. OIL WORLD: Oil World Weekly. ISTA Mielke GmbH, Hamburg (2018). https://www.oilworld.biz/t/tpublications/weekly. Accessed on 15 July 2019
5. Tripathy, S., Tadza, M.Y.M., Thomas, H.R.: Soil-water characteristic curves of clays. Can. Geotech. J. **51**(1–15), 2014 (2014)

Study on Bio-cementation of Ex-coal Mining Soil as a Road Construction Material

A. M. Indriani, Tri Harianto, A. R. Djamaluddin, and A. Arsyad

Abstract Hundreds of thousand hectares of ex-coal mining land are left without good management and use. This abandoned mine soil can be reused. This soil is composed of coal, sand and clay. Microstructural analysis shows that this material contains the chemical elements of quartz, titanium dioxide and clay minerals. This soil has been analyzed to determine its engineering properties and potentials for use in civil engineering construction. In this study, this soil was treated with bio-cementation to improve geomechanical properties by adding 3% and 6% Bacillus subtilis bacteria with bacterial culture for 6 days. Direct shear testing is done to see the increase in CBR and shear strength. The results showed that the CBR and shear strength of the soil increased with the addition of 6% bacteria and a curing time of 28 days then 3% bacteria in the same curing time.

Keywords Bio-cementation · Mine soil · Bacillus subtilis

1 Introduction

Indonesia has hundreds of thousands of hectares of ex-coal mining land that has been left without proper management by the government or mining companies. After analysis, the soil in these former mining areas can still be used as a road construction material. However, this soil needs to be improved in terms of engineering properties such as CBR and shear strength values. Turning the ex-coal mining soil to a cement-like material using bacteria (bio-cementation) is considered one of the bio-geotech technologies that are developing [1]. For geotechnical applications, microbial grouting is mostly utilized to strengthen the soils through increasing soil strength and stiffness through the biomineralization of $CaCO_3$ crystals that act as a cementing agent, binding soil particles together inside the soil matrix [2].

A. M. Indriani (✉) · T. Harianto · A. R. Djamaluddin · A. Arsyad
Hasanuddin University, South Sulawesi, Indonesia
e-mail: marini_sabrina@yahoo.com.sg

H. Hazarika et al. (eds.), *Advances in Sustainable Construction and Resource Management*, Lecture Notes in Civil Engineering 144,
https://doi.org/10.1007/978-981-16-0077-7_19

Recent advances in soil stabilization using calcite-induced microbial precipitation techniques ($CaCO_3$) (MICP) have been reported by Mujah et al. [4]. Microbial-induced calcium carbonate precipitation (MICP) deposition is a naturally occurring biological process in which microbes produce inorganic material as part of their basic metabolic activity. This technology has been explored and promising with potential in a variety of technical applications [3].

However, coal-mined soils contain heavy metals, so special attention needs to be paid in the process of applying bio-cementation using bacteria. Bio-cementation using Bacillus subtilis bacteria has been studied extensively focusing on sandy and clay soils, no one has yet done research on soils containing coal.

2 Materials and Methods

2.1 Soil Type

The ex-coal mining soil is obtained from Balikpapan, Province Kalimantan Timur, Indonesia. To provide detail of materials used in the experimental study, laboratory investigation program was carried out to evaluate the basic properties and mechanical properties of the untreated soil and stabilized soil, in this case, soil containing 15% of coal treated by microbially induced calcite ($CaCO_3$) precipitation (MICP).

Based on the sieve analysis, this soil is a sand because 0% retained from the sieve No. 4 and only 2% of the material passed from the sieve No. 200. Based on the USCS classification system, the soil is categorized as poorly sand (SP) with coefficient of uniformity(Cu = 1.8) and coefficient of curvature (Cc—2.86). The mechanical properties are shown in Table 1.

Table 1 Mechanical properties

Test		Test Results	Unit
Engineering properties of sample soil:			
Standard Proctor			
a	Maximum dry density, (γd)	1.77	g/cm^3
b	Optimum moisture content (OMC)	6.4	%
Direct Shear			
a	Cohesion (c)	5.8	kPa
b	Internal friction angle (f)	36	
California bearing ratio—Unsoaked (CBR)			
a	Unsoaked	34	%
b	Soaked	13	%

Microstructure analysis using X-ray diffraction (XRD) shows that minerals contained in soil with 15% coal are dominated by 92.5% silicon oxide $ -alpha quartz low (SiO_2) as much as 92.5% and 7.5% ilmenite (Fe O_3 Ti).

2.2 Preparation of Bacteria and Cementation Solution

The urease-producing bacteria used were *Bacillus subtilis.* B. subtilis is a Gram-positive, rod-shaped bacterium that forms heat-resistant, dormant spores. It is commonly found in the soil (Piggot, 2009) and can grow in a minimal medium containing only essential salts and carbon, nitrogen and phosphorus sources. *B. subtilis* grows at temperatures ranging from 10° to 55 °C, with the fastest growth rates at about 42° C. At 37° C, *it* grows with a doubling time of about 30 min in a rich medium.

In this experimental, *Bacillus subtilis* was cultured in B4 medium, which contained urea (20 g/L), nutrient broth (3 g/L), $NaHCO_3$ 2 · (12 g/L), $CaCl_2 \cdot 2H_2O$ 4 · (14 g/L) and NH4Cl (10 g/L) dissolved in distilled water. The cementation solution used in this study consisted of 0.25 M of urea and 0.25 M of calcium chloride. Variation of culture times is made in two variations, that is 3 days and 6 days.

2.3 Soil Treatment Procedures

Optimum condition for effective $CaCO_3$ precipitation to evaluate the optimum combination of culture and Bacillus sublitis concentrations that produces the most effective $CaCO_3$ crystals, a series of different culture and Bacillus sublitis concentrations, as listed in Table 2, were used in the experiment.

Bacteria-free soil samples for direct shear test were prepared at the optimum moisture content, i.e., 6.4% as shown in Table 1, on the other hand, for the samples containing bacteria, the required amount of bacteria solution (in place of some water), as shown in Table 2. Seventy two samples were prepared for direct shear testing on different days. All test samples were made in PVC mold and kept in appropriate condition until the test days.

Direct shear specimens were remolded according to standard method ASTM D-3080 being 6 cm in diameter and 2 cm high. Specimens were tested at curing times of 3 days, 7 days, 14 days and 28 days (Fig. 1).

Table 2 Combination bacteria culture and concentration used in the current study

Bacillus sublitis	Sample					
Culture	3 days			6 days		
Concentration	3%	4.5%	6%	3%	4.5%	6%

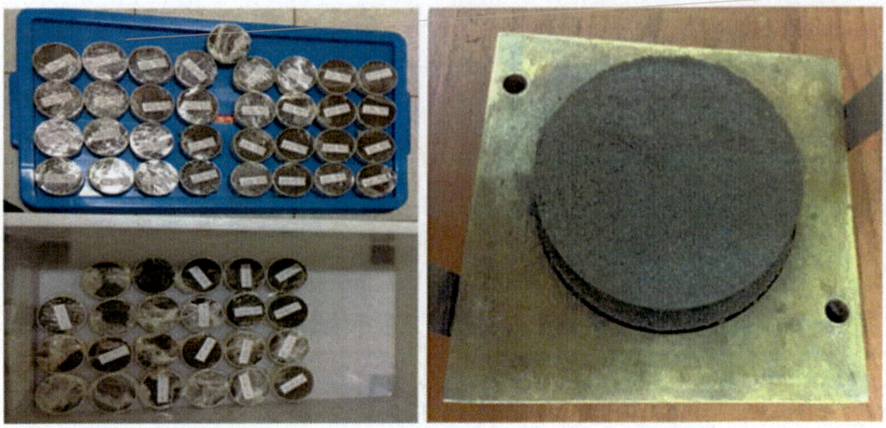

Fig. 1 Sample preparation

2.4 Microstructure Analysis

The bonding behavior of $CaCO_3$ crystal microstructural characteristics and the corresponding strength of bio-cemented ex-mine soil post-treatment were investigated using scanning electron microscopy (SEM).

3 Experimental Result

3.1 Effect of Bacterial Concentration

Precipitation was observed at a temperature of 30–32 °C of the growth medium. Curing is done in the direct shear test carried out for a period of 3, 7, 14 and 28 days to see changes due to the MICP process. From the results of sample testing, it can be seen that the shear stress for 15% of coal-contaminated soil-stabilized MICP using Bacillus subtilis bacteria gives excellent results. The effects of different concentrations and culture time of bacteria on the shear stress of the MICP samples are shown in Fig. 2 and Table 3.

From Fig. 2 of the relationship between normal stress and shear stress MICP 3 days and 6 days culture, it can seem that the value of shear stress continues to increase with longer curing time.

Figure 3 shows the increased cohesion resulting from the stabilization process with MICP compared to the untreated soil. There were also differences in the change in cohesion values between the MICP stable soil 3 days of culture and 6 days of culture. The addition of Bacillus subtilis in 3 days of culture, the more the concentration of bacteria added to the former coal mining soil, the higher the value of soil cohesion.

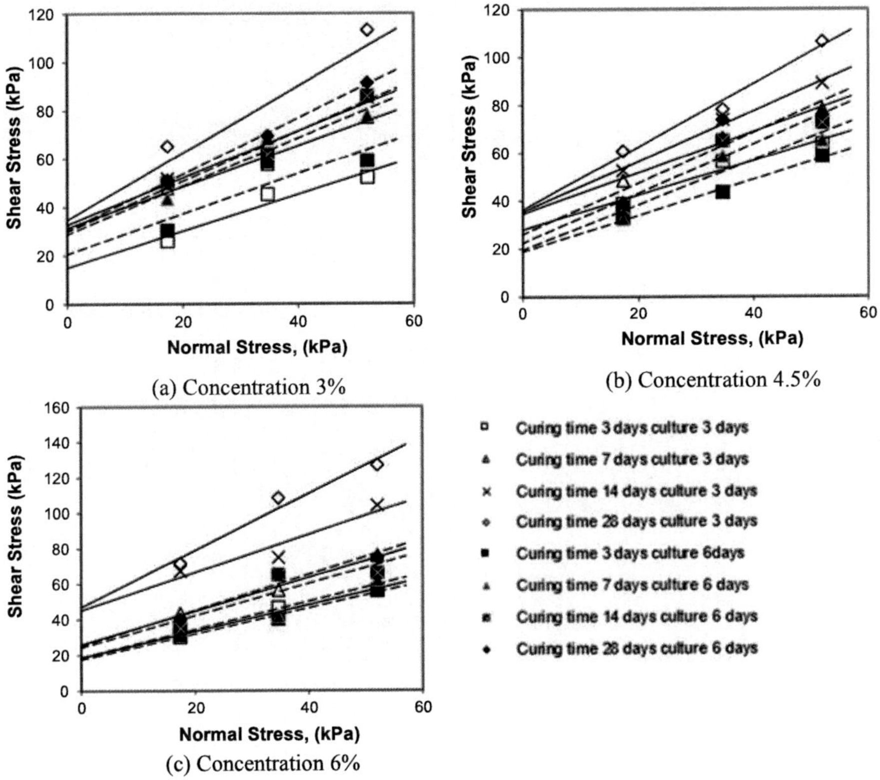

Fig. 2 Normal stress and shear stress relationship after MICP treated

Table 3 Shear stress parameters

MICP Bacillus subtilis		Curing time (days)	Parameters	
Culture (days)	Concentration (%)		*c* (kPa)	Φ (°)
3	3	28	36.2	53
	4.5		37.1	53
	6		47.1	58
6	3		30.4	49
	4.5		26.1	47
	6		25.4	45

The longer the curing time, the higher the cohesion value compared to untreated soil. But the opposite happened when the addition of Bacillus subtilis was to the culture for 6 days, the addition of bacterial concentration decreased the cohesion value. Even when compared with untreated soil, giving Bacillus subtilis bacteria culture for 6 days still increased the cohesion value.

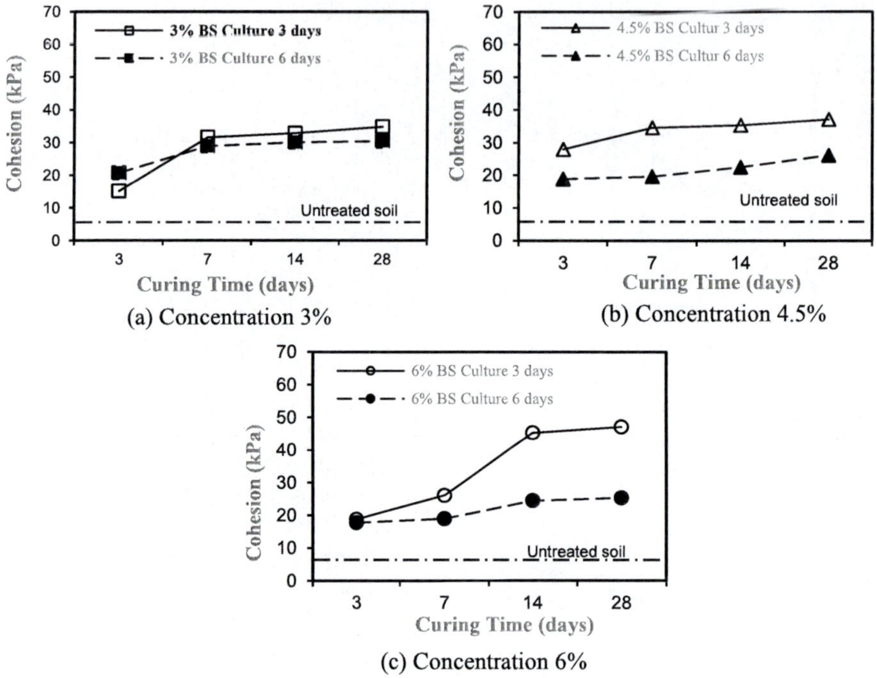

Fig. 3 Curing time and cohesion (c) relationship after MICP treated

The internal friction angle (ϕ) also increased. On the addition of concentrations of 3% and 4.5% with a curing period of 3, 7 and 14 days, the culture time of 6 days and the internal friction angle (ϕ) increased better than 3 days culture time, but at the curing period of 28 days, the value was lower than the 3 days culture time. At the addition of a concentration of 6% Bacillus subtilis, a culture time of 3 days is better than a culture time of 6 days. It is shown in Fig. 4.

This is consistent with the results of a previous study conducted by Cheng et.al (2017) which states that increasing the concentration of bacteria increases the soil UCS value due to the increase in $CaCO_3$ produced as shown in Fig. 5, where it is seen that increasing the concentration of bacteria increases the deposition of $CaCO_3$ so that the value of shear strength also increased.

3.2 *Effect of Bacterial Culture*

Culture has a very big influence on the soil stabilization process using the MICP method. From the test results using the 3-day culture of the Bacillus subtilis bacteria gave better results compared to the use of the 6-day culture. This is because in the 3-day culture, the Bacillus subtilis bacteria were still exponential phase and still close

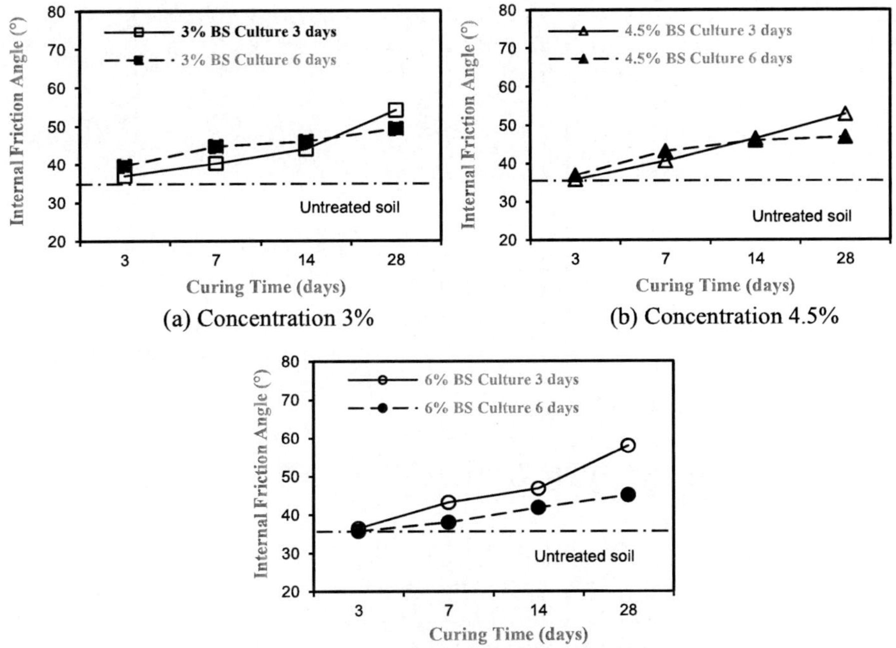

(a) Concentration 3%

(b) Concentration 4.5%

(c) Concentration 6%

Fig. 4 Curing time and internal friction angle (ϕ)relationship after MICP treated

Fig. 5 Relationship between increasing UCS value and $CaCO_3$ deposition due to increased bacterial concentration (Cheng et al. 2017)

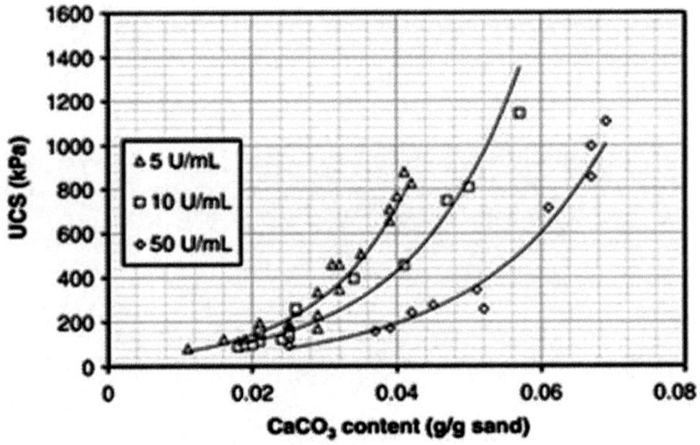

to the stationary phase while the 6 days of culture the Bacillus subtilis bacteria have entered the death phase so that the MICP process is no longer effective (Fig. 6).

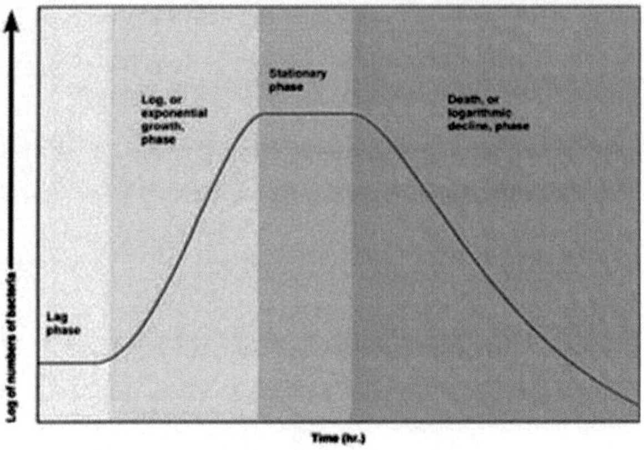

Fig. 6 Bacterial growth phase

3.3 *Microstructure Analysis*

The link between the $CaCO_3$ crystal microstructural characteristics and the corresponding strength of bio-cemented sand post-treatment was investigated using scanning electron microscopy (SEM) (**HITACHI FLEXSEM 100**). The bonding behavior between the host grain and the structure of the $CaCO_3$ crystals, as well as the evolution of the effective $CaCO_3$ crystal morphology, was examined. The precipitation of large $CaCO_3$ crystals would increase the area of contact for greater shearing resistance, thus contributing to a high shear strength value; however, the relative size between the CaCO3 crystals and sand particles is more important than the absolute size of the crystals themselves because the crystals need to be large enough to fill in the contact points of the different sand grain sizes [4] (Fig. 7).

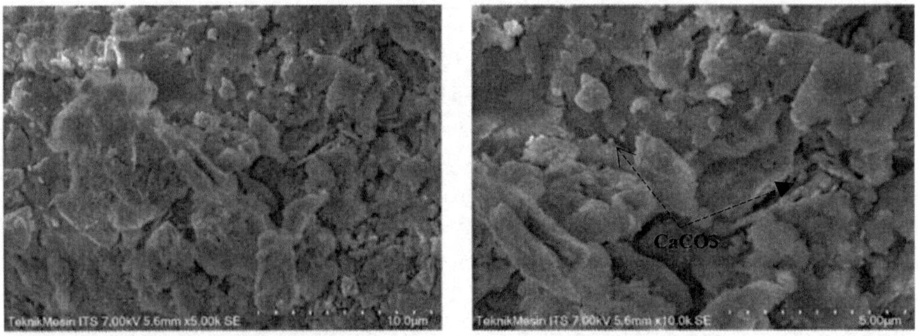

Fig. 7 Microstructure of soil sample

4 Conclusion

This paper has shown that the bacterium Bacillus subtilis can be used in the MICP process in coal-contaminated soils. After treatment, the shear strength value increased by more than 290% compared to untreated soil. That increasing the concentration of bacteria increases the soil shear strength. Culture has a very big influence on the soil stabilization process using the MICP method.

References

1. Al Qabany, A., Soga, K.: Effect of chemical treatment used in MICP on engineering properties of cemented soils. Bio- and Chemo-Mech. Process. Geotech. Eng. Geotechnique Symp. Print **2013**(4), 107–115 (2013). https://doi.org/10.1680/bcmpge.60531.010
2. Cheng, L., Cord-Ruwisch, R., Shahin, M.A.: Cementation of sand soil by microbially induced calcite precipitation at various degrees of saturation. Can. Geotech. J. **50**(1), 81–90 (2013). https://doi.org/10.1139/cgj-2012-0023
3. Dhami, N.K., Reddy, M.S., Mukherjee, M.S.: Biomineralization of calcium carbonates and their engineered applications: a review. Frontiers in Microbiology, 4(Oct), 1–13 (2013). https://doi.org/10.3389/fmicb.2013.00314
4. Mujah, D., Cheng, L., Shahin, M.A.: Microstructural and geomechanical study on biocemented sand for optimization of MICP process. J. Mater. Civ. Eng. **31**(4), 1–10 (2019). https://doi.org/10.1061/(ASCE)MT.1943-5533.0002660

Utilization of Coal Gangue for Earthworks: Sustainability Perspective

Mohammed Ashfaq, M. Heera Lal, and Arif Ali Baig Moghal

Abstract In recent years, there is a growing trend to utilize industrial residues/wastes with the aim of conserving natural resources. However, the environmental impacts in the form of carbon emission associated with their utilization ought to be predetermined prior to their promotion as a sustainable alternative to natural materials. This study aims to quantify the environmental impacts associated with the application of coal gangue (CG) in earthworks by performing carbon footprint analysis (CFA) and cost analysis (CA). An ongoing project of reinforced earthwork construction undertaken by the Government of Telangana, India, has been considered for the CFA and CA of coal gangue utilization. Prior to the CFA and CA, the feasibility of using CG for reinforced earth wall was ascertained by studying its geotechnical characteristics. Additionally, CFA was also performed to quantify the carbon emission associated with the disposal activity of unused CG. Results revealed that CG exhibited favorable geotechnical properties to enable its applications in earthworks. The CFA results indicate that the procurement and haulage of raw materials accounted for maximum carbon emissions and utilization of CG can eliminate 361 CO_{2eq} (kg) associated with its disposal. Further, the CA revealed that CG utilization in earthworks results in Re 3333/m^3 reduction in cost of construction. Furthermore, the results of the study revealed that the utilization of CG can lead to a significant decrease in the carbon footprints by eliminating the carbon emission associated with disposal of CG, thus creating a positive impact on the environment.

Keywords Coal gangue · Carbon footprint · Sustainability · Reinforcement · Retaining wall

1 Introduction

Climate change is a serious problem which must be addressed with holistic solution. A primary contributor to climate change crisis is the increase in greenhouse

M. Ashfaq · M. Heera Lal (✉) · A. A. B. Moghal
Department of Civil Engineering, NIT, Warangal 506004, Telangana, India
e-mail: mhl@nitw.ac.in

H. Hazarika et al. (eds.), *Advances in Sustainable Construction and Resource Management*, Lecture Notes in Civil Engineering 144,
https://doi.org/10.1007/978-981-16-0077-7_20

gases (GHG) in the atmosphere, caused mainly by fossil fuel consumption. With the global population growing at an unparalleled rate, demand for commercial energy has escalated proportionately. In India, fossil fuels are the major contributors in power generation with a share of more than 65% of total power output. Among the fossil fuels, coal is the primary source of energy with more than 550 million tons of annual generation. With an estimated reserve of more than 300 billion tons, the trend of relying on coal as a major source of fuel in power generation will persist. Another major contributor to greenhouse gas emissions is the built environment, accounting for more than 40% of global carbon dioxide emissions [1–3]. The built environment acts as an interface between the human society and ecosystem, and any disruption in its operation can have a multifaceted impact on them. The geotechnical engineering works have traditionally been one of the principal contributors to the built environment with huge consumption of both natural and man-made resources like soils, cement, concrete, industrial and mining wastes. In order to attain sustainable development, identification and development of economic and environmentally viable engineering products become paramount. It is generally considered that the utilization of industrial and mining wastes as an alternative to natural soils is an effective method to reduce financial and environmental burden associated with their procurement and operation. The coal mining activity involves release of waste at different levels of its execution and the largest among such wastes is CG, which accounts for 10–15% of total coal produced [4]. Though there are considerable reports available on the characteristics of coal combustion residues like coal ash and fly ash, very little information has been established on the properties of CG [5–10]. The CG is a heterogeneous well-graded solid waste produced during the mineral processing and washing phase of coal mining with traces of coal on it which vary with mining methodology and the prevailing geology [11, 12]. With the current practice of loose dumping, CG has been one of the largest industrial residues in terms of land occupied which may cause potential subsidence failure at the underground mining sites [13]. CG exhibits higher concentrations of trace metal elements and under rainfall-induced submergence, and these contaminants may be prone to leaching into the surrounding environments. The capitalization of this coal mining residue can have manifold benefits such as: (i) minimization of disposal costs, (ii) conservation of dwindling natural resources and (iii) mitigation of environmental impacts [11, 14–17]. The presence of silica and alumina along with the traces of clay minerals and residual coal fractions has led to the utilization of CG in various applications such as power generation, fertilizer production, brick production, cement and concrete production, as a backfill material and a landfill liner material [18–22]. Though the utilization of CG has increased in recent times, the volume of its generation is still huge. In this context, attempts should be made to explore new avenues for the safe disposal and utilization of CG. Bulk utilization of CG can be accomplished by using them in various civil engineering applications. More precisely, its application in geotechnical engineering as a potential embankment or fill material can prove to be a prudent attempt. However, prior to its utilization, clear understanding of geotechnical properties of CG becomes essential. Further, evaluation of environmental impacts associated with the utilization of coal in geotechnical can validate it as a sustainable alternative to the natural soils. An efficient tool in this regard is to estimate the environmental impacts in terms of energy consumption and carbon emissions, and there is sharp rise in application of life cycle analysis (LCA) in civil engineering works.

In recent times, a much simpler tool of LCA known as carbon footprint analysis (CFA) has been developed with the primary focus on CO_2 emissions, which contributes to more than 90% of total greenhouse gas emissions (GHG). Application of CFA in engineering practice eliminates complexities and simplifies the arduous task of quantifying all the greenhouse gases. Hence, in the present study, an attempt is made to perform comprehensive characterization of CG to validate their applicability in various civil engineering applications. Further, efforts have been made to perform CFA to quantify carbon emissions involved in the utilization of CG as a fill material in the construction of reinforced earth wall and its disposal.

2 Materials and Methodology

In the present study, CG procured from Kakatiya coal mines (18° 25′ 53.2″N; 79° 51′ 30.8″E), Bhupalpally, Telangana, India. The CG procured is pulverized using jaw crusher, and only 4.75 mm passing fraction has been considered for geotechnical characterization to eliminate the boundary effects associated with varying sizes. The physical properties of CG presented in Table 1 were determined by performing tests in accordance with their respective ASTM standards [Refer Table 1]. The compaction tests were performed for both standard and modified compaction scenario following procedure laid by ASTM D698 [23] and ASTMD 1557 [24], respectively. To find the shear parameters of CG, direct shear test was performed on CG in accordance with the procedure laid down by ASTM D3080 [25]. Prior to direct shear test, the samples were compacted to 95% of maximum dry density using static compactor, and the strain rate applied is 0.35 mm per minute. Further, to understand the influence of confining pressures on stress–strain behavior of CG unconsolidated undrained (UU), triaxial compression test was performed on coal gangue (with varying fines content) in accordance with procedure laid down by ASTM D2850 [26]. In order to avoid the influence of moisture fluctuations on the shear behavior, a fixed optimum moisture content of 17% was employed for all the coal gangue combinations. The confining

Table 1 Physical properties of coal gangue

Properties	Value	ASTM Standard
Specific gravity	2.57	ASTM D854-14 [30]
UCS classification	SP	ASTM D2487-17 [31]
Liquid limit (%)	28	ASTM D4318-17 [32]
Plastic limit (%)	NP	ASTM D4318-17
Plasticity index	NP	ASTM D4318-17
pH	7.17	ASTM D4972 [33]

pressure (CP) was varied from 50 to 150 kPa with an increment of 50 kPa. The range of confining pressure was fixed based on the previous research on coal tailings by Indraratna et al. [16] and Heitor et al. [27]. The hydraulic conductivity of CG was evaluated by using rigid wall permeameter in accordance with ASTMD 5856 [28]. The California bearing ratio (CBR) test on CG was carried out in accordance with the specifications laid down by ASTM D1883 [29]. The thorough understanding of mineralogy assists in the comprehension of mechanism involved in the variation in geotechnical property of CG, thus X-ray diffraction analysis was carried out on crushed (200 mm sieve passing) fraction of CG. From the clear and distinct peaks in X-ray diffractograms of CG presented in Fig. 1, it can be comprehended that the minerals present in CG are crystalline in nature. Further, it was observed that quartz (SiO_2) is the major mineral present in CG. However, the results (Fig. 1) also confirm the presence of clay fraction in the form of Kaolinite mineral which is a distinguished feature of CG compared to other coal combustion residues like fly ash and pond ash.

Previous researchers like Bouzza and Heerten [34] and Damians et al. [35] have established that methodology adopted in the evaluation of carbon emissions associated with the geotechnical works like MSE wall construction, subgrades and landfill liner works. The CFA has been performed for two cases, i.e., utilization of CG as a subgrade material and the disposal process of CG.

Further, the following boundary conditions have been adopted in CFA:

- Transport of labor from and to the construction site.
- Production of fuel and its haulage to the construction site.
- Post construction maintenance of the project.
- Manufacture of equipments and vehicles used in the project.

The exclusions made in the current study are in accordance with guidelines provided in ISO 14040 [36] and ISO 14044 [37]. The carbon emissions coefficient (CC) of a material represents the estimated CO_2 emissions involved in its production. The CC of the material considered in the current study was obtained from inventory of

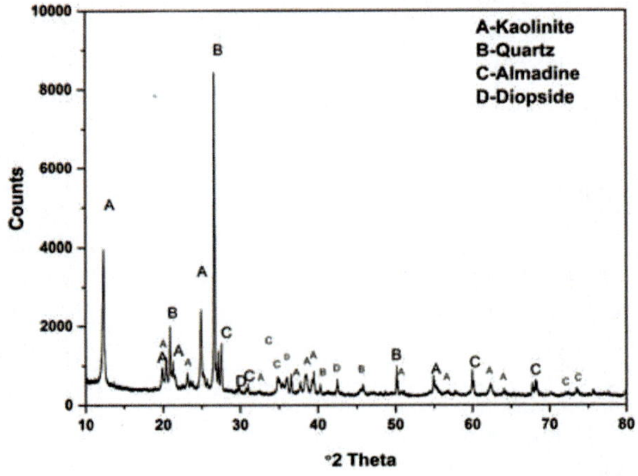

Fig. 1 X-ray diffractograms of crushed coal gangue

Hammond and Jones [38] and Ashfaq et al. [39]. The payload capacity, fuel economy and engine capacity of the vehicles used in the study were sourced from Davis et al. [40] and Shillaber et al. [41].

3 Results and Discussion

The physical properties of CG are presented in Table 1. Based on gradation curve, coarse coal gangue is classified under poorly graded sand (SP). The shear strength parameters take paramount importance especially in the construction of embankments and backfills. Further, it is observed that CG has favorable shear strength parameters in terms of friction component. Being non-plastic and cohesionless, CG derives all its shear strength from frictional component. It is worthwhile to note that in spite of its low dry density, CG has exhibited higher frictional angle than sandy soils. This can be attributed to the angularity and rough texture of CG which provide higher grade of interlocking in the shearing process. The positive shear strength parameters of CG provide a viable substitute to the natural geomaterial. Similar results were established by earlier characterization studies on CG by Ashfaq et al. [42].

3.1 Geotechnical Characterization

The following geotechnical properties of CG were assessed to confirm its applicability in earthworks. The grain size distribution (GSD) curve of CG is presented in Fig. 2, and the results are compared with other industrial residues like coal tailings,

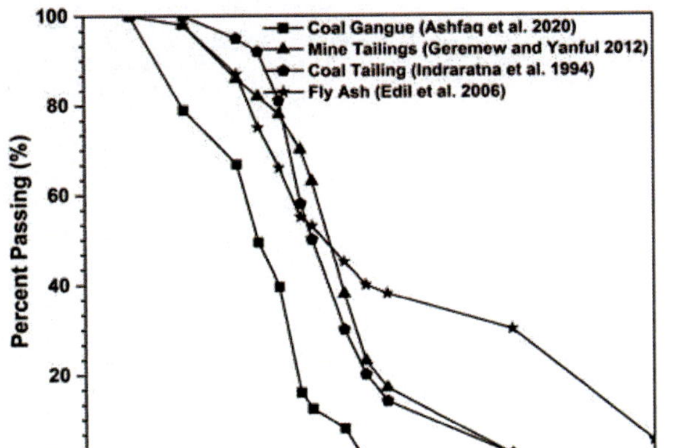

Fig. 2 Grain size distribution curve of coal gangue

fly ash and mine tailings. From the GSD curves, it can noted that the coarse fraction in CG is relatively more compared to other industrial residues. However, due to inherent heterogeneity, the GSD of CG exhibited more uniform gradation with varying particle size. The heterogeneity of CG can prove advantageous if applied as a fill material due to its relatively greater potential to dissipate pore water accumulation and thereby can eliminate the failure of hydraulic structures.

Compaction

The compaction behavior of CG was studied for both standard and modified proctor compaction method. The tests were conducted over varying contents (10–24%). From the compaction results presented in Fig. 3, it can be observed that the maximum dry density (MDD) and optimum moisture content (OMC) of CG are 1.99 (g/cm^3), 16% and 2.31 (g/cm^3), 18% for standard and modified compaction, respectively. The relatively lower MDD value of CG compared to natural soils (non-cohesive) can be attributed to its low specific gravity (2.57). In coherence with natural soils and other industrial wastes, the MDD and OMC values of CG for modified case are relatively higher. From both the compaction curves (refer Fig. 3), it can be noted that the distinct peak and rapid fall post peak were observed for both standard and modified compaction curves. The observed MDD values of CG are akin to silty sands which exhibit drainage characteristics. Further, the relatively lower MDD values of CG facilitate its greater workability as a fill material with relatively lower lateral earth pressure compared to conventional fill materials.

Shear Strength

The thorough understanding of shear strength properties is essential for application of CG as a fill material. The results obtained from DST test showed that the shear strength parameters, i.e., cohesion (C) and angle of frictional resistance (φ) of CG are 26 kN/m^2 and 39°, respectively. It can be noted that the being non-plastic in nature, CG has favorable shear strength which is primarily derived from frictional

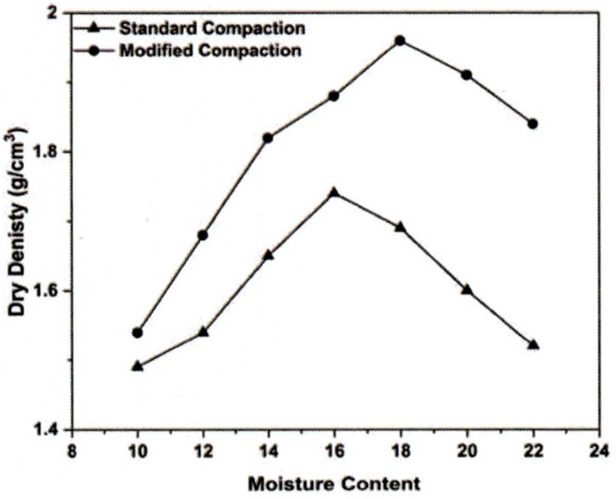

Fig. 3 Moisture content versus dry density variation of CG

component. The greater frictional resistance can be attributed to effective interlocking of CG particles attained by inherent angularity and rough texture of CG surface. Further, in order to study the effect of confining stress (CS) on the stress–strain behavior of CG, unconsolidated undrained triaxial test was performed and the results are presented in Fig. 4. From the results, it can be noted that the increase in CP resulted in variation in the stress–strain behavior of coal gangue. At a CS of 50 kPa, the maximum deviator stress (DS) of CG was observed as 280 kPa at axial strain of 7%. The DS value linearly increased with CS with the attainment of peak DS value at axial strain of 14%. With the increase in CS from 50 to 100 kPa, the maximum DS value increased by 125% which highlights the role of CS on stress–strain behavior of CG. The increase in CS has not only increased the corresponding DS value but also has delayed the attainment of peak DS from 7 to 14% axial strain. Irrespective of variation in CS value, the linear increase and subsequent fall post peak DS value demonstrate the perfect plastic deformation behavior of CG. Further, it also highlights the fact that the strain-softening behavior of CG is unaltered even at higher CS. This distinct stress–strain behavior of CG can be attributed to dense interlocking arrangement of CG particles which yields higher frictional resistance with relatively higher void ratio.

Hydraulic Conductivity

The understanding of hydraulic conductivity behavior is essential for the application of CG as a fill material in earthen dams and embankments. The hydraulic conductivity (HC) of CG was observed to 6.4×10^{-4} cm/s which is comparable to silts and silty sands. The observed HC values of CG are relatively lower compared to coal combustion residues like fly ash and pond ash. The lower HC can be attributed to the presence of traces of coal and clay fraction (refer Fig. 1). From the HC values of CG, it can be inferred that CG meets the IS 1498 specifications for fill material in embankments [1].

Fig. 4 Variation in stress–strain behavior of CG with confining stress

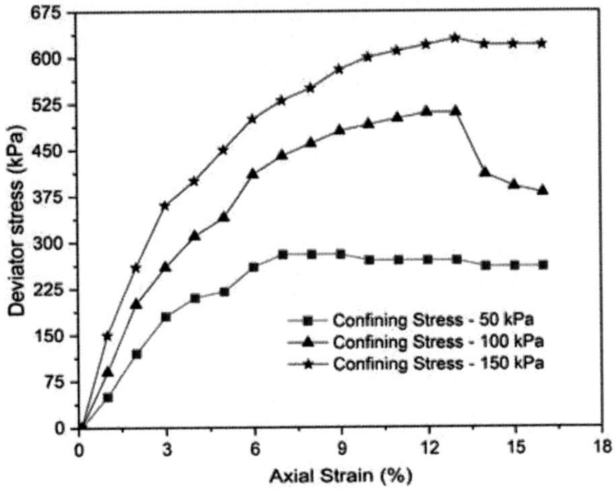

Table 2 Existing correlations for the estimation of resilient modulus based on CBR value

Correlation model	References
$M_R = 10$ CBR	Heukelom and Foster [44]
$M_R = 38$ CBR$^{0.711}$	Green and Hall [45]
$M_R = 18$ CBR$^{0.64}$	Lister and Powell [46]
$M_R = 21$ CBR$^{0.65}$	Ayres [47]
$M_R = 17.6$ CBR$^{0.64}$	AASHTO T.307 [48]
$M_R = 3$ CBR	Edil et al. [49]

California Bearing Ratio

In the present study, the CBR tests were performed to assess the feasibility of using CG as a subgrade material in pavement applications. From the tests, it was observed that the unsoaked and soaked CBR values of CG are 19% and 18%, respectively. It was observed that the loss in CBR values in soaked condition is negligible which highlights the fact that loss in bearing ratio due to capillary forces (under soaking) is minimal. Further, it can be comprehended that the observed CBR values of CG satisfy the design requirements for subgrade material as specified by IS 1498 [43]. However, the stabilization of CG with traditional additives like lime and cement can enhance its CBR value further which is always desirable for any subgrade material.

The resilient modulus (MR) is one of the essential parameters in the design of pavement structures, and hence an attempt was made to the estimate the MR values based on the observed CBR values. Many empirical correlations were proposed by previous researchers to estimate MR from CBR values which were analyzed (Table 2) to select the best possible model for the estimation of MR value.

Since most of the models over predicted M_R value, the correlation proposed by AASHTO was adopted in the present study due to its wider acceptance and application. The predicted M_R values for unsoaked and soaked case are 116 MPa and 111 MPa, respectively. In coherence with the observations made for CBR values, the highest M_R values were reported for unsoaked case. From the observed CBR and M_R values, it can be inferred that the CG can be utilized as a subgrade material.

3.2 *Carbon Footprint Analysis (CFA)*

To assess the environmental impacts associated with the utilization of CG as a fill material in embankment and its disposal, CFA was performed for both the cases of its utilization and disposal.

Case I. CFA of coal gangue application in earthworks. The potential environmental impacts envisaged with utilization of CG in geotechnical engineering applications are evaluated by performing CFA on its application as a fill material in highway embankment. For the present study, the extension of Karimnagar-Ramagundam highway (18° 38′ 33.3″N 79° 23′ 24.7″E) has been considered for CFA analysis. CFA has been performed in four stages. In the first stage, the embodied carbon of materials used is computed. Embodied carbon refers to emissions involved in the production/formation of raw material. The CC of materials emissions is taken from data established by Hammond and Jones [38] and Ashfaq et al. [39]. The second stage deals with the emissions related to the loading and transport of materials. The vehicles adopted for loading and transport are pickup excavator and heavy duty dumper of 10 tons and 25 tons capacity, respectively. The third stage involves the estimation of emissions related to the site operation. The site operations include loading of in situ soil/CG with a pickup excavator followed by spreading/placing of backfill of CG with bulldozer. Compaction of the placed backfill is done using a vibratory roller. Since emissions in the saturation of backfill (using water jets) are relatively insignificant due to absence of machinery/vehicle operations, their quantifications have not been included in the CFA. The final stage of CFA involves the estimation of CO_2 emissions related to the disposal/transportation of waste generated during the MSE wall construction. Due to the brevity, the fuel economy of each vehicle type and other sub-calculations such as determination of quantity of each material based on their respective unit weights are excluded from the study.

From the results presented in Figs. 5 and 6, it is evident that the raw material procurement and haulage contribute to the bulk of emissions which may be due to the import of backfill to the construction site. The embodied carbon, i.e., the carbon emissions in the generation/manufacture of coal is much higher compared to conventional or natural fill materials. This is due to the fact that coal mining involves the deep excavation involving application of huge mechanical and manual forces. On the contrary, the embodied carbon of CG is almost zero or negligible which is attributed to its mode of generation, i.e., embodied emissions are defined

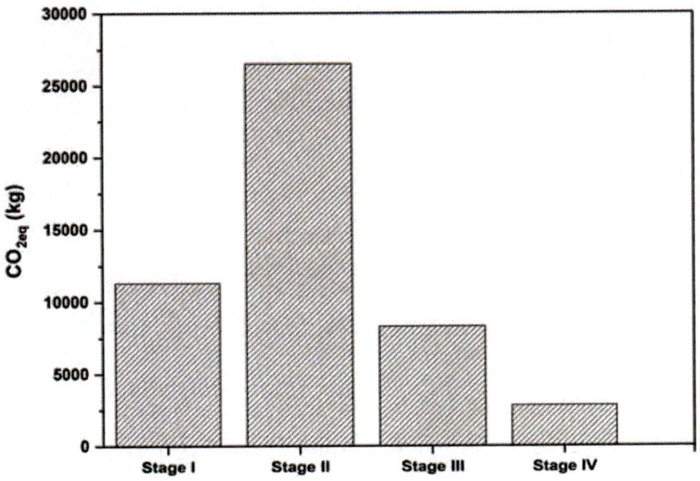

Fig. 5 Staged CO_2 emissions in coal gangue application and disposal scenario

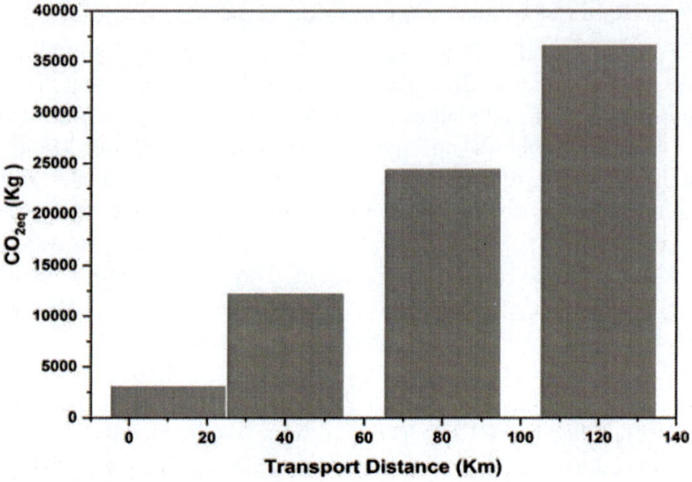

Fig. 6 Variation in CO_2 emissions with haulage distance

as the total emissions associated with various stages of manufacturing/production of product, the coal gangue being a by-product, and the embodied emissions for it are non-existent. The emissions involved in transport (Fig. 5) are inevitable and remain the same irrespective of the variation in fill material. The transport has a predominant effect on the overall emissions, and it is evident in Fig. 5. This is further explained in Fig. 6 which depicts the variation in the overall emissions with the increase in transport distance. The effect of haulage distance is clearly manifested with an exponential increase in emissions. It was observed that with the 10 km increase in distance of transport, the emissions increased by 3200 CO_{2eq} (kg) from 50 km. This further emphasizes the point that utilization of CG as an alternative to natural soils can substantially reduce carbon emissions, especially if the construction activity is taking place in the vicinity of mining sites or location(s) which are relatively at shorter distance compared to the conventional fill materials. In other words, it can be said that the viability of utilizing CG as an alternative fill material is greatly determined by its ease of procurement to the pavement construction site. Further, the influence of using stabilizers like lime and cement to enhance the mechanical properties of CG is beyond the scope of the current study. It is generally established that the usage of cement and lime further enhances the overall carbon emissions due to their very high carbon emissions coefficients. The high CC of lime and cement is due to their mode of manufacture which involves calcination at very high temperature using fossil fuels as a source of power.

Case II. CFA of coal gangue disposal. In this case, CFA was performed with the objective of estimating the CO_2 emissions related to the disposal activity of CG. Globally, there is no uniform mode of disposal for CG, for the brevity of study, the disposal method considered is specific to Bhupalpally mine site, Telangana State, India. Currently, CG is periodically collected from the mine site and is loosely

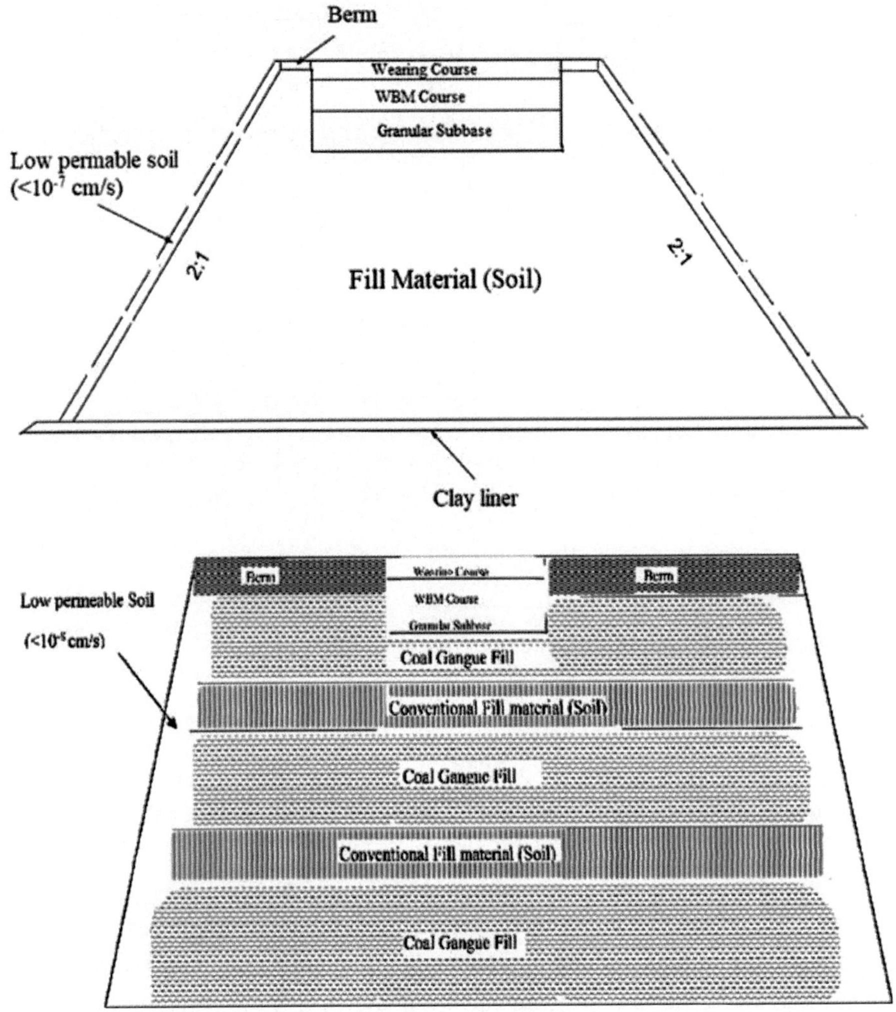

Fig. 7. Cross section of conventional and coal gangue fill

dumped in the nearby vacant lands located at a distance of 3 km (18° 27′ 23.8"N 79° 50′ 29.3"E). In this case, the carbon emissions are calculated for the embodied carbon, procurement/loading and haulage of CG. Pickup excavator of capacity 10 tons/Liter is used for loading of CG. For the transportation of CG to the dumping site, heavy duty dumper of 25 tons/liter capacity is employed. The detailed calculations involved in CFA of case II are presented in Table 3.

For both the cases, the embodied carbon of the materials accounted for the maximum amount of carbon emissions which indicates that the emissions rise proportionately with the increase in raw materials. Though the haulage distance of CG is relatively small, its enormous volume contributed to its greater emissions. It is

Table 3 CFA of coal gangue disposal

Process	Quantity of waste (kg)	Vehicle/machine	Capacity (kg) × 10^3	Distance (km)	Volume of fuel (L)	CO_{2eq} (kg)
Embodied carbon	29,510	–	–	–	–	206
Procurement	29,510	Pickup excavator	10	0	23	96
Haulage	29,510	Heavy duty dumper	25	3	14	59
The total CO_{2eq} (kg) emissions = 361						

inferred that application of CG in pavement application can result in a net savings of 361 CO_2eq (kg). The value obtained is specific to the Bhupalpally mining site, and it is subject to variation with the mining location. The mode of disposal and distance of disposal site are other influencing factors which determine the overall savings in carbon emissions.

3.3 *Cost Analysis (CFA)*

The economic viability is the primary factor which influences the decision of utilizing any novel material in earthworks. For the proposed embankment, the cost analysis was performed to evaluate the impact of CG utilization on the overall cost of construction. Further, in order to compare the results with the existing conventional fill materials, similar analysis was extended to local available soil. The cross sections of both the scenarios are presented in Fig. 7. Since the overall cost of construction directly increases with the dimensions of and also to facilitate the comparison of results, the dimensions of embankment for both the fill scenarios are assumed the same. However, the volume of fill varies owing to the difference in unit weight of the fill material. In the current study, the volume of CG fill is relatively greater compared to local soil due to lower unit weight. Since, the costs of CG procurement are not defined in Ministry of Road and Transport and Highways (MORTH) specifications, an assumed value equivalent to the average cost of procuring for mining/industrial wastes has been considered. It is noted that the overall costs of embankment with local soil are 36% higher compared to embankment with CG as fill material which is primarily due to the nominal cost of procurement for CG. This difference in cost of construction substantially increases with the dimensions of embankments. Further, it was also observed that the distance of haulage of material greatly influences the costs of construction. Similar observations were also made for the overall carbon emissions associated with haulage of CG. For every 5 km increase in haulage distance, the CG procurement costs increase by 22%. Thus, for situations in which the haulage distance is less than 23 km, CG as fill material is more economical alternative option over the conventional fill material (local soil). On the whole, the CA revealed that

CG utilization in earthworks results in Re 3333/m^3 reduction in cost of construction. Further, in addition to the reduction in overall project costs, CG utilization as a fill material provides the following benefits:

i. Conservation of scarce natural resource (soil).
ii. Avoids the rapid fluctuations in ground water table.
iii. Preservation of biodiversity of local environment.

Similar benefits were noted by Kumar and Patil [50] and Pant et al. [51] for coal combustion residues.

4 Conclusions

In the present paper, geotechnical properties of CG were examined to ascertain the feasibility of its application in pavements. The CFA was carried out for the construction of pavement. Further, CFA related to the disposal of CG was carried out. Following conclusions were drawn from the study:

- Geotechnical characteristics of CG revealed that it has favorable properties for its application as fill material.
- The emissions related to the embodied carbon of the raw materials were observed to be substantial.
- The emissions proportionately increased with transportation distance of CG which accounted for the maximum CO_2eq emission.
- In the current study, the net savings due to utilization of CG were observed to be 361 CO_2 (kg).
- The CA revealed that CG utilization in earthworks results in Re 3333/m^3 reduction in cost of construction.

Acknowledgements The authors are indebted to Singareni Collieries Company limited authorities for rendering timely help by permitting the procurement of coal gangue from Kakatiya coal mines, Bhupalpally, Telangana State, India.

References

1. Ashfaq, M., Heeralal, M., Moghal, A.A.B.: Characterization studies on coal gangue for sustainable geotechnics. Innov. Infrastruct. Solut. **5**(1). article no 15 (2020).
2. Ashfaq, M., Heeralal, M., Moghal, A.A.B.: Characterization of Heavy metals from coal gangue. (2021). In: Latha Gali, M., Raghuveer Rao, P. (Eds.) Problematic Soils and Geoenvironmental Concerns. Lecture Notes in Civil Engineering, vol. 88. Springer, Singapore. https://doi.org/10.1007/978-981-15-6237-2_8.

3. Ashfaq, M., Heeralal, M., Moghal, A.A.B.: Static and Dynamic leaching studies on coal gangue. In: Reddy K.R., Agnihotri A.K., Yukselen-Aksoy Y., Dubey B.K., Bansal A. (eds) Sustainable Environmental Geotechnics. Lecture Notes in Civil Engineering, vol 89. Springer, Cham. https://doi.org/10.1007/978-3-030-51350-4_28.
4. Liu, B., Liu, Z.L.: Recycling utilization patterns of coal mining in China. Resour. Conserv. Recycl. **54**, 1331–1340 (2010)
5. Pandian, N.S.: Fly ash characterization with reference to geotechnical applications. J. Indian Inst. Sci. **84**, 189–216 (2004)
6. Trivedi, A., Sud, V.K.: Collapse behavior of coal ash. J. Geotech. Geoenviron. Eng. **130**(4), 403–415 (2004)
7. Prakash, K., Sridharan, A.: Beneficial properties of coal ashes and effective solid waste management. Pract. Periodica. Hazard., Toxic, Radioact. Waste Manag. **13**(4), 239–248 (2009)
8. Sivapullaiah, P.V., Moghal, A.A.B.: CBR and strength behavior of class F fly ashes stabilized with lime and gypsum. Int. J. Geotech. Eng. **5**(2), 121–130 (2011)
9. Sivapullaiah, P.V., Moghal. A.A.B.: Lime leachability and CBR behavior of class F fly ashes. In: Proceedings of a Conference on 14th Asian Regional Conference on Soil Mechanics and Geotechnical Engineering, Hong Kong, China, pp. 44–49 (2011b)
10. Moghal, A.A.B.: A state-of-the-art review on the role of fly ashes in geotechnical and geoenvironmental applications. J. Mater. Civ. Eng. **29**(8). 04017072(1–14) (2017)
11. Indraratna, B., Rujikiatkamjorn, C., Chiaro, G.: Characterization of compacted coal wash as structural fill material. In: Geo-Congress March 25–29, 2012 Oakland, California, United States (2012)
12. Guo, S. (2017). Trace elements in coal gangue: a review. In: Al-Juboury A.I. (Eds.) Chapter 6 in Contributions to Mineralization, Intech Open, pp. 127–144.
13. Gao, Y., Huang, H., Tang, W., Liu, X., Yang, X., Zhang, J.: Preparation and characterization of a novel porous silicate material from coal gangue. Microporous Mesoporous Mater. **217**, 210–218 (2015)
14. Goh, A.T.C., Tay, J.: Municipal solid-waste Incinerator fly ash for geotechnical applications. J. Geotech. Eng. **119**(5), 811–825 (1993)
15. Kamon, M., Katsumi, T.: Civil engineering use of industrial waste in Japan. In: Proceedings of the International Symposium on Developments in Geotechnical Engineering, Bangkok, Thailand, pp. 265–278 (1994)
16. Indraratna, B., Gasson, I., Chowdhury, R.N.: Utilization of compacted coal tailings as a structural fill. Can. Geotech. J. **31**, 614–623 (1994)
17. Lim, T.T., Chu, J.: Assessment of the use of spent copper slag for land reclamation. Waste Manage. Res. **24**, 67–73 (2006)
18. Jablonska, B., Kityk, A.V., Busch, M., Huber, P.: The structural and surface properties of natural and modified coal gangue. J. Environ. Manage. **190**, 80–90 (2017)
19. Wang, J.M., Qin, Q., Hu, S.J., Wu, K.N.: A concrete material with waste coal gangue and fly ash used for farmland drainage in high groundwater level areas. J. Clean. Prod. **1121**, 631–638 (2016)
20. Wu, H., Wen, Q., Hu, L., Gong, M., Tang, Z.: Feasibility study on the application of coal gangue as landfill liner material. Waste Manage. **63**, 161–171 (2017)
21. Cheng, Y., Hongqiang, M., Hongyu, C., Jiaxian, W., Jing, S., Zonghui, L., Mingkai, Y.: Preparation and characterization of coal gangue geopolymers. Constr. Build. Mater. **187**, 318–326 (2018)
22. Dondrob, K., Koshy, N., Wen, Q., Hu, L.: Synthesis and characterization of geopolymers from coal gangue, fly ash and red mud. In: Proceedings of the 8th International Congress on Environmental Geotechnics, vol. 1. ICEG 2018. Environmental Science and Engineering. Springer, Singapore (2018).
23. ASTM: D698-12e2: Standard test methods for laboratory compaction characteristics of Soil Using Standard Effort (12,400 ft-lbf/ft^3 (600 kN-m/m^3)). ASTM International, West Conshohocken, PA, USA (2012)

24. ASTM: D1557-12e1: Standard Test Methods for Laboratory Compaction Characteristics of Soil Using Modified Effort (56,000 ft-lbf/ft^3 (2700 kN-m/m^3)). ASTM International, West Conshohocken, PA, USA (2012)
25. ASTM: D3080-11: Standard Test Methods for Direct Shear Test of Soils Under Consolidated Drained Conditions. ASTM International, West Conshohocken, PA, USA (2011)
26. ASTM: D2850-15: Standard Test Methods for Unconsolidated-Undrained Triaxial Compression Test on Cohesive Soils. ASTM International, West Conshohocken, PA, USA (2015)
27. Heitor, A., Indraratna, B., Kaliboullah, C.I., Rujikiatkamjorn, C., McIntosh, G.W.: Drained and Undrained Shear behavior of compacted coal wash. J. Geotech. Geoenviron. Eng. **142**(5), 04016006:1–10.
28. ASTM: D 5856-15: Standard Test Methods for Measurement of Hydraulic Conductivity of Porous Material Using A Rigidwall, Compaction-Mold Permeameter. ASTM International, West Conshohocken, PA, USA (2015)
29. ASTM: D1883–16: Standard Test Methods for California Bearing Ratio (CBR) of Laboratory-Compacted Soils. ASTM International, West Conshohocken, PA, USA (2016)
30. ASTM: D854-14: Standard Test Methods for Specific Gravity of Soil Solids by Water Pycnometer. ASTM International, West Conshohocken, PA, USA (2012)
31. ASTM: D2487-11: Standard Practice for Classification of Soils for Engineering Purposes (Unified Soil Classification System). ASTM International, West Conshohocken, PA (2017)
32. ASTM: D4318-17: Standard Test Methods for Liquid Limit, Plastic Limit and Plasticity Index of Soils. ASTM International, West Conshohocken, PA, USA (2012)
33. ASTM: D 4972-13: Standard Test Methods for pH of Soils. ASTM International, West Conshohocken (2013)
34. Bouzza, A., Heerten, G.: Geosynthetic applications—sustainability aspects in Handbook of Geosynthetic Engineering, Shukla S.K. (Eds.) Default Book Series Second edition, pp. 387–396 (2012).
35. Damians, I.P., Bathurst, R.J., Adroguer, E.G., Josa, A., Lloret, A.: Environmental assessment of earth retaining wall structures. Environ. Geotech. **4**(6), 415–431 (2017)
36. ISO (International Organization for Standardization): ISO 14040:2006: Environmental Management—Life Cycle Assessment–Principles and Framework. ISO, Geneva, Switzerland (2006)
37. ISO: ISO 14044:2006: Environmental Management—Life Cycle Assessment—Requirements and Guidelines. ISO, Geneva, Switzerland (2006)
38. Hammond, G., Jones, C.: Inventory of Carbon and Energy (ICE) Version 2.0. Sustainable Energy Research Team (SERT), University of Bath, Bath (2011). https://ghgprotocol.org/Third-Party-Databases/Bath-ICE
39. Ashfaq, M., Heeralal, M., Moghal, A.A.B., Murthy, V.R.: Carbon footprint analysis of coal gangue in geotechnical engineering applications. Indian Geotech. J. **50**, 646–654 (2019)
40. Davis, S.C., Diegel, S.W., Boundy, R.G.: Transportation Energy Data Book: Edition 31. Rep. No. ORNL-6987, U.S. Department of Energy (2012). https://tedb.ornl.gov/
41. Shillaber, C.M., Mitchell, J.K., Dove, J.E.: Assessing environmental impacts in geotechnical construction: Insights from the fuel cycle. In: Proceedings of Geo-Congress 2014, Geo-Characterization and Modelling for Sustainability, Geotechnical Special Publication-234 ASCE Reston, VA, pp. 3516–3525 (2014).
42. Ashfaq, M., Heeralal, M., Moghal, A.A.B.: Effect of coal Gangue particle size on its leaching characteristics. Geotech. Spec. Publ. **319**, 107–114 (2020)
43. IS 1498: Indian Standard Classification and Identification of Soils For General Engineering Purposes. Bureau of Indian Standards, New Delhi (1970)
44. Heukelom, W., Foster, C.: Dynamic testing of pavements. J. Soil Mech. Found. Div. **86**(1), 1–28 (1960)
45. Green, J., Hall, J.: Nondistructive Vibratory Testing of Airport Pavement. Technical Report S-75-14 (1975) Online source: ark:/67531/metadc304084
46. Lister, N.W., Powell, D.: Design practices for pavements in the United Kingdom. In: Proceedings of the 6th International Conference on the Structural Design of Asphalt Pavements, Ann Arbor, MI, USA(1987).

47. Ayres, M.: Development of a Rational Probabilistic Approach for Flexible Pavement Analysis. University of Maryland (Publisher), College Park, MD, USA (1997)
48. AASHTO T307: Determining the Resilient Modulus of Soils and Aggregate Materials. American Association of State Highway and Transportation Officials, Washington, DC, USA (2012)
49. Edil, T.B., Acosta, H.A., Benson, C.H.: Stabilizing soft fine-grained soils with fly ash. J. Mater. Civ. Eng. **18**(2), 283–294 (2006)
50. Kumar, S., Patil, C.B.: Estimation of resource savings due to fly ash utilization in road construction. Resour. Conserv. Recycl., **48** (2), 125–140 (2006)
51. Pant, A., Raman, G.V., Datta, M., Gupta, S.K.: Comprehensive assessment of cleaner, sustainable and cost-effective use of coal combustion residue (CCR) in geotechnical applications. J. Clean. Prod. **271**, 122570 (2020)

Physical and Numerical Modeling of Disaster Mitigation Techniques

A Consideration on Numerical Model for the Relationship Between Evaporation Efficiency and Volumetric Water Content

Yuta Jikuya, Kazunari Sako, and Shinichi Ito

Abstract It is important to obtain the amount of evaporation from the slope surface to analyze slope stability after rainfall. The bulk method is one of the evaporation estimation methods. Using this method, the amount of evaporated water from bare ground surfaces can be estimated from the general meteorological data. The evaporation efficiency, β, is an important parameter of the bulk method and is influenced by the volumetric water content in unsaturated soil. The molecular diffusion distance, $F(W_v)$, is also an important parameter to calculate β. $F(W_v)$ is influenced by the type of surface soil, the void ratio, and the volumetric water content, etc. The functional form of $F(W_v)$ was obtained from experiments, and the experimental relationships between the volumetric water content and $F(W_v)$ for loam and sand were explained by Kondo et al. In this paper, a numerical model for the molecular diffusion distance, $F(W_v)$, derived from soil parameters such as grain size and void ratio is proposed, and β obtained from our proposed numerical model is compared with the results of laboratory tests by Sako et al. As a result, the applicable scope of the numerical model and the necessity of improvement of the numerical model are discussed.

Keywords Evaporation · Evaporation efficiency · Bulk method · Unsaturated sandy soil

1 Introduction

Countermeasures for rainfall-induced slope failure are classified roughly into structural measures and non-structural measures. It takes a lot of time and money to implement structural measures at all dangerous points, so it is desirable to enhance non-structural measures. In non-structural measures, evacuation advisories and traffic regulations are conducted. However, conventional evacuation advisories and traffic regulations focus on setting standards for issuance based on rainfall intensity, and

Y. Jikuya (✉) · K. Sako · S. Ito
Kagoshima University, Kagoshima 890-0065, Japan
e-mail: k7665614@kadai.jp

H. Hazarika et al. (eds.), *Advances in Sustainable Construction and Resource Management*, Lecture Notes in Civil Engineering 144,
https://doi.org/10.1007/978-981-16-0077-7_21

there are no clear standards for cancelation of issuance. To set standards for cancelation of issuance, it is necessary to assess the slope stability after rainfall. It is important to obtain the amount of evaporation to analyze the slope stability after rainfall.

The bulk method is one of the evaporation estimation methods. In this method, if the exchange speed, g_a, and the evaporation efficiency, β, are known in advance, the amount of evaporated water from bare ground surfaces can be estimated from the general meteorological data. The molecular diffusion distance, $F(W_v)$, is an important parameter to calculate β. $F(W_v)$ is influenced by the type of surface soil, the void ratio, and the volumetric water content, W_v, etc. [1].

In this paper, a numerical model for the molecular diffusion distance, $F(W_v)$, derived from soil parameters such as grain size and void ratio is proposed, and the validity of the numerical model is discussed. The numerical model is based on the numerical model for the seepage behavior of unsaturated soil proposed by Sako et al. [2]. Firstly, the outline of the bulk method and the numerical model proposed by Sako et al. [2] is described. Especially, the relationship between the volumetric water content and the molecular diffusion distance (W_v–$F(W_v)$ relation) is explained. Secondly, the outline of our proposed numerical model is described, and the evaporation efficiency, β, is calculated using the numerical model. Comparing with the results of laboratory experiments by Sako et al. [3], the applicable scope of the numerical model and the necessity of improvement of the numerical model are discussed.

2 Outline of the Bulk Method

The estimation method of evaporation using the bulk method (β-method) is expressed as the following equation:

$$E = \rho g_a \beta \{q_{SAT}(T_s) - q_a\} \tag{1}$$

where E is the evaporation per unit area [kg m^{-2} s^{-1}], ρ is the air density [kg m^{-3}], g_a is the exchange speed [m s^{-1}], β is the evaporation efficiency [−], $q_{SAT}(T_s)$ is the saturation specific humidity at the soil surface temperature, T_s [kg kg^{-1}], and q_a is the specific humidity of the air [kg kg^{-1}]. Almost all parameters of the bulk method can be derived from the general meteorological data. However, the exchange speed, g_a, and the evaporation efficiency, β, have to be derived from the laboratory experiments. β is surface moisture availability and has a value in the range between 0 and 1 depending on the soil moisture (e.g., $\beta = 0$ at dry condition, $\beta = 1$ at saturated condition). Kondo et al. [4] have proposed the equation of the relationship between the volumetric water content and the evaporation efficiency (W_v–β relation) obtained from the experiments. Using $F(W_v)$ which is a function of the volumetric water content, W_v, β is expressed as the following equation:

Table 1 $F(W_v)$ constants for a typical type of soil

Type of soil	F_1 [m]	F_2	$W_{v,\,SAT}$
Loam	2.16×10^2	10.0	0.490
Sand	8.32×10^5	16.6	0.392
Fine sand	7.00×10^3	11.2	0.397

Source Kondo et al. [4]

$$\beta = \frac{1}{1 + \{g_a F(W_v)\}/D_{atm}} \tag{2}$$

where $F(W_v)$ is the molecular diffusion distance [m], and D_{atm} is the molecular diffusion coefficient for water vapor [m^2 s^{-1}]. The molecular diffusion distance, $F(W_v)$, is a parameter with a length dimension and expresses the distance of the flow of water vapor from the water surface in the soil pore to the land surface. The functional form of $F(W_v)$ is obtained from experiments and varies depending on the type of soil. Kondo et al. [1] have proposed the equation relating the volumetric water content to the molecular diffusion distance (W_v–$F(W_v)$ relation) obtained from the experiments. The equation of the W_v–$F(W_v)$ relation is expressed as the following equation:

$$F(W_v) = F_1\left(W_{v,SAT} - W_v\right)^{F_2} \tag{3}$$

where $W_{v,\,SAT}$ is the saturated volumetric water content [m^3 m^{-3}]. Table 1 shows the $F(W_v)$ constants, F_1 and F_2 for typical types of soil. When W_v is large, $F(W_v)$ is quite small. This is because water exists in large pores near the ground surface, and water vapor is easily transported to the ground surface. When W_v is small, $F(W_v)$ increases with a decrease in W_v. This is because water exists only in small complicated pores, and the distance for the water vapor to reach the ground surface from the pore is relatively large.

In this paper, the numerical model for the molecular diffusion distance, $F(W_v)$, derived from soil parameters such as grain size and void ratio is proposed.

3 Outline of the Numerical Model for the Seepage Behavior of Unsaturated Soil

Unsaturated soil is composed of three phases (viz. the solid phase (soil particles), the liquid phase (pore water) and the gas phase (pore-air)). Figure 1a shows a soil element with a few soil particles. This element can be modeled as shown in Fig. 1b. The voids are represented as a pipe with a diameter (D_v) and an inclination angle (θ). The soil particles are represented as other impermeable parts of the model minus the space occupied by the pipe. The element shown in Fig. 1a is called elementary

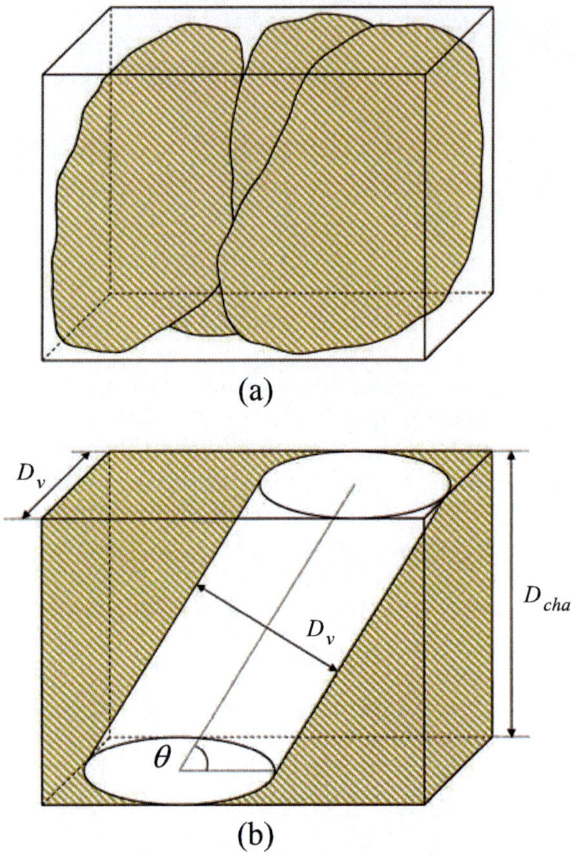

Fig. 1 **a** Elementary particulates; and **b** elementary particulate model

particulates, and the model shown in Fig. 1b is called an elementary particulate model. The volume of the elementary particulate model, V_e, and that of the pipe, V_p, are expressed as follows.

$$V_e = D_v \cdot \left(\frac{D_v}{\sin\theta} + \frac{D_{cha}}{\tan\theta} \right) \cdot D_{cha} \tag{4}$$

$$V_p = \pi \cdot \left(\frac{D_v}{2} \right)^2 \cdot \frac{D_{cha}}{\sin\theta} \tag{5}$$

where D_v is the diameter of the pipe, θ is the inclination angle of the pipe, and D_{cha} is the height of the elementary particulate model. In this paper, the height of the elementary particulate model, D_{cha}, is set to 10% passing particle size, $D_{s,10}$ (viz.

$D_{cha} = D_{s,10}$). It was found from previous research work (Akai [5]) that $D_{s,10}$ is one of the factors that influence the permeability coefficient.

The shape of voids in soil is complex because the shape and size of soil particles are random, and the structure of their assembly is irregular. Therefore, the diameter of the pipe, D_v, and the inclination angle of the pipe, θ, are regarded as random variables. Consequently, in this model, the probability density functions for D_v and θ are introduced. This model assumes that water enters in order from small pores and exits in order from large pores. Furthermore, the maximum diameter of pipes filled with water is defined as d. From the above, the void ratio, e, the degree of saturation, S_r, and the volumetric water content, W_v, are expressed through the following equations:

$$e = \int_{0}^{\infty} \int_{-\pi/2}^{\pi/2} \frac{V_p}{V_e - V_p} \cdot P_d(D_v) \cdot P_c(\theta) d\theta dD_v \tag{6}$$

$$S_r = \frac{1}{e} \int_{0}^{d} \int_{-\pi/2}^{\pi/2} \frac{V_p}{V_e - V_p} \cdot P_d(D_v) \cdot P_c(\theta) d\theta dD_v \tag{7}$$

$$W_v = \frac{1}{1+e} \int_{0}^{d} \int_{-\pi/2}^{\pi/2} \frac{V_p}{V_e - V_p} \cdot P_d(D_v) \cdot P_c(\theta) d\theta dD_v \tag{8}$$

where $P_d(D_v)$ is the probability density function for the diameter of the pipe, $P_c(\theta)$ is the probability density function for the inclination angle of the pipe, and d is the maximum diameter of pipes filled with water.

4 Numerical Model for the Molecular Diffusion Distance, $F(W_v)$

4.1 Outline of the Numerical Model for the Molecular Diffusion Distance, $F(W_v)$

The molecular diffusion distance, $F(W_v)$, is a parameter with a length dimension. The numerical model for the molecular diffusion distance, $F(W_v)$, is proposed using the model pipe shown in Fig. 2. The volume of pore water in the surface soil per unit volume, V_w, is modeled by the pipe with diameter D_p. In this model, the depth of the surface soil is assumed as 2 cm deep from the ground surface based on the experiment [3]. The maximum length of the model pipe is defined as L_{SAT}, and the length of the model pipe, L, is determined depending on the volumetric water content, W_v, in the surface soil. The difference between L_{SAT} and L is $F(W_v)$. $F(W_v)$ means

Fig. 2 Model pipe

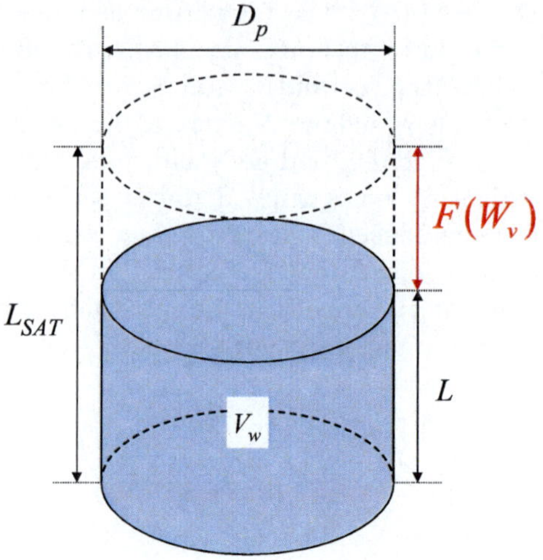

the transport length of water vapor from the water surface in the soil pore to the land surface. $F(W_v)$ is expressed as follows:

$$\begin{aligned} F(W_v) &= L_{SAT} - L \\ &= \left\{ \frac{e}{1+e}(1 - S_r) \right\} \bigg/ \left(\frac{D_p^2}{4}\pi \right) \end{aligned} \tag{9}$$

where e is the void ratio, S_r is the degree of saturation, and D_p is the diameter of the model pipe. The scheme of the model pipe to calculate the relationship between the volumetric water content and the molecular diffusion distance (W_v–$F(W_v)$ relation) is also shown in Fig. 2. To calculate the W_v–$F(W_v)$ relation by this scheme, it is necessary to determine the diameter of the model pipe, D_p. D_p is examined in the next section.

4.2 Examination of the Diameter of the Model Pipe

Table 2 shows the equations used to examine the diameter of the model pipe, D_p. In Case 1, the maximum diameter of pipes filled with water, d, is used as D_p, because it seems that d is the diameter of the pipe that mostly affects the molecular diffusion distance, $F(W_v)$. In Case 2, the mean value of the diameter of pipes filled with water is used as D_p. The probability density functions for the diameter of the pipe, $P_d(D_v)$, are corrected using the cumulative distribution functions for the diameter of the pipe, and D_p is calculated. In Case 3, the integration with respect to the product of the diameter of pipes filled with water and its probability is used as D_p. Unlike

Table 2 Equations used to examine D_p

Case 1	$D_p = d$	The maximum diameter of pipes filled with water
Case 2	$D_p = \int_0^d D_v \cdot \frac{P_d(D_v)}{\int_0^d P_d(D_v)dD_v} \mathrm{d}D_v$	The mean value of the diameter of pipes filled with water
Case 3	$D_p = \int_0^d D_v \cdot P_d(D_v)\,\mathrm{d}D_v$	The integration with respect to the product of the diameter of pipes filled with water and its probability

Case 2, $P_d(D_v)$ is not corrected using the cumulative distribution functions for the diameter of the pipe.

Using the equations shown in Table 2, the relationship between the volumetric water content and the molecular diffusion distance (W_v–$F(W_v)$ relation) was calculated. The diameter of the model pipe, D_p, was examined by comparing the calculation results with the experimental results by Sako et al. [3]. Figure 3 shows W_v–$F(W_v)$ relation obtained from a calculation using the equations shown in Table 2. The experimental results by Sako et al. [3] are also shown in Fig. 3. Toyoura sand (void ratio, $e = 0.785$) was used as the soil sample. In all cases, the calculation results are larger than the experimental results. It seems that the overestimation of the calculation results is caused by the calculation method of the volume of pore water in the surface soil. Therefore, it is necessary to improve the volume of pore water to be modeled. The improved model is proposed in the next section.

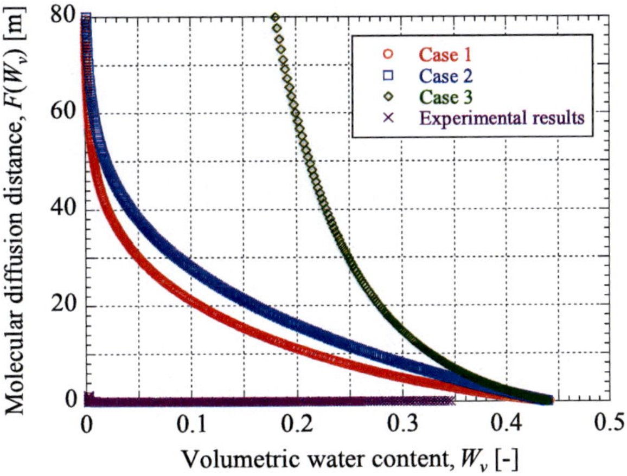

Fig. 3 Calculation results showing W_v and $F(W_v)$ relation

4.3 Improvement of the Numerical Model for the Molecular Diffusion Distance, $F(W_v)$

In this section, the improved model for the molecular diffusion distance, $F(W_v)$, is proposed. It seems that the molecular diffusion distance, $F(W_v)$, shown in Eq. (9) is calculated as the sum of multiple pores. It is the reason why the calculation results shown in Fig. 3 were overestimated. In the improved model, $F(W_v)$ is calculated as the mean value in pores filled with water. The volume ratio of the surface soil per unit volume and the elementary particulate model is used to improve the volume of pore water. In the improved model, the volume of pore water in the surface soil used in Eq. (9) is corrected using the volume of the elementary particulate model, V_e. Then, $F(W_v)$ is expressed as the following equation.

$$F(W_v) = \frac{\frac{e}{1+e}(1 - S_r)}{\frac{D_p^2}{4}\pi} \times \int_0^d \int_{-\pi/2}^{\pi/2} V_e \cdot P_d(D_v) \cdot P_c(\theta) d\theta dD_v \quad (10)$$

where d is the maximum diameter of pipes filled with water, V_e is the volume of the elementary particulate model, $P_d(D_v)$ is the probability density function for the diameter of the pipe, $P_c(\theta)$ is the probability density function for the inclination angle of the pipe, D_v is the diameter of the pipe, and θ is the inclination angle of the pipe. Figure 4 shows the relationship between the volumetric water content and the molecular diffusion distance (W_v–$F(W_v)$ relation) obtained from a calculation using the equations shown in Table 2. The experimental results by Sako et al. [3] are also shown in Fig. 4. It is found from Fig. 4 that the tendency of the calculation results for Case 3 is the closest to the experimental results. Therefore, it seems that the equation for Case 3 is the most appropriate as the diameter of the model pipe,

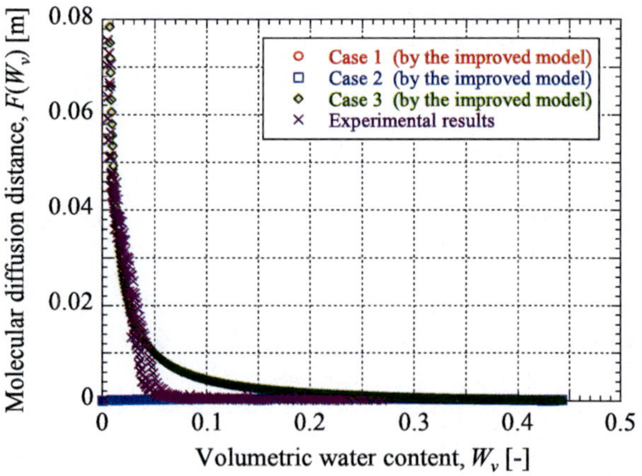

Fig. 4 Calculation results showing W_v and $F(W_v)$ relation (by the improved model)

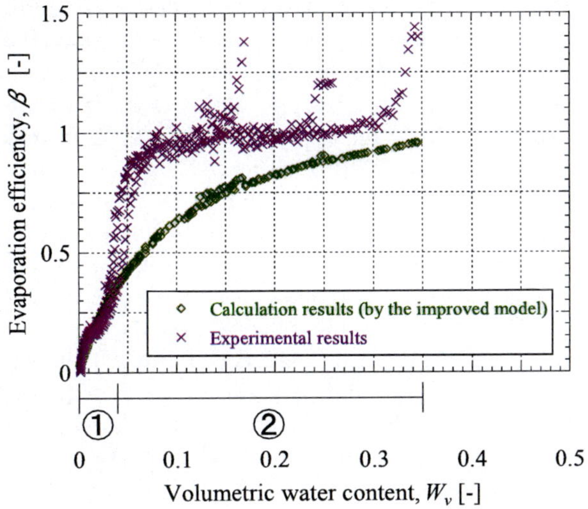

Fig. 5 Calculation results showing W_v and β relation

D_p. However, it is necessary to examine the validity of using the equation for Case 3.

5 Relationship Between the Volumetric Water Content and the Evaporation Efficiency

Figure 5 shows the relationship between the volumetric water content and the evaporation efficiency (W_v–β relation) obtained from the calculation using our proposed numerical model. From the results shown in Fig. 4, the equation for Case 3 was used as the diameter of the model pipe, D_p. The experimental results by Sako et al. [3] are also shown in Fig. 5. In the range of ① in Fig. 5, the tendency of the calculation results is almost the same as the experimental results. However, in the range of ② in Fig. 5, the tendency of the calculation results is different from the experimental results, because the calculation results of the molecular diffusion distance, $F(W_v)$, and the experimental results are different in terms of the volumetric water content, W_v, corresponding to the range of ② in Fig. 5. Therefore, it is necessary to improve the numerical model further.

6 Conclusions

The numerical model for the relationship between evaporation efficiency and volumetric water content was discussed in this paper. Firstly, a numerical model for the molecular diffusion distance, $F(W_v)$, derived from soil parameters such as grain size and void ratio was proposed as shown in Eq. (9). Three equations on the diameter of the model pipe, D_p, were also proposed. Comparing with the experimental data and the calculation data, the applicability of these equations for Eq. (9) was discussed. As a result, all calculation results were overestimated. Then, the improved equations on the molecular diffusion distance, $F(W_v)$, were proposed again, and the improved $F(W_v)$ using three equations on D_p was calculated. The equation for Case 3 resulted in a better result than the others. However, it is necessary to examine the validity of using the equation for Case 3. Finally, the relationship between the volumetric water content and the evaporation efficiency (W_v–β relation) was calculated using $F(W_v)$ derived from Eq. (10) and the D_p expressed by Case 3. In the range of low volumetric water content, the tendency of the calculation results showed good agreement with the experimental data. However, in the range of high volumetric water content, the tendency of the calculation results underestimated the experimental results. It is necessary to improve the numerical model further to be in good agreement with the experimental data.

References

1. Kondo, J., Saigusa, N., Sato, T.: A parameterization of evaporation from bare soil surfaces. J. Appl. Meteor. **29**, 385–389 (1990)
2. Sako, K., Kitamura, R.: A practical numerical model for seepage behavior of unsaturated soil. Soils Found. **46**(5), 595–604 (2006)
3. Sako, K., Moriiwa, M., Satomi, T.: Experimental consideration on evaporation efficiency β of unsaturated sandy soil surface. In: Proceedings of the 15th Asian Regional Conference on Soil Mechanics and Geotechnical Engineering, No.JPN-29 (2015).
4. Kondo, J.: Mizukankyo no kishogaku (Meteorology of water environment), Asakurashoten 197–198 (1994) (in Japanese)
5. Akai, K.: Events on Seepage, Soil Mechanics. In: Mogami, T. (Ed.) Gihodo, pp. 99–102 (1969) (in Japanese)

A Finite Difference Scheme for the Richards Equation Under Variable-Flux Boundary

Yasuhide Fukumoto, Fengnan Liu, and Xiaopeng Zhao

Abstract The Richards equation is a degenerate nonlinear partial differential equation which serves as a model for describing a flow of water through saturated/unsaturated porous medium under the action of gravity. This paper develops a numerical method, with a mathematical support, for the one-dimensional Richards equation. Implicit schemes based on a backward Euler format have been widely used, but have a difficulty in insuring the stability, because of the strong nonlinearity and degeneracy. A linearized semi-implicit finite difference scheme that is faster than the backward Euler implicit schemes is established, the stability of this scheme is proved by adding a small perturbation to the coefficient function, and an error estimate is made. It is found that there is a linear relationship between the discretization error in a certain norm and the perturbation strength.

Keywords Richards equation · Porous media · Finite difference scheme · Stability · Error estimate

1 Introduction

Prediction of movement of underground water is important in many branches of science and engineering, in particular, for natural disaster prevention. In case of a heavy rainfall, a large amount of water enters into the ground and, depending on its amount, a threat arises of river floods and/or landslides. They cause family tragedies, and, in the large, regional economic and human losses. Recently, in Asian regions including Japan, occurrence of river floods and landsides is the rule rather than the

Y. Fukumoto (✉)
Institute of Mathematics for Industry, Kyushu University, Fukuoka, Japan
e-mail: yasuhide@imi.kyushu-u.ac.jp

F. Liu
School of Mathematics and Physics Science, Dalian University of Technology, Panjin, China

X. Zhao
College of Sciences, Northeastern University, Shenyang, China

H. Hazarika et al. (eds.), *Advances in Sustainable Construction and Resource Management*, Lecture Notes in Civil Engineering 144,
https://doi.org/10.1007/978-981-16-0077-7_22

exception. Even in regions of mid latitudes, heavy rains or showers attack with a larger frequency. This trend is more so as years pass by. Typhoons and hurricanes also have a tendency to grow strong, accompanied by concentrated rains as well as strongly blowing wind, perhaps due to the temperature rise of the sea surface. In case a heavy rain lasts for a long time, landslides may take place in regions at the foot of a mountain and under a cliff. Prevention of such natural disasters associated with heavy rains is an issue of the highest demand for saving human lives. A real-time prediction for possible occurrence of such threats is strongly demanded, but the present status of numerical methods is poor. A rapid numerical calculation, backed by some sound mathematical basis, is yet to be feasible. Only a fully implicit difference scheme is mathematically supported, but the implicit numerical scheme heavily costs numerical resources.

Assuming that the unsaturated soil is a homogeneous isotropic porous medium (a fairly strong assumption), one can use an involved nonlinear degenerate diffusion equation to describe the fluid motion in an unsaturated zone, under the action of gravity force directed vertically downwards. This diffusion equation is called the Richards equation [1]. There are three alternative forms of the Richards equation, in the language of either the pressure head $h[L]$ or the moisture content $\theta[L]$ for the dependent variable. The constitutive relationship between $\theta = \theta(t, z)$ and $h = h(t, z)$ allows the conversion from one to another. The three forms, called the h-based, θ-based and mixed are, respectively,

- h-based

$$C(h)\frac{\partial h}{\partial t} - \nabla \cdot K(h)\nabla h - \frac{\partial K}{\partial z} = 0 \tag{1}$$

- θ-based

$$\frac{\partial \theta}{\partial t} - \nabla \cdot D(\theta)\nabla \theta - \frac{\partial K}{\partial z} = 0 \tag{2}$$

- Mixed

$$\frac{\partial \theta}{\partial t} - \nabla \cdot K(h)\nabla h - \frac{\partial K}{\partial z} = 0 \tag{3}$$

where the real-valued functions $C(h) = d\theta/dh$, $K(h)$ and $D(\theta) = K(h)/C(h)$, respectively, denote the specific moisture capacity function $[1/L]$, the unsaturated hydraulic conductivity $[L/T]$ and unsaturated diffusivity $\left[L^2/T\right]$. K(h) describes the ease with which water can move through pore spaces; it depends on the intrinsic permeability of the material, degree of saturation and density and viscosity of the fluid [2]. z denotes the vertical coordinate and is assumed to be positive in the upward direction, and the porous medium is assumed to be isotropic.

In this investigation, one-dimensional phenomena is considered. The first mathematical analysis for the Richards equation dates back to 1991. Gilding [3] supposed that the Richards equation can be expressed as a general advection–diffusion equation,

$$\theta_t = (a(\theta))_{zz} + (b(\theta))_z \tag{4}$$

where $(a(\theta))_{zz}$ and $(b(\theta))_z$ are the diffusion and advection terms, respectively. The author gave some pure mathematical analysis of the above equation.

Recently, Broadbridge et al. [4–6] constructed a class of solvable models by imposing some restrictions to the coefficient functions and provided exact solutions to them. Specifically, they posed

$$D(\theta) = \frac{a}{(b-\theta)^2} \text{ and } K(\theta) = \frac{\lambda}{2(b-\theta)} + \beta + \gamma(b-\theta) \tag{5}$$

where λ, β, γ, a and b are positive constants, and the boundary condition of constant flux

$$K(\theta) - D(\theta)\frac{\partial \theta}{\partial z} = r(\text{constant}) \text{ at } z = 0 \tag{6}$$

The other research direction on the Richards equation is to seek approximate solutions by numerical methods. Since the Richards equation is not easy to guarantee the regularity of its solution, there are few numerical theories regarding stability and error estimates. For our purpose of numerical analysis, one employs the empirical form of the coefficient function reported, based on the measurement, by Haverkamp et al. [7]:

$$\theta(h) = \frac{\alpha(\theta_s - \theta_r)}{\alpha + |h|^\beta} + \theta_r \text{ and } K(h) = K_s\frac{A}{A + |h|^\gamma}, \tag{7}$$

where $\alpha = 1.611 \times 10^6$, $\theta_s = 0.287$, $\theta_r = 0.075$, $\beta = 3.96$, $K_s = 0.00944$, $A = 1.175 \times 10^6$ and $\gamma = 4.74$. Haverkamp et al. [7] posed the following initial condition $h(z, 0) = -61.5\,\text{cm}$ and the boundary conditions $h(40\,\text{cm}, t) = h_{\text{top}} = -20.7\,\text{cm}$ and $h(0, t) = h_{\text{bottom}} = -61.5\,\text{cm}$ (the vertical dimension is assumed positive upward), and studied the numerical approximation for the Richards equation by using the simple backward Euler time-marching algorithm. Other than this, there are some papers which studied the numerical solutions for the Richards equation through adaptive time stepping [8, 9], iterative methods [8, 10, 11], for spatial discretization, finite element or finite volume methods [12–14] and finite difference methods [15–18].

The Richards equation is a strongly nonlinear degenerate PDE, and hence one should be cautious about the stability of its numerical solution. However, there are few papers dealing with the proof of stability. Radu et al. [19], assuming some regularity properties of the classical solution and imposing some restrictions on the coefficient functions, analyzed the stability of a fully implicit finite element scheme by applying the Kirchhoff transformation and integral in time. This type of analysis is applicable only to finite element schemes and others whose analysis rests on a continuous inner product. Finite difference schemes require a different treatment. Moreover, the order of the error can be improved.

In this paper, a linearized finite difference scheme for the Richards equation in the bounded domain [0, L], subject to the boundary conditions

$$h(0,t) = h_{\text{bottom}} = h_0 < 0 \text{ and } h(L,t) = h_{\text{top}} = h_{\text{L}} < 0 \tag{8}$$

and the initial condition

$$h(z,0) = h_0 < 0 \tag{9}$$

is established. There are three assumptions, without losing the generality, on the nonlinear functions $K(\text{s})$ and $C(\text{s})$ for s $\in R$ need to be stated:

(A1) $\left[K'(s)\right]^2 \le c_0 K(s),\ K'(s) \le c_1,\ C'(s) \le c_0',\ K''(s) \le c_1'$;
(A2) There exists $s_1,\ s_2,\ p \in R^+$ such that

$$K(s_1) - K(s_2) = (s_1 - s_2)K(s_1,s_2) \text{ and } \frac{[K(s_1,s_2)]^2}{K(s_2)} \le c_3\left(|s_2|^p + 1\right)$$

(A3) There exists a lower bound and an upper bound for $K(\text{s})$.

It is worth pointing out that Haverkamp et al.'s data satisfy (A1)–(A3). By making a rigorous analysis, one gives a mathematical proof to the stability of the present linearized finite difference scheme and pursues the convergence order of error. In order to overcome the degeneracy of $C(h) = \mathrm{d}\theta/\mathrm{d}h$, a positive perturbation ε is added to it, leading to the modified h-based Richards equation of the form

$$C_\varepsilon(h)\frac{\partial h}{\partial t} - \left[K(h)h_z\right]_z - K'(h)h_z = 0, \tag{10}$$

where $C_\varepsilon(s) = C(s) + \varepsilon$. With this modification, the degeneracy of the Richards equation is avoided, by maintaining the assumptions (A1)–(A3).

Augmenting the Richards equation with the perturbation, one has to work out how it influences the computational error. By making the error estimate for the perturbed equation, in a way parallel with the original h-based Richards equation, the convergence order $\mathrm{O}\left(\varepsilon^{3/4} + \Delta t^{3/4} + \Delta z^2/\Delta t^{1/4}\right)$ in the discrete L^∞-norm is obtained, showing that there is a linear relationship between the discretization error and $\varepsilon^{3/4}$. For the numerical approximation of the Richards equation, the present method is different from that of Radu et al. [19] and turns out to achieve its improvement. It should be emphasized that this scheme is computationally faster than other implicit schemes (including that of Raduet al.) and hence may be suitable for the longtime behavior of the infiltration process in unsaturated porous media.

2 Finite Difference Approximation

The purpose of this paper is to construct a linearized finite difference scheme, to prove its stability and to give the convergence order of error.

Let $\Delta z = L/M$ be the uniform step size in the spatial direction for a positive integer M, and

$$\overline{\Omega_h} = \{x_i = i\Delta z | i = 0, 1, \ldots, M\}.$$

Suppose that $\Delta t = T/N$ denotes the uniform step size in the temporal direction for any positive integer N. For a function w on $\overline{\Omega_h}$, denote $w_i = w(x_i)$ and define the difference operators

$$\nabla_h w_i = \frac{w_{i+1} - w_{i-1}}{2\Delta z}, \quad \nabla_h^+ w_i = \frac{w_{i+1} - w_i}{\Delta z}, \quad \nabla_h^- w_i = \frac{w_i - w_{i-1}}{\Delta z},$$

and

$$\Delta_h w_i = \frac{w_{i-1} - 2w_i + w_{i+1}}{\Delta z^2}, \quad 1 \le i \le M - 1.$$

The discrete L^2 inner product satisfies

$$\langle w, v \rangle = \Delta z \left\{ \frac{1}{2}(w_0 v_0 + w_M v_M) + \sum_{i=1}^{M-1} w_i v_i \right\}, \quad w, v \in \overline{\Omega},$$

and the corresponding discrete L^2-norm

$$\|v\|_h = \langle v, v \rangle^{1/2}.$$

The discrete H^1 seminorm $|\cdot|_{1,\mathrm{h}}$ and discrete maximum norm $|\cdot|_{\infty,\mathrm{h}}$ are defined, respectively, by

$$|v|_{1,h} = \left[\Delta z \sum_{i=1}^{M-1} (\nabla v_i)^2 \right]^{1/2}, \quad v \in \overline{\Omega},$$

$$|v|_{\infty,h} = \sup_i |v_i|, \; i = 1, 2, \ldots, M, \; v \in \overline{\Omega},$$

Let $H_i^n = \mathrm{h}(z_i, t_n)$ for $z_i = i\Delta z, i = 0, 1, \ldots, M, t_n = n\Delta t, \; n = 0, 1, \ldots, N$. Define $\delta_{\mathrm{t}} H^n = \left(h^{n+1} - h^n\right)/\Delta t$. The approximate solution $H^n \in \overline{\Omega_{\mathrm{h}}}$ for the problem (1)–(3) is a solution of the following scheme:

$$C_\varepsilon\left(H_i^n\right)\delta_t H_i^n - \nabla_h\left(K\left(H_i^n\right)\nabla_h H_i^{n+1}\right) = K'\left(H_i^n\right)\nabla_h H_i^{n+1}. \tag{11}$$

Here $C_\varepsilon(s) = C(s) + \varepsilon$, with a positive constant ε, and one adopts the following boundary conditions

$$H_0^n = h_0, \quad H_M^n = h_L, \tag{12}$$

and the initial condition

$$H_i^0 = \mathrm{h}_0, \quad 1 \le i \le M. \tag{13}$$

2.1 Existence and Uniqueness

The existence and uniqueness of solutions for the scheme (11)–(13) will be examined in the following.

Theorem 2.1 Suppose that (A1)–(A3) hold and that $K(s)$ and $C(s)$ are sufficiently smooth. Then, there exists a unique solution H^n for the scheme (11)–(13).

Proof Mathematical induction is used to prove the existence of solution for the scheme (11)–(13). To begin with, suppose that $H^0, \ldots, H^n$ exist. Then, for a fixed n and $1 \le i \le M$, (11) is rewritten as

$$C_\varepsilon(H_i^n)(H_i^{n+1} - H_i^n) - \Delta t[\nabla_h(K(H_i^n)\nabla_h H_i^{n+1}) + K'(H_i^n)\nabla_h H_i^{n+1}] = 0. \tag{14}$$

Let g, which is continuous, be a function defined by

$$g(V) = C_\varepsilon(H^n)(V - H^n) - \Delta t[\nabla_h(K(H^n)\nabla_h V) + K'(H^n)\nabla_h V]. \tag{15}$$

After taking an inner product of $g(V)$ with V, a simple calculation shows that

$$\begin{aligned}
\langle g(V), V\rangle_h \ge & \sum_{i=1}^{M-1}\Big[\Big(\frac{\varepsilon}{2} - \frac{c_0}{2}\Delta t\Big)V_i^2 - \frac{\varepsilon}{2}|H_i^n|^2\Big] \\
& + \frac{1}{2}\Delta t\Big[\Big(\frac{\varepsilon}{2} - \frac{c_0}{2}\Delta t\Big)V_0^2 - \frac{\varepsilon}{2}|H_0^n|^2\Big] \\
& + \frac{1}{2}\Delta t\Big[\Big(\frac{\varepsilon}{2} - \frac{c_0}{2}\Delta t\Big)V_M^2 - \frac{\varepsilon}{2}|H_M^n|^2\Big].
\end{aligned}$$

Consequently, for $\Delta t < \varepsilon/c_0$ and $\|V\|_h^2 = \varepsilon\|H^n\|_h^2/(\varepsilon - c_0\Delta t) + 1$, one obtains $\langle g(V), V\rangle_\mathrm{h} \ge 0$. Hence, from the Browder fixed point theorem, the existence of V^* for $g(V^*) = 0$ can be deduced. The fixed point $H^{n+1} = V^*$ satisfies the scheme (11)–(13). As regards the uniqueness, the proof is made along the same line with the error estimate. The latter is described at some length in Sect. 2.3, and one may leave out the former proof.

2.2 Stability

The stability of solutions for the scheme (11)–(13) is stated and is proved in the following.

Theorem 2.2 Suppose that (A1)–(A3) hold and that $K(s)$ and $C(s)$ are sufficiently smooth. Then,

$$\left|H^n\right|_{\infty,h} + \left\|H^n\right\|_{1,h} \le c_1,$$

where c_1 is a positive constant independent of Δt and Δz.

Proof The boundedness of the L^2-discrete norm will be considered firstly. If $\left\|H^{n+1}\right\|_{1,\mathrm{h}} \le \|H^n\|_{1,\mathrm{h}}$, one obtains the result straightforwardly. Alternatively, taking the inner product of (11) with H^{n+1}, it yields

$$\left\langle C_\varepsilon(H^n)\frac{H^{n+1}-H^n}{\Delta t}, H^{n+1}\right\rangle_h + \left\langle K(H^n)\nabla_h H^{n+1}, \nabla_h H^{n+1}\right\rangle_h \\ - \left\langle K'(H^n)\nabla_h H^{n+1}, H^{n+1}\right\rangle_h = 0, \quad 0 \le n \le N-1. \tag{16}$$

After some manipulation, one shows that

$$\left(\frac{\varepsilon}{2\Delta t} - \frac{c_0}{2}\right)\left\|H^{n+1}\right\|_h^2 + \frac{1}{2}\left\langle K(H^n)\nabla_h H^{n+1}, \nabla_h H^{n+1}\right\rangle_h \le \frac{\varepsilon}{2\Delta t}\left\|H^n\right\|_h^2, \tag{17}$$

resulting in

$$\left\|H^{n+1}\right\|_h^2 \le \frac{\varepsilon}{\varepsilon - 2c_0\Delta t}\left\|H^n\right\|_h^2, \tag{18}$$

and

$$\left\langle K(H^n)\nabla_\mathrm{h} H^{n+1}, \nabla_\mathrm{h} H^{n+1}\right\rangle_\mathrm{h} \le \frac{\varepsilon}{\Delta t}\left\|H^n\right\|_\mathrm{h}^2. \tag{19}$$

After repeated application, one obtains, for $\Delta t < \varepsilon/2c_0$,

$$\left\|H^{n+1}\right\|_h^2 \le \left(\frac{\varepsilon}{\varepsilon - 2c_0\Delta t}\right)^{n+1}\left\|H^0\right\|_h^2. \tag{20}$$

For the H^1 discrete seminorm of solutions for (11)–(13), assume that, at the outset, that $\left|H^j\right|_{1,\mathrm{h}} (j = 1, 2, \ldots, n)$ is bounded. Then, $K(H^n) \ge \bar{c}_0$. Let $\left|H^{n+1}\right|_{1,\mathrm{h}} > |H^n|_{1,\mathrm{h}}$, or alternatively, the boundedness of $\left|H^{n+1}\right|_{1,\mathrm{h}}$ can be obtained straightforwardly. Taking the inner product with $H^{n+1} - H^n$, one obtains

$$\left\langle C_\varepsilon(H^n)\frac{H^{n+1}-H^n}{\Delta t}, H^{n+1}-H^n\right\rangle_h + \left\langle K(H^n)\nabla_h H^{n+1}, \nabla_h(H^{n+1}-H^n)\right\rangle_h$$
$$-\left\langle K'(H^n)\nabla_h H^{n+1}, H^{n+1}-H^n\right\rangle_h = 0, \quad 0 \le n \le N-1. \tag{21}$$

After some calculation, one finds that

$$\frac{\varepsilon}{2\Delta t}\left\|H^{n+1}-H^n\right\|_h^2 + \left(\frac{\bar{c}_0}{2} - \frac{c_1^2}{2\varepsilon}\Delta t\right)\left\|\nabla_h H^{n+1}\right\|_h^2 \le \frac{\bar{c}_0}{2}\left\|\nabla_h H^n\right\|_h^2, \tag{22}$$

which implies that

$$\left\|\nabla_h H^{n+1}\right\|_h^2 \le \frac{\varepsilon\bar{c}_0}{\varepsilon\bar{c}_0 - c_1^2\Delta t}\left\|\nabla_h H^n\right\|_h^2, \tag{23}$$

for $\Delta t < \varepsilon\bar{c}_0/c_1^2$. After repeated application, one is led to

$$\left\|\nabla_h H^{n+1}\right\|_h^2 \le \left(\frac{\varepsilon\bar{c}_0}{\varepsilon\bar{c}_0 - c_1^2\Delta t}\right)^{n+1}\left\|\nabla_h H^0\right\|_h^2. \tag{24}$$

Note that (24) means the boundedness assumption of $\left|H^j\right|_{1,h}$ $(j = 1, 2, \ldots, n)$ is reasonable. Combining (20) and (24), by using Sobolev's embedding theorem, the proof is complete.

2.3 Error Estimates

One makes the error estimates for the scheme (11)–(13) in this subsection.

Theorem 2.3 Suppose that (A1)–(A3) hold and that $K(s)$ and $C(s)$ are sufficiently smooth. Let $h(x, t)$ be the solution of the problem (1), (8) and (9), with a sufficient regularity. Then, for a sufficiently small Δt, the solution of the difference scheme (11)–(13) converges to h (x, t), with the rate of convergence $O\left(\varepsilon^{3/4} + \Delta t^{3/4} + \Delta z^2/\Delta t^{1/4}\right)$ in the discrete L^∞-norm.

Proof Define the net function: $h_i^n = h(x_i, t^n), 1 \le i \le M, 0 \le n \le N$. By using the Taylor expansion, one obtains for $1 \le i \le M, 0 \le n \le N$,

$$\begin{cases} C\left(h_i^n\right)\delta_t h_i^n - \nabla_h K\left(h_i^n\right)\nabla_h h_i^{n+1} - K'\left(h_i^n\right)\nabla_h h_i^{n+1} = r_i^n, \\ h_i^0 = h_0, \\ h_0^n = h_0,\ h_M^n = h_L, \end{cases} \tag{25}$$

where r_i^n is the truncation error of the difference scheme (11)–(13). Hence, there exists a positive constant c_2 satisfying

$$\max_{i,n}|r_i^n| \le c_2(\Delta t + \Delta z^2),\ 1 \le i \le M,\ 0 \le n \le N-1.$$

Define the error $e_i^n = h_i^n - H_i^n,\ 1 \le i \le M,\ 0 \le n \le N$. Subtracting (11) from (16), one arrives at

$$\begin{aligned}[C(h^n)\delta_t h^n - C_\varepsilon(H^n)\delta_t H^n] - \nabla_h[K(h^n)\nabla_h h^{n+1} - K(H^n)\nabla_h H^{n+1}] \\ -[K'(h^n)\nabla_h h^{n+1} - K'(H^n)\nabla_h H^{n+1}] = r^n,\end{aligned} \tag{26}$$

and

$$e_i^0 = 0. \tag{27}$$

The inner product of (17) with e^{n+1} it yields

$$\|e^n\|_h = O(\varepsilon + \Delta t + \Delta z^2). \tag{28}$$

Next, taking the inner product of (17) with $e^{n+1} - e^n$, one has

$$\|\nabla_h e^n\|_h = O\left(\frac{\varepsilon}{\Delta t^{1/2}} + \Delta t^{1/4} + \frac{\Delta z^2}{\Delta t^{1/2}}\right). \tag{29}$$

Combining (19) and (20) and using Sobolev's embedding inequality, one arrives at

$$\|e^n\|_{\infty,h} \le \|e^n\|_h^{1/2}\|\nabla_h e^n\|_h^{1/2} = O\left(\varepsilon^{3/4} + \Delta t^{3/4} + \frac{\Delta z^2}{\Delta t^{1/4}}\right). \tag{30}$$

3 Numerical Results

The data are pocked out from the report of Haverkamp et al. [7]:

$$\theta(h) = \frac{\alpha(\theta_s - \theta_r)}{\alpha + |h|^\beta} + \theta_r, \quad K(h) = K_s\frac{A}{A + |h|^\gamma}. \tag{31}$$

Their derivatives are

$$C(h) = \theta'(h) = \frac{\alpha(\theta_s - \theta_r)\beta|h|^{\beta-1}}{(\alpha + |h|^\beta)^2},\ K'(h) = \frac{K_s A\gamma|h|^{\gamma-1}}{(A + |h|^\gamma)^2}. \tag{32}$$

Moreover, $\alpha = 1.611 \times 10^6$, $K_s = 0.00944$, $A = 1.175 \times 10^6$ and $\gamma = 4.74$. These data were obtained from the measurement and were used to illustrate an infiltration process in soil, with initial condition $h(z, 0) = -61.5$ cm and the boundary conditions $h(40\,\text{cm}, t) = h_{\text{top}} = -20.7$ cm and $h(0, t) = h_{\text{bottom}} = -61.5$ cm.

3.1 Error Analysis

In order to test the error estimate, one considers the following nonhomogeneous equation:

$$C(h)\frac{\partial h}{\partial t} - \nabla \cdot K(h)\nabla h - \frac{\partial K}{\partial z} = g(z, t), \tag{33}$$

with

$$h(x, 0) = -1.02z - 20.7, \tag{34}$$

and

$$h(0, t) = h_{\text{bottom}} = -61.5\,\text{cm}, \quad h(L, t) = h_{\text{top}} = -20.7\,\text{cm}. \tag{35}$$

In order for (33) to have an exact solution $h = -1.02z - 20.7 + t(z - 40)/4T$, the right-hand side of (33) has to take

$$\begin{aligned} g(z,t) = {} & \frac{5808800331328389z(z-40)\left(\frac{51z}{50} - \frac{tz(z-40)}{4T} + \frac{207}{10}\right)^{74/25}}{171798691184T\left[\left(\frac{51z}{50} - \frac{tz(z-40)}{4T} + \frac{207}{10}\right)^{99/25} + 1611000\right]} \\ & - \frac{5546t}{\left(\frac{51z}{50} - \frac{tz(z-40)}{4T} + \frac{207}{10}\right)^{237/50} + 1175000} \\ & - \frac{3613000706430075\left(\frac{t(z-20)}{2T} - \frac{51}{50}\right)\left(\frac{51z}{50} - \frac{tz(z-40)}{4T} + \frac{207}{10}\right)^{187/50}}{68719476736\left[\left(\frac{51z}{50} - \frac{tz(z-40)}{4T} + \frac{207}{10}\right)^{237/50} + 1175000\right]^2}. \end{aligned}$$

Denote h_i to be the exact solution of (1), (8) and (9), and H_i to be the numerical solution of the scheme (11)–(13). One makes an analysis of the error $e_i = |h_i - H_i|$ and particularly of the relation between the error and the perturbation.

Let $\Delta t = 0.0018$, $\Delta z = 0.04$. Figure 1 shows the variation trend of the error with different perturbations at 360 s. It is clear that as the perturbation is smaller,

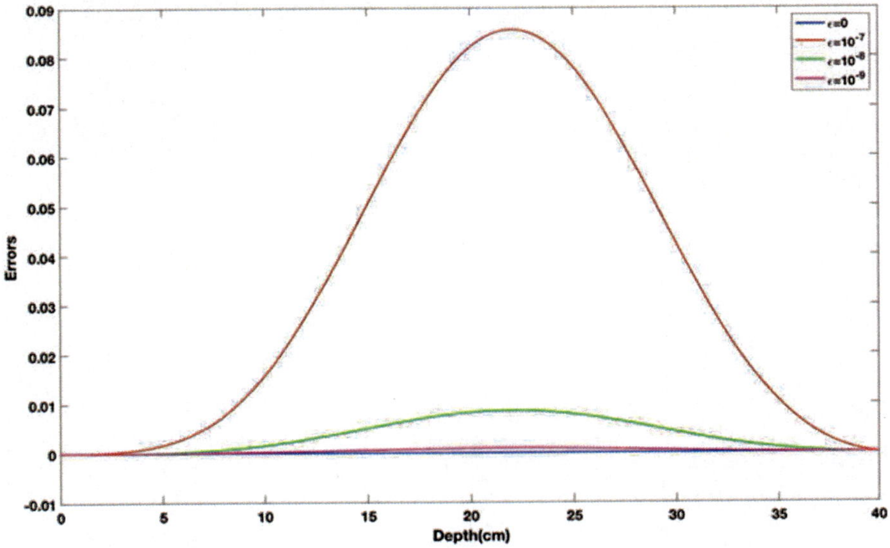

Fig. 1 Variation trend of errors with different perturbations at 360 s

the error becomes smaller. The best choice is $\varepsilon = 0$. This is because the data of Haverkamp et al. do not suffer from the degeneracy of the equation.

Figure 2 indicates that the relation between the error in the discrete L^∞-norm and the perturbation is linear, i.e., $|e^n|_{\infty,\mathrm{h}} = O(\varepsilon)$, which turns out to be better than the result of Theorem 2.3.

To consolidate this linear behavior quantitatively, Table 1 lists the relationship between the error in the discrete L^∞-norm and the perturbation. It clearly exhibits $|e^n|_{\infty,\mathrm{h}} \approx O(\varepsilon)$, which seems better than the result of Theorem 2.3. When the perturbation is switched off ($\varepsilon = 0$), the error attains the minimum value 4.1027×10^{-5}. By taking $\varepsilon = 0$, one can look into the relation between error estimate and mesh spacing. As read off from Table 2, supposing that $\Delta t^{3/4}/\left(\Delta z^2/\Delta t^{1/4}\right)$ is a fixed value, one finds that both the convergence order of Δt and Δz^2 are 1, which is also better than the result of Theorem 2.3.

3.2 *Pressure Head and Moisture Content*

To illustrate the variation trend of the pressure head $h(t, z)$ in time and depth, numerical calculation is performed based on (10), by leaving out the perturbation, and the result is shown in Fig. 3.

The moisture content is obtainable from the relation between the moisture content and the pressure head

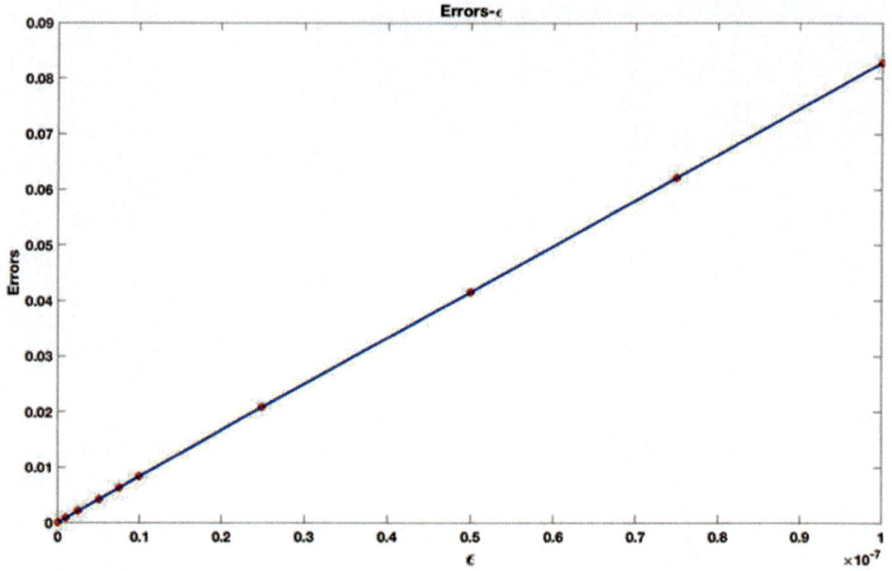

Fig. 2 Errors in the discrete L^{∞}-norm and the perturbation

Table 1 Relation between errors in discrete L^{∞}-norm and the perturbation

Perturbation	L^{∞}-norm of error	Convergence order
1.00×10^{-7}	8.5500×10^{-2}	–
0.75×10^{-7}	6.4200×10^{-2}	0.9988
0.50×10^{-7}	4.2800×10^{-2}	1.0000
0.25×10^{-7}	2.1400×10^{-2}	1.0000
1.00×10^{-8}	8.6000×10^{-3}	0.9953
0.75×10^{-8}	6.4000×10^{-3}	1.0078
0.50×10^{-8}	4.3000×10^{-3}	0.9922
0.25×10^{-8}	2.2000×10^{-3}	0.9773
1.00×10^{-9}	8.7052×10^{-4}	1.0109
0	4.1027×10^{-8}	–

Table 2 Relation between error estimate and mesh spacing

Δt	Δz	$\lvert e \rvert_{\infty,h}$	Convergence order of $\Delta t + \Delta z^2$
0.04000	0.2500	9.2224×10^{-4}	–
0.02000	0.1760	4.5885×10^{-4}	1.0049
0.01000	0.1250	2.2876×10^{-4}	1.0029
0.00500	0.0880	1.1427×10^{-4}	1.0010
0.00250	0.0625	5.7105×10^{-5}	1.0005
0.00125	0.0440	2.8543×10^{-5}	1.0010

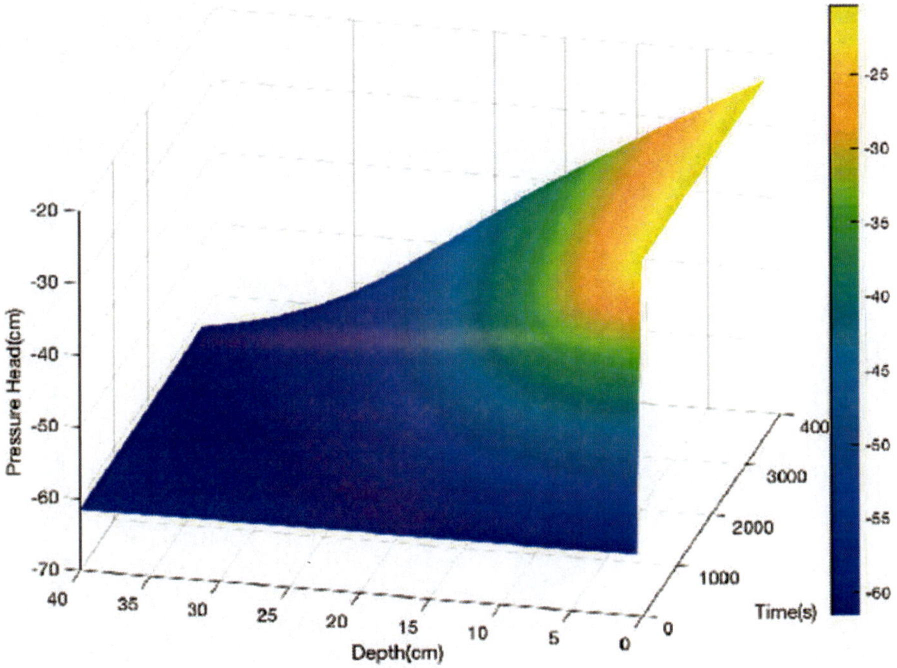

Fig. 3 Variation trend of pressure head in depth and time

$$\theta(h) = \frac{\alpha(\theta_s - \theta_r)}{\alpha + |h|^{\beta}} + \theta_r = \frac{1.616 \times 0.212 \times 10^6}{1.611 \times 10^6 + |h|^{3.96}} + 0.075 \tag{36}$$

Thus, the variation trend of moisture content in depth and time is described in Fig. 4.

4 Conclusion

The Richards equation is a degenerate nonlinear PDE that models a flow through saturated/unsaturated porous media. Research on its numerical methods was conducted in many fields. However, it is difficult to obtain stability of a numerical scheme because of the strong nonlinearity and degeneracy. This paper has proposed a linearized semi-implicit finite difference scheme and analyzed its stability by adding a small positive perturbation ε to the coefficient function of the Richards equation. It is worth pointing out that our linearized semi-infinite finite difference scheme is much more efficient that the existing backward Euler implicit schemes. By the numerical experiment, it is found that there is a linear relationship between the discretization error and ε better than our theoretical estimate. This observation suggests that the proof in Sect. 3.1 is

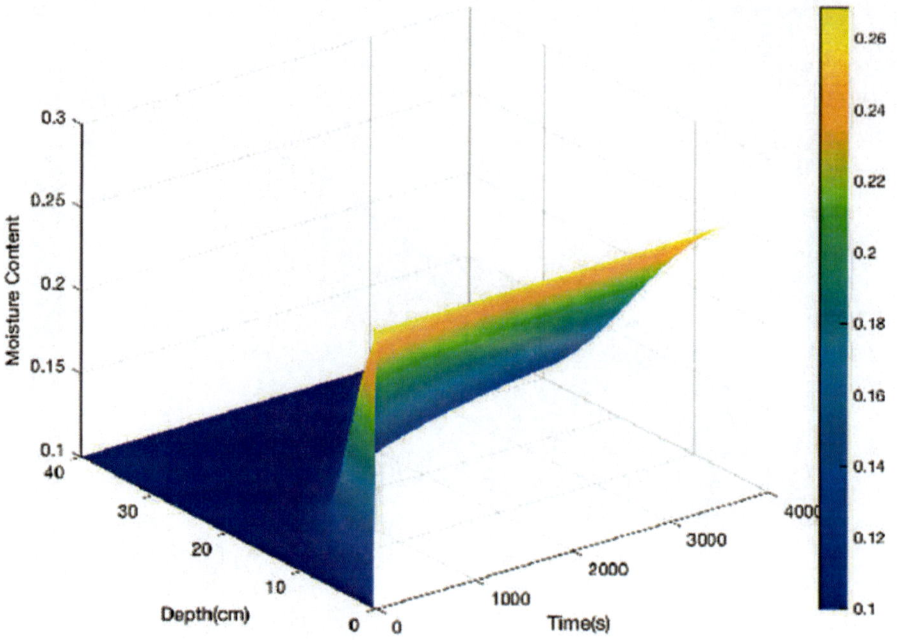

Fig. 4 Variation trend of moisture content in depth and time

not strict enough and has a room to be improved. A more strict error analysis and developing a two-dimensional numerical scheme, etc., are left for future tasks.

Acknowledgements YF was supported by a Grant-in-Aid for Scientific Research from the Japan Society for the Promotion of Science (grant no. 19K03672), FL was supported by Fundamental Research Funds for Central Universities (no. DUT19RC(4)038), and XZ was supported by China Postdoctoral Science Foundation (no. 2015M581689).

References

1. Fengnan, L., Fukumoto, Y., Zhao, X.: A linearized finite difference scheme for the Richards equation under variable-flux boundary conditions. J. Sci. Comput. **83**, 16 (2020)
2. Berardi, M., Vurro, M.: The numerical solution of the Richards equation by means of method of lines and ensemble Kalman filter. Math. Comput. Simul. **125**, 38–47 (2016)
3. Gilding, B.: Qualitative mathematical analysis of the Richards equation. Transp. Porous Med. **6**(5–6), 651–666 (1991)
4. Broadbridge, P., Edwards, M.P., Kearton, J.E.: Closed-form solutions for unsaturated flow under variable flux boundary conditions. Adv. Water Resour. **19**, 207–213 (1996)
5. Broadbridge, P., Daly, E., Goard, J.: Exact solutions of the Richard equation with nonlinear plant-root extraction. Water Resour. Res. **53**, 9679–9691 (2017)
6. Broadbridge, P., Triadis, F., Hill, J.M.: Infiltration from supply at constant water content: an integrable model. J. Eng. Math. **64**, 193–206 (2009)

7. Haverkamp, R., Vauclin, M.: A note on estimating finite difference interblock hydraulic conductivity values for transient unsaturated flow problems. Water Resour. Res. **15**, 181–187 (1979)
8. Kavetski, D., Binning, P., Sloan, S.W.: Adaptive time stepping and error control in a mass conservative numerical solution of the mixed form of Richards equation. Adv. Water Res. **24**, 595–605 (2001)
9. Williams, G.A., Miller, C.T.: An evaluation of temporally adaptive transformation approaches for solving the Richards equation. Adv. Water Res. **22**, 831–840 (1999)
10. Knabner, P.: Finite element simulation of saturated-unsaturated flow through porous media. In: Deuflhard, P., Engquist, B. (eds.) Large Scale Scientific Computing. Progress in Scientific Computing, vol. 7, pp. 83–93. Birkhäuser, Boston (1987)
11. Celia, M.A., Bouloutas, E.T.: A general mass-conservative numerical solution for the unsaturated flow equation. Water Resour. Res. **26**, 1483–1496 (1990)
12. Berganaschi, L., Putti, M.: Mixed finite elements and Newton-type linearizations for the solution of the Richards equation. Int. J. Numer. Methods Eng. **45**, 1025–1046 (1999)
13. Baca, R.G., Chung, J.N., Mulla, D.J.: Mixed transform finite element method for solving the non-linear equation for flow in variably saturated porous media. Int. J. Numer. Methods Fluids **24**, 441–455 (1997)
14. Eymard, R., Gutnic, M., Hillhorst, D.: The finite volume method for Richards equation. Comput. Geosci. **3**, 259–294 (1999)
15. Clement, T., Wise, W.R., Molz, F.J.: A physically based, two-dimensional, finite-difference algorithm for modeling variably saturated flow. J. Hydrol. **161**, 71–90 (1994)
16. Kumar, C.: A numerical simulation model for one-dimensional infiltration. ISH J. Hydraul. Eng. **4**, 5–15 (1998)
17. Szymkiewicz, A., Helmig, R.: Comparison of conductivity averaging methods for one-dimensional unsaturated flow in layered soils. Adv. Water Resour. **34**, 112–125 (2011)
18. Romano, N., Brunone, B., Santini, A.: Numerical analysis of one-dimensional unsaturated flow in layered soils. Adv. Water Resour. **21**, 315–324 (1998)
19. Radu, F., Pop, I.S., Knabner, P.: Order of convergence estimates for an Euler implicit, mixed finite element discretization of the Richards equation. SIAM J. Numer. Anal. **42**, 1452–1478 (2004)

Centrifuge Modelling Influence of Various Integration Schemes of Retaining Walls on Seismic Behaviour Using Tilting Table Test

Kazuya Sano, Kazuya Itoh, Tsuyoshi Tanaka, Naoaki Suemasa, and Takeharu Konami

Abstract More than 4000 retaining walls were damaged during the Kumamoto earthquake in 2016. Among them, approximately 60% were masonry retaining walls. There are mainly two types of masonry retaining walls in Japan. One is an air masonry retaining wall, which is unfilled with mortar, and the other is an integrated stone masonry retaining wall, which is filled with mortar. In addition, there is another type of retaining wall which is filled with mortar only on its surface so that the entire masonry retaining walls are not integrated and many of this type of retaining walls were damaged during the Kumamoto earthquake. The purpose of this study is examining how the integration of retaining walls affects their earthquake resistance. In this report, in order to confirm it from the deformation behaviour during an earthquake, we modelled the different types of retaining walls and conducted centrifuge tilting experiments. A series of centrifuge model tests were carried out under 20G, where a model retaining wall ground was tilted using a tilting table, a horizontal seismic force during an earthquake was reproduced, and the deformation behaviours of different types of model retaining walls during an earthquake were confirmed. Three types of model retaining walls made of plaster were reproduced including an air masonry retaining wall which is not integrated, a simple reinforced air masonry retaining wall which is integrated only on its surface, and a gravity retaining wall.

K. Sano (✉) · K. Itoh · T. Tanaka · N. Suemasa
Tokyo City University, Tokyo, Japan
e-mail: g2081613@tcu.ac.jp

K. Itoh
e-mail: itok@tcu.ac.jp

T. Tanaka
e-mail: ttanaka@tcu.ac.jp

N. Suemasa
e-mail: nsuemasa@tcu.ac.jp

T. Konami
OKASANLIVIC.CO., LTD., Tokyo, Japan
e-mail: konami@okasanlivic.co.jp

H. Hazarika et al. (eds.), *Advances in Sustainable Construction and Resource Management*, Lecture Notes in Civil Engineering 144,
https://doi.org/10.1007/978-981-16-0077-7_23

Keywords Masonry retaining walls · Centrifuge modelling · Image analysis

1 Introduction

In recent years, a lot of housing retaining walls have been damaged or even collapsed by several earthquakes in Japan. Particularly, more than 4000 retaining walls were damaged during the Kumamoto earthquake in 2016. Among them, approximately 60% were masonry retaining walls. There are mainly two types of masonry retaining walls in Japan. One is an air masonry retaining wall which is unfilled with mortar and is not allowed by the current Japanese act on regulation of residential land development. The other is an integrated stone masonry retaining wall, which is filled with mortar and is allowed by the current Japanese act on regulation of residential land development. In addition, there is another type of retaining wall which is filled with mortar only on its surface so that the entire masonry retaining walls are not integrated and many of this type of retaining walls were damaged during the Kumamoto earthquake [1]. The purpose of this study is to examine how the integration of retaining walls affects the earthquake resistance. In this report, in order to confirm it from the deformation behaviour during an earthquake, modelling the three different types of retaining walls: an air masonry retaining wall, a reinforced air masonry retaining wall and a gravity retaining wall which is allowed by the current Japanese construction law, we conducted centrifuge tilting experiments under 20G.

2 Outline of Experimental Equipment

2.1 Small Centrifugal Model Experimental Device

In this experiment, a small centrifugal model experimental device, owned by Tokyo City University, was used. Performance values and the photos of the device are shown in Table 1 and Fig. 1, respectively.

Table 1 Performance values of small centrifugal device

Radius of rotation	350 mm
Revolution per minute	100–510 rpm
Maximum acceleration	100 G
Mass	100 kg
External dimensions	1500 × 1450 mm

1) Outside of small centrifugal model experiment device

2) Platform inside small centrifugal model experiment device

Fig. 1 Small centrifugal model experiment device

2.2 Tilting Table

A tilting table [2] used for this report is shown in Fig. 2. We can tilt the table from 0 to 30° and load a horizontal seismic force $k_h = \tan\theta$ (k_h = horizontal seismic force, θ = tilting angle) on the modelled retaining walls.

Fig. 2 Tilting table device

3 Experimental Set-Up

3.1 Modelling of Retaining Walls

A soil container with 300 mm in width, 200 mm in height and 120 mm in depth was used for this experiment. All the sides of the container are made of aluminium except for one side, which is made of acrylic to observe the inside through it.

In this experiment, in order to confirm how the integration of retaining walls affects the earthquake resistance, three types of model retaining walls were reproduced on 1/20 scale by 3D printer: an air masonry retaining wall which is not integrated, a simple reinforced air masonry retaining wall which is integrated only on its surface and gravity retaining wall which is entirely integrated. The shape of the air masonry retaining wall is shown in Fig. 3. We used two types of retaining wall blocks whose height is 20 mm and widths are 19 mm and 38 mm, respectively, and a retaining wall base with 20 mm in height and 14 mm in width and those are made of plaster. The reinforced air masonry retaining wall was reproduced by putting 14 g of paper clay on the back of the wall. The diagram of the reinforced air masonry retaining wall is shown in Fig. 4. The diagram of the gravity wall, which is entirely integrated with PVC, is shown in Fig. 5. The surfaces of the retaining walls were installed 3 mm away from both sides of the soil container to reduce the friction, and sponge tape was attached between them to prevent the sand from spilling. Figures 6, 7 and 8 show the retaining wall blocks and the base, the reinforced air masonry retaining wall and gravity retaining wall used for this experiment, respectively. Silica sand No. 7 was used for the backfill in all the cases, which were made by air pluviation with a relative

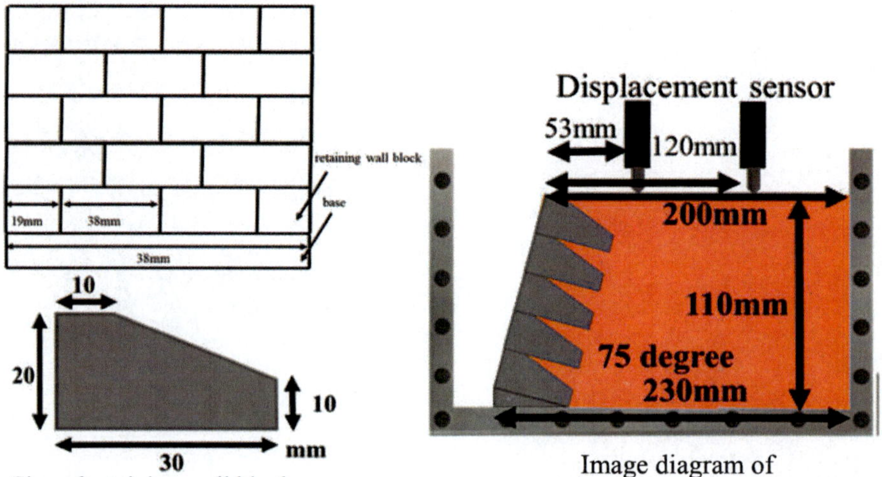

Size of retaining wall blocks

Image diagram of modelled air masonry retaining

Fig. 3 Air masonry retaining wall

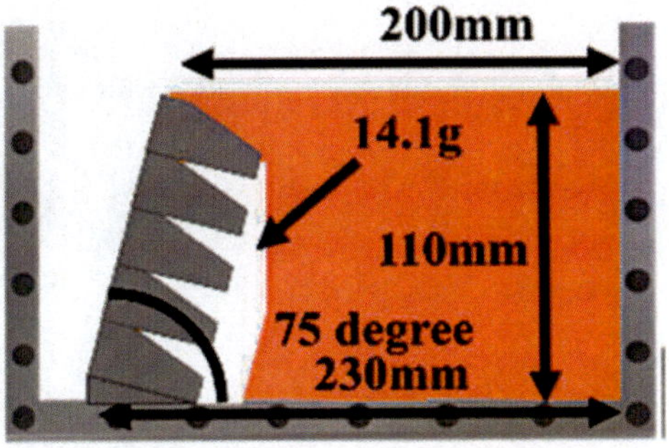

Fig. 4 Image diagram of reinforced air masonry retaining wall

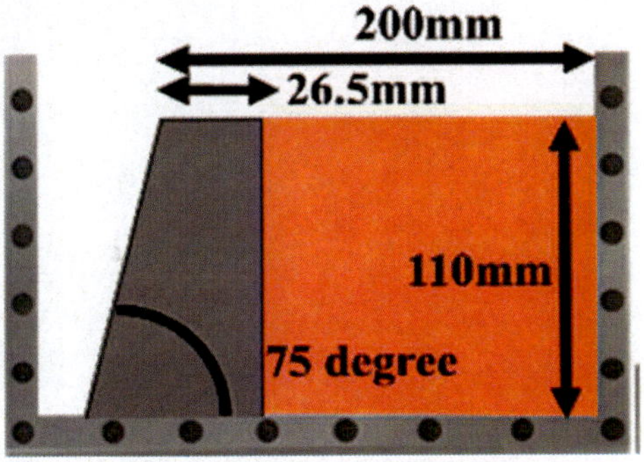

Fig. 5 Image diagram of gravity retaining wall

Fig. 6 Retaining wall blocks and base

Fig. 7 Reinforced retaining wall blocks and base

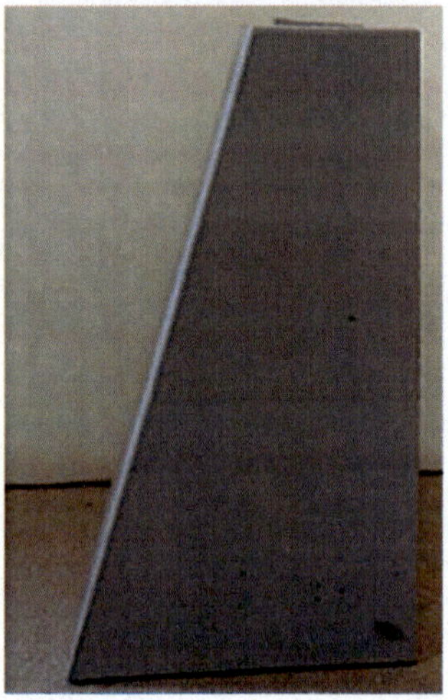

Fig. 8 Gravity retaining wall

density of 60%, 110 mm in height and 75° in tilt angle. The photos of the modelled retaining walls were shown in Figs. 9, 10 and 11, respectively.

3.2 Experimental Procedures

The state before the experiment is shown in Fig. 12. The modelled retaining wall was placed on the tilting table, where a HDV camera (GoproHEROsession5) was installed outside of the soil container for image analysis. The camera also tilts with the soil container. In addition, displacement sensors were attached at two points for measuring vertical displacement. One of the displacement sensors was attached at

Fig. 9 Modelled air retaining wall

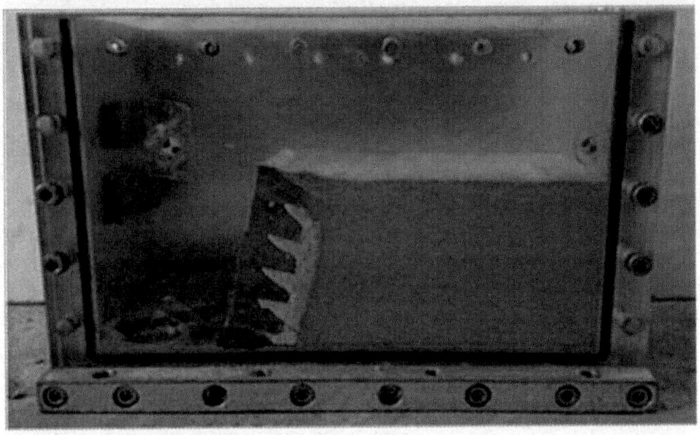

Fig. 10 Modelled reinforced air retaining wall

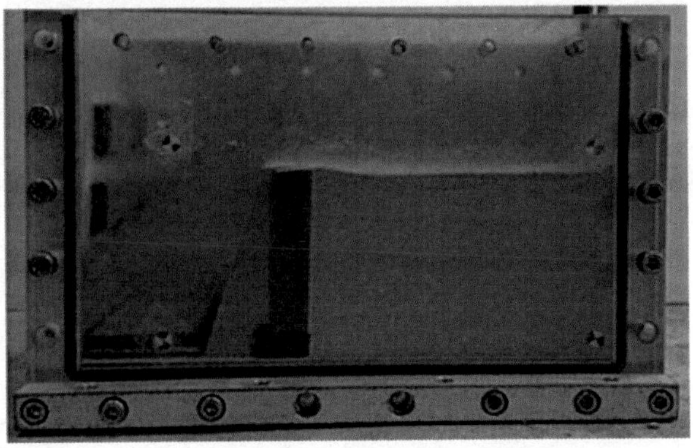

Fig. 11 Modelled gravity retaining wall

the point A, 53 mm away from the top of slope, and the other was attached at the point B, 120 mm away from the top of slope. In this experiment, a 20G centrifugal force was applied and 3 min after the force reached 20G, and a tilting experiment was conducted by changing the tilting angle from 0° to 25° under the condition of 0.16°/s in tilting speed.

Fig. 12 State before tilting experiment

4 Experimental Result

The displacement data at the points A and B are shown in Figs. 13 and 14, respectively. As for the air masonry retaining wall both in Figs. 13 and 14, after the displacement

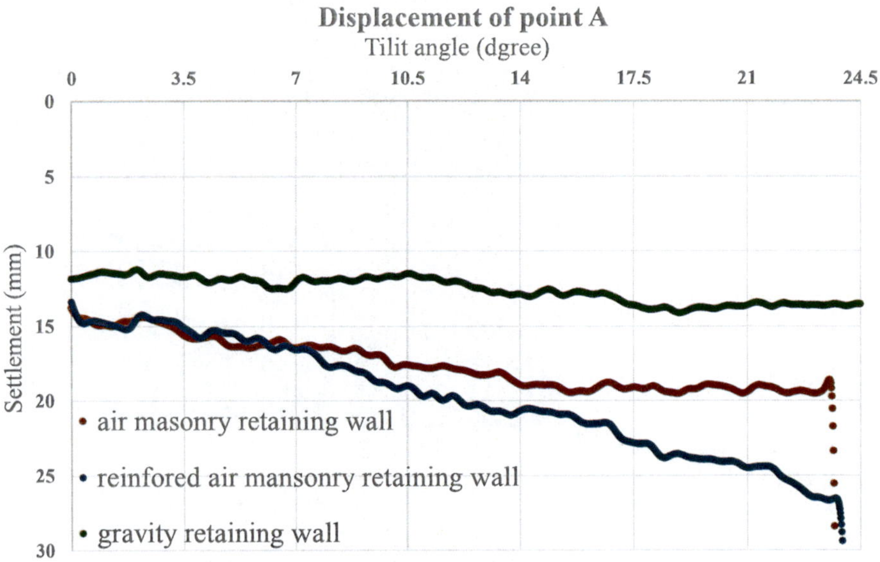

Fig. 13 Displacement data at point A

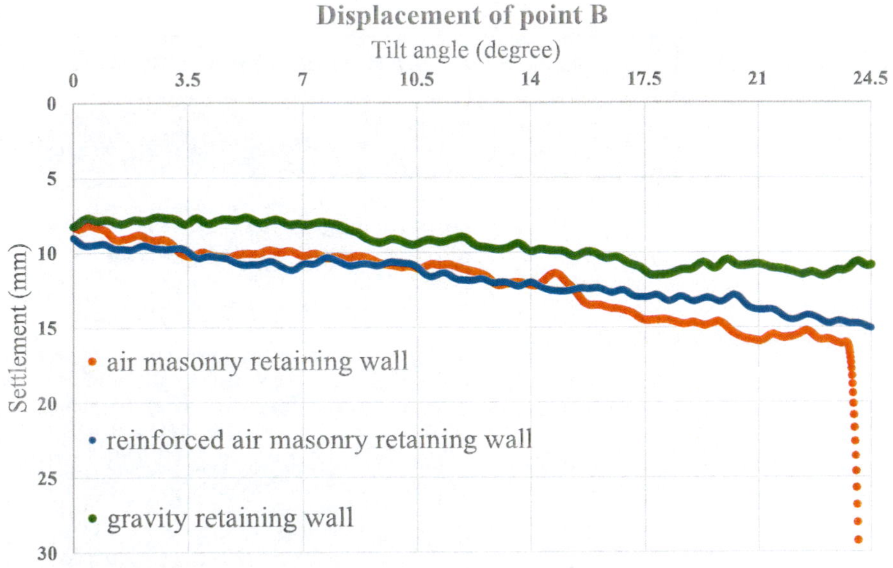

Fig. 14 Displacement data at point B

amount gradually increased according as the tilting angle increased, sudden rise was observed when the tilting angle turned to approximately 24° and which means the collapse of the retaining walls. Displacement of the point A of the reinforced air masonry retaining wall had been gradually increasing until 7°, and the displacement slope started more increasing after 7°. It could indicate that when tilting angle reached a certain angle, reinforcement was pushed to backfill. The tilting angle, when collapse occurred in the retaining wall, was a little bit bigger than that in the no-reinforced air masonry retaining wall. Finally, the displacements in both cases in the gravity retaining wall were microscopically small so that the collapse did not occur.

Next, we also conducted image analysis using the data obtained from the HDV camera for confirming the collapse behaviour of backfill. For image analysis, an image analysis software "TN-SKIP" [3] was used, which can analyse deformation, strain, etc., from the difference between the shade of the colour of the image before and after the deformation. Image analysis was categorized into four patterns, from 21° to 23.6° of tilt angle in all the retaining walls. The results of image analysis are shown in Table 2, where an increase in the maximum shear strain is shown by the change from cold colour to warm colour.

Table 2 Result of image analysis of air masonry retaining wall

Degree (°)	kh	Air masonry retaining wall
21→22°	0.40	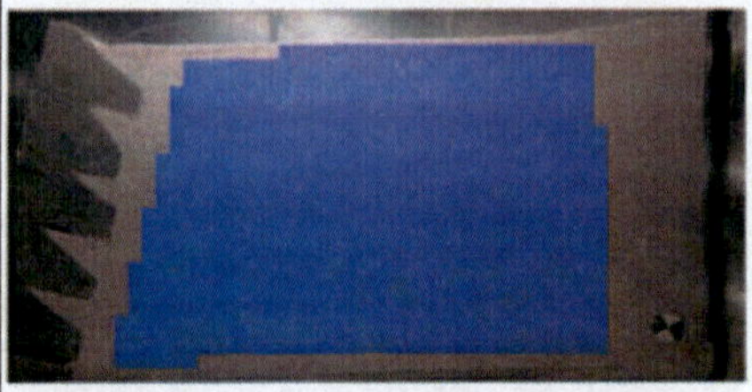
21→23°	0.42	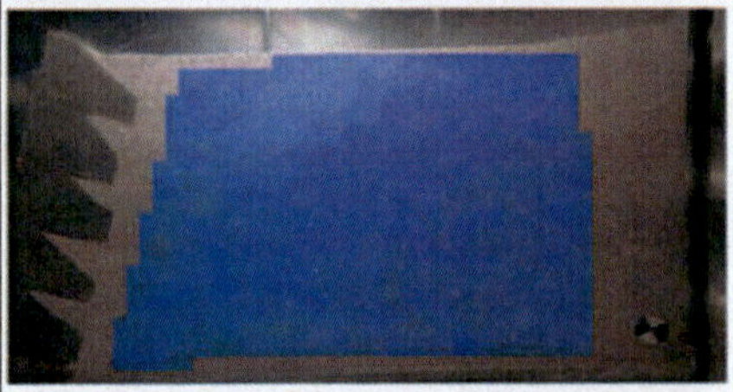
21→23.4°	0.43	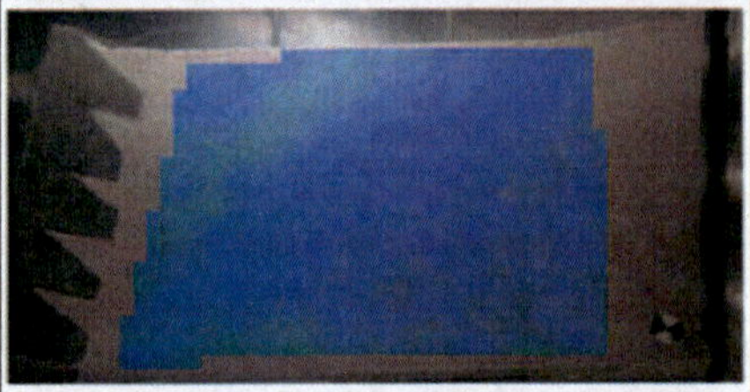
21→23.6°	0.44	

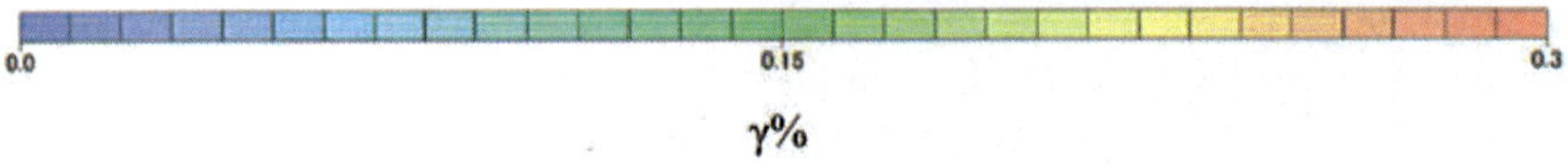

In Table 2, sliding line was not confirmed as 22°. Secondly, in 23° and 23.4°, faintly sliding line was confirmed. Finally, in 23.6°, clear sliding line was confirmed from the back of the first layer of retaining walls blocks to the centre of top of slope. In addition, in Table 3, clear sliding line was observed from the back of the base to the displacement sensor located at point A, and, that is, the tilt angle was higher

Table 3 Result of image analysis of reinforced air masonry retaining wall

Degree (°)	kh	Reinforced air masonry retaining wall
21→22°	0.40	
21→23°	0.42	
21→23.4°	0.43	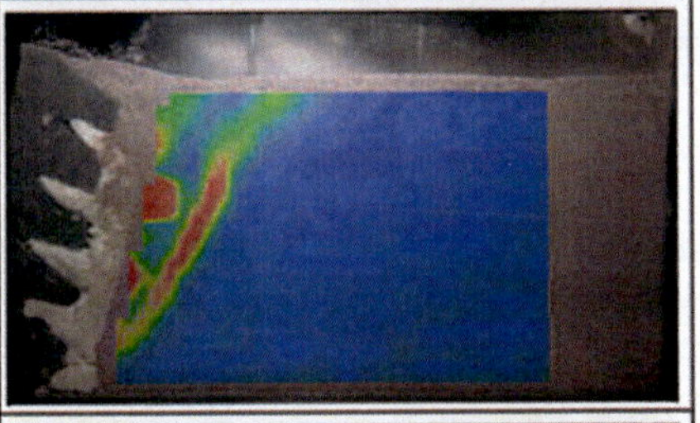
21→23.6°	0.44	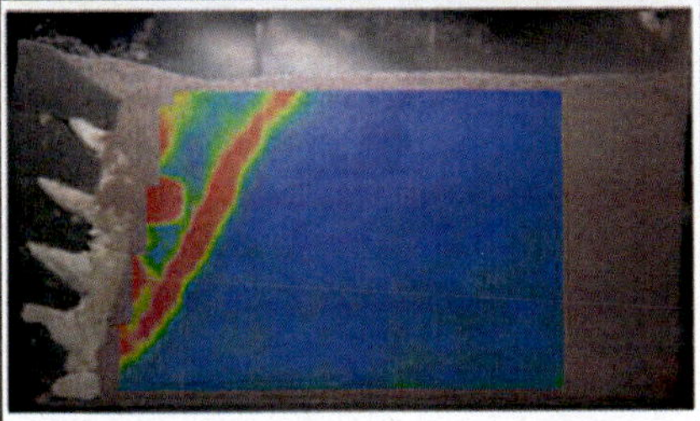

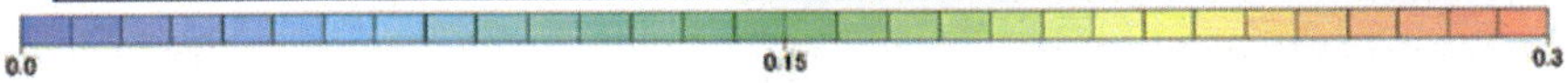

γ%

in the reinforced air masonry retaining wall than that in the air masonry retaining wall. From these results, the following two things are presumed. One is that collapse occurred in each retaining wall block in the air masonry retaining wall so that low liner shear strain was widely transmitted in the backfill. The other is that, in case of the reinforced air masonry retaining wall, collapse occurred in an integrated state so that high liner shear strain was transmitted in a limited area. On the other hand, in the gravity retaining wall, liner shear strain was not confirmed. This is presumed that accordingly as the reinforced area increased, the resistance of soil pressure increased. That is the reason why deformation happened in the gravity retaining wall (Table 4).

5 Conclusions

In this report, we modelled the three different types of retaining walls which are an air masonry retaining wall, a reinforced air masonry retaining wall and a gravity retaining wall and conducted centrifuge tilting experiments to examine how the integration of retaining walls affects the earthquake resistance.

As a result, when the air masonry retaining wall collapsed, low liner shear strain occurred in a broad area, while when the reinforced air masonry retaining wall collapsed, high liner shear strain occurred in a narrow area. In addition, it was presumed from this research that, if the material for the reinforcement was low stiffness, there was a tendency that the backfill pushed the reinforced wall and facilitated ground deformation during an earthquake. In the gravity retaining wall, liner shear strain was not confirmed. It is because the resistance against soil pressure increased accordingly as the reinforcement area increased, and as a result, the wall weight increased.

Table 4 Result of image analysis of gravity wall

Degree (°)	kh	Gravity retaining wall
21→22°	0.40	
21→23°	0.42	
21→23.4°	0.43	
21→23.6°	0.44	

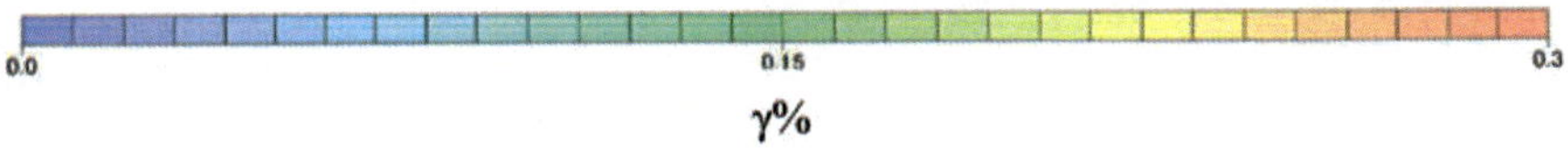

References

1. T Kiyota; Geotechnical damage caused by the 2016 Kumamoto earthquake Japan. Int. J. Geoeng. Case Hist. **4**(2), 78
2. N Itoh; Centrifuge simultaion of earthfill slope failure by inclined load of the 43rd Japanese Geotechnical Research Presentation 2008 (in Japanese)
3. Sreng, S.: Centrifugal loading tests of adjacent foundations and their FE-analysis using a new elasto-plastic model, Physical Modelling in Geotechnics-6th ICPMG '06

Effect of Bearing Pressure on Liquefaction-Induced Settlement in Layered Soils

Yutao Hu, Hemanta Hazarika, Siavash Manafi Khajeh Pasha, Stuart Kenneth Haigh, and Gopal Santana Phani Madabhushi

Abstract Earthquake-induced liquefaction causes extensive damage to infrastructure. Soil liquefaction-induced effects can account for a significant proportion of damage such as the settlement of existing buildings on liquefiable soils. The influence of structures on the behavior of liquefiable soil is complex. In this paper, the effect of bearing pressure on liquefaction-induced settlement in layered soils was studied. Two 1 g shaking table tests, with different model buildings and bearing pressures, were conducted to establish the ultimate settlement of different structures in liquefiable layered soils. The results showed that as expected the settlement of the heavier building was larger than that of the lighter one. However, the settlement ratio of two buildings was smaller than the bearing pressure ratio. This study will be used as a benchmark for further testing on improving performance through the use of soil reinforcement methods.

Keywords Liquefaction · Bearing pressure · Settlement · 1 g shaking table test

1 Introduction

Earthquake-induced liquefaction has been responsible for tremendous damage in historical earthquakes around world. This damage produced high economic and social costs mainly related to demolition, retrofitting and rebuilding of the affected structures. According to the results compiled by the Ministry of Land, Infrastructure, Transport and Tourism, in the 2011 Tohoku Earthquake, about 27,000 houses were damaged due to liquefaction. Although no damage was observed to walls or

Y. Hu (✉) · H. Hazarika
Kyushu University, Fukuoka 819-0382, Japan
e-mail: hu.yutao.594@s.kyushu-u.ac.jp

S. M. K. Pasha
IMAGEi Consultant Corporation, Tokyo 102-0083, Japan

S. K. Haigh · G. S. P. Madabhushi
University of Cambridge, Cambridge CB2 1PZ, UK

H. Hazarika et al. (eds.), *Advances in Sustainable Construction and Resource Management*, Lecture Notes in Civil Engineering 144,
https://doi.org/10.1007/978-981-16-0077-7_24

windows, some houses still experienced settlement and rotation during the earthquake [6]. Liquefaction-induced damage was also observed over a wide area following the 2016 Kumamoto Earthquake. Several areas in the southern part of Kumamoto City experienced liquefaction causing damage to residential houses due to differential settlement [3].

During earthquakes, an increase in pore water pressure may occur in saturated sandy soils as a result of suppressed soil contraction and the consequent transfer of stress to the pore fluid. Unlike denser sands which tend to dilate during shearing, looser sands tend to compress when a load is applied, at least after some strains are imposed. When the fully saturated soil is compressed, the pore water pressure tends to increase and attempts to flow to areas with lower pore water pressure (usually upward from the deeper layer to the ground surface). However, if the generated pore water pressure is high enough to induce liquefaction, the nearly undrained condition may result in partial or total effective stress loss. In consequence of this phenomenon, the strength and stiffness of the soil decreased, and the bearing capacity of the soil is reduced dramatically. If the bearing capacity decreases to such an extent that it is insufficient to hold the structure on such soils, a great subsidence takes place.

A number of researchers have studied liquefaction-induced settlement based on data from earthquakes. Liu and Dobry [5] focused on the variables of footing width and the depth of liquefiable soil, which is the primary basis of current practice. Coelho et al. [2] noted that soil beneath a shallow foundation in liquefiable soil tends to dilate owing to high static shear stresses which hence suppresses liquefaction at high bearing stresses leading to lower than expected settlements. Analysis of both new and previously published case histories by Bertalot et al. [1] supported this identification of footing bearing pressure as a significant additional variable. This study also suggested that high foundation bearing pressure may result in reduced settlement in the case of liquefaction of the foundation soil. However, the data of historical earthquakes vary significantly in local site factors and interpretation, which means that further findings would be better derived from controlled model testing.

As the influence of buildings on the behavior of saturated loose sands is complex, this study focuses on one specific factor. The main aim of the present work is to study the influence of buildings with different bearing pressures, on settlement in liquefiable soil. Two 1 g shaking table tests were conducted, in which bearing pressures were set in different values. The pore water pressure inside liquefiable soils and acceleration of the foundation were measured, analyzed and compared, as well as the settlement of the models. Results showed a considerable influence of the bearing pressure on the observed settlement.

2 Model Test

2.1 1 g Shaking Table Test

The two 1 g shaking table tests were conducted using the shaking table facility at the geo-disaster laboratory of Kyushu University. Models were constructed in a transparent Plexiglas container with dimensions of 180 cm × 40 cm × 85 cm. In order to reduce boundary effects, before each test the Plexiglas sides of the container were cleaned. Toyoura sand was used as foundation soil in these tests. The properties of the material are shown in Table 1. The saturation process could be performed by percolating water gradually and uniformly through three water inlets from the bottom of the container.

Soil–structure–fluid interaction can be simulated using the scaling law proposed by Iai [4]. Since this research involves liquefaction-induced damage to structure, it is the most suitable similitude relationship, and therefore this similitude law was used in this research. Throughout these tests, a geometrical scaling factor of 1:32 was set based on this law, as shown in Table 2.

Table 1 Properties of Toyoura sand

Property	Value
G_s	2.640
Dry density (g/cm^3)	1.506
e_{max}	0.976
e_{min}	0.611
D_{50} (mm)	0.160

Table 2 Scaling law used in the tests

Parameter	Scaling law (prototype/model)	Value
Length	n	32
Density	1	1
Stress and pressure	n	32
Time	$n^{0.75}$	13.5
Frequency	$n^{-0.75}$	0.074
Displacement	$n^{1.5}$	181.0
Acceleration	1	1

2.2 Model Preparation

Cross-sectional views of the tests performed are shown in Fig. 1. Different types of transducer were employed to measure acceleration, pore water pressure and displacement at different positions, as shown in this figure. The pore pressure transducers (PPTs) were fixed in place to monitor the pore water pressure in the exact locations. The laser micro-displacement transducers (LMDTs) were set at the middle of short sides on the top of the model buildings.

Case 1: A shallow foundation with a bearing pressure of 0.3 kPa was represented by a rectangular block of plexiglass material, with a cross-sectional area of 385 mm × 240 mm at model scale. A dense layer of Toyoura sand (D_r = 90%), representing

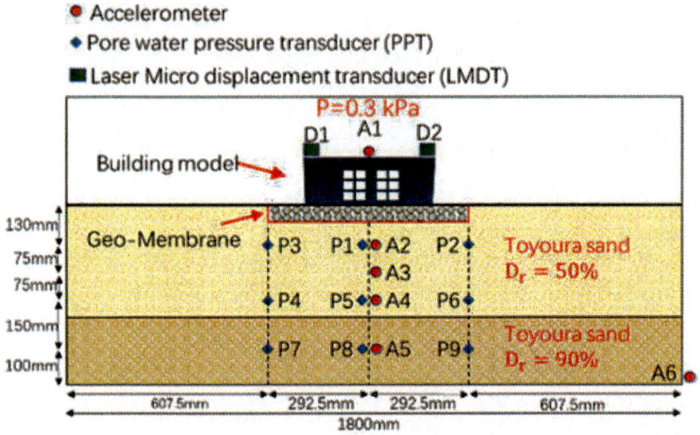

(a) Case 1: light-weight building

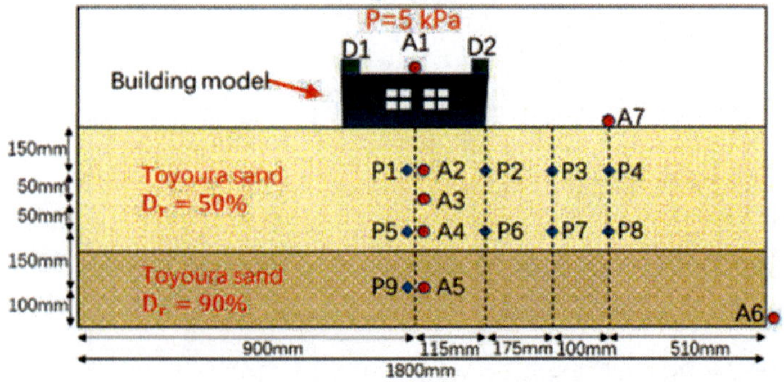

(b) Case 2: heavy-weight building

Fig. 1 Cross-sectional views of shaking table tests layout. **a** Case 1: lightweight building, **b** Case 2: heavyweight building non-liquefiable ground, was constructed using both dry deposition and tamping techniques. The depth of dense layer was 200 mm. The upper liquefiable layer (D_r = 50%) with depth in 330 mm was constructed only using dry deposition technique. To give enough bearing pressure, a 30-mm-thick geomembrane was set beneath the building. A sinusoidal acceleration of 350 Gal with frequency of 10 Hz and duration of 30 s was applied to the model.

Case 2: A shallow foundation of a heavy structure with a bearing pressure of 5 kPa was represented by a rectangular block of brass material, with cross-sectional area of 230 mm × 100 mm in model scale. A dense layer of Toyoura sand (D_r = 90%) representing non-liquefiable ground was constructed using both dry deposition and tamping techniques. The depth of dense layer was 200 mm. As no geomembrane has been used, the upper liquefiable layer (D_r = 50%) was 300 mm in depth, constructed only using dry deposition technique. Since the bearing pressure was about 17 times than in Case 1, due to the limitation of test equipment, sinusoidal acceleration of 300 Gal with frequency of 4 Hz and duration of 10 s was applied to the model.

3 Results

3.1 Time Histories of Settlement

The settlement time histories at each side of the two buildings, recorded by displacement transducers D1 and D2, are presented in Fig. 2. As shown in the figure, differential settlement happened to both buildings. The lighter building (0.3 kPa) settled only 4 mm at D1 side and 0.4 mm at D2 side. The heavier building (5 kPa) settled about 36 mm at D1 side and 32 mm at D2 side. Despite the bearing pressure ratio being about 16 between the two buildings, the ratio of maximum settlement is only 9. This may be due to the excess pore water pressures not reaching the full liquefaction levels below the heavier building as will be discussed in Sect. 3.2.

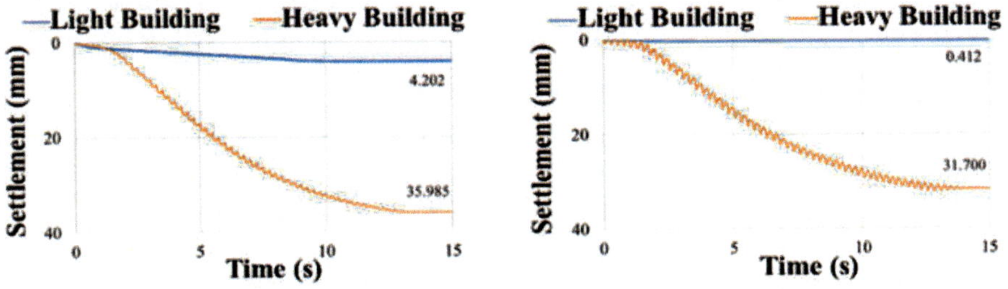

(a) Settlement recorded by D1 (b) Settlement recorded by D2

Fig. 2 Time history of settlement at each side of model. **a** Settlement recorded by D1, **b** settlement recorded by D2

Table 3 Settlement and rotation

Building	Sensor	Settlement	Rotation (anti-clockwise)
Light building	D1	4.202 mm	0.564°
	D2	0.412 mm	
Heavy building	D1	35.985 mm	1.066°
	D2	31.700 mm	

Differential settlement resulted in rotation of the two buildings. The settlement and angles of rotation are shown in Table 3. As shown in the table, the lighter building has experienced rotation of 0.564°. This may be due to the effect of inertia force.

3.2 *Excess Pore Water Pressure Ratios*

The settlements of the buildings discussed above were induced by generation of excess pore water pressures and consequent liquefaction of the loose sand (D_r = 50%). As the two buildings have different bearing pressures, it would be more prudent to plot the excess pore pressure ratio defined as $R_u = u_{expp}/\sigma'_{vo}$. These R_u plots are shown in Fig. 3 for PPTs directly beneath both buildings (P1 and P5 shown in Fig. 1) and those to one side of the buildings (P2 and P6 shown in Fig. 1). These transducers were located at the same depth of loose layer beneath the building in the two cases.

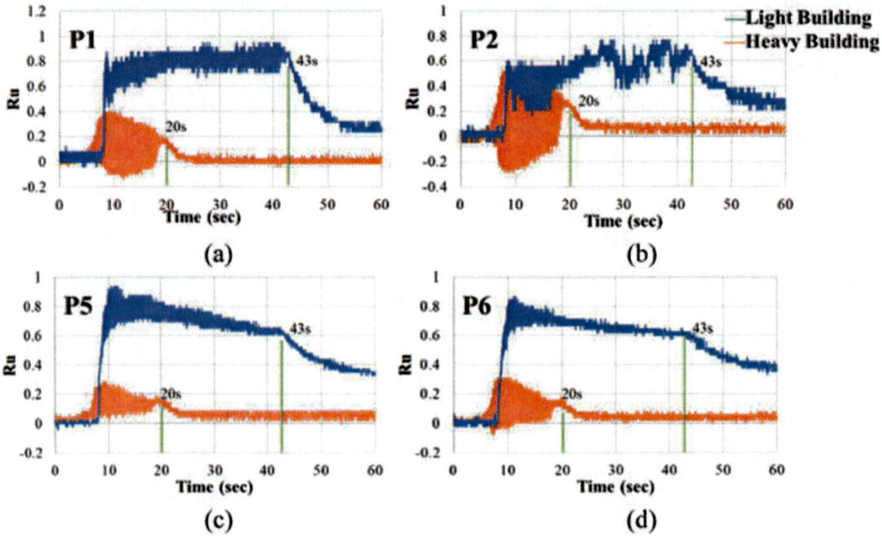

Fig. 3 Time history of excess pore water pressure ratio

(a) The settlement of light building

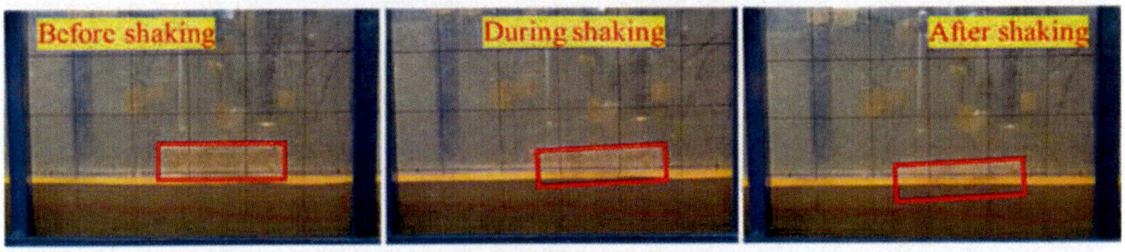

(b) The settlement of heavy building

Fig. 4 Settlement of buildings. **a** Settlement of light building, **b** settlement of heavy building

In this figure, P1 and P5 do not reach the full liquefaction value of 1 beneath the heavier building due to the dilatant effects of the static shear identified by Coelho et al. [2]. The deeper PPTs (P5 and P6 shown in Fig. 1) also show higher values than for the lighter building suggesting probable softening behavior of the soil. This is hypothesized as the main reason for a decreased settlement ratio is described in Sect. 3.1.

3.3 Building Settlements

Figure 4 shows photographs of cross-sectional views of two buildings during the shaking. In this figure, the lighter building showed insignificant settlement, while the heavier building sank until the original ground level and disappeared after the liquefaction event. The rotation of the heavier building was also more obvious than lighter one as shown in the figures.

3.4 Accelerations

Figure 5 shows the time history of acceleration for the light building in Case 1 under the input excitation with amplitude of 350 Gal. As can be seen from the figure, ground motion is amplified through the upward propagation within the soil from the bottom to the ground surface (A5–A2 shown in Fig. 1). The acceleration time histories also

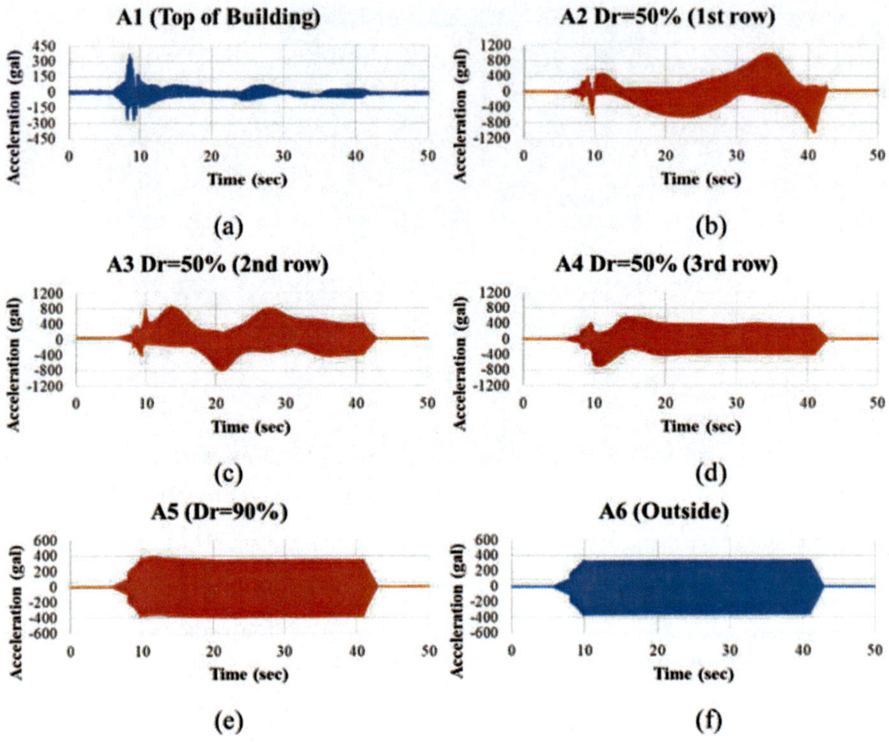

Fig. 5 Time histories of acceleration in light building case

showed spikes in the liquefiable layer (A2, A3 and A4 shown in Fig. 1). Larger acceleration spikes were observed as the density of soil increased due to subsequent shaking, which may be attributed to soil softening and re-stiffening induced by high level of excess pore water pressure.

Figure 6 shows the time history of acceleration for the heavy building in Case 2 under the input excitation with amplitude of 300 Gal. Larger accelerations were observed in shallower depth (A2, A3 and A4 shown in Fig. 1) in comparison with the base acceleration (A5 shown in Fig. 1).

4 Conclusions

In this paper, two buildings, one lighter (0.3 kPa) and one heavier (5 kPa) were tested on identical, layered soil profiles. Some of the important conclusions derived based on this study are as follows.

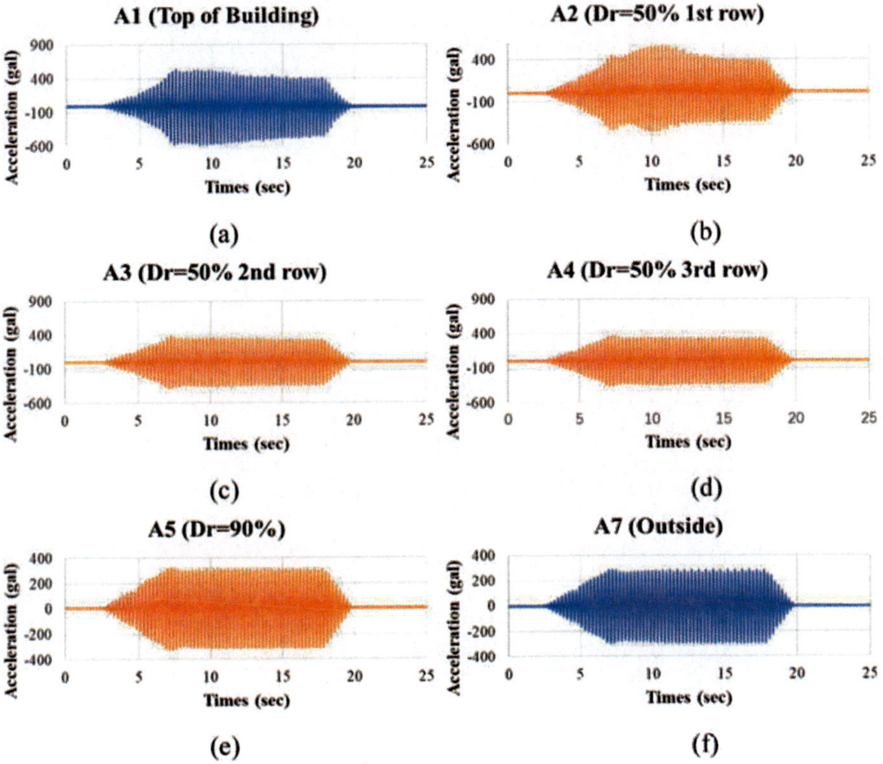

Fig. 6 Time histories of acceleration in heavy building case

(1) Although R_u value in the heavier building case is lower than that in the lighter building case, due to the weight of the buildings, the settlement of heavy building is higher than the light one.
(2) Both two buildings experienced rotation caused by differential settlement.
(3) The high level of excess pore water pressure induced softening in the soil in light building case, resulting in a disassociation between input motion and acceleration of soil.
(4) The settlement ratio of the two buildings, as well as the rotation pressure, was smaller than the bearing pressure ratio.

Several mitigation techniques have been introduced to reduce the effects of liquefaction on structures, particularly to control the excessive settlements. However, very few of these mitigation techniques have been developed for existing buildings that rest on liquefiable soils. Sustainable techniques are also needed to address this, and such techniques should be implementable for existing structure. These experimental results form the benchmark for further testing on reinforcement and the consequent improved performance.

Acknowledgements The authors would like to acknowledge the financial support provided by Kyushu University under Progress 100 project. Special thanks go to Mr. Yuichi Yahiro, technical assistant of Geo-disaster Laboratory of Kyushu University, for his help and support while conducting experiments.

References

1. Bertalot, D., Brennan, A.J., Villalobos, F.: Influence of bearing pressure on liquefaction-induced settlement of shallow foundations. Geotechnique **63**(5), 391–399 (2013)
2. Coelho, P.A.L.F., Haigh, S.K., Madabhushi, S.P.G., O'Brien, T.S.: Post-earthquake behaviour of footings employing densification to mitigate liquefaction. ICE Proc.—Gr. Improv., **11**(1), 45–53 (2007).
3. Hazarika, H., Kokusho, T., Kayen, R.E., Dashti, S., Fukuoka, H., Ishizawa, T., Kochi, Y., Matsumoto, D., Furuichi, H., Hirose, T., Fujishiro, T., Okamoto, K., Tajiri, M., Fukuda, M.: Geotechnical damage due to the 2016 Kumamoto earthquake and future challenges, Lowland Technol. Int. **19**(3), 189–204 (2017).
4. Iai, S.: Similitude for shaking table tests on soil-structure-fluid model in 1g gravitation field. Soils Found. **29**(1), 105–118 (1989)
5. Liu, L., Dobry, R.: Seismic response of shallow foundation on liquefiable sand. J. Geotech. Geoenviron. Eng. **123**(6), 557–567 (1997)
6. Yasuda, S., Harada, K., Ishikawa, K., Kanemaru, Y.: Characteristics of liquefaction in Tokyo Bay area by the 2011 Great East Japan Earthquake. Soils Found. **52**(5), 793–810 (2012)

Evaluation of Ground Anchor Residual Tension by Vibration Method

Kazuki Nawa, Atsushi Yashima, Yoshinobu Murata, Keizo Kariya, Hideki Saito, Kunio Aoike, Yoshinori Sone, and Mitsuru Yamazaki

Abstract A new method of non-destructive evaluation for residual tensile load of ground anchor has been developed. If the tendon tension part of a ground anchor is approximated by a "string" or "beam", the frequency of its free vibration is determined by line density and tendon length of the PC steel and the operating tension. Although the free vibration of tendon tension part cannot be directly excited because it is in the ground, it is confirmed by a physical model experiment that the free vibration can be excited by applying a power at the extra length of anchor head. A series of field experiments was conducted by using a small vibrator and an accelerometer. A swept-frequency vibration was applied to the extra length of anchor head, and the vibration waveform was measured at the same position. The observed waveform was analyzed by running spectrum analysis in order to find a resonance frequency which is used to calculate the residual tension load.

Keywords Ground anchor · Residual tension · Vibration · Resonance frequency

1 Introduction

Ground anchoring is one of the popular methods for maintaining the stability of slope. The ground anchoring method was introduced in Japan in 1957. Construction materials for ground anchors are easy to procure, and construction is also simple compared with the preventive pile method. Many anchors are constructed every year to stabilize cut slopes and prevent landslides. Periodic inspection of ground anchor is important to ensure the slope stability.

K. Nawa (✉) · A. Yashima · Y. Murata · K. Kariya
Gifu University, Gifu 5011193, Japan
e-mail: y4523019@edu.gifu-u.ac.jp

H. Saito · K. Aoike · Y. Sone
OYO Corporation, Saitama 3310812, Japan

M. Yamazaki
Central Nippon Highway Engineering Nagoya, Nagoya 4600003, Japan

H. Hazarika et al. (eds.), *Advances in Sustainable Construction and Resource Management*, Lecture Notes in Civil Engineering 144,
https://doi.org/10.1007/978-981-16-0077-7_25

There are many types of anchoring methods, as anchors have been improved based on experiences on the field. The fixed type anchor heads are classified into wedge-fixed type and nut-fixed type. The wedge–nut-fixed type is a kind of wedge-fixed type that combines wedge and nut. Figure 1 shows slopes with different type anchors and enlarged head portions.

The anchors are constructed in natural ground. Therefore, the decrease in function caused by a variety of factors such as the corrosion of steel and the effects of earth pressure is of primary concern. Initially, ground anchoring had been adopted as a temporary prevention method. Therefore, the protection against corrosion had been insufficient, particularly on anchor heads boundary of tendon-free part and anchor-fixed length. The ingress of water and air also contributes to the corrosion of anchors made of steel. By the first half of the 1980s, anchor failures caused by corrosion had been seen. We often observed jumping and lifting of the anchor head caused by rupture of the tendon-free part. For these reasons, in 1988, the standard document was revised [1]. After that, the anchors protected against corrosion started to be adopted [2].

When anchors are installed, load cells are set to some of the anchors. After that, the anchor tension is measured frequently to manage the anchor performance. After

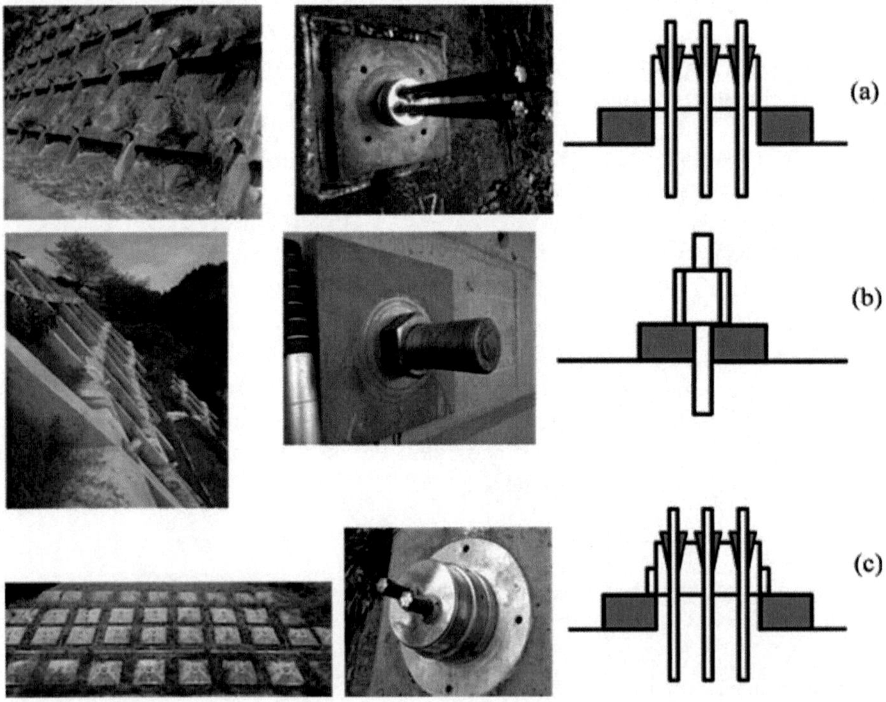

Fig. 1 Slope with different type anchors, **a** wedge-fixed type, **b** nut-fixed type and **c** wedge–nut-fixed type

confirming the convergence of anchor tension, the frequency of measuring is reduced. Daily inspections and periodic inspections are conducted mainly visually. When an abnormality is observed, there is a concern that the cut slope may be destabilized due to the change in the anchor tension. The residual tensile load of anchor is usually confirmed by the liftoff tests, which typically require time and cost. In order to overcome this shortcoming of the liftoff test, recently, more advanced hydraulic jacks (called SAAM Jack) exclusively for the purpose of liftoff tests have been developed [3]. The functions of the test equipment have been improved dramatically [4].

Even if the function of one anchor deteriorates, it does not always lead to instability of the entire slope. Therefore, it is important to evaluate the health of the overall slope by monitoring the residual tension of a large number of anchors [5]. Therefore, much more convenient as well as less expensive monitoring technique has been long awaited.

In this research, a non-destructive evaluation for residual tension load of ground anchors was developed. In the developing process, the tendon tension part of a ground anchor is assumed to be a "string" or "beam". Therefore, the frequency of its free vibration can be determined by line density and tension length of the PC steel and the operating tension. A series of field experiments were conducted by using a proposed measuring equipment. The observed tendon tension values were compared with those measured by liftoff test.

2 Physical Phenomenon Focused

We first assume that the tendon tension part of a ground anchor is approximated by a "string" fixed at both ends. If the tendon tension part is thin and long enough, the assumption of "string" is considered appropriate. Based on this assumption, the resonance frequency, f, of the free vibration of "string" in Fig. 2 can be determined by line density, μ (density per unit length), tension length of the PC steel, L and the operating tension, T by Eq. (1). Therefore, the residual tension can be determined by Eq. (2).

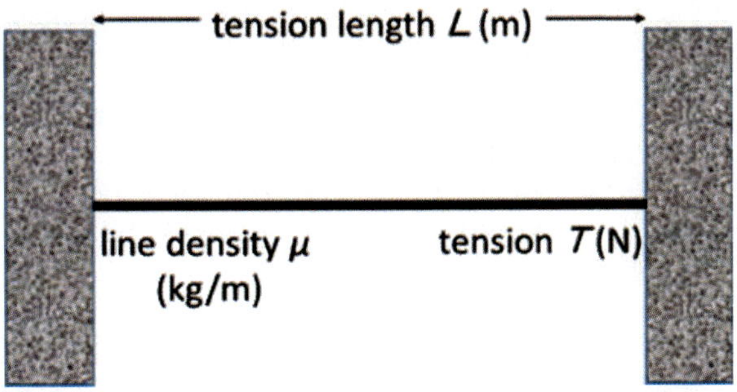

Fig. 2 "String" model fixed at both ends

$$f = \frac{1}{2L}\sqrt{\frac{T}{\mu}} \tag{1}$$

$$T = 4L^2 f^2 \mu \tag{2}$$

On the other hand, if the tendon tension part is thick and short, the tendon tension part of a ground anchor should be approximated by a "beam". For the beam bending vibration problem, the resonance frequency, f, is not dependent on the operating tension, T. We can introduce the new index ξ defined by Eq. (3) to classify the behavior of "string" and "beam" [6]. The thicker and shorter the tendon tension part is, the smaller the index ξ is.

$$\xi = \sqrt{\frac{T}{EI}}L \tag{3}$$

in which E is Young's modulus and I is the **moment** of inertia. The operating tension, T, can be calculated based on the value of ξ except when its value is less than 3.

3 Model Test and Discussion

In order to confirm the free length portion approximated as "string", we carried out large-scale experiments with three different type of anchors such as wedge-fixed type, nut-fixed type and wedge–nut-fixed type. The basic properties of anchors are summarized in Table 1. The lengths of tendon tension parts of different anchors are about 7 m. Its length is sufficiently long compared to the diameter of each steel bar. The experimental apparatus with 4000 kN reaction force capacity is shown in Fig. 3. The center hole jack was equipped at one end, and the accelerometer with vibrator was mounted on the extra length part of another end of anchor. The small target was attached on the center part of tendon tension part to monitor the vibration acceleration of the tendon part of anchor directly by using the laser Doppler vibrometer. The schematic view of the measuring system around the anchor head and the center of tendon part is also shown in Fig. 3.

The resonance frequency vibration of tendon tension part with the length of about 7 m of different anchor type was first generated by hitting the tendon part with a rubber hammer. The free vibration generated by tapping with a rubber hammer was

Table 1 Basic properties of anchors used in large-scale experiment

Head type	Diameter (mm)	Tendon length (mm)	Density/length (kg/m)
Wedge-fixed	12.7	6835	2.322
Nut-fixed	21.8	7172	2.480
Wedge–nut-fixed	15.2	7178	3.303

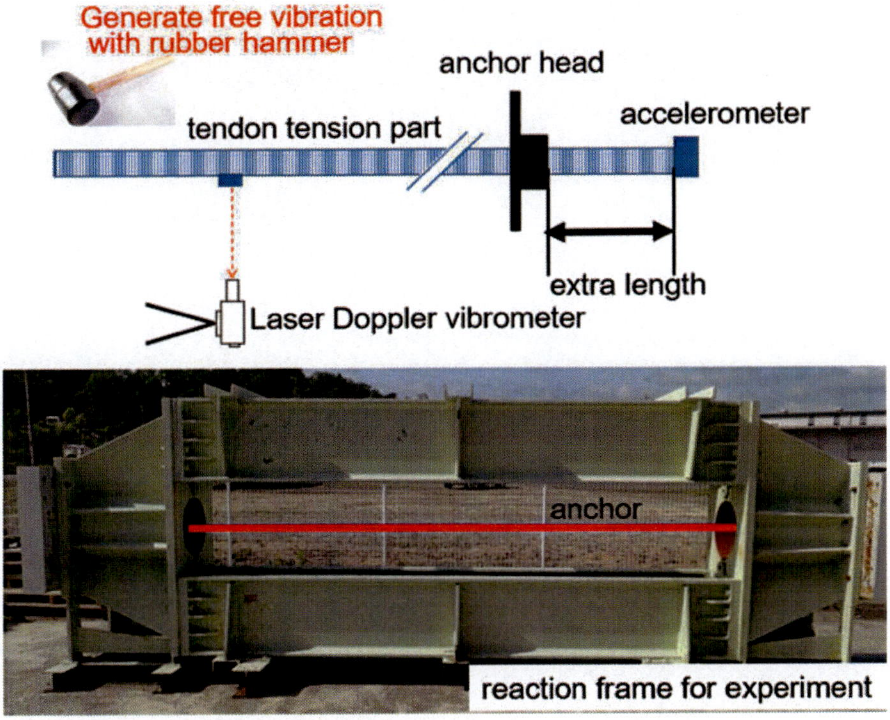

Fig. 3 Experimental apparatus with 4000 kN reaction force capacity and measuring system

monitored by the LD. The comparison between observed resonance frequencies and theoretical resonance frequencies calculated by Eq. (1) is shown in Fig. 4. It is found that the observed resonance frequency for all types of anchors coincides very well with the theoretical resonance frequency over a wide range of tension load. This means that the free length portion can be approximated as "string".

4 Field Measurement and Discussion

Anchor studies in this paper are the wedge-fixed type VSL and nut-fixed type SEEE ones which are frequently used in Japan.

Firstly, we tried to generate the free vibration of the tendon tension part by hitting the anchor head with a hammer. However, the free vibration was not excited. Therefore, we added the vibration perpendicular to the extra length of the anchor head by attaching the small vibrator as shown in Fig. 5.

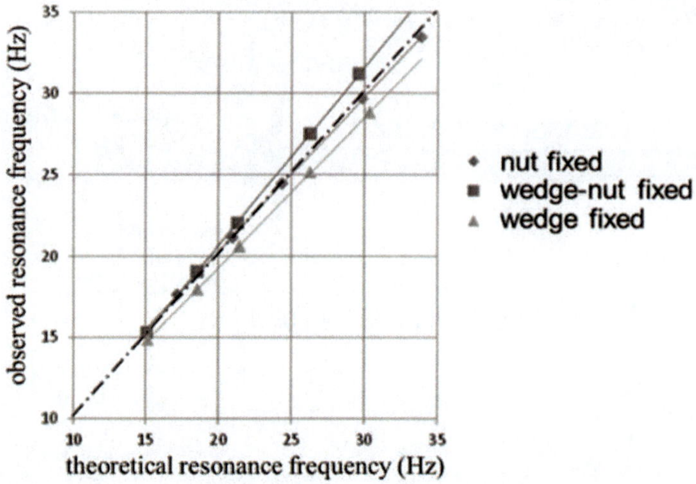

Fig. 4 Comparison between observed and theoretical resonance frequencies

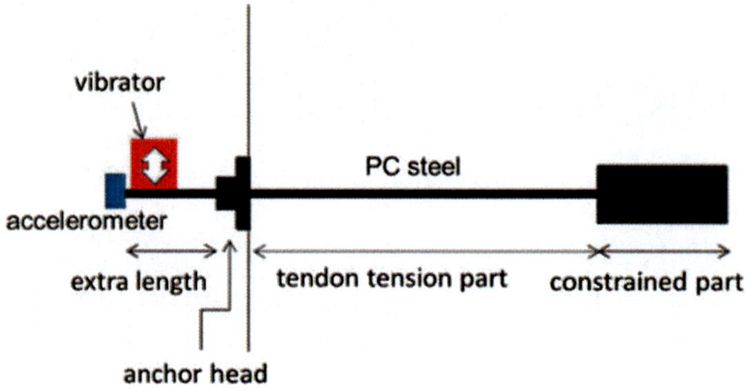

Fig. 5 Vibrator and accelerometer at extra length of anchor head

4.1 Application to VSL Anchor

Targeted anchors have four PC steel bars, 3.096 kg/m of line density and 4.0 m of tendon tension part. The designed tension load was 341.9 kN for four anchors measured in this study. The residual tensile loads for four anchors obtained by the liftoff tests are 255 kN, 309 kN, 392 kN and 517 kN, respectively. Based on these residual loads, the resonance frequencies of the tendon tension part were estimated by Eq. (1) between 36 and 51 Hz.

A small size vibrator was attached on the extra length of anchor head as shown in Photo 1. A swept-frequency sinusoidal vibration from very low to high frequency was applied to the extra length as shown in Fig. 6. Figure 7 shows the applied

Photo 1 Small size vibrator attached on the extra length of anchor head

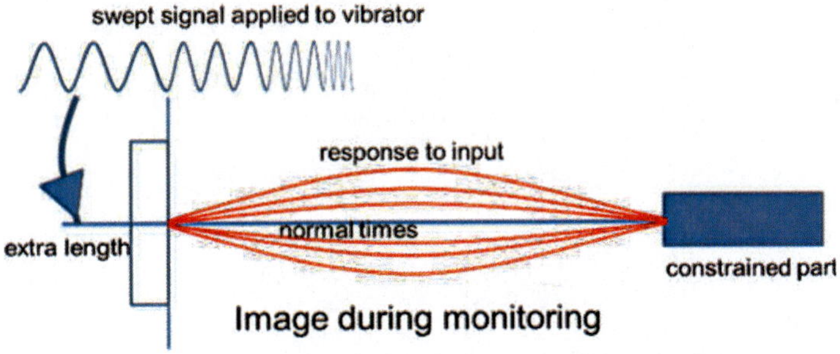

Fig. 6 Swept-frequency sinusoidal vibration from very low to high frequency

time history of swept-frequency vibration from 10 to 200 Hz for 50 s with 60 s of measuring time. The vibration waveform was measured at the same position with a small accelerometer as shown in Fig. 7. The acceleration was measured at 1 ms intervals and recorded through 16 bit A/D convertor.

When the applied frequency coincides with the resonance frequency of the tendon tension part, the response amplitude is expected to become large. We performed the running spectral analysis of the measured acceleration time history. Figure 8 shows an example of the analyzed result. The left figure in Fig. 8 shows the running spectrum. The luminance in the figure indicates the intensity of spectrum amplitude for each 0.512 s.

In the right figure in Fig. 8, the Fourier spectrum for each time step is described. From this analysis, we can easily understand that there are two peaks of the spectrum amplitude at 37 and 80 Hz. The former corresponds to the resonance frequency of the tendon tension part. On the other hand, the latter corresponds to the resonance frequency of the extra length of the anchor head which behaves as a cantilever.

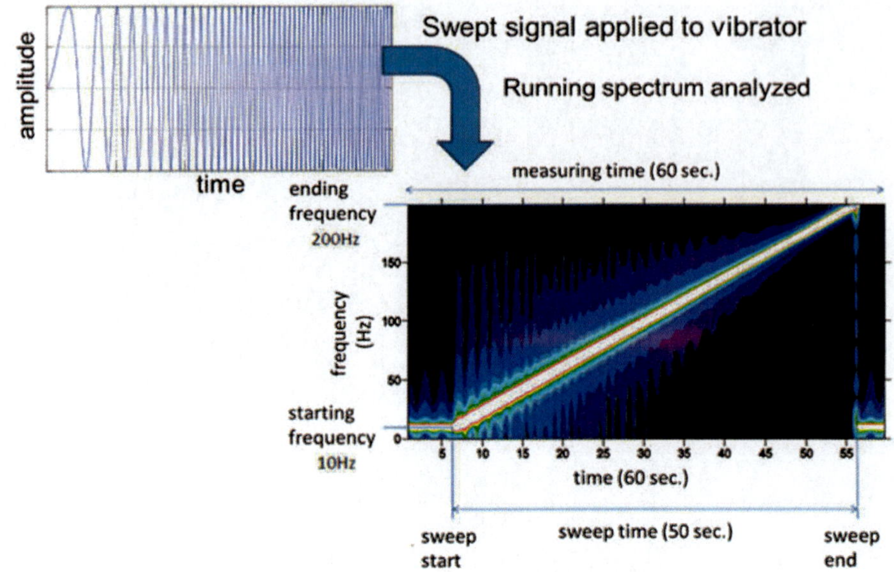

Fig. 7 Applied time history of swept-frequency vibration from 10 to 200 Hz for 50 s and vibration waveform measured

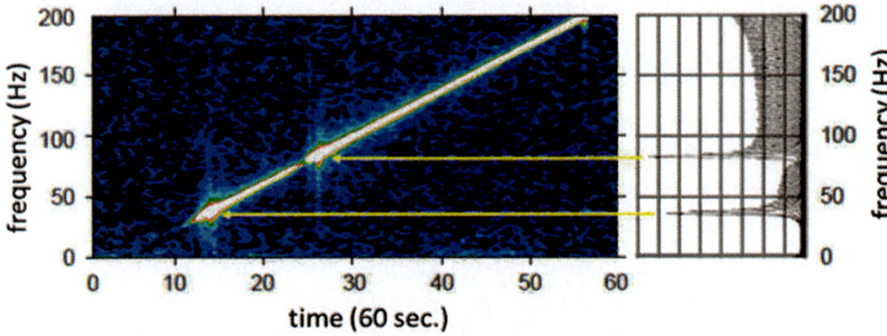

Fig. 8 Running spectrum and Fourier spectrum for each time step

The anchors targeted in this study have four PC steel bars. Figure 9 shows the running spectra for all four steel bars. We understand that the resonance frequency for each steel bar is different for wedge-fixed type VSL anchor. The resonance frequency of the tendon tension part is easily estimated based on the residual tensile load.

The comparison between the residual tensile load calculated by Eq. (2) and the measured residual tensile load by the liftoff test is summarized in Fig. 10. As each anchor has four extra length at the anchor head, the residual tensile load is calculated for each extra length. Therefore, the horizontal axis of the liftoff result is described

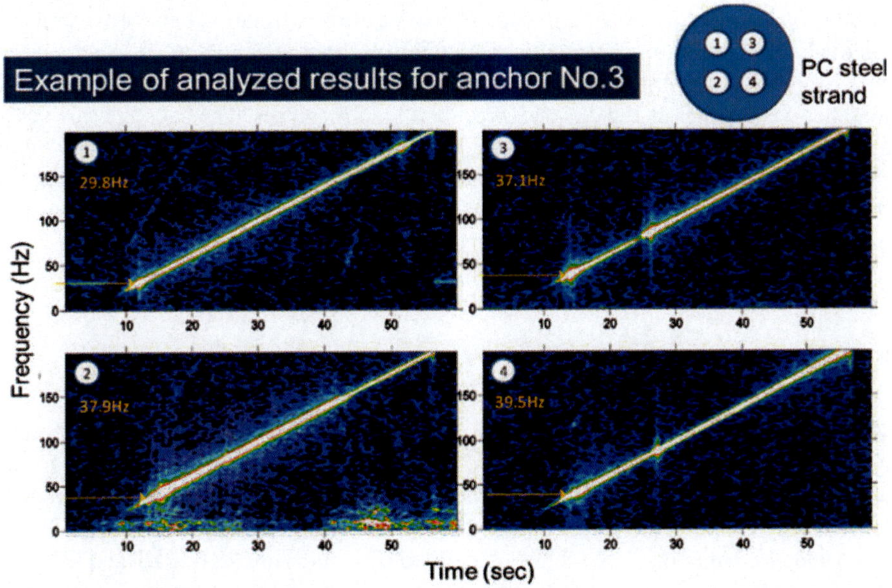

Fig. 9 Running spectra for all four steel bars

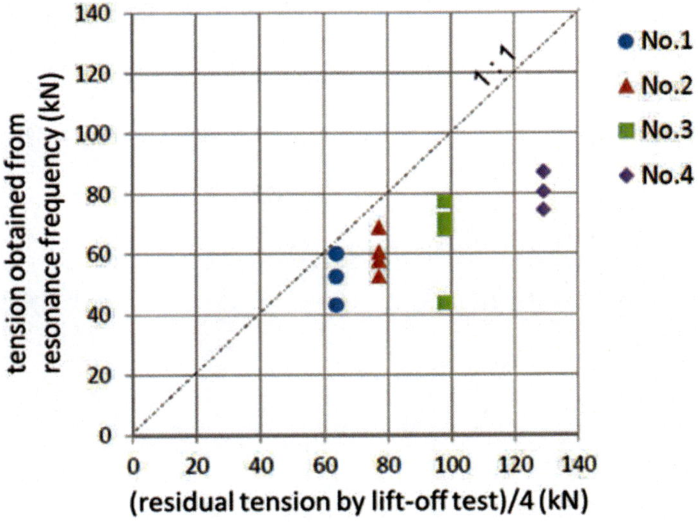

Fig. 10 Comparison between the residual tensile load calculated by Eq. (2) and the measured residual tensile load by the liftoff test

as one-fourth of the real residual load. There is a clear correlation between estimated residual loads in this study and observed ones by liftoff tests.

However, the divergence between the estimated residual load and the observed load becomes larger as the observed residual load increases. The cause of this deviation is under consideration. There are possibilities that the free length of tendon part is different from the design one and the length of the tendon tension part is insufficient to approximate it as "string". There needs to be further experiments and field measurements to answer these questions.

4.2 Application to SEEE Anchor

Targeted nut-fixed type anchors have 3.04 kg/m of line density for F70UA and 4.09 kg/m for F100UA. The tension part lengths are 4.0 m for F70UA and 10 m for F100UA, respectively. Photo 2 shows the location of anchors targeted. The designed tension loads were 317.0 kN or 360.4 kN. The liftoff tests were carried out for No.1, No.2 and No.3 anchors. The indices ξ defined by Eq. (3) are 28.0, 54.1 and 56.8 for three anchors. These values are large enough to approximate the tendon tension part as "string".

The spectrograms of extra length parts for three anchors are shown in Fig. 11. The residual tensile loads calculated by Eq. (2) are compared with measured ones by liftoff tests are also summarized in Fig. 11. The residual tensile load is found to be able to estimate based on the resonance frequency of the extra length.

Photo 2 Nut-fixed type anchor with relatively long tendon part

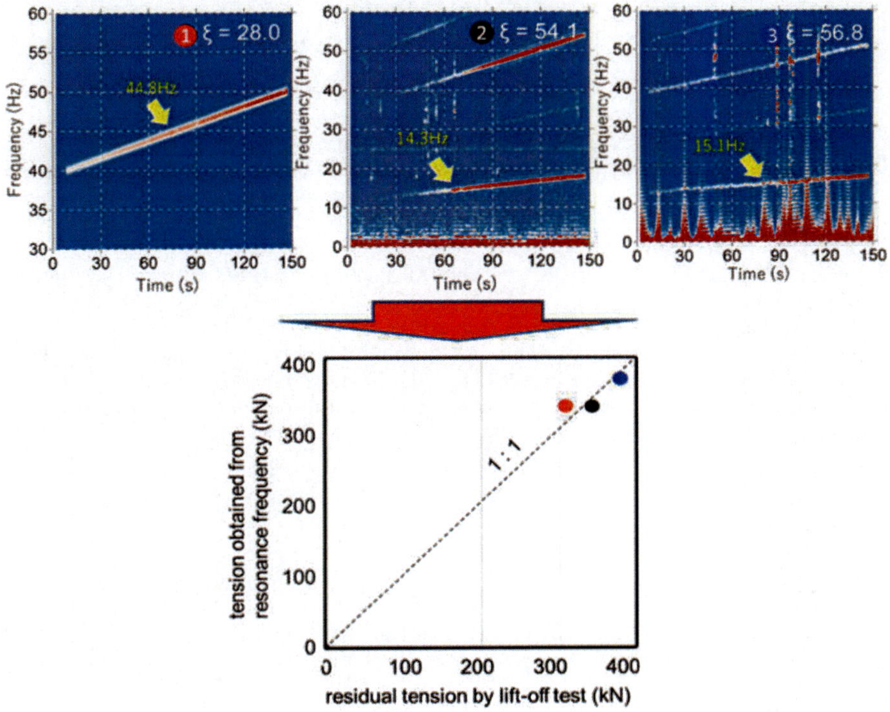

Fig. 11 Spectrograms of extra length parts for three anchors and comparison between the residual tensile load calculated by Eq. (2) and the measured residual tensile load by the liftoff test

5 Conclusions

We have proposed a new method of non-destructive evaluation for residual tension load of the ground anchor by the vibration method based on the assumption of "string" or "beam" of tendon tension part.

The free vibration of the tension part can be excited by the vibration perpendicular to the extra length of the anchor using a small vibrator. Based on the proposed technique, it is easily possible to obtain the magnitude of the residual tensile load of the ground anchor at least the wedge-fixed type VSL and nut-fixed type SEEE anchors.

Acknowledgements This work was partially supported by the JSPS (Grant No. 18K18875). We would like to thank Central Nippon Expressway Co., Ltd. for their cooperation in on-site measurement.

References

1. Okubo, K., Takemoto, M., Amano, H.: On the Ground Anchor in Expressway. Japan Road Association Miscellany, 27, 15 (2007)
2. Japanese Society of Soil Mechanics and Foundation Engineering: Ground anchor design and construction standards, the commentary, pp. 13–15 (1988)
3. Sakai, T., Yokota, D., Takemoto, M., Fujiwara, Y., Tsunekawa, Y.: The development of compact and lightweight maintenance jack and the investigation as the plane of ground anchor tension. Found. Work **38**(9), 79–82 (2010)
4. Sakai, T., Tsunekawa, Y., Yokota, D., Takemoto, M., Fujiwara, Y.: The evaluation of investigation of ground anchor slope as the plane by using SAAM jack. In: Proceedings of 65th Annual Scientific Society of Civil Engineers Conference, pp. 123–124 (2010) (in Japanese)
5. Hara, T., Tomioka, N., Soga, A., Yashima, A.: Probabilistic safety assessment of anchored slope. In: Proceedings of Geo-Risk 2017, ASCE, pp. 578–587 (2017)
6. Shinke, T., Hironaka, K., Zui, H., Nishimura, H.: Practical formulas for estimation of cable tension by vibration method. J. JSCE **294**, 25–32 (1980). (in Japanese)

Jet Grouted Columns with Horizontal Slab as Ground Improvement Towards Liquefaction Mitigation

Myat Myat Phyo Phyo, Hemanta Hazarika, Hiroaki Kaneko, and Tadashi Akagawa

Abstract A new countermeasure method of jet grout columns with horizontal slab in liquefaction mitigation is introduced in this study. In order to investigate the effectiveness of the proposed mitigation method, numerical studies on unimproved and improved ground were separately performed. Additionally, the validations of the numerical model geometry with boundary conditions and parameters used in UBC3D liquefaction model were conducted prior to the analyses to confirm the reliability of the numerical results. The effectiveness of the soil improvement method was assessed by comparing the changes in excess pore water pressure ratio, acceleration as well as distribution of shear stress and shear strain in the liquefiable soil before and after ground improvement. The results show that the new liquefaction mitigation method offers optimistic effect on control of shear deformation and excess pore water pressure.

Keywords Liquefaction prevention · Ground improvement · Jet grouting · Numerical simulation

1 Introduction

Recent and historical earthquakes have provided countless examples of the infrastructures damaging due to liquefaction. In Japan, many residential houses suffered severe damages induced liquefaction followed by strong earthquakes Niigata earthquake (1964), Hyogoken-Nanbu Earthquake (1995), Tohoku Earthquake (2011), etc. In order to consider remediation measures on vulnerable housing in liquefaction prone regions, researches and efforts are emphasized on the necessity of soil remediation against liquefaction. Many countermeasure methods, such as gravel drain

---

M. M. P. Phyo (✉) · H. Hazarika
Kyushu University, Fukuoka 819-0385, Japan
e-mail: mashwephyo@gmail.com

H. Kaneko · T. Akagawa
Japan Foundation Engineering Co., Ltd., Tokyo 151-0072, Japan

H. Hazarika et al. (eds.), *Advances in Sustainable Construction and Resource Management*, Lecture Notes in Civil Engineering 144,
https://doi.org/10.1007/978-981-16-0077-7_26

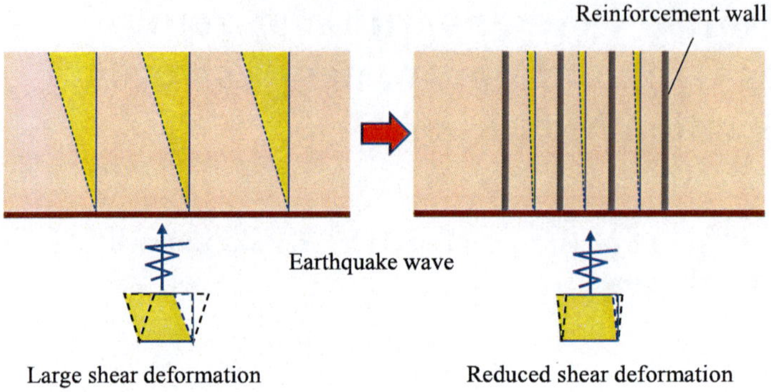

Fig. 1 Liquefaction mitigation mechanism of grid form wall

method, sand compaction pile method, deep mixing method and jet grouting, have been proposed to mitigate liquefaction. In Japan, the lattice-shaped cement mixing method is a well-known method to mitigate the liquefaction. This method uses intersected cement-treated columns to form grid shape in loose soil to restrict shear deformation and generation of excess pore pressure against earthquake as shown in Fig. 1. It was developed since in early 1990s but the effect of this improvement method was first evaluated in the Hyogoken-Nanbu earthquake in 1995. Since then, many numerical analyses, physical model tests and field tests have been performed to investigate the performance of the grid type, interaction between improved ground and surrounding ground (e.g. Kitazume [5]). Due to good results from numerical analyses and field tests, the grid form liquefaction mitigation technique has been frequently used in construction sites. As an example, this method was adopted to mitigate soil liquefaction damage in the land reclamation area, Urayasu City, where residential houses suffered severe damage due to liquefaction during the 2011 Tohoku earthquake [4].

Nevertheless, the effectiveness of jet grout columns in liquefaction prevention is particularly depending on many factors, such as improvement area ratio, wall stiffness and grid spacing. Based on the previous researches together with observations and experiences in the fields, ground improvement area ratio is predominantly effective in increasing the potential of liquefaction mitigation [8], [16]. However, the behaviours of the high-modulus jet grouting, interaction between improved ground and surrounding soil, the response of shear stress and shear strain distribution in improved ground are not fully explored yet.

In this study, a new countermeasure method, using small diameter jet grout columns with additional horizontal slabs, is introduced. This method is more beneficial compared to the conventional approach because it uses closely spaced jet grout wall ($L/H = 0.2$) to defend against liquefaction. In addition, the entire liquefaction prone layer is improved by jet grouting columns in both vertical and horizontal directions to restrain the shear deformation during an earthquake. In order to measure the effectiveness of the new method in reducing liquefaction risk, numerical analyses using an effective stress-based model UBC3D-PLM were performed. Validations for model geometry with boundary conditions and soil parameters used in liquefaction model were conducted prior to the analyses, to confirm the reliability of the numerical prediction. The effectiveness of closely spaced jet grout wall with horizontal slab in reduction of liquefaction was evaluated by varying the excess pore water pressure ratio, acceleration as well as distribution of shear stress and shear strain in the liquefiable soil.

2 Numerical Modelling

In this study, numerical simulations, using PLAXIS 2D, were performed to measure the effect of jet grout columns with horizontal slab in liquefaction mitigation. Beforehand, verification of numerical model geometry with boundary conditions and the parameters used in UBC3D liquefaction model were carried out to obtain the reliability of the numerical results.

2.1 Finite Element Model Validation

Numerical finite element (FE) modelling provides a significant aid in the determination of the dynamic response due to its inherent capability of detailed simultaneous prediction of stress distribution and displacements in the system. However, the reliability of a numerical prediction is largely influenced by soil constitutive model, boundary conditions and mesh distribution. Especially, consideration for boundary conditions is the most important in dynamic modelling since different wave propagation problems will have different boundary criteria. Therefore, simple verification for finite element model was conducted by comparing soil amplifications function obtained from the numerical and theory. Hence, a numerical analysis for a single layer of homogeneous linear elastic soil bed of 36 m thick was performed. Elasticity modulus, Poisson's ratio and unit weight used in the model were selected as 6000 MPa, 0.2 and 22 KN/m^3, respectively. To create a mesh distribution, 15 nodes triangular plain strain elements were used. The vertical boundaries were modelled with tied degree of freedom while compliance based was selected at the base as suggested in the site response and liquefaction evaluation by Brinkgreve [2]. The applied ground motion at the base of the model was the earthquake data of the Loma Prieta earthquake (1989) with the maximum peak ground acceleration of 0.3 g. The history of acceleration time was depicted in Fig. 2.

Amplification function in the numerical analysis is determined as the ratio of Fourier spectrum of the surface motion to the corresponding component of the bedrock motion. With the purpose of verification, the theoretical amplification factor of linear elastic model was calculated using Eq. 1 as proposed by Roesset [12], wherein H is the layer thickness, v_s is shear wave velocity, D is damping ratio, and f

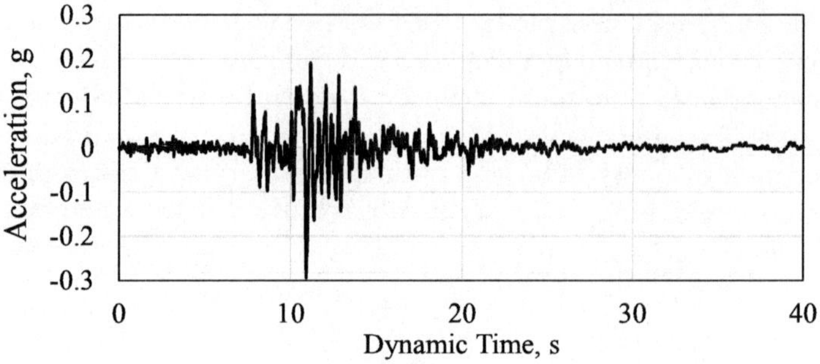

Fig. 2 Seismic input motion applied in the analyses

is frequency of the soil during vibration, respectively. As observed from Fig. 3, the amplification function obtained from numerical analysis is reasonably matching with the theoretical amplification function. Based on the results, the boundary concept of the model seems to yield quite adequate results regarding reproduction of soil amplification function.

$$A(f) = \frac{1}{\sqrt{\cos^2\left(2\pi \frac{H}{V_s} f\right) + \left(2\pi \frac{HD}{V_s} f\right)^2}} \tag{1}$$

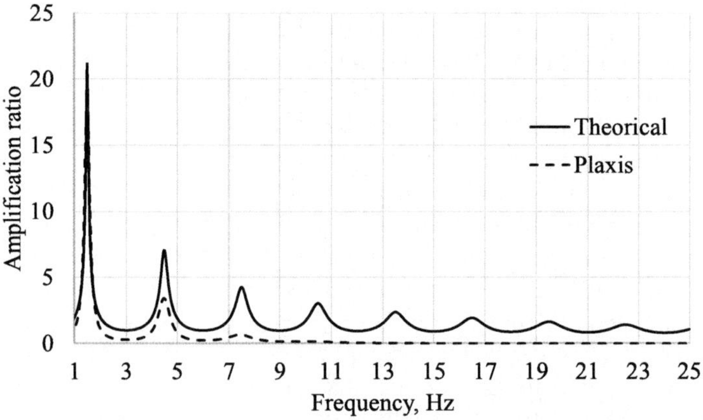

Fig. 3 Comparison of amplification functions obtained from theoretical and numerical

2.2 UBC3D-PLM Model Calibration and Parameter Selections

The UBC3D-PLM model is an effective stress model based on classical plasticity theory with a hyperbolic hardening rule to measure the development of excess pore water pressure and capture the onset of liquefaction under dynamic loading. The UBC3D-PLM model is developed by Tsegaye [15] and Petalas et al. [9] which is an extension of the UBCSAND model. The model utilizes the primary and secondary yield surfaces to account the effect of soil densification and predict the smooth transition into the liquefaction state under undrained loading. Furthermore, a soil densification rule is implemented to better predict the evolution of pore pressures during cyclic loading. Even though UBC3D-PLM is an advanced model, it is relatively simple to apply due to its reasonable number of parameters that can be extracted from laboratory or in situ tests. The liquefaction parameters are connecting with the corrected clean sand equivalent SPT blow count measurements $(N_1)_{60}$. However, the selection and calibration of parameters play a significant role to obtain reliable results.

In this study, the constitutive parameters were calibrated based on the curve fitting of the cyclic resistance ratios of sand of different relative densities, reproduced by means of cyclic direct simple shear test implemented in soil test facility of PLAXIS. With reference to Fig. 4, the comparison of cyclic strength curves obtained from the numerical and the semi-empirical methods are well agreed. The soil parameters used in UBC3D-PlM model and the generated deformation characteristics for D_r 50% are shown in Table 1 and Fig. 5, respectively. However, the UBC3D-PLM model is not advisable to use in static analysis since the parameters used in model are designated to evaluate liquefaction in loose soils and suitable only for dynamic calculation (Plaxis 2D Material Models Manual [10]). Therefore, hardening soil model was used in

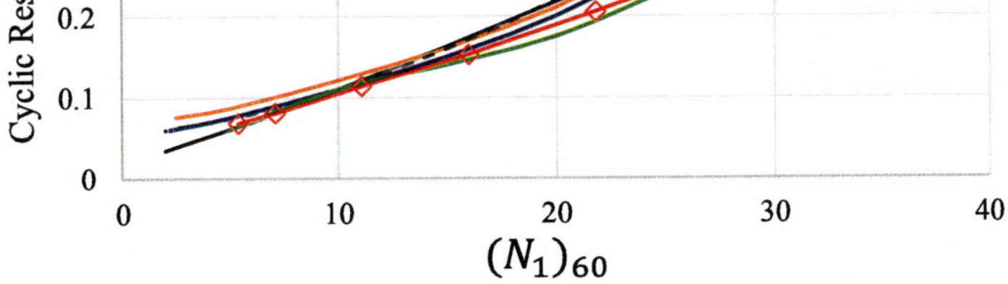

Fig. 4 Cyclic strength curve from UBC3D-PLM in comparison with empirical curves

Table 1 Parameters of liquefiable soil used in the UBC3D-PLM model

Parameter	Symbol	Value	Method/formula
Peak friction angle	ϕ'_p	35°	CD Test
Constant volume friction angle	ϕ'_{cv}	33°	CD Test
Elastic shear modulus number	K^e_G	967.67	$21.7 \times 20 \times (N_1)^{0.333}_{60}$
Elastic bulk modulus number	K^e_B	677.37	$0.7 \times k^e_G$
Plastic shear modulus number	K^p_G	458.40	$k^e_G \times (N_1)^2_{60} \times 0.003 + 100$
Elastic shear modulus index	me	0.5	Default
Elastic bulk modulus index	ne	0.5	Default
Plastic shear modulus index	np	0.4	Default
Failure ratio	R_f	0.77	$1.1 \times (N_1)^{-0.15}_{60}$
Densification factor	f_{dens}	0.45	Curve fitting
Post liquefaction factor	f_{post}	0.02	Curve fitting
Corrected standard penetration test	$(N_1)_{60}$	11.1	$D_r{}^2/15^2$

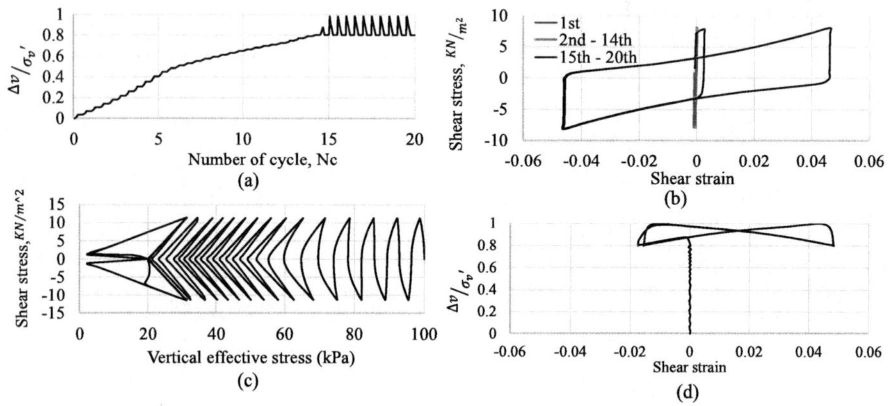

Fig. 5 Deformation characteristics generated for D_r 50% from PLAXIS soil test simulation **a** excess pore water pressure ratio developed with number of cycles, **b** shear stress versus shear strain curve, **c** shear stress versus vertical effective stress, **d** excess pore water pressure ratio versus shear strain

initial static phase to generate the stress state correctly for liquefiable soil prior to the dynamic phase.

2.3 *Numerical Simulation for Unimproved and Improved Ground Models*

In order to determine the performance of high-modulus jet grouted columns against liquefaction, the numerical simulations for unimproved and improved ground were conducted. As for the unimproved case, hypothetical three layers of soil profile were utilized. The water table was assumed to be coincided with the ground level. Herein, the response of soil liquefaction was measured at the 10-m-thick loose sand layer with respective relative density of $D_r = 50\%$, which was overlying 25 m of clay. The underlain layer was assumed to be a bedrock where the above-mentioned earthquake motion was imposed. Numerical simulations were conducted in three modelling phases. For the liquefiable sand layer, hardening soil model was used at the first initial and second plastic nil step phases to generate initial stress correctly. Thereafter, UBC3D-PLM model was applied at third dynamic stage in order to measure the development of excess pore water pressure and capture the onset of liquefaction of the loose sand layer under earthquake loading. The constitutive model and parameters used in numerical analyses for sand, clay and bed rock layer are shown in Tables 1, 2 and 3, respectively.

On the other hand, 0.6 m thickness of jet grout walls (10 m in height) was installed with 2 m interval with respect to spacing ratio 0.2 for the improved case. In addition, the slabs with 1 m thickness were horizontally added at every 1 m interval as shown in Fig. 6. Thereby, the loose sand layer was improved in both vertical and horizontal directions to restrain shear deformation during the earthquake. Mohr–Coulomb

Table 2 Parameters used in hardening soil model

Parameter	Symbol	(Unit)	Clay	Sand
Unit weight	γ_{sat}	(kN/m^3)	18	20
Effective cohesion	c'_{ref}	(kN/m^2)	13	0
Effective friction angle	ϕ'	(°)	22	35
Dilatancy angle	Ψ	(°)	-	1
Secant modulus	E_{50}^{ref}	(kN/m^2)	5436	20,380
Tangent stiffness for primary oedometer loading	E_{oed}^{ref}	(kN/m^2)	5436	20,380
Unloading/ reloading stiffness	E_{ur}^{ref}	(kN/m^2)	16,310	61,130
Power of stress-level decency	m		0.8	0.5

Table 3 Parameters of bedrock used in linear elastic model

Parameter	Symbol	(Unit)	Values
Unit weight	γ_{sat}	(kN/m^3)	22
Young's modulus	E	(MN/m^2)	6,000
Poisson's ratio	υ		0.2

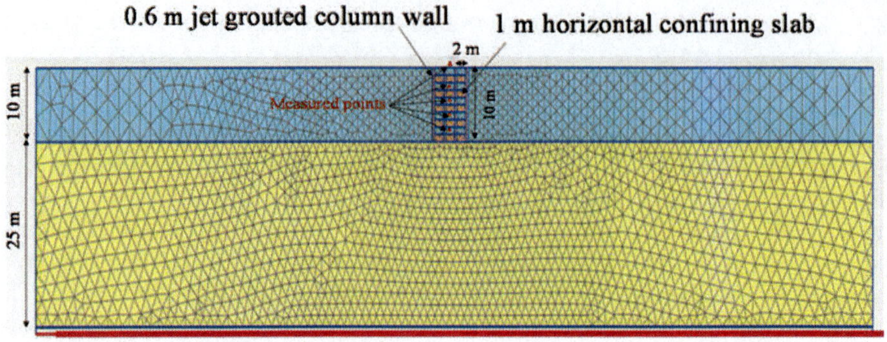

Fig. 6 Finite element model of the improved case

model was applied in jet grout columns modelling. Elasticity modulus, Poisson's ratio and undrained shear strength of the columns were selected as 1000 MPa, 0.2 and 1 MPa, respectively.

3 Result and Discussion

The aim of the study is to evaluate the effect of liquefaction mitigation by using closely spaced jet grout columns with the horizontal slab in liquefiable ground. Thus, numerical cases of soil improvement with and without were separately measured. The analysis results were evaluated based on the effect of jet grout columns on the distribution of excess pore water pressure ratio, acceleration, shear stress and shear strain in the liquefiable soil layer.

3.1 Acceleration

The surface acceleration behaviour becomes relevant in the evaluation of the soil improvement technique as soil softening can significantly affect to the building damage. The comparison of the surface acceleration responses obtained from the unimproved case and improved cases is presented in Fig. 7. With respect to Fig. 7, the peak accelerations of improved case and unimproved case were observed as 0.23 and 0.18 g, respectively. As a noticeable difference, the occurrence of soil acceleration attenuation behaviour was found in the unimproved case. It means that the soil acceleration appeared to decrease upon the onset of liquefaction (around dynamic time at 13 s) due to the reduction of soil strength and stiffness and increase in hysteretic damping. However, there was no trace of acceleration decreasing due to liquefaction in improved case. In other words, the newly proposed liquefaction countermeasure method can control the soil stiffness reduction against seismic loading.

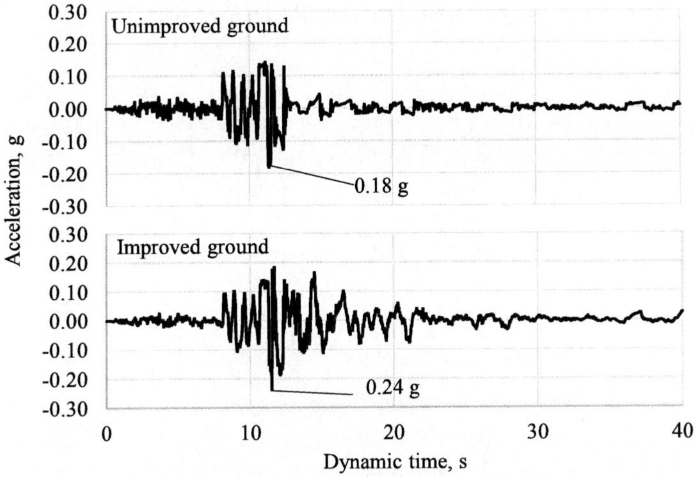

Fig. 7 Surface acceleration time histories measured from unimproved and improved cases

3.2 Excess Pore Water Pressure Ratio

Excess pore water pressure ratio is an indicator of liquefaction occurrence, which is calculated by means of vertical effective stress at the end of the dynamic calculation and initial effective vertical stress prior to the dynamic stage in UBC3D PLM [2]. The corresponding layer can be determined as a complete liquefied layer when the excess pore water pressure ratio is 1. The comparison results of excess pore water pressure distribution with dynamic time at different depths for unimproved and improved cases were indicated in Fig. 8. Based on the comparison results, it can be observed that the proposed liquefaction countermeasure method is effective in excess pore water pressure control.

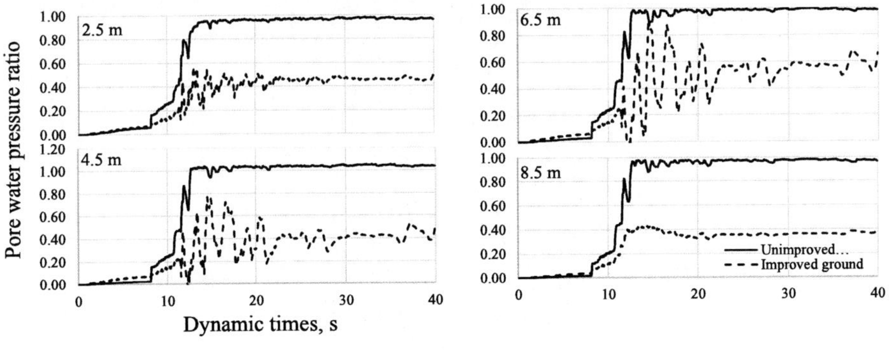

Fig. 8 Comparison results of PWP ratio measured from unimproved and improved cases

3.3 Shear Stress and Shear Strain Distribution

The effect of jet grout columns on shear stress and shear strain distribution in liquefiable soil was interpreted by calculating shear stress reduction ratio (R_d) and shear strain by Rayamajhi et al. [11]. Here, R_d is the ratio of the shear stress reduction coefficients, $r_{d\,\text{improved}}$ of the improved soil over that $r_{d\,\text{unimproved}}$ of the unimproved soil. Basically, the shear stress reduction coefficient, r_d, is derived from the simplified procedure of cyclic stress ratio (CSR) proposed by Seed and Idress [13], as shown ins Eq. 2.

$$\text{CSR} = \frac{\tau_s}{\sigma'_v} = 0.65\left(\frac{a_{max}}{g}\right)\left(\frac{\sigma_v}{\sigma'_v}\right)r_d \tag{2}$$

where τ_s is the average cyclic shear stress of the soil, σ'_v is the effective vertical stress of soil, σ_v is the total vertical stress of soil, a_{max} is the maximum ground surface acceleration, r_d is the shear stress reduction coefficient. The dynamic analyses provide the values of τ_{max} and a_{max} for unimproved and improved cases, from which the shear stress reduction coefficients, $r_{d,}$ could be computed based on Eq. 3.

$$r_d = \frac{(\tau_{max})_d}{(\tau_{max})_r},\ (\tau_{max})_r = \gamma' z a_{max} \tag{3}$$

$(\tau_{max})_d$ = maximum shear stress
Z = effective unit weight of soil.

The computed shear stress reduction ratios (R_d) were presented in Fig. 9. In other words, it shows the contribution of high-modulus column to reduce shear stress in liquefiable sand. A lower R_d indicates a lower potential of liquefaction in ground.

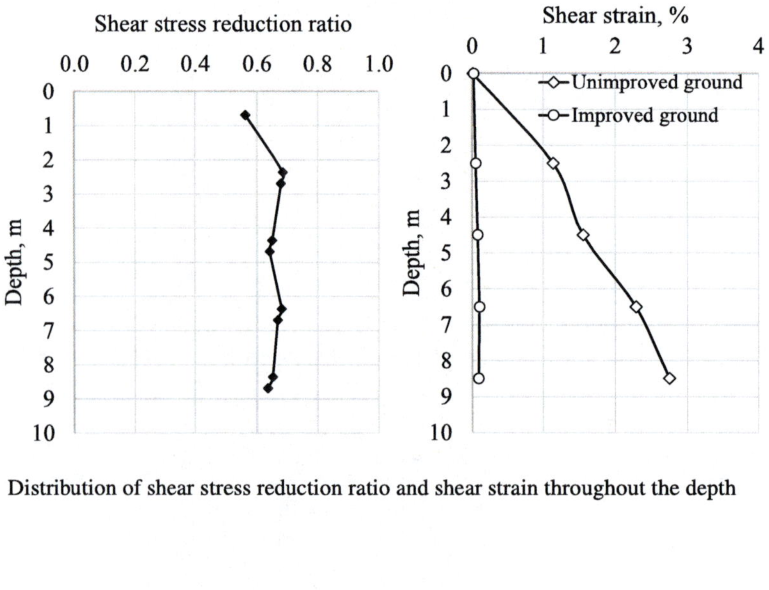

Fig. 9 Distribution of shear stress reduction ratio and shear strain throughout the depth

The range of shear strain distribution throughout the depth in unimproved case was noticed to be very wide compared to the improved case. It can be determined that the shear deformation of loose soil layer was effectively controlled by the vertical and horizontal reinforcement during the earthquake.

4 Conclusion

As an initial assessment for liquefaction prevention, the study on the effectiveness of jet grout columns with horizontal slab method was performed by means of numerical simulations. In order to assess the effectiveness of the proposed method, the seismic responses of the excess pore water pressure ratio, acceleration as well as distribution of shear stress and shear strain in the liquefiable soil were evaluated. The analyses result shows that the jet grout columns with horizontal slab method are effective in liquefaction prevention and advanced effective stress UBC3D-PLM can assist the seismic response assessment of the soil.

References

1. Baez, J.I.: A Design Model for the Reduction of Soil Liquefaction by Using Vibro-Stone Columns. University of Southern California, Los Angeles (1995)
2. Brinkgreve, R.B.J: Site response analysis and liquefaction evaluation. www.plaxis.nl (2015).
3. Galavi, V., Petala, A., Brinkgreve, R.B.J.: Finite element modelling of seismic liquefaction in soils. Geotech. Eng. J. SEAGS & AGSSEA **44**(3), 55–64 (2013)
4. Ishii, I., Towhata, I., Hiradate, R., Tsukuni, S., Uchida, A., Sawada, S., Yamauchi, T.: Design of grid wall soil improvement to mitigate soil liquefaction damage in residential areas in Urayasu. J. Jpn. Soc. Civ. Eng. **5**, 27–44 (2017)
5. Kitazume, M.: A parametric study on evaluation of stability of column type DM improved ground. In: Proceeding of International Symposium on Geoenvironmental Engineering, ISGE2009, vol. 9, pp. 920–925 (2009)
6. Kulemeyer, R., Lysmer, J.: Finite element method accuracy for wave propagation problems. J. Soil Mech. Found. Div. ASCE **99**(5), 421–427 (1973).
7. Martin, J.R., Olgun, C.G., Mitchell, J.K., Durgunolu, H.T.: High modulus for liquefaction mitigation. J. Geotech. Geoenviron. Eng. **130**(6), 561–571 (2004)
8. Namikawa, T., Koseki, J., Suzuki, T.: Finite element analysis of lattice-shaped ground improvement by cement-mixing for liquefaction mitigation. Soil Found. **47**(3), 559–576 (2007)
9. Petalas, A., Galavi, V., Brinkgreve, R.B.J.: Validation and verification of a practical constitutive model for predicting liquefaction in sands. In: Proceedings of the 22nd European Young Geotechnical Engineers Conference, Gothenburg, pp. 167–172. Sweden (2012).
10. Plaxis 2D—Material Models Manual 2018. Plaxis B.V, Delft, Netherlands (2018).
11. Rayamajhi, D., Nguyen, T.V., Ashford, S.A., Boulanger, R.W., Lu, J., Elgamal, A., Shao, L.: Numerical study of shear stress distribution for discrete column in liquefiable soil. J. Geotech. Geoenviron. Eng. ASCE **140**(3), 04013034 (2014).
12. Roesset, J.M.: Fundamentals of soil amplification. In: Hansen, R.J. (ed). Seismic Design for Nuclear Power Plants, pp. 183–244 (1970)
13. Seed, H.B., Idriss, I.M.: Simplified procedure for evaluating soil liquefaction potential. J. Soil Mech. Found. Div. ASCE, **97**(9), 1249–1273 (1971)

14. Toki, S., Tatsuoka, F., Miura, S., Yoshimi, Y., Yasuda, S., Makihara, Y.: Cyclic undrained triaxial strength of sand by a cooperative test program. Soils Found. **26**, 117–128 (1986)
15. Tsegaye, A.: Plaxis liquefaction model, external report. PLAXIS knowledge base. www.plaxis.nl (2010).
16. Takahashi, H., Kitazume, M., Ishibashi, S.: Effect of deep mixing wall spacing on liquefaction mitigation. In: Proceedings of the International Conference on Physical Modelling in Geotechnics, vol. 1, pp. 585–590. Hong Kong (2006).
17. Yamashita, K., Hamada, J., Yamada, T.: Field measurements on piled rafts with grid-form deep mixing walls on soft ground. Geotech. Eng. J. SEAGS AGSSEA **42**(2), 1–10 (2011)

Mechanical Behaviors of Assembled Multi-step Cantilever-Retaining Walls for High Embankments by Numerical Simulation Method

Zhaoying Li and Shiguo Xiao

Abstract Assembled multi-step cantilever-retaining walls are a new type of light-retaining structure with low carbon, which can be suitable for high fill earthworks and quickly installed in situ. In order to fully support practical design of the novel-retaining structure, mechanical behaviors of the structure under strip surcharge on the top surface of the backfill in a high embankment example are determined using numerical simulation method via Flac3D program. Analysis results show that the profile of the lateral earth pressure on the vertically assembled walls exhibits multi-segment polyline mode. Apart from the highest step wall member, there is obvious reduction effect of the earth pressure at the top of each wall member. The increases of width of heel plate of wall member, internal friction angle and unit weight of the back-fill are helpful to improve the overall stability of the wall-slope system and reducing lateral displacement of the whole wall. Moreover, considering easy construction of the structure in practice, relatively more steps under allowable conditions are also commonly recommended for the retaining structure.

Keywords Multi-step cantilever-retaining walls · Numerical simulation method · High embankments · Earth pressure · Overall stability

1 Introduction

Cantilever-retaining wall is one of light-retaining structures and often used for retaining embankments in the case of strict land restriction or low bearing capacity of foundation soil. However, height of the traditional cantilever-retaining wall is usually limited. For example, it should not be more than 6 m according to China Norm [1]. So, the traditional wall cannot be suitable for high embankments. In view of this,

Z. Li
Department of Geological Engineering, Southwest Jiaotong University, Chengdu 610031, China

S. Xiao (✉)
Key Laboratory of High-Speed Railway Engineering, Ministry of Education, Southwest Jiaotong University, Chengdu 610031, China
e-mail: xiaoshiguo@swjtu.cn

H. Hazarika et al. (eds.), *Advances in Sustainable Construction and Resource Management*, Lecture Notes in Civil Engineering 144,
https://doi.org/10.1007/978-981-16-0077-7_27

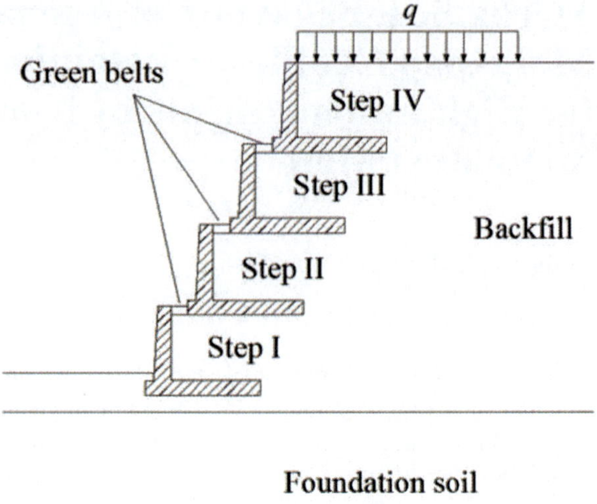

Fig. 1 Sketch map of multi-step cantilever-retaining walls

assembled multi-step cantilever-retaining walls are proposed in this paper to retain embankments with higher height such as 8 m, 10 m and much more (see Fig. 1). The new type of light-retaining structures can be constructed by assembling prefabricated members, each of which is just a traditional cantilever-retaining wall with smaller height.

Some researchers studied static performance of traditionally single cantilever-retaining wall [2–5]. Also, dynamic behaviors of the single cantilever wall are analyzed by some scholars [6–8]. Although there are some investigations on the traditional cantilever-retaining wall, mechanical behaviors of the new retaining structures are hardly reported in the literature. In order to fully study earth pressure on each wall, wall displacements and wall-slope overall stability, numerical simulation method via Flac3D program is used to solve the problem. Further, several influencing factors are analyzed, and the optimum type of the proposed retaining structures is provided in a practical high embankment of highway.

2 Numerical Modeling

The numerical model of the proposed structure used to retain fill earthworks can be established using FLAC3D program. Figure 2 shows a calculation model of four-step cantilever-retaining walls. The walls are usually prefabricated with C30 to C40 concrete in practice, so they are considered as elastic bodies in the model. The backfill, foundation soil and clay in green belts are modeled using perfectly elastic–plastic constitutive model with the associated flow rule and Mohr–Coulomb yield criterion. Strength parameters at the interfaces between the walls and backfills are assumed as 0.7 times of the soil ones. The distance from the toe of the lowest-step wall member to the left boundary with horizontal constraint is assumed as the fill height H, and

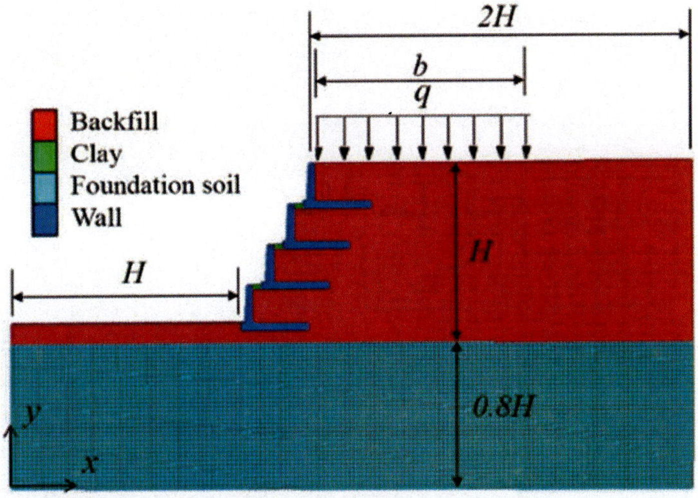

Fig. 2 Generally numerical simulation model of a slope retained by the wall

the distance from the top of the highest-step wall member to the right boundary with horizontal constraint is assumed as $2H$. Depth of the foundation soil is assumed as $0.8H$, and the bottom boundary is fixed in horizontal and vertical directions. The model with 1 m thickness out of plane can be meshed using eight-node hexahedron elements. The displacements of all nodes in the model are constrained in the direction perpendicular to the plane.

3 Computation Results of an Example

A practical embankment with 8.65 m height in a highway engineering in Sichuan Province is retained by the four-step cantilever-retaining walls precast with C35 concrete. The foundation soil consists of gravel soil, and well-compacted coarse sand is layered on it. The height of each wall member is 2.15 m, the total height of multi-step walls is 8 m, and 30-cm-thick clay layers partially exist between the walls connected by a series of discrete concrete blocks. The main physical and mechanical parameters in the example are shown in Table 1, which is obtained by field investigations and laboratory tests. Strip surcharge on the road surface is $q =$ 30 kPa with distribution width $b = 7.5$ m. According to the numerical modeling method mentioned above, the numerical model of the example is shown in Fig. 3, which has 10,372 elements and 21,230 nodes.

Table 1 Physical and mechanical parameters in the example

	Elastic modulus (MPa)	Poisson's ratio	Cohesion (kPa)	Internal friction angle (°)	Unit weight (kN/m^3)
Walls and Blocks	31,500	0.2	–	–	25
Clay	20	0.45	28	11	18
Backfill	30	0.32	0	35	20
Foundation soil	30	0.32	0	35	19

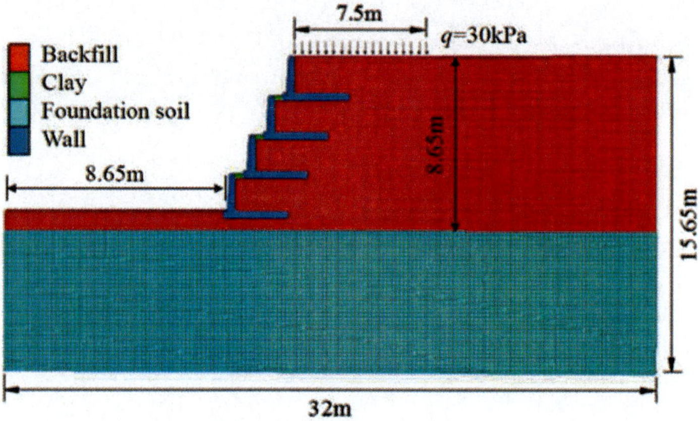

Fig. 3 Numerical model of the practical embankment retained by the wall

3.1 Earth Pressure

Distribution of the lateral earth pressure on the wall back is shown in Fig. 4, which indicates profile of the earth pressure on the whole wall is nearly multi-segment parabola regardless of small fluctuations due to shielding action of the nearby heel plates inserted in the filling soil. For each wall member, the maximum of the earth pressure occurs at its middle-lower part, but the minimum one lies basically at its top.

3.2 Wall Displacements

As shown in Fig. 5, distribution of the horizontal displacement of the whole wall exhibits parabola mode with the maximum 7.01 mm at the bottom of the step II wall

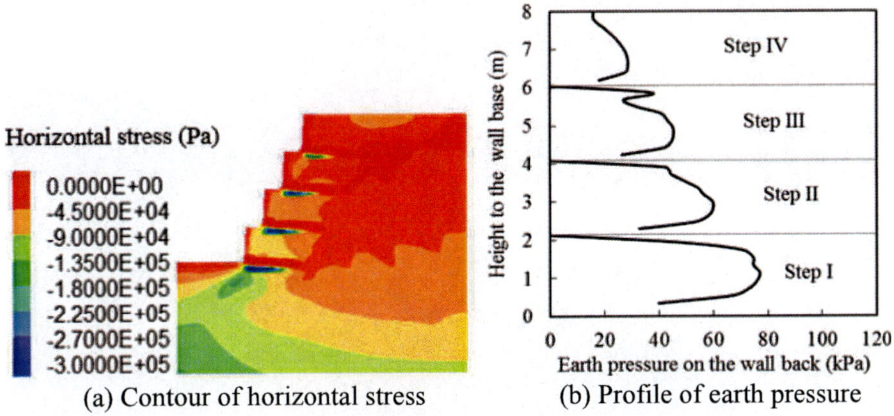

Fig. 4 Distribution of lateral earth pressure on the wall back

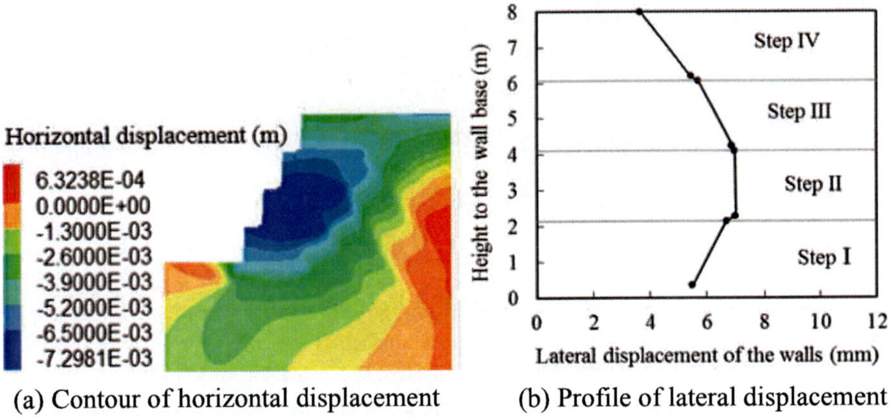

Fig. 5 Distribution of lateral displacement of the walls

member. In fact, movement of the whole wall is composite mode composed of translation and rotation. The upper three wall members rotate substantially toward the interior of the backfill, and the lowest wall member rotates slightly toward the opposite direction, while the whole wall translates toward the exterior of the embankment.

3.3 Wall-Slope Overall Stability

According to the maximum shear strain increment contour (see Fig. 6), critical slip band of the wall-slope system is passing by the heel of the lowest-step wall member with factor of safety $F_s = 1.38$. To some extent, one can simply assume a circular

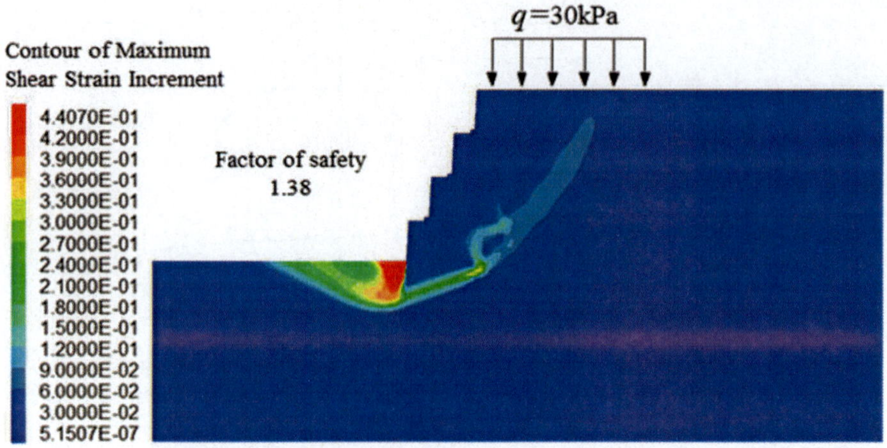

Fig. 6 Critical slip surface of the retained slope

slip surface passing by the lowest heel to be the critical slip surface of the wall-slope system.

4 Parameter Study and Structure Optimization Configuration

Some factors including number of wall members together with geometric parameters and physical properties of each member have important effect on the mechanical behaviors of the assembled walls. Based on the practical embankment example mentioned above and the control variable method, the influences of number of members, unit weight and internal friction angle of the backfill and width of the heel plate on mechanical behaviors of the wall are discussed as follows.

4.1 Number of Members

Table 2 shows influence of number of vertically assembled wall members on the wall-slope overall stability in the practical example. It can be seen that the factor of safety gets the minimum and maximum values under 5 and 2 wall members, respectively, which means type III is the comprehensively optimum structure form considering easy operation during prefabrication and installation of these members under the requirement of design factor of safety not less than 1.35. In fact, type III was used in the practical embankment engineering mentioned above, and it works well, which means the analysis results are acceptable.

Table 2 Factor of safety under various number of wall members

Type	Number of wall members	F_s
I	2	1.88
II	3	1.55
III	4	1.38
IV	5	1.29

4.2 Unit Weight of Backfill

Figure 7 shows the variation of lateral earth pressure distribution on the wall back with unit weight of the backfill in the case of four-step wall. The earth pressure on the wall back is increasing with the unit weight of the backfill, but their distribution modes under various unit weights are similar. Additionally, the maximum lateral displacement of the wall is increasing with the unit weight (see Table 3). However, the factor of safety of the wall-slope system is not decreasing as the unit weight increases, which is different from the general gravity-retaining wall. The reason lies that the factor of safety is influenced by not only earth pressure on the back of each wall member but also self-weight of the local backfill over each heel plate. In fact, the local backfill over each heel plate contributes resistant action for the wall member

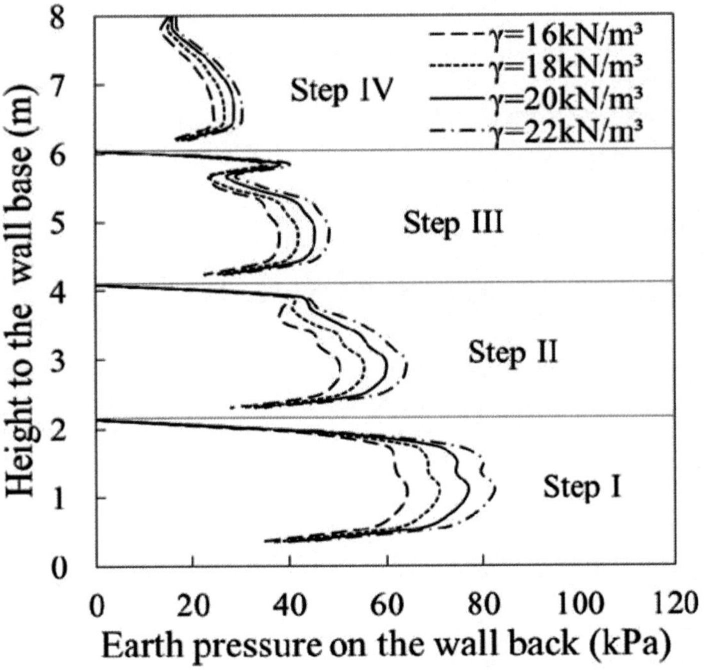

Fig. 7 Distribution of earth pressure on the wall back under various unit weights of backfill

Table 3 Responses of wall-slope system under various unit weights of backfill

Unit weight (kN/m^3)	16	18	20	22
Maximum lateral displacement of the whole wall (mm)	6.15	6.84	7.01	7.22
Resultant of earth pressure on the wall back (kN/m)	265.3	289.2	313.9	334.2
Factor of safety	1.38	1.38	1.38	1.39

Table 4 Responses of wall-slope system under various internal friction angles of backfill

Internal friction angle (°)	31	33	35	37
Maximum lateral displacement of the whole wall (mm)	16.86	9.91	7.01	4.69
Resultant of earth pressure on the wall back (kN/m)	332.4	315.0	313.9	311.1
Factor of safety	1.21	1.29	1.38	1.48

by equivalently adding self-weight of the whole wall. Although the earth pressure on the wall is increasing to some degree with the unit weight, the resistant effect of raising self-weight of the wall is predominant.

4.3 *Internal Friction Angle of Backfill*

As the internal friction angle of backfill increases, the earth pressure on the wall back is slightly decreasing (see Table 4). However, the maximum lateral displacement of the wall is decreasing obviously with the increase of the internal friction angle. Besides, the factor of safety of the wall-slope system is apparently increasing with the internal friction angle as expected.

4.4 *Width of Heel Plate of Wall Member*

Table 5 shows the maximum lateral displacement of the wall and earth pressure on the wall back are decreasing, respectively, with the increase of width of the heel plate of wall member. The results can be explained according to the shielding action of the earth pressure caused by the nearby heel plates inserted in the filling soil. Moreover,

Table 5 Responses of wall-slope system under various widths of the heel plate

Width of the heel plate (m)	1.62	2.12	2.62	3.12
Maximum lateral displacement of the whole wall (mm)	13.39	8.32	7.01	6.18
Resultant of earth pressure on the wall back (kN/m)	333.4	319.4	313.9	305.0
Factor of safety	1.23	1.32	1.38	1.45

the factor of safety of the wall-slope system is increasing as the width of the heel plate increases. The reason lies in two aspects including the reduction of the earth pressure and the increase of equivalent self-weight of the whole wall with the width of the heel plate.

5 Conclusions

(1) Assembled multi-step cantilever-retaining walls can be utilized for high fill earthworks, which can maintain the retained slope to be of high stability.
(2) Profile of the lateral earth pressure on the vertically assembled walls step by step exhibits multi-segment parabola mode. Apart from the highest-step wall member, there is obvious reduction effect of the earth pressure at the top of each wall member.
(3) Stability of the slope retained vertically by less wall members is better than that by more members. But the latter can be much more easily constructed in practice. So, relatively more steps under allowable conditions in practice are commonly recommended for the retaining structure.
(4) The increases of width of heel plate of the wall member, internal friction angle and unit weight of the backfill are the benefits for improving the overall stability of the wall-slope system and reducing lateral displacement of the whole wall. So, lengthening the heel plate to feasible extent in design and considerably improving degree of compaction of the backfill in construction are strongly suggested.

References

1. Code for design on retaining structures of railway subgrade (TB 100025–2006). China Railway Publishing House, Beijing (2014)
2. Bentler, J.G., Labuz, J.F.: Performance of a cantilever retaining wall. J. Geotech. Geoenviron. Eng. **132**(8), 1062–1070 (2006)
3. Chugh, A.K., Labuz, J.F.: Numerical simulation of an instrumented cantilever retaining wall. Can. Geotech. J. **48**(9), 1303–1313 (2011)
4. Greco, V.R.: Active earth thrust on cantilever walls in general conditions. Soils Foundations **39**(6), 65–78 (1999)
5. Ranjbar, K.A., Ganjian, N., Askari, F.: Stability analysis and design of cantilever retaining walls with regard to possible failure mechanisms: an upper bound limit analysis approach. Geotech. Geol. Eng. **35**(3), 1079–1092 (2017)
6. Cakir, T.: Backfill and subsoil interaction effects on seismic behavior of a cantilever wall. Geomech. Eng. **6**(2), 117–138 (2014)
7. Giri, D.: Pseudo-dynamic methods for seismic passive earth pressure behind a cantilever retaining wall with inclined backfill. Geomech. Geoeng. **9**(1), 72–78 (2014)
8. Osouli, A., Zamiran, S.: The effect of backfill cohesion on seismic response of cantilever retaining walls using fully dynamic analysis. Comput. Geotech. **89**, 143–152 (2017)

Microscopic Analysis of the Influence of Pore Size Distribution on SWCC Using Extended DDA

Longxiao Guo and Guangqi Chen

Abstract Soil–water characteristic curve (SWCC) is one of the most important properties of unsaturated soil, and pore size distribution (PSD) curve is commonly used to predict SWCC. In this study, an extended discontinuous deformation analysis (DDA) method is used to reveal the influence of PSD on SWCC from the micromechanism. First, the performance of the extended DDA method is validated by the experimental SWCC. And DDA microsimulation result is in good agreement with the experimental result. Then, the relationship between PSD curve and SWCC is analyzed based on the DDA simulation results. The microsimulation results show that PSD influences SWCC by changing the capillary radius in soil. Air-entry value decreases with the percentage of large pore increase. However, the effect of PSD on SWCC tends to diminish at the residual zone because the micropores do not show much difference between two models. This study extends DDA in unsaturated soil analysis and provides support for PSD-based SWCC prediction from the micromechanism.

Keywords Discontinuous deformation analysis · Unsaturated soil · Soil–water characteristic curve · Pore size distribution

1 Introduction

Unsaturated soil refers to the soil whose saturation is between 100 and 0. Comparing to saturated soil, unsaturated soil widely exists in nature, which in particular covers most of the arid and semiarid areas [1]. Unsaturated soil is regarded as three-phase material that contains soil particles, water, and air, and it embraces a wide range of soil types such as expansive soils, loess, and residual soil. They all have the basic characteristics, that is, there is matric suction in the soil. The presence of matric

L. Guo (✉) · G. Chen
Department of Civil and Structural Engineering, Kyushu University, Fukuoka 819-0395, Japan
e-mail: guo.longxiao.134@s.kyushu-u.ac.jp

G. Chen
e-mail: chen@civil.kyushu-u.ac.jp

H. Hazarika et al. (eds.), *Advances in Sustainable Construction and Resource Management*, Lecture Notes in Civil Engineering 144,
https://doi.org/10.1007/978-981-16-0077-7_28

suction makes soil behavior more complicated, and it does not adhere to the behaviors of saturated soil [2]. The swelling of expensive soil [3] and the collapsibility of loess that are closely related to matric suction can lead to a severe situation for engineering practice. However, saturated soil mechanics did not involve these important issues. According to the numerous researches and experiments on the unsaturated soils, the basic trend can be assured: The properties of shear strength and compressibility of unsaturated soil have a relatively positive relationship with the matric suction [4]. All the work has shown the importance of the role of matric suction. Therefore, unsaturated soil mechanics as an extension of saturated soil mechanics focuses more on the water-related properties, and one of the core issues is matric suction.

Soil–water characteristic curve (SWCC) is known as a constitutive relationship between water content and matric suction of soil. SWCC is not only providing a conceptual understanding of the energy state of the water phase in the soil but also is used to predict the permeability, shear strength of unsaturated soil [5]. For engineering practice, most engineering problems that happened in the unsaturated soil zone involve in matric suction loss such as rainfall, irrigation, and snowmelt. A good understanding of matric suction can help us in engineering practice in the unsaturated zone. For example, the landslide is a common natural disaster and causes casualties and property damage. Especially, rainfall-induced landslides can occur in a wide variety of climatic, geologic, and topographic settings, which are basic in wholly or partly the unsaturated zone. According to a recent survey, about half of the 40 most destructive landslide disasters worldwide in the past century resulted from prolonged or intense rainfall [6]. Although the volume of individual shallow landslides is often small, typically less than 1000 m^3, extensive areas are often affected [7]. For example, a shallow landslide triggered by rainfall in Yan'an, northwestern China, July 2013. It caused 8135 slope failure, 10,000 living caves damaged, and 45 fatalities. The mechanism analysis of landslides on long-term rainfall should consider the influence of matric suction. Water content increase in the soil would cause matric suction loss. Consequently, the factor of safety of slope will decrease, and then, slope failure occurred. Therefore, how matric suction changes with water content is the key issue for unsaturated soil research.

PSD is a pore characteristic that is commonly used to predict SWCC [8]. Traditionally, the research on the PSD and SWCC is based on the experiments, and the pressure plate apparatus, tensiometer techniques [9], etc., are the commonly used methods. With the development of computer technology, the numerical method can be an alternative approach for analyzing the SWCC. Discrete element method (DEM) [10] is increasingly used in the study of unsaturated soil at microscale [11]. DEM simulations can obtain detailed internal information such as coordination number and force chain efficiently [12]. Recently, DDA [13], a member of the family of DEM, has been employed in the studies of unsaturated soil. DDA has the advantages in microscale soil analysis, such as it has a variety of particle shapes that can be used to form complex soil microstructure. The simulation results have indicated that the extended DDA is a reliable and applicable numerical method to explore unsaturated soil behaviors [14].

In this research, the authors analyze the influence of PSD on SWCC from the micromechanism using the extended DDA method. First, the extended DDA is briefly introduced. Then, the experimental SWCC of Toyoura sand is used to validate the performance of extended DDA. Finally, the influence of PSD on SWCC is simulated and analyzed quantitatively. This study extends DDA in unsaturated soil analysis and provides support for PSD-based SWCC prediction from a micromechanism.

2 Methodology

2.1 *Fundamental Principle of Discontinuous Deformation Analysis (DDA)*

2D DDA, proposed by Shi (Shi, 1985), is a promising numerical method to simulate large deformation and displacement problems. It has been wildly used for modeling landslides, rockfall, and tunnel construction. DDA elements are the discrete blocks which are the deformable objects of any shape, and each block i has six unknown variables: rigid-body displacement (u_0, v_0), rigid-body rotation r_0, normal strain (ε_x, ε_y), and shear strain γ_{xy}, and it can be written as the following deformation matrix **D**:

$$\mathbf{D} = (u_0, v_0, r_0, \varepsilon_x, \varepsilon_y, \gamma_{xy})^{\mathrm{T}} \tag{1}$$

The displacement (u, v) of an arbitrary point (x, y) within a block is described as follows:

$$\begin{pmatrix} u \\ v \end{pmatrix} = \begin{pmatrix} 1 & 0 & -(y-y_0) & (x-x_0) & 0 & (y-y_0)/2 \\ 0 & 1 & (x-x_0) & 0 & (y-y_0) & (x-x_0)/2 \end{pmatrix} \begin{Bmatrix} u_0 \\ v_0 \\ r_0 \\ \varepsilon_x \\ \varepsilon_y \\ \varepsilon_{xy} \end{Bmatrix} = \mathbf{TD} \tag{2}$$

where (x_0, y_0) is the centroid of the block.

DDA is an energy-based implicit method, and it is established based on minimizing the total potential energy of the system. The simultaneous equilibrium equations of block system can be expressed as follows:

$$[\mathbf{M}]\{\ddot{\mathbf{D}}\}+[\mathbf{C}]\{\dot{\mathbf{D}}\}+[\mathbf{K}]\{\mathbf{D}\}=\{\mathbf{F}(t)\} \tag{3}$$

where [**M**] is the mass matrix; [**C**] is damping matrix; [**K**] is stiffness matrix; $\{\ddot{\mathbf{D}}\}$, $\{\dot{\mathbf{D}}\}$, and {**D**} are the acceleration vector, velocity vector, and displacement vector, respectively; and {**F**(t)} is the external force vector.

Assume that there are *n* blocks in a block system, the global equilibrium equation of the blocks is presented as follows:

$$\begin{bmatrix} \mathbf{K_{11}} & \mathbf{K_{12}} & \cdots & \mathbf{K_{1n}} \\ \mathbf{K_{21}} & \mathbf{K_{22}} & \cdots & \mathbf{K_{2n}} \\ \vdots & \vdots & \ddots & \vdots \\ \mathbf{K_{n1}} & \mathbf{K_{n2}} & \cdots & \mathbf{K_{nn}} \end{bmatrix} \begin{bmatrix} \mathbf{D_1} \\ \mathbf{D_2} \\ \vdots \\ \mathbf{D_n} \end{bmatrix} = \begin{bmatrix} \mathbf{F_1} \\ \mathbf{F_2} \\ \vdots \\ \mathbf{F_n} \end{bmatrix} \tag{4}$$

where K_{ij} is a 6×6 coefficient submatrix; F_i is a 6×1 loading submatrix on block *i*; $i, j = 1, 2, \ldots, n$ and *n* is the number of blocks in the defined block system.

2.2 The Extended DDA Method and Typical SWCC

Since DDA has advantages for analyzing the material like granular aggregates, a newly developed algorithm is incorporating with the conventional DDA program to simulate the mechanical behavior of unsaturated soil [14]. The fundamental problem of unsaturated soil simulation is to exert the matric suction among soil particles, meanwhile, calculate the capillary water volume under certain matric suction. In this simulation, soil particles are idealized as DDA blocks, and the matric suction is exerted among blocks based on Young–Laplace equation. The algorithm and its implementation based on the original DDA method and the capillary mechanics have been accomplished. In extended DDA, there are three main parts of the program implementation: 1. Determine potential contractile skin of particle; 2. Determine reasonable capillary water distribution among the multi-particle system; 3. Add matric suction and surface tension on the particle. Extended DDA can faithfully reflect the mechanism of the capillary action in the soil, which makes it possible to simulate SWCC microscopically.

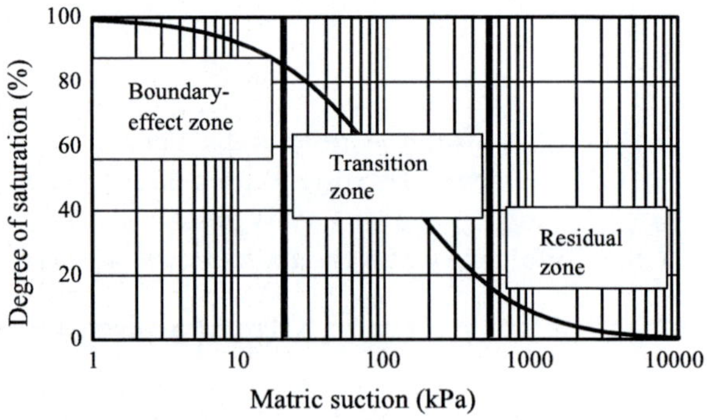

Fig. 1 Typical soil–water characteristic curve (SWCC)

A typical SWCC is shown in Fig. 1, which mainly divides into three zones: boundary-effect zone, transition zone, and residual zone. Correspondingly, unsaturated soil can be divided into three stages with the water filling in the pores, namely the boundary effect stage, transition stage, and residual stage. The transition zone is mainly discussed in unsaturated soil mechanics. In this zone, capillary water generally expands into a continuous water body in soil, and both water and air are continuous. Thus, it is also called water–air bi-open state for this zone, and capillary action provides the main part of matric suction. In this research, it is assumed that the matric suction is subjected to capillary mechanics.

3 Simulation of SWCC

3.1 Validation Case: SWCC of Toyoura Sand

Toyoura sand is a kind of naturally formed sand, and it is used as a standard material for experimental tests in Japan. An experimental SWCC of Toyoura sand is employed in this research. Abe [15] (1994) conducted SWCC tests on a Toyoura sand specimen by the pressure plate method. Physical parameters of Toyoura sand and DDA model are shown in Table 1. This experimental test result is adopted to validate the extended DDA method. The micro-DDA model and the important physical parameters are shown in Fig. 2 and Table 1, respectively. Grain size distribution (GSD) curve of Toyoura sand and DDA models are shown in Fig. 3.

Table 1 Physical parameters of Toyoura sand and DDA model

Material	Specific gravity, Gs	Median diameter, D50 (mm)	Uniformity coefficient, Uc	Coefficient of curvature, Cc
Toyoura sand	2.65	0.162	1.55	0.86
DDA particle	2.65	0.163	1.55	0.86

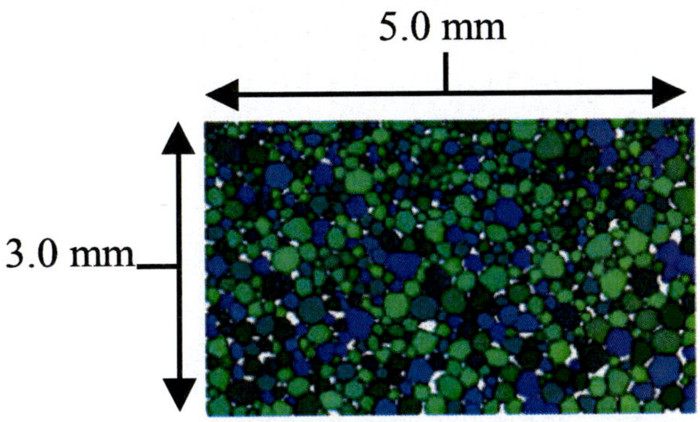

Fig. 2 Microscopic DDA model

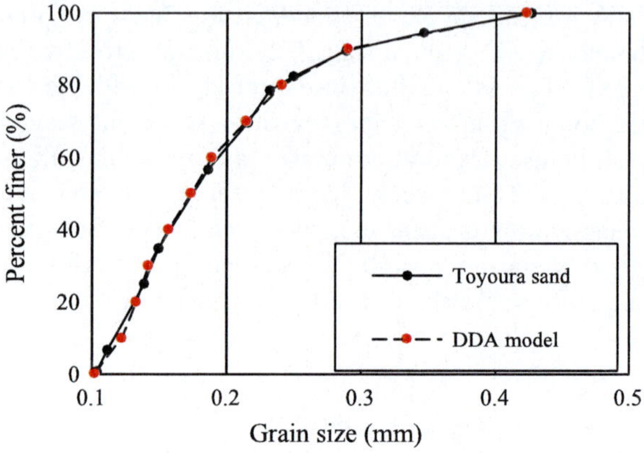

Fig. 3 GSD curve of Toyoura sand and DDA model

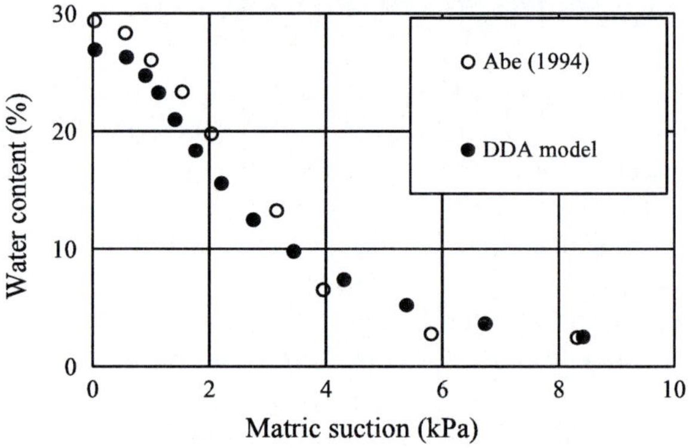

Fig. 4 Comparison between experimental result and DDA simulation result for SWCC of Toyoura sand

Figure 4 shows the comparison of the SWCC of experimental results and DDA simulation results. The simulation results have a good agreement with Abe's experimental SWCC. With the increase of the matric suction, water content sharply decreases within a suction range from 0 to 4 kPa, then gently decreases within a suction range from 4 to 8 kPa. Considering the various errors, although there are slight differences in the AEV between the test results and simulation results, the simulation results can quantitatively reflect the characteristics of SWCC. The comparison indicates that the extended DDA is qualified for the microanalysis of the SWCC of sandy soil.

3.2 Microscale Analysis on the Influence of PSD on SWCC

To analyze the influence of PSD on the SWCC, models with the same DDA blocks and the different PSD are built. As shown in Fig. 5, two models are made up of 600

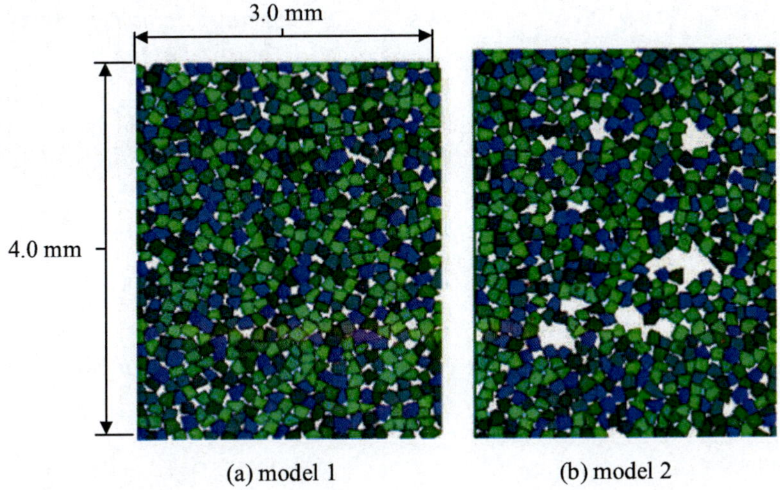

Fig. 5 Two micro-DDA models with different PSD

DDA blocks. Grain size distribution (GSD) curve and PSD curve of DDA models are shown in Figs. 6 and 7, respectively. The GSD of two models is the same. The minimum pore size of four models is around 0.01 mm, and the maximum pore size

Fig. 6 GSD curves of two DDA models

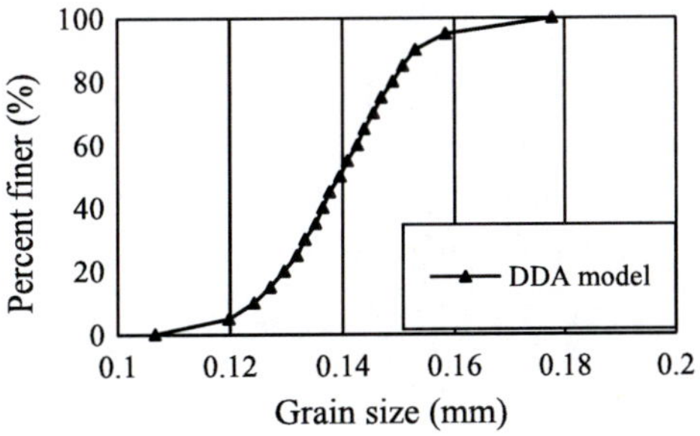

Fig. 7 PSD curves of two DDA models

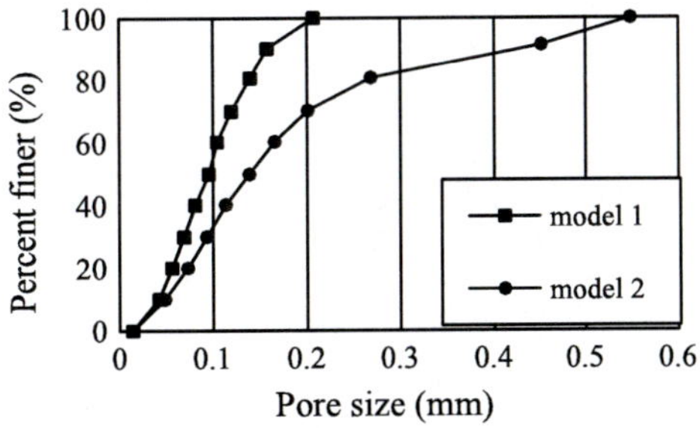

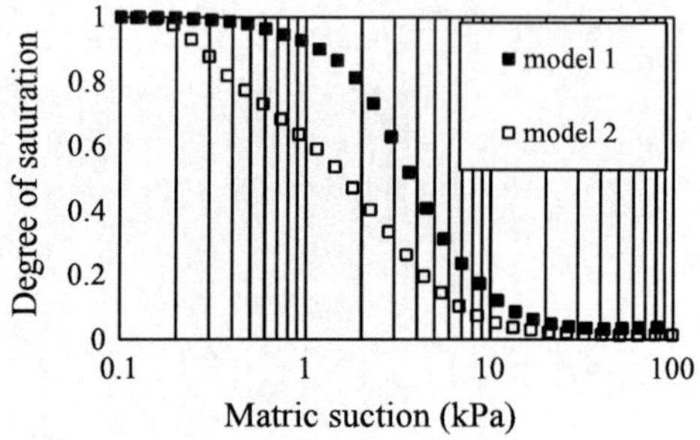

Fig. 8 SWCCs of two DDA models

is 0.21 mm, and 0.54 mm, respectively. The simulated SWCCs of two models are shown in Fig. 8.

4 Conclusions

This paper has analyzed the relationship between PSD and SWCC from the micromechanism based on the extended DDA method. The simulation results show that the PSD is a main variable factor of SWCC. There is a positive relationship between the slope of PSD curve and SWCC in the transition zone. Air-entry value decreases with the percentage of large pore increase. However, the micropores do not show much difference between two models, that is, the effect of PSD on SWCC tends to diminish at the residual zone. According to capillary mechanics, the capillary radius directly determines the value of matric suction, and PSD indirectly affects matric suction by changing the capillary radius of the soil. This study extends DDA in unsaturated soil analysis and provides support for PSD-based SWCC prediction from the micromechanism.

References

1. Fredlund, D.G., Rahardjo, H.: Soil mechanics for unsaturated soils. John Wiley & Sons (1993)
2. Mitchell, J.K., Soga, K.: Fundamentals of soil behavior, vol. 3. John Wiley & Sons New York (2005)
3. Li, Z., et al.: Study on SWCC of unsaturated expansive soil. Yantu Lixue (Rock Soil Mech.) **27**(5), 730–734 (2006)
4. Fredlund, D., Morgenstern, N.R., Widger, R.: The shear strength of unsaturated soils. Can. Geotech. J. **15**(3), 313–321 (1978)
5. Lu, N., Likos, W.J.: Suction stress characteristic curve for unsaturated soil. J. Geotech. Geoenviron. Eng. **132**(2), 131–142 (2006)
6. Sidle, R., Ochiai, H.: Processes, prediction, and land use Water resources monograph. American Geophysical Union, Washington (2006)
7. Lu, N., Godt, J.W.: Hillslope hydrology and stability. Cambridge University Press (2013)

8. Zhai, Q., et al.: Role of the pore-size distribution function on water flow in unsaturated soil. J. Zhejiang Univ.-Sci. A **20**(1), 10–20 (2019)
9. Lourenço, S., et al.: Development of a commercial tensiometer for triaxial testing of unsaturated soils. Unsaturated Soils **2006**, 1875–1886 (2006)
10. Cundall, P.A., Strack, O.D.: A discrete numerical model for granular assemblies. Geotechnique, **29**(1), 47–65 (1979)
11. Jiang, M., Leroueil, S., Konrad, J.: Insight into shear strength functions of unsaturated granulates by DEM analyses. Comput. Geotech. **31**(6), 473–489 (2004)
12. Scholtès, L., et al.: Micromechanics of granular materials with capillary effects. Int. J. Eng. Sci. **47**(1), 64–75 (2009)
13. Shi, G.H., Goodman, R.E.: Two dimensional discontinuous deformation analysis. Int. J. Numer. Anal. Methods Geomech. **9**(6), 541–556 (1985)
14. Guo, L., et al.: A method for microscopic unsaturated soil-water interaction analysis based on DDA. Comput. Geotech. **108**, 143–151 (2019)
15. Abe, H.: Experimental study on the estimation of mechanical properties for unsaturated soils. Ph. D. dissertation, University of Tokyo (1994)

Physical Model Tests for Newly Developed Breakwater Foundation Subjected to Earthquake and Tsunami

Babloo Chaudhary, Hemanta Hazarika, Akira Murakami, and Kazunori Fujisawa

Abstract Many breakwaters damaged due to earthquake and tsunami in the past. For example, several breakwaters collapsed by the 2011 off the Pacific Coast of Tohoku Earthquake and subsequent tsunami. It was found that these breakwaters damaged mainly due to the failure of their foundations. Therefore, countermeasures are urgently needed to be developed for breakwater foundation in order to make the breakwater safe against earthquake and tsunami. Recently, new countermeasures were developed by the authors for breakwater foundation in order to make it resilient against an earthquake and tsunami. This paper deals with evaluation of effectiveness of the developed foundation model by conducting shaking table tests and tsunami overflow tests. As reinforcing countermeasures, steel sheet piles and gabions are provided in the breakwater foundation. To see the performance of the developed model, comparisons are made between the developed foundation model and conventional foundation. Through the tests, it was found that the reinforced foundation performed well in reducing damage of the breakwater caused by the earthquake and tsunami.

Keywords Breakwater · Resilient foundation · Earthquake and tsunami

1 Background

Breakwater, an offshore structure, is mainly build for protection of ports and harbors from destructive effects of sea waves, typhoons and even tsunamis. It provides calm sea water for smooth movement of ships and other harbor activities at the seaport. It

B. Chaudhary (✉)
National Institute of Technology Karnataka (NITK) Surathkal, Mangalore 575025, India
e-mail: babloomit@gmail.com

H. Hazarika
Kyushu University, Fukuoka 819-0395, Japan

A. Murakami · K. Fujisawa
Kyoto University, Kyoto 606-8502, Japan

H. Hazarika et al. (eds.), *Advances in Sustainable Construction and Resource Management*, Lecture Notes in Civil Engineering 144,
https://doi.org/10.1007/978-981-16-0077-7_29

was found during past few decades that breakwaters are vulnerable to earthquake and tsunami. Many breakwaters collapsed by the 2011 off the Pacific Coast of Tohoku Earthquake and subsequent tsunami. The world deepest breakwater at Kamaishi port (Iwate Prefecture, Japan) was one of them which collapsed and failed to block the incoming tsunami waves. Foundation failure was the main reason of the collapse of the breakwater [1–5]. The rubble mound was scoured by the tsunami. Moreover, the caissons slid over the mound and sank into the sea. Due to the failure of the breakwaters, the tsunami could enter into the Kamaishi Port. Thus, the tsunami created deep devastation there. There were other breakwaters in Japan which were also collapsed mainly due to the failure of their foundations during the earthquake and tsunami [6–10]. Therefore, stability of foundation is very important to safety of breakwater during an earthquake and tsunami. Stability of breakwater is utmost important for safety of costal activities and human lives during such compound geo-disaster caused by an earthquake and subsequent tsunami.

Japan is one of the most earthquake-prone countries in the world. The predicted forthcoming mega-earthquakes, such as Tokai, Nankai and Tonankai earthquakes, are matter of great concerned for researchers and policy makers. These mega-earthquakes periodically occur about every 100–150 years. Not only in Japan but also in various other parts of the world, there are many earthquake prone coastal areas which are susceptible to be attacked by strong earthquakes, and the earthquakes can generate tsunami in the future. In order to prevent or at least reduce devastating damage triggered by earthquake and tsunami in the future, breakwater foundation should be enough strong to withstand earthquakes and tsunamis-induced forces. Therefore, it utmost important to developed countermeasure for breakwater foundation which can make the breakwater earthquake resistant and tsunami resilient.

Some research works have been done to develop countermeasures for breakwater foundation against tsunami. For example, Oikawa et al. [11] installed a row of steel piles in harbor side mound to increase stability of breakwater foundation against tsunami. Maruyama et al. [12] suggested to provide armor units (concrete blocks) on harbor side rubble mound in order to prevent any foundation failure caused by tsunami. Ueda et al. [13] proposed to cover seaside mound by membranes which can prevent seepage through the mound caused by water level difference during tsunami. Kikuchi et al. [14] installed wall of piles to reduce the impact of tsunami forces. However, all these countermeasures were suggested for enhancing the stability of breakwater foundation during tsunami only. These countermeasures were not aimed to mitigate the earthquake (that precedes a tsunami) induced damage of breakwater foundation.

A tsunami is generally preceded by a strong earthquake. Pore water pressure is generated in foundation soil by the earthquake which can cause deformation of foundation soils. The deformation of seabed foundation can result in settlement of the breakwater. Due to this settlement, height of the breakwater decreases. In this condition, if a tsunami strikes the breakwater, the breakwater may not block the tsunami waves, and the tsunami can easily overflow the breakwater. Therefore, earthquake-induced settlement of the breakwater should be prevented. This phenomenon makes tsunami as a compound disaster caused by both the earthquake and tsunami (not caused by only tsunami). Therefore, for an effective and practical solution for this problem, countermeasure must be effective against both the earthquake and tsunami. To see the effects of earthquake, that precedes tsunami, centrifuge model tests were conducted, and it was found that the earthquake-induced settlement of the breakwater was about 2 m [15]. To the end, new methodology was developed by Chaudhary et al. [16] to determine stability of a breakwater against combined effect of an earthquake and tsunami. Some reinforcing countermeasures were also developed for breakwater foundation against earthquake and subsequent tsunami [17–29]. New countermeasures have been developed by the authors for breakwater foundation to make the breakwater resilient against both the earthquake and tsunami. The authors used sheet piles and gabions as countermeasure elements, and developed a reinforcing model for breakwater foundation. The developed foundation model can reduce damage of breakwater caused by earthquake and tsunami.

2 Newly Developed Foundation Model

As countermeasures, the authors developed a new reinforcing foundation model which can provide resiliency to a breakwater against both the earthquake and subsequent tsunami. In this model, both sides of rubble mound were covered by several gabions as shown in Fig. 1. The purpose of putting the gabions over the mound is to protect the mound from scouring caused by the tsunami. It can also reduce lateral displacement of caisson due to friction between rubble mound and gabions during an earthquake. In addition, steel sheet piles are installed in the foundation (see Fig. 1). These sheet piles behave as cut off walls, and can prevent seepage beneath the caisson during the tsunami. The sheet pile can also restrict lateral deformation of seabed soils during the earthquake. It can also prevent deformation of rubble mound during the earthquake. In this way, deformation of seabed soils and rubble mound

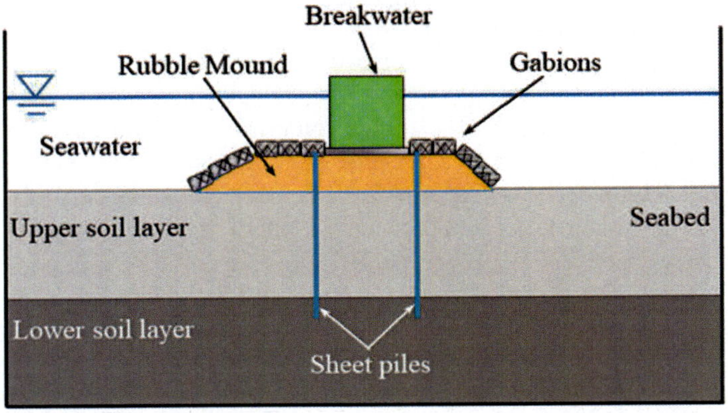

Fig. 1 Reinforcing foundation model for resilient breakwater against earthquake and tsunami

can be prevented or reduced up to great extent. Thus, settlement and horizontal displacement of breakwater can be reduced during the earthquake and tsunami. Such multifunctional measures can prevent any bearing capacity failure of the foundation during the earthquake and tsunami. In the following sections, effectiveness of the reinforcing model is described against earthquake and tsunami by conducting physical model tests. Shaking table tests and tsunami overflow tests were conducted to examine the performance of the developed model against earthquake and tsunami, respectively.

3 Shaking Table Tests

Shaking table tests were conducted for model breakwater for different earthquake ground motions. The acceleration, pore water pressure, water pressure, settlement and horizontal displacement of the caisson were monitored during the experiments. Through these tests, the reinforcing effects of the developed model under earthquake loadings were examined.

3.1 Model Description

The breakwater at Miyazaki port (Miyazaki Prefecture, Japan), which is likely to be affected by the predicted Nankai earthquake, was chosen as a prototype. About 2 m seismic subsidence is predicted in the Kochi area of Shikoku Island, Japan, by the Nankai earthquake [15]. The seismic subsidence might be due to widespread tectonic movement associated with the movement of the earthquake fault. Prototype to model ratio was 64. The breakwater consists of three box type caissons which were spanning across the soil box. The caissons were made of aluminum and filled with silica sands and lead balls to adjust weight (specific gravity 2.4) and center of gravity. One main caisson was at center, and two dummy caissons were on both sides of the main caisson. The seabed was made of two layers of Toyoura sands. The relative density of lower layer was 90%, and that of the upper layer was 60%. The rubble mound was made of crushed stones. The gabion was made of the crushed stones and wrapped with steel wire mesh. The dynamic characteristics of the sands and gravels against cyclic loading were determined through element testing [25, 26]. Sheet pile was made of steel plate of 3.2 mm thickness. Properties of foundation materials are listed in Table 1.

Table 1 Properties of foundation materials

Properties	Toyoura sands	Crushed stones
Dry density (g/cm^3)	1.506	1.416
Max. dry density (g/cm^3)	1.639	2.773
Min. dry density (g/cm^3)	1.336	1.781
Uniformity coefficient (U_c)	1.7	1.5
Mean grain size, D_{50} (mm)	0.16	4.7
Plastic limit (I_p)	NP	NP

3.2 Experimental Setup

Two models of breakwater foundation were used in the study (i) Conventional breakwater foundation (unreinforced foundation) and (ii) Breakwater foundation reinforced with gabions and two rows of sheet piles (reinforced foundation). Instrumentations for shaking table test are shown in Fig. 2. Displacement gauges were used to monitor settlements (at two locations) and horizontal displacements (at two locations) of the caisson. In addition, accelerations (of the seabed soils, mound and caisson), water pressure (on the caisson) and pore water pressure (in the foundation ground) were monitored at different locations of the models during the tests.

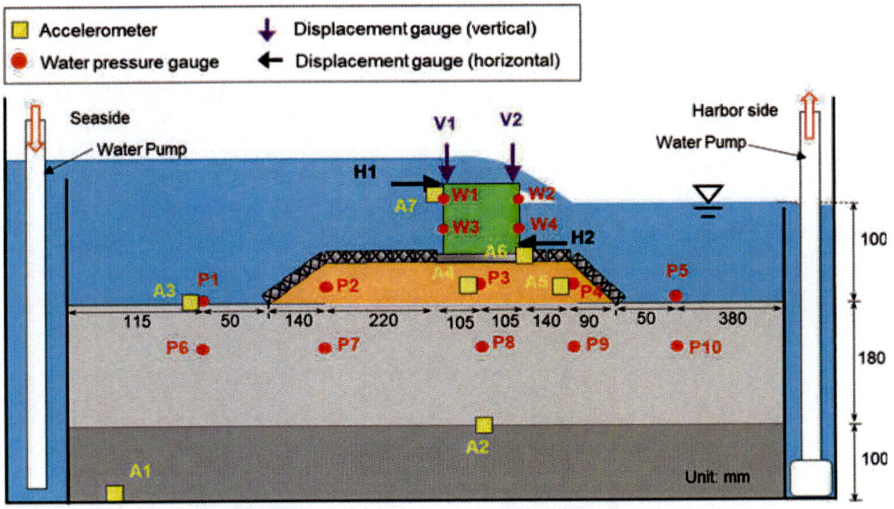

Fig. 2 Experimental setup and instrumentations for shaking table tests

3.3 Earthquake Loadings

For earthquake loadings, one foreshock and mainshock were used. Earthquake loadings were provided in the form of sinusoidal acceleration waves at the base of the soil box. Frequency (15 Hz) and duration (8 s) were same for the mainshock and foreshock. However, amplitude of the acceleration was different for the main shock (0.3 g) and foreshock (0.1 g). The parameters of earthquake chosen in such a way that in proto type, the earthquake will be somewhat similar to the 2011 off the Pacific Coast of Tohoku Earthquake. Sufficient time gap was provided between the foreshock and mainshock in order to dissipate excess pore water pressures in the foundation ground. It will be similar to the real ground event. Generally, a tsunami preceded by an earthquake.

3.4 Results and Discussions

To see the effectiveness of the reinforced model, comparisons are made between the unreinforced and reinforced foundation of breakwater. Average settlement and Average horizontal displacement are calculated based on data measure at V1 and V2, and H1 and H2, respectively.

Average settlement of the caisson for the reinforced and unreinforced foundation is shown in Fig. 3 during the main shock. It can be seen that the reinforced foundation shows considerably less average settlement compared to the unreinforced foundation. Lateral flow of seabed soil was one of the main reasons for settlement of the caisson

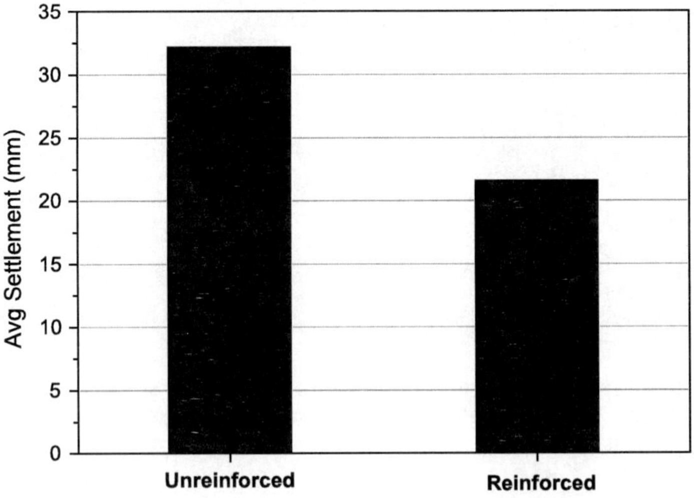

Fig. 3 Average settlement of the caisson during the mainshock of the earthquake

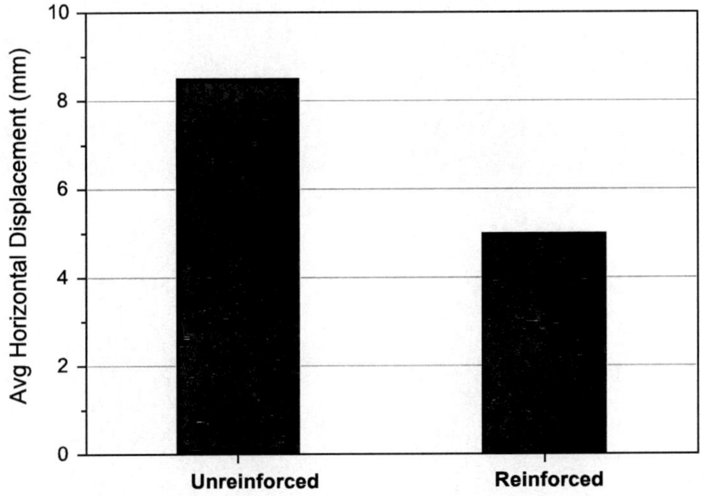

Fig. 4 Average horizontal displacement of the caisson during the mainshock of the earthquake

during the earthquake for the unreinforced foundation. However, the lateral flow of the seabed soils was resisted by the sheet piles, and hence comparatively less settlement of the caisson was observed for the reinforced foundation during the earthquake. It can be seen that the average settlement is 32.2 mm for the unreinforced foundation, which reduces to 21.6 mm for the reinforced foundation.

Average horizontal displacement for the reinforced and unreinforced foundations during the main shock (0.3 g) is shown in Fig. 4. Significant reduction in average horizontal displacement of the caisson can be seen in the case of reinforced foundation compared to the unreinforced foundation. Gabions are placed beside the breakwater. Due to mass of the gabion and friction (between the gabion and the rubble mound), the gabions could resist the lateral movement of the caisson during the earthquake. Another reason for reduction of horizontal displacement was lateral flow of foundation soils beneath the caisson, and it was resisted by the sheet piles of the reinforced foundation. Therefore, horizontal displacement of the breakwater could reduce significantly. It can be seen that the average horizontal displacement of the caisson is 8.5 mm for the unreinforced foundation, and it decreases to 5.0 mm for the reinforced foundation. Hence, the reinforced foundation could significantly reduce both average settlement and average horizontal displacement of the caisson, and provided resiliency to the breakwater against earthquake.

4 Tsunami Overflow Tests

The tsunami overflow tests were conducted on the same deformed models those obtained after the earthquake loadings. Water pumps were used to conduct the

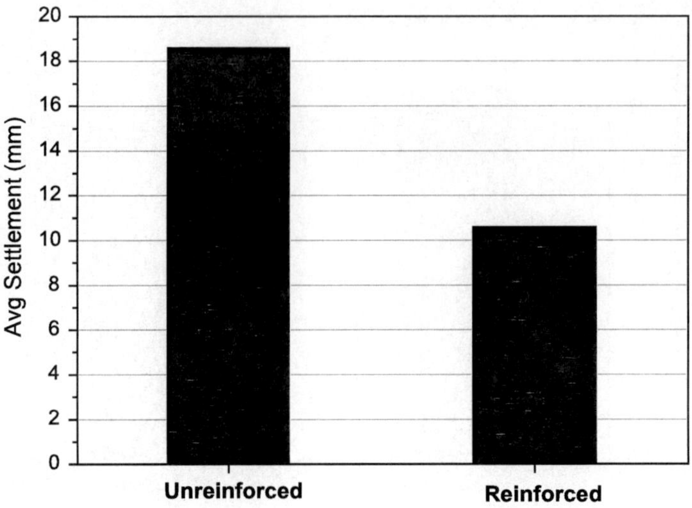

Fig. 5 Average settlement of the caisson during the tsunami overflow tests

tsunami overflow. The height of the tsunami wave was 140 mm (8.96 m $\approx$ 9 m in the prototype scale) above the seawater level on the seaside. The tsunami overflow was continued for 25 min.

The average settlement of the caisson during the tsunami overflow tests is shown in Fig. 5. It can be seen in the figure that the reinforced foundation considerably reduces the average settlement compared to the unreinforced foundation. The average settlement is 18.6 mm for the unreinforced foundation, and it decreases to 10.6 mm for the reinforced foundation. For the unreinforced foundation, the harbor side rubble mound was scoured by the overflowing tsunami. In addition, seepage and pipping were also observed in the foundation. These factors caused for the settlement for unreinforced foundation. However, for the reinforced foundation, the sheet piles behaved as cut off walls, and prevented the seepage. The gabions reduced scouring of the mound. Hence, less settlement of the breakwater was recorded for the reinforced foundation compared to the unreinforced foundation during the tsunami overflow tests.

5 Conclusions

New reinforcing countermeasures have been developed by the authors. In order to evaluate the effectiveness of the developed model, shaking table tests and tsunami overflow tests were conducted. The comparison has been made between the developed model and conventional breakwater foundation for different earthquake ground motions and tsunami overflows. The following conclusions could be derived based on this study.

1. Lateral flow of the seabed soils was the main reason of settlement of the caisson during the earthquake. The sheet piles of the reinforced foundation could resist such deformation. However, scouring, seepage and pipping of soils were the main causes of damage of the breakwater during the tsunami. The gabions and sheet pile were found effective to reduce scouring and seepage, respectively.
2. The reinforced foundation of the breakwater performed well in reducing settlement and horizontal displacement of the breakwater compared to the unreinforced foundation during the earthquake and tsunami.

Acknowledgements This study was funded by the Japan Iron and Steel Federation under priority themes research grant. The authors express their deep gratitude for the financial support. Special thanks go to Associate Professor Kiyonobu Kasama of Kyushu University for his generosity to allow using the shaking table testing facilities of Geo-disaster Prevention Laboratory.

References

1. Arikawa, T., Sato, M., Shimosako, K., Tomita, T., Yeom, G.S., Niwa, T.: Failure mechanism and resiliency of breakwater under tsunami. Technical Note No. 1269, Port and Airport Research Institute, Japan (2013). (in Japanese)
2. Arikawa, T., Sato, M., Shimosako, K., Hasegawa, I., Yoem, G.S., Tomita, T.: Failure mechanism of Kamaishi breakwater due to the great east Japan earthquake tsunami. In: Proceedings of the 33rd Conference on Coastal Engineering, Santander, Spain, (2012)
3. Takahashi, H., Sassa, S., Morikawa, Y., Takano, D., Maruyama, K.: Stability of caisson-type breakwater foundation under tsunami-induced seepage. Soils Found. **54**(4), 789–805 (2014)
4. Takahashi, S., Kuriyama, Y., Tomita, T., Kawai, Y., Arikawa, T., Tatsumi, D., Negi, T.: Urgent survey for 2011 Great East Japan earthquake and tsunami disaster in ports and coasts part I (tsunami). Technical note 1231, Port and Airport Research Institute, Japan, 1–9 (2011)
5. Kasama, K., Zen, K., Kasugai, Y.: Stability evaluation for the caisson-type composite breakwater under tsunami condition. Japanese Geotechn. Soc. Spec. Publ. (the 15th Asian Regional Conference, Fukuoka, Japan) **2**(72), 2465–2468 (2015)
6. Kazama, M., Noda, T.: Damage statistics (Summary of the 2011 Off the Pacific Coast of Tohoku Earthquake damage). Soils Found. **52**(5), 780–792 (2012)
7. Hazarika, H., Kasama, K., Suetsugu, D., Kataoka, S., Yasufuku, N.: Damage to geotechnical structures in waterfront areas of northern Tohoku due to the March 11, 2011 tsunami disaster. Indian Geotechn. J. **43**(2), 137–152 (2013)
8. Hazarika, H., Kataoka, S., Kasama, K., Kaneko, K., Suetsugu, D.: Composite ground disasters caused by the earthquake and tsunami in Aomori, Iwate Prefecture, northern Japan. Geotechn. Eng. J. Spec. Issue on 2011 Great East Japan Earthquake **7**(1), 13–23 (2012). (in Japanese)
9. Hara, T., Okawara, M., Osumi, T., Yamanaka, M., Ishihara, Y., Tsunekawa, Y., Okamura, M., Uzuoka, R.: Damages in south-central coastal area of Iwate prefecture in 2011 off the Pacific Coast of Tohoku Earthquake. Japan. Geotechn. J. **7**(1), 25–36 (2012)
10. Sugano, T., Nozu, A., Kohama, E., Shimosako, K., Kikuchi, Y.: Damage to coastal structures. Soils Found. **54**(4), 883–901 (2014)
11. Oikawa, S., Kikuchi, Y., Kawabe, S., Mizuno, R., Moriyasu, S., Tanaka, R., Takenaka, S.: An experimental study for reinforcing construction of breakwater with structural steel. In:

Proceedings of the Special Symposium of Japanese Geotechnical Society on the Great East Japan Disaster, Tokyo, Japan, pp. 703–709 (2014). (in Japanese)
12. Maruyama, S., Matsumoto, A., Hanzawa, M.: Experimental study on stability of armor units for harbor-side rubble mound of composite breakwater against tsunami. J. Jap. Soc. Civ. Eng. Ser. B3 (Ocean Eng.) **68**(2), I_7-I_12 (2012). (in Japanese)
13. Ueda, K., Iai, S., Tobita, T.: Centrifuge model tests and large deformation analyses of a breakwater subject to combined effects of tsunami. Soil Dynam. Earthq. Eng. **91**, 294–303 (2016)
14. Kikuchi, Y., Kawabe, S., Takenaka, S., Moriyasu, S.: Horizontal loading experiments on reinforced gravity type breakwater with steel walls. Japan. Geotechn. Soc. Spec. Publ. (the 15th Asian Regional Conference, Fukuoka, Japan) **2**(35), 1267–1272 (2015)
15. Central Disaster Management Council: Report of the 16th Committee Meeting. Investigation Committee on Tonankai Earthquake and Nankai Earthquake. Cabinet Office, Government of Japan (2003). (in Japanese)
16. Chaudhary, B., Hazarika, H.: Centrifuge modelling for stability evaluation of a breakwater foundation subjected to an earthquake and a Tsunami. Ocean Eng. **148**, 169–181 (2018)
17. Chaudhary, B., Hazarika, H., Ishibashi, I., Abdullah, A.: Sliding and overturning stability of breakwater under combined effect of earthquake and tsunami. Ocean Eng. **136**, 106–116 (2017)
18. Chaudhary, B., Hazarika, H., Murakami, A., Fujisawa, K.: Development of resilient breakwater against earthquake and tsunami. Int. J. Geomech. ASCE **19**(1), 1–17 (2019)
19. Chaudhary, B., Hazarika, H., Murakami, A., Fujisawa, K.: Geosynthetic-sheet pile reinforced foundation for mitigation of earthquake and Tsunami induced damage of breakwater. Geotext. Geomembr. **46**, 597–610 (2018)
20. Chaudhary, B., Hazarika, H., Murakami, A., Fujisawa, K.: Countermeasures for enhancing the stability of composite breakwater under earthquake and subsequent Tsunami. Acta Geotech. **13**(4), 997–1017 (2018)
21. Chaudhary, B., Hazarika, H., Nishimura, N.: Effects of duration and acceleration level of earthquake ground motion on the behavior of unreinforced and reinforced breakwater foundation. Soil Dynam. Earthq. Eng. **98**, 24–37 (2017)
22. Chaudhary, B., Hazarika, H., Murakami, A., Fujisawa, K.: Mitigation of earthquake induced damage of breakwater by geogrid reinforced foundation. Mar. Georesour. Geotechnol. **36**(7), 827–840 (2018)
23. Chaudhary, B., Hazarika, H., Pasha, S.M.K.: Countermeasures for breakwater foundation subjected to foreshocks and main shock of earthquake loading. Marine Georesources Geotechnol. **36**(3), 308–322 (2017)
24. Chaudhary, B., Hazarika, H., Nishimura, N.: Effects of reinforcement on the performance of breakwater foundation subjected to earthquake loadings. Int. J. Geotech. Eng. **11**(2), 186–197 (2017)
25. Chaudhary, B., Hazarika, H., Monji, N., Nishimura, K., Ishikura, R., Kasama, K.: New reinforcing method for improving the bearing capacity of breakwater foundation against earthquake and tsunami. Japan. Geotechn. Soc. Spec. Publ. **2**(35), 1273–1278 (2016)
26. Hazarika, H., Hara, T., Nishimura, K., Yamasaki, N., Monji, N., Chaudhary, B., Ishikura, R., Kasama, K.: Fundamental study on seismic resistant behavior of caisson type breakwater foundation reinforced by steel sheet pile and gabion. J. Jap. Assoc. Earthq. Eng. **16**(1), 184–204 (2016)
27. Hazarika, H., Chaudhary, B., Monji, N., Ishikura, R., Kasama, K., Hara, T., Yamazaki, N., Noda, T., Yamada, S.: Resilient breakwater foundation against level II earthquake and tsunami. In: Proceedings of the 6TH International Geotechnical Symposium on Disaster Mitigation in Special Geoenvironmental Conditions, Chennai, India, pp. 35–46, (2015) (Keynote)
28. Hazarika, H., Nishimura, K., Chaudhary, B.: Model testing on resilient solution for breakwater protection against tsunami. Japan. Geotechn. Soc. Spec. Publ. (International workshop on Geotechnics for resilient infrastructure- The second Japan-Indian Workshop in Geotechnical Engineering) **3**(2), 40–44 (2016)

29. Chaudhary, B., Hazarika, H., Monji, N., Nishimura, K., Yasufuku, N., and Ishikura, R.: Behavior of breakwater foundation reinforced with steel sheet piles under seismic loading. In: Hazarika, H., Kazama, M., Lee, W.F. (eds.) Geotechnical Hazards from Large Earthquakes and Heavy Rainfalls, pp. 449–461(2016)

Simulation of Crack Initiation and Propagation Using the Improved DDA

Mingyao Xia and Guangqi Chen

Abstract This paper analyses the rock fracture problem of Brazil disc split tests by using an improved DDA method. By incorporating a new rock fracture law and the discretization approach, the improved DDA can be used to simulate crack propagation during the rock fracture process in a two-dimensional plane stress state, such as the crack propagation in Brazil disc split tests. A numerical example of simulating the Brazil disc split tests for verification has been computed using the proposed DDA method. The simulated result agrees well with those obtained from the experiment, which proves the effectiveness and accuracy of the proposed DDA method. Meanwhile, the Brazil disc split test under different loading rates is performed systematically to investigate the external influences on rock fracturing. The result shows that the improved DDA can be well used for simulating the two-dimensional dynamic rock fracture problems in engineering.

Keywords Crack propagation · Brazil disc split test · Discontinuous deformation analysis

1 Introduction

Crack initiation and propagation in engineering materials can lead to structural failure. It is important to have good knowledge of material fracture problems, such as the initiation of a crack in the rock and propagation path; it may follow during its growth. The fracture problem in engineering materials like rock and concrete is one of the key factors influencing risk and safety assessment in disaster prevention, such as rockfalls, earthquake-induced landslides, rock blasting, tunnel collapse, and so on.

M. Xia (✉) · G. Chen
Department of Civil and Structural Engineering, Kyushu University, Fukuoka 819-0395, Japan
e-mail: xia.mingyao.629@s.kyushu-u.ac.jp

G. Chen
e-mail: chen@civil.kyushu-u.ac.jp

H. Hazarika et al. (eds.), *Advances in Sustainable Construction and Resource Management*, Lecture Notes in Civil Engineering 144,
https://doi.org/10.1007/978-981-16-0077-7_30

However, the modelling of crack initiation and propagation is a difficult and complicated problem in computational rock mechanics due to the difficulties of fracture mechanics for crack initiation and propagation. Especially, for the continuum-based numerical methods like the finite difference method (FDM), the finite element method (FEM), and the boundary element method (BEM), the solution process always encounters difficulties when the stress field is complicated. Discontinuous deformation analysis (DDA) method is a kind of discrete element method which is developed for modelling of large deformation and rigid body movements of discrete particles without analyzing the fracture analysis. However, it can also be an alternative approach for continuity to discontinuity problems when the fracture mechanics in DDA is constituted appropriately. Compared with continuum-based numerical methods, DDA is also a more attractive method for rock fracture problems due to its intrinsic feature of block discontinuity at the contact boundaries. The DDA, proposed by Shi [1], has been widely used for modelling the large deformation and rigid body movements such as landslides [2, 3], rockfall [4], and seismic response [5, 6]. To solve blast-induced rock mass failure, Ning et al. [7] improved the DDA method in modelling rock fracturing via advanced discretization within the original DDA framework. Jiao et al. [8] proposed a new two-dimensional contact constitutive model for the DDA method to simulate the fragmentation of jointed rock, and the improved DDA is successfully applied to reveal the triggering mechanism of a rockslide. The researcher indicated that DDA could be applied greatly to prevent the disaster if the DDA method can better simulate the rock fracture problem.

This paper focuses on analyzing the rock fracture problem by using an improved DDA (I-DDA) method. The Brazil disc split test is simulated using I-DDA to verify the effectiveness of I-DDA on simulation of rock fracturing problems. The simulated results are in agreement well with the experimental solutions, which indicate that the I-DDA is capable of simulating crack initiation and propagation in engineering materials.

2 Methodology

2.1 Discontinuous Deformation Analysis (DDA)

The basic theory of original two two-dimensional DDA consists of six basic variables including three rigid body displacements and three strain components: (u_0, v_0) is the rigid body translation, r_0 is the rigid body rotation, and (ε_x, ε_y, γ_{xy}) are the normal and shear strain components. The displacement (u, v) of an arbitrary point (x, y) within a block is re-described as follows:

$$\begin{pmatrix} u \\ v \end{pmatrix} = \begin{pmatrix} 1 & 0 & -(y-y_0) & (x-x_0) & 0 & (y-y_0)/2 \\ 0 & 1 & (x-x_0) & 0 & (y-y_0) & (x-x_0)/2 \end{pmatrix} \begin{Bmatrix} u_0 \\ v_0 \\ r_0 \\ \varepsilon_x \\ \varepsilon_y \\ \varepsilon_{xy} \end{Bmatrix} = \mathbf{TD} \tag{1}$$

where (x_0, y_0) is the centroid of the block.

The simultaneous equilibrium equations of block system are then derived by minimizing the total potential energy of the system, which can be expressed as follows:

$$[\mathbf{M}]\{\ddot{\mathbf{D}}\}+[\mathbf{C}]\{\dot{\mathbf{D}}\}+[\mathbf{K}]\{\mathbf{D}\}=\{\mathbf{F}(t)\} \tag{2}$$

where $[\mathbf{M}]$, $[\mathbf{C}]$, and $[\mathbf{K}]$ are the mass matrix, damping matrix, and stiffness matrix, respectively; $\{\ddot{\mathbf{D}}\}$, $\{\dot{\mathbf{D}}\}$, and $\{\mathbf{D}\}$ are the acceleration vector, velocity vector, and displacement vector, respectively; and $\{\mathbf{F}(t)\}$ is the external force vector.

With those formulations, the DDA method can be used to simulate large deformation and rigid body movements of discrete particles. Assuming damping is not considered in the calculations, according to the updated Lagrange description, the displacement vector at the beginning of the current time step is zero, and the simultaneous equilibrium equations derived by minimizing the total potential energy Π produced by the forces and stresses can be rewritten as follows.

$$\begin{bmatrix} \mathbf{K}_{11} & \mathbf{K}_{12} & \cdots & \mathbf{K}_{1n} \\ \mathbf{K}_{21} & \mathbf{K}_{22} & \cdots & \mathbf{K}_{2n} \\ \vdots & \vdots & \ddots & \vdots \\ \mathbf{K}_{n1} & \mathbf{K}_{n2} & \cdots & \mathbf{K}_{nn} \end{bmatrix} \begin{bmatrix} \mathbf{D}_1 \\ \mathbf{D}_2 \\ \vdots \\ \mathbf{D}_n \end{bmatrix} = \begin{bmatrix} \mathbf{F}_1 \\ \mathbf{F}_2 \\ \vdots \\ \mathbf{F}_n \end{bmatrix} \tag{3}$$

where K_{ij} is a 6×6 coefficient submatrix; F_i is a 6×1 loading submatrix on block i; $i, j = 1, 2, \ldots n$, and n is the number of blocks in the defined block system.

2.2 The New Tensile Fracture Law

Since the original DDA is based on the maximum normal tensile stress criterion, the occurrence of tensile failure is only related to the normal tensile stress. However, the maximum normal tensile stress criterion has a limitation in simulating tensile behaviour for various materials, and Zhang and Eckert [9] proposed a new unified tensile failure criterion, which has a great advantage in describing tensile behaviour of various materials. Here, the new tensile failure criterion is incorporated to improve the performance of DDA, which is expressed as follows.

$$\frac{\sigma^2}{\sigma_0^2} + \frac{\tau^2}{\tau_0^2} \geq 1, (\tau_0 > 0, \sigma_0 > 0) \tag{4}$$

σ_0 and τ_0 are tensile strength and shear strength, whereas σ and τ are normal stress and shear stress, respectively. This criterion is not only related to tensile stress but also affected by shear stress. It is based on the theory of distortion energy. The ratio $\alpha = \tau_0/\sigma_0$ was defined to reflect the differences of materials. For example, the shear strength of the soil is related to the value of cohesion c and friction angle φ, but the tensile strength of the soil is very small, which is closed to zero. Generally, the tensile strength of soil cannot be considered, and the value α is considered to be very large, while for granite, α can be taken as 2 according to shear and tensile strength from the experiment.

3 Simulation of Brazil Disc Split Test

To investigate the applicability of the proposed DDA method on the simulation of rock fracturing problems in Brazil disc split test, the failure process of rock disc under diametrical loadings is simulated. As shown in Fig. 1, the disc has a diameter of 0.1 m and a notch with a length of 0.03 m. The Brazil disc was discretized into 2690 triangular blocks by using DISTMESH, a mesh generator by Per-Olof Persson [10].

A loading force is applied through plates at the end of the loaded diameter for the disc. The loading point is located at the centroid of the loading plate, which linearly increases from 0.1 ms as shown in Fig. 2.

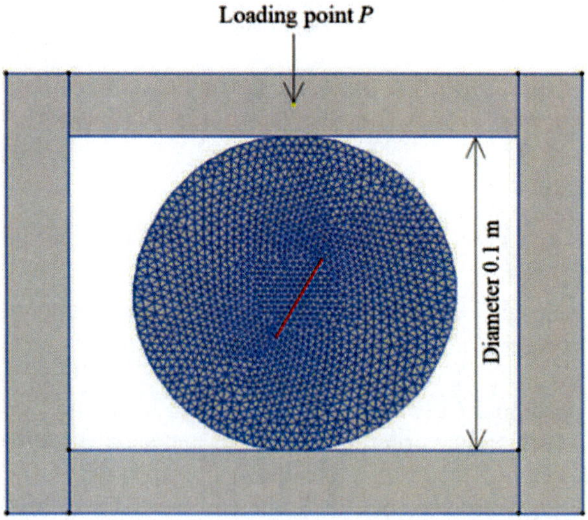

Fig. 1 Model of simulated Brazil disc split test

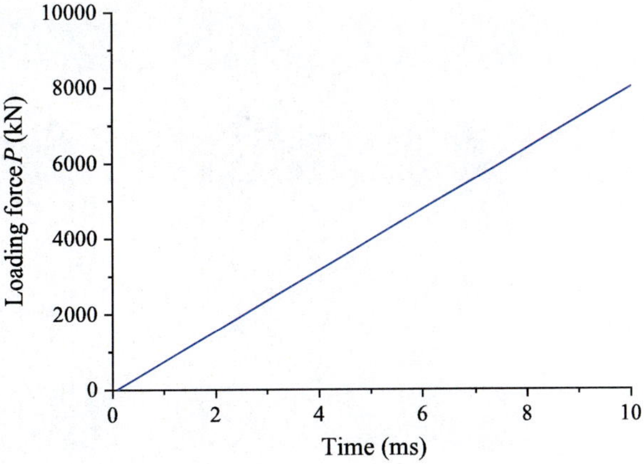

Fig. 2 Linearly increasing force at loading point *P*

Table 1 Test conditions

Joint type	Friction angle (°)	Cohesion (c/MPa)	The ratio α
Artificial joint	35	40	2
Crack	0	0	0

The material parameters based on Ning's researches [7] for the rock disc are adopted as follows: density $\rho = 2500$ kg/m^3, Young's modulus $E = 10$ GPa, and the Poisson's ratio $v = 0.2$. The joint mechanical parameters are shown in Table 1. Generally, there are two types of joints that form the edges of the triangular blocks, i.e. real crack and artificial joints. The grid inside the disc is virtual joints, which will develop into a real joint when the failure criterion is satisfied. The red line is the real crack with no strength, which donates a pre-existed crack and newly generated crack in the simulation. Note that the joint parameters for blocks along the initial crack are set as zero in this model; thus, a crack will be developed without loading at first to represent the pre-existed crack.

The simulated result of crack initiation and propagation in Brazil disc is obtained using the proposed DDA method. Figure 3 shows a comparison between the crack propagation path observed experimentally [11] and that simulated numerically, both the crack propagation paths of experimental and numerical methods tend to be parallel to the loading direction and to approach the loading points. The simulated result of the crack path in Brazil disc agrees well with the experiment results, which prove the accuracy of I-DDA on solving the fracture problems. It is therefore concluded that the proposed DDA method is capable of modelling of Brazil disc split test.

With the simulated result of Brazil disc split test in two-dimensional plane stress conditions, the compression induces tensile stresses normal to the vertical diameter which can be calculated. According to the ASTM [12], the suggested formula for calculating the splitting tensile strength σ_t (MPa) based on the Brazilian test is

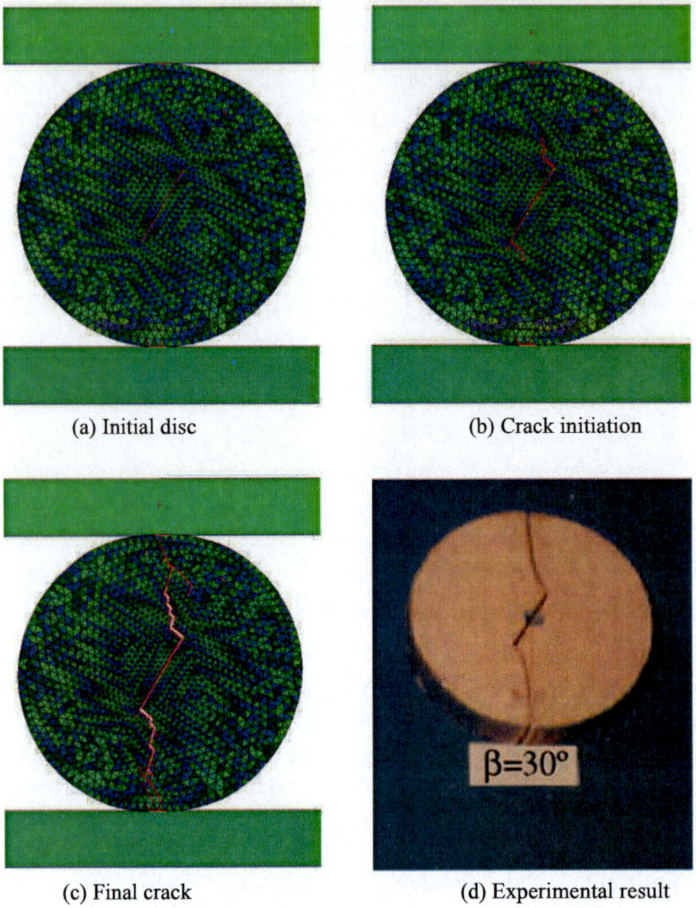

Fig. 3 Simulated result of Brazil disc split test

$$\sigma_t = \frac{2P_{\max}}{\pi DL} \tag{5}$$

where σ_t is the Brazil tensile strength, $P_{\max}$ is the maximum load, and D and L are the diameter and thickness of the specimen, respectively. As shown in Fig. 4, the tensile stress corresponding to the critical stretch calculated by Eq. (5) is 1.6 MPa, where the maximal load is $P_{\max} = 259$ kN.

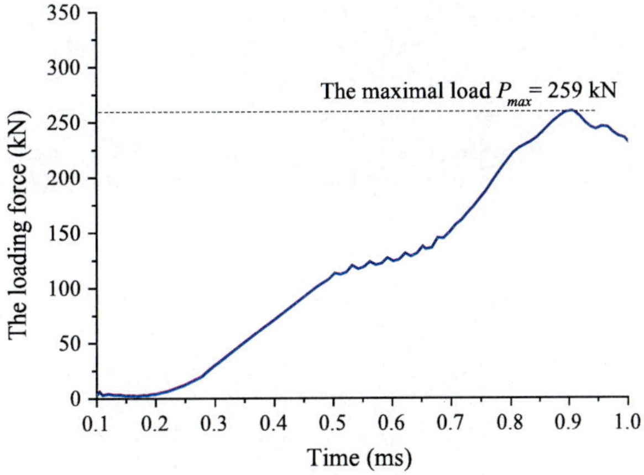

Fig. 4 Dynamic loading stress–time curve

4 Effect of the Loading Rate

The dynamic fracture behaviour of Brazil disc under different loading rates is distinctively different from that in static tests. To study the dynamic mechanical behaviour of Brazil disc, the rate dependency of crack damage stress is investigated by using the I-DDA. Figure 5 shows the dynamic loading history for the compression process of the Brazil disc split test. As the loading force increases, the slope of compression stress increases as well. The slope of the compression stress is then determined and used as the loading rate for the dynamic test. Here, to study the effect of the loading rate on the dynamic tensile strength, the same sample is loaded up to peak stress at different loading rates, and dynamic stress–time curves are obtained, as shown in Fig. 5.

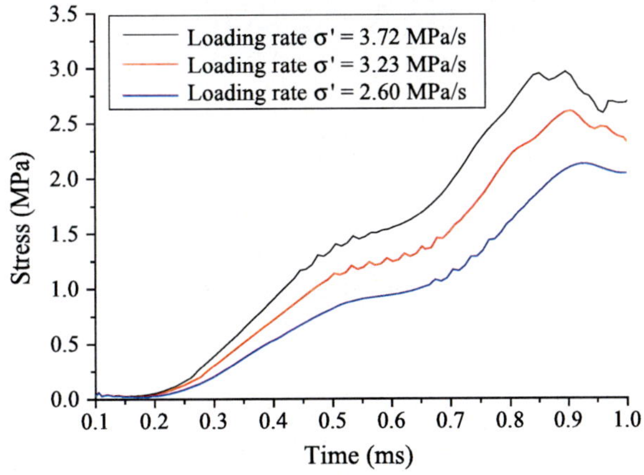

Fig. 5 Dynamic loading stress–time curve

It can be seen from simulated results that the compression-induced tensile stresses increase with respect to time at different loading rates. It should be noted that the stress listed herein is the compression stress of the disc rather than the tensile stress perpendicular to the loading axis. As expected, the stress–time behaviours exhibit a general increase at increased loading rates for rocks. As the loading rates increase from 2.60 MPa/s to 3.72 MPa/s, the obtained maximal loadings $P_{\max}$ are 2.12 MPa, 2.60 MPa and 2.96 MPa, respectively. The corresponding tensile strengths calculated from the suggested formula Eq. (5) are 1.66 MPa, 2.06 MPa, and 2.37 MPa. The calculated tensile strength under different loading conditions is defined as dynamic tensile strength. With the increasing loading rates, the simulated dynamic tensile strength increases. The effect of loading rates on dynamic tensile strength is consistent with the result of the previous researches on dynamic properties that the increased loading rates will improve the failure tensile strength of the rocks [13].

5 Conclusions

This paper analysed the rock fracture problem of Brazil disc in the two-dimensional stress field using the improved DDA method. By incorporating a new rock fracture law and the discretization approach, the improved DDA has been used to simulate crack initiation and propagation of the Brazil disc split test in a two-dimensional plane stress state. The simulated result agrees well with those obtained from the experiment, which proves the effectiveness and accuracy of the proposed DDA method. Meanwhile, the Brazil disc under different loading rates has been investigated to study the dynamic mechanical behaviour of Brazil disc. The simulated result using this improved DDA indicates that the dynamic tensile strength increases with the increased loading rate, which agrees with the dynamic properties of the rocks that the increased loading rates will improve the failure tensile strength of the rocks. With improvement, the DDA can be used to analyse the rock fracture problems in the two-dimensional stress state.

References

1. Shi, G.H.: Discontinuous Deformation Analysis: A New Numerical Model for the Statics and Dynamics of Block Systems. PhD thesis, Berkeley, University of California (1988)
2. Zhang, Y., Chen, G., Zheng, L., et al.: Effects of near-fault seismic loadings on run-out of large-scale landslide: a case study. Eng. Geol. **166**, 216–236 (2013)
3. Zhang, Y., Xu, Q., Chen, G., et al.: Extension of discontinuous deformation analysis and application in cohesive-frictional slope analysis. Int. J. Rock Mech. Min. Sci. **70**, 533–545 (2014)
4. Wu, J.H., Ohnishi, Y., Nishiyama, S.: A development of the discontinuous deformation analysis for rock fall analysis. Int. J. Numer. Anal. Meth. Geomech. **29**(10), 971–988 (2005)
5. Peng, X., Yu, P., Zhang, Y., et al.: Applying modified discontinuous deformation analysis to assess the dynamic response of sites containing discontinuities. Eng. Geol. **246**, 349–360 (2018)

6. Zhang, Y., Fu, X., Sheng, Q.: Modification of the discontinuous deformation analysis method and its application to seismic response analysis of large underground caverns. Tunn. Undergr. Space Technol. **40**, 241–250 (2014)
7. Ning, Y., Yang, J., An, X., et al.: Modelling rock fracturing and blast-induced rock mass failure via advanced discretisation within the discontinuous deformation analysis framework. Comput. Geotechn. **38**(1), 40–49 (2011)
8. Jiao, Y.Y., Zhang, X.L., Zhao, J.: Two-dimensional DDA contact constitutive model for simulating rock fragmentation. J. Eng. Mech. **138**(2), 199–209 (2012)
9. Zhang, Z.F., Eckert, J.: Unified tensile fracture criterion. Phys. Rev. Lett. **94**(9), 094301 (2005)
10. Persson, P.-O., Strang, G.: A simple mesh generator in MATLAB. SIAM Rev. **46**(2), 329–345 (2004)
11. Al-Shayea, N.A.: Crack propagation trajectories for rocks under mixed mode I-II fracture. Eng. Geol. **81**(1), 84–97 (2005)
12. ASTM: Standard test method for splitting tensile strength of intact rock core specimens. ASTM International, West Conshohocken, USA (2008)
13. Li, X.F., Li, X., Li, H.B., Zhang, Q.B., Zhao, J.: Dynamic tensile behaviours of heterogeneous rocks: the grain scale fracturing characteristics on strength and fragmentation. Int. J. Impact Eng **118**, 98–118 (2018)

Stability Assessment on Expressway Embankment by Automatic Survey System

Kosuke Nakashima, Atsushi Yashima, Yoshinobu Murata, and Keizo Kariya

Abstract Collapses of expressway embankments due to large earthquakes and heavy rainfalls have been reported. Those collapses often closed important transportation routes and resulted in serious damages in communities. Inappropriate groundwater treatment and slaking/weathering of embankment materials contribute to these collapses. In order to diagnose road embankment soundness, we have to visualize the inside the embankment. The automatic technology for surveys and evaluations of embankments by using surface wave logging and electric resistivity logging has been proposed. We performed initial diagnostics before the opening of the expressways to evaluate embankment materials and stability of the embankment, especially in the vicinity of the stiff structures, such as box culverts and around the boundary between embankment and cut section. The desired target inspection speed, 500 m per h (mean speed), was achieved. The stiffness of embankment was continuously obtained for many sections with different fill materials as the initial reference data for future inspection. Based on the comparison between initial reference data and those obtained in the future, it is possible to select the optimal countermeasures to ensure the continuous stability of the road embankment.

Keywords Stability assessment · Road embankment · Surface wave logging · Electric resistivity logging

1 Introduction

There have been many reports on collapses of road embankments due to heavy rainfalls and large earthquakes as shown in Photo 1. Those collapses resulted in serious damages in local communities. These collapses were also caused by inappropriate

K. Nakashima (✉)
Nippon Expressway Research Institute Company Limited, 1948508 Machida, Tokyo, Japan
e-mail: k.nakashima.ag@ri-nexco.co.jp

A. Yashima · Y. Murata · K. Kariya
Gifu University, 5011193 Gifu, Japan

H. Hazarika et al. (eds.), *Advances in Sustainable Construction and Resource Management*, Lecture Notes in Civil Engineering 144,
https://doi.org/10.1007/978-981-16-0077-7_31

Photo 1 Road embankment failure due to heavy rain [1] and large earthquake [2]

groundwater treatment and slaking/weathering of embankment materials. Therefore, our engineers have to develop the inspection technique to investigate the stability of embankments [1, 2].

There are also engineering problems in the maintenance of road embankment summarized as follows:

1. Quality control of compaction is only performed during construction stage. Initial inspection of outside appearance is only performed just after the construction.
2. If pavement subsidence occurs or lateral movement of embankment itself is observed, the countermeasure is determined by boring investigation or monitoring. Its inspection is expensive and takes long time.
3. Disasters occur without warning by earthquakes or heavy rains.

Yamabe et al. [3] and Kato et al. [4] investigated dynamic properties of expressway embankment from in situ tests. They concluded that the shear wave velocity structure was totally different in the respective embankment sections with different fill materials. Figure 1 shows the relationship between the mean value of the standard penetration test (SPT) blow number (N) for varying embankment fill material and the respective shear wave velocity (Vs). It is found from Fig. 1 that for the embankment sections constructed with crushed sedimentary rocks, the measured shear wave velocities were higher than 300 m/s. On the other hand, for the embankment section constructed with weathered rocks or clayey materials, the measured minimum shear wave velocity was lower than 200 m/s. It is found that the stiffness of the embankment is completely different depending on the fill materials.

In this paper, the automation technique for check and diagnosis of pavement as well as embankment structure is proposed by using surface wave logging and electric resistivity logging. These two inspection techniques are not new. We are developing automation technique to save time and cost for the inspection. We performed initial diagnostics before the opening of the expressways to evaluate embankment materials and stability of the embankment, especially in the vicinity of the stiff structures, such as box culverts and around the boundary between embankment and cut section. The

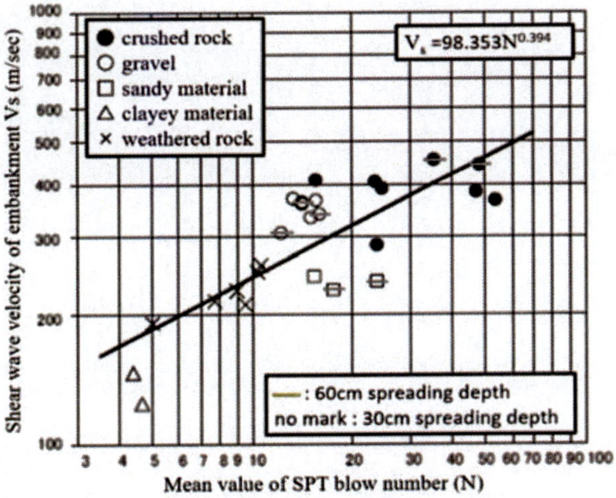

Fig. 1 Relationship between SPT blow number for fill material and its shear wave velocity [3]

desired target inspection speed, 500 m per h (mean speed), was achieved by the developed measuring system. The stiffness of embankment was continuously obtained for many sections with different fill materials as the initial reference data for future inspection. Based on the comparison between initial reference data and those obtained in the future, we can propose remedial works for pavement and embankment.

2 New Exploration System Developed

The two-dimensional surface wave exploration can evaluate shear wave velocity structure in ground down to half of logging length with numerous accelerometers. The exploration cost is, however, very high and takes one hour for 200 m long inspection. The electric resistivity exploration can evaluate groundwater table and soil particle size with the logging speed of 2000 m/h. The system, however, cannot follow on curved road during inspection.

In order to improve the inspection speed and operability of a current geophysical survey technique [5], we proposed a newly designed automation logging system. In order to achieve the stable movement of towed accelerometers and electric resistivity logging system, the carrying carts were designed and manufactured (Photos 2 and 3). The operability on a curved road was significantly improved.

The remaining problem on the two-dimensional surface wave logging is the noise induced by cars passing by. The very quiet environment is usually needed to measure an impact signal by the hammer accurately. However, the measured acceleration signal by the hammer is not large enough compared to the noise induced by cars passing by. We need to increase the signal/noise (S/N) ratio to obtain sufficient

Photo 2 Carrying carts for accelerometers

Photo 3 Carrying carts for electric resistivity logging system

surface wave signals. In order to overcome the traffic noise problem, we decided to use the impact power induced by the FWD test, which was found to be large enough.

The proposed system can speed up the integrated geophysical survey technique by using the two-dimensional surface wave method and electric resistivity method. Photo 4 shows the FWD vehicle towing the carrying carts for accelerometers and electric resistivity logging system. This proposed system can also evaluate the pavement and subgrade stiffness by the FWD test and the stiffness and stability of the subgrade and embankment body by the two-dimensional surface wave logging and electric resistivity logging at the same time.

Photo 4 FWD vehicle towing the carrying carts for accelerometers and electric resistivity logging system

3 Case Studies

We performed initial diagnostics before the opening of the three expressway sections to evaluate embankment materials and stability of the embankment.

3.1 Pre-opening Tokai-Hokuriku Expressway

The two-dimensional surface wave logging was carried out for the embankment and cut sections of the pre-opened Tokai-Hokuriku expressway in section about 1 km indicated by an arrow in Photo 5. The shear wave velocity distribution between the road surface and GL −16 m for the embankment and cut sections of 650 m is shown in Fig. 2. The fast shear wave velocity is observed in cut section. On the other hand, the slower shear wave velocity is observed in the embankment. The quite low

Photo 5 Embankment and cut sections of the pre-opened Tokai-Hokuriku expressway

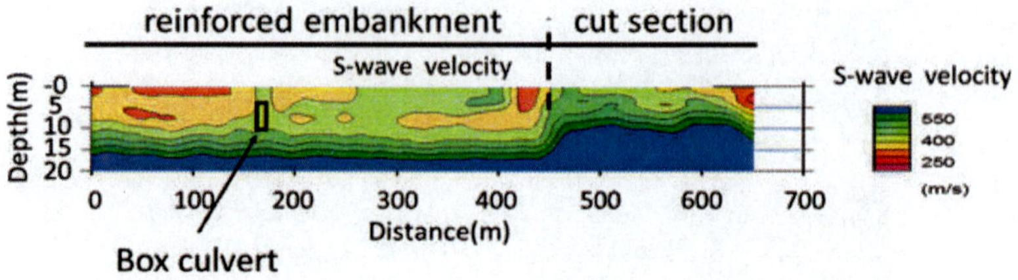

Fig. 2 Shear wave velocity distribution for the embankment and cut sections

shear wave velocities are in particularly recognized around the boundary between embankment and cut section. The low shear wave distribution is also observed around the underground box culvert.

It is well known that the densification of the fill materials is very difficult not only around the boundary between embankment and cut section but also around the boundary between embankment and stiff structure such as box culvert by the vibrating compactor. There will be a possibility of the seepage of groundwater around the boundary between embankment and cut section in the future. The groundwater and soft fill material easily accelerate the decrease in the strength of fill material. This will lead to the instability of road embankment. These accumulated data will be utilized as the initial reference ones for understanding the change in the internal mechanical condition of the embankment by future inspections.

3.2 *Pre-opening New Tomei Expressway*

The 55-km long New Tomei expressway was opened to traffic on February 13, 2016. Prior to this, we carried out surface wave logging for 15 km long embankment sections (totally 30 km for both lanes). An example of embankment and cutting sections is shown in Photo 6. The mean inspection speed of surface wave logging by the modified carts was more than 500 m/h. The mean record beyond the desired target speed was achieved.

The distribution of shear wave velocity of embankment sections along the New Tomei expressway is summarized in Fig. 3. From the inspection data, for the embankment sections constructed with crushed hard sedimentary rocks, the measured minimum shear wave velocity is found to be much higher than 300 m/s. On the other hand, for the embankment sections constructed with decomposed granite (sandy material), the measured minimum shear wave velocity is lower than 200 m/s.

The shear modulus is proportional to the square of shear wave velocity. Therefore, the stiffness of embankment with crushed hard sedimentary rocks is much higher than that of embankment with decomposed granite (sandy material). This finding agrees with the conclusion by Yamabe et al. [3] and Kato et al. [4] as shown in Fig. 1.

An example of details of shear wave velocity structure of embankment and cut sections is shown in Fig. 4. The surface wave logging was carried out not only on

Photo 6 Embankment and cut sections of New Tomei expressway

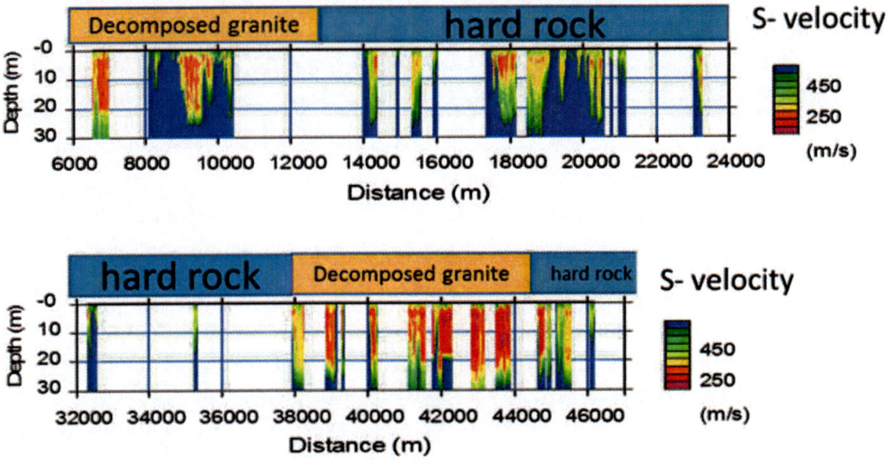

Fig. 3 Distribution of shear wave velocity along New Tomei expressway

the pavement surface but also on the horizontal berm of the embankment slope. In this figure, the shear wave velocity distribution in ground is not only shown, but also plan and longitudinal views are drawn. We can understand the high shear wave velocity in the cutting section. On the other hand, the low shear wave velocity in the embankment section is easily recognized.

The base course in the measured section is constructed by reinforced slab. Therefore, the stiffness of road pavement is very high. From the shear wave velocity distribution along measurement lines 1 and 2, it is found that the shear wave velocity in road surface area including pavement is higher than that in the embankment body.

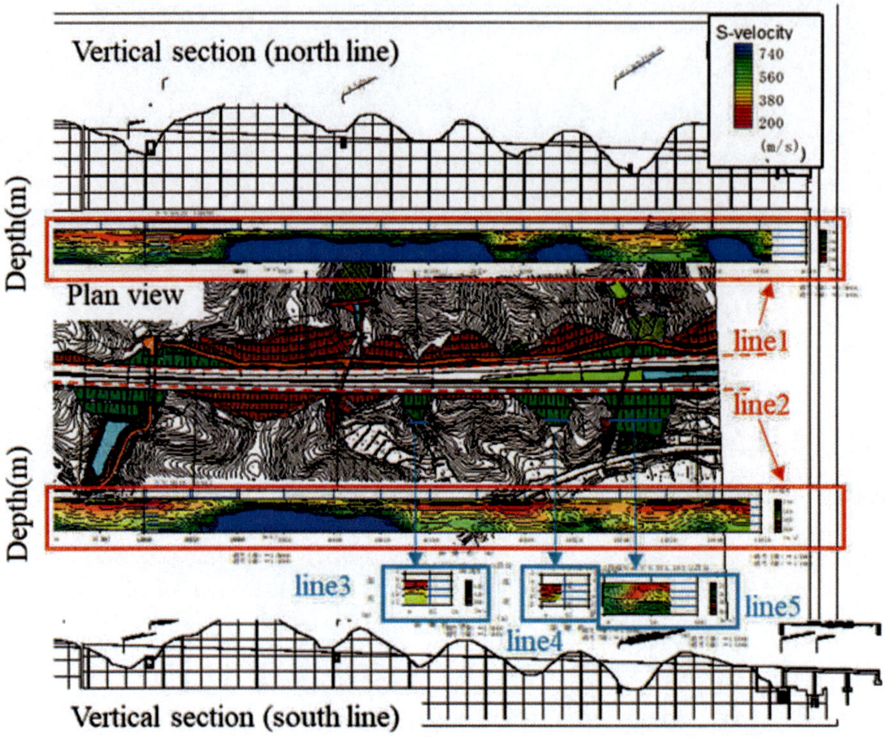

Fig. 4 Shear wave velocity of embankment and cut sections along New Tomei expressway

This shear wave velocity structure is found to reflect local structure of pavement appropriately. It is also confirmed that the shear wave velocity distribution can be obtained with enough precision by the two-dimensional surface wave exploration even if the road surface is covered by a relatively stiff pavement.

On the other hand, it is found from the shear wave velocity distribution along measurement lines 3, 4 and 5 on the horizontal berm of the embankment slope that the shear wave velocity near ground surface of the embankment slope is quite low. This is due to the densification difficulty during construction of embankment slope.

The expressway embankment is constructed using a compaction control by density criteria. However, due to the non-uniformity of embankment materials and differences in construction conditions, the density in the embankment varies considerably. Therefore, it is necessary to understand the non-uniformity of embankment materials and differences in construction conditions to assess the soundness of the completed embankment.

Photo 7 Embankment section of New Meishin expressway

3.3 Pre-opening New Meishin Expressway

The 23 km long New Meishin expressway opened to traffic on March 17, 2019. Prior to this, we carried out surface wave logging for 8.5 km long embankment sections (totally 17 km for both lanes). An example of embankment and section is shown in Photo 7. The height of embankment is much lower than that of New Tomei expressway. The mean inspection speed of surface wave logging by the modified carts was about 1200 m/h by using FWD vehicle. The mean record beyond the desired target speed was easily achieved.

An example of shear wave velocity structure of west lane is shown in Fig. 5. In this figure, the shear wave velocity distribution in both embankment and natural ground is shown. Almost all of embankments were constructed with crushed hard rocks. From the inspection data, for the embankment section constructed with those hard rock materials, especially from tunnel construction site, the measured minimum shear wave velocity was found to be higher than 300 m/s. This means that the embankment is quite stiff. On the other hand, for the embankment constructed with crushed Pleistocene soft rock materials close the cut section, the measured minimum shear wave velocity was more or less 200 m/s. The stiffness of embankment with crushed soft rocks is found to be low. This finding also agrees with the conclusion as shown in Fig. 1.

The electric resistivity distribution in a certain embankment section is shown in Fig. 6. The groundwater table is easily estimated from the resistivity structure in ground to be at around GL −4 m. By combining the stiffness of the embankment and groundwater table, the stability of the embankment can be assessed.

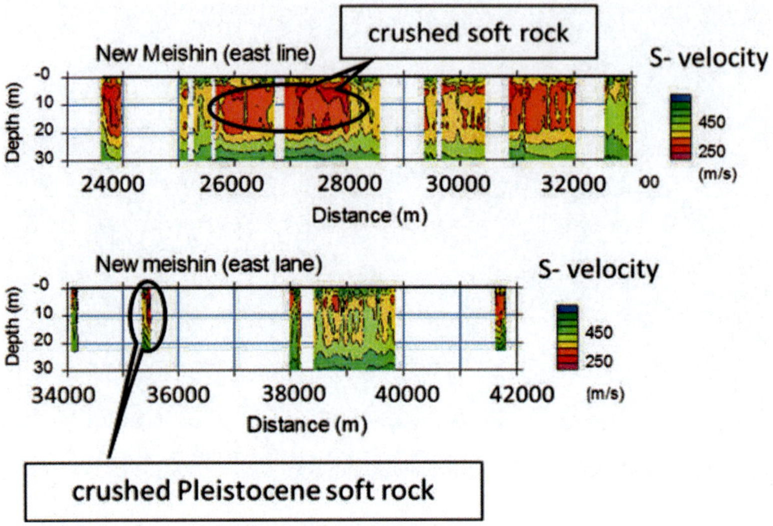

Fig. 5 Distribution of shear wave velocity along New Meishin expressway

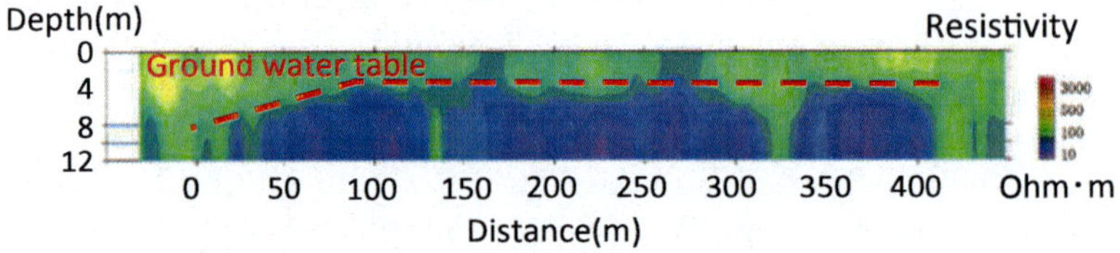

Fig. 6 Estimation of groundwater table by electric resistivity

4 Conclusions

The automation system for check and diagnosis of pavement as well as earthfill structure by using surface wave logging and electric resistivity logging was designed and manufactured. The desired target inspection speed, 500 m per h (mean speed), was achieved. The stiffness (Vs) of embankment was continuously obtained for many sections with different fill materials as the initial reference data for future inspection. Based on the comparison between initial reference data and those obtained in the future, we can propose remedial works for pavement and embankment. We can also assess the stability of embankment under the variation of shear wave velocity and groundwater table. Based on the proposed inspection technologies and remedial works, it is possible to provide sustainable road services.

Acknowledgements We would like to thank Central Nippon Expressway Co., Ltd. for their cooperation in on-site measurement.

References

1. Shima, S.: Damage report to Iwakuni and Miyajima, Civil Engineering magazine. Jap. Soc. Civ. Eng. **90**(11), 48–50 (2005). (in Japanese)
2. Central Nippon Expressway Co., Ltd.: Report on earthquake disaster of Tomei expressway Makinohara area, Exploration Committee Report (2009) (in Japanese)
3. Yamabe, S., Mishima, N., Mayuzumi, H., Nagao, K., Kitamura, Y., Kato, Fujioka, K.: Dynamic properties of highway embankment from in-situ test –field condition survey. In: Proceeding 39th Annual Meeting of JGS, pp. 1311–1312 (2004). (in Japanese)
4. Kato, Y., Nagao, K., Kitamura, Y., Fujioka, K., Mishima, N., Mayuzumi, H., Yamabe, S.: Dynamic properties of highway embankment from in-situ test –N-value and Vs-. In: Proceeding 39th Annual Meeting of JGS, pp. 1313–1314 (2004). (in Japanese)
5. Inazaki, T., Hayashi, K.: Utilization of integrated geophysical surveying for safety assessment of levee systems, Geotechnical Engineering magazine. Jap. Geotechn. Soc. **58**(8), 34–37 (2010). (in Japanese)

Stability of Reinforced Soil Quay Wall Subjected to Combined Action of Earthquake and Tsunami

Manu K. Sajan and Babloo Chaudhary

Abstract Reinforced soil quay walls are used as shore protection systems. Generally, horizontal layers of geogrids are provided as reinforcement in the backfill soil of the quay wall. These structures are internally stabilized by mobilized tensile strength of reinforcements. A quay wall can be subjected to tsunami load and earthquake load simultaneously. This condition occurs when an earthquake aftershock reaches the quay wall structure at the same time of a tsunami impact. Therefore, a combined analysis of quay walls subjected to earthquake and tsunami at the same time is necessary. In this study, horizontal slice method is used to evaluate the stability of the reinforced soil quay wall subjected to earthquake and tsunami. The failure surface is generated by optimizing the angle of failure plane of each slice, so that the mobilized tensile force on the reinforcement is maximum. Thus, the generated failure surface could justify the actual failure surface. It was observed that normalized force acting on the reinforcement is considerably increased under the combined effect of earthquake and tsunami. Stability of the wall is evaluated by varying several parameters, such as acceleration coefficient of earthquake motion, internal friction angle of soil, inclination and height of the quay wall, height of seawater level and height of tsunami waves, to find out the effect of these parameters on normalized reinforcement strength.

Keywords Earthquake · Reinforced soil quay wall · Tsunami

1 Introduction

Reinforced earth retaining structures are made by incorporating reinforcing elements inside the retained soil mass [1]. These reinforced soil retaining structures are generally designed to withstand loads acting in both vertical and horizontal directions by mobilizing the friction between reinforcement and soil. The loads acting in vertical

M. K. Sajan (✉) · B. Chaudhary
National Institute of Technology Karnataka, Surathkal, India
e-mail: manuksajan@gmail.com

H. Hazarika et al. (eds.), *Advances in Sustainable Construction and Resource Management*, Lecture Notes in Civil Engineering 144,
https://doi.org/10.1007/978-981-16-0077-7_32

direction include surcharge and self-weight, whereas the loads acting in horizontal direction include earth pressure and earthquake loads.

Quays are onshore structures build to facilitate the mooring of ships. In the case of reinforced soil quay walls, these structures are subjected to hydrodynamic forces and tsunami loads [2]. It was observed during the 2011 off the Pacific coast of Tohoku earthquake (the Great East Japan earthquake) that although the tsunami load was not considered and actual earthquake load was higher than designed load, more than 90% of the reinforced earth walls were not damaged [3]. This clearly indicates that the reinforced soil quay walls are efficient in withstanding tsunami loads.

During the 2011 off the Pacific coast of Tohoku earthquake, there were many aftershocks after the main shock of the earthquake which continued for several hours. It was reported that the Ofunato port in Japan was hit by an aftershock of magnitude, Mw = 7.9, and tsunami of height 3.2 m simultaneously at 15:15 (JST) [4]. This simultaneous impact of earthquake and tsunami might have caused the disastrous damages on breakwaters at the Ofunato port [4]. Therefore, it is relevant to analyze the stability of reinforced quay walls under the combined action of both earthquake and tsunami.

The analytical studies on combined action of earthquake and tsunami have been studied on breakwater structures [4]. Also, the effect of combined action of earthquake and tsunami on conventional type water front retaining wall has been studied earlier [5]. However, the effect of these forces on reinforced retaining walls has not been studied so far. Though [6–8] had done studies on the reinforced earth walls subjected to seismic loads using pseudo-static method, the effects of tsunami were not considered. Ahmad and Choudhury [9] proposed the seismic effect on reinforced earth wall using pseudo-dynamic method. The dynamic approach requires inputs of complex soil properties and wave parameters. Hence, as an easier and widely practiced method, the pseudo-static analysis has been adopted in this study. In the studies, related to the 2011 off the Pacific coast of Tohoku earthquake [10], it was suggested that the coastal structures such as quay walls should be designed by incorporating the structural measures to defend tsunami.

The numerical analysis of reinforced earth structures under seismic loads has been studied widely [6, 11–13], whereas the combined effect of the seismic load and tsunami load has not been evaluated yet. In the modified method of slices analysis for reinforced slopes under seismic loads, horizontal slices were considered instead of vertical. Therefore, the calculation of reinforcement forces became much simpler, since the force transfer between reinforcements of nearby slices has been avoided [14]. This new approach of determining the internal stability of reinforced slopes has been further studied by Nouri et al. [12], where pseudo-dynamic approach by incorporating horizontal slice method was done. The method evaluated the internal stability of reinforced soil system by determining the total tensile strength required for the reinforcement. Even though the horizontal slice method is widely acceptable for its simplest approach, the shear forces between two adjacent slices are not considered. Apart from that, the method generates the failure surface using optimization technique. This helps to develop a polylinear failure surface which gives an economical design as compared to linear failure surface [6].

The present study is thus focusing on the analytical approach to evaluate the stability of the reinforced soil quay walls by using pseudo-static method. The combined effects of hydrostatic forces, hydrodynamic forces, seismic loads and tsunami forces on the reinforced soil quay walls are considered to generate the most critical scenario. However, experimental tests of this study have to be done on later stage.

2 Method of Analysis

The seaward side of the quay wall is subjected to hydrodynamic forces along with tsunami forces. The tsunami force is assumed to impart a pulling effect on the reinforcements as the wave drawdown occurs. This is considered as a critical situation, where the tensile strength of reinforcement should be sufficiently large enough to withstand the tsunami load. The waterfront structure is also assumed to be subjected to surcharge loads during its operation. The quay wall system with a facing of concrete panels has been considered. The tsunami wave height (h_t) was varied to a maximum of 10 meters above the mean sea level. The objective of this study is to determine the tensile strength of the reinforcement by adopting horizontal slice method.

In the horizontal slice method, backfill soil of the reinforced quay wall is discretized into different horizontal slices [14]. From the studies of [15], an optimum number of slices required for a 5 m high reinforced earth wall was found out as 20 numbers. The increment in number of slices beyond this value yielded almost same values of total tensile strength of the reinforcement (Σtj). A similar trend was observed under the application of tsunami and earthquake loadings. Therefore, the number of slices considered in this study was limited to 20 numbers. The reinforcements were provided such that each slice holds one reinforcement at its center. For the simplicity in evaluation, the force transfers between adjacent horizontal slices were neglected [6]. Different forces acting on the quay wall system during the critical loading condition detailed in the below paragraphs are depicted in Fig. 1. The study is limited to the condition of quay wall with dry backfill by assuming an impermeable fascia.

2.1 Seismic Forces

The pseudo-static method was used to incorporate seismic effects on the wall system. In the study, M-O method was used. The method considers seismic forces as an inertia force acting on the failure wedge [16, 17]. The inertia forces were considered in both vertical and horizontal direction as Q_v and Q_h, respectively. Considering the weight of each slice as W_i, the seismic inertial forces can be expressed as below.

$$Q_h = k_h W_i \tag{1}$$

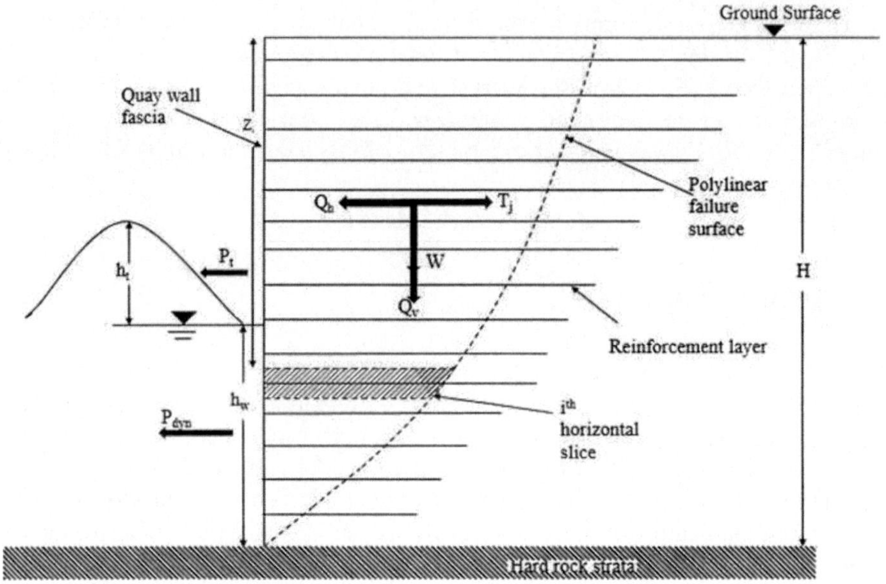

Fig. 1 Forces acting on the reinforced soil quay wall during earthquake and tsunami (drawdown condition)

$$Qv = k_v W_i \tag{2}$$

This method has the limitation of determining the appropriate pseudo-static coefficients. Furthermore, the efficacy of the method in calculating the inertial forces in liquefiable soils is not appreciable. Even though the pseudo-static analysis is not an accurate depiction of actual seismic loading on the soil mass, the simplicity in concept and calculation makes this method a useful tool in estimation of earthquake loading on reinforced earth walls.

2.2 *Tsunami Forces*

Tsunamis are large waves originated in the deep sea when epicenter of an earthquake occurs in the seabed. These waves will be of generally lower amplitude and longer wavelength in the deep sea, whereas when the waves approach the shore, the height of the wave drastically increases. When tsunami loads are considered acting on the quay wall, two cases have to be accounted. The first is the load transferred to the quay wall under the direct impact of a tsunami wave. Later is the pull-out force acting on the quay walls during the drawdown of the tsunami waves. The drawdown action would cause a destabilizing effect on the quay wall system, whereas the upward movement of tsunami wave would contribute more toward stabilizing the quay wall

[18]. Therefore, considering the critical condition, the direction of tsunami load was taken to be away from the quay wall system toward seaward side. The tsunami wave pressure acting on the quay wall (Pt) can be determined based on the Japanese design guidelines for coastal structures as follows [4].

$$P_t = \frac{1}{2}(P_1 + P_2)H \tag{3}$$

where

$$P_1 = \alpha_f \gamma_w (h_t + h_w) \tag{4}$$

$$P_2 = \alpha_f \gamma_w (h_t + h_w) - H \tag{5}$$

2.3 Hydrodynamic Pressures

The seismic event causes dynamic response in the water on seaward side. The outboard water thus imposes a hydrodynamic load (Pdyn) which can be evaluated from the solution defined by Westergaard [19].

$$P_{dyn} = \frac{7}{12} k_h \gamma_w h_w^2 \tag{6}$$

The hydrodynamic force and tsunami force are trying to pull the quay wall toward the sea. This effect was evident in the study on the water front geotechnical structures during 2011 off the Pacific coast of Tohoku earthquake [20]. It was observed that the structures failed toward to the outboard direction. Therefore, the direction of net forces was toward the seaside.

2.4 Equilibrium Conditions

The equilibrium condition of each slice wes considered for determining the tensile force acting upon the reinforcement in the slice. Referring to Fig. 2, the vertical forces at limiting state of equilibrium can be expressed as below [1].

$$N_i = \frac{V_i - V_{i+1} + (1 + k_v)W_i - cb_i \sin(\alpha_i)}{\tan\phi \sin\alpha_i + \cos\alpha_i} \tag{7}$$

where c is cohesion, b_i is the slanting width of ith slice, V_i and V_{i+1} are inter-slice vertical forces estimated as a summation of overburden pressure calculated as,

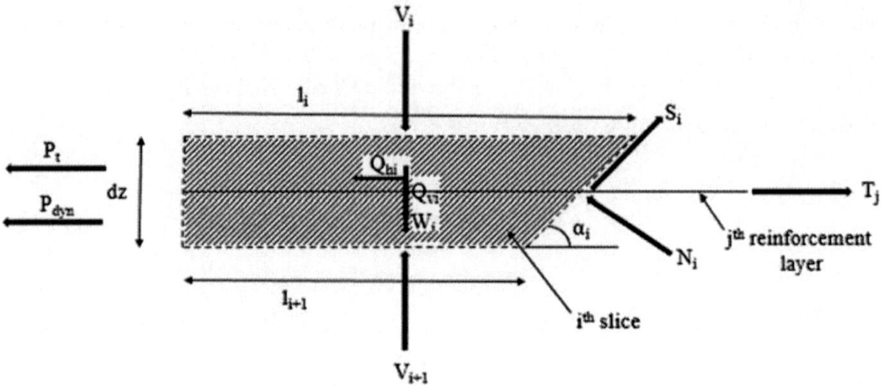

Fig. 2 Forces acting on *i*th horizontal slice

$$V_i = (1 + k_v)\gamma h_i l_i \tag{8}$$

l_i and l_{i+1} are the lengths of horizontal surfaces at top and bottom of slice, whereas b_i is the slanting length of the *i*th slice at the failure surface and α_i is the angle made between the failure surface and horizontal. While generating the failure surface, the slices are considered from the bottom of the reinforcing wall, in which the length of bottom of first slice was considered to be zero. In other words, the bottommost slice is given the shape of a triangle. Considering the limiting equilibrium of horizontal forces, the total tensile force on the reinforcement can be determined as,

$$\sum_{j=1}^{m} T_j = \sum_{i=1}^{n} N_i \sin\alpha_i + \sum_{i=1}^{n} W_i k_h + P_{dyn} + P_t - \sum_{i=1}^{n} S_i \cos\alpha_i \tag{9}$$

where the variables N_i and S_i are the normal and shear forces acting on the sliding face of the slice.

$$S_i = cb_i + N_i \tan\phi \tag{10}$$

The value of N_i is taken as zero instead of negative values. The value of T_j is maximized using the generalized reduced gradient (GRG) nonlinear optimization function in excel software such that the failure angles of each slices (α_i) are optimized to get the maximum value of T_j. The GRG nonlinear optimization function thus calculates the slope of objective function based on input values. Here, the objective function is summation of all T_j values determined for each layer. The input values are chosen as values of α_i. The maximum value of the α_i of each slice in the reinforced wall was restricted to be within 90° [7]. However, the exit angle was not fixed as 90°. Also, the minimum value of α_i was not fixed based on the M-O method. The restriction of α_i values based on the minimum angle of failure wedge proposed in the M-O method is based on the condition that ϕ-$\beta \geq \psi$, where β is the inclination of backfill surface to the horizontal and the value of ψ can be calculated as,

$$\psi = \tan^{-1}\left(\frac{k_h}{1 - k_v}\right). \tag{11}$$

This condition restricts the results, when higher values of k_h and lower values of ϕ are chosen. As an example, when the value of k_h is 0.5 and $k_v/k_h = 0.5$, for horizontal backfill surface, the calculated ϕ should be greater than 33.7°. Similarly, when value of k_h is considered as 0.6, the calculated ϕ should be greater than 40.6°. Thus, to avoid this restriction on ϕ values, the lower limit of α_i value was only limited to be positive real number. The wall friction is neglected, since the effect of wall friction in reinforcement force is negligible [7].

3 Results and Discussions

The normalized parameter K has been introduced to depict the effect of different factors on the reinforcement force [13].

$$K = \frac{\sum_{i=1}^{m} T_j}{0.5\gamma H^2} \tag{12}$$

The expression for K resembles the coefficient of lateral earth pressure in retaining wall design. The results obtained were plotted in Figs. 3, 4, 5 and 6. The parameters that were considered in the stability analysis include $H = 5$ m, $\gamma_d = 18$ kN/m^3, $\gamma_w = 10$ kN/m^3, $\phi = 15^{0-3}5^0$, $k_h = 0^-0.8$, $k_v/k_h = 0.5$, surcharge $q = 50$ kN/m^2, $h_w = 2$ m and $h_t = 0–10$ m.

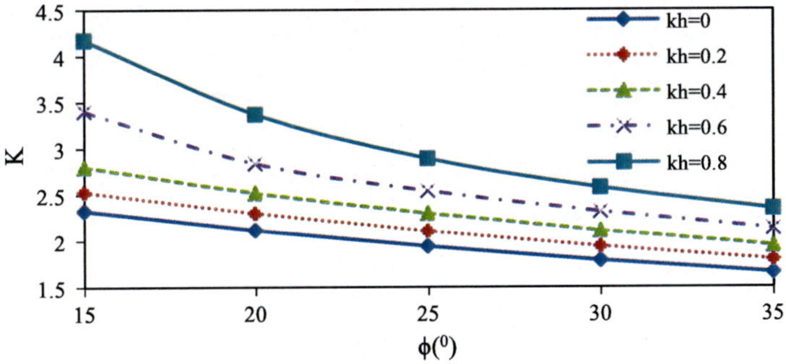

Fig. 3 Variation of K with ϕ for different values of k_h ($c = 0$, $q = 50$ kN/m^2, $\beta = 90°$ and $k_v = 0.5$ k_h)

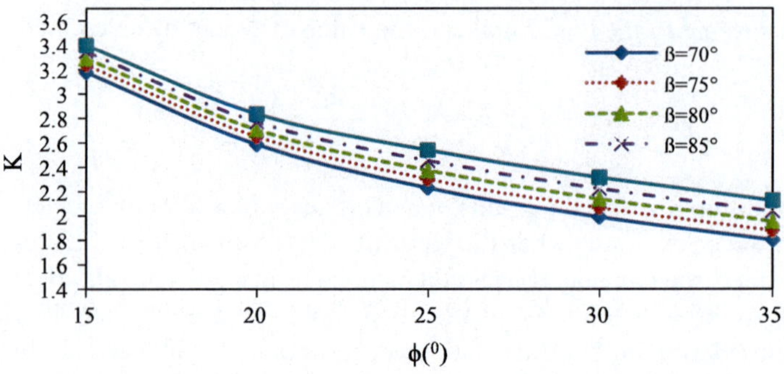

Fig. 4 Variation of K with ϕ for different values of β ($q = 50$ kN/m^2, $k_h = 0.6$, $c = 0$ and $k_v = 0.5$ k_h)

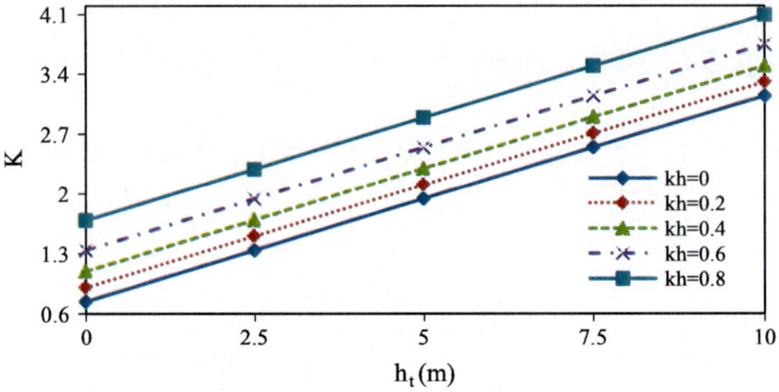

Fig. 5 Variation of K with h_t for different values of k_h ($c = 0$, $\phi = 35°$, $q = 50$ kN/m^2 and $k_v = 0.5$ k_h)

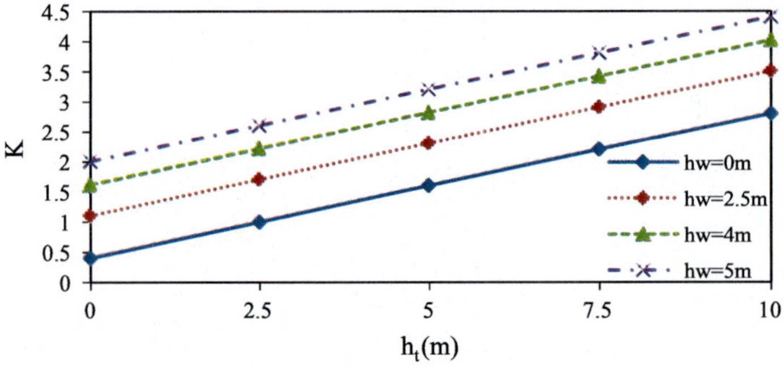

Fig. 6 Variation of K with h_t for different values of h_w ($c = 0$, $\phi = 35°$, $q = 50$ kN/m^2, $k_h = 0.6$ and $k_v = 0.5$ k_h)

3.1 *Effects of Horizontal Seismic Acceleration Coefficient (K_h)*

As shown in Fig. 3, the higher values of ϕ tend to reduce the K values. This decreasing trend in K is observed to be linear for lower values of k_h. The increase in k_h value has imposed larger forces on reinforcements, which necessitates the requirement of reinforcements of higher strength, to resist large earthquake loads. The angle of internal friction ϕ is an indicative of the capability of a soil to withstand shear forces. At higher values of ϕ, the soil transfers comparatively lesser force to the reinforcement. Thus, all the curves are observed to be having a negative slope.

3.2 *Effects of Inclination Angle of Quay Wall Fascia (β)*

Referring to Fig. 4, the inclination in the retaining wall fascia has reduced the reinforcement force. When the angle of internal friction in soil increases, the curves for various angle of inclination of fascia were showing a negative slope, which was evident. For further analysis, the fascia angle was chosen as 90°, since it imposed the maximum force on reinforcements.

3.3 *Effects of Height of Tsunami Waves*

The increase in height of tsunami waves (h_t) and higher seismic acceleration have imposed higher forces on the reinforcement layers. This effect is highlighted in Fig. 5. It is noticeable that there is a considerable hike in the reinforcement force as the tsunami height increases. Even at lower tsunami waves, there will be significant effect on the quay wall if the seismic impact is higher, whereas the effect of k_v is negligible in the calculation of reinforcement force. Therefore, in this study, the value of k_v has been chosen as 50% of k_h. The outboard water height level has also an effect on the reinforcement force as highlighted in Fig. 6. The increase in the level of water in outboard of quay wall to 5 m has imposed a 157% increase in the reinforcement strength values for the tallest tsunami impact of 10 m. The reinforcement force value for $\phi = 35°$ soils at a horizontal seismic acceleration of 0.6 without tsunami loading was 0.39 when there was no outboard water. Eventually, when the outboard water height was increased to 5 m, the force on reinforcement was increased by 5.1 times even without tsunami loading. Therefore, a close monitoring on the average height of water in the outboard is inevitable in the design of reinforced quay wall.

4 Conclusions

The stability analysis of reinforced soil quay walls, under the combined action of earthquake and tsunami, has been performed using the horizontal slice method and pseudo-static method. It was observed that factors such as horizontal seismic acceleration coefficients, height of tsunami waves, inclination of the quay wall fascia and internal friction angle of backfill soil play very important roles in the stability of the wall. Hence, these factors should be taken into careful consideration during the design process. Soils with higher shear strength tend to transfer lesser amounts of force to the reinforcements under critical conditions. Even at lower seismic acceleration, the influence of tsunami loading on reinforcement strength was considerably high.

Thus, the paper highlights the importance of incorporating the combined loading of earthquake and tsunami in the design calculation of reinforced earth water front structures. This study will ultimately contribute to a superior design protocol of waterfront structures under critical loading conditions.

References

1. Vidal, H.: La terre Armée. Annales de l'Institut Technique de Batiment et de Travaux Publics, France (1966)
2. Yang, K.-H., Wu, J.T.H., Chen, R.-H., Chen, Y.-S.: Lateral bearing capacity and failure mode of geosynthetic-reinforced soil barriers subject to lateral loadings. Geotext. Geomembr. **44**, 799–812 (2016)
3. Kuwano, J., Miyata, Y., Koseki, J.: Performance of reinforced soil walls during the 2011 Tohoku earthquake. Geosynthetics Int. **21**, 179–196 (2014)
4. Chaudhary, B., Hazarika, H., Ishibashi, I., Abdullah, A.: Sliding and overturning stability of breakwater under combined effect of earthquake and tsunami. Ocean Eng. **136**, 106–116 (2017)
5. Choudhury, D., Ahmad, S.M.: Stability of waterfront retaining wall subjected to pseudo-static earthquake forces. Ocean Eng. **34**, 1947–1954 (2007)
6. Ahmad, S.M., Choudhury, D.: Seismic internal stability analysis of waterfront reinforced-soil wall using pseudo-static approach. Ocean Eng. **52**, 83–90 (2012)
7. Gali, M., Murali Krishna, A.: Seismic response of reinforced soil retaining wall models: Influence of backfill relative density. Geotext. Geomembr. **26**, 335–349 (2008)
8. Sayed, S., Dodagoudar, G.R., Rajagopal, K.: Reliability analysis of reinforced soil walls under static and seismic forces. Geosynthetics Int. **15**, 246–257 (2008)
9. Ahmad, S.M., Choudhury, D.: Pseudo-dynamic approach of seismic design for waterfront reinforced soil-wall. Geotext. Geomembr. **26**, 291–301 (2008)
10. Raby, A., Macabuag, J., Pomonis, A., Wilkinson, S., Rossettob, T.: Implications of the 2011 great East Japan Tsunami on sea defence design. Int. J. Disaster Risk Reduction. **14**, 332–346 (2015)
11. Kerry Rowe, R., Skinner, G.D.: Numerical analysis of geosynthetic reinforced retaining wall constructed on a layered soil foundation. Geotext. Geomembr. **19**, 387–412 (2001)
12. Nouri, H., Fakher, A., Jones, C.J.F.P.: Development of Horizontal Slice Method for seismic stability analysis of reinforced slopes and walls. Geotext. Geomembr. **24**, 175–187 (2006)
13. Abdelouhab, A., Dias, D., Freitag, N.: Numerical analysis of the behaviour of mechanically stabilized earth walls reinforced with different types of strips. Geotext. Geomembr. **29**, 116–129 (2011)

14. Shahgholi, M., Fakher, A., Jones, C.J.F.P.: Horizontal slice method of analysis. Geotechnique **51**, 881–885 (2001)
15. Chandaluri, V.K., Sawant, V.A., Shukla, S.K.: Seismic stability analysis of reinforced soil wall using horizontal slice method. Int. J. Geosynthetics Ground Eng. **1**, 1–10 (2015)
16. Okabe, S.: General theory of earth pressure and seismic stability of retaining wall. Japan. Soc. Civ. Eng. **10**, 1277–1323 (1924)
17. Mononobe, N., Matsuo, H.: On the determination of earth pressures during earthquakes. In: Proceedings of World Engineering Congress, vol. 9, pp. 77–185. Japan (1929)
18. Kramer, S.L.: Geotechnical Earthquake Engineering, 1st edn. Prentice-Hall, United States of America (1996)
19. Westergaard, H.M.: Water pressures on dams during earthquakes. Trans. ASCE. **98**, 418–433 (1933)
20. Hazarika, H., Kasama, K., Suetsugu, D., Kataoka, S., Yasufuku, N.: Damage to Geotechnical Structures in Waterfront Areas of Northern Tohoku Due to the March 11, 2011 Tsunami Disaster. Indian Geotech. J. **43**, 137–152 (2013)

The Influence of Coupled Horizontal and Vertical Components of Strong Ground Motions on the Ground Response Analysis: Experimental and Numerical Approaches

Tsubasa Maeda, Hazarika Hemanta, and Yoshinori Kato

Abstract The large vertical ground motion during the 2008 Iwate-Miyagi Nairiku earthquake showed the importance of not only the horizontal ground motion but also the vertical motion in strong ground motion prediction and response analysis. In this study, considerations about the interaction behavior between the horizontal motion and the vertical motion during a strong earthquake were carried out by one-dimensional elasto-plastic earthquake analyses and cyclic shear box tests. It could be concluded from the results that the change of confining pressure caused by vertical motion affects the horizontal behavior, and strong horizontal motion produced vertical motion in the soil layer above the groundwater level.

Keywords Vertical motion · Seismic response analysis · Dilatancy · Elasto-plastic · Cyclic shear

1 Introduction

Earthquakes have caused a lot of economic and human damages since centuries. The 2016 Kumamoto earthquakes were a series of earthquakes including a magnitude (from Japan Meteorological Agency) 7.3 main shock, which occurred on April 16, 2016 (01:25 JST), beneath Kumamoto City, Kumamoto Prefecture, Japan, at an epicentral depth of about 12 km (32°45.2′ N, 130 °45.7′ E) and a foreshock earthquake with a magnitude (from Japan Meteorological Agency) 6.5 on April 14, 2016 (21:26 JST), at an epicentral depth of about 11 km (32° 44.5′ N, 130° 48.5′ E) [1, 2].

The earthquakes caused significant structural damage and huge loss lives in Kumamoto Prefecture. The Fire and Disaster Management Agency of Japan (FDMA) reported death of 270 people, severe injuries to 1184 people, and minor injuries to

T. Maeda (✉) · H. Hemanta
Kyushu University, Fukuoka 819-0395, Japan
e-mail: maeda.tsubasa.704@s.kyushu-u.ac.jp

Y. Kato
Maizuru College, National Institute of Technology, Kyoto 625-8511, Japan

H. Hazarika et al. (eds.), *Advances in Sustainable Construction and Resource Management*, Lecture Notes in Civil Engineering 144,
https://doi.org/10.1007/978-981-16-0077-7_33

1553 people. The structural damage amounted to 8657 houses completely destroyed, 3,4491 houses partially destroyed, and 155,095 houses damaged but habitable. The fire triggered from the earthquake destroyed 15 houses. In addition, more than 3 billion USD has been estimated for the civil infrastructure losses. Reconstruction cost is estimated to be around 6–8 billion USD [3, 4]. Further, more than 4000 aftershocks have been recorded by the Meteorological Agency of Japan since April 14, 2016 until November 30, 2017. The earthquake resulted in substantial damage to infrastructure including buildings, cultural heritage of Kumamoto castle, roads and highways, slopes and river embankment due to earthquake induced landslides and debris flows, and fault induced ground subsidence. At a surprisingly limited extent, liquefaction occurred only in a few districts of Kumamoto City and in the port areas [3, 5].

On the other hand, the Iwate-Miyagi Nairiku earthquake with a magnitude 7.2 occurred on June 14, 2008 (08:43 JST) in South Iwate Prefecture, Japan, at an epicentral depth of about 8 km (39° 01.7′ N, 140° 52.8′ E) [6]. A peak ground acceleration (PGA) in the earthquake was observed with 38.66 m/s^2 vertical acceleration, larger than the horizontal acceleration, at IWTH25 of KiK-net [7]. Generally, horizontal motion and vertical motion are handled separately in the creation of strong motion waveforms used in seismic design. When a strong earthquake occurs, it is possible that a large vertical acceleration occurs, so studying the vertical motion mechanism and the changes to horizontal motion by vertical motion is important in the evaluation of strong earthquakes. Aoi et al. [8] proposed that the vertical motion at IWTH25 occurred because particles are separated for each particle by large horizontal motion. Also, Morio et al. [9] presumed that the soil shallower than the groundwater level became elasto-plastic, so that strong vertical motion occurred because of dilatancy. It was also shown that the character of the waveform at IWTH25 could be predicted by analysis using the elasto-plastic constitutive law. However, detailed research about the vertical motion occurrence mechanism is needed because vertical motion has many unknown details. Therefore, in this study, cyclic direct shear tests and one-dimensional seismic response analysis were performed to clarify the basic coupled behavior, such as the occurrence of vertical motion caused by strong horizontal motion and the effect of vertical motion on horizontal motion.

2 Iwate-Miyagi Nairiku Earthquake

Figure 1 shows borehole and surface acceleration observed at IWTH25 during the 2008 Iwate-Miyagi Nairiku earthquake. UD2 and EW2 are surface acceleration, and UD1 and EW1 are borehole acceleration. Many characteristics of the observed waveform at IWTH25 have been shown in previous research [10].

1. In UD2, the maximum upward acceleration was 38.66 m/s^2, and the maximum acceleration of the three components composition was 40.22 m/s^2, which is the largest earthquake recorded.

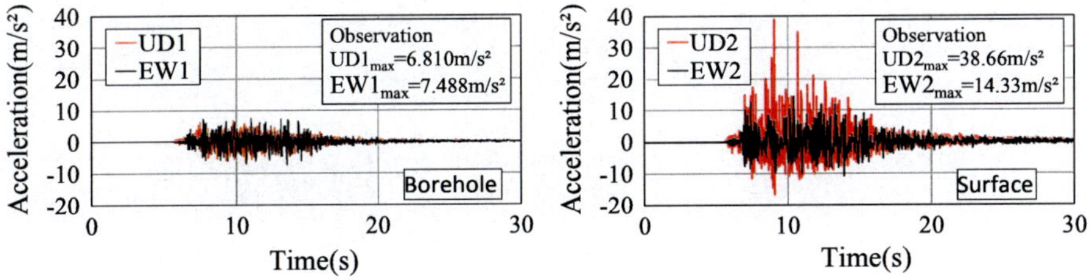

Fig. 1 Observed waveform on IWTH25

2. In UD2, the maximum downward acceleration was 17.03 m/s^2, which was less than half the upward acceleration, and the downward acceleration was small compared with the upward acceleration. Also, in the whole waveform, it shows asymmetry, and the downward acceleration peaked at 9.80 m/s^2.
3. The duration of the upward PGA was pulses shorter than that of the downward acceleration.
4. At the time when pulse upward acceleration was recorded, there were many peaks and valleys in the horizontal motion.

There is no vertical asymmetry in the underground records, and there are few times when the vertical and horizontal motion peaks and valleys coincide. From this, it is considered that the above features (1) through (4) occurred during the propagation of the ground surface from 260.4 m underground. Reproduction was attempted by using the one-direction seismic response analysis program SHAKE for the observed large acceleration record at IWTH25. The soil condition in Table 1 was decided upon based on the soil column at IWTH25. In Table 1, V_s is the S-wave velocity, and V_p is the P-wave velocity. In the analysis, the damping constant h is 5%, and the strain dependency is not considered. Figure 2 shows the horizontal acceleration waveform (EW2) and vertical acceleration waveform (U D2) on the surface obtained by SHAKE. In the horizontal direction, the acceleration in the analysis result is larger than the acceleration observed because of linear analysis. However, the analysis result for vertical acceleration is clearly smaller than that in the observation; also, in the

Table 1 Geologic data (IWTH25)

Layer number	Thickness (m)	Depth (m)	Mass Density ρ (t/m^3)	Vs (m/s)	Vp (m/s)
1	6	−6	1.9	430	850
2	28	−34	1.9	530	1770
3	30	−64	2.6	680	2310
4	48	−112	2.6	1120	2310
5	64	−176	2.6	1780	4010
6	28	−204	2.6	1380	2620
7	56.4	−260	2.6	1810	3180
8	–	–	2.6	1810	3180

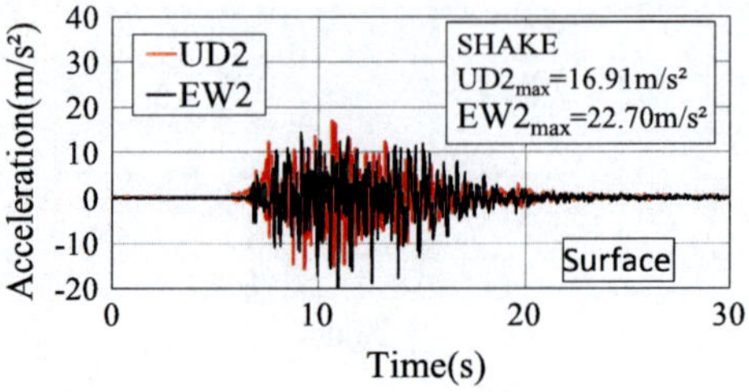

Fig. 2 Analysis waveform at surface

vertical direction, asymmetry does not appear. The vertical motion occurring from horizontal motion could not be considered by SHAKE, because it could not analyze the interaction of horizontal and vertical motion. Also, as above, the relationship to horizontal and vertical waveform could not be reproduced. The results obtained here, and the characteristics of (4) above suggest the possibility that vertical and horizontal motion interact in real ground.

3 Interaction of Horizontal and Vertical by Cyclic Direct Tests

Section 2 showed the possibility that the waveform of the interaction of horizontal and vertical motion can be observed in a previously observed seismic wave. During an earthquake, in laterally layered ground, seismic wave propagation corresponds to changes in that horizontal motion is cyclic shear and vertical motion is vertical stress or strain. Accordingly, the interaction of horizontal and vertical was examined by cyclic direct shear box tests using soil samples.

3.1 Effect of Vertical Motion on Horizontal Motion

First, an experiment was conducted on the effect of vertical motion on horizontal motion. It is generally known that there is a close relationship between effective stress and soil behavior. During an earthquake, the effective stress changes because of the vertical motion, which may affect the horizontal behavior. Therefore, the following experiment was performed. The experiment was carried out using over-compacted clay (σ_c = 1200 kN/m^2) prepared with water content w = 20% and dry density ρ_d = 1.77 g/cm^3. Using cyclic direct shear tests, a cyclic shear displacement with a period of 20 s and a half amplitude of 0.1 mm, as shown in Fig. 3, was added, and an experiment was conducted in which only a half-wave was input as the third wave in accordance with the third wave. The half-wave stress was increased so that the initial stress at the start of each shear doubled. The initial stresses at the start of shear were 20 and 100 kN/m^2. The over-compaction ratios were OCR = 60 and 12, respectively.

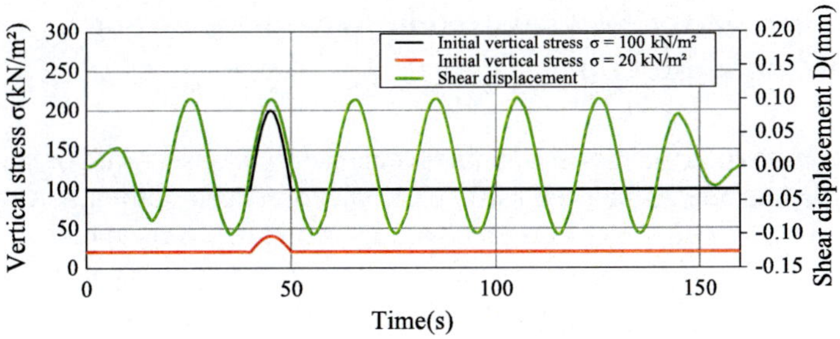

Fig. 3 Input waveform of test

Figure 4 shows the stress path, and Fig. 5 shows the relationship of shear stress and shear displacement. Figure 4 shows that, when vertical stress increased, shear stress also increased. This shows that typical interaction by effect of vertical motion on horizontal motion. Figure 5 shows that, when vertical stress increased, shear stress increased, which was not related to the initial vertical stress. Therefore, it is thought that, when vertical stress increases, shear stiffness changes. Because of the shear stiffness change resulting from vertical motion, the horizontal motion characteristics change.

Fig. 4 Stress path of test

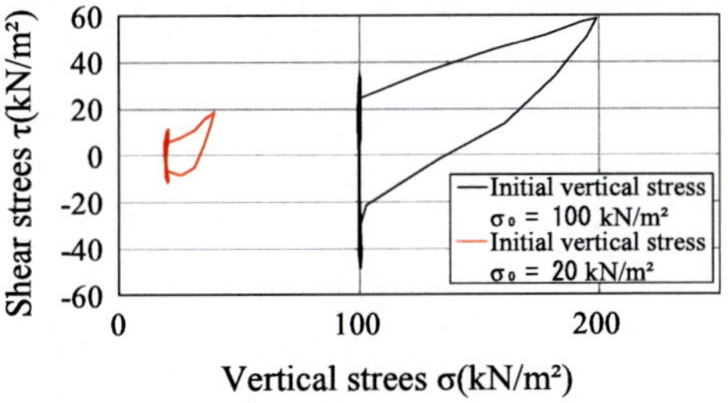

Fig. 5 Relationship between shear stress and shear strain

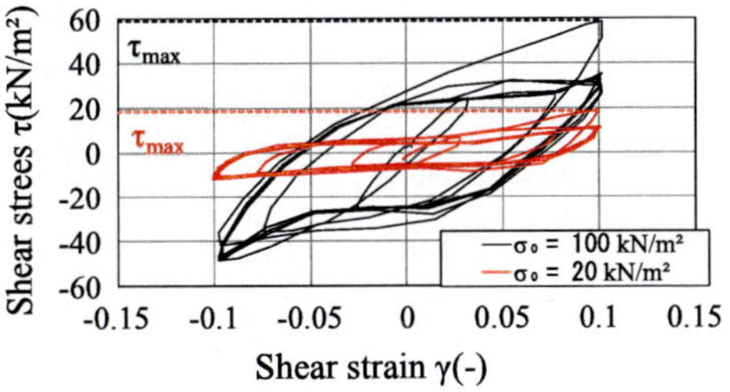

3.2 *Occurrence of Vertical Motion Because of Strong Horizontal Motion*

In granular materials, such as soil, volume changes occur as a result of shear deformation. It is considered that the same phenomenon occurs during a strong earthquake and that, when a strong horizontal motion is input, a vertical motion is caused by this phenomenon. Therefore, an experiment using only shear under a constant volume condition was performed, and the occurrence of vertical motion resulting from horizontal motion was considered.

A constant volume cyclic direct shear box test was performed on the same soil sample as described in Sect. 3.1. Before shear loading, consolidation was performed with initial stress $\sigma_0 = 20$ kN/m^2 until the volume change became zero. After that, as shown in Fig. 6, shear stress was gradually increased, and this was repeated until the soil sample broke. The speed at which the shear stress was applied was set so that one wave was completed in 120 s.

Figure 7 shows the stress path in the experiments. This shows typical interaction by occurrence of vertical motion because of strong horizontal motion. The vertical axis is vertical stress, and the horizontal axis is the history of shear stress. Vertical stress occurs by interaction despite only shear stress being input. Also, vertical motion occurs at twice the frequency of horizontal motion. It is thought that this phenomenon is caused by the dilatancy characteristic. The soil sample was over-compacted clay, so the vertical stress increased when the volume increased during constant volume tests. During earthquakes, the soil surface can be considered as a constant volume

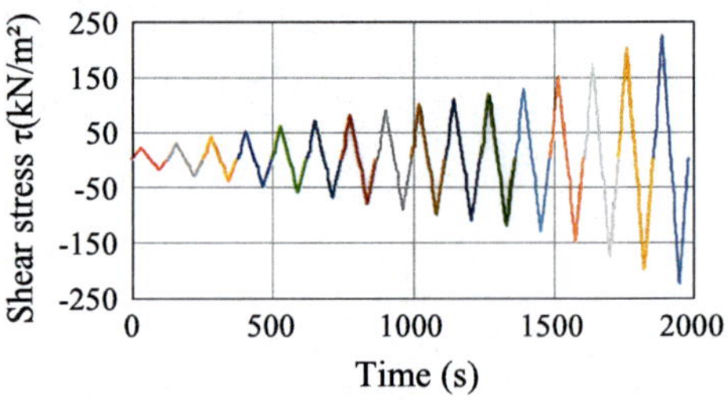

Fig. 6 Input waveform of box shear test

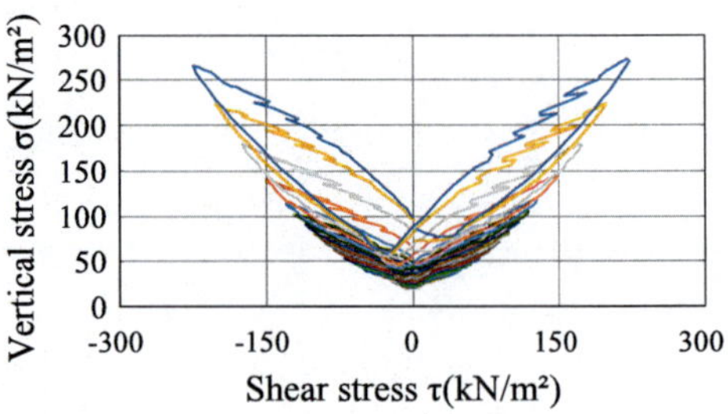

Fig. 7 Stress path of test

condition, so it is thought that vertical motion is excited at twice the frequency of horizontal motion because of dilatancy.

4 Elasto-Plastic Earthquake Response Analysis

The effects on horizontal motion of vertical motion and on vertical motion of horizontal motion were examined experimentally. It is necessary to study these phenomena by earthquake response analysis to provide strong earthquake prediction. Therefore, earthquake response analysis using a simplified one-dimensional model was conducted to deepen the consideration of the phenomena confirmed by experiments.

4.1 Effect on Horizontal by Vertical Motion

First, the analysis method and conditions are explained. A 40-m-thick one-dimensional ground was modeled by arranging 40 square elements of 1 × 1 m, and elasticity and elasto-plastic earthquake response analysis was performed using the analysis program FLIP [11]. Element numbers were assigned from top to bottom. The node numbers were incremented by 1 downward, with 1 in the upper left, and 41 in the upper right. The following values were set: mass density of ground $\rho =$ 1.8 t/m^3, S-wave velocity $V_s = 200$ m/s, P-wave velocity $V_p = 400$ m/s, cohesion $c = 30$ kN/m^2, friction angle $\varphi = 35°$, and Poisson ratio $\nu = 0.333$. The initial stress of the ground was set by self-weight analysis, and the self-weight analysis and dynamic analysis were set so that the horizontal and vertical displacements of two nodes at the same depth were equal. Rayleigh damping $\alpha = 0.6653$ and $\beta = 0.00{,}056$ were used. The step time dt = 0.001, the integration method was Wilson's θ method, and the nonlinear convergence calculation was done using the tangential stiffness method. $\theta = 1.4$ was set. As shown in Fig. 8, the horizontal and vertical motions with a frequency of 5 Hz and amplitude of 10.00 m/s^2 were only half-waves: ① only horizontal motions were input and ② both horizontal motion and vertical motion

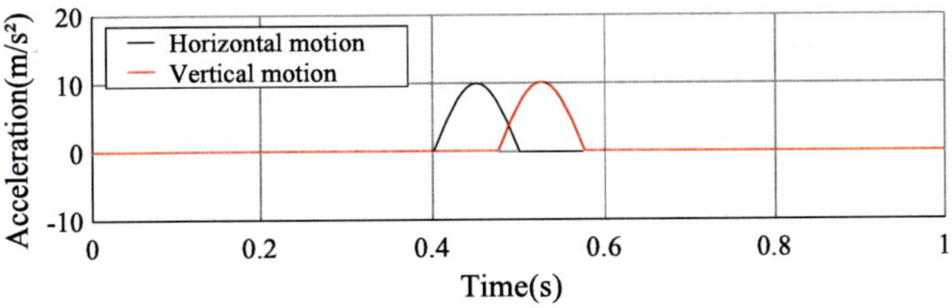

Fig. 8 Input waveform of analysis

were input. There were three analysis patterns: elastic (CASE 1) and elasto-plastic analysis (CASE 2) were performed for ① and elasto-plastic analysis (CASE 3) for ②.

The acceleration waveform at node 25 (depth 25 m) is shown in Fig. 9. From the top, the results are for CASE 1, CASE 2, and CASE 3. In CASE 1, the input horizontal motion is propagated as it is, and it can be seen that the wave reflected on the ground surface is propagated at approximately 0.8 s. Looking at CASE 2, it can be seen that there is vertical motion that could not be seen during CASE 1, even though only horizontal motion was input. Furthermore, the peak of horizontal motion is larger in CASE 3 with vertical motion. This phenomenon is thought to be caused by changes in horizontal motion resulting from changes in effective restraint pressure and shear stiffness due to vertical motion. Figure 10 shows the stress path of the element at 25 m underground. The vertical axis is the square root of the secondary stress invariant J_2, and the horizontal axis is I_1, the sum of the three principal stress components. The failure line is shown in the figure, but, in the elasto-plastic analysis, the input of large shear (horizontal motion) changes the effective stress along the failure line. It can be seen that this has occurred. From the above, it was possible to express the change

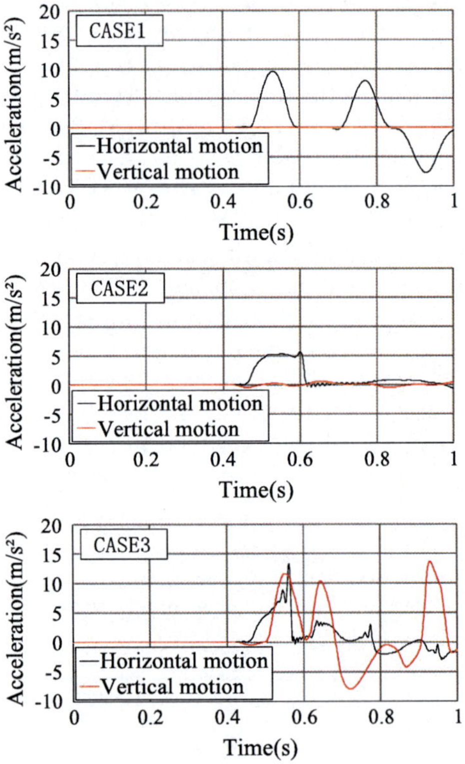

Fig. 9 Response waveform (depth of 25 m)

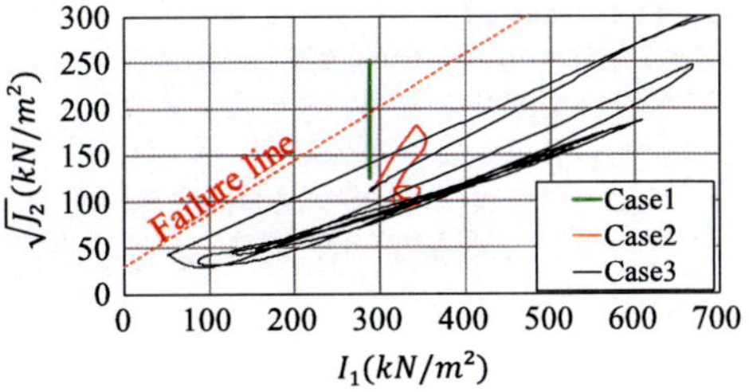

Fig. 10 Stress path of analysis

in horizontal behavior resulting from vertical motion by performing elasto-plastic analysis considering the dependence of shear stiffness on the restraint pressure.

4.2 Occurrence of Vertical Motion Caused by Strong Horizontal Motion

Here, the analysis conditions and methods are explained. The analysis area had a layer thickness of 10 m, and square elements of 1 × 1 m were arranged vertically. A program improved by Morio et al. [9] was used to introduce arbitrary constitutive laws into Fortran subroutine in the Zienkiewicz and Shiomi [12] two-phase effective stress analysis program. The yield surface is the Drucker–Prager model in Eq. (1), and the plastic potential is an unrelated flow law model in which φ in Eq. (1) is replaced with the dilatancy angle ψ.

$$f = \alpha I_1 + \sqrt{J_2} - \kappa \tag{1}$$

$$\alpha = \frac{2\sin\phi}{\sqrt{3}(3-\sin\phi)}, \kappa = \frac{6c\cos\phi}{\sqrt{3}(3-\sin\phi)}$$

Note that α and κ shown here are stress states in the triaxial compression test. The boundary conditions and analysis parameters are almost the same as the analysis in Sect. 4.1. The method used Newmark's β method, and the nonlinear convergence calculation used the initial stress method. The input seismic wave had a maximum acceleration of 5-Hz sinusoidal wave (horizontal motion) of 1.00 m/s^2. The vertical motion was considered because of the difference of the plastic potential surface by changing the dilatancy angle ψ.

Figure 11 shows the relationship between the maximum acceleration and the dilatancy angle on the ground surface. It can be seen that the maximum horizontal acceleration and the maximum vertical acceleration increased as the dilatancy angle increased.

This means that, when the dilatancy angle is reduced, negative dilatancy occurs, and the surface ground shows a strong plastic response. It is also shown that, when the

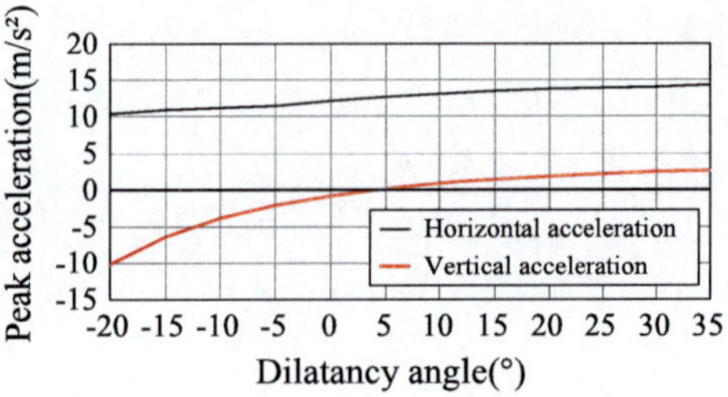

Fig. 11 Relationship between peak acceleration and dilatancy angle

dilatancy angle was increased, positive dilatancy occurred, and the plastic response was not as high as that of negative dilatancy. Figure 12a and b shows the surface waveforms at $\psi = -20°$ and 35°, respectively. It can be seen that, when the dilatancy angle was negative, downward acceleration occurred at the peak of horizontal motion, and, when the dilatancy angle was positive, upward acceleration occurred. In both cases, the frequency of vertical motion was twice that of horizontal motion. Figure 13 shows the stress path at a depth of 5 m at $\psi = 35°$. The vertical axis is the average principal stress, and the horizontal axis is the shear stress. A time of 0 s is indicated by a black circle, and a time of 4 s is indicated by a red circle. The shear stress

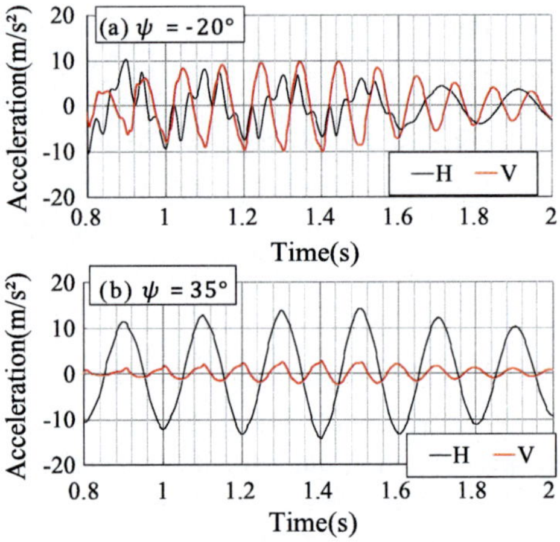

Fig. 12 Response surface acceleration of analysis

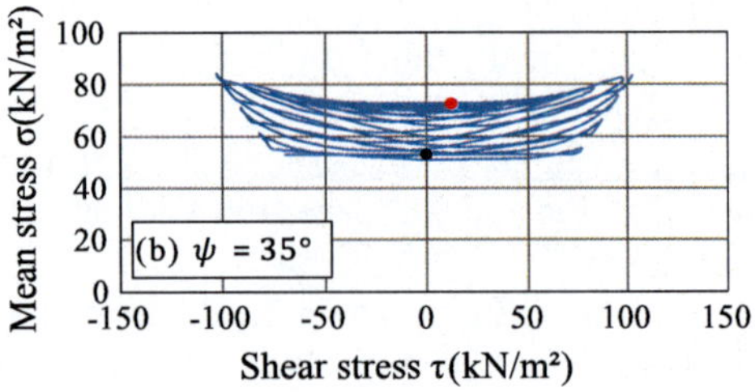

Fig. 13 Stress path of analysis

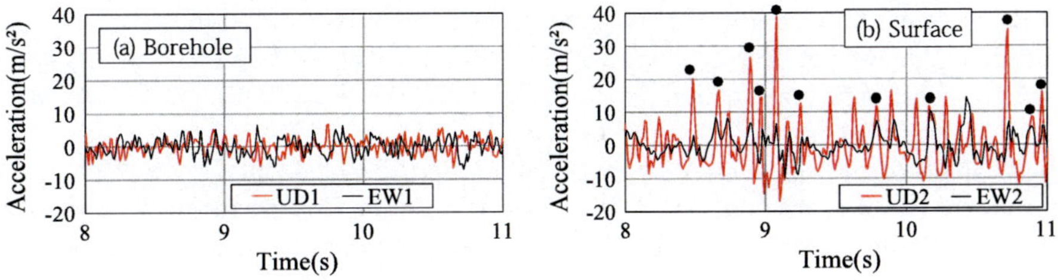

Fig. 14 Expanded view of IWTH25 (8~11 s)

changed as a result of horizontal motion, and the vertical stress increased because of repeated shearing. This is consistent with Fig. 7, which is the result of the experiment in Sect. 3.2. These results indicate that the occurrence of vertical motion resulting from the input of strong horizontal motion can be reproduced by seismic response analysis that introduces an elasto-plastic constitutive law that can express dilatancy characteristics. This shows that it can be expressed by changing the plastic potential surface.

5 Estimating Cause of Large Motion of Observed Earthquake

Based on the experiments and analysis results presented so far, an attempt is made to explain the observed waveform shown in Sect. 2. Figure 14 is enlarged views of observation records of IWTH25 for 8–11 s. In addition, (a) is borehole data, and (b) is surface data. In the observation data (a), there is no clear relationship between horizontal motion and vertical motion. However, in the surface observation waveform of IWTH25 in Fig. 14b, there are many points (black circles) where the positive peak of vertical motion and the peak of horizontal motion coincide. This phenomenon is the same as that in the case of positive dilatancy explained in Sect. 4.2. In particular, when looking at the vicinity of 9–10 s, it can be confirmed that the vertical motion occurs at twice the frequency of the horizontal motion, which is not different from the experimental results. From these facts, even in the actual ground motion, it is considered that vertical motion occurs due to the dilatancy characteristics of the surface ground below the groundwater level.

6 Conclusions

In this study, experiments and analyses were conducted on the coupled behavior of horizontal motion and vertical motion, and the results were considered. First, the Iwate-Miyagi Nairiku earthquake was analyzed, and it was shown that the influence of the interaction behavior of horizontal motion and vertical motion was observed, and that it could not be reproduced by one-dimensional seismic response analysis without considering interaction. Next, cyclic direct shear tests and one-dimensional seismic response analysis were performed.

As a result, it was shown by experiments that the horizontal motion changes due to the vertical motion, and that it can be expressed by performing an elasto-plastic earthquake response analysis considering the confinement dependency. In addition, the occurrence of vertical motion due to horizontal motion resulting from volume change (dilatancy characteristics) caused by the shear deformation of the ground can be confirmed by experiments and can be expressed by elasto-plastic seismic response analysis introducing constitutive laws that considered dilatancy characteristics. Based on these results, the Iwate-Miyagi Nairiku earthquake was evaluated. The earthquake at IWTH25 had a positive dilatancy. It can be explained the observed waveform at IWTH25.

Acknowledgements We acknowledge the support provided by the National Research Institute for Earth Science and Disaster Prevention (NIED) for the use of strong motion data.

References

1. Japan Meteorological Agency (JMA). http://www.data.jma.go.jp/svd/eqdb/data/shindo/Event.php?ID=198162. Last accessed 06 Nov 2020
2. Japan Meteorological Agency (JMA). http://www.data.jma.go.jp/svd/eqdb/data/shindo/Event.php?ID=197810. Last accessed 06 Nov 2020
3. Hazarika, H., Kokusho, T., Kayen, R.E., Dashti, S., Fukuoka, H., Ishizawa, T., Kochi, T., Matsumoto, D., Furuichi, H., Hirose, T., Fujishiro, T., Okamoto, K., Tajiri, M., Fukuda, M.: Geotechnical damage due to the 2016 Kumamoto Earthquake and future challenges. Lowland Technol. Int. **19**(3), 203–218 (2017)
4. Fire and Disaster Management Agency (FDMA). https://www.fdma.go.jp/disaster/info/items/kumamoto.pdf. Last accessed 06 Nov 2020
5. Japan Meteorological Agency (JMA). https://www.jma.go.jp/jma/kishou/books/gizyutu/135/ALL.pdf
6. Japan Meteorological Agency (JMA), http://www.data.jma.go.jp/svd/eqdb/data/shindo/Event.php?ID=172967. Last accessed 06 Nov 2020
7. Yamada, M., Mori, J., Heaton, T.: The slapdown phase in high-acceleration records of large earthquake. Seismol. Res. Lett. **80**(4), 559–564 (2009)
8. Aoi, S., Kunugi, T., Fujiwara, H.: Trampoline effect in extreme ground motion. Science **322**, 727–730 (2008)
9. Morio, S., Kato, Y., Fujii, T.: Elasto-plasticity consideration of the ground on the large acceleration time histories at two earthquake observation sites. Jap. Assoc. Earthq. Eng. **17**(4), 4_30–4_49 (2017)
10. Ohmachi, T., Inoue, S., Mizuno, K., Yamada, M.: Estimated cause of extreme acceleration records at the KiK-net IWTH25 station during the 2008 Iwate-Miyagi Nairiku earthquake. Jap. Assoc. Earthq. Eng. **11**(1), 1_32–1_47 (2011)

11. Iai, S., Matsunaga, Y., Kameoka, T.: Strain space plasticity model for cyclic mobility. Rep. the Port and Harbour Res. Inst. **29**(4), 27–56 (1990)
12. Zienkiewicz, O.C., Shiomi, T.: Dynamic behavior of saturated porous media: the generalized Biot formulation and its numerical solutions. Int. J. Numer. Anal. Meth. Geomech. **8**, 71–96 (1984)

Verification of Seismic Deformation of Road Embankment Laying on the Coast Using Numerical Analysis

Kentaro Kuribayashi, Tadashi Hara, and Shuichi Kuroda

Abstract Following the 2011 Off-Pacific Coast of Tohoku Earthquake, embankments have again attracted much attention as a tsunami restraining measure. Because of alluvial deposit widely distributed along the coast, liquefaction of the ground often obstructed the traffic function. In this paper, the characteristics of the sedimentary layer of the coast are sorted out from the viewpoints of landform division and grain size analysis results for a local coast in Kochi Prefecture. Furthermore, effective stress analysis is conducted at various liquefaction ground condition represented in the coastal area. Using the results of this analysis, the relationship between the characteristics of the ground and embankment deformation is established. Finally, a simple evaluation method for road embankment settlement is suggested using the results of the effective stress dynamic analysis which is often conducted for the design of road embankments with a large-scale earthquake in mind.

Keywords Liquefaction · Performance design · Road embankment · Numerical analysis

1 Introduction

In the 2011 Off-Pacific Coast of Tohoku Earthquake, the road embankments for coastal roads functioned to restrain tsunami greatly contribute to evacuation in the immediate aftermath of the disasters and to the swift transportation of relief goods for areas which were inundated by tsunami [1]. In view of such facts, the use of road embankment as a tsunamic restraining measure at the time of a major earthquake has been examined, especially for high standard coastal roads [1, 2].

K. Kuribayashi (✉) · S. Kuroda
Eight-Japan Engineering Consultants Inc, Okayama, Japan
e-mail: Japankuribayashi-ke@ej-hds.co.jp

T. Hara
Kochi University, Kochi, India

H. Hazarika et al. (eds.), *Advances in Sustainable Construction and Resource Management*, Lecture Notes in Civil Engineering 144,
https://doi.org/10.1007/978-981-16-0077-7_34

Typical examples of damage to banking structures by a major earthquake are slip failure of the banking on top of sloping ground or soft ground and step damage due to sinking of the foundation ground and embankment body at a connecting section with such structures as the back of an abutment and road crossing culvert crossing a road embankment. Coastal areas where alluvia made up of coastal deposits and river deposits are widely distributed cause particular concern for tsunami damage. In fact, many cases of damage due to liquefaction of the foundation ground have been reported in the past earthquakes [3, 4].

Meanwhile, the idea of performance design has been adopted for banking structures in recent years to the extent that the checking of post-earthquake deformation of an embankment is now required when Level 2 seismic motion is detected [5]. For calculation of the deformation volume of an embankment above liquefied ground, effective stress dynamic analysis using the two dimensions finite element method is popularly used. However, as this analysis and associated ground survey are both time and labour-consuming, the actual design usually selects a representative cross section of which the deformation volume is expected to be large for such checking using the analysis results for the selected cross section. As multiple above-mentioned connecting sections with road crossing structures tend to exist not only in the selected cross section but also in the entire design section, there is a high likelihood that the deformation volumes at these connecting sections differ from the settlement volume at the embankment crown obtained from the analysis results for the representative cross section. Because of this, a more reliable method is required to determine the settlement volume of an embankment along the longitudinal direction of a road, including at those sites where a crossing structure is constructed.

In this study, the construction of a road embankment in a coastal area is assumed. The sedimentation conditions of alluvium on the coast where damage by liquefaction and tsunami is expected to occur are sorted out, and numerical analysis reflecting such conditions is conducted to determine the tendency of the post-earthquake settlement volume at the embankment crown.

2 Selection of Representative Grounds

2.1 Grain Size Characteristics of Alluvia Forming Local Coasts of Kochi Prefecture

For the selection of representative grounds, the local coasts of Kochi Prefecture were considered to be typical candidates because these coasts were expected to experience massive damage by strong seismic motion and tsunami associated with an anticipated Nankai Trough earthquake in the future [6]. From among these coasts, some coasts were then selected based on verification of the existence of certain landform divisions which are likely to have formed an alluvium (flood plain; coastal plain; delta; natural levee; reclaimed land) using a land condition map [7]. This was followed by analysis of the tendency in terms of the mean grading texture (hereinafter referred to as "mean grading texture") using boring data for each local coast [8]. The mean grading texture at each local coast is the mean value of the representative grading texture which is calculated by Eq. (1) for each boring site for which grain size distribution data is available. The mean grading texture at each local coast shows the proportions of four grain types (gravel, sand, silt and clay).

$$G = \sum_{i=1}^{n} \left(G_i \times \frac{t_i}{t} \right) \tag{1}$$

where

- G mass percentage by soil type at each boring site
- G_i mean mass percentage by soil type in each soil layer
- t_i thickness of each soil layer
- t total thickness of all layers.

Figure 1 shows local coasts with landform divisions forming alluvium based on the land conditions map [7] and graphs of the mean grading texture as calculated in Eq. (1).

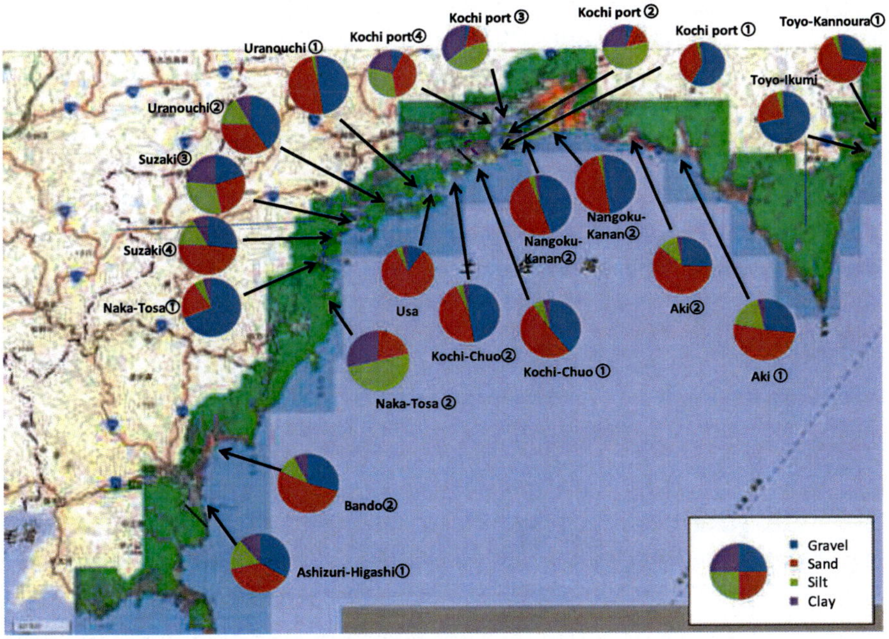

Fig. 1 Proportions of average grain size at each alluvia forming local coast (Background map published by Geospatial Information Authority of Japan [7, 8])

Coastal areas in the study area see soil layers mainly formed by sand and gravel. In general, river deposits near a river mouth contain much silt [9]. In Kochi Prefecture, however, coastal areas and mountain areas are close to each other, and many areas experience the conspicuous meandering of rivers [10], resulting in the frequent coastal presence of a sand and gravel layer formed by the flooding of a river.

In contrast, alluvium with dominant silt and clay and without much gravel is formed in some coastal areas. All of these coastal areas showing such a tendency are local coasts with a rias-type submergence topography where not only river deposits but also coastal deposits are likely to be accumulated [11]. It is believed that a conspicuous sand and gravel layer mentioned above is not formed in these areas because of reasons mentioned above in addition to the relatively small size of the river flowing through the plain.

The above findings clearly show that most alluvial land in coastal areas considered for the study has prominently sandy ground, sandy silt ground or sand and gravel ground consisting of river deposits and coastal deposits. This suggests that a significant impact of liquefaction by an earthquake must be taken into consideration when road embankments are constructed in these coastal areas.

3 Selection of Representative Grounds

The representative grounds to be selected are those for which there is concern regarding the occurrence of liquefaction, the grading texture of which resembles that of local coasts of Kochi Prefecture and for which cyclic undrained triaxial test results are publicly available. Figure 2 shows mean size distributions, which is circle graphs at local coasts shown in Fig. 1, and the grain size accumulation curves of the subject representative grounds for analysis which is described later. Table 1 shows the conditions for specimens which were collected at each representative ground and used for the cyclic undrained triaxial test.

1. Sandy ground

Specimens from the Ogata District located at the Bando ② local coast were selected to represent sandy ground which is the most common ground for local coastal areas in Kochi Prefecture [12]. These specimens were collected by the undisturbed sampling method in the coastal plain of the district in question and are sandy soil containing much sand. The sampling operation targeted two types of sandy ground which are loose sandy ground where specimens were collected from 3 m below the loosely sedimented ground surface and compacted sandy ground where specimens were collected from 8 m below the ground surface.

2. Sandy silt ground

There have been many cases where ground with a high fine-grain faction is damaged by liquefaction [13]. Specimens from Kanzaki River which is a Class A river flowing through Chiba Prefecture were used as alternative specimens representing sandy silt

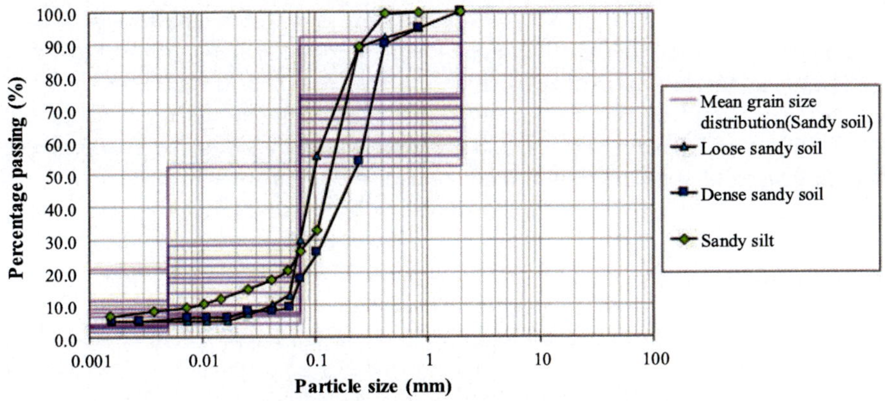

(a) Sand, sandy silt

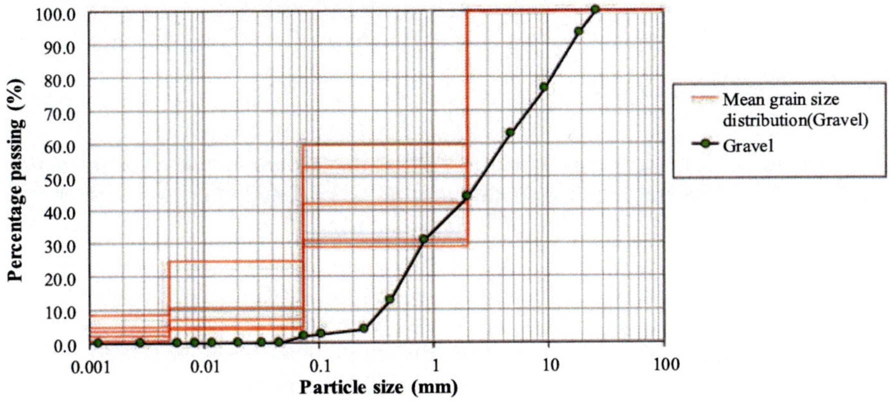

(b) Gravel

Fig. 2 Grain size accumulations of sample in each representative ground

Table 1 Conditions of sample in each representative ground

Representative ground	Radial stress (σ') (kPa)	Dry density (ρ_d) (g/cm^2)	Condition of soil specimen
Loose sandy soil	49	1.53	Undistributed (frozen sampling)
Dense sandy soil	98	1.27	Undistributed (frozen sampling)
Sandy silt	49	1.4	Undistributed (frozen sampling)
Gravel	98	1.98	Distributed and reconstructed

ground with a slightly higher fine-grain fraction than sandy ground [14]. These specimens were collected by the undisturbed sampling method from an old river channel near the present Kanzaki River and are classified to represent sandy silt ground. The sampling site was a confirmed liquefaction site during the 2011 Off-Pacific Coast of Tohoku Earthquake. It is characterized by low plasticity despite the fact that the soil has a fine-grain fraction of over 30%. This type of silty ground with less compaction produces a volume change similar to that of sandy ground at the time of liquefaction [15].

3. Gravelly ground

Specimens from the mouth of Niyodo River located at the Kochi-Chuo ② local coast were selected to represent gravel-dominant ground typically represented by local coasts in Kochi Prefecture [16]. These specimens were collected by the disturbed sampling method from the flood plain of Niyodo River.

4 Analysis Conditions

Effective stress dynamic analysis using the two dimensions finite element method was conducted, targeting the foundation ground and a road embankment to analyse the tendency of the deformation volume of a road embankment due to liquefaction caused by an earthquake. This analysis was based on the results of the cyclic undrained triaxial test conducted with specimens which displayed the representative grain texture in coastal areas. The programme used for the effective stress dynamic analysis was LIQCA [17, 18] which is capable of taking the drainage settlement after an event of liquefaction into consideration and which has been used for the residual deformation analysis of banking structures on sandy ground or gravelly ground at the time of an earthquake.

4.1 Element Simulation for the Representative Grounds

The physical properties of the ground which are compatible with the results of the cyclic undrained triaxial test for each representative ground were established through element simulation. Figure 3 shows the liquefaction strength curves based on the said test results and element simulation results. Figure 3 also includes the physical properties used for the reproduction analysis by LIQCA for banked levees along Yodo River which were damaged by the Southern Hyogo Earthquake in 1995 and element simulation results for comparison [19].

In Fig. 3, the liquefaction strength is slightly higher for the element simulation results than the test results in the region of a smaller number of cycles in particular. It is pointed out by the existing literature that the results of the cyclic undrained triaxial test tend to evaluate the liquefaction strength lower than the actual ground

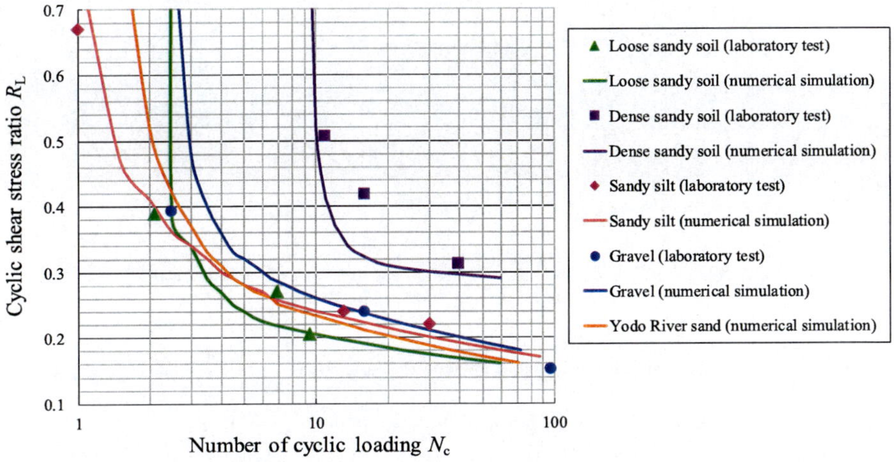

Fig. 3 Liquefaction strength curves in each representative ground

strength due to disturbance to the specimens and certain impacts of the test method [20]. As this tendency is more prominent in the range of a smaller number of cycles, it is judged that the element simulation results generally reproduced the liquefaction strength of the actual ground. For compacted sandy ground, the test results were not available for the range of 10 or less cycles.

4.2 Analytical Model

The analytical model is composed of banking and foundation ground (liquefied layer; non-liquefied layer). Figure 4 shows the analytical model, and Table 2 shows the analysed physical properties of each layer. The thickness of the liquefied layer is altered from 3 to 5 m and further to 8 m to analyse the relation between the liquefied layer thickness and the settlement volume. The cross section of the road embankment consists of a banking height of 5 m, crown width of 18 m and slope gradient of 1–1.8 as the standard cross-section conforming to the Road Construction Guidelines [7]. The groundwater level is set at GL −1.0 m in each case of analysis.

The boundary conditions for this analytical model are a viscous boundary for the basal plane and 1000 m wide pseudo-free ground on both end sides of the model. The entire model was subject to Rayleigh damping of the type proportional to the initial stiffness as this is popularly used in LIQCA, and a stiffness coefficient of α_1 was set at 0.003 [21].

The physical properties of the embankment were set based on the results of experiments with the existing road embankments, assuming sand and gravel materials of which the degree of compaction (*Dc*) is 95% [22–24]. The physical properties of the liquefied layer were set based on the element simulation results. For the value

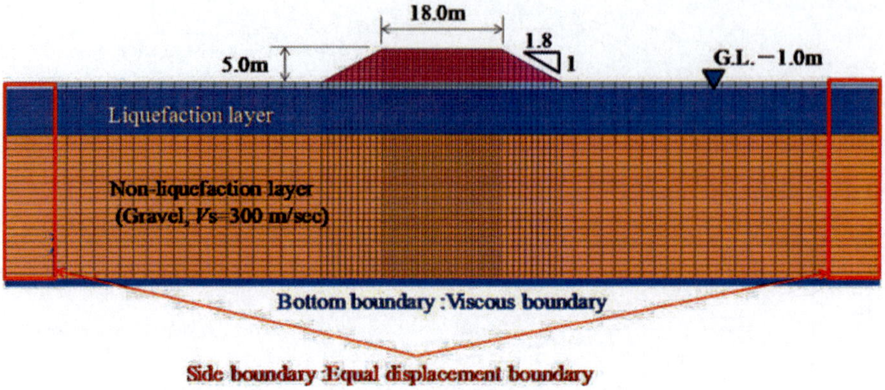

Fig. 4 Numerical model (LIQCA, thickness of liquefaction layer = 8 m)

of the shear wave velocity (Vs) used for calculation of the shearing rigidity of the ground, the values of the past tests implemented with representative grounds were used [12, 13, 16].

A non-liquefied layer was introduced under the liquefied layer to prevent impacts of the reflected wave from the base plane of the analytical model on the behaviour of the liquefied layer. The assumed non-liquefied layer was sand and gravel ground of $Vs = 300$ m/s which is equivalent to engineering foundation ground in the Specifications for Highway Bridges [25]. This non-liquefied layer was set as the elastic model to suppress the damping of input seismic motions by the non-liquefied layer.

Moreover, in order to consider the settlement of the embankment crown in the compressive dissipation process in the aftermath of the liquefaction of the foundation ground, post-earthquake compressive dissipation analysis was conducted, inheriting the situation of stress after the dynamic analysis. The compaction calculation period was set at approximately 52 h, and the residual deformation volume of the embankment was checked at the stage where the excess pore water pressure of the liquefied layer was sufficiently decreased.

4.3 Input Seismic Motions

The input seismic motions analysed were three Type II seismic motion waves of Level 2 seismic motions in Type I ground in the Specifications for Road Bridges which are widely used for the earthquake-resistant design of road embankments against Level 2 seismic motions as shown in Table 3 and Fig. 5.

The past examples of effective stress dynamic analysis based on the conventional microstrain theory point out the possibility of exaggerating the deformation of an embankment with seismic motions of which the duration is long, such as Type I seismic motions [26]. Therefore, Type I seismic motions were excluded from the

Table 2 List of input properties

		Elasto-plastic model						Elastic model
		Embankment (D c=95%)	Loose sandy soil	Dense sandy soil	Sandy silt	Gravel	Yodo river sand	Non liquefaction layer
Unit weight	γt(kN/m^3)	19.0	19.0	19.0	17.5	19.6	19.1	21.0
Permeability	k(m/s)	1.0×10^{-5}	3.8×10^{-5}	1.75×10^{-5}	1.75×10^{-5}	7.5×10^{-4}	1.3×10^{-5}	2.5×10^{-3}
Initial void ratio	e0	0.70	1.00	1.04	0.56	1.00	0.80	0.75
S wave velocity	Vs (m/s)	250	160	220	120	250	180	300
Initial shear modulus	Gmax(kN/m^2)	1.21×10^{s}	4.96×10^4	9.38×10^4	2.57×10^4	1.25×10^{s}	4.45×10^4	1.93×10^5
Cohesion	c′(kN/m^2)	0.00	0.00	0.00	0.00	0.00	0.0	0.00
Initial friction angle	Φ'(°)	40.00	32.60	38.50	34.00	33.00	39.0	40.00
Compression index	λ	3.41×10^{-3}	4.72×10^{-3}	7.00×10^{-3}	8.00E−03	7.00×10^{-3}	2.5×10^{-2}	
Swelling index	κ	3.41×10^{-4}	4.72×10^{-4}	7.00×10^{-4}	8.00E−04	7.00×10^{-4}	2.5×10^{-3}	
Pseudo overconsolidation ratio	OCR*	1.000	1.000	1.800	1.400	1.200	1.000	
Normalized initial shear modulus	G0/σm′	1913	1519	1436	787	1915	681	
Phase transformation stress ratio	Mm	0.909	0.909	0.944	0.838	0.909	0.980	
Failure stress ratio	Mf	1.336	1.072	1.283	1.122	1.087	1.300	
Hardening parameter	B0	4296	4000	4500	3000	5000	2500	

(continued)

Table 2 (continued)

		Elasto-plastic model						Elastic model
		Embankment (D c=95%)	Loose sandy soil	Dense sandy soil	Sandy silt	Gravel	Yodo river sand	Non liquefaction layer
	B1	43	20	50	30	50	25	
	Cf	0.00	0.00	0.00	0.00	0.00	0.00	
Reference value of plastic strain	γP*r	0.000	0.0070	0.0060	0.0060	0.0040	0.0020	
Reference value of elastic strain	γE*r	0.000	0.024	0.036	0.060	0.040	0.020	
Parameter od dilatancy	D0	0.000	0.800	0.800	0.500	0.500	1.000	
	n	0.000	6.000	7.200	3.000	4.000	4.000	

Table 3 Input earthquake motions

Input earthquake motion			Reference past earthquake
Type II	Inland earthquake	Type II-1	The Hyogo-ken Nanbu Earthquake in 1995
		Type II-2	
		Type II-3	

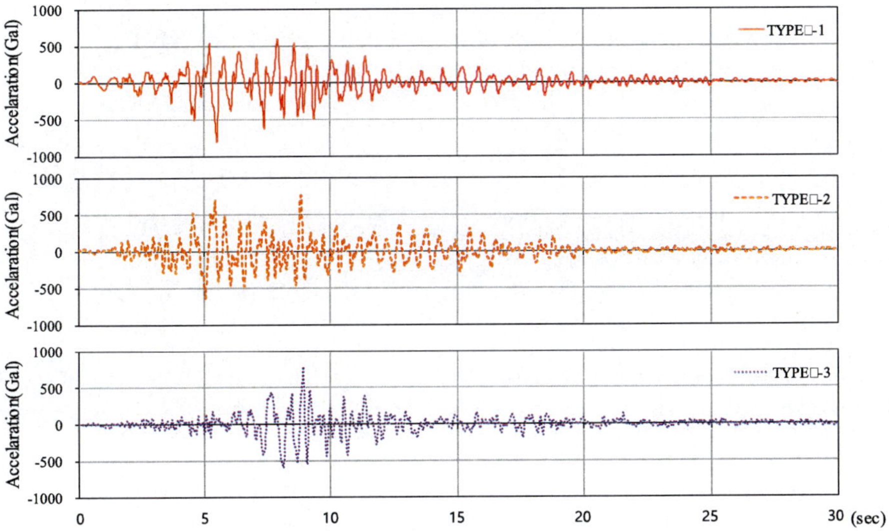

Fig. 5 Input earthquake motions

scope of the present analysis which is effective stress dynamic analysis based on the microstrain theory.

5 Analysis Results

5.1 Deformation Mode of Representative Cases

As examples of the analysis results, Figs. 6, 7, 8 and 9 show a residual deformation diagram and vertical displacement contour diagram after the completion of compression with Type 2-II-I-1 seismic motion which caused the largest deformation volume of the embankment among Type II seismic motions.

When the deformation modes of the liquefied layer thickness of 3, 5 and 8 m are compared with one another, deformation is concentrated at both the slope shoulders of the embankment in the case of a thinner liquefied layer, and the deformation mode is similar to that of a slip of the embankment itself. In those cases of a thicker liquefied

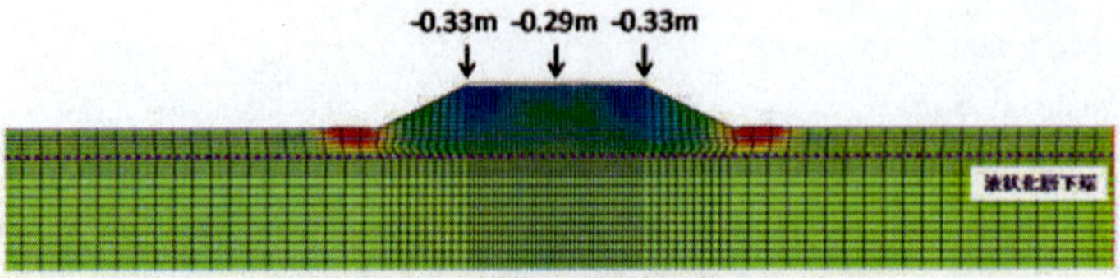

(a) Thickness of liquefaction layer = 3m

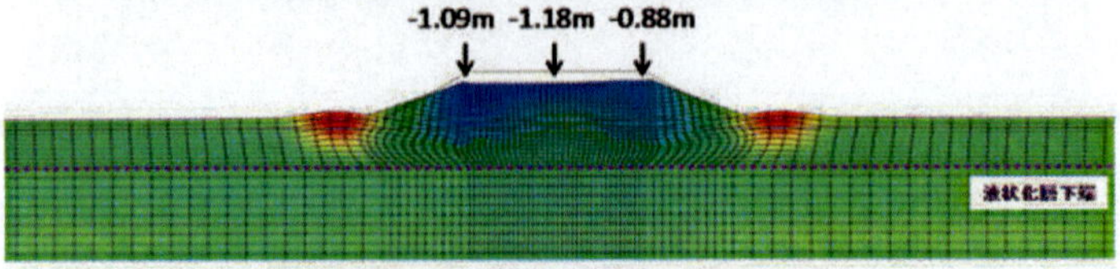

(b) Thickness of liquefaction layer = 5m

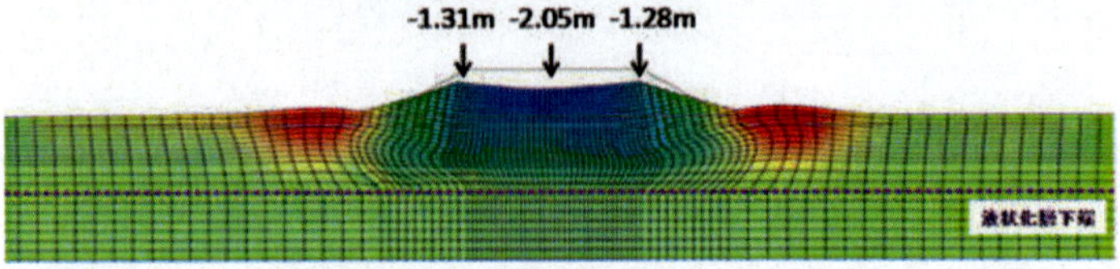

(c) Thickness of liquefaction layer = 8m

Fig. 6 Residual deformation and vertical displacement contour diagrams (ground: loose sandy soil, input ground motion: Type II-1, after dissipation of excess pore water pressure)

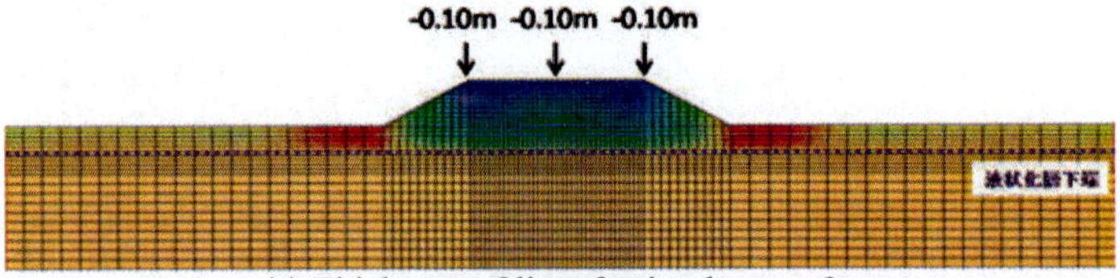

(a) Thickness of liquefaction layer = 3m

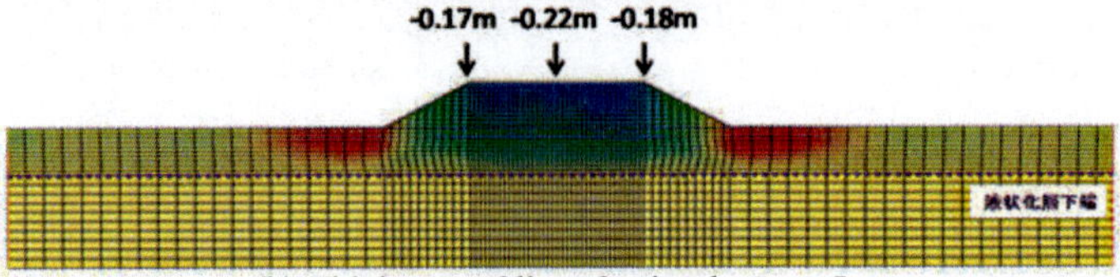

(b) Thickness of liquefaction layer = 5m

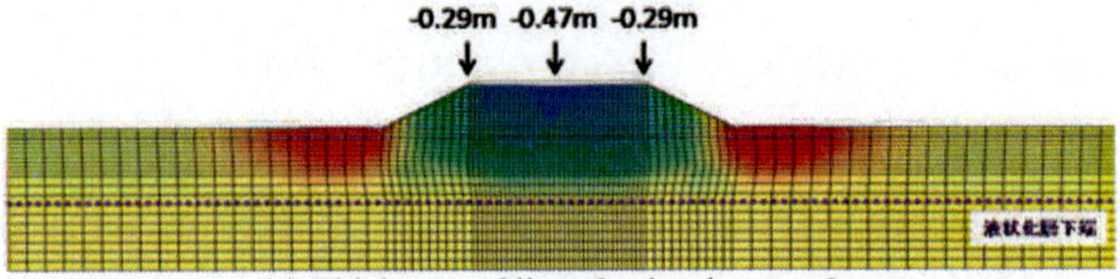

(c) Thickness of liquefaction layer = 8m

Fig. 7 Residual deformation and vertical displacement contour diagrams (ground: dense sandy soil, input ground motion: TYPEII-1, after dissipation of excess pore water pressure)

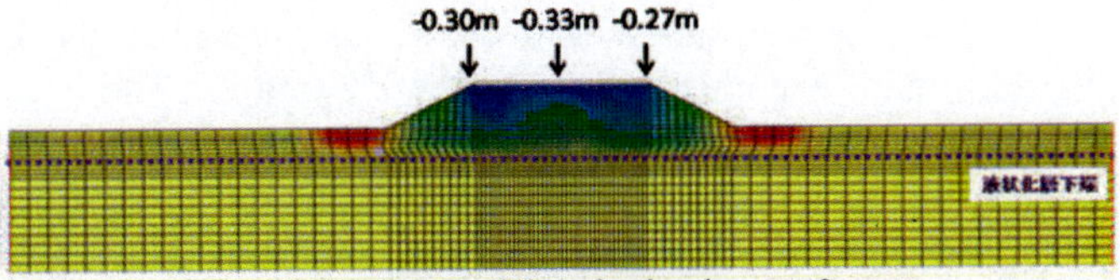

(a) Thickness of liquefaction layer = 3m

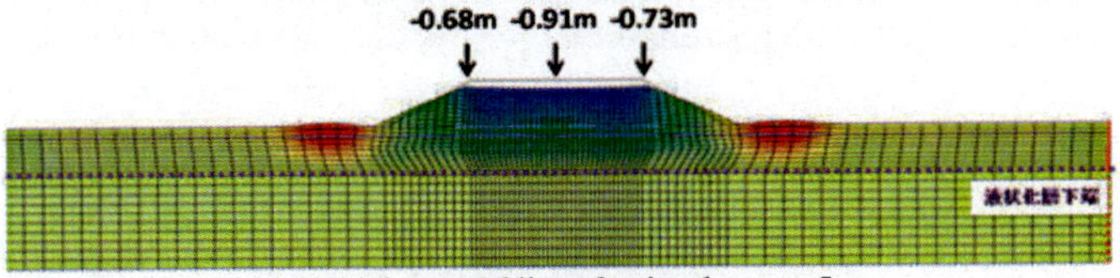

(b) Thickness of liquefaction layer = 5m

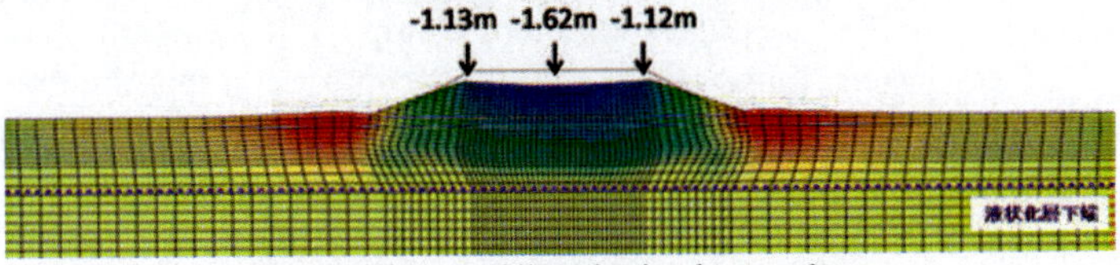

(c) Thickness of liquefaction layer = 8m

Fig. 8 Residual deformation and vertical displacement contour diagrams (ground: sandy silt, input ground motion: TYPEII-1, after dissipation of excess pore water pressure)

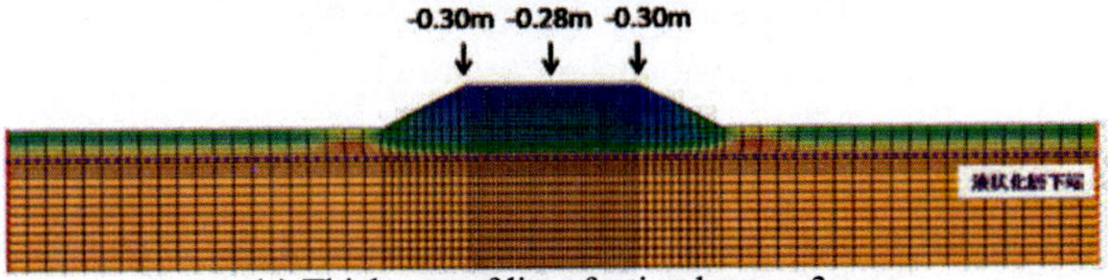

(a) Thickness of liquefaction layer = 3m

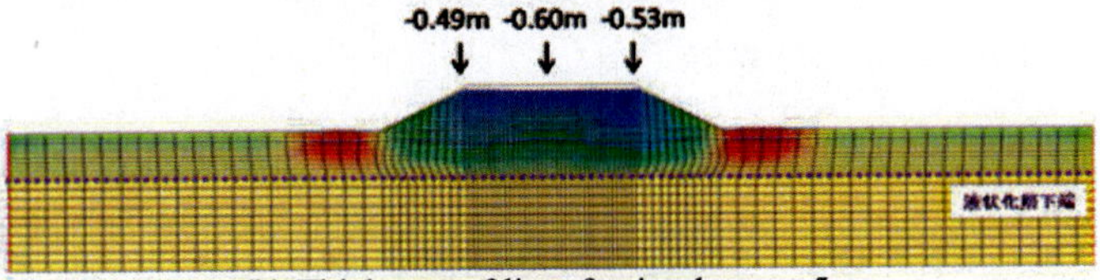

(b) Thickness of liquefaction layer = 5m

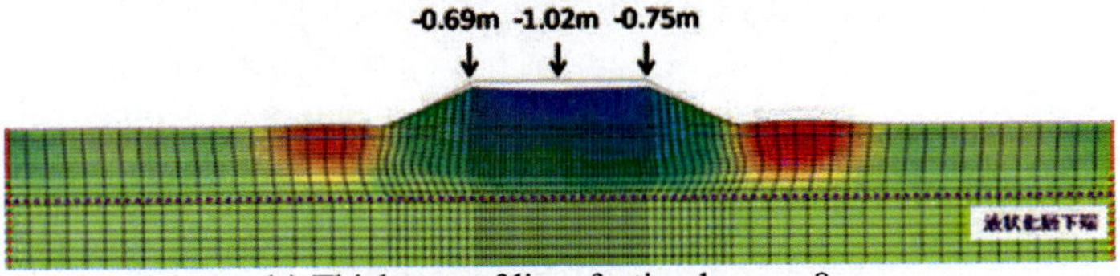

(c) Thickness of liquefaction layer = 8m

Fig. 9 Residual deformation and vertical displacement contour diagrams (ground: gravel, input ground motion: TYPEII-1, after dissipation of excess pore water pressure) layer, the soil flow towards both side ends of the embankment because liquefaction of the foundation ground becomes highly noticeable and the entire embankment sank with major settlement observed at the centre of the embankment. As the past cases of damage to the embankment due to liquefaction also show the settlement mode where the center of the embankment sank, leaving the slope shoulders less affected, the embankment deformation mode based on the present analysis results roughly conforms to that of the actual damage. Even though the deformation volume varies depending on the type of representative ground, the tendency of deformation mode is similar for all types of ground. In the case of compacted sandy ground, the embankment deformation is generally suppressed as the initial shearing rigidity and liquefaction strength are greater than those of loose sandy ground. In the case of sandy silt ground, as the liquefaction strength is greater than that of sandy ground in the range of two or more cycles, the settlement volume of the embankment crown is smaller than in the case of loose sandy ground. It can be said that the embankment crown settlement volume of both sandy ground and sandy silt ground depends on the characteristic of liquefaction strength at the target site rather than differences of the soil properties. Sand and gravel ground show similar liquefaction strength to that of sandy silt ground, and the embankment crown settlement volume is approximately half of that of loose sandy ground. The inferred reasons for this are that the liquefaction strength of sand and gravel ground is greater than that of loose sandy ground in the range of a smaller number of cycles, and that, the higher permeability coefficient of sand and gravel ground creates the condition where the excess pore water pressure caused by an earthquake is likely to dissipate quickly.

5.2 Analysis of Tendency of Embankment Crown Settlement Volume in the Face of Liquefaction Characteristics of Foundation Ground

Based on the results of a series of analyses, the relationship between the difference between the liquefied layer and non-liquefied layer and the embankment crown settlement volume is considered next. Figure 10 shows the correlation between the P_L values of banking structures damaged by liquefaction of the foundation ground in the past and the embankment crown settlement rate (= embankment crown settlement volume/embankment height). The examples of actual cases of damage referred to in this report are the results of an experiment using a centrifugal model as well as actual cases of embankments damaged by liquefaction due to earthquakes (1964 Niigata Earthquake through 1995 Southern Hyogo Earthquake) in the past as arranged by Naganawa et. al. [27] in addition to actual cases of damaged embankments due to the 2011 Off-Pacific Coast of Tohoku Earthquake (Type I seismic motion) and 1996 Southern Hyogo Earthquake (Type II seismic motion) [28–30] for which the damage was analysed by effective stress dynamic analysis. Here, the P_L value corresponding to each embankment settlement rate was calculated by the method used by Iwasaki

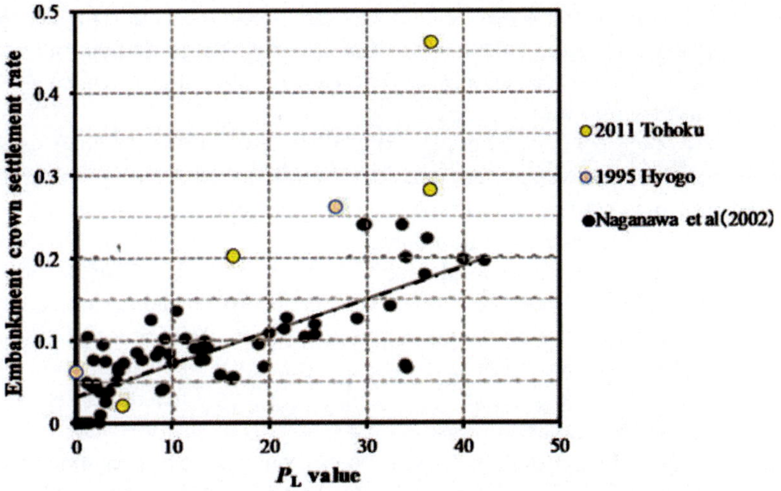

Fig. 10 Correlation diagram between the P_L value and the embankment crown settlement rate (examples of the past earthquakes)

et. al. [31] in those added cases.

$$p_L = \int_0^{20} (1 - F_L) \cdot (10 - 0.5z)\mathrm{d}z \quad (2)$$

where

P_L liquefaction potential index
F_L liquefaction resistance factor
z depth (m).

In contrast, the same value was calculated by applying the hyperbolic function for the weighting of the F_L value for calculation of the P_L value in those cases arranged by Naganawa et. al. [27], resulting in a smaller weighting factor than that of the method used by Iwasaki et. al. [31] in the range where the F_L value is larger than 0.2. It must, therefore, be noted that in the range of a small P_L value, it is possible for the P_L value in cases arranged by Naganawa et. al. [27] to tend to be slightly smaller than those of the added cases.

Figure 10 shows the positive correlation between the P_L value obtained from actual cases of damage and the model experiment and the embankment crown settlement rate. In Fig. 10, although various ground conditions are believed to be plotted, the embankment crown settlement rate is similar for all types of ground up to a P_L value of approximately 20. Although Naganawa et. al. [27] did not distinguish between Type I and Type II seismic motions, there is no significant difference in terms of the settlement rate depending on a different type of seismic motion even if the results of the present study are added. Meanwhile, when the P_L value becomes higher, there is much dispersion among the calculated embankment crown settlement rates. There is a case where the rate reaches nearly 0.5 around a P_L value of 40 plus, widely diverting from the correlation curve of Naganawa et. al. [27]

A similar arrangement to that in Fig. 10 was conducted using the analysis results of this report. The idea was to verify the correlation between the R_L value and the embankment crown settlement volume for five types of ground consisting of four types of selected representative ground and the sandy soil of Yodo River for which reproduction analysis of the actual damage had been conducted. The P_L value was calculated using the F_L value, in turn calculated based on the Specifications for Road Bridges [25]. The design lateral seismic factor required for calculation of the F_L value was set based on the ground type determined by the shear wave velocity (Vs) as well as layer thickness, and the local correction coefficient was set at the A1 area, meaning the strictest conditions.

Figure 11 shows the correlation between the P_L value and the embankment crown settlement rate (ratio of the embankment crown settlement volume to the embankment height) for liquefied ground. In this figure, the settlement volume at the end of seismic motion and the settlement volume following the dissipation of the excess pore water pressure were plotted for each type of ground. For the embankment crown settlement volume, the largest value among the analysis results for the settlement volume at the embankment crown was employed. Figure 10 shows that the higher the P_L value is, the more diverse the settlement volume at the embankment crown is due to the specific conditions of the liquefied ground. The embankment crown settlement rate varies between 0.15 and 0.45 when the P_L value is around 40. This situation resembles the tendency shown in Fig. 10. When the focus is placed on each type of ground, the P_L value and embankment crown settlement rate have a positive correlation in

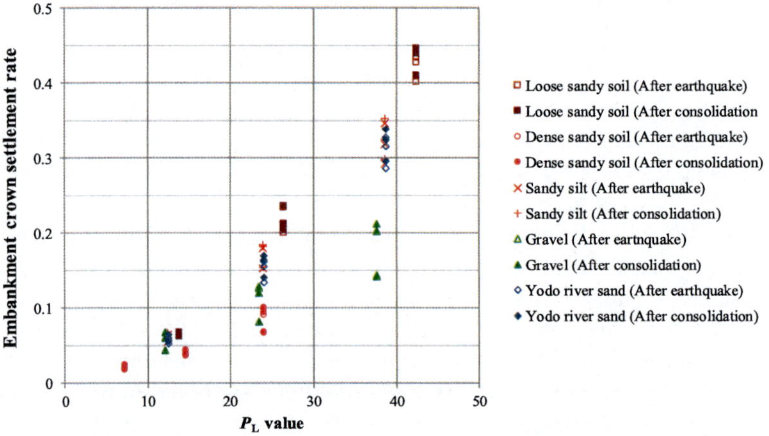

Fig. 11 Correlation diagram between the P_L value and the embankment crown settlement rate (results of analysis)

general within the scope of this study although dispersion due to a difference in seismic motion can be observed.

Such a finding qualitatively shows that dispersion around a P_L value of 40 as observed in actual cases of damage includes an influence of the foundation ground, suggesting the possibility of determining more accurate embankment crown settlement volumes when the ground conditions are known.

6 Conclusions

This study analysed the tendency between the degree of liquefaction of the foundation ground and the embankment crown settlement volume through a parametric study using two dimensions effective stress dynamic analysis of actual cases of damaged embankments as well as model grounds. From the study, the following conclusions are made:

1. As a result of arranging the grain size characteristics of coastal grounds using local coasts of Kochi Prefecture as examples, sandy ground was the most common ground type, followed by sand and gravel ground and sandy silt ground.
2. As a result of effective stress dynamic analysis reflecting the results of the cyclic undrained triaxial test, there is generally a positive correlation between the P_L value and the embankment crown settlement volume when evaluating the same type of ground individually even though dispersion of the settlement volume due to difference of the liquefied ground increases with a larger P_L value.

The results of this study suggest the possibility of interpolating the embankment crown settlement volume along the longitudinal direction of a road which cannot be evaluated solely by the results of conventional effective stress dynamic analysis of the representative cross section in the seismic performance evaluation of a road embankment by using boring data for the target road section. However, this method is believed to be effective only under the condition that there is a uniform spread of the liquefied layer as in the case of the road section targeted by this study.

References

1. Japan Geotechnical Society: Problems and Countermeasures about geo disaster in earthquakes ~ Lessons and Suggestions based on the 2011 off the pacific coast of Tohoku earthquake, pp. 4–10. Japan (2011)
2. Strategy meeting about Nankai Trough earthquake in Chubu area: Generic strategy of earthquake disaster prevention in Chubu area, pp. 34–38. Japan (2011)
3. JSCE: Reconnaissance report on the 1983 Nihonkai-chubu earthquake, pp. 185–266. Japan (1986)
4. Joint editorial committee for the report on the Great east japan earthquake disaster: Report on the Great east japan earthquake disaster, pp. 175–348. Japan (2012)

5. Japan Road Association: The standard of road embankment. Japan (2010)
6. Kochi prefecture: The 3rd document of the technique subcommittee about earthquakes and tsunami in Kochi prefecture. Japan (2013)
7. Geospatial Information Authority of Japan: Digital Topographic Map 25,000. Japan (2016)
8. Digital Map of Geospatial Information Authority of Japan, http://maps.gsi.go.jp/. Last accessed 31 Oct 2018
9. Saito, Y., Ikehara, K.: Sediment discharge of Japanese rivers and sedimentation rate and carbon content of marine sediments around the Japanese Islands, Chishitsu News, No. 452, pp. 59–64. Japan (2013)
10. Kochi Office of River and National Highway of Shikoku Regional Development Bureau. MLIT: River maintenance plan of Niyodo river. Japan (2012)
11. Sawamura, T.: Geological and topographical features and the disaster along the Tosa Bay Coast, Study Report of Kochi University, vol. 20, pp. 194–218. Japan (1971)
12. Hara, T., Sakabe, A., Kuroda, S., Kuribayashi, K., Heiguchi, M.: An analysis of influence factor against deformation characteristics by dissipation of excess pore water pressure after liquefaction. In: Proceedings of 2016 Conference on Geotechnical Engineering of JGS Shikoku Branch, pp. 157–158. Japan (2016)
13. Hara, T.: Relationships between geotechnical disaster and structures, Wood industry, vol. 66, no. 11, pp. 492–497. Japan (2011)
14. Hara, T., Sakabe, A., Numata, A., Mimura, K.: Liquefaction examination of shallow layer of Log piled ground by triaxial test. In: Proceedings of the 49th Japan National Conference on Geotechnical Engineering, pp. 1603–1604. Japan (2014)
15. Adachi, M., Yasuhara, K., Fukushima, M.: Liquefaction strength and post-liquefaction volumetric change of non-prastic slit. J. JSCE, **535**(III-34), 145–154. Japan (1996)
16. Kuribayashi, K., Hara, T., Shinagawa, D., Kuroda, S.: A study on seismic performance design of road embankment along the coast. In: Proceedings of the 19th Symposium on Performance-based seismic design method for bridges, pp. 313–318. Japan (2016)
17. Oka, F., Yashima, A., Shibata, T., Kato, M., Uzuoka, R.: FEM-FDM coupled liquefaction analysis of a porous soil using an elasto-plastic model. Appl. Sci. Res. **52**, 209–245. Japan (1994)
18. Oka, F., Kodaka, T., Kotani, Y.: Dynamic deformation analysis of river dike during earthquakes. In: Proceedings of the 37th Japan National Conference on Geotechnical Engineering, pp. 1291–1292. Japan (2002)
19. Uzuoka, R.: Analytical study on the mechanical mehavior and prediction of soil liquefaction and flow, pp. 193–206. Japan (2001)
20. Kiyota, T.: Liquefaction strength and small strain properties of in-situ frozen and reconstituted sandy soils, pp. 187–235. Japan (2007)
21. LIQCA Liquefaction Geo-Research Institute: Manual for LIQCA2D17 and LIQCA3D17. Japan (2017)
22. Kawajiri, S., Miura, M., Takahashi, M., Torii, N., Kawaguchi, T., Shibuya, K.: Deformation and strength characteristics of compacted soil over a wide strain range. J. JSCE A2(Appl. Mech.), **13**, 355–362. Japan (2010)
23. Kato, Y., Nagao, K., Kitamura, Y., Fujioka, K.: Dynamic properties of highway embankment from in-situ test ~N-value and Vs. In: Proceedings of the 39th Japan National Conference on Geotechnical Engineering, pp. 1313–1314. Japan (2004)
24. Fukuoka, T.: A problem about expansion to practical use of standard penetration test. Soil Found. **4**(2), 11–17. Japan (1956)
25. Japan Road Association: Specifications for highway bridges PART5 Seismic design. Japan (2017)
26. Toshizawa, M., Sakai, H., Uzuoka, R.: Applicability of effective stress analysis to seismic assessment for river dike during a long-time duration earthquake. J. Struct. Eng. **55A**, 415–420. Japan (2009)
27. Naganawa, T., Agiwa, H., Machida, F., Morimoto, I., Yamamoto, Y.: A study of liquefaction damage of soil structures. In: Proceedings of the 57th Japan Society of Civil Engineers Annual meeting, pp. 1033–1034. Japan (2002)

28. Tohoku Regional Development Bureau, MLIT: The 4th document of the technique subcommittee about restoration of Kitakami river dike. Japan (2011)
29. Japan Geotechnical Society: Dynamic analysis of ground ~ Basics to Application, pp. 120–123. Japan (2007)
30. Japan Institute of Country-ology and Engineering: Analytical method of seismic deformation of river dike, JICE Report, No.102001, pp. A1_1–A1_40. Japan (2002)
31. Iwasaki, T., Tatsuoka, F., Tokida, K., Yasuda,S.: Estimation of degree of soil liquefaction during earthquakes. Soil Found. **28**(4), 23–29. Japan (1980)

Characterization of Construction Materials

Behavior of Unsaturated Silty Soil Due to Change in Water Content and Suction

Ibrar Ahmed, Jiro Kuwano, Adnan Anwar Malik, and Dipak Raj Shrestha

Abstract To reduce and mitigate rainfall-induced problems, there is a need to improve understanding of the failure mechanism of unsaturated soil due to changes in water content in the soil body. Natural slopes, man-made slopes and embankments are generally made of unsaturated soils. In this regard, the objective of this study is to investigate the behavior of unsaturated silty soil due to change in water content and suction. The change in the water content of the sample represents the effect of rainfall. Moreover, this study aimed at understanding the effect of the degree of compaction on the strength and deformation characteristics of unsaturated silty soil. In the experiments, two series of laboratory element tests on double-cell triaxial machine carried out on DL clay (silty soil) in which water content varied from dry to wet of optimum moisture content, and the degree of compaction varied to study the effect at diverse overburden pressures. The test series is conducted in constant water content condition, and the measurements are closely monitored throughout the test course (initial suction to shear) to observe the changes in effective stress. The observed changes in the parameters are also presented in three dimensions to show the behavior of soil under the influence of more than two parameters. From this research, it is examined that with the increase in degree of compaction, volume change behavior transformed from compressive to dilative. Moreover, the strength of the silty soil increased with the increase in suction, i.e., due to decrease in water content, and vice versa.

Keywords Unsaturated soil · Suction · Compaction · Triaxial test

I. Ahmed (✉) · J. Kuwano · A. A. Malik · D. R. Shrestha
Saitama University, Saitama-shi 338-0826, Japan
e-mail: ahmed.i.345@ms.saitama-u.ac.jp

J. Kuwano
e-mail: jkuwano@mail.saitama-u.ac.jp

A. A. Malik
e-mail: adnanmalik@mail.saitama-u.ac.jp

D. R. Shrestha
e-mail: dipakrajshrestha@gmail.com

H. Hazarika et al. (eds.), *Advances in Sustainable Construction and Resource Management*, Lecture Notes in Civil Engineering 144,
https://doi.org/10.1007/978-981-16-0077-7_35

1 Introduction

A firm and stable subgrade is essential for the pavement structure's long-term bearing performance under vehicle loading. Some of the common problems of subgrade that affect the pavement stability are subgrade subsidence, subgrade slope collapse and ruts. These problems occurred due to insufficient bearing capacity of subgrade. The shear strength properties of the subgrade (compacted clayey soil) are affected by soil type, water content, degree of compaction and state of stress. The water content and degree of compaction are the main factors that affect the shear strength of a specific subgrade compacted soil [1]. In the case of compacted clayey soil, the suction and internal friction angle decreased with the increase in compaction and water content, and the cohesion peak appeared around the optimum moisture content [2]. In unsaturated soils, the cohesion and internal friction angle linearly decreased with the increase in water content. The decrease in cohesion due to water content is more evident than the internal friction angle [3]. The generation of pore water pressure during loading may cause geotechnical problems like slope failure [4]. The behavior of soil with a known initial state, a boundary condition and type of loading (due to any mechanical process) can be studied experimentally via the triaxial test [5, 6].

Melinda [7] investigated the shear strength and deformation characteristics of the residual soil during water infiltration that leads to failure. Rasool et al. [8] studied the behavior of unsaturated silty soil under constant shear stress and matric suction plane. The results indicated that with the increase of confinement (confining pressure), the water infiltration decreased due to the reduction in void ratio. The results also showed that a failure surface is a unique plane which is not affected much by drainage conditions.

Natural slopes, man-made slopes and embankments are generally made of unsaturated soils. In case of unsaturated soils, the important state variables are mean effective stress (p'), shear stress (q), void ratio (e) and one more variable among the following: suction (s), water content (w) and degree of saturation (S_r). In the above-mentioned studies, the research focused on the shear phase only to investigate the effect of various stress paths, water content variations during the constant net stress/shear stress/matric suction planes and variation of water content in the sample on the shear strength of unsaturated soils. The present study investigates the effect of water content and suction on the behavior of unsaturated silty soil (DL clay) in all the phases, i.e., initial suction, axis translation, consolidation and shear. The effect of degree of compaction on the behavior of unsaturated silty soil is also investigated. Finally, different stress variables are plotted in three dimensions (3D) to explain the behavior of unsaturated soil in all the phases of the experiment under the influence of more than two parameters.

2 Experimental Setup

Double-cell triaxial test apparatus having a strain control loading system was used in this research, as shown in Fig. 1. The advantage of using double-cell triaxial test apparatus is in its precise measurement of volume change [9] which is complex in unsaturated soils due to the presence of pore water pressure and pore air pressure. Pore water pressure was controlled and monitored from the bottom of the pedestal, whereas pore air pressure was controlled and monitored at the top cap. A membrane filter of high air entry value 420 kPa is used at the pedestal which does not allow the flow of air, while polytetrafluoroethylene (PTFE) sheet is used in the top cap, which does not allow the flow of pore water. A low-capacity differential pressure transducer (LCDPT) is used to measure the volume change behavior. Linear variable differential transducer (LVDT) was used to measure the axial deformation. The data/signals were acquired by all the transducers and then shifted to the amplifier, which minimizes the noise and increases the signal voltage. These analog signals were converted into digital signals by A/D board, and finally, a software (Digit Basic) was used, which presents the data in the form of physical values. The schematic diagram of the equipment is shown in Fig. 1.

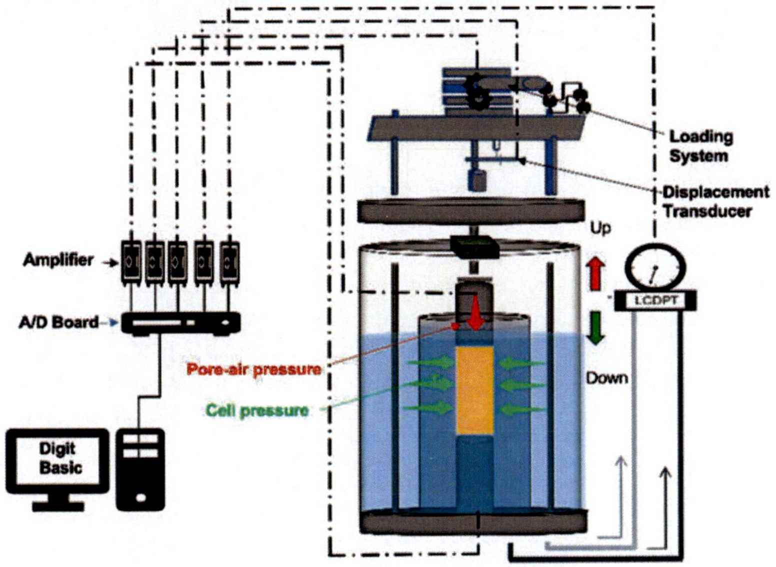

Fig. 1 Schematic illustration of double-cell triaxial test equipment

3 Material and Methodology

To study the behavior of unsaturated silty soil, DL clay (commercial name of soil) is used in this study, because its suction value stabilizes in a shorter interval of time so that each test could be performed in a stipulated time frame. According to Japanese Geotechnical Society (JGS), it is classified as having medium–low compressibility (ML) and composed of 90% silt and 10% clay which shows that its grain size is larger than average clay. This silty soil has a relatively uniform grain size distribution with a mean grain size D50 approximately 0.03 mm. The optimum moisture content of the soil was 20%, the maximum particle size was 0.039 mm, and maximum dry density was 1.55 g/cm^3. The appearance of freshly and freely deposited DL clay looks yellowish brown. The physical properties of DL clay are summarized in Table 1.

3.1 Sample Preparation and Test Phases

In this research, the samples (5 cm × 10 cm) were prepared with different initial water content (having the same dry density). Each sample was compacted in five layers by static compaction to achieve the uniform density (in element test; a slight change in the sample properties affects the test results). Two series of tests were performed in this study. In the first series of tests (Test Series # 1), three samples of DL clay were prepared. The water content was varied from dry to wet side of optimum moisture content. Moreover, two more samples were prepared for the second series of tests (Test Series # 2) with a higher degree of compaction to check the behavior of unsaturated silty soil at higher overburden pressure. The sample properties are shown in Table 2.

In this study, the tests were carried out in four phases, i.e., PI, PII, PIII, PIV, under constant water condition which replicates the situation in which excess pore air pressure dissipates rapidly, while pore water pressure dissipates with time [10] at confining stress of 100 kPa (see Fig. 2). High confinement was considered because, at this confinement, the top cap fully rests on the sample without any gap so that

Table 1 Physical properties of DL clay

Properties	Unit	Value
Density of soil particle, s	g/cm^3	2.654
Consistency	–	NP
Maximum dry density, d_{max}	g/cm^3	1.55
Optimum moisture content	%	20
Maximum particle size, d_{max}	mm	0.039
Coefficient of permeability, k_s	m/s	10^{-7}

Table 2 Sample properties

Test	Dry density (g/cm^3)	Degree of compaction (%)	Water content, *w* (%)	Void ratio, *e*	Remarks
1	1.26	81	15	1.07	Test Series # 1
2	1.26	81	20	1.07	
3	1.26	81	25	1.07	
4	1.37	88	10	1.01	Test Series # 2
5	1.37	88	15	1.01	

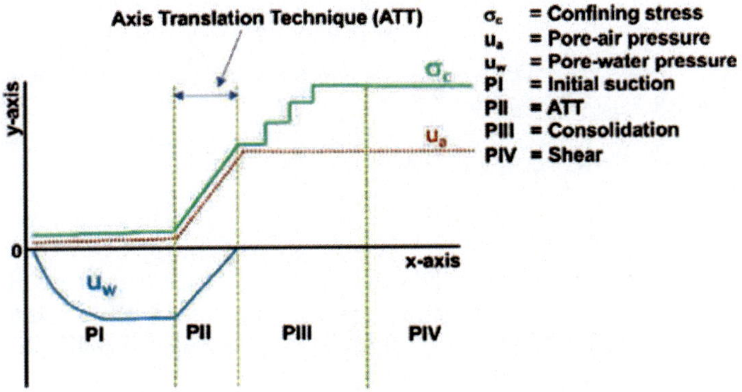

Fig. 2 Schematic diagram of test phases

the clear behavior of sample can be studied. In phase I (PI), the initial suction was measured by placing the sample on the pedestal. It was observed that suction value increased and then stabilized at a specific value and time for different water contents (see Fig. 3). It was also observed that the sample at higher water content has taken less time to reach the stabilized value of suction. In phase II (PII), as shown in Fig. 2, the axis translation technique (ATT) was applied to keep the pore water pressure positive so that no voids were developed inside the pedestal, which affects the reading of pore water pressure. The important thing that should be monitored during axis translation is that the pore air and cell pressures should be increased simultaneously, and it must be equivalent to the decreasing amount of pore water pressure. Otherwise, the effective stresses will increase and may affect the stresses inside the sample. In phase III (PIII), isotropic consolidation was carried out until the observed volume change from low-capacity differential pressure transducer (LCDPT) becomes constant. The consolidation process will occur after the application of the confining pressure if the pore fluids are allowed to drain. On the other hand, the consolidation process will not occur if the pore fluids are not allowed to drain (maintained in undrained condition) [10]. In phase IV (PIV), the sample was sheared up to 15% of axial strain, at the strain rate of 0.05%/min as per JGS standard [11]. All the phases are presented in Fig. 2.

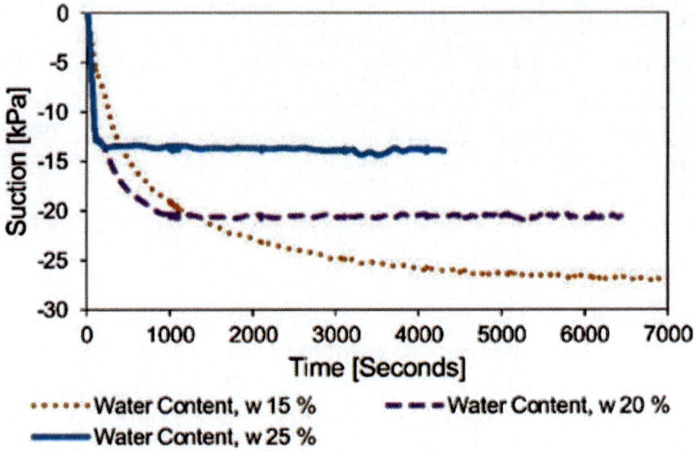

Fig. 3 Measurement of initial suction with time

Figure 4 showed the behavior of deviatoric stress (q), mean effective stress (p') and suction (s). This figure shows the result of Test 1 of the first series (Test Series # 1) having a dry density of 1.26 g/cm^3, the water content of 15%, and the suction against 15% water content is 25 kPa, while maximum deviatoric stress achieved is 272 kPa and mean effective stress is 202 kPa.

Mean effective stress is given by,

$$p' = \frac{\sigma_a' + 2\sigma_r'}{3} = (p - u_a) + s \cdot S_r \tag{1}$$

Effective axial stress is given by,

$$\sigma_a' = \sigma_{\text{anet}} + s \cdot S_r = (\sigma_a - u_a) + s \cdot S_r \tag{2}$$

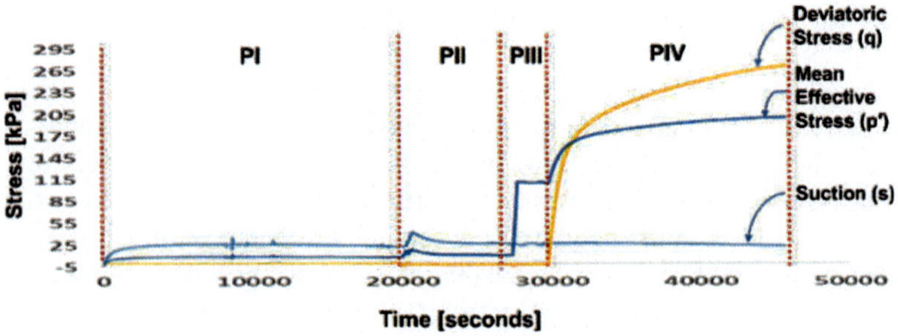

Fig. 4 Different test stages/phases

Effective radial stress is given by,

$$\sigma_r' = \sigma_{r\text{net}} + s \cdot S_r = (\sigma_r - u_a) + s \cdot S_r \tag{3}$$

$$q = \sigma_a - \sigma_r \tag{4}$$

where p' is mean effective stress, $\sigma_a{}'$ is effective axial stress, $\sigma_r{}'$ is effective radial stress, p is mean stress, u_a is pore air pressure, s is suction, S_r is degree of saturation, $\sigma_{a\text{net}}$ is net axial stress, σ_a is axial stress, $\sigma_{r\text{net}}$ is net radial stress, σ_r is radial stress, and q is deviatoric stress.

In PI and PII phases (see Fig. 4), during the initial suction measurement and axis translation, no axial and radial stresses were applied, while mean effective stresses are due to degree of saturation and suction. However, during isotropic consolidation, an equal amount of desired confining stress and axial stress is applied, which resulted in zero deviatoric stress (q), while mean effective stress is increased. During the shear, confinement is kept constant while axial stresses are increased as a result of which deviatoric stress and mean effective stress are increased as shown in Fig. 4 (PIV). For clear understanding/behavior of important parameters like deviatoric stress (q), mean effective stress (p') and suction (s) are also plotted in three dimensions as shown in Fig. 5.

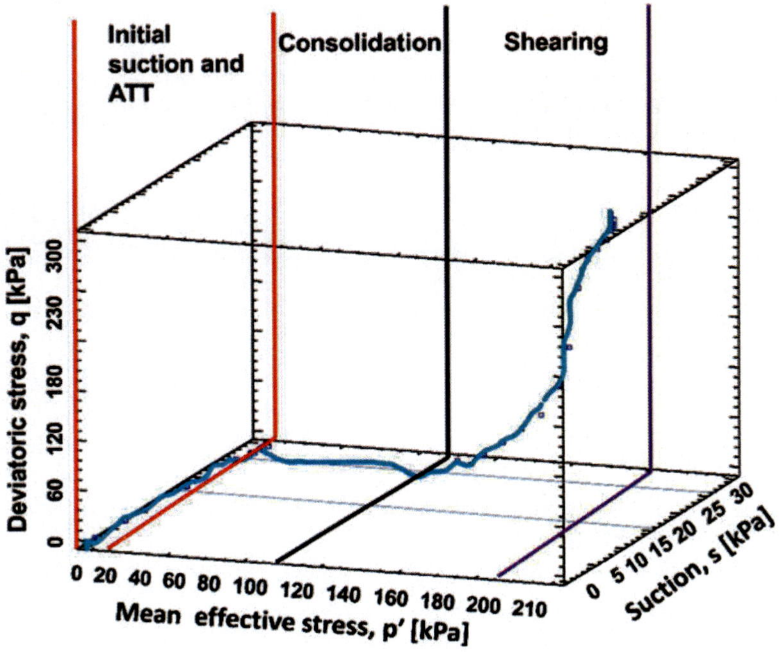

Fig. 5 Behavior of parameters during test phases (Test Series # 1)

4 Results

During the Test Series # 1 (degree of compaction, 81%) and Test Series # 2 (degree of compaction, 88%) on unsaturated silty soil (DL clay), the strength and deformation characteristics are emphasized. The shearing phase of the test is used to describe the mechanical properties of the material. In Test Series # 1, during the shearing, it is observed that deviatoric stress is more toward the dry side of optimum moisture content (OMC, 20%), while it is less on the wet side of OMC (see Fig. 6). This is because on dry side of optimum moisture content, the matric suction is more while matric suction is less on the wet side as shown in Fig. 7. Similar trend is observed in Test Series # 2, i.e., higher deviatoric stress is observed at lower water content.

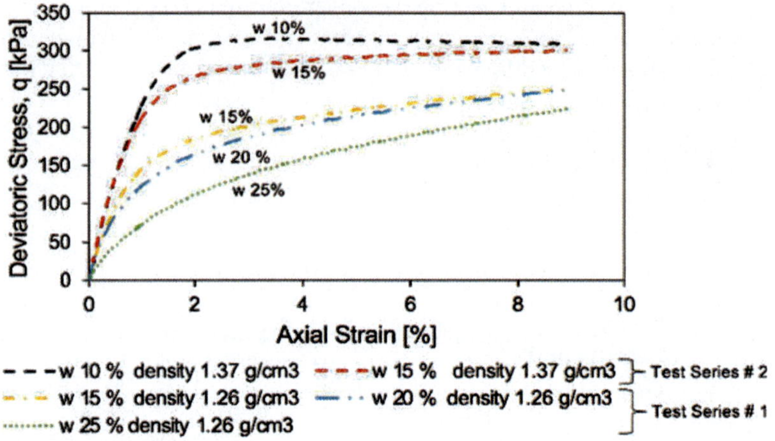

Fig. 6 Axial strain versus deviatoric stress

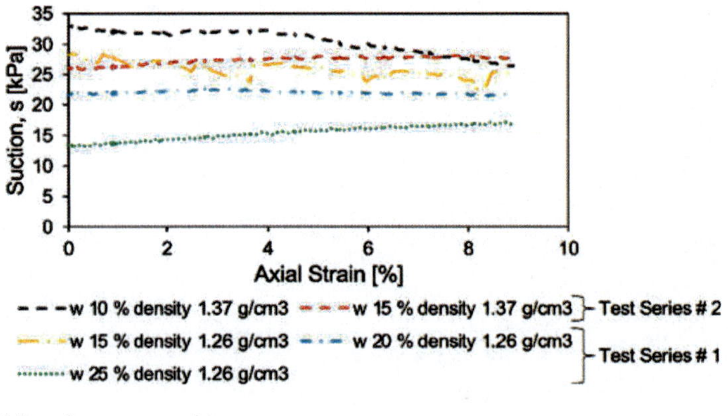

Fig. 7 Axial strain versus suction

Moreover, the deviatoric stress is more in Test Series # 2 than in Test Series # 1 due to higher degree of compaction.

The test results indicated that the sample at the degree of compaction of 81% (Test Series # 1) showed compressive behavior during shearing phase, while at the degree of compaction of 88% (Test Series # 2), it showed dilative behavior during shearing phase as shown in Fig. 8. The experimental result (Test Series # 1) further showed that with the increase of water content, the sample showed more compression and void ratio is decreased accordingly, while the degree of saturation is increased due to reduction in the void ratio as shown in Figs. 8 and 9, whereas on dilation side (Test Series # 2), with the increase in water content, the sample showed less dilation and degree of saturation is decreased slightly due to dilative behavior as shown in Figs. 8 and 9.

The change in void ratio (e), suction (s), and deviatoric stress (q) in all the phases, i.e., PI to PIV, under Test Series # 1 and Test Series # 2 with the initial water

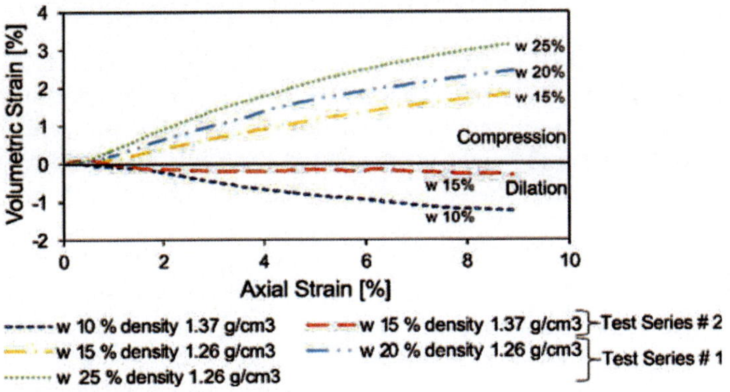

Fig. 8 Axial strain versus volumetric strain

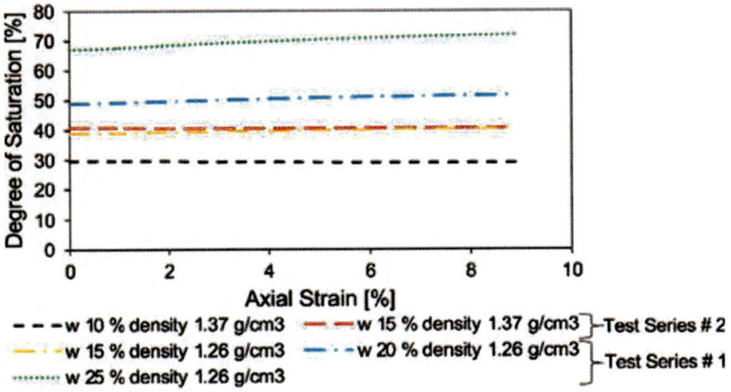

Fig. 9 Axial strain versus degree of saturation

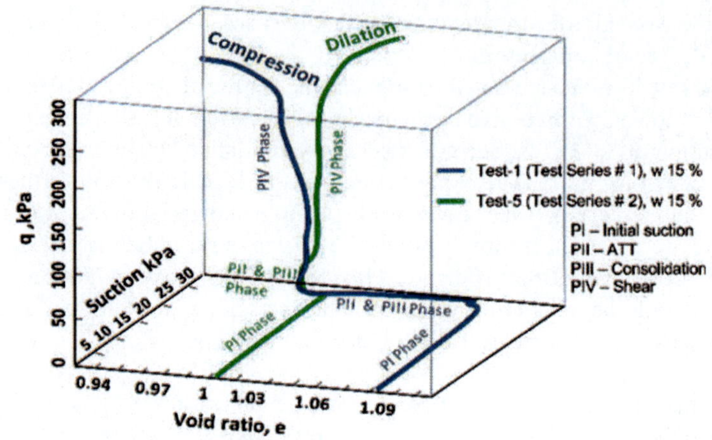

Fig. 10 Relationship between q, e and suction

content of 15% is plotted in three dimensions (3D) as shown in Fig. 10. The 3D representation indicates the compressive and dilative behavior of the sample (during shearing) prepared at the degree of compaction of 81% (Test Series #1) and 88% (Test Series # 2). The void ratio (e) at the end of consolidation phase (PIII) is quite close in both the test series, but it is completely opposite at the end of shearing phase (PIV) as shown in Fig. 10. The change in volumetric strain and void ratio through all the testing phases for Test Series # 1 is presented in Figs. 11 and 12. During the shearing phase (PIV), the volumetric strain increases, and void ratio decreases from the initial value. The change in volumetric strain and void ratio through all the testing phases for Test Series # 2 is presented in Figs. 13 and 14. During the shearing phase (PIV), the volumetric strain decreases, and void ratio increases from the initial value. It is also observed that after the consolidation phase (PIII), the void ratio in

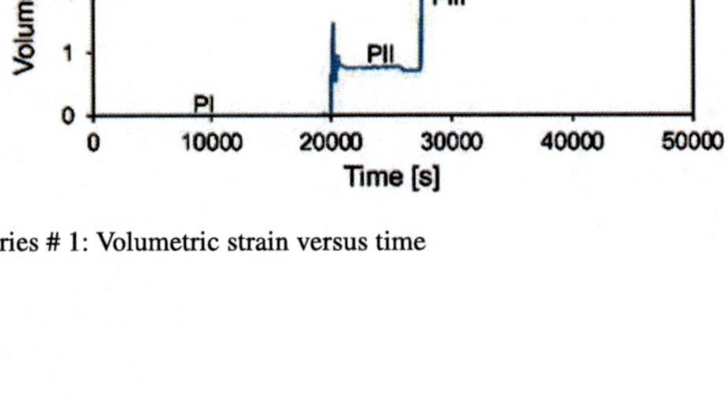

Fig. 11 Test Series # 1: Volumetric strain versus time

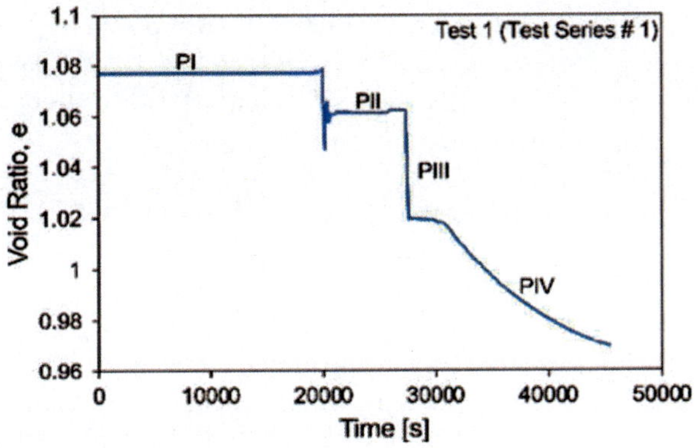

Fig. 12 Test Series #1: Void ratio versus time

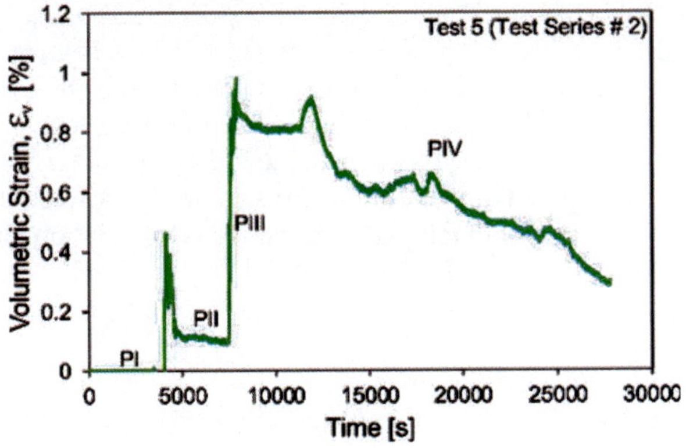

Fig. 13 Test Series # 2: Volumetric strain versus time

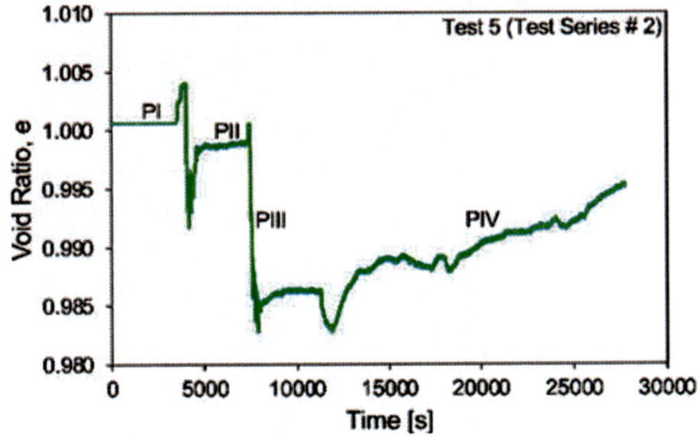

Fig. 14 Test Series # 2: Void ratio versus time

test series # 1 (1.019) is close to test series # 2 (0.985) but during shearing phase, the sample showed compressive behavior in test series # 1 and dilative behavior in test series # 2 as shown in Figs. 12 and 14. This indicates the influence of the initial state (particles arrangement) of the soil.

5 Conclusion

From this research, it is observed that during the axis translation phase if the increase in the pore air pressure and cell pressure is not synchronized, then the suction and effective stresses will increase, and it will affect the later phases (consolidation, shear) of the test.

It is concluded that with the increase of degree of compaction, strength increases, and volume change behavior clearly changed from compressive to dilative. On the compression side, with the increase of water content, the sample showed more compression, and the degree of saturation is increased with the decrease in void ratio. On dilation side, with increase in water content, the sample showed dilation, and the degree of saturation is slightly decreased with the increase in void ratio. The samples which are prepared at same initial water content (15%) for both the test series showed quite a close void ratio after the end of consolidation phase (PIII), but during shearing phase, the sample showed compressive behavior in test series # 1 and dilative behavior in test series # 2. This indicates the influence of the initial state (particles arrangement) of the soil.

The results also indicated that with the increase in suction of the sample, the strength increases and vice versa. It would be the primary reason that the increase in water content in the slopes of unsaturated soil results in the non-stability of the slopes.

References

1. Huzhu, Z., Hanbing, L., Jing, W., Weizhi, D.: Investigation of the effect of water and degree of compaction on the shear strength of clay soil material. Funct. Mater. **24**(2), 290–297 (2017)
2. Cokca, E., Erol, O., Armangil, F.: Effects of compaction moisture content on the shear strength of an unsaturated clay. Geotech. Geol. Eng. **22**, 285–297 (2004)
3. Shen, C.N., Fang, X., Wang, H.W., Sun, S.G., Guo, J.F.: Effect of suction, moisture content and dry density on shear strength of remolded unsaturated soils. Rock Soil Mech. **30**(5), 1347–1351 (2009)
4. Thu, M.T., Rahardjo, H., Leong, E.C.: Critical state behavior of a compacted silt specimen. Soils Found. **47**(4), 749–755 (2007)
5. Terzaghi, K., Peck, R.B.: Soil Mechanics in Engineering Practice, 2nd edn. Wiley, New York (1967)
6. Bishop, A.W., Henkel, D.J.: The Measurement of Soil Properties in the Triaxial Test, 2nd edn. Edward Arnold Publishers Ltd., Michigan (1964)
7. Melinda, F., Rahardjo, H., Han, K.K., Leong, E.C.: Shear strength of compacted soil under infiltration condition. J. Geotech. Geo-Environ. Eng. **130**(8), 807–817 (2004)

8. Rasool, AM., Kuwano, J., Tachibana, S.: Behavior of compacted unsaturated soil in isotropic compression, cyclic and monotonic shear loading sequences in undrained condition. In: 6th International Symposium on Deformation Characteristics of Geomaterials, pp. 267–274. IOS Press, Buenos Aires (2015)
9. Mendes, J., Toll, D.G., Evans, F.: A double cell triaxial system for unsaturated soils testing. In: Mancuso, C., Jommi, C., D'Onza, F. (eds.) Unsaturated Soils: Research and Applications, pp. 5–10. Springer, Heidelberg (2012)
10. Fredlund, D.G., Rahardjo, H.: Soil Mechanics for Unsaturated Soils. Wiley, New York (1993)
11. Japan Geotechnical Society Standards (vol. 2), 5.5 c& e, 527-2009

Cyclic Strength and Deformation Characteristics of Sand Containing Fines with Plasticity

Hideo Nagase, Akihiko Hirooka, Keigo Fukumoto, Keiichiro Miyaji, and Naoya Kawano

Abstract The strength and deformation characteristics of sandy soil with fines have not been sufficiently clarified, when the fine content ratio is high, but the plastic index is low, or the plastic index is high, but the void ratio is also high. The "ALID" program has been used to simulate the deformation of river levees during earthquakes. In some cases, the results of deformation obtained by analysis and determined in situ in the ground during earthquakes do not coincide entirely, because the properties of such soil with fines are unclear. In the present study, several series of cyclic loading tests and monotonic loading tests after cyclic loading were performed, in order to investigate the strength and deformation characteristics of sandy soil with several plasticity indexes, fine content ratios and void ratios. Consequently, the effects of the properties on the strength and deformation characteristics are clarified using the liquefaction strength and the shear modulus after liquefaction.

Keywords Liquefaction strength · Fine content ratio · Plasticity · Shear modulus

1 Introduction

In Urayasu City during the 2011 Great East Japan Earthquake and in Kumamoto City during the 2016 Kumamoto Earthquake, severe liquefaction occurred and wooden houses sustained wide and extensive damage in sandy soil deposits with fines. The occurrence of liquefaction and damage to structures built on such soil deposits and their liquefaction properties have often been studied. However, when the fine content

H. Nagase (✉) · A. Hirooka
Kyushu Institute of Technology, 1-1 Sensui-cho, Tobata-ku, Kitakyushu, Fukuoka, Japan
e-mail: nagase@civil.kyutech.ac.jp

K. Fukumoto
Mitsui Consultants Co., Ltd, Kyushu branch office, 2-14-1 Hakataeki-higashi, Hakata-ku, Fukuoka, Fukuoka, Japan

K. Miyaji · N. Kawano
Kyushu Institute of Technology, 1-1 Sensui-cho, Tobata-ku, Kitakyushu, Fukuoka, Japan

H. Hazarika et al. (eds.), *Advances in Sustainable Construction and Resource Management*, Lecture Notes in Civil Engineering 144,
https://doi.org/10.1007/978-981-16-0077-7_36

ratio is high, but the plastic index is low, or the plastic index is high, but the void ratio is high, the strength and deformation characteristics of sandy soil with fines have not been sufficiently clarified. The "ALID" program has been used to simulate the deformation of river levees during earthquakes. In some cases, the results of deformation obtained by analysis and determined in situ in ground during earthquakes do not coincide entirely, because the properties of such soil with fines are not clear. In the present study, several series of cyclic loading tests and monotonic loading tests after cyclic loading were performed, in order to investigate the strength and deformation characteristics of sandy soil with several plasticity indexes, fine content ratios and void ratios.

2 Sample and Test Procedures

Grain size distribution curves of Toyoura sand and the five kinds of mixed samples used in the tests are shown in Fig. 1. The test conditions and physical properties of the mixed samples are indicated in Tables 1 and 2, respectively. Toyoura sand and the mixed samples, which are constituted from Toyoura sand, DL clay and bentonite were used as the samples in the tests. DL clay is a non-plastic silt, and bentonite is a high plastic clay. DL means "drift less," and DL clay does not contain many particles smaller than 10 m.

In the cyclic loading test, the hollow cylindrical torsional shear test apparatus was used. The size of the specimen was 10 cm in outer diameter, 6 cm in inner diameter and 10 cm in height. In all the tests, the specimens were saturated and consolidated under isotropic condition with an initial effective confining stress of 98 kPa. A cyclic shear stress ratio of double amplitude shear strain DA = 7.5% at the 20th cycle,

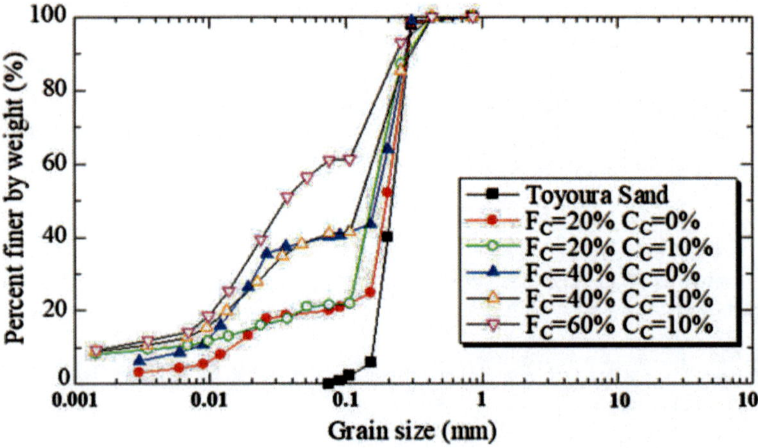

Fig. 1 Grain size distribution curves

Table 1 Test conditions

	FC (%)	CC (%)	Dr (%)
Case A	0	0	40
Case B	20	0	55
Case C		10	55
Case 1	40	0	60
Case 2		10	60
Case 3		10	37
Case 4	60	10	73

Table 2 Physical properties of mixed samples

	ρ_s (g/cm^3)	ρ_{dmax} (g/cm^3)	ρ_{dmin} (g/cm^3)	Liquid limit W_L (%)	Plastic limit W_P (%)	Plasticity index (I_p)
FC = 40% CC = 0%	2.632	1.966	1.374	NP	NP	–
FC = 40% CC = 10%	2.584	1.792	1.303	29.4	17.8	11.6
FC = 40% CC = 10%	2.660	1.776	1.226	32.1	20.4	11.8

which is called the liquefaction strength ratio R_{L20}, was obtained through the cyclic loading tests. After cyclic loading test using 20 cycles, the undrained condition was retained, and static undrained tests were also conducted at a strain speed of 5%/min.

3 Liquefaction Characteristics

3.1 Liquefaction Strength Characteristics

Figures 2 and 3 indicate the typical time histories of raw data on the shear stress ratio τ/σ'_0, the shear strain and the excess pore water pressure ratio $\Delta u/\sigma'_0$ obtained in Cases 3 and 4. The value of $\Delta u/\sigma'_0$ gradually increased during cyclic loading, although the value did not attain 1.0 at the 20th cycle. On the contrary, the shear strain also gradually increased during cyclic loading and attained 25–35% at the 20th cycle. It can be considered that this behavior took place because these specimens

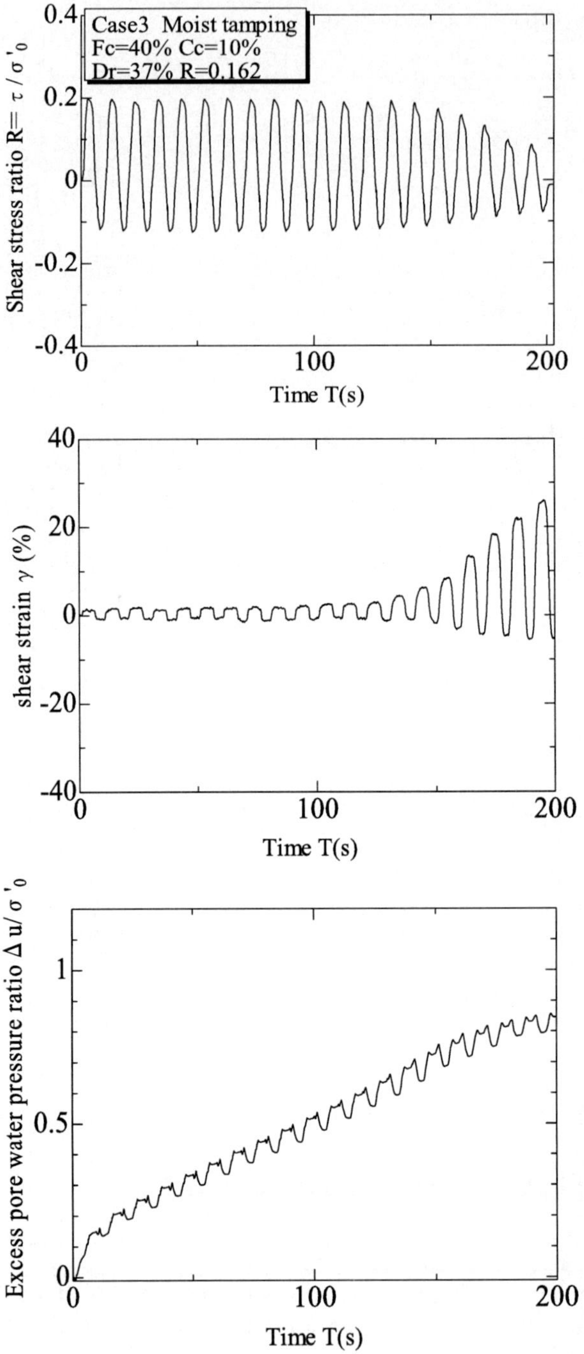

Fig. 2 Time histories (Case3)

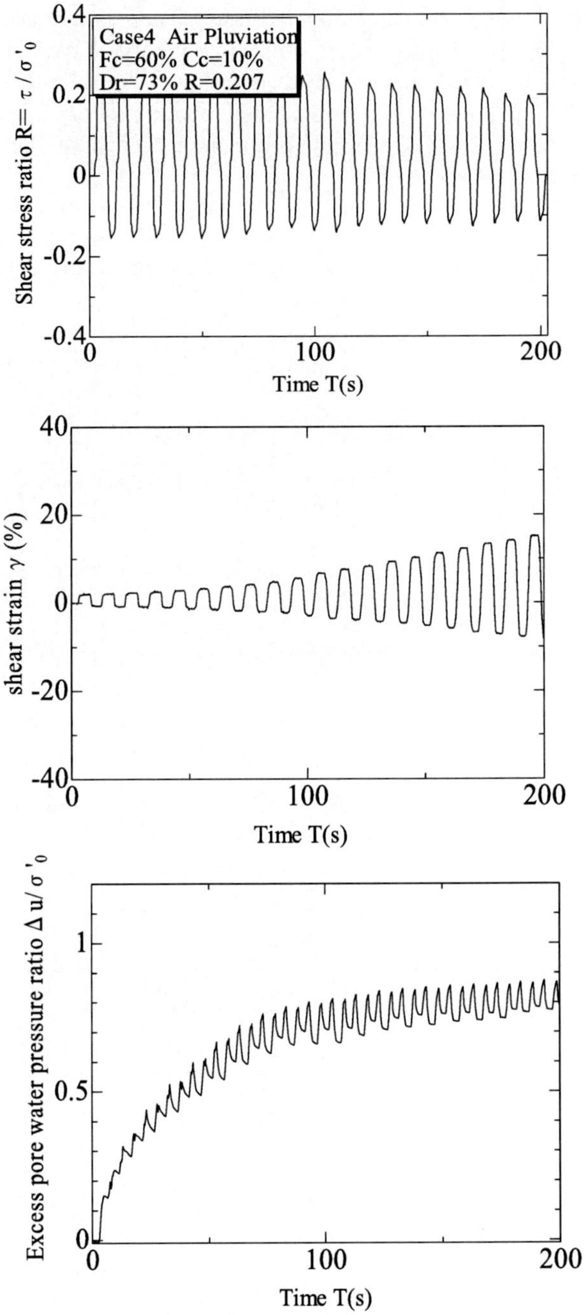

Fig. 3 Time histories (Case4)

were cohesive soils with a plasticity index of 11 to 12, and the viscose resistance was mobilized during cyclic loading. Therefore, this type of behavior should be called the "cyclic softening" phenomenon which is different from liquefaction. And the increment trend of the shear strain was observed more rapidly in Case 4 than in Case 3. Notably, the specimen in Case 3 indicated a phenomenon similar to liquefaction more precisely than that in Case 4.

Figures 4 and 5 show the liquefaction strength curves obtained in Case A and Cases 1, 2 and 3, and Case C and Cases 2 and 4, respectively. In the comparison between Cases 3 and 2 with a fine content ratio FC = 40% and a plasticity index I_p with a value nearly equal to 10, the liquefaction strength ratio R_{L20} decreased as the relative density D_r decreased. In the comparison between Cases 3 and 1, the value of R_{L20} in Case 3 was slightly larger than that in Case 1, because the cohesive resistance

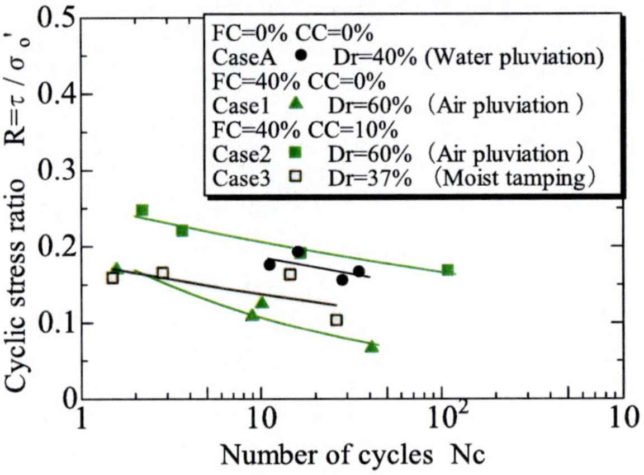

Fig. 4 Relationships between cyclic stress ratio and number of cycles

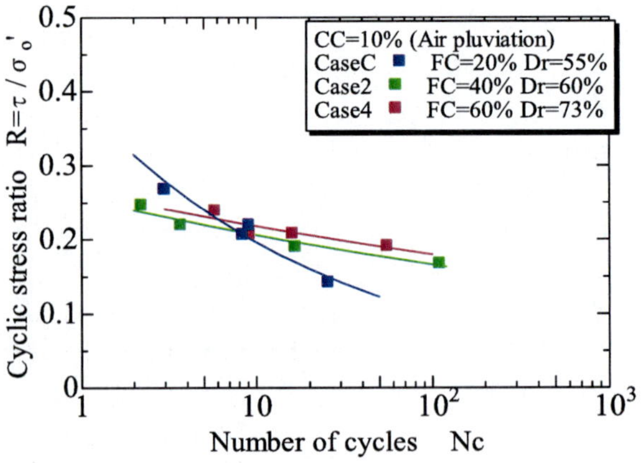

Fig. 5 Relationships between cyclic stress ratio and number of cycles

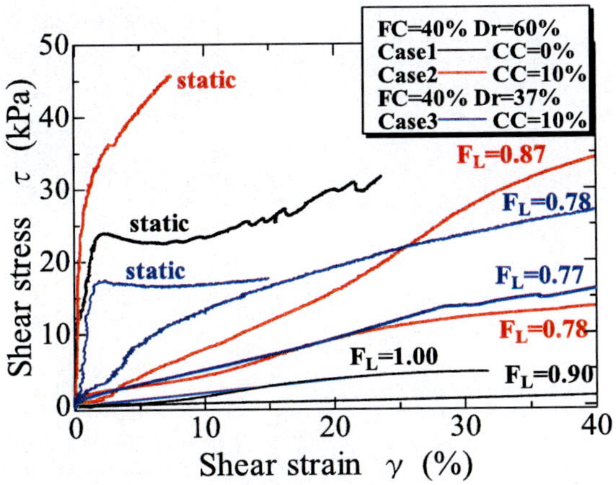

Fig. 6 Relationships between shear stress and shear strain

was mobilized between soil particles. On the contrary, when the value of FC was changed under the condition of the clay content ratio CC = 10%, the inclination of the liquefaction strength curve in Case C with FC = 20% was steeper than those in Case 2 with FC = 40% and Case 4 with FC = 60%. In the latter cases, change in the liquefaction strength curves was negligible. It should be noted that a large increase in the liquefaction strength could not be obtained in cases with $I_p = 10$ and FC 40%, in which the fines dominate the skeleton of soil particles.

3.2 Deformation Characteristics After Liquefaction

Figures 6 and 7 show a comparison between the shear stress versus shear strain relationships obtained by the static shear tests performed after the cyclic loading tests in Cases 1, 2 and 3, and Cases 1, 2 and 4, respectively. In Cases 3 and 2 with FC = 40% and $I_p = 10$, similar behaviors were observed in the relationships. In the comparison between the relationships in Cases 2 and 4, those curves also have a similar trend. The deformation characteristics of liquefied sandy soils hardly depend on the density of the specimen and the value of FC, if the value of $I_p = 10$.

The relationships between the reduction ratio of shear modulus $G_1/G_{0,i}$ and the factor of safety against liquefaction F_L are plotted in Fig. 8, according to reading the data in Figs. 6 and 7. Figure 9 shows a schematic illustration of shear stress versus shear strain after liquefaction. G_1 and G_2 mean the secant shear modulus after liquefaction in the region of a very small shear resistance and the secant shear modulus after a very large shear strain in the region of a recovered shear resistance, respectively. $G_{0,i}$ means the secant shear modulus at a shear strain of 0.1%, obtained from the shear test without cyclic loading. The results in Fig. 8 show that the plotted data is classified into two groups of Cases A, B, 1 with CC = 0% and Cases C, 2, 3,

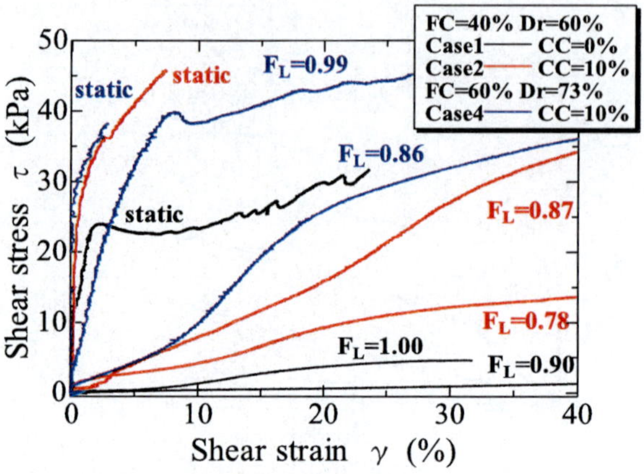

Fig. 7 Relationships between shear stress and shear strain

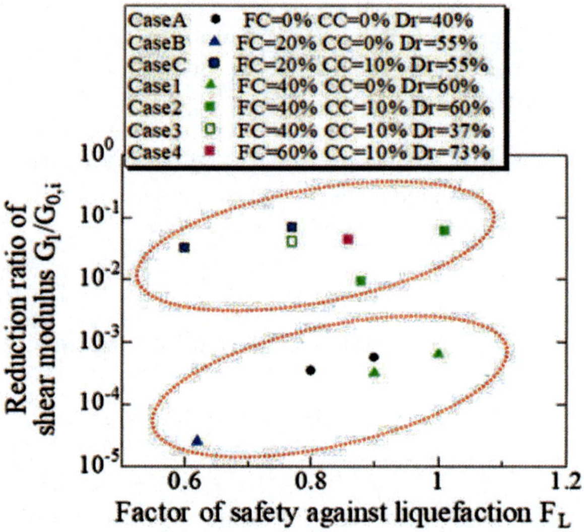

Fig. 8 Relationships between reduction ratio of shear modulus $G_1/G_{0,\mathrm{i}}$ and factor of safety against liquefaction F_L

4 with CC = 10%, and the value of $G_1/G_{0,i}$ decreases as the value of F_L decreases in each group. Furthermore, the reduction ratio of the shear modulus due to cyclic loading is two digits smaller in the former group than in the latter group. It can be considered that the reduction of the shear modulus was not remarkable in the latter group, because the cohesive resistance in the specimen with CC = 10% was mobilized in the tests after cyclic loading.

Figure 10 shows the chart of the shear modulus ratio G_1/σ'_0 versus the factor of safety against liquefaction F_L relationships used in the analysis of "ALID" for

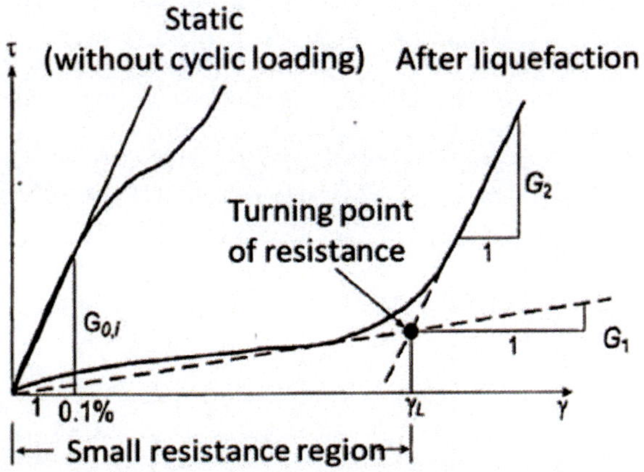

Fig. 9 Schematic illustration of shear stress versus shear strain after liquefaction

liquefaction-induced deformation. The data obtained by this study were plotted in Fig. 10, in order to compare the values of the preferred relationships called the Yasuda and Inagaki's formula with the values obtained from Figs. 6 and 7. Table 3 indicates the test condition of FC, CC and Dr, the values of the liquefaction strength ratio obtained from Figs. 4 and 5, and the values of R_L read from Fig. 10, corresponding to the plotted points. The data from Cases A, B, 1, 2, 4 show the same tendency as the values of the Yasuda and Inagaki's formula, although the data from Cases C and 3 underestimate the relationships, compared with the values of their formula.

4 Conclusions

In the present study, the following behaviors were observed in the cyclic loading tests and the static loading tests after liquefaction.

(1) The liquefaction strength of the mixed samples decreased as the relative density decreased in the case of FC = 40%, CC = 10% and $I_p = 10$, and dependency on the value of FC is negligible, if the value of FC is larger than 40%.
(2) The excess pore water pressure ratio did not perfectly attain 1.0 and the shear strain increased by larger than 20% in the cyclic loading tests on the specimens with CC = 10%. This behavior was called "cyclic softening" different from "liquefaction."
(3) The shear modulus G_1 was hardly influenced by the values of Dr and FC, if the value of $I_p = 10$.
(4) The shear modulus ratio versus the factor of safety against liquefaction relationships showed two tendencies in the test data, where one is the same as the Yasuda and Inagaki's formula and the other is the tendency to underestimate the value of their formula.

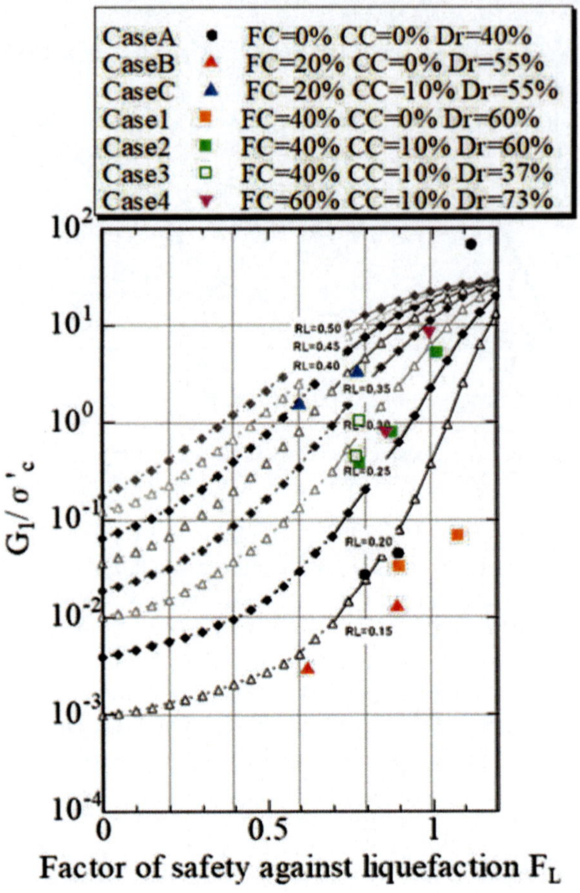

Fig. 10 Chart of shear modulus ratio G_1/σ_0' versus factor of safety against liquefaction F_L relationships

Table 3 Test conditions of FC, CC, Dr, and values of liquefaction strength ratio

	FC (%)	CC (%)	Dr (%)	R_{L20}	R_L
Case A	0	0	40	0.172	0.15
Case B	20	0	55	0.172	0.13
Case C	20	10	55	0.160	0.35
Case 1	40	0	60	0.109	0.12
Case 2	40	10	60	0.190	0.23
Case 3	40	10	37	0.162	0.25
Case 4	60	10	73	0.207	0.24

References

1. Miyaji, K., Kawano, N., Nagase, H., Hiro-oka, A., Fukumoto, K.: Liquefaction characteristics of sand containing fine grain with plasticity. In: Proceedings of the 54th Japan National Conference on Geotechnical Engineering, pp. 379–380 (2019) (in Japanese)
2. Miyaji, K., Nakano, T., Nagase H., Hiro-oka, A.: Liquefaction characteristics of sand containing fine grain with two different kinds of plasticity. In: Proceedings of the 53rd Japan National Conference on Geotechnical Engineering, pp. 479–480 (2018) (in Japanese)
3. Yasuda, S., Yoshida, N., Adachi, K., Kiku, H., Ishikawa, K.: Simplified evaluation method of liquefaction-induced residual displacement. J. Jpn. Assoc. Earthquake Eng. **16**(10), 31–50 (2016) (in Japanese)
4. Nagase, H., Hirooka, A., Fukumoto, K., Miyaji, K., Shichijyo, R.: Cyclic strength and deformation characteristics of volcanic ash soils deposited in Kumamoto district. In: Proceedings of the Technical Forum on Mitigation of Geo-disasters in Asia, pp. 45–50 (2019)
5. Nagase, H., Shimizu, K., Hiro-oka, A., Tanoue, Y., Saitoh, Y.: Earthquake-induced residual deformation of Ariake clay deposits with leaching. Soil Dyn. Earthquake Eng., 209–220 (2006)

Effect of Mineral Additives on the Strength Characteristics of a Laterite Soil

Tri Harianto and Widya Dwi Utami

Abstract The road access opening, frequently through several soil conditions, do not comply with the construction requirements. Thus, it needed materials that could protect the low bearing capacity of subgrade and strengthen the pavement layers from a load of the vehicle's wheel. This study aimed is to analyze the performance of the subbase layer, which consists of zeolite-stabilized laterite soil using water glass as an activator. The mechanical characteristics of the unconfined compressive strength (UCS) and California Bearing Ratio (CBR) value were investigated in this study. The soil sample was prepared with a zeolite percentage of 4, 8, 12, 16, 20%, and combining with 2% of water glass. Prior to the test, the soil samples set to the maximum dry density (MDD) and optimum moisture content (OMC) condition. The result of the mechanical characteristics of the stabilized soil showed that the higher UCS and CBR value was observed compared to untreated soil. The mechanism of the improvement of stabilized soil is also discussed.

Keywords Soil strength · Zeolite · Water glass

1 Introduction

The laterite soil derived from a wide variety of rocks weathering under strongly oxidizing and leaching conditions and rich in iron oxide. Laterite soil generally found in the humid climate, and it forms in a tropical and subtropical region. The climate (temperature, precipitation, leaching, and capillary rise), topography (drainage), vegetation, parent rock (iron-rich rocks), and time of these primary factors are the factors in the laterite formation.

Lateritic soils are indigenous materials that are cheaper and abundantly available as construction materials. Previous works reported the treatment of reclaimed asphalt pavement (RAP) with coal fly ash, RAP with new aggregates, and lateritic soil with forage ash. The results show that the CBR of lateritic soil improved with up to 6%

T. Harianto (✉) · W. D. Utami
Hasanuddin University, Perintis Kemerdekaan, 90245 Makassar, Indonesia
e-mail: triharianto@unhas.ac.id

H. Hazarika et al. (eds.), *Advances in Sustainable Construction and Resource Management*, Lecture Notes in Civil Engineering 144,
https://doi.org/10.1007/978-981-16-0077-7_37

forage ash treatment [1]. Laboratory testing results also show that the stabilization of lateritic soil improved California bearing ratio (CBR) and resilient modulus (Mr) values significantly [2–4].

Recently, the utilization of fiber in cement-stabilized soil increased the strength of the treated soil significantly [5, 6]. The ductility significantly improves without changing the compressive strength by the existence of fiber in the cement–soil mixture. The existence of fiber in the cement–soil mixture can suppress the development of crack formation during the unconfined compressive strength (UCS) test [7]. Moreover, the utilization of cement has severe environmental impacts, as it involves using vast amounts of fossil fuels as well as being responsible for the emission of more than 5% of all the carbon dioxide released worldwide [8]. Therefore, an attempt to reduce using cement in soil stabilization is increasing.

The utilization of mineral additive has received attention for potential applications in the soil stabilization method. The application of local material content (i.e., zeolite and water glass) as an additive to the lateritic soil has not widely applied. Therefore, the potential application of zeolite and water glass as an additive to improve the strength of laterite soil was investigated in this study.

2 Materials and Methods

2.1 Materials

The lateritic soil used in this study was collected from a borrow pit as disturbed samples at Sangkaropi, Toraja, Indonesia. The lateritic soil was crushed using a hand hammer, from its lump state to smaller sample sizes able to pass through a 0.425-mm aperture sieve following ASTM C702-98 (2003). The natural moisture content of the lateritic soil was determined to be 32% and classified as a silt with high plasticity (MH).

The stabilization agent used in this study was zeolite with water glass (sodium silicate) as an activator. Zeolite is crystalline alumino-silicates consisting of three-dimensional frameworks of SiO_4 and AlO_4 linked through oxygen bridges. The zeolite was crushed state to smaller sample sizes able to pass through a 0.075-mm aperture sieve following ASTM C702-98 (2003).

2.2 Methods

The lateritic soil (LS) and zeolite water glass-stabilized laterite (ZW-LS) soil were tested to determine the index properties, particle-size distribution, soil classification, specific gravity, and compaction characteristics following procedures outlined in standard ASTM codes. The unconfined compression test (UCT) and the California bearing ratio (CBR) unsoaked tests were carried out in accordance with the procedure outlined in ASTM D2166 and D1883-07e2, respectively. Lateritic soil was stabilized with zeolite by concentrations of 4, 8, 12, 16, and 16% and 2, 4, and 6% of water glass solution. All the specimens were subjected to curing for 7, 14, and 28 days. Prior to the test, the soil specimens were set to the maximum dry density (MDD) and optimum moisture content (OMC) condition. An optimum proportion is determined during the preliminary mix design tests.

3 Results and Analysis

3.1 Laterite Soil Characteristics

The soil sample was air dried for one day prior to testing in order to simulate the fields' condition. This procedure conducted due to the different environment usually affected the index properties of the soil samples [9]. The physical and mechanical properties of soil sample are summarized in Table 1. The engineering properties of the soil sample were determined in accordance with American Society for Testing and Materials (ASTM, 1992).

Table 1 Properties of the lateritic soil

Designation	Value	Unit
A. Physical properties		
Specific gravity	2.68	
Nature water content	32	%
Soil classification		
a. USCS	MH	
b. AASTHO	A-5	
Atterberg limit		
a. Liquid limit (LL)	59	%
b. Plastic limit (PL)	49	%
c. Plasticity index	10	%
B. Mechanical properties		
Standard proctor test		
a. Optimum moisture content (OMC)	25	%
b. Maximum dry density (MDD)	14.8	KN/m^3
Unconfined compression test (UCT)		
Compressive strength	97.1	KN/m^2
California bearing ratio (CBR)		
Unsoaked	19.0	%

3.2 Unconfined Compressive Strength

The variation of UCS of ZW-stabilized LS with curing time for 0, 7, 14, and 28 days is shown in Fig. 1. Based on these results, there is a general increase in the UCS values with increasing zeolite and water glass content. Generally, the increase of UCS value was found to 3–25 times of magnitude (300–2500 N/m^2) compared to untreated soil (97.1 kN/m^2). The highest increase of UCS value was found for 20% zeolite and 6% of water glass. This behavior tends to be similar to other zeolite and water glass content. In a mixture of 20% zeolite and 6% water glass (28 days curing time), an increase of soil strength is almost 25 times of magnitude compared to untreated LS. Increasing the strength of ZW-stabilized LS is strongly influenced by the content of minerals present in the zeolite and water glass.

The variation of the UCS value of zeolite-stabilized LS with various water glass content is shown in Fig. 2. All the UCS tests are conducted with 0, 7, 14, and 28 days. Based on these results, there is a general tendency of the increase in the UCS values with increasing zeolite and water glass content. Generally, the significant improvement of UCS value is for up to 7 days. The highest increase of UCS value was found for 6% of water glass. This behavior tends to be similar to other zeolite and water glass content. In a mixture of 20% zeolite and 6% water glass (28 days curing time), an increase of soil strength is almost 25 times of magnitude compared to untreated LS. The improvement magnitude of UCS in all various mixing for 7 days

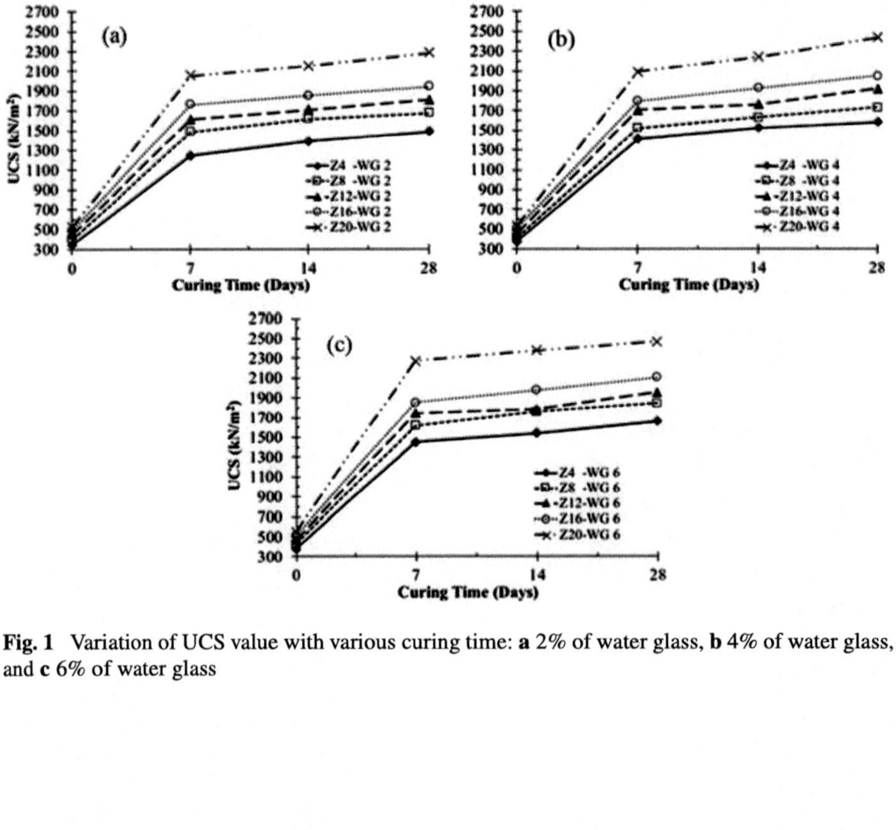

Fig. 1 Variation of UCS value with various curing time: **a** 2% of water glass, **b** 4% of water glass, and **c** 6% of water glass

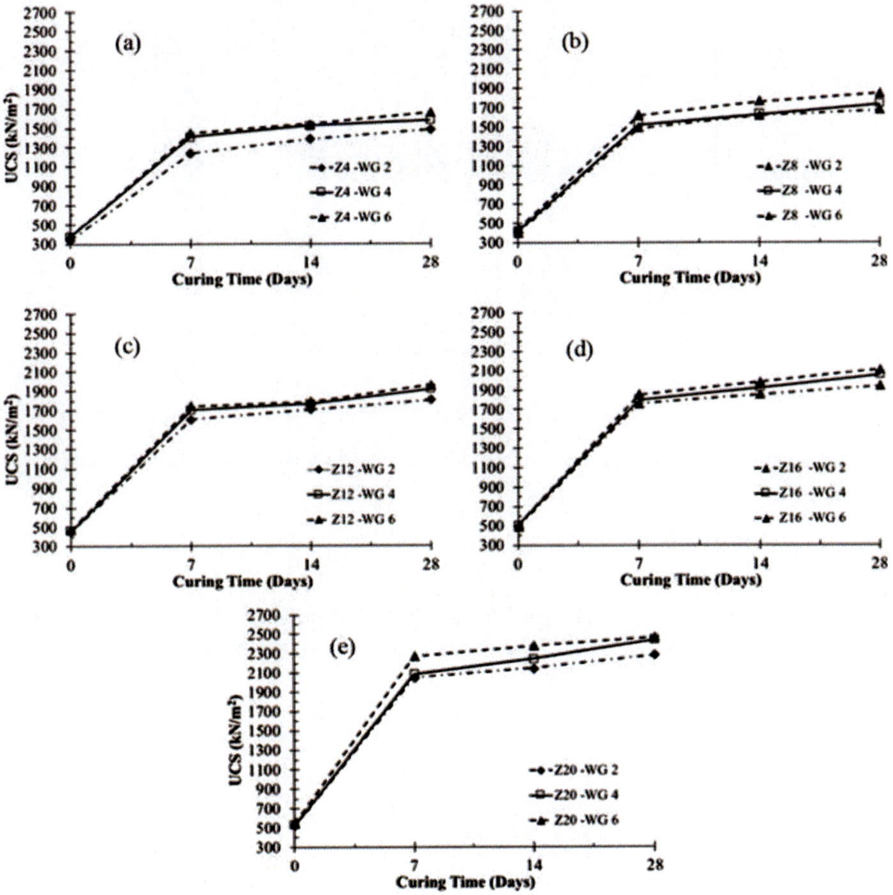

Fig. 2 Variation of UCS value with various zeolite content and curing time: **a** 4% of zeolite, **b** 8% of zeolite, **c** 12% of zeolite, **d** 16% of zeolite, and **e** 20% of zeolite

curing time as a primary cementing process is shown in Fig. 3. It can be seen that for the higher water glass content, the increase of the magnitude of UCS improvement is higher due to the higher amount of zeolite–water glass hydration products. Therefore, the water glass has a significant effect on the zeolite-stabilized LS improvement.

The increase in the UCS mainly attributed to the reactions between soil and additive materials (zeolite and water glass), including pozzolanic reaction. This products bind the ZW and LS particles together into a strong matrix [10]. The high silica and alumina in the zeolite reacted with mineral in Lateritic soil. This is because fine-grained soil particle composition (SiO_2, Al_2O_3) also participate in cementation reaction [11, 12].

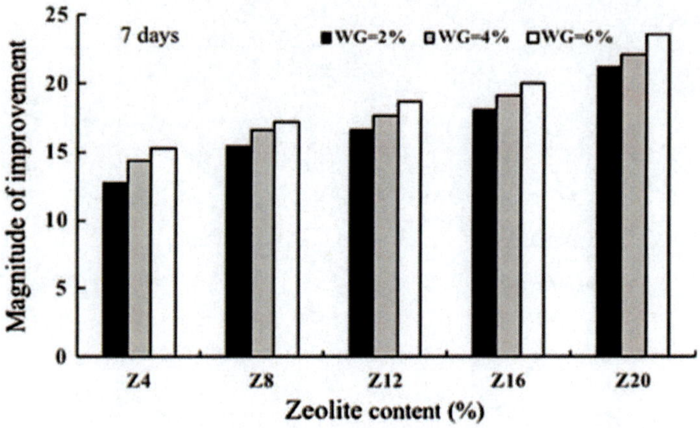

Fig. 3 Magnitude of UCS improvement of zeolite-stabilized LS with water glass activator

3.3 California Bearing Ratio

The variation of CBR (unsoaked) of ZW-stabilized LS mixes with various curing time is shown in Figs. 4 and 5. The CBR value of the ZW-stabilized LS increased with

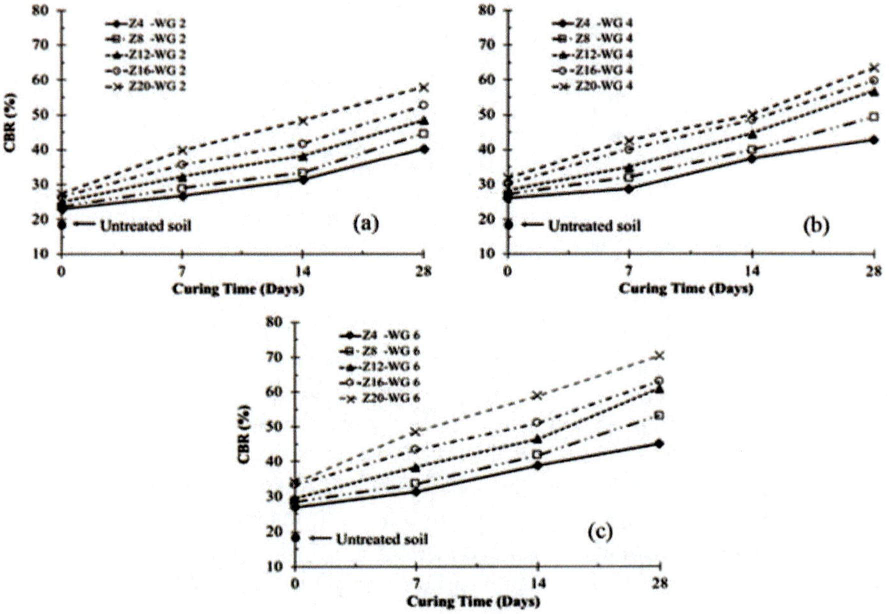

Fig. 4 Variation of CBR value with various curing time: **a** 2% of water glass, **b** 4% of water glass, and **c** 6% of water glass

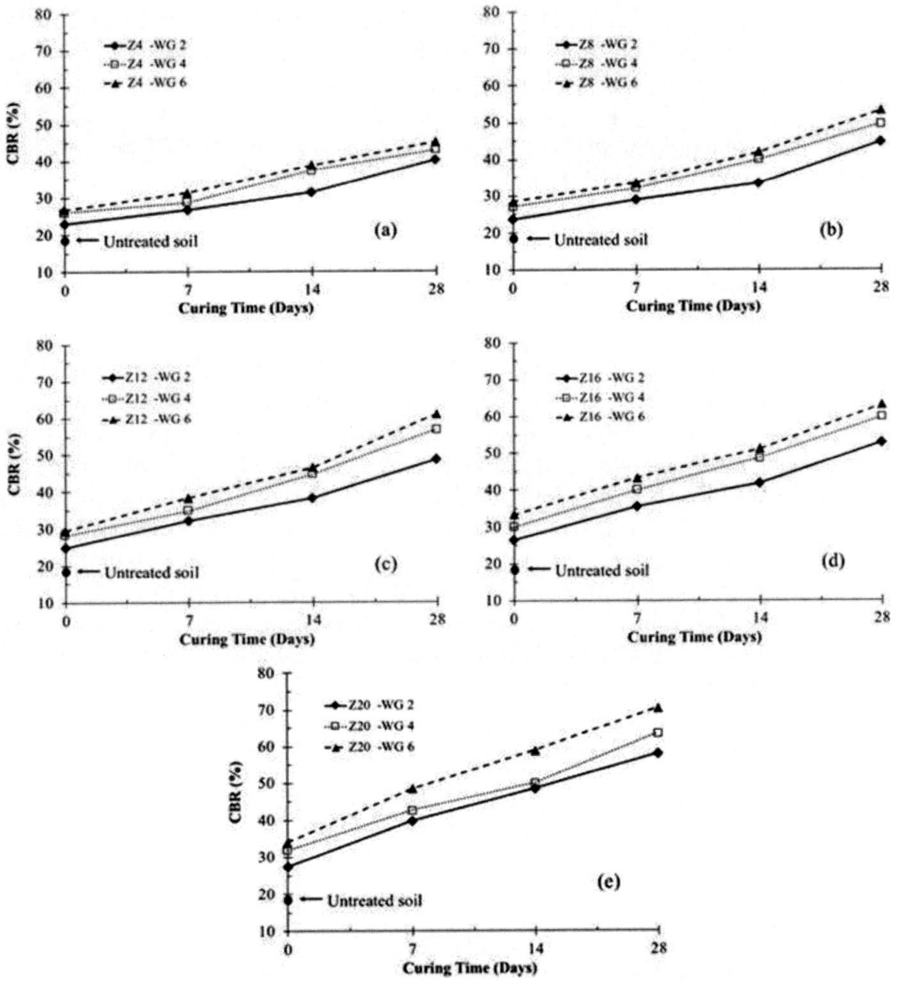

Fig. 5 Variation of CBR value with various curing time: **a** 4% of zeolites, **b** 8% of zeolite, **c** 12% of zeolite, **d** 16% of zeolite, and **e** 20% of zeolite

increased curing time. Increasing zeolite content also increases the CBR value, and the highest CBR value was found for 28 days. The CBR value of ZW-stabilized LS mixes for 28 days curing time shows that the CBR value increased three times greater of 2% water glass and almost four times of 6% water glass compared to the untreated LS. These values met the requirement of Indonesian National Standard (SNI-03-3438-1994) for subgrade and subbase layer of pavement foundation. The Indonesian general specification recommended that a minimum CBR value of subgrade and subbase are 6 and 20%, respectively.

The reason for this improvement is due to the pozzolanic reactions of zeolite with laterite soil. This results in the agglomeration of soil particles and causes an increase in strength gain. Overall, it is observed that the addition of water glass on the zeolite-stabilized LS leads to an improvement of CBR value. This increase reflected an improvement of bearing capacity of the zeolite-stabilized LS mixed with water glass. The bearing capacity improvement is a result of the stiffening of the soil by effect of the cement hydration [12]. Moreover, the increasing of CBR value of the cement-treated soil is in good agreement with the previous study by [13] and [14].

4 Conclusions

The present study was conducted to investigate and evaluate the effectiveness of water glass as an activator mixed with zeolite-stabilized laterite soil. The zeolite and water glass were mixed with the laterite soil and compacted at various zeolite and water glass content. Lateritic soils are indigenous materials that are cheaper and abundantly available as construction materials. The UCS value increased significantly with the addition of **zeolite** and water glass compared to the untreated soil. The high silica and alumina in the zeolite reacted with mineral in lateritic soil. This is because fine-grained soil particle composition (SiO_2, Al_2O_3) also participates in cementation reaction. The CBR value of the ZW-stabilized LS increased with increased curing time. Increasing zeolite and water glass content also increase the CBR value, and the highest UCS and CBR value was found for 28 days. These values met the requirement of Indonesian National Standard (SNI-03-3438-1994) for subgrade and subbase layer of pavement foundation. The Indonesian general specification recommended that a minimum CBR value of subgrade and subbase are 6 and 20%, respectively.

References

1. Olugbenga, O.A., Nurudeen, T.B., Adewale, A.C.: Effects of forage ash as stabilizing agent in lateritic soil for road. Innov. Sci. Eng. **1**, 1–8 (2011)
2. Edil, T.B., Benson, C.H.: Sustainable Construction Case History: Fly Ash Stabilization of Road-Surfaced Gravel. World of Coal Ash, Covington, KY (2007)
3. Hatipoglu, B., Edil, T.B., Benson, C.H.: Evaluation of base prepared from Rd. surface gravel stabilized with fly ash. In: Proceedings of Geo-Congress 2008: Geotechnics of Waste Management and Remediation, ASTM, West Conshohocken, PA, pp. 288–295 (2008)
4. Osinubi, K.J.: Influence of compaction delay on the properties of cement stabilized lateritic soil. J. Eng. Res. **6**(1), 13–25 (1998)
5. Harianto, T., Du, Y., Hayashi, S., Suetsugu, D., Nanri, Y.: Geotechnical properties of soil-fibre mixture as a landfill cover barrier material. Geotech. Eng. J. Southeast Asian Geo Soc. **39**(3), 137–143 (2008)
6. Tang, C., Shi, B., Gao, W., Chen, F., Cai, Y.: Strength and mechanical behaviour of short polypropylene fiber reinforced and cement stabilized clayey soil. Geotext. Geomemb. **25**, 194–202 (2007)
7. Liu, C., Starcher, R.D.: Effects of curing conditions on unconfined compressive strength of cement and cement-fiber-improved soft soil. J. Mater. Civ. Eng. **25**(8), 1134–1141 (2013). https://doi.org/10.1061/(ASCE)MT.1943-5533.0000575

8. Provis, J., Deventer, J.V.: Alkali Activated Materials: State of the Art Report. RILEM TC 224-AAM, Springer, Netherlands (2014)
9. Moh. Z.C., Mazher, M.F.: Effects of method of preparation on index properties of lateritic soils. In: Proceedings of the 7th International Conference on Soil Mechanics and Foundation Engineering, Mexico City, Mexico, vol. 1, pp. 23–25 (1969)
10. Li, L., Benson, C.H., Edil, T.B.: Sustainable construction case history: fly ash stabilization of recycled asphalt pavement materials. J. Geotech and Geo. Eng. **26**(2), 177–187 (2008)
11. Mariri, M., Ziaie Moayed, R., Kordnaeji, A. Stress-strain behavior of loess soil stabilized with cement, zeolite and recycled fiber. J. Mater. Civil Eng. **31**(12), 04019291-1–04019291-10 (2019)
12. Harianto, T., Sitepu, F., Jasruddin: Strength improvement of cement stabilized soil by binder mineral additive. Lowland Technol. Int. J. **21**(2), 90–97 (2019)
13. Osula, D.O.A.: Evaluation of admixture stabilization for problem laterire. J. Transp. Eng. **115**(6), 674–687 (1989)
14. Mengue, E., Mroueh, H., Lancelot, L., Eko, R.M.: Mechanical improvement of a fine-grained lateritic soil treated with cement or use in road construction. J. Mater. Civ. Eng. **29**(11), 040172061–040172122 (2017). https://doi.org/10.1061/(ASCE)MT.1943-5533.0002059

Effect of Spatial–Temporal Discretization Order on the Selection of Lattice Boltzmann Forcing Strategy in Convective Flow Simulation Within Internal Geometrical Arrangement of Concrete Structure

Aditya Dewanto Hartono, Kyuro Sasaki, and Ronald Nguele

Abstract This article reports on comprehensive assessment regarding conjoined effect of discretization order of the continuous Boltzmann equation and discrete forcing scenario in the feasibility of lattice Boltzmann method (LBM) as an interlinked hydrothermodynamics solver. A two-dimensional natural convection phenomena within the geometrical pore arrangement of the concrete structure was modeled as a synergetic embodiment of the discrete lattice fluid and thermal counterparts. Four different combination strategies of LBM implementation were carefully examined, elucidating theoretical and numerical segments, in order to expose any plausible discrepancy in the retrieved steady-state solutions. It was found that the numerical outcomes from distinct LBM strategies return equivalent results upon few selected key physical properties, despite incisive difference that emerges in the theoretical aspect. Excellent agreement with classical computational techniques was observed for all considered treatment options, underlining validity of our strategies. This study represents a further step toward clarification of the convoluted issue regarding proper selection of discretization order and forcing scheme in LBM simulation.

Keywords Lattice Boltzmann method · Natural convection · Discretization order of Boltzmann equation · Forcing schemes · Concrete structure

1 Introduction

Thermal expansion characteristics of the upholding steel within the concrete has been identified as the plausible reason for internal microdefect generation. This is particularly evident in tropical region of the globe in which excessive rainfall

A. D. Hartono (✉) · K. Sasaki · R. Nguele
Resources Production and Safety Engineering (REPS) Laboratory, Department of Earth Resources Engineering, Kyushu University, Fukuoka 812-0395, Japan
e-mail: adityadewanto@gmail.com

H. Hazarika et al. (eds.), *Advances in Sustainable Construction and Resource Management*, Lecture Notes in Civil Engineering 144,
https://doi.org/10.1007/978-981-16-0077-7_38

is accompanied with intense exposure of sunlight. Liquid aqueous phase from the rainfall was absorbed by the concrete structure and reside within its geometrical pore arrangement. During sweltering daylight, convective flow emanates within the porous interior configuration of the solid ambient. Thermal expansion of steel in superfluous heat flux region of the domain has potential to induce internal fracture. In the long term, this inner microdefect would proliferate; thus stimulating structural instability that provokes undesirable and detrimental ramifications upon the steadiness of construction.

Lattice Boltzmann method (LBM) has been recognized as a powerful numerical technique for simulating thermo-hydrodynamics manifestation [6, 10]. Due to its unique features that offer plentiful advantages over the quintessential numerical techniques [7], the popularity of LBM upon researchers has escalated during the last few years. LBM allows investigators to systematically elaborate on numerous systems that once considered arduous; thus providing an opportunity to elucidate advanced physical phenomena through numerical approach. LBM has been adopted in diverse problems related to numerical heat transfer and computational fluid dynamics, encompassing straightforward to challenging configurations.

One particular phenomena of interest is natural convection. Such physical configuration is eminent upon researchers due to its extensive real-world applications. Numerous investigations had embarked on evaluating the phenomena using classical numerical methods. Among them, Mayne et al. [9] successfully adapted finite element scenario to simulate thermally driven cavity problem, covering broad range of Rayleigh numbers. They justified that the h-adaptive finite element method was suitable for modeling such system with high accuracy while maintaining proper computational time. Similar configuration was studied by Yu and Tian [15] using finite difference. The authors concluded that second-order accuracy solutions were attained by adapting the stream function–velocity formulation. A number of studies have employed LBM to simulate the interlinked hydrothermal impartment within a closed enclosure. Both standard [12] and sophisticated [6, 14] LBM approaches have been fulfilled with encouraging feedbacks.

Predominantly, convective flow modeling using LBM requires assimilation of external force term into the native form of the discrete Boltzmann expression. Miscellaneous strategies to incorporate forcing term into LBM have been proposed [7]. In this context, Mohamad and Kuzmin [11] carried out thorough assessment upon different forcing schemes in LBM. Three of the most popular forcing scenarios were selected and tested for natural convection modeling in closed- and open-ended square domain. Their evaluation suggested that the diverging LBM forcing schemes basically exhibit equivalent performances. Subsequently, the impact of discretization order of the continuous Boltzmann equation upon the efficacy of LBM simulation was accentuated [13]. The annotation demonstrated that the selection of discontinuation order of Boltzmann equality possesses certain influences on the numerical outcomes. Nevertheless, such entailments were affirmed as case-dependent.

Despite the availability of few treatises that engaged on investigating individual effects of forcing models and discretization order on LBM's functionality, no single attempt exist in elaborating collective ramification of the two aforementioned critical elements. Existing reports on natural convection modeling using LBM mainly focused on practical aspects, incorporating dissimilar expressions of forcing and discontinuation strategies. This situation is disadvantageous upon the newcomers in LBM, presumably stimulating puzzling interpretation about the particular subject. Therefore, this work aims to undertake comprehensive investigation pertinent to the conjoint contribution of forcing schemes and discretization order of Boltzmann equality upon the feasibility of LBM as hydrothermodynamics solver. The buoyancy-driven natural convective flow within a two-dimensional equilateral rectangle was selected as the appropriate dynamical phenomena to evaluate the issue. The deliberation covers numerical implementation upon the considered combination of forcing-discretization schemes, corroborated by succinct explanation of the apposite theoretical grounding. Attention was given to expose plausible discrepancy between the treatments upon few tangible physical quantities. Concurrently, the accuracy and validity of every manifestations were justified by performing careful comparison with former numerical studies, covering finite difference (FDM), finite volume (FVM), finite element (FEM), and advanced lattice Boltzmann (LBM) techniques.

The current report is hierarchically designated as follows. Section 2 highlights substantial aspects of LBM related to the synergetic flow of heat and mass in a plain ambient. Emphasize was given on elaborating options upon discretization procedure of the continuous Boltzmann equality as well as introducing the prevalent forcing schemes in LBM. The subsequent passage explicates natural convection arrangement as the chosen numerical benchmark case. Thereupon, Sect. 4 poses numerical results together with germane discussions and comments. Finally, Sect. 5 ratifies some concluding remarks of the current research article.

2 Fundamentals of Lattice Boltzmann Method for Simultaneous Heat and Mass Transfer

One predominantly eminent applications of lattice Boltzmann method (LBM) in modeling complex flow phenomena is the concurrent movement of fluid and thermal quantities [6, 10]. In contrast from the traditional numerical procedures, LBM regards the dynamic elements as an agglomeration of their constituent molecules manifested commensurately by the particle distribution function (PDF). As a collective epitome, the PDF contains important messages regarding flow conditions. Therefore, vital information concomitant to the behaviors of the moving materials is recorded by means of tracking the evolution of the related PDF. In this vein, the prevalent dynamical equations, consisting the continuity, momentum (or Navier–Stokes), and heat formulation, are solved indirectly using discrete Boltzmann expression. Continuum parameters are recovered through rigorous analytical assessment of the moments of the PDF [13].

In this work, the flowing materials were distinguished into different mesoscopic embodiments. In the so-called double distribution functions (DDF) approach [6, 10], hydrothermal delineations were connected through the external buoyancy force expression [7]. As a consequence of the DDF technique, two sets of PDF prevail as a relevant embodiment of each corresponding flowing materials; one PDF resembles hydrodynamics flow and the other portrays thermal dissemination. Hereinafter, f and g were declared as the PDF representations of the mass (alongside momentum) and thermal populations, respectively.

To facilitate the deliberation, separate expositions of LBM implementation upon each category of the flowing substances were administered. Discontinuation procedures of the commensurate Boltzmann formulas for each individual circulating materials were introduced first, emphasizing on establishment of distinct discrete Boltzmann manifestations. Thereupon, key concepts to employ LBM for each counterpart of the flowing materials were prescribed. Subsequently, inclusion of external forcing terms into the prevalent discrete Boltzmann formulation was discussed. Following the feedbacks of the previous explications is the clarification of emerging combinations between discretization and forcing arrangements, which is the mainstay of the current investigation. The section ended with succinct annotations regarding macroscopic manifestations upon the aforementioned combinations between the two central elements.

2.1 *LBM for Mass and Momentum Transport*

Discrete Boltzmann Representation for Mass and Momentum Fraction. The continuous Boltzmann formula primarily describes evolution of the corresponding PDF due to the inherent microscopic collisions which take place among its constituent particles. In the context of mass and momentum transport, the PDF is considered as the commensurate lattice realization of the associated elements. General expression of the Boltzmann equality occupies the following remark [7]:

$$\frac{\partial f}{\partial t} + \xi_\alpha \frac{\partial f}{\partial x_\alpha} + F_\alpha \frac{\partial f}{\partial \xi_\alpha} = \Omega(f). \tag{1}$$

Here, ξ_α and Ω denote microscopic velocity of the particles and collision operator, respectively. Contributions from subsidiary physical stimulus upon the primary dynamical element of the flow are imposed within Boltzmann expression as the external force density vector, portrayed by F_α. In order to be reconcilable for computational purposes, Eq. (1) needs to be modified into its equivalent discrete quantity.

Proportionately, parameterization of Eq. (1) can be executed following different truncation order of expansion upon the Boltzmann equality, establishing distinct entailments of the well-known lattice Boltzmann equation (LBE) [7, 13] following the single-relaxation-time (SRT) mechanism. Such dichotomy elicited two expositions of LBE, each corresponds to the first and second truncation order of Boltzmann formulation in velocity and spatial–temporal spaces:

$$f_i(x_\alpha + \xi_{i\alpha}\Delta t, t + \Delta t) - f_i(x_\alpha, t) = -\frac{\Delta t}{\tau_\nu}\left[f_i(x_\alpha, t) - f_i^{\mathrm{eq}}(x_\alpha, t)\right] + R_i(x_\alpha, t)\Delta t \tag{2}$$

$$\overline{f}_i(x_\alpha + \xi_{i\alpha}\Delta t, t + \Delta t) - \overline{f}_i(x_\alpha, t) = -\frac{\Delta t}{\bar{\tau}_\nu}\left[\overline{f}_i(x_\alpha, t) - f_i^{\mathrm{eq}}(x_\alpha, t)\right] + R_i(x_\alpha, t)\Delta t\left[1 - \frac{\Delta t}{2\overline{\tau}_\nu}\right]. \tag{3}$$

Concurrently, the following definitions were introduced:

$$\overline{f}_i = f_i + \frac{\Delta t}{2\tau_\nu}\left[f_i - f_i^{\mathrm{eq}}\right] - \frac{R_i \Delta t}{2} \tag{4}$$

$$\overline{\tau}_\nu = \tau_\nu + \frac{\Delta t}{2}. \tag{5}$$

In this vein, f_i specifies the discrete mode of PDF manifestation upon the flowing mass and momentum, while R_i depicts the forcing arrangement. Thorough exposition upon the latter property would be given in the upcoming segment of the article. Furthermore, $\xi_{i\alpha}$ designates the discontinuous velocity of particles and τ_ν implies the infinitesimal relaxation time of the propagating unit. Parameter f_i^{eq} notifies equilibrium distribution function, which theoretically implies stability condition of particular PDF with the absence of molecular collisions. The sophisticated form of the equilibrium distribution function fulfilled the Maxwell–Boltzmann (MB) statement [7]. Discrete counterpart of the MB formulation occupies central role in LBM and is defined as:

$$f_i^{\mathrm{eq}} = w_i \rho\left[1 + \frac{\boldsymbol{\xi}_i \cdot \boldsymbol{u}}{c_s^2} + \frac{(\boldsymbol{\xi}_i \cdot \boldsymbol{u})^2}{2c_s^4} - \frac{\boldsymbol{u} \cdot \boldsymbol{u}}{2c_s^2}\right] \tag{6}$$

Physical properties ρ and c_s define the macroscopic fluid density and lattice speed of sound, respectively, while w_i portrays the weighting factors for each corresponding discrete velocity directions.

Fundamental expositions which connect LBM and macroscopic flow properties were obtained from the definition of PDF moment expressions correspond to discontinuation order of the Boltzmann equality [7, 13]. Relaxation time parameter for each commensurate discretization orders is associated to its pertinent macroscopic fluid kinematic viscosity (ν). Table 1 outlines the underlying relationships between LBM parameter and its accompanying tangible properties.

Table 1 Fundamental relations between LBM and tangible fluid properties correspond to spatial–temporal discretization order of the Boltzmann equation

Spatial–temporal discretization order	Relations of macroscopic to LBM parameters
First order	$\rho = \Pi^{\text{eq}} = \sum_i f_i^{\text{eq}} = \sum_i f_i$ (7)
	$\rho u_\alpha = \Pi_\alpha^{\text{eq}} = \sum_i \xi_{i\alpha} f_i^{\text{eq}} = \sum_i \xi_{i\alpha} f_i$ (8)
	$\nu = \rho c_s^2 (\tau_\nu - \frac{\Delta t}{2})$ (9)
Second order	$\rho = \Pi^{\text{eq}} = \sum_i f_i^{\text{eq}} = \sum_i \overline{f}_i + \frac{\Delta t}{2} \sum_i R_i$(10)
	$\rho u_\alpha = \Pi_\alpha^{eq} = \sum_i \xi_{i\alpha} f_i^{eq} = \sum_i \xi_{i\alpha} \overline{f}_i + \frac{\Delta t}{2} \sum_i \xi_{i\alpha} R_i$(11)
	$\nu = \rho c_s^2 (\overline{\tau}_\nu - \frac{\Delta t}{2})$ (12)

Forcing Schemes in LBM Formulation. LBM defines inclusion of external influences into the flowing system through the forcing term. This unique feature is one of the primary aspects that renders LBM favorable among practitioners. The distinctive mechanism allows smooth inclusion of additional physics to the mathematical remark of LBM. Within the continuous Boltzmann framework, this additional stimulus is manifested by F_α (or equivalently $\boldsymbol{F}$). Following similar parameterization notion in LBM, variable F_α needs to be discretized as well. The procedure prompts discrete counterpart of forcing configuration depicted by R_i.

Miscellaneous scenarios have been proposed in order to advocate proper embodiment of R_i. In this work, we particularly interested in two of the most prominent schemes in discretizing additional impetus into LBM configuration. The first scenario was based on involvement of forcing moments down to the first-order fraction. Meanwhile, the other strategy entangled forcing moments up to the second-order term. The former expository was introduced by Luo [8], while the latter was proposed by Guo et al. [3, 4]. Both remarks were represented accordingly as

$$R_i = w_i \frac{\boldsymbol{\xi}_i \cdot \boldsymbol{F}}{c_s^2} \tag{13}$$

$$R_i = w_i \left[\frac{\boldsymbol{\xi}_i - \boldsymbol{u}}{c_s^2} + \frac{(\boldsymbol{\xi}_i \cdot \boldsymbol{u})\boldsymbol{\xi}_i}{c_s^4} \right] \cdot \boldsymbol{F} \tag{14}$$

For the purpose of examination, the two aforementioned forcing expressions were declared as the configuration of interest to be integrated with the discrete LBE formulation.

Combination Scenario of Discretization Order and Forcing Schemes. Spatial and temporal discretization of the continuous Boltzmann equation generates discontinuous LBEs, as were displayed by Eqs. (2) and (3). Likewise, parameterization of the additional stimulus term produces discrete forcing schemes, described by Eqs. (13) and (14). These outcomes were occupied to proclaim plausible combination between

Table 2 Possible combinations of forcing and ST discretization

Combination strategy	Spatial–temporal discretization order	Forcing model
I	First order (Eq. 2)	Luo (Eq. 13)
II	First order (Eq. 2)	Guo (Eq. 14)
III	Second order (Eq. 3)	Luo (Eq. 13)
IV	Second order (Eq. 3)	Guo (Eq. 14)

the two critical factors. Table 2 summarizes commensurate combination strategy to be assessed in the current examination. Hereinafter, the term combination strategy refers to the definition documented in Table 2.

LBM Arrangement for Mass and Momentum Fraction. Two fundamental stages mark the distinctive characteristic of LBM, namely collision and streaming [7]. Both phases occur during the same computational time step. Collision step takes the form of the discrete LBE remark depicted in Eq. (2) or (3). Adapting meager mathematical manipulations to both expositions produce the following expressions for collision step:

$$f_i^* = f_i - \frac{\Delta t}{\tau_\nu}\left[f_i - f_i^{\text{eq}}\right] + R_i \Delta t \tag{15}$$

$$\overline{f}_i^* = \overline{f}_i - \frac{\Delta t}{\overline{\tau}_\nu}\left[\overline{f}_i - f_i^{\text{eq}}\right] + R_i \Delta t\left[1 - \frac{\Delta t}{2\overline{\tau}_\nu}\right]. \tag{16}$$

After collision, the particles propagate to other lattice position within the flow domain. This is the streaming phase of LBM, represented as:

$$f_i(x_\alpha + \xi_{i\alpha}\Delta t, t + \Delta t) = f_i^*(x_\alpha, t) \tag{17}$$

$$\overline{f}_i(x_\alpha + \xi_{i\alpha}\Delta t, t + \Delta t) = \overline{f}_i^*(x_\alpha, t) \tag{18}$$

In this vein, the movement of particles follows certain velocity scheme that was firmly established in LBM's theorem. In this work, the D_2Q_9 LBM velocity scheme was employed for mass and momentum transport. Figure 1 illustrates the adopted velocity pattern. In this two-dimensional flow configuration, the discrete hydrodynamics particles have nine possible directions of movements within the associated lattice domain, including stationary position.

For this particular arrangement, the velocity of particles $\boldsymbol{\xi}_i$ and the weighting factor w_i occupy the following mathematical statements [6, 10]:

$$\boldsymbol{\xi}_i = \begin{cases} (0, 0); & i = 0 \\ \left(\cos\left[(i-1)\frac{\pi}{2}\right], \sin\left[(i-1)\frac{\pi}{2}\right]\right)c; & 1 \le i \le 4 \\ \left(\cos\left[(2i-9)\frac{\pi}{4}\right], \sin\left[(2i-9)\frac{\pi}{4}\right]\right)\sqrt{2}c; & 5 \le i \le 9 \end{cases} \tag{19}$$

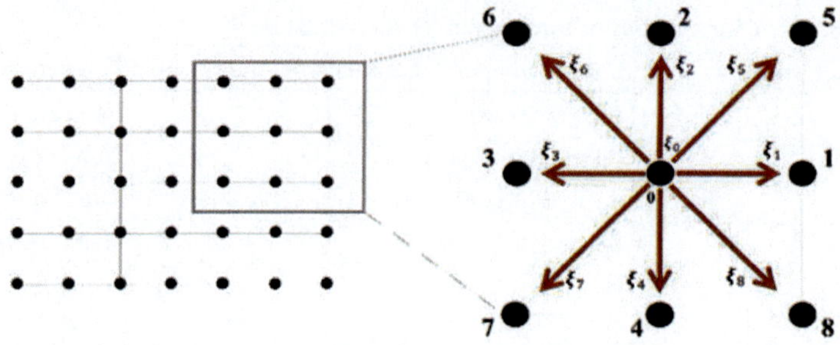

Fig. 1 D_2Q_9 velocity arrangement for mass and momentum propagation

$$w_i = \begin{cases} \frac{4}{9}; & i = 0 \\ \frac{1}{9}; & 1 \le i \le 4 \\ \frac{1}{36}; & 5 \le i \le 9. \end{cases} \tag{20}$$

Here, c represents the lattice speed and is defined as $c = \frac{\Delta x}{\Delta t}$.

Macroscopic Representations and Residual Terms. As numerical hydrodynamics solver, LBM has remarkable capacity to return the dynamical fluid representation at macroscopic level. Basically, LBM returns the prominent mass and momentum formulations in the following general expressions:

$$\frac{\partial \rho}{\partial t} + \frac{\partial}{\partial x_\alpha}(\rho u_\alpha) = \Psi \tag{21}$$

$$\frac{\partial}{\partial t}(\rho u_\alpha) + \frac{\partial}{\partial x_\beta}(\rho u_\alpha u_\beta) = -\frac{\partial}{\partial x_\beta} P + \nu \frac{\partial}{\partial x_\beta}\left[\frac{\partial u_\alpha}{\partial x_\beta} + \frac{\partial u_\beta}{\partial x_\alpha}\right] + F_\alpha + \Upsilon + \Xi. \tag{22}$$

Here, the continuity and Navier–Stokes equations are represented by Eqs. (21) and (22), accordingly. In the latter equality, P and ν typify fluid pressure and kinematic viscosity, respectively.

Nevertheless, since LBM was based on continuous Boltzmann equation which describes the behavior of gas, additional macroscopic terms beyond the typical fluid mechanics portrayal were recovered. These ancillaries appeared in both mathematical remarks of hydrodynamics. Parameter Υ denotes inherent residual fraction due to adjusted expansion in Boltzmann equation [7]. The quantity depends on the selected discretization order of the Boltzmann equation and is depicted as

$$\Upsilon = \begin{cases} -\left(\tau_\nu - \frac{\Delta t}{2}\right)\frac{\partial}{\partial x_\beta}\frac{\partial}{\partial x_\gamma}(\rho u_\alpha u_\beta u_\gamma), & \text{first-order discretization} \\ -\left(\overline{\tau}_\nu - \frac{\Delta t}{2}\right)\frac{\partial}{\partial x_\beta}\frac{\partial}{\partial x_\gamma}(\rho u_\alpha u_\beta u_\gamma), & \text{second-order discretization.} \end{cases} \tag{23}$$

Table 3 Residual terms in the recovered continuity (Ψ) and momentum (Ξ) equation

Combination strategy	Ψ	Ξ
I	$-\frac{\Delta t}{2}\frac{\partial}{\partial x_\gamma}F_\gamma$	$\left(\tau_\nu - \frac{\Delta t}{2}\right)\frac{\partial}{\partial x_\beta}\left[u_\alpha F_\beta + F_\alpha u_\beta\right] - \frac{\Delta t}{2}\frac{\partial F_\alpha}{\partial t}$
II	$-\frac{\Delta t}{2}\frac{\partial}{\partial x_\gamma}F_\gamma$	$-\frac{\Delta t}{2}\frac{\partial}{\partial x_\beta}\left[u_\alpha F_\beta + F_\alpha u_\beta\right] - \frac{\Delta t}{2}\frac{\partial F_\alpha}{\partial t}$
III	0	$\left(\overline{\tau}_\nu - \frac{\Delta t}{2}\right)\frac{\partial}{\partial x_\beta}\left[u_\alpha F_\beta + F_\alpha u_\beta\right]$
IV	0	0

Subsequently, parameters Ψ and Ξ characterized the supplementary fractions of the recovered continuity and Navier–Stokes formulas, respectively. Both expositions rely upon combination between discretization order of Boltzmann equality and forcing scheme. Based on the touted combination strategies displayed in Table 2, the corresponding expressions for Ψ and Ξ were encapsulated in Table 3.

2.2 *LBM for Thermal Transport*

Discrete Boltzmann Representation for Thermal Fraction. Tantamount LBM principles prevail for the thermal counterpart of the flowing materials. Nevertheless, different from the mass and momentum, LBM procedure for heat propagation does not incorporate external impetus term. Hence, instead of Eq. (1), the underlying Boltzmann equality for heat circulation is explicated by

$$\frac{\partial g}{\partial t} + \xi_\alpha \frac{\partial g}{\partial x_\alpha} = \Omega(g). \tag{24}$$

Here, f was replaced with g as the associated representation of thermal particles in LBM. Employing appropriate adjustments to Eqs. (2) and (3) based on remark in Eq. (24), the fundamental LBE for heat transport was generated as

$$g_i(x_\alpha + \xi_{i\alpha}\Delta t, t + \Delta t) - g_i(x_\alpha, t) = -\frac{\Delta t}{\tau_c}\left[g_i(x_\alpha, t) - g_i^{eq}(x_\alpha, t)\right] \tag{25}$$

$$\overline{g}_i(x_\alpha + \xi_{i\alpha}\Delta t, t + \Delta t) - \overline{g}_i(x_\alpha, t) = -\frac{\Delta t}{\overline{\tau}_c}\left[\overline{g}_i(x_\alpha, t) - g_i^{eq}(x_\alpha, t)\right]. \tag{26}$$

The relaxation time for heat circulation was now depicted by τ_c. Parameter g_i^{eq} specifies thermal equilibrium distribution function and is defined as

$$g_i^{eq} = w_i T\left[1 + \frac{\boldsymbol{\xi}_i \cdot \boldsymbol{u}}{c_s^2} + \frac{(\boldsymbol{\xi}_i \cdot \boldsymbol{u})^2}{2c_s^4} - \frac{\boldsymbol{u} \cdot \boldsymbol{u}}{2c_s^2}\right] \tag{27}$$

where T indicates temperature of the ambient.

In this framework, since LBM manifestation for thermal flow does not assimilate supplementary forcing mechanism in its formulation, distinct portrayals of thermal LBM formulated in Eqs. (25) and (26) are principally equivalent. Hence, instead of binary delineations, single representation of LBE for thermal counterpart was adopted. Hence, the appropriate LBE remark for heat transfer is depicted by Eq. (25). Concomitant macroscopic parameters, such as temperature and thermal diffusivity, were recovered by the following exegesis:

$$T = \Pi_c^{\text{eq}} = \sum_i g_i^{\text{eq}} = \sum_i g_i \tag{28}$$

$$D = c_s^2\left(\tau_c - \frac{\Delta t}{2}\right) \tag{29}$$

LBM Arrangement for Thermal Fraction. For thermal counterpart, collision and streaming steps were prescribed correspondingly as follows:

$$g_i^* = g_i - \frac{\Delta t}{\tau_c}\left(g_i - g_i^{eq}\right) \tag{30}$$

$$g_i(x_\alpha + \xi_{i\alpha}\Delta t, t + \Delta t) = g_i^*(x_\alpha, t) \tag{31}$$

To describe heat propagation, the D2Q5 lattice scheme was applied into the thermal LBE. Figure 2 demonstrates the particular velocity pattern. Different from hydrodynamics mechanism, the thermal particles have five possible ways to propagate within the lattice domain. Likewise, the quiescent condition was taken into consideration. The concomitant velocity of thermal particles and the weighting fractions were prescribed as:

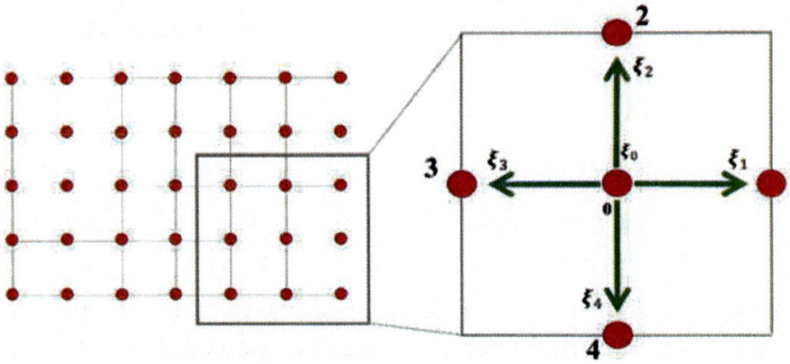

Fig. 2 D2Q5 lattice arrangement for thermal propagation

$$\xi_i = \begin{cases} (0, 0); & i = 0 \\ \left(\cos\left[(2i-1)\frac{\pi}{4}\right], \sin\left[(2i-1)\frac{\pi}{4}\right]\right)\sqrt{2}c; & 1 \le i \le 4 \end{cases} \tag{32}$$

$$w_i = \begin{cases} \frac{2}{6}; & i = 0 \\ \frac{1}{6}; & 1 \le i \le 4 \end{cases} \tag{33}$$

Macroscopic Representation and Residual Terms. Proper LBM strategy for thermal flow simulation returns the appropriate heat equation at macroscopic standpoint, as is described by the following formula:

$$\frac{\partial T}{\partial t} + \frac{\partial}{\partial x_\alpha}(T u_\alpha) = D\frac{\partial^2 T}{\partial x_\alpha^2} + \Lambda + \Phi. \tag{34}$$

Here, D specifies thermal diffusion coefficient. As was the case for hydrodynamics, thermal LBM arrangement produced tangible residual fractions, as well. Parameter Λ denotes the inherent ancillary term due to Boltzmann discretization and is defined as:

$$\Lambda = \frac{D}{c_s^2}\frac{T}{\rho}\frac{\partial}{\partial x_\alpha}\left(F_\alpha - \frac{\partial}{\partial x_\beta}P\right). \tag{35}$$

On the other hand, Φ portrays additional factor which depend upon the combination between discretization order of the Boltzmann equality for mass and momentum fractions and forcing scenario. In other words, residual fractions from hydrodynamics term have profound influences upon the thermal material, as well. Table 4 summarizes supplementary expressions in the recovered heat formulation.

Table 4 Residual terms in the recovered heat equation (Φ)

Combination strategy	Φ
I	$\frac{D}{c_s^2}\frac{\partial}{\partial x_\alpha}\frac{T}{\rho}\left[\left(\tau_\nu - \frac{\Delta t}{2}\right)\frac{\partial}{\partial x_\beta}\left[u_\alpha F_\beta + F_\alpha u_\beta\right] - \frac{\Delta t}{2}\frac{\partial F_\alpha}{\partial t}\right]$
II	$\frac{D}{c_s^2}\frac{\partial}{\partial x_\alpha}\frac{T}{\rho}\left[-\frac{\Delta t}{2}\frac{\partial}{\partial x_\beta}\left[u_\alpha F_\beta + F_\alpha u_\beta\right] - \frac{\Delta t}{2}\frac{\partial F_\alpha}{\partial t}\right]$
III	$\frac{D}{c_s^2}\frac{\partial}{\partial x_\alpha}\frac{T}{\rho}\left[\left(\overline{\tau}_\nu - \frac{\Delta t}{2}\right)\frac{\partial}{\partial x_\beta}\left[u_\alpha F_\beta + F_\alpha u_\beta\right]\right]$
IV	0

3 Problem Description and Boundary Conditions

3.1 Problem Description: Natural Convection

Natural convection flow was selected as the appropriate physical configuration to assess our combination treatments. In this work, we exclude the effect of porous media within the solid structure and focus on the basic configuration that constitutes the flow system. Within this frame of mind, an equilateral rectangle structure was generated as the prototype ambient of the flow. The enclosure was filled with a stationary fluid. The vertical walls were subject to different thermal conditions in which the hot (T_h) and cold (T_c) temperatures were imposed to the left and right side of the cavity, respectively. Meanwhile, the top and bottom walls were perfectly insulated. Figure 3 illustrates the two-dimensional arrangement of convective flow considered in this study. Boussinesq approximation was employed to the system, rendering the external stimulus form as buoyancy force. Hence, the external force term possesses the following remark:

$$\boldsymbol{F} = \rho \boldsymbol{g} \beta (T - T_0) \tag{36}$$

In which $\boldsymbol{g}$, β, and T_0 prescribe the gravitational acceleration, thermal expansion coefficient, and reference temperature, respectively.

A rigorous review on dimensional analysis revealed key dimensionless group that substantiates the free convective transport. The principal dimensionless properties were listed as follows:

$$\mathrm{Pr} = \frac{\nu}{D}; \quad \mathrm{Ra} = \frac{g\beta \Delta T l^3}{\nu D}; \quad \Theta = \frac{T - T_0}{\Delta T}; \quad u^* = \frac{ul}{D}; \quad t^* = \frac{tD}{l^2} \tag{37}$$

where Pr and Ra designate the Prandtl and Rayleigh number, respectively; Θ, u^*, and t^* specify the dimensionless temperature, velocity, and time, accordingly.

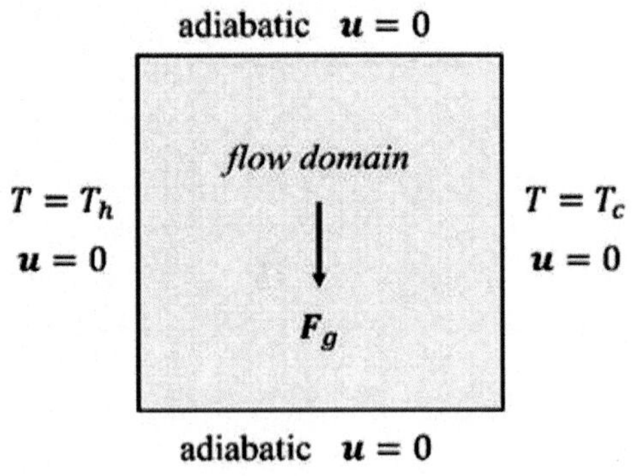

Fig. 3 Problem description of natural convection flow

3.2 Boundary Conditions

Boundary treatments are customarily defined in terms of macroscopic quantities of the propagating units. This reality is in contrast with LBM approach which regards the flowing materials from statistical physics standpoint. Therefore, direct application of continuum boundaries as in the accustomed numerical methods is undoubtedly irreconcilable. In order to be parallel with LBM notion, tangible boundary properties have to be decomposed into their corresponding particles counterpart. Such endeavor need to be performed in a manner that satisfies the fundamental conservation laws [7].

Few techniques have been proposed in order to employ boundary conditions into LBM. In this work, the non-equilibrium-bounce-back (NEBB) [7, 16] and the anti-bounce-back (ABB) [7, 14] procedures were imposed upon the hydrodynamics and thermal constituents of the flow, respectively.

Boundary Conditions for Mass and Momentum Fraction. Based on the established flow configuration displayed in Fig. 3, the following NEBB treatments were imposed to each of the pertinent walls of the domain.

$$\text{Left boundary:}\begin{cases} \rho_{\text{left}} = f_0 + f_2 + f_4 + 2(f_3 + f_6 + f_7) - \frac{1}{2c}(F_x\Delta t) \\ f_1 = f_3 - \frac{1}{6c}(F_x\Delta t) \\ f_5 = f_7 - \frac{1}{2}(f_2 - f_4) - \frac{1}{6c}(F_x\Delta t) - \frac{1}{4c}(F_y\Delta t) \\ f_8 = f_6 + \frac{1}{2}(f_2 - f_4) - \frac{1}{6c}(F_x\Delta t) + \frac{1}{4c}(F_y\Delta t) \end{cases} \tag{38}$$

$$\text{Right boundary:}\begin{cases} \rho_{\text{right}} = f_0 + f_2 + f_4 + 2(f_1 + f_5 + f_8) + \frac{1}{2c}(F_x\Delta t) \\ f_3 = f_1 + \frac{1}{6c}(F_x\Delta t) \\ f_7 = f_5 + \frac{1}{2}(f_2 - f_4) + \frac{1}{6c}(F_x\Delta t) + \frac{1}{4c}(F_y\Delta t) \\ f_6 = f_8 - \frac{1}{2}(f_2 - f_4) + \frac{1}{6c}(F_x\Delta t) - \frac{1}{4c}(F_y\Delta t) \end{cases} \tag{39}$$

$$\text{Top boundary:}\begin{cases} \rho_{\text{top}} = f_0 + f_1 + f_3 + 2(f_2 + f_5 + f_6) + \frac{1}{2c}(F_y\Delta t) \\ f_4 = f_2 + \frac{1}{6c}(F_y\Delta t) \\ f_7 = f_5 + \frac{1}{2}(f_1 - f_3) + \frac{1}{4c}(F_x\Delta t) + \frac{1}{6c}(F_y\Delta t) \\ f_8 = f_6 - \frac{1}{2}(f_1 - f_3) - \frac{1}{4c}(F_x\Delta t) + \frac{1}{6c}(F_y\Delta t) \end{cases} \tag{40}$$

$$\text{Bottom boundary:}\begin{cases} \rho_{\text{bottom}} = f_0 + f_1 + f_3 + 2(f_4 + f_7 + f_8) - \frac{1}{2c}(F_y\Delta t) \\ f_2 = f_4 - \frac{1}{6c}(F_y\Delta t) \\ f_5 = f_7 - \frac{1}{2}(f_1 - f_3) - \frac{1}{4c}(F_x\Delta t) - \frac{1}{6c}(F_y\Delta t) \\ f_6 = f_8 + \frac{1}{2}(f_1 - f_3) + \frac{1}{4c}(F_x\Delta t) - \frac{1}{6c}(F_y\Delta t) \end{cases} \tag{41}$$

Boundary Conditions for Thermal Fraction. The ABB boundary statements were adopted for the vertical hot and cold walls, as follows:

Left boundary:

$$g_1 = -g_3^* + 2w_3 T_h \tag{42}$$

Right boundary:

$$g_3 = -g_1^* + 2w_1 T_c \tag{43}$$

Meanwhile, the insulated top and bottom walls were adhered to the central difference exposition. The corresponding mathematical remarks for adiabatic horizontal walls were as follows:

Top boundary:

$$g_i^t = \frac{4}{3} g_i^{t-1} - \frac{1}{3} g_i^{t-2} \tag{44}$$

Bottom boundary:

$$g_i^b = \frac{4}{3} g_i^{b+1} - \frac{1}{3} g_i^{b+2} \tag{45}$$

where t and b typify the top and bottom boundary lattice position, respectively.

4 Results and Discussions

In the current segment of the article, the feedback from convective flow simulation using LBM was presented. Comprehensive evaluation for every combination strategies between discretization order and forcing schemes was highlighted. Initial assessment was to substantiate the adopted LBM configurations. This was performed by verifying numerical outcomes with predecessor works. Thereupon, behavior of natural convection flow within solid concrete was exhibited. Such embodiment typically reflects moderate movements of the dynamical unit. Hence, simulation was performed for a small interval of low Ra. Following this exposition is discussions regarding the effect of different LBM combination treatments, which is the mainstay of current examination.

Assimilation of the NEBB boundary conditions demands utilization of relatively moderate lattice nodes in order to maintain numerical stability [7]. This condition generally reflects in unfavorable computational requirement. Hence, in order to shed the particular handicap, we employ the graphical processing unit (GPU) architecture in our study. The parallel nature of GPU-based computing allowed significant reduction in computational cost for numerical endeavors which incorporate fairly advanced number of discrete lattice nodes. In this work, GPU programming was developed within Python environment using the open-source Cupy library [1] which allows high-performance matrix computations within GPU framework.

4.1 Numerical Validation

To confirm the efficacy of LBM performance in convective flow modeling, we compared the numerical outcomes of our exercises with published literature. Natural convection modeling within a differentially heated equilateral domain with $\mathrm{Pr} = 0.71$ and $\mathrm{Ra} = 10^4$ was selected as the benchmark configuration. Simulation was performed using dimensionless counterparts of the associated flow properties. Hence, the outcomes were exhibited as non-dimensional units. Simulation was undertaken for all combination treatments between discretization order and forcing arrangement listed in Table 2 up to steady-state condition, which was defined to be the situation in which the following dimensionless velocity and temperature remarks were satisfied:

$$\frac{\sum_{i,j}\left|\boldsymbol{u}_{ij}^{*n+1}-\boldsymbol{u}_{ij}^{*n}\right|}{\sum_{i,j}\left|\boldsymbol{u}_{ij}^{*n+1}\right|}\le\varepsilon \text{ and } \frac{\sum_{i,j}\left|\Theta_{ij}^{n+1}-\Theta_{ij}^{n+1}\right|}{\sum_{i,j}\left|\Theta_{i,j}^{n+1}\right|}\le\varepsilon. \tag{46}$$

Here, the value of ε was taken to be 10^{-9}. Figure 4 illustrates numerical outcomes for key flow parameters under the prescribed conditions. Basically, all considered combinations returned similar denouements. The flow profiles were in excellent agreement with outlines in previous investigations [6, 9, 10, 12, 14, 15].

To further substantiate the occupied configurations, critical flow properties were conscientiously evaluated against the classical computational methods, such as finite difference (FDM) [2], finite volume (FVM) [5], and finite element (FEM) [9]. Concurrently, we compared the key parameters with advanced multiple-relaxation-times (MRT) LBM arrangement [14], as well. Table 5 and Table 6 outline the corresponding assessments. Therein, parameters u and v max indicate the maximum x velocity at the vertical center line and the maximum y velocity at the horizontal center line of the flow domain, respectively. Excellent conformity was observed for all combination strategies considered, justifying the capacity of our LBM scenarios for natural convection modeling.

4.2 Convective Flow Simulation Within the Concrete

Inherent porous structure of the solid concrete allows the aqueous liquid material from the rainfall to be absorbed and reside within the geometrical pore arrangement of the

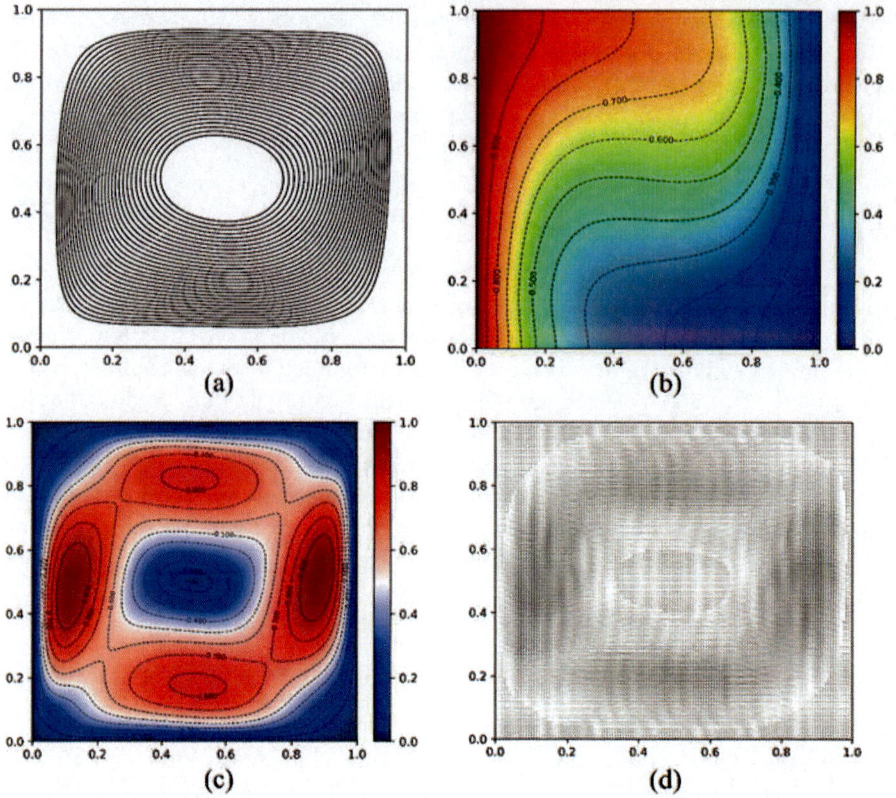

Fig. 4 LBM simulation for natural convection flow with Pr = 0.71 and Ra = 10^4. The figure demonstrates essential flow features, including **a** streamlines, **b** isotherms, **c** circulating speed, and **d** vector field profiles

Table 5 Benchmark solutions of natural convection simulation for Pr = 0.71 and Ra = 10^4, comparing present study with FDM [2] and FVM [5]

Physical parameters	Present study (combination strategy)				FDM [2]	FVM [5]
	I	II	III	IV		
Average Nu at the hot wall	2.24200	2.24199	2.24198	2.24197	2.238	2.4475
u max	16.16355	16.16313	16.16325	16.16313	16.178	16.1802
v max	19.58472	19.58428	19.58461	19.58448	19.617	19.6295
Location of u max	0.8206	0.8206	0.8206	0.8206	0.8230	0.82551
Location of v max	0.1196	0.1196	0.1196	0.1196	0.1190	0.12009

Table 6 Benchmark solutions of natural convection simulation for Pr = 0.71 and Ra = 10^4, comparing present study with FEM [9] and advanced LBM [14]

Physical parameters	Present study (combination strategy)				FEM [9]	MRT LBM [14]
	I	II	III	IV		
Average Nu at the hot wall	2.24200	2.24199	2.24198	2.24197	2.2593	2.2448
u max	16.16355	16.16313	16.16325	16.16313	16.1798	16.1833
v max	19.58472	19.58428	19.58461	19.58448	19.6177	19.6282
Location of u max	0.8206	0.8206	0.8206	0.8206	0.8235	0.8232
Location of v max	0.1196	0.1196	0.1196	0.1196	0.1195	0.1189

domain. During scorching day, the outer fraction of the concrete structure was struck by intense sunlight, constituting dissimilar temperature condition between inner and outer portion of the solid object. This situation catalysts convective movement of the ambient substance within the domain.

In this work, convective flow simulation within the geometrical pore structure of the solid ambient was conducted using the prescribed LBM strategies. The left border side of the cavity was presumed as the heated fraction of the concrete material which temperature condition was set to be 45 °C, while the opposite right margin was considered as the inner part of the structure with average temperature value of 20 °C. In order to mimic liquid phase, proper adjustment was applied to the associated Prandtl number, which value was set to be Pr = 8. The simulation was performed incorporating three modest conditions of Rayleigh number, assimilating interval of $10^2 \leq \text{Ra} \leq 10^4$. Computation was carried out using dimensionless properties. Steady-state criteria were adhered to the conditions stated in Eq. (46). Numerical responses were demonstrated as streamlines and isotherms, commensurately displayed in Figs. 5 and 6.

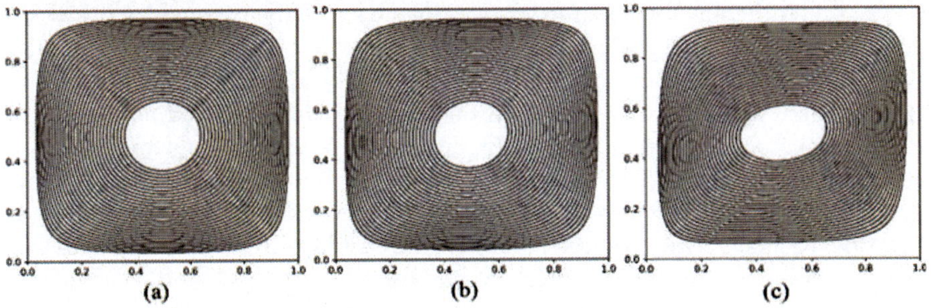

Fig. 5 Streamlines for natural convection flow involving liquid water (Pr = 8) within geometrical arrangement of the solid concrete for **a** Ra = 10^2, **b** Ra = 10^3, and **c** Ra = 10^4

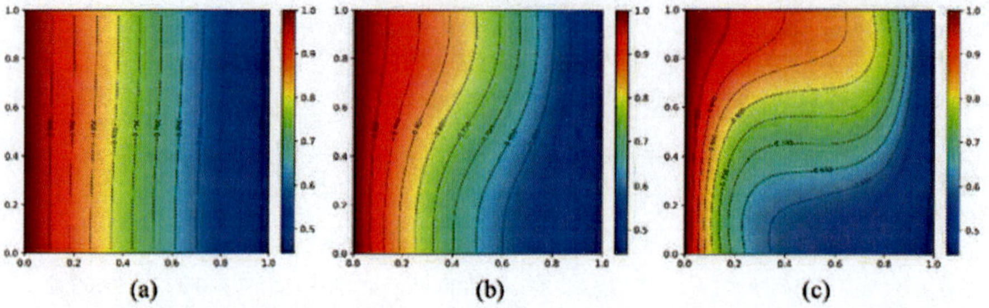

Fig. 6 Isotherms for natural convection flow involving liquid water (Pr = 8) within geometrical arrangement of the solid concrete for **a** Ra = 10^2, **b** Ra = 10^3, and **c**Ra = 10^4

The hot wall at the left boundary transfers heat to the adjacent fluid, escalating its temperature while simultaneously reducing density. Due to buoyancy assist, the warmer portion of the fluid raises toward the upper segment of the enclosure. Contrasting situation prevails at the left margin of the rectangular domain, in which the fluid particles were cooled down by the imposed cold temperature condition. Continuous ascending and descending movements of the fluid at the opposing vertical walls stimulate global circulation within the cavity, emanating convective transport. In addition to its capacity as the primary flowing unit, fluid also contributes as the media for heat transfer within the domain. Likewise, the thermal fraction also delivers influence to the fluid behavior by changing its density. Hence, natural convection comprises two interlinked physical elements that proceed synergistically within the domain of interest. These behaviors are well-captured in Figs. 5 and 6.

For Ra = 10^2, the mode of thermal dissemination was mainly dominated by conduction. This exposition is illustrated in Fig. 6a, where temperature impartment exhibits vertical profile. With escalating Ra, the fluid circulates faster, thus promoting the contribution of convection in heat transport. This situation is well-documented in Fig. 6b, c where the capacity of fluid as thermal carrier has been underlined for Ra = 10^3 and Ra = 10^4, respectively.

To evaluate the influence of discretization order and forcing schemes, numerical results from different combination strategies were carefully examined. Critical flow quantities, such as average Nusselt number (Nu) at the hot wall, maximum x-velocity at the vertical midline (u max), maximum y-velocity at the horizontal center line (v max), and computational time until convergence were selected and compared for different combination treatments. Figure 7 demonstrates the outcomes for such endeavor. A closer inspection of Fig. 7 reveals key findings of this treatise. Numerical denouements of the three physical quantities of the flow indicate that all combination strategies basically return equivalent results. The existing differences were extremely small, thus inconsequential for practical purposes. Figure 7a, b, and c captures the notable features for three key properties, including average Nu at the hot wall, u max and v max, correspondingly.

At a first glance, these behaviors seem to be in contradiction with theoretical exposition which clearly demonstrates the presence of residual remarks in the retrieved

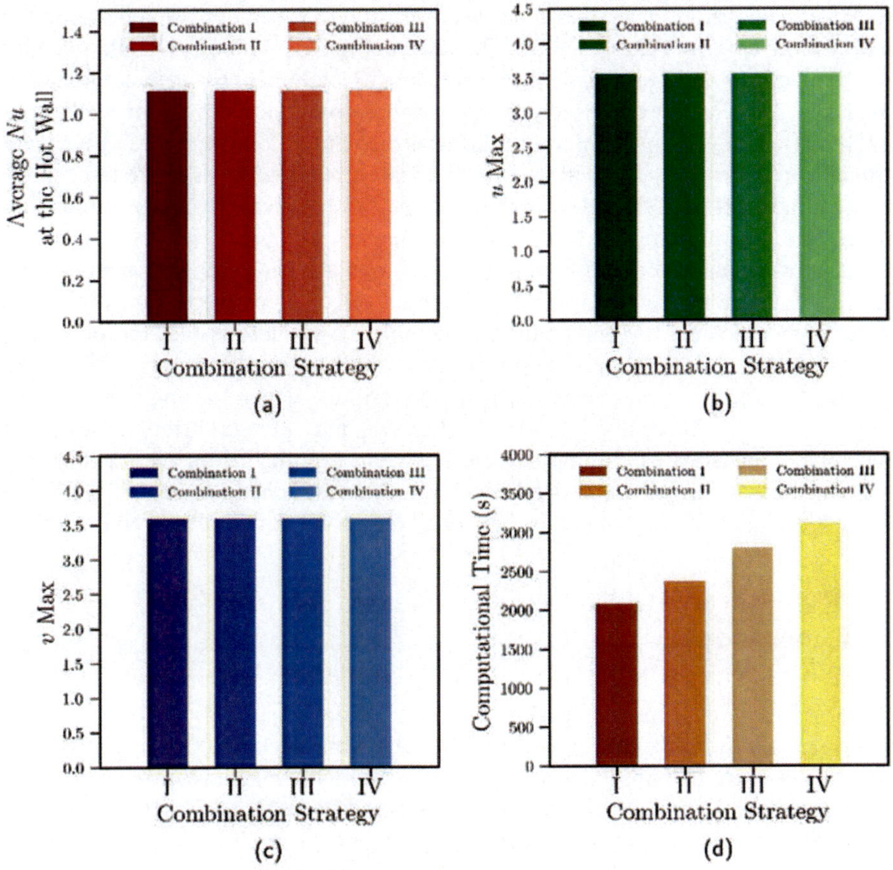

Fig. 7 Comparison of numerical outcomes among different combination strategy of discretization order and forcing scenario for key simulation parameters, such as **a** average Nu at the hot wall, **b** u max, **c** v max, and **d** computational time

continuum expressions, as were succinctly documented in Tables 3 and 4. Nevertheless, this situation is not unexpected. A possible explanation for this is that those additional terms only possess infinitesimal contribution during dynamical process of convective flow. A review of Tables 3 and 4 exposes strong dependence of residual terms on the applied buoyancy force $\boldsymbol{F}$. The employed Boussinesq approximation necessitates small deviation in fluid density throughout the simulation. Consequently, the variation of $\boldsymbol{F}$ in spatial and temporal coordinates was inconsiderable, contributing upon exiguous values for residual expressions. Hence, for convective flow, all combination strategies considered in this study were essentially producing equivalent performance. These results are in agreement with those obtained by Mohamad and Kuzmin [11] in their work.

Nevertheless, significant difference was observed in the computational demand to achieve convergence solution. As is illustrated in Fig. 7d, numerical simulation with combination I and II generally exhibits lower computational cost than combination III and IV. This was due to the increasing complexity of the accompanying mathematical remarks of the second-order discretization and Guo forcing scheme, which were implemented in strategy III and IV. Intricate matrix manipulations related to combination strategy III and IV reflect in the necessity for extra calculation procedures in the algorithm, which translate in higher computational demand.

As a final remark, the information gained from our study is capable in answering the nebulous knowledge regarding the unified effect of discretization order and forcing scenario in LBM simulation. A notable caveat here is that the outcomes of our investigation might not be transferable to other flow phenomena. Since the residual fractions in the retrieved macroscopic formulas depend strongly on the variation of external force, distinct flow mechanisms may pursue different numerical behaviors. Further research should be carried out to explore the impact of discretization order and forcing schemes in other complex transport phenomena outside natural convection. Such attempt would be beneficial to expand our feedbacks in this work.

5 Conclusions

This article has presented comprehensive deliberations regarding the conjoint effect of Boltzmann discretization order and external forcing configuration upon LBM's capacity as numerical hydrothermodynamics solver. Natural convection phenomena within a thermodynamically closed enclosure were selected as the benchmark system to test the contribution of the two critical elements. The particular configuration was accounted as the suitable prototype for the more general convective flow that occurs within internal pore fraction of the concrete structure. The D_2Q_9 and D_2Q_5 velocity schemes were adopted as the concomitant arrangements for the fluid and thermal constituents of the flowing units, respectively. Four distinct combinations of discretization order and forcing arrangements emanated from assimilation of two plausible options for both physical quantities. Numerical simulation was performed using each combination options incorporating modicum interval of low Rayleigh number.

Key physical quantities were carefully evaluated and recorded as the appropriate simulation responses. Numerical outcomes for critical physical quantities showed excellent agreement with former studies, confirming the validity of our models. The findings reveal that the four considered combinations display equivalent performances in terms of numerical accuracy and stability, despite incisive theoretical evidence of the presence of residual remarks in the restored macroscopic formulations. We argue that this behavior is due to the inherent nature of natural convection flow, which includes inconsiderable variations of the external stimulus form both spatially and temporally. This condition translates into inconsequential contribution from the residual fractions to the retrieved solutions during computational processes. Nevertheless, the four combinations demonstrated significant discrepancy in computational cost. The more sophisticated mathematical description of second-order discretization and Guo forcing scenario reflect in higher computational demand, stimulating longer simulation time for combination III and IV. Hence, when computation cost is of main concern, combinations I and II are the most felicitous strategies to be applied in natural convection simulation using LBM. Otherwise, all considered strategies were suitable to be implemented as LBM configuration in modeling convective flow system.

Acknowledgements We are indebted to Dr. Fajril Ambia of SKK Migas, Indonesia, for his fruitful suggestions upon development of the computer codes for this work. Mr. Hartono thanks the Japanese Ministry of Education, Culture, Sports, Science, and Technology (MEXT) for financial support toward the fulfillment of this study.

References

1. CuPy Homepage: https://cupy.chainer.org/. Last accessed 19 Feb 2020
2. Davis, D.V.: Natural convection of air in a square cavity a benchmark numerical solution. Int. J. Numer. Meth. Fluids **3**, 249–264 (1983)
3. Guo, Z., Zheng, C., Shi, B.: discrete lattice effects on the forcing term in the lattice Boltzmann method. Phys. Rev. E **65** (2002)
4. Guo, Z., Zhao, T. S.: Lattice Boltzmann model for incompressible flows through porous media. Phys. Rev. E **66** (2002)
5. Hortmann, M., Peric, M., Scheuerer, G.: Finite volume multigrid prediction of laminar natural convection benchmark solutions. Int. J. Numer. Meth. Fluids **11**, 189–207 (1990)
6. Jami, M., Moufekkir, F., Mezrhab, A., Fontaine, J.P., Bouzidi, M.: New thermal MRT lattice Boltzmann method for simulations of convective flows. Int. J. Thermal Sci. **100**, 98–107 (2016)
7. Krüger, T., Kusumaatmaja, H., Kuzmin, A., Shardt, O., Viggen, E.M.: The Lattice Boltzmann Method. Principles and Practice. Springer International Publishing, Switzerland (2017)
8. Luo, L.S.: Theory of lattice Boltzmann method: lattice Boltzmann models for non ideal gases. Phys. Rev. E **62**(4), 4982 (2000)
9. Mayne, D.A., Usmani, A.S., Crapper, M.: h-Adaptive finite element solution of high Rayleigh number thermally driven cavity problem. Int. J. Numer. Meth. Heat Fluid Flow **10**, 598–615 (2000)
10. Mezrhab, A., Moussaoui, M.A., Jami, M., Naji, H., Bouzidi, M.: Double MRT thermal lattice Boltzmann method for simulating convective flows. Phys. Lett. A **374**, 3499–3507 (2010)
11. Mohamad, A.A., Kuzmin, A.: A critical evaluation of force term in lattice Boltzmann method, natural convection problem. Int. J. Heat Mass Transf. **53**, 990–996 (2010)
12. Shu, C., Peng, Y., Chew, Y.T.: Simulation of natural convection in a square cavity by Taylor series expansion- and least squares-based lattice Boltzmann method. Int. J. Mod. Phys. C **13**(10), 1399–1414 (2002)
13. Silva, G., Semiao, V.: First- and second-order forcing expansions in a lattice Boltzmann method reproducing isothermal hydrodynamics in artificial compressibility form. J. Fluid Mech. **698**, 282–303 (2012)
14. Wang, J., Wang, D., Lallemand, P., Luo, L.S.: Lattice Boltzmann simulations of thermal convective flows in two dimensions. Comput. Math. Appl. **65**, 262–286 (2012)

15. Yu, P.X., Tian, Z.F.: Compact computations based on a stream-function-velocity formulation of two-dimensional steady laminar natural convection in a square cavity. Phys. Rev. E, 036703 (2012)
16. Zou, Q., He, X.: On pressure and velocity boundary conditions for the lattice Boltzmann BGK model. Phys. Fluids **9**, 1591 (1997)

Effects of Water Absorption and Retention Performance of Paper Sludge Ash in Combination with Cement to Stabilize Dredged Soil

Phan Nguyen Binh, Kimitoshi Hayano, Mochizuki Yoshitoshi, and Hiromoto Yamauchi

Abstract Due to its ability to absorb and retain water, paper sludge ash, a cinder generated from the incineration process of paper sludge, is high potential in used as complement to cement in stabilizing dredged clay soil. In this study, an attempt is made to bring out the effectiveness of combining paper sludge ash (PS ash) and cement in stabilizing dredged clay through a series of unconfined compression tests. Based on the investigated water absorption and retention performance of PS ash, a new parameter, unabsorbed water/cement ratio, W^*/C, which is the ratio by weight of the clay–water content that was unabsorbed and unretained by PS ash to cement content, was used to assess the strength development of PS ash–cement-treated clay. It is revealed that the strength of the stabilized clay is governed by the parameter W^*/C. The combination of cement and PS ash gives a treatment effect comparable to that of cement and even better for high water content clay. Experimental results show that it is high potential for PS ash–cement-treated clay to be used in soft soil stabilization.

Keywords Paper sludge ash–cement-treated clay · Water absorption and retention · Unconfined compressive strength

1 Introduction

In recycling of dredged clay, a common treatment used to achieve considerable strength is adding stabilizing agents, such as cement, which leads to economic and

---

P. N. Binh (✉) · K. Hayano
Graduate School of Urban Innovation, Yokohama National University, Yokohama 240-8501, Japan
e-mail: binh8789@gmail.com

M. Yoshitoshi
Sustainable Eco Corp, Tokyo 235-0033, Japan

H. Yamauchi
DOMI Environmental Solutions, Tokyo 299-0268, Japan

H. Hazarika et al. (eds.), *Advances in Sustainable Construction and Resource Management*, Lecture Notes in Civil Engineering 144,
https://doi.org/10.1007/978-981-16-0077-7_39

environmental issues. To overcome these issues, many studies on the use of sustainable alternative stabilizing agents and/or complements generated from industries have been conducted to reduce the input of cement, such as using fly ash, bassanite, and so on [1, 2].

Paper sludge ash (PS ash), a cinder generated from the incineration process of paper sludge, is such an alternative sustainable material for dredged soil stabilization. Due to the porous structures with many complex voids, PS ash can immediately improve stability of dredged clays by absorbing and retaining excess clay–water [3–6]. However, since PS ash does not excessively solidify as high as cement [3], the improvement effect achieved by using only PS ash is lower than using only cement. Besides, it has been found that the cementation bond strength increases as the clay-water/cement ratio decreases [7]. Therefore, it suggests that by combining PS ash and cement in stabilizing dredged clays, the excess water in clays can be reduced and therefore increasing the clay–water/cement ratio, resulting in enhancing the improvement effect of cement.

This paper deals with a fundamental experimental study to show the effectiveness of a combination of PS ash and cement in stabilizing dredged clay. A series of unconfined compression tests (UCTs) was conducted on both cement-treated clay and PS ash–cement-treated clay to access improvement effects achieved on them. The ability to absorb and retain water of PS ash was also investigated. Based on the test results, a new parameter, unabsorbed clay–water/cement, identified as the ratio by weight of clay–water content that is unabsorbed and unretained by PS ash to cement content, is proposed to describe the strength development of PS ash–cement-treated clay.

2 Materials and Specimen Preparation Method

2.1 Materials

Ao clay, commonly used in laboratory in Japan, was used in this study. The clay is categorized into CL according to the Unified Soil Classification System with the following physical properties: specific density, $\rho_s = 2.716$ g/cm^3; liquid limit, $w_L = 40.7\%$; plastic limit, $w_P = 23.7\%$; plastic index, $I_P = 17.0$. Since the purpose of this study is to stabilize dredged soft soils having water content higher than the liquid limit, the water content of Ao clay was adjusted to 81.4% ($2w_L$) before adding stabilizing agents. In this study, a commercial type of PS ash, named PS ash-based stabilizer, generated by insolubilizing heavy metals in the original PS ash, was used with ordinary Portland cement to stabilize the Ao clay. The particle size distributions of Ao clay and PS ash-based stabilizer are shown in Fig. 1.

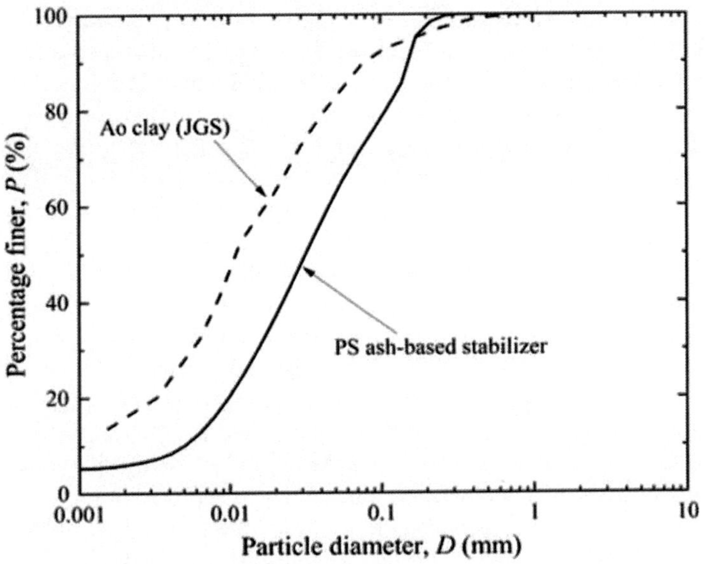

Fig. 1 Particle size distributions of Ao clay and PS ash-based stabilizer

2.2 *Water Absorption and Retention Performance of PS Ash, W_{ab}*

In this study, the water absorption and retention performance of PS ash, W_{ab}, identified as the ratio of the weight of water absorbed and retained by a given amount of PS ash to the weight of that PS ash. The value of W_{ab} was determined by the method whose principle is shown in Fig. 2a. Accordingly, an electromagnetic sieve shaker

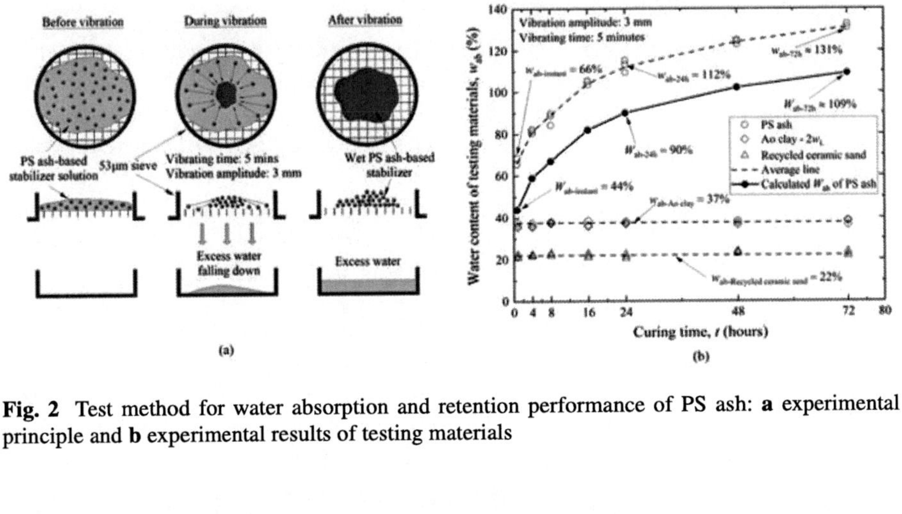

Fig. 2 Test method for water absorption and retention performance of PS ash: **a** experimental principle and **b** experimental results of testing materials

was used to generate a three-dimensional vibration in a stainless sieve to separate PS ash particles that absorbed and retained water from unabsorbed and unretained water in PS ash solution. Due to the vibration, these PS ash particles retained on the sieve, while the unabsorbed and unretained water passed through the sieve. The water content of these PS ash particles, w_{ab}, represents the water absorption and retention performance of PS ash.

In the experiments, the PS ash was first soaked in distilled water using a PS ash/water ratio of 0.25 and cured for various periods before conducting the experiment. The water absorption and retention performance of Ao clay and recycled ceramic sand were also investigated by the same method. Ao clay particles were mixed with distilled water to achieve a water content of 81.4% ($2w_L$) while recycled ceramic sand was mixed with distilled water using a dry material/water ratio of 0.25. Experimental results (Fig. 2b) indicate that the water content, w_{ab}, of retained PS ash increases with the increase in curing time. Meanwhile, the w_{ab} value of recycled ceramic sand is constant at 22% though it has no ability to absorb and retain water. This suggests that even under the impact of vibration, a certain amount of free water was still trapped in the voids between recycled ceramic sand particles due to suction. It is also found that w_{ab} value of Ao clay is constant at approximately 37%, which is higher than that of the recycled ceramic sand. This is because the Ao clay has ability to absorb and retain water by electrochemical bond of clay particles while recycled ceramic sand has not. Same performance may occur in case of PS ash so that the actual W_{ab} of PS ash should be less than the w_{ab} obtained in Fig. 2b by a corresponding amount of water retained by suction. In this study, the actual W_{ab} of PS ash at a given time is approximately calculated by subtracting a value equal to w_{ab} of recycled ceramic sand from corresponding w_{ab} of PS ash obtained in Fig. 2b and is represented by black solid symbol line in Fig. 2b.

2.3 Mixture Design Selection

It has been reported that the cementation bond strength of cement-treated clay is governed by the clay–water/cement ratio, W/C [7]. Same performance is expected to occur in case of PS ash–cement-treated clay. In order to access the differences in strength developments of PS ash–cement-treated clay and cement-treated clay, several mixture ratios were selected to achieve various values of W/C in both cases of stabilized clay. 14 cases of cement-treated clay were conducted with the value of W/C ranged from 0.25 to 3.0. To easily prepare the samples, 14 mixing cases of PS ash–cement-treated clay were adopted with differences in ratio by weight of materials as shown in Table 1. By varying the ratio by weight between stabilizers, various W/C values of PS ash–cement-treated clay were obtained. Due to the water absorption and retention performance of PS ash, the mobility of PS ash–cement-treated clay is expected to be lower than that of cement-treated clay at a same W/C. Therefore, the investigated W/C range of PS ash–cement-treated clay was selected to be wider than that of cement-treated clay.

Table 1 Mixing cases of PS ash–cement-treated clay

Case no.	1	2	3	4	5	6	7
Stabilizers/dry clay	1.5			1.25		1.0	
PS ash/cement	0.2	0.5	0.9	0.1	0.5	0.1	0.3
W/C	0.65	0.81	1.03	0.72	0.98	0.9	1.06
Case no.	8	9	10	11	12	13	14
Stabilizers/dry clay	0.75			0.5			
PS ash/cement	0.1	0.3	0.8	0.5	1	2	3
W/C	1.19	1.41	1.95	2.44	3.26	4.87	6.51

2.4 Specimen Preparation Method

The specimens of stabilized clays, used for UCTs, were prepared by tamping method based on the Japanese Geotechnical Society Standard, *JGS 0821–2009*. Dry Ao clay was first mixed with distilled water to be a homogenous slurry. Then, based on the soil treatment type, different stabilizers were added into the slurry. Whenever a stabilizer was added, the overall mixture was mixed for 10 min. After mixing, the final mixture was transferred into plastic molds in layers of 30 g. The dimensions of the plastic molds were 5 cm in diameter and 10 cm in height. To completely remove the air bubbles inside the specimens, the mold was dropped 20 times from a height of 2 cm per layer of mixture. The weight of mixture layers, times, and height of dropping were selected to ensure that the compaction effect in the tamping is the same for all mixing ratios though the initial density of each mixing case is different due to the differences in water content of the mixtures. UCT specimens then were cured in seal in room temperature (20 ± 1 °C).

Based on the determined values of W_{ab}, two mixing methods were performed in case of PS ash–cement-treated clay, including *Method 1—mixing cement and PS ash simultaneously with clay slurry* and *Method 2—adding cement one day after mixing PS ash with clay slurry*. For the *Method 2*, mixture of PS ash and clay slurry was cured in seal bag for 24 h in room temperature (20 ± 1 °C) before adding cement.

In concrete technology, the 3-day compressive strength can gain about 50% of 28-day strength, which is traditional characterized for concrete strength. In this study, the selected W/C range is not too high, same performance was expected to be obtained in stabilized clays; therefore, all specimens in this study were cured in 3 days for the first stage of study.

3 Experiment Results and Discussions

3.1 *Influence of W/C to Strength Development of Stabilized Clays*

The relation between unconfined compressive strength, q_u, and W/C of cement-treated clay, summarized in Fig. 3a, shows that at W/C greater than about 0.5, the strength of cement-treated clay is governed by W/C as Miura et al. reported [7], in which, with the decreases of W/C, the increases of unconfined compressive strength are observed. However, the trend ceases to follow at very low W/C. The compressive strength of cement-treated clay drops dramatically when the W/C is less than about 0.5. It suggests that due to the lack of water, it was difficult to mix the clay and cement thoroughly; therefore, in some cases, cement cannot completely react with water and give lower unconfined compressive strength. This performance is similar to that in concrete technology shown in Fig. 3b though the materials used as aggregates are different. The reason is that the strength developments of both concrete and clays stabilized by cement are directly governed by the cement hydration instead of the aggregate type. Based on the results shown in Fig. 3a, it can be confirmed that without fully compaction, the optimum W/C for stabilizing clay by cement is about 0.5.

A unique q_u–W/C relation of cement-treated clay at W/C of 0.5 and greater has been obtained as follows:

$$q_u = \frac{a}{\left(1 - e^{-b(W/C)}\right)^c} \text{ (MPa)} \tag{1}$$

where q_u is the unconfined compressive strength at a specific age, W/C is the water/cement ratio by weight, and a, b, and c are empirical constants. Here, $a =$

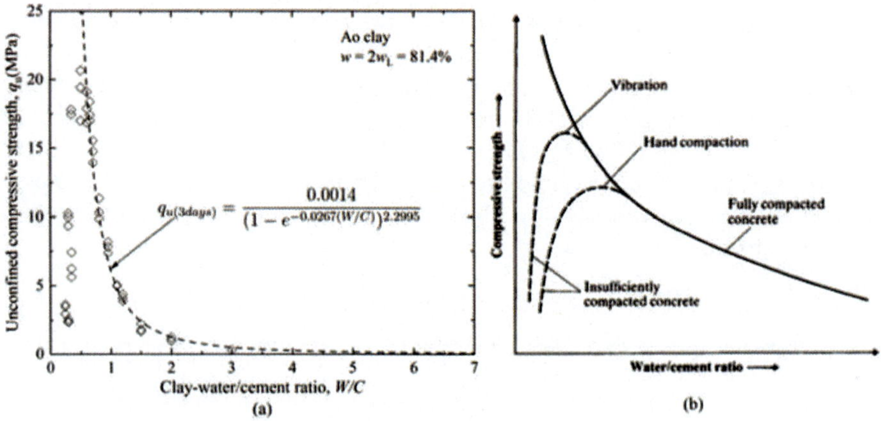

Fig. 3 Relations between compressive strength and water/cement ratio of **a** cement-treated clay and **b** concrete [8]

0.0014, $b = 0.0267$, and $c = 2.2995$ are obtained for the cement-treated clay subjected to 3 days curing.

Figure 4 summarizes the q_u–W/C relation of PS ash–cement-treated clay and compares it to the same relation of cement-treated clay represented by a dotted line. Experimental results of PS ash–cement-treated clay do not seem to follow the obtained q_u–W/C relation of cement-treated clay. At the same value of W/C, there is a large scattering in q_u even though the W/C was greater than 0.5. It reveals that the water absorbing and retaining ability of PS ash has changed the amount of clay–water that could react with cement. Therefore, to assess accurately the effect of PS ash in stabilizing clay, the water absorption and retention performance of PS ash should be considered. The results in Fig. 4 also indicate that there are differences in improvement effect between the two mixing methods. For example, q_u produced by *Method 2* is higher than that of specimens produced by *Method 1* at high W/C. Instead, the q_u of specimens produced by *Method 2* is lower than that of specimens produced by *Method 1* at low W/C, because due to the lack of water, cement cannot completely react with water.

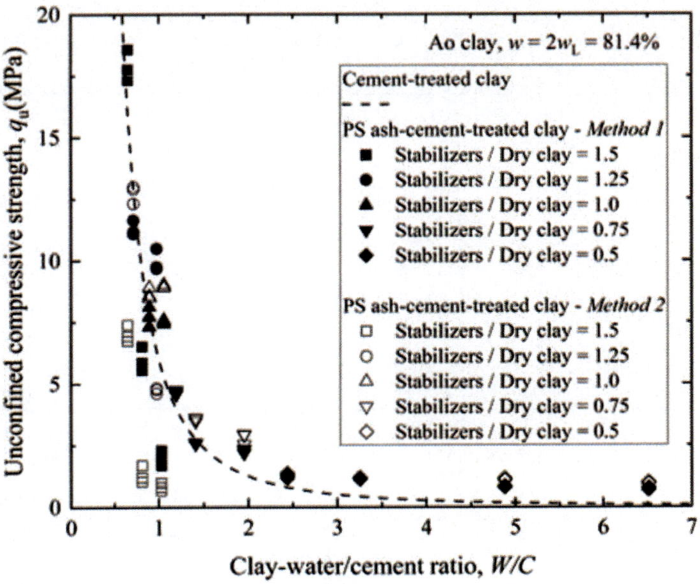

Fig. 4 q_u–W/C relation of PS ash–cement-treated clay

3.2 *Influence of Water Absorption and Retention Performance of PS Ash to Strength Development of PS Ash–Cement-Treated Clay*

Based on the investigated water absorption and retention performance of PS ash, W_{ab} (Fig. 2b), the alternative parameter W^*/C, identified as the ratio by weight of clay-water that is unabsorbed by PS ash to cement content, was used to determine the effects of absorption and retention performance of PS ash on strength development of PS ash–cement-treated clay. Without using PS ash, W^*/C is equal to W/C; therefore, W^*/C can also be used in the assessment of strength development of the cement-treated clay. The values of W^*/C corresponding to the determined W_{ab} are shown in Table 2.

Figure 5a shows the q_u–W^*/C relation of PS ash–cement-treated clay produced by both *Method 1* and *Method 2* and compares them to the same relation of cement-treated clay. It can be obtained that the q_u–W^*/C relation of PS ash–cement-treated clay is similar to those of the cement-treated clay and concrete as in Fig. 3, compared

Table 2 W^*/C of mixing cases of PS ash–cement-treated clay

Case no.	1	2	3	4	5	6	7
W/C	0.65	0.81	1.03	0.72	0.98	0.9	1.06
W^*/C ($W_{ab\text{-}instant}$)	0.56	0.59	0.64	0.67	0.76	0.85	0.93
W^*/C ($W_{ab\text{-}24\,h}$)	0.47	0.36	0.22	0.63	0.53	0.81	0.79
Case no.	8	9	10	11	12	13	14
W/C	1.19	1.41	1.95	2.44	3.26	4.87	6.51
W^*/C ($W_{ab\text{-}instant}$)	1.15	1.28	1.60	2.22	2.82	4.00	5.19
W^*/C ($W_{ab\text{-}24\,h}$)	1.1	1.14	1.23	1.99	2.36	3.08	3.81

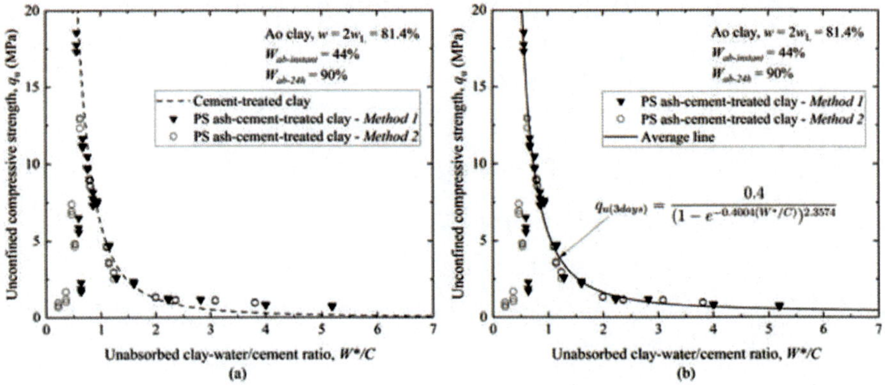

Fig. 5 q_u–W^*/C relation of PS ash–cement-treated clay: **a** compared with cement-treated clay and **b** q_u–W^*/C relation express formula

to the results shown in Fig. 4. As same as those of the cement-treated clay, the optimum compressive strengths of PS ash–cement-treated clay in both two mixing methods can be achieved at around W^*/C of 0.5. It also be obtained that there is no difference in strength development of PS ash–cement-treated clay between *Method* 1 and *Method* 2. The longer the curing time after the PS ash is added, the lower W^*/C is achieved at the time of cement addition and, therefore, resulting in higher value of q_u. However, when W^*/C was less than 0.5, the lack of water makes the mixture not to be thoroughly mixed or completely compacted, which contributed to low q_u values.

Based on the results shown in Fig. 5a, the q_u–W^*/C relation of PS ash–cement-treated clay at W^*/C of 0.5 and greater of both two methods can be expressed as one formula in the same form of Eq. 1 as follows:

$$q_u = \frac{x}{\left(1 - e^{-y(W/C)}\right)^z} \text{ (MPa)} \quad (2)$$

where q_u is the unconfined compressive strength of PS ash–cement-treated clay at a specific age, W^*/C is the ratio by weight of clay–water that is not absorbed by PS ash to the cement content, and *x, y,* and *z* are empirical constants. Here, $a = 0.4$, $b = 0.4004$, and $c = 2.3574$ are obtained for the PS ash–cement-treated clay subjected to 3 days curing. (Fig. 5b).

Figure 6 illustrates the comparison in q_u–W^*/C relations between PS ash–cement-treated clay and cement-treated clay at W^*/C of 0.5 and greater, by experimental results (Fig. 6a) and by express equations (Fig. 6b). As shown in Fig. 6, the effect of combining PS ash and cement was different from that of cement on the values of q_u depending on the range of W^*/C. At low W^*/C, q_u of the PS ash–cement-treated clay was lower than q_u of the cement-treated clay. The trend gradually reversed with the increase in W^*/C. At high W^*/C, q_u of the PS ash–cement-treated clay became

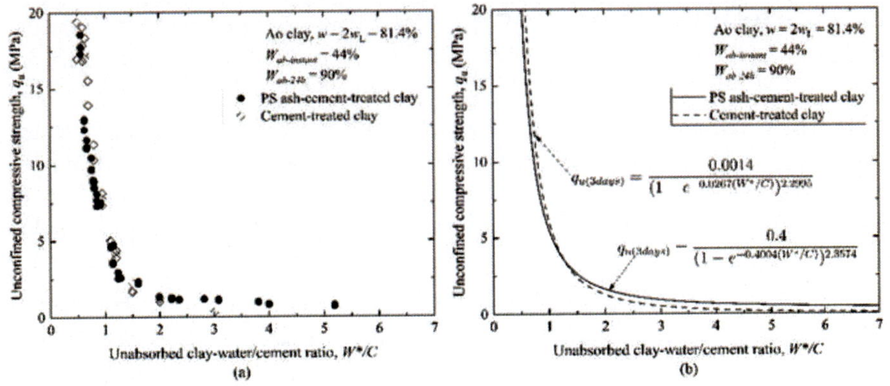

Fig. 6 Comparison in q_u–W^*/C relations of stabilized clays: **a** experimental results and **b** express equations

higher than q_u of the cement-treated clay. This phenomenon can be explained by the contribution of PS ash in stabilizing clay.

At high W^*/C, the mixture is liquid enough to thoroughly mix and therefore cement particles in the PS ash–cement-treated clay mixture could completely react with water, as well as cement particles in the cement-treated clay mixture. Additionally, cement cannot immediately react with all unabsorbed clay–water; therefore, PS ash could further reduce W^*/C of the PS ash–cement-treated clay mixture by continuing to absorb clay–water during the curing time. As a result, the total bond strength of cement in PS ash–cement-treated clay was higher than that in cement-treated clay. Furthermore, because the chemical composition of PS ash is similar to cement [9], PS ash can also react with water and contribute to the bond strength and, therefore, contributed to the overall strength of the PS ash–cement-treated clay. Consequently, at higher W^*/C values, q_u of PS ash–cement-treated clay is higher than that of cement-treated clay.

On the other hand, at low W^*/C, the mixture is not liquid enough to be mixed thoroughly. Consequently, cement particles in the PS ash–cement-treated clay mixture could not react with water as well as cement particles in the cement-treated clay mixture. Moreover, because the value W^*/C was low, most of the unabsorbed and unretained clay–water instantly reacted with cement so that PS ash could not further reduce W^*/C. Thus, the bond strength of cement in the PS ash–cement-treated clay was lower than that in the cement-treated clay and q_u of the PS ash–cement-treated clay is lower than q_u of the cement-treated clay at low W^*/C values.

In most applications, the cement content in cement-treated clays varies from 50 to 200 kg per cubic meter of clay [10], equivalent to W/C of approximately 5 to 25, corresponding to W^*/C greater than 3.0 where q_u of the PS ash–cement-treated clay is higher than that of cement-treated clay; therefore, it is high potential to combine PS ash and cement in soil stabilization in practice.

4 Conclusions

This study aims to evaluate the possibility of combining PS ash and cement for dredged clay soil stabilization. A series of UCTs was conducted on both cement-treated clay and PS ash–cement-treated clay to evaluate the strength development as well as to access the potential of using a combination of PS ash and cement in stabilizing dredged clay soil. From the results, key conclusions can be made as follows.

1. Due to the ability to absorb water of PS ash, parameter W/C cannot explain the strength development PS ash–cement-treated clay. Therefore, it is important to consider the water absorption and retention performance of PS ash to accurately assess the effect of PS ash in stabilizing clay.
2. It reveals that the newly proposed parameter W^*/C, which is the ratio by weight of clay–water that is unabsorbed by PS ash to the cement content, can explain the

strength developments of both cement-treated clay and PS ash–cement-treated clay. It indicates that the q_u–W^*/C relationships of both stabilized clays are similar.
3. Due to the ability to continue absorbing water during the curing time of PS ash, depending on the value of W^*/C, the effect of using a combination of PS ash and cement in stabilizing dredged clay may be better or worse than using only cement. At high W^*/C, PS ash–cement-treated clay has higher q_u than that of cement-treated clay. In contrast, q_u of PS ash–cement-treated clay is lower than q_u of cement-treated clay at low W^*/C.
4. Experimental results after 3 days of curing indicate that for most applications, the combination of PS ash and cement can give comparable improvement effect to cement and even better in stabilizing dredged clay soil. However, further studies will be required to confirm the statement.

References

1. Horpibulsuk, S., Phetchuay, C., Chinkulkijniwat, A., Cholaphatsorn, A.: Strength development in silty clay stabilized with calcium carbide residue and fly ash. Soils Found. **53**(4), 477–486 (2013)
2. Kamei, T., Ahmed, A., Ugai, K.: Durability of soft clay soil stabilized with recycled Bassanite and furnace cement mixtures. Soils Found. **53**(1), 155–165 (2013)
3. Mochizuki, Y.: Study on subjects and applicability for mud improvement due to mixing with paper sludge ash. Jpn. Geotech. Soc. Special Publ. **4**(5), 105–108 (2016)
4. Kumara, G.H.A.J.J., Tani, K., Mochizuki, Y.: Stability of embankments filled with dredged clay improved by paper sludge ashes. In: 11th JSCE International Summer Symposium. Tokyo (2009)
5. Mochizuki, Y., Yoshino, H., Saito, E., Ogata, T.: Soil improvement due to mixing with paper sludge ash. In: 1st International Conference on Sustainable Construction: Waste Management. Singapore (2004).
6. Mochizuki, Y., Yoshino, H., Saito, E., Ogata, T.: Effects of soil improvement due to mixing with paper sludge ash. In: Proceeding of China-Japan Geotechnical Symposium, pp. 1–8 (2003)
7. Miura, N., Horpibulsuk, S., Nagaraj, T.S.: Engineering behavior of cement stabilized clay at high water content. J. JSCE (Soils Found.) **41**(5), 33–45 (2001)
8. Neville, A.M.: Properties of concrete, 4th edn. Prentice Hall, Upper Saddle River (2000)
9. Kawai, S., Hayano, K., Yamauchi, H.: Fundamental study on curing effect and its factor on the strength deformation characteristics of PS ash-based improved soil. J. JSCE (C: Geotechnics) **74**(3), 306–317 (2018) (in Japanese)
10. Chiu, C.F., Zhu, W., Zhang, C.L.: Yielding and shear behaviour of cement-treated dredged materials. Eng. Geol. **103**(1–2), 1–12 (2009)

Fundamental Study of Adsorption Thin Layers for Safe Storage of Heavy Metal Contaminated Soil

Osamu Otsuka, Ryo Nishizato, Minoru Okuno, Naoto Watanabe, Tsutomu Matsuo, Tomomichi Obuchi, Takashi Mizobuchi, Yoichi Kitamura, Eriko Kikawa, Ken Sato, and Hidekazu Shimohara

Abstract In Japan, arsenic is contained in excavated muck. Arsenic is both highly toxic and soluble, and the Japanese government has set environmental standards to prevent arsenic from penetrating groundwater. Industries are obliged to take countermeasures to prevent the permeation of excavated muck containing arsenic into groundwater. Arsenic levels should not exceed values established in environmental standards, regardless of whether the pollution is from human or natural sources. Currently, countermeasures such as the use of adsorption layers are applied to prevent heavy metals from penetrating groundwater. The adsorption layer method involves mixing chemical reagents that adsorb heavy metals, with non-contaminated soil on site to build an adsorption layer. The contaminated soil is then embanked on the adsorption layer. If leachate containing arsenic is generated, it adheres to the adsorption layer when it flows out of the soil, preventing it from penetrating groundwater. The arsenic adsorption effect is diminished if the on-site mixing of chemicals and non-contaminated soil is incomplete. Therefore, to ensure optimal performance of the adsorption layer, the use of high-quality "pre-made adsorbent layers" is considered. It is difficult to construct pre-made adsorbent layers in a manner that corresponds to the landform. Newly developed pre-made adsorbing layers can be adapted to the shape of the landform, which reduces the material disposal rate and subsequent environmental impact. The adsorption thin layer was manufactured and exposed to hot air welding tests and construction tests. It can be easily arranged, both on flat surfaces and slopes.

Keywords Adsorption layer method · Adsorption thin layer · Arsenic · Heavy metal · Soil treatment

---

O. Otsuka (✉) · R. Nishizato · M. Okuno · N. Watanabe · T. Matsuo
KFC Ltd., 1-19 Ookuwa, Kazo, Saitama, Japan
e-mail: ootsuka.osamu@kfc-net.co.jp

T. Obuchi · T. Mizobuchi · Y. Kitamura
Toppan Printing Co., Ltd., 1-5-1 Taito, Taito-ku, Tokyo, Japan

E. Kikawa · K. Sato · H. Shimohara
ES General Laboratory Co., Ltd., 1-8-1 5jou, Nakanumanishi Higashi-ku, Sapporo, Hokkaido, Japan

H. Hazarika et al. (eds.), *Advances in Sustainable Construction and Resource Management*, Lecture Notes in Civil Engineering 144,
https://doi.org/10.1007/978-981-16-0077-7_40

1 Introduction

Arsenic (As) is a highly toxic and soluble element. Due to this, arsenic in contaminated soil can potentially permeate and diffuse into groundwater by leachate such as rainwater, posing increased hazards to health. In Japan, arsenic is found in excavated mucks across the country [1]. The Japanese government has set environmental standards to limit the adverse effects of arsenic on public health. Even if arsenic originates from natural sources, companies have an obligation to take measures to prevent the permeation of arsenic in the excavated soil into groundwater. The adoption of countermeasures against contamination by heavy metals has resulted in an increase in the cost of tunnel construction, which generates a large amount of excavated soil.

The current countermeasures against heavy metal (e.g., arsenic) contamination include water shielding containment and adsorption layers. Water shielding containment is a method to separate the contaminated soil and water by enclosing the contaminated soil with an impermeable membrane to prevent leaching of arsenic from contaminated soil by rainwater. An impermeable membrane, which is a factory product, is laid at the site, and separate membranes are joined so that water does not leak from the gap between the membranes (see Table 1). Heavy machinery is not required for its construction. If rainwater enters from the seams of the impermeable membrane, it is expected that contained heavy metals could dissolve from contaminated soil. Therefore, leachates must meet environmental standards before

Table 1 Characteristics of three different countermeasures against heavy metal contamination (○: Advantages, ×: Disadvantages)

	Water shielding containment	Adsorption thin layer	Adsorption layer method
Pre-made	○	○	×
Construction by human power	○	○	×
Construction for slope	○	○	△
Leachate discharge to the environment	×	○	○
Overview	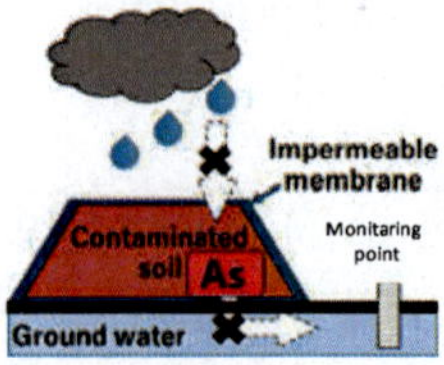	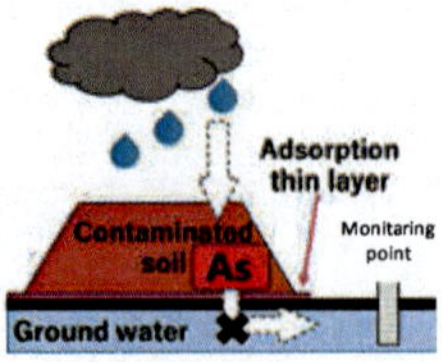	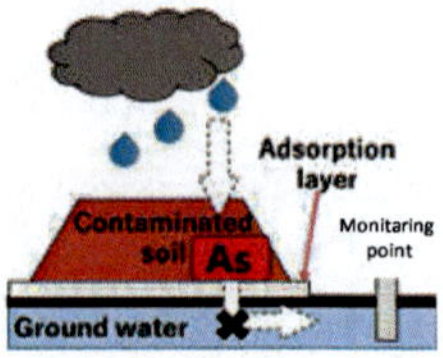

being discarded. Consequently, water shielding containment requires a leachate purifier system. The installation and operation of a leachate purifier system entail cost increases. To prevent cost increases, an adsorption layer method was developed that does not require purification of the leachate.

In the adsorption layer method, a layer is constructed from a reagent which adsorbs the relevant heavy metal(s) [2, 3]. Heavy metals in the leachate can be adsorbed by the adsorbent by passing the leachate through the adsorption layer. The leachate that has passed through the adsorption layer can then be disposed of without further treatment because the heavy metals have been removed [4]. To build an adsorption layer, the adsorbent is mixed on-site with non-contaminated soil, by using heavy machinery. Contaminated soil is then embanked on top of the adsorption layer. Poor mixing of non-contaminated soil and adsorbent reduces the adsorption capacity of the adsorption layer. In recent years, an adsorption sheet method has been proposed to ensure the quality of the adsorption layer. In this method, a pre-made adsorption layer is used [5, 6]. This adsorption sheet method is expected to combine the advantages of water shielding containment and the adsorption layer method. However, it is difficult to construct the adsorption sheet in a way that conforms to the landform, and the material loss rate tends to be high. Finally, the heavy metal adsorption thin layer (adsorption thin layer) was developed as a new "pre-made adsorption layer" that could be cut and joined to adapt its shape to the landform. This feature of adapting to the landform results in a lower material disposal rate, contributing to a reduction in environmental impact. The adsorption thin layer was manufactured with a zero-valent iron-based adsorbent as a "pre-made adsorption layer," and joining and construction of the layer were carried out in this study.

2 Overview of the Heavy Metal Adsorption Thin Layer

The adsorption thin layer comprises of hydrophilic nonwoven fabric layered on a uniformly packed layer containing a zero-valent iron adsorbent (3 kg/m^2) (Fig. 1).

Since the amount of adsorbent filled affects the arsenic adsorption capacity of the adsorption thin layer, we measured the filled adsorbent weight (Fig. 2). All the measured products contained more than the guaranteed minimum amount of adsorbent. Further, the $+3\sigma$ and -3σ deviations from the average weight were calculated to be 3.8 kg and 3.1 kg, respectively. The latter value is still greater than the guaranteed minimum amount of adsorbent. Statistically, the probability that the weight of adsorbent in the adsorption thin layer lies in the range $\pm 3\sigma$ from the average is 99.7%. Therefore, the probability that the weight of adsorbent in the adsorption thin layer is greater than the guaranteed minimum will be greater than 99.7%. These results indicate that the process of manufacturing adsorption thin layers produces adsorption layers of high quality.

A zero-valent iron-based adsorbent with high arsenic adsorption capacity was selected as the filler. Zero-valent iron is expected to have insolubilization effects on both As (III) and As (V) by adsorption ($H_2AsO_3^- + Fe\text{–}OH \rightarrow FeH_2AsO_3, HAsO_4^{2-}$

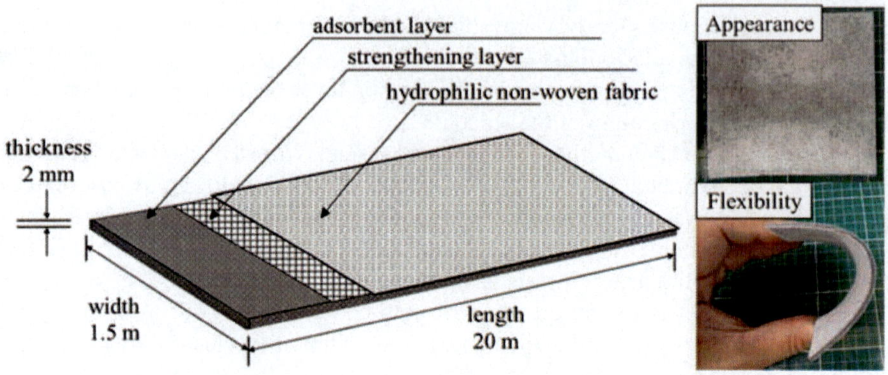

Fig. 1 Composition, appearance, and flexibility of a heavy metal adsorption thin layer

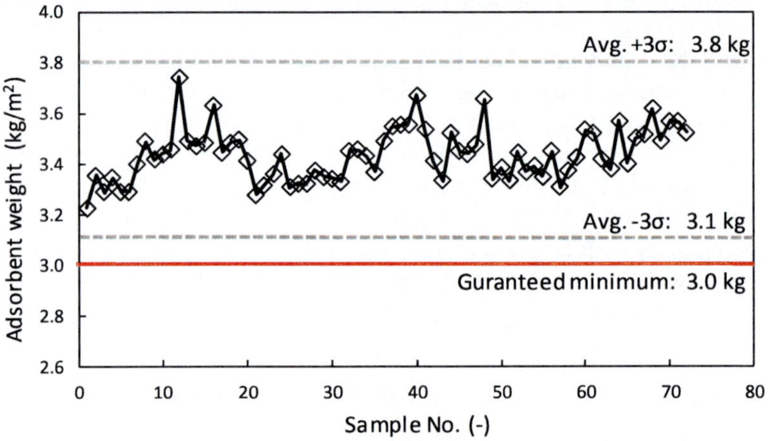

Fig. 2 Variation of the amount of adsorbent in products manufactured with an adsorption thin layer

$+ Fe\text{–}OH + H^+ \rightarrow FeHAsO_4^- + H_2O$) and co-precipitation ($Fe^{3+} + HAsO_4^{2-} \rightarrow FeAsO_4 + H^+$) [7–9]. Furthermore, because the adsorption layer is very thin (with a total thickness of only 2 mm), it can be cut with a utility knife and is very flexible.

3 Materials and Methods

3.1 Up-Flow Percolation Test

Up-flow percolation tests were performed with reference to ISO 21268-3 [10]. A cylindrical column of 100 mm in height and 50 mm in diameter was packed with

285 g of silica sand (grain size: No. 5). The adsorption thin layer was cut with a utility knife to a diameter of 50 mm and placed on the upper part of the column. A concentration of 0.03 mg/L of As (V) standard solution was passed through at a flow rate of 98 ml/h. A solution having a liquid–solid ratio of 0.2–8.4 was passed through the adsorption thin layer and collected as a test solution. At the same time, an As (V) standard solution was also collected as a test solution. The concentration of arsenic (V) in the test solution was determined by ICP-MS (7800 ICP-MS; Agilent technologies).

3.2 *Hot Air Welding Test*

Adsorption thin layers were cut to a width of 300 mm and length of 1.5 m as test specimens. In order to join two test specimens, the ends were overlapped by 100 mm and joined using a hot air automatic welding machine (Twiny T Leister Technologies) (Fig. 4a). The welding speed was 1.5 m/min and the welding temperature was set between 400 and 500 °C in the laboratory at ambient temperature of 20 °C. The tensile strength of the weld was measured in accordance with test methods for geotextiles (JIS L1908).

3.3 *Construction Test*

An outline of the adsorption thin layer construction method is shown in Fig. 6. Protective mats with a thickness of 10 mm were laid above and below the adsorption thin layer to prevent damage from external forces. The adsorption thin layers were joined by hot air welding.

3.4 *Heavy Machinery Running Test*

To simulate the uneven conditions on a real surface, gravel was sprinkled on the ground. The adsorption thin layer and protective mats were constructed on a flat surface, and 500 mm of protective soil was filled. Twenty reciprocates were run above the protective soil using a general-purpose small turning machine (311 D RR; Caterpillar Japan LLC) weighing approximately 12,000 kg. After the runs, the protective soil was removed to check whether there was any damage to the adsorption thin layer.

4 Results and Discussion

4.1 Up-Flow Percolation Test

An up-flow percolation test was performed to examine the arsenic adsorption capacity of the adsorption thin layer in conditions simulating heavy rain. The transition of arsenic concentration is shown in Fig. 3. The arsenic concentration in the inflow solution was constant at 0.03 mg/L from the beginning to the end of the test. On the other hand, the arsenic concentration in the outflow solution (which had passed through the adsorption thin layer) was consistently less than 0.01 mg/L. The standard arsenic concentration allowed in groundwater, as determined by the Japanese government, is 0.01 mg/L. Therefore, it was suggested that the arsenic that was initially present in the solution that had passed through the adsorption thin layer was removed in sufficient amounts, and the leachate can therefore be discarded without treatment. In the up-flow percolation test, the total amount of arsenic introduced into the column was 67.2 μg, of which 53.2 μg was adsorbed on the adsorption thin layer. The adsorption rate was calculated to be 79%.

To simulate heavy rain, up-flow percolation tests were carried out with the inflow rate being four times larger than that of the ISO21268-3. If the flow rate is reduced to simulate light rain, the time of contact of the arsenic with the adsorbent in the adsorption thin layer is increased and the adsorption performance is expected to improve. Zero-valent iron also can adsorb and insolubilize various heavy metal ions (e.g., Se, Pb, and Cr) by chemical reduction and co-precipitation [11, 12]. Therefore, it is reasonable to expect that the adsorption thin layer can adsorb heavy metals as well.

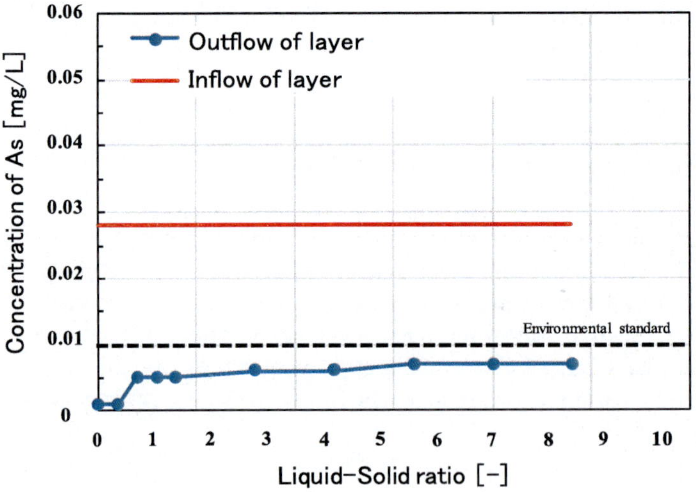

Fig. 3 Arsenic concentration in eluate

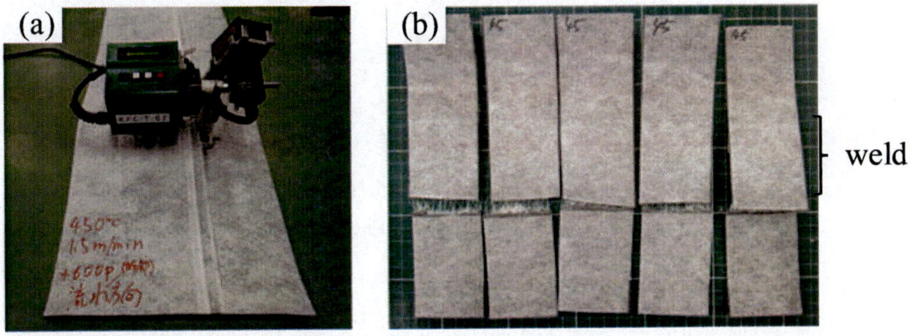

Fig. 4 Photographs of **a** hot air welding test; **b** weld tensile test

4.2 Hot Air Welding Test

When adsorption thin layers are not adequately joined when they are being laid, the leachate may pass between separate parts of the adsorption thin layer and penetrate groundwater. Therefore, it was verified whether separate parts of the layer had been properly joined by hot air welding. It was found that the test specimens were welded properly in the laboratory at a welding temperature of 400 °C or higher. Upon examining the tensile strengths of the welds, it was found that all test specimens were securely joined because the base material broke in preference to the weld (Fig. 4b).

However, the tensile strength decreased remarkably once the temperature exceeded 450 °C. The tensile strength of the weld was strongest when welding was performed at 450 °C; the average tensile strength was 380 N/50 mm (Fig. 5).

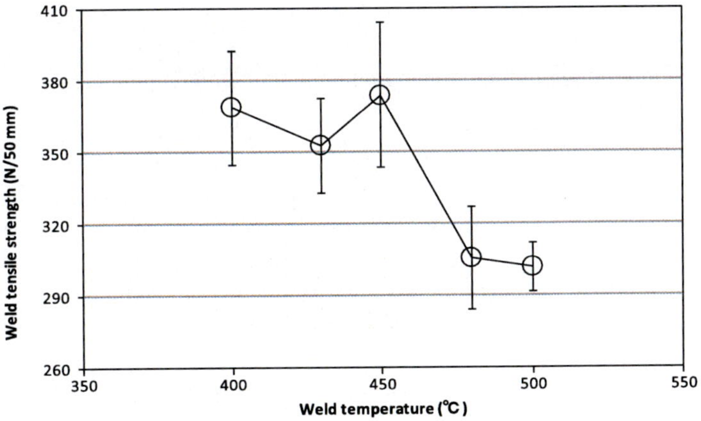

Fig. 5 Effect of welding temperature on tensile strength

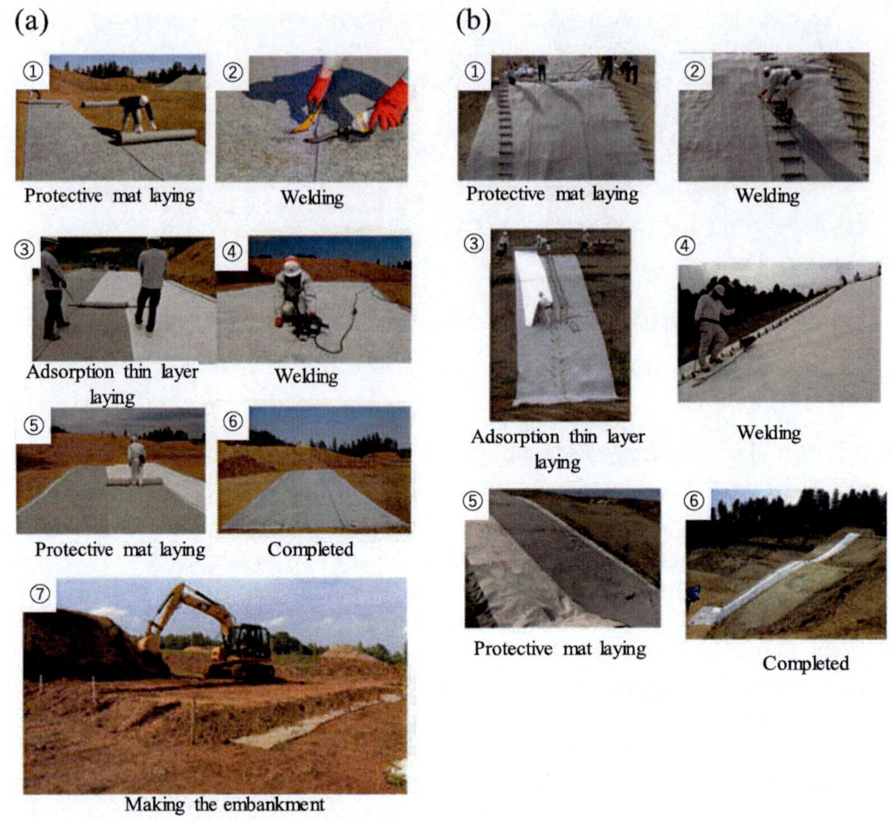

Fig. 6 Photographs of construction test for **a** flat surface and **b** slope

4.3 Construction Test

It was verified whether the produced adsorption thin layer can be constructed without breakage. The cone index of the ground at the construction site was found to be 1,200 kN/m^2 or more by a cone penetration test. The adsorption thin layer was laid according to the construction method described. On a flat surface, an adsorption thin layer measuring 114 m^2 (5.7 × 20 m) can be manually constructed without breakage (Fig. 6a). In addition, the adsorption thin layer was safely installed without injury or accident. Adsorption thin layers can also be welded at the construction site (Fig. 6a ④).

After testing the layer on a flat surface, it was laid on a slope with an angle of 21°–46°. Even on a slope, 119 m^2 (7.0 × 17 m) of adsorption thin layer can be installed manually without breakage (Fig. 6b). The adsorption thin layer was also safely installed in slope conditions without injury or accident. Adsorption thin layers can be welded in slope conditions at the construction site (Fig. 6b ④). The adsorption

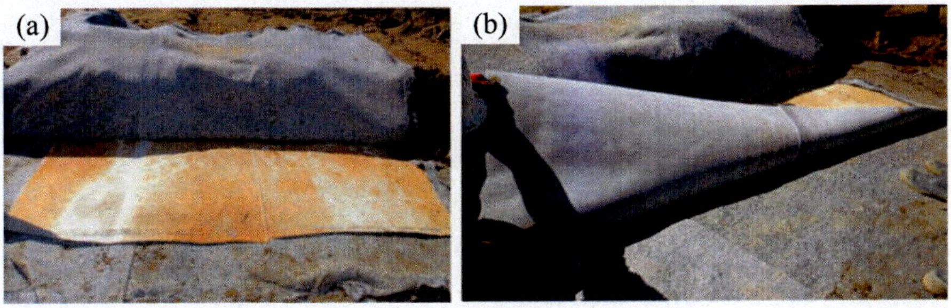

Fig. 7 Photographs of heavy equipment running test for **a** the adsorption thin layer and **b** mats at the bottom

layer method is difficult to construct on a slope. It is expected that adsorption thin layers will allow the construction of adsorption layers even on terrain on which earlier methods have been difficult to apply.

4.4 Heavy Machinery Running Test

In making an embankment of arsenic contaminated soil, a heavy machine reciprocates on the protective soil. Heavy machinery running tests were conducted to confirm that the adsorption thin layer was not damaged by the action of the heavy machine. After the adsorption thin layers were constructed on a flat surface, 500 mm of soil was piled up to protect the mat. The protective soil subsided to a maximum of 205 mm under heavy machine running. However, the heavy machinery did not directly touch the protective mat because the subsidence was less than the protective layer height. The results are shown in Fig. 7. The surface of the adsorption thin layer was only discolored to brown by moisture in the soil, and neither the adsorption thin layer nor the protective mat was damaged. From this result, it was indicated that the adsorption thin layer had sufficient strength to sustain actual use.

5 Conclusions

The adsorption thin layer was devised as a countermeasure against heavy metal contamination with low environmental impact. A high-quality adsorption layer was successfully processed into a sheet by factory manufacture. It can reduce the material disposal rate as it can be cut and joined. Up-flow percolation tests confirmed that arsenic was adequately removed from the solution passing through the adsorption thin layer. Therefore, the arsenic adsorption capability of zero-valent iron was verified, even when the adsorption layer was processed into a sheet. Furthermore, it was confirmed by construction tests that it was possible to install the layers on both flat surfaces and slopes. From the above results, the heavy metal adsorption thin layer method is expected to become a new heavy metal contaminated soil countermeasure with low environmental impact. The adsorption thin layer has an arsenic removal capability equal to or greater than that of the adsorption layer method, while retaining the ease of construction of the impervious membrane.

Acknowledgements We are very grateful to Dr. Hideki Yoshida, Associate Professor at Muroran Institute of Technology for his valuable operation of the laboratory tests. We would also like to offer our thanks to Ms. Ayako Gomi for her technical assistance with in creating the drawings.

References

1. Imai, N., Terashima, S., Ohta, A., Mikoshiba, M., Okai, T., Tachibana, Y., Togashi, S., Matsuhisa, Y., Kanai, Y., Ueoka, A., Taniguchi, M.: Geochemical map of Japan. Geol. Survey Jpn., 209 (2004)
2. Bruce, A.M., Mathew, L.H., Christopher, A., Jory, A.Y.: Arsenic (III) and arsenic(V) reactions with zero valent iron corrosion products. Environ. Sci. Technol. **36**, 5455–5461 (2002)
3. Jia-lin, M., Inui, T., Katsumi, T., Takai, A., Kuninishi, K., Hayashi, S.: Performance of sorption layer using Ca/Mg immobilizing agent against natural contamination. In: Proceedings, 10th Asian Regional Conference of IAEG (2015)
4. Tabelin, C., Igarashi, T., Yoneda, T., Tamamura, S.: Utilization of natural and artificial adsorbents in the mitigation of arsenic leached from hydrothermally altered rock. Eng. Geol. **156**, 58–67 (2013)
5. Daisuke, I., Daiki, T., Toshihumi, I., Takushirou, T. and Kazuhiro, M.: Performance of heavy metal adsorption mat against soil with naturally occurring heavy metals. In: Abstract, Japan Society of Civil Engineers, Miyagi, Japan, pp. 391–392 (2016)
6. Pimolpun, K., Pitt, S.: The study of competitive adsorption of heavy metal ions from aqueous solution by aminated polyacrylonitrile nanofiber mats. Energy Procedia **56**, 142–251 (2014)
7. Sasaki, K., Nakano, H., Wilopo, W., Miura, Y., Hirajima, T.: Sorption and speciation of arsenic by zero-valent iron. Colloids Surf. A: Physicochem. Eng. Aspects **347**(1–3), 8–17 (2009)
8. Eljamal, O., Sasaki, K., Hirajima, T.: Numerical simulation for reactive solute transport of arsenic in permeable reactive barrier column including zero-valent iron. Appl. Math. Model. **35**(10), 5198–5207 (2011)
9. Wilopo, W., Sasaki, K., Hirajima, T., Yamanaka, T.: Immobilization of arsenic and manganese in contaminated groundwater by permeable reactive barriers using zero valent iron and sheep manure. Mater. Trans. **49**(10), pp. 2265–2274 (2008)
10. ISO 21268-3:2019 Soil quality—Leaching procedures for subsequent chemical and ecotoxicological testing of soil and soil-like materials—Part 3: Up-flow percolation test (2019)
11. Das, S., Lindsay, M.B.J., Essilfie-Dughan, J., Hendry, M.J.: Dissolved selenium (VI) removal by zero-valent iron under oxic conditions: influence of sulfate and nitrate. ACS Omega **2**(4), 1513–1522 (2017)
12. Azzam, A., El-Wakeel, S., Mostafa, B., El-Shahat, M.: Removal of Pb, Cd, Cu and Ni from aqueous solution using nano scale zero valent iron particles. J. Environ. Chem. Eng. **4**(2), 2196–2206 (2016)

Influence of Sand on the Behavior of Buffer Material Used for the Nuclear Waste Disposal

Padavala Hari Krishna and Adla Nandini

Abstract Clay–sand mixtures are being planned for use as buffer materials for high-level nuclear waste disposal. These mixtures are attracting greater attention as buffer materials because they offer properties of low volumetric shrinkage and high swelling. The influence of sand on the behavior of clay–sand mixtures was investigated by many researchers through a series of laboratory experiments. The study illustrates the role of coarser fraction in controlling swelling and shrinkage of clay–sand mixtures. In the present study, experiments were conducted using oedometer tests to investigate swelling and shrinkage properties at different sand contents by experimentation. The clay–sand mixtures were prepared with sand contents of 0%, 10%, 30%, and 50% by weight of clayey soil. The study showed a decrease in swelling potential and volumetric shrinkage with increase in the amount of sand. An increase in dry unit weight and a decrease in respective moisture content by an increase in the amount of sand were observed in the compaction tests. In addition, attempts were also made to find the influence of varying proportions of coarse sand and fine sand mixtures on swelling and shrinkage properties of these mixtures.

Keywords Buffer material · Clay–sand · Atterberg limits · Swelling potential · Volumetric shrinkage

1 Introduction

Disposal of high-level radioactive nuclear waste in deep geological repository aims to provide a long-term and permanent waste management solution. International consensus (IAEA 2003) has proposed geological disposal as the safest option [1]. In general, isolation of the waste from biosphere depends on passive multibarrier systems. These typically consist of the natural geological barrier provided by the repository host rock and an engineered barrier system. As a part of engineered barrier

P. H. Krishna (✉) · A. Nandini
Department of Civil Engineering, National Institute of Technology Warangal, Warangal, Telangana, India
e-mail: phari@nitw.ac.in

H. Hazarika et al. (eds.), *Advances in Sustainable Construction and Resource Management*, Lecture Notes in Civil Engineering 144,
https://doi.org/10.1007/978-981-16-0077-7_41

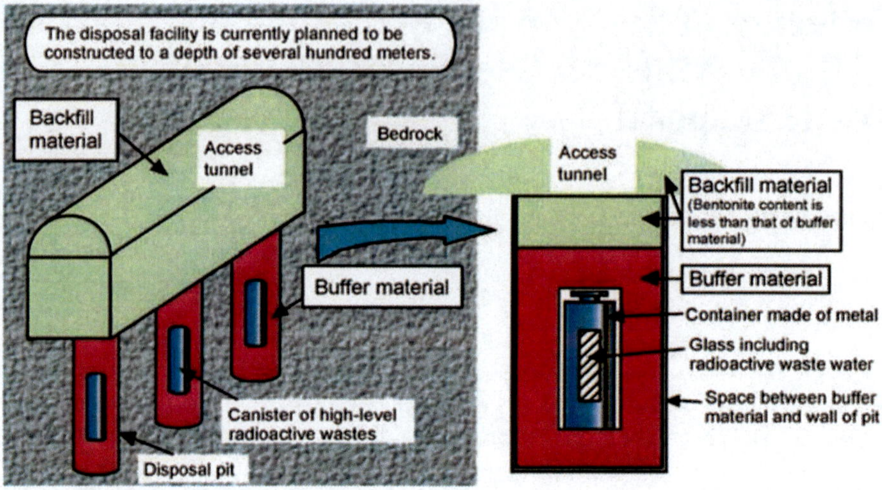

Fig. 1 An example of disposal facility and pit for high-level radioactive wastes in Japan (vertical emplacement type, after Ogata et al. [2])

system, bentonite and bentonite–sand mixtures are used as buffer material in between waste canister and host rock. Black cotton soil has characteristics of self-patching and restoration, so it is suitable to be used as an effective barrier material for isolating the nuclear wastes from groundwater as shown in Fig. 1 [2]. But to reduce the problem of shrinkage cracks and to increase the strength and stability, bentonite is blended with coarser particles for use in nuclear waste disposal units [3, 4].

The buffer material should primarily be capable of maintaining its integrity at elevated temperatures (up to 150 °C) and in the saline groundwater that may exist in disposal vault [5]. On drying, it should not shrink away from the repository host rock. In addition, its bearing capacity must be sufficient to support the waste container and to minimize settlement [3]. The dry density of in situ compacted pure clayey soils can be low; however, by mixing sand to these soils, the attainable dry density can be increased. In general, unless compacted to very high densities (that is greater than 2.0 Mg/m^3 pure Na bentonite) pure clays may not meet the above requirements [6]. Generally, the dry density of pure clays will be around 1.3 Mg/m^3, when compacted in situ. Such high densities could be attained by pre-compacting the clay with sand to improve the strength properties and thus their dry density can be increased [7, 8].

Apart from this, as the groundwater resets after construction of the disposal repository, the buffer material will swell due to water uptake. Swelling may be desirable as in many hydraulic containment applications such as: (1) soil barriers in landfill liner and cover systems and (2) barriers for disposal of radioactive waste [9, 10]. And experimental studies on the swelling capacity of sodium bentonite and its sand mixtures have been taken by many researchers [11–14]. Swelling and volumetric shrinkage due to varied temperatures are the main controlling factors in the stability of the underground structures and affect the coefficient of permeability [15].

Therefore, prediction of the swelling potential of black cotton soil–sand mixtures due to water uptake is an important technical issue and volumetric shrinkage due to elevated temperature both in theory and in engineering application is to be resolved for the deep disposal of nuclear waste. As such the shrinkage and swelling characteristics are not well addressed with respect to the coarse sand and fine sand content separately.

Toward that goal, the present study is aimed at investigating the swelling potential and volumetric shrinkage at maximum dry densities of black cotton soil–sand mixtures with varying amount of fine and coarse sand contents (0, 10, 30, and 50%) through experimentation.

2 Materials and Testing Program

2.1 Materials

The material used in this study to represent the buffer material was a laboratory mixture of black cotton soil and graded silica sand. The black cotton (BC) soil used in this study has been collected from the northern side of NIT-Warangal campus [16]. The soil is collected by open excavation, from depth of 1–1.5 m from natural ground level. This soil has highly expansive Montmorillonite mineral causing it to have a very high potential for swelling and shrinkage. The properties of the soil used in the present investigation are given in Table 1.

Table 1 Properties of the natural soil

Parameter	Value
Specific gravity	2.658
Gravel (%)	1
Sand (%)	4
Silt (%)	27
Clay (%)	68
Liquid limit (%)	104
Plastic limit (%)	22
Shrinkage limit (%)	13
IS classification	CH
Maximum dry density (Mg/m^3)	1.53
Optimum moisture content (%)	24.8
Free swell index (%)	209

Locally available Godavari river coarse sand and fine sand are taken for this study. The mixtures were prepared with the proportion of coarse and fine sand ranging in 0, 10, 30, 50% of total weight of mixture.

2.2 *Testing Program*

Atterberg Limits

Liquid limit (fall cone test) [17], plastic limit test [17], shrinkage limit test [18], specific gravity test [19], compaction test [20], and free swell index [21] were conducted according to the IS: 2720 standards for each mixture.

Swell Potential and Volumetric Shrinkage

The swell potential of each mixture was performed separately as per IS: 2720.The term swell potential (Sp) is generally used to indicate the amount of vertical swell (expressed as percent of initial sample thickness) obtained under a particular surcharge (usually 1 psi or 6.9 kPa), where Sp is swelling potential in percent. Swelling potential or volume change is defined as the ratio of increase in thickness to the initial thickness of the soil sample compacted at optimum moisture content (OMC). It was measured in the conventional oedometer cell as shown in (Fig. 2) performed on compacted soil samples. The soil sample was compacted at OMC and placed in a consolidation ring and was soaked under a token surcharge of 6.9 kPa, [22] so that the sample undergoes free swell. The prepared specimen was inundated with water and full swell attained within 72–96 h. The percentage swell of the specimens during soaking was measured as swell potential.

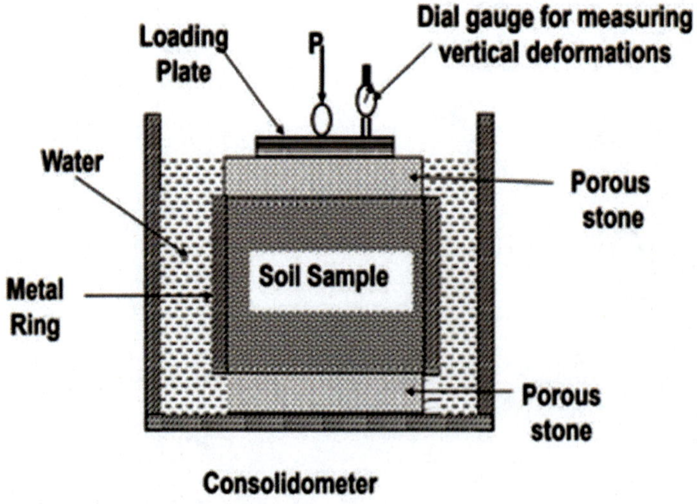

Fig. 2 Consolidometer to determine swelling potential

Similarly, volumetric shrinkage test was performed as per IS 2720—Part6, 1972. Volumetric shrinkage is defined as the decrease in volume, expressed as a percentage of the soil mass when dried, of a soil mass when the water content is reduced from a given percentage to the shrinkage limit.

A positive value represents volume shrinkage while a negative value represents volume expansion. A high positive value represents high volume shrinkage, which may cause sink mark or void.

In the present study, the clay–sand mixture is oven dried and sample is prepared at its optimum moisture content and is statically compacted in a mold having diameter 6 cm and height 2 cm in UTM (Fig. 3). Then, the sample is removed from mold and kept under room temperature. And the next day, these samples are transferred to the oven.

Thus, the sample is dried in oven until the weight of specimen is constant. Thus, the samples are taken out of oven and change in volume of specimen due to drying of the soil mixture after it reached equilibrium to its original volume on the first day of preparation of samples is volumetric shrinkage of the specimen. Average of volumetric shrinkage of both samples is considered for accuracy.

Fig. 3 Oedometer mold placed in UTM

3 Results and Discussions

3.1 Atterberg Limits

Liquid Limit: The liquid limit of clay–sand mixtures was found and their variations were summarized in Table 2. As can be observed from this table, it is found that liquid limit decreases with increase in sand content. This may be mainly due to the fact that the "liquid limit" is considered as purely as the water holding capacity of the soil, then the surface area plays a primary role. The inter particle forces in the non-clay fraction are negligible, and the individual particles in this fraction have negligible amounts of adsorbed water associated with them. Therefore, with an increase in the percentages of sand, lower values of liquid limit are observed.

Plastic Limit: The plastic limit values of the clay–sand mixtures are also given in Table 2. As can be observed from this table, it is clearly seen that plastic limit decreases with increase in sand content. This may be mainly due to decrease in the plasticity nature of mixture as particular amount of clay is to be replaced by sand grains, having no plasticity.

Shrinkage Limit: When the soil is essentially with no sand sized particles, maximum shrinkage limit was obtained. With the inclusion sand particles, the relative grain size distribution gets improved, resulting in decreased shrinkage limit. Researchers [23] have proved that shrinkage limit of a soil is the result of packing phenomenon and is primarily controlled by the relative grain size distribution of the soil. With proper proportioning of the coarser fractions in the mix, a lower shrinkage limit can be obtained which can be even less than that obtained for cohesive soils (the minimum value of the shrinkage limit is obtained for 10% sand and 90% clay mix which is less than that for native clay soil. From Table 2, it is observed that shrinkage limit with the addition of 10% sand to BC soil is decreased by about 8% and there after it is clearly seen that increase in the proportion of the coarser fraction results in an increase in the shrinkage limit. Thus, the shrinkage limit is increased by 15% with

Table 2 Atterberg limits of BC soil with varied proportions of fine and coarse sand

S. No.	Properties	BC soil	BC soil + 10% FS	BC soil + 30% FS	BC soil + 50% FS	BC soil + 10% CS	BC soil + 30% CS	BC soil + 50% CS
1	Liquid limit (%)	104	84	62	50	88	65	55
2	Plastic limit (%)	22	21.5	19.5	18	–	–	–
3	Plasticity index	82	62.5	42.5	32	–	–	–
3	Shrinkage limit (%)	13	14	15	16	–	–	–

the addition of 30% fine sand to BC soil and increased by 23% with the addition of 50% fine sand to BC soil.

3.2 Compaction Studies

Considering the use of compacted clay–sand mixtures as buffer materials for waste disposal, studying compaction characteristics has a great significance. Tests are conducted for varying proportions of fine sand and coarse sand with black cotton soil and the results are given in Table 3. As can be observed from this table, the maximum dry density value of clay–sand mixture is increasing with an increase in the percentage of sand, which is in conformity with the observations by other researchers [24].

Similarly, the optimum water content of the specimen decreases with increasing fine sand content, as the surface area of black cotton soil is more than that of clay–fine sand mixtures, as it requires more water to lubricate during compaction to attain a considerable dry density. It is observed that the maximum dry unit weight of the specimen increases with increase in size of sand due to higher specific gravity of coarse sand grains.

Fine sand mixtures have high optimum water contents compared to coarse sand as surface area of fine sand mixtures is more compared to coarse sand it needs more water to attain a considerable dry density. Optimum moisture content is increased by about 6% by adding 10% fine sand to black cotton soil compared to coarse sand which resulted in an increase by 8.5%. Similar kinds of results are observed for other percentages of clay–sand mixtures.

The higher values of maximum dry density and lower values of optimum moisture content for all the clay–sand mixtures indicate possible higher compaction of these buffer materials in the field, which surely increases the performance of the disposal facilities of nuclear waste.

Table 3 Maximum dry density and optimum moisture content values of BC soil with fine and coarse sand

S. No.	Properties	BC soil	BC soil + 10% FS	BC soil + 30% FS	BC soil + 50% FS	BC soil + 10% CS	BC soil + 30% CS	BC soil + 50% CS
1	Maximum dry density (Mg/m^3)	1.53	1.64	1.71	1.75	1.66	1.72	1.83
2	Optimum moisture content (%)	24.81	20.0	19.4	18.27	18.5	17.0.45	16.0

Table 4 Free swell index limits of soil mixtures

Soil type	FSI (%)	Degree of expansion
BC soil	209	Very high
BC soil + 10% sand	130	High
BC soil + 30% sand	35	Moderate
BC soil + 50% sand	10	Low

3.3 Swelling and Shrinkage Properties

Free Swell Index (FSI)

Free swell index is increase in the volume of soil, without any external constraints, on submergence in water. The FSI values of native BC soil and clay–sand mixtures were determined as per IS: 2720 (Part 40)—1977 and the values are given in Table 4.

As can be seen from this table, black cotton soil has FSI value of 209%, indicating very high degree of expansion and it is observed that with the addition of 10% sand FSI value was found as 130%, although decreased by about 38%, which still indicates high degree of expansion, while the addition of higher percentages of sand could reduce the FSI values which indicates the degree of expansion as low.

Swell Potential

Swell potential of a soil depends on clay content. And the variation of the swell potential with time for the untreated black cotton soil and different types of clay–sand mixtures for both fine sand and coarse sand are presented in Figs. 4 and 5, respectively. It can be observed from these figures, that the swell potential varies differently with respect to time period for swelling.

The increase in swell potential in the initial period of time of 10 min is very less, whereas the value is increasing exponentially after the first 10 min up to 1000 min.

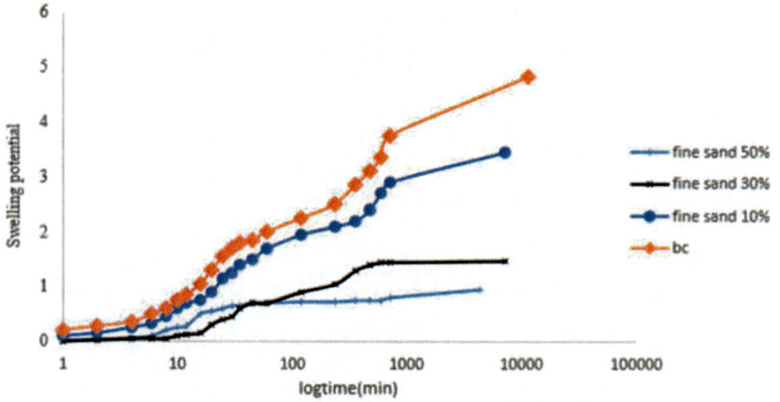

Fig. 4 Swell potential (%) versus log time graph for BC with fine sand mixture

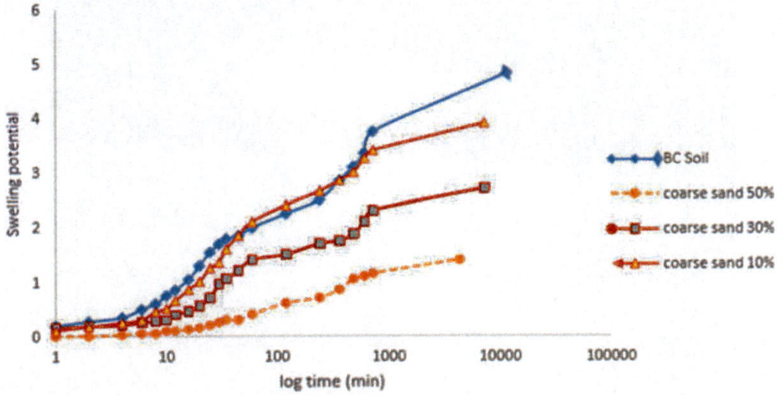

Fig. 5 Swell potential (%) versus log time graph for BC with coarse sand mixture

Table 5 Swell potential for soil mixtures

Soil type	BC soil	BC Soil + 10% FS	BC soil + 30% FS	BC soil + 50% FS	BC Soil + 10% CS	Bc soil + 30% CS	BC soil + 50% CS
Swell potential (%)	4.82	3.45	1.475	0.95	3.90	2.70	1.40

And after 1000 min, again the slope of the swell potential for both types of clay–sand mixtures are given in Table 5.

It can be observed from this table that the swell potential value decreases with an increase in the percentage of sand portion and decreases by a maximum of 80% with find sand and by around 70% with coarse sand.

Swell potential of BC soil as described earlier falls in high degree of expansion category. Thus, it is undesirable to use in construction of buffer. Comparatively fine sand decreases the swelling behavior of BC soil than coarse sand.

Volumetric Shrinkage

Volumetric shrinkage is determined for soil mixtures at maximum dry density. Sample is prepared and compacted in the mold and placed in UTM explained previously and the volumetric shrinkage is determined for various proportions of fine sand and coarse sand content with BC soil.

In order to get more accurate results, two soil specimens are used for these shrinkage tests and the average value of shrinkage for these two samples is considered for understanding the volumetric shrinkage of clay–sand mixtures. Figure 6 shows the two samples of untreated black cotton soil before and after shrinkage procedure.

The soil samples of clay–fine sand mixtures (10, 30, and 50%) before and after shrinkage are presented in Figs. 7, 8, and 9, respectively. Similarly, the soil samples of clay–coarse sand mixtures (10, 30, and 50%) before and after shrinkage are presented in Figs. 10, 11, and 12, respectively.

Fig. 6 Shrinkage of BC soil

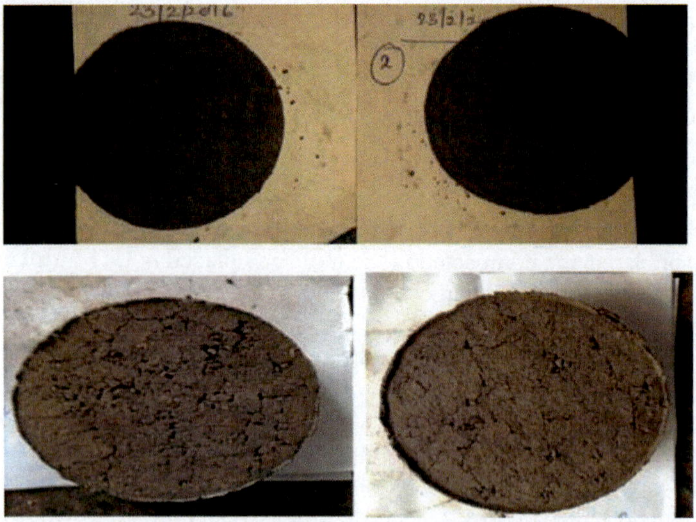

Fig. 7 Shrinkage of BC soil with 10% fine sand

Fig. 8 Shrinkage of BC soil with 30% fine sand

The average ultimate volumetric shrinkage values and the time taken for the shrinkage of all these categories of clay–sand mixtures along with that of original untreated black cotton soil are given in Table 6.

As can be observed from the above table, in general the volumetric shrinkage values of clay–sand mixtures are less than that of untreated black cotton soil. And the values of clay–fine sand mixtures are found to be less than that of clay–coarse sand mixtures, which are similar to that of swell potential values of these mixtures. There is a maximum decrease of volumetric shrinkage by about 6.6% with fine sands when compared to 4.7% decrease with the addition of coarse sands. The percentage decrease in the volumetric shrinkage of clay–fine sand mixtures is more than that of clay–coarse sand mixtures, for all the proportions tried in the present study.

Lesser values of volumetric shrinkage for all the clay–sand mixtures clearly indicate possible higher performance of these soils over the original black cotton soil in the functioning of containment zones of nuclear wastes.

Fig. 9 Shrinkage of BC soil with 50% fine sand

Fig. 10 Shrinkage of BC soil with 10% coarse sand

4 Conclusions

The following conclusions are drawn from the present study.

1. The plasticity index values of the clay–sand (both fine and coarse sand) mixtures were decreased by considerable percentage than those of original black cotton soil.

Fig. 11 Shrinkage of BC soil with 30% coarse sand

2. The maximum dry density values of clay–sand mixtures are higher than that of original clayey soil while the optimum moisture content values are lesser than that of original black cotton soil. These findings will clearly indicate the possibility of higher compaction potency for the clay–soil mixtures when compared to clayey soils.
3. With increasing sand content in both the types of sands, lesser values are observed for both swell potential and volumetric shrinkage.
4. Out of the two types of sands used in the present investigation, fine sand is preferably used over coarse sand in clay-sand mixtures for the usage of buffer materials, considering the higher reduction in volumetric shrinkage. Although the swell potential and maximum dry density of former is less than the latter.

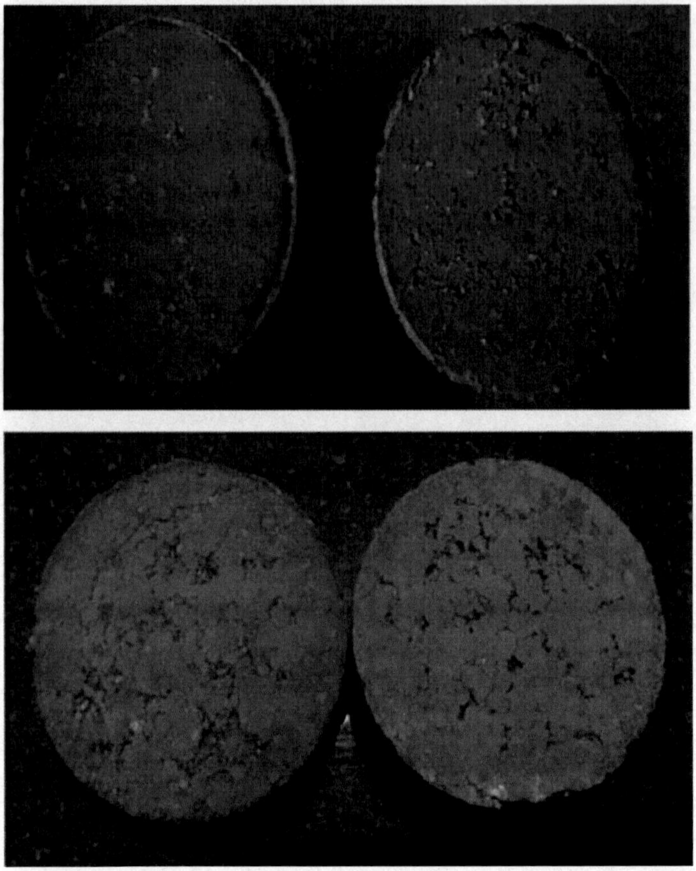

Fig. 12 Shrinkage of BC soil with 50% coarse sand

Table 6 Volumetric shrinkage and time taken for shrinkage values

Soil type	BC soil	BC soil + 10% FS	BC soil + 30% FS	BC soil + 50% FS	BC soil + 10% CS	BC soil + 30% CS	BC soil + 50% CS
Volumetric shrinkage (%)	13.16	11.67	6.52	4.53	12.58	10.46	6.95
Time taken (days)	27	24	23	20	26	23	21

5. As lesser values of volumetric shrinkage and higher values of swell potential are preferred for buffer material usage in the containment portions of nuclear waste disposal based on the functioning of containment zones of nuclear wastes, out of all the mix proportions investigated, 10% fine sand and clay mixture is found to be most optimum for this purpose.

Further studies preferably in the field are required before recommending these clay–sand mixtures as buffer materials used for nuclear waste disposal.

References

1. IAEA (International Atomic Energy Agency): Scientific and technical basis for the geological disposal of radioactive wastes. IAEA Technical Reports Series No. 413. IAEA, New York (2003)
2. Ogata, N., Kosaki, A., Ueda, H., Asano, H., Takao, H.: Execution techniques for high level radioactive waste disposal: IV. Design and manufacturing procedure of engineered barriers. J. Nuclear Fuel Cycle Environ. **5**(2), 103–121 (in Japanese with English abstract) (1999)
3. Dixon, D.A., Gray, M.N., Thomas, A.W.: A study of the compaction properties of the potential clay-sand buffer mixtures for use in nuclear fuel waste disposal. Eng. Geol. **21**, 241–255 (1985)
4. Ada, M.:. Performance assessment of compacted bentonite/sand mixtures utilized as isolation material in underground waste disposal repositories. Thesis report. The Graduate School of Natural and Applied Sciences of Middle East Technical University (2007).
5. Fritz, P., Barker, J.F., Gale, J.: Isotope hydrology at the Stripa test site. In: Proceedings of N.E.A. Workshop, International Stripa Project, Stockholm (1982)
6. KBS-III: Final storage of spent nuclear fuel (Ill-barriers). SKBF/KBS, Swedish Nuclear Fuel Supply Company, Stockholm 1983
7. Mitchell, J.K.: Fundamentals of Soil Behaviour, pp. 323–324. Wiley, Toronto (1976)
8. Howell, J.L., Shackelford, C.D., Amerand, N.H., Roslyn, T.: Stern. compaction of sand-processed clay soil mixtures, Geotech. Test. J. (1997)
9. Pusch, R., Homark, H., Karnland, O.: Microstructural impact on the conductivity of smectitie buffer clays. In: Proceedings of 9th International Clay Conference, Strasbourg, pp. 127–137 (1990)
10. Pusch, R.: Waste disposal in rock. Dev. In: Geotechnical Engineering, p. 76. Elsevier, Amsterdam (1995)
11. Nayak, N.V., Christensen, R.W.: Swelling characteristics of compacted expansive Soils. Clays Clay Miner. **19**(4), 251–261 (1974)
12. Mollins, L.H., Stewart, D.I., Cousens, T.W.: Predicting the properties of sand-bentonite mixtures. Clay Miner. **31**, 243–252 (1996)
13. Komine, H., Ogata, N.: Experimental study on swelling characteristics of sand-bentonite mixture for nuclear waste disposal. Soils Found. **39**(2), 83–97 (1999)
14. Sun, D., Cui, H., Sun, W.: Swelling of compacted sand-bentonite mixtures. Appl. Clay Sci. **43**, 485–492 (2009)
15. Sivapullaiah, P.V., Sridharan, A.: Hydraulic conductivity of bentonite sand mixtures. Can. Geotech. J. **36**, 406–413 (2000)
16. IS: 1498: Indian Standard Methods of test for soils: "classification and Identification of soils for engineering purposes" (1970)
17. IS: 2720 Part 5: Indian Standard Methods of test for soils: determination of liquid & plastic limit (1985)
18. IS: 2720 Part 6: Indian Standard Methods of test for soils: "determination of shrinkage factors" (1972)

19. IS: 2720 Part 3/Sec 1: Indian Standard Methods of test for soils: determination of Specific gravity of fine-grained soils (1980)
20. IS 2720: Part 8: Indian Standard Methods of test for soils: "determination of water content-dry unit weight relation using heavy compaction" (1983)
21. IS: 2720 Part XL: Indian Standard Methods of test for soils: "determination of free swell index of soils" (1977)
22. Seed, H.B., Woodward, Jr., R.J., Lundgren, R.: Prediction of swelling potential for compacted clays. J. Soil Mech. Found. Div. Am. Soc. Civ. Eng. (SM3) **88**, 53–87 (1962)
23. Sridharan, A., Prakash, K.: Shrinkage limit of soil mixtures. Geotech. Test. J. **23**, 3–8 (2000)
24. Calabar, A.F., Mustafa, W.S.: Behaviour of sand-clay mixtures for road pavement subgrade. Int. J. Pavement Eng. **18**, 714–726 (2017)

Swell-Compressibility Behavior of Geopolymer Blended Expansive Clays

T. V. Nagaraju, B. M. Sunil, and Babloo Chaudhary

Abstract This paper presents the influence of GGBS-based geopolymer on swell-compressibility characteristics of oven-dry, expansive clay passing 4.75 mm sieve. One-dimensional swell-consolidation tests were conducted on the expansive clay passing through 4.75 mm sieve to which GGBS was added at 0, 5, 10, 15, and 20% by dry weight of the clay. Rate of heave, swell potential, swelling pressure, and linear shrinkage were evaluated. Rate of heave and swell potential decreased significantly with increase in GGBS content. The paper also explores the microstructure behavior and surface texture of the GGBS-based geopolymer–clay blends using the scanning electron microscopy (SEM). The SEM results revealed that the higher Si/Al compounds increase the dense phase of geopolymer products. Moreover, geopolymer synthesis can contribute to the bulk utilization of industrial by-products.

Keywords Expansive clays · Geopolymers · Microstructure · Swell · Shrinkage

1 Introduction

Expansive clays are problematic by virtue of their swelling and shrinkage behavior with variation of moisture content [1]. Expansive clay as foundation soil or as a sub-grade soil leads to severe distresses and cause damages for both substructure and superstructure [1, 2]. To counteract these distresses caused by the expansive clays, many ground improvement techniques were in practise by means of chemical modification, mechanical modification, and hydraulic modification [2–4].

Chemical amelioration techniques are quite successful to enhance the properties of expansive clays. Moreover, the availability of gravel soils for the purpose of mechanical stabilization is not always possible in the field which increases the potentiality of the chemical stabilization. Furthermore, increasing industrialization generates large quantities of industrial products such as fly ash, pond ash, ground granulated blast furnace slag (GGBS), silica fume and rice husk ash (RHA). Earlier,

T. V. Nagaraju (✉) · B. M. Sunil · B. Chaudhary
National Institute of Technology Surathkal, Surathkal, Karnataka 575025, India
e-mail: varshith.varma@gmail.com

H. Hazarika et al. (eds.), *Advances in Sustainable Construction and Resource Management*, Lecture Notes in Civil Engineering 144,
https://doi.org/10.1007/978-981-16-0077-7_42

cement and lime stabilization has been intensively used in treating the expansive clays [4–6]. However, the manufacturing process of ordinary Portland cement emits larger quantity of CO_2. On the other hand, lime is also energy intensive. These numerous studies [7–9] showed that cement and lime are quite successful in counter acting the swell-shrink behavior, improving the strength characteristics and reducing the plasticity nature of the clay.

There are industrial by-products as mentioned earlier can be utilized as a partial replacement for treating expansive clays in the construction. These pozzolanic materials were helpful in improving the strength characteristics and swell–shrink behavior [10]. The main reaction that occurs in the soil pozzolanic materials is flocculation (soil particles combined together to form flocks) and cementation (minerals in clay particles reacts with Si and Al ions in the pozzolanic material to produce cementitious compounds) [11].

In current research, geopolymers were used in soils as a stabilizer where large quantities of industrial by-products can be utilized as a precursor. The use of geopolymers in the soils with addition of precursor can develop geopolymer products [12–14]. The geopolymer products were developed based on the reaction between the alkali-activated solution and presence of Si and Al ions in the precursor [15]. Moreover, hardening of geopolymer blends occurs due to the condensation between the Si and Al ions and there hardening is more rapid than the presence of Si ions alone [16, 17]. UCS results of lateritic soil significantly increased with alkali activation, and they meet the strength requirements of pavements having both light and heavy volume traffic [18].

The main objective of the present work is to analyze the effects of geopolymers on the swell–compressibility behavior. One-dimensional swell consolidation tests were conducted on the geopolymer blended expansive clays for determining the rate of heave, swell potential, and swelling pressure. In addition to swell–compressibility behavior, subsequent scanning electron microscopy (SEM) analysis was carried for the geopolymer–soil blends with varying precursor content.

2 Materials and Methods

2.1 Materials

The experimental investigation was carried on the expansive clay having the free swell index 145%. The soil sample was collected at a depth of 0.5 m below the ground surface from Bhimavaram, Andhra Pradesh, India. The precursor used in this study is GGBS which was obtained from the Vespra Pvt. limited, Vijayawada, India. The properties of an expansive clay and GGBS were tabulated in Table 1. In this study, geopolymers were used to chemically alter the properties of expansive clays. Alkali-activated solution was prepared using the sodium hydroxide solution (NaOH)

Table 1 Properties of an expansive clay and GGBS

Property	Silica fume content				
	0%	5%	10%	15%	20%
Heave (mm)	1.84	1.54	1.42	1.36	1.32
Swell potential (%)	9.2	7.7	7.1	6.8	6.6
Swelling pressure (kPa)	110	140	215	–	–
Rebound (mm)	0.32	0.24	0.18	0.12	0.07
Linear shrinkage (%)	8	4	3	–	–

of 10 molarity and sodium silicate gel (Na_2SiO_3), and their ratios Na_2SiO_3/NaOH were maintained as 2.5.

2.2 Methods

Two series of experimental investigations were conducted. In the first investigation, swell–compressibility behavior of the geopolymer blends with varying GGBS content was studied. In the second investigation, SEM analysis was carried for the geopolymer–soil blends.

The swell-consolidation tests were conducted on the untreated expansive clay and expansive clay treated with GGBS-based geopolymer with varying GGBS content as 5%, 10%, 15%, and 20% by dry weight of the soil. The oven-dry expansive clay passing through 4.75 mm sieve was treated with the GGBS-based geopolymers with varying GGBS content, and the premixed geopolymer blended soil was cured for the 5 days. The prepared geopolymer–soil blend (passing through 4.75 mm) was weighed corresponding to the γ_d chosen (13 kN/m^3) and the volume of the consolidometer ring (diameter = 60 mm; height = 20 mm) for the performance of one-dimensional swell-consolidation tests. The blended sample was statically compacted in the consolidometer ring in four layers each of 5 mm thickness so as to ensure a uniform γ_d. A filter paper and a porous stone were placed at each end of the sample, and this unit was placed in the consolidometer after positioning the loading pad. This assembly was mounted on the loading frame, and the samples were allowed to undergo free swell by inundation for three days (exactly 72 h) under a surcharge of 5 kPa. The dial readings were continuously monitored as the samples underwent swelling. After equilibrium heave, which would generally be reached in three days in oedometer specimens of 20 mm thickness, the samples were subjected to consolidation under increased vertical stresses.

Swell potential ($S\%$) was determined as the ratio of increase thickness (ΔH) to the original thickness (H), expressed as $S\%$. And swelling pressure (p_s) was determined from e-log p curves as the pressure corresponding to the initial void ratio.

Other series of investigation, scanning electron microscopy (SEM) test was carried for the aforementioned geopolymer–soil blends, to understand about microstructural behavior of geopolymer–clay blends.

3 Results and Discussion

3.1 Effects of Geopolymer on Swell–Compressibility

Figure 1 shows the rate of heave of untreated clay and clay treated with the GGBS-based geopolymers with variation of GGBS content. The untreated clay attained an equilibrium heave of 1.84 mm in the end of 3 day (4320 s). The rate of heave of the GGBS-based geopolymer blends decreased significantly with increasing precursor content from 0 to 20%. This could be attributed to the presence of geopolymers that influence the expanding lattice structure of montmorillonite mineral and also development of geopolymer products. Another curious point which may be highlighted in respect of rate of heave of geopolymer blends was there was no initial hike in the rate of heave as compared to the untreated expansive clay. This is due to the geopolymer blends that were pretreated which not allow the water into the clay pockets.

Figure 2 shows that the swell potential decreased with increase in GGBS content in the GGBS-based geopolymers. The reduction of swell potential was from 9.2 to 6.6% when precursor (GGBS) content increased from 0 to 20%. The improvement in the swell potential values was due to the reduction of surfaces forces within the clay particles.

The e-log *p* data curves of the untreated clay and clay are treated with the GGBS-based geopolymers with varying precursor content. The e-log p curves clearly indicated that swelling pressure (p_s) could not be determined when GGBS content is

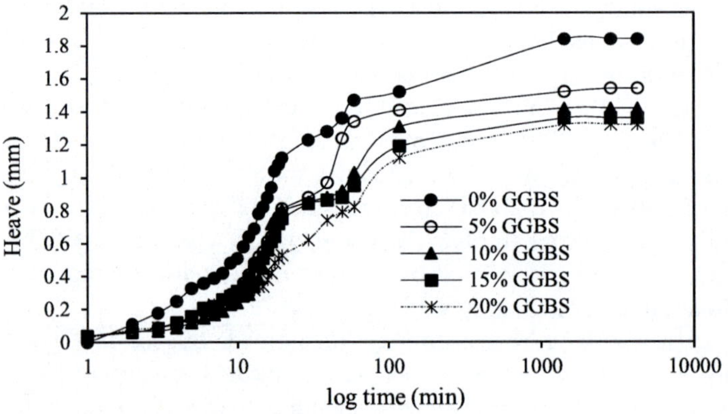

Fig. 1 Rate of heave of clay for different percentage of GGBS

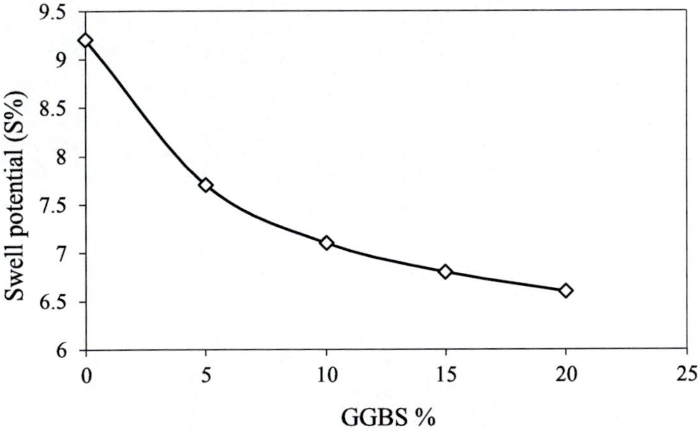

Fig. 2 Effects of precursor content on swell potential

more than the 15% (see Fig. 3). At higher precursor contents, efficient and effective geopolymer products would develop dense matrix in the blends, which helps in not vent of any water molecules, and therefore, the expanding lattice of montmorillonite structure is counteracted. Even the GGBS-based geopolymer soil bended samples in an inundation time of three days in the oedometer not show any change in swelling pressure due to the strong geopolymer products. Rebound and linear shrinkage was decreased with increasing GGBS content in the geopolymer–clay blends. This can be attributed due to the strong geopolymer products that offer resistance against the both axial and lateral displacement of clay grains (Fig. 4).

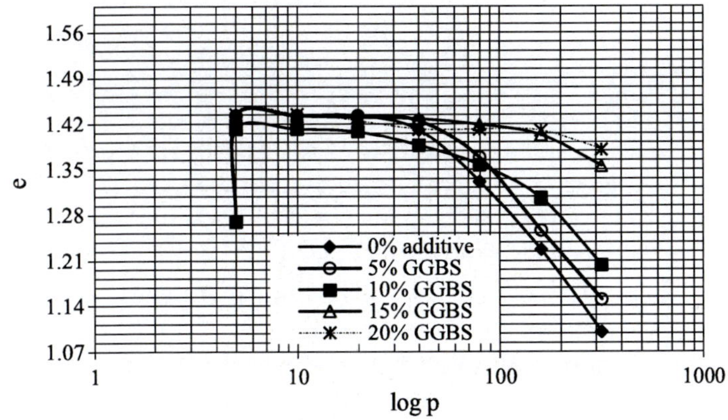

Fig. 3 E-logp curves of geopolymer–clay blends

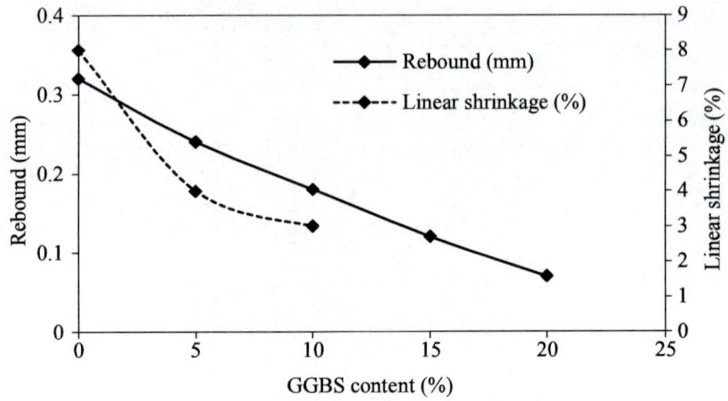

Fig. 4 Effects of precursor content on rebound and linear shrinkage

3.2 *Effect of Geopolymer on Microstructure*

Figure 5a shows that the untreated expansive clay having open texture. Geopolymer-treated SEM micrographs revealed that the dense compacted geopolymer products exists when the precursor content more than 10% in the blends, leading to an improvement in the swell–compressibility behavior. The dense matrix is formed due to the reaction of Si ions and alkali-activated gel, tends to formation of CSH gel highlighted in Fig. 5c. The surface texture in Fig. 5d clearly shows pockets on the surface due to the possibility of higher concentration leaching of Al and Si oxides from GGBS (20%) by alkali dissolution of the precursor. Further, allows the formation of Si–O–Si bonds.

4 Conclusions

This paper presents the influence of GGBS-based geopolymers on expansive clays. The GGBS-based geopolymers are formed by activating GGBS, which is rich in combination of silica (40%) and alumina (13.5%). The following conclusions can be drawn from the foregoing experimental study:

1. Rate of heave and swell potential of the GGBS-based geopolymer clay decreased with increasing GGBS content in the blends.
2. Swelling pressure (p_s) of the GGBS-based geopolymers increased significantly with increasing precursor content. However, swelling pressure could not be determined when the precursor content is more than the 15% in the geopolymer clay blends owing to the development of efficient geopolymer products.
3. SEM micrographs revealed that the geopolymer–clay blends having strong bonds due to the leaching of Si and Al ions in the alkali dissolution of the precursor.

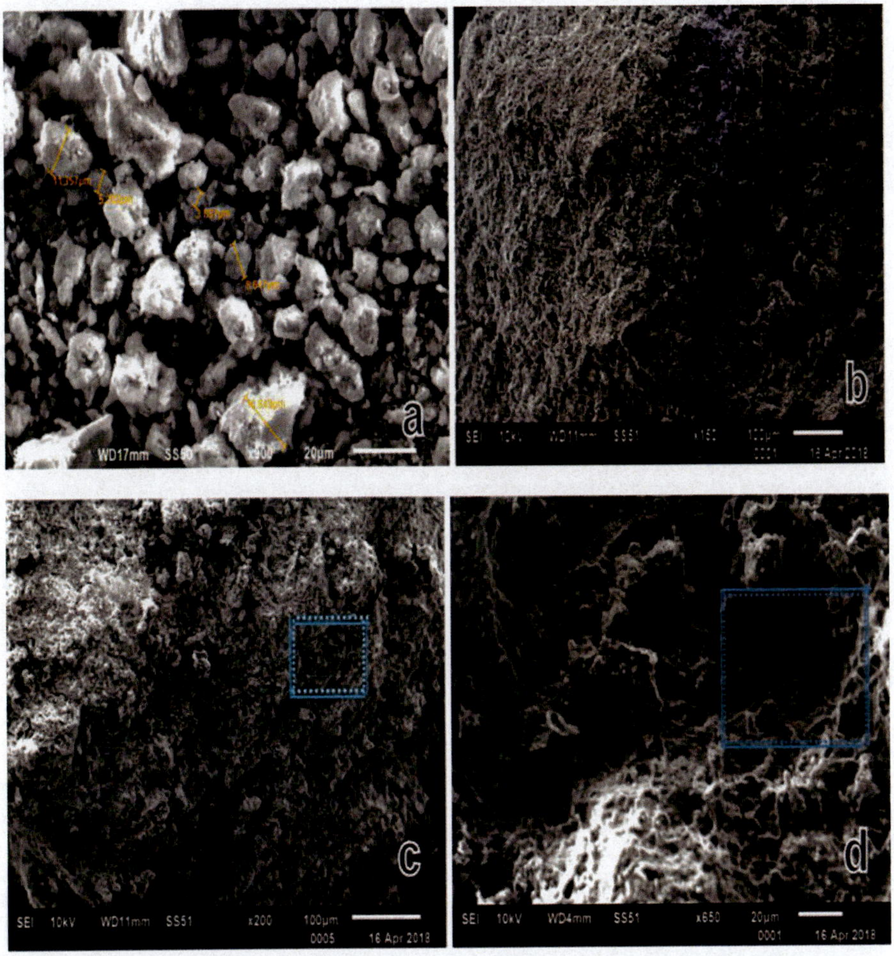

Fig. 5 SEM micrographs of geopolymer–clay blends with varying GGBS content **a** untreated expansive clay, **b** 10% GGBS, **c** 15% GGBS, and **d** 20% GGBS

4. Alkali activation of GGBS can be helpful in bulk utilization of GGBS in soil stabilization, whereas GGBS content alone cannot give better results as compared with alkali-activated GGBS [10].

References

1. Chen, F.H.: Foundations on Expansive Soils. Elsevier Scientific Publishing Co., Amsterdam (1988)
2. Nelson, D.J., Miller, J.D.: Expansive Soils: Problems and Practice in Foundation and Pavement Engineering. Wiley, New York (1992)
3. Rajagopal, K.: The geosynthetics for sustainable construction of infrastructure projects. Indian Geotech. J. **47**(1), 2–34 (2017)
4. Nagaraju, T.V., Satyanarayana, P.V.V.: Geotechnical aspects of various constructions along the canal embankments. In: Tyagaraj, T. (ed.) Ground Improvement Techniques and Geosynthetics, Chap. 16, pp. 143–150. LNCE. Springer Nature, Berlin. ISBN: 978-981-13-0558-0 (2019)
5. Cokca, E.: Use of class C fly ash for the stabilization of an expansive soil. J. Geotech. Geoenviron. Eng. **127**(7), 568–573 (2001)
6. Yadu, L., Tripathi, R.K.: Effects of granulated blast furnace slag in the engineering behaviour of stabilized soft soil. Procedia Eng. **51**, 125–131 (2013)
7. Sherwood, P.T.: Soil Stabilization with Cement and Lime, pp. 14–55. HMSO Publications Center, London (1995)
8. Nalbantoglu, Z.: Lime Stabilization of Expansive Clay, Expansive Soils—Recent Advances in Characterization and Treatment. In: Al-Rawas, A.A., Goosen, M.F.A. (eds.) (2006)
9. Phanikumar, B.R., Nagaraju, T.V.: Engineering behaviour of expansive clays blended with cement and GGBS. Ground Improvement **171**(3), 167–173 (2018). https://doi.org/10.1680/jgrim.17.00054
10. Phanikumar, B.R., Nagaraju, T.V.: Swell and compressibility of GGBS–clay mixes in lumps and powders: effect of 4% lime. Indian Geotech. J. **49**(2), 161–169 (2019) ISSN 2277-3347
11. Chittoori, B., Puppala, A.: Quantitative estimation of clay mineralogy in fine-grained soils. ASCE J. Geotech. Geoenviron. Eng. **137**(11), 997–1008 (2011)
12. Ghadir, P., Ranjbar, N.: Clayey soil stabilization using geopolymer and Portland cement. Constr. Build. Mater. **188**, 361–371 (2018)
13. Miao, S., Shen, Z., Wang, X., Luo, F., Huang, X., Wei, C.: Stabilization of highly expansive black cotton soils by means of geopolymerization. J. Mater. Civ. Eng. **29**(10) (2017)
14. Bilondi, M.P., Toufigh, M.M., Toufigh, V.: Experimental investigation of using a recycled glass powder-based geopolymer to improve the mechanical behavior of clay soils. Constr. Build. Mater. **170**, 302–313 (2018)
15. Nagaraju, T.V., Prasad, C.D.: New prediction models for compressive strength of GGBS-based geopolymer clays using swarm assisted optimization. In: Adv. Comput. Methods Geomech. 367–379 (2020)
16. Zhang, M., Guo, H., El-Korchi, T., Zhang, G., Tao, M.: Experimental feasibility study of geopolymer as the next-generation soil stabilizer. Constr. Build. Mater. **47**, 1468–1478 (2013)
17. Kim, Y., Dang, M.Q., Do, T.M.: Studies on compressive strength of sand stabilized by alkali-activated ground bottom ash and cured at the ambient conditions. Geo-Engineering **7**, 15 (2016)
18. Phummiphan, I., Horpibulsuk, S., Phoo-ngernkham, T., Arulrajah, A., Shen, S.-L.: Marginal lateritic soil stabilized with calcium carbide residue and fly ash geopolymers as a sustainable pavement base material. J. Mater. Civ. Eng. **29**(2) (2017)

Swelling Behavior and Permeability of Polymerized Bentonite with Due Monomers

Nie Jixiang and Chai Jinchun

Abstract To improve the anti-cationic solution capability of geosynthetic clay liners (GCLs), new polymerized bentonites (PBs) were produced for potential use as cores of GCLs. PBs were produced using free radical polymerization method with due monomers, acrylic acid (M_1) and acrylamide (M_2), and potassium persulfate as initiator (I). The natural bentonite (UB) used was a sodium type one. The adopted pH was 6, $I/(M_1 + M_2)$ ratio of 0.005, and $(M_1 + M_2)$/UB ratio of 0.05 (0.05 PB) and 0.1 (0.1 PB). The results of the X-ray diffraction (XRD) indicate that PBs produced in this study can be classified as micro-composite. The swelling capacity of the PBs was investigated by free swelling index (FSI) tests. In deionized water, 0.1 M to 0.6 M NaCl and 0.03 to 0.06 M $CaCl_2$ solutions, the values of FSI of PBs are higher than that of UB. For PBs, FSI increased with increase of polymer/bentonite ratio. In deionized water, FSI of 0.1 PB was more than 5 times that of the UB. The values of permeability of the PBs and UB with 0.6 M NaCl solution were deduced from oedometer test results. At the same void ratio, the permeability of 0.1 PB was more than one order lower than that of the UB. For the 0.1 PB, for void ratio up to 7.5, the value of k was less than 10^{-10} m/s. Therefore, 0.1 PB can be used as core material of GCLs to be used under higher concentration of Na^+ environment.

Keywords GCL · Polymerized bentonite · Free swelling index

1 Introduction

Geosynthetic clay liners (GCLs) are widely used as landfill liners to prevent waste leachate flowing into the surrounding ground [1, 2]. GCLs are factory-manufactured consisting of a thin layer of sodium bentonite sandwiched between two geotextiles

N. Jixiang
Department of Civil Engineering and Architecture, Saga Univ, Saga, Japan

C. Jinchun (✉)
Honjo-machi, Saga 840-8502, Japan
e-mail: chai@cc.saga-u.ac.jp

H. Hazarika et al. (eds.), *Advances in Sustainable Construction and Resource Management*, Lecture Notes in Civil Engineering 144,
https://doi.org/10.1007/978-981-16-0077-7_43

that are bonded by needle punching or stitching [3]. Sodium bentonite (UB) has low permeability to water (typically <10^{-10} m/s) [4–6] and excellent swelling ability with water [7–9], which can self-heal certain damage holes on the GCLs. However, for cationic solutions, the permeability of UB increases [10, 11] and the self-healing capacity decreases [12]. Therefore, under an environment of higher concentration of cation solution, GCLs cannot effectively act as a barrier. In the past decades, researchers have modified UB with polymers to improve its swelling properties [13, 14]. The results of the studies confirmed that the polymerized bentonite (PB) had higher swelling capacity and lower permeability in cationic solutions than the corresponding UB [15]. While for higher concentration cationic solutions, such as 0.6 M NaCl and 0.6 M $CaCl_2$ solutions, the literature reported PBs still cannot meet the requirement for using as a core material of GCLs.

In this study, a new PBs was produced with two monomers using free radical polymerization method. The swelling properties and permeability of the newly produced PBs were measured by free swelling index (FSI) tests and consolidation tests, respectively, and the results are reported. The discussions were made on its potential use in GCLs.

2 Materials Used and Polymerization Method

2.1 Bentonite

Natural sodium bentonite from Wyoming, USA was used. To remove impurity material, the natural sodium bentonite was stirred for 30 min in deionized water and static precipitated for 48 h. Then extract the upper slurry from the container as untreated sodium bentonite (UB). The dried UB will be used for producing PB.

2.2 Polymerization Method

Acrylic acid ($C_3H_4O_2$) (M_1) and acrylamide (C_3H_5NO) (M_2) as monomers, potassium persulfate as initiator (I), and deionized water as the solvent were used in the process of polymerization. The conditions adopted for polymerization are M_1/M_2 ratio of 0.5, $I/(M_1 + M_2)$ ratio of 0.5%, $(M_1 + M_2)$/UB ratio of 5% (0.05 PB) and 10% (0.1 PB). Free radical polymerization method was used with following steps:

1. Prepare sodium bentonite slurry. 10 g dried powdered UB was mixed into 200 ml deionized water with a ratio of 1:20, by stirring it for 1 h and keeping it more than 12 h in static state to form a homogeneous sodium bentonite slurry.
2. Prepare monomer mixture solution. M_1/M_2 ratio was set as 1/2. 0.333 g M_1 was dissolved into 30 ml deionized water, which was neutralized with 1 M NaOH solution to make pH of 6.0 (±0.05). Then 0.666 g M_2 was added into

Table 1 Specific gravity, plastic limit, and liquid limits of the bentonite

Properties	UB	0.1 PB
Liquid limit w_l (%)	716.0	1005.0
Plastic limit w_p (%)	46.0	40.6
Specific gravity G	2.75	2.61

the neutralized solution and stirred evenly for 1 min to form mixed monomer solution.

3. Prepare composite slurry. The mixed monomer solution was mixed into the sodium bentonite slurry and stirred for 20 min to form a homogeneous mixture slurry. Then 0.005 g potassium persulfate was dissolved in 20 ml deionized water, which was added into the mixture slurry and continued to keep stirring for 10 min. After that the composite slurry was formed.
4. Polymerize bentonite. The composite solution was vacuumed in airtight container using 0.1 MPa for 2 h to remove O_2. The whole composite slurry was exposed into the oven at 100 °C to polymerize for 24 h. The PB was made into powder passing through 75 μm sieve. Liquid limits (w_l), plastic limits (w_p), and specific gravities (G) of the UB and the 0.1 PB are listed in Table 1. The specific gravity of PB is a little smaller than that of UB, and it may be attributed that the structure of UB is modified by polymers.

3 X-Ray Diffraction Analysis

The X-ray diffraction (XRD) tests were conducted using a SHIMADZU XRD -6000 for the 0.1 PB and UB. The 0.1 PB or UB was filled into the specimen container of 20 mm in diameter and 1.25 mm in height. The voltage and current of the X-ray tubes used were 40 kV and 30 mA, respectively. Figure 1 shows that the position

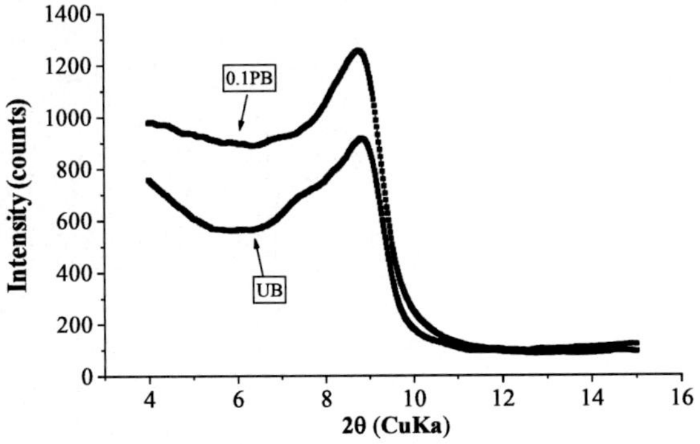

Fig. 1 XRD patterns of UB and 0.1 PB

of the diffraction peak of 0.1 PB is almost the same as the UB, which indicates the interlayer spacing of sodium bentonite crystal was not modified by the polymerization. Generally, according to the state of the long chains of the polymers insertion into the silicate layer of sodium bentonite, clay–polymer composite structure can be classified into three types: (1) phase-separated (micro-composite or conventional composite); (2) intercalated; and (3) exfoliated structures, which include one micro-composite structure and two nanocomposite structures [16, 17]. The result in Fig. 5 shows that the 0.1 PB belong to phase-separated composite.

For phase-separated structure, the polymer cannot be embedded between the silicate sheets of bentonite, and the bentonite particles are coated by the polymer. Therefore, it is inferred that for 0.1 PB, polymers did not enter the interlayer of sodium bentonite crystals and only wrapped around the particles of the sodium bentonite.

4 Swelling Properties of the PB and UB

The swelling capacity of the UB and PBs was evaluated using free swelling index (FSI) tests which were conducted following the ASTM D5890 [18]. 2 g dried of PBs or UB was slowly and evenly dropped into the desired solution of 100 ml at a rate of 0.1 g per 10 min. The final volume of the swelled sample was measured after 24 h. Each case was repeated three times to check the repeatability of the tests, and the adopted values of FSI are mean values.

The FSI values of UB and PBs are shown in Fig. 2 for three solutions, namely deionized water, 0.6 M NaCl, and 0.03 M $CaCl_2$. For all three solutions, the PBs had higher FSI value than that of UB. Also, for the PBs, FSI increased with polymer/bentonite ratios. In deionized water, FSI value of 0.1 PB is more than 5 times that of the UB. With cation solutions, FSI value of 0.1 PB is approximately 1.6 times that of UB in 0.6 M NaCl and 1.2 times that of UB in 0.03 M $CaCl_2$ solution.

Figure 3 shows the FSI values of 0.1 PB and UB in NaCl solutions with different concentrations of 0.1 M, 0.3 M, and 0.6 M. For all the concentrations, 0.1 PB had higher swelling capacities than that of UB. With the increase of NaCl concentration, FSI value was reduced. However, percentage difference between the values of FSI of the 0.1 PB and UB increased, i.e., the reduction of FSI with increase concentration of NaCl was lower for 0.1 PB than that for UB. For the three concentrations, FSI values of 0.1 PB meet the minimum requirement ($\geq$24 ml/2 g) to be used in GCLs. However, in 0.6 M NaCl solution, FSI value of UB is 17 ml/2 g which is lower than 24 ml/2 g. Therefore, 0.1 PB has a better swelling ability to resistance the high concentration Na^+ ion.

FSI values of PB in $CaCl_2$ solutions are shown in Fig. 4. 0.1 PB had a high FSI value than that of UB in 0.03 M $CaCl_2$ solution. However, PB almost had the similar FSI value as that of UB in 0.06 M $CaCl_2$ solution.

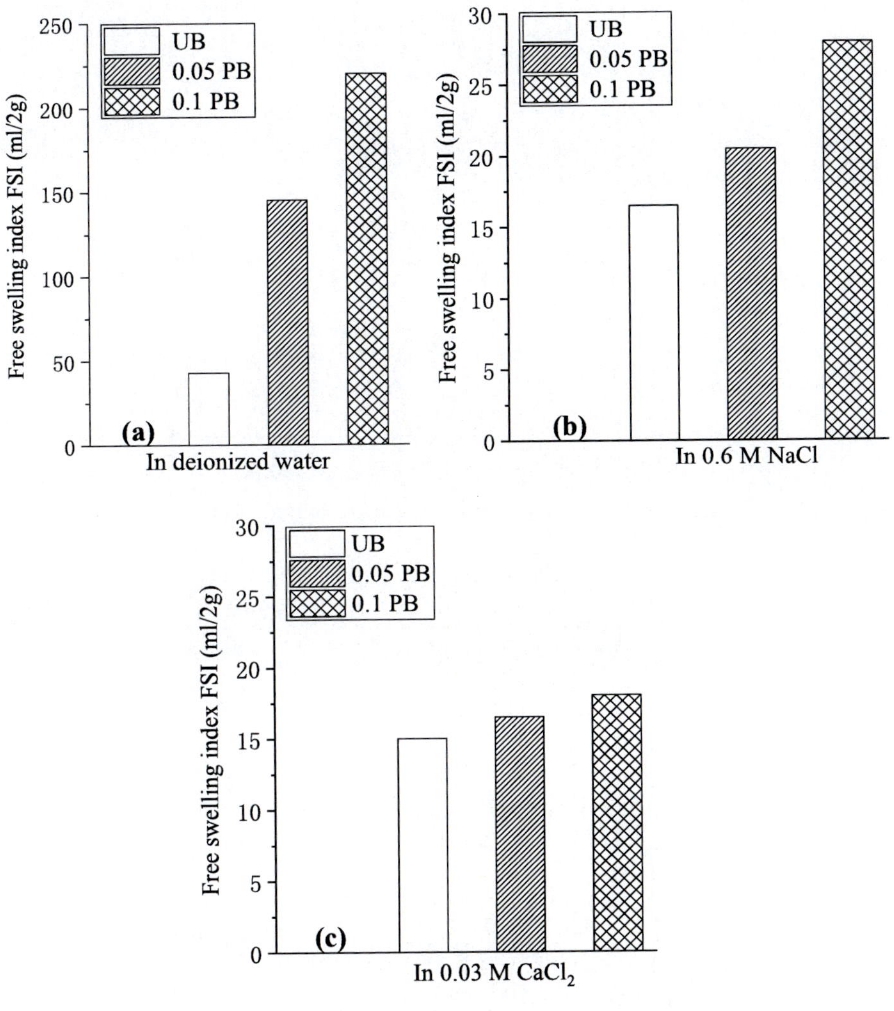

Fig. 2 Comparing FSI values of PB and UB with **a** deionized water, **b** 0.6 M NaCl, and **c** 0.03 M $CaCl_2$ solution

5 Permeability of the PB and UB

The permeability (k) was evaluated from odometer test results. The odometer tests were conducted following ASTM D2435 [19]. There are publications in the literature that show that for low permeability materials, the value of k deduced from oedometer test results is comparable with directly measured k [20, 21].

Dry powder of the PB or UB was thoroughly mixed with the 0.6 M NaCl liquid at a liquid content of two times its liquid limit and cured it for at least 24 h. After that, the slurry was pre-consolidated under a vertical effective stress of 20 kPa in an odometer

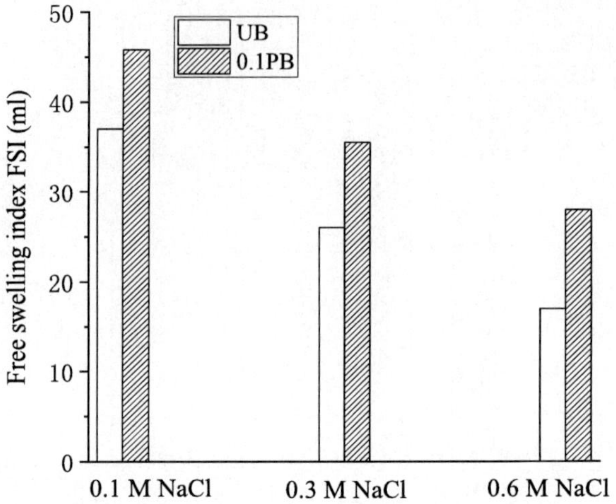

Fig. 3 FSI values of PB and UB with different concentration in NaCl solutions

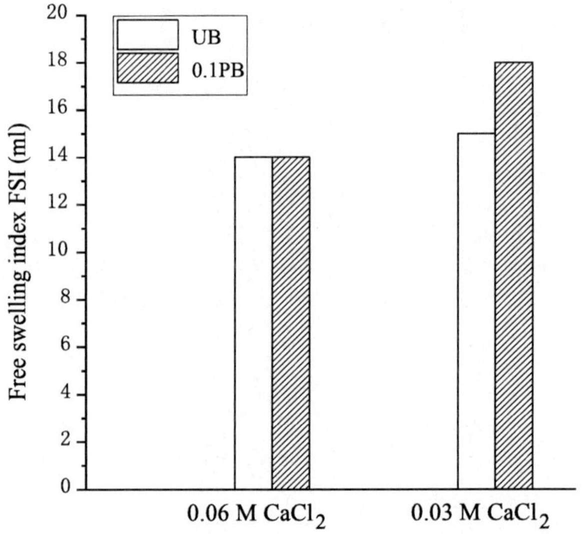

Fig. 4 FSI values of PB and UB with different concentration in $CaCl_2$ solutions

ring 60 mm in diameter and 60 mm in height until the rate of vertical deformation was less than 0.003 mm/h. The consolidation test was carried out with vertical effective stress from 20 to 1280 kPa, by doubling the stress for each subsequent step. The initial water content (w_i), initial void ratio (e_i), and compression index (C_c) of the 0.1 PB and the UB are given in Table 2.

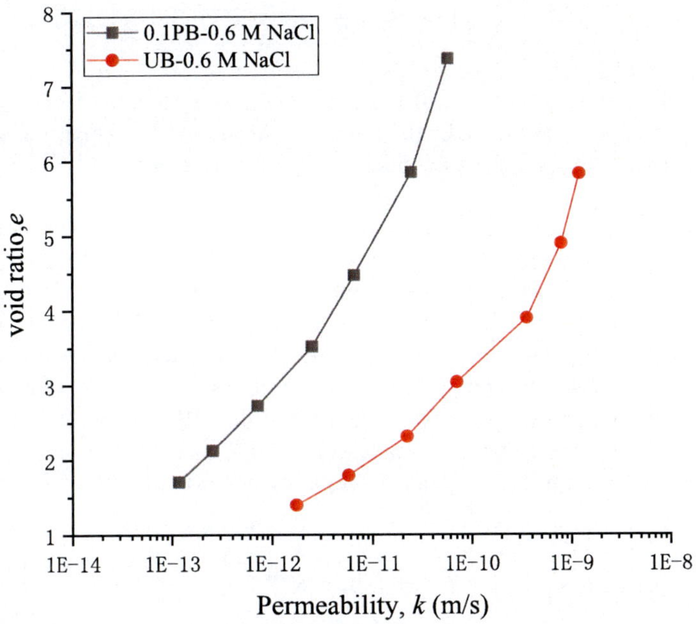

Fig. 5 Relationship of permeability (k) and void ratio (e)

Table 2 Initial water content (w_i), initial void ratio (e_i), and compressibility index (C_c) of the UB and 0.1 PB in 0.6 M NaCl solution

	w_i	e_i	C_c
UB	249.50	6.62	2.07
0.1 PB	451.03	8.09	2.31

The obtained values of permeability are presented in a relationship between void ratio (e) and permeability (k) as shown in Fig. 5. The 0.1 PB samples had obviously lower values of k than that of the UB in 0.6 M NaCl solution. For the 0.1 PB, for void ratio up to 7.5, the value of k was less than 10^{-10} m/s. It suggests that the 0.1 PB can be used as a barrier material where higher concentration of Na^+ ion solutions exist.

6 Conclusions

New polymerized bentonites (PBs) were produced using free radical polymerization method with due monomers, acrylate acid (M_1) and acrylamide (M_2), and potassium persulfate as initiator (I). Other conditions adopted were: pH of 6, $I/(M_1 + M_2)$ ratio of 0.005, and $(M_1 + M_2)$/UB ratio of 0.05 (0.05 PB) and 0.1 (0.1 PB). Properties of the PBs and UB were investigated using free swelling index (FSI) tests, consolidation tests, and XRD analyses. Based on the test results, the following conclusions can be drawn.

1. The results of X-ray diffraction (XRD) analysis show that for the PBs produced, polymer chains did not enter the interlayer of the sodium bentonite crystals. It is postulated that the PBs produced in this study can be classified as micro-composite.
2. In deionized water, 0.1 M to 0.6 M NaCl and 0.03 to 0.06 $CaCl_2$ solutions, the values of FSI of PBs are higher than that of UB. For PBs, FSI increased with increase of polymer/bentonite ratio. In deionized water, FSI of 0.1 PB was more than 5 times the UB. In 0.6 M NaCl solution, FSI of 0.1 PB was more than 24 ml/2 g, which means that the 0.1 PB can be used as the core of GCL.
3. With 0.6 M NaCl solutions, at the same void ratio, the permeability of 0.1 PB was more than one order lower than that of the UB. For the 0.1 PB, for void ratio up to 7.5, the value of *k* was less than 10^{-10} m/s. Therefore, 0.1 PB can be used as a barrier material under higher concentration of Na^+ environment.

Acknowledgements This work has been supported by Grants-in-Aid for Scientific Research (KAKENHI) of Japan Society for the Promotion of Science (JSPS) with a grant number of 17K06558.

References

1. Komine, H., Ogata, N.: Predicting swelling characteristics of bentonites. J. Geotech. Geoenviron. Eng. ASCE. **130**(8), 818–829 (2004). https://doi.org/10.1061/(ASCE)1090-0241(2004)130:8(818)
2. Katsumi, T., Ishimori, H., Ogawa, A., Yoshikawa, K., Hanamoto, K., Fukagawa, R.: Hydraulic conductivity of nonprehydrated geosynthetic clay liners permeated with inorganic solutions and waste leachates. Soils Found. **47** (Katsumi, Ishimori et al. 2007), 79–96 (2007). https://doi.org/10.3208/sandf.47.79
3. Shackelford, C., Benson, C., Katsumi, T., Edil, T., Lin, L.: Evaluating the hydraulic conductivity of GCLs permeated with non-standard liquids. Geotext. Geomemb. **18**(2–4), 133–161 (2000). https://doi.org/10.1061/(ASCE)GT.1943-5606.0001844
4. Alther, G.R.: The role of bentonite in soil sealing applications. Bull. Assoc. Eng. Geol. **19**, 401–409 (1982). https://doi.org/10.2113/gseegeosci.xix.4.401
5. Kolstad, D., Benson, C., Edil, T.: Hydraulic conductivity and swell of nonprehydrated GCLs permeated with multispecies inorganic solutions. J. Geotech. Geoenviron. **130**(12), 1236–1249 (2004). https://doi.org/10.1061/(ASCE)1090-0241(2004)130:12(1236)
6. Benson, C., Oren, A., Gates, W.: Hydraulic conductivity of two geosynthetic clay liners permeated with a hyperalkaline solution. Geotext. Geomemb. **28**(2), 206–218 (2010). https://doi.org/10.1016/j.geotexmem.2009.10.002
7. Prongmanee, N., Chai, J.C.: Effect of shape of damage hole on self-healing capacity of GCL. Geosynth. Eng. J. **32**, 59–64 (2017). https://doi.org/10.5030/jcigsjournal.32.59
8. Scalia, J., Benson, C., Bohnhoff, G., Edil, T., Shackelford, C.: Long-term hydraulic conductivity of a bentonite-polymer composite permeated with aggressive inorganic solutions. J. Geotech. Geoenviron. Eng. **140**(3) (2014). https://doi.org/10.1061/(ASCE)GT.1943-5606.0001040

9. Jo, H.Y., Katsumi, T., Benson, C.H., Edil, T.B.: Hydraulic conductivity and swelling of nonprehydrated GCLs permeated with single-species salt solutions. J. Geotech. Geoenviron. Eng. ASCE **127**(7), 557–567 (2001). https://doi.org/10.1061/(ASCE)1090-0241(2001)127:7(557)
10. Razakamanantsoa, A.R., Barast, G., Djeran-Maigre, I.: Hydraulic performance of activated calcium bentonite treated by polyionic charged polymer. Appl. Clay Sci. **59**, 103–114 (2012). https://doi.org/10.1016/j.clay.2012.01.022
11. Chai, J.C., Sari, K., Hino, T.: Effect of type of leachate on self-healing capacity of geosynthetic clay liner. Geosynth. Eng. J. **28**, 93–98 (2013). https://doi.org/10.5030/jcigsjournal.28.93
12. Bohnhoff, G.L., Shackelford, C.D.: Improving membrane performance via bentonite polymer nanocomposite. Appl. Clay Sci. **86**, 83–98 (2013). https://doi.org/10.1016/j.clay.2013.09.017
13. Di Emidio, G., Van Impe, W., Mazzieri, F.: A polymer enhanced clay for impermeable geosynthetic clay liners. In: Proc., 6th Int. Conf. on Environmental Geotechnics, Balkema, Rotterdam, Netherlands, pp. 963–967 (2010). https://hdl.handle.net/1854/LU-1079370
14. Tian, K., et al.: Polymer elution and hydraulic conductivity of bentonite–polymer composite geosynthetic clay liners. J. Geotech. Geoenviron. **145**(10) (2019). https://doi.org/10.1061/(ASCE)GT.1943-5606.0002097
15. Di Emidio, G., Mazzieri, F., Verastegui-Flores, R.D., Van Impe, W., Bezuijen, A.: Polymer-treated bentonite clay for chemical-resistant geosynthetic clay liners. Geosynth. Int. **22**, 125–137 (2015). https://doi.org/10.1680/gein.14.00036
16. Alexandre, M., Dubois, P.: Polymer-layered silicate nanocomposites: preparation, properties and uses of a new class of materials. Mater. Sci. Eng. R Rep. **28**(1–2), 1–63 (2000). https://doi.org/10.1016/S0927-796X(00)00012-7
17. Ruiz-Hitzky, E., van Meerbeek, A.: Clay mineral- and organoclay–polymer nanocomposite. In: Bergaya, F., Theng, B.K.G., Lagaly, G. (eds.), Handbook of Clay Science. Elsevier, Amsterdam; Boston, pp. 141–245 (2006). https://doi.org/10.1016/S1572-4352(05)01018-4
18. ASTM: Standard test method for swell index of clay mineral component of geosynthetic clay liners. ASTM D5890. West Conshohocken, PA: ASTM (2006). https://doi.org/10.1520/D5890-19
19. ASTM: Standard test methods for one dimensional consolidation properties of soils using incremental loading. ASTM D2. West Conshohocken, PA: ASTM (2018). https://doi.org/10.1520/D2435_D2435M-11R20
20. Bohnhoff, G.L., Shackelford, C.D.: Consolidation behavior of polymerized bentonite-amended backfills. J. Geotech. Geoenviron. Eng. **140**(5), 04013055 (2014). https://doi.org/10.1061/(ASCE)GT.1943-5606.0001079
21. Quang, N.D., Chai. J.C.: Permeability of lime-and cementtreated clayey soils. Can. Geotech. J. **52**(9), 1221–1227 (2015). https://doi.org/10.1139/cgj-2014-0134.

Suitability of Mechanically Biologically Treated Waste for Landfill Covers

T. G. Parameswaran, N. Anusree, P. Sughosh, G. L. SivakumarBabu, and T. K. K. Chamindu Deepagoda

Abstract Landfills are one of the major sources of atmospheric methane (CH_4), which causes global warming. The microbial oxidation of methane in engineered covers is considered a potent option for the mitigation of methane emissions from landfills or sites containing wastes of low methane generation rates. Studies have shown that microbial oxidation of CH_4 in landfill cover soil is enhanced in the presence of organic matter-rich substrates. Hence, in this study, the methane oxidation potential for a mechanically separated, biologically treated, and anaerobically digested waste (MBT waste) is being investigated. Column experiments were devised for the study; the results of which indicate that the oxidation potential of the material is far above the reported average values. This suggests that the material could be satisfactorily used as a cover material in landfills. The gas concentration profiles obtained from the experiments were validated with the numerical model which gave matching results. This suggests that the methodology adopted to model the methane oxidation in the numerical model is satisfactory and could be used for future investigations.

Keywords Methane migration · TOUGH3 · Landfill covers · Methane oxidation

T. G. Parameswaran (✉) · G. L. SivakumarBabu
Department of Civil Engineering, Indian Institute of Science, Bangalore 560012, India
e-mail: parameswarantg@gmail.com

G. L. SivakumarBabu
e-mail: glsivakumar@gmail.com

N. Anusree
Center for Sustainable Technologies, Indian Institute of Science, Bangalore 560012, India
e-mail: anusreen.129@gmail.com

P. Sughosh
Department of Civil Engineering, Jawaharlal Nehru National College of Engineering, Shimoga 577204, India
e-mail: sughosh.p@gmail.com

T. K. K. C. Deepagoda
Department of Civil Engineering, University of Peradeniya, Peradeniya 20400, Sri Lanka
e-mail: chamindu78@yahoo.com

H. Hazarika et al. (eds.), *Advances in Sustainable Construction and Resource Management*, Lecture Notes in Civil Engineering 144,
https://doi.org/10.1007/978-981-16-0077-7_44

1 Introduction

Landfills are one of the major sources of atmospheric methane (CH_4), which causes global warming. The greenhouse potential of CH_4 is 21 times higher than that of CO_2 [7]. Fugitive and residual emissions of methane are found to have reduced through microbial methane oxidation in landfill cover soils. Recently, a number of researchers have recognized the potential for designing soil covers and biofilters that maximize CH_4 oxidation [3, 10]. More recently, research has focused on methods to increase CH_4 oxidation by improving landfill covers by adding organic-rich materials such as sludge and composts [8].

As per the latest report by Central Pollution Control Board of India, 127,486 tonnes of MSW is being generated per day as a result of rise in the economy [9]. It is reported that around 70–90% of Municipal solid waste (MSW) generated is disposed in open dumps and in landfills [6]. Unscientific methods of waste disposal would lead to air, soil, and groundwater pollution and thus in several countries, waste treatment units are installed which act as an intermediate treatment facility that takes a place before landfilling.

Studies have shown that methane oxidation in covers could be sufficiently enhanced through the addition of organic-rich materials which have sufficient porosity to assure the required supply of oxygen coupled with a satisfactory retention time [5]. Compost from waste treatment units, compost with wood chips or compost–soil mixtures are the commonly used biocover materials [2, 12].

Pariatamby et al. [7] investigated the methane oxidation efficiency with the addition of different organic wastes in different combinations through batch studies. The study concluded that addition of organic wastes at optimum ratio and combination with compost as landfill cover material will have a significant effect on methane emission reduction. Zulfiany et al. [13] investigated the characteristics of compost produced by waste treatment unit (WTU) in City of Depok for its potential as a biocover material. The results of the column study indicated that the compost could be sufficiently used for a biocover material. Huber et al. [4] tested the methane oxidation rates of 30 different compost materials (of different age and input materials) with laboratory column setups. The study suggested that bulk density, the organic matter, total organic carbon, nutrient content, water holding capacity, and electrical conductivity had a relevant impact on the oxidation efficiency of the investigated materials.

Literature reports various studies on compost (or waste) materials for its usability as a biocover in landfills. Even then, it is to be understood that expense of these materials is high so they are best suited for application in hotspots [1] or biowindrows [11]. Mechanically biologically treated (MBT) waste may be a viable alternative material for biocovers, since it is readily available waste in landfills and is less costly than compost. Thus in this study, the potential of mechanically biologically treated (MBT) waste material is checked for biocovers through column experiments and the results are validated through numerical modeling.

2 Materials and Methods

The MBT waste used in this study was collected from a municipal solid waste landfill in Mavallipura, India, and was anaerobically digested for 470 days. The MBT waste reaching Mavallipura landfill site was subjected to mechanical separation (involving a series of shredding and screening) and biological processes before disposing into landfills. The residual obtained was anaerobically digested which was used for the current study.

Column setups were utilized for the study. The sections were made of polyvinyl chloride, with 80 cm height and a 15.4 cm diameter. Every section had an aggregate of five side inspecting ports situated at 5–10 cm interims along the length of the column. An extra examining port was set on the outlet, to gauge the concentration of CH_4 inside the headspace region. The MBT waste was filled in the column to a height of 60 cm at a bulk density of 745 kg/m^3. An inlet gas 99.5% CH_4 was fed into the base of the segments at a stream rate of 25 mL/min simulating the range of fluxes encountered at landfills. Atmospheric air was passed over the headspace of column at a stream rate of 100 mL/min (to maintain natural oxygen concentration) after it went through the humidification chamber to give adequate O_2 important to CH_4 oxidation. Inlet flow rates were checked persistently utilizing rotameters. The flow rate was verified using bubble flow meters. The column experiment was conducted at room temperature. A 5 cm thick gas distribution layer composed of gravel was installed at bottom of column to allow a uniform flow of gas to the base. Above this layer, a geonet was put in to retain the MBT waste in position. Then the column was filled with MBT waste over the gravel layer in layers, with each layer subjected to light, uniform compaction. The schematic diagram and the laboratory column setup for the study are given in Fig. 1.

Gas samples from column were sampled by a pressure-lock syringe through rubber septa and were analyzed a gas chromatograph. In the gas chromatograph, O_2 and N_2 were separated using a molecular sieve and detected using a gas chromatograph equipped with a thermal conductivity detector (TCD). The gas chromatograph settings for oven temperature were 50 °C and with carrier gas as hydrogen. The peaks obtained were quantified with Peak ABC integration software and the gas sample concentrations were determined by comparing with standard gases.

3 Experimental Results

A gradual increment in methane oxidation rates was observed in column for 30 days. A decrease in the amount of oxygen and simultaneous increase in carbon dioxide concentration is seen in the column which is due to the action of methanotrophic bacteria. Figure 2 shows the variation of the methane oxidation potential with time.

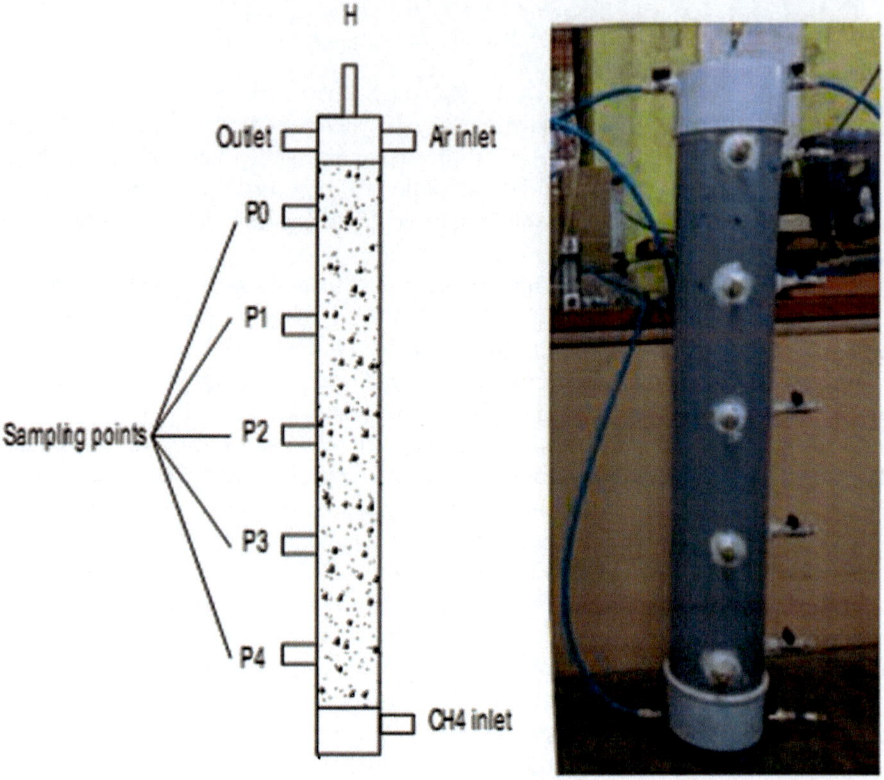

Fig. 1 Schematic diagram and laboratory setup of the column

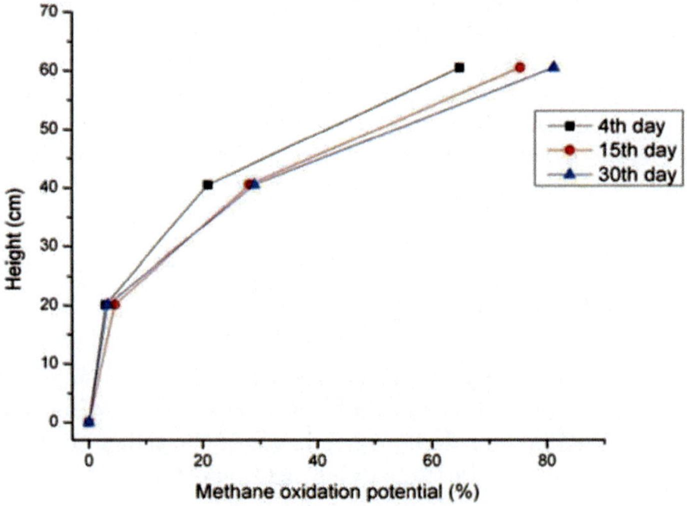

Fig. 2 Variation of methane oxidation potential with time

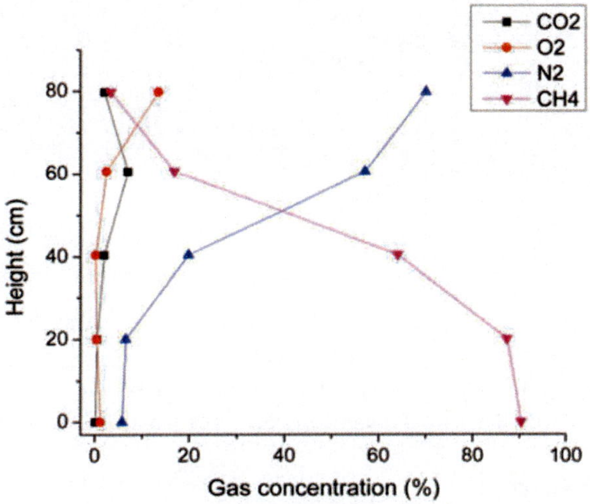

Fig. 3 Gas concentration profiles at the end of 30 days

The concentration values reached a steady state at the end of 30 days. The vertical gas concentration profile obtained for MBT waste in the column for 30 days is as given in Fig. 3.

Experimental results show a methane oxidation efficiency of 81% (calculated as the ratio of difference of bottom and top methane concentrations to the bottom concentration) for the MBT waste which is high enough to limit the atmospheric methane flux to a safe level.

4 Numerical Modeling

Numerical modeling was carried out with TOUGH3 EOS7CA to validate the experimental results. The details of the simulator are given in Table 1, governing equations in Table 2, and the properties considered in the simulation are as given in Table 3.

Since TOUGH3 EOS7CA does not have a module for oxidation, sinks were introduced in various elements in the column which extracts methane at a specific rate. As the oxygen diffusion is more at the top and less at the bottom, variable methane extraction rates were specified along the depth of the column. The oxidation capacity from the experiments is as shown in Table 4.

In the column study, as the methane concentrations were measured by collecting only the gas samples, in the numerical model only the methane present in the gaseous phase is taken for validation even though the migration occurs in both aqueous and gaseous phases. The methane concentration profile observed from the experiments and that obtained from the numerical simulation are as plotted in Fig. 4.

Table 1 Details of the simulator

Characteristics	Parameters chosen
Module used for simulations	EOS7CA; integral finite difference method
Solution of equations	Newton–Raphson iterations
Air composition	79% N_2, 21% O_2 by volume
Gas solubility	As per Henry's law
Mass transport mechanisms	Advection and Fickian molecular diffusion
Advective transport	As per multiphase extension of Darcy's law
Heat transport	Conduction, convection, and radiative transfer
Real gas mixture properties	Peng Robinson (1976)
Tortuosity value	Millington-Quirk model (1980)
Relative permeability calculation	Van Genuchten–Mualem model (Mualem, 1976; van Genuchten, 1980)
Capillary pressure function	Van Genuchten function (1980)
Components	Air, water, non-condensable gas (NCG), gas tracer, and brine
Phases	Two phases—gaseous and aqueous phases
Side boundaries	Neumann boundary condition (no flux)
Atmosphere	Dirichlet boundary condition (by assigning large volume element)

It can be seen that the methane gas profiles obtained from the numerical model is well correlated with the experimental results.

5 Conclusion

The methane oxidation efficiency in conventional covers is estimated to exist within a range of 0 to 10% of the liberated methane. Research on the topic of methane oxidation in covers promises possible values of even 100%, 40% being estimated as the average value. Since an oxidation potential of 81% was obtained for MBT waste in the column study, they can satisfactorily be used as an effective cover material. Also, as the methane concentration profile accounting the methane oxidation in the numerical model is well correlated with that from the experimental study, the methodology adopted for modeling the methane migration in TOUGH3 EOS7CA is satisfactory enough to be used for future purposes.

Table 2 Governing equations in the simulator

Eq. no	Equation	Description
1	$d_{0\beta} = \varphi\tau\rho_\beta d_\beta^\kappa$	Effective diffusion coefficients in soil
2	$\frac{\mathrm{d}}{\mathrm{d}t}\int\limits_V M^\kappa V_n = \int\limits_V F^\kappa \cdot n\mathrm{d}\Gamma_n + q^\kappa \mathrm{d}V_n$	Mass and energy balance equation
3	$M^\kappa = \varphi\sum_\beta S_\beta\rho_\beta X_\beta^\kappa$	Mass accumulation
4	$F_\beta = \rho_\beta u_\beta = -k\frac{k_{r\beta}\rho_\beta}{\mu_\beta}(\nabla P_\beta - \rho_\beta g)$	Phase flux
5	$F_{\mathrm{adv}}^\kappa = \sum_\beta X_\beta^\kappa F_\beta$	Component flux (advective)
6	$f_\beta^\kappa = -\varphi\tau\rho_\beta d_\beta^\kappa \nabla X_\beta^\kappa$	Component flux (diffusive)
7	$d_g^\kappa(P,T) = d_g^\kappa(P_0,T_0)\frac{P_0}{P}\left[\frac{T+273.15}{273.15}\right]^\theta$	Temperature dependence of gas diffusion coefficients (Vargaftik, 1975; Walker et al., 1981)
8	If $S_l < S_{ls}$, $k_{rl} = \sqrt{S^*}\left\{1-\left(1-[S^*]^{1/\lambda}\right)^\lambda\right\}^2$ If $S_l \geq S_{ls}$, $k_{rl} = 1$ and If $S_{gr} = 0$, $k_{rg} = 1 - k_{rl}$ If $S_{gr} > 0$ $k_{rg} = \left(1-\hat{S}\right)^2\left(1-\hat{S}^2\right)$ Where $S^* = \left(\frac{S_l - S_{lr}}{S_{ls} - S_{lr}}\right)$ and $\hat{S} = \left(\frac{S_l - S_{lr}}{1 - S_{lr} - S_{gr}}\right)$	Relative permeability (van Genuchten Mualem, 1980)
9	$P_{\mathrm{cap}} = -P_0\left(\left[S^*\right]^{-1/\lambda} - 1\right)^{1-\lambda}$; $P_{\max} \leq P_{\mathrm{cap}} \leq 0$	Capillary pressure function (van Genuchten, 1980)
10	$P_g^\kappa = K_H X_w$	Henry's law

Table 3 Properties considered for the simulation

Property	Values considered
Porosity	0.5
Absolute permeability (m^2)	5×10^{-13}
Degree of saturation	0.54
λ	0.11
S_{lr}	0.1
S_{ls}	1
S_{gr}	0.01
$1/P_0$	8.5×10^{-4}
$P_{\max}$	10^{10}

Table 4 Oxidation rates used for the simulation

Depth (cm)	Oxidation rate (kg/s)
60.5	1.173×10^{-7}
40.4	1.715×10^{-7}
20	5.921×10^{-8}

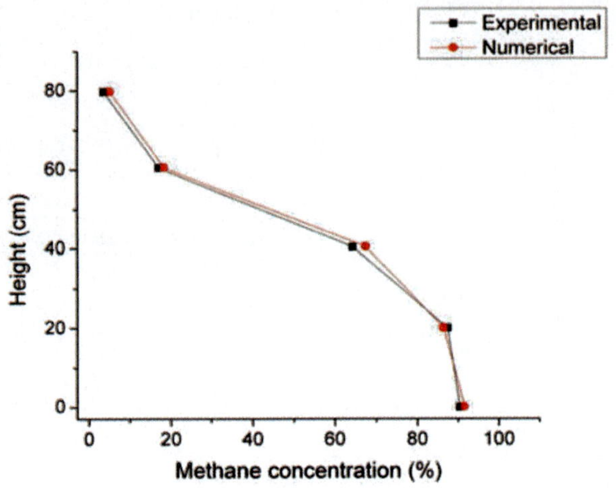

Fig. 4 Methane concentration profiles

References

1. Abichou, T., Chanton, J., Powelson, D., Fleiger, J., Escoriaza, S., Lei, Y., Stern, J.: Methane flux and oxidation at two types of intermediate landfill covers. Waste Manage. **26**(11), 1305–1312 (2006)
2. Barlaz, M.A., Green, R.B., Chanton, J.P., Goldsmith, C.D., Hater, G.R.: Evaluation of a biologically active cover for mitigation of landfill gas emissions. Environ. Sci. Technol. **38**(18), 4891–4899 (2004)
3. Feng, S., Ng, C.W.W., Leung, A.K., Liu, H.W.: Numerical modelling of methane oxidation efficiency & coupled water-gas-heat reactive transfer in a sloping landfill cover. Waste Manage. **68**, 355–368 (2017)
4. Huber, H.M., Tintner, J., Böhm, K., Lechner, P.: Scrutinizing compost properties and their impact on methane oxidation efficiency. Waste Manage. **31**(5), 871–883 (2011)
5. Humer, M., Lechner, P.: Design of a landfill cover layer to enhance methane oxidation-results of a two year field investigation. In: Proceedings Sardinia, vol. 1, pp. 541–550 (2001, October)
6. Kumar, A., Sharma, M.P.: Estimation of GHG emission and energy recovery potential from MSW landfill sites. Sustain. Energy Technol. Assessments **5**, 50–61 (2014)
7. Pariatamby, A., Cheah, W.Y., Shrizal, R., Thamlarson, N., Lim, B.T., Barasarathi, J.: Enhancement of landfill methane oxidation using different types of organic wastes. Environ. Earth Sci. **73**(5), 2489–2496 (2015)
8. Qin, L., Huang, X., Xue, Q., Liu, L., Wan, Y.: In-situ biodegradation of harmful pollutants in landfill by sludge modified biochar used as biocover. Environ. Pollut. **258**, 113710 (2020)

9. Rana, R., Ganguly, R., Gupta, A.K.: Physico-chemical characterization of municipal solid waste from Tricity region of Northern India: a case study. J. Mater. Cycles Waste Manage. **20**(1), 678–689 (2018)
10. Rannaud, D., Cabral, A., Allaire, S.E.: Modeling methane migration and oxidation in landfill cover materials with TOUGH2-LGM. Water Air Soil Pollut. **198**(1–4), 253 (2009)
11. Scheutz, C., Fredenslund, A.M., Chanton, J., Pedersen, G.B., Kjeldsen, P.: Mitigation of methane emission from Fakse landfill using a biowindow system. Waste Manage. **31**(5), 1018–1028 (2011)
12. Stern, J.C., Chanton, J., Abichou, T., Powelson, D., Yuan, L., Escoriza, S., Bogner, J.: Use of a biologically active cover to reduce landfill methane emissions and enhance methane oxidation. Waste Manage. **27**(9), 1248–1258 (2007)
13. Zulfiany, E.N., Kristanto, G.B., Hartono, D.M.: Assessment of compost from waste treatment unit as biocover material for Cipayung landfill at City of Depok. In: International Conference on Advances in Civil, Structural, Environmental, and Biotechnology, pp. 13–17. The IRED, Kuala Lumpur (2014)

Synergy Between Poly Vinyl Alcohol and Bentonite in Drilling Fluid Formulation: An Experimental Study

Danielle Poungui, Kyuro Sasaki, Yuichi Sugai, and Ronald Nguele

Abstract The formulation of a drilling fluid is mostly defined by the type of formation to drill. Ensuring drilling mud's effectiveness before throughout a drilling process implies monitoring its physico–chemical properties, including cake filtration. Cake filtration of a drilling mud can be either good or poor (loss filtration or fluid drag effect) depending on the mud swelling efficiency. A drilling mud made of bentonite (or Barite) still needs the assistance of appropriate additives to carry it to its optimum performance. The present study aims at monitoring the cake filtration of a water-based drilling mud using polyvinyl alcohol (PVA), which is known and used for its swelling abilities. The mud samples were formulated by blending two different types of PVAs (standard & synthetic PVAs) with deionized water at various concentrations (0.0%–0.3%) and 7% of bentonite. pH and density of muds were measured before filtration tests. The cake filtration was evaluated by injecting 10 mL of mud samples at constant pressure into a stainless steel Millipore sigma filter holder 90 mm at a constant pressure of 1.2 MPa. Results show that the addition of PVA increases the mud's acidity with the concentration of polyvinyl alcohol. The observed higher reliable content of the samples with standard PVA was confirmed by their higher density values, leading to lower filtrate productions than the muds samples with synthetic PVA.

Keywords Polyvinyl alcohol-bentonite dispersions · Cake filtration · Polymer · Drilling engineering · Rheology

1 Introduction

Water-based drilling fluid or WBDF is the general term associated to drilling mud (DF) which is formulated by addition of water (base fluid) and reactive solids as additives. To date, WBDF has still many disadvantages reported including formation damage, low cake properties, and high fluid loss [1]. The mud filtration stage begins

D. Poungui (✉) · K. Sasaki · Y. Sugai · R. Nguele
Kyushu University, Fukuoka 819-0395, Japan
e-mail: daniellepoungui@mine.kyushu-u.ac.jp

H. Hazarika et al. (eds.), *Advances in Sustainable Construction and Resource Management*, Lecture Notes in Civil Engineering 144,
https://doi.org/10.1007/978-981-16-0077-7_45

after forming a stable internal filter cake, which only allows the mud filtrate to invade the formation while setting the stage for the further evolution of the external filter cake. The fluid drag effect causes the filter cake to undergo compaction by allowing the smaller particles to penetrate deeper through it [2]. It is further reported that several properties among which formation permeability can alter the filtration performance of a typical DF [3].

The crossflow of the mud against the borehole, which also altered the filtration performance, is directly related to rheological properties of DFs. Therefore, ensuring a successful and safe drilling process implies monitoring and controlling physio–chemical characteristics of drilling muds. Extensive studies on polyvinyl alcohol (PVA), a water-soluble synthetic polymer, have unveiled its swelling abilities [4]. This feature has been exploited with to overcome in drilling fluid-related issues such as reducing torque and drag, controlling fluid loss, improving wellbore stability, and drilling performance [1].

For example, İşçi and Turutoğlu (2011) showed that DFs, prepared using both bentonite and sepiolite, could be stabilized by PVA, exhibiting thereby excellent rheological properties [5]. Although a wealth of information has been reported in the synergy between PVA and clays, few works have tackled the extent to which these synergistic interactions would alter the filtration of DF [6].

There, herein is aimed to evaluate the influence of PVA addition on the DF filtration. Several DFs were formulated from bentonite clays and aqueous PVAs. A series of filtration tests were conducted at 25 °C to investigate the influence of not only the type of PVA but also the concentration within the drilling mud. Mass of the filtrate as well as mud cake was selected as control parameters.

2 Materials and Methods

2.1 *Materials*

Bentonite clay was selected as primary clay for DF formulation. The polymers were obtained from Mitsubishi chemicals. The properties of the polymers used in this study are outlined in Table 1.

Bentonite and the polymers were used as received without any form of purification.

Table 1 Properties of PVA polymer[a]

Code	Charge/type	Hydrolysis type
PVA-1	Non-ionic/Standard	Fully hydrolyzed
PVA-2	Cationic/Modified	Partially hydrolyzed

Table 2 Composition of the formulated drilling muds

		No PVA	Standard PVA	Modified PVA
		DF_0	DF_1	DF_2
Bentonite		7 wt.%		
PVA	a		0.1 wt.%	
	b		0.2 wt.%	
	c		0.3 wt.%	

2.2 *Methods*

2.2.1 Drilling Fluid Formulation

The formulation of sample mud was performed following Eq. (1),

$$DFx\ (a, b, c) = \text{Bentonite} + (DI + x(a, b, c)\%PVA) \tag{1}$$

In total, seven samples were prepared using Eq. (1), with DF prepared without PVA been the control (DF_0). Table 2 outlines the compositions of the DF as per formulated in this work.

Sodium bentonite was added to the base fluids after ensuring the homogeneity of water–polymer mixtures by stirring. All drilling muds contained the same amount of water and bentonite. The samples were left for 12 h settling hours before performing any measurements.

2.2.2 Drilling Fluid Filtration

Prior the filtration, both the density and the pH of DF were recorded. Approximately, 10 mL of the investigated DF was transferred to the filtration cup of a stainless steel Millipore sigma filter holder within which was pre-installed a weighed filter paper (Φ 2.70 μm).

The cell was closed and a constant pressure of 1.2 MPa was then applied onto the muds by injection of Helium gas. The effluent fluid (filtrate) was recorded every hour for a total of 5 h. The mud cake mass was obtained at the end of each filtration test as the mass difference between the filter paper weight before and after filtration.

3 Results and Discussion

3.1 Cake Filtration

Figure 1 shows the results of cake filtration in respect of polymer additions.

Though the injecting pressure was set to be constant throughout the filtration, there was oscillation of pressure during filtration process. The increase of pressure was the accumulation of helium gas in the filtration area as it was also breaking the gas–solid/liquid interface and penetrating the mud. Under an applied pressure, fine particles trickle through the cake and block the pores of the cake or the filter medium [7]. The filtrate therefore goes through open pores of the cake before being produced, resulting in gas ejection and pressure reduction.

However, the rate of filtrate loss over time and the amount of filtrate production are an equally important indicator of the solid–liquid ratio and porous nature of the mud, directly related to the good isolation. The excess of filtrate production obtained during filtration tests of samples DF0 and DF1/2 at 0.1% PVA indicates a large amount of liquid content in the said samples. The least polymer addition in the samples has a negligible effect on the inhibition performance of the muds. The lesser inhibition and expansion of clay particles of the more deficient hydration in the mud make it easier to produce. Such cases should be avoided in situ to prevent loss fluid and view drilling problems related to it. On the other hand, when it was found that adding PVA, especially at a concentration of 0.3 wt%, a lower volume in the filtrate is obtained. Under such circumstances, it requires more time and more force for the gas to find ways thought the mud, thus reducing the gas's sweeping efficiency and a significant reduction of liquid filtration into the formation. Making the muds with 0.2% PVA concentration has relatively better filtrate production. A rational polymer concentration provides a better clay dispersion and solid–liquid proportions required for good filtration and convenient mud cake thickness after filtration.

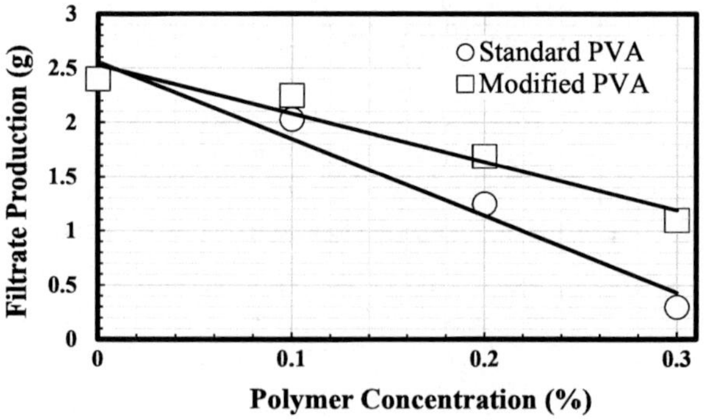

Fig. 1 Cake filtration readings

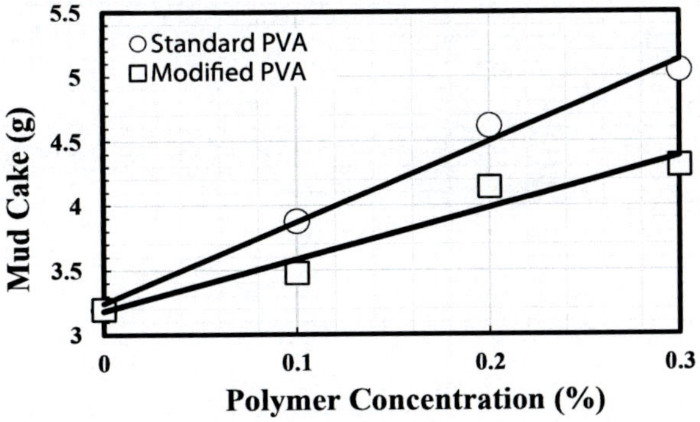

Fig. 2 Mud cake mass as function of PVA addition

3.2 *Mud Cake Mass*

Figure 2 shows the results of the remaining mud cake as function of the PVA concentration.

A positive correlation between the mud cake and the polymer concentration could be seen with a larger volume obtained for standard PVA. For example, adding 0. 2wt% of standard PVA yields 4.6 g of cake mud, which was 11% higher compared to the DF formulated from modified PVA. Recalling the results conveyed from Fig. 1, the mud samples with cationic modified polymer showed more liquid content than those with standard polymer.

In general, during inhibition, water is gradually squeezed out from the upper and lower layers of clay with an increment of the adsorption degree, thereby reducing clay of hydration [8]. PVA has the ability to maintain a certain amount of refined grains in the mud mainly by bridging role and forming a stable space grid structure with clay particles, creating a compact mud cake on the borehole wall [7]. The lower mass of cake obtained from DF2 could be imputed to a selective adsorption of clay particles for cations, especially the cations with lower hydration [8]. An internal filter cake is formed when well-sized mud particles deposit in and around the wellbore, in high concentration [2] Also, the fact that the cake thickness increases with the concentration in polymer is due to static filtration conditions, which consists of the dead-end filtration in filter press and accumulation of suspended particles on the filter face [1].

3.3 *Influence of pH and Density on Cake Mud Filtration*

3.3.1 Contribution Mud Acidity on Mud Cake

Figure 3 shows the influence of the mud acidity of mud filtration.

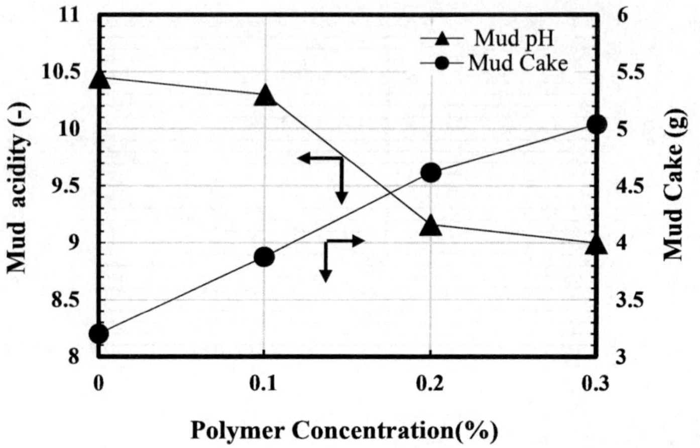

(a) Muds formulated from standard PVA

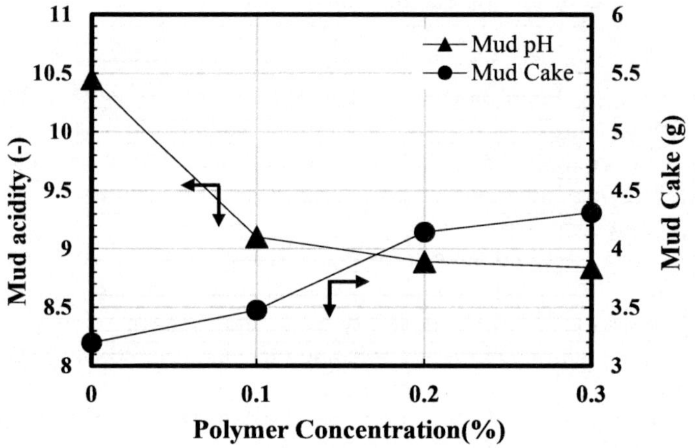

(b) Muds formulated from modified PVA

Fig. 3 Influence of pH and relation with mud cake

The acidity of the mud decreases monotonically with the addition of PVA-OH in drilling muds. Considering that the optimal pH for better water-based drilling mud performance ranges between 8 and 10 [3], the polymer has a regulating effect on the drilling mud.

A Michigan State University's journal discussing about soil pH stated that, the main elements affecting soil pH are calcium, magnesium and potassium. On their own, these elements keep pH on the alkaline side. Soil pH is related to how well the soil hangs onto these elements. sand typically has a lower pH than clay since water moves through sand faster than clay [9]. When added to drilling mud, the polymer sticks to clay particles surface through hydrogen bonding, which is the case of standard PVA (DF_1, Fig. 3a). However, this effect is most expected from clays with higher cation exchange capacities and expandable clays that allow water to access the interlamellar space which houses the exchangeable cations[10]. With the presence of a charged moiety on the PVA ($NH4^+$), a competitive adsorption with between the different silicate ions ($SiO_3{}^{2-}$, $SiO_4{}^{4-}$) of the bentonite clay is prompted. The opposite charged ions packed and are held together by forces of electrical attraction with each cation attracting all neighboring anions [11]. This causes a decrease in mud cake mass.

In a similar work, Gamal et al. showed that increasing the pH from 9 to 10 resulted in increasing the filtrated volume by 33% [12]. They argued that altering the pH of the aqueous solutions, within which the bentonite clay is mixed, has an impact on the filtration properties for the mud.

3.3.2 Contribution Mud Density on Mud Cake

Figure 4 Established the relationship between the mud cake and the mud density

It was found that density of the mud increases with the polymer concentration of that cake mud. Those with a high concentration of polymer were denser and exhibited a gelling-like behavior as it shows in Fig. 1. Increasing the polymer concentration promotes a highly charged solution from which stronger electrostatic interactions exist.

Muds, formulated from polymers (DF_1 or DF_2), showed higher inhibiting performance, herein expressed by the density, due to polymer high molecular weight of the polymer. A plausible reason could be the electrostatic attraction of polymer with clay particles and other adjacent clay particles creates bridge flocculation made by two clay particles with their surrounding polymer extensions when they approached each other [13].

However, there are suspensions in which flocculation occurs while the electrostatic forces are less important (DF_0 for example). In that event, the adsorption can cause repulsion between solid particles and the attraction of counter ions into the particles [14]. This can explain why samples with standard polymers bearing lesser charges particles, thus hosting lesser reactions, have higher swelling. As a result of these interactions, the reduction of water content in the solution as new clusters are formed among clay particles leading to much more resistance against fluid flowing [15]. Therefore, the polymer mud could prevent rock dispersion and keep the stability of the borehole.

3.4 Conclusions

This experiment was a synergistic study of the interactions between polyvinyl alcohol (PVA) and clay particles as well as the influence of pH and density on cake filtration. Based on the results obtained above, a better filtration rate is obtained when PVA is

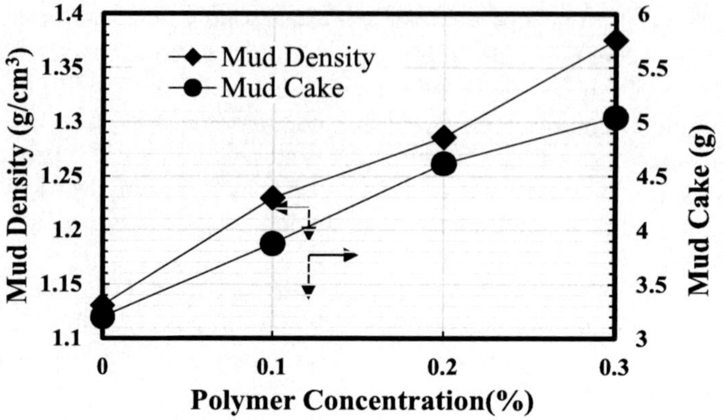

(a) Muds formulated from standard PVA

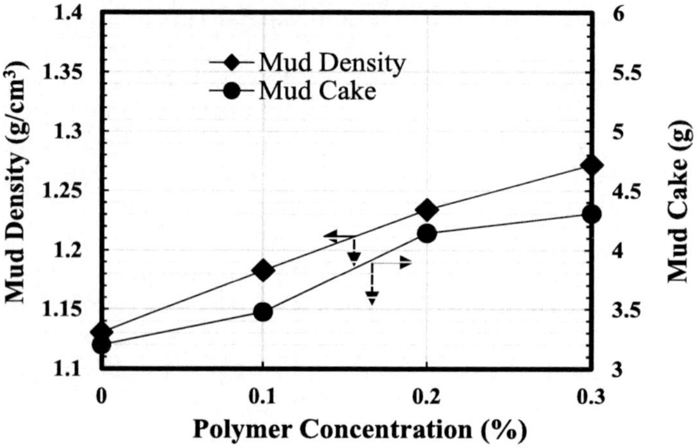

(b) Muds formulated from modified PVA

Fig. 4 Mud density in respect of polymer concentration

added as additive in the mud. Using standard PVA, a larger mud cake was obtained owing to hydrogen bonding between the silicate moieties of the bentonite and the hydroxyl group of PVA.

The addition of cationic PVA promotes a better filtration due to the competitive adsorption between the charged sites of the polymer and that of the bentonite clay. The concentration at which the polymer is added to the mud is a determinant factor of mud filtration performance. Enabling us to know at which polymer concentration drilling problem such as fluid loss and fluid frag can be avoided. A maximum polymer concentration can lead the properties of the designated mud to exceed the range of acceptability of significant drilling mud. The addition of cationic polymer can somehow improve the performance of DF.

References

1. Saboori, R., Sabbaghi, S., Kalantariasl, A., Mowla, D.: Improvement in filtration properties of water-based drilling fluid by nanocarboxymethyl cellulose/polystyrene core–shell nanocomposite. J. Pet. Explor. Prod. Technol. **8**(2), 445–454 (2018)
2. Ezeakacha, C., Salehi, S., Hayatdavoudi, A.: Experimental Study of Drilling Fluid's Filtration and Mud Cake Evolution in Sandstone Formations, no. December, 2016
3. Australian Drilling Industry Training Committee Limited. (ed.).: The Drilling Manual, 5th edn. Taylor & Francis Group, Boca Raton (2015)
4. Caenn, R., Chillingar, G.V.: Petroleum Drilling Fluids: State of the Art (1996)
5. Işçi, E., Turutoğlu, S.I.: Stabilization of the mixture of bentonite and sepiolite as a water based drilling fluid. J. Pet. Sci. Eng. **76**(1–2), 1–5 (2011)
6. Taghdimi, R., Karimi, G.: Modification of Nano Clay Systems: An Approach to Stabilizing Drilling Fluids, vol. 829, pp. 818–824 (2014)
7. Walzel, P.: Filtration-Fondamentals. Ullmann's Encycl. Ind. Chem. 25.1–25.40 (2012)
8. Qi, L.: The application of polymer mud system in drilling engineering. Proc. Eng. **73**, 230–236 (2014)
9. Goldy, R.: Understanding Soil pH Part I—MSU Extension (2011)
10. Hatch, C.D., Wiese, J.S., Crane, C.C., Harris, K.J., Kloss, H.G., Baltrusaitis, J.: Water Adsorption on Clay Minerals As a Function of Relative Humidity: Application of BET and Freundlich Adsorption Models (2012)
11. Shan, W., Tao, S., Fu, F., Yue, W., Zhao, Z.: Research on the drilling fluid technology for high temperature over 240°C. Proc. Eng. **73**(May), 218–229 (2014)
12. Basfar, S., Al-Majid, A., Gamal, H., Elkatatny, S.: Effect of pH on rheological and filtration properties of water-based drilling fluid based on bentonite. Sustainability **11**(23), 1–13 (2019)
13. Işçi, S., Ünlü, C.H., Atici, O., Öngör, N.: Rheology and structure of aqueous bentonite-polyvinyl alcohol dispersions. Bull. Mater. Sci. **29**(5), 448–456 (2006)
14. Brostow, W., Lobland, H.E.H., Pal, S., Singh, R.P.: Polymeric Flocculants for Wastewater and Industrial Effluent Treatment, vol. 31, pp. 157–166 (2009)
15. Işci, S., Günister, E., Ece, Ö.I., Güngör, N.: The modification of rheologic properties of clays with PVA effect. Mater. Lett. **58**(12–13), 1975–1978 (2004)

Advancement in Sustainable Construction Techniques

Bearing Capacity Determination of Jointed-Timber Piles in the Saga Lowland

Sailesh Shrestha, Yuki Matsumoto, Shunsuke Moriyama, and Norihiko Miura

Abstract The pile load tests were conducted in the soft clay with high compressibility and lower shear strength with depths until 10–30 m in the Saga lowland, Japan. Generally, single-timber piles (STP) were used as the pile foundation of the small-scale buildings in this area. However, differential settlements were experienced in these buildings. To overcome the problems, the tip of the timber piles was required to reach supporting stratum at greater depth. In order to reach the stratum, the timber piles were connected with several joints. Until then, there were methods to calculate the bearing capacity of STP. However, there was no detailed study to describe the bearing capacity and its mechanism of the jointed-timber piles (JTP). Based on the several pile load test results, the bearing capacity evaluation method has been proposed for the JTP in terms of bearing capacity coefficients α and β. It was confirmed that the second limit resistance forces were higher than the bearing capacity evaluated with the proposed method, including the reduction effect of joints, length/diameter ratio.

Keywords Jointed-timber piles · Second limit resistance force · Bearing capacity · Saga lowland

1 Introduction

Timber is one of the construction resources which is sustainable in the context of Japan. To preserve the woods in a well-balanced condition, the matured trees whose capacity to absorb CO_2 is lower can be harvested and replaced by planting trees. When the new trees grow, they absorb and store the CO_2 emissions that are driving

S. Shrestha (✉) · Y. Matsumoto · S. Moriyama
Kyushu Piling Co. Ltd., Yanagawa City, Fukuoka 832-0082, Japan
e-mail: sailesh@alumni.ait.asia; office@qp-pile.com

N. Miura (✉)
Laboratory of Soft Ground, Saga 840-0811, Japan
e-mail: miuran@viola.ocn.ne.jp

H. Hazarika et al. (eds.), *Advances in Sustainable Construction and Resource Management*, Lecture Notes in Civil Engineering 144,
https://doi.org/10.1007/978-981-16-0077-7_46

global warming. The trees can be planted in a vast amount, and the surplus trees can be used for engineering purposes.

Single-timber piles (STP) were used as the foundation of the small-scale buildings in the Saga lowland, coastal area of the Ariake sea, Japan (Fig. 1). Here, the STP means the pile without any joint and is only one piece of the timber. Since the depth of the soft clay/soft ground in the Saga lowland varied from 10 to 30 m with high compressibility and low shear strength, due to the short length of the STP piles, the problem of differential settlement of the building foundation occurred. The construction code of conduct became very strict in this area with regard to the settlement. To reduce the settlement of the building, it is necessary to use the longer length of timber piles to reach the supporting strata at greater depth with higher standard penetration test, SPT (N) values. The maximum length of each timber stake used as the pile foundation should be less than or equal to 12 m. Due to transport restrictions because of the narrow road to the site, the average length of the timber stakes used is 3, 4, 5, and 6 m. Therefore, to meet the design length reaching the bearing stratum, the timber stakes are jointed with joint reinforcement (steel pipes or rebar) between them. The conventional method is available for calculating the bearing capacity of the STP. However, there is no method update to calculate the bearing capacity of the jointed-timber pile (JTP). In this paper, an attempt has been made to understand the mechanism and estimate the bearing capacity of the JTP.

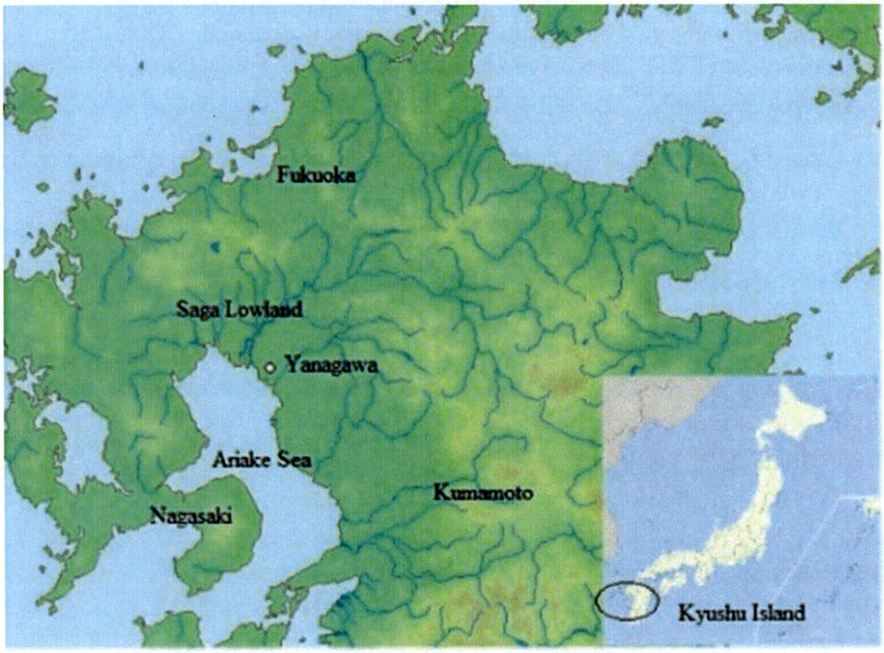

Fig. 1 Saga lowland

Fig. 2 Jointed-timber pile installation

JTP consists of top concrete or preservative-treated timber combined above the groundwater level and timber stake under the ground water level as a composite pile to reduce the settlement and increase the bearing capacity of the soft ground in Saga lowland (Fig. 2). From the results of maximum pile load or second limit resistance force $\left(R'_u\right)$ from the 54 number of pile load tests in 12 different locations in the Saga lowland, the bearing capacity evaluation method for the design of JTP has been proposed in which the bearing capacity coefficients α and β is based on the results of the Swedish Weight Sounding (SWS) test. The proposed method carries out a large number of loading tests. The results from the proposed method for JTP were compared with the results from the STP in terms of the relation between the shear resistance (τ) and $\overline{N'_f}$ values. Further, the bearing capacity of the pile in terms of allowable compressive force ($_LR_a$) was also compared with the second limit resistance force $\left(R'_u\right)$ and the bearing capacity determined using the coefficients α and β (R_u).

2 Field Tests

Field tests were conducted in the 12 different locations in the Saga lowland. The locations and the test periods of all the sites and the range of the pile's depth into the ground have been shown in QP Code 2 2018, the manual of Kyushu Piling co. Ltd, certified by the General Building Research Corporation of Japan (GBRC) [1]. The range of length of the timber piles was 2.5–23.5 m.

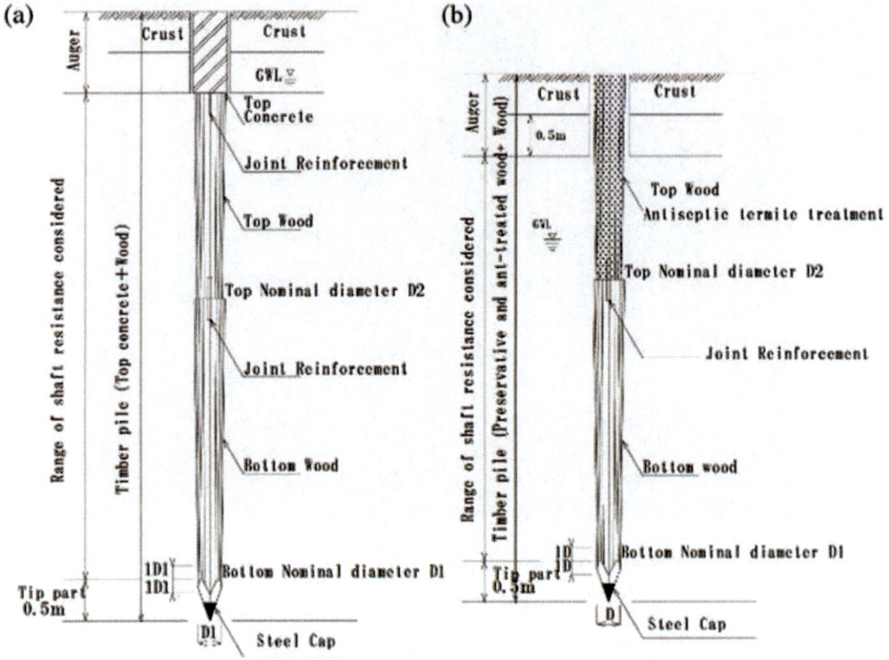

Fig. 3 Schematic of Jointed-timber pile **a** Top concrete as a cap, **b** treated timber as a cap

2.1 *Field Installation*

Pile driving equipment (Pulsonic 4B, pushing force 10–16 tonnes, 20–60 Hz) is used for driving the timber pile deep into the ground. It is reported in the literature "the timber piles remained healthy for hundreds to thousands of years as long as they are used in a reducing environment below groundwater level." Above the groundwater level, timber piles should be treated with a preservative to prevent degradation of the wood from insect attack, for example, termites and white ants. Figure 3 demonstrates the layout of the timber piling adopted in this JTP method. Timber logs are connected through the steel pipes joints. Figure 3a shows the configuration in which the timber stake is covered with the top concrete. In contrast, Fig. 3b shows the timber stake covered with preservative-treated timber above the groundwater level.

2.2 *Ground Survey Tests*

The Saga lowland's ground profile is reported in the literature [2]. Before installing the timber piles, the ground is surveyed by the Swedish weight sounding (SWS) test. In the current pile load test locations, the standard penetration test (SPT) was also performed to assure the ground properties. The Japanese Geotechnical Society (JGS) specifies the SWS test target ground to a maximum depth of about 10 m. It is due to the influence of shaft friction near the surface layer and deep layer. However, from construction experience, the accuracy of the SWS survey depth of 10 m or more is not significantly impaired on the soft ground along the Saga lowland. The test site's geotechnical properties are described in the QP code 2 2018 [1] document, but here, the ground profile of only one site E has been illustrated in Fig. 4. In Fig. 4, the SWS value of the topsoil 0.0–1.9 m depth was 0.9 after loosened by the auger. To reduce the lateral ground displacement, an auger is used to loosen top hard soil. Therefore, in the design, the SWS value of the topsoil was ignored.

Fig. 4 Ground profile

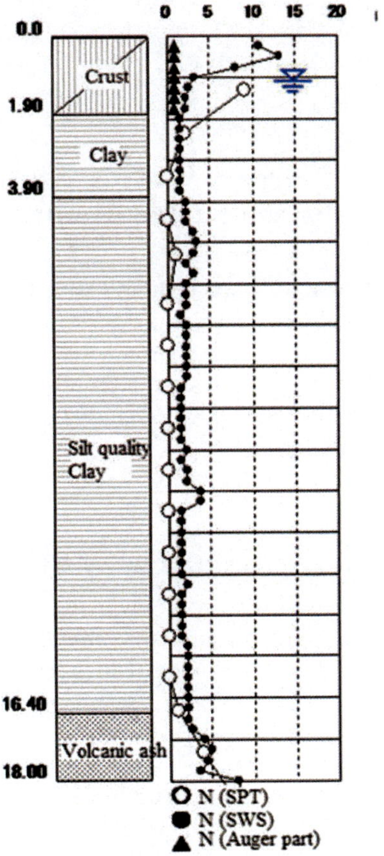

2.3 Procedure for Static Axial Compressive Load Test

A total of 54 axial compressive load tests from site A to site L was carried out following the Japanese Geotechnical Society Standards [3] [JGS1811-2002]. The loading method is a static one-step loading method. In the one-step loading method, the test was carried out for 30 min in each step and took 5–7 h to apply the final load thoroughly. The loading stages were eight or more, and the loading speed was the planned maximum load/number of loading stages. The applied load increment was maintained using a calibrated hydraulic jack, and the vertical displacement of the pile head was measured using four dial gauges attached to two reference beams. The displacements obtained from the four dial gauges attached on opposite sides of the pile head were averaged to determine the corresponding head displacement of the pile. For each load increment, the pile head displacement was recorded at elapsed times of 1, 2, 5, 10, 15 min, and every 15 min.

3 Proposed Bearing Capacity Evaluation Method for the Jointed-Timber Piles

A total of 54 axial compressive load tests (referred also as pile load tests) from Site A to Site L was performed in 12 different locations in Saga lowland. In this section, the bearing capacity evaluation method of jointed-timber piles is established at first. The safety of using the proposed bearing capacity evaluation method is confirmed by comparing it with the second limit resistance force $\left(R_u'\right)$ obtained from the pile load test results.

3.1 Determination of Bearing Capacity Evaluation Method

According to the notification of [4], there is a provision for the bearing capacity calculation with the equation as follows:

$$R_u = \alpha \overline{N_\mathrm{p}'} A_\mathrm{p} + \beta \overline{N_\mathrm{f}'} \pi DL \tag{1}$$

where β = soil resistance factor at the periphery according to local experience, α is the coefficient related to the tip support force, A_p = area of the end pile at the tip, $\overline{N_\mathrm{p}'}$ and $\overline{N_\mathrm{f}'}$ are the average value of the results from the SWS test, D = diameter of the pile, and L = length of the pile. To successfully use Eq. (1) in the field, if the N values are known from the ground survey tests, then the value of the α and the β should be estimated. The value of α and β vary depending on the location. Since the ultimate load measured when a load is applied to the head of the pile is the sum of the tip support force and the peripheral friction force, various combinations can be

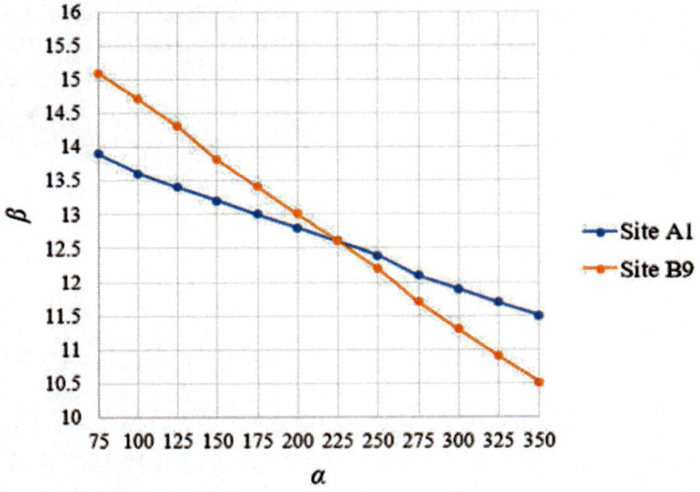

Fig. 5 Relationship between α and β

considered between the coefficient α and the coefficient β. For each of the 54 tests that were confirmed, the value of α is taken as the width of the interval of 25 from 75 to 350 [4]. Then the value of β is back-calculated from Eq. (2).

$$\beta = \frac{R'_u - \alpha \overline{N'_p} A_p}{\overline{N'_f} * \pi DL} \tag{2}$$

where R'_u = second limit resistance force obtained from the pile load test. The values of the α and β for all the sites are tabulated in QP code 2 2018 [1]. The intersection point of the lines with minimum β values was taken (Fig. 5). Figure 5 shows an inflection point at $\alpha = 225$ and $\beta = 12.6$. Thus, the values of $\alpha = 200$ and $\beta = 12.0$ was set for safe evaluation method. Substituting the value of the α and β in Eq. (1), the obtained equation is as follows:

$$R_u = 200\overline{N'_p} A_p + 12.0\overline{N'_f} \pi DL \tag{3}$$

which is proposed for the design of the JTP. According to the [4] of Japan rule, the long-term bearing capacity of the pile is considered as one-third of the total value as depicted below:

$$\text{Long term allowable bearing capacity, } R_a = 1/3 R_u \tag{4}$$

where R_a = allowable bearing capacity (kN), R_u = design bearing capacity from Eq. 3 (kN). The allowable bearing capacity obtained from Eq. (4) must be higher than the design load. Further, the analysis of the bending failure of the piles is essential. The load–moment (M–N) interaction curve can interpret the bending failure of the pile.

In this research, the author did not include the bending moment investigation because the pile head is free, i.e., the pile head are not connected to the footing of the building, and in the case of earthquake or eccentrical load, there is no considerable bending moment in the piles.

3.2 *Results from Pile Load Tests*

Table 1 gives the value of the second limit resistance force of pile load tests at two different locations, Site E and Site L, and Fig. 6 shows the head displacement relationship under load. The other location test results are included in the QP code 2 2018 [1]. The second limit resistance force is defined as the maximum load capacity of the pile when the tip displacement of the timber pile reaches 10% of the nominal diameter or when the indentation resistance reaches the maximum within the range of 10% or less, as defined by JGS. If the tip displacement could not be measured, the value of tip displacement is equivalent to the head displacement. In Fig. 6a, the pile with a length of 5.5 m has a maximum load of 76 kN, and with a length of 16.5 m

Table 1 Second limit resistance force from the pile load test

Test number	Length of pile (m)		Nominal diameter (mm)		Second limit resistance R'_u (kN)	Value calculated by proposed method			
						R_p (kN)	R_f (kN)	R_u (kN)	R'_u/R_u
E1	5.5		160		76	10.2	54.1	64.3	1.18
E2	9.4		140		112	4.6	88.3	92.9	1.21
E3	4.8	16.5	210	178	285	8.2	205.9	214.1	1.33
	6.0		180						
	5.7		150						
L1	5.8	23.5	220	185	249	8.8	150.8	159.6	1.56
	6.0		200						

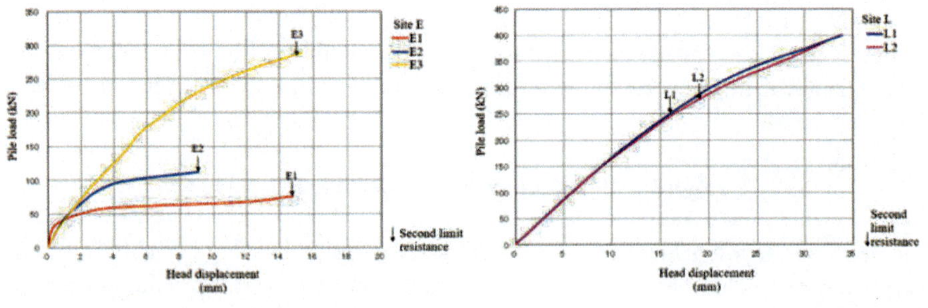

Fig. 6 Pile head displacement to pile load

has a maximum load of 285 kN. It proves that the timber pile has considerable shaft resistance and varies significantly with the pile's length.

3.3 Validation of the Proposed Bearing Capacity Evaluation Method

The design bearing capacity (R_u) of each test pile was calculated by Eq. (3) in Sect. 3.1. It was then compared with the second limit resistance force, R'_u directly obtained from the pile load test. Table 1 shows the second limit resistance $\left(R'_u\right)$ ratio from the pile load test to the design bearing capacity (R_u) calculated by Eq. 3. Here, in Fig. 7 four sites have JTP, whereas eleven sites have STP. Further, Fig. 7 shows the relation between these. According to the proposed bearing capacity evaluation method, the second limit resistance force of the timber pile was 1.3 times than the estimated design bearing capacity by Eq. (3). The proposed approach is safe to be used in the field. This method can be used for both JTP and STP.

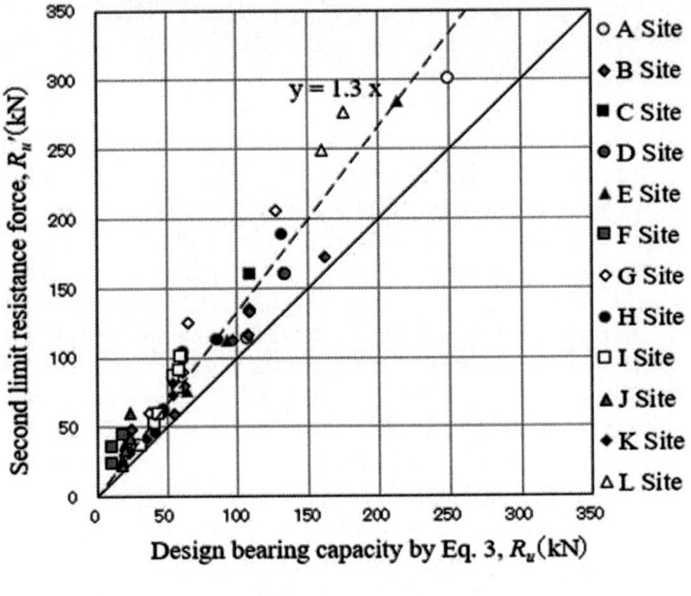

Fig. 7 Ratio of second limit resistance force $\left(R'_u\right)$ to the design bearing capacity (R_u)

4 Performance Confirmation of the Jointed-Timber Piles

The timber piles are connected with joint reinforcement to meet the design length. In this case, it is necessary to consider the decrease in timber bearing capacity regarding the shear strength and the compressive strength aspects.

4.1 Shear Resistance of the Timber Piles with or Without Joints

Since the bearing capacity due to peripheral friction is dominant to the tip bearing capacity (Table 1), the shear resistance acting at the periphery (τ) has been investigated to confirm the single-timber pile and jointed-timber pile's bearing capacity mechanism. Here,

$$\tau = R_f/\pi DL \tag{5}$$

where τ = Shear resistance of pile acting at the periphery (kN/m^2).

Figure 8a represents the shear resistance of all the timber piles from short to long, i.e., without joints (STP) and with joints (JTP) from length 2.5 to 23.5 m, including overall Sites from A to L. The upper limit of overall tests data hints $\tau = 16.9\ \overline{N_f'}$ whereas the lower limit hints $\tau = 12.8\ \overline{N_f'}$. The value of $\tau = 16.9\ \overline{N_f'}$ can be compared with the value of $\tau = 17.3\ \overline{N_f'}$ in Fig. 8b, which is the relationship between τ and $\overline{N_f'}$ when the timber piles are connected with the joints in between. Therefore, it can be judged that there is no decrease in the bearing capacity of the pile due to the joints. The results of the jointed-timber pile bearing capacity in terms of shear resistance have no reduction compared with the results of the single-timber pile. Hence, it is seen that the bearing capacity mechanism is same for both STP or JTP. In the legend at the bottom right corner, *n* means the number of joints of the pile. The length of the pile for the Site E and Site L is in Table 1. Site H has the two piles of length 9.5 and 14.5 m, and Site I has a pile of length 9.5 m. It can be seen, Figs. 5 and 8a, the results agree with each other. In Fig. 5, the value of β was 12.6, which was the minimum slope intersection from all the sites. Also, it can be seen in Fig. 8a, the line of τ vs $\overline{N_f'}$ with a minimum slope is 12.8. Further, previously as it was decided to use the value of $\beta = 12.0$, the line corresponding to it, $\tau = 12.0\ \overline{N_f'}$ lies lower than all the lines, which means, $\beta = 12.0$, is a safe value for design.

4.2 Compressive Strength of the Jointed-Timber Pile

The calculation of the allowable compression force of the timber pile is based on Eq. (6).

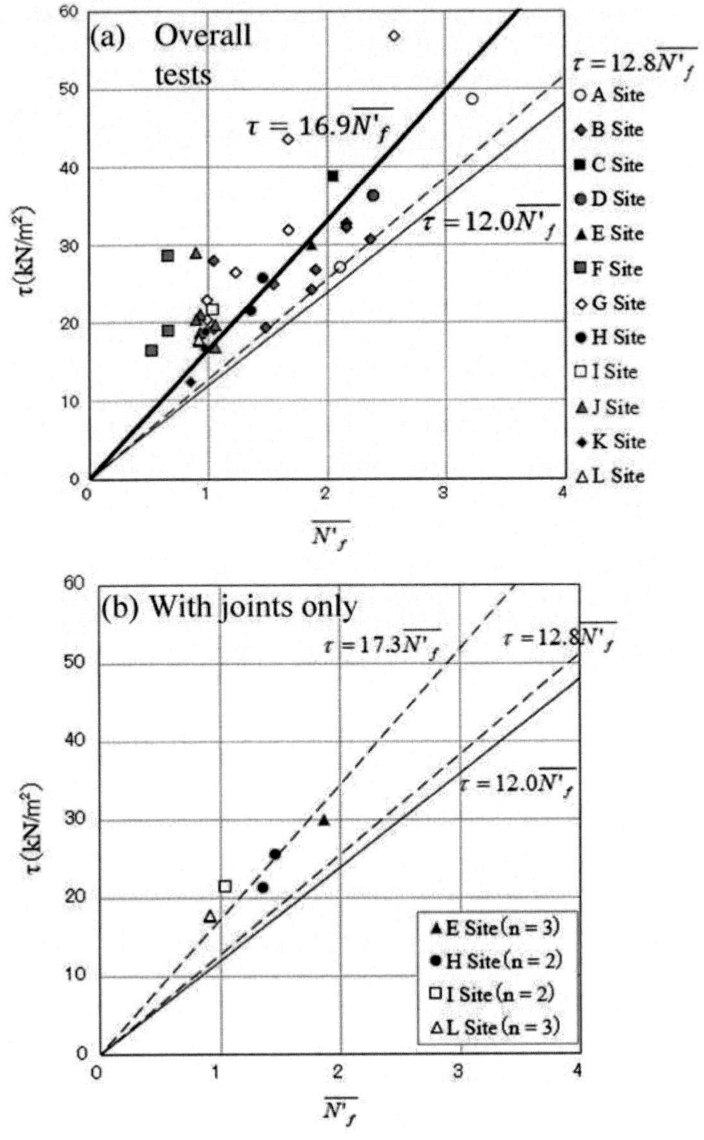

Fig. 8 Relationship between the τ and N value at the periphery of the pile $\left(\overline{N'_f}\right)$

$$\text{Long term, } {}_{L}R_{\text{a}} = A'_{\text{p}} \cdot {}_{L}\, f_{\text{c}} \tag{6}$$

${}_{L}R_{a}$ is the long term allowable compression force (kN) of the timber pile, A'_{p} is the nominal diameter cross-section area (m^2); ${}_{L}f_{\text{c}}$ is the long-term allowable compressive strength (kN/m^2) of the timber pile, where ${}_{L}f_{\text{c}} = 0.7 \times F_{\text{c}} \times (1.1/3)$. Here, 0.7 is the moisture content influence coefficient, F_c is the compressive strength, according to the [5] notification no. 1524. Equation 6 is a limit state design considering the ultimate limit state, where the maximum load is considered beyond the elastic limit.

4.3 *Reduction in the Compressive Strength of the Timber Pile Due to the Joints*

It is necessary to consider the reduction in compressive strength due to the joints in JTP. In the case of a maximum length of 23 m, three pieces of wood, and top concrete or top wood with preservative-treated timber are joined, so the maximum number of joints is three. Because of safety, a reduction rate of 0.1 for top concrete to timber joint (the joint is rebar) and 0.15*n* for timber to timber joints (joint is steel pipes) is considered. Further, it is necessary to consider the reduction rate *b* as:

$$b = (L/D - 60)/100 \tag{7}$$

where the length (*L*) is the total length of the pile in m, and the diameter (*D*) is the diameter at the top of the pile. Thus, the material final allowable compression force considering the two reduction rates (1) number of joints (2) L/D ratio is as follows:

With top concrete (long term)

$$_{L}R_{\mathrm{a}} = 0.7F_{\mathrm{c}}.(1.1/3)\cdot(1 - 0.15n + 0.05 - b)\cdot A'_{\mathrm{p}} \tag{8}$$

Without top concrete (long term)

$$_{L}R_{\mathrm{a}} = 0.7F_{\mathrm{c}}.(1.1/3)\cdot(1 - 0.15n - b)\cdot A'_{\mathrm{p}} \tag{9}$$

where A'_{p} is the wood cross-sectional area (m^2) based on the nominal diameter considering weighted average, *n* is the number of joints. The reduction rate of 0.1 has been included in Eq. 8. The long-term allowable compression force $_{L}R_a$ determined using Eq. (8), and the R_{a} determined using Eq. (4) was verified by comparing with the 1/3 of the second limit resistance (R'_u) as shown in Table 2.

Table 2 Comparison of allowable bearing capacity (R_{a}) and compression force ($_{L}R_{\mathrm{a}}$)

Test number	Length of pile (m)	Nominal diameter (mm)	Top diameter (mm)	$(1 - 0.15n + 0.05 - b)$	$_{L}R_{\mathrm{a}}$ (kN)	$R_{\mathrm{a}} = 1/3R_u$ (kN)	$R'_u/3$ (kN)
E3	16.5	178	250	0.48	62.5	71.4	95.0
L1	23.5	185	250	0.20	28.1	53.2	83.1
L2	23.4	202	250	0.20	33.6	58.4	92.2

The maximum allowable compressive force (${}_{L}R_{\mathrm{a}}$) is smaller than the long-term allowable bearing capacity (R_{a}) from the proposed method in all the cases. Therefore, in the field construction for the safety, the smaller value of the ${}_{L}R_{\mathrm{a}}$ or R_{a} was taken as the final design bearing capacity. Further, both the values are smaller than the value of 1/3 of the second limit resistance force value $\left(R'_{u}\right)$. Therefore, it is verified that the force equation (Eqs. 8 and 9) gives a safe value. Even though the timber pile in the field tests was able to sustain the high load, the reason behind using the ${}_{L}R_{\mathrm{a}}$, i.e., the minimum of the ${}_{L}R_{\mathrm{a}}$ or R_{a} is for the purpose of safe design.

5 Conclusions

Single-timber piles (STP) were used in the Saga lowland, Japan, as a pile foundation for the small-scale buildings. However, the problem of differential settlement of the building foundation occurred. To reduce the settlement and increase the bearing capacity of the highly compressible and low shear strength soft clay, timber piles had joint reinforcement between them to reach the bearing stratum at a greater depth. These piles with a joint in between them were called jointed-timber piles (JTP). Until now, the conventional bearing capacity method only explains the STP. However, no method is available to evaluate the bearing capacity of the JTP and its mechanism. To understand clearly and propose a bearing capacity evaluation method for JTP, 54 different pile load tests were conducted at 12 different locations in the Saga lowland, and the following conclusions were achieved:

(1) The proposed bearing capacity evaluation method is safe to calculate the bearing capacity of the JTP since its calculated value (R_{u}) is lower than the value of the second limit resistance force $\left(R'_{u}\right)$ obtained from the pile load tests for all the Sites A to L.

(2) The proposed method is successful in satisfying the criteria of the STP. There is no decrease in the bearing capacity of the JTP in terms of the shear resistance (τ) compared with the bearing capacity of the STP.

(3) Two reduction rates (1) number of joints (n) (2) length-to-diameter ratio (L/D) has been implemented in the evaluation of the bearing capacity due to the compressive strength of the timber pile. Finally, the minimum value of the bearing capacity determined from the long-term allowable compressive force (${}_{L}R_{\mathrm{a}}$) or from the proposed method (calculated from the bearing capacity coefficients α and β) were chosen for the safe design. Both these values were smaller than the second limit resistance force $\left(R'_{u}\right)$ from the loading tests.

Acknowledgements The authors would like to thank the president of the Kyushu Piling Co. Ltd. Mr. Shujiro Matsumoto for always inspiring to do the research and thank Mr. S. Yoshida for completing the report to GBRC.

References

1. Kyushu Piling Co. Ltd.: QP Code 2 2018. Building Technology Performance Certification Evaluation Summary Report, GBRC 12-18, p. 155 (2018)
2. Chai, J.-C., Shrestha, S., Hino, T.: Failure of an embankment on soil-cement column-improved clay deposit investigation and analysis. J. Geotech. Geoenviron. Eng., ASCE, **145**(9) (2019)
3. Japanese Geotechnical Society: Standard for Method for Static Axial Compressive Load Test of Single Piles. JGS1811-2002 (2002)
4. Ministry of Land: Infrastructure, Transport and Tourism. Notification No. 1113. http://www.mlit.go.jp/en/index.html (2001)
5. Ministry of Land: Infrastructure, Transport and Tourism. Notification No. 1524. http://www.go.jp/en/index.html (2007)

Experiment of a Hard Rock Excavator by Using an Edge Excavation

Yudai Ihara, Takahisa Shigematsu, Shinichi Kawamura, Yoshihiro Ohnishi, Sho Miyanaga, and Muhamad Afif Bin Osman

Abstract Blasting method is a conventional method for hard rock mass excavation. Recently, the use of blasting method is getting difficult due to environmental problems, such as noise pollution and ground vibration. Mechanical excavation method is an alternative for the blasting method; however, the present excavation method has two problems: the low efficiency due to the wear of cutting bits and the high cost to exchange cutting bits. Thus, through this research, we are trying to innovate the edge excavation method in order to solve the two problems. The purpose of this study was to develop a rock excavator attached to backhoe by using multistage edge excavation method. We carried out the experiments by using model excavator to find out its efficiency and force acting on the rock excavator. In this study, we conducted a basic experiment by using a displacement controlling method to determine the load required to control a model excavator. The specimen used was high-strength mortar specimen. In the experiment, a model excavator was attached to the experimental device, and the speed was set to be constant by displacement control. Excavation was performed by rotating the specimen. The vertical force F_z, torque T, excavation depth z, and soil volume m, the excavation time t were measured. As a result, the load required for the model excavator was verified and the capability of the experiment to be performed through load controlling was confirmed.

Keywords Multistage type edge excavation · Edge excavation · Displacement control · Disk cutter bit

Y. Ihara · T. Shigematsu (✉) · S. Kawamura · S. Miyanaga · M. A. B. Osman
National Institute of Technology, Kure 737-8506, Japan
e-mail: Sigematu@kure-nct.ac.jp

Y. Ohnishi
Ehime University, Matsuyama 790-8577, Japan

H. Hazarika et al. (eds.), *Advances in Sustainable Construction and Resource Management*, Lecture Notes in Civil Engineering 144,
https://doi.org/10.1007/978-981-16-0077-7_47

1 Introduction

Blasting method is a conventional method for hard rock mass excavation. Recently, the use of blasting method is getting difficult due to environmental problems, such as noise pollution and ground vibration. Mechanical excavation method is an alternative for the blasting method; however, the present excavation method has two problems, the low efficiency due to the wear of cutting bits and the high cost to exchange cutting bits. Thus, through this research, we are trying to innovate the edge excavation method in order to solve the two problems. Figure 1 shows the outline of the plane excava-

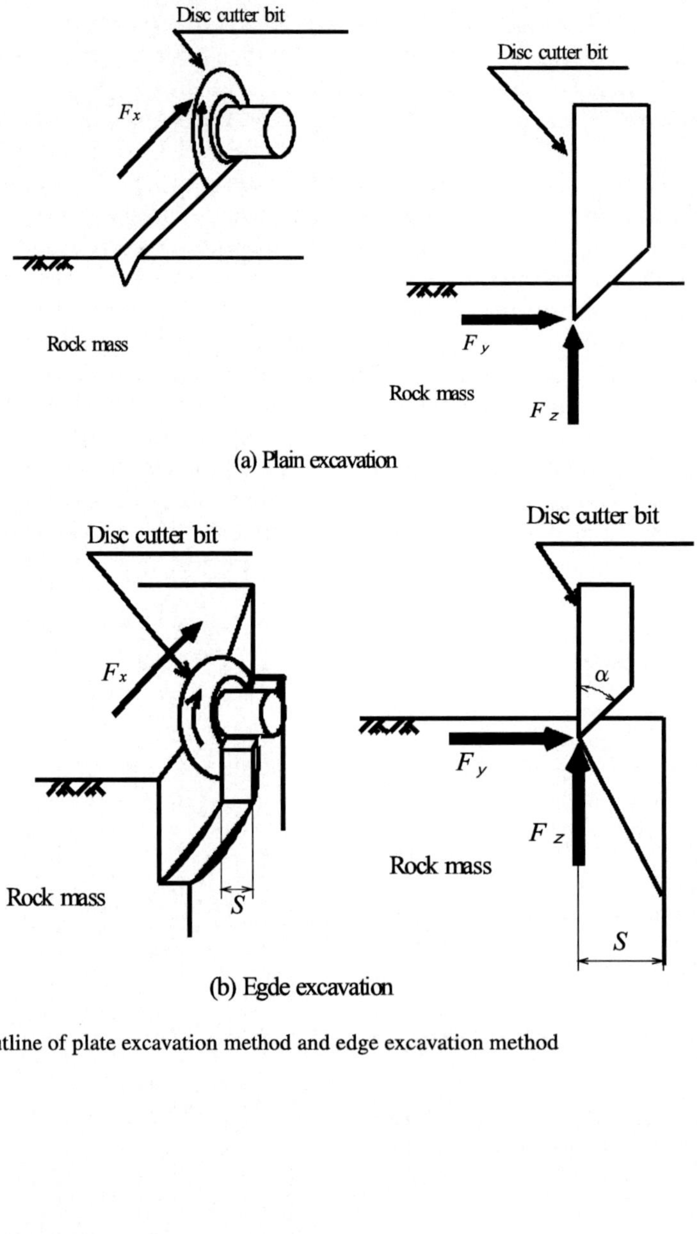

Fig. 1 Outline of plate excavation method and edge excavation method

tion and the edge excavation. Based on other research so far, almost all excavation methods have adopted the plain excavation method, in which the excavation of the plain rock mass was carried out under the condition of one degree of freedom. The use of edge excavation method was considered to be more effective than the plain excavation method. The failure mechanisms of the conventional plane excavation method by using disc cutter bits in tunnel-boring machines are, as mentioned by Snowdon et al. [1], both the compressive failure and the tensile failure in the rock when the penetration of an adjacent disc cutter bit is made. On the other hand, the edge excavation method experiences shear and tensile failure in stepped rock, which is shaped by the wedge action of a ripping-type disc cutter bit, in the condition of two degrees of freedom. A specific cutting energy is calculated as the ratio of an excavation power of a disc cutter bit to the amount of cutting debris per unit time [1]. The specific cutting energy of the edge excavation method with a ripping-type disc cutter bit is much smaller than that of the conventional plane excavation method involving a tunnel-boring machine. Muro et al. [2–5] have reported that the measured specific cutting energies of the edge excavation method were one-tenth energies of the conventional plane excavation method for several kinds of rock. Instead of a point attack bit, disc cutter bits are often used during excavations because the disk cutter bits can spin around and avoid the strong impact of cutting loads, while the point attack bits receive the impact completely and directly.

The purpose of this study was to develop a rock excavator attached to backhoe by using multistage edge excavation method. By using this technology to manufacture a backhoe, we will be able to respond to unmanned dismantling of nuclear power plantsand other sites where increased production is expected in the future. Previously, experiments were conducted with the model excavator to find out the efficiency and the force acting on the rock excavator [6]. Further experiments conducted with the "model excavator" showed that the accumulated sediment caused the decrease of the excavation efficiency [7]. Furthermore by changing the position of the model excavator from vertical to horizontal setting, the removal efficiency of excess sediment had remarkably improved. This improvement provides the following benefits: (1) increase of excavation speed, (2) decrease of torque, (3) increase of overall excavation efficiency.

In this study, we conducted a basic experiment by using a displacement controlling method to determine the load required to control a model excavator. The vertical force Fz, torque T, excavation depth z, and soil volume m, the excavation time t were measured.

2 Experiment Method and Model Excavator

2.1 Model Excavator

Photo 1 and Fig. 2 show the outline of model excavator. The model excavator was attaches with four disk bits under the steel pedestal. Four disk bits were arranged respectively according to the placement interval of cutter, cutter 1 was 45 mm outside the center of the specimen, cutter 2 was 24 mm outside the cutter 1, cutter 3 was 25 mm outside the cutter 2, and cutter 4 was 25 mm outside the cutter 3. Disk cutter bits size was configured, cutter 1 was φ100 mm, cutter 2 is φ90 mm, cutter 3 and cutter 4 was φ100 mm. The model excavator had three stages. First stage was cutter 1 and cutter 2 contact at the same time and plain excavate. First stage formed 2 free side. After that, second stage excavator by cutter 3, and finally, third stage excavator

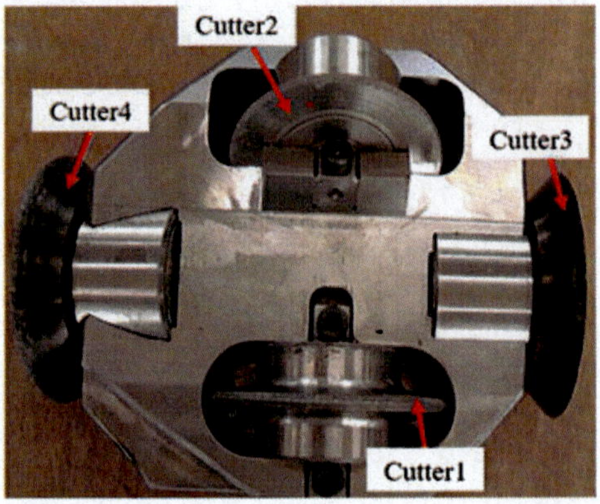

Photo 1 Model excavator

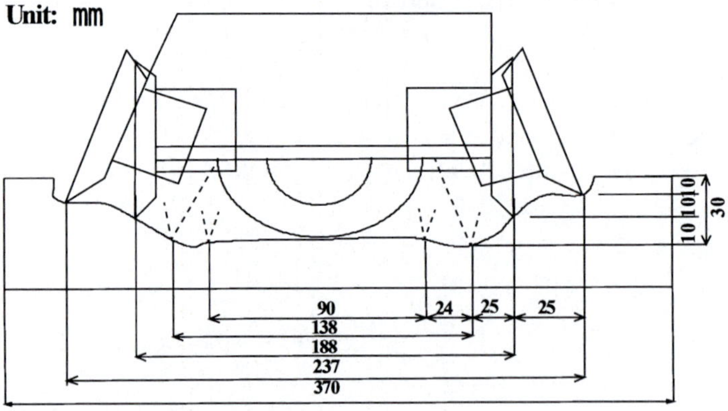

Fig. 2 Model excavator arrangement

by cutter 4. Also, cutter 2 and cutter 4 deployed diagonally due to lower peripheral friction.

2.2 Model Machine

Figure 3 shows the outline of model machine, Photo 2 shows the specimen after the excavation. Model machine apparatus consisting of a worm jack, a turntable of 400 mm diameter, a model excavator, and an axial torque transducer. The apparatus had a height of 1750 mm, a width of 700 mm, and a depth of 700 mm. The turntable was driven by a 1.5 kw motor via a reduction gear; the rotation speed was kept at

Fig. 3 Outline of model machine

Photo 2 Specimen after the excavation

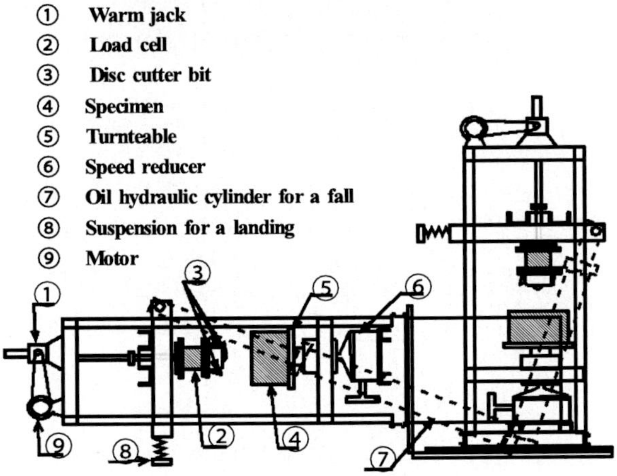

Table 1 One axis compressive strength of high-strength mortar specimen

Specimen numbers	One axis Compressive strength of high-strength mortar specimen (N/mm^2) (Strength after three month)	Average (N/mm^2)
①	161.6	153.5
②	147.2	
③	151.7	

2.0 r.p.m by using a screw rod. Displacement was controlled with a worm jack. It was made possible to move up and down the apparatus operating the worm jack. By changing the position of model excavator from vertical to horizontal settings, removal of the excess sediments was remarkably improved. The axial torque transducer placed on top of the excavator measured the thrust F_z, torque T. The displacement sensor was used to measure the penetration depth z. Vertical forces F_z, excavation depth z, and torque T relative to excavation time t were measured by a data logger that recorded a single point at 0.02 s. The model machine can be tilted using the hydraulic cylinder. It was possible to roll up the device 90° due to discharge excavated soil efficiently by expanding and contracting the hydraulic cylinder beside the machine.

2.3 Specimen

Table 1 shows the one axis compressive strength of high-strength mortar specimen. Specimen uses high-strength mortar specimen (axis compressive strength 153.5 N/mm^2) with water binder ratio $W/B = 17\%$. The shape of each specimen use columnar form, the height was 172 mm, and the diameter was 365 mm.

2.4 Displacement Control Experiment

Displacement control which was to excavation with a constant excavation depth z uses a worm jack which can set a certain amount of penetration. Set speed v_{set} was set to 5 patterns (0.0223, 0.0302, 0.038, 0.0459, 0.0537 mm/s) and each set of speed was performed 3 times except for set speed $v_{set} = 0.0537$ mm/s was performed 2 times.

2.5 Method

The specimens were excavated up to excavation depth of $z = 35$ mm while moving toward the model excavator at a constant speed. At the same time, the turntable revolution was set at 2 r.p.m. The specimens were first set to steel moulds and tighten with M8 hex bolt before fix to the turntable by using M10 hex bolt. The excavation time, t, vertical force F_z, excavation depth z, torque T, excavated soil volume m where measured through data collection machine every 0.02 s. The machine was connected to ground wire to prevent noise from occurring.

3 Experimental Results and Discussion

3.1 Excavation Time

In this chapter, the results of high strength mortar specimen and set speed $v_{\mathrm{set}} = 0.0223$ mm/s will be considered as an example.

(1) **Vertical force F_z and excavation time t**

Figure 4 shows the relationship between vertical force F_z and excavation time t. The feature of plain excavation was that the excavation stage can be roughly divided into two stages. After the initial edge excavation in which a large amount of separation was generated and excavation was performed, the steady edge excavation is performed in which a small amount of separation was generated and then excavated. As a result, the vertical force F_z increases from the start of the experiment to around 340 s. This can be considered as plain excavation. The reason was that immediately after the start of the experiment, the cutter cuts into the specimen, and the excavated plain tends to become unstable as the two free plains were formed. Then, in the vicinity of 540 s, the phenomenon that the vertical force F_z increased again due to the contact of the cutter 3 was observed. This was considered to correspond to the initial end plain excavation. There was a place where the fluctuation range of the value becomes

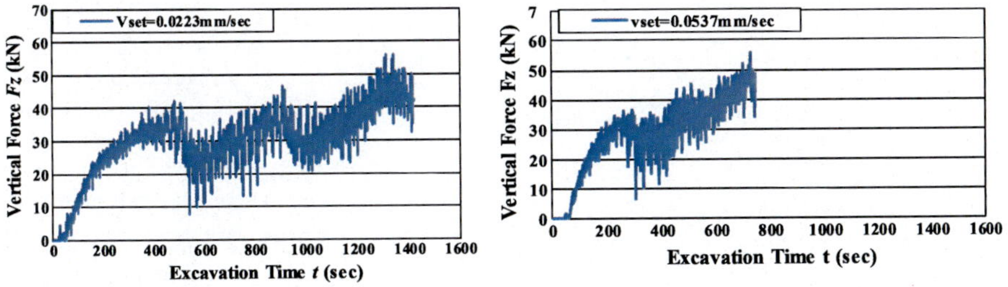

Fig. 4 Relationship between the force F_z and the excavation time

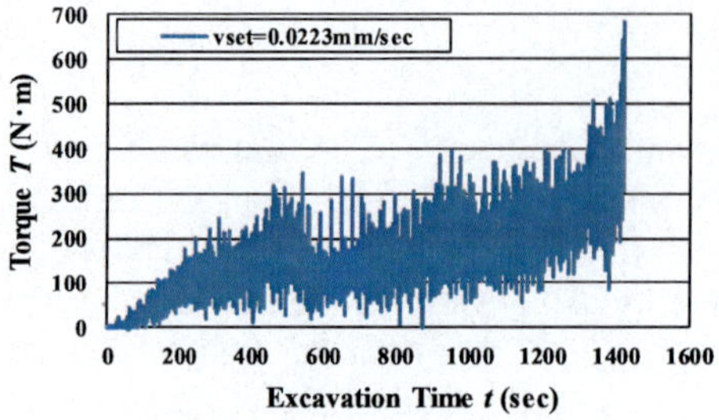

Fig. 5 Relationship between the torque T and the excavation time t

small due to the stable plain excavation. This was due to the steady edge excavation. In addition, there was a tendency for the increase of F_z during plain excavation immediately after the start of the experiment to be larger than in the initial edge excavation. In addition, these tendencies were similar at other set speeds v_{set}.

(2) **Torque T and excavation time t**

Figure 5 shows each slowest speed and fastest speed the relationship between torque T and excavation time t. Right after the start of the experiment, the torque T increases rapidly as well as the vertical force F_z. It was considered that as the cutter cuts into the specimen, torque T increased rapidly due to the unstable excavation plain. In addition, the fluctuation ranged of the torque T value became shorter near 400 s. As a possible factor, the torque T was considered to be stable due to the fact that the cutter cuts sufficiently as the excavation progresses and the excavation plain stabilizes. In addition, regarding the tendency of the graph, the same tendency was observed at other set speed.

(3) **Excavation depth z and excavation time t**

Figure 6 shows each slowest speed and fastest speed the relationship between the excavation depth z and the excavation time t. Experiment was conducted until the excavation depth z reached 35 mm as an experimental condition. Therefore, the excavation time t was measured at the moment the cutter hits the specimen to the time

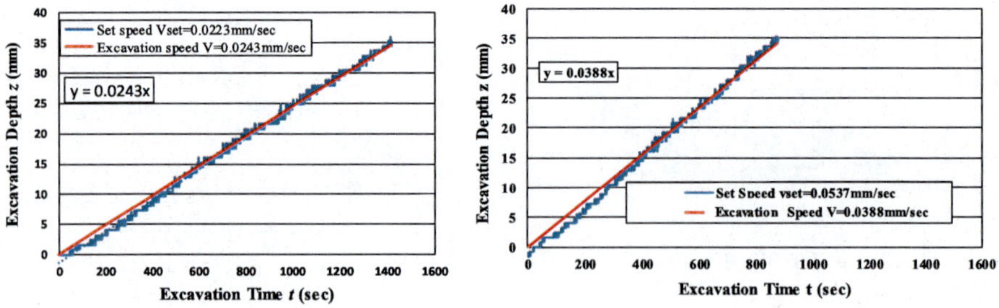

Fig. 6 Relationship between the excavation depth z and the excavation time t

when excavation depth reached 35 mm. In addition, it was confirmed that the excavation had occur with the excavation speed v at a constant fluctuation because the experiment was performed by the displacement control performed by the displacement control. In addition, even with other set speeds v_{set}, there was a tendency for excavation to proceed around the set speed v_{set}. In addition, even with other set speeds v_{set}, there was a tendency for excavation to proceed around the set speed v_{set}.

3.2 *Relationship with Excavation Speed*

(1) **Maximum vertical force F_{zmax}**

Figure 7 shows the relationship between the maximum vertical force F_{zmax} and the excavation speed v. By obtaining the maximum vertical force F_{zmax}, it became an index for judging how much vertical force F_z should be applied when performing an experiment in load control in order to obtain a certain excavation speed v. To determine the maximum vertical force F_{zmax}, the maximum vertical force from the start of the experiment to the end of the experiment can be defined as the maximum vertical force F_{zmax}. As a result, the maximum vertical force F_{zmax} tended to increase as the excavation speed v increased. In addition, even when comparing the same measurement times for different set speeds v_{set}, it was confirmed that the higher the excavation speed v, the larger the vertical force F_z.

(2) **Maximum torque T_{max}**

Figure 8 shows the relationship between the maximum torque T_{max} and the excavation speed v. By obtaining the maximum torque T_{max}, the maximum torque T_{max} applied to the cutter during excavation can be known, and it can be used as an index to judge how much torque T was required for the excavation. The maximum torque T_{max} was

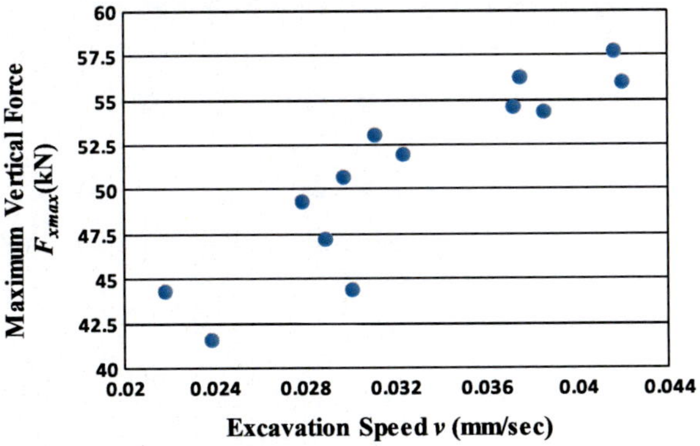

Fig. 7 Relationship between the maximum vertical force F_{zmax} and excavation speed v

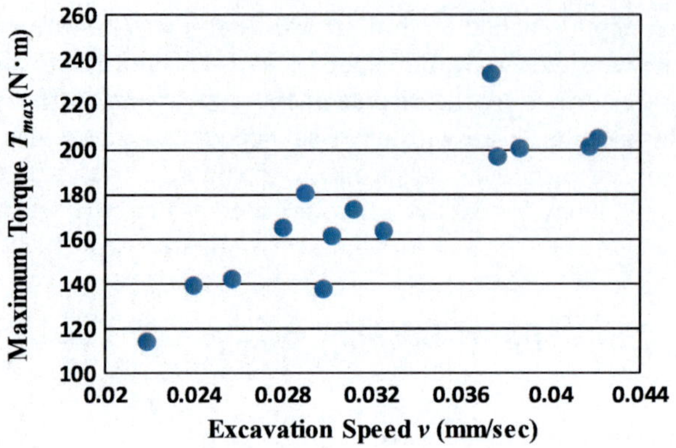

Fig. 8 Relationship between the maximum vertical torque $T_{\max}$ and excavation speed v

determined from the maximum value from the start of the experiment to the end of the experiment. As a result, an increasing tendency ware seen as the excavation speed v increased. As a factor, the vertical force F_z and the torque T increase as the excavation speed v increases. In addition, even if the same excavation time T at different set speeds v_{set} was compared, the torque T was large when the excavation speed v is fast, confirmed.

(3) **Excavation amount per revolution V_{round}**

Figure 9 shows the relationship between the excavation volume per revolution V_{round} and the excavation speed v. It was possible to judge the excavation efficiency by obtaining the excavation volume V_{round} per one rotation of the maximum vertical force F_z. The calculation method for the excavation volume V_{round} per revolution was calculated from the mass per unit of the specimen. The number of revolutions of the specimen during the experiment was obtained by dividing the excavation time T

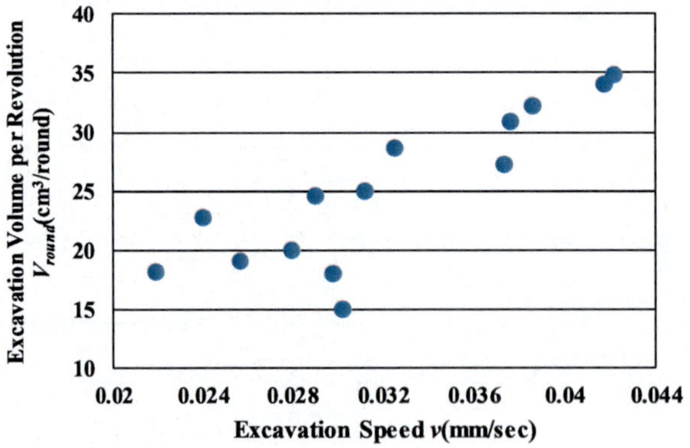

Fig. 9 Relationship between the excavation volume per revolution V_{round} and excavation speed v

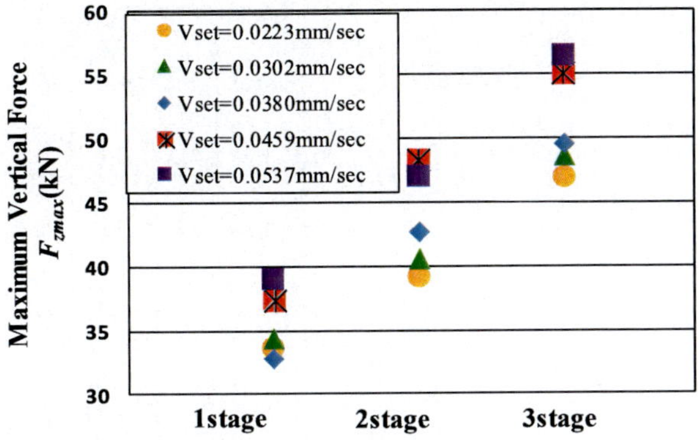

Fig. 10 Relationship between the maximum vertical force F_{zmax} for each stage and set speed v_{set}

by the rotation speed of 2 r.p.m. The excavated volume was divided by the number of revolutions of the specimen to obtain the excavation volume per revolution V_{round}. As a result, as the excavation speed v increased, the excavation volume per revolution V_{round} tended to increase. However, the amount of soil volume m was about the same value, and no change in the amount of soil volume m was observed at the excavation speed v. This was because the experiment was performed with displacement control, and the faster the excavation speed v, the shorter the excavation time T and the smaller the divided value.

3.3 *Maximum Vertical Force* $\mathbf{F}_{zmax}$ *for Each Stage and Set Speed* $\mathbf{V}_{set}$

Figure 10 shows the relationship between the maximum vertical force F_{zmax} for each stage and set speed v_{set}. As a result, it was found that the set speed v_{set} and the maximum vertical force F_{zmax} are not correlated. Looking at the overall tendency, there was a tendency that three set speeds v_{set} with small set speed v_{set} and two set speeds v_{set} with large set speed v_{set} were collected. Moreover, the difference in the maximum vertical force F_{zmax} was noticeable as the number of steps increased.

3.4 *Difference Between Set Speed* $\mathbf{V}_{set}$ *and Excavation Speed* $\mathbf{V}$

Figure 11 shows the relationship between the set speed v_{set} and the excavation speed V. The 45° line is drawn to show the set speed v_{set} and the excavation speed V in a one-to-one relationship. There was a tendency to move away from the 45° line as the set speed v_{set} increased. As a possible cause, it was considered that the reaction force applied to the cutter increased due to the increase of vertical force F_z due to

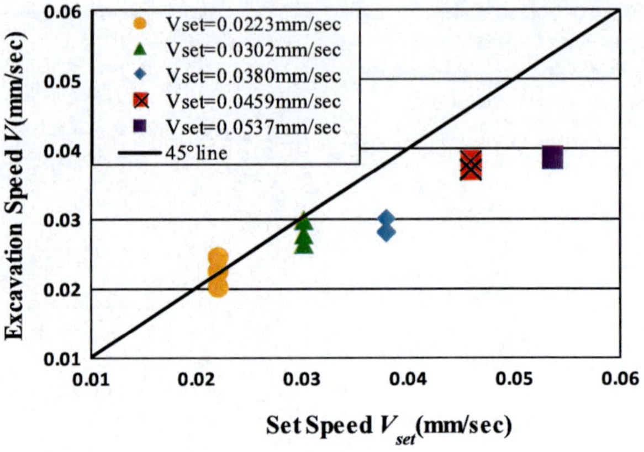

Fig. 11 Relationship between the set speed v_{set} and the excavation speed V

the use of the high-strength mortar specimen, and the vertical force F_z exerted by the warm jack reached its limit. Therefore, it is necessary to carry out an experiment again with lower W/B. Also, since the deviation for each set speed v_{set} was small, it was possible to predict the excavation speed V to some extent.

4 Conclusion

(1) The vertical force F_z rapidly increases until the initial plain excavation and steady plain excavation occurs. After that, the vertical force F_z decreases, and the initial end face excavation can be repeated again confirmed. A similar tendency was observed for the torque T. The torque T tended to decrease rapidly during centering.

(2) As the set speed v_{set} increased, the error in the excavation speed increased, so it is necessary to perform an experiment with a design with a lower water-binder ratio *W/B*.

(3) In the displacement control experiment, the same tendency was seen in vertical force, torque, and excavation depth for any set speed v_{set}, and it can be predicted. Therefore, since the force could be grasped by the displacement control by the new model excavator, it can be applied in the load control.

Acknowledgements This work was supported by JSPS Grant-in-Aid for scientific research Grant Number JP17K06525.

References

1. Snowdon, R.A., Ryley, M.D., Temporal, J.: A study of disc cutting in selected British rock. Int. J. of Rock Mech. Min. Sci. Geomech. Abstrs., **19**, 107–121 (1982)
2. Muro, T., Tsuchiya, K., Kohno, K., Waka-bayashi, Y.: Experimental considerations for the steady excavation performance of a disc cutter bit in an edge part of mortar specimen. J. Geotech. Eng., JSCE, 687 (III-56), 37–47 (2001) (in Japanese)
3. Muro, T., Tsuchiya, K., Kohno, K.: Steady state edge excavation property of a disc cutter bit. In: Proceedings of the 6th Asia-Pacific Conference of the International Society for Terrain Vehicle Systems, 49–58, Bangkok, Thailand, 2001
4. Muro, T., Tsuchiya, K., Kohno, K.: Experimental considerations for steady state edge excavation un-der a constant cutting depth for a mortar specimen using a disk cutter bit. J. Terrramech. **39**(3), 143–159 (2002)
5. Muro, T., Tsuchiya, K. Kohno, K.: Steady state edge excavation of rock material in a multiple tunneling machine. J. Constr. Manag. Eng., JSCE, **777** (VI-65), 109–124 (2004) (in Japanese)
6. Shigematsu, T., Muro, T., Terao, N., Oda, N., Hanaoka, T.: Development of a hard rock excavator by using an edge excavaton. In: Proceedings of the 16th International Conference International Society for Terrain-Vehicle Systems, pp. 276–280, Sapporo, Japan, 2008
7. Shigematsu, T., Kitaoka, I., Tatsuro, M., Oda, N., Kawamura, S.: Efficiency tests of a model trench excavator by using multistage type edge excavation. J. Geotech. Eng., JSCE, **69**(2), 121–128 (2013) (in Japanese)

Full-Scale Testing of a Structure on Improved Soil Replaced with Rubber–Gravel Mixtures

Athanasios Vratsikidis, Angelos Tsinaris, Anastasios Kapouniaris, Anastasios Anastasiadis, Dimitris Pitilakis, and Kyriazis Pitilakis

Abstract We present the results of an extensive large-scale experimental campaign on the dynamic response of rubber–gravel mixtures as an innovative seismic isolation material. In the first series of experiments, the foundation soil immediately below the prototype structure of EUROPROTEAS was replaced only with gravel to serve as benchmark tests, while in the following tests two rubber–gravel mixtures with increasing rubber content per mixture weight were used. The experimental campaign included free- and forced-vibration tests. A large number of instruments of various types (accelerometers, seismometers, shape-acceleration arrays, and laser sensors) were installed on the structure, in the foundation soil and at the adjacent soil surface in order to obtain a well-instrumented 3D set of recordings to study the response of the structure and wave propagation in soil media. In this study, we seek to investigate the isolation capability of the rubber–gravel mixtures under dynamic loading. Our primary goal is to assess the effect of the rubber content of the improved foundation soil in the stiffness and the damping of the soil-structure system.

Keywords Experimental geotechnical engineering · Soil–rubber mixtures · Seismic isolation · Large scale · Field test

1 Introduction

Over the last years, the disposal of a large volume of used tires has become a severe environmental problem. A recent trend is the use of recycled tires in civil engineering projects [1, 2]. An application in earthquake engineering consists in placing a shallow and resilient layer of rubber–soil mixture (RSM) underneath a structural foundation, an improvement method known as geotechnical seismic isolation [3–5]. This emerging application is mainly based on theoretical and experimental studies that have proven that granulated RSM has attractive mechanical and dynamic properties [6–8].

A. Vratsikidis (✉) · A. Tsinaris · A. Kapouniaris · A. Anastasiadis · D. Pitilakis · K. Pitilakis
Aristotle University of Thessaloniki, Thessaloniki, Greece
e-mail: avratsik@civil.auth.gr

H. Hazarika et al. (eds.), *Advances in Sustainable Construction and Resource Management*, Lecture Notes in Civil Engineering 144,
https://doi.org/10.1007/978-981-16-0077-7_48

In this paper, we present an extensive experimental campaign conducted on the large-scale prototype structure of EUROPROTEAS located in the EUROSEIS-TEST experimental facility (https://euroseisdb.civil.auth.gr). The experiments were conducted after replacing the foundation soil with three different rubber–gravel mixtures (RGM). The standard gravelly soil derived from coarse uniform gravel quarry, while the synthetic material (rubber) is derived from recycled used tires and, after appropriate processing, can be found in the form of various sizes. Free- and forced-vibration testing was performed over a wide range of load levels and input frequencies in order to investigate the response of the soil-structure system, evaluate the foundation soil damping behavior under the effects of different rubber content per mixture weight, and to determine the optimum rubber content.

2 EUROPROTEAS Facility

2.1 Structure

EUROPROTEAS prototype structure was particularly designed to promote soil-structure interaction phenomena (Fig. 1). It consists of three identical reinforced concrete slabs, each having a mass of 9 Mg, one representing the surface foundation and the other two the superstructure mass. The roof mass is supported by four steel columns QHS 150 × 150 × 10 mm which are connected with steel X-braces L 100 × 100 × 10 mm in all directions to ensure the symmetry of the structure [9, 10].

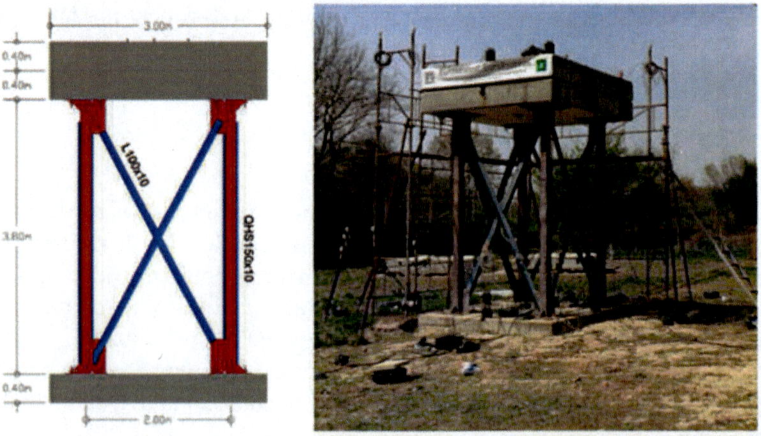

(a) (b)

Fig. 1 **a** A 2D sketch and **b** a photograph of the EUROPROTEAS structure (https://euroseisdb.civil.auth.gr)

The outer dimensions of the structure are 3 × 3 × 5 m, while its total weight is approximately 28.5 Mg. The X-bracings and upper roof slab are removable allowing four different configurations of structure's mass and stiffness covering a wide range of natural frequencies between 1.78 and 13.06 Hz. For the experiments described herein the configuration of EUROPROTEAS involved bracing in all four sides and two reinforced concrete slabs on the roof.

2.2 *Foundation Soil*

The foundation soil stratigraphy and its dynamic properties are well documented from extended geophysical and geotechnical tests reported in earlier studies [11, 12]. In the series of experiments presented herein, the uppermost 0.5 m of the foundation soil was replaced with RGM backfills.

Prior to the modification of the foundation soil, a preliminary study was carried out at the Laboratory of Soil Mechanics, Foundations and Geotechnical Earthquake Engineering of the Aristotle University of Thessaloniki to define the physical properties of the mixtures of gravel with granulated rubber at a percentage of 0, 10, and 30% per mixture weight to be used for the foundation soil improvement. The small strain dynamic response of RGMs is characterized by low values of stiffness and high values of damping. Additionally, the stiffness degradation and the damping increase exhibit a more linear behavior in the medium and high strain range as the rubber content increases [13–15].

According to [16], the relative ratio of the mean grain size of the soil–rubber mixture is a determinant parameter for the behavior of the mixture. Herein, we studied an uniform quarry gravel with angular particles fraction of mean grain size D_{50} equal to 20.76 mm as primary physical material (*G*), whereas a granular fraction of rubber of mean grain size D_{50} equal to 3.27 mm was studied as primary synthetic material (*R*). The rubber grain is tire-derived materials and has been included in ASTM and CEN standards in recent decades [17, 18]. The specific gravity of soil solids, G_s, was determined according to the ASTM D854-02 specification [19] and was found equal to 2.67 for the gravel fraction and 1.10 for the rubber fraction, respectively.

According to [20, 21], it is seen that in a range of the axial deformation up to 20%, the examined specimens of pure gravel fraction as well as the mixtures of gravel with rubber content exhibited "loose sand behavior" without reaching positive values of volumetric strain. The majority of the examined specimens present a contractive behavior with a tendency of reduction as the axial deformation increases, regardless of the uniformity of the sample, the particle size, and the relative ratio $D_{50,s}/D_{50,r}$ of the mean diameter of its granules. However, as the rubber percentage in the mixture increases or as the level of the applied envelope stress increases the specimens exhibit a more intense contractive behavior, which means that the addition of the rubber in the mixture leads to a reduction of the dilation angle values. This is because of the high compression and contraction of the soil structure during the consolidation stage

Table 1 Properties and classification of natural and synthetic examined materials

Properties	Gravel	Rubber
Material ID	G	R
G_s	2.67	1.10
D_{10} (mm)	14.69	2.07
D_{30} (mm)	18.80	2.67
D_{50} (mm)	20.76	3.27
D_{60} (mm)	21.67	3.56
C_u	1.48	1.72
C_c	1.11	0.97
e_{max}	1.385	1.616
e_{min}	0.844	1.075
USCS classification	GP	Granulated rubber

due to the high level of the applied radial stress and the high deformability of the rubber granules. Therefore, during the failure stage, the rubber grains, as part of the sample solid structure, have a small margin of deformation left, while the solid structure, in which the soil grains are participated, is partly rearranged.

Table 1 summarizes the physical characteristics of the natural soil and the synthetic material used in this research, while Fig. 2 shows the grain size distribution of the above materials that were used in this research and affect the mechanical properties of the examined samples. The above parameters were determined according to the specification ASTM C136 [22]. The classification of the used physical soils and synthetic materials were performed adopting the D2487-00 [23] and D6270-98 [17] of ASTM specifications respectively.

Three soil pits having dimensions 3.2 × 3.2 × 0.5 m were excavated (Fig. 3). The first soil pit was filled only with gravel to serve as a benchmark foundation type. The other two pits were filled with RGM of different rubber content by mixture weight. Table 2 summarizes the properties of the RGM (G/R), for different percentages of rubber content (0%, 10%, and 30%) per mixture weight (~ 0%, 25%, and 75% per volume mixture, respectively), used as foundation material in the experimental program. For the definition of the properties, two samples were taken from the bottom and the top layer of each mixture. The same values of the properties were estimated in both samples; however, there was a discrepancy in the determination of the relative density of the third mixture, which had a higher value at the bottom layer despite the fact that the compaction tests conducted were identical and the energy of the compaction was the same. We believe that this is attributed to the poor compaction of the foundation soil and that proper layering of mixtures consisted of such materials is an important factor to achieve uniform relative density.

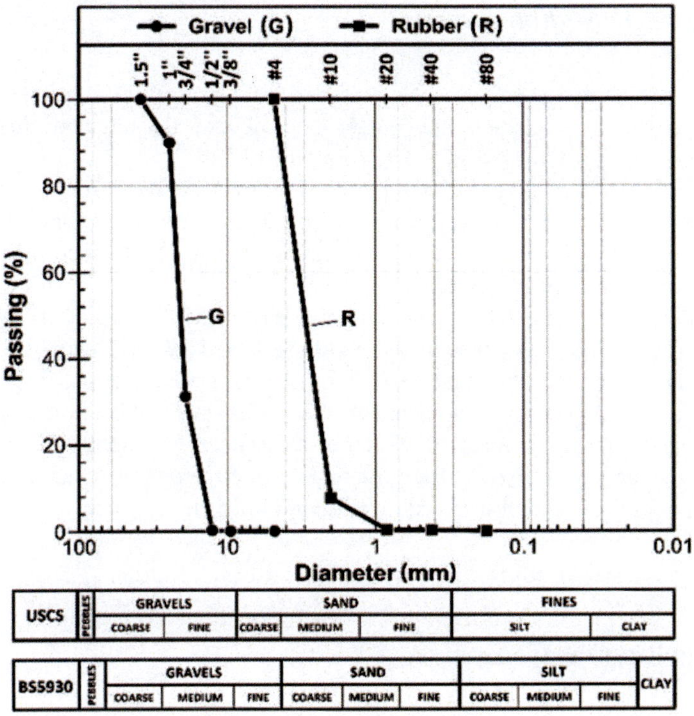

Fig. 2 Sieve analysis of the gravel and rubber materials used in the experiments

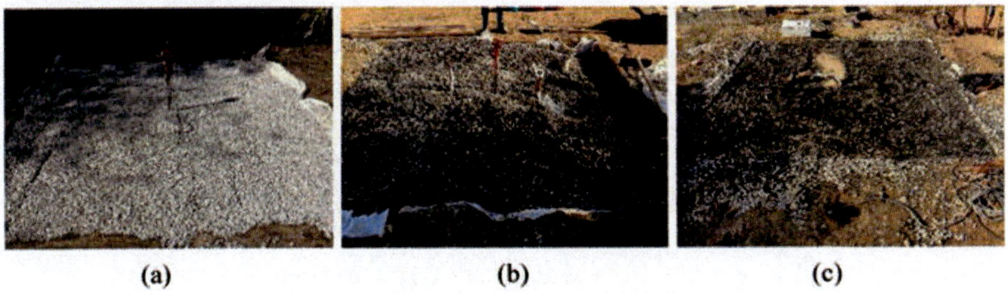

Fig. 3 Three soil pits filled with the rubber–gravel mixtures, **a** G/R 100/0, **b** G/R 90/10, and **c** G/R 70/30

Table 2 Foundation soil types used in the experiments

Foundation ID	Rubber by mixture weight (%)	$D_{50,r}/D_{50,s}$	G_s	D_r (%)	γ_d (kN/m^3)
G/R 100/0	0	–	2.67	97.81	16.2
G/R 90/10	10	0.16	2.51	98.49	15.2
G/R 70/30	30		2.19	58.52–71.03	11.8

2.3 Instrumentation Layout

A dense instrumentation scheme was designed in order to fully monitor and record the response of the structure, the foundation soil, and the adjacent soil in three dimensions (Fig. 4).

The structure was instrumented with nine triaxial accelerometers. Five of them were mounted on the roof, whereas the foundation was instrumented with four accelerometers. Moreover, laser sensors were installed to record the vertical displacement of the foundation slab.

Eleven seismometers were installed on the soil surface buried at the level of the foundation base. Eight of them were placed along the axis of loading, while the others were placed on the perpendicular axis. A 1.2 m shape-acceleration array equipped with 8 triaxial MEM sensors having an intermediate distance of 0.15 cm was installed immediately below the geometrical center of the structure to capture the response of the RGM mixture. Additionally, four uniaxial accelerometers were buried in specific locations under the foundation in order to fully monitor the response of the foundation soil.

3 Experimental Campaign

3.1 Free-Vibration Experiments

We carefully designed the free-vibration experiments to capture the response of the structure for a wide range of pull-out force amplitudes. The forces were applied to the superstructure by a wire rope of 14 mm diameter. The one end of the wire rope was clamped at a 3 ton reinforced concrete counterweight buried in the soil 20 m away from EUROPROTEAS (Fig. 5a). The other end of the rope along with a load cell that measured the applied tension force was fastened on a special steel hook installed on the top roof slab (Fig. 5b). Tension was applied by a pulling hoist and when the desired level of force was reached, the wire rope was cut loose and the structure oscillated freely until rest. In total, 15 pull-out force amplitudes varying between 1.35 and 15.20 kN were applied (Table 3).

3.2 Forced-Vibration Experiments

An eccentric mass vibrator system was installed at the geometrical center of the roof slab as a source of harmonic excitation (Fig. 6a). The axis of the produced harmonic force was applied along the main axis of the structure forming an angle of approximately 30 degrees with the magnetic north–south direction (Fig. 4). The shaker has four pairs of plates (A, B, C, and D) in four different sizes that can be used

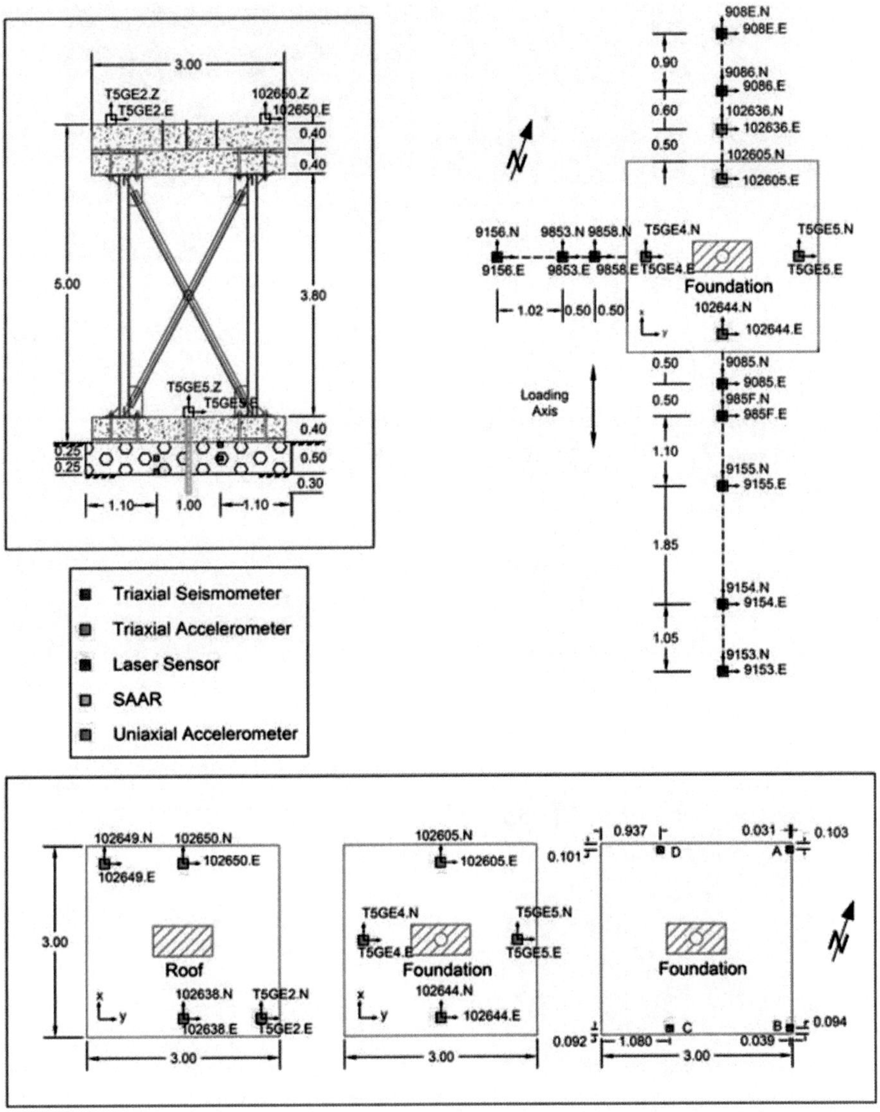

Fig. 4 Cross section of the foundation soil and structure instrumentation (top left), plan view of the foundation and soil surface instrumentation (top right), and plan view of the roof slab instrumentation (bottom left), the foundation slab instrumented with the accelerometers (bottom center) and with the laser sensors (bottom right)

Fig. 5 a Pulling hoist clamped on the counterweight and b the load cell attached to the roof slab

Table 3 Summary of the free-vibration experiments per each foundation soil mixture

Foundation soil type	G/R 100/0	G/R 90/10	G/R 70/30
Experiment ID	Force (kN)		
A	1.60	1.90	1.35
B	10.40	5.40	2.90
C	15.20	10.60	6.28
D	2.10	3.50	9.80
E		7.50	15.00
F		2.80	

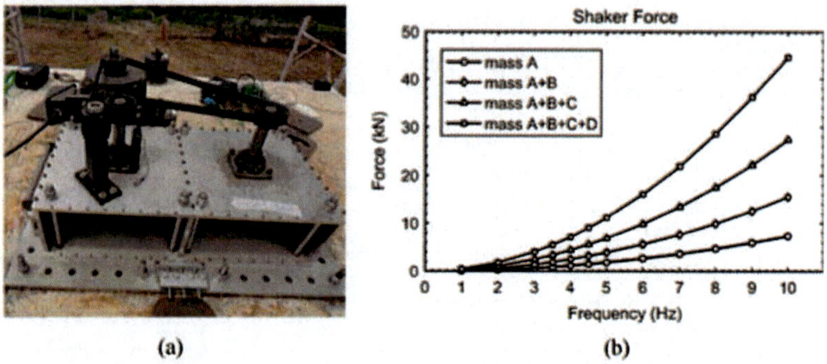

Fig. 6 **a** Eccentric mass shaker installed on the upper roof slab and **b** the force-frequency relationship governing its function

to adjust the vibrator's eccentricity. The amplitude of the output force can be adjusted according to the eccentricity and the operating frequency (Fig. 6b). The performed forced-vibration experiments covered a wide range of excitation amplitude varying between 0.07 and 28.50 kN in a frequency range of 1–10 Hz (Table 4).

Table 4 Summary of the forced-vibration experiments

Experiment ID	Mass/Plates	Eccentricity (kg m)	Frequency range (Hz)	Force amplitude (kN)
A	A	1.85	1–10	0.07–7.30
B	A + B	3.93	1–10	0.15–15.50
C	A + B + C	6.93	1–10	0.30–27.30
D	A + B + C + D	11.31	1–8	0.50–28.50

4 Experimental Results

4.1 Period Elongation

The maximum amplitude of the acceleration recorded at the roof versus the period of the input signal during the forced-vibration Experiments B and D for each foundation soil mixture are reported in Fig. 7. The natural period of the soil-structure system for the benchmark test, in which the foundation soil is replaced only by gravel, is estimated at 0.25 s in both the experiments. The effect of the 10% rubber content per mixture weight does not seem to significantly affect the natural period of the system which remains approximately the same. However, it appears to slightly increase the damping of the system as the values of the recorded roof acceleration for the same input frequencies are reduced. On the other hand, in the case of the G/R 70/30 foundation soil mixture, the period is clearly shifted to 0.4 s implying a strong effect of the increased rubber content in the stiffness of the soil mixture-structure system.

Similar conclusions are drawn by looking at the transmissibility functions (Fig. 8) of the three different soil-structure systems, which are estimated according to [24]. The acceleration response factor R_d is calculated as the ratio of the acceleration recorded at the roof to the shaker force normalized by the superstructure mass. The acceleration response function is decreased in the case of the G/R 70/30 foundation soil mixture indicating an increase in the value of damping. This is attributed to

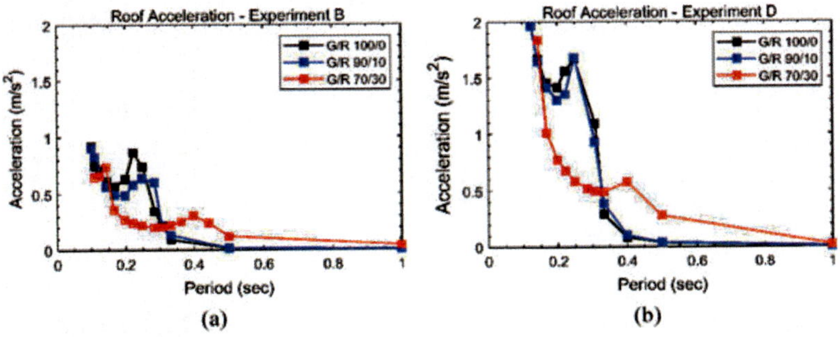

Fig. 7 Maximum amplitude of the acceleration recorded at the roof during the forced-vibration **a** experiment B and **b** experiment D at different excitation frequencies

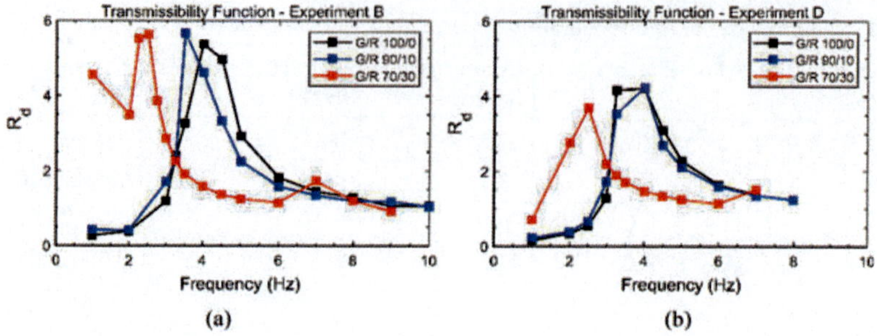

Fig. 8 Transmissibility functions estimated for the forced-vibration **a** experiment B and **b** experiment D

the strong effect of the increased rubber content in the mixture. Additionally, the acceleration response function is decreased for large amplitude forces for the same foundation soil mixture (e.g., in Experiment D) indicating an increase in the damping due to possible nonlinearities introduced in the response of the system.

4.2 *Roof Response Decay*

In Fig. 9a, we compare the decay of the amplitude of the acceleration recorded at the roof of the structure during the two free-vibration experiments labeled A in the case of the foundation soil mixture G/R 100/0 and G/R 70/30 (Table 3). The amplitude of the applied pull-out force in both experiments was approximately 1.5 kN. The acceleration values normalized by the initial maximum recorded acceleration in each test are presented in Fig. 9b. A decrease of more than 70% in the amplitude is noticed after three cycles of oscillation in both tests. This pronounces the large amount of energy dissipated in only the first few cycles of oscillation. However, the rate of

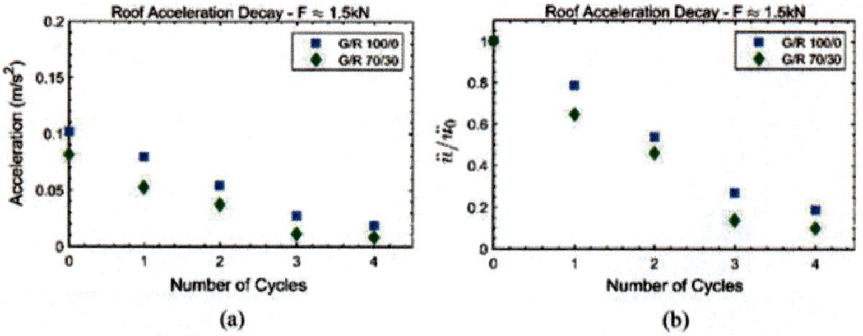

Fig. 9 Decay of the roof response **a** as it was recorded and **b** normalized by the initial acceleration value with respect to the number of cycles of oscillation during the free-vibration experiment A (Table 3) for G/R 100/0 and G/R 70/30

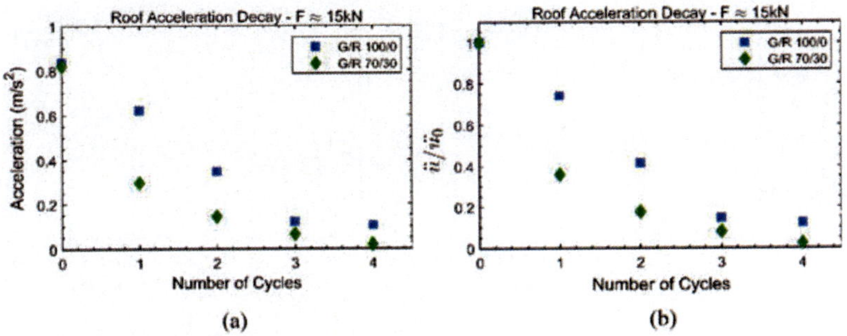

Fig. 10 Decay of the roof response **a** as it was recorded and **b** normalized by the initial acceleration value with respect to the number of cycles of oscillation during the free-vibration experiment C in the case of the G/R 100/0 mixture and experiment E in the case of G/R 70/30 mixture (Table 3)

the decrease of the acceleration seems to be slightly greater in the case of the G/R 70/30 mixture indicating the contribution of the rubber content in the damping of the system.

While the rate of the decrease of the response appears almost similar in the case of the gravelly foundation soil irrespectively of the amplitude of the applied pull-out force, we notice a much steeper decline when the rubber content is present and the amplitude of the pull-out force is much greater (Fig. 10). A decrease in the acceleration by 50% is shown in the first cycle of oscillation, whereas in the next cycle the response is almost negligible demonstrating the increased damping of the system due to the increased rubber content.

4.3 Adjacent Soil Motion Decay

We investigated the wave propagation in the soil and specifically the decay of the amplitude of motion with increasing distance from the structure when it was subjected to harmonic sinusoidal forces during the forced-vibration Experiment D, in which the greatest forces are applied at the roof of the structure (Table 4).

In the case of the G/R 100/0 foundation soil mixture, the amplitude of the response recorded at the soil surface is greater when the structure is excited close to its resonant frequency indicating that the foundation soil and the structure respond to the loading as a whole system (Fig. 11). A reduction in the motion by 80% is noticed at a distance equal to B/6 (B = foundation width) compared to the response recorded at the top of the foundation, whereas the amplitude of the response can be considered negligible at a distance equal to 2B/3 irrespectively of the excitation frequency. In the presence of the rubber in the foundation soil, the decrease is greater at the same distances as the foundation soil contributes to the dissipation of a bigger portion of energy (Fig. 12). At a distance equal to B/6, the decrease of the recorded velocity is approximately

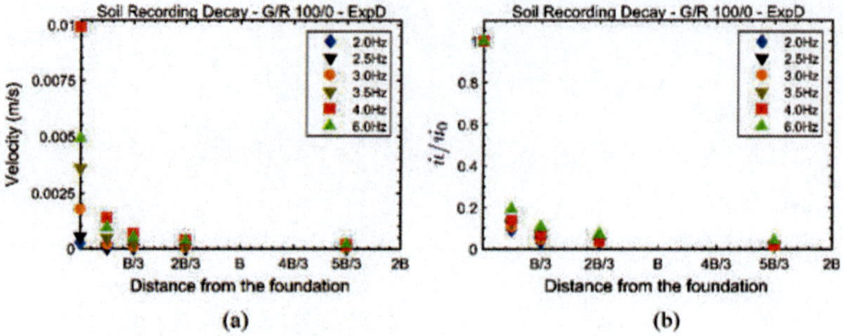

Fig. 11 Decay of the recorded soil surface response **a** as it was recorded and **b** normalized by the velocity value recorded on the foundation at increasing distance from the foundation in forced-vibration experiment D in the case of the G/R 100/0 foundation soil

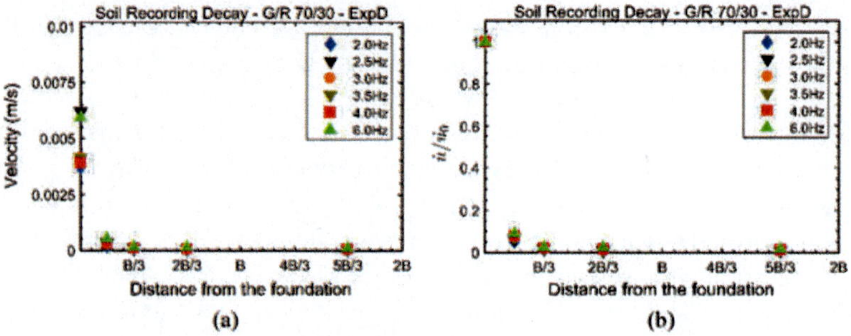

Fig. 12 Decay of the recorded soil surface response **a** as it was recorded and **b** normalized by the velocity value recorded on the foundation at increasing distance from the foundation in forced-vibration experiment D in the case of the G/R 70/30 foundation soil

90%, while at distance equal to B/3 the amplitude of the motion is decreased by almost 98% irrespectively of the excitation frequency.

5 Conclusions

We performed a series of free- and forced-vibration experiments to study the behavior of RGM as an alternative and innovative foundation soil improvement. We specifically focused on the effect of the rubber content per mixture weight by investigating the dynamic response of the prototype structure of EUROPROTEAS founded on three different RGM mixtures. The results pronounced that a 10% rubber content per mixture weight has a slight and almost negligible effect in the response of the structure. On the other hand, an increase in the rubber content to 30% per mixture weight affects significantly the predominant frequency and the damping of the soil-structure system. Additionally, it is evident that there is a significant decline in the amplitude of the recorded response at the adjacent soil surface indicating a greater dissipation of energy in the foundation soil.

Acknowledgements This study was performed in the framework of the European project "Seismology and Earthquake Engineering Research Infrastructure Alliance for Europe—SERA-TA—H2020 (Grant Agreement 730900)."

References

1. Edeskär, T., Westerberg, B.: Tire shreds used in a road construction as a lightweight and frost insulation material. In: Oriz de Urbina, G., Guomans, H. (eds.) 5th International Conference on the Environmental and Technical Implications of Construction with Alternative materials, pp. 293–302, INASMET, San Sebastian (2003)
2. Humphrey, D.: Tire derived aggregate as lightweight fill for embankments and retaining walls. In: Hazarika, H., Yasuhara, K. (eds.) Proceedings of the International Workshop on Scrap Tire Derived Geomaterials-Opportunities and Challenges, pp. 59–81, Yokosuka, Japan (2007)
3. Tsang, H.-H.: Seismic isolation by rubber-soil mixtures for developing countries. Earthq. Eng. Struct. Dynam. **37**(2), 283–303 (2008)
4. Senetakis, K., Anastasiadis, A., Trevlopoulos, K., Pitilakis, K.: Dynamic response of SDOF systems on soil replaced with sand/rubber mixture. In: Proceedings of the ECOMAS Thematic Conference on Computation Methods in Structural Dynamics and Earthquake Engineering, Rhodes, Greece (2009)
5. Pitilakis, K., Trevlopoulos, K., Anastasiadis, A., Senetakis, K.: Seismic response of structures on improved soil. In: De Roeck, G., Degrande, G., Lombaert, G., Muller, G. (eds.) Proceedings of the 8th International Conference on structural dynamics (EURODYN2011), pp. 72–81, Leuven, Belgium (2011)
6. Anastasiadis, A., Senetakis, K., Pitilakis, K.: Small-strain shear modulus and damping ratio of sand-rubber and gravel-rubber mixtures. Geotech. Geol. Eng. **30**(2), 363–382 (2012)
7. Anastasiadis, A., Senetakis, K., Pitilakis, K., Gargala, C., Karakasi, I.: Dynamic behavior of sand/rubber mixtures. Part I: Effect of rubber content and duration of confinement on small-strain shear modulus and damping ratio. J. ASTM Int. **9**(2), 221–247 (2012)
8. Tsinaris, A., Anastasiadis, A., Pitilakis, K.: Mechanical behavior of lightweight soil/rubber mixtures in a wide shear strain range. In: Proceedings of the 16th European Conference on Earthquake Engineering (16ECEE), Thessaloniki, Greece (2018)
9. Pitilakis, D., Rovithis, E., Anastasiadis, A., Vratsikidis, A., Manakou, M.: Field evidence of SSI from full-scale structure testing. Soil Dyn. Earthq. Eng. **112**, 89–106 (2018)
10. Vratsikidis, A., Pitilakis, D.: Full-scale free- and forced-vibration experiments at the Euro-Proteas SSI facility: experimental data exploitation. In: Proceedings of the 16th European Conference on Earthquake Engineering (16ECEE), Thessaloniki, Greece (2018)
11. Pitilakis, K., Raptakis, D., Lontzetidis, K., Tika-Vasillikou, T., Jongmans, D.: Geotechnical and geophysical description of EUROSEISTEST, using field, laboratory tests and moderate strong motion records. J. Earthq. Eng. **3**(3), 381–409 (1999)
12. Raptakis, D., Chávez-García, F.J., Makra, K., Pitilakis, K.: Site effects at Euroseistest-I. Determination of the valley structure and confrontation of observations with 1D analysis. Soil Dyn. Earthq. Eng. **19**(1), 1–22 (2000)
13. Anastasiadis, A., Senetakis, K., Pitilakis, K., Gargala, C., Karakasi, I.: Dynamic behavior of sand/rubber mixtures. Part I: Effect of rubber content and duration of confinement on small-strain shear modulus and damping ratio. J. ASTM Int. **9**(2), 1–19 (2012)
14. Pistolas, G.A., Anastasiadis, A., Pitilakis, K.: Dynamic behavior of granular soil materials mixed with granulated rubber: influence of rubber content and mean grain size ratio on shear modulus and damping ratio for a wide strain range. Innov. Infrastruct. Solutions **3**(47) (2018)

15. Pistolas, G.A., Pitilakis, K., Anastasiadis, A.: A numerical investigation on the seismic isolation potential of rubber/soil mixtures. Earthq. Eng. Eng. Vibr. **19**, 683–704 (2020)
16. Senetakis, K.: Dynamic properties of granular soils and mixtures of typical sands and gravels with recycled synthetic materials. Ph.D. dissertation, Department of Civil Engineering, Aristotle University of Thessaloniki Greece (in Greek) (2011)
17. ASTM Standard Test Methods for Use of Scrap Tires in Civil Engineering Applications: D6270-98. Annual Book of ASTM Standards, ASTM International (1998)
18. CEN: Materials Produced from End of Life Tyres-Specification of Categories Based on Their Dimension(s) and Impurities and Methods for Determining Their Dimension(s) and Impurities, European Committee for Standardization, CEN/TS 14243:2010, Brussels (2010)
19. ASTM Standard Test Methods for Specific Gravity of Soil Solids by Water Pycnometer: D854-02. Annual Book of ASTM Standards, ASTM International (2002)
20. Pistolas, G.A.: Experimental and numerical investigation of the implementation of recycled materials mixtures in the foundation of structures for the improvement of seismic behavior. Ph.D. dissertation. Department of Civil Engineering, Aristotle University of Thessaloniki, Greece (in Greek) (2015)
21. Tsinaris, A.: Experimental and numerical investigation of improving the seismic response of retaining structures with backfill of mixtures of lightweight materials. Ph.D. Dissertation. Department of Civil Engineering, Aristotle University of Thessaloniki, Greece (in English & Greek) (2018)
22. ASTM: Standard Test Method for Sieve Analysis of Fine and Coarse Aggregates: C136-06. Annual Book of ASTM Standards, ASTM International (2006)
23. ASTM Standard Practice for Classification of Soils for Engineering Purposes (Unified Soil Classification System): D2487-00. Annual Book of ASTM Standards, ASTM International (2000)
24. Chopra, A.: Dynamics of Structures: Theory and Applications to Earthquake Engineering, 2nd edn. Prentice Hall Inc., New Jersey (2001)

Interdependence and Rationality Between Sustainable Indicators and Criteria—A Fuzzy AHP Approach

Suchith Reddy Arukala and Rathish Kumar Pancharathi

Abstract The rapid urbanization is likely to impose tremendous pressure on the available natural resources due to their depletion, seriously affecting the Social, Environmental, Economic and Technological (SEET) indicators. The identification of criteria and indicators to achieve Sustainable Construction (SC) is a challenging task in terms of balancing and interrelating them. The present study while investigating the most significant criteria that contribute to SC, establishes inter-relationship between them and SEET indicators by utilizing the Fuzzy Analytical Hierarchy Process (AHP), a Hybrid Multi-Criteria Decision Making (MCDM) method. Based on Delphi Technique (DT) and Relative Importance Index (RII), eight criteria have been identified including Water Efficiency, Materials and Waste Management, Health and Well-being, Energy Efficiency, Sustainable Sites, Social Welfare, Transportation, and Management. The findings of the study reveal that the criteria Materials and waste management and Energy Efficiency have attained the highest relative weights of 13.96 and 12.63 respectively. Similarly, among SEET indicators, the Environmental and Technological indicators have secured 30.15 and 28.52 relative weights respectively. This well-established inter-relation between indicators and criteria will facilitate the decision-makers/stakeholder to understand the degree of performance between sustainable criteria and indicators for achieving sustainable buildings. In addition to this, a computerized building assessment tool which can facilitate the formulation of guidelines by policy-makers was also developed.

Keywords Sustainable construction · Environment · Building assessment tool · Green buildings · Multi-criteria decision method · Fuzzy · AHP

S. R. Arukala (✉)
Department of Civil Engineering, KITS Warangal, Warangal, Telangana, India
e-mail: asr.nitwarangal@gmail.com

R. K. Pancharathi
Department of Civil Engineering, NIT Warangal, Warangal, Telangana, India
e-mail: rateeshp@gmail.com

H. Hazarika et al. (eds.), *Advances in Sustainable Construction and Resource Management*, Lecture Notes in Civil Engineering 144,
https://doi.org/10.1007/978-981-16-0077-7_49

1 Introduction

Compared to developed countries, developing countries have got a new trend of accepting green building guidelines [1–3]. Every country is in the process of developing its own rating system or guidelines to achieve an overall sustainable built environment. For example, prominent assessment tools like the Leadership in Energy and Environmental Design (LEED) scheme in the US, and the Comprehensive Assessment System for Building Environmental Efficiency (CASBEE) in Japan. Based on the LEED revision, India introduced the Indian Green Building Council (IGBC) assessment method in the year 2000. The Green Rating for Integrated Habitat Assessment (GRIHA) in India [4] and the Building Research Establishment's Environment Assessment Method (BREEAM) in Australia also uses the country-specific format of Norway, Sweden, Spain, and the Netherlands. The assessment method developed created for one nation or region might not be directly applicable to others, because, a number of factors may prevent the transfer of currently available environment assessment tools to other nations [5]. Some of these factors include regional differences, climatic conditions etc.

1.1 Need for Sustainability in Indian Built Environment

From a case study conducted by Gesellschaft für Internationale Zusammenarbeit (GIZ), it was acknowledged that developing countries need further action and development in policy, regulation, and strengthening the regulatory authorities and reinforcing the existing laws towards sustainability. There is lack of awareness on sustainability/green practices and their benefits [6]. Most of the issues of sustainability are interrelated in existing methods, and the focus is mainly given to environmental aspects [7]. This clearly specifies disregard for the economic, social, and technological aspects of sustainability, which could further lead to ecological imbalance and thereby, miss the real goal of sustainable development. For example, though GRIHA has attempted to make green building assessment tools, it was not able to fully incorporate the social, economic and cultural elements in the sustainability assessment criteria [8]. The existing building assessment tools are hence, limited to uni-dimensional sustainability. Some of the assessments were based on a Triple Bottom Line (TBL) approach i.e., Environment, Social, and Economic dimensions [9, 10]. An in-depth study of the literature indicated that the Technological component's significance could be enhanced by incorporating recent technological advances in sustainability in the construction sector. Technological advances have always been the cornerstone in mitigating the unavoidable side effects of development and in surpassing the limits/constraints dictated by the other indicators of sustainability. For instance, a shift from working stress method to limit state method in the design philosophy led to thinner and more economical sections without compromising the safety and durability, and introduction of steel columns and steel beams in lieu of stonewalls as structural materials made the towering skyscrapers possible. The technological dimension can be incorporated by rejuvenating ideas of Reuse, Recycle, Reduce, Renew, and Regenerate (5R's) into implementable solutions to the existing TBL to achieve sustainable construction [3]. In other words, various methods and approaches are required for benchmarking the threshold values and targets to transform a theoretical concept into practical implementation. Secondly, policies and guidelines for proper governance are needed. Finally, it is essential to recognize the advantages of supporting techniques and technology for achieving sustainable harmony in the construction industry. The present study encompasses the Social, Environmental, Economic, and Technological (SEET) aspects in achieving sustainable construction. This study is a continuation of the author's work [3] to quantify the building performance. The present study concentrates on quantifying the interdependency between SEET indicators and criteria using the Fuzzy Analytical Hierarchy Process (FAHP), a Hybrid Multi-Criteria Decision Making (MCDM) method considering local context, climate conditions, culture, topography and ethical aspects prevailing in developing countries.

1.2 Fuzzy Numbers and Linguistic Terms

Lofti Zadeh (1965), introduced the Fuzzy set theory in order to make decisions for problems dealing with vagueness, subjectivity, and impreciseness [11]. Consider a TFN defined by $\tilde{A} = (a, b, c)$, where $\mu_A(x)$ is the degree of belonging or membership value of the element in the universe of discourse. The fuzzy triangular scale adopted in the study is shown in Table 1.

$$\mu_A(x) = \begin{cases} \frac{x-a}{b-a}, \\ \frac{c-x}{c-b}, \\ 0 \end{cases} \begin{bmatrix} a \leq x \leq b \\ b \leq x \leq c \\ \text{otherwise} \end{bmatrix} \tag{1}$$

Table 1 Linguistic terms and corresponding triangular fuzzy numbers

Saaty scale	Definition (level of importance)	Fuzzy triangular scale
1	Equal	(1, 1, 1)
3	Weak	(2, 3, 4)
5	Fair	(4, 5, 6)
7	Strong	(6, 7, 8)
9	Absolute	(9, 9, 9)
2 4 6 8	Intermediate values	(1, 2, 3) (3, 4, 5) (5, 6, 7) (7, 8, 9)

Consider two fuzzy numbers $\widetilde{A_1}$ and $\widetilde{A_2}$, where $\widetilde{A_1} = (a_1, b_1, c_1)$ and $\widetilde{A_2} = (a_2, b_2, c_2)$ whose operations of addition, multiplication, division, and reciprocal are defined by Eqs. 2 to 5.

$$\widetilde{A_1} \otimes \widetilde{A_2} = (a_1 + a_2,\ b_1 + b_2,\ c_1 + c_2) \tag{2}$$

$$\widetilde{A_1} \otimes \widetilde{A_2} = (a_1 \otimes a_2,\ b_1 \otimes b_2,\ c_1 \otimes c_2) \text{ for } a_1 > 0,\ b_1 > 0,\ c_1 > 0 \quad (i = 1, 2) \tag{3}$$

$$\widetilde{A_1} / \widetilde{A_2} = (a_1/c_2,\ b_1/b_2,\ c_1/a_2) \text{ for } a_1 > 0,\ b_1 > 0,\ c_1 > 0 \ \ (i = 1, 2) \tag{4}$$

$$\tilde{A}_1^{-1} = (1/c_1, 1/b_1, 1/a_1) \text{ for } a_1 > 0,\ b_1 > 0,\ c_1 > 0 \tag{5}$$

1.3 Analytical Hierarchy Process (AHP)

The AHP is a qualitative and quantitative MCDM technique used to evaluate the relationship between two or more components [12], in a logical way. The decision problem is analyzed at various levels of hierarchy structure, to enable them to be evaluated independently. The human perception involved with uncertainty and ambiguity can be resolved with the fuzzy logic concept. In the present study, to establish the interrelationship among the criteria, indicators, and criteria to indicators, the Fuzzy Analytical Hierarchy Process (FAHP) has been employed in decision making.

2 Comparison of Existing Building Assessment Tools

Based on the credibility and recognition of the rating system, four assessment tools were selected, compared, and analyzed for similarities and dissimilarities present in the developed, and developing countries. Though these rating systems seem to have some criteria in common (names), they differ in their meaning and understanding. This is mainly due to varied climate, culture, region, awareness, practices and assessment method. In addition to this, the rating systems are not unique in nature, dimension and do not comply with the requirements. The assessment tools in the developed countries i.e., LEED and BREEAM being the most prominent and globally established assessment tools in the domain of sustainable construction are considered in the present work. Also, assessment tools used in developing countries like GRIHA and IGBC have been considered for comparison. The specific purpose to compare these tools is to check whether these assessment criteria and attributes are transferrable and adaptable to suit the circumstances for developing countries like India. Table 2 compares and summarizes, the components, features, and criteria of

Table 2 Criteria comparison of existing Assessment tools

Criteria	Attributes	Factors	BREEAM	LEED	IGBC	GRIHA
Sustainable site and ecology	Construction site	Selection of site	√	√	√	√
		Protection of site	√	√	√	√
	Ecological value	Land contamination	√	√	√	√
		Mitigating ecological impact	√	√	√	√
		Balancing site ecology	√	√	√	√
		Protecting biodiversity	√	√	√	√
	Transport	Ease of accessibility	√	√	√	√
		Developing density	√	√	√	X
		Intercommunity network	√	√	√	X
		Safety of pedestrian	√	√	X	X
		Car parking facility	√	√	√	X
Energy efficiency	Energy performance	HVAC	√	√	√	√
		Rate of ventilation	√	√	√	X
		Internal and external lighting	√	√	√	√
		Provision of hot water	√	√	√	√
		Heat transmission	√	√	X	X
		Renewable technology on energy	√	√	√	√
		Monitoring energy	√	√	√	X
		Energy saving	√	√	√	√
		CO_2 Strategy	√	√	√	X
Water efficiency and water management	Water	Reducing the consumption of water	√	√	√	√
		Harvesting water	√	√	√	√
		Recycling of water	√	√	√	√

(continued)

Table 2 (continued)

Criteria	Attributes	Factors	BREEAM	LEED	IGBC	GRIHA
		Innovative water recycling technology	√	√	√	√
		Water conservation technique	√	√	√	√
		Water irrigation technique	√	√	√	√
		Groundwater recharge	√	√	√	√
Material	Material category	Low impact environment material	√	√	√	√
		Use of non-renewable resources	√	√	X	X
		Material reuse	√	√	√	X
		Using innovative technology for non- structure	√	√	√	X
		Insulating component	√	√	X	X
		Material finishing	√	√	√	X
		Local resources utility	√	X	√	X
		The efficiency of material over LC	√	√	X	X
Pollution and risk	Emissions and disaster	Global warming potential for refrigerant	√	√	√	X
		Noise pollution	√	√	√	X
		Preventing pollution leaks	√	X	√	X
		Water pollution	√	√	X	X
		Effect of heat island	√	√	√	X
		Source of NO_x emission	√	√	X	X
		Carbon emission	√	√	√	X
		Fire safety	√	√	X	√
		Natural Disaster	√	√	X	X

(continued)

Table 2 (continued)

Criteria	Attributes	Factors	BREEAM	LEED	IGBC	GRIHA
Indoor environment quality	Noise and acoustics	Level of noise emitting	√	√	X	√
		Insulation to sound source	√	X	X	√
		Absorption of sound acoustics	√	X	X	√
	Lighting and illumination	Active lighting	√	√	√	√
		Lighting control	√	√	√	√
		Open view	√	√	√	X
		Measuring and control on glaring	√	√	X	X
		Level of illumination	√	√	√	X
		Daylight factor	√	√	√	X
	Ventilation	Natural ventilation	√	√	√	√
		Type of ventilation	√	√	√	√
		Supply of purified and fresh air	√	√	√	√
		Air monitoring sensor	√	√	√	X
		Monitoring on carbon emission	√	√	√	
	Contamination level	Unstable compounds	√	√	√	√
		Pollution of electromagnetic waves	X	X	X	X
		Level of microbiological content	√	√	X	X
	Thermal comfort	Controlling zone	√	√	X	X
		Heating, cooling, humidity, vaporcontrol, and comfort	√	√	√	√

the existing assessment tools (BREEAM, LEED, IGBC, and GRIHA) to understand the depth of each of the criteria and their related attributes. The tick mark symbol '√' represents that the criteria are included in the respective assessment tool, whereas cross mark 'x' represents that it does not. Some of the criteria, which contribute to building sustainability, are neither included in IGBC nor GRIHA. Similarly, the attributes, which are included in IGBC, are not included in GRIHA and vice-a-versa, for example, ventilation, CO_2 emissions, and material efficiency. Also, some criteria like topographical consideration, climatic conditions, local context, and regional variations are not at all considered.

For instance, energy is considered as a key category for all assessment methods and is given the highest possible points. BREEAM measures Building Energy Performance (BEP) along with CO_2 emission reductions with the target of net-zero emissions. On the other hand, LEED emphasizes reduction of energy costs for BEP rather than CO_2 emissions, which is in line with the standards of the American Society of Heating, Refrigerating and Air-conditioning Engineers (ASHRAE). LEED mainly focuses on renewable energy utilization for measuring BEP and energy optimization. However, energy monitoring and enhanced commissioning are not considered in the GRIHA rating system. The three assessment methods (LEED, BREEAM, and GRIHA), evaluate most of the major water quality and quantity parameters. Indoor water use reduction, potable water use reduction, water recycle and reuse, wastewater treatment and efficient landscaping are the common criteria considered in all the three rating tools. Water leak detection and water metering are considered important criteria, but they are not considered in GRIHA for water efficiency evaluation. Waste related criteria and their parameters are integral to all three-assessment methods. Within this broad criterion, waste management and recycling emerge as the most important parameters, due to their importance in minimizing the negative impacts of waste generation for humans and the surrounding environment. Construction materials, is another important element of the environmental assessment method due to the impact of material consumption on building users and the environment. BREEAM and LEED emphasize sourcing of raw materials, but it is not considered in GRIHA. Renovation of abandoned buildings is considered in LEED, but not considered in GRIHA evaluation criteria for assessment. Indoor environmental quality (IEQ) is considered as a key objective for all building assessment methods. BREEAM and GRIHA include this category under Health and Well-being section. LEED assesses this category through low-emitting materials, indoor air quality, and quality views. Similarly, BREEAM assesses this category through visual comfort, the impact of refrigerants and noise pollution. GRIHA assesses this category through air quality; low-VOC paints, sanitation/safety facilities, but at the same time omits visual comfort, quality views, and hazards in its criteria. Tobacco smoke control, pollution, thermal comfort, and air quality are commonly considered in all three assessment methods. Light pollution reduction and joint use of facilities are considered in LEED but not considered in GRIHA. All the tools evaluated in this study offer credits to encourage and support sustainability measures. BREEAM considers Management as a separate category for its assessment, while LEED distributes management parameters across several assessment categories. BREEAM covers sustainable management principles more comprehensively than LEED. Transportation is considered as a separate category in LEED and BREEAM. LEED addresses transportation through factors Location and Transportation criterion with 16 possible points whereas, BREEAM, assesses the same with 13 possible points. From the observations, it was found that the criterion Transportation and Management was not considered in GRIHA for environmental assessment.

3 Methodology

The identification of criteria related to sustainable construction from various sources including existing building rating systems was carried out to address Agenda 21 and UN initiatives towards sustainability. The methodology to assess the relative weights of criteria and attributes and establish interrelationship among them is shown in Fig. 1. The significant criteria that can assess the sustainable performance of a building are identified based on the existing assessment tools, guidelines, and policies. Based on the comparison of tools like BREEAM, LEED, IGBC, and GRIHA, the criteria and sub-criteria are assessed and checked for the possibility of transferring and adopting to developing countries. While diagnosing the similarities and differences in various existing tools, the study emphasized on the suitability of potential and possible criteria to be considered. Further, the diagnosed criteria are refined and screened out using Delphi Technique (DT) to reach a consensus decision. Based on the comparative discussion carried in the paper by the same authors ([3, 5]), the significant criteria and attributes are adopted. The identified list is then refined and was utilized to develop priorities and weights through quantitative research methods and MCDM techniques.

To determine the relevant attributes for building assessment, data collected from the DT has been evaluated using the Relative Importance Index (RII) based on Eq. 6.

$$\text{RII} = \frac{\sum_{i=1}^{N} P_i R_i}{N \times n} \tag{6}$$

where

RII Relative Importance Index;
P_i Respondent's rating;
R_i Number of respondents placing identical weighting/rating;
N Sample size;
n Highest attainable score.

The attributes whose RII value is more than or equal to 0.7 has been screened out for selecting the most significant parameters to assess the performance of the building. The ultimate criteria and attributes for sustainable building assessment are determined as shown in Table 3.

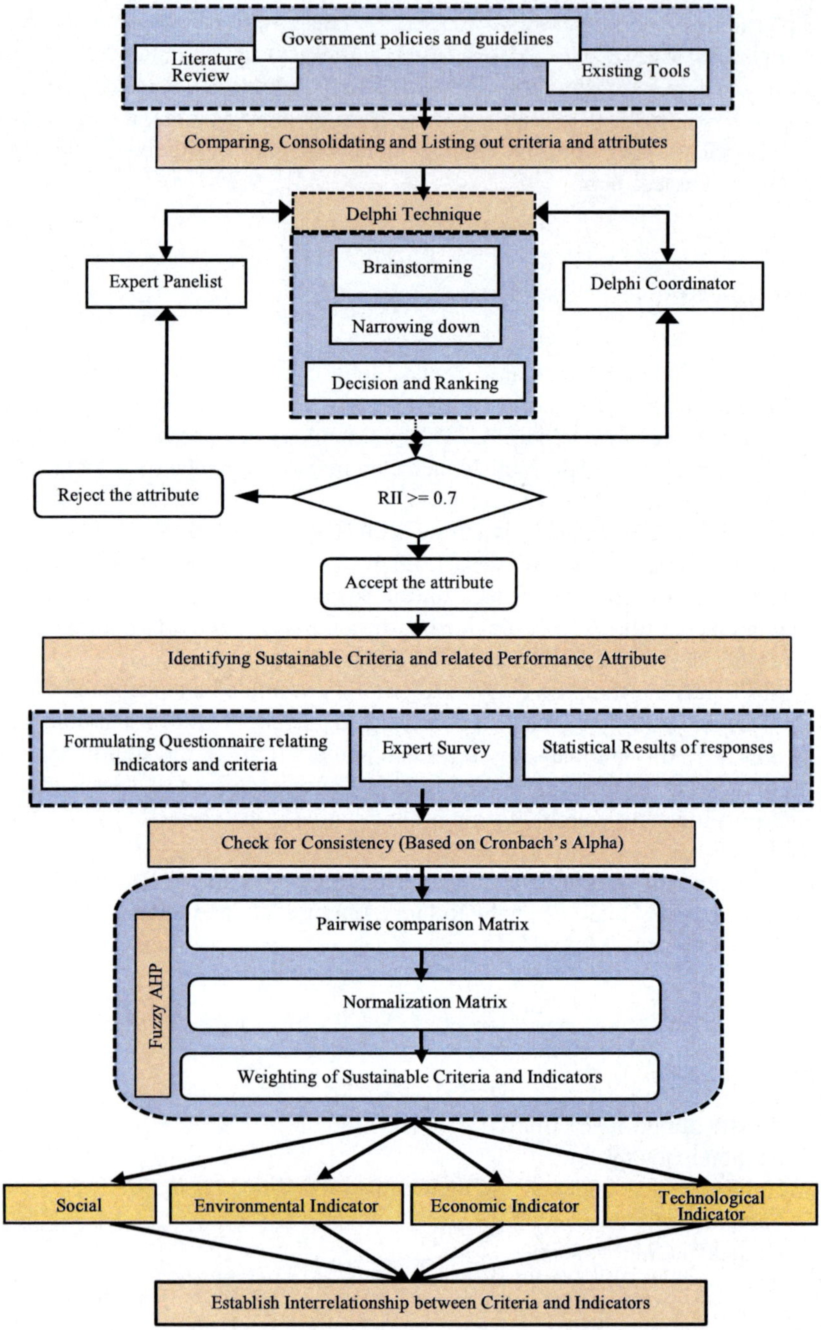

Fig. 1 Methodology to assign relative weights and establish interdependency for criteria and indicators

Table 3 Most prominent sustainable criteria and sub-criteria for developing countries

Code	Criteria	Attributes
WE	Water efficiency	Water monitoring and leak detection (A1) Building water use reduction (A2) Recycle and reuse of water (A3) Rainwater management (A4) Reduction in landscape water requirement (A5)
MW	Materials and waste management	Low-energy materials (B1) Regionally available materials (B2) Recycled and re-use materials (B3) Responsible sourcing (B4) Efficient waste management (B5)
HW	Health and well-being	Water quality & water pollution (C1) Outdoor & indoor noise levels (C2) Sanitation/Safety facilities & Accessibility (C3) Minimize ozone depletion (C4)
EE	Energy Efficiency	Renewable energy production (D1) Energy-efficient appliances (D2) Energy monitoring (D3) Reduction in energy consumption associated with interior lighting (D4) Adequate Daylight (D5) Energy-efficient vertical transportation systems (D6)
SS	Sustainable sites	Site selection (E1) Protect or restore habitat (E2) Heat island reduction (E3) Open space (E4) Light Pollution (E5) Efficient ventilation (E6) Conservation of soil surrounding the building (E7)
SW	Social welfare	Knowledge and Awareness towards sustainability (F1) Local Economic Development (F2) Design for durability (F3)
T	Transportation	Public transport accessibility (G1) Use of Bicycles (G2) Proximity to amenities (G3) Environmentally friendly pavements at the building site (G4) Reduced parking footprint (G5)
M	Management	Managing the balance between the building and its immediate surrounding (H1) Managing fire prevention facilities (H2) Preventing the reckless dumping of polythene products at the building site (H3)

To determine the priority weights of criteria towards each of the indicators, pairwise comparison of the criteria and indicators is performed for each individual judgments and then averaged. To evaluate the relative weights of criteria and indicators, the present study collected data from a structured questionnaire survey responses (147 no's) from all the stakeholders of the construction industry, each belonging to categories such as Academicians, Engineers, Designers, Architects, Consultants, Contractors, and Others. Further, the study utilized the FAHP for assigning the relative weights to criteria and indicators.

4 Data Collection, Results and Discussion

The respondents were invited to assess the level of importance of criteria and indicators by assigning a score on the seven (7) point Likert scale and converted to a fuzzy triangular scale as suggested by [13]. A score of '1' indicates as 'not important' whereas, '7' indicates 'highly important'. This icludes professionals from all domains of civil engineering. Among the 147 responses, incomplete data and data which is not reliable to be considered have been neglected. As required from population size, 96 reliable and complete responses were selected for further investigation. The consistency of the data is checked using Cronbach's alpha coefficient calculated using the following Eq. 7.

$$\alpha = \frac{N \times c}{v + (N - 1) \times c} \tag{7}$$

N is the number of items, *c* is the average inter-item covariance among the items, and *v* is the average variance. In general, the alpha score of more than 0.70 is considered acceptable [14]. In the present study, the Cronbach's alpha was calculated for four different groups (Social, Environmental, Economic, and Technological), from the information provided by 58 valid respondents. In all the groups, the '**α**' values were found to be more than 0.80. Thus, the data provided was found to be reliable and was of good quality. The procedure adopted by the authors [5, 15] was used to evaluate the relative weights using Fuzzy AHP. To de-fuzzy, the obtained relative weights, center of the area method is proposed. The relative weights for sustainable criteria and indicators for four no's of 8 × 8 matrices were evaluated for 58 respondents (Fig. 2). The average of individual priority weight is evaluated, using arithmetic mean operation. Similarly, the relative weights of sustainable indicators (SEET) with respect to sustainable criteria; where eight no's of 4 × 4 matrices for 58 respondents are performed and the final weights are obtained as shown in Fig. 3. The interrelated weights are obtained by multiplying the relative weights of sustainable indicators and criteria (Fig. 4).

From Fig. 2, considering Social indicator, the criteria Sustainable Sites (SS) has attained the highest weight (i.e., 13.93) and overall rank sixth among all the 32 criteria (say 8 criteria × 4 indicators). Following this, the criteria, Transportation (T)

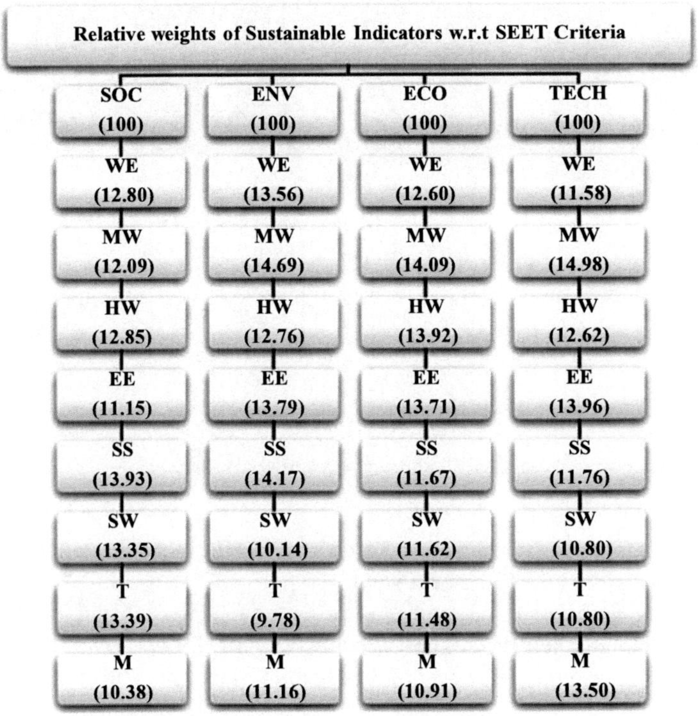

Fig. 2 Relative weights of sustainable criteria w.r.t sustainable indicators

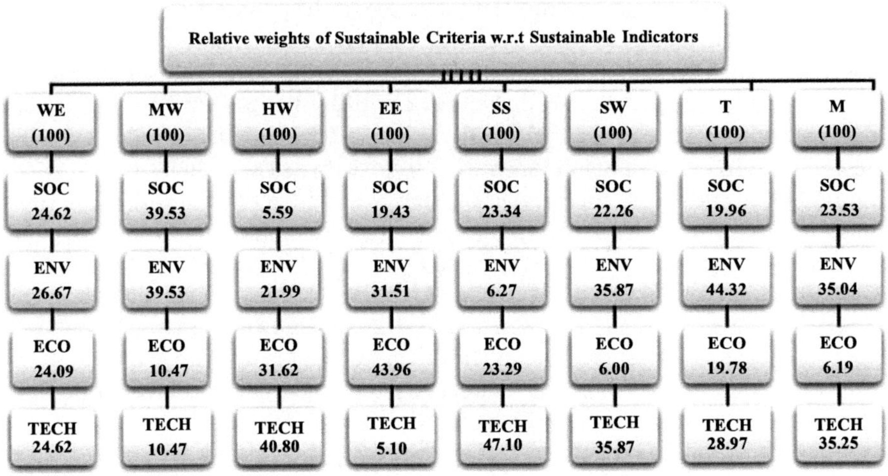

Fig. 3 Relative weights for sustainable indicators w.r.t sustainable criteria

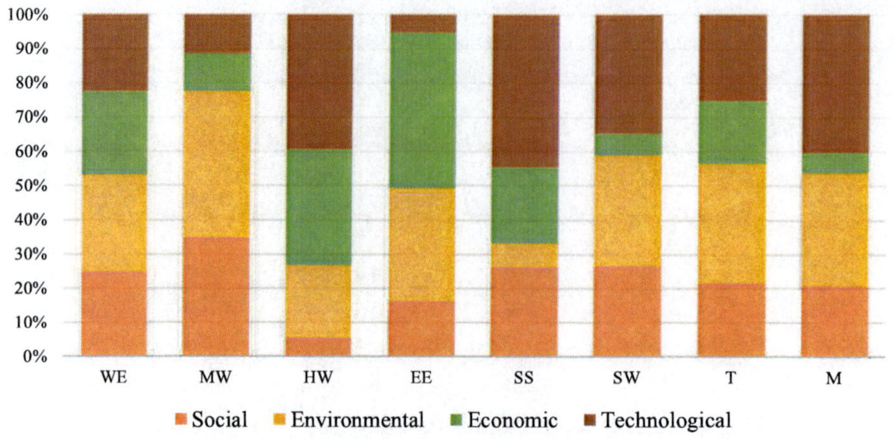

Fig. 4 Interrelated weights for SEET indicator w.r.t sustainable criteria

and Social Welfare (SW) has achieved a nearby weight of 13.39 and 13.35 with an overall ranking of 12 and 13 respectively.

Considering Environmental indicator, the criteria, Materials and Waste Management (MW) has attained the highest weight (i.e., 14.69) and overall rank second among all the 32 criteria. The criteria, Sustainable Sites (SS) and Energy Efficiency (EE), achieved a weight of 14.17 and 13.76 with an overall ranking of eight and three respectively. Considering the Economic indicator, the criteria MW has attained the highest weight (i.e., 14.09) and an overall rank of fourth among all the 32 criteria. Next, to it, Health and well-being (HW) and Energy Efficiency (EE) have achieved a weight of 13.92 and 13.71 with an overall ranking of seven and nine respectively. Similarly, in the Technological indicator, MW and EE have attained the highest weight (i.e., 14.98 and 13.96) and ranked first and fifth among 32 criteria respectively.

From Fig. 3, it can be observed that the criterion MW has a major role in creating social justice in the built environment. It also reveals that with the use of efficient, non-pollutant, and eco-friendly vehicles, there will be a reduction in the emissions and pollutants, further leading to reduced environmental impacts. Similarly, it can be noticed that the use of EE materials and technologies will benefit the user over a period of time. The initial cost may be high, but the cost to benefit ratio would be very low, due to a reduction in operational and maintenance costs. With proper implementation of guidelines and policies with respect to sustainable design principles of the buildings (technological indicator), the criteria, SS, eventually provides the source to attain the allotted weight.

From Fig. 4, it can be noticed that for assessing the performance of a building towards sustainability, the criteria EE has the highest interrelationship weight (6.03) corresponding to Economic indicator, Secondly, MW corresponding to Environmental indicator has a higher weight (5.81). Similarly, SS in Technological indicator (5.54), MW in Social (4.78), and Management (M) in Technological (4.76) indicator.Considering the interrelated weights of criteria and indicators, the average weights obtained by pooling along with the criteria with regard to Social, Environmental, Economic and Technological indicators (i.e., 2.77, 3.67, 2.66 and 3.45) is taken as a cut off value to observe the effective or most significant criteria responsible for indicator performance. Then, the normalized weights are calculated for criteria and indicators. It is noteworthy to observe that the Technological indicator incorporated along with the Triple-Bottom line (Environmental, Social and Economic) has the highest weight (28.4) among SEET indicators. Similarly, Sustainable sites (SS) and Materials and Waste Management (MW) have achieved higher weights (16.91 and 15.56 respectively) among the 8 criteria.

5 Conclusions

Based on the comparison of LEED, BREEAM, IGBC, and GRIHA assessment tools and guidelines undertaken, the adaptability of various criteria and their corresponding attributes in developing countries like India were observed considering regional variation, culture, heritage, climatic conditions, and topographical aspects. This enabled to explore and bring out the similarities and dissimilarities that exist in the building assessment tools of developed and developing countries.

- The study defines 37 attributes broadly under eight major criteria that are most appropriate for assessment of sustainable performance for construction in developing countries. These criteria include Water Efficiency (WE), Materials and Waste Management (MW), Health and Well-being (HW), Energy Efficiency (EE), Sustainable Sites (SS), Social Welfare (SW), Transportation (T), and Management (M) using Delphi technique and Relative Importance Index.
- The relative weights of criteria and indicators along with the interdependency are calculated using the Fuzzy Analytical Hierarchy Process, a subjective pairwise comparison.
- Among all the criteria, MW has attained the highest relative weight of 13.96 and subsequently, EE attained 13.15 using Fuzzy Analytical Hierarchy Process. It is noteworthy to observe that both these criteria belong to the Technological indicator.
- Among SEET indicators, the Environmental indicator has secured the highest relative weight of 30.15 and the Technological indicator is next to this with a weight of 28.52. The importance of Technological indicator which was not finding a deserved place in the Triple bottom line approach. In this way, the study brought out the significance of the proposed 'Technological' indicator and encouraged the Quadra-Bottom Line approach in implementing and achieving sustainable construction.

Thus, the findings facilitate the incorporation of innovative ideas and implement the concepts of 5R's into sustainable design principles. This interdependency obtained in the study between criteria and indicators facilitates the scope for the development of a framework for a sustainable building assessment in developing countries like India. These criteria facilitate policymaking, guidelines, and development of a comprehensive green building rating tool.

References

1. Mori, K., Christodoulou, A.: Review of sustainability indices and indicators: Towards a new City Sustainability Index (CSI). Environ. Impact Assess. Rev. **32**(1), 94–106 (2012)
2. Du Plessis, C.: Agenda 21 for: Sustainable construction in developing countries—a discussion document, p. 91 (2002)
3. Reddy, A.S., Raj, P.A., Kumar, P.R.: Developing a Sustainable Building Assessment Tool (SBAT) for Developing Countries—Case of India, pp. 137–148 (2018)
4. Vyas, G.S., Jha, K.N.: Identification of green building attributes for the development of an assessment tool: a case study in India. Civ. Eng. Environ. Syst. **33**(4), 313–334 (2016)
5. Reddy, A.S., Kumar, P.R., Raj, P.A.: Quantitative assessment of sustainable performance criteria for developing a sustainable building assessment tool (SBAT). In: International Conference on Sustainable Infrastructure 2019: Leading Resilient Communities through the 21st Century. Reston, VA: American Society of Civil Engineers, pp. 689–702 (2019)
6. Economic Policy Forum: Promoting sustainable and inclusive growth in emerging economies: Green Buildings, pp. 1–62 (2014)
7. Teng, J., Mu, X., Wang, W., Xu, C., Liu, W.: Strategies for sustainable development of green buildings. Sustain. Cities Soc. **44**(September 2018), 215–226 (2019)
8. Ubarte, I., Kaplinski, O.: Review of the sustainable built environment in 1998–2015. Eng. Struct. Technol. **8**(2), 41–51 (2016)
9. Raouf, A.M.I., Al-ghamdi, S.G.: Building information modelling and green buildings: challenges and opportunities, vol. 2007 (2019)
10. Jiang, Q., Liu, Z., Liu, W., Li, T., Cong, W.: A principal component analysis based three-dimensional sustainability assessment model to evaluate corporate sustainable performance. J. Clean. Prod. **187**, 625–637 (2018)
11. Nikravesh, M., Kacprzyk, J., Zadeh, L.: Forging new frontiers: fuzzy pioneers I (2007)
12. ALwaer, H., Clements-Croome, D.J.: Key performance indicators (KPIs) and priority setting in using the multi-attribute approach for assessing sustainable intelligent buildings. Build. Environ. **45**(4), 799–807 (2010)
13. Reddy, A.S., Kumar, P.R., Raj, P.A.: Developing a material sustainable performance score (MSPS) to select an alternative cementitious material. Cem. Wapno Bet. **24**(1), 68–79 (2019)
14. Tavakol, M., Dennick, R.: Making sense of Cronbach's alpha. Int. J. Med. Educ. **2**, 53–55 (2011)
15. Diabagaté, A., Azmani, A., El Harzli, M.: Selection of the best proposal using FAHP: case of procurement of it master plan's realization. Int. J. Electr. Comput. Eng. **7**(1), 353–362 (2017)

Introduction of the Recycling and Reusing Method of Screw Steel Pile EAZET

Shuo Teng and Tadashi Maejima

Abstract EAZET pile is a type of screw steel pile which consists of a pile shaft and a helical plate attached to the bottom. It is commonly known as an environmentally friendly pile method. It generates no surplus soil nor muddy water during installation by screwing the pile into the ground with its helical plate at pile tip. On the other hand, by rotating reversely, it can be pulled out and retrieved, thus also be recycled and reused if necessary. This article presents the recycling and reusing procedure and criteria of screw steel pile EAZET based on experience. For safe and smooth retrieval, capable machine should be chosen carefully in advance. The capacity of the machine and the torque during retrieval should be constantly noticed, especially when the retrieval is conducted long time after installation. The reason is that the ground disturbed by the pile installation has recovered, where it requires larger torque for pulling out. Retrieved piles will be examined via visual check, dimension check, and bending test of the pile shaft. After that, they will be divided into reusable members and non-reusable members. Reusable members can be repaired for future reuse. On the other hand, non-reusable members are recycled as reproduction materials for components in recycling factory. Recyclable and reusable features of EAZET enable it to be further adopted by temporary structures besides its original market. It can contribute to the transformation from economical society to recycling society where the efficient usage and recycling are emphasized.

Keywords Screw steel pile · Recycle · Reuse · Environmental friendly method · Sustainable society

S. Teng (✉) · T. Maejima
Asahi Kasei Construction Material Corporation, Tokyo, Japan
e-mail: shuo.tb@om.asahi-kasei.co.jp

T. Maejima
e-mail: maejima.tc@om.asahi-kasei.co.jp

H. Hazarika et al. (eds.), *Advances in Sustainable Construction and Resource Management*, Lecture Notes in Civil Engineering 144,
https://doi.org/10.1007/978-981-16-0077-7_50

1 Introduction

In recent years, steel rotary penetrating piles with helical plates attached to the pile tip are frequently used. Taking the environmental impact and the increasing numbers of urban constructions into consideration, demands have been growing for small-sized machines which can enter confined areas, and environmentally friendly method which generates no surplus soil, low vibration, and low noise. Screw steel pile method satisfies these requirements quite well and is considered as a suitable method which meets the requirements of pile for buildings. Besides practical uses in building construction area, screw steel piles have also been gradually adapted for temporary structures and tower structures due to its easiness of retrieval and high uplift capacity achieved by helical plates. This article introduces the installation method and management, and some construction cases along with several points of attention.

2 Construction Plans for Installation and Retrieval

The installation and retrieval flows are shown in Fig. 1.

2.1 Planning Stage

For installation, selection of the machine is important. It is ideal to select machine when taking the pile specification, difficulty of penetration into the ground, sufficient embedment depth and site space, into consideration.

For retrieval, besides the considerations above, the piles' installation condition should also be noticed. The longer time is after installation, larger force would be required for rotating the piles reversely since the ground around the piles would recover and apply larger resistance and friction. Moreover, since it is always assumed that the whole pile be retrieved, the torsional failure should be avoided considering the efficiency. Therefore, the upper limit of the rotational torque is determined by the torsional strength of pile shaft; thus machines with torque control device should better be selected. As a result, it is possible that the machines for installation and retrieval are different due to the pile specifications, installation periods, etc.

2.2 Preparation Stage

For higher retrieval efficiency, it is necessary to reveal 500 mm down from the foundation bottom and have metal fitting attached in advance for retrieval. If damage

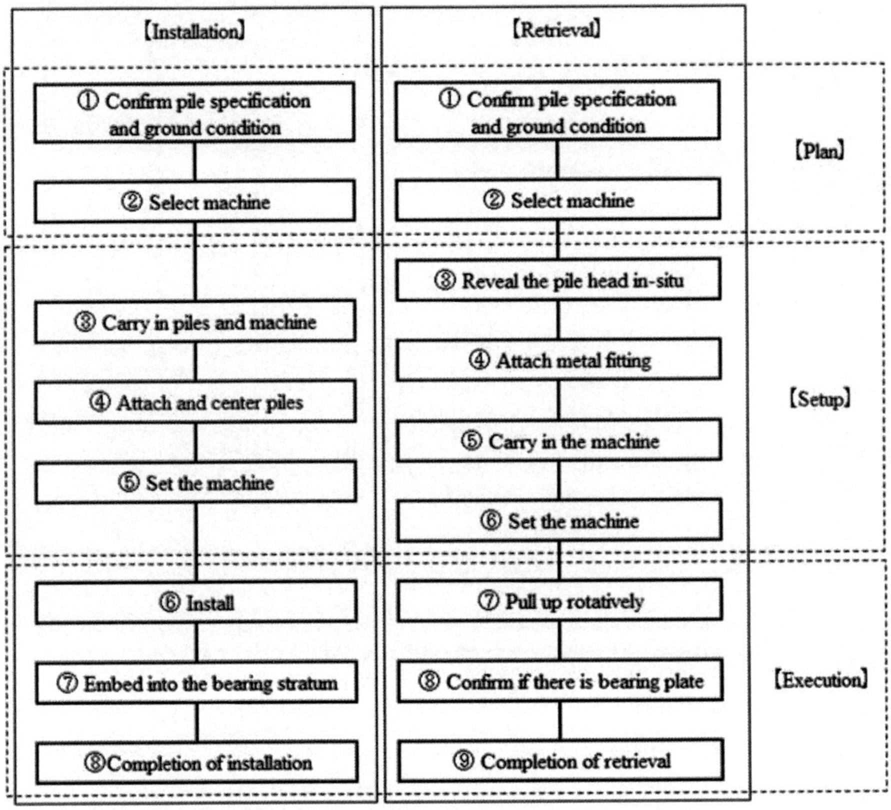

Fig. 1 Execution flow of installation and retrieval

is discovered in the steel shaft, the deformed part shall be cut and removed before pulling out for safety matters. While rotating reversely and pulling out, piles can be cut at the same time according to later use.

2.3 Installation and Retrieval Stage

The piles shall be installed vertically into the bearing stratum with adequate embedment depth. The installation accuracy, the running torque, and crowd force should always be noticed and compared to the soil investigation results during the installation.

After a period of time after installation, since the ground disturbed by the pile installation has recovered, it requires larger torque initially for pulling out and the retrieval becomes more difficult. Noticing the torsional strength of the pile, forward and reverse rotation shall be conducted to reduce the friction at pile head and transmit

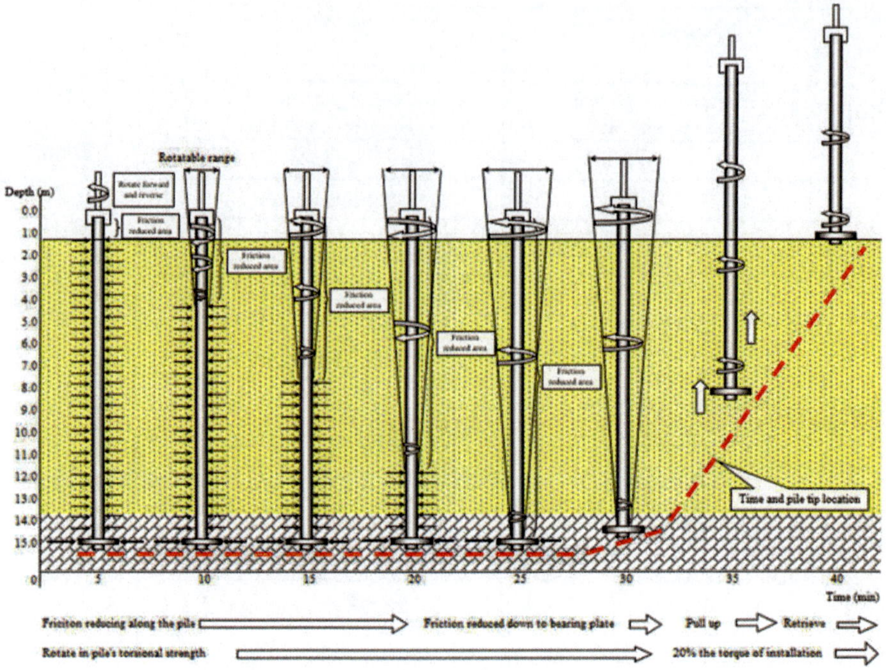

Fig. 2 Diagram for retrieval

the rotational force down to the pile tip (Figs. 2 and 3). Situation where the pile was broken as shown in Fig. 4 may occur. In this case, despite the forward and reverse rotation be conducted to the pile head, the friction along the pile shaft still cannot be reduced. This results in longer retrieval time or impossibility of retrieval at all.

3 Management Items for Installation and Retrieval

Management items for construction are shown in Fig. 5.

Regarding the installation process, both material quality and construction quality shall be managed.

As for material quality management, management value and management method are established for the soundness of pile shaft and bearing plate, the size of pile diameter and length, etc.

As for construction quality management, items such as location and verticality can be raised. Generally speaking, it is desirable to keep the offset within 100 mm and inclination within 1/100.

Since rather larger rotation torque is applied to the pile, installation condition should be noticed one by one to avoid damage. Meanwhile, the installation resistance,

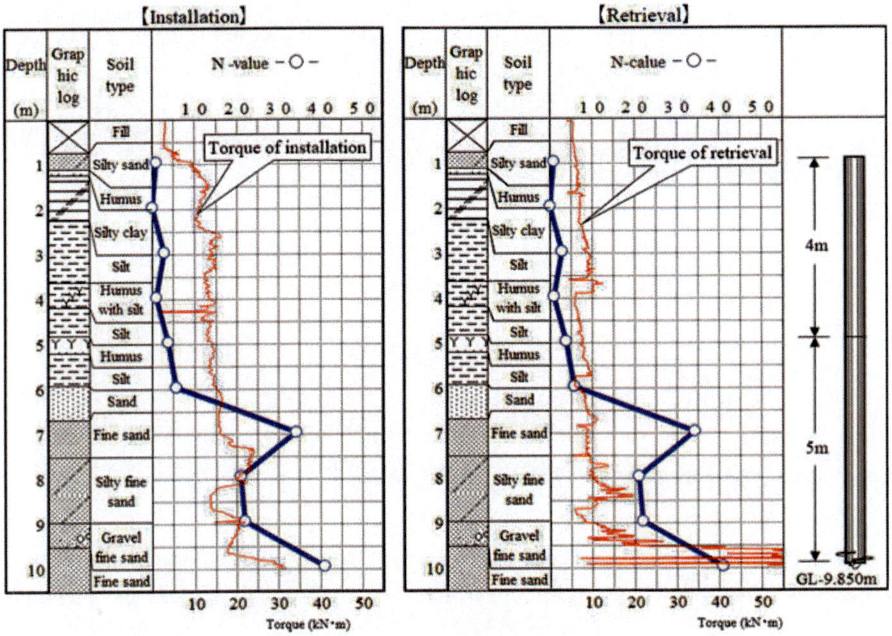

Fig. 3 Comparison of torque of installation and retrieval

Fig. 4 Torsional failure

soil condition, bearing stratum level, and embedment depth should be compared to the soil investigation results comprehensively.

Management of joint is also significant. Joints can be divided into welded joints and mechanical joints. For the former ones, items such as the skill of the engineer,

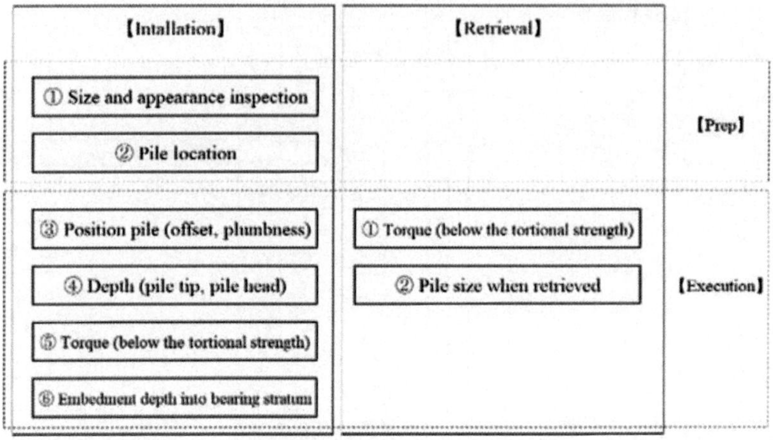

Fig. 5 Management items for installation and retrieval

shape of welding groove, weather, and selection of the welding rods can be considered important. For the latter ones, the bending, compression, tension performance, and assemble process management can be given.

Regarding the retrieval process, the vital issue is to keep the torsion force down within the torsional strength of the pile. Moreover, it may happen that the construction record has not been reserved, or the construction did not exactly follow the construction plan thus the pile length is unknown. Therefore, it is necessary to make sure that the whole pile and the bearing plate have been successfully retrieved.

4 Execution Method for Retrieval

In the case of retrieval long time after installation, construction machine and method shall be selected based on ground condition and pile specification (Fig. 6).

Since the dismantling of the upper structure may cause the decrease of ground strength thus the machine may overturn, the ground contact pressure shall be noticed. If the capacity is not adequate, it becomes impossible to retrieve by rotationally pulling out. The method shall be reconsidered. Besides rotation reversely, using casing to forcibly reduce the skin friction can also be adopted.

The selection of the method can be discussed according to the embedment depth into the bearing stratum. The record and condition of the installation shall be confirmed and then choose a more efficient one.

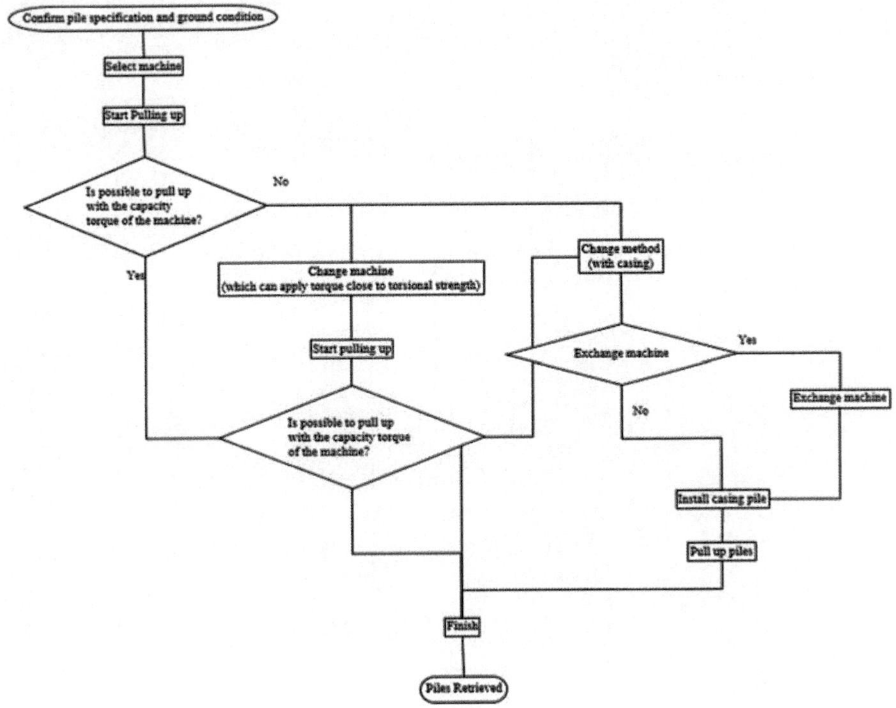

Fig. 6 Method and machine selection flow for retrieval

5 Case Study

5.1 Retrieval of Large Piles

Construction Overview. EAZET was adopted because this project required method with least impact on natural environment and possibility of removal and EAZET satisfied perfectly with appropriate cost and construction period.

The piles were installed from May 2003 to July 2004 and were removed from February 2006 to August 2006. The piles were 114.3–609 mm for shaft diameter, 250–800 mm for bearing plate diameter, 5–18 m for length, and STK400, STK490 for material steel. The machines were selected based on pile specifications and ground conditions.

Ground Overview. The main composition of the ground is gravel from the surface to GL −4 m below, and clayish gravel for bearing stratum with N-value more than 50. The groundwater was from GL −5 to −7 m. Since the site was rather large (installed with more than 2000 piles), the bearing stratum depth was uneven from GL −10 to −30 m (Fig. 7).

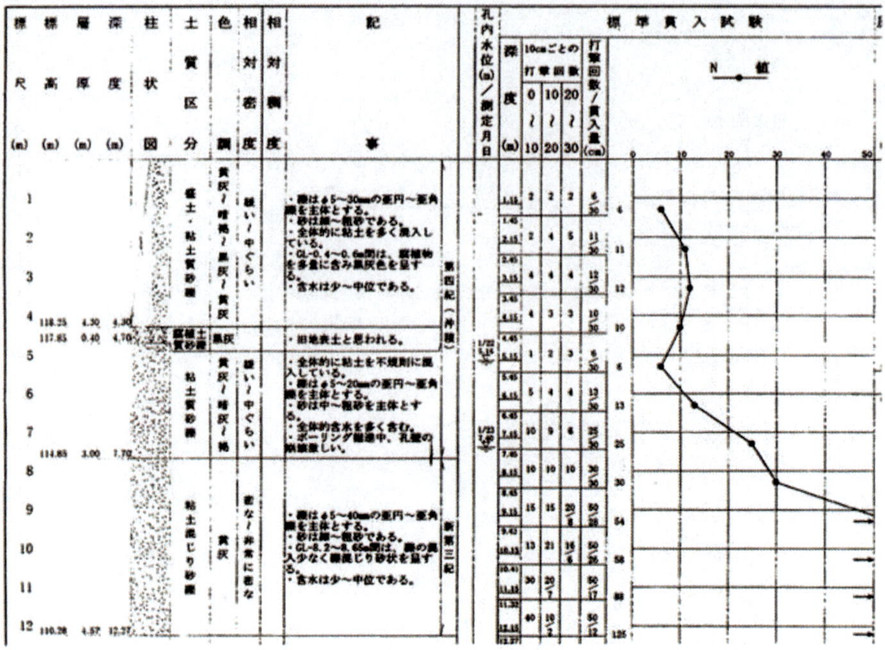

Fig. 7 Boring log

Retrieval Plan. For piles with diameter no more than 400 mm, same machine as installation was used. On the other hand, for piles with diameter no less than 400 mm and ratio of length by diameter more than 25, full-slewing machine was chosen (Table 1).

Retrieval Execution. Nearly 2000 piles were retrieved and became precious experience for us (Fig. 8). The understandings gained from this experience are presented as follows:

Table 1 Machine selection

Diameter (mm)	Machine type	Capacity (kN m)	Torque during installation (kN m)
114.3–267.4	Small machine (high torque)	60–100	60–100
318.5–355.6	Small machine (high torque)	60–100	100–250
	Small machine (super high torque)	100–250	
400–609.6	Small machine (super high torque) Full-slewing machine	100–400	250–400

(a) Retrieved piles

(b) Site view

(c) Various machines

(d) Various machines

Fig. 8 Site view

1. Regarding piles with length by diameter ratio no less than 20, due to the repetition of rotation forward and reverse, the torsion performance may act. Therefore, the friction resistance should be reduced gradually from the pile head to tip to ensure the execution efficiency.
2. Regarding piles with length by diameter ratio no more than 20, piles can be pulled out with only half rotation times of installation because the torsional force can be transmitted directly to the bearing plate.
3. As long as the friction resistance can be reduced down to the pile tip, piles can be pulled out with 20% the torque of installation.

To summarize, execution efficiency was mainly affected by the time spent on reducing the friction. It may be considered effective to the plan for auxiliary work such as pre-boring around the pile.

5.2 Reuse of Piles with Verification Bending Test

Introduction. Screw pile method requires no cement nor muddy water so it is easier to retrieve without damage to the piles. Retrieved piles can be divided into those which can be repaired to meet the performance regulation, and those which should be recycled.

Site Overview. At this site, screw piles used for temporary structures were retrieved after 6 months of construction. Performance test have been conducted afterward.

The diameter of pile shaft was 267.4 mm, thickness was 6.0 mm, diameter of bearing plate was 500 mm, and the length was 9 m (5 m + 4 m). The piles were supported by volcanic sand layer around G.L. −9 m. The boring log of the ground and the torque data collected are shown in Fig. 9.

Considering all the data and restrictions of the site, small machine was selected. Since the period after installation was half year, and the N-value of beating stratum was about 20, the retrieval work was rather smooth (Fig. 10).

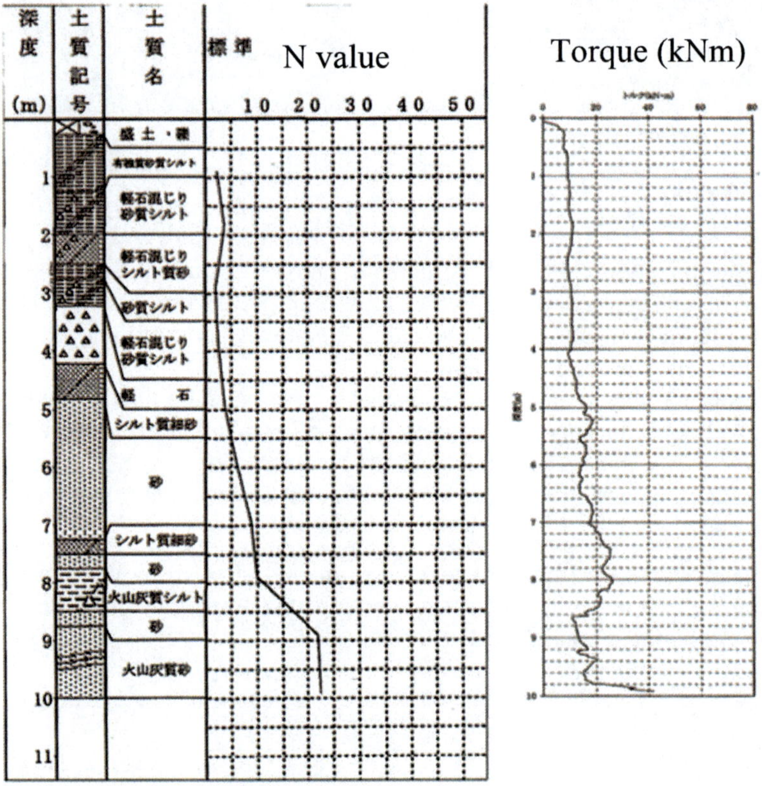

Fig. 9 Comparison of boring log and torque data

(a). Revealed pile head.

(b). Set pile head cap.

(c). Reverse rotation.

(d). Joint cutting.

(e). Pulling out.

(f). Retrieved piles.

Fig. 10 Site view

Performance Verification Tests. Performance verification tests were carried out to retrieved piles in terms of the following three aspects.

1. Visual inspection
2. Size inspection
3. Performance verification via bending tests.

As for visual inspection and size inspection, although small scratches can be seen on the pile surface due to the installation, obvious damage could not be discovered. Only the drilling blade attached to the bearing plate was worn away slightly. The size did not change from that before installation.

Meanwhile, considering the corrosion and other possible problems, bending tests were conducted to the pile shaft. The piles specifications are shown in Table 2, where the calculations are based on Eq. 1. Measurement is displayed in Fig. 11, and loading pattern is presented in Table 3, respectively. Actual condition during experiment is shown in Fig. 12.

$$M = \frac{1}{8}W(2L_1 - L) + \frac{P}{4}(L_1 - 1)$$
$$P = \frac{8M - W(2L_1 - L)}{2(L_1 - 1)} \quad (1)$$

Table 2 Pile properties

Length L (m)	Weight W (kN)	Second moment of area (mm^2)	Yield strength σ (N/mm^2)	Short-term allowable moment M (kN m)	Short-term allowable load P (kN)
4.0	1.52	42.11	235	74.01	112.9

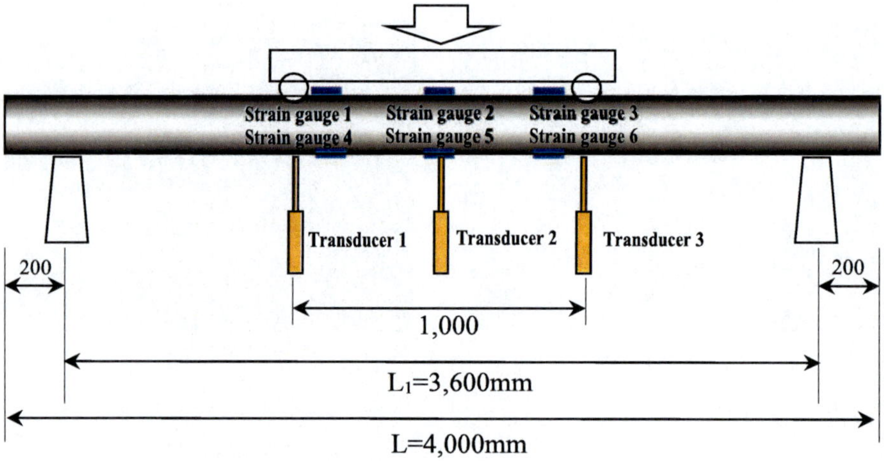

Fig. 11 Experimental program

Table 3 Load cycle (example)

Pile No.	Load cycle (kN)
1	0–28.3 (P/4)–56.5 (P/2)–84.7 (3P/4)–112.9 (P)–141.1–169.4 (Ultimate)–P_u

(a). Overall view.

(b). Measurement equipment.

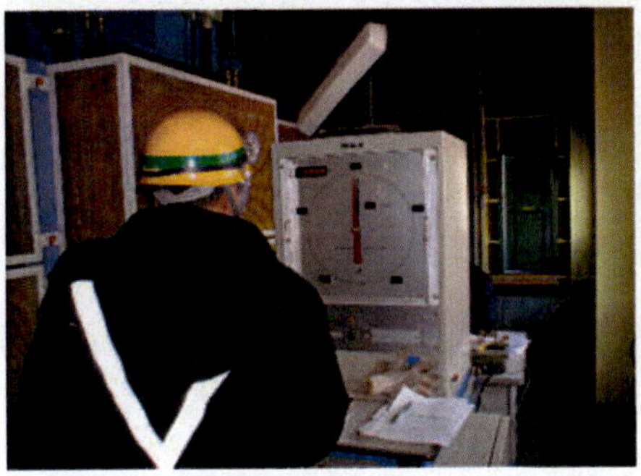

(c). Loading machine.

(d). Loading condition.

Fig. 12 Loading test condition

The relationship of load–displacement and load–strain at the center is shown in Figs. 13 and 14, respectively. The maximum load reached 178 kN. In addition, the strain at the short-term allowable load was 1050 $\mu\varepsilon$. Therefore, it seemed that the performance was the same as newly manufactured, thus satisfied as reusable members.

This site provided good condition for easy retrieval, while it may still prove that there is high possibility for retrieved piles to be reused.

5.3 *Verification Corrosion Test*

Since in the two cases above, the piles were retrieved after a relatively short period of time (two years and half a year, respectively), it may be doubtful about the adaptability for long-term projects. Therefore, quality of steel piles 25 years after installation underground has been verified.

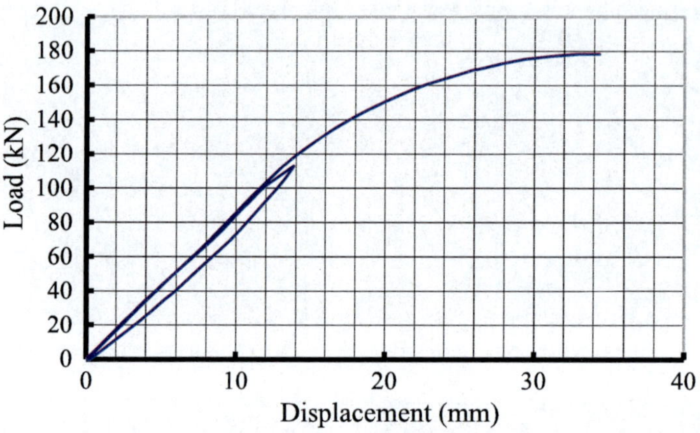

Fig. 13 Load–displacement relationship

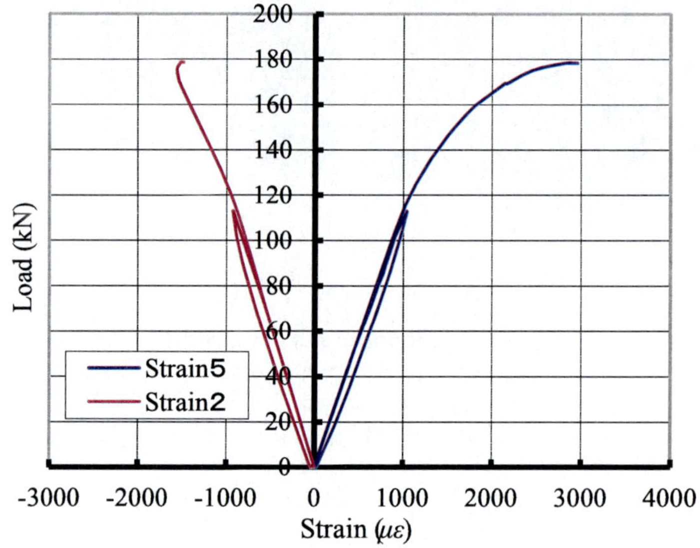

Fig. 14 Load–strain relationship

Introduction. In order to investigate the corrosion condition of steel piles underground for long term, steel piles installed underground for 25 years have been pulled out and inspected. The groundwater was considered not changing exceedingly since it could not been discovered after digging down for 2 m. Furthermore, the pH value of the ground was 4.52–7.16, which is a friendly environment for corrosion.

The pile length was 5 m with 2 m of leading pile and 3 m of extension pile. The piles' diameter was 114.5 mm, and thickness was 4.5 mm for leading piles and

(a). Piles.

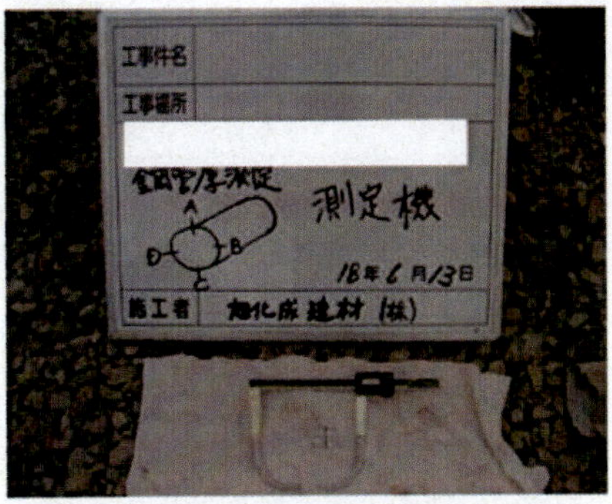

(b). Special venire caliper.

(c). Cut piles.

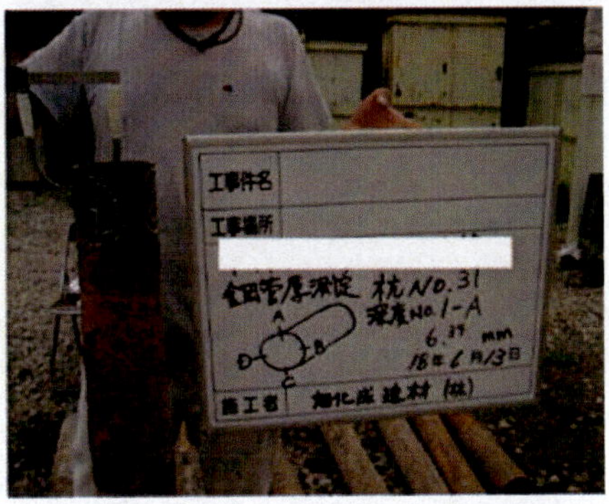

(d). Measurement situation.

Fig. 15 Site view

6.0 mm for extension piles. Pile size was inspected and compression tests have been conducted to these piles to verify the strength. Figure 15 shows the situation of piles and measurement.

Pile Size Measurement. After retrieval, the 5 m piles were cut into 1 m length for further investigation. Thickness of four locations of each cross section was measured. Measurement locations were chosen including the minimum thickness in each cross section. Moreover, locations with severe corrosion along the pile were additionally measured in each piece.

The results are presented in Tables 4 and 5.

Comparing Tables 4 and 5, it can be seen that except for a few locations with large corrosion allowance more than 2 mm, the corrosion allowance along the pile was considerable. It can be determined that there piles generally qualified as foundation members after 25 years installation.

As mentioned in session 5.2, retrieved pile can be divided into two types: ones that can be reused as construction members directly, and ones that should be melted

Table 4 Regular measurement

Specimen	Corrosion allowance (mm)						
	1	3	9	27	29	31	Ave
1	0.15	0.05	0.24	0.03	0.08	0.25	0.133
2	0.15	0.25	0.24	0.10	0.20	0.09	0.170
3	0.10	0.33	0.12	0.14	0.09	0.22	0.168
4	0.04	0.27	0.19	0.57	0.18	0.15	0.231
5	0.12	0.09	0.14	0.44	0.12	0.23	0.191
Ave	0.111	0.199	0.186	0.255	0.135	0.188	0.179

Table 5 Severe corrosion measurement

Specimen	Corrosion allowance (mm)						
	1	3	9	27	29	31	Ave
1	2.42		0.60		0.21		1.08
2	2.06	2.21	0.97	1.24			1.62
3				0.33	0.42		0.38
4	0.24					1.02	0.63
5		0.15				0.64	0.40
						Ave	0.82

and turned into steel materials. It has been known in this corrosion test that only few locations would be damaged while most parts of the pile are satisfying. Therefore, it provides an idea that if retrieved pile can be repaired locally and then reused, it would be more cost efficient and environmentally friendly. This kind of method has not been established and should be a future task.

Compression test. Piles were cut into pieces in 200 mm length and compression tests were carried out (Figs. 16 and 17). Part of the results is shown in Table 6.

The specimens failed due to buckling and the compressive strength greatly exceeded the design value.

6 Summary

More experiences of pile retrieval and reuse/recycle should be gained and then establish complete standards and manuals for management the whole steps. Moreover, method of partial repair for reuse should also be established and realize ideal pile recycle system as illustrated by Fig. 18. Retrieved piles can be reused directly, can be reused under partial repairment, or be recycled as steel material and turned into new members.

Fig. 16 Compression test

Fig. 17 Specimens after test

As a result, besides many other advantages over other methods, screw pile method can greatly contribute to environmental-friendly construction and sustainable society. Further spread can be expected in many aspects.

Table 6 Compression test results

Specimen	Volume (cm^3)	Density (g/cm^3)	Compressive strength (kN)	Design value (kN)
1–2	338.8	7.23	566.5	453
3–5	352.5	7.01	649.0	322
3–2	328.4	7.19	569.0	453
27–3	300.8	7.35	558.0	453
31–4	319.2	6.77	531.5	322

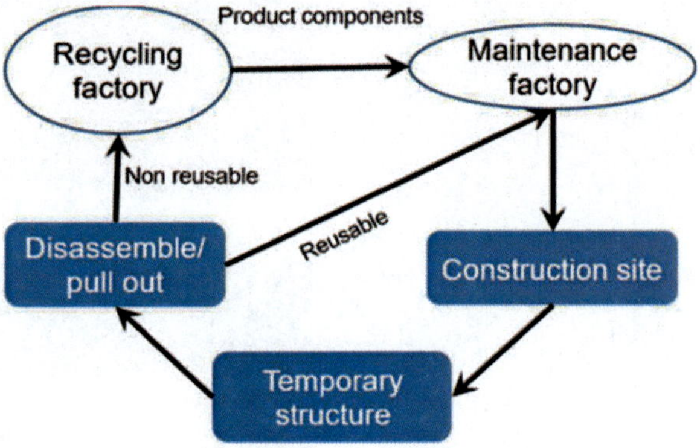

Fig. 18 Ideal pile recycle system

Mechanical Behavior of Cement-Treated Soils with Nanosilica—A Green Binder

Gizem Aksu and Tugba Eskisar

Abstract This study deals with the mechanical properties of cement-treated soils with the addition of nanosilica particles. Different industries discovered the beneficial uses of nanoparticles; however, the potential use of nanoparticles in geotechnical engineering is a subject that is still unclear and needs some investigation. Nanosilica particles could act as a green binder for geotechnical site works. Therefore, in this study, the effect of nanosilica material in cement-treated clay and cement-treated clayey sand specimens were examined. Index properties, compaction characteristics and unconfined compressive strength of nanosilica added cement-treated soils were assessed. Tests were carried out on specimens with 0%, 5%, and 10% cement by dry weight of soil with 0%, 0.3%, 0.5%, and 0.7% nanosilica content to evaluate the strength properties of the specimens. The curing periods were 7 and 28 days. The results of the preliminary tests showed that, the inclusion of nanosilica resulted in a limited increase in the optimum moisture content of the specimens and the maximum dry density of the specimens had a limited decrease at 0.7% nanosilica content. The performance achieved by the addition of nanosilica particles was very remarkable compared to the specimens that were prepared only with cement. Clayey sand specimens with 5% cement and 0.5% nanosilica had a similar compressive strength with specimens that had 10% cement in 28 days. Consequently, inclusion of nanosilica particles in soils is a promising development to reduce the cement amount in cement-treated soils.

Keywords Nanosilica · Green binder · Unconfined compressive strength

1 Introduction

Improvements in nanotechnology made the production of nanomaterials possible and different industries discovered the beneficial uses of nanoparticles. Nanoparticles are new materials which effects are less known in civil engineering applications. ASTM

G. Aksu · T. Eskisar (✉)
Ege University, Izmir 35100, Turkey
e-mail: tugba.eskisar@ege.edu.tr

H. Hazarika et al. (eds.), *Advances in Sustainable Construction and Resource Management*, Lecture Notes in Civil Engineering 144,
https://doi.org/10.1007/978-981-16-0077-7_51

E2456 [1] defines nanoparticles as ultrafine particles with lengths in two or three dimensions greater than 1 nm and smaller than 100 nm and which may or may not exhibit a size-related intensive property.

Addition of nanosilica material may decrease the need of using cement in the soil stabilization and expedite the duration of the stabilization [2]. When nanosilica is used with cement, the hydration process is accelerated due to the high surface energy of the nanosilica [3]. The hydration products cover the particles in the soil matrix leading to some physical changes including particle filling and nucleation effects [4, 5]. The hydration products are better distributed, resulting in a denser soil matrix.

A few studies were done on the use of nanosilica particles in geotechnical engineering. In these studies, addition of nanosilica has improved the characteristics of the soils dramatically [6, 7].

Bahmani et al. [6] assessed the effect of nanosilica material on consistency limits, compaction, permeability, and compressive strength of cement admixed soils. The addition of nanosilica positively affected the compaction properties, and the permeability of the soil specimens. 0.4% nanosilica addition resulted in the highest strength of the cement admixed soil.

Ren and Hu [8] examined the physical and mechanical properties of nanosilica added silty clay. They performed specific gravity tests, Atterberg limits tests, unconfined compression tests, and worked on the frost heave response of specimens. Increasing the nanosilica rate increased the plastic limits, liquid limits and the unconfined compressive strengths, but decreased the effects of frost heave. The specific gravity was found to be stable in any nanosilica content.

Bahmani et al. [9] investigated the effects of size and replacement ratio of 15 nm nanosilica on physical characteristics of cemented soil. Unconfined compression tests were conducted on 7–14–28 day specimens with replacement ratios from 0.2% up to 1% of nanosilica particles. 15 nm nanosilica had an accelerated impact on the specimens.

Chobbasti and Kutanaei [7] conducted tests to observe the mechanical characteristics of cemented sandy soil. Cement content was 6% by weight of the sandy soil, and nanosilica was 0%, 4%, 8%, and 12% by weight of cement in their study. Curing periods were for 7–28–90 days. Unconfined compression tests and compaction tests showed that adding optimum percentages of nanosilica enhanced mechanical properties of the soil.

Lv et al. [10] reported the stress–strain development of nanosilica added loess soil. The sizes of nanoparticles were 10 nm, 29 nm, and 100 nm. The addition of nanosilica was found to improve the mechanical properties of loess soil effectively. Stiffness and strength increased with increases in nanosilica content. 10 nm nanosilica exhibited strain hardening behavior ending up with a brittle failure at a nanosilica content of 7%.

The potential use of nanoparticles in geotechnical engineering is a subject that needs more investigation. As a member of nanoparticles family, nanosilica particles could act as a green binder for geotechnical site works. The aim of the study was to examine the effect of 15 nm size nanosilica particles in cement-treated clay and cement-treated clayey sand specimens. Atterberg tests, compaction tests, and unconfined compression tests were held. Nanosilica content effect on cement admixed soils wasreported for soil specimens which were cured for 7 and 28 days.

2 Materials and Methods

Clay and clayey sand used in this study were classified according to USCS. Clay was mainly composed of kaolinite minerals and classified as ML, and clayey sand was classified as SM-SC. Index properties of the soils were determined according to ASTM D6913 [11] and ASTM D4318 [12] standards and these properties are given in Table 1.

Nanosilica particles with a size of 15 nm was used. The purity of this material was 99.5%. Technical details of nanosilica are presented in Table 2. Ordinary Portland cement (OPC) was preferred as a stabilizing agent to treat the specimens.

Specimens with 0%, 5%, and 10% cement by dry weight of soil, and with 0%, 0.3%, 0.5%, and 0.7% nanosilica content by dry weight of soil were prepared to evaluate the strength properties of the treated and untreated soils.

While determining the Atterberg limits, liquid limit tests, and plastic limit tests were carried out in accordance with ASTM D4318 [12] standard. First of all, a dry mixture of cement and ground was prepared. Nanosilica was mixed with a high speed

Table 1 Geotechnical properties of the soils

Property	Clayey sand	Clay
USCS symbol	SM-SC	ML
Liquid limit	22	49
Plastic limit	15	31
Plasticity index	7	19
Percent passing No. 200	40	100

Table 2 Technical properties of the nanosilica particles used in this study

Property	Value
Purity	99.5%
Color	White
Average size (nm)	15
Specific surface (m^2/g)	150–550
Bulk density (g/cm^3)	<0.1
Specific gravity (g/cm^3)	2.2
Elements (%)	Fe Ca Ti Na
	0.002 0.007 0.012 0.003

mixer (120 rpm) for 5 min in water and then added to the soil–cement mixture. During the experiments, liquid limit of the specimen was increased gradually until the liquid limit range was achieved. In the plastic limit test, the experiments were carried out at single water contents, just after adding the well mixed nanosilica–water. Since the specimens contained cement, both liquid limit and plastic limit tests have been completed in less than 10 min considering that cement hydration would take place in a short period of time. The same mixing technique was applied to the preparation of unconfined compression test specimens.

Standard Proctor energy was applied while determining the compaction characteristics and during the preparation of cylindrical specimens for unconfined compression testing. Compaction tests were performed according to ASTM D 698 [13]. The tests were completed in less than 30 min to prevent the hardening of the mixture. The unconfined compressive test specimens had a diameter of 5 cm and a height of 10 cm keeping the H/D ratio as 2. The curing periods of the specimens were chosen as 7 and 28 days to observe the early strength gain. Figure 1 shows some of the clay specimens which were admixed with cement and/or nanosilica and cured for 7 days.

The stress–strain behavior of the specimens were attained by unconfined compression tests. The unconfined compression tests were in conformity with ASTM D2166 [14]. The loading rate was 1.42 mm/min during the tests and 10% of axial displacements were tracked during testing.

Fig. 1 Cement and/or nanosilica admixed clay specimens with a curing period of 7 days

3 Results and Discussion

3.1 Atterberg Limits

The effects of the intrusion of 0%, 5%, and 10% cement on liquid limit (LL), plastic limit (PL), and plasticity index (PI) of the soils are shown in Fig. 2. Liquid and plastic limits increased with the addition of cement in clay specimens. Locat et al. [15] and Chew et al. [16] explained that in clay soils aggregation and cementation of particles created larger clusters in the clay soil matrix. Locat et al. [17] added that the entrapped water within intra-aggregate pores, increased the apparent water content without truly affecting inter-aggregate reactions. In clayey sand specimens, the liquid limit did not change with the increment of cement content, but the plasticity index increased at a cement content of 5% and slightly decreased at a cement content of 10%. At a cement rate of 10%, cementitious products might be accumulating onto the surfaces of the flocculated clay clusters and the larger voids in the clay–sand medium due to the presence of sand particles enhancing a lower plasticity index.

The Atterberg limits of cement admixed soils along with the inclusion of nanosilica at rates of 0%, 0.3%, 0.5%, and %0.7 are presented in Fig. 3. There was an apparent distinction of Atterberg limits between nanosilica-treated cement admixed clay and clayey sand soils. The plastic limits of the clay specimens were around 35 and 37. The reason for this observation could be attributed to the non-plastic nature of the nanosilica particles not being able to change the plasticity of the clay soil. The liquid limit of clay specimens with 5% cement increased while the liquid limit of clay specimens with 10% cement decreased with the rate of nanosilica added to the specimens. As the plastic limits were almost stabilized, the plastic limit values followed the same trend with the liquid limit values.

On the contrary, liquid limits of nanosilica-treated cement admixed clayey sand specimens were between 24 and 25. Only 0.7% nanosilica-treated 5% cement

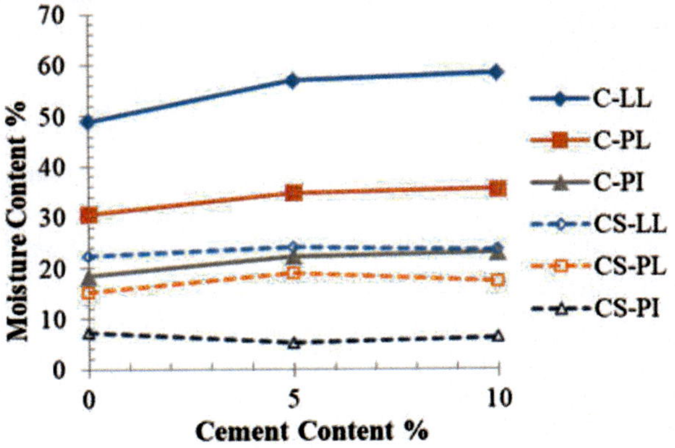

Fig. 2 Atterberg limits of cement admixed clay (C), and clayey sand (CS) specimens

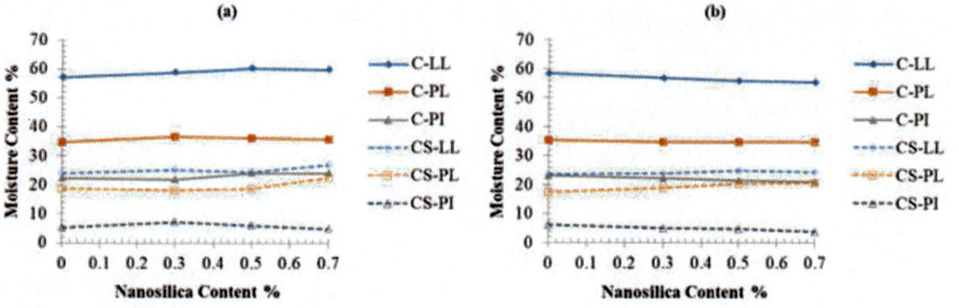

Fig. 3 Atterberg limits of nanosilica-treated specimens, **a** with 5% cement, **b** with 10% cement (C denotes clay, CS denotes clayey sand)

admixed clayey sand had a liquid limit of 27 (Fig. 3a). Plastic limit values of the clayey sand specimens slightly increased with the nanosilica content. 0.3% of nanosilica inclusion to the 5% cement admixed clayey sand resulted in the peak plasticity index (PI = 7) of the test group followed by a decrement of plasticity index. Increment of nanosilica content dramatically decreased the plasticity index of 10% cement admixed clayey sand specimens. The lowest value of plasticity index was achieved as 4 in 0.7% nanosilica-treated 10% cement admixed specimens (Fig. 3b).

The decrement of the plasticity index could be a result of better packing of particles with the support of nanoparticles in the soil medium. Also, Qing et al. [18] and Bahmani et al. [6] reported that the nanoparticles could be covered by very thin layers of water molecules and soil matrix is therefore able to plasticize with less amount of water.

3.2 Compaction Tests

Untreated clay and clayey sand specimens had optimum water contents of 29.7% and 13.1%, respectively. As cement was added to both clay and clayey sand, it was concluded that the maximum dry unit weight (or maximum dry density—MDD) and the optimum water content of the specimens decreased (Fig. 4). This finding was compatible with the results of Otoko [19] and Faria et al. [20].

Figure 5 shows the optimum water contents and the maximum dry unit weights (or maximum dry density—MDD) of the nanosilica-treated cement admixed soil specimens. When nanosilica particles were added, the maximum dry unit weight and the optimum water content increased in comparison to solely cement admixed clay (Fig. 5a). A similar observation was made for clayey sand specimens (Fig. 5b). However, the optimum nanosilica content was specific to the amount of cement content, regardless of the soil type. For example, for the specimens treated with 10% cement, the optimum nanosilica content was 0.5%, but for the specimens treated with 5% cement, the optimum nanosilica content was 0.3%. This finding showed that in order to achieve effective compaction, nanosilica content should be proportional to

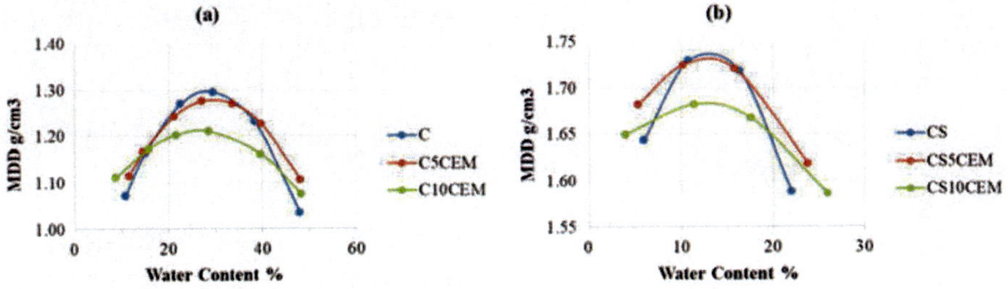

Fig. 4 Compaction curves of untreated and cement admixed **a** clay, **b** clayey sand

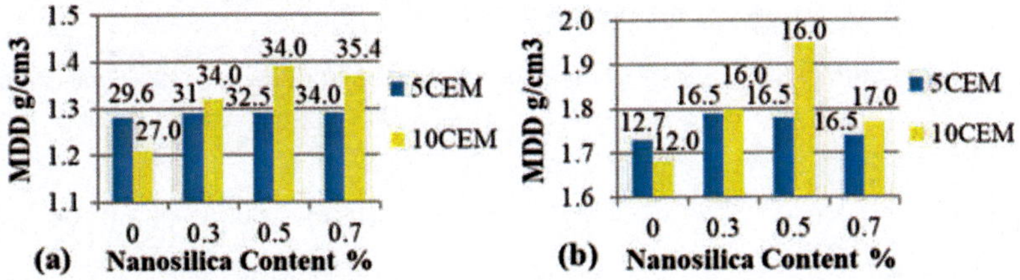

Fig. 5 Compaction characteristics of **a** clay, **b** clayey sand specimens (Numbers above the bars denote the optimum water content of specimens)

the cement content. Proper rate of cement and nanosilica existence further accelerated the dissolution of C_3S and formation of C–S–H with its activity being inversely proportional to the size [21].

3.3 *Unconfined Compression Tests*

Untreated specimens were tested to achieve the unconfined strength values as the control group. Clay specimens had a strength of 205 kPa and clayey sand specimens had a strength of 148 kPa. Exemplary stress–strain curves of nanosilica-treated 5% cement admixed clay specimens are shown in Fig. 6. These specimens were cured for 7 days.

It was seen that untreated clay was very ductile and with the addition of 5% cement, the behavior switches to a brittle one with a two-fold increment of unconfined compressive strength in 7 days. The addition of nanosilica to the cement admixed clay specimens further increased the unconfined compressive strength. This increase could be achieved only by choosing the optimum nanosilica content, for example, 0.5% nanosilica addition gave the best performance among 5% cement-treated clay specimens. The addition of more nanosilica reduced the unconfined compressive strength of the specimens. At this stage, 0.5% nanosilica was capable of adequately wrapping around the clay particles and filling the inter-particle voids, the presence of more nanosilica caused a disintegrated soil structure with a lower compressive

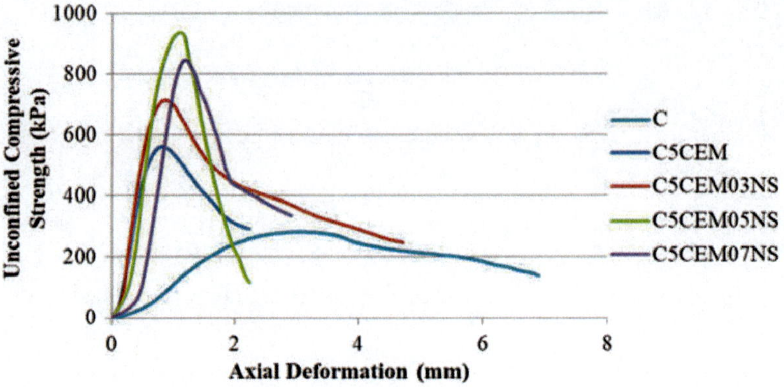

Fig. 6 Stress–strain curves of nanosilica-treated 5% cement admixed clay specimens cured for 7 days

strength. It is also noteworthy to express that more the nanosilica content more the ductile behavior.

Table 3 shows the results of unconfined compressive tests after curing for 7 and 28 days.

There was a remarkable strength gain when nanosilica was added to the specimens. Although a relatively higher strength could be achieved by cement treatment, this effect could be doubled by the presence of nanosilica particles. Also, it is possible to use less amount of cement with nanosilica to reach the similar strength value obtained by the addition of higher amount of cement. When 28 days performances of the clay specimens were considered, 5CEM07NS specimens had higher strength compared to 10CEM specimens. Clayey sand specimens could easily enhance a strength over 3000 kPa with a combination of 5% cement and nanosilica in 28 days.

Table 3 Unconfined compressive strength (UCS) of the specimens

Specimen name	Clay (UCS-kPa)		Clayey sand (UCS-kPa)	
	7 days	28 days	7 days	28 days
5CEM	563	1192	1007	3069
5CEM03NS	714	1287	1846	3212
5CEM05NS	938	1699	1895	3713
5CEM07NS	845	1610	2030	2645
10CEM	927	1413	3014	3831
10CEM03NS	1374	2243	3195	4817
10CEM05NS	1475	1974	3041	4462
10CEM07NS	1151	2298	3609	5566

Another fact is the strength development between 7 and 28 days specimens was two times more for 0.7% nanosilica added cement-treated clay specimens. The performance of nanosilica could be better achieved for clayey sand specimens. 5566 kPa was achieved for 10CEM07NS in clayey sand specimens in 28 days. The reason could be due to the variety of particle sizes in the soil medium could create interlocking of the particles and allowing cementitious contents and nanosilica particles to take place in chemical reactions forming a stiff body.

4 Conclusions

Index properties, compaction characteristics and, strength development of nanosilica added cement-treated specimens were evaluated in this study. The mechanical properties of cement-treated clayey sand soil are reported for the first time with this study. It should be emphasized that higher strength values could be obtained in clayey sand soils compared to the literature on clay or sand soils [6–9].

There was an apparent distinction of Atterberg limits between nanosilica-treated cement admixed clay and clayey sand soils. The plastic limits of the clay specimens were constant, while the liquid limit of the clayey sand specimens was constant with the increment of nanosilica particles. The liquid limit of clay specimens with 5% cement increased while the liquid limit of clay specimens with 10% cement decreased with the rate of nanosilica added to the specimens. It is seen that with the addition of nanosilica particles, optimum water content increased and maximum dry unit weight decreased under a standard compactive effort. The strength development was evident in 7 days specimens and further increase of strength was observed in 28 days specimens. 5566 kPa was achieved for 10CEM07NS in clayey sand specimens in 28 days. Nanosilica particles also contributed to the ductility of the soil–cement mixtures. In order to find the governing nanosilica content upon the unconfined compressive strength, long-term behavior should be explained, and for this reason, the performances of 90 and 150 days cured specimens are planned to be tested in a future study.

With this study, it is also shown that nanosilica particles are promising new materials that could be used in soil improvement applications as a green binder, and further inspection of nanosilica will help better understanding of this new material and more application areas would emerge.

References

1. ASTM E2456: Standard terminology relating to nanotechnology. West Conshohocken, PA (2012)
2. Choobbasti, A.J., Vafaei, A., Kutanaei, S.S.: Mechanical properties of sandy soil improved with cement and nanosilica. Open Eng. **5**, 111–116 (2015)

3. Farzadnia, N., Ali, A.A.A., Demirboga, R.: Development of nanotechnology in high performance concrete. Adv. Mater. Res. **364**, 115–118 (2012)
4. Lin, K.L., Chang, W.C., Lin, D.F., Luo, H.L., Tsai, M.C.: Effects of nano-SiO_2 and different ash particle sizes on sludge ash-cement mortar. J. Environ. Manage. **88**, 708–714 (2008)
5. Hou, P., Wang, K., Qian, J., Kawashima, S., Kong, D., Shah, S.P.: Effects of colloidal nanoSiO_2 on fly ash hydration. Cement Concr. Compos. **34**, 1095–1103 (2012)
6. Bahmani, S.H., Farzadnia, N., Asadi, A., Huat, B.B.K.: Stabilization of residual soil using SiO_2 nanoparticles and cement. Constr. Build. Mater. **64**, 350–359 (2014)
7. Choobbasti, A.J., Kutanaei, S.S.: Microstructure characteristics of cement-stabilized sandy soil using nanosilica. J. Mech. Geotech. Eng. **9**, 981–988 (2017)
8. Ren, X., Hu, K.: Effect of nanosilica on the physical and mechanical properties of silty clay. Nanosci. Nanotech. Lett. **6**, 1010–1013 (2014)
9. Bahmani, S.H., Farzadnia, N., Asadi, A., Huat, B.B.K.: The effect of size and replacement content of nanosilica on strength development of cement treated residual soil. Constr. Build. Mater. **118**, 294–306 (2016)
10. Lv, Q., Chang, C., Zhao, B., Ma, B.: Loess soil stabilization by means of SiO_2 nanoparticles. Soil Mech. Found. Eng. **54**, 409–413 (2018)
11. ASTM D6913: Standard test methods for particle-size distribution (gradation) of soils using sieve analysis. West Conshohocken, PA (2017)
12. ASTM D4318: Standard test methods for liquid limit, plastic limit, and plasticity ındex of soils. West Conshohocken, PA (2017)
13. ASTM D698: Standard test methods for laboratory compaction characteristics of soil using standard effort (12 400 ft-lbf/ft^3 (600 kN-m/m^3)). West Conshohocken, PA (2012)
14. ASTM D2166: Standard test method for unconfined compressive strength of cohesive soil. West Conshohocken, PA (2016)
15. Locat, J., Berube, M.A., Choquette, M.: Laboratory investigations on the lime stabilization of sensitive clays: shear strength development. Can. Geotech. J. **27**, 294–304 (1990)
16. Chew, S.H., Kamruzzaman, A.H.M., Lee, F.H.: Physicochemical and engineering behavior of cement treated clays. J. Geotech. Geoenviron. Eng. **130**(7), 696–706 (2004)
17. Locat, J., Tremblay, H., Leroueil, S.: Mechanical and hydraulic behaviour of a soft inorganic clay treated with lime. Can. Geotech. J. **33**, 654–669 (1996)
18. Qing, Y., Zenan, Z., Deyu, K., Rongshen, C.: Influence of nano-SiO_2 addition on properties of hardened cement paste as compared with silica fume. Constr. Build. Mater. **21**(3), 539–545 (2007)
19. Otoko, G.R., Battistelle, R.A.G., Neves, C.: On the economic use of cement in soil stabilization. Int. J. Eng. Technol. Res. **2**, 1–7 (2014)
20. Faria, O.B., Battistelle, R.A.G., Neves, C.: Influence of the addition of "synthetic termite saliva" in the compressive strength and water absorption of compacted soil cement. Ambient Constr. **16** (2016)
21. Aggarwal, P., Singh, R.P., Aggarwal, Y.: Use of nanosilica in cement based materials—a review. Cogent Eng. **2**(1), 1–11 (2015)

Structural Stability Against Earthquake and Tsunami Using Environmentally Sustainable Materials

Kiran Hari Pradhan, Hemanta Hazarika, and Yasuhide Fukumoto

Abstract Tsunami activated by earthquake is known to be one of the most powerful natural disasters. For instance, in March 11, 2011, tsunami activated by earthquake led to several compound disasters in Japan. Many coastal protection structures such as seawalls and breakwaters were found to be damaged seriously. In order to protect coastal structures from such kind of devastating damage in the future, it is essential to take proper countermeasures. A tire retaining wall in Iwate prefecture located just about 150 m away from a completely collapsed sea wall was found to be neither scoured nor damaged by tsunami, which indicates that the flexible tire structure has a great potential to reduce the tsunami impact force as compared to heavy and rigid structures. Therefore, main object of this research was to demonstrate how effectively tire structure will function against the earthquake and tsunami to protect sea wall from scouring on the foundation and resulting damage. A new model for simulation of tsunami impact force has been developed in Geotechnical Engineering Laboratory of Kyushu University. In which absorption of tsunami impact force and the dispersion effect of the tire structure was examined. Field experiments were also performed with various types of plants cultivated inside the soil-filled tires to see how the greening effect could be maintained. Results showed a better performance of the sea wall when protected with tires placed behind the sea wall. Also, field test showed that the greening effect could be maintained by cultivating suitable plants inside the tires.

Keywords Impact force · Scouring · Sea wall and tire retaining wall

1 Introduction

Tsunami activated by earthquake is known to be one of the most powerful natural disaster which causes serious damage to many coastal protection structures. For example, tsunami generated by Tohoku earthquake of magnitude Mw-9.0 followed

K. H. Pradhan (✉) · H. Hazarika · Y. Fukumoto
Kyushu University, 744 Motooka, Fukuoka, Japan
e-mail: kiranharip@gmail.com

H. Hazarika et al. (eds.), *Advances in Sustainable Construction and Resource Management*, Lecture Notes in Civil Engineering 144,
https://doi.org/10.1007/978-981-16-0077-7_52

several compound disasters in coastal area of Japan in March 11, 2011. Many coastal protection structures such as seawalls and breakwaters were found to be damaged seriously. But, surprisingly a waste tire retaining wall which was located just about 150 m away from a completely collapsed sea wall was found to be neither scoured nor damaged by tsunami. To know the reason behind that, many site investigations and laboratory tests were conducted by Hazarika et al. [1]. This case study revealed that the flexible material like tire can resist the earthquake and tsunami forces due to advantageous physical and mechanical characteristics that such material possesses.

According to destruction mechanism due to earthquake and tsunami, the damage of the sea wall is mainly due to scouring at the back of the structures by drainage impact force and undertow (back wash) impact force of tsunami. Hence, to protect sea walls from earthquake and tsunami, construction of tire retaining wall behind sea wall as shown in Fig. 1 is considered to be one of the effective measures.

In this research, a new anti-tsunami technique is developed, in which tires filled with soils are used behind the sea walls to reduce the impact of tsunami force, and hence prevent the scouring of soils and foundations of sea walls. Main purpose of this research is to demonstrate how effectively the tire will function against the earthquake and tsunami to protect concrete sea wall.

The property of tire can also be explained by Newton law, i.e., the force of impact will be less when the time interval will be large. Because of flexible nature of tires, the impact force will be acting for a longer time, which resulted in a lesser force acting on the structure.

$$F = mv/\Delta t \tag{1}$$

A new model for simulation of tsunami impact force (length 120 cm, height 100 cm, width 30 cm) was developed in Geotechnical Engineering Laboratory of Kyushu University and experiments were conducted for the sea wall by putting tires behind the wall. Performances of the tire retaining wall under the impact force of

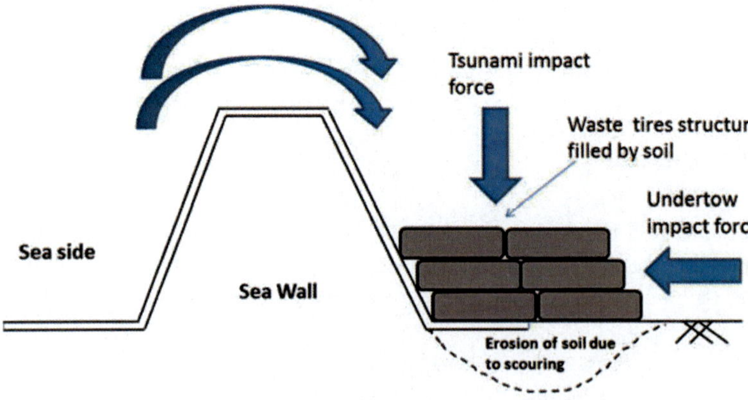

Fig. 1 Protection of seawall by waste tire structure

water were observed. The schematic diagram for the simulated condition used in the research is shown in Figs. 2 and 3. There is a hinged gate on the top of embankment model to reproduce the overflow phenomena by water reservoir. Water Falling Test (WFT) is conducted for the sea wall with tires behind it.

Figure 2 shows the simulated condition used in the research. Several testing configurations have been conducted for different condition of tire placement and the results were compared with the case of without using tire behind the sea wall. Numerical

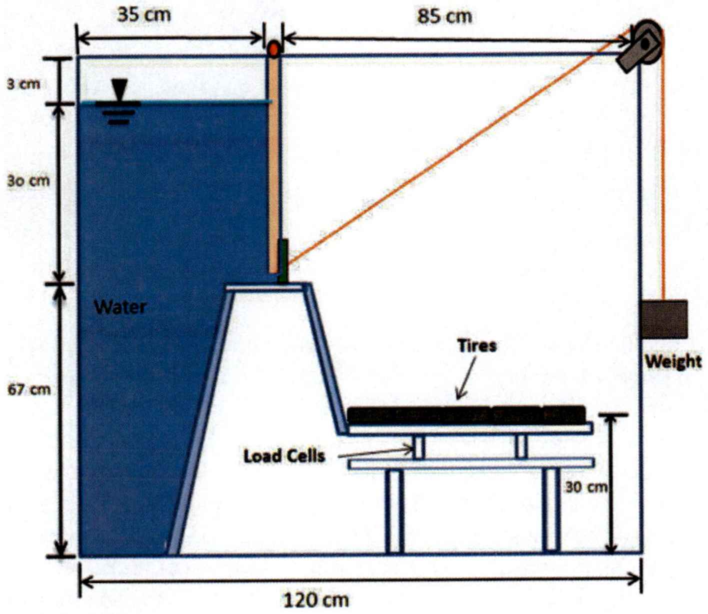

Fig. 2 Model for simulation of tsunami impact force

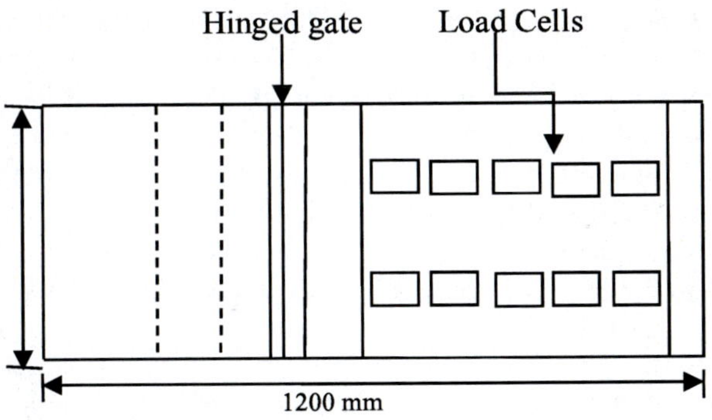

Fig. 3 Top view of water falling test apparatus

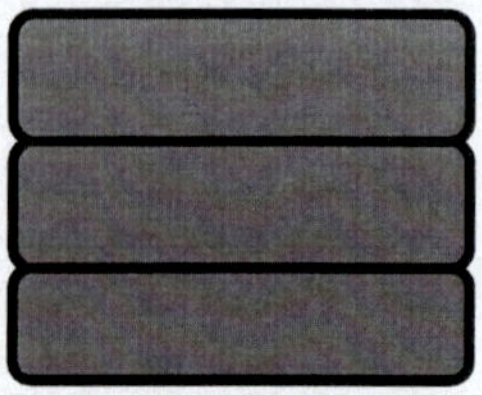

Fig. 4 Real tire

Fig. 5 Cultivation of plants inside real tire

simulation results show that a sea wall with tires placed behind it could withstand the tsunami impact force better.

Furthermore, the Impact force reducing property of flexible material due to its internal mechanism was also verified by conducting numerical analysis based on Isakovitch theory where effects due to interactions between internal molecules of the tire material are studied when it will be disturbed by some external energy. For this, one study case had been selected, that is Thermal attenuation and dispersion of sound in a periodic emulsion presented by Yasuhide Fukumoto et al. [2]. He investigated the attenuation and dispersion of sound waves in suspensions and emulsions caused by the thermal-transport process to constitute the effective compressibility of the system.

The striking feature is that a drastic change in attenuation occurs when the thermal conductivity of the particle is substantially larger than that of the ambient fluid. It means that the flexible material like the tire has some property to absorb external energy and transmit it to the other form of energy. The results obtained through this numerical study explains the results obtained through laboratory testing. Because according to the simulation analysis of the model tire experiment the impact wave was found to be smoothed than the cases without using tire. Thus, the mechanism of impact force reduction was clarified through this study.

In addition to the existing tire retaining wall, the effectiveness of greenery and scenic beauty is also the main concern of this research. In order to preserve the environment, field experiments were also performed in which various types of plants were cultivated inside the soil-filled tires to see how the greening effect could be maintained to preserve the environment as shown in Figs. 4 and 5.

2 Water Falling Test (W.F.T.)

Model tires (85 mm outer diameter and 21 mm thickness) filled with Masado soil compacting to dry density of 1.43 gm/cm^3 as shown in Figs. 6 and 7 are used and the tire samples are subjected by water impact force. Performance of the tire retaining wall under the impact force of water was observed. Then reduction of water impact force is calculated by connecting with data loader.

Several testing configurations have been conducted for different condition of tire placement so that the performance could be easily evaluated and the results were compared with the case of without using tire behind the sea wall. Following are the different cases tire specimen for the experiment.

Case 1: Without tire, Case 2: 2 Layer of tires without soil Case 3: 3 Layer of tires without soil, Case 4: 2 Layer of tires with soil, Case 5: 3 Layer of tires with soil, Case 6: 3 Layer of stepped tires with soil (Fig. 8).

Fig. 6 Tire model (Hollow)

Fig. 7 Tire model (filled with soil)

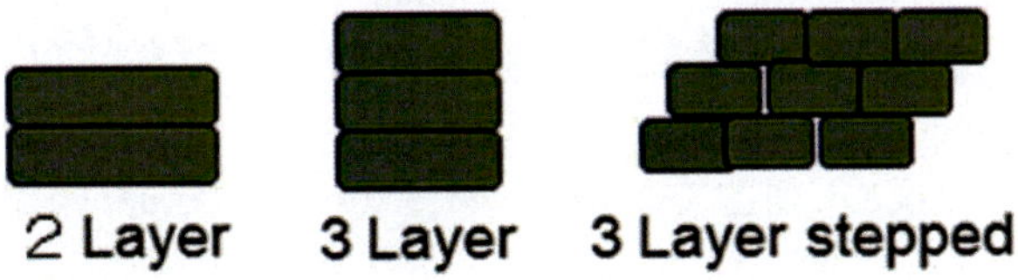

Fig. 8 Different condition of tire placement for WFT

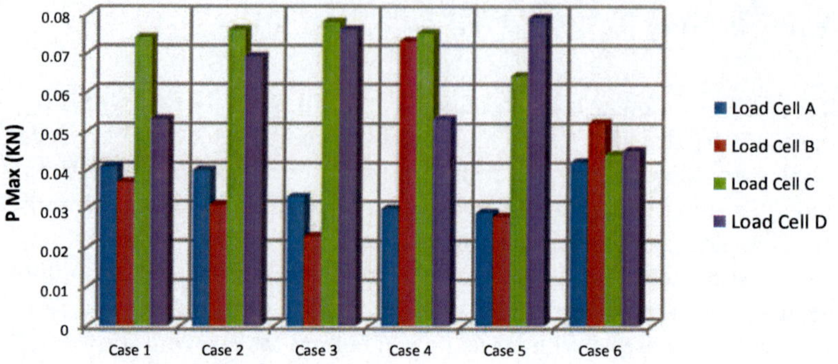

Fig. 9 Maximum value of impact force for each case

Table 1 Result of drainage impact test in KN

Load cell	Case 1	Case 2	Case 3	Case 4	Case 5	Case 6
A	0.041	0.04	0.033	0.03	0.029	0.042
B	0.037	0.031	0.023	0.073	0.028	0.052
C	0.074	0.076	0.078	0.075	0.064	0.044
D	0.053	0.069	0.076	0.053	0.079	0.045

3 Test Result

Figure 9 shows water impact force recorded by each load cell A, B, C, and D for each case. As seen from the figure, impact force was reduced gradually in Case 2 and Case 3 for load cell A and B in compare to Case 1 when the layer is increased for hollow tire. Similarly impact force was reduced gradually in Case 4 and Case 5 for load cell A and B when the layer is increased for tire filled with soil. Likewise, impact force was reduced significantly in case of 6 for Load cell C and D in compare to Case 1 (Table 1).

4 Cultivation of Plants in the Tire Structure

Kirinsou and Dichondra salt-tolerant plant has been used which is considered to be capable of withstanding the salinity condition. The plants were cultivated inside the tires filled with soil maintaining dry density of 1.43 gm/cm^3 to see how effectively plants could maintain green and beautiful environment. Following cases of specimen has been carried out for the vegetation experiment and observed the possible growth.
Case 2: Tire two-stage, Case 3: ire three-stage
Case 4: 3 gear tire (stepped) (Fig. 10).

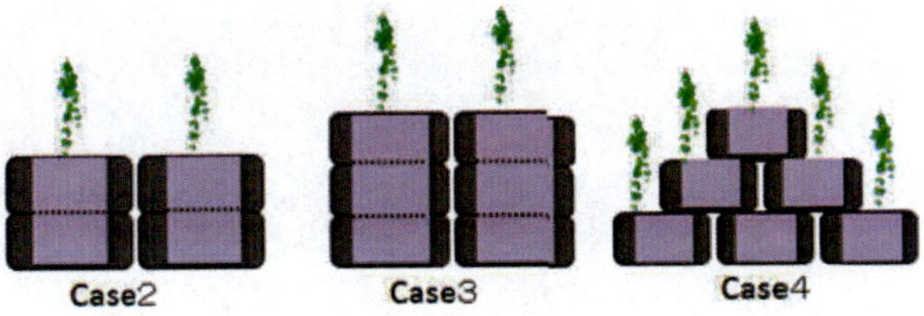

Fig. 10 Different condition of tire placement for cultivation

Both the plants were cultivated on July 2013 and monitored regularly until September 2013. As per visual inspection both Decondra and kirinsou plants were found to be growing satisfactorily from the month of July 2013 to September 2013. Figures 11 and 12 are photos of Decondra and kirinsou plants for Cases 2, 3, and 4 taken at the end of July 2013, respectively. Similarly, Figs. 13 and 14 are photos for Decondra and kirinsou plants for Cases 2, 3, and 4 taken at the end of September 2013, respectively. As seen photos from Figs. 11 and 13, Decondra plant is growing satisfactorily from the month of July 2013 to September 2013. Similarly, as seen photos in Figs. 12 and 14. Kirinsou plant is also found to be growing satisfactorily from the month of July 2013 to September 2013.

Fig. 11 Decondra in different cases at the end of July 2013

Fig. 12 Kirinsou in different cases at the end of July 2013

Fig. 13 Decondra in different cases at the end of Sept. 2013

2 layer 3 layer 3 layer slope

Fig. 14 Kirinsou in different cases at the end of Sept. 2013

5 Conclusion

Results from aforementioned experiments to reduce the tsunami impact force by using flexible materials such as tires filled with soil, showed a better performance. However, accuracy should be improved by increasing the number of load cells. Also, field test shows that the greening effect could be maintained by planting trees inside the tires and could be one of the effective methods for recycling of waste tire. Finally, instead of thinking about rigid and heavy structure, developing and implementing appropriate measures that can protect the infrastructures from such future devastating earthquakes and tsunami is today's environmentally sustainable technologies.

References

1. Hazarika, H., Hara, T. Furuichi, H.: Soil-Structure Interaction during Earthquake and Tsunami—Two Case Studies from the Latest Disaster in Japan Eighteenth International Conference on Soil Mechanics and Geotechnical Engineering, pp. 131–142 (September, 2013)
2. Fukumoto, Y., Izuyama, T.: Therma lattenuation and dispersion of sound in a periodicemulsion, pp. 4905–4921 (October, 1992)

Tip Shape Effect on Screw Pile Installation and Ultimate Resistance

Adnan Anwar Malik, Yahya Ndoye, and Jiro Kuwano

Abstract In the field of deep foundations, the screw pile is one of the new ways of construction that can support infrastructure against extreme weather events. Moreover, its eco-friendly installation mechanics that involve quick installation, minimum noise and vibration; no wastage of material and reusability makes it more efficient than conventional piling techniques such as concrete piles which involves noise, consumption of natural resources, and wastage of materials. However, the use of screw pile in the dense ground with large helix is still a concern as higher torque and pushing force is required, which sometimes exceeds the limits of machinery. In this regard, the present study focused on the effect of helix tip shape on the installation torque and force under dense ground conditions. In addition to this, the effect of tip shape on ultimate resistance also studied. To investigate these effects, model-scale pile load tests are conducted. Three kinds of pile tip shapes, i.e., flat, cone, and cutting-edge ends are considered in this study. Toyoura sand is used as a model ground. Ground disturbance due to pile installation also monitored at various distance from the edge of the pile. To check the effectiveness of the screw pile, its results compared with the results of the straight pile having the same shaft diameter under similar ground conditions. The test results indicated that the overall performance of the screw pile with cutting-edge is better than other tip shape piles, and also better than straight shaft pile in terms of installation force, torque, surrounding ground disturbance, and ultimate resistance.

Keywords Screw pile · Tip shape · Straight shaft pile · Pile installation · Pile ultimate resistance · Ground disturbance

A. A. Malik (✉) · J. Kuwano
Saitama University, Saitama, Japan
e-mail: adnanmalik@mail.saitama-u.ac.jp

Y. Ndoye
Giken Ltd. (Formerly Master Student at Saitama University), Tokyo, Japan

H. Hazarika et al. (eds.), *Advances in Sustainable Construction and Resource Management*, Lecture Notes in Civil Engineering 144,
https://doi.org/10.1007/978-981-16-0077-7_53

1 Introduction

The frequency of extreme weather events such as heavy downpours and floods, heat-waves, and typhoons increasing every passing year and many scientists believe that the reason behind is climate change. Climate change is closely related to industrialization, and it has serious impacts on existing and future infrastructure, including high costs for adaptation, maintenance, and potential negative impacts on transit [11]. To protect the infrastructure from extreme weather events, researchers are intensively working to find new ways of construction which should be sustainable and environmentally friendly. In the field of deep foundations, the screw pile is one of the new ways of construction that can support infrastructure against extreme weather events. Its eco-friendly installation mechanism that involves quick installation, minimum noise and vibration, no wastage of material and reusability makes it more efficient than conventional piling techniques such as concrete piles which involve noise, consumption of natural resources and wastage of materials. It consists of a circular or square shaft with one or more helices affixed to the shaft [5].

The installation of screw pile is done by a combination of pressing and torque. To minimize the ground disturbance, it is better to install the screw pile with high torque, low-speed torque motor [9]. The tip end of the central shaft can be opened or closed. The installation of screw pile affects the ground density, and this effect is less in cohesive soils but more in loose to medium dense sand [1]. The performance of screw pile depends on soil type, pile dimension, screw geometry, soil strength, shaft diameter, installation depth, load direction, installation torque, and installation force [3–5, 10, 12]. The helix bending deflection/deformation also affects the performance of the screw pile with a single helix if it is more than a critical value [6]. The helix deformation that significantly affects the ultimate resistance of the screw pile can be observed from the graphical method such as Decourt method [8].

Both screw pile and straight shaft pile (installed by pressing method) are environmentally friendly driven piles. The screw pile having a close-end with single helix perform better than straight shaft pile with close-end under similar shaft diameter and ground conditions; however, to obtain the similar ground conditions, pile installation was not considered [7]. The effect of installation on the performance of driven pile is very important because it disturbs the surrounding ground. This disturbance might increase or decrease the density of the ground [2]. The use of screw pile in the dense ground with large helix is still a concern as higher torque and pushing force is required which sometimes exceeds the power limits of the machinery. Moreover, the disturbance in the surrounding ground due to its installation is also a concern, especially in urban areas, as no soil is removed. In the dense ground, the parameters such as helix pitch, helix diameter, helix thickness, pile tip shape significantly affect the applied torque and pushing force. In this regard, the present study focused on the effect of helix tip shape on the installation torque, pressing force, and surrounding ground disturbance. In addition to this, the screw pile ultimate resistance was also investigated. To achieve the objectives, model scale pile load tests were conducted.

In order to show the effectiveness of screw piles, its results were compared with a straight shaft-driven piles (installed by pressing).

2 Testing Equipment

A steel container having a diameter of 1000 mm with a depth of 1100 mm was used to prepare the model (see Fig. 1a). Dry Toyoura sand was used to develop the model ground. A model pile having shaft diameter (D_S) of 21.7 mm with a length of 600 mm was used (see Fig. 1b). The used model pile can be converted into a screw pile by changing the toe-end with a helix (see Fig. 1b). In case of screw pile, helix diameter (D_H) of 43 mm with a pitch (P_H) of 24 mm (measured from the center of the top edge to center of the bottom edge of the helix) was used. Three different types of tip shapes, i.e., flat end, cone end, and cutting-edge end were used (see Fig. 1c).

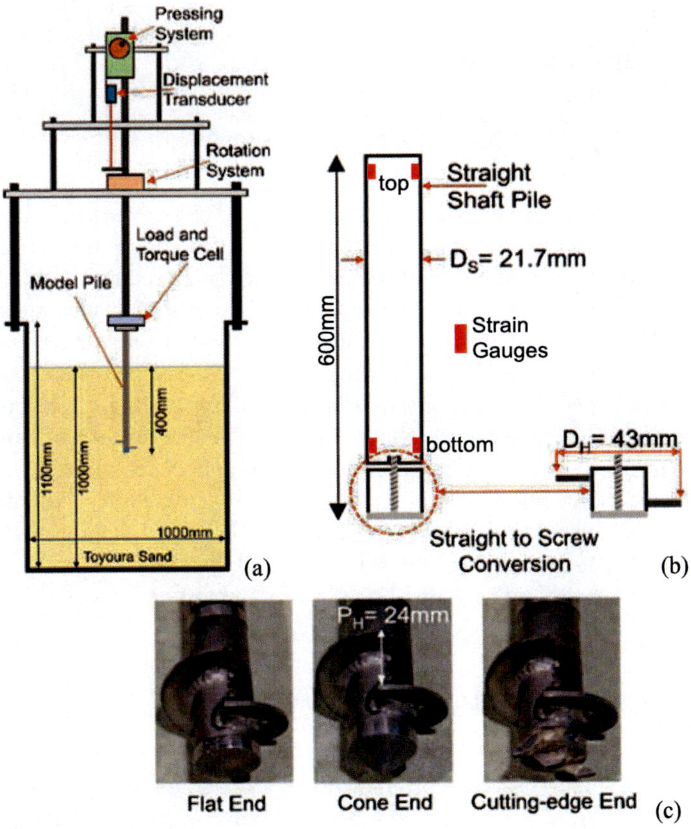

Fig. 1 **a** Model container **b** Model pile **c** Model pile tips

3 Testing Procedure

The model ground was prepared by compacting Toyoura sand in five layers, each having a thickness of 200 mm after compaction. The relative density of each layer was approximately 80%. After preparing the ground, the initial condition of the model ground was checked by inserting a 3 mm thin steel needle with the help of hand-held load cell. The penetration load (PL) was recorded during the insertion of the needle. The needle was inserted up to 450 mm depth at distances of $1D_H$ (1 times helix diameter), $2D_H$, $3D_H$ and $4D_H$ from the edge of the pile. After that, the screw pile was installed by pressing and rotation. The pressing and rotation were adjusted in such a way that one pitch penetration (24 mm) achieved in one rotation of helix. In case of straight shaft pile, only pressing was used to install the pile. All piles were installed in the same manner. The pile was installed up to 400 mm depth. After the installation of the pile, the ground condition was again checked by inserting a thin steel needle at the same locations. Then, a pile load test was conducted.

4 Results and Discussions

4.1 Screw Pile Tip Shape Effect on Installation Effort

The screw piles with flat end, cone end, and cutting-edge end were installed by using the following parameters: Pressing rate = 15 mm/min, Rotation = 0.625 rotation/min, Penetration of pile per rotation = 24 mm/rotation (1 pitch/rotation).

In all the tests, the pressing rate and rotation were adjusted in such a way that one pitch penetration ($P_H = 24$ mm) achieved in one rotation of helix. It is observed from the test results that the installation load is more for flat-end pile than cone-end and cutting-edge pile (see Fig. 2a). The installation load for cutting-edge and cone-end screw piles are similar to each other (see Fig. 2a). The test results indicated that the installation torque is more or less similar for the flat-end, cone-end, and cutting-edge end screw piles up to installation depth of 150 mm (3 times helix diameter, $3D_H$) (see Fig. 2b). After that, the cutting-edge screw pile showed higher installation torque than the other two types (see Fig. 2b). It is presumed that the impact of cutting-edges become more prominent at high depth that's why the torque increased.

4.2 Effect of Tip Shape on Surrounding Ground Disturbance

The initial condition of the model ground was checked by inserting 3 mm steel needle at a distance of $1D_H$, $2D_H$, $3D_H$ and $4D_H$ from the edge of the pile. The load required

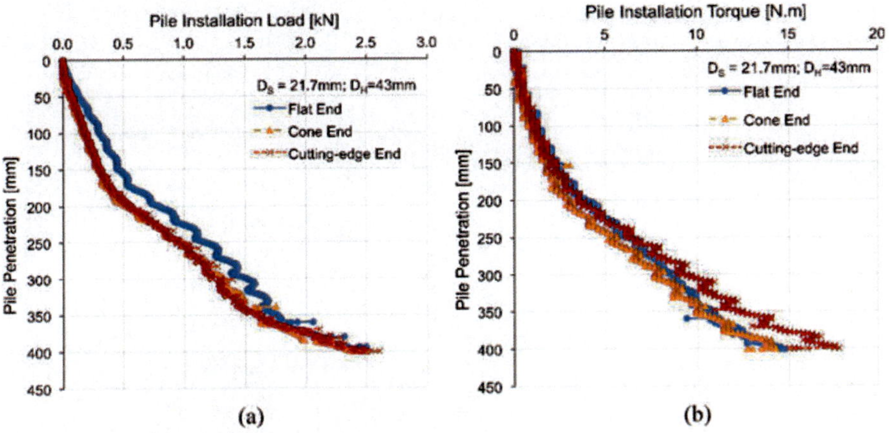

Fig. 2 **a** Pressing load **b** Installation torque

to insert the thin steel needle into the model ground was recorded by using a hand-held load cell. It observed that the penetration load after preparing the model ground for all test scenarios is almost similar (see Fig. 3).

After the pile installation, the model ground condition was again checked by inserting 3 mm steel needle at a distance of $1D_H$, $2D_H$, $3D_H$, and $4D_H$ from the edge of the pile. The comparison of penetration load before and after screw pile installation for the three types of end shapes, i.e., flat end, cone end, and cutting-edge end is shown in Fig. 3. The results indicated that in all test cases, the penetration load measured after the installation of the screw pile reduced from the initial penetration load (see Fig. 3). In all the cases, the reduction in penetration load (loosening of the ground) after the installation of the pile is high between 200 mm and 400 mm of depth, i.e., $4.6D_H$ to $9.3D_H$. Whereas, the ground disturbance below the pile tip, i.e., below 400 mm is considerably less (see Fig. 3). The screw pile with cutting-edge produced less ground disturbance than flat-end and cone-end screw piles up to $2D_H$ distance from the edge of the central shaft. Moreover, it showed more densification below the pile tip than the other screw piles at $1D_H$ distance. It is also observed that with the increase in distance away from the pile shaft edge, i.e., $1D_H$ to $4D_H$ the ground disturbance reduced.

4.3 *Screw Pile Tip Shape Effect on Ultimate Resistance*

The pile load tests were conducted on screw pile having a flat end, cone end, and cutting-edge end. In the load tests, 2 mm/min of loading rate was used. The skin resistance was obtained by subtracting top strain gauge load with bottom strain gauge load (refer, Fig. 1b). Whereas, end bearing resistance was obtained by subtracting skin resistance from top gauge load. The test results indicated that the total pile

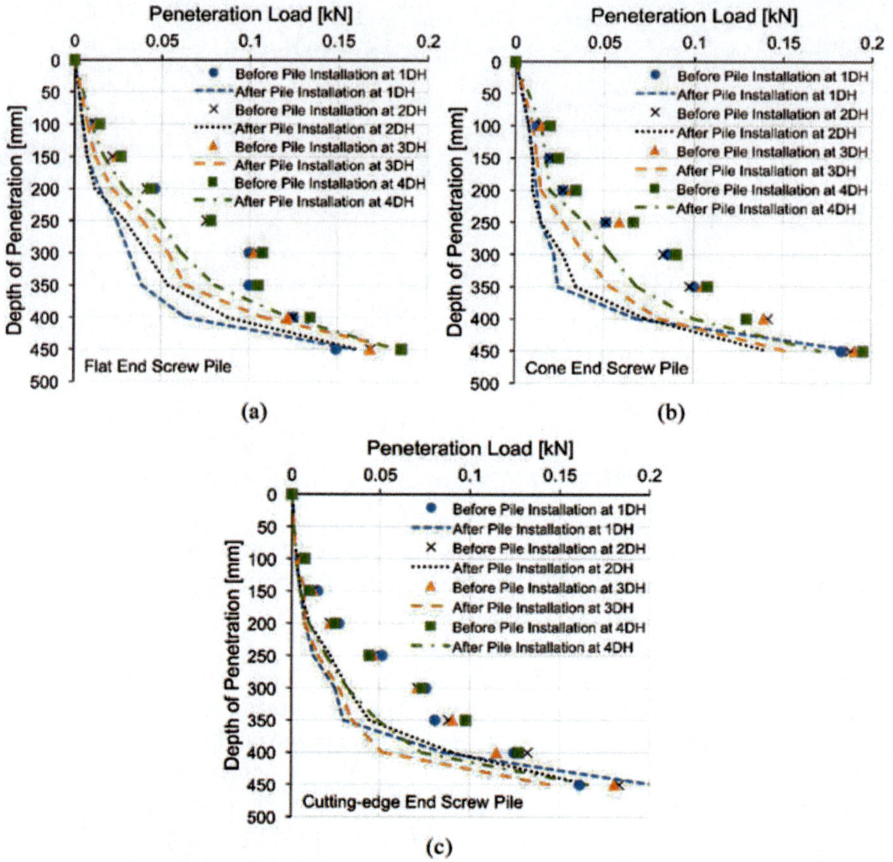

Fig. 3 **a** Before and after penetration load for flat-end screw pile **b** For cone-end screw pile **c** For cutting-edge end screw pile

resistance (end bearing resistance + skin resistance) of cutting-edge pile and cone end pile is more than the flat-end pile but similar to each other (see Fig. 4a). The skin resistance of cutting-edge pile is greater than cone-end pile and flat-end pile (see Fig. 4b). The overall end bearing resistance of cone-end pile is more than the flat-end and cutting-edge pile. However, at the initial loading stage, i.e., up to 2.5 kN, the end bearing resistance of cutting-edge pile and cone-end pile is similar (see Fig. 4b).

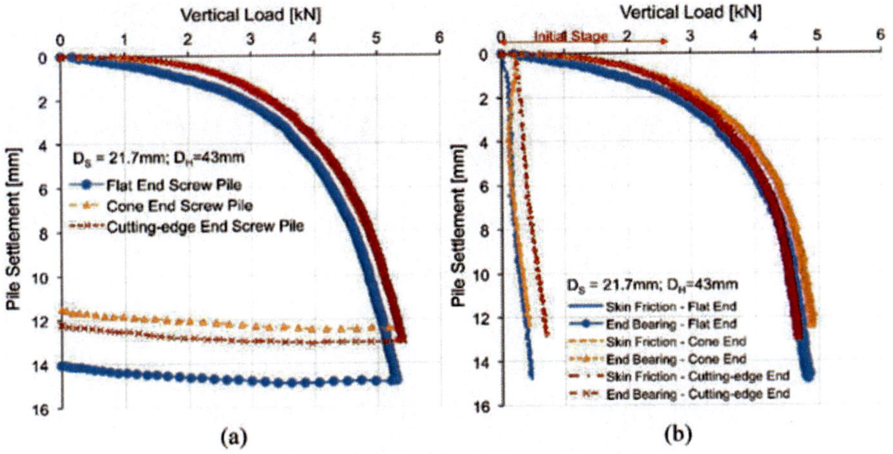

Fig. 4 **a** Ultimate load-settlement behavior **b** End bearing and skin resistances

4.4 Comparison Between Screw Pile and Straight Shaft Pile

Pile load tests were conducted to showcase the effectiveness of the screw pile over straight shaft pile. The pressing method was used to install the straight shaft pile having flat end and cone end. Whereas, straight shaft pile with cutting-edge was installed by pressing and rotation. The pressing rate and rotation were similar for both types of piles. The shaft diameter of screw pile and straight shaft pile was similar, i.e., $D_S = 21.7$ mm. In case of screw pile, the helix diameter (D_H) was two times the shaft diameter (D_S), i.e., $2D_S = 43$ mm. The ground was compacted in a dense state with a relative density of 80%.

Installation Effort. In the case of flat-end and cone-end screw pile and straight shaft piles, the pressing load required to install the piles are similar although the helix diameter of the screw pile is two times the shaft diameter of straight shaft pile (see Fig. 5). These results indicated that less installation effort is needed to install a screw pile with a larger effective tip area, and the larger effective tip area contributes to bearing resistance.

In case of cutting-edge screw pile and straight shaft pile having similar shaft diameter, the piles were installed by pressing and rotation. The test results indicated that by including rotation with the cutting-edge end, the installation load required to install the straight shaft pile is reduced by 30% in comparison with screw pile at maximum installation depth (see Fig. 6a). Moreover, the installation torque is reduced by 46% in comparison with the screw pile at maximum installation depth (see Fig. 6b). These results showed that in the case of straight shaft pile with cutting edge, the inclusion of rotation can reduce the installation load and torque.

Ground Disturbance. In this comparison, the ground penetration ratio (GPR) was estimated by taking the ratio between penetration load (PL_A) measured after the

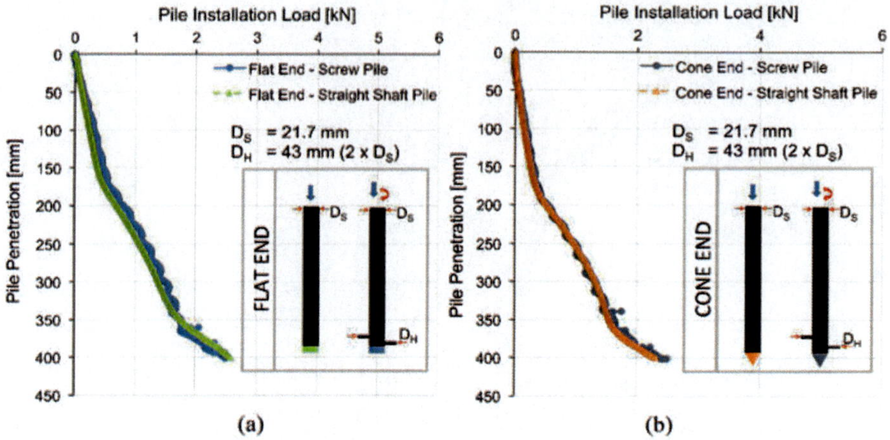

Fig. 5 **a** Comparison of installation load for flat-end piles **b** For cone-end piles

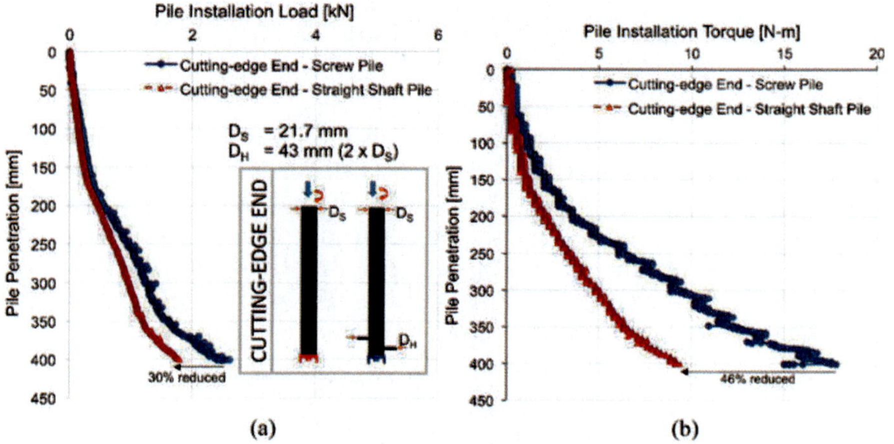

Fig. 6 **a** Comparison of installation load for cutting-edge piles **b** Installation torque for cutting-edge piles

installation of the pile to the penetration load (PL_B) measured before the installation of the pile, i.e., GPR $= PL_A/PL_B$.

The surrounding ground disturbance due to pile installation at a distance of $1D_H = 2D_S$ from the edge of pile shaft can be divided into three zones, i.e., near ground surface (Z-I), the middle zone (Z-II), and below pile tip or helix (Z-III). In all the cases, the screw pile showed lesser ground disturbance than straight shaft pile near the ground surface, i.e., up to 100 mm depth although helix diameter is two times straight shaft pile (see Fig. 7a). The ground disturbance below the pile tip or helix (at depth 450 mm) is considerably small for both type of piles; however, the screw pile showed densification in all the test cases (see Fig. 7a). The densification is highest

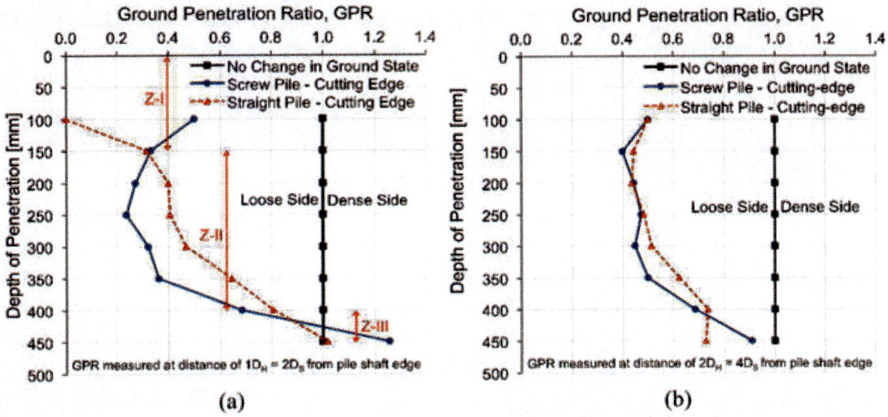

Fig. 7 **a** Comparison of ground disturbance for cutting-edge screw and straight piles at a distance of $1D_{\mathrm{H}} = 2D_{\mathrm{S}}$ **b** For cutting-edge piles at a distance of $2D_{\mathrm{H}} = 4D_{\mathrm{S}}$

in case of cutting-edge screw pile. The surrounding ground disturbance due to pile installation at a distance of $2D_{\mathrm{H}} = 4D_{\mathrm{S}}$ is more or less similar to each other for all the test cases (see Fig. 7b). The results indicated that although screw pile helix diameter is two times the shaft diameter of straight shaft pile, the ground disturbance is close to straight shaft pile which clearly shows the benefit of using screw pile.

Ultimate Resistance. In all the test cases, i.e., flat end, cone end, and cutting-edge end, ultimate resistance of screw pile is higher than straight shaft pile (see Fig. 8). It is also observed that the ultimate resistance of the screw pile with cutting-edge and cone ends is two times the ultimate resistance of straight shaft pile (ultimate resistance

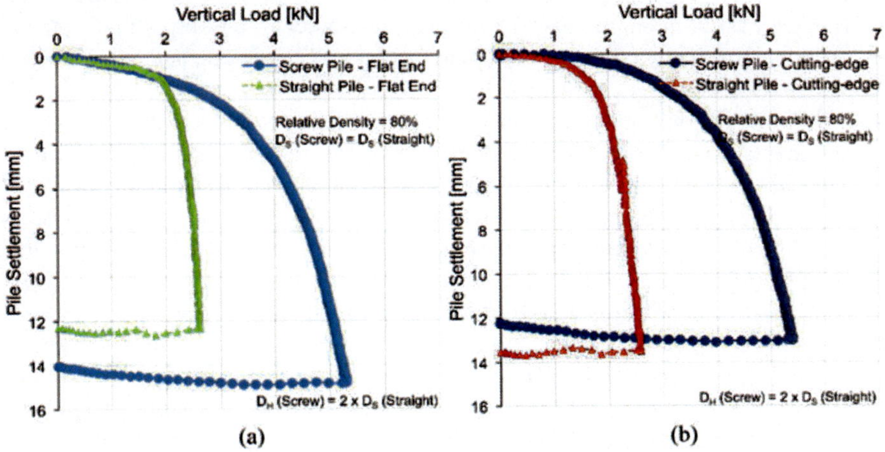

Fig. 8 **a** Comparison of load-settlement response of screw and straight shaft pile with flat ends **b** For cutting-edge ends

considered at settlement equals to 10% of helix diameter). Moreover, in the case of screw pile, cutting-edge, and cone-end piles showed higher ultimate resistance than flat-end pile.

5 Conclusions

In this study, the effect of the pile tip shape on the installation effort and ultimate resistance of screw pile was investigated under dense ground condition. The end tip shapes used in this study are flat end, cone end, and cutting-edge end. Screw pile having a shaft diameter of 21.7 mm with a helix diameter of 43 mm was used. Straight shaft pile having a similar shaft diameter with screw pile was also used to compare the performance of both types of piles. Following are the main outcomes of this study;

- In case of screw piles, cutting-edge and cone end screw piles required less installation load to install the pile up to desired embedment depth than flat-end screw pile. The installation torque was more or less similar for the flat-end, cone-end, and cutting-edge end screw piles up to installation depth of $3D_{\mathrm{H}}$. After that, the cutting-edge screw pile showed higher installation torque than the other two types.
- The penetration resistance was reduced in all the cases of screw piles, i.e., flat end, cone end, and cutting-edge end after the pile installation, which indicated the loosening of the ground. The ground disturbance (loosening) below the pile tip was considerably less than the ground disturbance along the pile shaft in all the tip shape cases of screw pile. The screw pile with cutting-edge produced less ground disturbance than flat-end and cone-end screw piles up to $2D_{\mathrm{H}}$ distance from the edge of central shaft. Moreover, it showed more densification below the pile tip than the other screw piles at $1D_{\mathrm{H}}$ distance away from the edge of the central shaft.
- The test results indicated that the ultimate resistance of cutting-edge screw pile and cone-end screw pile was more than the flat-end screw pile but similar to each other.
- In the comparison of the screw and straight shaft piles with flat and cone ends, the pressing load required to install the piles were similar, although the helix diameter of the screw pile was two times the shaft diameter of straight shaft pile. Moreover, in comparison of the screw and straight shaft piles with cutting-edge ends, the inclusion of rotation in straight shaft pile installation reduced the installation load by 30% and torque by 46% from screw pile case.
- In all the tip shape cases at $1D_{\mathrm{H}}$ distance from the edge of pile shaft, the screw pile showed lesser ground disturbance than straight shaft pile near the ground surface, i.e., up to 100 mm depth, although helix diameter of screw pile was two times straight shaft pile. The ground disturbance below the pile tip (at depth 450 mm) was considerably small for both types of piles. The surrounding ground disturbance due to pile installation at a distance of $2D_{\mathrm{H}} = 4D_{\mathrm{S}}$ was more or less similar to each other for all the test cases.

- In all the test cases, i.e., flat end, cone end, and cutting-edge end, ultimate resistance of screw pile was higher than straight shaft pile. Moreover, the highest ultimate resistance observed for screw piles with cutting-edge and cone ends, i.e., two times the ultimate resistance of straight shaft pile.

References

1. Adams, P.A.: Helical pile application and design. PhD Engineer, LLC, pp. 1–15 (2011)
2. Aziz, A.A.R., Salman, M.M.: The effect of improvement surrounding soil on driven pile friction capacity. Al-Nahrain J. Eng. Sci. **20**(1), 36–48 (2017)
3. Garnier, J., Gaudin, C., Springman, S., Cullingan, S.M., Goodings, D., Konig, D., Kutter, B., Phillips, R., Randloph, M.F., Thorel, L.: Catalogue scaling laws and similitude questions in geotechnical centrifuge modeling. Int. J. Phys. Mod. Geotechn. **3**, 1–24 (2007)
4. Ghaly, A., Hanna, A., Hanna, M.: Installation torque of screw anchors in sand. Soils Found. **31**(2), 77–92 (1991)
5. Li, W., Deng, L.: Axial load tests and numerical modeling of single-helix piles in cohesive and cohesionless soils. Acta Geotech. **14**, 461–475 (2019)
6. Malik, A.A., Kuwnao, J., Tachibana, S., Maejima, T.: Effect of helix bending deflection on load settlement behaviour of screw pile. Acta Geotech. **14**, 1527–1543 (2019)
7. Malik, A.A., Kuwnao, J., Tachibana, S., Maejima, T.: End bearing capacity comparison of screw pile with straight pipe pile under similar ground conditions. Acta Geotech. **12**(2), 415–428 (2017)
8. Malik, A.A., Kuwnao, J., Tachibana, S., Maejima, T.: Interpretation of screw pile load test data using extrapolation method in dense sand. Int. J. Geomate **10**(1), 1567–1574 (2016)
9. Perko, H.A.: Helical Piles: A Practical Guide to Design and Installation. Wiley, New Jersey (2009)
10. Sakr, M.: Relationship between installation torque and axial capacities of helical piles in cohesionless soils. Can. Geotech. J. **52**(6), 747–759 (2014)
11. Schweikert, A., Chinowsky, P., Espinet, X., Tarbert, M.: Climate change and infrastructure impacts: comparing the impact on roads in ten countries through 2100. Procedia Eng. **78**, 306–316 (2014)
12. Tsuha, C.H.C., Aoki, N.: Relationship between installation torque and uplift capacity of deep helical piles in sand. Can. Geotechn. J. **47**(6), 635–647 (2010)

Climate Change Independent Natural Disasters

Application of Deep Mixing Method for Mitigation of Potential Seismic and Hurricane-Induced Hazards

Rakshya Shrestha, Nozomu Kotake, and David Yang

Abstract This paper describes the application of Deep Mixing Method (DMM) for the mitigation of potential hazards induced by natural disasters such as hurricane-induced ground flooding and seismic ground shaking for two projects in the USA. The two projects completed in 2014 and 2016 are used as case examples to present the design, construction, and quality control of the DMM work for hazard mitigation. DMM was installed in the first project called P-17A primarily to reinforce soft soils within and below a levee to provide resistance against lateral loads that will be induced from high flood generated by hurricane. The second project called Perris Dam included the use of DMM to improve potentially liquefiable soils within the foundation of an earthen dam primarily to limit the deformation of the dam during a seismic event.

Keywords DMM · Seismic remediation · Hurricane protection · Dam · Levee

1 Introduction

Deep Mixing Method (DMM) is an in situ soil mixing method whereby soil is improved by injecting cement grout through single or multiple augers that simultaneously mix the soil, forming in-place soil–cement columns, walls or panels that reinforce soft or loose ground. DMM can produce soil-cement elements in versatile configurations for mitigation of potential liquefaction and lateral spreading, improvement of bearing capacity and control of settlement for a wide variety of construction projects including wharf, petroleum refinery, roadway, dam, and levee. For hazard

R. Shrestha (✉) · N. Kotake · D. Yang
JAFEC USA, Inc, 2025 Gateway Place, Suite 180, San Jose, CA 95110, USA
e-mail: rshrestha@jafecusa.com

N. Kotake
e-mail: nozomu@jafecusa.com

D. Yang
e-mail: davidyang@jafecusa.com

H. Hazarika et al. (eds.), *Advances in Sustainable Construction and Resource Management*, Lecture Notes in Civil Engineering 144,
https://doi.org/10.1007/978-981-16-0077-7_54

mitigation, the main objective is to obtain a reinforced ground with improved strength and reduced compressibility compared to the native ground.

This paper presents two projects in the USA where DMM was installed to improve soft and loose soils within and below the levee and dam foundation respectively: (1) P-17 A Project and (2) Perris Dam Project. At P-17A, DMM panels were installed through the levee and soft foundation soils and keyed into competent stratum to provide shear resistance to the levee against lateral loads from high flood water generated by hurricane and to transfer the vertical load of the raised levee to the bearing stratum for long-term settlement control. At Perris Dam, DMM was installed to strengthen the loose and potentially liquefiable sandy soils to limit the seismic deformation of the dam during an earthquake event and to support a new stabilization berm at the downstream toe of the dam. This paper will review a brief background for each of these projects, discuss the subsurface soil conditions and present the design, construction, and quality control of the DMM work for these levee repair and dam remediation projects.

2 Project Backgrounds and Design Approaches

2.1 P-17A Project, Louisiana

In August 2005, Hurricane Katrina caused major flooding of the Mississippi river and damaged the flood protection system in southeast Louisiana causing $125 billion (USD) in damage and over 1200 deaths. This damage was aggravated by Hurricane Rita in September 2005 causing an additional 120 fatalities and a loss of $18 billion. These storms created catastrophic damage along the coast of Louisiana leading to one of the largest disaster-related migration in the US history. In an effort to minimize further damage and to enhance the protection level, a number of levees located on the west bank of the Mississippi river in Louisiana were incorporated into the Hurricane Protection project by the United States Army Corps of Engineers (USACE). DMM techniques were then utilized by the USACE in several of these levee repair projects. The P-17A project which was executed in 2014 is one of such projects. This is one of the largest DMM levee projects of the USACE. The largest DMM levee project in the US is called LPV-111, which was completed in 2012 and where 8.9 km of levee was treated by 1.3 million cu. m of DMM [5].

DMM ground improvement was installed to reinforce soft soils within and below the levee for emergency repair of a 2.2 km long section of a levee along the west bank of Mississippi River in Louisiana. The main purposes of DMM ground improvement in this project were to provide shear resistance to the levee against lateral loads from high flood water generated by hurricane and also to serve the function of supporting additional vertical load created by the increased height of the levee for hurricane protection. Consequently, to control the long-term settlement of the remediated levee, which might otherwise occur due to the compression of soft foundation soils.

Subsurface Conditions

Subsurface investigations performed along the P-17A levee identified three main soil strata: soft to medium stiff lean clay (CL), fat clay (CH) and medium dense to dense clayey/silty sand (SC-SM). Clay layers were present to approximately 7.5 m below the ground surface while sandy layers were prevalent below the clay layers. Traces of organic fat clay layers were encountered intermittently. Fortunately, these layers were not abundant enough to cause concern.

DMM Design and Layout

DMM panels, comprising of a series of overlapping soil–cement panels were oriented in transverse direction to the levee axis as shown in Fig. 1a. The panels were designed to provide shear resistance to the levee against lateral loads from high flood water generated by hurricane. These panels were installed through the levee and soft foundation soils and keyed into competent stratum. Depth of these panels varied between 8.9 and 14 m below the working platform. A cross section is shown in Fig. 1b. DMM panel width varied between 5.2 m (narrow section) to 16.4 m (wider section) as shown in Fig. 1a. To satisfy the width requirement, DMM panels installed in the narrow section included two overlapping elements (one element consisted of six columns) where the three edge columns of each element fully overlapped at the center. The wider section included up to five elements overlapped with the consecutive elements installed. The center to center spacing between the panels was maintained at 3.0 m and the total treatment ratio of DMM was 21%. Treatment ratio is defined as the area of the soil that is mixed divided by the total area of soil within the zone to be treated. 1822 elements were installed along the levee alignment amounting to approximately 41,000 cubic meters of DMM treatment.

2.2 *Perris Dam Project, California*

Perris Dam is an earth fill dam constructed between 1970 and 1974 with an embankment length of 3.5 km, a maximum crest height of 38.4 m, and a capacity of 162 million cubic meters of water. It is located in Riverside County in California, 5 miles from San Jacinto fault zone, which is capable of moment magnitude 7.5 earthquake. The foundation at Perris Dam consists of up to 90 m of fine sandy alluvial soils. The Perris Dam Foundation Study Report, prepared by California Department of Water Resources [4] indicated that thin sandy layers and pockets in the dam foundation were susceptible to potential liquefaction and severe loss of strength during a large earthquake event. The liquefaction of these foundation soils could potentially result in slope failure of the embankment dam and uncontrolled release of water from the Lake Perris reservoir endangering the people living downstream of the dam. A reservoir water level restriction of 7.6 m below the maximum water surface elevation had been in place since 2005 to maintain safe operation of Perris Dam and Lake until the remediation is implemented to control the crest deformation between 0.9 and 1.5 m

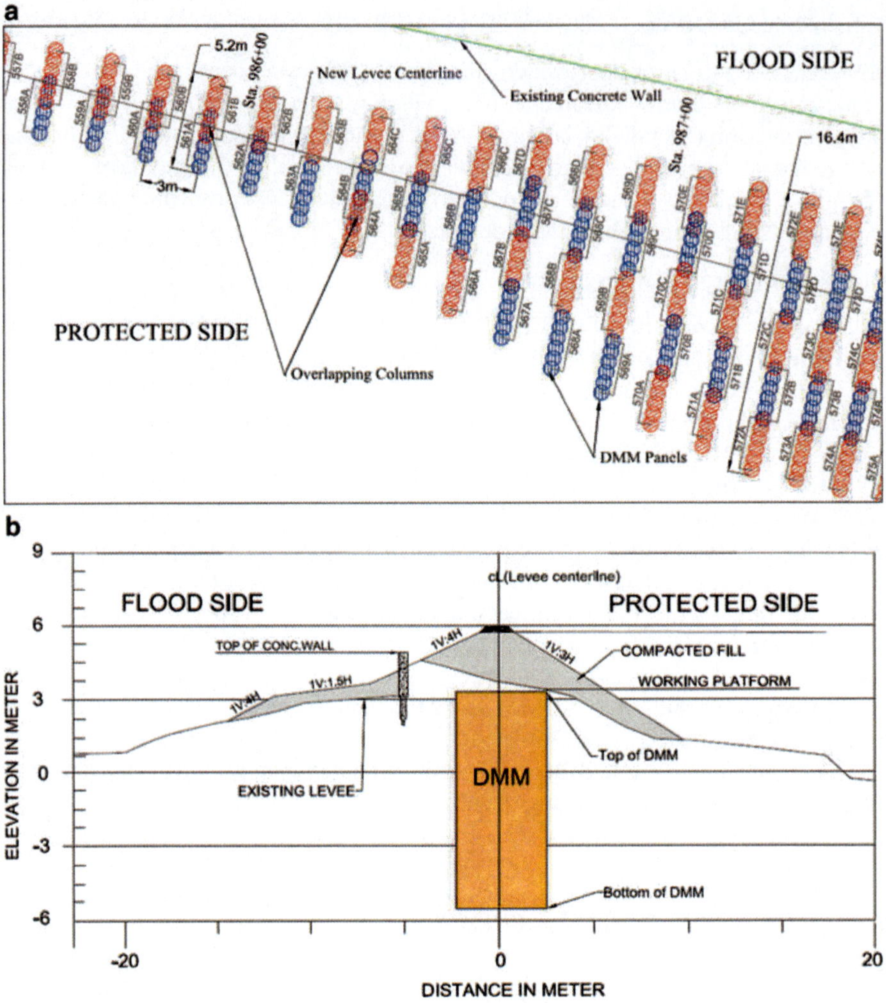

Fig. 1 **a** Layout of DMM at P-17A Project. **b** Cross section of DMM at P-17A Project

during the design earthquake event. DMM was selected as the most cost-effective remediation alternative in conjunction with a new downstream berm to reinforce these foundation soils and limit the seismic deformation of the dam.

DMM ground improvement was installed approximately 1.3 km along the downstream toe of the dam in a cellular pattern as shown in Fig. 2a. The purpose of DMM was also to connect the new compacted downstream berm to the stronger soils at depth as shown in Fig. 2b to reduce the deformation of the dam embankment during earthquake. The overall purpose of the Perris Dam Project was to upgrade the seismic safety of the dam to meet the new seismic standards and enhance public's

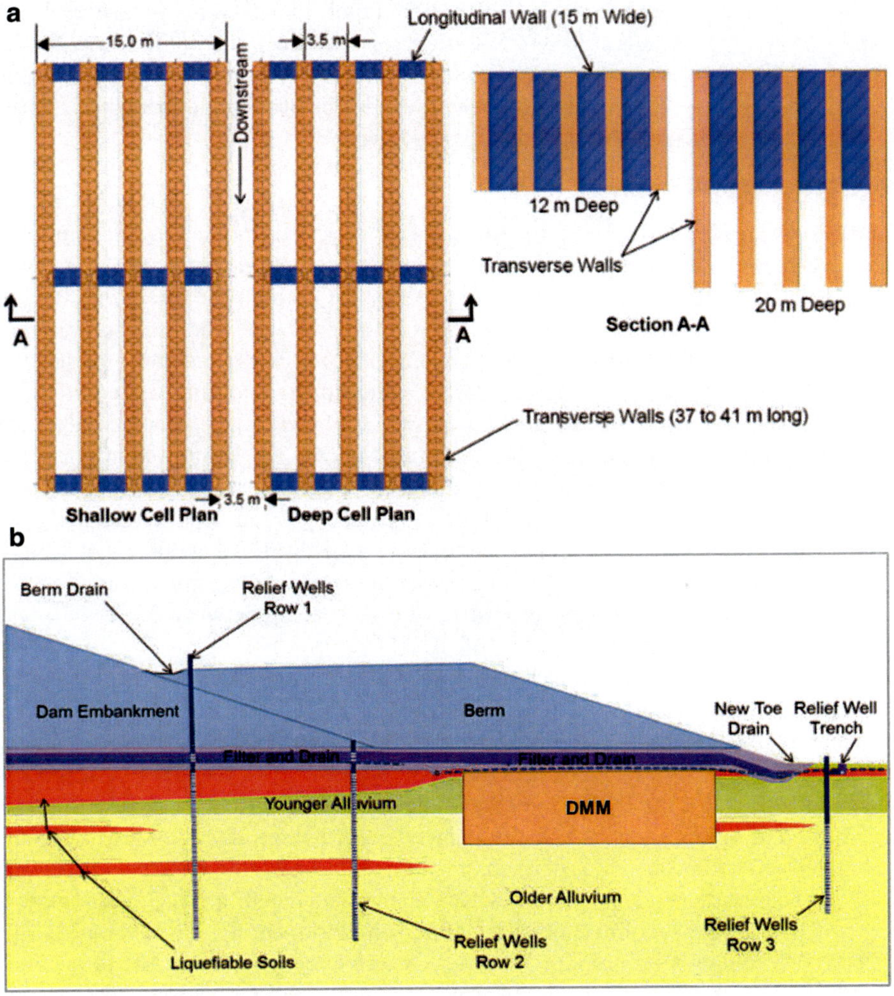

Fig. 2 **a** Typical DMM cell layout at Perris Dam **b** DMM remediated section at Perris Dam [2]

safety. Perris Dam is believed to be the largest application of DMM for dam remediation in the USA to date. In addition to the DMM work performed to treat the alluvial soils beneath the footprint of the berm, the remediation works included extension of the existing blanket drain, construction of new drain line, new toe drain, and relief well system as shown in Fig. 2b.

Subsurface Conditions

Perris Dam is founded predominantly on alluvial soils which primarily consists of silty sands and clayey sands with occasional lenses of poorly graded sand, gravel and sandy silt. This alluvium extends down to 90 m. Beneath the alluvium lies granitic bedrock. As mentioned earlier, the DWR 2005 report indicated the presence of thin sandy layers and pockets in the dam foundation that were susceptible to potential liquefaction and severe loss of strength during a large earthquake event. Most of the liquefiable soils are located in the upper 8 meters of the foundation with isolated liquefiable pockets at deeper depths of up to 18 m.

DMM Design and Layout

DMM cells, each 15 m wide in the longitudinal direction (parallel to the dam axis) and 37–41 m wide in the transverse direction (perpendicular to the dam axis) were installed. Each cell consisted of five transverse walls (T-walls) and three-longitudinal walls (L-walls) as shown in Fig. 2a. DMM installed in cellular configuration are proven to be effective for liquefaction prevention through the reduction of seismic induced strain and excessive pore water pressure during and after the earthquake. At some stations along the alignment, DMM cells were installed in an alternate sequence of shallow and deep cells as seen in Fig. 2a. The shallow portion of the cells were designed to reinforce the liquefiable sandy layers located within 12 m below the ground surface while the portion extended to 20 m depths were designed to reinforce the isolated liquefiable sandy pockets below 12 m depth. A total of 72 cells each spaced at 3.5 m from the adjacent cells were installed amounting to a total volume of 246,000 cubic meters of DMM. The c/c spacing between the T-walls was also 3.5 m.

T-walls were designed to resist the major seismic force along the perpendicular direction of the dam and L-walls were designed to brace the T-walls and resist minor seismic force along the longitudinal direction of the dam. The total treatment ratio of the DMM cells was 42%, out of which the 38% was from the T-walls. FLAC models were developed for the DMM remediated dam sections and dynamic analysis was performed to estimate the seismically induced deformation. It was assumed that the untreated soil within the DMM cells contain liquefiable soils. Based on the analyses, it was determined that under a seismic event with moment magnitude of 7.5, a treatment ratio of 38% for the T-walls (major lateral force direction) and a design average DMM unconfined compressive strength (UCS) of 3.5 MPa were required to limit the seismic deformation of the dam to 0.8 m and 1.0 m. These deformations were calculated using the assumptions of elastic DMM strength and cracked DMM strength, respectively [2]. The corresponding vertical crest settlements were calculated as 0.5 m and 0.7 m, respectively. The maximum allowable seismic crest deformation required by DWR was 0.9–1.5 m to release the reservoir water level restriction and to return dam to normal operation.

3 DMM Acceptance Criteria

Design of the DMM panels, walls or cells used for the projects included both geometric design and material design. The geometric design included treatment layout and treatment depth as discussed in Sect. 2. The material design included the required strength of the soil-cement product. Based on the geometric and material designs, a set of acceptance criteria was established for the DMM construction. These acceptance criteria were included in the project specifications and served as a foundation for DMM construction and quality control.

For P-17 A project, DMM was designed to achieve a UCS of 0.7 MPa (100 psi) at 28 days of installation of the element. It was required that 90% of the cored samples tested from each DMM element had to be greater than 0.7 MPa at 28 days. Perris Dam project was designed to achieve an average UCS of 3.5 MPa (500 psi) at 28 days. The average was determined from UCS results of cored samples obtained from three consecutive DMM elements that were cored to full depth. It was also required that 90% of the core samples tested from these three elements had to be greater than 2.1 MPa (300 psi). Such strength acceptance criteria based on average and percentile strength requirements are selected to cope with the inherent variation of the sub-surface soils and soil–cement product and to truncate these natural variations. Although UCS of the core samples governed the acceptance criteria of the product, criteria such as recovery of the core runs and uniformity of the soil-mixed product such as the presence of unmixed soil lumps were also evaluated for both projects before the product was finally accepted. To ensure that DMM elements were located according to the plans, horizontal and vertical tolerances and overlaps between the adjacent columns and elements were monitored and continuity of the elements for the full depth was confirmed.

4 DMM Construction

4.1 Materials and Equipment

Materials used for the projects included cement and water. No additional admixtures were used for both projects. Cement comprising of 25% Type II Portland cement and 75% slag was used for the P-17A project and Portland cement Type II was used for Perris Dam project. Fresh water was used to prepare the grout in both projects. Equipment required for the projects included the grout plant and the mixing rig. The grout plant consisted of cement silos, grout mixer, agitator, water tank, and high-pressure pumps. Grout produced by the plant was pumped to the agitator tank from which in turn grout was pumped to the mixing rig. Two sets of equipment were used for the Perris Dam project and the mixing rigs were four-axis auger equipment as shown in Fig. 3. Each axis was 1.52 m in diameter. The mixing rig used in P-17A project was a six-axis auger equipment with each axis being 0.91 m in diameter.

Fig. 3 DMM mixing equipment at Perris Dam Site

4.2 *Quality Control Program*

The quality control program included installation of field validation test elements, real-time monitoring of mixing parameters during installation, and sampling and testing of the completed DMM. Real-time monitoring of the mixing parameters such as penetration and withdrawal speed of the mixing shaft, rotation speed of the mixing blades, grout injection volume and average amperage of the mixing motors was performed using a computer operated quality control system for every element installed. These parameters were recorded at every 1-foot interval along the depth of the element and daily quality control reports were prepared.

The quality of the product was verified by coring and testing a certain specified percentage of the total DMM elements installed. Each core was logged, evaluated for the uniformity of the soil–cement mix and selected representative specimens from the recovered cores (Fig. 4a.) were prepared and sent to the laboratory for strength testing. Figure 4b. shows DMM test elements installed during the field validation phase at Perris Dam project. Soils surrounding these test elements were excavated to expose the installed DMM to ensure that a smooth and intact DMM top was obtained.

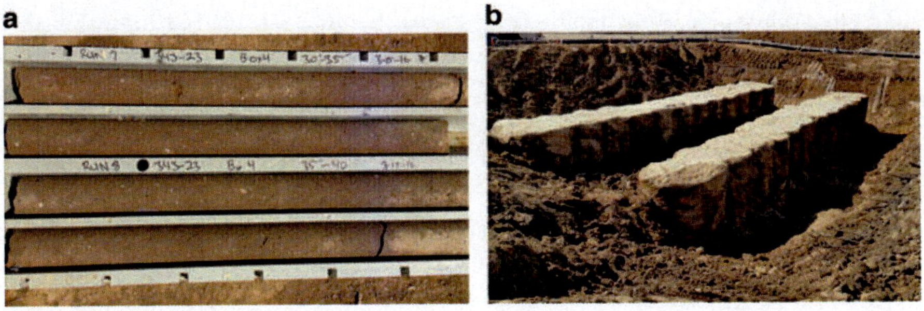

Fig. 4 **a** DMM Core runs **b** Exposed DMM test panels at Perris Dam site

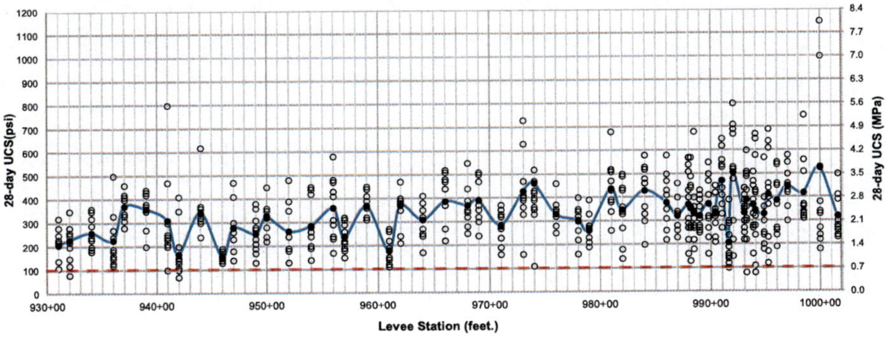

Fig. 5 28-day UCS of the cored samples at P-17A Project

4.3 Strength Data

A total of 55 full-depth cores were retrieved and 557 cored samples were tested for UCS at P-17A project. All 557 test results are plotted in Fig. 5. The average UCS obtained per cored element is also plotted and connected by blue solid line. It can be seen in Fig. 5 that the test results showed a wide range of distribution with few values falling below 0.7 MPa (red horizontal dashed line). The design strength requirement (as described in Sect. 3, DMM Acceptance Criteria) was to maintain 90% of the UCS data to be above 0.7 MPa. The cumulative distribution curve of the UCS is presented in Fig. 7a. for the entire project shows that, only 1% of all the UCS data was less than 0.7 MPa (red vertical dashed line) or 99% of the UCS data was greater than 0.7 MPa, showing that the design strength requirement was adequately satisfied. For Perris Dam project, 330 full-depth cores were obtained, and 2647 cored samples were tested. Figure 6 plots the distribution of average 28-day UCS across the DMM treatment zone along the dam profile. It can be seen from the figure that the average 28-day UCS attained the required average value of 3.5 MPa. The design strength requirement was also to maintain 90% of the UCS data of the project to be above 2.1 MPa. The cumulative distribution curve of the UCS data presented in Fig. 7b.

shows that only 0.1% of the total UCS tests was below 2.1 MPa (red vertical dashed line) or 99.9% of the total tests was above 2.1 MPa. This UCS data obtained from the field retrieved cored samples after 28-day of curing also showed that the actual average UCS was about 1.3 times higher than the design average UCS of 3.5 MPa.

Some low strength elements were encountered; this occurred mainly in the areas with high ground water level and high seepage. Although the specified strength was adequately met, few low UCS values were still obtained, and the data showed variation.

This can be attributed to the inherent variation in the properties of the in situ soils and ground conditions, presence of small unmixed clay lumps during the mixing process and perhaps due to the sample disturbance (invisible fissure cracks) induced in the process of coring, preparing, transporting, and testing that induced further material variability. The coefficient of variation of the UCS values obtained was 43% for P-17A and 27% (relatively less variation) for Perris Dam project.

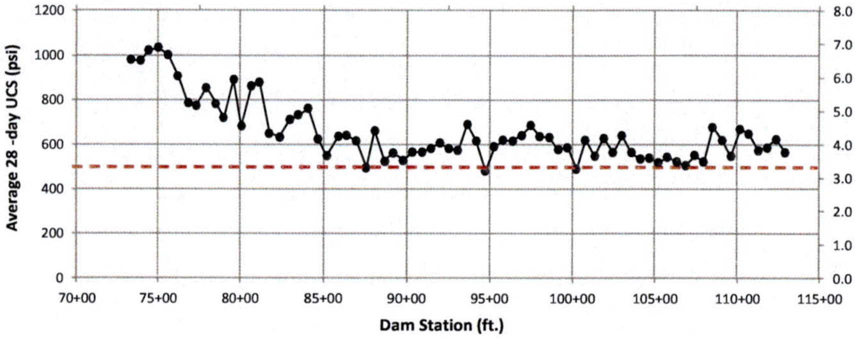

Fig. 6 Average 28-day UCS of the cored samples at Perris Dam Project

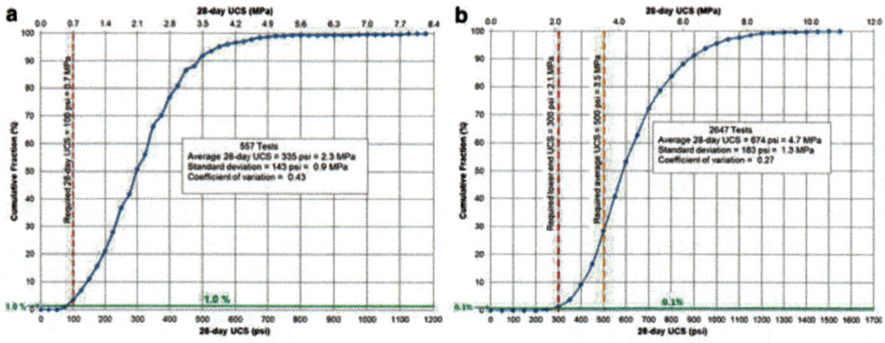

Fig. 7 **a** Cumulative frequency plots for 28-day UCS data at P-17A. **b** at Perris Dam

5 Concluding Remarks

This paper is an illustration of important application of DMM for natural hazard mitigation including flood control of levee and seismic mitigation of earthen dam. DMM was used to increase the strength of soft and liquefiable soils in order to provide foundation support, increase stability, and provide resistance toward vertical and lateral loads that would be induced by the potential geotechnical hazards. Despite some inevitable natural variability, careful monitoring of the quality control system enabled maximum control over the DMM installation parameters and resulted in highly quality soil–cement product. Strength values obtained from the core test results showed that the specified strength requirements for both projects were successfully attained, and the design purpose was met.

Acknowledgements The authors would like to recognize the hard work and professionalism exhibited by the owners, designers, general contractors, project and field engineers, superintendents, and the workers involved in both projects without which the successful completion of these large-scale mitigation projects would not have been possible.

References

1. Cali, P., Lelong, B., Bruce, D., Valagussa, S., Beckerle, J., Gardener, J., Filz, G.: Overview of Deep Mixing at Levee LPV 111. In: 4th International Conference on Grouting and Deep Mixing (DFI), New Orleans (2012)
2. Chen, W., Friesen, S., Driller, M.: Seismic remediation of Perris dam. In: Proceedings of Deep Mixing Conference, San Francisco (2015)
3. Chen, W., Driller, M., Friesen, S: Seismic Remediation of Perris Dam using CDSM. In: Proceedings of 37th Annual USSD Conference, Anaheim, CA (2017)
4. DWR: Perris Dam Foundation Study, California Department of Water Resources (2005)
5. Shrestha, R., Griffin, R., Cali, P., Kafle, S., Watanabe, M., Yang, D.: Deep mixing for levee repair at hurricane protection project, P-17A, Louisiana, USA. In: International Conference on Deep Mixing (DFI), San Francisco (2015)

Characteristics of Surficial Mass Movements on Cut Slopes of the Prime National Highway 1 (PNH-1) in Bhutan

Kiyoharu Hirota, Tomohiro Nishimura, Takeshi Kuwano, and Tomoharu Iwasaki

Abstract This article presents characteristics of surficial mass movements after cutting a slope through the Prime National Highway 1 (hereinafter abbreviated as PNH-1) in Bhutan. The surficial mass movement, in the form of slope failures, affects the ability of traffic to move along the PNH-1. The geology in the area of the PNH-1 consists of metamorphic rocks. There are several types of surficial mass movement in slopes, for instance, rock/soil slide, debris flow, and rock falls. Slope failure has occurred due to the geological discontinuities, the boundary of strata, joints, and so on. From the observation of more than 100 slope failures along the PNH-1, we classified the surficial mass movements into three types: "Denudation," "Erosional flow," and "Slide." Among these three types, "Denudation" and "Erosional flow" destabilize the surface layer of the slope after a while after cutting, while "Slide" tends to collapse on the slope within a short time after cutting. According to Varnes' landslide classification, surficial mass movements such as "Denudation" and "Erosional flow" are categorized as "Flows," and "Slide" are categorized as "Slides."

Keywords Surficial mass movement · Cut slope · Geological discontinuity

1 Introduction

This paper describes the characteristics of surficial mass movement as slope failures observed on the Primary National Highway 1 (hereafter referred to as PNH-1) in Bhutan. Surficial slope failures on the cut slope in Bhutan are mainly affected by the widening of the PNH-1. The slope failures are more concentrated in the rainy season between June and September. In Japan, over 70% of slope disasters are caused by surficial mass movements [1]. During slope observations on the PNH-1, we considered that more than 80% of slope disasters were due to the surficial mass movement. Therefore, they were not recognized as large-scale landslides during on-site observations. The ratio of the number of occurrences for the surficial mass movement is an

K. Hirota (✉) · T. Nishimura · T. Kuwano · T. Iwasaki
Kokusai Kogyo Co. Ltd., 2 Rokubancho, Chiyoda-Ku, Tokyo 102-0085, Japan
e-mail: sbhirota@gmail.com

H. Hazarika et al. (eds.), *Advances in Sustainable Construction and Resource Management*, Lecture Notes in Civil Engineering 144,
https://doi.org/10.1007/978-981-16-0077-7_55

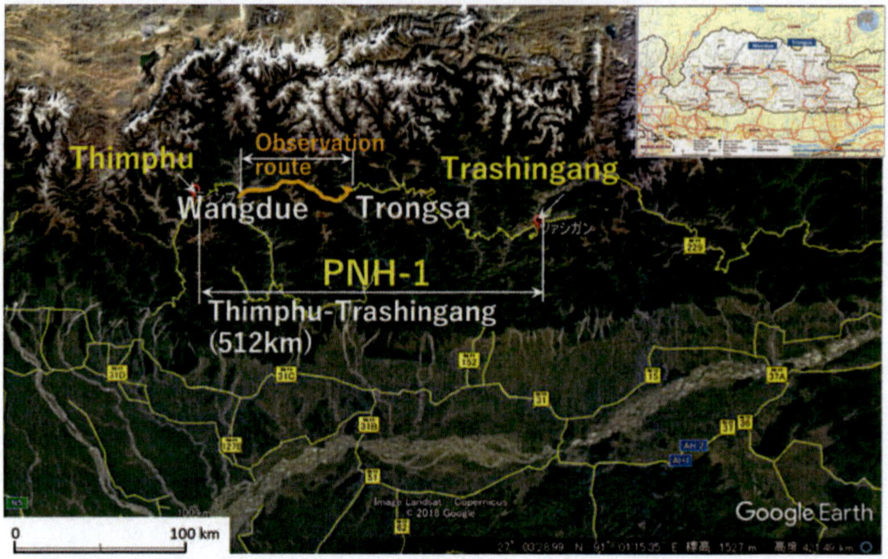

Fig. 1 Observed route (*Image Source* Google Earth)

interesting issue both in Bhutan and in Japan. The issue of surficial mass movements is a common problem and methods to prevent road disasters after cutting the slope is the same in both countries. In this paper, for the initial research focusing on the surficial mass movement, the typological classification of slope failures and the main factors for elucidating the mechanism of the surficial mass movements are discussed.

2 Investigated Slopes on the PNH-1

2.1 Investigated Slopes on the National Highway

The observed route for the cut slope is on the part of the PNH-1 between Wangdue and Trongsa heading east from the capital Thimphu (Fig. 1). The distance between the two towns of Wangdue and Trongsa is around 100 km.

2.2 How to Select Cut Slopes for Observation

The section observed for this project is a part of the PNH-1 from Wangdue (the Lobeysa Regional Office) to Trongsa (the Trongsa Regional Office) where two regional offices of the Department of Road, Bhutan, are located. This section was selected as it contains newly cut slopes due to widening of the road. For the selection of specific cut slopes, sites were selected using maps from the previous road works project, Google Earth images, and on-site observations of each cut slope.

2.3 Basic Features of the Investigated Area

Geological Setting. The investigation route has five units. The GHlml (Lower metasedimentary unit), the GHlo (Orthogneiss unit), the GHlmu (Upper metasedimentary unit) in the Greater Himalayan Zone, the Pzc (Chekha Formation), and the Pzu (Maneting Formation) in the Tethyan Himalayan Zone from the lower level in geologic time (Fig. 2). The Greater Himalayan Zone consists mainly of metamorphic rocks originated in amphibole and granitic rocks, while the Tethyan Himalayan Zone consists mainly of metamorphic rocks that originated from sedimentary rocks.

Mass movements as slides with geological discontinuity on cut slopes are particularly likely to occur in granitic rocks of the GHlo and schist of the GHlmu distribution areas.

Precipitation. Bhutan's rainy season is from June to September. Data are shown for two towns, Wangdue and Trongsa, from 2015 to 2017 (Fig. 3a, b). In the four-month period from November to February of the following year, the monthly precipitation is almost always below 30 mm at the two sites.

Comparing the annual precipitation data between Wangdue and Trongsa, it is about 600 mm per year in Wangdue and 1500 mm per year in Trongsa. The annual precipitation in Trongsa is more than double that in Wangdue (Table 1). Moreover, in Trongsa, the rainfall exceeds 100 mm beginning in May (or April depending on the year) and continues until September. In Bhutan, the rainy season generally is from June to September. Precipitation data is consistent with this rainy season.

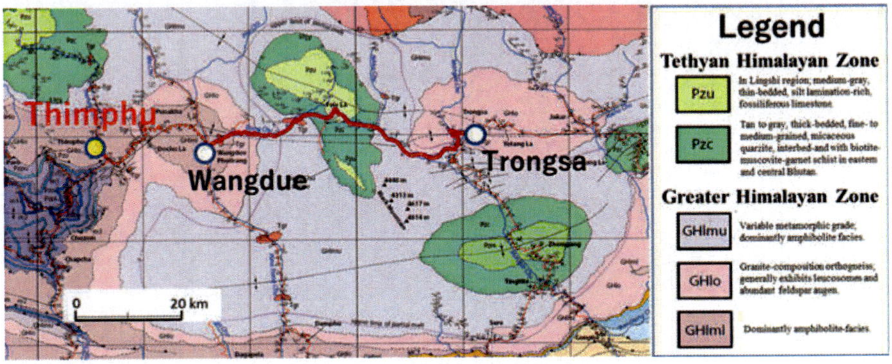

Fig. 2 Geologic map of the investigated area (modified from Long [2])

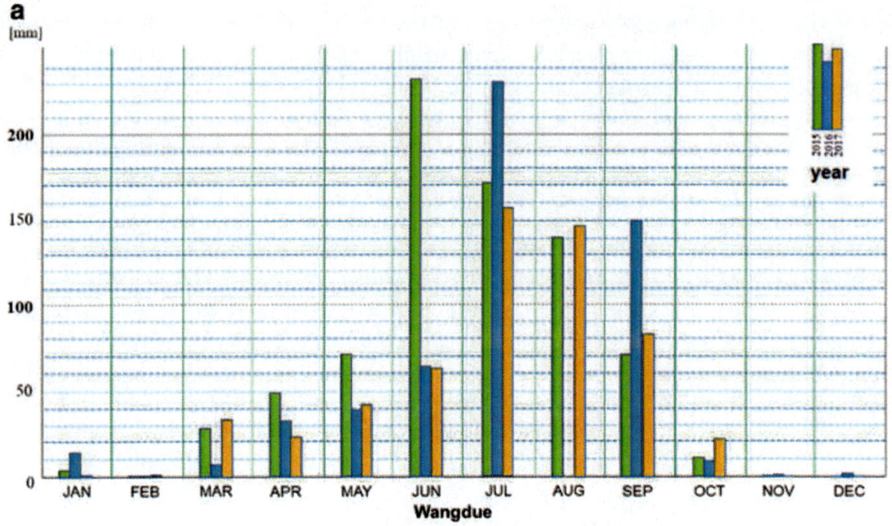

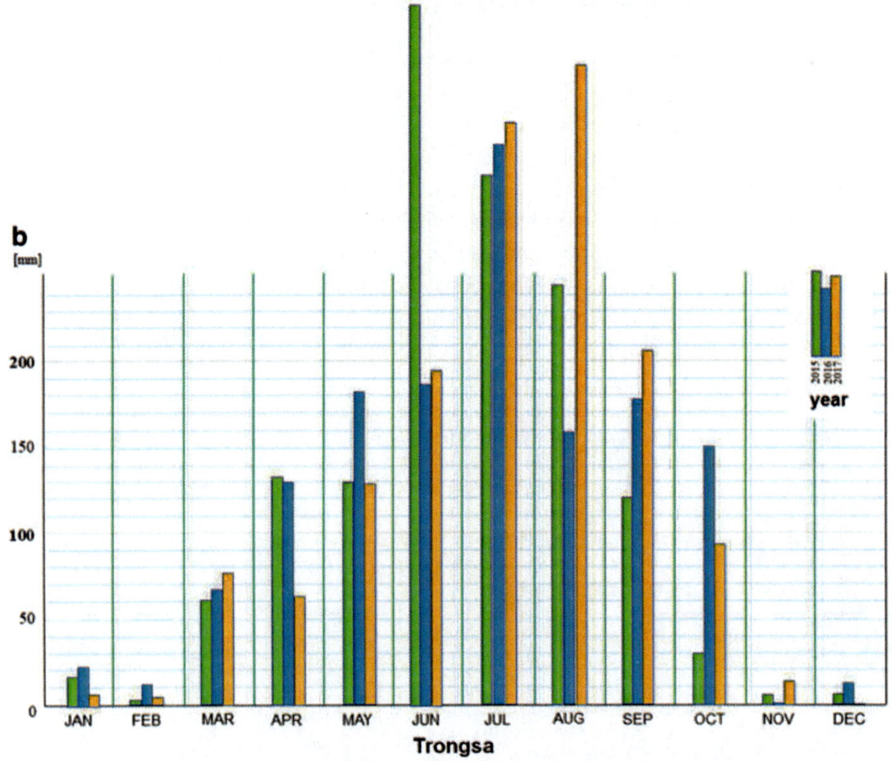

Fig. 3 **a** Precipitation graph for Wangdue, Bhutan (2015–2017). (*Source* Department of Roads in Bhutan: hereafter referred to as DoR data). **b** Precipitation graph for Trongsa, Bhutan (2015–2017) (*Source* DoR data)

Table 1 Precipitation data for Wangdue and Trongsa (2015–2017)

	Wangdue									Precipitation (mm)			
Month	Jan	Feb	Mar	Apr	May	June	Jul	Aug	Sep	Oct	Nov	Dec	Sum
2017	0.6	0.6	33.6	22.9	42	62.9	**156**	146.1	82.9	22	0	0	569.6
2016	14.2	0	7.2	32.3	38.4	64	**229.1**	–	149	8.9	1	1.9	546+
2015	4	0	28.6	48.8	71	**230.7**	170.5	139.1	71.1	10.7	0.1	0	774.6
	Trongsa												
2017	6.6	3.9	76.4	62.9	127.6	194	337.5	**371.1**	204.3	92.9	13.5	0	1490.7
2016	22.7	11.5	66.6	128.9	180.8	186.5	**324.3**	157.8	175.5	149.9	1.2	12.1	1418.9
2015	17.1	2.5	60.3	132.2	128.3	**406.4**	306.6	243.7	119.5	29.5	5.6	5.9	1457.9

Edited from the DoR source data

3 Results and Discussion

3.1 Classification of Surficial Mass Movement in Cut Slopes

Over 100 cut slopes have been observed in the observation section. For every 100 slopes observed, 26 sites will be classified as "Denudation," 56 will be "Erosional flow" sites, and the remaining 16 will be "Slide" sites. The following are the three types of surficial mass movements in cut slopes based on the current PNH-1 observation results.

1. Denudation: Movement of surficial materials (Fig. 4a, b). There are three denudation processes: erosion, weathering, and mass wasting.
2. Erosional flow: Collapse of different geological boundaries or portions with different erosion resistance boundaries (Fig. 4c, d).
3. Slide: Sliding due to geological discontinuity such as joint, schistosity, and fault. Collapse from a stable base rock boundary regardless of soils or rocks (Fig. 4e, f). Slides are prominently observed in the granitic rocks and schist distribution areas along the PNH-1 in the study area.

3.2 Main Causes for Each Type of Slope Failure

Surficial mass movements are almost always related to geological discontinuity which becomes the boundary of sliding plane in Bhutan (Fig. 5). Of the kinds of geological discontinuities, the most common structures are observed are "bedding plane" and "schistosity" in Bhutan.

Denudation: Denudation occurs due to water from geological boundaries with different susceptibility to weathering with an example of this being between coarse-grained soils and fine-grained soils (Fig. 4a, b). As surface weathering progresses, it can be observed as surficial erosion and gully erosion on the slope.

Fig. 4 Examples of slope failures on cut slopes (**a** Denudation, **b** Denudation (enlarged view of "a"), **c** Flow, **d** Erosional flow, **e** Slide (soils), **f** Slide (rocks))

Fig. 5 Geological discontinuity on cut slopes (**a** Slide (soils), **b** Slide (rocks); white arrows: geological discontinuity)

Erosional flow: This occurs when materials mainly weathered from gneissose granite, deposit in the valley and flow down to the road due to precipitation (Fig. 4c). Materials on the slope move after weathering at the geological boundary. Also, materials weathered and/or deposited on the cut slope will move downward.

Slide: In the case of "Slide (soils)," we can find geological discontinuity between soils and rocks (Fig. 5a). The slip surface as geological discontinuity at "Slide (rocks)" slopes can be clearly seen (Fig. 5b). We found that geological discontinuities greatly affected to the slope failure on the cut slope at the sites observed in Bhutan.

Besides the classification of mass movement, it is important to define the relationship between stable and unstable slopes. By defining both Denudation and Erosional flow in the slope, it helps the government of Bhutan to plan which countermeasures to use against unstable slopes.

Observed geological boundaries and joints consist of geological discontinuities; however, it is not clear what the trigger point of slope failure is on geological discontinuity. In order to know which boundaries where Rock Slides and/or Soil Slides occur, it is necessary to recognize where the geological discontinuity is. Clarifying the cause and mechanism of slope failure and predicting the extent of damage will be important for future road maintenance and implementing slope countermeasures.

3.3 Correspondence with Varnes' Landslide Classification

According to the typology of slope failures observed, "Denudation" and "Erosional flow" are included in "FLOWS" as the type of movement in Varnes' Landslide Classification (Varnes 1978). Three types of surficial mass movements caused slope failure are "Denudation," "Erosional Flow," and "Slide" which classified into the two categories of "FLOWS" and "SLIDE" in the Varnes' Landslide Classification (Table 2).

Surficial mass movement is almost included in "TRANSLATIONAL SLIDES," on the contrary there are few "ROTATIONAL SLIDES" in cut slopes along the PNH-1 in Bhutan.

4 Conclusion

The type of surficial mass movement is divided into three types, "Denudation," "Erosional flow," and "Slide" which quickly causes instability due to cutting the slope.

Denudation is surficial erosion on the cut slope where it consists of talus and/or weathered soils. At first, the gully erosion occurs on slopes, and slope failure begins in small blocks. Erosional flow occurs at the cut slope where talus/detritus deposits on the slope and in the valley. A slide occurs in the rock slope with geological

Table 2 Abbreviated version of Varnes' Landslide Classification

Type of movement		Type of material		
		Bedrock	Engineering soils	
			Predominantly coarse	Predominantly fine
Falls		Rock fall	Debris fall	Earth fall
Topples		Rock topple	Debris topple	Earth topple
Slides	Rotational	Rock slide	Debris slide	Earth slide
	Traslational			
Lateral spreads		Rock spread	Debris spread	Earth spread
Flows		Rock flow	Debris flow	Earth flow
		(deep creep)	(soil creep)	
Complex		Combination of two or more principal types of movement		

surficial mass movements: yellow areas are modified from Jan Novotný [3]

discontinuities. Mainly, slides are prominently observed in the granitic rocks and schist distribution areas along the PNH-1.

The factors causing surficial mass movements consist of three key factors: "Geological discontinuity" as an inner factor, "Precipitation" as an outer factor under the natural force, and "Cut slope" as an outer factor due to road building (Fig. 6). The outer factor "Precipitation" influences all other surficial mass movements. The period affected by precipitation is during the rainy season from June to September. Furthermore, the combination of "geological discontinuity" and "cut slope" causes "Slide" in the short term, especially the rainy season.

Acknowledgements We conducted this investigation as a part of "*The project for Capacity Development on Countermeasures of Slope Disaster on Roads in Bhutan*" in partnership with the Japan International Cooperation Agency (JICA). We would like to express our gratitude to the members of the JICA Headquarters and JICA Bhutan Office, our counterpart of the Department of Roads in Bhutan, and many other people in Bhutan for their support during this project.

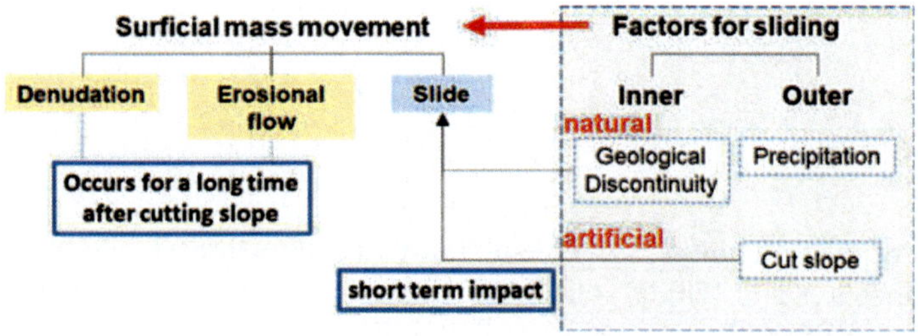

Fig. 6 Relationship between "surficial mass movement" and "factors for sliding"

References

1. Sasaki, Y.: Proposal of systematic investigation procedure for surface slope failure by soil survey, 1–20. Symposium of Kansai Br. of Japan Landslide Society (in Japanese) (2012)
2. Long, S., MacQuarrie, N., Tobgay, T., Grujic, D., Hollister, L.: Geologic map of Bhutan. J. Maps (2011)
3. Jan Novotný: Vearns Landslide Classification (2013). http://www.geology.cz/projekt681900/vyukove-aterialy/2_Varnes_landslide_classification.pdf 2019/11/29

Estimation of Liquefaction Area Using the Screw Driving Sounding Data in Kumamoto

Kohei Tsuji, Naoaki Suemasa, and Tsuyoshi Tanaka

Abstract On April 2016, the Kumamoto earthquake occurred. The earthquake caused serious liquefaction damage in the long and narrow area of about 5 km north and south between the Shirakawa River and the Midori River that flow west through Kumamoto City. In This area is called "Ekijyoka no obi (Liquefaction belt)," and many houses have sunk or tilted. The screw driving sounding test has some advantages including simpler test system and better cost efficiency for penetration in comparison with standard penetration test (SPT) and cone penetration test (CPT). For these reasons, the screw driving sounding test is often used in residential land surveys in Japan, and there is a lot of existing data in this area. In this paper, the spatial information on the Holocene geological situation of this neighborhood was generated made in this area spatial information using the screw driving sounding test data, and the phase change of the liquefied area, and the non-liquefied area was compared. As a conclusion, the spatial extent of the liquefied soil layer was estimated.

Keywords Earthquake · Sounding · Liquefaction

1 SDS Machine and Method

The concept of the SDS test system using the plasticity theory analogy model was proposed by Suemasa et al. (2005).

The plasticity theory analogy model was originally proposed for a shallow foundation problem by Nova and Montrasio (1991).

Figure 1 shows an automatic machine for the SDS.

A monotonic loading system is used for the SDS, where the number of load steps is increased to 6 while the rod is always rotated at a constant rate during the test.

K. Tsuji (✉)
Geotechnical Engineering Laboratory, Japan Home Shield Corporation, Sumida-ku, Tokyo, Japan
e-mail: kohei.tsuji@j-shield.co.jp

N. Suemasa · T. Tanaka
Department of Urban and Civil Engineering, Tokyo City University, Setagaya-ku, Tokyo, Japan

H. Hazarika et al. (eds.), *Advances in Sustainable Construction and Resource Management*, Lecture Notes in Civil Engineering 144,
https://doi.org/10.1007/978-981-16-0077-7_56

Fig. 1 SDS test machine

The step loads are 0.25, 0.38, 0.5, 0.63, 0.75, 0.88, and 1 kN in order, and the load is increased every revolution of the rod. Measurement items in the test are maximum torque (Max.*T*), average torque (Av.*T*), minimum torque of rod (Min.*T*), penetration amount (*L*), settlement rate (*V*), and the number of rotations of the rod (*N*). The data is measured every revolution of the rod. In the SDS, a set of the loading system is conducted at every 25 cm of settlement of the rod, and immediately after the last load, the rod is lifted up by 1 cm and then rotated to measure the rod friction.

The concept of estimating rod friction is shown in Fig. 2. The rod friction can be separated into a vertical component (W_f) and a horizontal component (T_f) as the rod penetrates into the ground while being rotated. A weighted load (W_a) and measured torque (T_a) are defined as follows,

$$W_a = W_f + W \tag{1}$$

$$T_a = T_f + T \tag{2}$$

where W and T are the load and torque at the screw point, respectively. The maximum shear stress acting on the rod surface is described as,

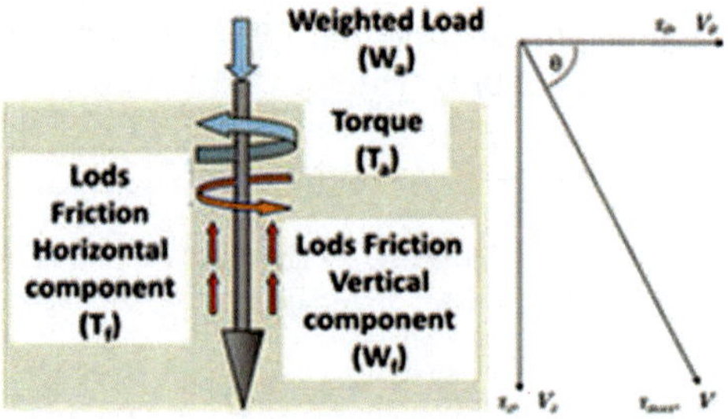

Fig. 2 Concept of a rods friction

$$\tau_{\max} = \frac{T_m}{2\pi r^2 \cdot L} \tag{3}$$

where T_m is torque against the rod friction measured at the end of a loading set, r is the radius of the rod, and L is the total penetrated length. Supposing that the combination rate (V) between rotation speed (V_θ) and settlement speed (V_z) is equal to the one between horizontal shear stress (τ_θ) and vertical shear stress (τ_z) on the surface of the rod, the formulas can be given as follows,

$$\tau_\theta = \tau_{\max} \cdot \cos\theta \tag{4}$$

$$\tau_z = \tau_{\max} \cdot \sin\theta \tag{5}$$

By substituting Eqs. (4), (5) into (6) and (7), the vertical and the horizontal components of the rod friction are obtained as follows,

$$T_f = 2\pi r^2 \cdot L \frac{v_\theta}{\sqrt{v_v^2 + v_\theta^2}} \cdot \frac{T_m}{2\pi^2 L} \tag{6}$$

$$W_f = 2\pi r \cdot L \frac{v_v}{\sqrt{v_v^2 + v_\theta^2}} \cdot \frac{T_m}{2\pi^2 L} \tag{7}$$

2 The Survey Site

The survey site is located in the Kumamoto Plain. The Kumamoto Plain is bounded by the outer rim of Aso on the east side and the Ariake Sea on the west side. Shirakawa flows to the north of the plain, and Kase and Midorikawa flow west to the south. The "liquefaction belt" where the survey site is located is an area extending from the Shirakawa River to the south side, branching out from the Shirakawa River, and continuing to the Kase River with a width of 50–100 m and a length of about 5 km. The "liquefaction belt" is the largest and most continuous in the Kumamoto Plain, located in a slight elevation zone mainly composed of sandy soil. Liquefaction damage was concentrated during the Kumamoto earthquake. Figures 3, 4, and 5 show the land condition map around the survey site. The survey site is a residential area that suffered liquefaction damage in the "liquefaction belt." Many structures around the surveyed residential land deformed and subsided due to liquefaction, and the houses in the residential land showed a relative settlement of the floor surface of more than 10 cm and a slope of about 10/1000 depending on the location. Figure 6 shows a photograph of the settlement around the study area.

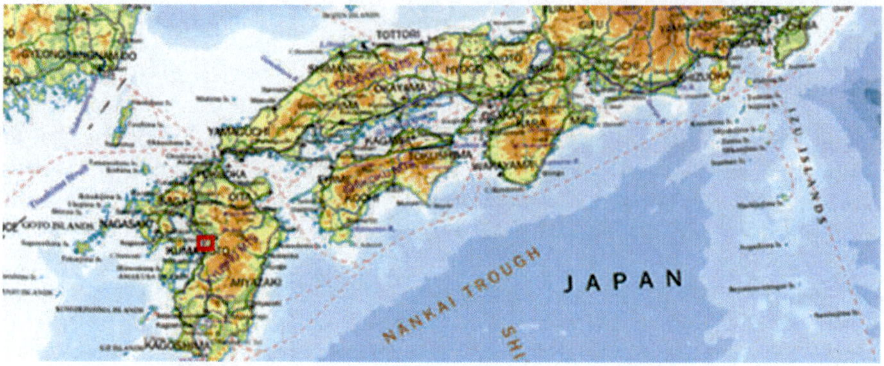

Fig. 3 Survey location map

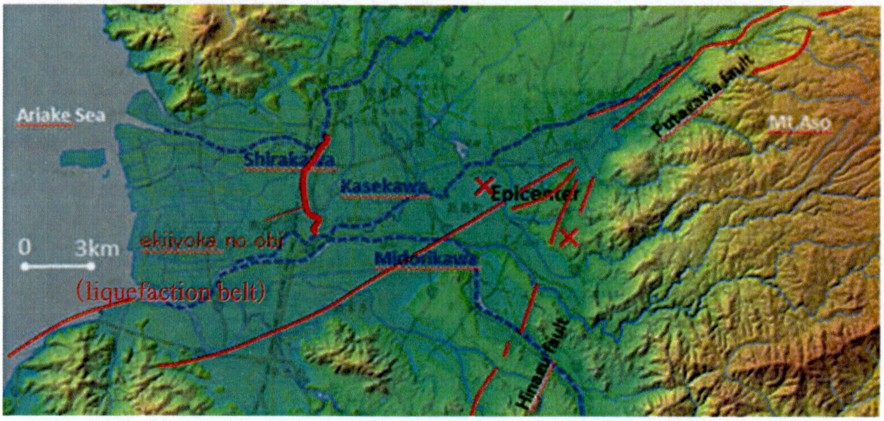

Fig. 4 Schematic of the survey site

3 Survey Boring and Screw Driving Sounding

The analysis parameters that are effective in preparing the soil classification and estimated soil profile from the test results obtained by SDS are described below. The penetration energy per revolution when the screw point penetrates into the ground is expressed by the following equation.

$$\mathrm{d}E = \pi T \mathrm{d}n_{ht} + W \mathrm{d}s_t \tag{8}$$

The total energy E is required to penetrate the screw point for a depth of 25 cm in each test section of the SDS, E, which is expressed as: Here which is the summation of the rotational energy due to torque ET and the penetration energy due to vertical load are ET and EW.

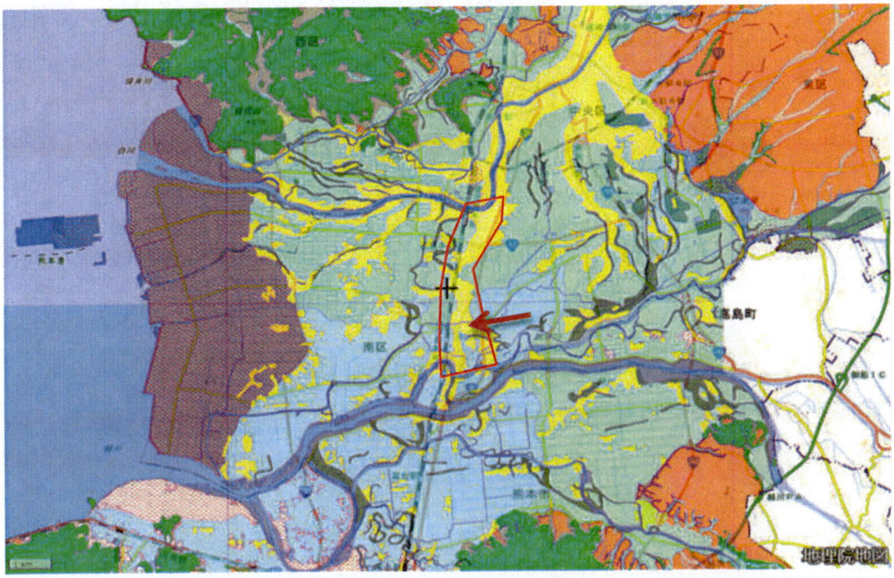

Fig. 5 Located in a slight elevation zone mainly composed of sandy soil (yellow zone)

Fig. 6 Damage situation

$$E = ET + EW \tag{9}$$

The total energy E is an index that indicates the hardness and tightness of the survey ground. ET/EW, which is the ratio of the rotational energy to the penetrating

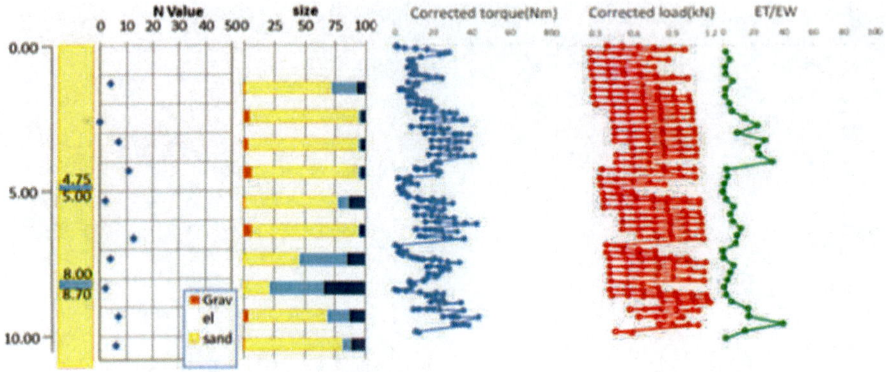

Fig. 7 Borehole survey and SDS results

energy, represents the torque contribution rate during the penetrating and is large in sandy ground.

Figure 7 shows an example of a graph of the borehole survey results accompanied with and a graph of the screw driving sounding (hereinafter called "SDS") results.

The SDS results includes corrected torque, corrected load, and ET/EW. As shown in the depth distribution of corrected torque and ET/EW, it is estimated that there is a thin layer of cohesive soil between sandy layers at GL-5.0 m and GL-8.0 m. The same stratum structure was observed in the result of boring. The depth distributions of corrected torque and corrected load values in SDS tend to be almost the same as that of the standard penetration test results. The SDS parameters including corrected torque and ET/EW can indicate that layers shallower than 5 m from the surface tend to liquefy.

4 Geological Section Estimated by the SDS

As shown in Fig. 8, there are about 150 sites that SDS was performed near the liquefaction belt. These results were used to create the section view.

From previous research results, it was found that ET/EW have a good relationship with FC, where Fc becomes less than 20% when ET/EW is larger than 10 as seen in Fig. 9. Therefore, ET/EW is one of the useful parameters to predict that a layer may liquefy. Figure 10 shows the resulting vertical cross sections to show the stratum structure in the liquefaction belt area. From the cross sections of ET/EW distribution shown in Fig. 10, it is found that there are multilayers of clay and sand depositing in the area outside the liquefaction belt, and sand dominantly deposited in the shallower part in the liquefaction belt.

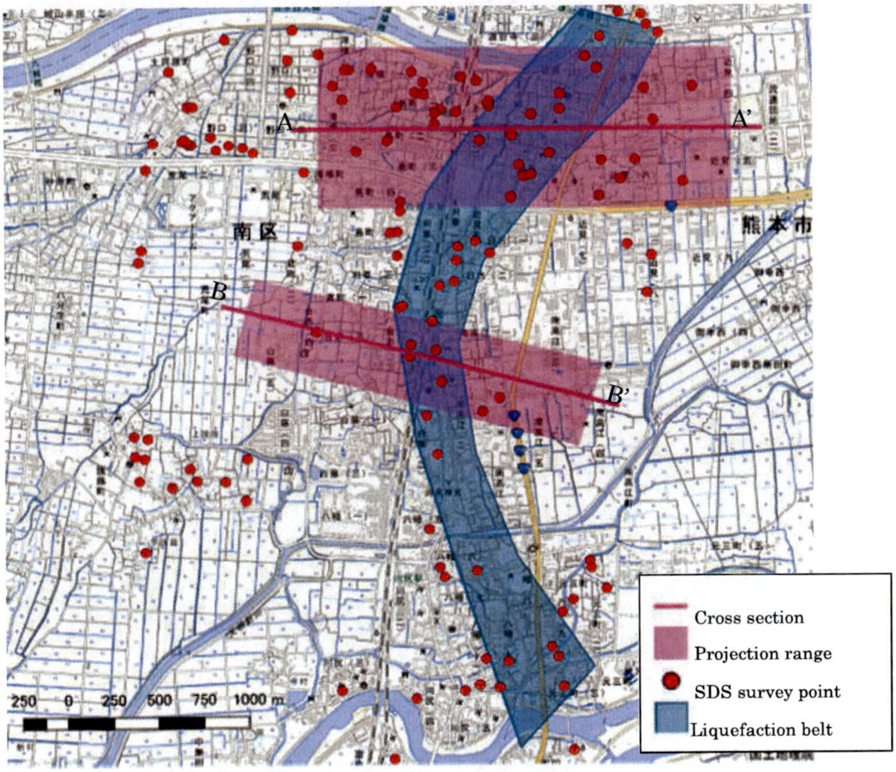

Fig. 8 Location of the SDS and the resulting cross section

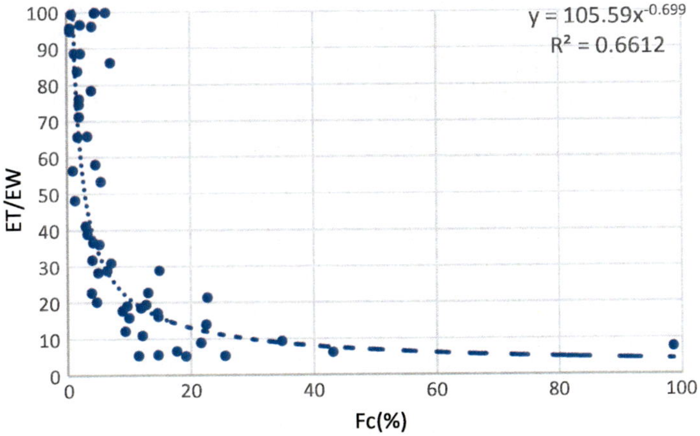

Fig. 9 Relationship between ET/EW value and Fc

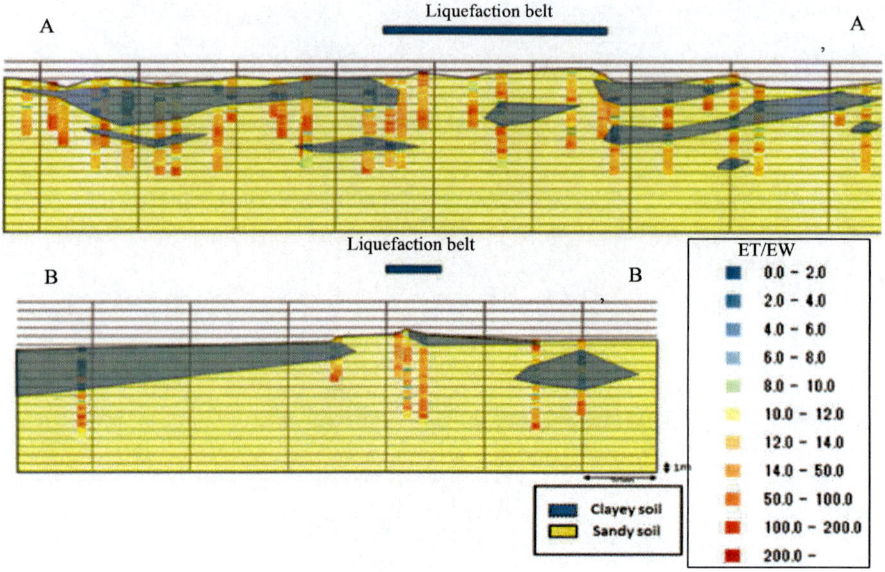

Fig. 10 Cross section

5 Conclusions

The conclusions obtained from the field investigation of the SDS are as follows

1. Accurate geological section can be created by comparison with SPT.
2. Alternate layers of sand and clay are distributed near the study area.
3. The soil layer in the vicinity of the survey site has a drastic vertical layer change and poor horizontal layer continuity.
4. There are multilayers of clay and sand depositing in the area outside the liquefaction belt, and sand is dominantly deposited in the shallower part in the liquefaction belt.

Acknowledgements I would like to thank Prof. Hazarika Hemanta for useful discussions and carefully proofreading the manuscript.

References

1. Murakami, S. (Fukuoka University): Geotechnical Society 2016 Kumamoto Earthquake Ground Disaster Investigation Team Liquefaction Group Report (24 June 2016)
2. Maeda, Y., Yamato, S., Sugano, Y., Sakai, G., Suemasa, N., Tanaka, T.: Evaluation of soil liquefaction potential by screw driving sounding test in residential areas, 6ICEGE, Christchurch, New Zealand (2015)

3. Tanaka, T., Suemasa, N., Yamato, S.: Assessment of Soil Characteristics by Screw Driving Sounding (19SEAGC & 2AGSSEA), Kuala Lumpur (2016)
4. Tsuji, K., Suemasa, N., Tanaka, T., Sugano, Y., Maeda, Y.: Comparison of SDS data before and after Kumamoto earthquake. In: 52nd Japan National Conference on Geotechnical Engineering (2017)

Deterministic Seismic Hazard Analysis of Ankleshwar City, Gujarat

Manali Patel, Chandresh Solanki, and Tejas Thaker

Abstract Earthquake disaster study as well as mitigation is important for the safe and economic design of structures. Ankleshwar is the major industrial hub of India, situated in Bharuch district, located at south-west coast of Gujarat and lies in seismic zone III as per Indian seismic zonation. In present study, the seismic risk analysis of Ankleshwar city is carried out considering the past earthquake records. Earthquake catalogue consisting of all earthquake events from 1819 to 2019 is prepared within 400 km radius of the Bharuch district. After identifying the seismic sources and declustering of earthquake catalogue, the shortest distances from each seismic source responsible for the tectonic activity have been calculated. For the Ankleshwar city, the peak ground acceleration values at rock level have been determined using predictive relationships. The analysis reported that the PGA values are ranging from 0.18 to 0.38 g with respect to maximum credible earthquake of magnitude (M_w) 5.8 generated from Son Narmada Fault. The basic design parameters arrived from the PGA model can be used for the seismic design of structures in the Ankleshwar city.

Keywords Deterministic seismic hazard analysis · Foreshocks · Aftershocks · Maximum credible earthquake · Peak ground acceleration

1 Introduction

Earthquake disaster study as well as mitigation is important for the safe and economic design of structures. Worldwide, India is considered as seismically more active country. Ankleshwar is the major industrial hub of India, situated in Bharuch district, located at south-west coast of Gujarat and lies in seismic zone III as per Indian seismic zonation (IS 1893:2016). In the present analysis, an area of 400 km radius around the

M. Patel (✉) · C. Solanki
S V National Institute of Technology, Surat, Gujarat, India
e-mail: davemanali99@gmail.com

T. Thaker
Pandit Deendayal Petroleum University, Gandhinagar, Gujarat, India

H. Hazarika et al. (eds.), *Advances in Sustainable Construction and Resource Management*, Lecture Notes in Civil Engineering 144,
https://doi.org/10.1007/978-981-16-0077-7_57

Bharuch city has been taken and the damage caused due to Bhuj earthquake in 2001 has been considered. In the present study, an attempt has been made to carry out the seismic risk analysis of Ankleshwar city considering the past earthquake records. The purpose of deterministic seismic hazard analysis (DSHA) is to characterize the different possible ground motions for a particular site by considering previous earthquake records. In present analysis, the sources which can generate earthquake ≥4 magnitude are considered.

2 Preparation of Earthquake Catalogue

Due to improper record of synthesized earthquake events, it is necessary to collect the ancient earthquake data from the several sources with the aim to prepare the suitable earthquake catalogue for the region. Earthquake catalogue consisting of all earthquake events from 1819 to 2019 is prepared within 400 km radius of the Bharuch city. The instrumental data were collected from the various sources such as United States Geological Survey (USGS), Global Instrumental Earthquake Catalogue (ISC-GEM), Institute of Seismological Research (ISR) and the historical data that were collected from the various researchers like Oldham [1], Chandra [2] and Thaker [3]. In the obtained data, there were different magnitude scales with many repetitions; however, to overcome this, enough attention is provided to decluster and convert the magnitudes in one scale. The present catalogue comprises of 292 events from the year 1819 to 2019 with various moment magnitudes (M_w) ranging from 2.0 to 7.8 from 1819 to 2019.

3 Completeness of Earthquake Catalogue

In the historical time (before 1958), due to unavailability of instruments, the earthquake events with lower magnitudes have not been detected, and hence, the complete earthquake catalogues were not available frequently. In 1985, Tinti and Mulargia [4] differentiated recorded seismicity with the real seismicity to find out the incompleteness period in the catalogues. Therefore, it is necessary to have the complete earthquake catalogue. The completeness period for various magnitude classes can be determined by two methods, namely visual cumulative (CUVI) method [4] and STEPP method [5], respectively. In the year 1973, Stepp [5] introduced a method to find the completeness of earthquake data in which he used the statistical estimation approach where the variance (σ) of the estimates of sample mean (λ) is inversely proportional to the number of observations in the sample. In the CUVI method, the cumulative number of events for each magnitude class is computed corresponding to each year. The earthquake data considered to be complete when the trend of the data is stabilized to approximate straight line.

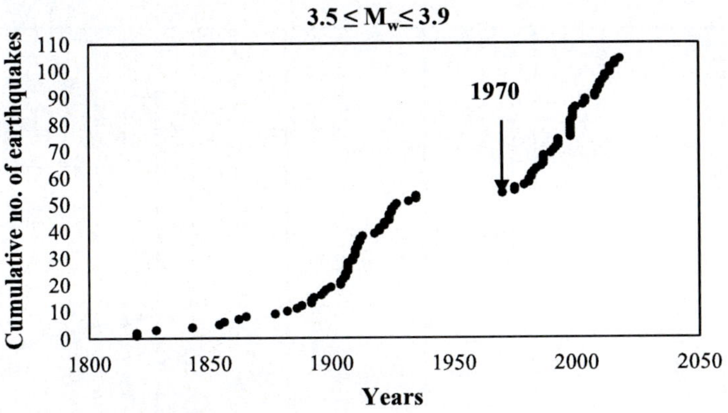

Fig. 1 Cumulative visual interpretation method

By using these two methods, the completeness period can be determined for different magnitudes (Figs. 1 and 2). Table 1 provides the information about the completeness period estimated by these two methods.

4 Regional Recurrence Relationship

The regional magnitude–frequency relationship is the key factor for the characterization of seismic activity.

For determining the seismicity parameters, usually Gutenberg and Richter relationship [6] is taken. In this study, the following equation has been considered:

$$\log_{10} N = a - bM_{\mathrm{w}} \tag{1}$$

where N is cumulative number of earthquakes $\geq$ a particular magnitude M_{w}/year, and 'a' and 'b' parameters define the seismicity of the area.

The obtained recurrence relationships from STEPP and CUVI methods (Fig. 3) are as follows:

$$\log_{10} N = -0.861M_{\mathrm{w}} + 3.81,\ R^2 = 0.993 \tag{2}$$

$$\log_{10} N = -0.932M_{\mathrm{w}} + 4.15,\ R^2 = 0.991 \tag{3}$$

From the present comparative analysis, it is found that the obtained seismicity parameters in both the methods are similar. Henceforth, the generalized equation for the study region was developed as follows:

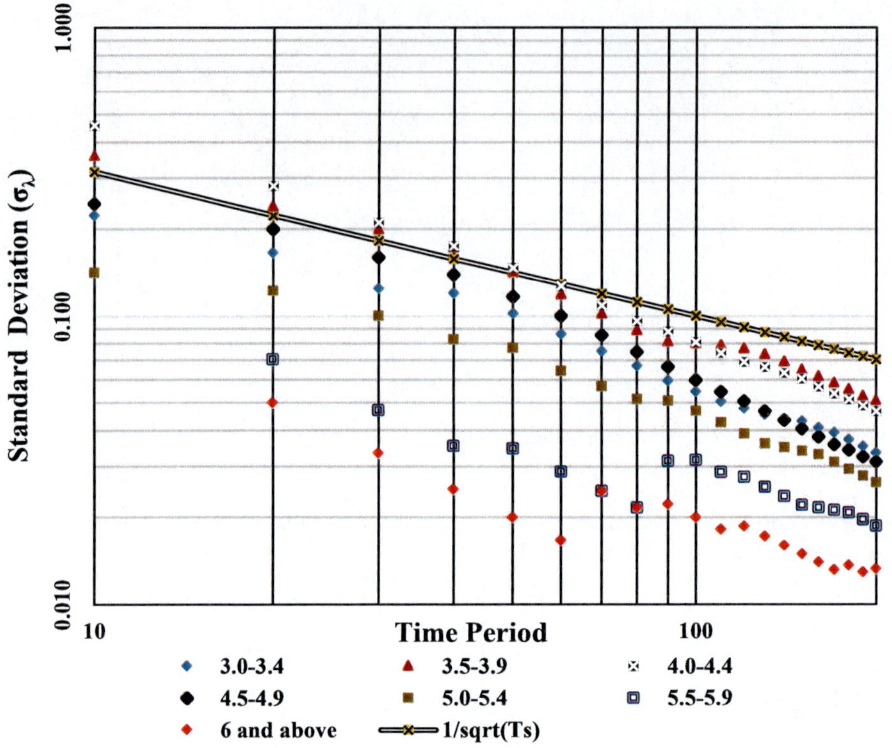

Fig. 2 STEPP method

Table 1 Completeness period by CUVI and STEPP methods

Magnitude class	CUVI method		STEPP (1973) method	
	Period	Years	Period	Years
3.00–3.49	1956–2019	63	1959–2019	60
3.50–3.99	1970–2019	49	1969–2019	50
4.00–4.49	1968–2019	51	1969–2019	50
4.50–4.99	1962–2019	57	1959–2019	60
5.00–5.49	1950–2019	69	1949–2019	70
5.50–5.99	1900–2019	119	1899–2019	120
6.00–6.49	1845–2019	174	1839–2019	180
6.5 and above	1819–2019	200	**	**

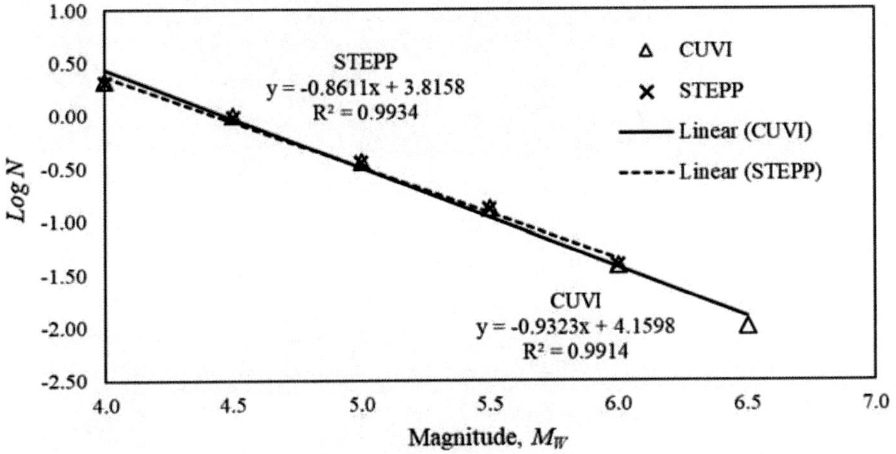

Fig. 3 Regional frequency–magnitude relationships

$$\log_{10} N = (-0.896 \pm 0.0356)M_w + (3.987 \pm 0.172) \tag{4}$$

For the computation of regional frequency-magnitude relationships, the CUVI and STEPP methods have been adopted.

5 Seismotectonic Model

In the current study, generated earthquake catalogue can be beneficial for the entire Peninsular India; however, damages occurred due to earthquakes are limited at particular distances. The generated earthquake catalogue lies between latitudes 19°N to 25° N and longitudes 69° E to 76° E covering all the magnitudes ranges between 2 and 7.8.

The seismotectonic atlas [7] was used to plot the faults which are the prime governing seismic sources. The SEISAT, 2000 [7] was developed by the Geological Survey of India (GSI). The seismic events with different magnitudes were superimposed on these plots where the 13 major faults were considered which are capable to generate peak ground acceleration greater or equal to 0.01 g. The seismotectonic model of Bharuch region is represented in Fig. 4.

6 Seismic Hazard Analysis

The assessment of strong ground motion at a particular site can be predicted by seismic hazard analysis. The advantage of seismic hazard analysis is to prepare the

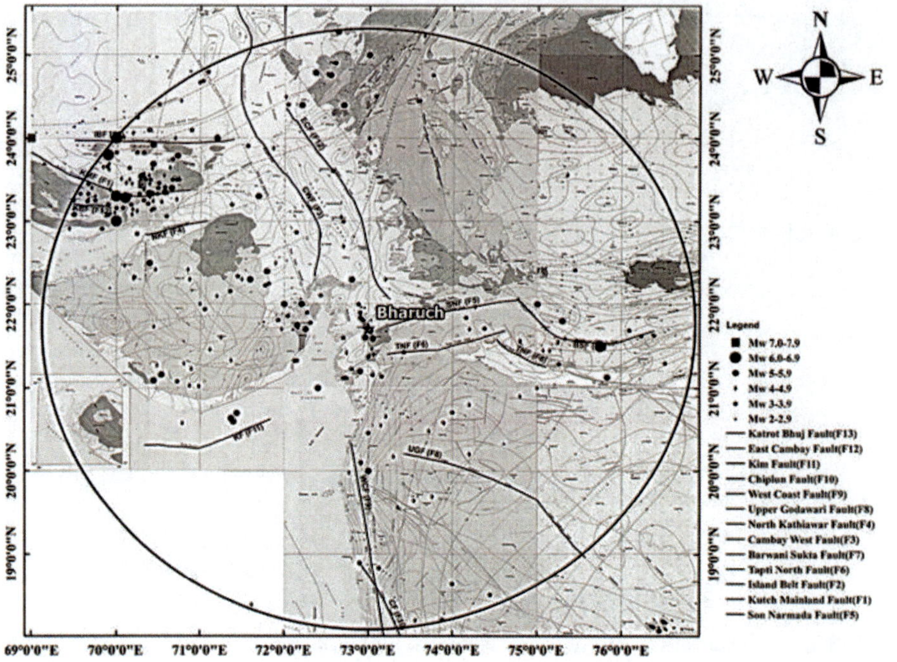

Fig. 4 Seismotectonic model for the study region

seismic zonation maps which can be useful for earthquake-resistant design of any structure. The various ground-shaking parameters can be used to describe the seismic severity of any specific site.

In comparison to peak velocity and displacement time histories, the use of peak ground acceleration has been widely adopted as the ground motion measurements. Majority of the codal provisions and the design procedures are based on PGA and normalized response spectra. Generally, the peak ground acceleration can be computed with two.

In general, the PGA of any specific site is computed using the probabilistic and deterministic seismic hazard analysis. The probabilistic analysis used the lowest to highest ground motions, whereas the deterministic analysis only considers the highest ground motions. Raghu Kanth and Iyengar have proposed attenuation relationship for western-central region, southern region and Peninsular region of India [8, 9].

In present study, the following attenuation relationship for Ankleshwar region is proposed:

$$\ln Y = C_1 + C_2(M - 6) + C_3(M - 6)^2 - \ln R - C_4 R + \ln \varepsilon \tag{5}$$

where $C_1 = 1.7236$, $C_2 = 0.9453$, $C_3 = -0.0725$ and $C_4 = 0.0064$ are the constants provided by Raghu Kanth and Iyengar [9] for the western-central portion of India, Y is associated with PGA or spectral acceleration at bed rock level in terms of g, M is the moment magnitude, R is refers to the hypo central distance and ε is the error connected with regression taken as 0.3439.

7 Results and Discussion

For the rigorous seismic hazard analysis, the Ankleshwar city was divided into 65 number of grids (3 km × 3 km). The hazard maps have been developed for the Ankleshwar city by considering the higher value of peak ground acceleration at 31 accessible locations (Fig. 5). The majority of the industries are situated in north-east part of Ankleshwar city. As per the present analysis, the peak ground acceleration values for this region varies between 0.29 and 0.38 g. However, it drops to 0.18–0.28 g for the western part of the city.

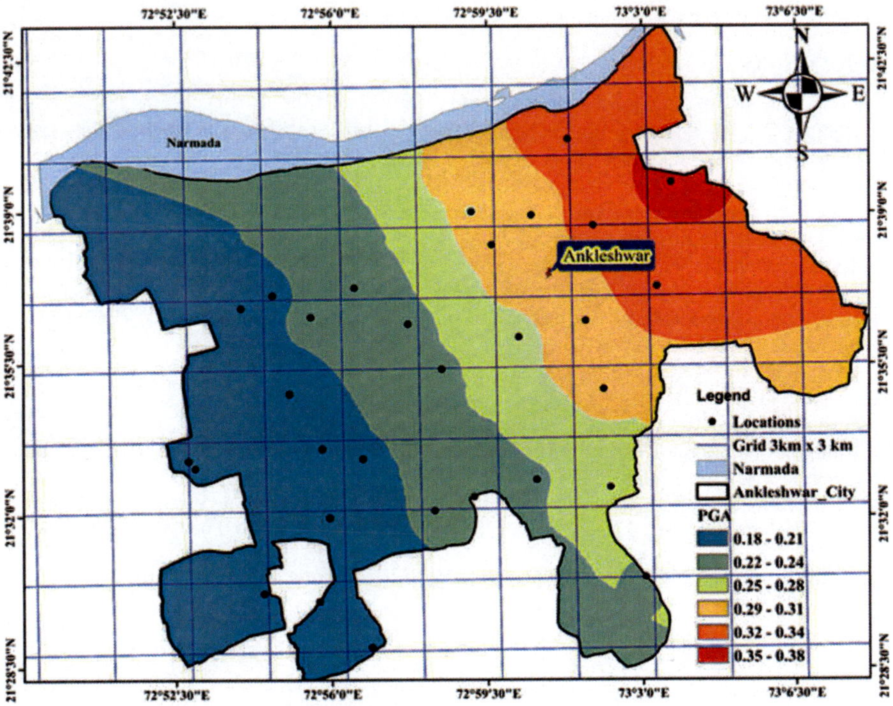

Fig. 5 Seismic hazard maps for Ankleshwar region

8 Conclusions

In present study, the deterministic seismic hazard analysis (DSHA) approach was used to assess the peak ground acceleration of Ankleshwar city which falls under moderate zone of earthquake, i.e., zone III (BIS 1893). To assess PGA values, the maximum credible earthquake (MCE) and past seismicity parameters were used, and the entire region was divided into grids of 3 km $\times$ 3 km. From the present analysis, it can be stated that the eastern part of the city is more severe with PGA values >0.29 g in comparison with western part of city where PGA values are ranging between 0.18 and 0.29 g. The BIS (1893:2016) has suggested generalized PGA value as 0.16 g for entire seismic zone III. The present analysis can be useful for the protection of existing structures and designing the new structures. As it is the preliminary analysis, further the probabilistic seismic hazard analysis, ground response analysis and liquefaction studies are required for effective consideration.

References

1. Oldham, T.A.: Catalogue of Indian Earthquakes, pp. 163–215. Geological Survey India (1883)
2. Chandra, U.: Earthquakes of PI—a seismotectonic study. Bull. Seismol. Soc. Am. **67**(5), 1387–1413 (1977)
3. Thaker, T.: Seismic Hazard Analysis and Microzonation Studies for Surat City and Surrounding Region. Indian Institute of Delhi (2012)
4. Tinti, S., Mulargia, F.: Completeness analysis of a seismic catalogue. Ann. Geophys. (1985)
5. Stepp, J.C.: Analysis of completeness of the earthquake sample in the Puget sound area and its effect on statistical estimates of earthquake hazard. In: International conference on microzonification, pp. 897–910 (1973)
6. Gutenberg, B., Richter, C.F.: Frequency of earthquakes in California. Bull. Seismol. Soc. Am. 185–188 (1944)
7. Seismotectonic Atlas of India. Geological Survey of India (2000)
8. Iyengar, R.N., Raghu Kanth, S.T.G.: Attenuation of strong ground motion in peninsular India. Seismol. Res. Lett. **75**, 530–540 (2004). https://doi.org/10.1785/gssrl.75.4.530
9. Raghu Kanth, S.T.G., Iyengar, R.N.: Estimation of seismic spectral acceleration in Peninsular India. J. Earth Syst. Sci. **116**, 199–214 (2007). https://doi.org/10.1007/s12040-007-0020-8

Large Strain Torsional Shear Tests on the Mechanism of Long-Distance Flow Slide in Palu, Central Sulawesi, Indonesia

Risqi Faris Hidayat, Takashi Kiyota, Muhammad Umar, and Hasbullah Nawir

Abstract The catastrophic flow slide in Palu city has proven that surface ground could laterally deform in several kilometers due to the earthquake motion even though the ground inclination is very gentle (1–5%). This paper aimed to understand the flow deformation behavior of Toyoura sand and flow sand samples collected from Sibalaya trench No. 3 in the modified hollow cylindrical torsional shear apparatus by using the water-inflow concept. Static liquefaction tests with constant shear stress are employed to identify the possibility of the specimen to flow under small initial static shear in drained conditions. In comparison, using the same field sand sample, an undrained cyclic test followed by monotonic loading is conducted to observe the specimen behavior under the undrained condition. It is observed that continuous flow behavior can be observed under small initial static shear stress with water-inflow concept. The continuous flow behavior of the sand specimen is defined by the initial density, which also affects the volume expansion behavior. The static liquefaction test using flow sand sample shows that this material could flow under the water-inflow mechanism. As a comparison, this material shows non-flow behavior in the typical undrained test, and even the specimen is liquefied.

Keywords Static liquefaction · Torsional shear test · Flow slides · Sulawesi earthquake · Confined aquifer

1 Introduction

Long-distance flow slide disaster was observed in several places in Palu city; Petobo, Balaroa, Jono Oge, Lolu, and Sibalaya after a 7.5 Mw intraplate earthquake hit Central Sulawesi and the surrounding place on September 28, 2018. The United

R. F. Hidayat (✉) · T. Kiyota · M. Umar
Institute of Industrial Science, The University of Tokyo, Tokyo, Japan
e-mail: risqifaris.hidayat@yahoo.com

H. Nawir
Department of Civil Engineering, Bandung Institute of Technology, Bandung, Indonesia

H. Hazarika et al. (eds.), *Advances in Sustainable Construction and Resource Management*, Lecture Notes in Civil Engineering 144,
https://doi.org/10.1007/978-981-16-0077-7_58

States Geological Survey [1] notified that the epicenter (0.256 S and 119.846 E) was located 70 km from the northern part of Palu city at the shallow depth of around 20 km. This earthquake also triggered a tsunami in the coastal areas of Palu, Donggala, and Mamuju.

A site investigation was conducted to reveal the mechanism of this long-distance flow failure. This ground movement could flow, massively, at the very gently sloping ground, ranging around 1–5% [2–4]. Hidayat et al. [2] discovered that sand ejecta, as the evidence of soil liquefaction, has been found in several places around the affected areas. They also reported that the freshwater pond inside the affected area could be observed, even two weeks after the earthquake. Kiyota et al. [3] described that the shallow groundwater in the affected areas was under pressure, based on the interview with the residents. They considered that the cause of this long-distance flow slide was a significant reduction in the effective stress and the shear strength of the surface ground by a large amount of groundwater that probably supplied from the confined aquifer layer or underground fault water.

Kokusho [5] reported, based on his experimental results, that when the water interlayer is formed below the sloped less permeable layer, a large deformation could be observed in the less permeable layer and its upper layers due to the gravity force while maintaining their shape. Okamura et al. [4], who conducted several trench surveys in the Sibalaya area, reported that in order to employ the water interlayer as the only factor to promote this long-distance flow slide, the water interlayer should be smooth and continuous all over the area, and this assumption is not always met with the investigation results. Thus, this water interlayer concept could not fully explain the occurrence of long-distance flow slides in Palu City.

In the typical laboratory undrained shear test, at the large strain, the strength and stiffness of sandy soil are eventually recovered, even in very loose sand [6]. These experimental results probably suggest that the long-distance flow slide is unlikely to occur at the gently sloped ground under the undrained condition. However, considering the inflow of the groundwater from the confined aquifer or underground fault water might supply abundant water to the liquefied layer, the significant positive dilatancy is allowed to occur. This condition reduces the shear resistance of the surface ground significantly, resulting in a long-distance flow slide even in the gently sloped ground.

This paper aims to understand the flow deformation behavior of clean sand (Toyoura sand) as well as flow sand samples collected from Sibalaya trench No. 3 [4] in the modified hollow cylindrical torsional shear apparatus [7] using the water-inflow concept. Static liquefaction tests with constant shear stress are employed to identify the possibility of the sand specimen to flow under small initial static shear in drained conditions. In comparison, using the same field sand sample, an undrained cyclic test followed by monotonic loading is conducted to observe the specimen behavior under the undrained condition.

2 Large Strain Torsional Shear Tests

2.1 Apparatus and Sample Preparation

A fully automated torsional shear apparatus with a hollow cylindrical specimen, developed in the Institute of Industrial Science, University of Tokyo [7], is employed in this study (Fig. 1). The specimen has 100 mm of outer diameter, 60 mm of inner diameter, and 200 mm of height. This apparatus could achieve double amplitude of shear strain ($\gamma_{\text{-DA}}$) levels exceeding 100% by using a belt-driven torsional loading system that is connected to an AC servomotor through electromagnetic clutches and a series of reduction gears. An external potentiometer with a wire and a pulley is employed to measure the large deformation. Specified shear stress amplitude is controlled by a data acquisition system connected to a computer, which monitors the outputs from the load cell and calculates the shear stress. The measured shear stress is corrected for the effects of the membrane force by utilizing an empirical equation for the same specimen size from Umar [8].

Toyoura sand ($G_s = 2.648$, $e_{\max} = 0.948$, $e_{\min} = 0.619$) and field sand from Sibalaya Trench No. 3 ($G_s = 2.639$, $e_{\max} = 0.948$, $e_{\min} = 0.519$) with 10% of fines content (non-plastic) are used in this study. This field sand sample is extracted from the trench investigation conducted by Okamura et al. [4] after the disaster. The grain size distribution of both samples is shown in Fig. 2.

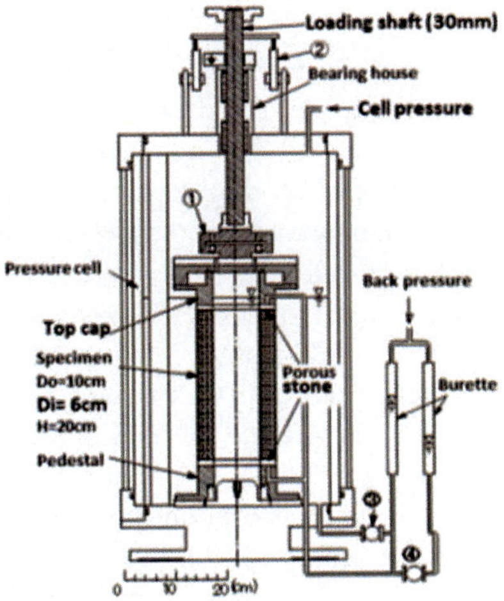

Fig. 1 Schematic diagram of hollow cylindrical torsional shear apparatus [8]

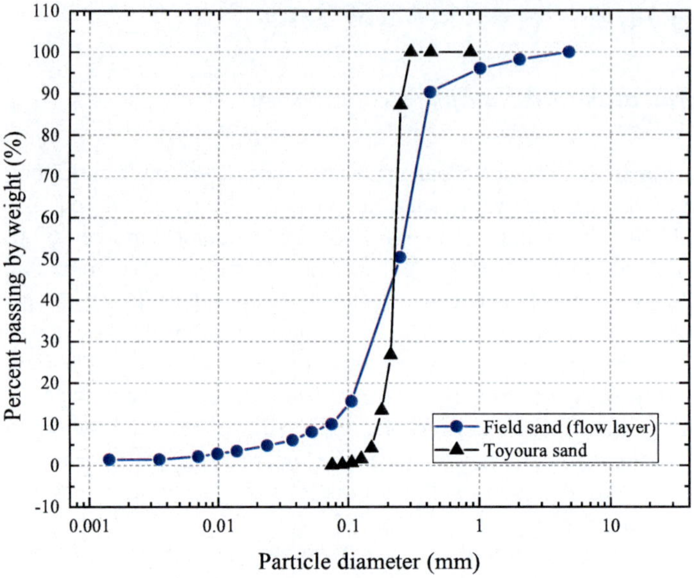

Fig. 2 Particle size distribution of Toyoura sand and field sample obtained from Trench No. 3 Sibalaya area

2.2 *Testing Procedures*

Static liquefaction test with water inflow under constant shear stress. In this study, a series of static liquefaction tests with constant shear stress is conducted in the modified torsional shear apparatus. All specimens were prepared by air pluviation method. The specimen was saturated by double vacuum method.

For Toyoura sand, the specimen was subjected to isotropic consolidation to mean effective stress (p') of 100 kPa with the back pressure of 200 kPa. After the consolidation, the initial static shear stress (τ_{static}) of 7.5 kPa was applied to the specimen under drained condition. For the field sample, it was consolidated to $p' = 50$ kPa and τ_{static} of 4 kPa was applied. Both testing conditions on the Toyoura sand specimen and field sample represent the ground inclination of 1.5–2%, according to the formula provided by Chiaro et al. [9].

After applying the τ_{static}, the p' was gradually reduced by increasing the back pressure with the rate of 5 kPa/min under constant τ_{static}. This reduction was continued until the specimen showed continuous flow behavior under the constant τ_{static}. During the reduction of the p', the height of specimen was not kept fixed. The test was terminated when the shear strain (γ_{zq}) reached more than 80%. The list of static liquefaction tests in the torsional shear apparatus is shown in Table 1.

Undrained cyclic loading followed by a monotonic loading test. This test was carried on the field sample. At first, the specimen was consolidated to $p'_{\text{ini}} = 50$ kPa and

Table 1 List of experiments on modified torsional shear apparatus

Code test	Relative density, Dr (%)	Static shear stress, τ_{static} (kPa)	Mean effective stress, p' (kPa)	B value	Material	Test
TS 1	23.4	7.5	100	0.96	Toyoura	Static liquefaction
TS 2	44	7.5	100	0.96	Toyoura	Static liquefaction
TS 3	50	7.5	100	0.96	Toyoura	Static liquefaction
TS 4	58	7.5	100	0.97	Toyoura	Static liquefaction
TS 5	73	7.5	100	0.96	Toyoura	Static liquefaction
TS 6	84	7.5	100	0.96	Toyoura	Static liquefaction
NS 1	75	4	50	0.94	Sibalaya	Static liquefaction
NS 2	84	4	50	0.94	Sibalaya	Undrained cyclic loading + undrained monotonic loading

τ_{static} of 4 kPa was also applied. After that, the specimen was subjected to undrained cyclic loading (τ_{cyclic}) of 20 kPa (double amplitude) until the specimen reached γ_{zq} more than 7.5% (double amplitude). From this state, the specimen was subjected to undrained monotonic loading until the γ_{zq} reached 90% of a single amplitude.

3 Experimental Results and Discussion

3.1 Flow Deformation Behavior Under Static Liquefaction Test with a Constant Shear Stress of Toyoura Sand

Figure 3 shows the time series of stress–strain development of the loose specimen (TS1). In this study, the reduction of mean effective stress (p') was promoted by the increase of the back pressure. As the p' was reduced, the volumetric strain (ε_{vol}) increased progressively. The progression of ε_{vol} induced the development of shear strain (γ_{zq}) in the specific increment. The γ_{zq} development of this specimen can mainly be classified into three failure states, notated as A, B, and C, that correspond to the flow deformation behavior.

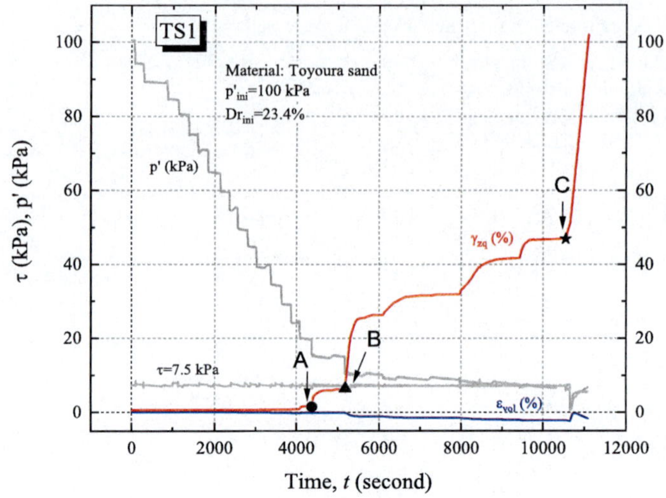

Fig. 3 Stress–strain development of loose specimen (TS1) in time series

For specimen TS1, at state A, γ_{zq} was remarkably developed from 1.6 to 6.4% at $p' = 15$ kPa. However, it gradually discontinued afterward. The subsequent reduction of p' induced the development of γ_{zq} (state B) from 6.4 to 25%, and it continued until γ_{zq} reached 47%. The continuity of γ_{zq} was suddenly followed by the sudden drop of both p' and τ (state C). After the sudden drop, both p' and τ were recovered.

The same observation could be investigated at dense specimen (TS6), as shown in Fig. 4. For specimen TS6, state A occurred at the $p' = 8$ kPa, where γ_{zq} developed

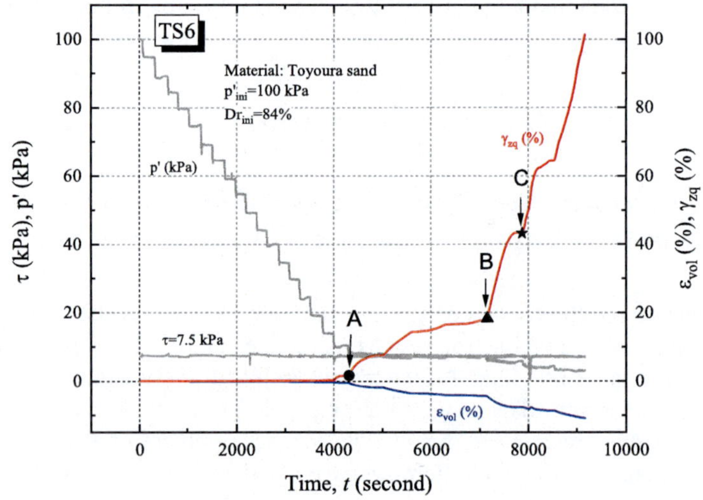

Fig. 4 Stress–strain development of loose specimen (TS6) in time series

from 2.3 to 18%. State B happened at the $p' = 5$ kPa, where the continuous development of γ_{zq} could be observed from 18 to 43%. After γ_{zq} reached 43%, state C could be observed where p' and τ drop abruptly. In this case, the recovery of stress in TS6 was faster than that of TS1.

The observed behavior of Toyoura sand at states A, B, and C corresponds to the development of flow behavior due to the water inflow. At state A, the specimen showed a primarily softening behavior which probably relates to the phase transformation of the specimen. Further reduction of p' resulted on a continuous development of shear strain, which was observed at state B. This behavior represents the flow behavior of the specimen under the constant shear stress. Finally, at state C, the unique phenomenon of the sudden drop of stresses could be observed. This stress drop might be interpreted as the sign of the loss of particle interlocking due to the large deformation of the specimen under drained condition. This behavior was followed by the stress recovery, representing the recovery of the soil strength. By comparing the stress recovery of TS1 and TS6, it is clear that stresses in TS1 took a longer time to recover than TS6, representing that loose specimen requires more time to recover the soil strength in order to mobilize the same acting shear stress.

After the stress recovery, the shape of the specimen was altered and became non-uniform. Figure 5 shows the change of the specimen shape during the experiment. Figure 5a, b corresponds to the specimen shape before the stress drop. The shape of the specimen was still uniform. Some membrane wrinkles were observed as the shear strain was increasing. Figure 5c shows the shape after the stress drop. It could be seen that the specimen was heavily distorted after the stress drop. Therefore, the behavior of the specimen after state C could be unreliable and biased and would be excluded in the results. Figure 5d presents the specimen shape at $\gamma_{zq} = 100\%$.

Figure 6 is the plot of the volumetric strain and shear strain of Toyoura sand. Sento et al. [10] and Yoshimine et al. [11] reported that the volume expansion of clean sand was dependent on the initial density of the specimen. The higher the initial density, the higher development of volumetric strain would be developed. The same results can be obtained in this study that specimen TS6, which is the densest specimen,

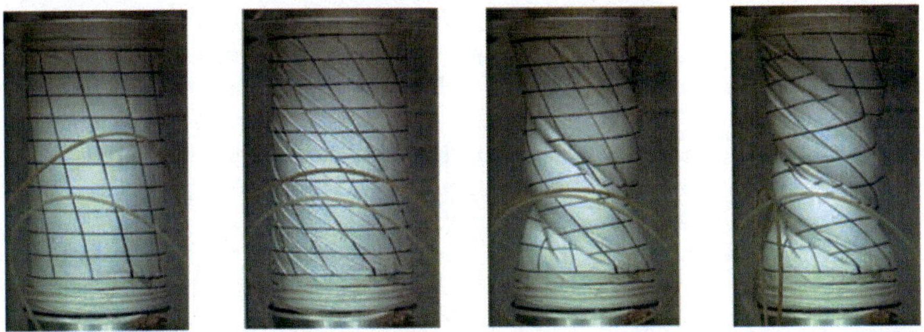

(a) γ_{zq}=5% (b) γ_{zq}=42% (c) γ_{zq}=65% (d) γ_{zq}=100%

Fig. 5 Specimen shape change of TS1 during static liquefaction test in torsional shear

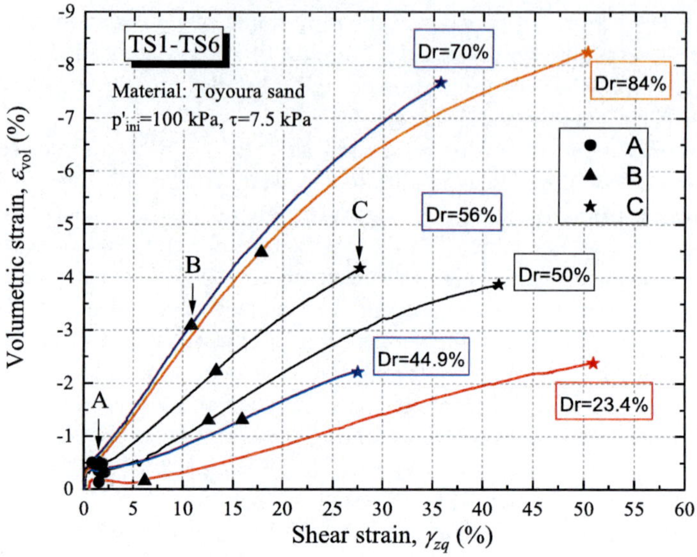

Fig. 6 Relationship of volumetric strain and shear strain of Toyoura sand

shows more development of volumetric strain (ε_{vol}). This result also represents that dense material needs more water to induce continuous flow states. The state A, B, C can be observed at all specimens in this study, regardless of their initial densities.

3.2 *Flow Deformation Behavior of in Situ Sandy Soil*

A static liquefaction test under constant shear stress in the torsional shear apparatus using a field sample was conducted to demonstrate whether the current flow disaster can be explained by the water-inflow mechanism. The τ_{static} was set to 4 kPa with the p' of 50 kPa, assuming that the ground inclination is around 1.5%. The field density of this material was estimated from the undisturbed sample, which was 1.485 g/cm^3. The initial relative density is calculated as 75%, stating that the field sample was in a dense state.

Figure 7 shows the stress and strain development of static liquefaction using this field material in the torsional shear apparatus.

It is identified that the specimen showed the failure state A, B, and C under this condition. The state A appeared at $p' = 6$ kPa. The state B, as the starting of continuous flow failure, was shown at $p' = 2$ kPa with the ε_{vol} of -4.1%, while the state C appeared when p' reached 1 kPa. Comparing with the specimen TS5, whose $Dr_{ini} = 73\%$ and flow at the $\varepsilon_{vol} = -3\%$, it can be inferred that this flow sample ($Dr_{ini} = 75\%$) significantly requires more water to produce continuous flow behavior. This

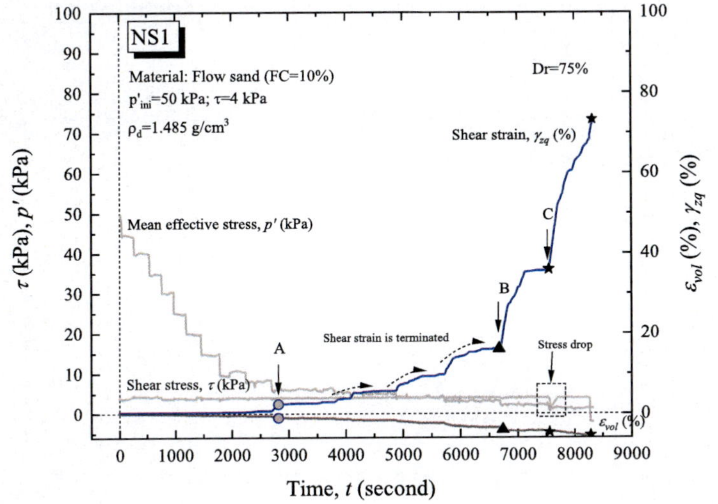

Fig. 7 Strains and stresses development in time series of static liquefaction with constant shear stress of field sample

result roughly gives an idea that the materials having fine content (FC = 10%) might need more water to produce flow behavior compared to the clean sand.

In order to investigate whether this phenomenon could occur under undrained conditions, an experiment of undrained cyclic loading followed by the undrained monotonic loading was conducted as a comparison to the abovementioned static liquefaction test. In this experiment, the field sand specimen was prepared by air pluviation method with a dry density of 1.56 g/cm^3 (Dr = 84%).

As the specimen was subjected to the undrained cyclic loading after the consolidation, the shear strain (γ_{zq}) of the specimen reached 10.2% (double amplitude), stating that the specimen is already liquefied ($\gamma_{zq} > 7.5\%$). From this state, the liquefied specimen was directly subjected to the undrained monotonic loading until the shear strain reached more than 80% (γ_{zq} > 80%). The effective stress path and the shear stress and shear strain relationship are shown in Figs. 8 and 9, respectively.

It is shown that during the monotonic loading, the shear stress kept increasing as the shear strain increased, showing the non-flow behavior at the large strain state (γ_{zq} > 85%). This result indicates that the soil liquefaction cannot solely trigger the flow failure in this field condition. However, it might be necessary to examine this concept on the rheological experiment, such as using a ring shear apparatus, to investigate the flow behavior of specimen under undrained conditions.

The static liquefaction test describes the flow behavior of the specimen in the static condition. Flow behavior observation under dynamic conditions might be necessary to get a comprehensive interpretation to explain the mechanism of the current disaster.

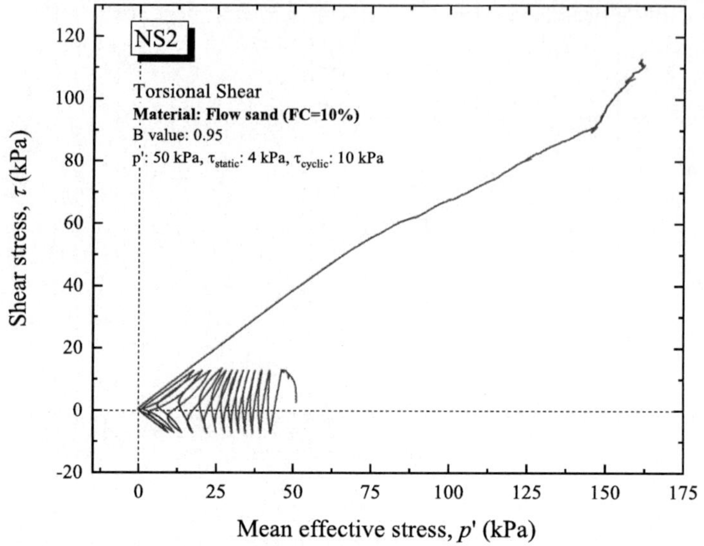

Fig. 8 Effective stress path of sandy soil under undrained cyclic loading test followed by monotonic loading

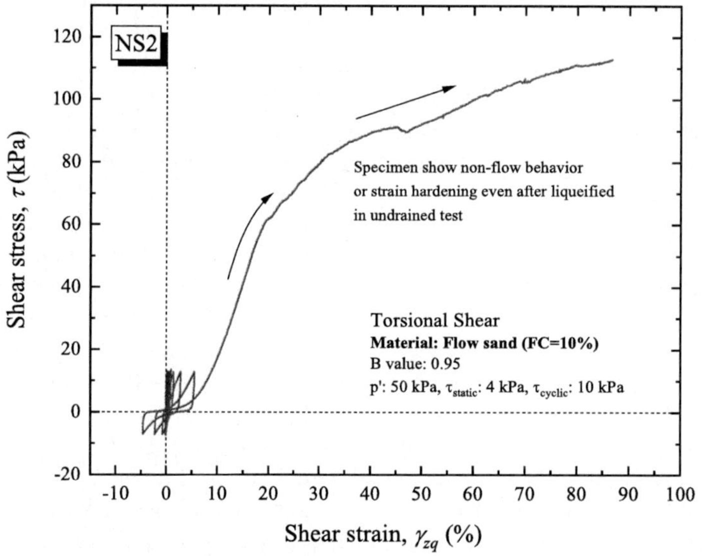

Fig. 9 Shear stress and shear strain relationship of sandy soil under undrained cyclic loading until shear strain (γ_{zq}) reached 10.2% (DA) followed by undrained monotonic loading

4 Conclusion

The flow behavior of Toyoura sand specimen can be observed in the torsional shear test under constant small initial static shear stress. Preliminary softening, continuous flow, and stress drop phenomena could be observed on each specimen, regardless of their initial densities. It is found that the dense specimen requires more water to produce the continue flow behavior. Furthermore, the static liquefaction test on flow sand from Trench No. 3 Sibalaya showed that this material could flow under the water-inflow mechanism. As a comparison, this material showed non-flow behavior in the typical undrained test; even the specimen was liquefied.

Acknowledgements This work was partly supported by JSPS KAKENHI Grant Number JP19KK0108 and grant funding from the JSCE Committee of Promoting Innovation in Infrastructure Management.

References

1. The United States Geological Survey, Earthquake Hazards Program, The United States Geological Survey (28 Oct 2018). https://earthquake.usgs.gov/earthquakes/eventpage/us1000h3p4/executive
2. Hidayat, R.F., Kiyota, T., Tada, N., Hayakawa, J., Nawir, H.: Reconnaissance on liquefaction-induced flow failure caused by the 2018 Mw 7.5 Sulawesi Earthquake, Palu, Indonesia. J. Eng. Technol. Sci. **52**(1), 51–65 (2020)
3. Kiyota, T., Furuichi, H., Hidayat, R.F., Nawir, H.: Overview of long-distance flow-slide caused by the 2018 Sulawesi earthquake, Indonesia. Soils Found. **60**(3), 722–735 (2020)
4. Okamura, M., Ono, K., Arsyad, A., Minaka, U.S., Nurdin, S.: Large-scale flowslide in Sibalaya caused by the 2018 Sulawesi earthquake. Soils Found. **59**(5), 1148–1159 (2020)
5. Kokusho, T.: Water film in liquefied sand and its effect on lateral spread. J. Geotech. Geoenviron. Eng. **125**(10), 817–826 (1999)
6. Kiyota, T., Koseki, J., Sato, T.: Relationship between limiting shear strain and reduction of shear moduli due to liquefaction in large strain torsional shear tests. Soil Dyn. Earthq. Eng. **49**, 122–134 (2013)
7. Kiyota, T., Sato, T., Koseki, J., Abadimarand, M.: Behavior of liqueified sands under extremely large strain levels in cyclic torsional shear tests. Soils Found. **48**(5), 727–739 (2008)
8. Umar, M.: Degradation of stress-strain properties of sand in undrained torsional shear tests. Ph.D. thesis, The University of Tokyo, Tokyo (2019)
9. Chiaro, G., Sato, T., Kiyota, T., Koseki, J.: Effect of initial static shear stress on the undrained cyclic behavior of saturated sand by torsional shear loading. In: 5th International Conference on Earthquake Geotechnical Engineering, Santiago, Chile (2011)
10. Sento, N., Kazama, M., Uzuoka, R., Ohmura, H., Ishimaru, M.: Possibility of post liquefaction flow failure due to seepage. J. Geotech. Geoenviron. Eng. **130**(7), 707–716 (2004)
11. Yoshimine, M., Nishizaki, H., Amano, K., Hosono, Y.: Flow deformation of liquefied soil under constant shear load and its application to analysis of flow slide of infinite slope. Soil Dyn. Earthq. Eng. **26**(2–4), 253–264 (2006)

One-Dimensional Soil Column Simulation on Water Film Formation During Earthquake

Divyesh Rohit, Hemanta Hazarika, Chengjiong Qin, Tsubasa Maeda, Takaji Kokusho, and Yuichi Yahiro

Abstract The 2018 Sulawesi earthquake of magnitude Mw 7.5 triggered large-scale flow slides and ground liquefaction in the Palu Valley at multiple locations on a ground of very gentle gradient. Findings of the post event forensic geotechnical investigation conducted by the authors of this research highlighted the presence of silty clay or sandy clay layers sandwiched between sandy or sandy gravel layers at multiple locations in the flow slide zones. These sandwiched layers tend to have low permeability as compared to overlying and underlying layers sandy strata. To study the contribution of these sandwiched layers on the triggering of flow slides, a preliminary research was conducted using 1D soil column with clay and non-plastic fines sandwiched between sand layers. The findings highlighted that the excess pore water pressure reached its peak value by an instant shock. A thin interlayer water called water film is easily formed below the clay and non-plastic fines seam on impact. The influence of water film on pore pressure dissipation and overall soil settlement was studied. Water film formed beneath the clay and non-plastic seam permits the soil layer above it to float, with continuous deformation in soil due to change in equilibrium condition.

Keywords Flow slide · Liquefaction · Pore water pressure · Water film

1 Introduction

The 2018 Sulawesi earthquake of magnitude (Mw) 7.5 triggered large-scale flow slides in Balaroa, Petobo, Jono Oge and Sibalaya areas of Palu city in the Sulawesi Island of the Indonesian archipelago. The extent of flow slides ranged from few

D. Rohit (✉) · H. Hazarika · C. J. Qin · T. Maeda · Y. Yahiro
Kyushu University, Fukuoka 819-0395, Japan
e-mail: rohit.divyesh.530@s.kyushu-u.ac.jp

T. Kokusho
Chuo University, Tokyo 192-0393, Japan

H. Hazarika et al. (eds.), *Advances in Sustainable Construction and Resource Management*, Lecture Notes in Civil Engineering 144,
https://doi.org/10.1007/978-981-16-0077-7_59

meters to few kilometers. The locations where the flow slides occurred had a very gentle ground slope of about 1–5% [1].

The authors along with other researchers conducted forensic geotechnical investigation at various flow slide sites in Palu city. The site in Jono Oge for example had flowing water sources on three sides, which caused a shallow water table (Fig. 1a). The presence of very loose soil layers at shallow depth of less than 5 m was highlighted in the results of portable dynamic cone penetration tests conducted during the survey (Fig. 1b). Also, the presence of low-permeable silty clay as well as clayey silt layers sandwiched between sand and sandy gravel layers was confirmed in Jono Oge (Fig. 1c).

According to past research, long-distance flow slides or long run out landslides are transformed from the metastable structure of a slope made of a large loose, cohesionless mass which is triggered by external or internal disturbance, for instance earthquake motion, blasting, collapse of cave, snowmelt or a landslide movement [2, 3]. Landslides when transformed into flow slides can have a run-up of long distance even on a ground with gentle gradient [4]. During a slope failure, the metastable structure of a slope gets damaged suddenly, which leads to rapid increase in pore water pressure (PWP) thereby transforming the failing mass into a debris sheet with considerably uniform thickness. And, if the underlying soil layer downhill is loose,

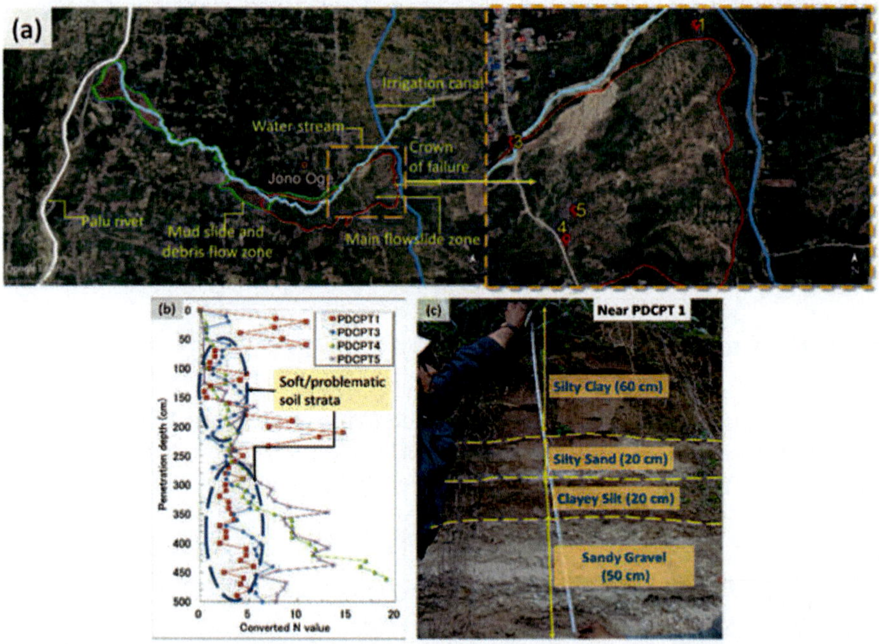

Fig. 1 Site investigation results of Jono Oge flow slide zone: **a** characteristics of failure zone (modified from Google Earth) and location of PDCPT tests; **b** results of PDCPT tests and **c** soil profile of the open trench near crown of the failure zone (modified from Hazarika et al. [1])

saturated and of low permeability, it will react to the unanticipated undrained load by causing structural collapse, resulting in the partial or whole liquefaction at the bottom of the debris sheet thereby decreasing the resisting force [5].

Recent studies have highlighted the formation of a thin layer of interlayer trapped water called water film below low-permeable clayey layers to be one of the probable factors contributing to triggering of such large-scale flow slides [1]. Kokusho [6, 7] conducted multiple tests on one-dimensional (1D) soil models as well shaking table (2D) tests to evaluate the phenomenon of void redistribution and water film formation in stratified soil deposits with the presence of low-permeable silt seams sandwiched between sandy layers. The studies revealed the formation of water film during liquefaction in stratified sandy soils with low permeability sandwiched layers. Also, the pore water pressure developed below the sandwiched layer is inhibited from dissipation, which forms a thin interlayer water film which is expulsed from the lower layers. The current research focuses on a preliminary study of excess pore water pressure inhibition using Toyoura sand as permeable loose sandy layer and pearl clay and non-plastic fines as sandwiched seam using 1D soil model. The settlement of soil layers at the surface and the interlayer interface is studied along with the pore water pressure buildup and dissipation at different heights for different interlayer conditions, as well as development and decay of water film.

2 1D Model Tests for Pore Water Pressure Development and Dissipation

To study how the presence of low permeability interlayer seam affects the pore water pressure dissipation in a soil layer and the development and decay of water film, 1D model tests were conducted in the laboratory by simulating the soil stratification conditions in which a clay or silt seam is sandwiched between sand layers.

2.1 1D Soil Model Setup

In this test, a saturated loose one-dimensional (1D) sand layer was prepared by air pluviation of Toyoura sand in a Lucite tube of 9.6 cm internal diameter and 100 cm height. The relative density of the soil was maintained throughout at approximately 50% by calibrating the height of fall. The soil column was saturated with water from bottom, which caused certain amount of settlement in the layer height. Pearl clay and non-plastic fines (<0.075 mm) sieved from available sand were used for making the sandwiched seams. The properties of the materials used are given in Fig. 2a. The thickness of clay and silt layer was maintained at 2 cm across the tests with uniform mass maintained for all the tests. The basic soil properties for pearl clay are sourced from Sun et al. [8]. The 1D setup was imparted with a shock using a suspended

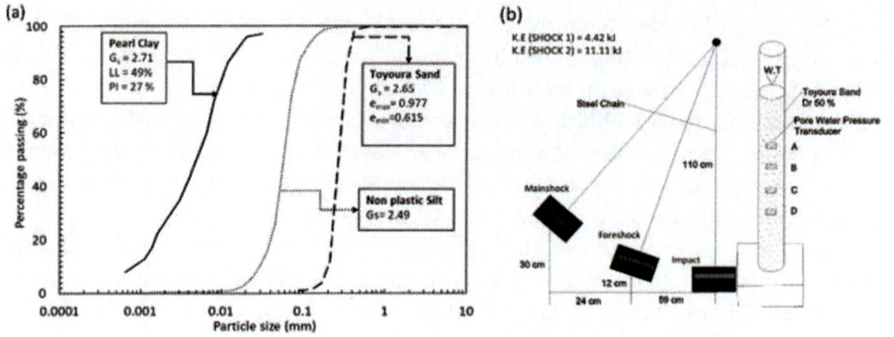

Fig. 2 1D impact tests: **a** properties of materials used and **b** mechanism of impact

concrete cylinder of mass 3.777 kg, 10 cm diameter and 20 cm in length. The model was imparted with two shocks, the second greater than the first one. The shock 2 was provided after the PWP completely dissipated post the shock 1. The amount of energy transferred to the soil model for all the cases during shock 1 and shock 2 was controlled by maintaining the distance and height of the hanging cylinder manually (Fig. 2b). Four typical test cases were considered for evaluating the influence of low-permeable interlayers seams between sand layers, which are as follows: Case I: only sand layer; Case II: single clay seam in between sand layer; Case III: two clay seams alternatively between sand layers; and Case IV: top clay seam and bottom seam with non-plastic fines between the sand layers (Fig. 3a–d). The pore water pressure transducers with nomenclature, A, B, C and D, are used to record the pore water pressure.

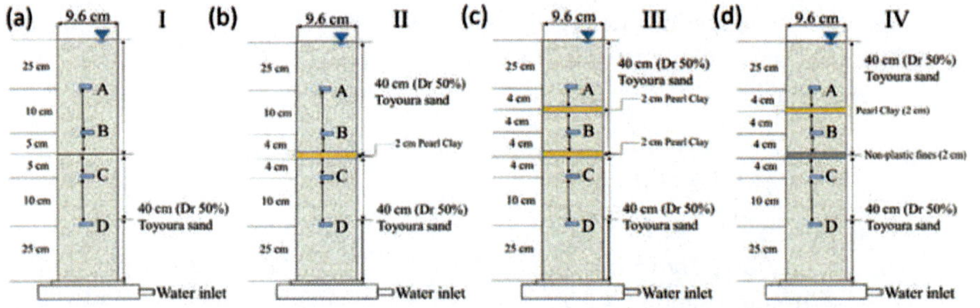

Fig. 3 1D experimental setup for the test cases: **a** case I; **b** case II; **c** case III and **d** case IV

3 Pore Water Pressure (PWP) Development and Dissipation After Impact

The influence of low permeability interlayers on the variation in pore water pressure, water film formation and the settlement in soil layer at different elevations is discussed in this section.

3.1 Trend for Excess Pore Water Pressure (EPWP)

To evaluate the influence of single and multiple low permeability seams on the overall stability of the soil column and the EPWP dissipation after the impact, the PWP is plotted against the elapsed time (secs) as a function of the overburden stress (R_u) (Fig. 4a–d). The PWP reaches its peak point as soon as the shock is imparted to the model. In Fig. 4a, the EPWP for all the transducers dissipates within 15 s. While in Fig. 4b, the EPWP pressure dissipates slowly for the PWP transducers C and D, which are below the clay seam due to the suppression of dissipation of EPWP by the clay layer. In Fig. 4c, the EPWP for the transducers B, C and D which are placed below the clay seams dissipates very slowly as compared to A for both shock 1 and shock 2, while the EPWP developed at B which is just below the upper clay seam dissipates in less time than at C and D, this can be attributed to the presence of additional lower clay seam above C and D which create additional obstruction for the migration of pore water from the bottom sand layer. Furthermore, in Fig. 4d, the EPWP at B and C dissipates slowly with an identical trend while the EPWP for at D, which is near the base of the column, shows a bit quicker dissipation as compared to B and C. Moreover, the inset plots in Fig. 4a–d show the peak R_u values (R_{umax}), and it can be observed from the plots that the R_{umax} values for Case II and IV show a similar trend for shock 1 and shock 2. The discrepancy in the trend for other cases may be attributed to the different degree of consolidation at different depths due to shock 1 and shock 2.

3.2 Variation of Pore Water Pressure with Time Across Soil Column Height

In Fig. 5, the vertical distribution of pore water pressure is across the soil column with respect to elapsed time after the impact. From the figure, the PWP reaches its peak immediately on impact, which later dissipates with respect to time. The PWP normalizes quickly near the surface, due to the free passage of water available for all the cases. Furthermore, the presence of clay and silt seam affects the development and dissipation PWP for the underlying layers. Also, the values of PWP at the bottom of the layer are higher in case of dual interlayer seams just after the impact. The

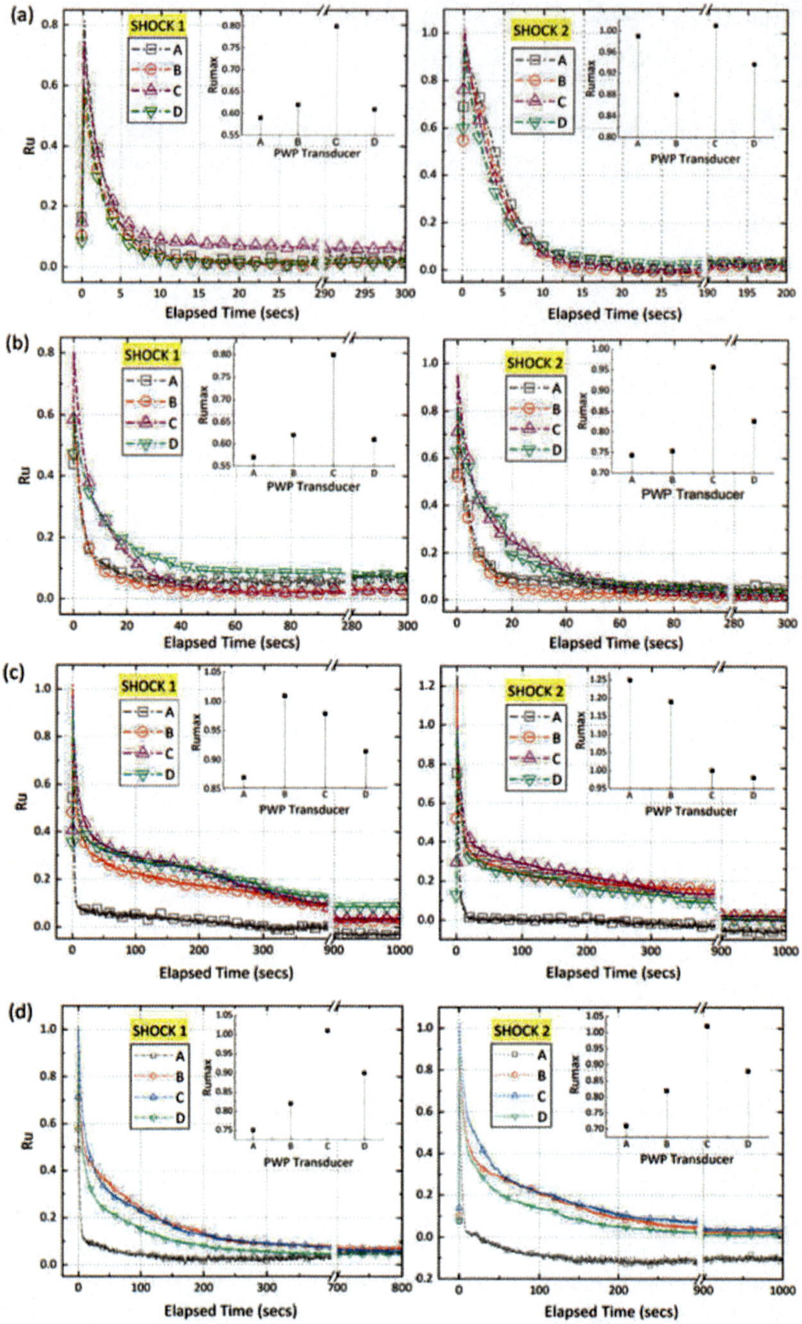

Fig. 4 Dissipation of excess pore water pressure with elapsed time for different PWP transducers as a function of overburden stress: **a** case I; **b** case II; **c** case III and **d** case IV

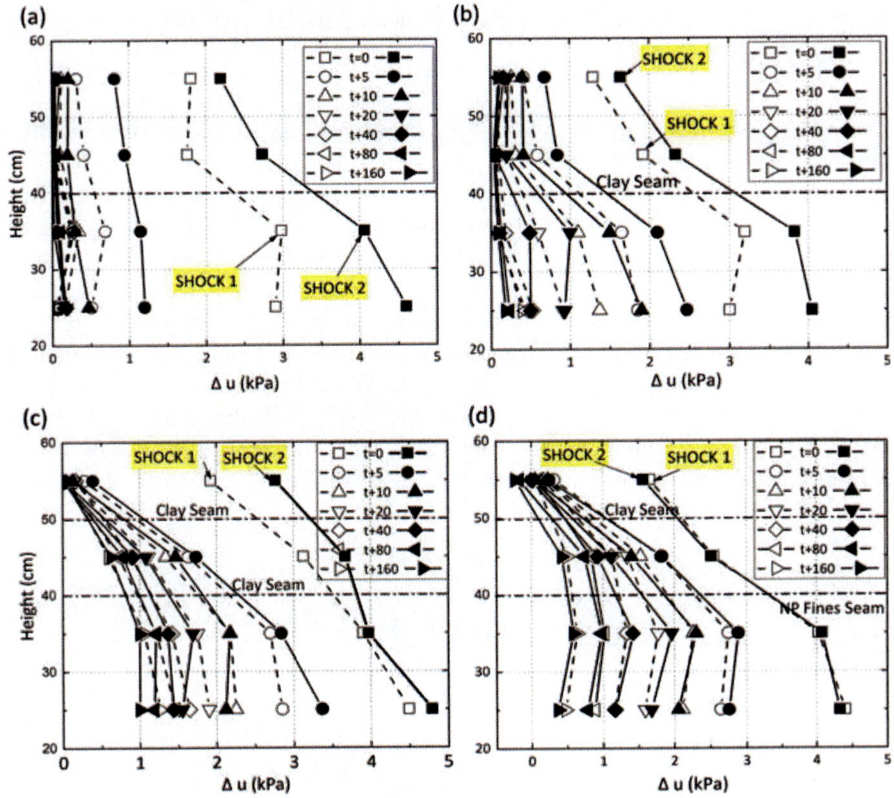

Fig. 5 Distribution of PWP dissipation with elapsed time across the height of the soil column: **a** Case I; **b** Case II; **c** Case III and **d** Case IV

presence of dual seams not only affects the development process of PWP but also the dissipation process, as observed that even after $t + 160$ s, the PWP for lower layers is higher for Case III and IV as compared to Case II. The permeability of interlayer seams also affects the dissipation as in Case III, where the values of PWP are higher after a time interval as compared to Case IV. As the mass hits the Lucite tube setup, the soil layers settle due to rearrangement and densification of loose soil particles. This re-arrangement of soil matrix affects the ability of the soil layer to permit the migration of pore water to adjacent layers. Further, if layers of low permeability are present in a soil stratum, the ability of the soil layers to migrate the excess pore water in case of a sudden impact or shear force is significantly affected. Low permeability soil can cause delay in the dissipation of EPWP pressure developed during dynamic or impact loading in soil which disturbs the equilibrium condition in soil to reach initial static shear stress after liquefaction, thereby further creating unstable condition in the soil which may trigger flow slides due.

4 Settlement of Soil Layer and Development and Decay of Water Film

Images from specific time interval of the test after the impact were extracted from the video acquisition of the tests to measure the thickness of the water film and settlement of the soil layer. The measuring tape fixed on the equipment was used as a reference thickness calculator. An open-source application [9] was used to measure the thickness of water film at millimeter scale from the images. Figure 6a, b depict the settlement of soil layer as well as development and decay of water film (read as WF here after) with elapsed time in soil layers at the top and bottom seams, while Fig. 6c represents the settlement in and surface layer. In Fig. 6a, the settlement trend is identical for both Case III and IV, whereas the WF thickness for Case III increases and then decreases in a more gradual manner. In Case IV, the thickness of WF is higher for top seam, while the water film forms under the lower seam just for a short period of time. Furthermore, in Fig. 6b, for Case I, there is no water film formation

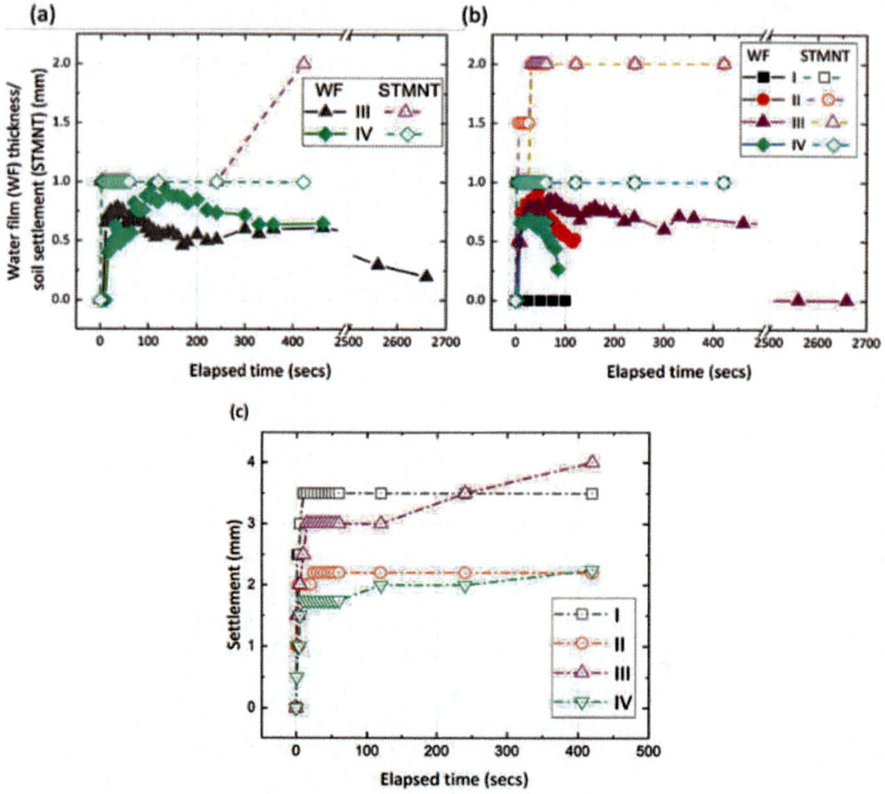

Fig. 6 Water film (WF) thickness and soil settlement (STMNT) (mm) against elapsed time (secs): **a** top seam; **b** bottom seam for cases (I–IV) and **c** at the surface of the soil column

due to the absence of any inhibiting layer. Also, the settlement in here is sudden, with the final settlement occurring within few seconds after the impact. For Case II, the WF thickness sees a gradual increase and then decrease, while the settlement also occurs gradually over a time period, similar to Case III. WF in Case III shows the most gradual development and decay among all. For Case IV, the WF develops and decays quickly as compared to other model cases due to presence of lower non-plastic fines interlayer. The settlement in this case imitates the trend of Case I. In Fig. 6c, the settlement at the surface of the soil column is shown for all the cases for the shock 2. The figure highlights that the soil settlement occurs sudden for Case I after the impact, while for Case II, the settlement is bit gradual and low like Case IV. For Case III, the settlement occurs gradually over the time as the EPWP dissipates with time and the WF decays.

The formations of water film for the Cases II, III and IV are shown in Fig. 7. Gradient color mode is applied to the images to make the water film more distinctive. Here, for Case II, the development of WF is simple due to presence of single clay seam. For Case III, the water film first develops under lower clay seam with higher thickness as compared to the upper seam, later when the WF below lower seam starts decaying, the thickness of WF below the upper seam increases. Furthermore, in Case IV where the lower seam is of non-plastic fines, the WF film decays very quickly, dissipating the PWP in the lower sand layer and migrating the pore water to the middle sand layer. The migrated pore water results in increase in EPWP in middle layer and thereby increase in WF thickness below the upper clay seam. In brief, the formation of water film inhibits the passage of pore water from lower layers to upper layers and leads to a slow dissipation of PWP resulting in delayed settlement of the overall soil layer.

5 Conclusions

1D soil model tests were conducted to simulate the formation of water film under low-permeable soil seams, which cause the delay in dissipation of pore water pressure, resulting in overall instability and delayed failure in soil layer. Following key inferences can be made from the test results:

1. The dissipation of excess pore water pressure developed due to impact loading is influenced by the vertical soil heterogeneities like presence of finer soil interlayer seams or layers with low permeability. The presence of multiple low interlayer seams can further complicate the dynamic response of saturated stratified soils by providing multiple inhibiting layers for dissipation of excess pore water pressure.
2. The permeability of the seams also affects the dissipation process as seen from the results of Case III and IV, and the non-plastic fines could dissipate excess pore water pressure in a lower time period as compared to pearl clay seam.

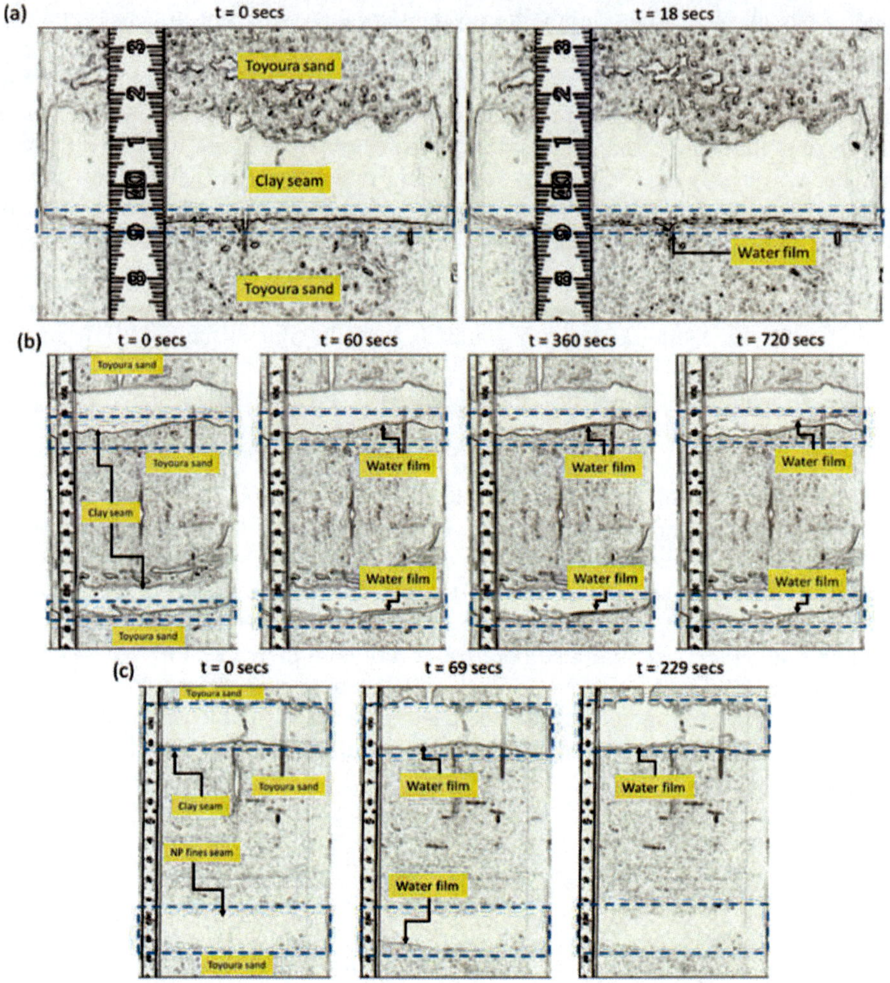

Fig. 7 Water film development and dissipation under the clay and non-plastic fines seam at different time interval (*t*): **a** Case II; **b** Case III and **c** Case III

3. The settlement of the soil layer is delayed by the development of water film in lower layers. Alternatively, the final settlement occurs after the water film has decayed fully.
4. Water film can form in soil even without liquefaction as observed for Case II.

This research provides a preliminary exposition on the influence of low permeability soil seams on the dissipation of excess pore water pressure and in turn dynamic stability of the stratified soils, further evaluation of the effect of overburden pressure on the development of water film is required.

Acknowledgements The authors would like to acknowledge the financial support for this research under JSPS (Japan Society for Promotion of Science) KAKENHI (Grant Number 20H02244).

References

1. Hazarika, H., Rohit, D., Pasha, S.M.K., Maeda, T., Masyhur, I., Arsyad, A., Nurdin, S.: Large distance flow-slide at Jono-Oge due to the 2018 Sulawesi Earthquake, Indonesia. Soils Found. (2020) (accepted)
2. Bjerrum, L.: Subaqueous Slope Failures in Norwegian Fjords, vol. 88, pp. 1–8. Norwegian Geotechnical Institute Publication (1971)
3. Hungr, O.: Flow slides and flows in granular soils. In: *Proceedings of International Workshop on Occurrence and Mechanisms of Flows in Natural Slopes and Earth Fills* (2003). https://ww.unina2.it/flows2003/flows2003/articoli/Hungr%20-Flows.pdf. Accessed 24 Sept 2020
4. Hutchinson, J.N.: A sliding-consolidation model for flow slides. Can. Geotech. J. **23**, 115–126 (1986)
5. Qiao, S.F., Clayton, C.R.I.: Flow slides run-out prediction using a sliding-consolidation model. Landslides **10**, 831–842 (2013)
6. Kokusho, T.: Water film in liquefied sand and its effect on lateral spread. J. Geotech. Geoenviron. Eng. **125**(10), 817–826 (1999)
7. Kokusho, T.: Mechanism for water film generation and lateral flow in liquefied sand. Soils Found. **40**(5), 99–111 (2000)
8. Sun, D.A., Matsuoka, H., Xu, Y.F.: Collapse behavior of compacted clays in suction-controlled triaxial tests. Geotech. Test. J. **27**(4), 362–370 (2004)
9. Neggers, J.: Jann5s/measuretool. https://github.com/Jann5s/measuretool. GitHub. Retrieved 28 Sept 2020

Risk Assessment of Landslide Disaster for Detecting Isolated Communities in Wakayama Prefecture After Nankai Trough Earthquake

Katsuki Hozumi, Hisakazu Sakai, and Yoshio Kajitani

Abstract Isolated communities after earthquakes face severe problems, such as rescue delays and a lack of daily necessities due to prolonged times required for these to reach isolated areas. It is crucial, especially in mountainous regions, to take countermeasures for slopes that would cause an isolation problem with high probability. Therefore, we proposed an estimation methodology for detecting the isolated communities in mountain region using the slope failure probabilities during earthquakes. In this process, the slope failure probability was calculated using disaster prevention Karte and stability survey tables recorded by every road administration in Japan as well as the peak ground velocity and slope angles. Subsequently, the interruption probability for each road was estimated according to the calculated slope failures. The validity of the proposed method was verified by comparing it with road regulations that came into effect after the 2004 Niigata-ken Chuetsu earthquake. The proposed method was applied to the emergency route in Wakayama Prefecture during a Nankai Trough mega scenario earthquake, and the road interruption probability was subsequently calculated. Furthermore, we showed a case study of detecting slopes that should be taken countermeasures with priority among the fragile slopes along two major routes and demonstrated the effectiveness of the proposed methodology.

Keywords Disruption estimation · Nankai trough earthquake · Isolation assessment · Slope failure · Road network

K. Hozumi (✉)
School of Engineering and Design, Hosei University, Shinjuku, Tokyo, Japan
e-mail: katsuki.hozumi.4c@stu.hosei.ac.jp

H. Sakai
Faculty of Engineering and Design, Hosei University, Shinjuku, Tokyo, Japan

Y. Kajitani
Faculty of Engineering and Design, Kagawa University, Takamatsu, Kagawa, Japan

H. Hazarika et al. (eds.), *Advances in Sustainable Construction and Resource Management*, Lecture Notes in Civil Engineering 144,
https://doi.org/10.1007/978-981-16-0077-7_60

1 Introduction

Isolating the damaged region after strong earthquakes causes severe problems, such as rescue delays and a lack of daily necessities, owing to the prolonged time required for accessing isolated areas. This is an ineluctable issue, particularly in mountainous areas. Earthquake disaster mitigation measures for slopes with a high risk of road blockage, inducing an isolation problem, have been promoted by administration officials. However, the budgets and labor power for slope failure countermeasures are limited, which makes the implementation of measures impossible for all slopes and becomes an obstacle to the implementation of isolation countermeasures. To determine the priority of mitigation measures, this study aimed to develop a methodology of detecting high-risk isolated regions by assessing the probabilities of road blockage and slope failure during an earthquake.

Although Ono and Hongo [1] evaluated the isolation risk using the slope and peak ground acceleration (PGA)-based slope failure probability evaluation formula proposed by Sakai et al. [2], they reported only the road blockage probability under a uniform seismic strength condition in the target region. Ono and Hibi [3] applied the methodology of Ono and Hongo to the 2004 Niigata-ken Chuetsu and the 2016 Kumamoto earthquakes and showed the utility of the method. Kondo et al. [4] evaluated the isolation risk in Wakayama Prefecture based on quite simple seismic damage charts for bridge damage, landslide and tsunami disaster. Many isolation risk studies regarding landslide have used approximate superficial slope features, such as the elevation and curvature. However, a slope stability is strongly influenced by not only slope configuration but also the ground condition of the slope, such as rock weathering, spring water and surface ground material. Therefore, slope ground condition should be introduced into more precise estimation of seismic slope stability. Konno et al. [5] conducted the regression analysis of seismic damage probability for artificial and natural slopes based on the road disaster prevention data and the peak ground velocity (PGV) as well as the slope topography data. The road prevention data enumerates the slope ground conditions strongly relating to slope stability as well as slope configuration. The authors also applied the regression results to the road network in Nagano Prefecture and estimated the isolation risk. In this study, we verified the availability of the regression method proposed by Konno et al. in case of another earthquake. We also conducted isolation assessment against a future scenario earthquake and attempted to detect fragile slopes to be effective countermeasures.

2 Methods

At first, we verified the availability applicability of the regression method proposed by Konno et al. by applying it to the slope failure during the 2004 Niigata-ken Chuetsu earthquake and the isolated region after the earthquake. Subsequently, we estimated isolation risk in Wakayama Prefecture against a Nankai Trough mega scenario earthquake and attempted to give priority to fragile slopes along two road routes in order not to yield an isolation region.

2.1 Evaluation of Slope Failure Probability

In this study, the slope failure probability assessment was performed based on the method of Konno et al.'s research [5]. Konno et al. conducted logistic regression analyses for the seismic slope failure probability of artificial and natural slopes based on the damaged and undamaged slope data of the 2014 earthquake in the northern Nagano Prefecture of Japan. The data include slope information listed in disaster prevention Karte and stability survey tables (hereinafter referred to as disaster prevention charts) as well as PGVs and slope angles. The accuracy of the disaster prevention chart items in the equations was improved by removing the explanatory variables with a confidence interval of 90% and p-values (significance probability) of 0.1 or higher. The formulas for artificial and natural slopes are expressed by Eqs. 1 and 2. Therefore, the real slope failure can be calculated using these equations.

Slope failure formula for slope:

$$\begin{aligned} y &= \frac{1}{1+\exp(-z_1)} \\ Z_1 &= 10.2 + 3.92\beta_3 - 1.99\beta_9 + 0.707\beta_{10} \\ &\quad - 1.26\beta_{12} + 0.061\beta_{13} + 0.233\beta_{14} \end{aligned} \tag{1}$$

Slope failure formula for natural slope:

$$\begin{aligned} y &= \frac{1}{1+\exp(-z_2)} \\ Z_2 &= -7.86 + 1.23\beta_3 + 1.27\beta_6 + 2.24\beta_7 + 3.10\beta_9 \\ &\quad - 1.11\beta_{10} - 0.933\beta_{12} + 0.024\beta_{13} + 0.157\beta_{14} \end{aligned} \tag{2}$$

y: Slope failure probability.

β_3: Clear knick line (convex); 0: No, 1: Yes.

β_6: Overhang; 0: No, 1: Yes.

β_7: Water collecting slope type; 0: No, 1: Yes.

β_9: Convex slope such as ridge tip; 0: No, 1: Yes.

β_{10}: Pumice stone or float stone (0: stable; 1: slightly unstable; 2: unstable).

β_{12}: Surface coverage (1: mainly woody; 2: complex (bare/grass/woody); 3: bare ground to grass).

β_{13}: Slope angles (°).

β_{14}: PGV (cm/s).

2.2 Disruption Evaluation of Road Networks

Road disruption occurs even if only one of the road slopes fails. Accordingly, road disruption does not occur only when all road slopes do not collapse. The disruption probability between two points of a road path can be calculated using Eq. 3 with the collapse probabilities of all slopes along the path.

In this paper, road disruption indicates a state wherein a road is impassable owing to earthquake-induced slope failure.

$$Pr = 1 - \prod_{i=1}^{n}(1 - y_i) \quad (3)$$

where P_r is the disruption probability for a road; y_i is the slope failure probability for each slope; n is the number of slopes between two points.

The validation was confirmed by estimating the recall, specificity and accuracy. Recall is the ratio of the correctly road disruptions predicted to all real road disruptions. Specificity is the ratio of the correctly road non-disruptions predicted to all real road non-disruptions. Accuracy is the ratio of the correctly predicted roads to the whole roads of subjects. With higher accuracy, the classification becomes better. Each probability can be calculated using Table 1 and Eqs. 4–6. In this study, the slope failure probability assessment predicted the probability. Next, we had to estimate the road disruption probability border that occurred over a certain probability and did not occur under a certain probability. We considered the following four conditions: (1) the sum of the disruption probabilities and no disruption probabilities are maximum; (2) minimum error discrimination (minimax method); (3) the probability of the drastic increase of road disruption frequency; (4) the average minimum error discrimination with the receiver operating characteristic (ROC) curve. In this study, the border was established based on the safest value of the four methods. The border of (1) is the probability when the total hit ratio calculated by Eqs. 4, 5 and 7 is maximum. The border of (2) is the total minimum calculated by Eqs. 4, 5 and 8 which is minimum. The border of (3) is the probability of the disruption frequency increasing with consideration to the estimated results. Moreover, (4) was estimated by the ROC curve, which is a graph made by the true positive fraction (TPF) and the false positive fraction (FPF). TPF is the fraction of positives predicted correctly and corresponding to the vertical axis in Fig. 1. FPF is the fraction of negatives misclassified as positives and corresponding to the horizontal axis in Fig. 1. The ROC curve is created by plotting these fractions. Additionally, area under the curve (AUC) stands for the area under the ROC curve and represents the hit rate. The closer AUC is to 1, the better the accuracy is. We calculated the balanced error rate (BER) that is simply the average number of road predictions in error divided by the total number of road predictions. The probability when BER is minimum is denoted as p-th, and p-th is considered as the border (Table 1).

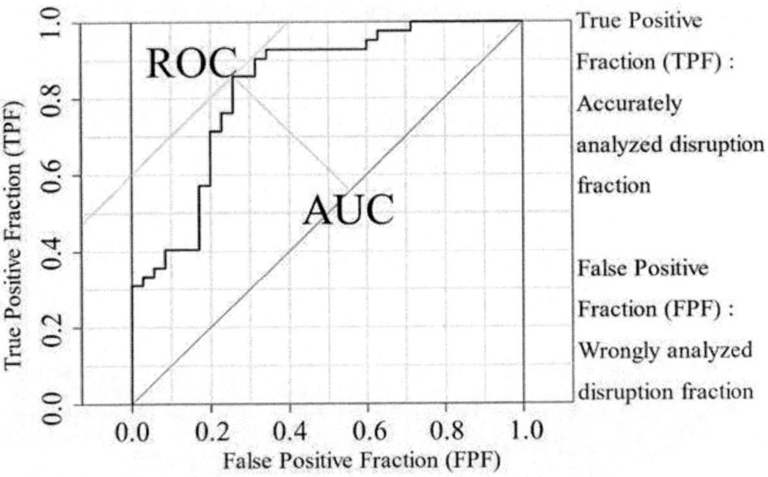

Fig. 1 Concept of the ROC curve

Table 1 Prediction results for estimating recall, specificity and accuracy

		Prediction	
		Disruption	Non-disruption
Real	Disruption	TP	FN
	Non-disruption	FP	TN

TP Disruptions were predicted, and disruptions were actually confirmed.
FP Disruptions were predicted, but disruptions were not actually confirmed.
FN Disruptions were not predicted, but disruptions were actually confirmed.
TN Disruptions were not predicted, and disruptions were not actually confirmed.

$$\text{Recall} = \frac{TP}{TP + FN} \quad (4)$$

$$\text{Specificity} = \frac{TN}{FP + TN} \quad (5)$$

$$\text{Accuracy} = \frac{TP + TN}{TP + FP + FN + TN} \quad (6)$$

$$\text{The total hit rate} = \text{Recall} + \text{Specificity} \quad (7)$$

$$\text{The total minimum} = \text{Disruption}|\text{Recall} - \text{Specificity}| \quad (8)$$

3 Validity of Risk Assessment Methodology

3.1 Slope Failure Probability Evaluation

The availability of seismic assessments for slope failures was verified based on the disaster data for the 2004 Niigata-ken Chuetsu earthquake. The slope failure probability predictions were performed for Ojiya City, Tokamachi City, Nagaoka City and a part of Uonuma City, where many landslides occurred during the earthquake. The slope failure prediction results are presented in Figs. 2 and 3. The value of the

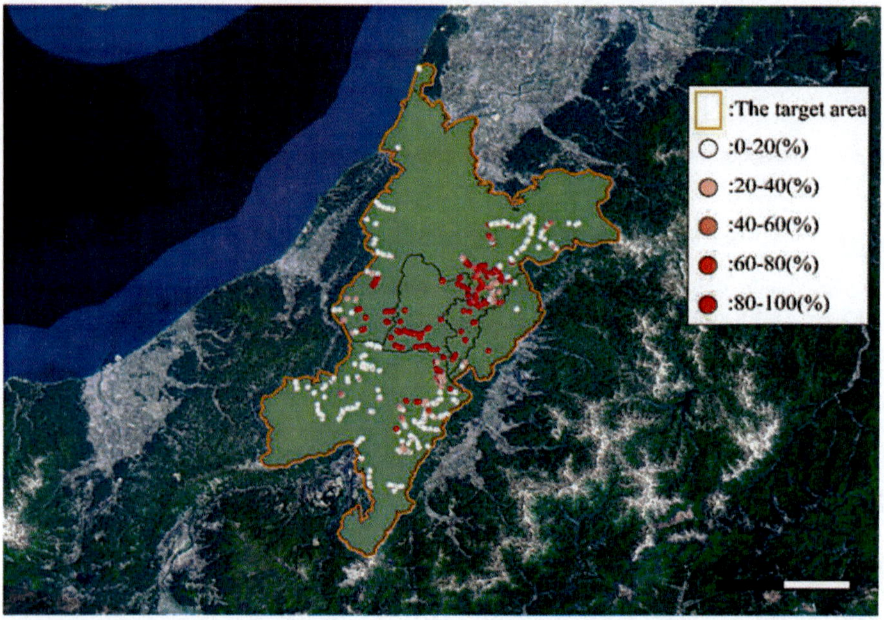

Fig. 2 Artificial slope estimation for Niigata Prefecture. This map is based on the Digital Map 200,000 (Map Image) published by the Geospatial Information Authority of Japan [6]

predicted slope failure was particularly high for Ojiya City and the former Yamakoshi area in Nagaoka City.

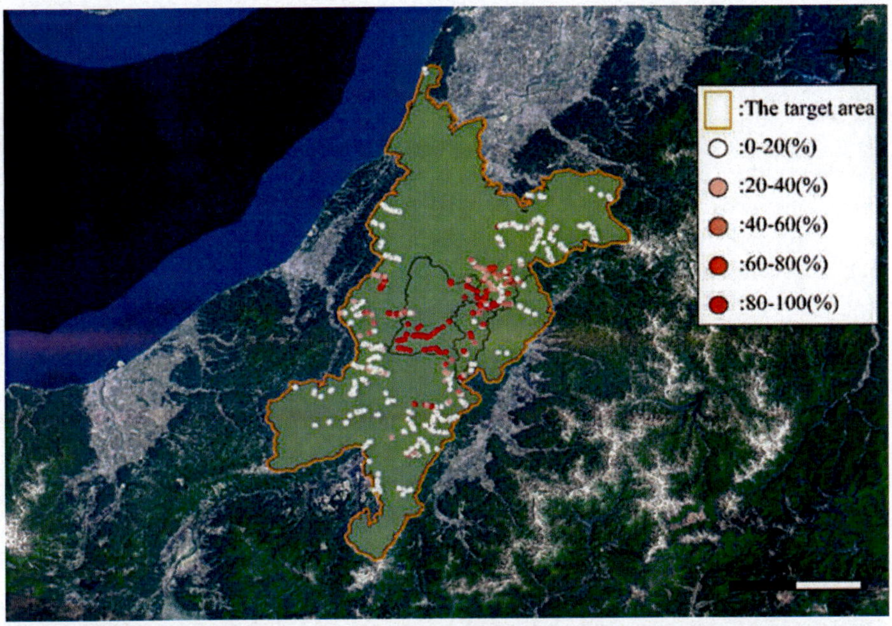

Fig. 3 Natural slope estimation for Niigata Prefecture. This map is based on the Digital Map 200,000 (Map Image) published by the Geospatial Information Authority of Japan [6]

3.2 Disruption Evaluation for Road Network and Fitting to the Previous Disaster Data

The effectiveness of the isolation evaluation method was investigated for Nagaoka City including the former Yamakoshi area, where isolation had occurred within a large area. First, the road disruption probability was estimated based on the slope failure probability. Comparing between the actual and estimated road disruption probability, we calculated the hit rate of the road blockage. When the road disruption probability border was set on the safest side of all predictions, the border value was calculated as 56.4%. The hit rate is shown in Table 2. The table indicates that this road disruption estimation process has a high accuracy of 79.2%.

Table 2 Hit rate of the estimated road disruption

Border	56.4%	Prediction			Results	
		Non-disruption	Disruption	Sum	Hit rate	
Actual	Non-disruption	25	10	35	Specificity	71.4%
	Disruption	6	36	42	Recall	85.7%
Results	Sum	31	46	77	Accuracy	**79.2%**

4 Detection of Isolated Region in Wakayama Prefecture

4.1 Prediction of Slope Failure Probability

The slope failure probability during the Nankai Trough mega scenario earthquake and the road network's sustainability after the earthquake were evaluated in Wakayama Prefecture, Japan. Herein, the detached territories were excluded. In this study, we set the border of disruption probability as 56.4% (accuracy: 79.2%). Figures 4 and 5

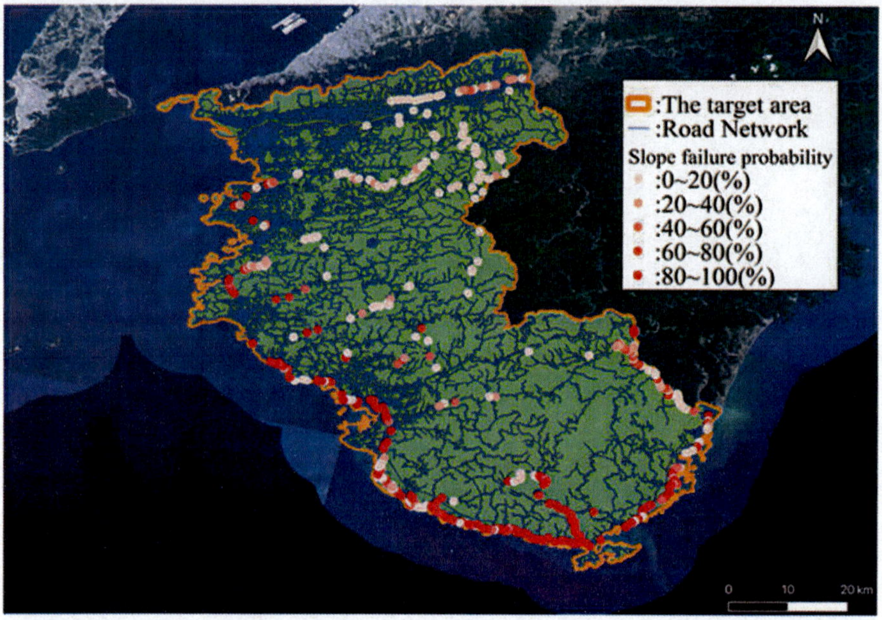

Fig. 4 Artificial slope estimation in Wakayama Prefecture. This map is based on the Digital Map 200,000 (Map Image) published by the Geospatial Information Authority of Japan [6]

show the slope failure estimation for artificial and natural slopes, respectively. The estimated slope failure at the inland of Wakayama Prefecture has a relatively low value. In contrast, the slopes along the coast area may fail with higher probabilities.

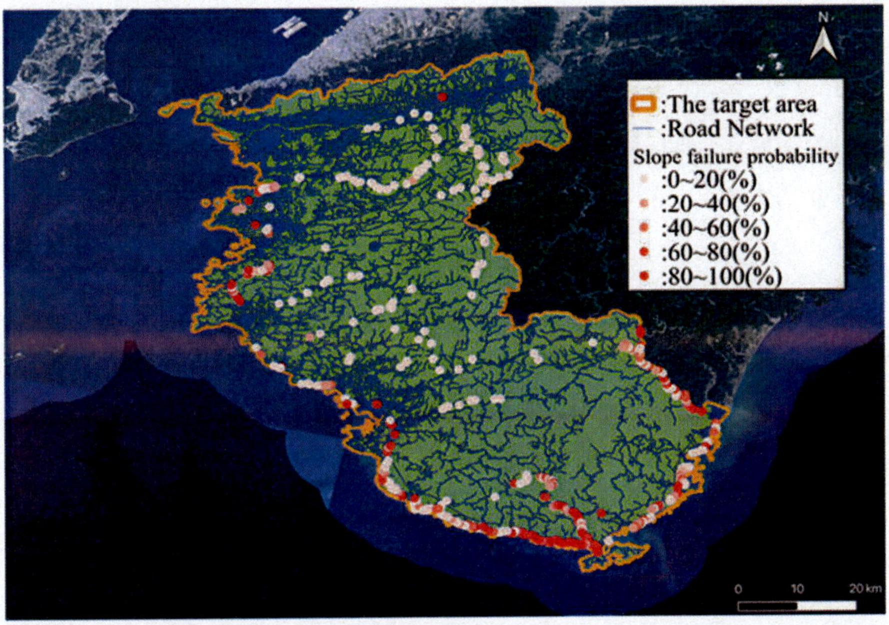

Fig. 5 Natural slope estimation in Wakayama Prefecture. This map is based on the Digital Map 200,000 (Map Image) published by the Geospatial Information Authority of Japan [6]

4.2 Disruption Evaluation of Road Network

Based on the above slope damage predictions, the road disruption probability was calculated. The estimated probability is shown in Fig. 6. The high low in the probability distribution is like the slope failure probability distribution.

4.3 Sample Priority Detection of Fragile Slopes to Countermeasure.

In order to verify the utility of the abovementioned methodology, we attempted to detect fragile slopes that have a high risk of a trigger of road blockage and should be countermeasures. The verification sample is for the two routes connecting the disaster rescue supply and the storage site, and the disaster prevention center of the Ryujin Administration Bureau in Tanabe City as shown in Fig. 7. The isolation risk results are shown in Table 3. From the table, the maximum slope failure probability was estimated as 65.1% in Route 1 and as 51.1% in Route 2. The road disruption probability was estimated as 94.5% in Route 1 and as 59.0% in Route 2. These probabilities clearly indicate that slope failure and road disruption are likely to occur. If the road disruption probabilities are greater than the border value (56.4%), road

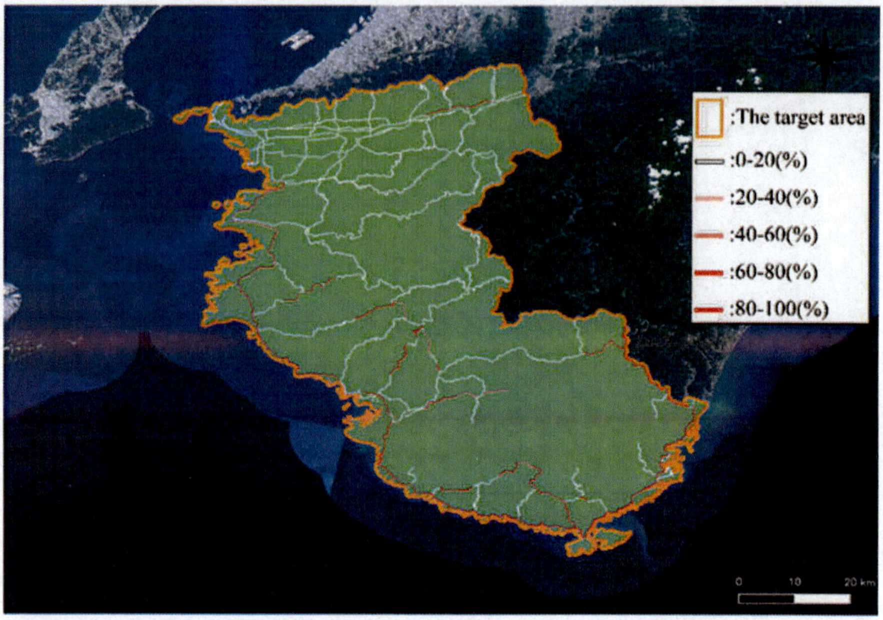

Fig. 6 Disruption probability in Wakayama Prefecture. This map is based on the Digital Map 200,000 (Map Image) published by the Geospatial Information Authority of Japan [6]

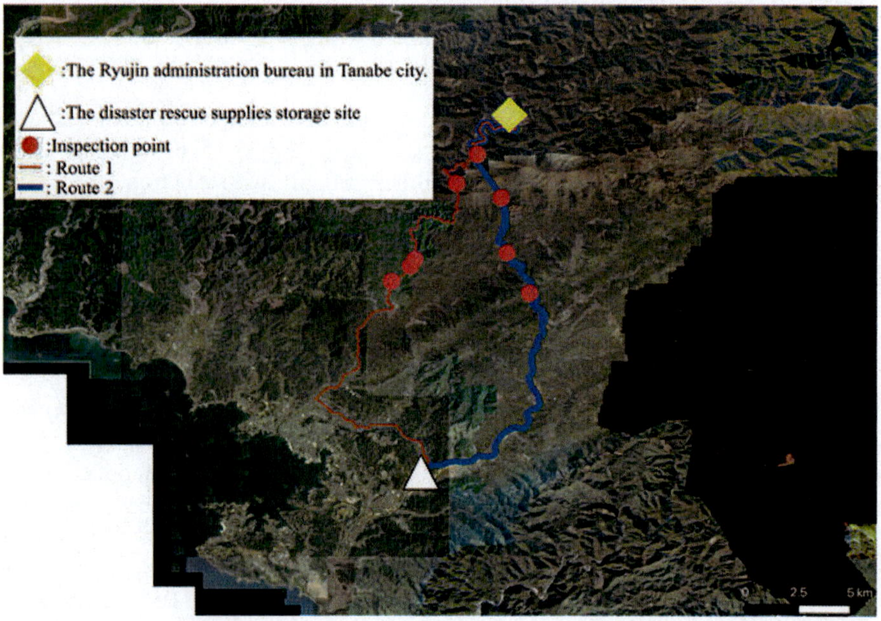

Fig. 7 Verification route. This map is based on the Digital Map 50,000 (Map Image) published by the Geospatial Information Authority of Japan [7]

Table 3 Isolation estimation results

	Before countermeasures		After countermeasures	
	Max slope failure probability (%)	Disruption probability (%)	Max slope failure probability (%)	Disruption probability (%)
Route 1	65.1	94.6	23.9	50.3
Route 2	51.1	59	12.6	29.6

disruption will likely occur with high probability. Subsequently, we verified whether countermeasures reduce the slope failure and the road disruption. "The countermeasures were conducted," which is defined as a "yes problem" in the disaster prevention charts and turns into a "no problem." In this case, the slope failure probability and road disruption probability were calculated. The maximum slope failure probability was reduced to 23.9% in Route 1 and reduced to 12.6% in Route 2. The road disruption probability also decreased to 50.3% in Route 1 and decreased to 29.6% in Route 2. These probabilities remain under the border we set. Therefore, preferential countermeasures should be implemented in Route 2.

5 Conclusions and Discussion

Isolated communities after earthquakes face severe problems, such as rescue delays and a lack of daily necessities due to prolonged times required for these to reach isolated areas. It is crucial, especially in mountainous regions, to take countermeasures for slopes that would cause an isolation problem with high probability. This research provided the estimating procedure of isolation risk caused by earthquake-induced slope failure. The procedure has consisted of the probabilities of slope failure during earthquake and road blockage after the earthquake. The slope failure probability was calculated by Konno's regression models for artificial and natural slopes. The regression model was based on the data listed in disaster prevention charts as well as slope configuration characteristics and PGV values.

At first, we applied the methodology to the 2004 Niigata-ken Chuetsu earthquake and verified the validation of the process. As the results, when the border of the road disruption rate was set as 56.4%, the predicted road disruption was very high accuracy of 79.2%. We confirmed the availability of the method for predicting road disruption after an earthquake.

Secondly, we estimated the slope failure probability during the Nankai Trough mega earthquake and road disruption probability after the earthquake in Wakayama Prefecture. The estimated slope failure probabilities can be utilized for predicting the isolation risk of some communities after the scenario earthquake.

Finally, we attempted to detect slopes that had a high risk of route disruption and should be countermeasures with priority. The result suggested the utility of the methodology of estimating road disruption to maintain the road networks efficiently after an earthquake.

Acknowledgements The authors would like to thank the following persons for their invaluable contributions. Y. Ono, Tottori University, provided his expertise on OSMnx and Python programming. K. Ootomo, Central Research Institute of Electric Power Industry, instructed a method of writing paper. R. Kawai, Regional Development Bureau of the Ministry of Land and Infrastructure, adviced on using disaster prevention charts.

Y. Suzuki and K. Watase, Wakayama Prefecture; Y. Nakajima and Y. Iguchi, Transport and tourism Kinki Regional Development, Bureau of the Ministry of Land and Infrastructure provided disaster prevention charts.

R. Konno and K. Nishida, graduate students of Hosei University, provided their helpful advice on research methods.

The Nankai Trough mega earthquake model study group in the cabinet office provided seismic waveform data. This study was supported by JSPS KAKENHI Grant Number JP 17K01343.

References

1. Ono, Y., Hongo, S.: An attempt to quantify the isolation possibility of mountainous areas based on slope failure probability during earthquake. In: The Editorial Committee on Journal of Japan Society of Civil Engineers, Japan Society of Civil Engineers 2018Annual Meeting, Proceedings of the Annual Meeting of Japan Society of Civil Engineers (JSCE) 2018, ser. A1, vol. 74, no.4, pp. I_313–I_319, Japan Society of Civil Engineers (JSCE), Tokyo, Japan (2018)
2. Sakai, H., Okumura, M., Shioaki, T., Kagawa, T., Hasegawa, K., Sawada, S., Tatano, H.: A fundamental study on simplified method for calculating earthquake induced slope collapse ratio. In: The Editorial Committee on Journal of Japan Society of Civil Engineers, Japan Society of Civil Engineers 2013Annual Meeting, Proceedings of the Annual Meeting of Japan Society of Civil Engineers (JSCE) 2013, ser. A1, vol. 69, no. 4, pp. I_142–I_147, Japan Society of Civil Engineers (JSCE), Tokyo, Japan (2013)
3. Ono, Y., Hibi, K.: Evaluation of isolation risk in mountainous region considering road blockages caused by earthquake-induced slope failures: Application to 2004 Niigata-ken Chuetsu and 2016 Kumamoto earthquakes. In: Journal of JAEE, vol.19, no.6, pp. 6_232–6_243, Tokyo, Japan (2019). https://doi.org/10.5610/jaee.19.6_232
4. Kondo, S., Terumoto, K., Ota, K., Kataie, Y., Takao, H., Kawata, Y.: A study of prediction isolated district map focused on road blockage. In: Bimonthly journal of Institute of Industrial Science, the University of Tokyo, vol. 62, no. 4, pp. 417–419, Tokyo, Japan (2010). https://doi.org/10.11188/seisankenkyu.62.417
5. Konno, R., Sakai, H., Kajitani, Y.: Reliability assessment of slopes and prediction of isolated areas in the northern area of Nagano prefecture using disaster prevention charts. In: The Editorial Committee on Journal of Japan Society of Civil Engineers, Japan Society of Civil Engineers 2017Annual Meeting, Proceedings of the Annual Meeting of Japan Society of Civil Engineers (JSCE) 2017, vol. 72, p. I_106, Japan Society of Civil Engineers (JSCE), Tokyo, Japan (2017)
6. Geospatial Information Authority of Japan: Digital Map 200000 (Map Image). https://maps.gsi.go.jp/#9/33.985502/135.585022/&base=ort&ls=ort&disp=1&vs=c1j0h0k0l0u0t0z0r0s0m0f0&d=m (1974–2008). Last accessed 2020/7/16
7. Geospatial Information Authority of Japan: Digital Map 50000 (Map Image). https://maps.gsi.go.jp/#9/33.985502/135.585022/&base=ort&ls=ort&disp=1&vs=c1j0h0k0l0u0t0z0r0s0m0f0&d=m (1974–2008). Last accessed 2020/7/16

Innovative and Sustainable Techniques in Geoengineering

DCM Columns and PVDs Hybrid Ground Improvement for Embankment Construction

Junfeng Ni and Jinchun Chai

Abstract A case history of embankment construction on soft clay deposit improved by a combination of deep cement mixing (DCM) columns and prefabricated vertical drain (PVDs) installation method was simulated and investigated by finite element analysis (FEA). The function of the DCM columns is to control the settlement and lateral displacement under the toe of the embankment, and the function of the PVDs is to accelerate consolidation rate for a zone under the central of the embankment. For this case, only the surface settlements were measured, and the simulations agree with the measured data well, while the results of FEA give relative large lateral displacements under the toe of the embankment with a value of about 370 mm. Further numerical investigation was carried out assuming part of the preloading load was vacuum pressure, with which to result in the similar final surface settlement. The numerical results indicate that with the vacuum pressure, the lateral displacement under the toe of the embankment as well as the maximum compressive stress in the columns can be reduced significantly. These results indicate that combination of DCM columns and PVDs installation with vacuum pressure as part of preloading pressure can be an effective and economic ground improvement method for embankment construction on soft clay deposit.

Keywords Embankment · Deep cement mixing · PVD · FEA

1 Introduction

Deep cement mixing (DCM) and preloading with installation of prefabricated vertical drain (PVD) are two commonly used soft ground improvement methods. DCM normally forms soil–cement columns in the ground. The columns can increase the

J. Ni (✉) · J. Chai
Department of Civil Engineering and Architecture, Saga University, 1 Honjo, Saga 840-8502, Japan
e-mail: nijunfeng007@163.com

J. Chai
e-mail: chai@cc.saga-u.ac.jp

H. Hazarika et al. (eds.), *Advances in Sustainable Construction and Resource Management*, Lecture Notes in Civil Engineering 144,
https://doi.org/10.1007/978-981-16-0077-7_61

strength and stiffness of the deposit in a relatively short time [1–3]. On a DCM column improved deposit, the ground deformation can be substantially reduced/controlled. However, DCM is a relative expensive ground improvement method comparing with preloading method. While preloading with PVD installation needs a longer time and will induce large ground deformation, it may affect the surrounding structures and sometimes it is not suitable in urban areas [4–6].

Combination of DCM and preloading with PVDs can be an effective and economic soft ground improvement technique for embankment construction. The idea is improving the deposit under the toes and part of an embankment slopes by DCM columns, which can limit/control the deformation of the adjacent areas, and improving the middle zone under the embankment by preloading with PVD installation to reduce the total cost of the ground improvement. The load for preloading can be embankment fill and/or vacuum pressure. Kajio reported a case history of a road embankment construction on a soft clay deposit using combined DCM column and preloading with PVD installation ground improvement technique [7]. To design this kind of hybrid ground improvement, there are still issues need to be investigated, such as the width to be improved by DCM, the area improvement ratio by DCM columns, compressive as well as tensile strength requirement for the columns.

In this study, firstly the field case reported by Kajio was simulated by finite element method (FEM), and the simulated surface settlements are compared with field measured data. Then, further numerical investigations were carried out assuming using both embankment fill and vacuum pressure as preloading pressures. Finally, discussions are made on the efficiency of the technique.

2 A Case History

2.1 Subsoil Profile

This case history was reported by Kajio [7]. The project was a section of Takeo bypass, Saga, Japan. At the site, the thickness of a soft clay deposit is around 10 m. From the ground surface, there is a sandy soil layer of 0.26 m thick underlain by a thick clay soil layer with about 10.17 m in thickness. Below the clay layer, there is stiff sand layer.

Some physical and mechanical properties of subsoils are summarized in **0** (data from Kajio) [7]. The measured underground water level was 1.1 m from the ground surface at the time of the site investigation (Fig. 1).

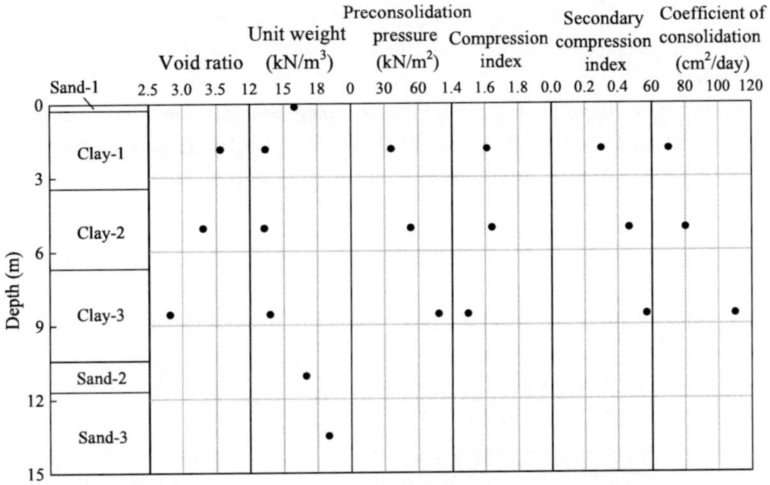

Fig. 1 Soil profiles at the project site

2.2 DCM Column Construction and PVD Installation

0 shows the cross-sectional and plan views of DCM columns and PVDs. Since L-type reinforced concrete retaining walls were constructed on both sides of the embankment, the top of the columns was 1.4 m below the ground surface with a length of 9.03 m. The diameter of columns was 1.6 m. The columns were constructed in a way that each time two connected columns were constructed simultaneously. The spacing between other adjacent two columns was 2.0 m (center to center), which results in an area improvement ratio, $\alpha = 55.7\%$ ($\alpha = A_c/A$, where A_c is the total cross-sectional area of the columns and A is the total cross-sectional area of the zone improved by the columns). The unconfined compressive strength (q_u) of the columns was 769 kPa.

The PVDs were installed from the ground surface and fully penetrated the soft clay soil layer with a length of 10.43 m. The spacing between adjacent PVDs was 1.0 m with a square pattern.

2.3 Embankment Construction and Filed Monitoring

The design height of the road was 3.0 m above the ground surface. To reduce residual settlement of the road, 2.0-m-thick additional fill was used during preloading period (with a total elapsed time of 238 days). Before pavement construction, the fill above the designed height of the road was removed. The cross section of the embankment is shown in **0**. The L-type retaining walls had a height of 4.4 m and base width of 3.0 m. The fill material for the embankment is decomposed granite and after compaction, the unit weight was about 20 kN/m^3. The top and bottom widths of embankment were 18.2 m and 24.2 m, respectively. The slope of the embankment above the retaining wall was 1:1.5 (V:H).

There is no detailed construction record at the section considered in this study, and it was assumed that 5.0-m-thick fill material was constructed within 50 days (0.1 m/day). The settlement was monitored at S1 (shown in **0**) which was 0.5 m above the original ground surface. The monitoring works begun after 0.5 m fill was constructed (Fig. 2).

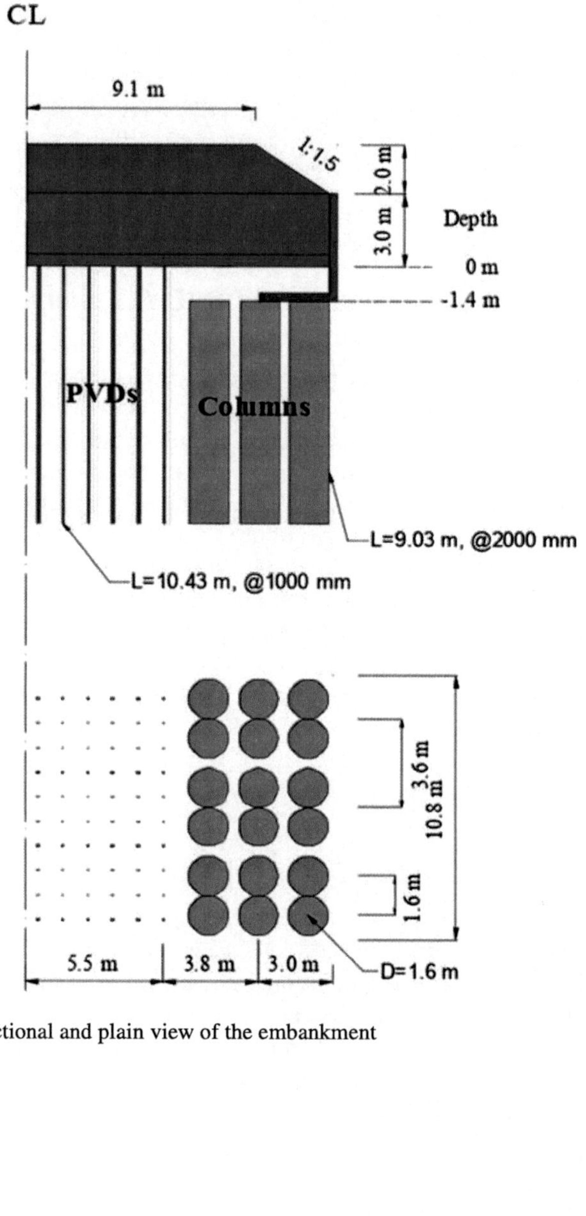

Fig. 2 Cross-sectional and plain view of the embankment

2.4 Finite Simulation

(1) Modeling the effect of PVDs.

Chai et al. proposed a simple method for modeling PVD-improved soil zones[8]. In the method, an equivalent vertical hydraulic conductivity (k_{ve}) of PVD-improved zone is calculated by the following equation:

$$k_{ve} = \left(1 + \frac{2.5l^2 k_h}{\mu D_e^2 k_v}\right) k_v \quad (1)$$

where D_e is the diameter of a PVD unit cell; k_v and k_h are vertical and horizontal hydraulic conductivities of the natural soil, respectively; and l is the drainage length of PVDs. μ is a parameter considering the effects of spacing, smear and the well resistance of PVDs, which can be expressed as follows [9]:

$$\mu = \ln \frac{n}{s} + \frac{k_h}{k_s} \ln(s) - \frac{3}{4} + \pi \frac{2l^2 k_h}{3q_w} \quad (2)$$

where $n = D_e/d_w$ (d_w is the equivalent diameter of a PVD), $s = d_s/d_w$ (d_s is the diameter of the smear zone), k_s is the hydraulic conductivity of the smear zone, and q_w is the discharge capacity of a PVD.

For this case history, $D_e = 1.13$ m (1.13 times spacing of square pattern), and $D_w = 0.06$ m. q_w of 100 m^3/year and d_s of 0.1 m were assumed. Besides, the value of k_s is assumed as $k_h/5$ for PVDs-improved zone. The calculated values of k_{ve} for corresponding soil layers are listed in Table 1.

(2) Mesh for finite element analysis.

Plain strain finite element analyses (FEA) were conducted to simulate this case history. The finite element mesh adopted is shown in **0**. Due to the symmetry, only half of the embankment was simulated with the model cross-sectional area of 30 m in depth and 60 m in the horizontal direction. Fifteen nodal triangular elements were employed in the model. For deformation boundary condition, the bottom was fully fixed in all directions but the top boundary was free. For both the left and right boundaries, the horizontal displacement was fixed. For the drainage boundary conditions, the left and right boundaries were impermeable, while the top and bottom were permeable.

In the plain strain analysis, the columns were modeled under equal EI condition (E is Young's modulus and I is moment of inertia of cross section). The equivalent plain strain column has a thickness of 1.11 m. At column and soils interface, joint elements were used.

Considering large deformation phenomenon, updating nodal coordinates option was adopted at the end of each phase. For establishing initial stress state in the

Table 1 Initial modeling parameters for numerical analysis

Depth m	Layer	E (kPa)	V	C′ (kPa)	¥ (°)	Yt (kN/m^3)	e	k_V (k_{ve}) $\times 10^{-4}$ m/day	kh $\times 10^{-4}$ m/day	λ*	K*	OCR
0–0.26	Sand-1	5000	0.10	2	35	16.00	1	100	100	–	–	–
0.26–3.43	Clay-1	–	0.15	5	30	13.40	3.566	0.56 (11.15)	0.84	0.153	0.028	2.57
3.43–6.68	Clay-2	–	0.15	5	30	13.30	3.310	0.99 (19.56)	1.49	0.165	0.047	1.80
6.68–10.43	Clay-3	–	0.15	5	30	13.80	2.820	2.69 (52.56)	4.04	0.17	0.065	1.83
10.43–11.68	Sand-2	8000	0.10	10	35	17.00	0.8	1000	1000	–	–	–
11.68–30	Sand-3	7.6×104	0.10	10	35	19.00	0.8	2500	2500	–	–	–
Embankment		1000	0.45	10	35	20.00	0.8	2500	2500	–	–	–
Columns		7.69×10^4	0.10	385	0	Same with corresponding soil				–	–	–
L-type Retaining wall		2×10^6	0.10	10000	0	Same with corresponding soil				–	–	–

deposit, the groundwater level of 1.0 m below the ground surface, at reset earth pressure coefficient, $K_0 = 0.5$ were assumed.

(3) Model parameters.

The clay soil layers were modeled by soft soil model (SSM) [10], while all sand layers were analyzed with Mohr–Coulomb model (MC). The adopted model parameters are given in Table 1. The void ratio (e), unit weight (y_t) and OCR are from Kajio [7]. The vertical hydraulic conductivity (k_v) was estimated from the reported value of coefficient of consolidation. In SSM, the value of λ^* is calculated by $\lambda^* = \lambda/(1 + e_0)$ and κ^* equals $\kappa/(1 + e_0)$, where λ and κ are slopes of virgin compression line and reloading/unloading line in $e - \ln p\prime$ plot (p' is consolidation pressure), respectively. Poisson's ratio (ν) was assumed to be 0.1 for sand and CDM columns and 0.15 for clays. The effective cohesion (c') was estimated based on the soil properties, and the effective internal friction angle (ϕ') was assumed as 30° for clays and 35° for sands. Young's modulus (E) for sand was assumed based on experience. The horizontal hydraulic conductivity (k_h) was assumed to be equal to vertical hydraulic conductivity (k_v) for sands but 1.5 times for clays [1]. The values of hydraulic conductivity in Table 1 are initial ones. During consolidation process, hydraulic conductivity varied with void ratio according to Taylor's equation [11].

$$k = k_0 \times 10^{(e-e_0)/C_k} \tag{3}$$

where e_0 is initial void ratio, k_0 is the initial hydraulic conductivity, and C_k is a constant. The adopted C_k value was 0.5 e_0 [12].

The void ratio, unit weight and hydraulic conductivity of the columns were assumed to be the same as the surrounding soils. Young's modulus (E) was evaluated to be 100 times q_u [13], and the tensile strength σ_t was set as $q_u/10$. The effective cohesion (c') was estimated to be 1/2 of q_u, by assuming the effective stress internal friction angle (ϕ') of zero. For embankment, E was assumed to be 1000 kPa and ν was assumed to be 0.45 [14]. The values of c' and ϕ' were assumed to be 10 kPa and 35°, respectively. For retaining wall, E was 2×10^6 kPa, and ν is 0.1. The value of c' is assumed as 10,000 kPa and ϕ' of zero.

Fully coupled consolidation analyses were carried out using Plaxis 2D V2018. Two cases were analyzed. Case 1 simulated the field case, and Case 2 is an assumed case to investigate the effectiveness of using vacuum pressure as a preloading pressure.

2.5 *Results and Discussions*

Only the surface settlements at S1 (see Fig. 3) were measured, and the measured and simulated (Case-1) results are compared in Fig. 4. It can be seen that the simulations have a good agreement with the measured data.

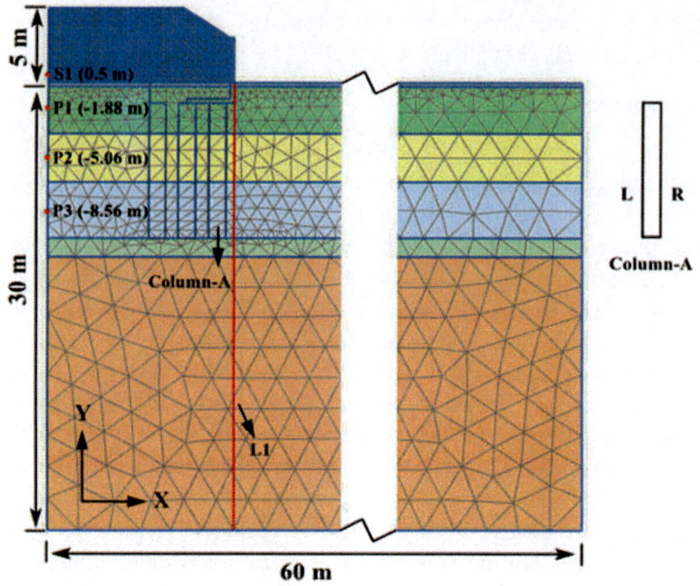

Fig. 3 Meshed model and interested points

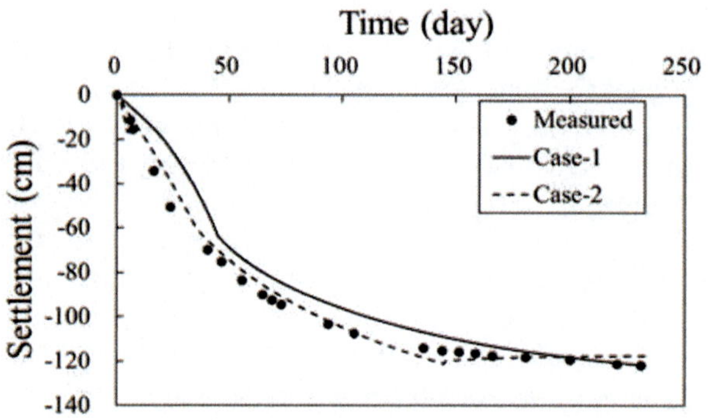

Fig. 4 Variation of settlement at S1 with elapsed time

Further, the surface settlement profile, excess pore water pressures at some selected points, lateral deformation of the columns and effective vertical stresses in the column were analyzed from the results of numerical analysis (Case-1).

The surface settlement profiles are shown in Fig. 5. It can be seen that larger settlement occurred on the PVD-improved zone, which is about three times of that on the DCM columns improved zone. It is also noted that heaving occurred in a zone from the toe of the retaining wall (12 m from the center) to about 20 m from the center. It is considered that the heaving is mainly caused by the outward deformation of the retaining, which will be presented later.

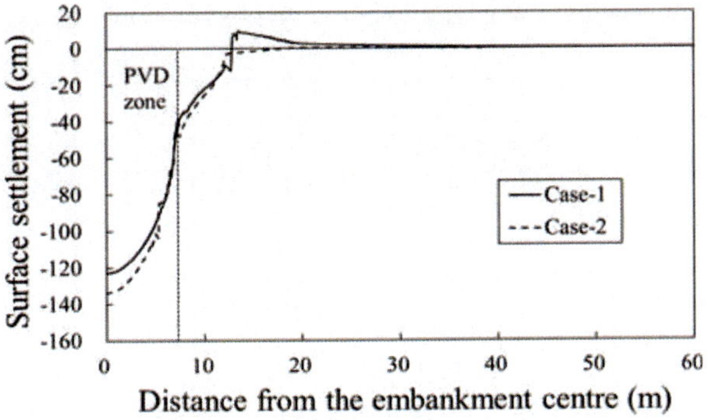

Fig. 5 Surface settlement with distance at 238th day

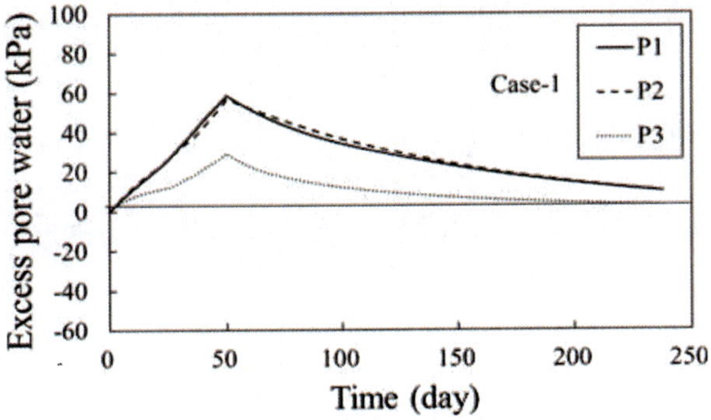

Fig. 6 Variation of excess pore water pressures with time

Figure 6 depicts the variation of excess pore water pressures, *u*, with time at P1, P2, P3 (see 0). The simulated maximum *u* was about 60 kPa, and most *u* were dissipated at the time of removing part of the fill.

Figure 7 presents the simulated lateral deformation profile at L1 (see 0) at 238 days elapsed time. The simulated maximum lateral deformation was 370 mm (Case-1), which occurred at the ground surface.

Figure 8 shows the effective vertical stresses at left (L) and right (R) sides of column A (see Fig. 3 for its location) under the embankment toe. There was no tensile stress occurred, and the maximum compressive stress was below the corresponding strength.

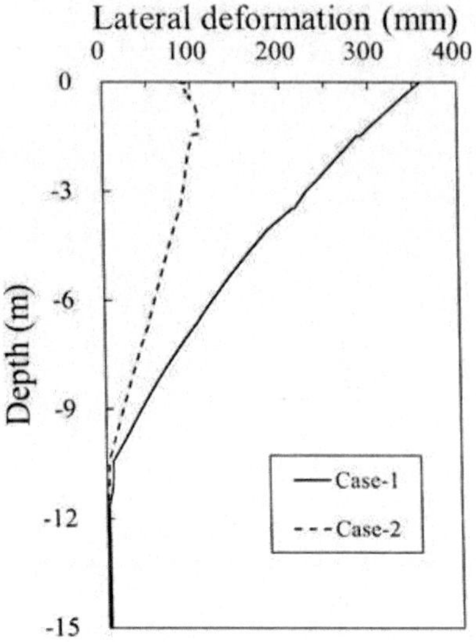

Fig. 7 Final lateral deformation at toe of embankment

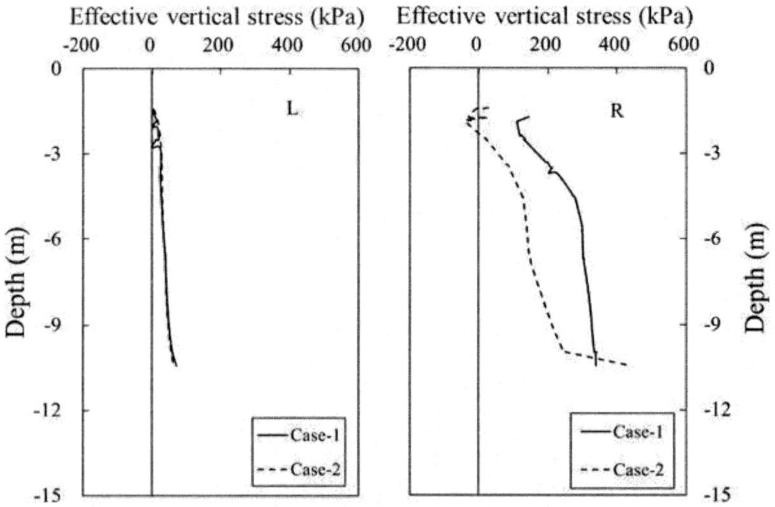

Fig. 8 Effective vertical stress with depth for column A

3 Further Numerical Investigation

3.1 *Assuming Using Vacuum Consolidation*

As shown in Fig. 7, the simulated lateral displacement is relatively large (Case-1). Since vacuum pressure is an isotropic consolidation pressure, which can reduce the outward lateral displacement of the treated area, further numerical investigation was carried out assuming a combined preloading pressure, a surcharge load plus a vacuum pressure. For the field case, about 1.0-m-thick extra fill was removed before the pavement construction. Therefore, it was assumed that instead of 2.0-m-thick extra fill, only apply 1.0-m-thick extra fill, and the embankment construction time was shortened to 40 days (filling rate of 0.1 m/day). The assumed vacuum pressure was 80 kPa, and duration of vacuum pressure application was 150 days (to result in the similar final settlement as the actual field case). Another modification of the test condition is to shorten the PVD installation depth to 9.43 m (leave 1.0-m-thick clay without PVD). The deposit was a double drainage one, and if PVDs fully penetrated the deposit, vacuum pressure will be lost into the bottom sand layer [15, 16].

3.2 *Results and Discussions*

As shown in Fig. 4, the simulated final settlement was about the same as that of case 1. The simulated surface settlement profile under the embankment is about the same as that of case 1, but no heaving away from the toe (Fig. 5). With vacuum pressure as the part of preloading pressure, the lateral displacement under the toe of the retaining wall reduced significantly (Fig. 7), and the simulated maximum vertical compressive stress in column A reduced also (Fig. 8).

The simulated excess pore water pressures variations at P1, P2 and P3 of case 2 are shown in Fig. 9. The simulated maximum absolute vacuum pressure at P1 was about 50 kPa at the time of stopping vacuum pump.

The above results indicate that using vacuum pressure as part of preloading pressure can be an effective and economic ground improvement method.

4 Conclusions

A case history of embankment construction on soft clay deposit improved by a combination of deep cement mixing (DCM) columns, and prefabricated vertical drain (PVDs) installation method was simulated and investigated by finite element analysis (FEA). DCM columns were constructed under both sides of the embankment slopes, and PVDs were installed in the center zone. At the site, the thickness of the clay deposit was about 10 m, and the embankment fill thickness was 5.0 m. Based

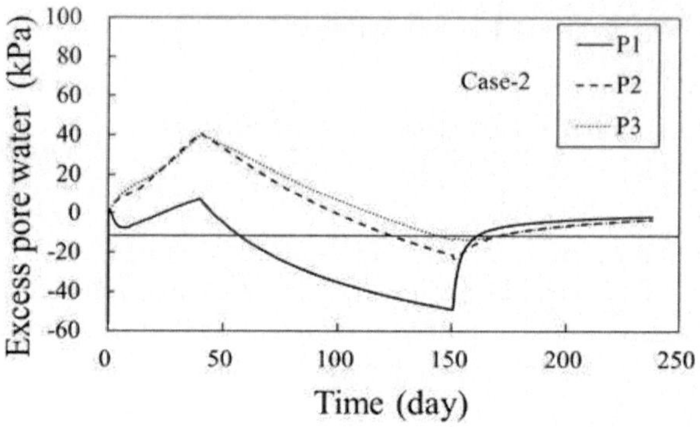

Fig. 9 Variation of excess pore water pressure with elapsed time in case 2

on the field measured settlements and the results of FEA, the following conclusions can be drawn.

(1) For this case, only the surface settlements were measured and the simulations agree with the measured data well. The results of FEA indicate that the stresses in the columns are lower than the corresponding strengths, and the design was safe, while the lateral displacement under the toe of the embankment was relatively large with a value of about 370 mm.
(2) Assuming the part of the preloading load was vacuum pressure, with which to result in the similar final surface settlement, the numerical results indicate that the lateral displacement under the toe of the embankment as well as the maximum compressive stress in the columns can be reduced significantly.
(3) It is suggested that combination of DCM columns and PVDs with vacuum pressure as part of preloading pressure can be an effective and economic ground improvement method for embankment construction on soft clay deposit.

References

1. Chai, J., Igaya, Y., Hino, T., Carter, J.: Finite element simulation of an embankment on soft clay—case study. Comput. Geotech. **48**, 117–126 (2013)
2. Chai, J.C., Miura, N., Koga, H.: Lateral displacement of ground caused by soil-cement column installation. J. Geotech. Geoenviron. Eng. **131**, 623–632 (2005)
3. Yapage, N.N.S., Liyanapathirana, D.S., Kelly, R.B., Poulos, H.G., Leo, C.J.: Numerical modeling of an embankment over soft ground improved with deep cement mixed columns: case history. J. Geotech. Geoenviron. Eng. **140** (2014)
4. Shen, S.-L., Chai, J.-C., Hong, Z.-S., Cai, F.-X.: Analysis of field performance of embankments on soft clay deposit with and without PVD-improvement. Geotext. Geomembranes **23**, 463–485 (2005)
5. Sinha, A.K., Havanagi, V.G., Mathur, S.: An approach to shorten the construction period of high embankment on soft soil improved with PVD. Geotext. Geomembranes **27**, 488–492 (2009)

6. Sinha, A.K., Havanagi, V.G., Mathur, S.: Inflection point method for predicting settlement of PVD improved soft clay under embankments. Geotext. Geomembranes **25**, 336–345 (2007)
7. Kajio, T.: Consideration on settlement control of surcharge embankment in road maintenance. In: 73rd Annu. Sci. Lect. Meet. JSCE, Hokkaido, Japan (2018)
8. Chai, J.C., Shen, S.L., Miura, N., Bergado, D.T.: Simple method of modeling PVD-improved subsoil. J. Geotech. Geoenviron. Eng. **127**, 965–972 (2001)
9. Hansbo, S.: Consolidation of Fine-grained Soils by Prefabricated Drains. Proc. Int. Conf. Soil Mech. Found. Eng. **3**, 677–682 (1981)
10. Vermeer, P.A., Neher, H.P.: A soft soil model that accounts for creep. In: Beyond 2000 Comput. Geotech. Ten Years PLAXIS Int. Proc. Int. Symp. Amsterdam, March 1999, pp. 249–261 (1999)
11. Taylor, D.W.: Fundamentals of soil mechanics. **66**, (LWW, 1948)
12. Tavenas, F., Jean, P., Leblond, P., Leroueil, S.: The permeability of natural soft clays. Part II: Permeability characteristics. Can. Geotech. J. **20**, 645–660 (1983)
13. Porbaha, A., Shibuya, S., Kishida, T.: State of the art in deep mixing technology. Part III: geomaterial characterization. Proc. Inst. Civ. Eng. Improv. **4**, 91–110 (2000)
14. Chai, J., Chun, Shrestha, S., Hino, T., Uchikoshi, T.: Predicting bending failure of CDM columns under embankment loading. Comput. Geotech. **91**, 169–178 (2017)
15. Ong, C.Y., Chai, J.C., Hino, T.: Degree of consolidation of clayey deposit with partially penetrating vertical drains. Geotext. Geomembranes **34**, 19–27 (2012)
16. Chai, J.-C., Carter, J.P., Hayashi, S.: Vacuum consolidation and its combination with embankment loading. Can. Geotech. J. **43**, 985–996 (2006)

Evaluation of Hybrid Pile Supported System for Protecting Road Embankment Under Seismic Loading

Chengjiong Qin, Hemanta Hazarika, Siavash Manafi Khajeh Pasha, Hideo Furuichi, Yoshifumi Kochi, Daisuke Matsumoto, Takashi Fujishiro, Shinichiro Ishibashi, Naoto Watanabe, and Shigeo Yamamoto

Abstract A 1 g shaking table testing program was conducted to assess the dynamic performance of an innovative cost-effective countermeasure technique for new and existing road embankment on a liquefiable foundation subjected to cyclic sinusoidal loading. The model was shaken with a scaled earthquake motion having peak base acceleration of 0.2 g and 0.3 g in the first and second events which simulate the foreshock and mainshock of the 2016 Kumamoto earthquakes. In a series of four different shaking table model tests, the dynamic response of the embankment was investigated first without, and then considering three countermeasure techniques: pile supported embankment, connected piles-geogrid system supported embankment, and hybrid pile system supported embankment. The mechanism of remediation and performance of each proposed countermeasure techniques are discussed

C. J. Qin (✉) · H. Hazarika
Kyushu University, Fukuoka 819-0395, Japan
e-mail: qinchengjiong@yahoo.co.jp

S. M. K. Pasha
IMAGEi Consultant, Tokyo 102-0083, Japan

H. Furuichi
Giken Ltd., Tokyo 135-0063, Japan

Y. Kochi
K's Lab Inc., Yamaguchi 753-0212, Japan

D. Matsumoto
Japan Foundation Eng., Co., Ltd., Fukuoka 815-0075, Japan

T. Fujishiro
Geo-disaster Prevention Institute, Kitakyushu 802-0003, Japan

S. Ishibashi
Nihon Chiken Co., Ltd., Fukuoka 812-0894, Japan

N. Watanabe
KFC Ltd., Nagoya 461-0048, Japan

S. Yamamoto
Chuo Kaihatsu Corporation, Fukuoka 814-0103, Japan

H. Hazarika et al. (eds.), *Advances in Sustainable Construction and Resource Management*, Lecture Notes in Civil Engineering 144,
https://doi.org/10.1007/978-981-16-0077-7_62

based on dynamic response of embankment. The effectiveness of each technique on preventing the development of pore water pressure and limiting the deformation of embankment was studied in this research. The hybrid pile supported system was found to be effective in maintaining the overall stability of embankment during earthquake loadings.

Keywords Shaking table test · Pile supported embankment · Hybrid pile retrofit technique · Vertical settlement · Liquefaction-induced lateral displacement

1 Introduction and Background of Study

A sequence of two strike-slip earthquakes of 6.5 magnitude foreshock and 7.0 magnitude main shock struck Kumamoto city of Japan on April 14, 2016 and April 16, 2016. Many highway embankments were damaged in Kyushu Expressway at Kumamoto Prefecture, Japan, due to the 2016 Kumamoto earthquakes [1]. The stratified cohesionless soil layers of different permeabilities and shear strength, such as sand-silt and volcanic soils systems, were evaluated thoroughly using field measurements and laboratory experiments and found to be common cause of liquefaction and formation of sand boils during the 2016 Kumamoto earthquakes [2]. Many road embankments in this region are founded on liquefiable loose soils which raises the need for development of innovative applicable and cost-effective deformation countermeasure technique for new and existing highway embankments. A more recent example of liquefaction-induced ground deformation is provided by [3] where severe damage was sustained by a highway embankment in Mashiki machi due to the 2016 Kumamoto earthquake.

The failure of the Mashiki machi embankment (Fig. 1) during the 2016 Kumamoto earthquake is an example of embankment failure caused by loss of foundation strength due to liquefaction. Furthermore, highway embankments are very important and strategic infrastructures which play a crucial role in resilience to natural disasters and providing road transport services in disaster relief after an extreme event [4].

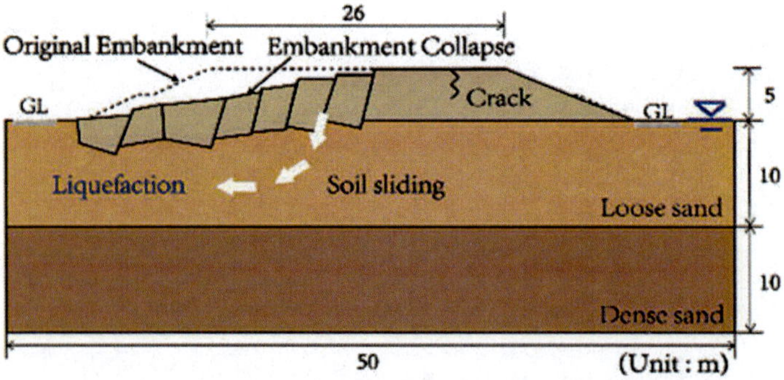

Fig. 1 Failure of road embankment during 2016 Kumamoto earthquake [4]

In order to prevent liquefaction-induced damage to similar highway embankments due to the future earthquakes, it is necessary to adopt appropriate deformation countermeasures to protect new and existing highway embankments and remediate the associate ground settlements. Therefore, developing deformation countermeasure techniques for highway embankments will be discussed in this research.

2 Previous Studies

Previous numerical and experimental studies (e.g., 1 g shaking table tests and centrifuge tests) have shown that liquefaction-induced failure of road embankments occurred mainly due to the liquefaction of foundation, excessive settlements, lateral spreading, and instabilities of slopes [5–7]. In most of reported cases of foundation failure, excess pore water pressure and associated shear strains were developed in liquified zones outside or below the embankment and extends inside the geo-structures. The large settlements developed shear failure zones and lateral spreading will eventually lead to localized failure or total collapse of embankments [8]. Among the common preventive measures, piles foundation is often being used to prevent plastic deformation in the ground, such as landslides, slope failure, or lateral flow. With the development of pile foundation method, the successful use of this method has been described by several investigators [9–12]. Previous numerical studies have suggested the 2D theoretical analysis of the lateral force acting on stabilizing piles considering plastic deformation in surrounding ground of piles [10, 11]. Poulos proposed a simple approach to design the stabilizing piles which can be used for reinforcing highway embankments. He reported that for safe design of piles (i) a relatively large diameter piles are required to resist shear and moment; (ii) the center-to-center spacings required typically range between 2 and 4 diameters. However, it was commonly acknowledged that 2D solutions were conservative to analysis the lateral force of pile in comparison with 3D solutions. However, a new approach for the displacement of the reinforced slope reinforced by a row of piles was developed deeply in 3D condition by [12]. Furthermore, restraining pile head from rotation was found to increase the maximum nominal shear strength and acting as monolithic structure leading to uniform deformation of piles [13]. Previous studies on beam-pile and anchor-pile supported embankments using 1-g shaking table tests have shown that the anchor-pile systems can yield better performance in supporting the embankment slopes than that of beam-piles [14].

3 New Deformation Countermeasure Technique

Despite many studies, some features of dynamic soil pressure are not yet fully understood, and in some cases, the engineering practice for seismic design in reinforcing the slope still largely relies on experience. However, it is necessary to develop stabilizing

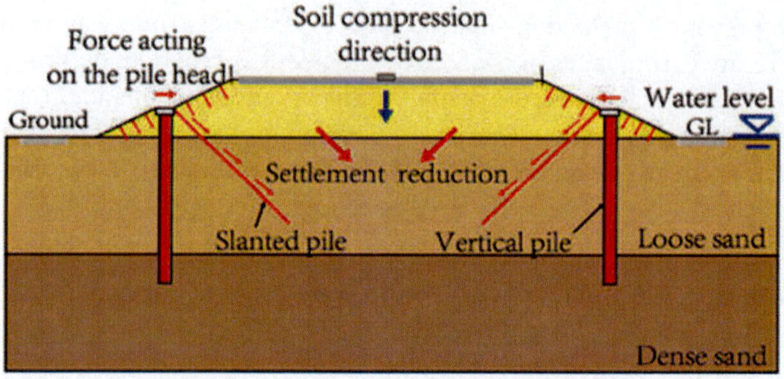

Fig. 2 Schematic view of model highway embankment (reinforced case)

structure for new and existing highway embankments to reduce large deformation during strong earthquake and prevent liquefaction-induced damage. A new deformation countermeasure technique of hybrid pile supported system is introduced as shown in Fig. 2. The vertical piles increase the stability of embankment slopes and foundation. In order to enhance the transmission of induced dynamic forces to vertical piles during shaking, they were connected to each other laterally using geogrid. A slanted pile was screwed to each vertical pile, to transfer load to the surrounding soil through the friction between slanted pile and soil. Effectiveness of the proposed technique against earthquake was evaluated through a series of 1 g shaking table tests.

4 Test Conditions

The model of the foundation and embankment was constructed in a rigid box, made of acrylic material, and confined by steel frames at the sides. The highway embankment in Mashiki machi [3], which was damaged during Kumamoto earthquake, was considered as prototype in the experiments. The prototype to model ratio was 50, and similitude relationship suggested by [15] was used to determine various parameters for the experiment, which is given in Table 1. Vertical piles were made of aluminum,

Table 1 Similitude for 1 g shaking table tests

Parameter	Scaling factor	Parameter	Scaling factor
Length	N	Displacement	$N^{1.5}$
Density	1	Permeability	$N^{0.75}$
Stress	N	Axial stiffness	$N^{3.5}$
Pore water pressure	N	Bending stiffness	$N^{4.5}$
Acceleration	1	Friction	1

and its dimension was determined according to scaling law and the axial stiffness and bending stiffness of the prototype model. Slanted piles were made of steel. The physical properties of the foundation materials are given in Table 2. Toyoura sand was used as foundation soil in these tests. A dense layer of sand (Dr = 90%) representing non-liquefiable ground was constructed using both dry deposition and tamping techniques. The upper liquefiable layer (Dr = 60%) was constructed only using dry deposition technique.

The frequency of input excitations for foreshocks and mainshocks was determined according to scaling laws and recorded acceleration data in KMMH16 station. Features of input excitations are given in Table 3. A mainshock that followed a foreshock was imparted to the model. A minimum interval of 15 min was considered between foreshock and mainshock tests to ensure full dissipation of developed excess pore water pressure during shaking (Fig. 3).

1 g shaking table tests were conducted for four different configurations of highway embankment and foundation as shown in Fig. 4: (Case 1) Unreinforced foundation of embankment, (Case 2) only two rows of piles to support embankment, (Case 3) two rows of combined piles and geogrid to support embankment, and (Case 4) hybrid pile system to support embankment. In the reinforced foundations, the center-to-center spacing is similar, as a 10/3 times pile diameter [9]. Instrumentations for the

Table 2 The properties of reinforcing materials

	Parameter	Unit	Prototype	Model (Propose)	Model (Practice)
Steel pile	Diameter	mm	1500	30	30
	Thickness	mm	6	1.2	3
	Length	mm	15,000	300	300
	Young's modulus	MN/m^2	206,000	206,000	73,000
	Bending stiffness	MN/m^2	3.2×10^3	3.67×10^{-3}	3.43×10^{-3}
Slanted Pile	Diameter	mm	90	1.8	3
	Length	mm	12,500	250	250
	Young's modulus	MN/m^2	206,000	206,000	206,000
	Bending stiffness	MN/m^2	1.33	1.1×10^{-7}	8.2×10^{-7}
Geo-grid	Tensile strength	kN/m	≥ 120	3	3.9
	Mass per unit area	g/m^2	720	720	700
	Aperture size	mm	50×50	1×1	2.5×2.5

Table 3 Characteristics of input motions

Input excitations	Properties		
	Amplitude of acceleration (Gal)	Frequency (Hz)	Duration (s)
Foreshock	200	10	10
Mainshock	300	10	10

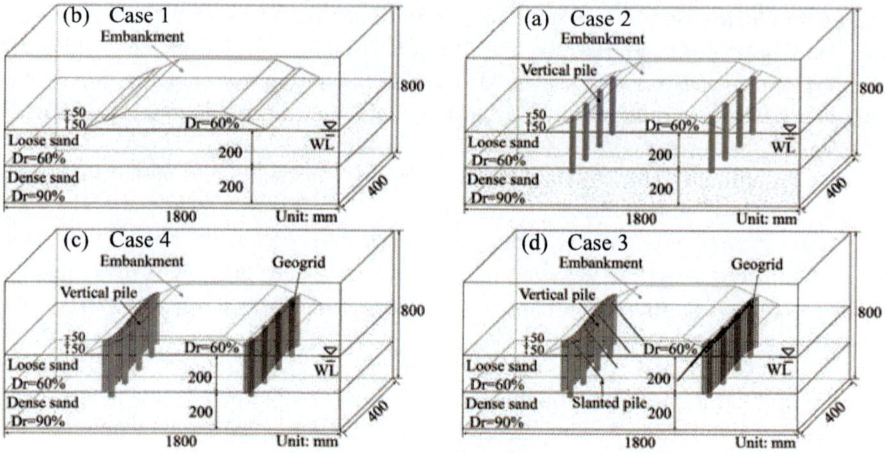

Fig. 3 Layouts of models: **a** Case 1—Unreinforced foundation of embankment, **b** Case 2—only two rows of steel piles reinforcement, **c** Case 3—two rows of combined geogrid and steel piles reinforcement, and **d** Case 4—hybrid vertical piles and slanted piles system reinforcement

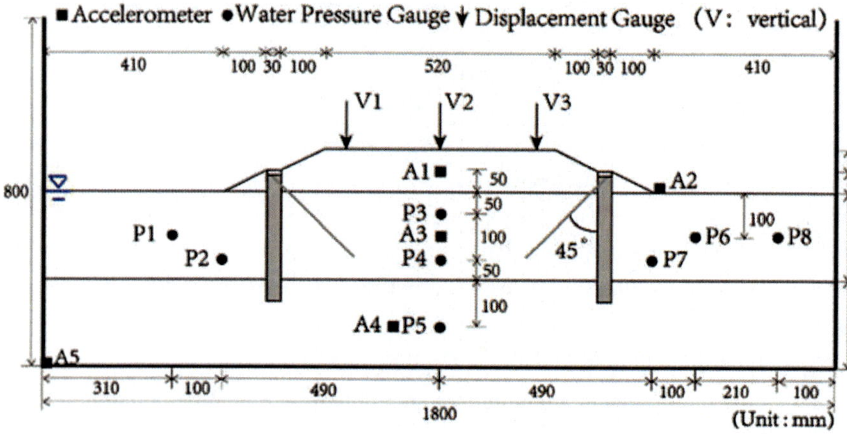

Fig. 4 Instrumentations for the shaking table tests

shaking table is shown in Fig. 5. Three displacement gauges were used to monitor settlements of embankment. Five accelerometers were installed at different locations within the soil profile to record acceleration responses. Eight pore water pressure gauges were used to monitor excess pore water pressure at different locations of saturated foundation.

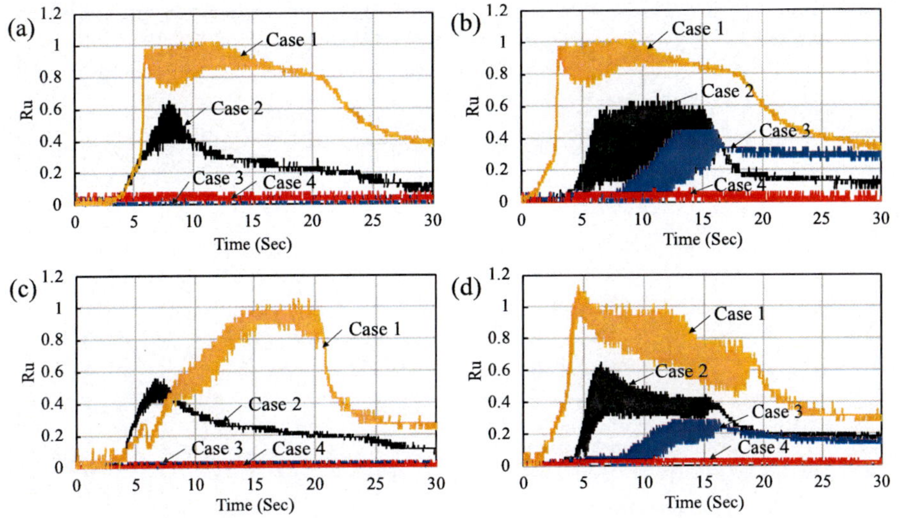

Fig. 5 Time histories of excess pore water pressure ratio ($R_u = \Delta u/\sigma'_3$): **a** Ru measured at P3 during foreshock, **b** Ru measured at P3 during mainshock, **c** Ru measured at P4 during foreshock, and **d** Ru measured at P4 during mainshock

5 Results and Discussion

Time histories of excess pore water pressure ratio ($R_u = \Delta u/\sigma'_v$) measured at P3 and P4 below the embankment during foreshock and mainshock are shown in Fig. 5. Vertical dashed lines define the boundaries where the excess pore water pressure starts to generate and fully dissipates. P3, P4, and P5 were installed at 50 mm, 150 mm, and 300 mm depth below the center of embankment, respectively. In the unreinforced foundation (Fig. 5a–d), instruments installed at the center of loose sand layer (P3 and P4) measured larger excess pore water pressure ratio values ($R_u = \Delta u/\sigma'_3$) in comparison with those of reinforced one during foreshock and mainshock. R_u value of 1.0 is achieved under the embankment of unreinforced case which is an indication of a state of full liquefaction. In contrast to the unreinforced embankment, the pile supported embankment can reduce the development of R_u. In the case of connected piles-geogrid supported embankment (Case 3) and hybrid pile system supported embankment (Case 4), R_u recorded a value of 0.1 (P3, P4) at the foundation which is significiantly lower than the value of 1.0 recorded in the Case 1 during foreshock. However, R_u measured at P3 and P4 started to reach the peak at $t = 7$ s and $t = 13$ s in Case 2 and Case 3, respectively; during mainshock, represented Case 3 can delay the development of R_u. There was no significant peak of R_u on P3 and P4 at Case 4 during foreshock and mainshock (Fig. 5).

Time histories of vertical displacement measured at V1, V2, and V3 are plotted in Fig. 6. In the cases which full state of liquefaction (R_u) is observed (Fig. 5), vertical displacement increased rapidly until the ending of foreshock and mainshock

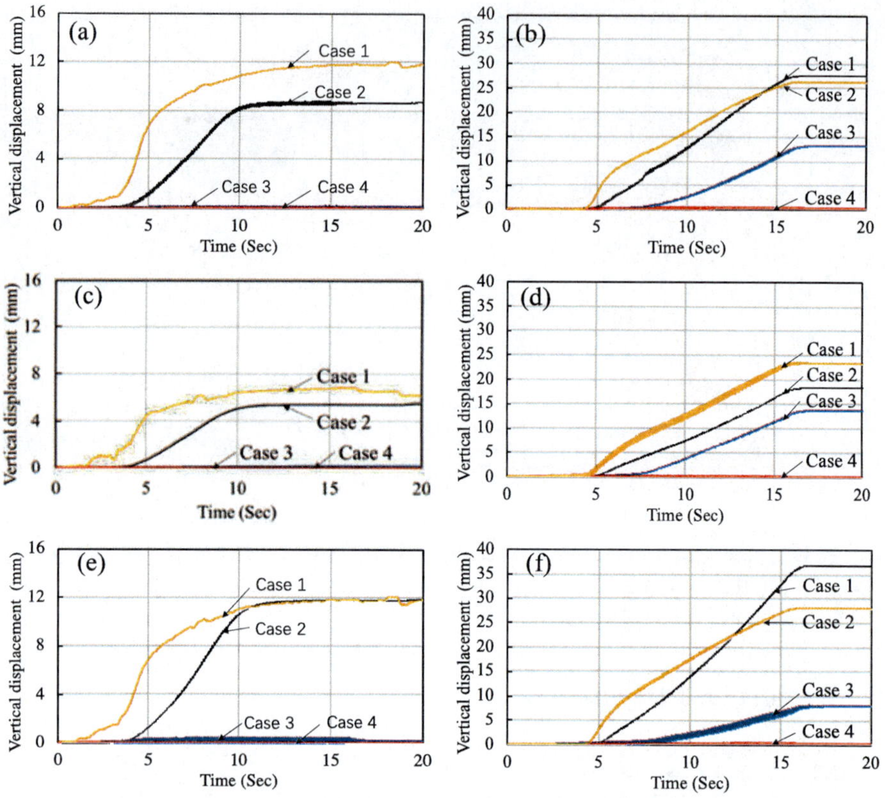

Fig. 6 Time history of vertical displacement: **a** settlement measured at V1 during foreshock **b** settlement measured at V1 during mainshock, **c** settlement measured at V2 during foreshock, **d** settlement measured at V2 during mainshock, **e** settlement measured at V3 during foreshock, and **f** settlement measured at V3 during mainshock

(Fig. 6). It was found that connected piles-geogrid system can only delay the accumulation of settlement during foreshock. No significant excess pore water pressure ratio and vertical displacement was recorded in Case 4 during foreshock and mainshock. Maximum vertical displacement measured at V2 installed at the center of embankment during foreshock and mainshock is shown in Fig. 7. The vertical displacement in Case 3 increases significantly during mainshock than foreshock. Connected piles-geogrid system can limit liquefaction-induced damage to earth-structure. Therefore, embankment can be still operated with a minimum maintenance following an earthquake as strong as foreshock in this study. In the case of reinforced embankment, vertical displacement decreased by 99% which indicates effectiveness of the hybrid pile supported system in stabilizing embankment slopes. It is worth mentioning that the slanted pile increase bearing capacity by increasing the friction between soil and pile during seismic loadings.

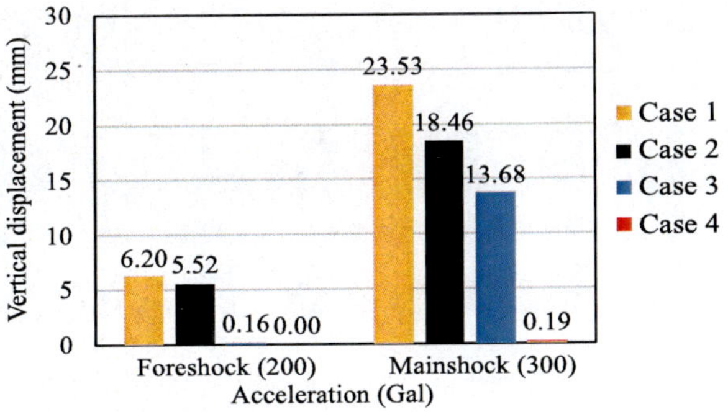

Fig. 7 Peak vertical displacement measured at V2 during foreshock and mainshock

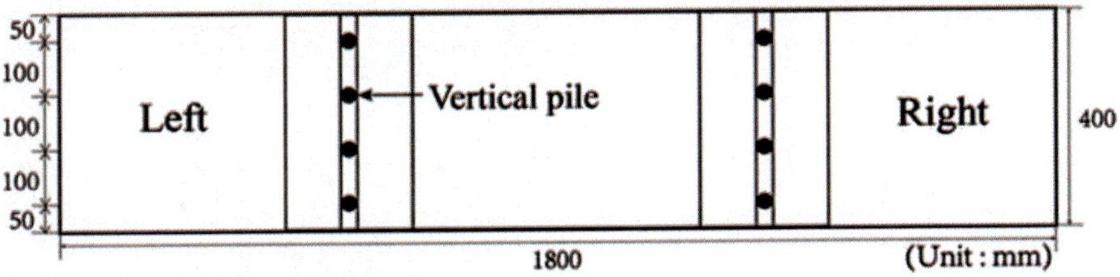

Fig. 8 Plan view of 1G shaking table test

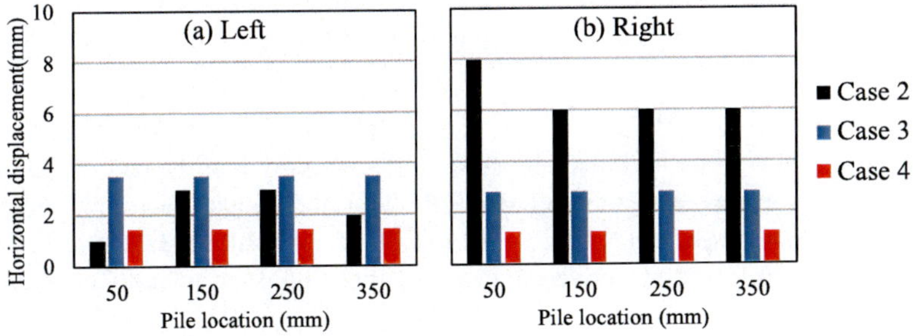

Fig. 9 Residual deformation of pile heads on left **a** and right **b** during foreshock

Residual deformation of the tips of the piles were measured at the end of each loading for all cases and average values are plotted in Figs. 9 and 10. As shown in Fig. 9, the average residual deformation of pile tips at the end of foreshock is less than 10 mm in Case 2, Case 3, and Case 4. Case 3 has similar residual deformation in comparison with Case 2 during foreshock. However, the residual deformation in Case 2 and Case 3 showed more significantly than hybrid pile supported system during mainshock (Fig. 10). Maximum values of residual deformation of tip of the piles are 46, 26, and 5 mm for Case 2, Case 3, and Case 4, respectively. Slanted piles

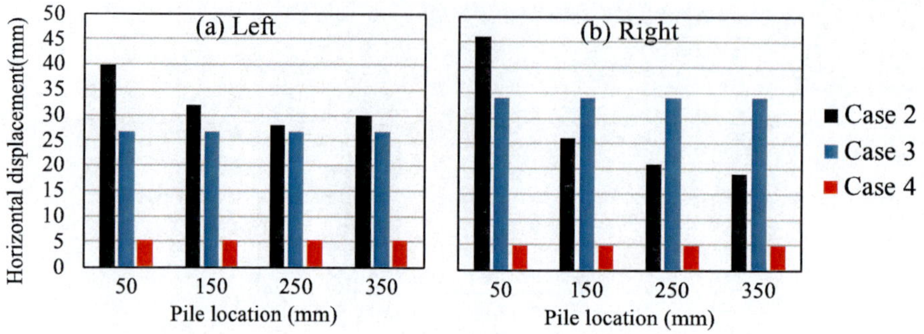

Fig. 10 Residual deformation of pile heads on left **a** and right **b** during mainshock

can increase the shear strength and restrict the horizontal deformation of pile heads effectively in Case 4 during foreshock and mainshock.

6 Conclusions

Performance of a highway embankment on a liquefiable foundation layer subjected to cyclic sinusoidal loading has been evaluated in this study. In addition, a new reinforcing technique for foundation of highway embankment was developed that can mitigate damage of embankment due to the failure of its foundation during the earthquake. The following conclusions could be drawn based on the result of shaking table tests:

1. In the case of embankment without any reinforcing technique, the excess pore water pressure generated in the loose saturated soil layer of the foundation during the earthquake reduces the soil strength. This led to excessive settlement and lateral displacement and finally instability of embankment. Therefore, for a sustainable and cost-effective solution against liquefaction-induced damage, a new countermeasure should be developed.
2. Considering the region around Kumamoto province with a high risk of liquefaction, several road embankments and foundations were broken by the liquefaction-induced damage. New countermeasure technique includes the connection by geogrid and slanted pile strengthened each vertical pile. The connection by geogrid can increase the maximum shear strength to limit the development of excess pore water pressure. Slanted piles can increase the bearing capacity to maintain the stability of road foundation and embankment. In addition, no significant R_u were evaluated to the effect of hybrid pile supported system during foreshock and mainshock.
3. Lateral deformation of embankment foundation is one of main reason of embankment failure during the earthquake. Time histories of vertical displacement and residual deformation of tips of the piles show that no significant deformation

during foreshock and mainshock. Hybrid pile supported system can prevent such lateral deformation, resulting in more stable embankment.

In order to find an effective reinforcing technique for highway embankment, more detail studies are needed. Research is also ongoing to evaluate the earthquake resistant characteristics of the proposed technique through the effective stress analysis using FEM.

Acknowledgements Financial aid for this study provided by NEXCO and affiliated organizations is gratefully acknowledged. The authors also express their sincere gratitude to Mr. Yuichi Yahiro, Technical Assistant at Geo-disaster Prevention Engineering Laboratory of Kyushu University for his technical support during the shaking table tests.

References

1. Hazarika, H., Kokusho, T., Kayen, R.E., Dashti, S., Fukuoka, H., Ishizawa, T., Kochi, Y., Matsumoto, D., Furuichi, H., Hirose, T., Fujishiro, T., Okamoto, K., Tajiri, M., Fukuda, M.: Geotechnical damage due to the 2016 Kumamoto earthquake and future challenges. Lowland Technol. Int. **19**(3), 189–204 (2017)
2. Ogo, K., Hazarika, H., Kokusho, T., Matsumoto, D., Ishibashi, S., Sumartini, W.O.: Analysis of liquefaction of volcanic soil during 2016 Kumamoto earthquake based on boring data. Lowland Technol. Int. **19**(4), 245–250 (2018)
3. West Nippon Expressway Company Limited: Kumamoto Earthquake Response Committee WG2 Report: About earthquake resistance performance of highway embankment (2017)
4. Tokida, K.: Seismic potential improvement of road embankment. In: Moustafa, A. (ed.) Advances in Geotechnical Earthquake Engineering: Soil Liquefaction and Seismic Safety of Dams and Monuments, pp. 269–296 (2012)
5. Lu, C.W., Oka, F., Zhang, F.: Analysis of soil–pile–structure interaction in a two-layer ground during earthquakes considering liquefaction. Int. J. Numer. Anal. Meth. Geomech. **32**(8), 863–895 (2008)
6. Maharjan, M., Takahashi, A.: Liquefaction-induced deformation of earthen embankments on non-homogeneous soil deposits under sequential ground motions. Soil Dyn. Earthq. Eng. **66**, 113–124 (2014)
7. Oka, F., Tsai, P., Kimoto, S., Kato, R.: Damage patterns of river embankments due to the 2011 off the Pacific Coast of Tohoku Earthquake and a numerical modeling of the deformation of river embankments with a clayey subsoil layer. Soils Found. **52**(5), 890–909 (2012)
8. Sasaki, Y., Tamura, K.: Failure mode of embankments due to recent earthquakes in Japan. In: Proceedings of the 4th International Conference on Earthquake Geotechnical Engineering, No. 1479, Thessaloniki, Greece (2007)
9. Poulos, H.G.: Design of reinforcing piles to increase slope stability. Can. Geotech. J. **32**(5), 808–818 (1995)
10. Ito, T., Matsui, T.: Methods to estimate lateral force acting on stabilizing piles. Soils Found. **18**(4), 43–59 (1975)
11. Ito, T., Matsui, T., Hong, W.P.: Extended design method for multi-row stabilizing piles against landslide. Soils Found. **22**(1), 1–13 (1982)
12. He, Y., Hazarika, H., Yasufuku, N., Han, Z., Li, Y.G.: Three-dimensional limit analysis of seismic displacement of slope reinforced with piles. Soil Dyn. Earthq. Eng. **77**, 446–452 (2015)
13. Hazarika, H., Watanabe, N., Sugahara, H., Suzuki, Y.: Influence of placement and configuration of small diameter steel pipe pile on slope reinforcement. In: Proceedings of the 19th

International Conference on Soil Mechanics and Geotechnical Engineering, pp. 2151–2154, South Korea (2017)
14. Ma, N., Wu, H.G., Ma, H.M., Wu, X.Y., Wang, G.H.: Examining dynamic soil pressure and the effectiveness of different pile structure inside reinforced slopes using shaking table tests. Soil Dyn. Earthq. Eng. **116**, 293–303 (2019)
15. Iai, S.: Similitude for shaking table tests on soil-structure-fluid model in 1g gravitational field. Soils Found. **29**(1), 105–118 (1989)

Evaluation and Short-Term Test on Potential Utilization of Ground Source Heat Pump for Space Cooling in Southeast Asia

Arif Widiatmojo, Yutaro Shimada, Isao Takashima, Youhei Uchida, Srilert Chotpantarat, Punya Charusiri, Juraluk Navephap, and Trong Thang Tran

Abstract The demand for air conditioning in Southeast Asia has increased in recent years owing to rapid economic growth in the region. On the other hand, power generation still relies on fossil fuels. A ground source heat pump (GSHP) is an alternative to reducing the energy required for cooling and heating in many countries. The use of a GSHP in a tropical climate like Southeast Asia is mainly for cooling (i.e., heat rejection), and the temperature differences between soil background and atmosphere are low throughout the year. This paper addresses the potential use of GSHP in Southeast Asia based on short-term performance data. Four GSHP systems have been installed in Thailand and Vietnam. Data gathered from the short-term operations provides important insights on GSHP performance. A comparison with an air source heat pump (ASHP) indicates that a GSHP installed at the Bangkok, using two vertical ground heat exchangers (GHEs), achieves a maximum of 33% electricity reduction. Similarly, GSHPs installed at the Saraburi site using horizontal GHEs achieves a maximum of 18.5% and 40.3% less electricity during highest and lowest temperature seasons, respectively, compared with an ASHP. The results also show that parallel–series arrangement improves the thermal performance than those of series–parallel arrangements. Considering the wide temperature climate range of Hanoi, a GSHP can also be used for heating purposes during a short winter period.

A. Widiatmojo (✉) · Y. Uchida
National Institute of Advanced Industrial Science and Technology, Koriyama 963-0298, Japan
e-mail: arif.widiatmojo@aist.go.jp

Y. Shimada
Tokyo Institute of Technology, Yokohama 226-8503, Japan

I. Takashima
Akita University, Akita 010-0851, Japan

S. Chotpantarat · P. Charusiri
Chulalongkorn University, Bangkok 10330, Thailand

J. Navephap
The National Geological Museum, Khlongluang 12120, Thailand

T. T. Tran
Vietnam Institute of Geosciences and Mineral Resources, Hanoi, Vietnam

H. Hazarika et al. (eds.), *Advances in Sustainable Construction and Resource Management*, Lecture Notes in Civil Engineering 144,
https://doi.org/10.1007/978-981-16-0077-7_63

While the short-term results discussed in this paper suggest the potential use of GSHP in Southeast Asia, further long-term analysis, cost evaluations and system optimization are required.

Keywords Ground source heat pump (GSHP) · Space cooling · Tropical country · Southeast asia

1 Introduction

The population of Southeast Asia was almost 640 million in 2016 and is expected to increase to 760 million by 2040, assuming 0.7% annual population growth. Urbanization is an important factor that affects total energy consumption [1–3]. The residential sector accounts for the second highest electricity demand after the industrial sector, growing by an average annual rate of 7.5% owing mainly the increasing number of appliances. Introduction of energy-efficient products can restrain the household energy demand. However, the market price of energy-efficient products, for instance, five-star-rated energy-efficient air conditioners, are higher than those of regular products [4, 5]. Figure 1 shows the relationship between GDP growth of Southeast Asian countries and the electricity demand. Strong correlation between rapid GDP growth and per-capita electricity consumption growth can be clearly seen in this figure, in comparison to OECD member countries and the rest of the world.

The Japan Refrigeration and Air Conditioning Industry Association (JRAIA) reported that in 2016, Vietnam and Thailand were the second and third countries with highest air conditioner demand in Southeast Asia, with a total of 1.98 million and 1.56 million units per year, respectively. Indonesia ranked first with 2.3 million units in the same year [6]. Annual regional air conditioner demand increased from 12.2 million units/year in 2011 to 16.4 million units/year in 2016, equivalent to an average of 6.1% increase per year. Data provided by the Ministry of Energy of Thailand indicate that the residential sector consumes 20.4% of national electricity with

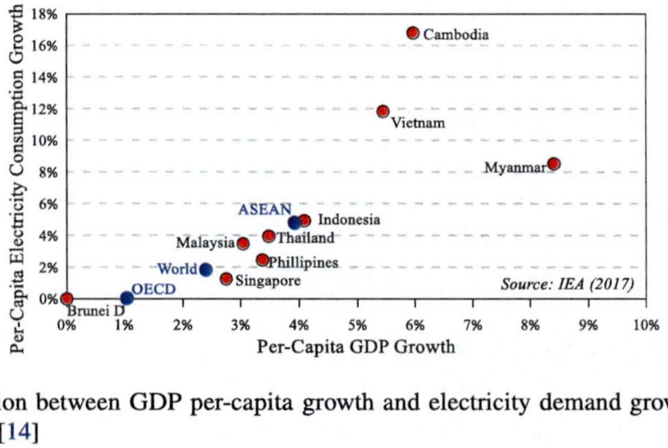

Fig. 1 Correlation between GDP per-capita growth and electricity demand growth of Southeast Asian countries [14]

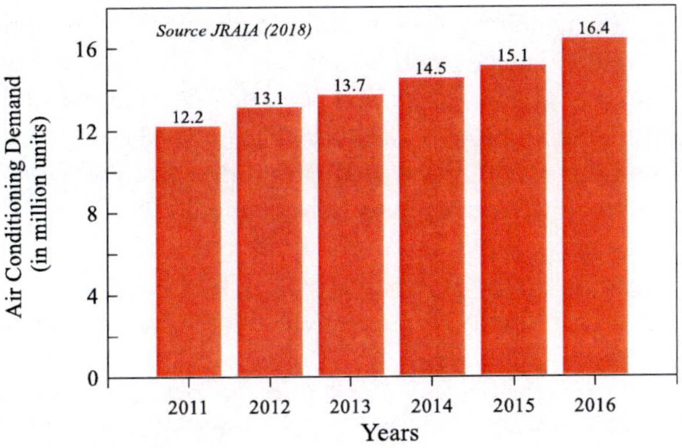

Fig. 2 Annual air conditioner demand in Asia (except Japan and China) [6]

46% and 17% for air conditioning and refrigeration, respectively [7]. Houses and residential areas account for 38% of electricity usage in Brunei Darussalam, representing 63% of total electricity customers, whereas cooling consumes 62% and 74% of building energy for residential and commercial infrastructure, respectively [8, 9]. Figure 2 shows the growth of air conditioner demand in Asia, with the exception of Japan and China.

Several studies have focused on climate change and the urban heat island (UHI) phenomenon. Arifwidodo and Chandrasiri showed the relationship between energy consumption and the cooling degree day (CDD) index, implying a significant influence of household energy consumption and UHI levels in Bangkok, Thailand [10]. They further reported that the UHI level in Bangkok is higher than other major cities including San Francisco, San Diego, and Shanghai. Other studies have also investigated the increasing UHI level in major Southeast Asian cities [11–13].

To counter this problem, the governments of Southeast Asian countries have considered a wide range of policies and commitments [14]. The Thai government began labeling the ratings of air conditioners in 1990. Thus, most air conditioner users in Thailand are supposed to be aware of the labeling method. Likewise, the government of Brunei Darussalam is currently developing energy labeling for air conditioners [15]. Governments of Southeast Asian countries have worked together to establish policy roadmaps to promote harmonization of energy performance standards for air conditioners [16].

Many studies and approaches have been implemented to evaluate potential electricity and CO_2 emission reduction [17–19], and others have pointed out limitations that may hinder the use of ground source heat pump (GSHP) [20–22]. GSHP offers an alternative for energy-efficient cooling and heating in many four-season countries. A GSHP system uses annually stable ground temperature as a heat source or sink, instead of outdoor air. The coefficient of performance (COP) of a GSHP system normally ranges from 3 to 5, whereas ASHP generally has COPs ranging from 2 to 4 [23].

Permchart and Tanatvanit performed field experiments to investigate the possibility of direct GSHP expansion and estimated a four-year payback period for the system [24]. Khedari et al. also addressed potential applications of ground-coupled direct expansion air conditioning in Thailand. The major disadvantage of this system is the ground circulation loop. In their study, high-pressure refrigerant flows directly through the ground loop without transfer to a secondary loop as in GSHP. The direct expansion requires underground burial of copper coils, which are highly corrosive. This reduces the ground loop lifetime, especially in a high-precipitation areas like in Southeast Asia. Additionally, high-pressure loops have a higher probability of leakage, which are difficult to maintain.

Bruelisauer et al. reviewed some alternatives to replace conventional ASHP in Singapore. According to their study, the most commonly used ASHP provides the worst thermal performance. The best-performing cooling alternatives are cooling towers using wet-bulb temperature, followed by a heat sink using a water body (e.g., river, lake and sea) and ground-based heat sink (i.e., GSHP) [25]. Giang et. al. performed energy consumption analysis from five high-story building in Hanoi. They found that during summer and winter, air-conditioning system of these buildings consumed 38% and 78% of the total electricity use, respectively [26].

Yasukawa et al. investigated regional groundwater temperature variation by measuring the vertical temperature variation of several observation wells in Thailand and Vietnam [27]. Vertical groundwater temperature profiles were measured from 55 and 31 wells in Thailand and Vietnam, respectively, with depths varying from 35 to 240 m. The location of the observation wells in the Chao-Phraya plain, Thailand, is concentrated in seven areas: Bangkok, East Bangkok, Kanchanaburi, Ayutthaya, South Nakhon Sawan, North Nakhon Sawan, and Phitsanulok-Sukhotai. In Vietnam, the location of the observation wells was focused in the vicinity of the Red River delta around Hanoi city. They observed a regional variation of subsurface temperature in Thailand and Vietnam of 3.4 and 2.0 °C, respectively. They also identified that in Nakhon Sawan, Phitsanulok, Kanchaburi, and Hanoi, the subsurface temperature was about 5 °C lower than the monthly mean maximum air temperature over a four-month period. In Bangkok and Ayutthaya, the average subsurface temperature was lower than the monthly mean maximum air temperature throughout the year. These findings indicate the potential use of the subsurface as a heat sink for cooling purposes. In their subsequent study, a GSHP system was installed in Kamphaengphet district, north of the Nakhon Sawan district and southwest of Phitsanulok [28].

2 Challenges of GSHP Applications in Southeast Asia

GSHP is widely used in four-season countries, and the number of installed systems and total capacities are increasing. However, there are barriers that limit the wide spread of GSHP use. These barriers can be divided into four major classifications: financial and market, technical, legislation and policy, and lastly, social acceptance [20].

In addition to common barriers, application of GSHP in Southeast Asia encounters obstacles as described below.

Technical factors:

- The use of GSHP is mainly for cooling. Imbalance between heat rejection and heat extraction can lead to long-term thermal build-up surrounding the GHE, which may decrease the thermal efficiency.
- Small temperature difference between background soil and atmosphere [27, 29]
- Lack of data on how GSHP performs under tropical climate conditions
- Lack of technical expertise on the proper sizing and installation of GSHP.

Financial and market factors are:

- High initial cost and considerably cheaper cost of ASHP
- Low regional electricity cost owing to government fuel subsidy
- No existing GSHP market

Legislation and Policy:

- No government incentives
- Government regulations on the use of renewable energy or energy-efficient appliances remain limited

Social acceptance:

- Lack of knowledge about GSHP technologies
- Lack of awareness on energy savings.

Considering these barriers, the Renewable Research Center of National Institute of Advanced Industrial Science and Technology (AIST) initiated collaboration with Chulalongkorn University, the Department of Mineral Resources (DMR) of Thailand, and the Vietnam Institute of Geosciences and Mineral Resources (VIGMR) of Vietnam to conduct research on potential applications of GSHP in tropical regions of Southeast Asia. The main goals are as follows:

- Provide GSHP performance analysis under tropical climate conditions
- Introduce GSHP technology to local communities
- Evaluate the cost required to install a GSHP system in comparison with ASHP
- Develop system optimization framework to evaluate the long-term sustainability and best possible scenario of GSHP application in Southeast Asia.

The aim of this paper is to present the initial results of GSHP test studies in Thailand and Vietnam, focusing on the first and second main goals stated above. The results presented here focus on the short-term data, while the medium-to-long-term performance data are still being gathered and will be published in another paper. From the short-term performance evaluation, we emphasize several points as follows:

Potential electricity reduction in comparison to ASHP

Potential performance and cost reduction by using shallow horizontal heat exchanger

Effect of different heat exchange fluid flow configuration to the thermal performance of GSHP

Potential application of GSHP for both cooling and heating during short winter period (in Hanoi).

3 GSHP System Installation

To date, four systems have been installed in Southeast Asia. Two systems in Thailand were established in cooperation with Chulalongkorn University, at the Bangkok campus and the Saraburi campus. Another system was installed in the Geological Museum of Thailand under joint research with the Department of Mineral Resources, Ministry of Natural Resources and Environment of Thailand. A GSHP system was also installed in Hanoi under cooperation with the Vietnam Institute of Geosciences and Mineral Resources. Several GSHP projects have been carried out in Southeast Asia in recent decade, excluding four projects discussed in this paper. The temperature sensors used in all systems are listed in Table 2 (Table 1).

3.1 Thailand

The study area is located in the upper and lower plains of the Chao Phraya Plain, Thailand. Seawater intrusion and some land subsidence were observed in the Bangkok area owing to excessive groundwater extraction in the area. The lower plain ranges ~200 km from south to north and ~175 km from east to west. The Chao Phraya

Table 1 Other GSHP projects in Southeast Asia (from 2008 ~)

No	Location	Date	GHE type	Current status	Remarks
1	Kamphaengphet (Thailand)	October 2006	1 × 57 m well double u-tube	Finished	First GSHP in SE-Asia [a]
2	Chiang Mai (Thailand)	March 2008	1 × 80 m well single u-tube	N/A	N/A
3	Katsertsart Univ., Bangkok (Thailand)	July 2010	Horizontal GHE	N/A	N/A
4	Bandung Institute of Technology, Bandung (Indonesia)	July 2013	Horizontal GHE	Inactive	Modified ASHP, 25% electricity reduction
5	West Java Mineral Resources Agency	March 2015	2 × 50 m double u-tube	Inactive	Modified ASHP

[a] Yasukawa et al.[28]

Table 2 Specification of temperature sensors

Sensor location	Sensor type	Accuracy (°C)
Indoor air temp	T-type thermocouple	+/−0.5
Atmospheric air temp	T-type thermocouple	+/−0.5
Heat transfer fluid inlet	Pt-100 platinum resistance	+/ −0.2
Heat transfer fluid outlet	Pt-100 platinum resistance	+/ −0.2
Ground temperatures	NTC thermistor	+/ −0.2

River extends from the Nakhon Sawan province to the Gulf of Thailand in the south. Hydrogeological data suggest that the Bangkok area has eight aquifers within 600-m depth, six of which extend to the north of the Pathum Thani and Ayuthaya provinces. Sorted from the top, the main four upper aquifers are the Bangkok, Phra Pradaeng, Nakhon Luang, and Nonthaburi aquifers. A detail about hydrogeological description is provided by Buapeng [30]. The location of three GSHP sites is shown in Fig. 3. According to the Koppen climate classification, the upper region of Thailand has a tropical monsoon climate, whereas the lower region (southern Thailand) has a wet climate/tropical savanna. The rainy season in the southwest area is from mid-May to mid-October, and the winter period in northeast Thailand occurs from mid-October to mid-February. In 2017, the annual precipitation in the Bangkok area was 2670 mm/year, which is higher than other areas, such as Lopburi province (1282 mm/year) and Pak Chong, Nakhon Ratchasima province (1400 mm/year). Monthly average low temperature, high temperature, and relative humidity recorded in Bangkok in a full year (2019) are presented in Fig. 4. The highest monthly temper-

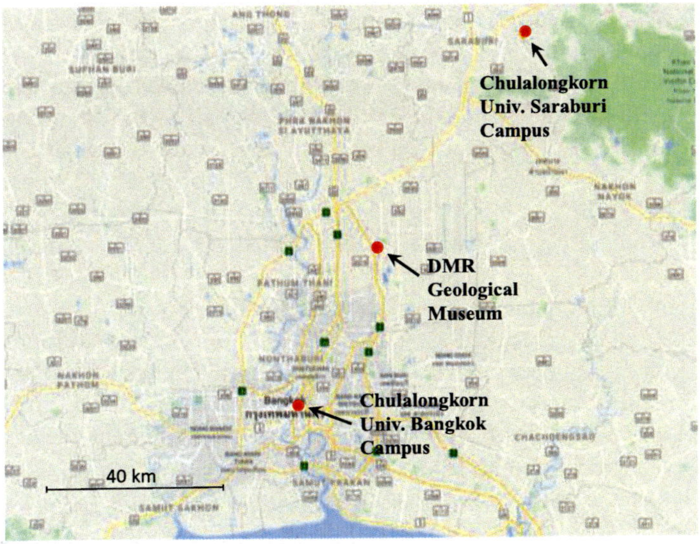

Fig. 3 Location of three GSHP sites in Thailand

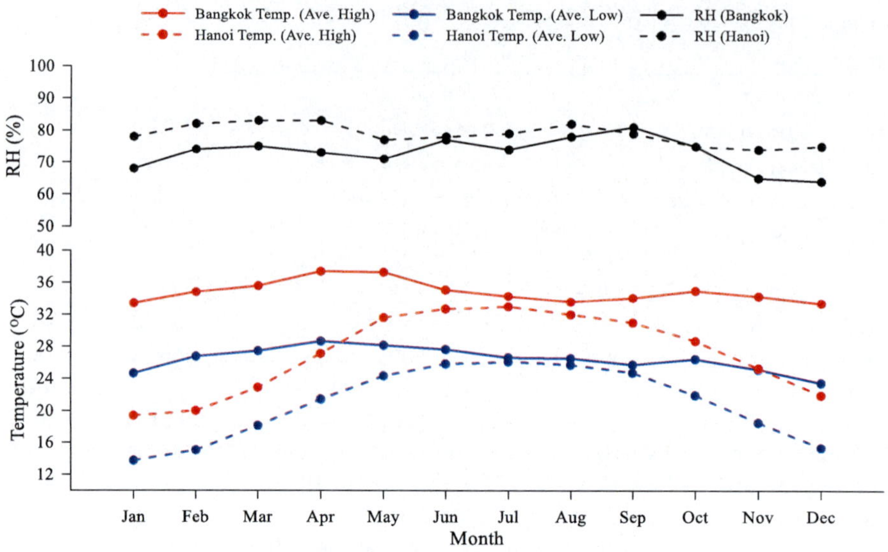

Fig. 4 Monthly average highest air temperature, lowest air temperatures, and relative humidity in Bangkok and Hanoi

ature occurred in April–May, while the lowest temperature occurred in December. On the other hand, highest and lowest relative humidity occurred in September and December, respectively.

Chulalongkorn University Bangkok Campus. The GSHP system at Chulalongkorn University, on the Bangkok campus, was installed in May 2014. The system has two vertical boreholes with a single U-tube arrangement. Both boreholes were supposed to be 50 m in depth; however, only one borehole was installed as planned and the other collapsed, leaving about 15 m depth for heat exchanger installation. The experimental room is located on the 2nd floor of the Parot Racha building, having a dimensions of 3 × 4.75 × 3.5 m. The building was previously used as an office and meeting place but is no longer used for regular official activities. For performance comparison analysis, a new ASHP was installed to replace the broken old ASHP (see Fig. 5). The cooling capacity of the ASHP is 3.42 kW with the rated cooling COP of 3.3.

Based on well temperature observation data gathered from 14 wells located in the Bangkok and East Bangkok areas, the average ground temperature from surface to 60-m depth ranges between 29.5 and 31.7 °C [28]. The geological data indicate that the uppermost aquifer (Bangkok aquifer) is located ±25 m from the surface with an average thickness of 10–15 m. The ground temperature at the GSHP site at 0–50 m depth ranges between 29 and 30.7 °C [31]. The geology within the Bangkok metropolitan area consists mainly of basement complex and alluvial deposits.

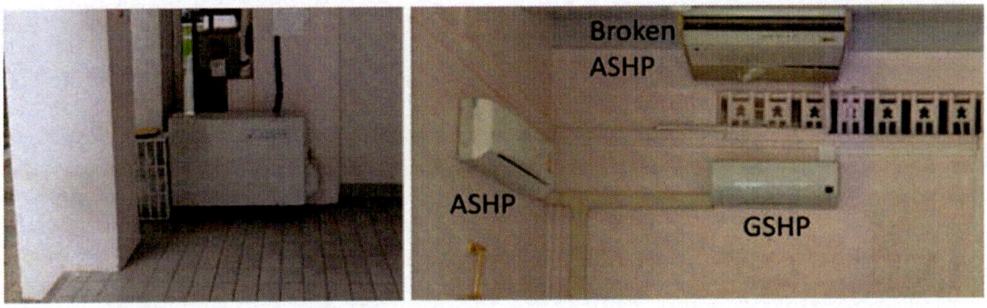

Fig. 5 Heat pump (left) and fan coil units (right). ASHP was installed for performance comparison analysis at Chulalongkorn University, Bangkok

Chulalongkorn University Saraburi Campus. On the Saraburi site, two GSHPs were connected in series with horizontal ground heat exchangers. An ASHP was also installed for performance comparison. The experiments were performed in an office in main building 1 of the Center of Fuels and Energy Laboratory. The experimental room had dimensions of $3 \times 8 \times 2$ m and was used on a daily basis. The building wall material is fiber cement board with thermal conductivity of $\lambda = 0.14$ Wm^{-1} K^{-1} and heat capacity $C = 600$ Jkg^{-1} K^{-1} [29]. The first work was performed in June 2015 with the installation of 200-m high-density polyethylene (HDPE) pipes, capillary mats, the main data acquisition system, and both GSHPs. The capillary mat heat exchanger is a sheet-like heat exchanger with 117 small capillary pipes, which enable a larger effective heat transfer area [32]. The final work was accomplished in November 2016 with installation of background soil temperature sensors and an additional 300-m HDPE pipe. In total, 500 m of HDPE piping in both slinky and helical configurations and two capillary mats, each with dimension of 5.6×0.9 m, were installed in a 73.7 m^2 footprint. One of the GSHPs is a modified ASHP by replacing the original heat exchanger with a plate heat exchanger, allowing heat exchange between the heat pump's refrigerant and ground loop circulation fluid. Heat pump specifications in the Saraburi site are listed in Table 3. Figure 6 shows the installation of capillary mat heat exchangers and slinky-configured HDPE pipes for the GSHP system. The schematic figure of the heat pumps and the ground heat exchanger arrangement can be seen in Fig. 7. For the detail of the installed system, readers can refer to the previously published paper about this GSHP system [29].

National Geological Museum.

The Golden Jubilee National Geological Museum is the largest geological museum in Thailand, located in Pathumthani province about 50 km north of Bangkok. The area is situated in the lower Chao Phraya basin. The hydrogeology of Pathumthani is formed by six main aquifers: Bangkok, Phra Pradaeng, Nakhon Luang, Nonthaburi, Sam Khok, and Phaya Thai. The topography of the basin is relatively flat with an elevation of 1–5 m above sea level. The regional slopes have an average gradient of 4%–10%. Many tributaries originate from the Chao Phraya River, most of which are perennial.

Table 3 Specification of heat pumps used in Chulalongkorn University, Saraburi campus (after Widiatmojo et al. [29])

Heat pump	Cooling capacity (kW)	Heating capacity (kW)	Inverter	Refrigerant	Energy rating[1]	Remarks
GSHP 1	4	5	Yes	R410a	-	Imported from Japan; equipped with inverter control
GSHP 2	N/A[2]	-	No	R410a	N/A[3]	Replaced with 7.03 kW Water-R410A plate heat exchanger; no inverter control
ASHP	3.5	-	Yes	R410a	5 stars	Equipped with inverter control

[1]Certification standard issued by Electricity Generating Authority of Thailand (EGAT); [2] the original cooling capacity was 3.6 kW; [3] the original energy rating was five stars (2011 standard certification)

Fig. 6 Installation of capillary mat heat exchanger (left) and slinky-configured HDPE pipe (right)

The system was installed in the museum's souvenir shop in March 2016, using a 2 × 50 m BHE with a double U-tube configuration. Figure 8 shows the installation site, borehole heat exchangers, and the museum's souvenir shop where the Fan Coil Unit (FCU) is installed. The heat pump is identical to those installed at another GSHP sites.

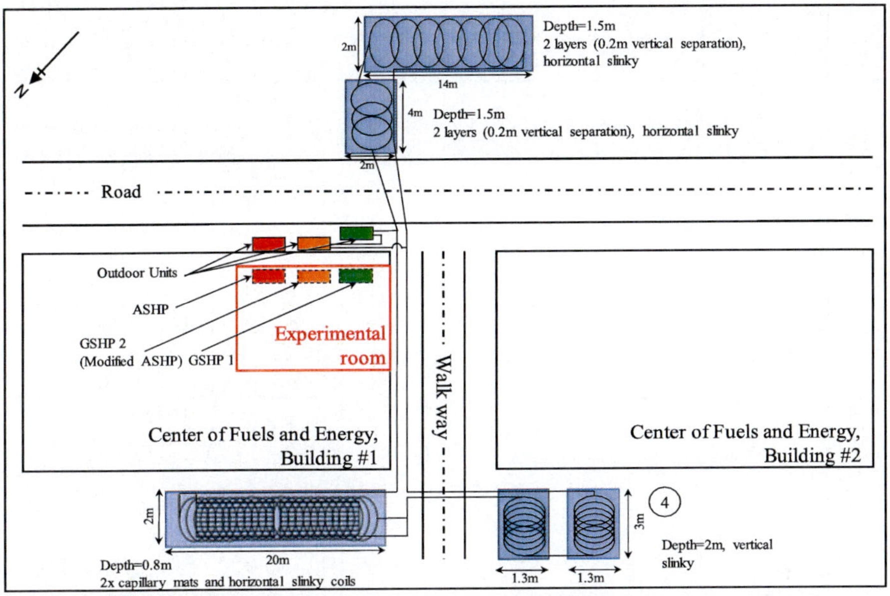

Fig. 7 Layout of shallow ground heat exchangers, pipe connection and heat pump arrangements in Saraburi site

Fig. 8 Installation of GHE into the borehole (middle); museum's souvenir shop (right)

3.2 *Vietnam (Hanoi)*

Hanoi is situated in the Red River delta in the northern region of Vietnam. The main aquifers in Hanoi province can be classified into Holocene or Pleistocene aquifers. The genesis of sedimentary deposits suggests that gravels and sands in channel structures are aquifers. On the other hand, alluvial ridge silts and clays form low permeability layers. Depending on the spatial correlations, these features form confined

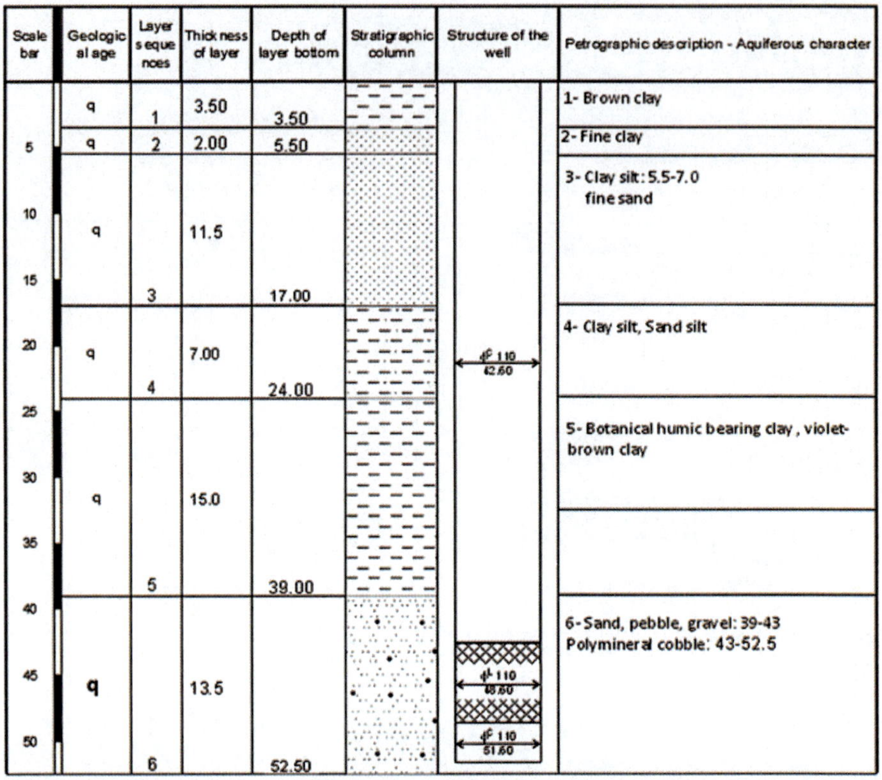

Fig. 9 Stratigraphic description of the Hanoi GSHP site

and unconfined characteristics. As can be seen in stratigraphic description (Fig. 9), a permeable layer is located at about 40–50 m depth.

The climate of Hanoi is categorized as humid subtropical with high amounts of precipitation. The average total annual precipitation is 1600 mm, with heaviest rainfall in June–September. The highest and lowest temperatures occurred in July and January, respectively. Monthly average data, shown in Fig. 4, indicate that Hanoi has lower average temperature compared to Bangkok. The annual pattern of temperature is also similar to those in four-season countries located at northern hemisphere. It also indicates that Hanoi has a wider temperature range compared with other regions in Southeast Asia due to its higher latitude location.

The GSHP system is installed in the director's room of the Vietnam Institute of Geoscience and Mineral Resources (VIGMR). The room is located on the 3rd floor with dimensions of $4.5 \times 4.5 \times 3$ m. The heat pump is connected in series into two vertical 50-m boreholes with a double U-tube configuration. The installation was completed in October 2016. Figure 10 shows the insertion of a GHE (HDPE pipes) into one of the boreholes from the director's room and the heat pump. As a part of our goal to introduce GSHP technology to local communities, a workshop was held in

Fig. 10 GSHP installation at VIGMR, Hanoi, Vietnam

February 2017 at VIGMR, Hanoi focusing on an introduction and practical analysis of GSHP data. The workshop was attended by university students, local enterprises, and officials from government agencies.

4 Thermal Response Test

The thermal response test (TRT) is a method to determine the average thermal conductivity of the soil medium surrounding the ground heat exchanger. Determination of the average thermal conductivity (also known as apparent thermal conductivity) depends upon the assumption of ideal line-source heat transfer from the borehole to the surrounding soil. To ensure the measured thermal conductivity adequately represents the average of value of grouting material, surrounding soil, and convective heat transfer due to groundwater flow (apparent thermal conductivity), the constant heating is typically applied for more than 1–2 days, depending on the soil conditions and heating rate [33].

The thermal conductivity approximation is based on the slope of the mean fluid temperature against logarithmic time, k (°C). According to line source theory, a slope can be evaluated as a function of apparent thermal conductivity:

$$k = \frac{Q}{4\pi L\lambda} \tag{1}$$

where Q (W) is the heating rate and L (m) is the borehole depth.

5 Discussion

5.1 Chulalongkorn University Bangkok Campus

Performance Evaluation. Several tests were performed to evaluate thermal performance and compare electricity reduction with a normal ASHP [31]. Intermittent tests for both GSHP and ASHP were carried out from June to September, during which the average room temperature was 24.9 and 24.5 °C for GSHP and ASHP, respectively. The difference here is attributed to the position of the indoor units and the room temperature sensor. Each test was completed in an average of 7 h daily (9am–4 pm). Figure 11 shows the relationship between average power consumption with outside temperature during operation. In general, GSHP consumes less electricity than ASHP for a given temperature. Based on the linear regression for both HPs, the gap between two lines increases along with the outdoor temperature, indicating that the GSHP can provide higher electricity reduction even at a higher outdoor temperature. Further calculation for low (28 °C) and high (33 °C) outdoor temperatures shows that the percentage of power consumption reduction is 33.3% and 24%, respectively.

The COP is the ratio of heat removed from the room to the work required:

$$\text{COP}_{\text{cooling}} = \frac{Q_T}{W_T} = \frac{Q_\text{R} - \varepsilon W_C}{W_T}, \quad \text{where} \begin{cases} \varepsilon = 1(\text{Cooling}) \\ \varepsilon = -1(\text{Heating}) \end{cases} \tag{2}$$

where Q_R (W) is the rate of heat extraction or rejection from/into the ground, Q_T (W) is the rate of heat added or removed from/into the building, and W_T (in W) is the total electrical power consumption, calculated as:

$$W_T = W_C + W_F + W_P \tag{3}$$

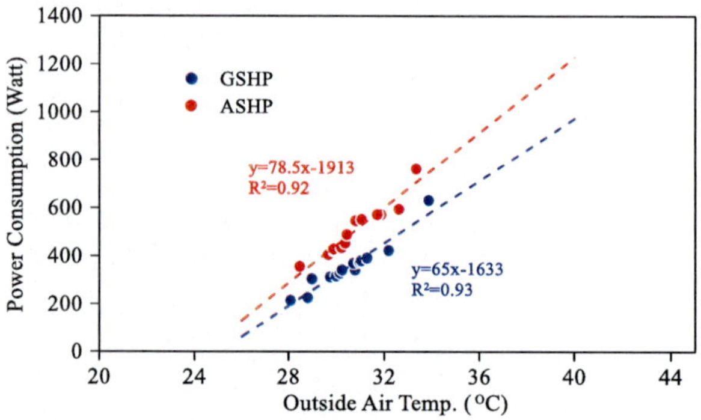

Fig. 11 Relationship between power consumption and outside air temperature for both ASHP and GSHP systems in Chulalongkorn University, Bangkok

$$Q_R = \varepsilon(T_{\text{out}} - T_{\text{in}})\rho c V_m, \quad \text{where}\begin{cases} \varepsilon = 1(\text{Cooling}) \\ \varepsilon = -1(\text{Heating}) \end{cases} \tag{4}$$

where W_C, W_F, and W_P are the electrical power for the compressor, fan, and circulation pump, respectively, while ρ (kg/m^3), c (J/(kgC), and V_m (m^3/s) are the density, specific heat capacity, and flow rate of the heat exchange fluid, respectively. Tout and Tin are the GHE fluid temperature leaving and entering the heat pump. Figure 12 shows the COP of GSHP during daily operation with the evaluated value of COP = 4.3. The observed COP fluctuations are attributed to daily outdoor temperature fluctuations and also the data acquisition interval (1 h).

Thermal Response Test in Bangkok GSHP Site. For this purpose, the 50-m borehole heat exchanger was temporarily disconnected from the heat pump to create a connection with the TRT apparatus. To the best of authors' knowledge, this TRT measurement was the first test to be carried out in Southeast Asia. Figure 13 shows

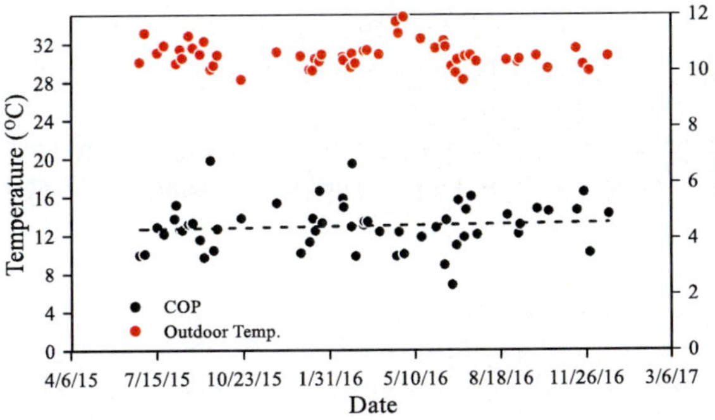

Fig. 12 Average daily coefficient of performance (COP) of GSHP system installed at Chulalongkorn University, Bangkok campus

Fig. 13 TRT measurement at the GSHP site at Chulalongkorn University, Bangkok Campus

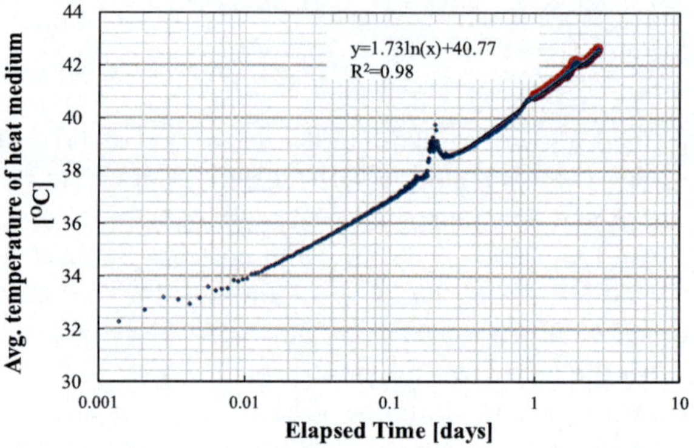

Fig. 14 Linear regression of mean fluid temperature and logarithmic elapsed time for the TRT measurement at Chulalongkorn University, Bangkok campus

the preparation and arrangement of TRT measurements at Chulalongkorn University, Bangkok campus. A constant heating rate was set to $Q = 2000$ W for 3 days, and the flow rate was set to an average of 14.86 L/min. The fluid used for circulation was mixture of propylene glycol water. Before applying the constant heat, the fluid was pre-circulated for about 1.5 h. During the pre-circulation, the stable background soil temperature was found to be 28.8 °C. This value is slightly below the average groundwater temperature from observation well surveys in Bangkok area [27]. The semi-log temperature showing fluid temperature against time with a slope of $k = 1.73$ is shown in Fig. 14. Substituting $Q = 2000$ W and $L = 50$ m into Eq. (2), the apparent thermal conductivity is determined as $\lambda = 1.82$ $\mathrm{Wm^{-1}\ K^{-1}}$. This value is relatively high for the current geological setting in the Bangkok area, likely owing to the intense groundwater extraction (pumping) in the area, which generates artificial groundwater flow [34].

5.2 *Chulalongkorn University Saraburi Campus*

The GSHP systems installed in Saraburi have two distinguished features compared with the other installed systems. First, the horizontal heat exchanger is cheaper than vertical ones because they do not require drilling for the borehole heat exchanger. The horizontal heat exchanger can be installed in a relatively shallow trench that may be excavated by manual (human) labor. Second, the Saraburi site has two GSHPs. GSHP 1 is similar to those installed in other Southeast Asian GSHP sites, while the other one, GSHP 2 is modification from an ASHP.

A comprehensive review on the short-term performance of both GSHPs during highest temperature season can be found in [29], where a two-month intermittent operation was performed for both GSHPs and a normal ASHP. The recorded temperature data and power consumption were then analyzed. GSHP 1 and GSHP 2 consumed

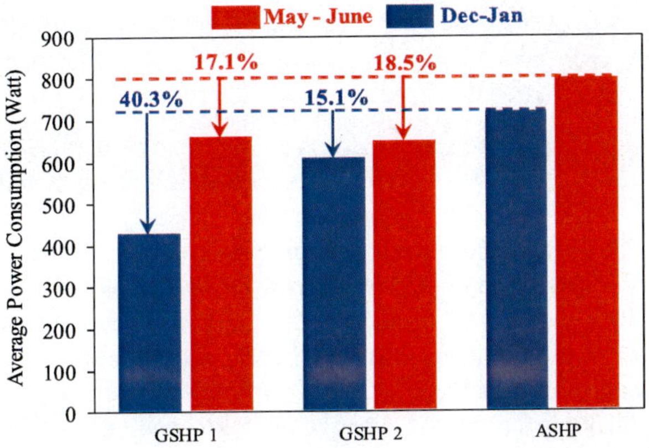

Fig. 15 Electricity reduction of GSHP 1 and GSHP 2 in comparison to normal air source heat pump (ASHP)

17.1% and 18.4% less energy, respectively, than that required by the normal ASHP. Intermittent test studies also have been carried out during the lowest temperature season (December–January). During these periods, GSHP 1, GSHP 2 and normal ASHP run, in total, 145.5 h, 170.83 h and 146.7 h, respectively. The GSHP 1 and GSHP 2 consumed an average of 428.8 W and 609.4 W of electricity, respectively, in contrast to 718.05 W for the ASHP. It corresponds to 40.3% and 15.1% electricity reductions for both GSHPs. It is important to note that the GSHP 2, which originally was an ASHP, provided less electricity reduction during low temperature season than during the high temperature season. On the other hand, GSHP 1 indicated significant electricity reduction during low-temperature season. This can be explained by the use of inverter, which control the speed of compressor during low thermal load, such as in low-temperature season. Figure 15 shows the comparison of average electricity consumption during both temperature seasons.

Economic evaluation based on NPV analysis shows that the payback period for the GSHP 2 (modified system) considering an annual electricity reduction of 18.4% is 15 years. Further, assuming if 40% electricity reduction can be achieved annually, the payback period reduces to 6.3 years [29].

The cost ratio for the ground heat exchanger installation to the cost for heat pump modification to the ground heat exchanger cost and installation is 0.6 for the case of GSHP 2. This cost ratio should be higher than 1 for the case of GSHP with the vertical heat exchanger.

The results of this study also pointed out the possible thermal stabilizing effect of rainwater infiltration into the ground. Ground temperature sensor installed in some location around the ground heat exchanger show temperature decrease following the rainfall period. This effect is expected to have significant role in thermal performance of GSHP, considering high amount of precipitation in tropical region [29].

5.3 National Geological Museum, Department of Mineral Resources

The GSHP system in the National Geological Museum operates regularly during the museum's hours (8 am–4 pm). The gift shop's staff is responsible to operate the system.

Initially, the GHE was configured in 'parallel–series' arrangement. In this configuration, two boreholes were connected in parallel, whereas two sets of U-tubes in each borehole were connected in series. The GHE configuration was modified in March 2019. In the modified version, two boreholes were connected in series and the U-tubes in each borehole were arranged in parallel, herein after 'series–parallel' arrangement. The purpose of this re-arrangement is to evaluate the best piping configuration suited to the system. To the best of author's knowledge, studies focusing on the evaluation of piping arrangement are scarce; furthermore, it strongly depends on the various factors, such as fluid pump capacity, apparent thermal conductivity, ground temperature, thermal load, borehole pipe arrangement and also depth and number of boreholes. In parallel–series configuration, heat exchange fluid is split into two flows, each of which only pass a single borehole heat exchanger twice (two loop each borehole). While in series–parallel configuration, fluid pass both boreholes once in parallel flows. Figure 16 illustrates both configurations from two different years, before the modification 2018 and after the modification 2019.

Figure 17 compares the COP of the GSHPs and the recorded outside air temperatures for both tests. Both tests were carried out in June 2018 and June 2019. In general, the COPs in parallel–series configuration are higher than the series–parallel.

During an elapsed time of ±400 min where both outside air temperatures are nearly similar, the parallel–series configuration provides higher COP. Likewise, over an elapsed time of ±350 min, the parallel–series configuration performs better despite the higher outside air temperature. Figure 18 shows the relationship between outside air temperature and COP over an elapsed time of ±400 min. It can be seen that the

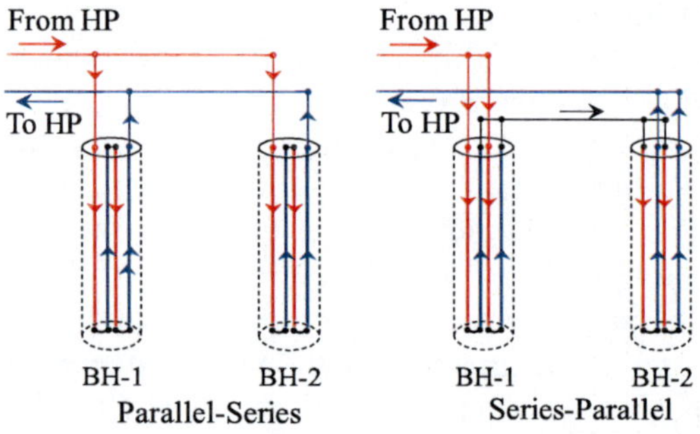

Fig. 16 GHE configurations in the National Geological Museum, Department of Mineral Resources, Thailand

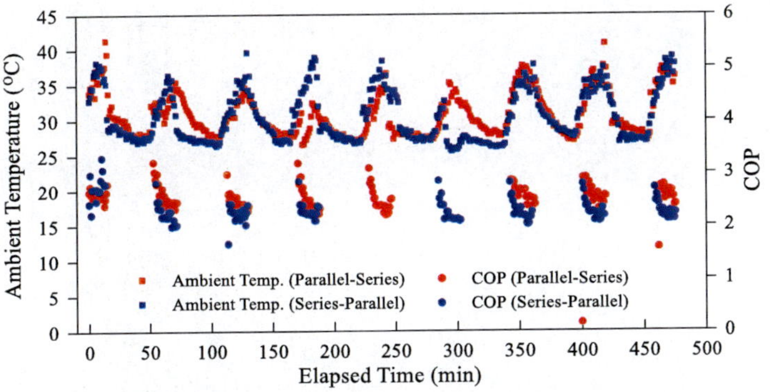

Fig. 17 Comparison of GSHP COPs and outside air temperature during the test period

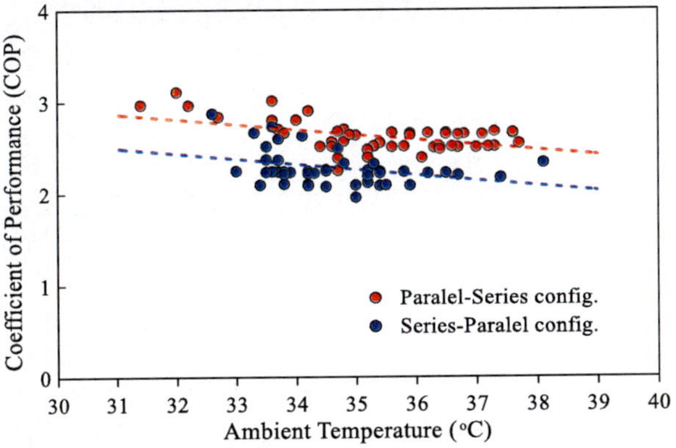

Fig. 18 Relationship between COP and outside air temperature for both configurations

parallel–series configuration provides better thermal performance. Linear regressions for both data indicate the COP average difference of 0.3.

We found that the average COP in series–parallel configuration is 2.30, 9.4% lower than the average COP using parallel–series (COP = 2.54). This corresponds to a reduction of average temperature differences between HP's fluid inlet–outlet and flow rates. Average temperature differences using the series–parallel configuration are 2.44 °C, 7.9% lower than the average value using the parallel–series configuration. Moreover, the average flow rates of heat exchange fluid indicate a 1.4% reduction from 24.17 Lt/min in 'parallel–series' to 23.84 Lt/min for 'series–parallel' configuration. The flowrate reduction suggests the increasing hydraulic resistance of the GHE after modification. Based on this short-term performance evaluation, parallel–series is considered as the most suitable configuration for this GSHP system.

Fig. 19 TRT measurement in the GSHP site at VIGMR, Hanoi

5.4 VIGMR (Hanoi) Site

Thermal Response Test (TRT). The TRT was carried out on August 19–21, 2018 using one of the 50-m boreholes. Because the GHEs in the Hanoi site use a double U-tube, the double loop was temporarily bypassed into a single-loop arrangement. The TRT measurement uses a similar apparatus to the one used in Bangkok site. Figure 19 shows the measurement and equipment used in TRT test at VIGMR, Hanoi.

The fluid was first circulated without applying heat. The background temperature was 27.2 °C, lower than the Bangkok site. The heating rate was set to an average of $Q = 1950$ W for 2 days, while the fluid circulation flow rate was set to 15 L/min. Figure 20 shows the semi-log plot of average fluid temperature against time, from which, using Eq. 1, the apparent thermal conductivity of $\lambda = 1.42$ $\mathrm{Wm^{-1}\ K^{-1}}$ was evaluated for the Hanoi GSHP site. Note that this value is lower than Bangkok, likely owing to the deeper aquifer, lower groundwater flow, and high portion of clay–silt sediment in the upper 50 m.

GSHP Performance. The GSHP system at the Hanoi site is operated based on the requirement of the room's occupant (director of VIGMR). The data are recorded continuously by the data logger and downloaded periodically by the staff. According to the data records and personal communication, the system is not used on daily basis and only if the outdoor temperatures are too hot. In the season with modest temperatures, the occupant prefers opening the windows for passive air circulation rather than using the GSHP.

To examine the GSHP performance during the hottest period in Hanoi, we carried out continuous test for two consecutive days. During the test, the temperature was set to 25 °C. The result presented in Fig. 21 shows that the highest outdoor temperature reached 36.6 °C on July 3, while the lowest was 28.7 °C on July 5. The average

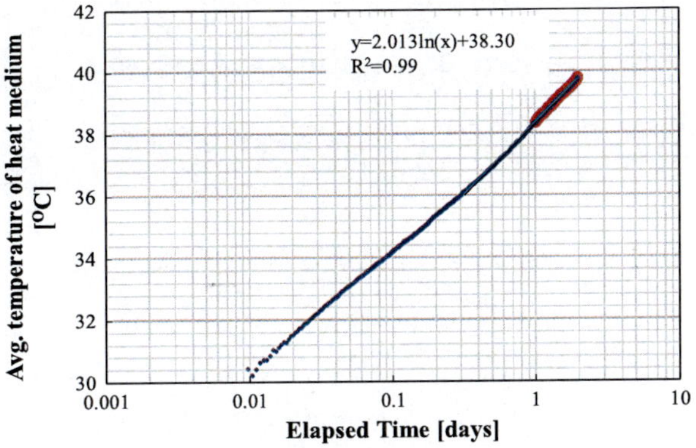

Fig. 20 Linear regression of mean fluid temperature and logarithmic elapsed time for the TRT measurement at VIGMR, Hanoi

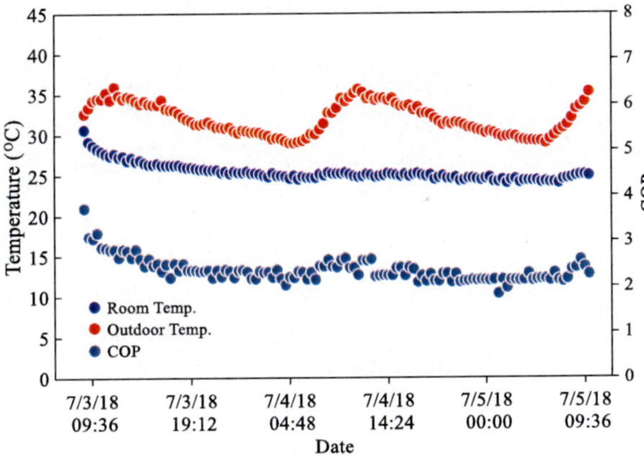

Fig. 21 Continuous cooling using GSHP installed in VIGMR (Hanoi)

outdoor temperature during the 2-day was 31.9 °C and the average COP was 2.34. During the test period, the COP declined in the beginning of the first day; however, the data do not indicate a direct relationship between COP and outdoor temperature variation.

As mentioned in Sect. 3.2, Hanoi has a unique climate with a wide temperature range compared with other major cities in Southeast Asia. A continuous heating test was also carried out by using the Hanoi GSHP. The test was completed from January 30 to February 1, 2018. During the test, the minimum and maximum recorded outdoor temperatures were 10.1 and 15.3 °C, respectively, with an average of 13.04 °C. The results are shown in Fig. 22. The average COP during the continuous test was 3.64. Fluctuations in room temperature and COP are clearly observed. Further examination into the power consumption data show that these fluctuations were likely due to the heat pump's inverter adjusting the compressor output. However, it must be

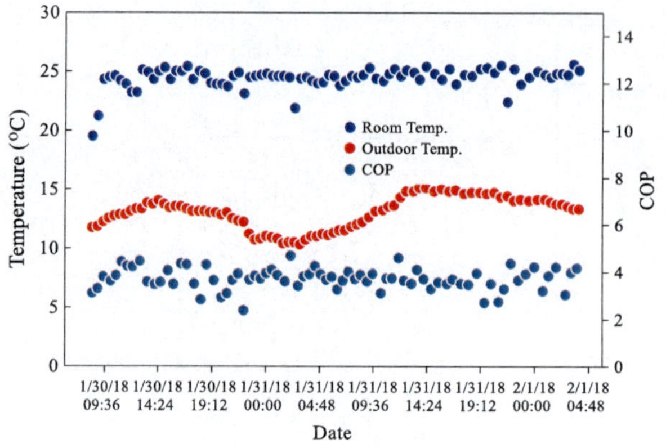

Fig. 22 Continuous heating using GSHP installed in VIGMR (Hanoi)

noted that this value only reflects the GSHP performance during the hottest season, while the annual performance evaluation must also be included during modest- and low-temperature seasons. The data for the Hanoi GSHP are limited because the system does not operate regularly, especially in the mild temperature other than June–September.

Comparing the COP of the Hanoi and Bangkok systems, the Hanoi system has lower COP, whereas the Bangkok site has favorable geological conditions for GSHP installation, reflected by the higher thermal conductivity ($\lambda = 1.82$ Wm^{-1} K^{-1}) compared with the Hanoi site ($\lambda = 1.42$ Wm^{-1} K^{-1}). The room at the Hanoi site is larger with considerably more activities (e.g. meetings and receptions), compared with the Bangkok site. On the other hand, the experimental room for the Bangkok site is a smaller and almost no regular daily activity.

6 Future Works

The short-term test results have shown an early insight on potential utilization of GSHP system in tropical climate of Southeast Asia. However, critical evaluations related to the system sustainability and possible system improvement are currently difficult due to the limited field test data. In order to perform such evaluations the following points are crucial to be considered in the next studies:

Improvement of data acquisition system, including temperature sensors measuring the heat pump component (i.e., refrigerant temperatures and pressures, ground temperature response, etc.).

Long-term data acquisition and analysis

Detailed economic–social analysis of GSHP utilization, combining the long-term performance data, economic growth and social behavior on energy utilization.

Potential influence of regional groundwater flow on the GSHP performance

Development of GSHP system model to evaluate the long-term performance and possible optimization of heat exchanger capacity, size, length, including the potential use of hybrid GSHP systems.

7 Conclusions

We performed operational tests of GSHPs in the tropical climate of Southeast Asia. A total of four systems were installed: three in Thailand and one in Vietnam. One of these systems uses shallow–horizontal heat exchangers, while the others using vertical (boreholes) heat exchangers. Short-term operational data from all system were obtained and evaluated. TRT tests were also carried out at the Bangkok and Hanoi sites.

From the short-term tests, the main findings are summarized as follows:

GSHP system in Chulalongkorn University, Bangkok, utilizes two 50-m-vertical heat exchangers in single U-tube configuration; however, one of the borehole has only approximately 20 m depth due to the collapse of the borehole's wall.

From the TRT measurement, the apparent thermal conductivity was found to be 1.82 Wm^{-1} K^{-1}, while the background soil temperature was 28.8 °C

From the short-term operational tests, the GSHP system in Bangkok wsd found to have 33.3% and 24% average electricity reduction compared to ASHP for the outdoor temperature of 28 and 33 °C, respectively, with the average COP of 4.3

Two GSHPs and an ASHP were installed in Chulalongkorn University, Saraburi campus. Both GSHPs were connected in series with horizontal (shallow) ground heat exchangers.

Short-term operational tests during high-temperature season indicated that GSHP 1 and GSHP 2 provided 17.1% and 18.4% electricity reductions, respectively, in comparison to the ASHP. On the other hand, during lowest temperature season, GSHP 1 and GSHP 2 consumed 40.3% and 15.1% less electricity than the ASHP.

A GSHP in National Geological Museum, Department of Mineral Resources (DMR), Thailand, has been installed, utilizing two 50-m-vertical heat exchangers in double U-tube configuration.

The GSHP performances under two different piping configurations, namely 'parallel–series' and 'series–parallel' were tested in June 2018 and June 2019, respectively.

The GSHP test in 'series–parallel' configuration was found to have 7.9% lower temperature difference between HP's fluid outlet–inlet, and 1.4% reduction GHE fluid flow rate due to increasing hydraulic resistance.

The average COPs are 2.54 and 2.30 for 'parallel–series and 'series–parallel', respectively (9.4% of average COP reduction).

A GSHP system has been installed at Vietnam Institute of Geoscience and Mineral Resources, Hanoi, Vietnam, utilizing two 50-m-vertical heat exchangers in double U-tube configuration.

From the TRT measurement conducted in this site, the apparent thermal conductivity was found to be 1.41 Wm^{-1} K^{-1}, with the background soil temperature 27.2 °C.

Unlike other regions in Southeast Asia, GSHP heating can also be used during the short winter period at in Hanoi, where the use of GSHP for heating is possible.

Two-day continuous tests for cooling and heating were carried out; the results show average COP of 2.34 and 3.64 respectively.

Despite the limited short-term performance analysis, the comparison of electricity consumptions against ASHP, such as in Saraburi and Bangkok site, suggests the potential energy reduction by utilizing GSHP.

To perform more detailed analysis, future studies will focus on the improvement of the data acquisition systems, long-term performance analysis, socio-economic analysis, potential influence of regional groundwater flow, as well as system modeling and optimization.

Acknowledgements The authors would like to thank to Ms. Sasimook Chokchai; Mr. Podchana Jamngoen of the Department of Geology, Chulalongkorn University; Mr. Narongsuk Sodsaard of the Faculty of Science, Chulalongkorn University, for providing assistant for system installation and data management; Prof. Hikari Fujii and Mr. Hiroyuki Kosukegawa of Akita University for discussion and support during TRT measurements; and Mr. Nguyen Tien Quang of VIGMR for data curation and system maintenance.

This research is financially supported by:

The Leading Initiative for Excellent Young Researcher (LEADER), Ministry of Education, Culture, Sport, Science and Technology, Japan

The National Research University Project, Office of Higher Education Commission (WCU-58-017-EN and NRU59-052-EN)

The Thammasat University Research Fund under the Research University Network (RUN) initiative

Coordinating Committee for Geoscience Program in East and Southeast Asia (CCOP), Development of Renewable Energy for Ground Source Heat Pump system (Sub-Project)

References

1. Le, V.T., Pitts, A.: A survey on electrical appliance use and energy consumption in Vietnamese households: case study of Tuy Hoa city. Energy Build **197**, 229–241 (2019). https://doi.org/10.1016/j.enbuild.2019.05.051
2. Bakirtas, T., Akpolat, A.G.: The relationship between energy consumption, urbanization, and economic growth in new emerging-market countries. Energy **147**, 110–121 (2018). https://doi.org/10.1016/j.energy.2018.01.011
3. Doan, V.Q., Kusaka, H.: Projections of urban climate in the 2050s in a fast-growing city in Southeast Asia: the greater Ho Chi Minh City metropolitan area. Vietnam. Int. J. Climatol. **38**, 4155–4171 (2018). https://doi.org/10.1002/joc.5559
4. Supasa, T., Hsiau, S.-S., Lin, S.-M., et al.: Household energy consumption behaviour for different demographic regions in Thailand from 2000 to 2010. Sustainability **9**, 2328 (2017). https://doi.org/10.3390/su9122328
5. Sriamonkul, W., Intarajinda, R., Tongsuk, N., Saengsuwan, S.: Life cycle cost analysis of air conditioning system for residential sector in Thailand. GMSARN Int. J. **5**, 131–138 (2011)

6. Japan Refrigeration and Air Conditioning Industry Association (JRAIA): World Air Conditioner Demand by Region (2018)
7. Ministry of Energy of The Kingdom of Thailand: Energy Statistic of Thailand (2018)
8. Ahmad, A., Othman, H.M.: Electricity consumption in Brunei Darussalam: challenges in energy conservation. Int. Energy J. **14**, 155–166 (2014)
9. (EDPMO): EDPMO (2013) Peer Review on Energy Efficiency in Brunei Darussalam
10. Arifwidodo, S., Chandrasiri, O.: Urban Heat Island and Household Energy Consumption in Bangkok. Elsevier B.V, Thailand (2015)
11. Yang, J., Santamouris, M.: Urban Heat Island and mitigation technologies in Asian and Australian Cities—impact and mitigation. Urban Sci. **2**, 74 (2018). https://doi.org/10.3390/urbansci2030074
12. Estoque, R.C., Murayama, Y., Myint, S.W.: Effects of landscape composition and pattern on land surface temperature: an urban heat island study in the megacities of Southeast Asia. Sci. Total Environ. **577**, 349–359 (2017). https://doi.org/10.1016/j.scitotenv.2016.10.195
13. Kardinal Jusuf, S., Wong, N.H., Hagen, E., et al.: The influence of land use on the urban heat island in Singapore. Habitat Int. **31**, 232–242 (2007). https://doi.org/10.1016/j.habitatint.2007.02.006
14. IEA: Southeast Asia Energy Outlook 2017 (2017)
15. Abas, A.E.P., Mahlia, T.M.I.: Development of energy labels based on consumer perspective: room air conditioners as a case study in Brunei Darussalam. Energy Rep. **4**, 671–681 (2018). https://doi.org/10.1016/j.egyr.2018.10.003
16. ASEAN-SHINE: Promotion of Higher Efficiency Air Conditioners in ASEAN: A Regional policy Roadmap (2015)
17. Bayer, P., Saner, D., Bolay, S., et al.: Greenhouse gas emission savings of ground source heat pump systems in Europe: a review. Renew. Sustain. Energy Rev. **16**, 1256–1267 (2012). https://doi.org/10.1016/j.rser.2011.09.027
18. Noorollahi, Y., Gholami Arjenaki, H., Ghasempour, R.: Thermo-economic modeling and GIS-based spatial data analysis of ground source heat pump systems for regional shallow geothermal mapping. Renew. Sustain. Energy Rev. **72**, 648–660 (2017). https://doi.org/10.1016/j.rser.2017.01.099
19. Soni, S.K., Pandey, M., Bartaria, V.N.: Ground coupled heat exchangers: a review and applications. Renew. Sustain. Energy Rev. **47**, 83–92 (2015). https://doi.org/10.1016/j.rser.2015.03.014
20. Karytsas, S., Choropanitis, I.: Barriers against and actions towards renewable energy technologies diffusion: a principal component analysis for residential ground source heat pump (GSHP) systems. Renew. Sustain. Energy Rev. **78**, 252–271 (2017). https://doi.org/10.1016/j.rser.2017.04.060
21. Garber, D., Choudhary, R., Soga, K.: Risk based lifetime costs assessment of a ground source heat pump (GSHP) system design: methodology and case study. Build. Environ. **60**, 66–80 (2013). https://doi.org/10.1016/j.buildenv.2012.11.011
22. Park, N., Jung, S., Park, H., et al.: Payback period estimation of ground-source and air-source multi heat pumps in Korea based on yearly running cost simulation. Int. Refregeration Air Cond. Conf. (2010)
23. Rosen, M.A., Dincer, I.: Exergy as the confluence of energy, environment and sustainable development. Exergy Int. J. **1**, 3–13 (2001). https://doi.org/10.1016/S1164-0235(01)00004-8
24. Permchart, W., Tanatvanit, S.: Study on using the ground as a heat sink for a 12,000-Btu/h modified air conditioner. World Acad. Sci. Eng. Technol. **3**, 120–123 (2009)
25. Bruelisauer, M., Meggers, F., Leibundgut, H.: Choosing Your Heat Sink for Cooling in Tropical Climates, pp. 1339–1345 (2012)
26. Giang, N.D., Ichinose, M., Sasaki, R., Tokuda, E.: Assesment of energy consumption as a performance index in high-rise buildings in Hanoi, Vietnam. Build. Perform. **8**, 25–38 (2017)
27. Yasukawa, K., Uchida, Y., Tenma, N., et al.: Groundwater temperature survey for geothermal heat pump application in tropical Asia. Bull. Geol. Surv. Jpn. **60**, 459–467 (2009). https://doi.org/10.9795/bullgsj.60.459

28. Yasukawa, K., Takashima, I., Uchida, Y.: Geothermal heat pump application for space cooling in Kamphaengphet, Thailand. Bull. Geol. Surv. Jpn. **60**, 491–501 (2009)
29. Widiatmojo, A., Chokchai, S., Takashima, I., et al.: Ground-source heat pumps with horizontal heat exchangers for space cooling in the hot tropical climate of Thailand. Energies **12**, 1274 (2019). https://doi.org/10.3390/en12071274
30. Buapeng Somkid: The Use of Environmental Isotopes on Groundwater Hydrology in the Selected Areas in Thailand (1990)
31. Chokchai, S., Chotpantarat, S., Takashima, I., et al.: A pilot study on Geothermal Heat Pump (GHP) use for cooling operations, and on GHP site selection in tropical regions based on a case study in Thailand. Energies **11**, 2356 (2018). https://doi.org/10.3390/en11092356
32. Widiatmojo, A., Gaurav, S., Ishihara, T., et al.: Experiments Using Capillary Mat as Ground Heat Exchanger for Ground Source Heat Pump Heating Application, pp. 363–378 (2019). https://doi.org/10.4236/epe.2019.1111024
33. Sarbu, I., Sebarchievici, C.: General review of ground-source heat pump systems for heating and cooling of buildings. Energy Build. **70**, 441–454 (2014). https://doi.org/10.1016/j.enbuild.2013.11.068
34. Onodera, S., Saito, M., Sawano, M., et al.: Effects of intensive urbanization on the intrusion of shallow groundwater into deep groundwater: examples from Bangkok and Jakarta. Sci. Total Environ. **404**, 401–410 (2008). https://doi.org/10.1016/j.scitotenv.2008.08.003

Experimental Study on Strengthening Near-Surface of Slopes Using Bio-grouting Technique

Sivakumar Gowthaman, Kazunori Nakashima, Hiromi Nakamura, and Satoru Kawasaki

Abstract Near-surface instability due to incessant rainfall events poses challenges to the maintenance of earth structures. Bio-grouting (also be referred to as microbial induced carbonate precipitation (MICP)) is a recently emerged soil improvement technique, revealing high potential for stabilizing near-surface of slopes. The technique promotes the cementation of embedded soil using calcium carbonate that precipitates biochemically. This paper presents a bench-scale experimental program, and the objectives were (i) to understand how the treatment protocols impact the strengthening of near-surface and (ii) to assess the profile of treated slope. For those, a series of slope models was treated by various experimental protocols using surface spraying technique. During the treatment, bacteria culture and cementation resources were sprayed in two subsequent phases. The findings suggest that the bio-grouting responses vary depending on volume of cementation solution supplied. High supply of cementation solution developed a highly nonuniform-treated profile compared with low supplies. Also, 1 mol/L concentration of cementation solution is found to be the optimum for the treatment, providing strong intergranular bridging. The spatial distribution of calcium carbonate showed the treated slope can be considered into three layers: surface-crust layer, cemented soil layer and uncemented soil, suggesting that the application technique may provide erosion protection via the crust formed along the outer surface of the slope and cemented soil material formed on the interior.

Keywords Microbial induced carbonate precipitation (MICP) · Slope near-surface · Surface spraying · Slope model · Calcium carbonate

S. Gowthaman (✉) · K. Nakashima · S. Kawasaki
Faculty of Engineering, Hokkaido University, Sapporo 060-8628, Japan
e-mail: gowtham1012@outlook.com

H. Nakamura
Hokkaido Office, East Nippon Expressway Company Limited, Sapporo 004-8512, Japan

H. Hazarika et al. (eds.), *Advances in Sustainable Construction and Resource Management*, Lecture Notes in Civil Engineering 144,
https://doi.org/10.1007/978-981-16-0077-7_64

1 Introduction

Stability of embankment slopes is always a crucial concern in the field of construction engineering, as the constancy of transportation structures are reliant on the stability of the slopes. As indicated in Fig. 1, the potential causes for slope instability ranges from deep-seated failures to sediment yield, suggesting that the stabilization needs to be considered by the means of enhancing (i) structural conditions and (ii) cover conditions [1]. Mass movements and failures occur along the weakest sub-surface, when the driving forces of a collective slope material exceed the resistive forces, without necessarily being influenced by water [2]. Shallow failures are often triggered during or immediately after prolonged rainfalls/ snow-melt, owing to the infiltration. The excess infiltration is found to be diminishing the effective stress, reducing the soil strength and may increase the failure potential in embankment slopes [3, 4]. As described in Fig. 1, mechanical methods are generally applied to enhance the structural stability and performance of the slopes.

Enhancing the cover conditions, on the other hand, is another important requirement to both natural and engineered slopes. Sediment yield is often reported to be a global threat [5, 6], occurring due to the complex interactions of sub-processes between the detachment and transport of surface materials. Human activities always have a major influence in natural slope processes. During constructions, the slopes are modified in a way to fit the construction requirements (e.g., clearance of vegetation and changes in topography); as the result, near-surface substrates are significantly disturbed [7]. Moreover, due to the direct impacts of climatic factors, near-surface substrates are tended to be weakened more rapidly [8], attaining high erodibility risk. Therefore, in terms of eliminating risks and saving economy, cover condition of the slope needs to be enhanced immediately after the construction processes.

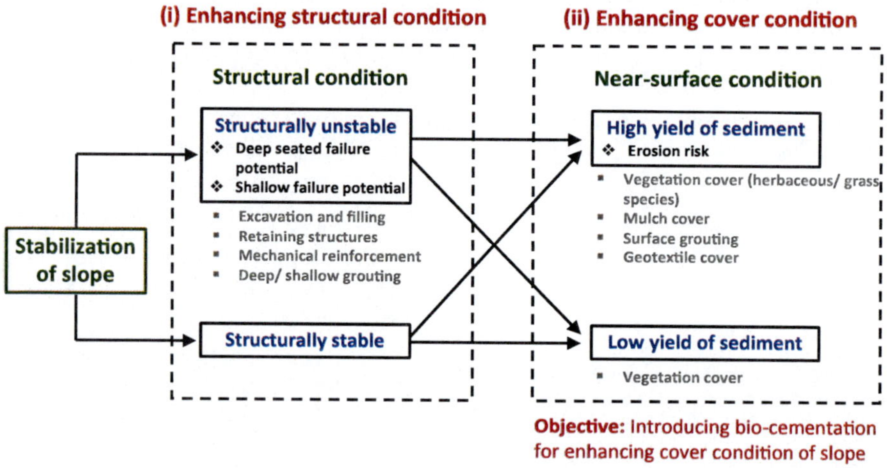

Fig. 1 Slope stabilization processes and objective of the research work

The available methods for enhancing cover condition of the slope are diverse, having both benefits and drawbacks. Seeding/ vegetation is the eco-friendliest method; however, their growth and survival are reported to be limited in cold and arid regions, and it generally takes quite a long time to effectively achieve the benefits [7, 9]. Grouting is the most popular and widely used method, but from the sustainable perspective, the use of cement and other synthetic binders are reported to be less preferred [10, 11], and there continues to be a need to explore new reliable techniques.

Bio-grouting (also be referred to as microbial induced carbonate precipitation (MICP)) is a potential bio-mediated soil improvement technique that relies on sustainable and environmentally friendly processes [10, 12–14]. Within the recent past, number of studies have demonstrated that the technique could potentially be used to solve many geotechnical challenges including liquefaction mitigation [15], slope soil stabilization [16, 17], fugitive dust control [18] and erosion control [7, 19]. In the bio-grouting process, the ureolytic bacteria hydrolyze the urea into ammonium and carbonate ions (Eq. 1), and the calcium carbonate bio-cement is finally produced within embedded soil while supplying calcium ions (Eq. 2), effectively enhancing the physical and mechanical characteristics of soil.

$$CO(NH_2)_2 + 2H_2O \xrightarrow{\text{ureolyticbacteria}} CO_3^{2-} + 2NH_4^+ \tag{1}$$

$$Ca^{2+} + CO_3^{2-} \rightarrow CaCO_3 \downarrow \tag{2}$$

The objectives of this research work are to understand how the treatment protocols impact the strengthening of near-surface and to assess the profile of treated slope. To achieve the first objective, a series of slope models was treated by various experimental protocols using surface spraying technique. Finally, the treatment was advanced to bench-scale model slope to assess the profile of treated slope and to demonstrate the feasibility of bio-grouting under open cold environmental conditions.

2 Materials and Methods

2.1 Slope Soil

One of the erosion-prone expressway slopes located in Hokkaido, Japan (the northernmost island of Japan, located in subarctic region, experiencing a cold climate), has been chosen for the investigation as the representative. The particle size distribution curve of the slope soil is presented in Fig. 2. Based on the Japanese Industrial Standard (JIS), the slope soil can be categorized as fine sand with the mean particle

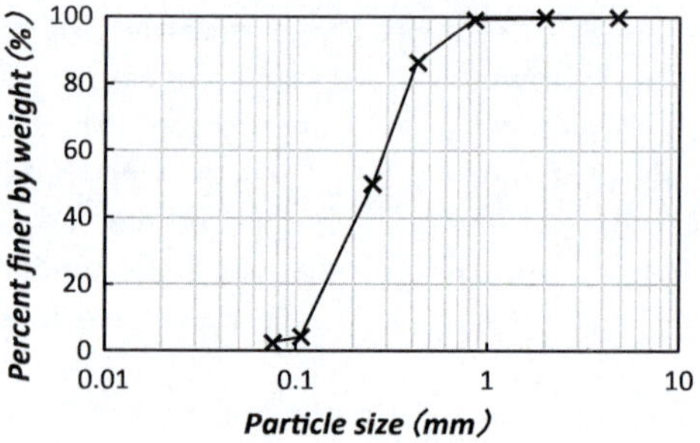

Fig. 2 Grain size distribution of representative slope soil investigated

diameter of 0.23 mm. The intrinsic carbonate content of the soil was found to be 0.22% by weight.

2.2 Slope Models and Test Cases

Slope models used in this study were essentially chosen to be two different scales (as shown in Fig. 3): small-scale and bench-scale. In the first set of experiments (Part 1), small-scale slopes were used to optimize the treatment process considering the effects of supply volume and resources concentration on treatment efficiency. Table 1 presents the detail of the cases considered. The effects of supply volume were studied by Cases 1–3. Cases 2 and 4 were used to evaluate the effects of concentration of cementation solution on MICP improvement. In the second set of experiment (Part 2, indicated as Case 5), a full bench-scale slope was given the

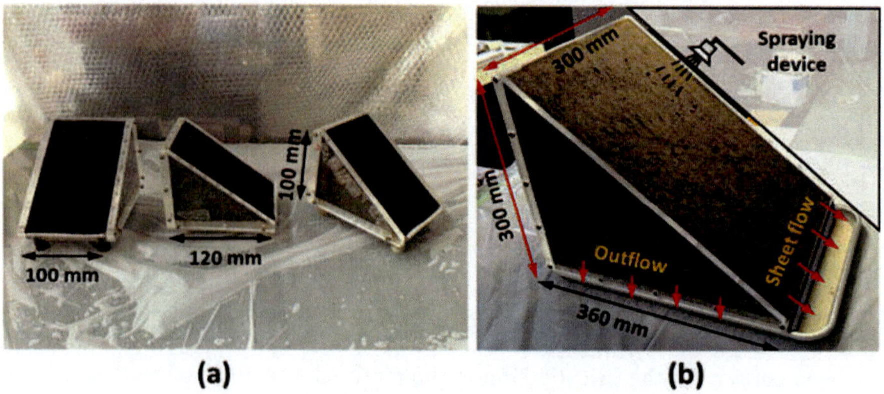

Fig. 3 Slope models used in this study: **a** small-scale and **b** bench-scale slopes

Table 1 Test cases and treatment conditions

Case No	Scale	Initial density (g/cm^3)	No. of bacteria sprays	No. of cementation sprays	Spray volume (per spray) (mL)	Concentration of cementation resources (mol/L)	Treatment duration (days)
Case 1	Small	1.58	2	14	$0.25 \times P_v$	1	14
Case 2	Small	1.52	2	14	$0.5 \times P_v$	1	14
Case 3	Small	1.52	2	14	$1 \times P_v$	1	14
Case 4	Small	1.51	2	14	$0.5 \times P_v$	0.5	14
Case 5	Bench	1.51	2	14	$0.5 \times P_v$	1	14

P_v: pore volume of the target near-surface zone

optimum treatments (based on the findings in Part 1) under an open environment, followed by the assessment of treated slope profile.

2.3 *Treatment Process*

The ureolytic bacteria (*Lysinibacillus xylanilyticus*), isolated from native slope soil, were used for to treat the slope models. The isolation process of the bacteria can be found from our previous work [20]. The bacteria were cultured in NH_4-YE (ATCC 1376) medium in shaking incubator at 25 °C and 160 rpm for 48 h until the growth (OD_{600}) achieved around 4.0 (48–72 h). The growth conditions were chosen based on the optimal enzymatic performance of the bacteria [17]. NH_4-YE medium consisted of dissolved ammonium sulfate, yeast extract and tris buffer of 10 g/L, 15.7 g/L and 20 g/L, respectively. The cementation solution consisted of calcium chloride, urea and nutrient broth at required concentration (in accordance with the test conditions, refer Table 1). In this study, two potential concentrations were investigated: 0.5 and 1 mol/L (Cases 2 and 4, respectively, refer Table 1), using 1:1 chemical ratio of $CaCl_2$: urea.

During the treatment, a spraying device (as indicated in Fig. 3) was used to introduce the bacteria and cementation solutions to the surface of the slopes in two subsequent phases. In the first-phase, the bacteria culture was sprayed, followed by the spraying of cementation solution in second-phase. Between the two phases, a time gap (1–2 h) was given for effectively immobilizing bacteria cells within the near-surface soils, as suggested in previous works [17, 21]. A moderate spray rate of around 45 mL/min was chosen herein, which is relatively in consistent with previous spraying applications [22], and the sprayer was held vertically to the slope surface during the spraying. Cementation spaying was performed every 24 h, whereas the bacteria culture was sprayed at the beginning (day 1) and mid (day 7) of the 14 days treatment. The additional information regarding the treatment conditions are presented in Table 1. As shown in Fig. 3b, the sheet flow solutions (i.e., the solutions flown over the surface) were collected during the treatment to quantify the efficiency, and the outflow solutions (i.e., the solutions which attained the outlet through infiltration) were used to monitor the internal chemical conditions (i.e., pH and Ca^{2+} concentrations).

2.4 Evaluation Methods

After the treatment, the slopes were allowed to cure for 72 h under the same environmental conditions. The assessment program consisted of needle penetration tests, measurement of calcium carbonate content and scanning electron microscopy (SEM). The needle penetration tests (SH-70, Maruto Testing Machine Company, Tokyo, Japan, see Fig. 4) were performed in accordance with JGS 3431–2012 [23], evaluating the unconfined compression strength (UCS) of treated near-surface. In each point, tests were performed in triplicate, and the average value was used for the representation.

For measuring the calcium carbonate content in cemented soil, a simplified device (shown in Fig. 5a) was used to measure the pressure of CO_2 (g) released when the cemented specimen was reacted with HCl in a closed controlled system [24]. Using the calibration curve developed between calcium carbonate and gas pressure (Fig. 5b), the carbonate content (%) was estimated as the ratio between mass of the precipitated $CaCO_3$ and mass of the soil before treatment. The SEM analysis was performed by using Miniscope TM 3000 (Hitachi, Tokyo, Japan).

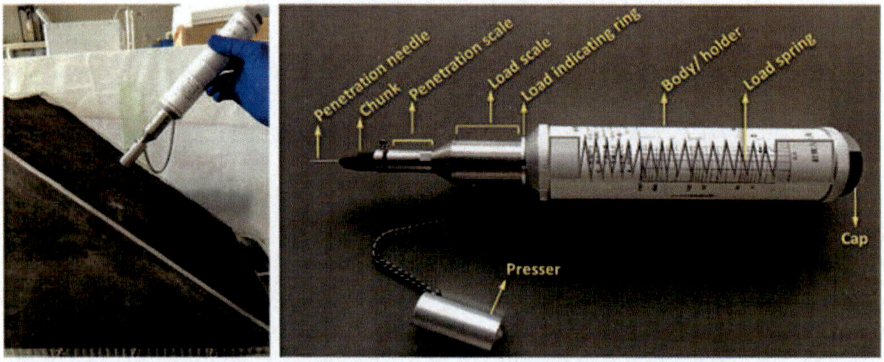

Fig. 4 Needle penetrometer (SH-70)

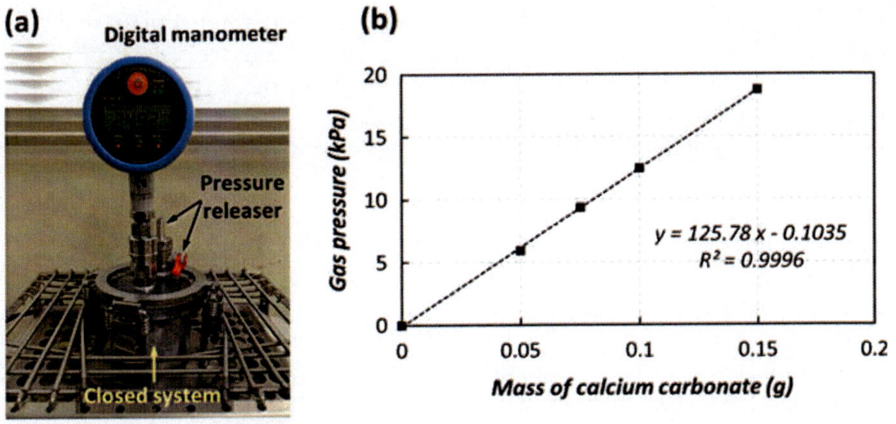

Fig. 5 Figures of the **a** device used to measure the $CaCO_3$ content and **b** calibration curve developed between mass of $CaCO_3$ and gas pressure

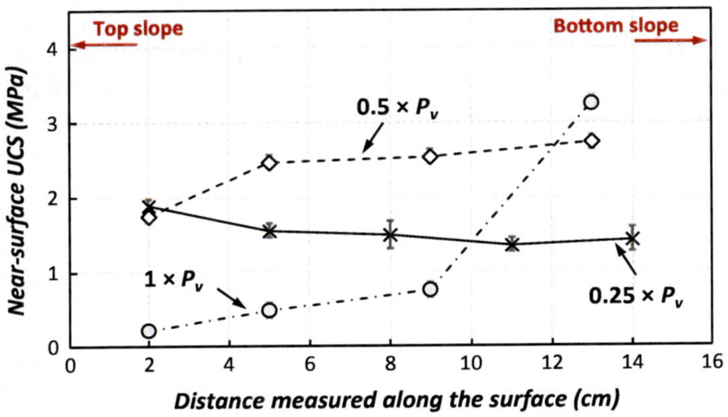

Fig. 6 Near-surface UCS of the slopes treated using different supply volumes

3 Results and Discussion

3.1 Effect of Spray Volume

The MICP treatment was designed to target the topmost 30 mm near-surface of the slope for enhancing the cover condition. Based on the pore volume of target near-surface zone (P_v), three volumes of cementation solution ($0.25 \times P_v$, $0.5 \times P_v$ and $1 \times P_v$) were systematically chosen (Cases 1–3, see Table 1). Figure 5 presents the near-surface UCS of the slopes treated using different supply volumes. It should be noted that the reference strength, i.e., the UCS of the untreated soil material was found to be around 50 kPa. From Fig. 5, it can be seen that the MICP treatment could enhance the UCS significantly, up to around 3 MPa. However, the uniformity of the treatment appears to be influenced by the supply volume. For the slope treated by $0.25 \times P_v$, UCS value slightly decreases toward the slope bottom, and that slightly increases when the spray volume is $0.5 \times P_v$. However, in both the cases, an acceptable uniformity could be seen in strengthening along the surface. On the other hand, the near-surface treated by $1 \times P_v$ of supply volume shows a heterogeneous distribution in UCS, varying between 0.2 and 3 MPa. The UCS value at the bottom slope is around 12 times higher than that of top slope. The similar tendency was observed in repeated experiments as well. Overall, from the results, the effective spray volume can therefore be suggested to be $0.5 \times P_v$.

Many previous studies have evidenced that the increase in number of cementation treatments increased the precipitation content of calcium carbonate in soil columns, hence enhanced the strength characteristics [25, 26]. Thus, before this work, it was expected in a similar way that the increase in supply volume might contribute to effective enhancement of near-surface. However, the observation has disproven the hypothesis, revealing that spraying high volume of solutions (i.e., $1 \times P_v$) results non-uniformity in cementation. Basically, supply of high bacteria culture was reported to have positive effect in MICP, which increases both the yield of calcium carbonate and crystal nucleation sites [27]. In slopes, high supply of cementation solution would tend to cause increased generation of sheet flow over the surface/ near-surface (further explained in subsequent paragraphs), which is suspected to be stimulating the transportation of bacteria cells toward bottom slope, leading to non-uniformity in surface UCS. However, the available results are deficient to fully demonstrate the above conclusion; therefore, further investigations need to be performed.

The inclined nature of slope naturally typically leads to the production of sheet flow, and that highly depends on the soil type (i.e., gradation) and slope gradient. At the beginning of every spraying, the generation of sheet flow was low; when the near-surfaces reach nearly the saturation stage, further spraying would partially turn into sheet flow. In all the cases, it can also be observed that with the increase in number of sprays, the generation of sheet flow increased, and the outflow decreased (refer Fig. 7). This could be explained by the occurrence of crystallization. The progression of calcium carbonate crystals decreased the permeability of the near-surface, which led to the reduction in infiltration, increasing the sheet flow generation. Nevertheless, the quantity that infiltrated into the surface (measured in terms of outflow) is considered to be contributing to the MICP, determining the efficiency of supply.

In the case of $0.25 \times P_v$, low quantity of cementation solution was sprayed every 24 h, which resulted less generation of sheet flow (Fig. 7a). On the other hand, in $1 \times P_v$ case, high quantity was sprayed every 24 h, resulted the development of high sheet flow during the treatment (Fig. 7c). For instance, on 14th spraying, the sheet flow was around 120 mL among total effective supply of 180 mL. In fact, the generation of sheet flow represents the leftover of non-utilized resources, which directly affects the economy and efficiency of the treatment. In order to relate the supply quantity and efficiency of the treatment, the "supply efficiency (SE)" is proposed herein as an index parameter (Eq. 3). Simply, the SE can be defined as the effective supply that contributes to the MICP among total supply. Herein, the infiltrated quantity (Eq. 4)

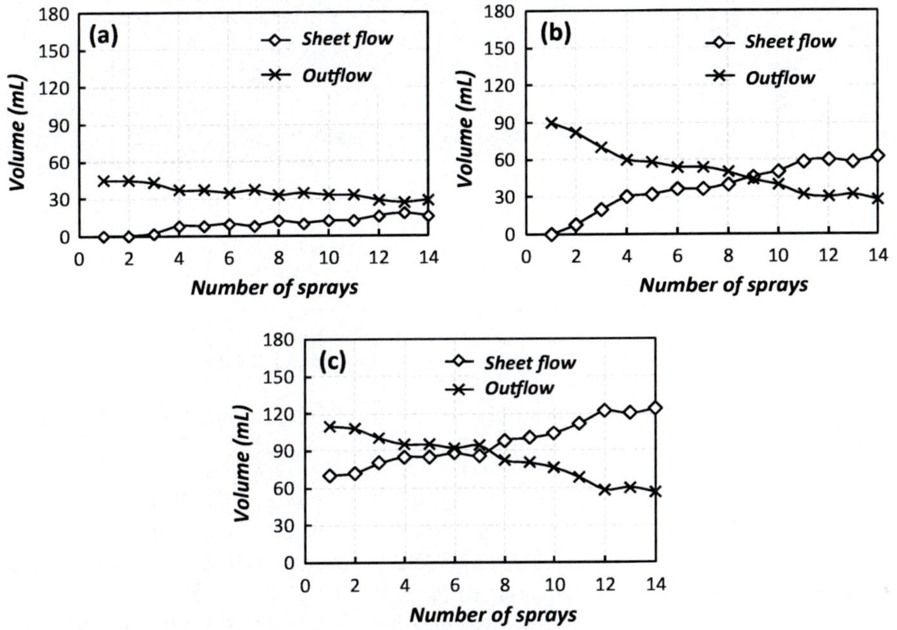

Fig. 7 Supply volumes converted to sheet flow and outflow in slope **a** $0.25 \times P_v$, **b** $0.5 \times P_v$ and **c** $1 \times P_v$

is considered as contributor. Accordingly, at any (*n*th) number of supplies,

$$\text{Supplyefficiency (SE)} = \frac{\text{Cumulativeofinfiltrationquantity}}{\text{Cumulativeofsupplyquantity}} \times 100\% \quad (3)$$

$$\text{Infiltrationquantity} = \text{Supplyquantity} - \text{sheetflow} - \text{loss} \quad (4)$$

The spray loss was found to be around 10% of the supply volume. The supply efficiencies by the end of the treatment were around 71%, 52% and 42%, respectively, for the cases of $0.25 \times P_v$, $0.5 \times P_v$ and $1 \times P_v$, suggesting that high supply lessens the supply efficiency for the slope soil considered in this research work. The efficiency might vary depending on the spray rate, position of spraying and gradation of the soil as well; therefore, future studies should also consider other factors impacting SE.

3.2 *Effect of Cementation Solution Concentration*

Figure 8 presents the comparison of near-surface UCS of the slopes treated with two different concentrations of cementation solution (0.5 and 1 mol/L). The observation reveals that the treatments using both concentrations were able to improve the surface

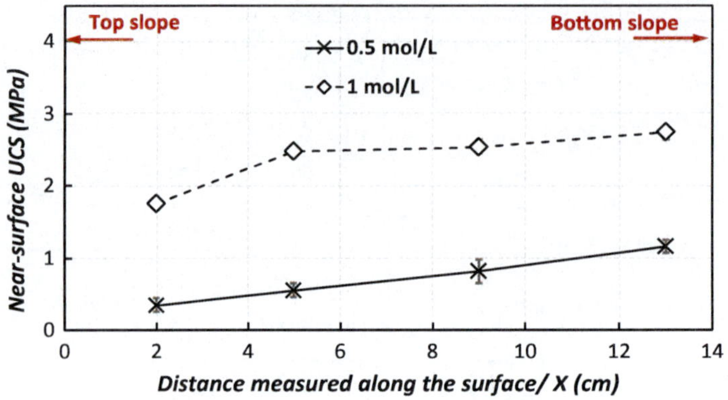

Fig. 8 Near-surface UCS of the slopes treated using 1 and 0.5 mol/L cementation solutions

conditions; however, the concentration of the cementation solutions significantly governs the MICP improvement of near-surface. The near-surface treated by 1 mol/L shows the UCS values between 1.8 and 2.8 MPa. On the other hand, 0.5 mol/L resulted relatively low strength, UCS between 200 kPa and 1 MPa.

The precipitation content of calcium carbonate governs the mechanical response of the MICP soil [20, 21, 25, 26]. When the slope is treated using 0.5 mol/L, relatively lower precipitation of calcium carbonate occurs, leading to the weaker particle connections (see Fig. 9a), which is unlikely to be desirable for the stabilization of slope near-surfaces. On the other hand, stronger intergranular bridging could be observed in treatment using 1 mol/L (see Fig. 9b), which can be considered to be the major contributor to erosion resistance.

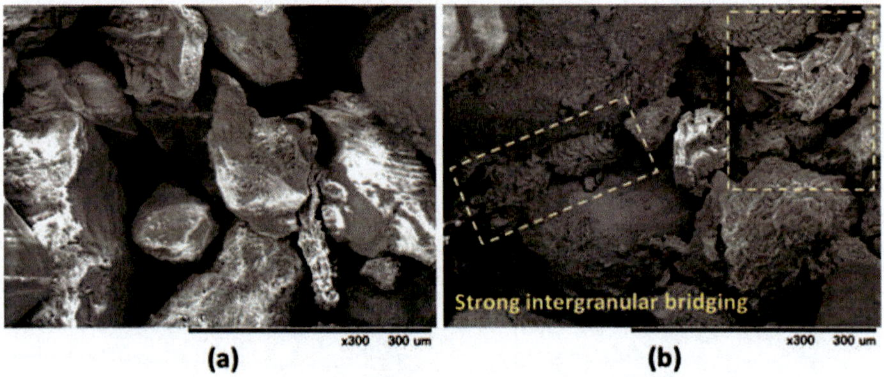

Fig. 9 SEM images of soils treated by **a** 0.5 mol/L and **b** 1 mol/L cementation solutions

3.3 Profile of Cemented Slope

Based on the outcomes of the series of small-scale slope model solidification tests, the optimum treatment conditions were derived and implemented to a full bench-scale slope model in an open (temperature uncontrolled) environment (Case 5, refer Table 1) in order to demonstrate the applicability in real-field, promoting the technique.

Considering the application in Hokkaido (located in cold subarctic region, experiencing cold climate), the slope was subjected to cold climate for the treatment. The temperature during the treatment was monitored in real time by the sensors attached to the slope surface, and the measurements are plotted in Fig. 10. It can be seen that the average temperature was around 15 °C within the first 192, but the temperature dropped to the lowest value of 5 °C for the next 48 h, and then once again that increased at latter periods. Overall, the treatment temperature ranged between 5 and 20 °C. Diurnal fluctuations also could be observed; in night times, the temperature dropped by 5–6 °C. As per the biological response of the bacteria [17], the bacteria could precipitate $CaCO_3$ even at low temperatures; however, the effective performance has been reported to be between 15 and 25 °C. Accordingly, it can be understood that the bacteria were effectively performing during around 70% of the treatment process.

It was also found that the pH of the outflow solutions was between 8.0 and 8.5 in most of the times, suggesting the occurrence of MICP reactions that turned the solutions to alkaline condition. The Ca^{2+} values indicated that the resources were utilized up to around 68%. Until the 11th number of treatments, the pH was found to be between 8.0 and 8.5; however, afterward, a drop could be seen, led the pH value to around 7.8. Similar tendency was also observed in Ca^{2+} measurements, i.e., rapid increase was observed in Ca^{2+} readings after 12th treatment. This could be probably attributed to the low temperature effect. When the temperature decreased below 10 °C, the MICP reactions were likely to be trivial within soil, resulted the decrease in pH and increase in leached Ca^{2+}.

Figure 11 presents the variation of near-surface UCS of the slope. It can be seen that relatively a marked improvement was achieved, and that ranged between 1.6

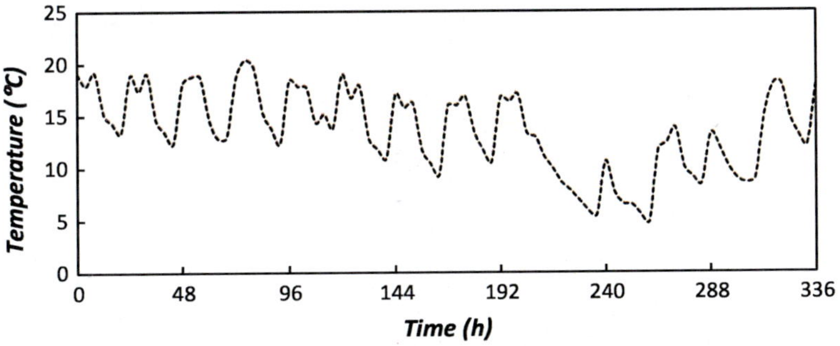

Fig. 10 Temperature measurement from sensors, during the treatment

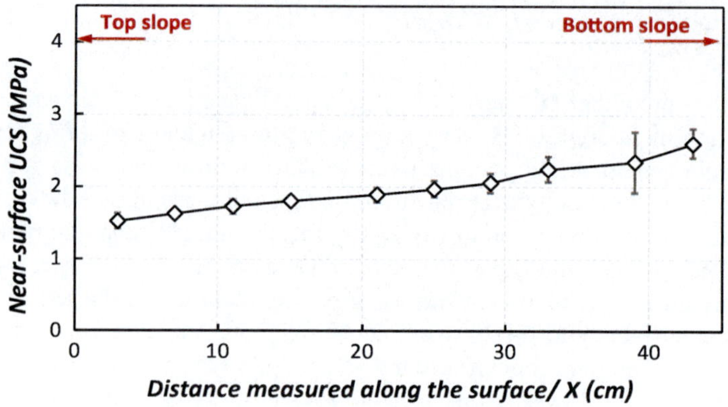

Fig. 11 Variation of near-surface UCS of bench-scale slope

and 2.8 MPa. Similar to that observed in small-scale slope (Case 2), higher UCS was observed at bottom slopes, which could probably be attributed to the flow lines (i.e., sheet flow) and slight transportation of bacteria cells toward bottom slope. Previous studies have suggested that the sand should be treated to a UCS value of around 1.5 MPa to prevent the dynamic and liquefaction damage [28]. Therefore, the near-surface UCS achieved herein is considered to be sufficient enough to withstand against the sediment yield and to provide a reliable cover condition to the expressway slopes.

Subsequently, the mold was dismantled at vertical assembly, and the cementation was assessed along the vertical direction. To evaluate the spatial distribution of calcium carbonate, samples were collected from different depth positions of the slope and tested. Figure 12a presents the distribution of calcium carbonate with the depth measured from the slope surface. It can be seen that high cementation was achieved at surface zone, and the cementation decreased with the increasing depth. This could be possibly due to the distribution of the bacteria cells; as the soil investigated herein

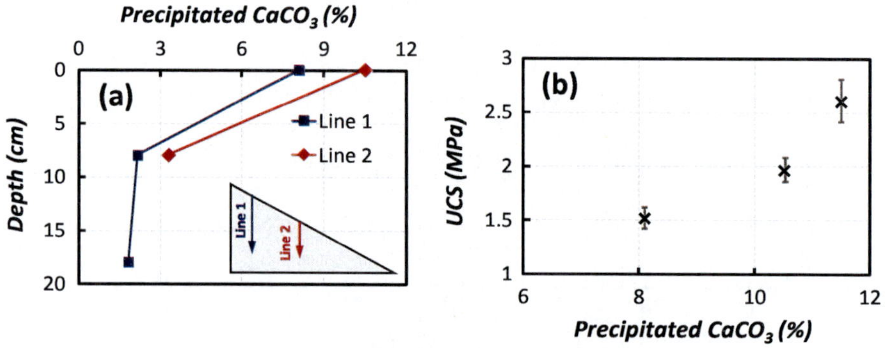

Fig. 12 **a** Distribution of $CaCO_3$ with the depth and **b** plot of UCS against $CaCO_3$ measurements

Fig. 13 Cemented crust layer and categorization of treated slope profile

was fine sand, higher bacteria cells would be filtered at near-surface zone compared to that of deeper zones. In Fig. 12b, the measured calcium carbonate content is plotted against the UCS values. The results indicate that the UCS was highly governed by the precipitated $CaCO_3$, exponentially increasing with the increase in $CaCO_3$, and the tendency reported herein is in a good agreement with the plot reported in previous works [20].

The assessment also showed that the surface crust layer was formed to a thickness ranging between 25 and 30 mm (see Fig. 13), and that primarily comprised of well-developed calcium carbonate bonds, wherein the calcium carbonate was found to be highly distributed, up to around 12%.

The assessment further demonstrated that the MICP-treated slope can be considered into three categories/ layers: (i) well-cemented surface crust, (ii) cemented soil layer and (iii) uncemented soil (as illustrated in Fig. 13). The near-surface crust was underlain by a cemented soil layer, and that was underlain by uncemented soils. The cemented soil layer was not be able to withstand the existed shape, as the formed calcium carbonate content was found to be low and was unable to facilitate strong connections. From the SEM analysis, the calcium carbonate crystals were found to be located on the surface of the grains. The thickness of this cemented soil layer was unable to measure exactly; but the measurement of calcium carbonate content of the soil sampled at various locations revealed that the thickness of cemented soil layer could be nearly around 40–60 mm. Below that, no considerable calcium carbonate deposits were found, remaining as same as untreated soil.

In the MICP treatment, the particle contacts are cemented by calcium carbonate, which leads to the aggregation of soil particles, resulting the increase in mean size of substrate. The surface crust layer achieved herein is a well-aggregated matrix and is unlikely to be considered as soil grains; hence, the probable mean particle size may be huge. Therefore, the crust formed on the surface can be expected to provide a protection through sort of armoring/ shielding mechanism and can be primarily responsible for surface erosion/ sediment yield resistance. The thicker the crust layer enable the better the cover condition, preserving the slope near-surface against the degradation processes. Moreover, the formation of calcium carbonate on the surface of the soils located in cemented soil layer is likely to increase the mean particle diameter. This suggests that the cemented soil layer (even slight crystallization on grain surface) can also be attributed to the increase in both internal and external erosion resistances. Therefore, the cemented soil layer can be potentially considered as secondary shield/ armor against erosion process.

Several studies witnessed that the MICP treatment responses are highly determined by the grain size distribution of the soil material [17, 27]. Coarse sands consist of large pores and facilitate high infiltration of sprayed solutions. On the other hand, the fine soils spaces pose limitations regarding the treatment depth. The less void spaces affecting the transportation of bacteria calls onto the soil and infiltration volume. Although sufficient MICP resources are supplied, the absence/ limited availability of the bacteria at depths leads to the formation of thin surface crust. As demonstrated, for the fine sand studied herein, the near-surface cover appears to be limited to the depth of around 30–60 mm. To expand the understanding on applicable range of MICP, similar works need to be performed in the future on different soil materials, and the efficiency needs to be correlated with the grain size distribution in terms of both resource utilization and effective treatment depth.

4 Conclusions

The bio-grouting responses have been found to be varying depending on volume of cementation solution supplied. High supply of solution develops a highly nonuniform profile compared with low supplies. For the near-surface stabilization of slope considered herein, 0.5 times the pore volume of target near-surface zone was found to be the optimum, and the corresponding supply efficiency appeared to be acceptable.

Treatment using 1 mol/L cementation solutions showed better response than that of 0.5 mol/L. When the slope is treated using 0.5 mol/L, relatively lower precipitation of $CaCO_3$ occurred (led to 200 kPa–1 MPa) compared to that achieved in treatment using 1 mol/L (led to 1.8–2.8 MPa). 1 mol/L treatment was found to provide strong intergranular bridging between soil particles.

The bench-scale model results revealed that the MICP treatment in an open environment, i.e., within 15–25 °C, insignificantly affected the microbial performance, demonstrating the feasibility for the applications in cold regions. The near-surface achieved the UCS of 1.6–2.8 MPa with acceptable uniformity. Measurement of calcium carbonate and assessment demonstrated that the treated slope profile can be considered into three layers: surface-crust layer, cemented soil layer and uncemented soil, suggesting that the application technique may enhance the cover condition of the slope via (i) the crust formed along the outer surface of the slope and (ii) cemented soil material formed on the interior.

References

1. Kumar, N., Das, D.: Nonwoven geotextiles from nettle and poly (lactic acid) fibers for slope stabilization using bioengineering approach. Geotext. Geomem. **46**, 206–213 (2018)
2. Guerra, A.J.T., Fullen, M.A., Jorge, M. do C.O., Bezerra, J.F.R., Shokr, M.S.: Slope processes, mass movement and soil erosion: a review. Pedosphere **27**, 27–41 (2017)
3. Harden, C.P., Scruggs, P.D.: Infiltration on mountain slopes: a comparison of three environments. Geomorphology **55**, 5–24 (2003)
4. Muntohar, A.S., Liao, H.-J.: Rainfall infiltration: infinite slope model for landslides triggering by rainstorm. Nat. Hazards **54**, 967–984 (2010)
5. Tang, C.S., Wang, D.Y., Shi, B., Li, J.: Effect of wetting-drying cycles on profile mechanical behavior of soils with different initial conditions. CATENA **139**, 105–116 (2016)
6. Zhang, X., Hu, M., Guo, X., Yang, H., Zhang, Z., Zhang, K.: Effects of topographic factors on runoff and soil loss in Southwest China. CATENA **160**, 394–402 (2018)
7. Jiang, N.-J., Tang, C.-S., Yin, L.-Y., Xie, Y.-H., Shi, B.: Applicability of microbial calcification method for sandy-slope surface erosion control. J. Mater. Civ. Eng. **31**, 04019250 (2019)
8. Dixon, N., Crosby, C.J., Stirling, R., Hughes, P.N., Smethurst, J., Briggs, K., Hughes, D., Gunn, D., Hobbs, P., Loveridge, F., Glendinning, S., Dijkstra, T., Hudson, A.: In situ measurements of near-surface hydraulic conductivity in engineered clay slopes. Q. J. Eng. Geol. Hydrogeol. **52**, 123–135 (2019)
9. Chirico, G.B., Borga, M., Tarolli, P., Rigon, R., Preti, F.: Role of vegetation on slope stability under transient unsaturated conditions. Procedia Environ. Sci. **19**, 932–941 (2013)
10. DeJong, J.T., Mortensen, B.M., Martinez, B.C., Nelson, D.C.: Bio-mediated soil improvement. Ecol. Eng. **36**, 197–210 (2010)
11. Gowthaman, S., Nakashima, K., Kawasaki, S.: A state-of-the-art review on soil reinforcement technology using natural plant fiber materials: past findings, present trends and future directions. Materials. **11**, 553 (2018)
12. Achal, V., Kawasaki, S.: Biogrout: A novel binding material for soil improvement and concrete repair. Front. Microbiol. **7** (2016)
13. Ivanov, V., Chu, J.: Applications of microorganisms to geotechnical engineering for bioclogging and biocementation of soil in situ. Rev. Environ. Sci. Biotechnol. **7**, 139–153 (2008)
14. Omoregie, A.I., Palombo, E.A., Ong, D.E.L., Nissom, P.M.: Biocementation of sand by *Sporosarcina pasteurii* strain and technical-grade cementation reagents through surface percolation treatment method. Constr. Build. Mater. **228**, 116828 (2019)
15. Montoya, B.M., DeJong, J.T., Boulanger, R.W.: Dynamic response of liquefiable sand improved by microbial-induced calcite precipitation. Géotechnique **63**, 302–312 (2013)
16. Gowthaman, S., Mitsuyama, S., Nakashima, K., Komatsu, M., Kawasaki, S.: Microbial induced slope surface stabilization using industrial-grade chemicals: a preliminary laboratory study. Int. J. GEOMATE **17**, 110–116 (2019)
17. Gowthaman, S., Iki, T., Nakashima, K., Ebina, K., Kawasaki, S.: Feasibility study for slope soil stabilization by microbial induced carbonate precipitation (MICP) using indigenous bacteria isolated from cold subarctic region. SN Appl. Sci. **1**, 1480 (2019)
18. Meyer, F.D., Bang, S., Min, S., Stetler, L.D., Bang, S.S.: Microbiologically-induced soil stabilization: application of *Sporosarcina pasteurii* for fugitive dust control. Geotech. Spec. Publ. 4002–4011 (2011).
19. Jiang, N.-J., Soga, K.: Erosional behavior of gravel-sand mixtures stabilized by microbially induced calcite precipitation (MICP). Soils Found. **59**, 699–709 (2019)
20. Gowthaman, S., Mitsuyama, S., Nakashima, K., Komatsu, M., Kawasaki, S.: Biogeotechnical approach for slope soil stabilization using locally isolated bacteria and inexpensive low-grade chemicals: a feasibility study on Hokkaido expressway soil. Japan. Soils Found. **59**, 484–499 (2019)
21. Feng, K., Montoya, B.M.: Influence of confinement and cementation level on the behavior of microbial-induced calcite precipitated sands under monotonic drained loading. J. Geotech. Geoenviron. Eng. **142**, 04015057 (2016)

22. Gu, J., Suleiman, M.T., Bastola, H., Brown, D.G., Zouari, N.: Treatment of sand using microbial-induced carbonate precipitation (MICP) for wind erosion application. Geotech. Spec. Publ. 2018-March, 155–164 (2018)
23. JGS 3431–2012: In: Japanese Standards and Explanations of Geotechnical and Geoenvironmental Investigation Methods, No. 1. JGS Publication (in Japanese), pp. 426–432. Tokyo, Japan (2012).
24. Fukue, M., Nakamura, T., Kato, Y.: Cementation of soils due to calcium carbonate. Soils Found. **39**, 55–64 (1999)
25. Cui, M.J., Zheng, J.J., Zhang, R.J., Lai, H.J., Zhang, J.: Influence of cementation level on the strength behaviour of bio-cemented sand. Acta Geotech. **12**, 971–986 (2017)
26. Gowthaman, S., Nakashima, K., Kawasaki, S.: Freeze-thaw durability and shear responses of cemented slope soil treated by microbial induced carbonate precipitation. Soils Found. (2020) (article in press).
27. Tang, C.S., Yin, L.Y., Jiang, N.J., Zhu, C., Zeng, H., Li, H., Shi, B.: Factors affecting the performance of microbial-induced carbonate precipitation (MICP) treated soil: a review. Environ. Earth Sci. **79**, 94 (2020)
28. Duraisamy, Y.: Strength and Stiffness Improvement of Bio-cemented Sydney Sand. Doctoral dissertation. University of Sydney, Sydney, Australia (2016)

Sustainable Improvement Technology for Soft Grounds—The Geo Drain SPD Method

Hironobu Iitsuka and Yoshiki Nara

Abstract We will introduce the effectiveness of Geo Drain SPD method, which is an advanced vacuum consolidation method that reduces usage of the amount of finite natural resource by using biodegradable materials, based on construction example and laboratory tests.

Keywords Soft ground · Vertical drain · Vacuum method · Sand · Biomass plastic

1 Introduction

A typical method for improving soft grounds on a large scale and economically is the vertical drain (VD) method. (the sand drain method and the PVD ≪ prefabricated vertical drain ≫ method are generically called the VD method.) The VD method is a method of reducing harmful residual settlement and uneven settlement after construction, by accelerating consolidation of viscous ground and by settling ground in advance.

The VD method has a long history, with the first achievement in Japan in the early 1950s. It spreads widely from now on and has become a reliable method which has important achievements. However, the VD method has a problem in its sustainability, as construction using the VD method requires natural resources—high-quality sand, embankment surcharge or petroleum-derived plastic—which is limited. Besides, petroleum plastic drain materials installed in the ground left indefinitely, and some cases have been reported to cause problems later.

Therefore, we'd like to introduce The Geo Drain SPD method. This method is the vacuum consolidation method, which is enabled to reduce consumption of natural resources and to resolve the problem of drains left in the ground.

H. Iitsuka (✉) · Y. Nara
Industrial Material and Products Team, Chikami Miltec Inc, Tokyo, Japan
e-mail: hiitsuka@c-miltec.co.jp

Y. Nara
e-mail: ynara@c-miltec.co.jp

H. Hazarika et al. (eds.), *Advances in Sustainable Construction and Resource Management*, Lecture Notes in Civil Engineering 144,
https://doi.org/10.1007/978-981-16-0077-7_65

2 The Transition of the Vertical Drain Method

2.1 From the SD Method to the PVD Method

The sand drain (SD) method, the packed drain method, and the PVD method (prefabricated vertical drain) are mentioned as the representative examples of the VD method. Each method is a method of accelerating consolidation and has the same principle of improvement: Firstly, install vertical drainage layer with high drainability in the viscous ground. Secondly, shorten the drainage distance of water in the ground discharged by the load of embankment surcharge.

The SD method was mainly used before; however, it has been recently replaced by the PVD method, and this shift is due to the depletion of sand resources and economics. Compared with the SD method, the PVD method requires smaller construction machines and shorter construction cycle; in short, it is more economical for constructions on land.

2.2 The Horizontal Drain (PHD)

In the VD method, it is necessary to discharge water discharged vertically from the ground in the horizontal direction. When this drainage doesn't work well, it will cause delay of consolidation. Smooth drainage is important for achieving the expected consolidation effect. In recent years, procurement of high-quality sand, which was used as sand mat in conventional method, has been difficult due to depletion caused by large-scale projects. Therefore, the plastic horizontal drain materials have been developed as a substitute for natural sand (Fig. 1; Table 1).

The PHD has the flexibility to follow settlement of ground, and they are dedicated to horizontal use, which are different from the PVD. The improvement of consolidation delay by horizontal drain is shown in the report of Fukuda et al. [1].

In addition, the report by Ikehata et al. shows the observation results of the dissipation of excess pore water pressure by a horizontal drain [2].

Fig. 1 Pictures of PHD and PHD construction status

Table 1 PHD specification

Item		Unit	SB drain specification
Material	Core		PVC
	Filter		Polyester nonwoven
Size	Thickness	mm	8.0
	Width	mm	305 ± 3.0
			205 ± 3.0
	Length	m	50
Core compressive strength		kN/m^2	>200
Permeability		m/s	$>5.0 \times 10^{-2}$
Filter permeability		m/s	$>1.0 \times 10^{-4}$

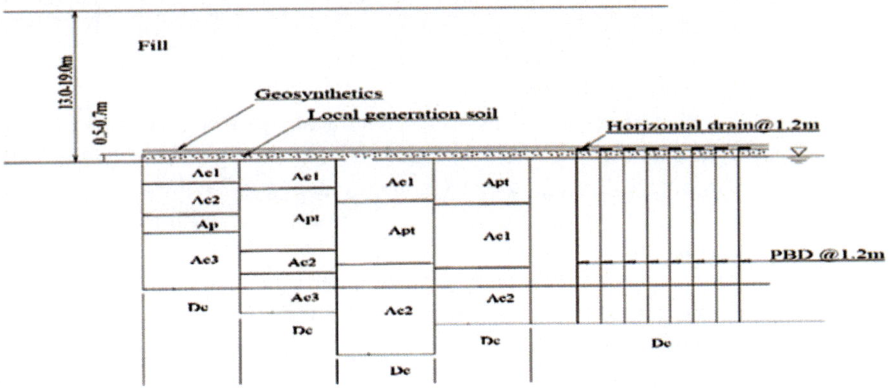

Fig. 2 Soil profile in object region [2]

The following is an example of measurement during embankment construction. It is a change with time of excess pore water pressure at the intersection of the vertical drain with the horizontal drain (Figs. 2 and 3).

In some parts, an excess pore water pressure of about 20 kPa was observed at the early stage of embankment. However, in other cases, generation of excess pore water pressure was not observed even when the embankment height reached a maximum of 19 m. Therefore, it is judged that the excess pore water pressure generated by the embankment is drained by the horizontal drain, and its effectiveness is shown.

2.3 *The Vacuum consolidation method*

The vacuum consolidation method was introduced to Japan in the 1960s, and its construction achievements have been increasing since the late 1980s. Its feature is to

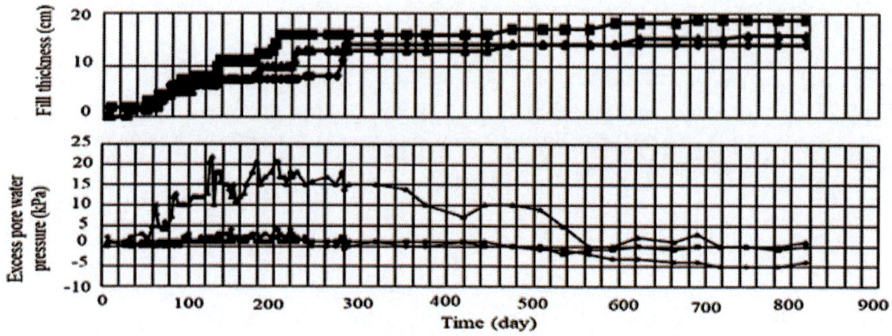

Fig. 3 Measured excess pore water pressure [2]

drain underground water by negative pressure instead of conventional surcharge and to accelerate consolidation. The vacuum equipment instead of surcharge is used to apply a load, and it is unnecessary to load and remove surcharge as a result, whole construction period can be shortened.

The general construction procedure is 1. Cover ground surface of improvement area with air-tight sheet, 2. Place PVD, PHD and drainage pipe under the sheet. After vacuum equipment starting, measuring instruments gauge settlement and negative pressure. It will stop when achieving improvement goals (residual settlement, consolidation density, etc.) set in advance.

Basically, the vacuum method is applicable where the PVD method could be applied. Depending on conditions of soil layer and surrounding area, however, auxiliary method might be required.

3 The Geo Drain SPD Method

3.1 Outline of the Method

The Geo Drain SPD method is an improved technology that is more eco-friendly and sustainable while using principles of conventional vacuum consolidation method.

The feature of this method is that firstly, natural materials such as embankment surcharge and sand for sand mat are not used, and secondly, biomass plastics products are can be used as drainage materials (PVD.) PHD is used instead of sand for sand mat. From PVD to PHD, and from pipes to vacuum equipment are directly connected and integrated by using PHD, which can be expected to reduce pressure loss. Perforated pipes, which are used as collecting pipe so far, are no longer needed with appearance of special jig connecting PHD and collecting pipes. Therefore, the risk of clogging of pipes is eliminated, and high vacuum is achieved (Fig. 4).

One vacuum equipment covers approximately 3000 m^2 of land. Vacuum equipment is a unit of necessary equipment and is easy to transport, install and remove.

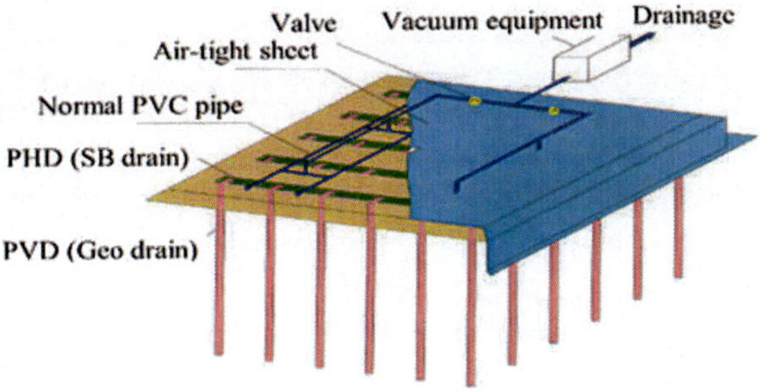

Fig. 4 Geo Drain SPD method outline [3]

After starting, it monitors measuring basically on automatic operation, and we can grasp settlement and pressure in real time.

Heavy equipment works such as excavation and transportation are disappeared, which will lead to reducing CO_2. In addition, use of biomass plastic products will resolve the problem of drains left not degradable in the ground. The biomass plastics will be introduced later.

3.2 *Construction Example*

One construction example of the Geo Drain SPD method is shown below. This case is a construction of land formation. The Geo Drain SPD method was adopted to resolve a lack of embankment surcharge and to shorten construction period. As for the construction conditions, installation length of PVD is 12 m(L), and installation interval is 1.5 m(d).

As for the ground condition, there is a humus layer approximately 5 m above and an alluvial clay layer approximately 7 m below it (Fig. 5).

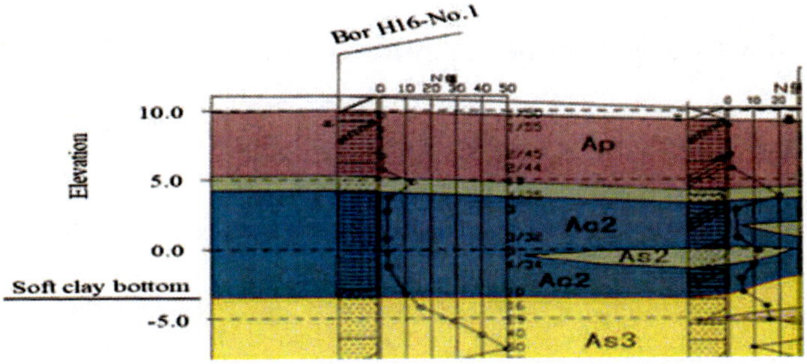

Fig. 5 Estimated soil layer cross section

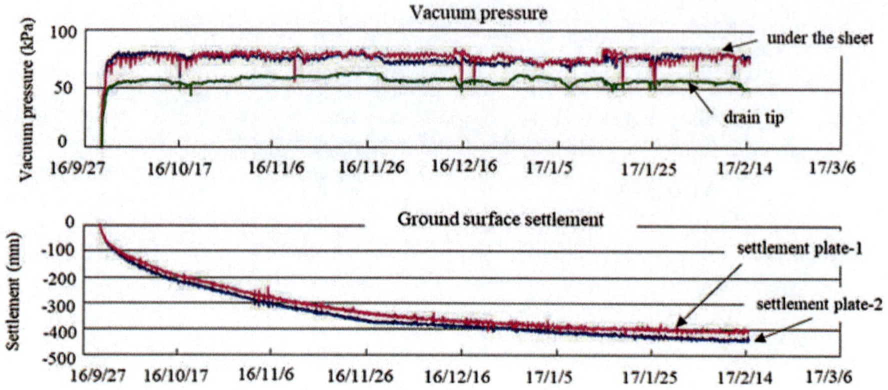

Fig. 6 Changes in vacuum pressure and settlement over time

Because of the presence of an intermediate sand layer, steel sheet piles were installed at the boundary of improvement area as a water shielding measure. We divided the improvement area into 13 lots(one lot is approximately 2500 m^2,) and one vacuum equipment covered one lot.

Settlement plates and vacuum gauges were installed in all lots in order to monitor settlement by settlement plate and pressure under the sheet or the end of the drain by vacuum gauge. The measurement was operated automatically, and multiple items were measured and monitored simultaneously to grasp the progress of consolidation. The final settlement was predicted by hyperbolic function based on the actual settlement data, and vacuum operation was set to stop when the residual settlement was 10 cm or less.

After vacuum operation was stopped, the moisture level was checked from the end of the sheet and dealt as necessary. As a result, a pressure of 60–80 kPa was acting under the sheet, and the settlement proceeded smoothly. The design pressure under the sheet was 60 kpa, though (Fig. 6).

4 The biomass-base-biodegradable PVD

4.1 The Features

"Biomass plastic" is a kind of polymer which is synthesized chemically or biochemically with raw material derived from organic resources. Sakai and Shirai proposed Synthesis method of PLA from food waste, and they propose social system which can reduce waste volume in city [4].

Many kinds of biomass plastics such as polylactic acid (PLA), polyhydroxyalkanoates (PHA), cellulose acetate (CA), Nylon11 (PA11), Nylon4 (PA4), bio-based-polybutylene succinate (bio-PBS), bio-based polyethylene (bio-PE) are already commercially produced and used in human life.

Biomass plastics are categorized "biodegradable" and "non-biodegradable."

PLA, PHA, CA and bio-PBS are categorized as biodegradable, while PA4, PA11, bio-PE are categorized as non-biodegradable.

Biodegradable plastics are catabolized by the action of microorganisms in the nature under certain conditions. Eventually, they are catabolized into carbon dioxide and water.

Biodegradable plastics have already been used in many application examples in which taken advantage of their features while they are still not popular in construction industry (https://www.jbpaweb.net/bp/bp_product.htm).

"Nonal drain," PVD for Geodrain SPD method, has same shape as conventional PVD derived from petroleum, and its performance and mechanical properties are also the same as conventional products.

We first tried to find suitable biomass plastic to develop biodegradable PVD. After trials with some raw materials, we found a suitable combination of biodegradable raw materials (data not shown).

Then we performed some test with actual clay soil in order to confirm Nonal drain can satisfy standard of Japan at Geo Research Institute (Osaka, Japan) [5]. We measured discharge capacity under similar condition with actual construction site that Nonal drain bent in clay soil. Through these tests, we could confirm that Nonal drain can be used as PVD (data not shown) (Fig. 7).

Table 2 shows specification of Nonal drain. These properties satisfy the requirements of public work in Japan.

After launch in 2005, Nonal drain was used for Tokyo port expansion, shopping mall construction, residential zone development, etc.

Nonal drain is an eco-friendly material derived from plant that will return to the nature after finish its role (Fig. 8). Nonal drain is decomposed after construction, and plants will grow by using water and carbon dioxide catabolized from Nonal drain by microorganisms. In short, Nonal drain is a sustainable PVD circulating in the nature.

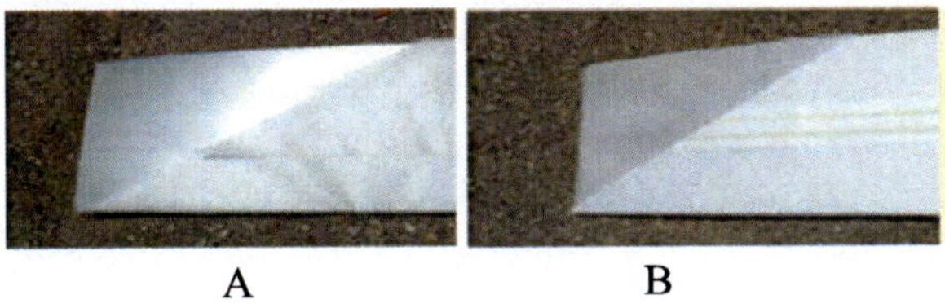

Fig. 7 Picture of PVD:A: Nonal drain, B: Geo drain (conventional PVD)

Table 2 Nonal drain specification

Item		Unit	Specification	
			Nonal drain	Geo drain
Material	Core		Biodegradable resin	Polyolefin resin
	Filter		Biodegradable nonwoven	Polyester nonwoven
Size	Thickness	mm	3.3 ± 0.5	3.5 ± 0.5
	Width	mm	98.5 ± 5.0	
	Length	m	250	
Tensile strength		kN/width	>2.0	
Permeability		m/s	$>5.0 \times 10^{-2}$	
Discharge capacity		m^3/year	>500	
Filter permeability		m/s	$> ,0 \times 10^{-4}$	

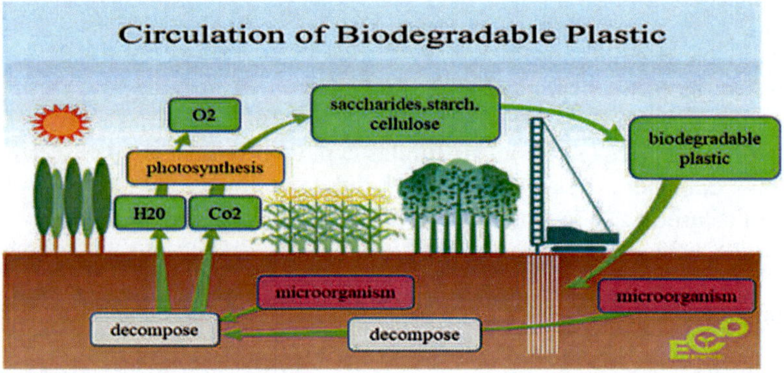

Fig. 8 Circulation diagram

4.2 Reduction of Oil Consumption

For conventional PVD materials, a large amount of petroleum-based resins are consumed for production. On the other hand, Nonal drain can reduce about 75% (w/w) of consumption of petroleum material because its raw material is mainly derived from plant. (Fig. 9) By using the Nonal drain, we can greatly reduce the amount of oil consumption for ground improvement project.

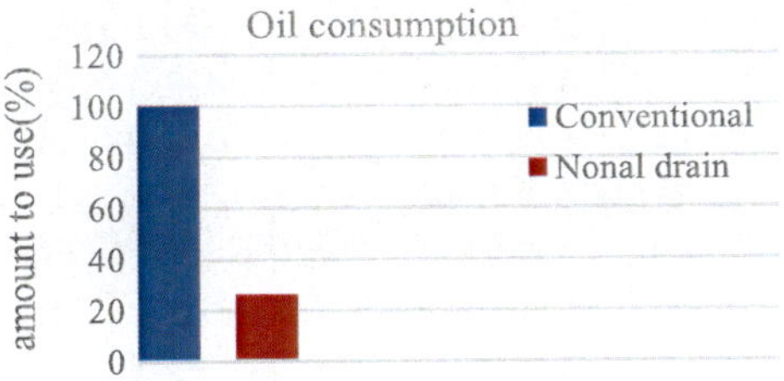

Fig. 9 Example of oil consumption reduction effect vertical axis shows rate of consumption of petroleum raw material (petroleum raw material/PVD, w/w); convention: conventional PVD, NND: Nonal drain

4.3 Countermeasure for Microplastics Issue

The problem of microplastics has become an important issue of the world in recent years. The microplastics are defined as dumped plastics that have flowed into the sea and rivers, have been subdivided by the effects of waves, ultraviolet rays, etc., and have become smaller than 5 mm. These plastic wastes are taken in by many organisms through the food chain. Microplastics became concern that they will have a harmful effect on wildlife and the human body.

3R (Reduce, Reuse, Recycle) is mentioned as a countermeasure for this problem. In 3R, "Reduce," which will reduce production and consumption, is the most effective measure. But PVD is usually left in the soil and cannot be recycled because it cannot be removed from soil after construction work. Therefore, biodegradable Nonal drain can reduce environmental stress from microplastic.

4.4 Biodegradation in Construction Site

We installed Nonal drain into the soil of construction site for residence zone development. Then we measured time course of biodegradation.

In order to measure rate of biodegradation, we used tensile strength as an indicator because biodegradation of polymer means reduction of molecular size resulting in embrittlement. In case biodegradation proceeds, tensile strength shall reduce accordingly.

As shown in Fig. 10, tensile strength gradually reduced and about 38% reduced at 28 months from installation. This result proved Nonal drain was degraded in the clay soil. After 28th month, we could not pick up specimen because of embrittlement.

Based on approximate calculation, Nonal drain seemed to reach the final stage of degradation about 72 months after installation.

Figure 11 shows the changes in coefficient of permeability after being left in the indoor case for 3 months, 6 months, 12 months, 24 months, respectively. At 24 months, the coefficient of permeability meets the product standard value [6].

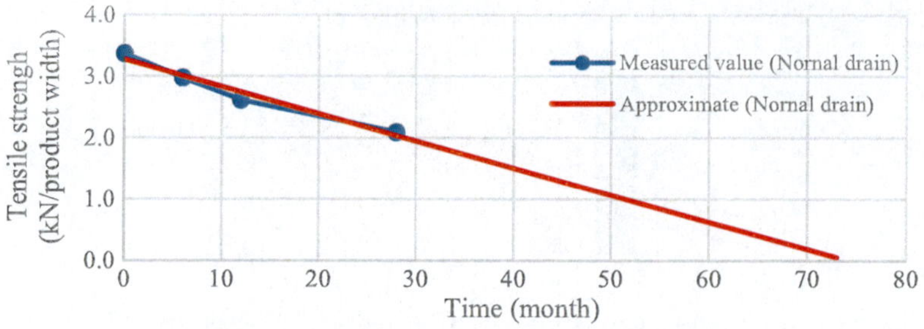

Fig. 10 Change in Nonal drain tensile strength over time vertical axis: tensile strength (at break), horizontal axis: elapsed time from installation

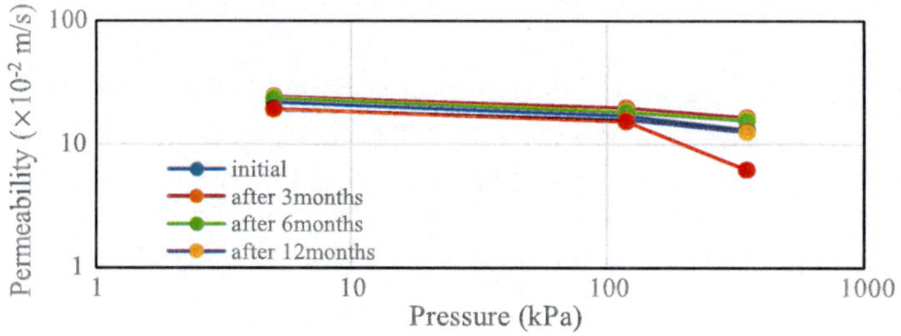

Fig. 11 Nonal drain permeability test result vertical axis: permeability, horizontal axis: pressure [6]

Although the rate of biodegradation is affected by some environmental factors, we expect these results can be reference for following investigation in anticipation of rate of biodegradation.

5 Problems

5.1 Surcharge

Currently, the vacuum pressure is about 60 to 70kpa. However, when the design load is required more, the embankment combined construction method of embankment on the sheet is used, and the embankment reduction by further increasing the vacuum is an issue.

5.2 *Bioplastic*

We have some subjects to improve about Nonal drain, e.g., control of degradation rate. Degradation rate of Nonal drain is thought to be affected by several factors in the nature because biomass plastic is derived from the nature. At this time, we cannot anticipate the degradation rate in each soil because of lack of sufficient data. We have to collect many data from many kind of soil conditions to solve this important subject.

Additionally, the decomposition of Nonal drain is accelerated by UV, high humidity, and high temperature, so it is necessary to take care about stock condition at the construction site.

6 Prospects

6.1 *Surcharge*

To reduce the amount of embankment by further increasing the vacuum pressure, measures to prevent air leakage and reduce pressure loss are necessary. We are examining sheet end treatment to maintain airtightness ant piping methods to reduce pressure loss and are proceeding with development for achieving high vacuum.

6.2 *Bioplastic*

We are trying to control degradation rate by modification of formula of Nonal drain. Future technology to control the rate make Nonal drain to be used easier and allow biomass plastic can be applied to several other purpose in construction industry.

Furthermore, when new technology which produces PLA from food waste will be operated, Nonal drain will be able to contribute development of sustainable society further.

7 Conclusion

In this report, we introduced Geo Drain SPD method which can reduce usage of sand and soil when soft ground improvement. This means we can reduce environmental stress from excavation and transportation. By using biomass-based-biodegradable plastics for PVD, we can reduce environmental problems such as microplastic problem and the crisis of oil depletion (Table 3).

Table 3 Method comparison

Method	Sand blanket	Embankment	Petroleum-derived plastic
Sand drain	Use	Use	
PVD	Use	Use	Use
Geo Drain	Not use	Use	Reduction
Geo Drain SPD	Not use	Reduction	Reduction

We propose with "Geo Drain SPD method" and "Nonal drain" a new sustainable construction method contributing SDGs concept.

Acknowledgements UNITIKA LTD., Geo Drain Association

References

1. Fukuda, Ito, Suwa, Nomura, Hashizume: Presentation of two layers horizontal drain model and design method for reduction of consolidation delay. Geosynthetics Papers **26** (2011)
2. Ikehata, Kodera, Nonaka, Okamoto, Miura, Muramatsu: Case study on soft subsoil measures by using the zonal horizontal drain and the plastic board drain that replaces sand mat. In: 57th Annual Scientific Lecture Meeting of JSCE (September 2002).
3. Geo Drain Association : Geo Drain SPD Method technical data (2019)
4. Sakai and Shirai: Total Recycle System for Municipal Food Waste for Poly-L-Lactic Acid Output, Haikibutsu-gakkai-shi 89–96, 15(2), 2014
5. Kamon and Miura: Plastic Bord drain kouhou-sono riron to jissai, ISBN:9784306024106, 2009
6. Geo Drain Association : Experiment report for confirmation of physical property change of plant drain material in soil (2020)

Climate Change Induced Disasters

Development of a Junior High School ESD Program to Raise the Disaster Prevention Awareness Based on the Experience of Heavy Rainfall in Lowland

Takashi Shimoyamada and Satoquo Seino

Abstract In this study, we developed a public junior high school ESD program to increase awareness of disaster prevention from heavy rain experiences. The theme of this program was "What is the phenomenon caused by heavy rain in our town? What are the issues to be improved?". In this program, we constructed an inquiry learning consisting of production activities and presentation activities. By this program, we taught to increase disaster prevention awareness of students. From the results of the questionnaire survey, awareness of disaster prevention was raised, and motivation to contribute to the community was recognized.

Keywords Disaster prevention education · Production · Presentation · Lowland

1 Introduction

In the lowlands covered by this study, water use and disaster risk were taken into account in the context of sustainability in relation to Sustainable Development Goals (Goal 11: Sustainable Cities and Communities) with recent climate change. Management balance has become an important issue. The school district targeted for this report (Kubota Town, Saga City, Saga Prefecture, Japan) is located in Kyushu and is included in the Saga Plain consisting of low-lying land (Figs. 1 and 2). Since the Holocene glacial retreat, the Saga Plain has continued to expand while heading south toward the Ariake Sea due to natural land reclamation and reclamation. The Saga Plain is a rich granary area, but it is difficult to secure agricultural water because the ratio of mountainous area to paddy area is small. Therefore, water from the Kase River and the Rokkaku River is being led by creeks that develop into every corner of the plain so that water resources can be reused efficiently.

T. Shimoyamada (✉)
Nabeshima Junior High School, 1-19-1 Nabeshima, Saga City 849-0937, Japan
e-mail: simo0818@icloud.com

S. Seino
Kyushu University, 744 Motooka, Nishi-Ku, Fukuoka 819-0395, Japan

H. Hazarika et al. (eds.), *Advances in Sustainable Construction and Resource Management*, Lecture Notes in Civil Engineering 144,
https://doi.org/10.1007/978-981-16-0077-7_66

Fig. 1 Location of Saga Prefecture

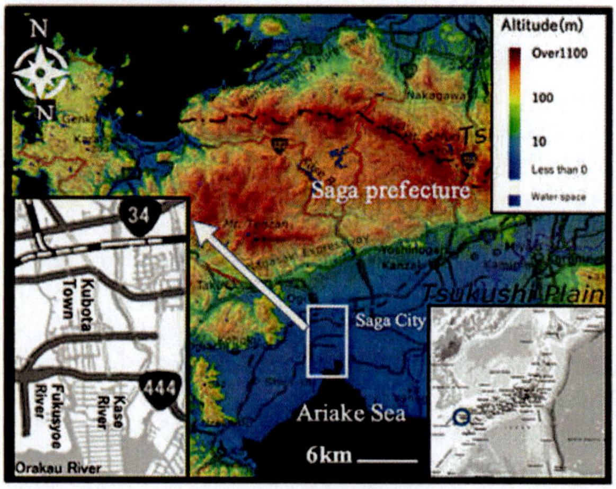

Fig. 2 The school district of the public school that conducted this study in the basin of a first-class river "Kase River" (Saga city). This map was based on the GSI Maps published by the Geospatial Information Authority of Japan

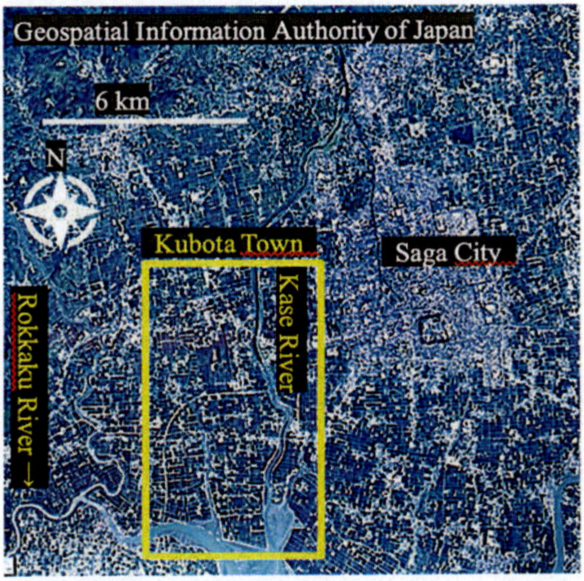

Fig. 3 Irrigation canal "Creek" spreading innumerably in the Saga Plain including Kubota Town

Agricultural water stored in innumerable creeks is pumped to each paddy field by agricultural pumps and returned to the creek (Figs. 3 and 4). The reduced water can be pumped again. Most of the origins of the creeks are those with water route or excavated for irrigation. After that, it was consolidated by the irrigation and drainage project and the field improvement project, and the creek became linear and efficient [1]. The water pumped from these creeks into the paddy fields continues to provide a rich harvest. On the other hand, during heavy rains, a large amount of precipitation temporarily gathers in the countless creeks, and the water overflowing from the creeks is eventually temporarily stored in the vast paddy fields, which may lead to roads and houses. It is thought that the influence is suppressed. Even in the case of heavy rain, the water of the creek that has lost its destination due to the closed gutter is drained directly to the river leading to the Ariake Sea by drainage gutter and the pumps of drainage pump stations in the area. However, the drainage from drainage gutter and this pump is affected by the water level of the river and the tide level of the Ariake Sea (the tide level at high tide exceeds 5 m and flows back through the river). This difficulty in draining, along with flooding due to rising water levels in rivers and floods due to levee breaks, poses a flood risk in the region.

Previous research [2–4] on disaster prevention education reported cooperation with the government and schools, disaster drills, and disaster prevention volunteer activities with the aim of improving disaster prevention resilience in the region. The public junior high school, which was the subject of this study, provided guidance on flood damage through evacuation drills and subject learning before the heavy rain in Saga. However, those instructions needed further improvement to raise awareness of

Fig. 4 Storage gate and pump for pumping installed in the irrigation channel "Creek"

disaster prevention in the school district. Therefore, in this study, for the presentation at the 2019 school cultural festival, we developed and implemented an ESD program related to SDGs Goal 11 from the disaster prevention side for all students who experienced the damage caused by heavy rain in Saga Prefecture.

The purpose of this survey is to raise awareness of disaster prevention among junior high school students who may be affected by heavy rain in the future. Through the ESD program we have developed, students can organize their experiences and observations of heavy rainfall damage and broaden their values on the disaster prevention aspects of Goal 11 of the SDGs. And it can increase students' awareness of disaster prevention. Students will characterize lowland areas with this ESD program consisting of multiple inquiry learning. For this reason, students will carry out production activities that utilize the experience of heavy rain damage of all school students and the hydrological and scientific knowledge of the area (Kubota Town). In addition, students will be able to make presentations summarizing the results of all exploration and learning activities and will be able to propose them as junior high school students' problems for improvement of school districts, together with local exchange activities. This study links the heavy rain disaster with domestic and international schools, school districts, and lowlands. In addition, it is one of the basin learnings that we have developed so far, which is a rare case of tackling the SDGs goal 11 from the disaster prevention side through an ESD program that incorporates production activities and presentation activities.

2 August 2019 "Torrential Rain in Saga Prefecture"

During the heavy rain in Saga in August 2019, Takeo City and Omachi Town were severely damaged by the inland water flood caused by the creeks in the Rokkaku River and Kase River Basin. Regarding the heavy rain, the weather of conditions of the Saga Local Meteorological Observatory "announces heavy rain special warning at 05:50 on August 28, 2019, in the northern and southern parts of Saga Prefecture (after that, Kiyama Town and Tara Town are added)", "All heavy rain special warnings were canceled at 14:55 on the 28th," and "The 72-h rainfall was 461.0 mm in Saga, which was the highest value in the history of observation".

In Kubota Town, Saga City (Fig. 2), which is the school district where the author used this study, the maximum rainfall per hour was 114.0 mm (Fukushoe Sanchoibi: August 28, 4:00 a.m. to 5:00 a.m.), the creek overflowed due to heavy rainfall with a cumulative rainfall of 471 mm on 8/26-8/31. Heavy rain on Saga and subsequent Typhoon No. 17 (Sep. 21-Sep. 23) confirmed severe damage to paddy rice and soybean and six underfloor floods. In addition, the main roads of the school district leading to the evacuation centers were temporarily flooded at multiple locations, which resulted in car breakdowns on closed roads and congestion on bypass roads. The deepest water level observed by students around their homes was above the knees, which was almost the same as the assumed water level (less than 0.5 m, underfloor flood) in the school district in "Saga City Water Hazard Map ". At our school, the area near the school gate was flooded for several hours after 5 a.m. on August 28. The road and parking lot in front of the main gate were flooded below the knee. The cause was the flood of a creek that crossed the road leading to the school gate from the intersection of the Saga Prefectural Outer Ring Road "Shisei Elementary and Junior High School". On the other hand, the gate on the opposite side of the west school did not flood.

During this heavy rain in Saga, which lasted for several days, heavy rain special alerts, flood alerts, lightning alerts, strong wind alerts, landslide disaster alerts, flood alerts, evacuation orders, and river flood information were issued day and night. It is probable that for several days, students and their families endured fears and anxieties, paying attention to various media for disaster and weather information in order to do their best to save lives. The water flooded near the main gate was discharged next day.

3 "Basin Learning Field" in the Kase R. and Rokkaku R. Basins

The Kase River basin (basin area 368 km^2) and the Rokkaku River basin (341 km^2), which are the target of this study, are the targets of the basin learning, which is our research activity. The junior high schools and school districts targeted by this survey are located on the right bank of the Kase River basin and relatively far from the

Rokkaku River (Fig. 3). The Saga Plain lies in the low plains extending southward from the Sefuri Mountains in the northern part of Kyushu (800–1000 m above sea level, about 50 km east–west, about 25 km north–south, Mt. Sefuri 1055 m) and extending to the Ariake Sea. This Saga Plain is located on the Saga Prefecture side of the Chikushi Plain. And the Saga Plain which consists of the basins of the Kase River and the Rokkaku River. In the Kase River basin, rainfall in the mountains spread out like a fan in the Saga Plain, many disasters occurred due to breakwaters and floods, and flood control and water use projects have been continued for a long time. In river basins, excavation and logging are required to improve drainage capacity in addition to improvement of levees and soil. In addition, the Rokkaku River starts from around Mt. Shinroku (447 m) in Takeo City, meanders the Shiroishi Plain including Takeo City and Omachi Town, and joins the Ushizu River and pours into the Ariake Sea. The Rokkaku River basin is a region heavily flooded by internal and external floods. And, it is a lowland that was heavily damaged by the heavy rain in Saga in the summer of 2019.

The area around the mouths of the Kase River and Rokkaku River has become lowlands due to natural reclamation and reclamation. Due to the soft geology and the tide of the Ariake Sea, it is lower than the sea level at high tide. For this reason, the discharge of innumerable tributaries and irrigation waterways (creeks) tends to deteriorate, and water damage is said to occur easily.

4 Setting and Verification of Research Hypotheses

Along with recent climate change, the balance between water use and disaster risk management has become an important issue not only for the school district but also for sustainability in the lowlands of the world. ESD stands for Education for Sustainable Development, and it is an education that starts from the problems facing the region and develops into the problems facing the world such as the modern environment, poverty, human rights, peace, and development. ESD is an interdisciplinary and comprehensive learning and activity that aims to create new values and actions that help solve problems and build a sustainable society. Furthermore, ESD is said to be an education that fosters leaders who build a sustainable society.

The hypothesis of this study was that the ESD program we developed would increase students' awareness of disaster prevention. In this program, in order to raise awareness of disaster prevention, we instructed public junior high school students in Kubota Town about production activities based on disaster experiences and presentation of learning results. And the theme of the program we developed was "What is the phenomenon caused by the heavy rains in our town? What are the issues to be improved?". This theme was set up for the 2019 school culture festival. In this ESD program, students learned the contents of Table 1 according to the plan shown in Table 2. Finally, we conducted a questionnaire survey to confirm the increased awareness of disaster prevention among students through this program.

Table 1 Inquiry learning of ESD program

	Learning outside the school	Production Activities	Exchange Activities
Inquiry learning	Heavy rain experience at home	Wall newspaper, hazard map, and model research poster	Exhibition announcement oral presentation

Table 2 Inquiry learning plan of ESD program

August 28, 2019	"Saga heavy rain disaster" observation of all students in each home
August 29–November 29	"Saga heavy rain disaster Wall Newspaper" production
September 30	"First questionnaire survey" for all junior high school students
October 15–25	Production for school cultural festival "Kubota Town Inland Water Hazard Map" "Kubota Town flood hazard map" model topographic map, "research poster" panel
October 26	Exhibition of products produced at school cultural festival
November 27	"Second questionnaire survey" for third-grade junior high school students
November 31 to December 1	"Kyushu "River" workshop in saga" oral presentation/exhibition announcement
February 1, 2020	"Saga City disaster prevention training" oral presentation/exhibition announcement

5 Hazard Map Creation and Discovery Based on Heavy Rain Experience

5.1 Discovered from the Plan "Kubota Town Inland Flood Map" Created

The first questionnaire (total of 217 people (84 in first grade, 67 in second grade, 66 in third grade), 201 in total) investigated the flood area and its depth that each student witnessed around their home. The inundation depth was recorded on a blank map with easy-to-understand expressions such as ankles, knees, and waist height. In addition, each student wrote down a closed road on a blank map. Then, the third grader aggregated the survey results of all students and summarized them in the "Kubota Town Inland Water Hazard Map" (Fig. 5, left).

At this time, the students and teachers looking down on the completed flat hazard map (Fig. 5, left) remembered past flooded fields and closed roads. And they discovered that there was an overlap between the past and present disaster areas.

Fig. 5 "Kubota Town Inland Flood Map" (left) produced by students in the third year of a junior high school and satellite images (right) regarding heavy rain in Saga on August 28, 2019

5.2 Solution by Three-Dimensional Topography Model "Kubota Town Inland Water Hazard Map"

In order to solve the cause of the overlap between the past and present disaster areas within the school district, we instructed the production of the three-dimensional topographic model Fig. 6 "Kubota Town Inland Water Hazard Map (coloring every 0.5 m above sea level: length 2.6 m $\times$ width 1.6 m)" by students. Figure 6 was created by superimposing Fig. 5 (left) and the school district contour map [5, 6]. Students also created a pyramid of quadrangular pyramids that are color-coded according to the depth of the flood (red is "knee depth", yellow is "ankle depth", and blue is "closed road"). Students fixed them in the representative position in Fig. 6.

Finally, students and teachers confirmed this completed the three-dimensional topographic model. And it turned out that the main cause of the overlap between past and present "flooded areas and closed roads" was a slight difference in elevation in the lowlands. The difference was visually between "2 m above sea level" and "more than 2 m above sea level".

5.3 Research Poster Production

We instructed third-year students to create a research poster outlining their research and learning outcomes. At that time, the homes of all the students, who were the observation points of heavy rain, were not evenly distributed in the school district,

Fig. 6 Three-dimensional topography model "Kubota Town Inland Water Hazard Map"

and there were biased observation records such as floods and road closures. So we got more accurate and scientific information about the district's characteristic floods. We instructed the students to add them to the study poster for comparison.

Therefore, we obtained the image taken by JAXA [7] at 12:11 on August 28 at Daichi 2 PALSAR-2 (high resolution 3 m resolution, HH polarization mode). And we instructed the students to utilize the image in their research posters. The orange part in Fig. 5 (right) is estimated to have been flooded, but this is an estimated value that includes changes in image brightness such as urban development and seasonal changes in farmland and changes that are not related to disasters. It is reported that this image is the result of analysis using past PALSAR-2 data of the Ministry of Land, Infrastructure, Transport and Tourism and flood estimation data.

5.4 Presentation of the Heavy Rain Damage Survey of the School District

The results of all the above work and surveys were widely reported by students at the 2019 School Cultural Festival and at the presentation opportunities in Table 3 in exhibition and oral presentations. The students who made these presentations were asked questions and advice from visitors such as cities and towns, government officials, experts, and school district residents. In addition, students were able to discuss with their attendees the impact of the slight difference in elevation in the lowlands on the inundation position and inundation depth and the suggestions for improvement from the junior high school students to the school district regarding the

Table 3 Presentation opportunities through oral presentations and exhibition announcement

School cultural festival	Public school in Kubota Town, Saga City
The 19th Kyushu "River" Workshop in Saga	Organizer: The 19th Kyushu "River" Workshop in Saga Executive Committee, Co-sponsor: Kyushu River Association Group Liaison Conference, Sponsor: Kyushu Regional Development Bureau, Ministry of Land, Infrastructure, Transport and Tourism
First Year of Reiwa Saga City Disaster Prevention Drill	Ministry of Land, Infrastructure and Transport Kyushu Regional Development Bureau Takeo River Office, Saga City/Kubota Town, and related organizations

disaster prevention aspect of SDGs Goal 11. These discussions provided valuable educational opportunities for students.

In this way, disseminating the results learned by junior high school students to the community can be considered as a form of social participation. As a citizen of the school district, the opportunity to announce the results will be a valuable educational opportunity to raise awareness of disaster prevention and foster positive humanity.

6 Questionnaire Survey and Results

The method of verifying the research hypothesis in this study was a questionnaire survey. Each survey consisted of questions that included options and free explanations. About a month after the production and display of the Saga Heavy Rain Disaster Wall Newspaper started (September 30), we conducted the first questionnaire survey to collect damage information and records from all students. 217 students: 84 in first grade, 67 in second grade, 66 in third grade, total 201 respondents). In addition, we conducted a second questionnaire on November 27, after the end of the school festival, for all the third-grade junior high school students (62 students) who produced according to Tables 1 and 2.

From these results (Fig. 7 1,2,3), we were able to clarify the situation around the student's house during heavy rain in Saga Prefecture. And more, from Fig. 7 4,5, it was found that students are interested in disaster volunteers in the event of a heavy rain disaster, measures for the next disaster, support, and reconstruction. Also, from the results of the first survey, it was found that about 47% of the students in the third grade of junior high school want to participate in volunteer activities in case of a disaster caused by heavy rain. In the second survey after the implementation of the ESD program, the third-year middle school students' volunteer activities motivation to participate increased to 76%. In addition, 79% of all third graders predicted heavy rain disasters in the school district in the future.

In addition, from the results of aggregating all student suggestions for open-ended questions in the first survey, Table 4 shows the improvement tasks that each student

Table 4 "Issues for improvement that can be done by me" based on the experience of heavy rain in Saga

In the second questionnaire for the third year of junior high school, the students made a free description of improvement suggestions for future flood damage. () shows the same number of proposals.
・Preparation of emergency food and drinking water (11) ・Prepare disaster prevention goods and disaster prevention backpacks (10) ・Prepared to evacuate at any time (10) ・Learn more about evacuation areas and dangerous areas (4) ・Confirmation of disaster contact network with family and evacuation site (3) ・Disaster volunteer activities (3) ・Evacuate early (3) ・Collect information and tell everyone (3) ・Determine the evacuation site with your family in advance (1) ・Preparing sandbags (1), deepening knowledge about disasters (1)

Table 5 Improvement problems that the third year of junior high school wants to propose to the administration based on the experience of heavy rain in Saga

In the second questionnaire for the third year of junior high school, the students made a free description of improvement suggestions for future flood damage. () shows the same number of proposals.
・Improvement and strengthening of levees (22) ・Evacuation centers, volunteers, preparation / storage of relief supplies (18) ・Prompt coverage of evacuation information and surrounding area (13) ・Update/detailing of hazard maps, installation of hazard maps in various places, and distribution to all households (11) ・Creating a town that predicted flooding/flooding, measures and preparation (9) ・Evacuation drills (corresponding to floods) and evacuation routes are known (8) ・Improvement of roads (drainage, size, difficulty in flooding) (8) ・Assistance with support money (6) ・Responding to disasters and floods and reviewing early recovery after a disaster (4) ・Review of areas and evacuation routes/places that are easily flooded (3) ・Drone delivery system when you can't go to shelters and isolated houses (1), Mountain cliff surface fixing (1), Tsunami countermeasure (1), Building remodeling (1), Abnormal weather as a countermeasure against sediment-related disasters (1), Safe evacuation of elderly people (1), Countermeasures against water infiltration from the bottom of food racks in stores (1)

can perform based on their experience of heavy rain in Saga. In addition, Table 5 was able to show the improvement issues that students would like to suggest to government agencies regarding disaster risk reduction (Fig. 7).

As mentioned above, junior high school students who experienced heavy rainfall in Saga learned the importance of balancing the water use of the creek with the flood crisis through this ESD program. Students were also able to think about the sustainable development of the school district toward SDGs Goal 11 from the perspective of disaster risk reduction easily.

7 Conclusions

In this research, for the purpose of improving disaster prevention awareness, we developed an ESD program consisting of 2D/3D hazard map creation activities based on experience and observation of heavy rain damage of junior high school students and activities to announce learning results.

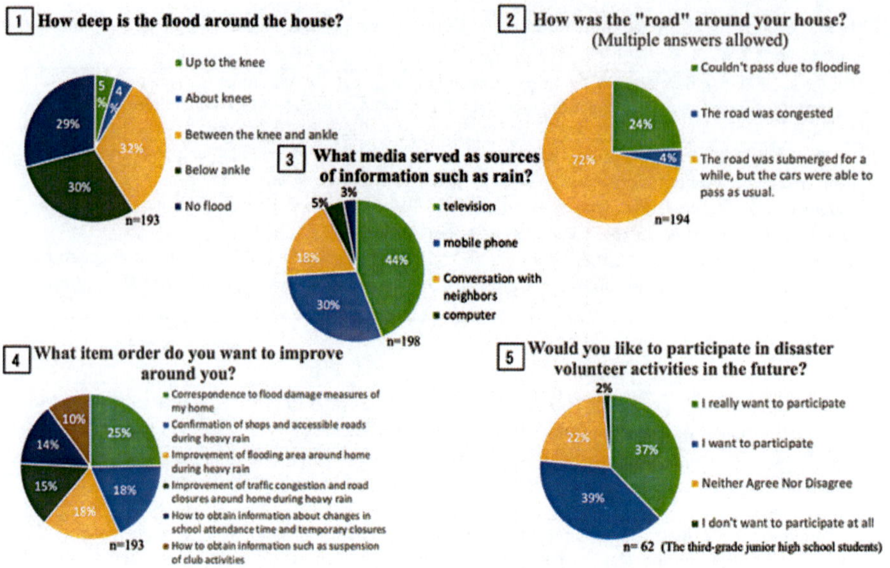

Fig. 7 Results of the questionnaire surveys on disaster prevention awareness

Through the implementation of this ESD program, junior high school students increased their interest in improving the area from a disaster prevention standpoint and were aware of disaster risk reduction in order to create a sustainable city that meets SDGs Goal 11. And, we were able to increase the students' willingness to participate in building a sustainable society with disaster risk reduction as one of the issues and disaster awareness.

Developed ESD programs may lead to change junior high school students' values and behaviors for balancing water use and flood risk management to protect our lives and societies living in the lowlands. This ESD program was also effective from the perspective of linking flood damage and water use balance common to lowlands in Japan and overseas with educational sites and regional development.

However, in the future, if the climate change becomes large, this ESD program alone will not be enough to protect lives from flood damage and as education aimed at building societies from the perspective of disaster prevention through sustainable development. It is necessary to incorporate various improvements for school education.

For this reason, in Japan, where there are many natural disasters, it is necessary to develop, accumulate, and update ESD programs for school education and social education in school districts and administrative units that have experienced characteristic disasters many times in the past. In order to raise disaster prevention awareness and resilience to characteristic disasters in various regions, the history of disasters in the region, disaster and archeological sites, hydrology, meteorology, earthquakes, volcanoes, geology, civil engineering, and infrastructure.

In addition to management, we think necessary to learn information and preparations such as hazard maps, timeline, various warnings, and evacuation methods. And more, we think it is necessary to train evacuation shelter management support experiences by junior and senior high school students, disaster volunteer experiences, home stockpiling preparation, and for the elderly, infants and foreigners.

References

1. Kyushu Agricultural Administration Bureau, Creek in Saga Plain (Saga Prefecture), https://www.maff.go.jp/kyusyu/seibibu/history/06.html. Last accessed 29 July 2020
2. Oita Prefectural Board of Education, Dis-aster Prevention Education Case Studies <1st>, https://www.pref.oita.jp/site/gakkokyoiku/2000435.html. Last accessed 29 March 2020
3. Oikawa, Y., Taguchi, K.: Marine education as disaster prevention -for sustainable symbiosis between the sea and people-, Curriculum Development of Marine Education: Research and Practice, The Research Center for Marine Education, Ocean Alliance, pp. 361–381. The University of Tokyo Japan Educational Press (2015)
4. Kimura, N., Yamashiki, Y., Kisić., I.: Awareness-raising of flood risk and building resilience among children in Zagreb, Croatia. Hydrol. Res. Lett. **8**(1), 64–70 (2015)
5. Tani, K.: Development and algorithm of a web site to draw contour lines of Japan using altitude tile data, Occasional paper of Department of Geography, Saitama **35**, 73–83 (2016)
6. Tani, K.: Web contour maker (ver1.08) using the elevation tiles of the Geographical Survey Institute, http://ktgis.net/service/webcontour/index.html. Last accessed 29 March 2020
7. Japan Aerospace Exploration Agency. https://www.eorc.jaxa.jp/earthview/2019/tp190829html. Last accessed 29 March 2020

River Education on Sustainability and Disaster Prevention at Elementary Schools in Island Regions

Hidefumi Hatashima and Satoquo Seino

Abstract The participation of elementary school students studying in the area is important for the sustainability of the area. This paper is a practice of an elementary school in Tsushima City, an island region of Japan. Two years ago, elementary schools have been conducting river education through collaborative learning. Through river education, elementary school students were able to understand that the river was clean, the extent of the brackish water area was discovered, the river business was understood, and salt damage was understood. We believe that river education will lead to disaster education as a result. Also, elementary school children want the water they use as drinking water to be clean forever, and they can imagine their future life from the perspective of disaster prevention. It was revealed that such elementary school students' understanding contributes to regional sustainability.

Keywords River · Sustainability · Nita river · Farmland development

1 Introduction

Creating a sustainable society requires scientific understanding of rivers and wise decision making. River education requires a comprehensive approach to science and society. The introduction of river education in Japan's primary education was amendment of River Law in 1997. In addition to the conventional flood control and utilization, improvement and conservation of river environment were stipulated in the legal purpose. Understanding of local rivers can be concretely imagining disasters in an emergency by environmental education during normal times.

In this paper, we discuss sustainability and disaster prevention education based on case of river education at Tsushima Municipal Nita Elementary School in Tsushima City, which has been struggling with water use and agriculture historically (see Fig. 1).

H. Hatashima (✉) · S. Seino
Graduate School of Engineering, Kyushu University, Fukuoka City, Japan
e-mail: h.hatashima@civi.kyushu-u.ac.jp

H. Hazarika et al. (eds.), *Advances in Sustainable Construction and Resource Management*, Lecture Notes in Civil Engineering 144,
https://doi.org/10.1007/978-981-16-0077-7_67

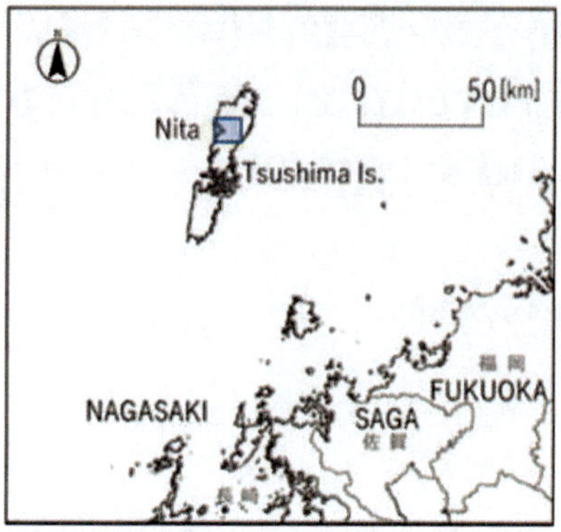

Fig. 1 Study site Nita District, Tsushima Island (From the National Geographic Institute)

2 Background of This Study

2.1 Previous Research

Since 2000, integrated studies have been newly established to promote cross-sectional and comprehensive learning and exploratory learning in the elementary school curriculum. Environmental education was also illustrated as a learning activity. In this learning, the place to learn expanded not only to the school but also to the area outside the school, and the utilization of local human resources increased. Children have more opportunities to learn knowledge from the community.

The importance of disaster prevention education has been pointed out since the 2011 Great East Japan Earthquake. The Ministry of Education, Culture, Sports, Science and Technology pointed out that in promoting disaster prevention education in schools, "By relating activities for learning the history and natural environment of the region, children and students will benefit the region from various perspectives" [1]. In addition, Teramoto stated in his geography report that disaster education requires the development of geographical imagination [2]. Teramoto pointed out the opportunity to walk experientially in the precincts to develop this ability.

From these previous studies, by setting opportunities for learning the history and environment of the region during integrated studies, exploring by walking and setting collaborative learning with the region, all knowledge of the region can be acquired. It is clear that it fosters geographical imagination and leads to disaster awareness. Furthermore, if river education that leads to regional sustainability can be developed at the time of integrated studies, it may be possible to develop disaster prevention education.

In this practice, the sustainability of the area is promoted by the river education, and the relation to the disaster prevention education is examined.

2.2 Realities of the Area

In the Seta area, which is located on the north side of the school district, the Ministry of Agriculture, Forestry and Fisheries has been engaged in a multi-functional agricultural function countermeasure project since 2014. Through this project, people in the Seta area have been able to cultivate abandoned land. They have also planted trees on both banks of the Nita River to improve the landscape of cultivated land and rivers. Furthermore, they have turned their attention to developing human resources for the next generation of agricultural engineers, as agriculture has supported regional and Japanese food production. Therefore, starting in 2018, the aim was to develop human resources for agricultural engineers, and cooperation with elementary schools began (see Figs. 2 and 3). At the beginning of April, two representatives from the Seta area came to our school. Then, in May, we and local representatives confirmed the program and the activities began. Hart discussed the participation of children and these learnings and also participated in agricultural administration [3].

Now we will explain the river education and relationship with the area before this learning began. River education teaches the function of water flowing in science classes. There is a possibility that it may develop from science learning to conduct research on aquatic life and water quality, but elementary school students did not learned that.

At Nita Elementary School, upper elementary school students are working on rice production. For more than 10 years, the Kaidokoro area has been cooperating with the school. In addition, they were learning about regional food culture through

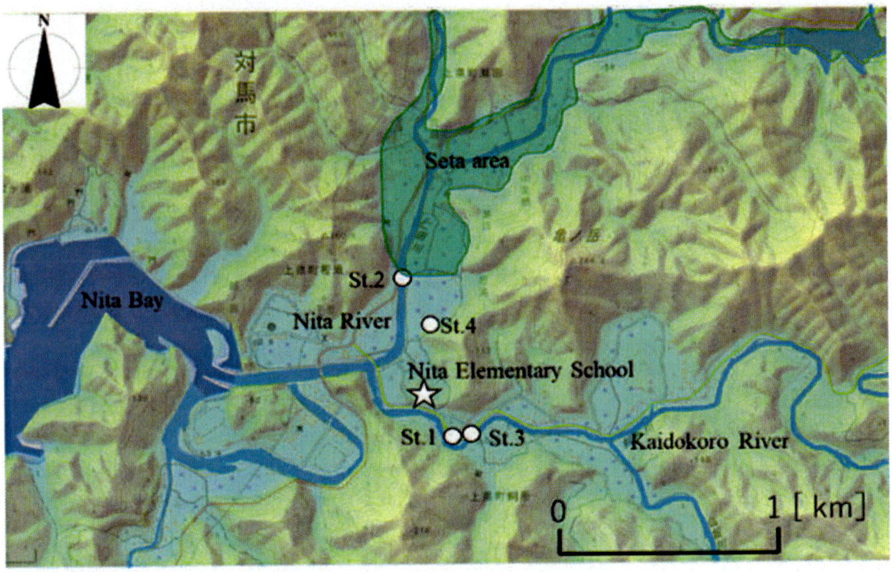

Fig. 2 Land in the Nita area (From the National Geographic Institute)

Fig. 3 Elementary school students receiving guidance on farm work on abandoned arable land [2019.5.25]

interaction with elderly care facilities in the Kashitaki area. In the lower grades of elementary school, the elderly men in the area taught students old games.

In other words, it can be said that elementary school students at Nita Elementary School learned about rivers in science lessons and had a lot of interaction with the community.

3 Study Site and Methods

Nita Elementary School, Tsushima City, Nagasaki Prefecture, which conducted this research, has 53 students and 9 teachers in 2018. It is located on a hill in the Nita area of Kamiagata Town, Tsushima City. In front of the school is the Kaidokoro River (stream extension 12.9 km, basin area 26.20 km), a tributary of the Nita River (stream extension 14.2 km, basin area 106.09 km), and the road along the river is a school route(see Fig. 2).

The Nita elementary school has been practicing river education since 2018. Promoting practice outside the school in collaboration with the community. In 2019, Tsushima City suffered flood disaster every 50 years, three times. In this area as well, there was no human injury, but there was damage caused by the calves being raised.

The analysis target of the educational contents is the review notes of the students and the lesson records of teachers during the practice in 2018 and 2019. In addition, knowledge of river life such as land use of the target river, agricultural land development, migratory fish, etc. was obtained by interview survey with residents.

4 Surveys

We will describe the practice in the Nita River basin at Nita Elementary School, the natural conditions there, and the development process. Tables 1, 2, 3, 4 and 5 show the elementary school students' awareness of the techniques of vegetable making and water circulation. The authors categorized the awareness of elementary school students.

4.1 Practice of Utilization of Abandoned Land

St.4 in Fig. 2 is a farmland developed jointly by the Seta area and elementary school students. From Table 1, elementary school students learned how to grow vegetables, such as erecting struts according to soil preparation and vegetable growth. From Table 2, it was also understood that the Kashitaki area where the developed farmland is located uses the water of Meboro Dam. It was also found that the river construction expanded the brackish water area and constructed a movable weir. The expansion of the brackish water area is also apparent from the fact that the goby classified in Table 3 lived in the brackish water area.

Elementary school students have learned not only the agricultural technique but also the fact that the cultivated land is taken from the dam, the nearby well water

Table 1 Understanding elementary school students in vegetable cultivation

Categorization			Students' expression
Vegetable cultivation skills	Preparation for cultivation	Revitalization of abandoned fields	Remove roots and stones from abandoned fields
		Agricultural action	Loosen the hardened soil in the field, sprinkle fertilizer, and mix it with a machine
			Ridges of the fields covered with vinyl protection
	Procedure		Soften the soils, put the seed, spray the water
			Water and fertilizer for harvest
			Along the growth of vegetables, band them to pillars by eight shape knot
	Tools		Utilization of bamboo as the pillar

Table 2 Understanding elementary school students in water awareness

Categorization		Students' expression
Awareness of water	Old water use	There was an old well in the area of Nita
		Wells become tide water and are rarely used
		The field is made near the pond
		In the old days, farmers pumped water from the river
		The water supply was formed, and the well water was not used
	Effects of river civil engineering	The water in the fields of the Kashitaki area comes from the Meboro Dam
		The tide rose by the river construction, and a movable weir was formed
		The water taken from the weir is sent to the fields in the Shimoharu area

Table 3 Understanding elementary school students in identification of goby

Categorization			Students' expression
Classification of goby	Body morphology	Overall morphology	Striped pattern with 5 fins
		Detail pattern of parts	Some cheeks are red
		Overall color	The body is pale
			The body is yellowish
		Detail color of parts	The belly is pale
			The eyes are blue or green
			There is no orange in the tail as seen in other goby
		Detail color with morphology	The mouth is red, and there is a bump
	Habitat	Overall	Inhabits shallow places
			Inhabits brackish waters
		Detail	Not on rocks, but on pebbles

Table 4 Understanding elementary school students in biotope making

Categorization		Students' expression
Biotope making	Land height	The land is low and not suitable for farmland
	Relationship with dam	The water of the biotope has disappeared
	Natural environment	I would like to add aquatic plants to the biotope, but I am not growing up in the Nita River now
		Frogs and the Japanese giant beetle lived in the biotope
	Relationship with dam	The water of the biotope has disappeared

Table 5 Understanding elementary school students in water quality survey

Categorization		Students' expression
Water quality survey	Understanding the meaning of values	The COD values of the Nita and Kaidokoro Rivers are different
		The Kaidokoro River has a COD value of 0, so it is clean because it does not require oxygen to decompose dirt
	River cleanliness	Kaidokoro River is cleaner than Nita River
		The COD values of the Nita and Kaidokoro Rivers are low, so it's beautiful
	Survey methodology	We examined the results in three groups and found no significant differences in COD values

cannot be used due to salt damage, that is, the influence of civil engineering works on the river rehabilitation water cycle.

A biotope was made next to the farmland shown at St4 in Fig. 2. As a practice to learn the management of aquatic ecosystem, we have been working on biotope making since June 2018. We investigated the aquatic plants that inhabited the Nita River, and transplanted hornwort that grew naturally in the wetlands of the Tanohama area of the school district to the biotope (see Fig. 4). As you can see from Table 4, in the fall, we were able to observe the representative creature of the paddy field. However, the biotope failed to retain water and was completely drained in December.

In 2019, we moved the cultivated land, which is a place for field activities of agricultural practice, near the school. Biotope making was supported by an environmental protection organization in 2019. The water retention capacity was improved, and the goby that inhabit the Nita River was released. Water does not drip even now.

Fig. 4 Elementary school students making biotopes on abandoned arable land [2019.6.21]

4.2 Survey of Aquatic Organism and Water Quality

We also worked on river environmental surveys. Surveys of river organisms were done by our elementary school students in 2018 (see Fig. 5). Students surveyed aquatic organisms at St. 1 in Fig. 2. Students were able to find goby, crab, and

Fig. 5 Elementary school students surveying aquatic organisms [2019.8.1]

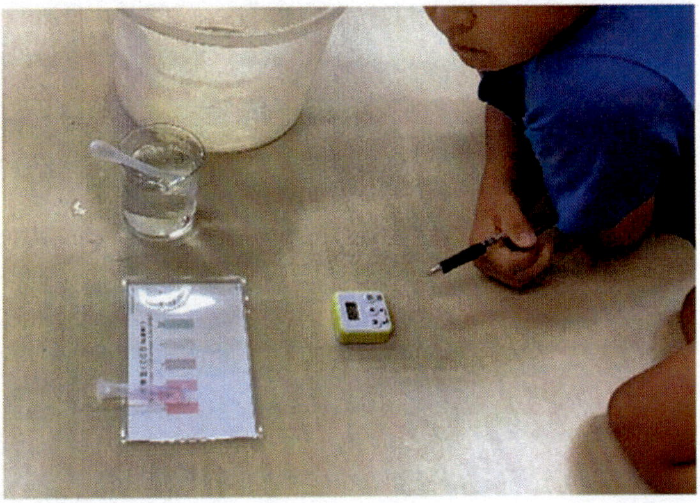

Fig. 6 Elementary school students collecting water for river water quality surveys [2019.8.3]

mollusks that live in brackish waters. In 2019, we began a water quality survey by working on a pack test to measure COD values (see Fig. 6). They surveyed aquatic organisms at St. 3 in Fig. 2. These activities were partnered with university research institutes and water quality research experts.

In the August study, we captured and observed dobsonflies, which are indicator organisms of the clear stream. In addition, COD values ranged from 0 to 2, and elementary school students were able to recognize that the quality of the river in the school district was good, together with the evaluation of experts. COD values up to 2 indicate clean water because it indicates rainwater and upstream water from rivers. COD values up to 2 indicate rainwater and upstream water of the river, 2 to 10 indicate downstream water of the river, and 20 indicates sewage and sewage. The Nita River has a COD value of 2, so it is clean water. These learnings solve a trend away from river mentioned by Nakamura [4].

4.3 Land Use and Farmland Development

The history of the Nita River basin was examined. The area is dotted with ruins from more than 7000 years ago. Being able to see the entire area and to confirm the residence by being a human being continuously through the times of hunting, gathering, and farming means that sustainable living had been established in this basin.

The Kashitaki area in the middle reaches was developed in Nitta in the 1700s. In the 1970s, 1980s, and 2000s, the area was developed as agricultural land (see Fig. 7).

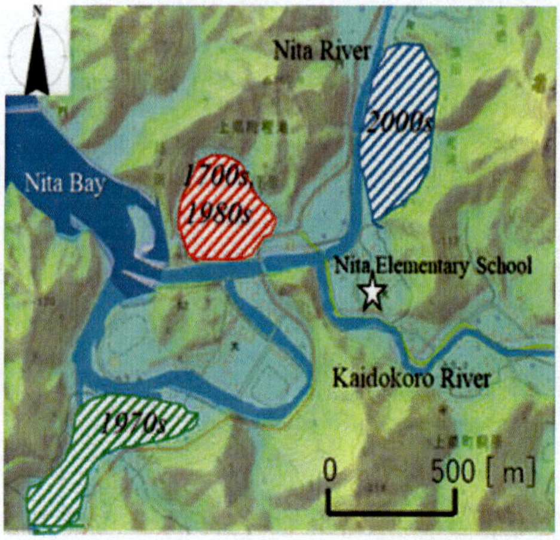

Fig. 7 State of farmland development in Nita area

However, agricultural land that has been permanently developed over 300 years is at a turning point. There was a shortage of successors due to the national reduction policy and group employment during the high economic growth period. Since these farmlands were originally floodplains of the alluvial plains that were deposited by the flood of the Nita River, they have been hit by floods every time of heavy rain.

The reality is that some land is not suitable for cultivation. Despite many years of farmland development, many are now abandoned.

4.4 Biota of Nita River Before River Civil Engineering Business Development

Currently, to prevent floods, the Meboro and Nita dams have been constructed in the Nita River, and the meandering estuary has been expanded with straightening the flow and straightening the flow. According to interviews with the residents, before the construction of these rivers, both sides of migratory fish, ayu, were seen in the entire Nita River, and Japanese eels and Mokuzu crabs, which are organisms in the brackish waters, were inhabited. It was clear that the Nita River was rich in nature, although there was damage from the flood, such as water plants growing in the upper stream. Even now, natural sweetfish inhabitants can be seen in the Nita and Kaidokoro rivers.

5 Result

Figure 8 is a flowchart showing this practice. First, we will discuss what kind of river education practices were connected from the actual conditions of the community and schools.

In this area, we are promoting a multifaceted agricultural low-functionality countermeasure project. River projects have been promoted in this area to prevent drought and floods. The school had multiple classes because it was connected to researchers and it is a small school. Therefore, I was able to study aquatic organism researches and water quality surveys by having researchers. In addition, I was able to make biotope and experience farming in abandoned cultivated land during the time of

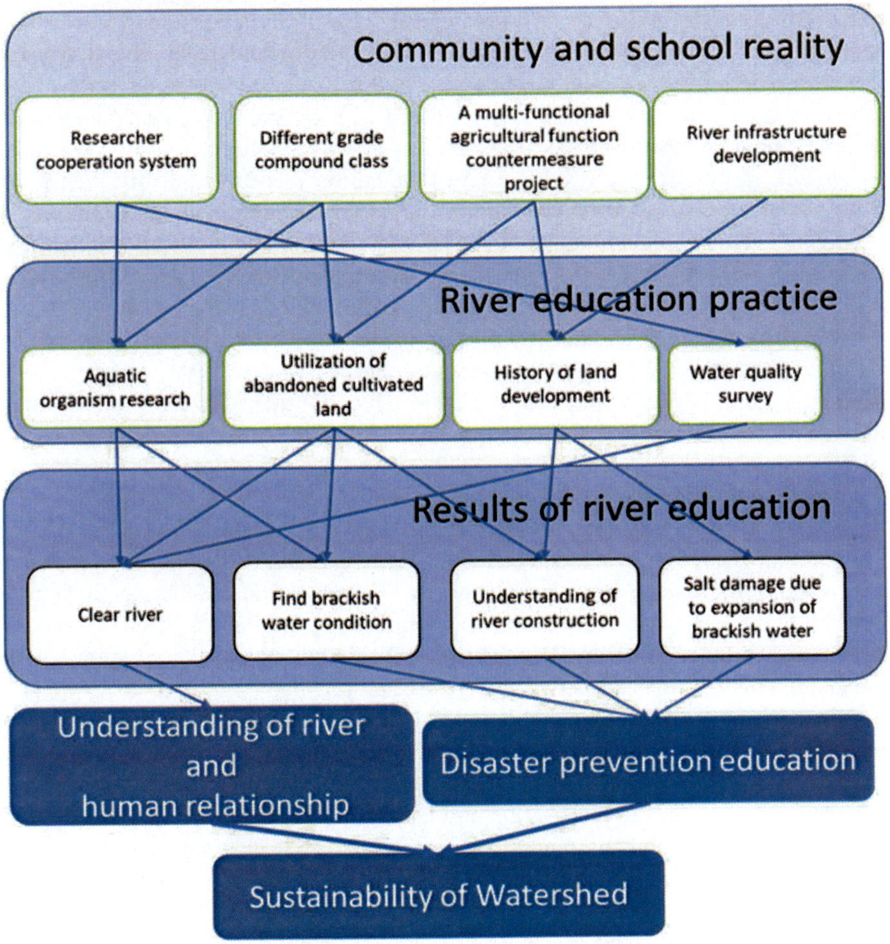

Fig. 8 River education and disaster prevention education, flow diagram of local sustainability

general studies and integrated studies shown in the curriculum guidelines. By participating in the multi-functional agricultural function countermeasure project, I was able to experience farming and know that farmland developed. Land development was to supplement the opening of cultivated land, and it was necessary to develop river infrastructure.

Next, we will discuss what kind of results were obtained by practicing river education. In the river survey, it became clear from the water quality survey that the river was a clear stream, and from the aquatic organism survey, the place that was thought to be fresh water was brackish water. Elementary school students will discover again that the river is clean because it takes advantage of the abandoned farmland to pass along the riverbank, which is also a school road. And they also understood that the water in the cultivated land came from distant dams, and that the nearby wells and rivers were brackish and could not be used for agriculture. They also understood that a dam was needed for agricultural use and a weir was needed to secure fresh water. Examination of the history of land development revealed that river works were necessary to prevent flood damage. However, due to the river infrastructure construction, the brackish water area has expanded, causing well water and salt damage to farmland.

We will summarize the results of river education. The extent of brackish water is closely related to sea level rise due to global warming and storm surges. If we can recognize how much brackish water is, we can think about where to live. It can be understood that the flood countermeasures are taken by the river infrastructure construction. However, we must understand that we cannot say that we are protecting everything by these works. The land where salt damage occurs is low in the first place. In other words, it is necessary to understand that it is not suitable for farmland or residential land. Through river education, the extent of brackish waters and the limits of river civil engineering projects are known, and the ability to decide whether or not to live can be developed, leading to disaster prevention education. It is this learning that fosters the geographical imagination through the town walk described above.

In other words, by grasping the size of the brackish waters and the status of river infrastructure development, it can be judged that land with salt damage is dangerous for disaster prevention. This is the reason why river education and disaster prevention education are related. By promoting river education, we want to leave a beautiful river that can be used as drinking water forever. Elementary school students can understand the relationship between the river and people.

Through the practice of collaborating with local residents, elementary school students found that river rehabilitation, land development and coordination with the environment, and farmland that had been cultivated by ancestors' hardships are being abandoned due to disaster prevention and changes in social conditions. Elementary school students learned the natural characteristics of local rivers and gained the experience of thinking from a historical perspective. Such educational programs are expected to develop human resources who can predict what will happen in the future due to climate change and sea level rise, and to develop it into regional disaster prevention education. River education has expanded the framework for collaborative learning through the cooperation of local human resources and the involvement of various research institutions, experts, and organizations. Elementary school students continue to the present by finding a viewpoint to explore the Nita River in a multifaceted and multifaceted manner. By utilizing abandoned cultivated land, river biological surveys, and water quality surveys, we were able to get a glimpse of the benefits and negative aspects of river civil engineering projects.

6 Conclusion

In the practice of river education, I will mention the educational effects. The first is that students have the ability to interact. Elementary school students had the opportunity to talk to people from various regions by utilizing abandoned farmland. I was able to interact with the researchers I met for the first time. It can be said that the ability to interact has improved by expanding the learning target. In other words, it can be said that elementary school students have the power to classify living things. Students were also able to add power to classify organisms by traits. It is a classification of goby. At first, they were classified according to the overall shape, but they were judged comprehensively based on the shape, color, habitat, etc. We believe that such educational effects are close to the OECD's key competency [5].

Tsushima, which is an island, has a mountainous area and an inlet, and has little flat land, so the industry was live trade and fishing. Only by studying the area of the elementary school ward, the development of farmland and the securing of water have continued to be promoted since the early modern period, but however there was a fight against flooding in lowlands, storm surge disaster, and salt runup. Even under such conditions, it was found that the residences were able to sustainably live on this island by devising their residences on hills and lowlands, industry for semi-agriculture and semi-fishing, and agriculture for the right areas. This kind of learning is UNESCO's quality education [6].

References

1. Ministry of Education: Promotion of disaster prevention education at schools. Primary Educ Mater **981**(6), 2–7, Toyokan Publisher (2019) (in Japanese)
2. Kiyosi, T.: Adaptation to school course of study of the protection against disasters and the part of the social studies. Geographical Rep **114**, 29–38, Aichi University of Education, (2012) (in Japanese)
3. Hart, R.A.: Children's participation: the theory and practice of involving young citizens in community development and environmental care, pp. 41–56, Hobunsya (2000) (in Japanese)
4. Shinichiro, N.: Creation of awareness to solve a trend away from rivers : a practice of river education in the Zenpukuji River, Tokyo, Policy Practice Stud **4**(1), 11–20 (2018) (in Japanese)

5. OECD. https://www.oecd.org/pisa/35070367. Last accessed 30 Aug 2020
6. UNESCO. https://en.unesco.org/themes/education-sustainable-development. Last accessed 30 Aug 2020

Instrumentation for Disaster-Affected Reinforced Slopes in Uttarakhand, India

Minimol Korulla, K. S. Beena, and Pankaj Kr. Mourya

Abstract Landslide at Birahi in Chamoli district of the Lesser Himalayas in India had affected traffic safety along National Highway NH-58. Geotechnical and geophysical investigations were done to evaluate the causes of failures and to develop remedial measures. A slide-zone area of overburden with boulders was identified which matches with the borehole data for analyzing the probable causes of failure. Remedial measures adopted for this unstable stretch including drainage network implementation, erosion control measures, nailing on the hillside with facia, and a very tall unique retaining structure, i.e., 25 m high reinforced soil system with flexible facia on the valley side of slide area are briefed in this paper. However, the focus of the paper is on the illustration of how displacement and pressure distribution behavior of composite reinforced soil system of 25 m high installed is getting monitored through field instrumentation. The maximum displacement values are calculated using the empirical formula suggested by FHWA-NHI-10-024 guidelines. The paper also mentions about the numerical analysis of reinforced soil structure using Plaxis software. The displacement values from the results of Plaxis software analysis are reported and compared with the values obtained from empirical formulas of FHWA-NHI-10-024 guidelines.

Keywords Composite soil reinforcement system · Field instrumentation · Numerical analysis

1 Introduction

In June 2013, a cloud burst damaged the beauty of the Devabhoomi (Land of Gods), Uttarakhand state in India with heavy floods in river Mandakini and its tributaries.

M. Korulla (✉) · K. S. Beena
School of Engineering, Cochin University of Science and Technology, Kochi, Kerala, India
e-mail: m.korulla@maccaferri.com

P. Kr. Mourya
Ministry of Road Transport and Highways, New Delhi, India

H. Hazarika et al. (eds.), *Advances in Sustainable Construction and Resource Management*, Lecture Notes in Civil Engineering 144,
https://doi.org/10.1007/978-981-16-0077-7_68

This led to the development of around 2300 landslides along the river Mandakini, making it the biggest disaster after the 2004 tsunami. After the natural disaster during June 2013, the Ministry of Road Transport and Highway of India decided to undertake the "Rehabilitation/Mitigation of landslide locations with slope protection and drainage improvement at various stretches on NH-58 in Chamoli district in the state of Uttarakhand."

Birahi, a Village in Chamoli at the Uttarakhand State of India, is situated at an altitude of 1200 m above sea level. It is situated on the riverbanks of Alaknanda and Birahi Ganga on National Highway NH58.

2 Problem Description and Causative Forces of Two Landslides at Birahi

The region of Birahi falls under seismic zone V. Presence of fault lines has made the area highly vulnerable to landslides and slips. At Birahi, two landslides were identified, one was at the hillside and the other was in between the roads of hairpin bend (refer Fig. 1). Construction operations during the widening of the road aggravated major destabilizing forces in this specific stretch. Cutting of hillside slope disturbed

Fig. 1 Two-slided portions with hill side and valley side unstable patches at Birahi

the equilibrium of the forces maintained in nature till then. Loss of vegetation cover aggravated the destabilizing forces and reduced the effect of restoring forces. Accumulation of colluviums on the valley side slope caused heavy slope and toe erosion, when it is encountered with surface runoff due to high torrential rains. This erosion disturbed the stability of the road considerably.

3 Geological, Hydrological and Geotechnical Investigations

Geological, hydrological, geotechnical and geo-morphological studies are essential for cause analysis of a landslide affected area. Local specialists and international experts are consulted during the analysis and solution framing of this project by the consultants involved. The rocks in the region are identified as highly jointed, weathered, and fractured quartzitic schist belonging to the Garhwal group. Surficial material (topsoil) of slope in between the roads consists of soil boulder mix and is prone to sliding. From the geotechnical investigation, it was observed that the slope of the unstable zone is around 50° with 1 m surficial instability thickness having cohesion of 3–5 kPa and an angle of internal friction of 30–32°.

4 Adopted Solution for Birahi Landslide

The solution adopted for the hillside and valley side is depicted in the drawing shown as Fig. 2.

On the hillside, the slope was left untreated after cutting during the construction of the road where the slide has occurred. It was stabilized by providing a drapery steel net system secured with surface anchors. The drapery system used was supplemented with a reinforced geosynthetic mat as a facia system along with self-drilling anchors of 25 mm diameter at 3 m spacing in the longitudinal and transverse direction. Here, a drapery net with surface anchors was used for surficial (topsoil) stabilization and self-drilling anchors were recommended for deep stabilization. Reinforced geosynthetic mat was used for surface erosion control as per the strength requirement in line with Indian National Highway Guidelines [1, 2]. On the valley side, for retention of 12m wide road, a composite reinforced soil system of 25m maximum height has been provided [3, 4]. A proper drainage network was provided on the hill side as well as valley side including edge drains and chute drains [5].

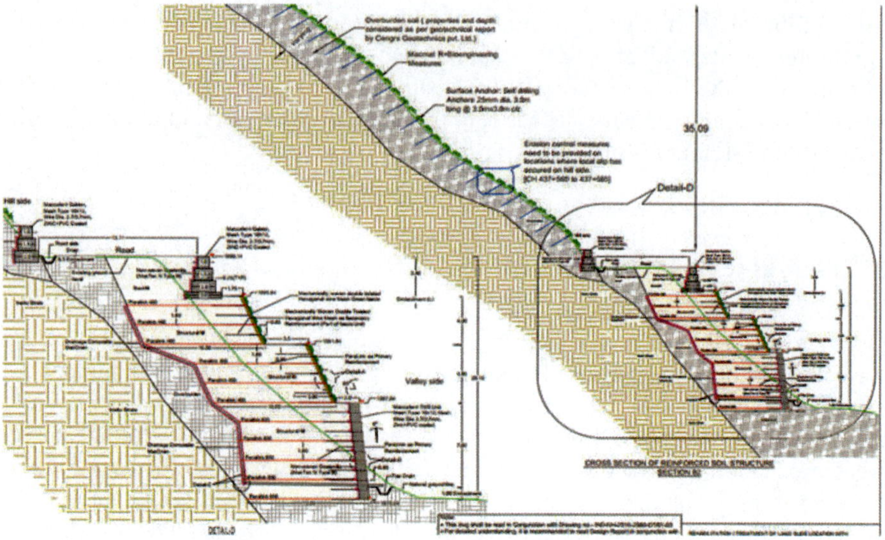

Fig. 2 Adopted landslide mitigation solution for unstable patches at Birahi

4.1 Characteristics of Unique 25-m-High Composite Reinforced Soil System for Road Retention in the Valley Side

The slope angle more than the friction angle with low cohesion value indicates that a geotechnical stabilization needs to be done as the first step before bringing back vegetation through erosion control measures for this slide affected stretch. The area where the rehabilitation measures were constructed is within first-degree seismic zone, i.e., zone-V and with heavy rainfall during monsoon time. Therefore, a flexible retention system that could withstand larger seismic activity (with a proven record) and a free-draining facing were the essential components of the selection criteria for the retaining system. The composite reinforced soil system technology for a height of 25 m maximum provided an opportunity to fulfill both requirements. The composite soil reinforcement system adopted consists of two types of reinforcements, a primary reinforcement with polymeric nature, i.e., a geogrid with high strength and secondary reinforcement with woven steel wire mesh of medium strength. At Birahi location which this paper is referring to, road retention is provided in the valley side by a combination of a reinforced soil system and a gabion gravity wall. Here, the bottom portion of the reinforced soil structure is made of gabion facing units with an integrated tail, and the top portion is made of green facia with an integrated tail. High-strength geogrids as primary reinforcement with higher spacing interacts with soil and develop adequate resistance against large lateral and vertical forces. The integrated tail of gabion/vegetated facia acts as secondary reinforcement and that will impart stability to the facia system against local forces.

5 Theoretical Deformation Predictions of Reinforced Soil Structure Using Plaxis Software and Comparison with Maximum Probable Values as Per FHWA-NHI-10-024 [6]

Composite reinforced soil system with gabion facia and vegetated facia was selected for road retention due to its flexibility, seismic performance, cost-effectiveness, speed of construction, and suitability for use with the local materials. Other than these merits, this system also has the following general and technical advantages in comparison to the conventional retaining structures like rubble masonry, reinforced cement concrete, etc. The flexibility of the system within acceptable limits helps to accommodate horizontal movements and differential ground settlement. The flexibility also helps to redistribute stresses and strains without any compromise in structural integrity when the wall is subjected to dynamic forces. To determine deformations and to examine the soil–structure interaction effect for 25 m high composite reinforced soil system, numerical modeling using Plaxis software was performed (refer Fig. 4). FHWA-NHI-10-024 [6] is defining a formula for determining the maximum lateral displacements for reinforced soil structure having both extensible and inextensible reinforcement (refer Fig. 3). The maximum lateral displacements for 25 m high reinforced soil system having 14 m base width and with an extensible reinforcement are shown below, and it comes out to be 466 mm in this case. The maximum deformation which is the outcome of Plaxis analysis is 264.7 mm which is less than the maximum probable horizontal movement defined by FHWA-NHI-10-024 guidelines [6].

Calculation as per FHWA-NHI-10-024 [6] is as follows.

$$\text{Maximum lateral displacement}(\delta_{\max}) = \delta_R * \frac{H}{75} \tag{1}$$

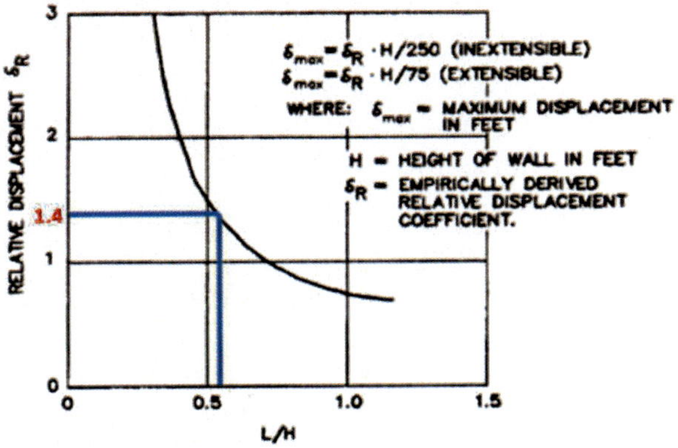

Fig. 3 Empirical curve for estimating lateral displacement for reinforced soil structures (reference—FHWA-NHI-10-024 [6])

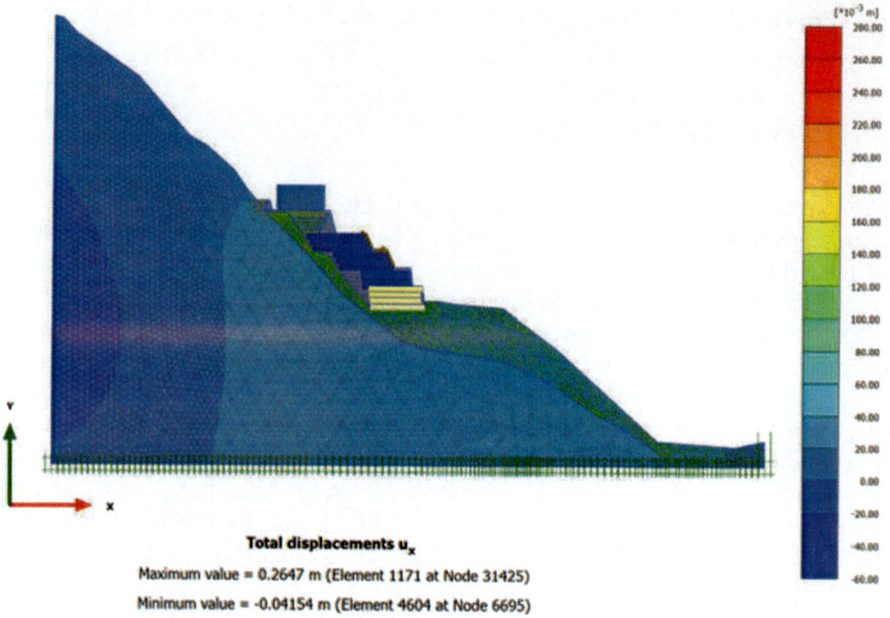

Fig. 4 Photograph showing lateral displacement output from Plaxis analysis for 25 m high reinforced soil structure

$$\frac{L}{H} = \frac{14(\mathrm{m})}{25(\mathrm{m})} = 0.56 \tag{2}$$

$$\delta_R \text{ from the graph for } \frac{L}{H} \text{ value of } 0.56 = 1.4 \tag{3}$$

$$\delta_{\max} = 1.4 * \frac{25(\mathrm{m})}{75} = 0.466\ \mathrm{m} = 466\ \mathrm{mm} \tag{4}$$

6 Actual Site Predictions Using Instrumentation and Monitoring

It is important to understand the behavior of reinforced soil structure of this magnitude (25 m) which is very much unique in nature via instrumentation and monitoring. Furthermore, due to the challenging site location of Birahi where the rainfall intensity is high, and which falls in seismic zone V with the high vulnerability of surface erosion and further landslides, makes instrumentation and monitoring work much essential for the overall mitigation and stabilization scheme. Hence to observe and safeguard the integrity of the structures adopted, suitable instruments for monitoring

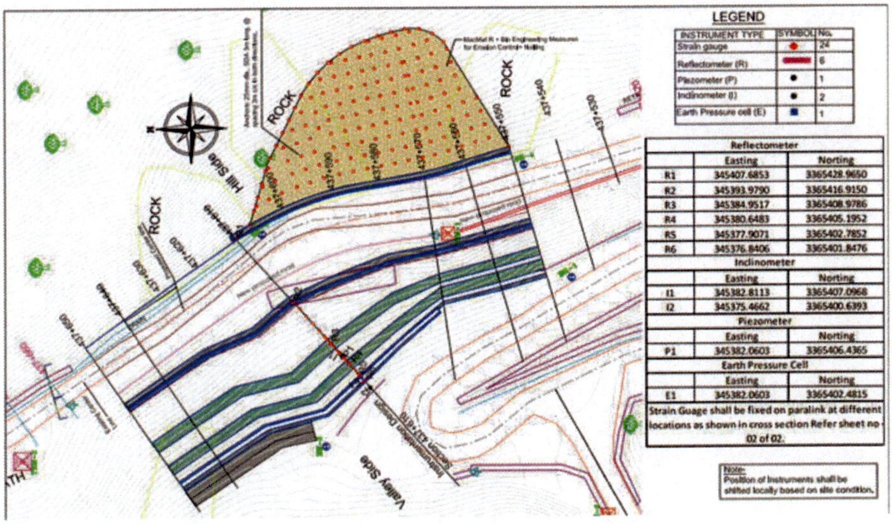

Fig. 5 Plan showing instrumentation details at Birahi location at Chainage 437+610

were proposed on the site during construction (Fig. 5). Major parameters that need to be monitored are as follows: (1) drainage behavior of the ground, (2) forces on polymeric reinforcement, (3) deterioration of facing and other elements, (4) performance of any structure supported by reinforced valley slopes, and (5) horizontal and vertical movements of slope surface and overall structure.

It was decided that field observations on local face deformations, horizontal movements, and vertical settlements of most critical structures shall be documented to validate the results from numerical analysis and instrumentation. It was also decided that post-construction behavior of critical structures needs to be monitored for a minimum of four years from the date of completion as per the verified completion certificate.

6.1 Different Types of Instruments Used for Monitoring Reinforced Soil System Behavior at Birahi Site

To monitor the different influencing parameters of reinforced soil system behavior and performance, the instruments, namely slope inclinometer, strain gauges, earth pressure cells, piezometers, and reflectometers, were installed at various positions (Fig. 6 and Table 1). Chainage 437+610 was selected for fixing the instruments. Strain gauges were fixed on the primary reinforcements, i.e., polymeric geogrids which are assumed as extensible reinforcements to observe the strains developed at different layers in the reinforced soil structure (Fig. 7). Moreover, to detect the pore water pressure, one piezometer is also located in a PVC pipe of diameter 4 in. which

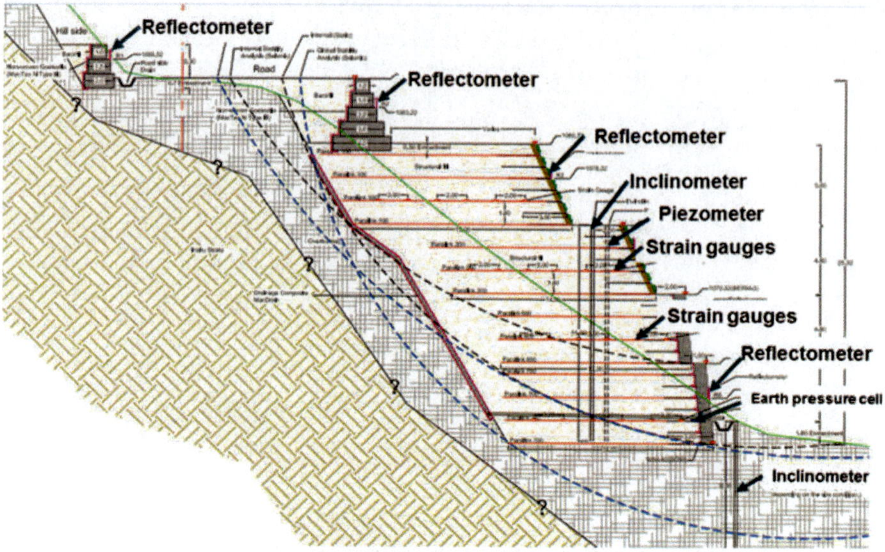

Fig. 6 Instrumentation scheme for Birahi site

Table 1 Different types of instruments installed and their general functions

Instrument	Function
Reflectometer	Deflection of facia elements in all three directions
Strain gauge	Strain developed in primary reinforcement
Slope inclinometer	Wall deformation
Earth pressure cells	Total pressure in earth fills and embankments
Piezometers	Pore water pressure measurement

Fig. 7 Fixing the strain gauges with plates on primary reinforcement (geogrid) at the site for valley side reinforced soil system

Fig. 8 Installation of inclinometer in the borehole (left-aligned) and reflectometer (right-aligned)

reached till the bottom layer of the valley side retention system. Earth pressure cells at the bottom-most retention systems at the backside of the facia were positioned to monitor the variation of the stresses where the shear failure is most anticipated. Apart from these, six reflectometers were also placed at the valley side retention system to observe the tilt of the structure with time (Fig. 8). Moreover, two inclinometers were put at the valley side to check the overall deformation i.e., both horizontal and vertical movements/displacements (Fig. 8). Figure 6 is showing the instrumentation scheme for reinforced soil structure at CH 437+610.

6.2 Monitoring Status

The instruments were installed during the initial phase of construction of the composite reinforced soil structure itself, i.e., in October 2017, and it continued till the construction was completed during April 2018. The process of monitoring is continuing at the site, and it is expected to continue.

7 Summary of Results and Conclusions

The maximum lateral displacements for 25 m high reinforced soil system having 14 m base width came out to be 466 mm as per the suggested calculation from FHWA-NHI-10-024 guidelines [6] for critical loading conditions and the maximum deformation which has come from Plaxis analysis is 264.7 mm. The inclinometer and reflectometer readings so far show that the structure has not shown any significant or noticeable deformations or lateral movements. Earth pressure readings for the bottom-most layer vary from 223 to 241 kPa which are quite less than the theoretically predicted maximum analytical values. The piezometer readings in the monsoon and non-monsoon months range between 11 and 35 kPa, which shows the structural backfill and the gabion facia due to the free-draining nature do not carry significant pore water pressure. The strain values recorded are negligibly small so far. Recording of data and reviewing shall be continued for a few more years to arrive at conclusions. The present trend in the data shows the structure is performing as per the intended use and as per the assumptions carried out during analysis and design. The small values of field observation figures and instrumentation results indicate that the mobilization of forces at the site has not reached near to critical load combination at any stage of construction or post-construction stage.

Acknowledgements Authors express the deep gratitude and appreciation toward the end client, consulting and contracting organizations involved in this project for the cooperation and understanding for the execution of this project, especially officials of PIU MoRTH (Project Implementation Unit-MoRTH Dehradun), NHIDCL, and the technical as well as field implementation team of Maccaferri Environmental Solution Pvt Limited, Gurgaon, RG Buildwel Engineers Ltd., and MAGOTEC. Special thanks to Mr. Jagat Narayan of NHIDCL and Col. PS Tamhan of MAGOTEC. The support and guidance from Mr. Vikramjiet Roy, Mr. Dipankar, Dr. Mahajan, Mr. Rupesh, Mr. Rudra, and Ms. Padmaja of Maccaferri are highly commendable.

References

1. IRC 56: Recommended for Practice for Treatment of Embankment and Roadside Slopes for Erosion Control. Indian Roads Congress, India (2011)
2. MoRTH: Specifications for Road and Bridge Works. Indian Roads Congress, India
3. BS 8006 Part1: Code of Practice for Strengthened/Reinforced Soils and other Fills (2010)
4. AASHTO: LRFD Bridge Design Specifications. American Association of State Highway and Transportation Officials, Washington, D.C. (2012)
5. IRC SP 42: Guidelines for Road Drainage. Indian Roads Congress, India (2014)
6. FHWA-NHI-10-024: Design and Construction of Mechanically Stabilized Earth Walls and Reinforced Soil Slopes. Federal Highway Administration, Washington, D.C. (2009)

Proactive Conservation of Main Tower of Bayon, Angkor, Against Global Warming

Yoshinori Iwasaki, Mitsuharu Fukuda, Mitsumasa Ishizuka, and Takeshi Nakagawa

Abstract This paper describes the special characteristics of the ancient temples in Angkor in terms of geotechnical points of view, which have never been mentioned in the past because simply the soils and foundations have never been studied. Japanese Government Team for Safeguarding Angkor (JSA) introduced the study of soil and foundation. The heritage structure of Bayon in Angkor stands upon shallow direct foundation. The high main tower stands upon manmade fill of 14 m in thickness without special supporting such elements as piles. The amazingly strong strength of the manmade fill that had made to support heavy stone structure is the one of the character-defining elements of the authenticity of the foundation of Bayon in Angkor Thom. If the sand fill shows the shear strength of the common sandy, the fill could not keep standing safely under the rainy season in the south Asian region of the squall with heavy rain for the last 700 years since thirteenth century. To preserve the main stone tower, the paper proposes a simple method to cover the surface of the fill of the mound with impermeable layer to prevent the failure of the sand filled mound from the anticipated long and heavy rain in the coming global warming.

Keywords Bayon · Sand fill mound · Direct foundation · Global warming

Y. Iwasaki (✉)
Geo Research Institute, 2-1-2 Otemae, Osaka 540-0008, Japan
e-mail: yoshi-iw@geor.or.jp

M. Fukuda
Taisei Geotech, Kurume, Fukuoka, Japan

M. Ishizuka
JSA/JASA Site Office, Angkor, Siem Reap, Cambodia

T. Nakagawa
JSA, Waseda University, Tokyo, Japan

H. Hazarika et al. (eds.), *Advances in Sustainable Construction and Resource Management*, Lecture Notes in Civil Engineering 144,
https://doi.org/10.1007/978-981-16-0077-7_69

1 Introduction

Geotechnical engineering plays one of the important roles to safeguarding cultural heritages and has made such key contribution as in restoration work for the inclined Pisa Tower. When the restoration of the Pisa Tower was discussed, only the inclination of the tower itself was considered as to be preserved. However, since the foundation is considered as one of the important elements of the heritage structure, the direct foundation of the Pisa Tower, which was the major reason to have caused the inclination, is considered as the character-defining elements of the authenticity as well as soft soil conditions.

In the past, foundations are generally considered as simply nothing but an element to support the heritage structures and not considered as a part of the heritage.

However, the recent trend of conservation of heritage indicates that a foundation system has become to be considered as one of the basic components of the heritage structure. In 2010, the international standard of ISO 13822 on assessment of structural safety was renewed and added an Annex "Heritage Structures" and stressed the importance of the foundation [3].

In this paper, the new concept of "authenticity of foundation" is discussed as well as the character-defining elements of authenticity that should be evaluated and restored.

2 Authenticity

2.1 Authenticity of Cultural Heritage

Authenticity was defined in the Venice Charter of 1964 as heritage composed from original material, original position, original design as well as original procedure (Venice Charter 1964) [1]. The concept of the Venice Chapter is called "anastylosis (Greek)," which means "take column back to original position." Anastylosis implies that original stone columns spread over a historical ruin shall be rebuilt at their original positions.

The principle of anastylosis was developed based on conservation of stone structure in Europe and does not give any values of heritage to such repaired materials as often seen in wooden structures in Japan.

Later in 1994, the concept of the authenticity was expanded by the Nara Document on Authenticity to cover various methods characterized by the region to which the heritage belongs. Region specific methods that were developed in some area are also accepted as the characteristic of authenticity [2].

Character-defining elements are defined as historic materials, forms, locations, spatial configurations, morphology, concept and details, structural design, uses, and cultural associations that contribute to the heritage value of a structure that shall be retained in order to preserve its heritage value.

2.2 *Authenticity of Cultural Heritage*

In 2005, ISO 13822 (Bases for design of structures—Assessment of existing structures) was reviewed for renewal. ISCARSAH (International Scientific Committee on the Analysis and Restoration of Structures of Architectural Heritage under ICOMOS) had proposed to include heritage structures in the standard and worked together for five years to develop. The ISO 13822 has been updated in 2010 and added an Annex-I (informative) Heritage Structure, which has expanded the heritage structure to include foundation as a part of the structures.

The Annex-I clearly states as in paragraph of I.5.3 Authenticity of foundation that "From the point of view of conservation, foundations are not different from the rest of the structure and should be assessed and rehabilitated taking into consideration their heritage value. This involves the requirement to identify their authenticity and character-defining elements." [3].

3 Bayon Temple in Angkor Thom

Bayon, as shown in Fig. 1, is the central temple of Angkor Thom constructed around in late twelfth to early thirteenth century.

Figure 2 shows section and plan view of the Bayon Temple. Japanese Government of Safeguarding Angkor Team (JSA) performed archaeological trenching study at the northeastern corner as "long trench" as shown in Fig. 2.

Fig. 1 Bayon temple

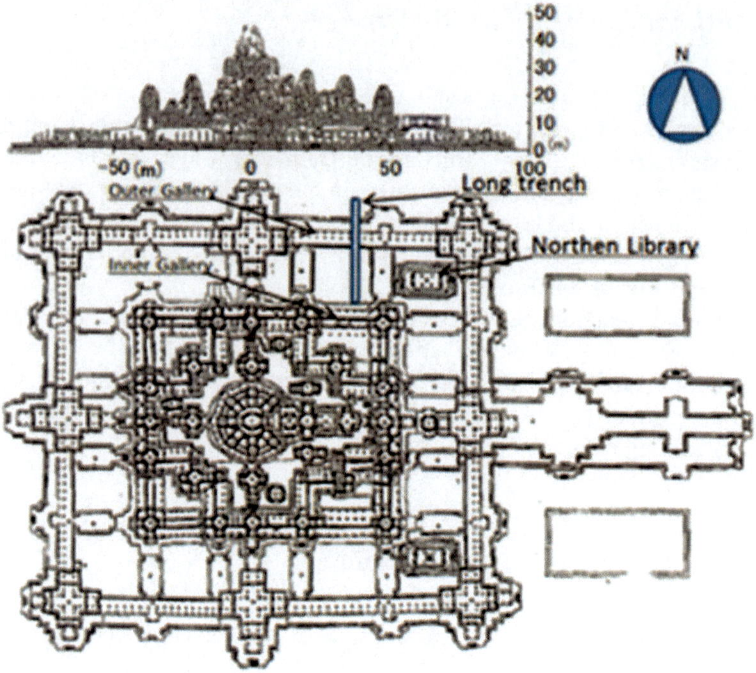

Fig. 2 Plan view and section

3.1 *Trenched Foundation*

The result of the long trench has revealed the following facts as shown in Fig. 3 [4].

1. The surface of the first terrace was covered with sandstone and laterite block, beneath which densely compacted sandy filled layer was identified.
2. Compacted sandy filled layer continued with and additional laterite block layer to the level of the original ground surface.
3. It was identified that the compacted sandy filled layer continues from the original ground level to 2–3 m in thickness.

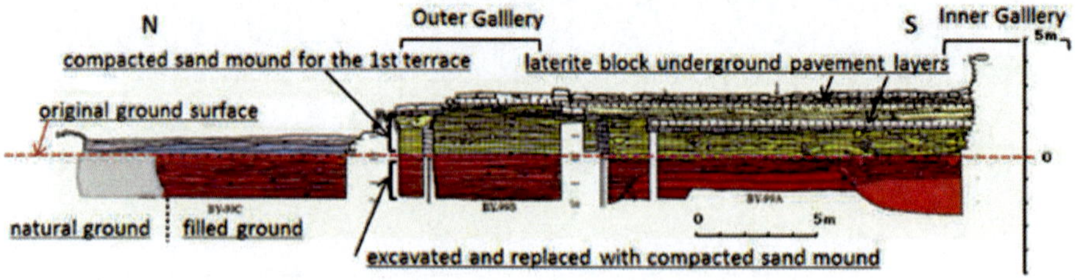

Fig. 3 Long trench of archaeological study

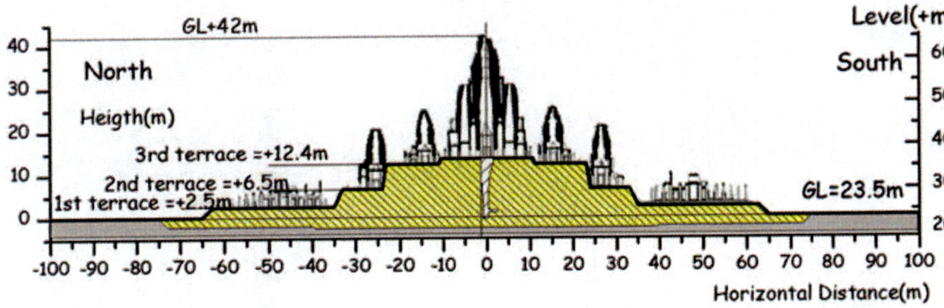

Fig. 4 Trenched foundation

4. The compacted sandy filled layer below the original surface was found to continue horizontally to about 10 m outside of the outer gallery.

The archaeological study suggests that the foundation work for the Bayon is "trenched foundation" and was systematically constructed by excavating the original ground 2–3 m vertically as well as about 10 m outside of the temple boundary of the outer gallery as shown in Fig. 4. The stone masonry structures were constructed upon terraces with three different levels of 2.95, 6.95, and 12.40 m.

4 Platform Foundation of Main Tower of Bayon Temple in Angkor Thom

At the center of the base of the main tower of Bayon, a vertical shaft was excavated in 1933 by French as shown in Fig. 5.

Stone masonry main tower of 31 m in height stands upon the platform of manmade fill of 14 m in thickness by direct shallow base stone as shown in Fig. 5. It is amazing that the tower has been standing for 800 years since the late twelfth early thirteenth century upon the manmade fill [5].

Several borings were carried out to study the platform foundation. The results of two borings are shown in Fig. 5 BYV-2009 at the vertical shaft and BYV-2010 at the edge of the foundation of the tower on the third terrace. Boring BYV-2009 at the vertical shaft shows very loose sand layer of SPT, $N < 4$, which shows no compaction efforts were paid by French conservator. Boring BYV-2010 shows laterite block of 6 m in thickness beneath the pave stone followed by densely compacted sandy fills.

The manmade filled layer showed very large SPT, N-values larger than 100–250 as shown in Fig. 6 with water contents. The sampled soil, when dry, is just like soft rock rather than sand.

A sample of the sandy fill was put into water as shown in Fig. 7. The stiff sandy fill collapsed in the water within 10 min (Iwasaki 2018).

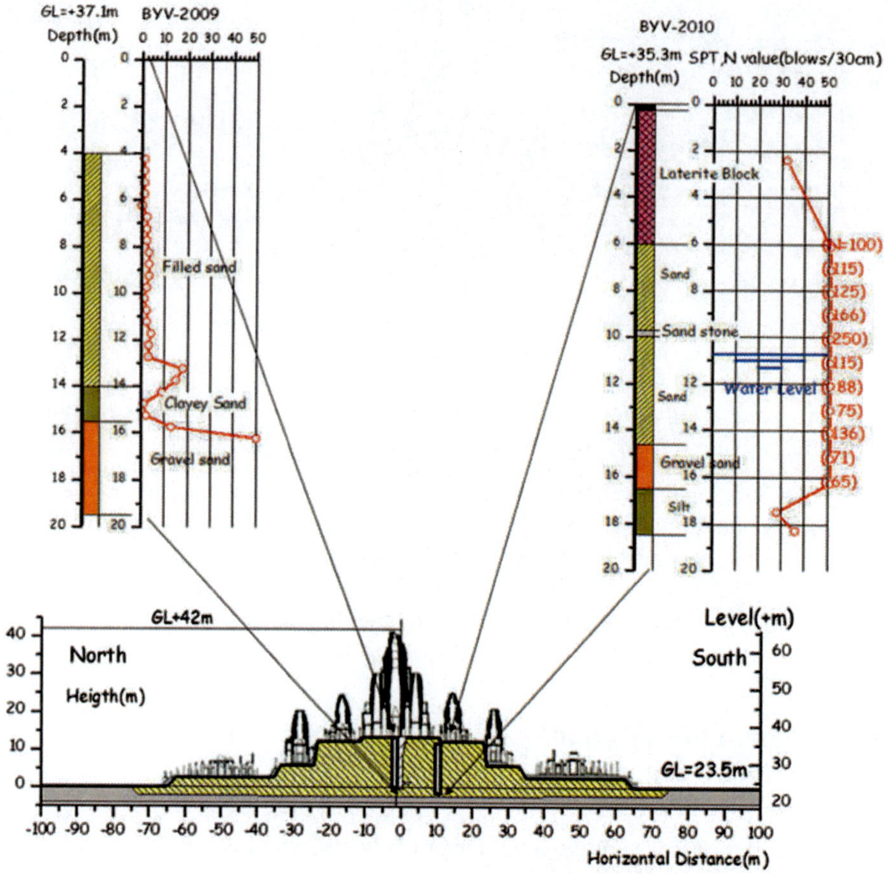

Fig. 5 Boring results at the platform foundation of central tower of Bayon

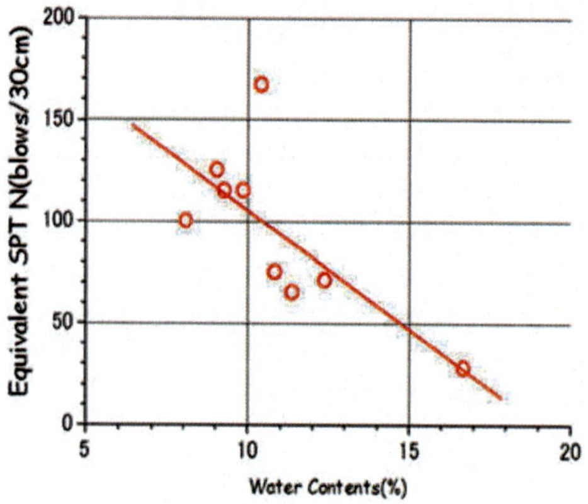

Fig. 6 SPT *N*-value

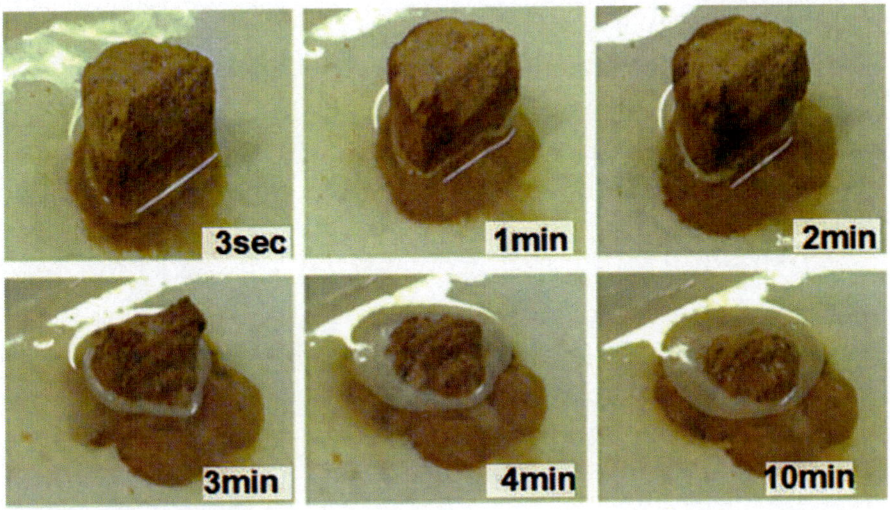

Fig. 7 Collapse of stiff sandy filled soil in water

4.1 *Characteristics of Filled Soils in Bayon Temple*

Grain size distributions of the samples of filled soils by the vertical boring within the platform foundation are plotted in Fig. 8 including additional horizontal boring in the platform, Northern and Southern Libraries, N1 Prasat Suor Prat and Baphuon Temple. Surprisingly, two kinds of clayey and sandy fills are clearly separately grouped with uniformity.

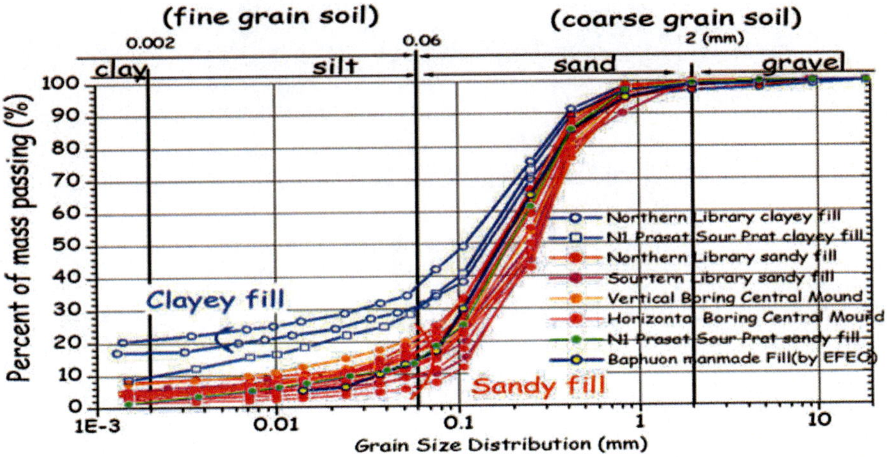

Fig. 8 Grain size distribution

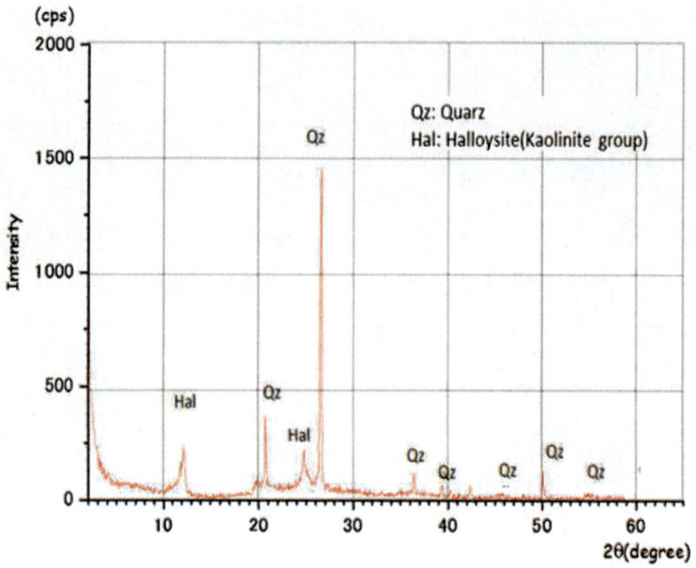

Fig. 9 X-ray diffraction analyses

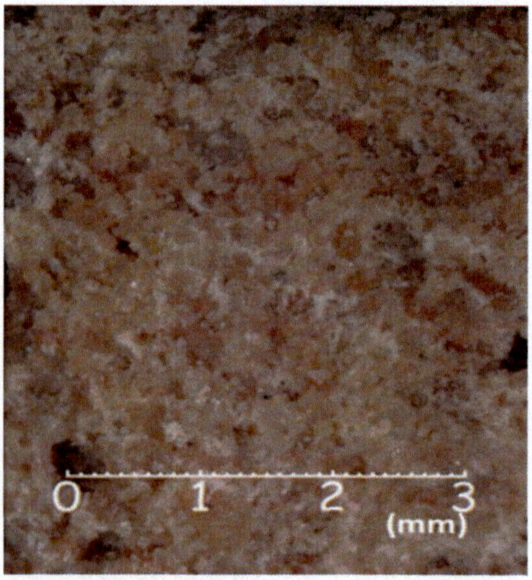

Fig. 10 Micrograph of sandy fill soil

X-ray diffraction analysis was performed to study the mineral component of the filled soil. Figure 9 shows the results of X-ray diffraction analysis. The major mineral is quartz with clay mineral of halloysite of kaolin group.

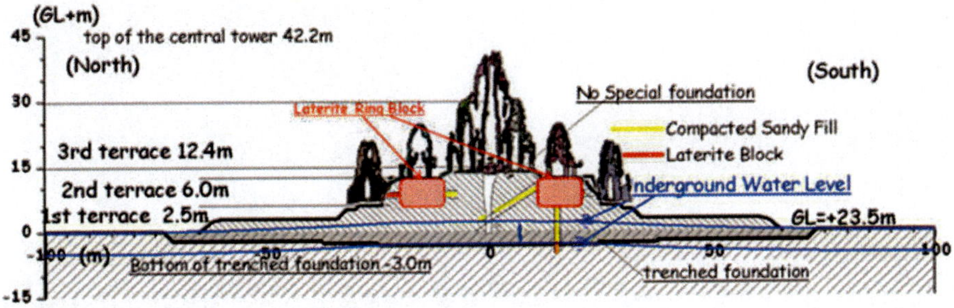

Fig. 11 Trenched foundation structure and laterite ring block in foundation

The micrograph sample of the sandy fill is also shown in Fig. 10. The rounded particles of sand of quartz are gathered filled with kaolin clay [5].

Based upon several borings, ring-shaped laterite body has been identified in the filled mound as shown in Fig. 11.

5 Proactive Countermeasures Against Failure of the Mound

In the history of Angkor, heavy rains had caused failures of filled mound and embankments.

One of the examples of such failure was the failure of northern east slope of Baphuon Temple (Figs. 12 and 13), constructed in 1060, near Bayon in the same Angkor Thom site during a heavy rain in June 1943 as shown in Fig. 14.

The characteristic of grain size distribution of the mound of Baphuon is shown in Fig. 8, which shows the distribution shape is the same as those of Bayon Temple.

This implies the original mound of the Baphuon has very high strength and had been stable for 900 years. The stability of high mound of 40 m in height with the steep angle of 45° of the manmade filled structure (shown in Fig. 13b) is never realized by the present common knowledge of the geotechnical engineering.

In 1960, EFEO (Ecole Francaise d'Extreme-Orient), French governmental organization had tried to reconstruct the failed mound by the compaction of the original soil. However, the fill height was reached at 5 m in height, the filled mound had failed for three times.

The clear mechanism of stability of the high mounds for several hundred years from tenth -thirteenth century to twenty-first century for such high mound of Bayon Temple and Angkor Wat in Angkor are not known and shall be studied step by step gradually with the restoration works.

Fig. 12 Baphuon temple in Angkor Thom

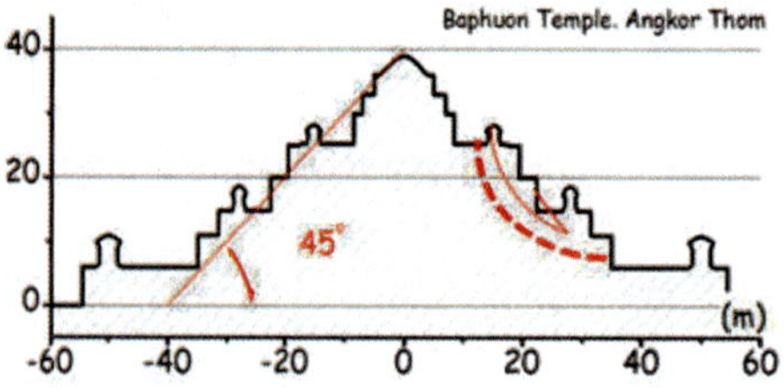

Fig. 13 Vertical section of Baphuon (east side)

5.1 *Safety Mechanism of High Rise Filled High Mound of Bayon for 700 years*

The effects of rainfall to the sandy filled mound were evaluated by monitoring the moisture change of mound. Moisture sensors were installed at the surface of the top third terrace at several depths as shown in Fig. 15.

Fig. 14 Failure northeastern corner of the Baphuon temple by a heavy rain

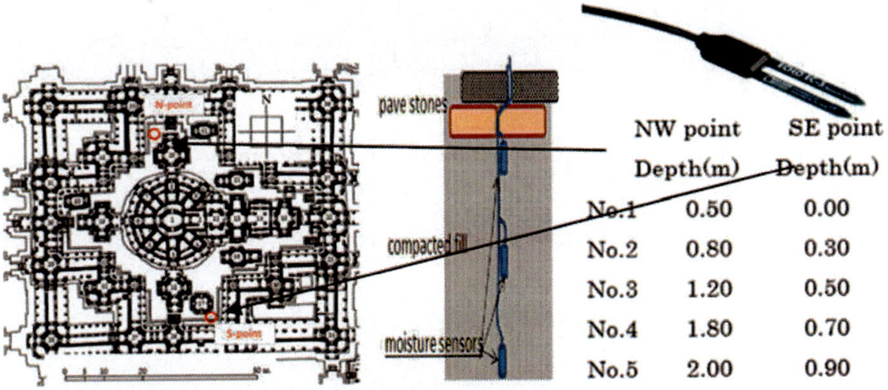

Fig.15 Installation of moisture sensors in the filled mound

An example of monitored results during May 27 to 28 in 2017 is shown in Fig. 16. The rainfall intensity was about 10–15 mm/10 min for about one and half hours of 80 mm in total. The rain was very heavy intensity associate with squall in monsoon area.

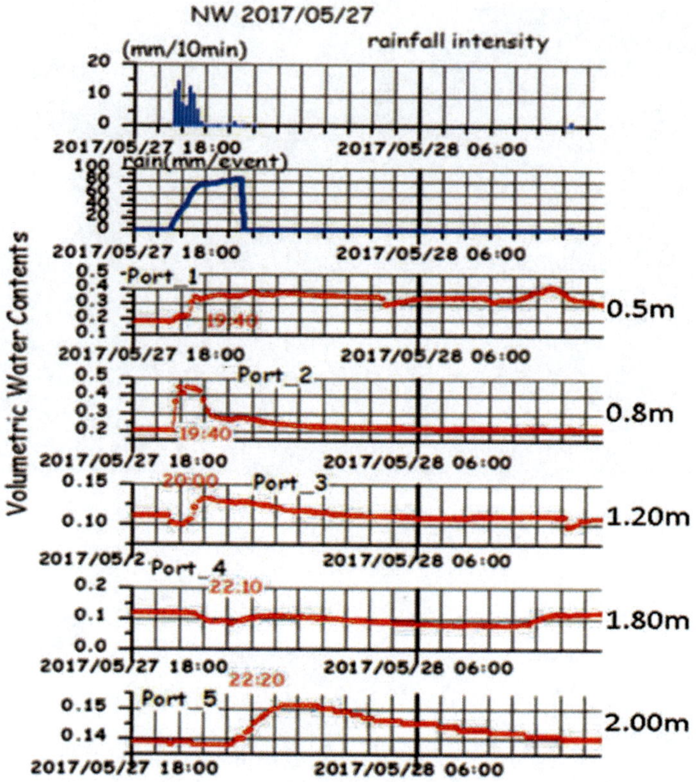

Fig. 16 Monitored results of water contents at several depths under the top surface

5.2 *Preparing to Cover Mound Surface with Impervious Soil Layer*

The common conventional way to protect the cultural heritage has been some reaction after failure. However, since the heavy and continuous rainfall is anticipated in Angkor region, which shall be inevitable to keep the sandy filled mound in the safe condition. The 10 to 20 cm of the top surface of the soil mound id being planned to be replaced with slake lime mixed soil, which keeps its strong strength under water and suitable to protect the sandy mound of the platform. Geotextile may be used for stabilizing the slaked lime mixed soil against tension cracks as shown in Fig. 17.

Observational procedure shall be applied for the restoration work. However, the observational method in geotechnical engineering is to compare the monitored data with some predicted calculation. In the restoration work for the cultural heritage, the monitored results shall be evaluated by engineering senses than numerical bases. If the water shielding is found not enough, additional steps to improve are to be applied.

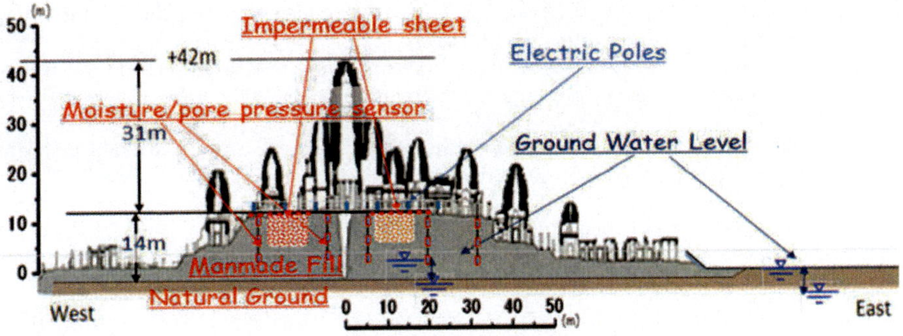

Fig. 17 Impervious layer and sheet with monitoring system

The conservation of cultural heritage is different from the conventional repairments of common structures. The major key difference is to improve the safety or to keep the authenticity as well as possible. The principles countermeasures for the heritage structures are based upon (1) Minimum intervention, (2) Use the same material and the method, and (3). Apply reversible method to be able to change if the results are not appropriate. The step-by-step procedures of try and confirm shall be applied to keep the character of the authenticity as well as possible to keep the integrity as the historical heritage to be retained.

6 Conclusions

The manmade disaster of the heavy rainfall by global warming shall bring manmade sandy filled foundation platform in Angkor. The recent activity of Japanese Government Team for Safeguarding Angkor has been successfully introduced geotechnical study on the monuments.

The results of the filled soil as very strong in dry but very weak in wet condition was identified as the unique character of the Angkor monuments as the characteristic elements of the authenticity to be preserved.

It is now being planned not only to preserve the authenticity of the amazing techniques of the ancient Khmer engineer but also to protect the great monuments of the human being based by proactive geotechnical engineering.

References

1. Angkor Charter: https://apsaraauthority.gov.kh/?page=detail&menu1=647&ctype=document&id=647&ref_id=6&lg=e (2014)
2. Nara Doc: www.icomos.org/charters/nara-e.pdf (1994)

3. ISO_STANDARD 13822-2010: Bases for Design of Structures—Assessment of Existing Structures, Annex I Heritage Structures (2010)
4. Narita, T., Nishimoto, S., Shimizu, M., Akazawa, Y.: Outline of excavations and investigation at the outer gallery of Bayon complex. Annual Technical Report 2000, JASA, Tokyo, pp. 3–22 (2000)
5. Iwasaki, Y., Fukuda, M.: Preservation of the main tower of Bayon temple, Angkor, Cambodia. In: Geotechnics and Heritage—Historic Towers, pp.191–227. CRC Press, Balkema (2018)

Geological and Hydrological Phenomena on Hazard Estimation

Groundwater Level Changes in the Coastal Construction Site of Coal-Fired Power Plant, Cilacap, Indonesia; Natural or Construction Effect?

Doni Prakasa Eka Putra, Rilo Restu Surya Atmaja, Fania An Nisaa, Kurnianto Dwi Setyawan, and Pramono Hadi

Abstract The construction of power plant with the capacity of 1000 MW in the coastal area of Cilacap, Central Java, Indonesia, started from October 2017 until 2019. The construction is started by excavating the area at least 8 m of depth to build the foundation and underground cooling tunnel. During the dry season in 2018, community who live around the construction site reported groundwater level depletion in their wells. The power plant authority argued that groundwater extraction for shrimp ponds along the beach is also responsible for the groundwater level depletion in the region. This study aims to determine whether groundwater level declining in this area caused by natural conditions or due to construction activities. Field observations were carried out both in the rainy and dry seasons in 2019. It consisted of groundwater level measurement, resistivity measurement, and inventorying the utilization of groundwater. Groundwater modelling used to simulate the impact of groundwater extraction and construction of the power plant towards the groundwater flow around the construction site. The results suggested that seasonal difference within a year affects to the declining of groundwater level of 1.06 m by average at the community's wells and the whole research area, while the dewatering practices by the power plant authority around the construction site of foundation and underground cooling tunnel were responsible to the declining of groundwater level of 1–2 m limited 500–700 m from the construction site.

Keywords Power plant · Dewatering · Coastal aquifer · Groundwater modelling

D. P. E. Putra (✉) · R. R. S. Atmaja · F. A. Nisaa · K. D. Setyawan
Department of Geological Engineering, Faculty of Engineering, Universitas Gadjah Mada, Grafika 2, Yogyakarta, Indonesia
e-mail: putra_dpe@ugm.ac.id

P. Hadi
Center of Environmental Study, Universitas Gadjah Mada, Lingkungan Budaya Sekip Utara Street, Yogyakarta, Indonesia

H. Hazarika et al. (eds.), *Advances in Sustainable Construction and Resource Management*, Lecture Notes in Civil Engineering 144,
https://doi.org/10.1007/978-981-16-0077-7_70

1 Introduction

Coal-fired power plant in Cilacap, Indonesia, increased its capacity by building a new power plant next to the existing power plant with the capacity of 1000 MW. The construction began in October 2017. The initial stage of construction started by building the foundation, basement, underground tunnel, and the power plant. The depth of excavation was carried out more than 8 m of depth. Raft foundation type was applied. Underground tunnel were built to drain sea water as cooling water to the power plant.

Because the location of power plant is close to the sea, the groundwater level in the area was very shallow, below 1 m. Therefore, during the excavation and construction, the contractor applied dewatering to avoid the water in the construction site.

Eventually, in the middle of 2018, problems occurred. Society who lives in the surrounding of power plant area complained about the depletion of their wells and the emergence of salt water in their wells. However, the power plant company argued that the pumping of seawater and groundwater for shrimp ponds by community industry is also responsible for the groundwater level declining and seawater intrusion.

Therefore, this study is important to identify the main cause of the problems using scientific analysis. This study aims to identify the cause of groundwater level depletion which was caused whether by natural phenomenon (seasonal difference) or purely due to construction activities.

Numerous research investigated the climatic and seasonal effect to the groundwater level depletion and resources [1–5], while other researches identify the influencing factors on groundwater drought and depletion by investigating the effect of climatic and anthropogenic factors [6].

Groundwater numerical modelling is widely used to understand the dynamics of an aquifer system, to characterize the underground water flow, and to simulate impacts of different management scenarios [7]. Groundwater models are adopted as tools to address salt intrusion problems in coastal aquifers [8–10], while groundwater head simulation using numerical modelling is also applied in archaeological deposits and tunnel construction [11]. Groundwater models are utilized to simulate over-extraction of groundwater [12, 13]. In this study, numerical groundwater models are used to understand the groundwater flow and the effects of construction and over-extraction of groundwater, coupled by seasonal measurement of groundwater level to understand to natural condition.

2 Study Area

The coal-fired power plant is located in the south coast of Cilacap, Central Java, Indonesia. Next to the power plant is Indian Ocean. Cilacap area is consisted of Basalt, Gabon Breccia Formation, Jampang Breccia Formation, Nusakambangan

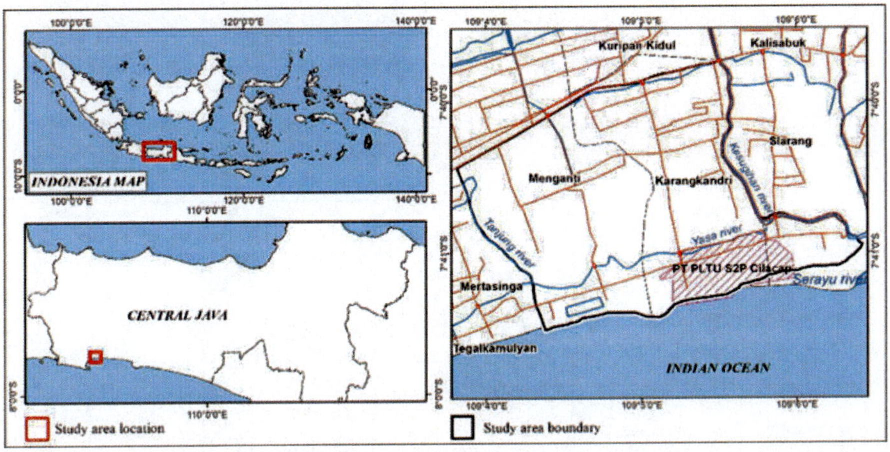

Fig. 1 Study area is located in the Cilacap regency, Central Java, Indonesia. Coal-fired power plant area is presented as purple hatch pattern next to Indian Ocean

Tuff Formation, Pamutuan Sandstone Formation, Kalipucang Limestone Formation, Halang Sandstone Formation, Coastal Deposits, and Alluvium [14–16].

The coal-fired power plant is located on coastal deposits comprised of sand and iron sand deposit. The study area is presented in Fig. 1. Study area is limited by river in the east and west. Kesugihan river in the east, Tanjung river in the west, and then Indian Ocean in the south. Settlement area is located at the north and east of power plant area.

3 Methodology

The methodology in this research is consisted of groundwater level measurement, resistivity measurement, inventorying the utilization of groundwater around the construction site, and groundwater flow modelling.

Groundwater level measurement conducted in the wet and dry seasons of 2019. The measurement in the wet season is carried on April 2019, while measurement in the dry season is carried on August 2019. The groundwater level measurement is conducted by measuring 125 dug wells and six bore wells in the settlements around the coal-fired power plant area. In addition, observation and flow rate measurement of river in 30 locations are conducted. The observation and measurement of groundwater level and river level are shown in Fig. 2.

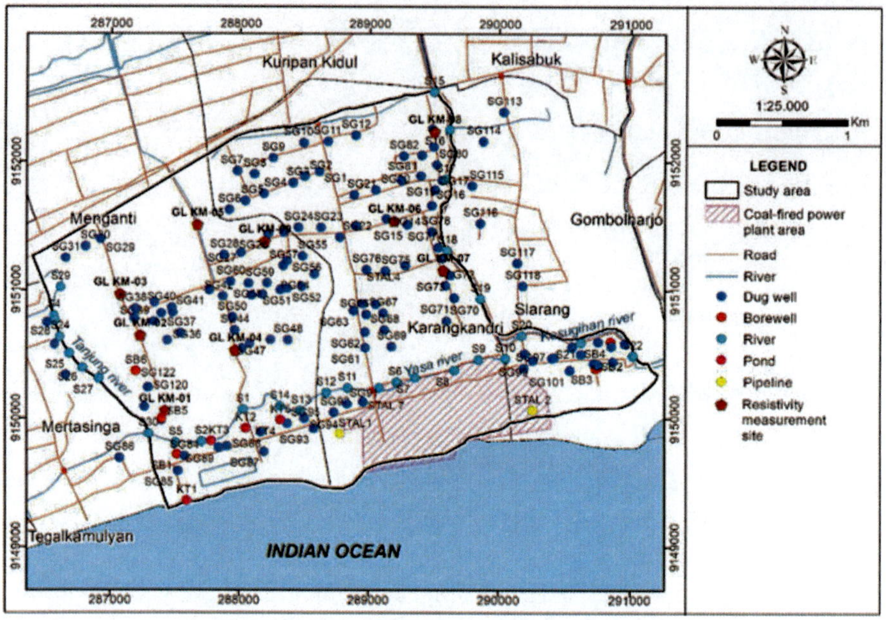

Fig. 2 Map of measurement of groundwater level, river, water body and resistivity

The resistivity measurement is conducted using 1D Schlumberger method in nine locations (see Fig. 2). The length of each measurement is 250 m. This length is assigned to provide subsurface interpretation up to 100 m of depth. The result of each points is correlated to understand the subsurface lithology and its distribution.

Observation and inventory of groundwater utilization by power plant and society are conducted to understand the estimation of groundwater pumping rate in the power plant area and its vicinity. Measurement of groundwater pumping discharge is conducted at five shrimp ponds and two channels.

Secondary data such as rainfall and temperature from BPS-Statistics Agency of Cilacap is collected to calculate the evapotranspiration and infiltration [17].

Groundwater modelling is carried out to simulate the impact of groundwater extraction and power plant construction practice towards the groundwater flow and groundwater level around the construction site. For the numerical model, Visual MODFLOW 3.1 applied to simulate groundwater flow and pumping activities.

Identification of seasonal effect and construction effect is conducted by comprehensively comparing groundwater level depth on two seasons and the result of groundwater modelling from different scenarios.

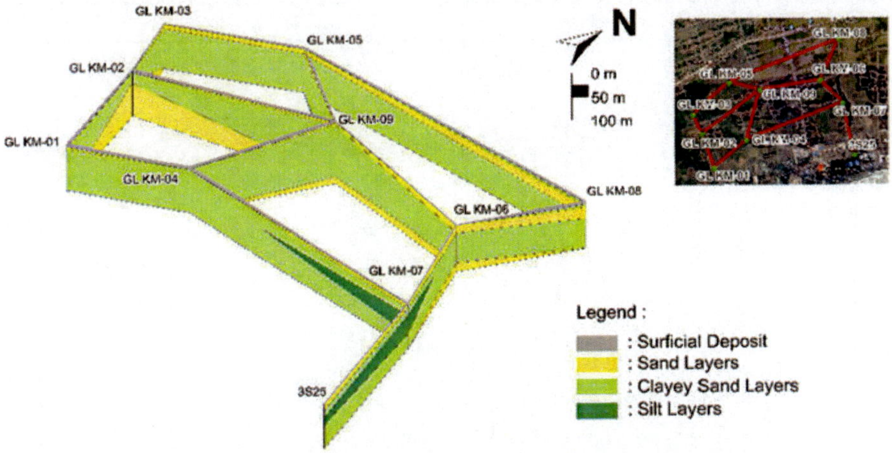

Fig. 3 Fence diagram of subsurface lithology. This diagram is generated by correlating nine locations of Schlumberger resistivity measurement

4 Results

4.1 Geology

Based on resistivity measurement on nine locations, the deposits in the research area are consisted of clayey sand, sand, and silt deposits (see Fig. 3). Loose deposits are distributed on the surface. Lenses of silt deposit layer are located in the southeast of study area. Lenses of sand deposit layer are getting thicker in the west part of study area. These two sand layers sandwich the clayey sand layer. The upper sand layer is getting thinner to the south, while clayey sand layer is comprised of dominantly the whole research area.

While surface geology map of study area is comprised of sand and silty sand deposit. Most of research area is covered by silty sand, while the coal-fired power plant is located on sand deposits (see Fig. 4). There are three aquifer layers in the research area represented by upper sand layer, clayey sand layer as middle layer, and lower sand layer. Those three aquifers are categorized as coastal aquifers due to its location.

4.2 Groundwater Utilization

The observation and inventorying of groundwater utilization discovered that daily need of water by each house is ranging between 200–1500 l/day, while each shrimp ponds by average utilize groundwater discharge of 225 m^3/day. Pumping rate of

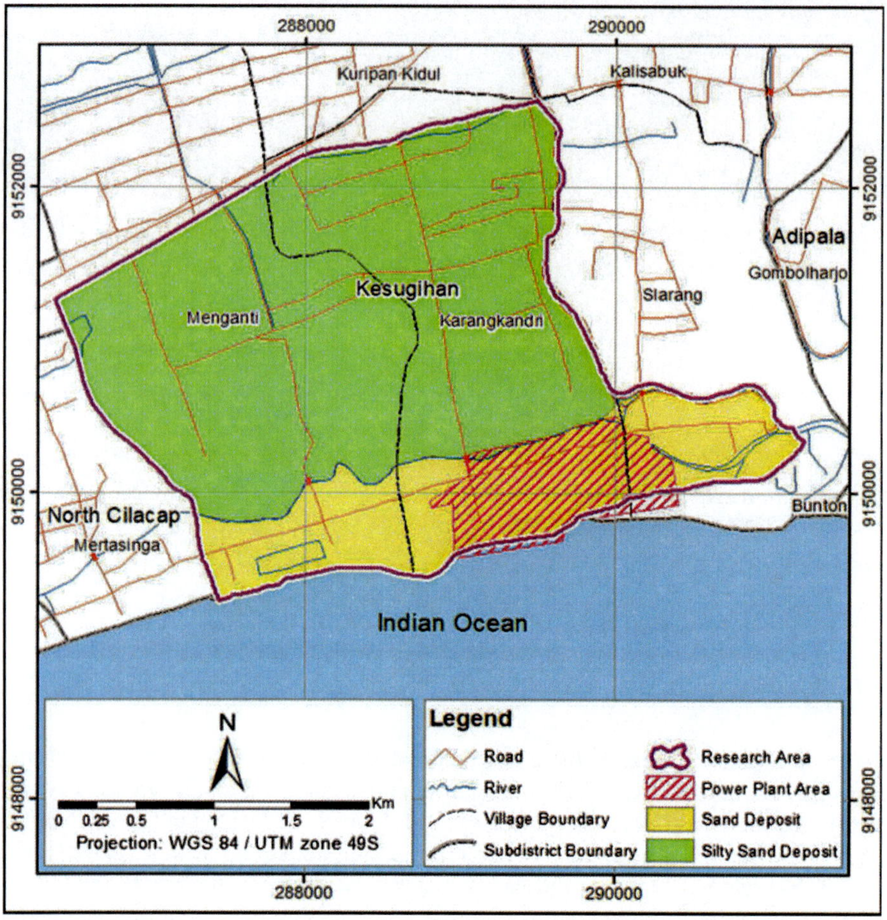

Fig. 4 Surface geology map of study area

groundwater during the construction of power plant is 100 l/s, while pumping rate of groundwater (dewatering) from excavation and underground tunnel construction is 1300 l/s.

4.3 Model Input

The top sand layer and surficial layer aquifer (K_1) have the hydraulic conductivity of 4.9×10^{-4} m/s. The K value of clayey sand aquifer is 5.0×10^{-5} m/s (K_2), while hydraulic conductivity of silt aquitard is 2.94×10^{-8} m/s (K_3). Lastly, sand layer aquifer at the lower depth hasthe hydraulic conductivity of 4.9×10^{-4} m/s (K_4) [18].

Based on rainfall data from 2009 to 2019 by BPS-Statistics Agency of Cilacap Regency, the average of annual precipitation in Cilacap Regency of 2746 mm/year [17]. This figure is resulted from rainfall monitoring data from year of 2009 until 2017. Calculation of evapotranspiration using Thornwaite (1948) method results in 1751 mm/year, while run-off resulted by calculation using Soil Conservation Service (1964) and CN number by Hudson (1993) is 787 mm/year. Finally, the infiltration value in the research area of 294 mm/year.

4.4 Groundwater Level

Groundwater level depth map of dry season of 2019 is shown in Fig. 5. Groundwater level depth in the west part is shallower than the east part. Groundwater level depth is ranging between 1.2–3.95 m even in dry wells, while groundwater level depth map of rain season of 2019 is presented in Fig. 6. Groundwater level depth in rainy season is ranging between 0.28–3.3 m. Mostly, groundwater level depth in rainy season is below 2 m. It can be concluded by comparing both map that groundwater level depth in dry season depleted in the whole research area including settlement area and power plant area. The average gap of groundwater level depth in two seasons is of 1.06 m.

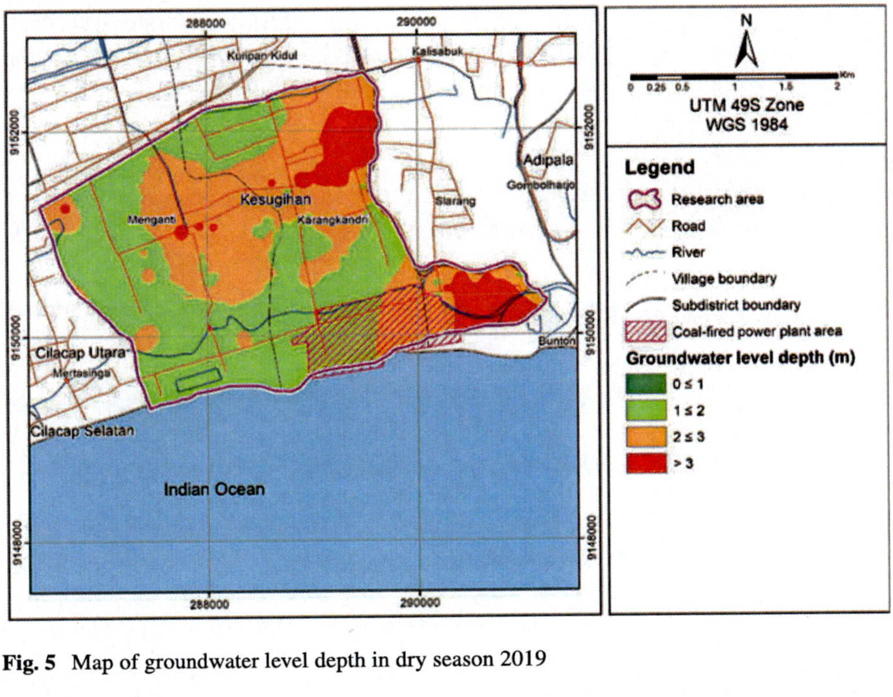

Fig. 5 Map of groundwater level depth in dry season 2019

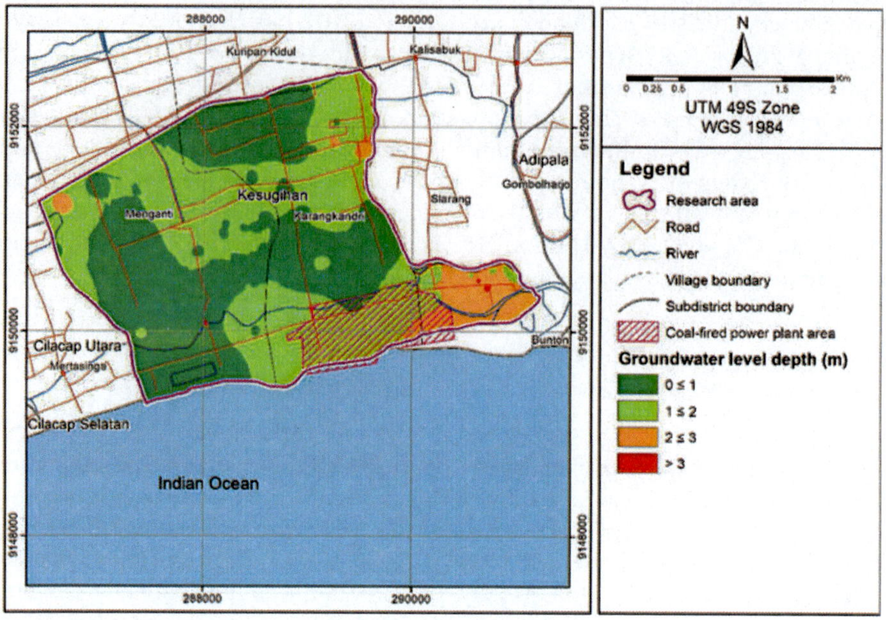

Fig. 6 Map of groundwater level depth in rainy season 2019

The depth of groundwater level in the construction site is ranging between 2 and 3 m, while settlement area in the east of power plant area has groundwater level depth of more than 3 m. The depletion of groundwater level depth significantly occurred in the centre east settlement and coastal area including power plant area and settlement in the east of power plant area.

4.5 Groundwater Flow Direction

Groundwater flow direction map of actual condition is generated by interpolating groundwater level elevation from wells and river water level (see Fig. 7). Groundwater flows from northwest to the southeast.

Groundwater level elevation in the research area ranging from −2 below sea level to 7 m above sea level. In the east part, groundwater flow sto the Kesugihan river as gaining stream type. However, in the downstream, Kesugihan river apparently supplies the aquifer, while in the west part, Tanjung river acts as losing stream and groundwater divider.

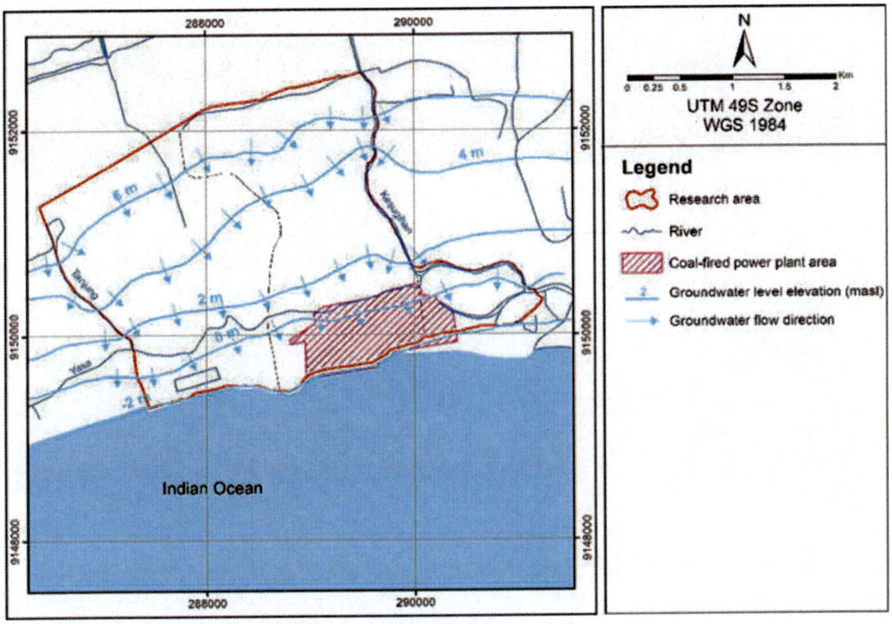

Fig. 7 Groundwater level elevation and flow direction map

4.6 Groundwater Simulation

4.6.1 Conceptual Model

According to hydrogeological unit and subsurface layers, study area is classified into four units including aquifer 1 (upper sand layer), aquifer 2 (clayey sand layer), and aquitard (silt layer), and aquifer 3 (lower sand layer). Aquifer 1 assigned to sand layer at the upper depth. Thickness of aquifer 1 is ranging from 20 to 50 m. Aquifer 2 of clayey sand is sandwiched by aquifer 1 (upper sand layer) and aquifer 3 (lower sand layer) with thickness of 40 m, while aquifer 2 is assigned to silt layer with thickness of 30 m.

4.6.2 Model Discretization

Visual MODFLOW works using block-centred finite difference method. The research area covers area of 16.17 km^2. Discretization of model area divided model area into 13,140 grids with the grid size of 50 $\times$ 50 m^2.

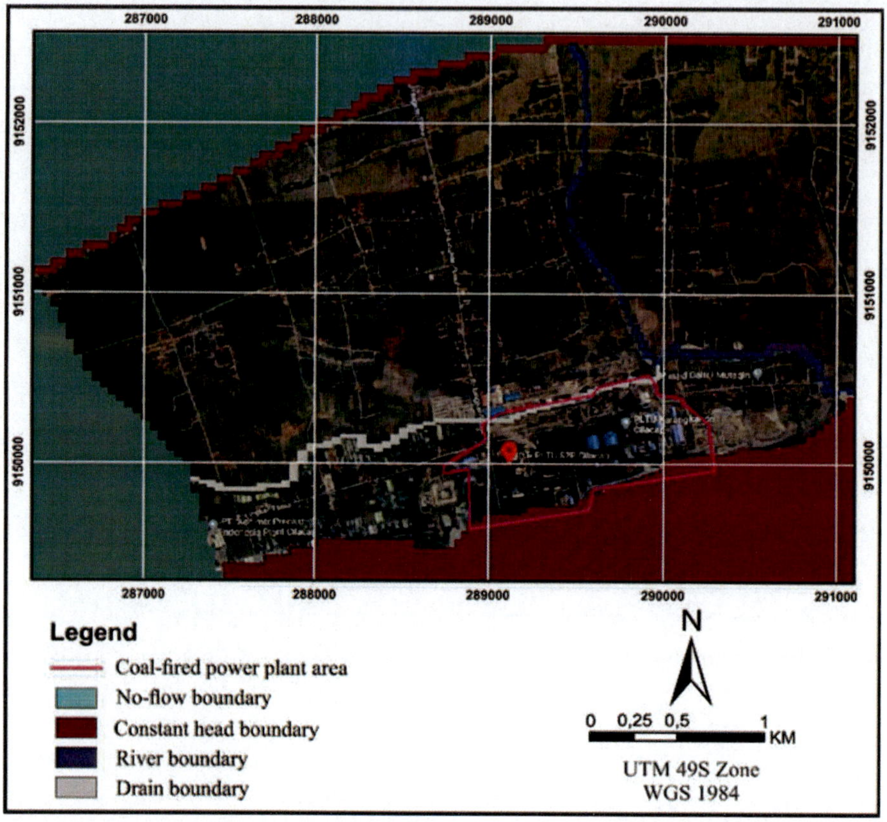

Fig. 8 Boundary condition for groundwater modelling

4.6.3 Boundary Condition

Tanjung river acts as groundwater divider then assigned as no-flow boundary, while Kesugihan river acts as gaining stream type assigned as river boundary. The constant head boundary assigned to the north and south. Constant head boundary in the north part assigned from observed value of groundwater level elevation, while Indian Ocean in the south is assigned as constant head boundary (see Fig. 8). Yasa river in the south part is assigned as drain boundary.

4.6.4 Calibration and Sensitivity Analysis

Calibration and parameter adjustment are required in modelling; thus, the calculated value of groundwater head is as close as possible with actual condition or observed value [19]. Manual calibration strongly connects with the sensitivity analysis, which

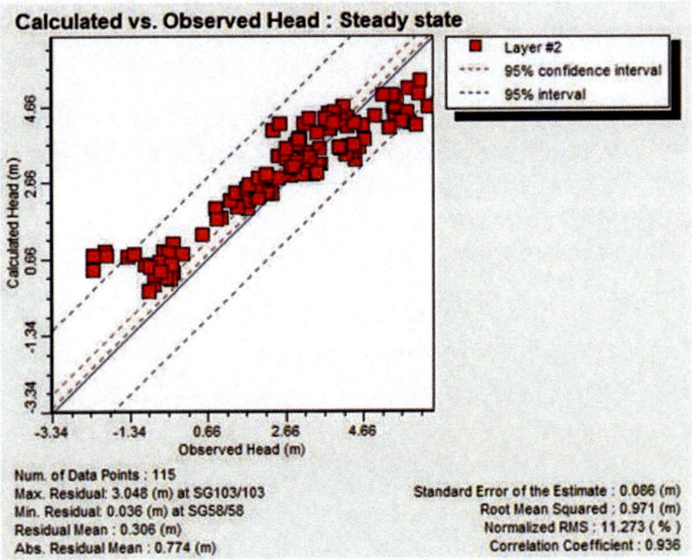

Fig. 9 Calibration graph

put into evidence the variation of the model response as result of variation of uncertain parameters [20].

The parameter adjusted during calibration was only hydraulic conductivity (K). The modified K is as follows: The K_1 sand layer and surficial layer aquifer is 9.0×10^{-4} m/s. The K_2 clayey sand aquifer is 4.96×10^{-4} m/s, while the K_3 silt aquitard is 5.88×10^{-7} m/s. Lastly, K_4 sand layer aquifer at the lower depth is 9.0×10^{-4} m/s.

The result of calibration graph is presented in Fig. 9. The result of calibration is "good", where coefficient correlation is 0.936 and the Standard Error of Estimate is 0.086 m.

4.6.5 Scenario 1: Actual Condition

Groundwater modelling is carried out on four simulation scenarios. The first scenario is the current condition representing the actual condition of the construction site and surrounding area. This means that the result of this model represents condition after all construction just finished. Therefore, underground cooling tunnel as impervious construction is assigned as no-flow boundary/inactive cells. Tunnel construction is located at the depth of 8 m from surface.

Groundwater head is ranging from 2 m below sea level in the beach until 7 m above sea level in the north. Groundwater head in the power plant area is ranging from 2 to 0 m (see Fig. 10).

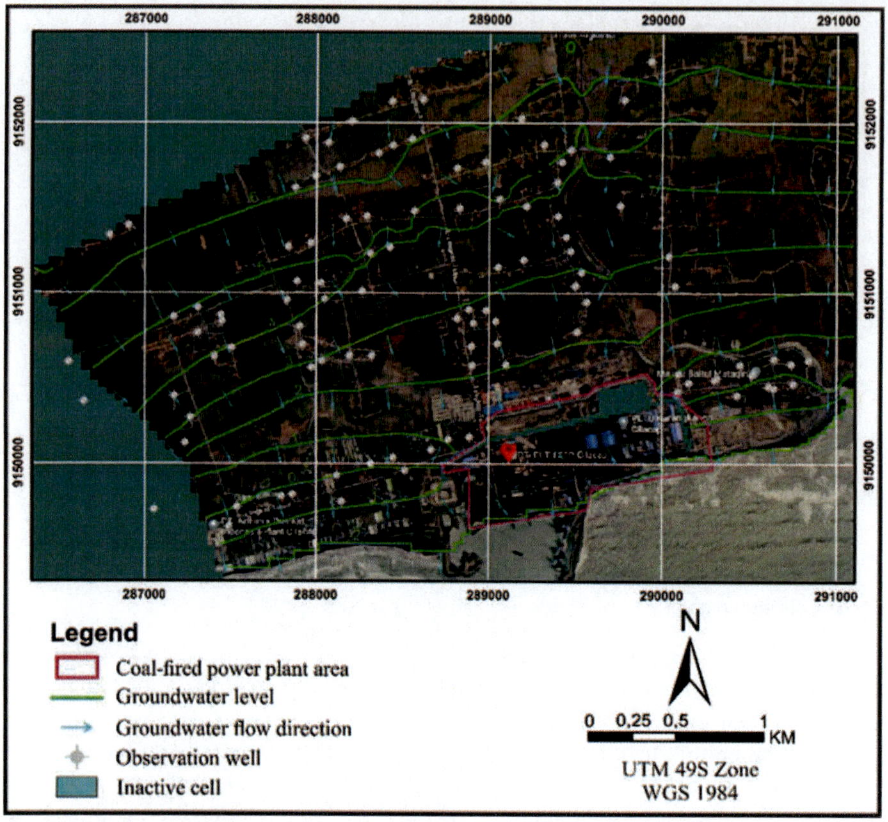

Fig. 10 Scenario 1. Simulation of actual condition

4.6.6 Scenario 2: Before Construction

The second scenario simulates the condition before the construction of basement and underground tunnel is carried out. This scenario is used as the basis of the next two dewatering scenarios.

Groundwater head is ranging between 0–7 m above sea level (see Fig. 11). Groundwater head in the power plant area is ranging from 0–1 m above sea level.

4.6.7 Scenario 3: Dewatering for Power Plant Construction

The third scenario simulates the dewatering practices when the basement of the power plant was constructed. The total pumping rate is 100 l/s. Groundwater head ranging from 2 m to 7 m above sea level (see Fig. 12). It is clearly seen that groundwater level

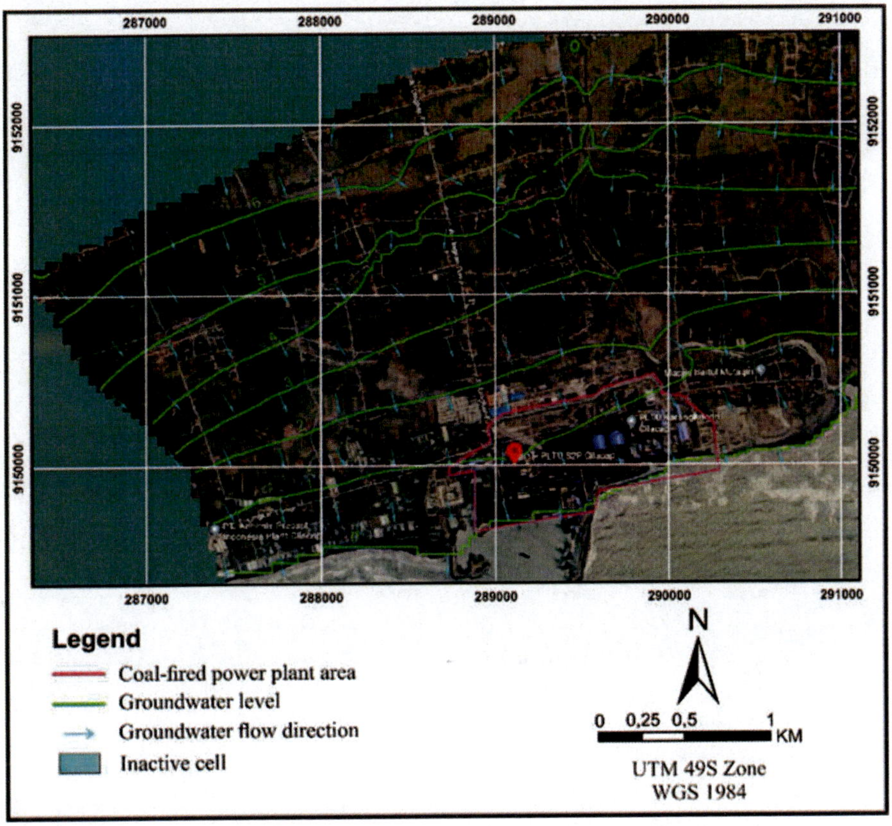

Fig. 11 Scenario 2. Simulation before construction

declined at the power plant area where dewatering practice occurred. The dewatering affected the declining of groundwater level up to 2 m and drew seawater to the construction area.

The results showed that the settlement area in the radius of 500 m from construction site had undergone a declining groundwater level up to 2 m, while the groundwater level in the area of settlement located in the far north from construction site not affected significantly by dewatering.

4.6.8 Scenario 4: Dewatering for Underground Cooling Tunnel Construction

The fourth scenario simulates the dewatering practice when the underground tunnel was constructed. The total pumping rate of 1300 l/s is distributed by 13 pumps (10 l/s for each well). Outside the power plant area, groundwater head is ranging between

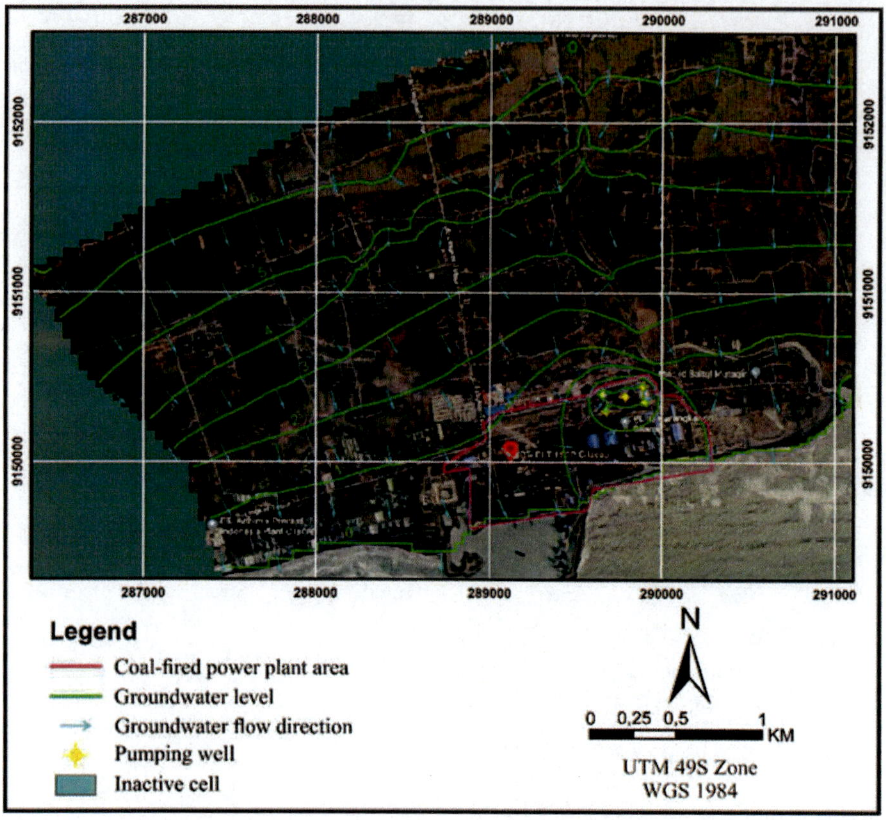

Fig. 12 Scenario 3. Dewatering simulation for foundation construction

0–7 m above sea level. Groundwater head in the settlement area is not affected by dewatering and visually not experienced any significant changing.

The result showed that the affected area of dewatering around power plant area is larger than the previous scenario (see Fig. 13). The declining of groundwater level occurred in the radius of 700 m from construction site up to more than metres. Dewatering practices drew seawater to the construction area.

5 Discussion

Identification of the cause of groundwater level changes is conducted by examining the difference of groundwater head at the wet and dry seasons and the difference of groundwater level before, during, and after (actual condition) the construction of power plant and underground tunnel.

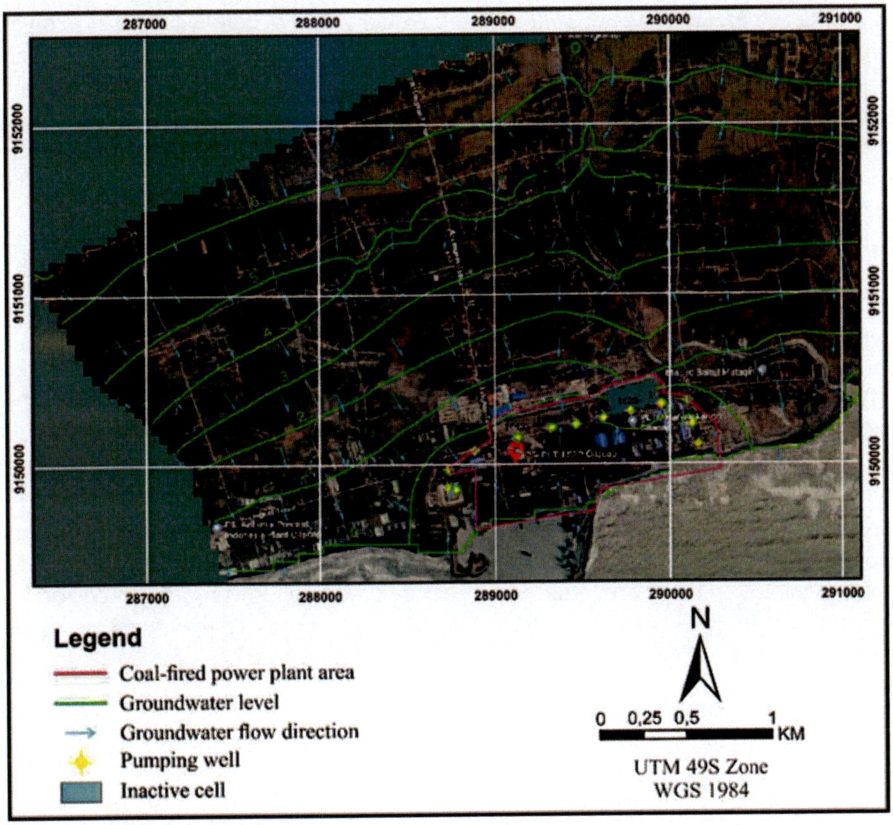

Fig. 13 Scenario 4. Dewatering simulation for underground cooling tunnel construction

The comparison of groundwater level in the dry season and wet season at the settlement area located in the north of construction site showed that the groundwater level depleted 0.05–2.2 m in the dry season. The maximum declining could be higher due to the dry condition of the wells. Thus, the groundwater level in the dry wells is not measured. The average of the groundwater level declining is 1.06 m. In the dry season, the declining of groundwater level occurred in almost whole research area.

Influence of anthropogenic factors to the changes of groundwater level is investigated by examining the difference of groundwater modelling result from various scenarios. Based on comparison of scenario model 1 (simulation after construction/actual condition) with scenario model 2 (simulation before construction), the groundwater level in the southern part and power plant area declined of 2 m. However, groundwater level in the middle and the northern part (settlement area) are constant or only very slight changes.

Based on dewatering scenarios, the affected radius of groundwater level declining is ranging from 500 to 700 m from the construction site. Groundwater level declined 1–2 m. Settlement area in the east of power plant area is affected by the dewatering. The results of dewatering scenarios showed that the groundwater level in the middle and northern part of research area (settlement area) is constant to very slight changes. It showed that dewatering practice did not affect the groundwater level in the settlement area in the middle and northern part of research area.

Based on discussion above, groundwater level changes in the settlement area in the middle and northern part of research area caused by natural effect specifically seasonal difference, while groundwater level declining in the power plant area, southern part of research area, and settlement in the southeastern part of research area is caused by dewatering and construction factors by power plant authority.

6 Conclusions

Based on groundwater modelling and comparison of groundwater level in the wet and dry season, the results suggested that seasonal difference affects the depletion of groundwater level of 0.05–2.2 m or 1.06 m by average at the community's wells and the whole research area.

While the dewatering practices by the power plant authority around the construction site of foundation and underground cooling tunnel were responsible to the declining of groundwater level of 1–2 m at the radius of 500–700 m from the construction site, the settlement in the east of power plant area is affected by the dewatering practices.

Lesson learned that natural factors specifically seasonal difference affect massively the groundwater level changes, while dewatering or construction factors affect locally or limited area.

Acknowledgements The authors would like to express gratitude to Centre of Environmental Study, Universitas Gadjah Mada, Indonesia, and coal-fired power plant of PT Sumber Segara Primadaya, Cilacap, Indonesia, who have supported this research.

References

1. Araki, M., Shimizu, A., Kabeya, N., Nobuhiro, T., Ito, E., Ohnuki, Y., Tamai, K., Toriyama, J., Tith, B., Pol, S., Lim, S., Khorn, S.: Seasonal fluctuation of groundwater in an evergreen forest, central Cambodia: experiments and two-dimensional numerical analysis. Paddy Water Environ, **6**, 37–46 (2008)
2. Singh, R. D., Kumar, C. P.: Impact of climate change on groundwater resources. In: 2nd National Ground Water Congress, pp. 332–350. New Delhi (2010)
3. Kumar, C.P.: Climate change and its impact on groundwater resources. Res. Inventy Int. J. Eng. Sci. **1**(5), 43–60 (2012)
4. Deane, D.C., Harding, C., Aldrige, K.T., Goodman, A.M., Gehrig, S.L., Nicol, J.M., Brookes, J.D.: Predicted risk of groundwater decline in seasonal wetland plant communities depend on basin morphology. Wetlands Ecol. Manage. **26**, 359–372 (2018)
5. Patil, N.S., Chetan, N.L., Nataraja, M., Suthar, S.: Climate changes scenarios and its effect on groundwater level in Hiranyakeshi watershed. Groundwater Sustain. Dev. **10**, 1–7 (2020)
6. Mustafa, S.M.T., Abdollahi, K., Verbeiren, B., Huysmans, M.: Identification of the influencing factors on groundwater drought and depletion in north-western Bangladesh. Hydrogeol. J. **25**, 1357–1375 (2017)
7. Zhuo, Y.: A critical review of groundwater budget myth, safe yield, and sustainability. J. Hydrol. **370**, 207–213 (2009)
8. Calvache, M.L., Pulido-Bosch, A.: Effects of geology and human activity on the dynamics of salt-water intrusion in three coastal aquifers in southern Spain. Environ. Geol. **30**, 215–223 (1997)
9. Sherif, M., Kacimov, A., Javadi, A., Ebraheem, A.A.: Modelling groundwater flow and seawater intrusion in coastal aquifer of Wadi Ham UAE. Water Resour. Manage. **26**, 751–774 (2011)
10. Kuorgialas, N.N., Dokuo, Z., Karatzas, G.P., Panagopoulos, G., Soupios, P., Vafidis, A., Manoutsoglou, E., Schafmeister, M.: Saltwater intrusion in an irrigated agricultural area: combining density-dependent modeling and geophysical methods. Environ. Earth Sci. **75**, 1–13 (2015)
11. Kitterød, N.O.: Archaeological deposits and tunnel constructions: simulations of groundwater head at the Old Wharf of Bergen Norway. Quat. Int. **368**, 51–67 (2015)
12. Alfaro, P., Liesch, T., Goldscheider, N.: Modelling groundwater over-extraction in the southern Jordan Valley with scarce data. Hydrogeol. J. **25**, 1319–1340 (2017)
13. Khan, S., Rana, T., Gabriel, H.F., Ullah, M.K.: Hydrogeologic assessment of escalating groundwater exploitation in the Indus Basin, Pakistan. Hydrogeol. J. **16**, 1635–1654 (2008)
14. Kastowo.: Peta Geologi Lembar Majenang, Jawa. Skala 1:100.000. Pusat Penelitian dan Pengembangan Geologi, Bandung (1975)
15. Asikin, S., Handoyo, A., Pratistho, B., Gafoer, S.: Peta Geologi Lembar Banyumas, Jawa. Skala 1:100.000. Pusat Penelitian dan Pengembangan Geologi, Bandung (1992)
16. Surono, S.: Peta Geologi Lembar Pangandaran, Jawa. Skala 1:100.000. Pusat Penelitian dan Pengembangan Geologi, Bandung (1992)
17. Cilacap, B.P.S.K.: https://cilacapkab.bps.go.id/subject/151/iklim-climate.html#subjekViewTab3. Last Accessed 28 May 2019
18. Putra, D.P.E.: Laporan Penelitian Air Tanah di Wilayah PLTU PT Sumber Segara Primadaya Cilacap dan sekitarnya, Kabupaten Cilacap. Jawa Tengah. Pusat Studi Lingkungan Hidup Universitas Gadjah Mada, Yogyakarta (2019)
19. Boonstra, J., De Ridder, N.A.: Numerical modelling of groundwater Basin: a user-oriented manual. International Institute for Land Reclamation and Improvement, Wageningen (1981)
20. Spitz, K., Moreno, J.: A practical guide to groundwater and solute transport modeling. Willey, New York (1996)

Landslide Susceptibility Assessment in Trenggalek, East Java, Indonesia: A Geological Overview

Wahyu Wilopo, Egy Erzagian, Diyaning Ratri, and Teuku Faisal Fathani

Abstract The number of landslide occurrences in Trenggalek tends to increase every year, and it represents a significant constraint on development, causing high levels of economic loss and substantial numbers of fatalities. A complete understanding of geological conditions is essential needed for spatial planning and to reduce the risk of lost lives and infrastructure damage. Surface mapping at regional scale comprised of lithology, geomorphology, geological structures, and land-use conditions was conducted in an area of 1261 km^2 to develop a detailed landslide susceptibility map. The surface mapping showed that landslides commonly occurred on steep slopes of the fault-controlled hills that having inclination ranging from 16° to 28°, in the areas covered by residual soils of Mandalika Breccia unit, with high geological structure density in the settlement areas. The high susceptibility zone covered 46,75% of the research area and was characterized by having slope inclination ranging from 8° to 45°, geological units of Mandalika Breccia and Arjosari Sandstone, and land uses of settlement, dry fields, and plantation. Undisturbed soil samples collected from the high susceptibility area exemplify the clay–silt material, and XRD analysis showed the dominant clay mineral composition of halloysite and smectite. Major ion components from precipitation water, seepage, and surface water samples were used to determine hydraulic connectivity in high susceptibility zone. Both seepage and surface water indicate sodium bicarbonate-type water similar to the chemical composition of the precipitation water. The landslide mechanism in the research area is mainly controlled by lithological conditions and triggered by high rainfall.

Keywords Landslide susceptibility · Surface mapping · Trenggalek · Indonesia

W. Wilopo (✉) · E. Erzagian · D. Ratri · T. F. Fathani
Universitas Gajah Mada, Yogyakarta, Indonesia
e-mail: wilopo_w@ugm.ac.id

H. Hazarika et al. (eds.), *Advances in Sustainable Construction and Resource Management*, Lecture Notes in Civil Engineering 144,
https://doi.org/10.1007/978-981-16-0077-7_71

1 Introduction

Landslides are natural disasters that are controlled by geological conditions, including slopes, soil or rock types, geological structures, slope hydrology, and land use [1]. Landslides can be triggered by human activities and/or natural processes, including cutting slopes, changing slope use, vibrations/earthquakes, and rain [2]. Global climate change has increased in the intensity of rain, which has resulted in many landslides with many casualties [3]. Therefore, many researchers carry out landslide mapping with various approaches, including landslide inventory-based probabilistic approach, heuristic approach, statistical approach, and deterministic approach [4]. Some researchers also use geomorphological approaches to map landslide susceptibility [5]. Utilization of all geological parameters in landslide susceptibility mapping will give better results than using only one parameter [6]. Geographical information system (GIS) applications are widely used to support landslide susceptibility analysis using various parameters and various approach methods [7–9]. The accuracy of the landslide vulnerability map is greatly influenced by the map scale, mapping units, and mapping methods [10].

The research area is located in Trenggalek Regency, East Java Province, Indonesia, as shown in Fig. 1. Based on Regional Geological Map of Pacitan, Tulungagung, and

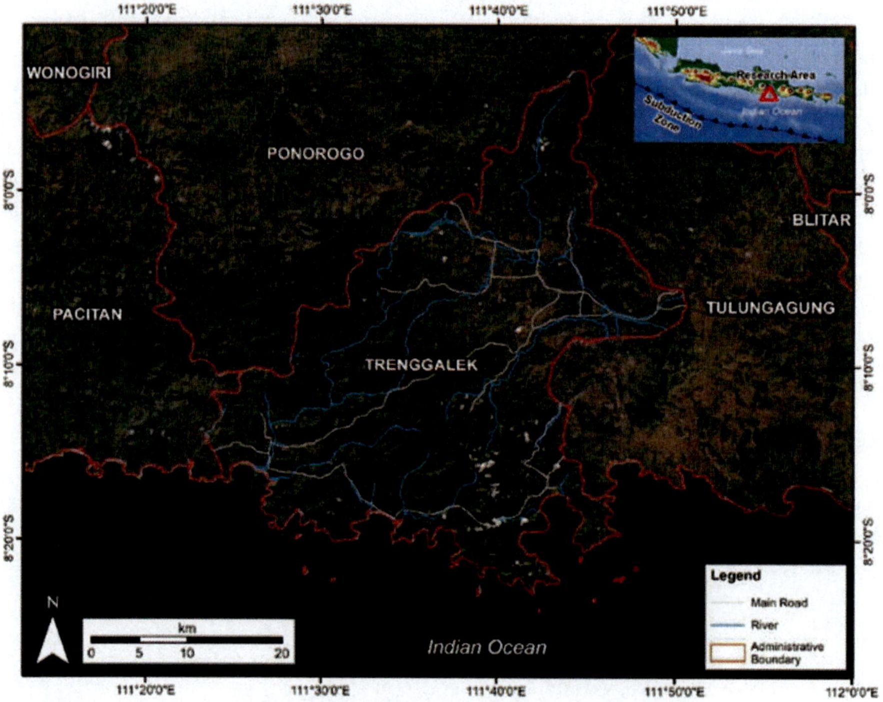

Fig. 1 Research area of Trenggalek Regency, East Java, Indonesia

Fig. 2 Typical landslides along road cut slopes in Trenggalek

Madiun Scale 1:100,000, the Trenggalek area mostly consists of Oligocene–Quaternary rocks [11]. The lithology is mostly weathered with low strength properties. The geomorphology of the Trenggalek area is the result of volcanism and is strongly affected by the geological structure, such as Puger Fault and Kambengan Fault. Land use in Trenggalek has been used for rice and dry fields (73.18%) and settlements (9.28%). Considering the relatively low population density, the development of public facilities in Trenggalek is quite feasible. However, many landslides have occurred and threatened the communities in the area, as shown in Fig. 2. Landslides have occurred in the past few years and are mostly triggered by the high rainfall, especially during the annual rainy season from October to April. Most of the Trenggalek Regency area is located on the steep slopes that were prone to a landslide. On the other hand, many residents living in disaster-prone areas of the landslide do not have sufficient knowledge regarding disaster mitigation because of the low level of education [12]. Therefore, to anticipate the impact of more significant losses and fatalities, it is necessary to research the geological conditions of Trenggalek Regency, including the landslide susceptibility.

This paper presents the result of surface geological mapping in the Trenggalek area at a regional scale comprised of lithology, geomorphology, geological structures, and land-use condition to develop a detailed landslide susceptibility map. Henceforward, this vulnerability map can be used for spatial planning purposes and to identify the main controlling factor of landslides. Lithological, geomorphological, geological structures, and land-use maps will be presented in this paper, and in the end, the main controlling factor of landslides will be highlighted.

2 Methodology

The methods used for this assessment are divided into two, namely field and laboratory investigations. Field investigations including investigations of lithology conditions, geomorphology, geological structure, precipitation and seepages water, land use, and susceptibility of the research area to landslide were conducted during surface engineering geological mapping at a regional scale.

Topographical maps analysis was conducted before fieldwork using a geographical information system (GIS) tool to support the field investigations. It combined with field observation of surface topography and active geomorphological processes to produce a geomorphological map. The research area was classified into several geomorphological units based on morphography, morphogenesis, and morphometry [13, 14].

Investigation of rock properties and geological structures was implemented, and the results were presented in a geological map and structural density map. The rocks classification genetic used were based on the classification in the regional geological map of Pacitan, Ponorogo, and Madiun area [11]. A structural density map was established upon lineaments and interpretation of geological structures found during field investigation. Field observation and interpretations of the satellite images of the existing land uses were conducted to develop a land-use map.

Slope inclination, lithology, structural density, and land-use data were used as parameters in the landslide susceptibility zonation. Each of the parameters is processed using weights of evidence method. The criteria used for landslide susceptibility zonation are shown in Table 1.

Table 1 Criteria of landslide susceptibility zonation

Parameter	Classification	Value	Weight (%)
Lithology	Mandalika breccia, Arjosari sandstone	3	52
	Mandalika lava, alluvial	2	
	Intrusion, Campurdarat limestone, Jaten conglomerate, Wuni volcanic breccia, Oyo calcareous sandstone, Wonosari limestone, Argokalangan volcanic breccia	1	
Slope inclinations	>25°	3	20
	15°–25°	2	
	<8°–15°	1	
Structural density	>1.5 km/km^2	3	20
	1–1.5 km/km^2	2	
	<1 km/km^2	1	
Land use	Settlement	3	8
	Rice and dry fields	2	
	Forest	1	

The lithological unit, slope inclination, structural density, and land-use intensity are given 52%, 20%, 20%, and 8% weights, respectively. Each parameter was then divided into several sub-parameters that were assigned a value from 1 to 3. The highest value represents the most susceptible parameter to cause a landslide, while the lower value represents the less susceptible parameter to cause a landslide. Then the score of each classification is obtained by multiplying the value by the weight. Based on the results of the calculation, the research area is then divided into three zones of low, intermediate, and high landslide susceptibilities.

As the presence of water on a slope is considered as one of triggering factors in landslides, field observation of water seepage, surface water, and precipitation water was also conducted, upon which major ion components from these water samples were used to determine hydraulic connectivity in high susceptibility zone. Soil sampling weathered rock was also carried out for XRD analysis, which aims to determine the presence of clay minerals that hold the potential to cause landslides.

3 Results and Discussion

3.1 Geological Condition

The research area consists of 11 lithological units and is divided into four geomorphological units. During the surface mapping, 62 landslides were identified. Most of the landslides were relatively small, having approximately a size of 3 m wide by 3 m high, although slightly more massive landslides that have approximately a size of 8 m wide by 8 m high were also observed.

Lithology. The lithology sequences from the oldest to the youngest unit are Mandalika lava, Mandalika breccia, intrusion, Arjosari sandstone, Campurdarat limestone, Jaten conglomerate, Wuni volcanic breccia, Oyo calcareous sandstone, Wonosari limestone, Argokalangan volcanic breccia, and alluvial. Most of the rocks in the field already weathered from slightly to completely decomposed. XRD analysis from weathered Mandalika breccia showed the dominant clay mineral composition of halloysite and smectite. Besides, minor clay minerals, such as kaolinite, plagioclase, and mica are also can be found. Halloysite is quite susceptible to changes in geotechnical characteristics. Smectite is expansive and high plasticity; therefore, it is easy to swell when infiltrated by water.

Landslides commonly occurred in the Mandalika breccia unit, which extends from the south to the north due to the absence of smectite mineral. Mandalika breccia has the widest coverage in Trenggalek, as shown in Table 2. Mandalika lava had several landslides more than Arjosari sandstone. However, Arjosari is more vulnerable due to its intense weathering and its numerous discontinuities.

Geological map of the research area is shown in Fig. 3a. Although the alluvial had no landslides occurrence, it still holds potential movement due to its loose material compositions. As the result of considerably different engineering characteristics

Table 2 Lithological unit of the research area

Unit	Area (%)	Number of landslides observed
Mandalika Lava	11.96	10
Mandalika Breccia	39.16	31
Arjosari Sandstone	9.93	6
Intrusion	1.69	0
Campurdarat Limestone	7.66	4
Jaten Conglomerate	2.37	2
Wuni Volcanic Breccia	1.77	3
Oyo Calcareous Sandstone	0.51	0
Wonosari Limestone	2	4
Argokalangan Volcanic breccia	2.08	2
Alluvial	20.81	0

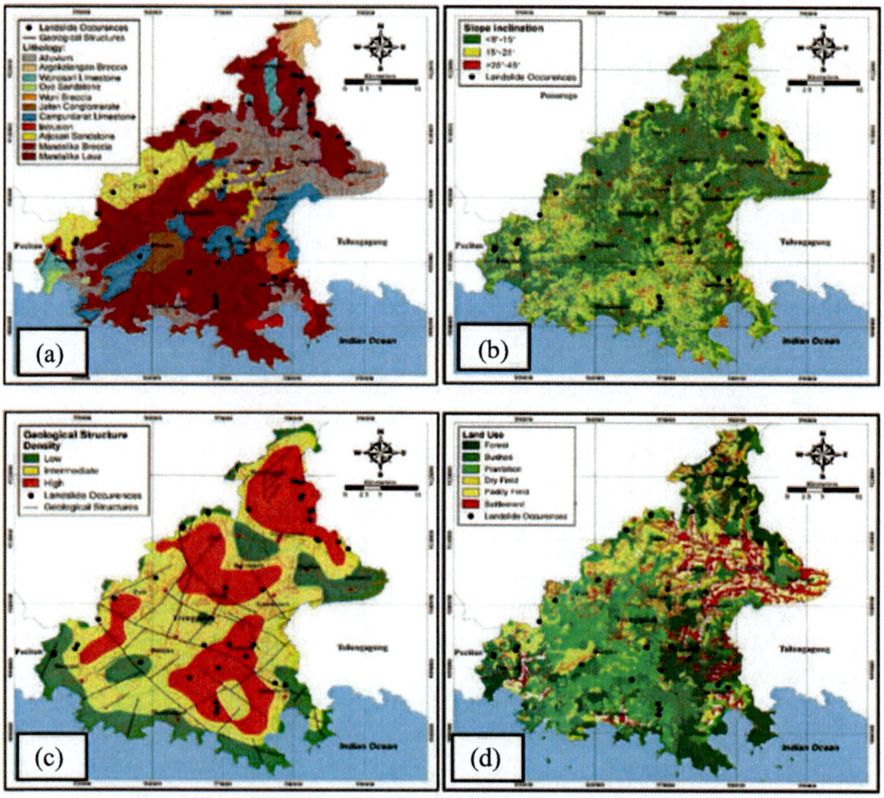

Fig. 3 **a** Geological map of the research area; **b** Slope inclination map of the research area; **c** Structural density map of the research area and **d** Land-use map of the research area

Table 3 Characteristics of the geomorphological unit

Unit	Area (%)	Slope inclination	Elevation (m)
Gentle slopes of the strong-eroded plain	15.19	2°–8°	0–120
Intermediate slopes of the fault-controlled hills	29.34	8°–19°	6–677
Steep slopes of the fault-controlled hills	40.16	16°–28°	60–1002
Steep slopes of the lava flow ridge	15.31	17°–26°	116–1164

of the geological units, the residual soils of the intrusion, Campurdarat limestone, Jaten conglomerate, Wuni volcanic breccia, Oyo calcareous sandstone, Wonosari limestone, and Argokalangan volcanic breccia have a relatively thin bed (around 0.5–1 m) and had relatively small landslides dimension.

Geomorphology. The geomorphological condition of the research area is divided into four units, as shown in Table 3.

The slope inclination map can be shown in Fig. 3b. The active surface geological processes in the steep slopes of the lava flow ridge and the fault-controlled hills were mainly erosion, while the active processes in the intermediate slopes of the fault-controlled hills and gentle slopes of the strong-eroded plain were mainly erosion and deposition.

During the field observation, 35 landslides were found in steep slope inclinations (>25°) of steep slopes of the fault-controlled hills and steep slopes of the lava flow ridge landforms, 25 were found in moderate slope inclinations (15°–25°), and two were found in gentle slope inclinations (<8°–15°) of intermediate slopes of the fault-controlled hills landform.

Geological Structures. Unlike other regions, the geological structure in Trenggalek had a significant role in the occurrence of landslides. The geological structure of either joints or faults could easily become a pathway for rainwater to weaken the rocks. The structural density map of the research area is shown in Fig. 3c.

Land-Use Condition. Land use in the study area consists of forest, bushes, plantation, dry field, paddy field, and settlement, as shown in Fig. 3d, which are then grouped into three categories, i.e., forest, rice and dry fields, and settlement. Rice and dry fields were the primary land use in the research area (73.18%), while the settlement area was only a small part of the research area (9.28%). Road cut slopes consisting of low strength soils and rocks in the research area were commonly unprotected and, therefore, prone to failure, particularly during rainy seasons.

3.2 *Landslide Susceptibility*

The result of the research is shown in the map of landslide susceptibility in Fig. 4. The high susceptibility zone covered approximately of 46.75% the research area and

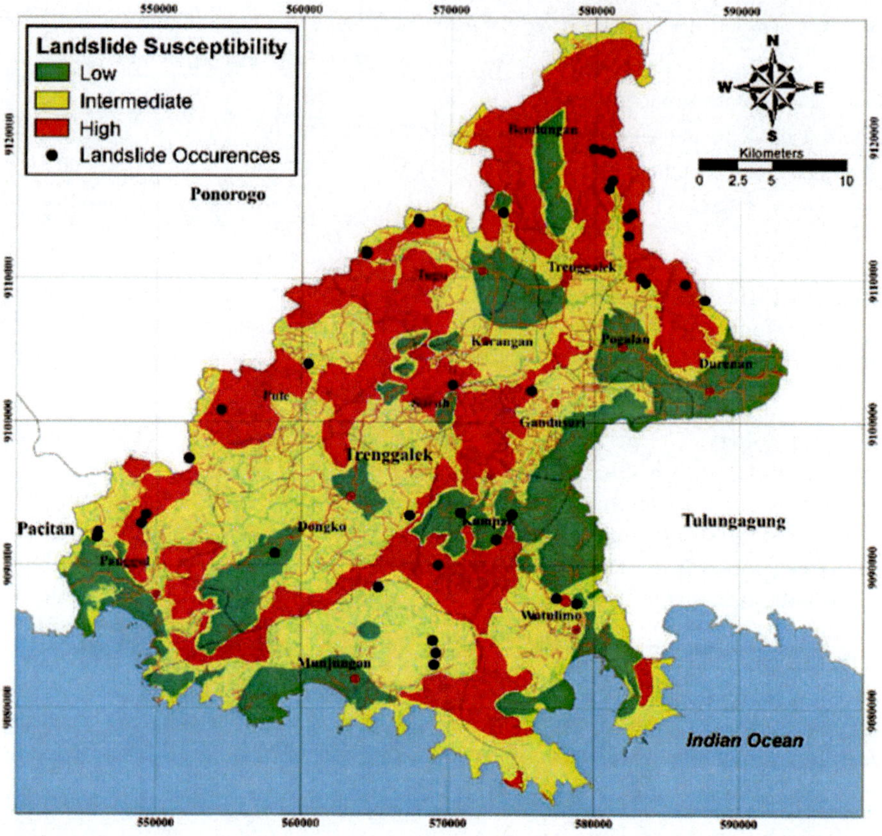

Fig. 4 Landslide susceptibility map of the Trenggalek Regency

characterized by having slope inclination ranging from 8° to 45°, having geological unit consists of breccia and sandstone, having high-density geological structures, and having land use in the form of settlement, dry fields, and plantation. The moderate susceptibility zone consisted of 36.71% of the research area and characterized by having slope inclinations ranging from 8° to 45°, having geological units of alluvial, breccia, sandstone and lava, having moderate to high-density geological structures and having land use in the form of plantation and settlement. The low susceptibility zone occupied 16.54% of the research area and characterized by having slope inclination ranging from 8° to 45°, having a geological unit of alluvial, limestone, intrusion, lava, and calcareous sandstone, having low-density geological structures and having land use in the form of forest and plantation.

The result of the study shown that landslides could have occurred in the slopes from 8° to 45° and mainly located in the fault-controlled hills landform. From the wide ranges of inclinations, it is known that the slopes inclinations were not the main controlling factor of landslides in the research area. The main controlling factors of landslides in the research area are lithological conditions in which weathered intensively and formed a reasonably thick soil, dominantly composed of clay–silt material. Dominant clay mineral composition of halloysite and smectite indicates young clay minerals that quite susceptible to the change of geotechnical properties [15].

Major ion components from precipitation water, seepage, and surface water samples are collected to determine hydraulic connectivity in the hydrological system in high susceptibility zone. There are six water samples taken from several villages to be analyzed with the results, as shown in Table 4.

Both seepage and surface water indicate sodium bicarbonate type water, which is similar to the chemical composition of the precipitation water. The enrichment of calcium and chloride elements on the surface water is a result of the interaction of rainwater with rock and soil. Nglinggis and Prambon areas consist of carbonate rocks, which are members of Wonosari Limestone. It shows that precipitation water mainly seeped through the joints and other discontinuities, making the rock prone to the weathering process. The presence of discontinuity planes also acts as a weak plane that facilitates the occurrence of the soil movement.

4 Conclusion

Surface mapping at 1:25,000 scale comprised of lithology, geomorphology, geological structures, and land-use condition was conducted in an area of 1261 km^2 to develop a detailed landslide susceptibility map. Landslides commonly occurred on steep slopes of the fault-controlled hills that having inclination ranging from 16° to 28°, in the areas covered by residual soils of Mandalika Breccia unit, with high geological structure density in the settlement areas. The high susceptibility zone covered 46.75% of the research area and characterized by having slope inclination ranging from 8° to 45°, having geological units of Mandalika Breccia and Arjosari Sandstone, and having land uses in the form of settlement, dry fields, and plantation. The rainfall water was confirmed as a source of groundwater in the slope that triggers a landslide according to their chemical composition. Therefore, it is necessary to control both surface and subsurface drainage on the slope.

Table 4 Water analysis in Trenggalek

Location	Water source	Analysis result								TDI	Water type
		Na^+	Ca^{2+}	Mg^{2+}	K^+	HCO_3^-	SO_4^{2-}	NO_3^-	Cl^-		
Trenggalek	Precipitation	0.48	0.16	0.04	0.03	0.59	0.13	0.01	0.01	1.44	Sodium bicarbonate
Nglinggis	Surface	2.91	2.22	0.89	0.15	3.37	0.54	0.24	0.50	10.82	Sodium-calcium bicarbonate
Prambon	Surface	0.74	0.47	0.20	0.03	0.79	0.15	0.01	0.01	2.40	Sodium-calcium bicarbonate
Prambon	Surface	1.13	0.94	0.61	0.03	2.08	0.23	0.01	0.06	5.07	Sodium-calcium bicarbonate
Nglinggis	Seepage	1.39	1.48	0.77	0.03	2.77	0.25	0.08	0.11	6.88	Calcium-sodium bicarbonate
Prambon	Seepage	1.30	0.86	0.58	0.03	1.98	0.13	0.07	3.27	8.19	Sodium-calcium chloride bicarbonate

Acknowledgements The research was supported by the Faculty of Engineering Universitas Gadjah Mada and Regional Disaster Management Agency of Trenggalek. Assistants of GAMA-InaTEK during the field and laboratory investigations are gratefully acknowledged.

References

1. Karnawati, D., Fathani, T.F., Wilopo, W., Setianto, A., Andayani, B.: Promoting the hybrid socio-technical approach for effective disaster risk reduction in developing countries. WIT Trans. Built Environ. **119**, 175–182 (2011)
2. Fathani, T.F., Karnawati, D., Wilopo, W.: An integrated methodology to develop a standard for landslide early warning systems. Nat. Hazards Earth Syst. Sci. **16**, 2123–2135 (2016). https://doi.org/10.5194/nhess-16-2123-2016
3. Haque, U., Silva, P.F.D., Devoli, G., Pilz, J., Zhao, B., Khaloua, A., Wilopo, W., Andersen, P., Lu, P., Lee, J., Yamamoto, T., Keellings, D., Hong, W.J., Glass, G.E.: The human cost of global warming: deadly landslides and their triggers (1995–2014). Sci. Total Environ. **682**, 673–684 (2019)
4. Van Westen, S.J., van Ach, T.W.J., Soeters, R.: Landslide hazard and risk zonation—why is it still so difficult? Bull. Eng. Geol. Environ. **65**, 167–184 (2005). https://doi.org/10.1007/s10064-005-0023-0
5. Van Westen, S.J., Rengers, N., Soeters, R.: Use of geomorphological information in indirect landslide susceptibility assessment. Nat. Hazards **30**, 399–419 (2003)
6. Segoni, S., Papafico, G., Luti, T., Catani, F.: Landslide susceptibility assessment in complex geological setting: sensitivity to geological information and insights on its parameterization. Technical note. Landslides 11 (2020). https://doi.org/10.1007/s10346-019-01340-2
7. Casagli, N., Catani, F., Puglisi, C., Delmonco, G., Ermini, L., Margottini, C.: An inventory-based approach to landslide susceptibility assessment and its application to the Virginio River Basin, Italy. Environ. Eng. Geosci. **X**(3), 203–216 (2004)
8. He, H., Hu, D., Sun, Q., Zhu, L., Liu, Y.: A landslide susceptibility assessment method based on GIS technology and an AHP-weighted information content method: a case study of Southern Anhui, China. ISPRS Int. J. Geo-Inf. **8**(266), 23 (2019). https://doi.org/10.3390/ijgi8060266
9. Psomiadis, E., Papazachariou, A., Soulis, K.X., Alexiou, D.S., Charalampopoulos, I.: Landslide mapping and susceptibility assessment using geospatial analysis and earth observation data. Land **9**(133), 26p (2020). https://doi.org/10.3390/land9050133
10. Erener, A., Duzgun, H.S.B.: Landslide susceptibility assessment: what are the effects of mapping unit and mapping method? Environ. Earth Sci. **66**, 859–877 (2012)
11. Samodra, H., Suharsono, Gafoer, S., Suwarti, T.: Geological Map of Pacitan, Tulungagung and Madiun Sheet, Scale 1: 100.000. Geological Research and Development Centre, Bandung, Indonesia (1992)
12. Wilopo, W., Fathani, T. F.: An integrated approach to community empowerment for disaster risk reduction; A case study in Plosorejo Village, Tawangmangu District, Karanganyar Regency. In: Proceedings of the National Seminar on Strengthening Indonesia's Resilience through Disaster Risk Reduction, Yogyakarta, Indonesia (2014) (in Indonesia language)
13. Lobeck, A. K.: Geomorphology, an Introduction to the Study of Landscapes, 731p. McGraw-Hill Book Company, Inc. (1939)
14. Verstappen, H. T.: Applied Geomorphology: Geomorphological Surveys for Environmental Development, 437p. Michigan University (1983)
15. Shirahata, H., Asahi, H., Oura, H.: Relationship between rock alteration and landslides in the Noboribetsu district, Southwest Hokkaido. J. Jpn. Soc. Eng. Geol. 28-2 (1987)

Riverbed Fluctuation Analysis of Small Rivers Equipped with Stream Barb Groins

Mizuki Sakai, Yugo Hashimoto, and Eiji Matsushita

Abstract Streambed degradation significantly affects riverbanks, causing various problems such as unstable flood control. Stream barbs are therefore, used for streambank stabilization; they are low-height masonry groins placed from the riverbank to the upstream side and are intended to mitigate riverbed degradation as a form of river flow control. In this study, the effect of stream barbs installed in 2018 in the Tagawa River in Nagano Prefecture, Japan, was investigated via riverbed fluctuation analysis using International River Interface Cooperative software. A terrain model for simulation was created using a computer-aided design river section view, and a hydrograph for May to December of 2018 was calculated from the stage discharge curve (H-Q curve). The Tagawa River was assumed to have a four-stage peak flow rate (maximum of 160 m^3/s), with a design discharge of 410 m^3/s. The analysis confirmed that regardless of the peak discharge scale, sedimentation of soil occurs at under 30–20 m^3/s corresponding to the gradual decrease curve of discharge. The fluctuation analysis results were compared with drone-captured images from January 2019, whereby the flow meandering conditions of watercourse formation, sedimentation of soil, and scouring sites were determined. Overall, the simulation results were confirmed to agree closely with the images. It was also confirmed that the new side bar generated via the barb construction is effective in maintaining the sediment and preventing the decline of the riverbank against flooding at the scale at which the sidebar was generated, although the scouring becomes deep during flooding.

Keywords Fluctuation analysis · Stream barb · Riverbed degradation

1 Introduction

Riverbed degradation has significant effects on riverbanks and causes various problems such as unstable flood control [1]. To address this issue, stream barbs, which are low-height masonry groins, are used for streambank stabilization and placed from

M. Sakai (✉) · Y. Hashimoto · E. Matsushita
National Institute of Technology, Nagano College, Nagano, Japan
e-mail: mizukis@nagano-nct.ac.jp

H. Hazarika et al. (eds.), *Advances in Sustainable Construction and Resource Management*, Lecture Notes in Civil Engineering 144,
https://doi.org/10.1007/978-981-16-0077-7_72

the riverbank to the upstream side; they are designed to mitigate riverbed degradation as a form of river flow control [2, 3]. A barb does not inhibit the flow of floods but gathers soil and sediment along the bank revetment, making a side bar. If the barb work is made lower from the riverbank to the tip with a side slope, then the side bar forms easily. Additionally, it is more effective to install the barb at an angle of 20°–30° from the riverbank to the upstream side [4, 5]. Many small- and medium-sized rivers are susceptible to more intense disasters with recent climate change. One effective mitigation measure is maintaining the riverbed height and stabilizing the riverbank by installing barbs.

Nagano Prefecture is one of Japan's leading alpine areas, with many rapid small- and medium-sized rivers that flow into larger rivers. The Tagawa River is a small river that flows from Shiojiri to Matsumoto city, Nagano Prefecture. It is located 626 m above sea level at lat. 36°10′15″ N, long. 137°57′58″ E, with a length of 18.4 km (Fig. 1). The middle basin of the Tagawa River has experienced problems related to riverbed degradation. River improvement projects were carried out in the analysis area in 2018. Three wooden mattress stream barbs on the left bank and two riprap works were constructed.

In this study, the effect of stream barbs installed in April 2018 at Tagawa River were investigated using the International River Interface Cooperative (iRIC) riverbed fluctuation analysis.

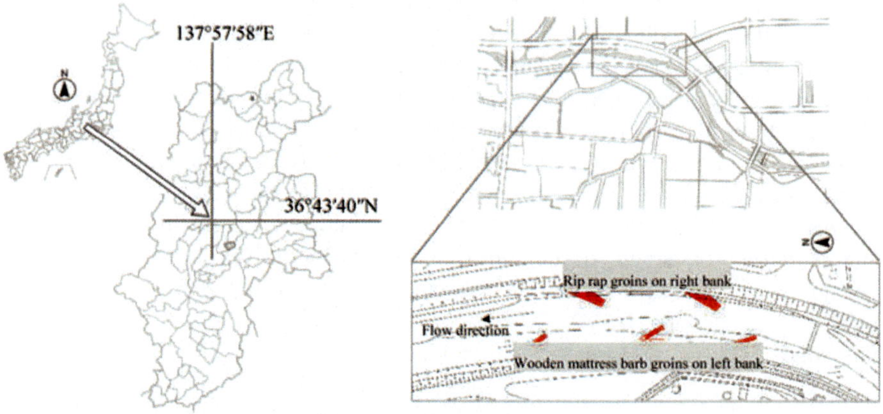

Fig. 1 Outline map of Tagawa River showing geographical location and construction drawing for the analysis area

2 Materials and Methods

2.1 Fluctuation Analysis Using iRIC

Fluctuation analysis was carried out using the iRIC software (ver. 2.3) with the Nays2DH solver. The iRIC analysis software for river flow and riverbed variation combines the Multi-dimensional Surface-Water Modeling System (MD_SWMS), developed by US Geological Survey (USGS), and River Interface Corporative (RIC Nays), developed by the Foundation of the Hokkaido River Disaster Prevention Research Center. In this application, Nays2DH was developed from Nays2D. Nays2DH is a planar two-dimensional solver developed to calculate river flow, riverbed change, and riverbank erosion.

Terrain Model Construction Method. The general steps of performing a simulation are creating a grid for calculation using river survey data and/or digital elevation model (DEM) data; setting simulation discharge, boundary conditions, roughness, and other parameters; and then finally running the simulation. However, the study area was a section of the river that consisted of completed construction works. Thus, the terrain model for simulation was created by five river section views of computer-aided design (CAD) drawings obtained from surveying and mapping for construction works. Terrain data were created using the cross-sectional view and the plane figure at the completion of the river improvement construction in the analysis section. Although the cross-sectional view shows five cross sections at 50-m intervals (total construction section length: 200 m), the terrain data had seven cross sections (analysis target area: 300 m) because the analysis section was extended by 50 m from the upstream and downstream ends of the construction section for stable analysis.

To import the terrain data into iRIC, it is necessary to create a.riv file that holds the coordinate values at points (obtained by dividing each cross section into 76 points, including changed gradient points) and the elevation values. These values were calculated using CAD software (JW_CAD). The analysis grid is used by the solver to perform calculations. One of the iRIC grid generation functions uses the river surveying data (.riv file) to generate the grid. In this analysis, the number of cross-grid divisions on the moving bed was 12, the cross-grid on the fixed bed was divided into six cross sections, and the division interval between cross sections was 1.3 m. The elevation on grid points was calculated by linear interpolation (using distance along the grid line) based on the.riv file. This interpolation method is appropriate because it was confirmed from the construction drawing that the bed slope was constant.

Survey Methodology of Riverbed Material. The grain size accumulation curve was obtained from an experiment using onsite soil samples taken in 2019. Five samples were collected from the analysis field. Though the grain size distribution of the riverbed material is generally examined via image analysis techniques, sieve analysis was used in this case to achieve an accurate grain size distribution abiding by the Japan Geotechnical Society standard [6]. The sample collected from a location between the upper and middle barbs of the left bank on the site was used for fluctuation analysis by iRIC. For this site, it was confirmed that 89.5% of the sample was under 75 mm in grain size. The resulting D_{50} was 30.2 mm.

Analytical Condition of Model and Hydraulic Parameters. The following three aspects were examined prior to this analysis: (1) whether to use single or mixed particle sizes for the bed material in the analysis, (2) whether to consider only the bed load or the bed load and suspended sand in the condition of sediment transportation, and (3) whether the particle size is less than 0.85 mm. The analysis conditions determined based on the examinations were (1) the bed material was to contain mixed particle sizes, (2) the particle size used was to be 0.85 mm or more, and (3) sediment transportation was to consider only the bed load. Additionally, the coefficient of roughness was set to 0.05 for the fixed bed and 0.03 for the moving bed, from the characteristics of the river in line with the practical guide manual issued by the Ministry of Construction. Two types of fluctuation analyses were performed under the following conditions.

Examination of Riverbed Fluctuation via Long-Term Runoff Analysis. The hydrograph for analysis was created from the hydrograph in the analysis section, obtained from the stage discharge curve (H-Q curve) for seven months, from May to December 2018. The average daily discharge of 0.45 m^3/s or more was extracted; this value (0.45 m^3/s) is the 275-day discharge confirmed by the discharge curve in the analysis section. This result was obtained on the assumption that the bed does not fluctuate below the 275-day discharge. Additionally, because large floods were confirmed two times (in July and October 2018), the average hourly discharge was extracted from those events. All the extracted data were arranged in the order of the time axis, and one flow was discharged every hour in the analysis. We compared the results obtained from the analysis of the model with the elevation data obtained from aerial images from a drone flight in January 2019. As a comparison method, two types of deposit amounts from the time of completion were obtained from the aerial images and analysis results, and the accuracy was confirmed from the difference between the two types of deposit amounts. Additionally, we confirmed the change of the particle size and the bed height during the floods from the analysis results. The amount of deposition was obtained by multiplying the area value of the analysis grid (calculated by Heron's formula) and the change in elevation from the time of completion.

Examination of Riverbed Fluctuation with Change in Discharge Scale. To clear the effects on sedimentation condition by discharge volume, four-stage peak flow rates (maximum 150 m^3/s) were assumed for the Tagawa River, where the design discharge was 410 m^3/s. The hydrograph for this analysis was created by adding two larger floods to the previous analysis. The gradual decrease curves and peak discharges of these floods were adjusted on the basis of the October hydrograph (peak discharge of approximately 60 m^3/s). We considered the situation of riverbed fluctuation in each flood from the viewpoints of the scale of peak discharges and the deposition tendency of the gradual decrease curves.

2.2 *Aerial Photography by Drone and Image Analysis Method*

After the completion of the river improvements in May 2018, two flood events occurred in the investigation area in July and October 2018. An aerial image was captured by a drone (DJI Phantom4) in January 2019 to confirm whether the riverbed movement was associated with the sediment movement due to the floods. The images were captured with a 75% overlapping rate in the direction of the air route and 60% overlapping rate between air routes. From a height of 40 m, 87 images were captured by automatic flight. One image was developed from these photographs using AgiSoft MetaShape. AgiSoft MetaShape software is used for automatic close-range photogrammetry data processing. This image contains the altitude data from the GPS function of the drone. Additionally, three-dimensional data were obtained by using survey coordinates. The image data were used as measured data to compare with the results of the model analysis.

3 Results and Discussion

3.1 *Results of Model Analysis and Aerial Images*

Comparison of Model and Image. The aerial photograph taken by the drone in January 2019 and the simulation results of the fluctuation analysis are shown in Fig. 2.

It was observed that sedimentation occurred between the upper and the middle barb at the left riverbank, and the riverbed was deepened by scour at the tip of the barb. Comparing the model results with drone images confirmed the close agreement in the sedimentary parts and scouring sites. However, the flow meandering condition of the watercourse formation of the model was stronger than that shown in the aerial image. Therefore, the lower sedimentary area of the second barb of the left bank in the model became smaller than that in the aerial image. The barbs on the left bank are permeable because they have wooden mattress groins. Although the aerial image shows the scour and sedimentation reflecting the characteristics of the actual permeability of the barbs, it is considered that the model result was different because of the impermeable groins setting. For the riprap groins on the right bank, a large particle size (e.g., 500 mm) was given and was analyzed as a moving bed. However, because the particle size distribution was set to be uniform, the particle movement due to the flood was not accurately reflected. Additionally, these groins were set with a large-size riprap with permeability; however, water flow on ripraps could not be reflected, and it was considered that the meandering condition became stronger. Although it was confirmed that the trend of riverbed elevation change could be

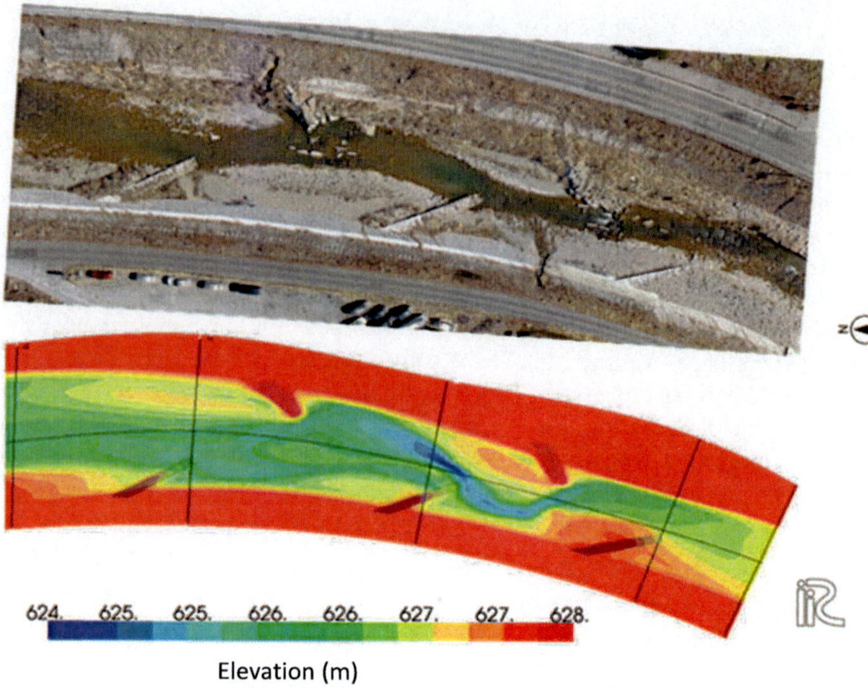

Fig. 2 Aerial image of analysis area taken in January 2019 and results of fluctuation analysis of riverbed

estimated, it is necessary to further modify the model conditions to accurately reflect the soil sedimentation volume as a revetment protection effect by the barbs.

Accuracy Check of the Model. We calculated the amount of soil sedimentation after completion for both the aerial image and the model. The result of the model was subtracted from the result of the aerial image, assuming the result of the aerial image as the true value. The positive and negative errors of the model result are shown in the occurrence frequency distribution graph on the left side of Fig. 3. In the right side of Fig. 3, the difference in sedimentation volume between the aerial image and the model simulation is shown for each mesh. Although the mesh size is

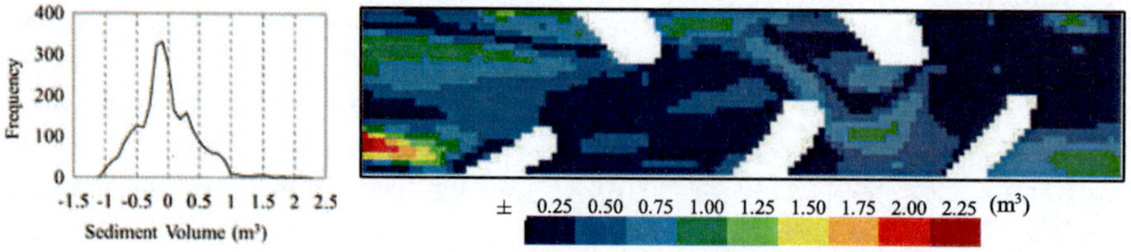

Fig. 3 Accuracy check from the viewpoint of difference in sediment volume between aerial image and model

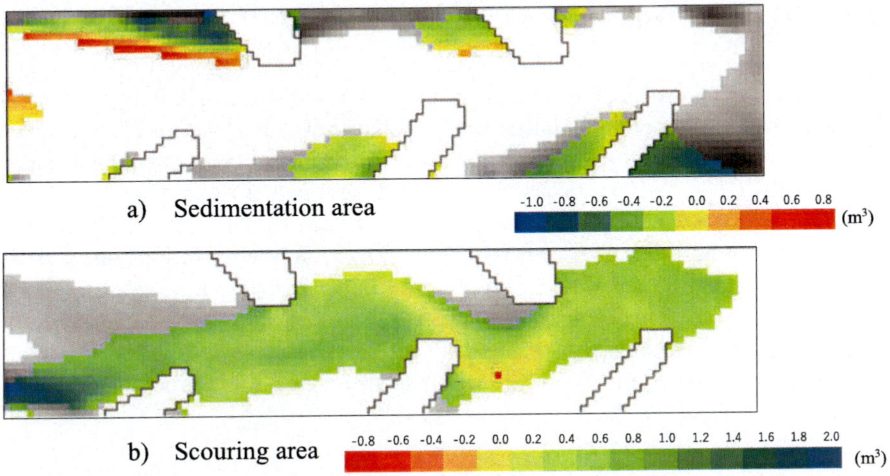

Fig. 4 Sedimentation and scouring trends of the aerial image and the model

different at each point, the analysis section is shown as a rectangle. The positive and negative errors are shown by color bars; and the error is smaller at the blue points and larger toward the red points. A total of 2614 mesh data were analyzed, of which 1567 results were in the range from -0.3 to $+0.3$. The accuracy of this data is confirmed by the prevalence of the blue color in the right figure; however, it was confirmed that the error was large in two places, the sedimentary areas between the riprap groins on the right bank and upstream of the first barb on the left bank.

In Fig. 4, the error distribution due to changes in riverbed elevation is shown for both the sedimentary part and the scouring part. The gray gradation part is the part where accumulation and scour were confirmed by the aerial image, but it was not captured by the model. In the color bar, the error is smallest at the yellow points. The areas where the model underestimated the elevation change are shown in blue (e.g. the scour is shallow in the scour part, and less deposition is in the sedimentation part), and areas of model overestimation are shown in red. In the sedimentary area, upstream sedimentation and sedimentation between the barb works are underestimated by the model, and downstream sedimentation near the water route is overestimated. In contrast, it was confirmed that although the scouring part showed a relatively close agreement between the aerial image and the model, the model underestimated the areas from the barb tips. It was suggested that one of the reasons for the errors in both the sedimentation area and the scour area was the meandering of the water route in the model. Although it was confirmed that the trend of riverbed elevation change could be estimated, it is necessary to further modify the model conditions to accurately reflect the soil sedimentation volume as a revetment protection effect by barbs. In particular, it should consider the permeability setting, the confirmation of the upstream and downstream bed slopes, and the sediment supply conditions.

3.2 Riverbed Deformation During the Flood

Time trends of riverbed deformation by the flood and mean diameter of riverbed materials before the flood (I), during peak discharge (II), and after the flood (III) are shown in Fig. 5. The left and right sides show the results of cross sections A and B, respectively. Cross sections A and B have both sedimentation and scouring areas, with a transition zone between these areas. The riverbed materials were composed of small- to medium-sized sediments (under 40 mm). In the transition zone of cross section A, the riverbed elevation decreased during the peak discharge of the flood. Simultaneously, the transition zone that forms the side bar near the scouring disappeared; the riverbed elevation of the scouring areas increased, and the grain size became smaller. After the flood, the side bar at the transition zone was expanded

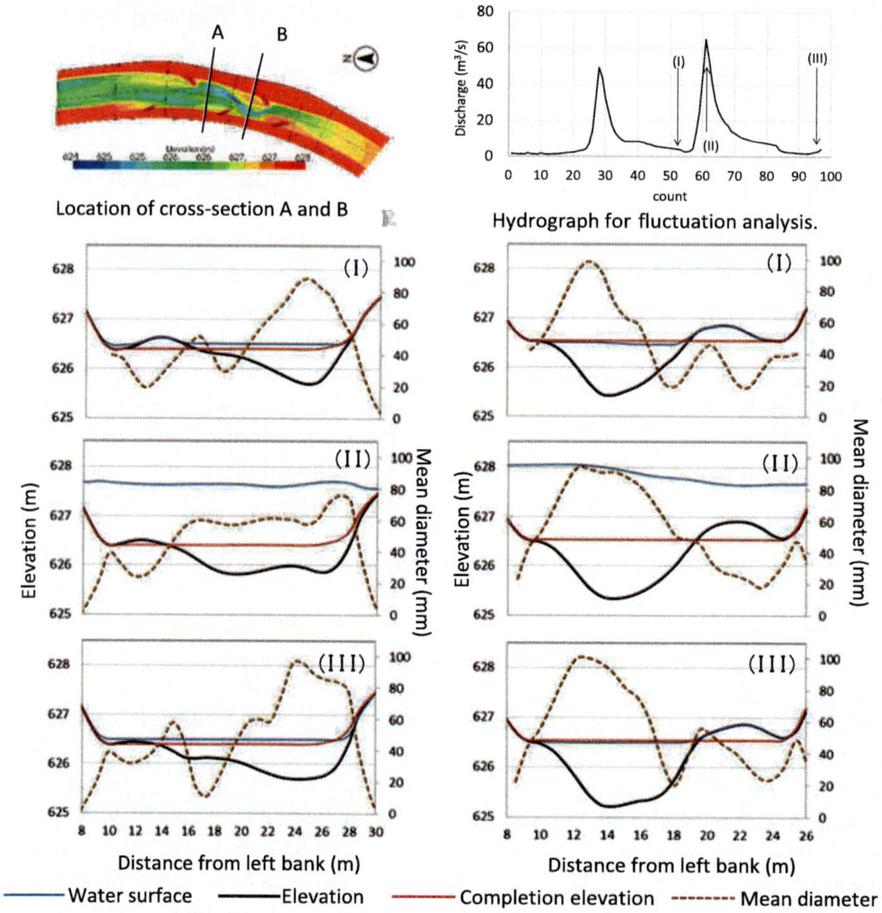

Fig. 5 Results of riverbed deformation by flood at cross sections A and B in the numerical simulation

mainly by small-diameter sand; it returned to a riverbed topography similar to that before the flood occurred. The grain size proportion also recovered to that before the flood. Sedimentary parts between barbs were composed of small-sized soils with an average grain size of 20–30 mm. During the floods, streams overflowed above the new sedimentary parts generated by previous floods; small- and medium-sized sand was re-deposited by the current flood. At a point 18.1 m from the left bank, the elevation decreased from 626.3 to 625.8 m due to the flood. However, re-deposition occurred at the same point, and the elevation recovered gradually to 626.1 m. The sedimentation areas remained during the floods, although their positions changed closer to the bankside. In cross section B, however, the transition zone generated by the previous flood was eliminated by the larger-scale flood and did not recover. The elevation of the sedimentation area reduced. The Shields number is a dimensionless number that explains the attractive force at a riverbed. The critical Shields number for D50 was calculated using an equation developed by Iwagaki [7]. That number for this analysis area is 0.05, which is the value at which the sediment will start to move. In cross section B, the Shields number always exceeds 0.05 during a flood. It was suggested that sedimentation did not occur during the gradual decrease curve. In cross section A, the Shields number became less than 0.05 at the discharge used for analysis. Therefore, it was considered that the movement of the sediment was small, and sedimentation was also promoted at the transition zone, as in the water route, by the sediment supply from the upstream cross section B. It was confirmed that the scouring depth and position of the sedimentation area were changed by the flood scale, but the composition of the mean diameter at both A and B before the flood shows a similar tendency to that after the flood.

Regardless of the discharge peak, it was suggested that the deposition tendency was repeated during the gradual decrease curve. To promote the sedimentation of soil near the bank revetment and re-deposit, it after flooding on the side bar, it is necessary to carefully consider the position of the barbs, the change in the Shields number depending on the flow velocity, and the condition of the upstream soil supply.

3.3 *Riverbed Fluctuation by Discharge Intensity*

A hydrograph of the four-stage peak discharge used for fluctuation analysis and the results of riverbed deformation with the elevation of the scouring areas (A, B, and C) and sedimentation areas (D and E) are shown in Fig. 6. Large scouring occurred at a peak discharge of 50 m^3/s near the tips of the barbs A and B. The riverbed was rendered the lowest at site B, with a decrease of approximately 1.5 m. In subsequent floods, the riverbed elevation temporarily decreased during the flood; then, re-deposition occurred at the time of the gradual decrease curve, and the riverbed did not decrease further. The lower elevation limit was approximately 625 m. The degradation tendency of site C is different from that of sites A and B; the large reduction in the riverbed due to one flood did not occur, and the riverbed gradually decreased as the flow rate increased. Temporary and rapid elevation changes were

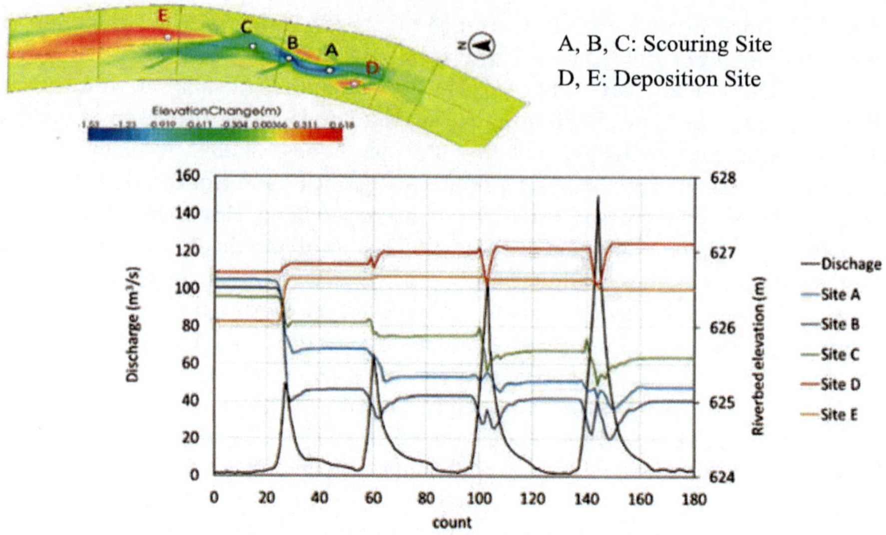

Fig. 6 Riverbed fluctuation of four-stage discharge intensity. The left axis shows the four-stage discharge used for numerical simulation, and the right axis shows the riverbed elevation of each point

confirmed to occur at the peak discharge at any site, which were suggested to be related to the change in stream velocity and bedload with an increase in discharge. Regardless of peak discharge, re-deposition of soil occurred below 30–20 m^3/s in the gradual decrease curve. In the sedimentary area, i.e., the first barb on the left bank at site D, the riverbed rose gradually as the discharge increased and temporarily decreased during floods; however, it tended to rise while redepositing during the gradual decrease curve. This result suggests that sedimentation beneath the barb is directly caused by repeated floods. At site E, far from the barb, a large amount of sedimentation occurred until the second peak (less 70 m^3/s), but the riverbed gradually decreased at the peak thereafter. It is necessary to investigate the effective distance and area from barbs and the tendency of flooding on a larger scale to determine whether sedimentation can be maintained.

4 Conclusions

Simulations using barbs agreed closely with aerial photographs taken by a drone showing actual flow and riverbed conditions. The new side bar generated by the construction of the barb was effective at maintaining the sediment and preventing the decline of the riverbank against a flood of the scale at which the side bar was generated, although the scouring became deep during flooding. However, the barb installed at the site where the riverbed was degraded had a higher Shields number and a lower sedimentation effect than the downstream barb. Thus, placing barbs on the upstream side of the problematic site is effective in revetment protection. Further research is required for greater-scale flooding and long-term effects from barbs.

References

1. Harada, M., Kayaba, Y.: Perspective on study of riverbed environment in the upper and middle reaches. Ecol. Civil Eng. **18**(1), 3–18 (2015) (in Japanese)
2. Takakoka, H., Harada, M., Ooishi, T., Kayaba, Y.: Characteristics of bed variation and flow diversity due to "barb" groins. In: Proceedings of Annual Meeting of Environmental Systems Research, pp. 103–108 (2013) (in Japanese)
3. Kyuka, T., Takebayashi, H., Fujita, M. Kimura, I., Shimizu, Y.: Flow, sediment transport and bed deformation around spur dikes on erodible and on inerodible bed fields. Ecol. Civil Eng. **19**(2), 181–201 (2017) (in Japanese)
4. U.S. Department of Agriculture, Natural Resources Conservation Service: Engineering Technical Note 23, Design of Stream Barbs (Version 2) (2005).
5. Harada, M. Takakoka, H. Oishi, T. Kayaba, Y., Fujita, Y.: Characteristics of bed deformation around submerged groins with various angle advances in river sediment research. In: Proceedings of the 12th International Symposium on River Sedimentation (ISRS2013), pp. 851–857 (2013)
6. Japanese Geotechnical Society Standards Laboratory Testing Standards of Geomaterials (vol. 1) (2015): JGS 0131 Test method for particle size distribution of soils (JIS A1204)
7. Iwagaki, Y., Tsuchiya, Y., Sakai, M.: Basic studies on the wave damping due to bottom friction. Coastal Eng. Jpn. **8**(1), 37–49 (1965)

The Mechanism of Riverbank Erosion Caused by Ship-Generated Waves Along Hau River's Entrance Navigation Channel, Southern Vietnam

Son Huu Nguyen, Tin Trung Huynh, Vinh Trong Bui, and Ngo Van Dau

Abstract From the past, earning living of the people in the lower of Mekong River has been acquainted with the ships and boats in the density river system. Gradually, waterway has become the key factor for area developing, intensity of commercial, and recreational navigation. Inland movement of the cargo ship brings a significant efficient on cargo shipping which impulses the regional economic development. However, impacts from inland movement should be considered seriously. The drawn followed by ship movement put a pressure on the riverbanks which causes turbulence and erosion potentially. This paper carried out several field measurements on bank structures and ship movements to assess the potential impacts of ship-induced wave on Hau River's entrance navigation. As results, the riverbanks could resist the external shear stress by maximum wave height from 40 to 60 cm.

Keywords Artificial channel · Ship-induced wave · Turbulence · Bank erosion

1 Introduction

Because of the high economic importance of the waterway traffic, Hau's River has become the key channel, which allows the large cargo ship to enter the inland ports. Due to the high sedimentation rate of Dinh An estuary, a new artificial route, which is shown in Fig. 1, has been provided nearby the existing route to ensure the safe movement of the vessels. Natural erosion occurs by many factors which are natural properties and human-being factors [2]. These natural properties are wind-wave, current, tidal, geological structure of the riverbank. Human activities contribute a significant impact on the bank loss including buildings along the riverbank, ship movement, dredging, etc.

Wind generated wave is the natural properties, which requires a long distance of surface water to affect the bank and have limited chances to cause erosion. In this paper, the authors have carried out several field measurements to determine the

S. H. Nguyen (✉) · T. T. Huynh · V. T. Bui · N. Van Dau
Ho Chi Minh City University of Technology—VNU-HCMC, Ho Chi Minh City, Vietnam
e-mail: nguyenhuuson@hcmut.edu.vn

H. Hazarika et al. (eds.), *Advances in Sustainable Construction and Resource Management*, Lecture Notes in Civil Engineering 144,
https://doi.org/10.1007/978-981-16-0077-7_73

Fig. 1 Maps of studied area

eroding resistant of the riverbanks of the navigation channel of Hau's River. The main reason of this paper is to determine the potential impacts of the ship movement to the bank's stability.

The navigation profoundly impacts on aquatic ecosystems, bank erosion, and deposition [1]. During the movement on the surface water, they create the hydrodynamic impacts, lead the pressure, and generate the front wave to the riverbank [3]. Ship-generated wave has been the most topic which took considerations of many researcher around the word. Relationship of ship movement and water disturbance has been developed by Havelock (1908). During the movement in unrestricted water body, the primary wave system is rather small, as transport of water to the sides is possible. Nowadays, it has experimentally determined the maximum wave height which is resisted by bank sediment from eroding [6]. The primary wave system is here depicted more accurately than the idealized state described earlier, with the crests displaced some distance from the ship bow and stern.

The wave system created by a vessel is distinctively different depending on several factors. First, the characteristics of ship-induced waves differ greatly depending on whether they are formed in deep or shallow and unrestricted or restricted waters [4]. Other factors which affect the wave system are ship dimensions and hull design [5]; movement velocity, and shallow water depth [5]. Depending on the river profile, water flows have different transport patterns, regarding for example the mean flow rate. In addition, the location of the shipping lane has also significantly affect on the ship-induced.

2 Methodology

2.1 *Field Investigation and Measurement*

The authors have carried out several field trips to measure the wave characteristics, sample the bank material composition. The borehole samples were picked up and experimental analysis in the laboratory. The analytical results from borehole sample allows the author to determine the resistance of the riverbank from external forces causing eroding. Location of site investigation is given in Table 1.

Meanwhile, to evaluate the potential impacts of the ship-generated waves, the on-site wave measurement using the AWH-Infinity Wave Height Meter has been conducted. The purpose of site monitoring is to measure the wind-wave and ship-generated wave at the project area. Chosen location for monitoring is (01) at the SK-01 (in the artificial estuary) and QCB-01 in Quan Chanh Bo channel (Table 1).

Table 1 Location of site investigation

No.	Sample ID	Position		Parameters measured
		Latitude (m)	Longitude (m)	
1	KT01L	666,152.15	1,058,543.58	Geological sample
2	KT01R	666,552.00	1,058,782.54	Geological sample
3	KT02L	664,959.66	1,059,654.87	Geological sample
4	KT02R	665,172.13	1,060,035.63	Geological sample
5	KT03L	662,805.12	1,061,624.06	Geological sample
6	KT03R	663,192.92	1,061,754.08	Geological sample
7	KT05L	660,985.96	1,063,402.70	Geological sample
8	KT05R	661,301.16	1,063,716.40	Geological sample
9	QCB01L	659,824.60	1,064,031.00	Geological sample
10	QCB01R	660,231.14	1,064,440.25	Geological sample
11	M1-2	658,145.00	1,064,558.00	Geological sample
12	M1-3	653,379.81	1,065,148.94	Geological sample
13	M1-4	653,854.96	1,064,870.03	Geological sample
14	HK1-1	649,795.53	1,064,424.92	Geological sample
15	HKT16-1	643,496.31	1,063,193.90	Geological sample
16	HKT26-1	650,294.24	1,064,241.05	Geological sample
17	HKT28-1	647,118.32	1,063,619.96	Geological sample
18	HKT23-1	645,748.05	1,063,663.12	Geological sample
19	QCB-01	643,872.50	1,063,038.77	Wave measurement
20	QCB-02	643,552.94	1,063,193.90	Wave measurement
21	SK-01	667,101.01	1,058,292.07	Wave measurement
22	SK-02	666,664.53	1,058,216.49	Wave measurement

Fig. 2 Borehole sample of riverbank material (left) and setting of site wave measurement using AWH-Infinity equipment (right)

2.2 Determine the Ship-Generated Wave

During the later nineteenth century, the first ship wave theory was developed by Kevin using Bernoulli equation [4]. The primary theory of ship-generated wave is given as Eq. 1.

$$\left(\frac{U^2}{2} + gz + \frac{p}{\rho}\right)_1 = \left(\frac{U^2}{2} + gz + \frac{p}{\rho}\right)_2 = \text{constant} \tag{1}$$

where U = ship velocity relative to water velocity (m/s); g = acceleration due to gravity (m/s^2); z = distance from reference level (m); p = pressure (Pa); and ρ = water density (kg/m^3). The maximum wave height created by a ship is the relationship of ship velocity (U) and the overall ship dimensions which can theoretically be estimated as:

$$H = \frac{U^2}{g}\beta(F_* - 0.1)^2\left(\frac{y}{L_s}\right)^{-1/3} \tag{2}$$

The bed shear stress followed by ship movement is dependent by orbital velocity (U_{orb}, m/s) and wave friction factor (f_w). The orbital velocity of the linear wave is defined by:

$$U_{\text{orb}} = \frac{\pi H}{T * \sinh(2\pi h/L)} \tag{3}$$

where U_{orb} = wave orbital velocity near the bed (m/s); H = wave height (m); T = wave period (s); L = wave length (m); and h = SWL depth (m). The wave friction factor is defined by the relationship of length scale $A = (U_{\text{orb}} * T)/2\pi$ (m) and roughness z_0 (m) as $f_w = 1.39\left(\frac{A}{z_0}\right)^{-0.52}$. The bed shear stress from oscillatory flow alone is defined by, among others, Voulgaris et al. (1995), modified by Soulsby (1997) and Althage (2010) [7] define the wave orbital velocity as:

$$\tau_w = \frac{1}{2}\rho f_w U_{\mathrm{orb}}^2 \quad (4)$$

3 Results

3.1 Bank Material Composition

Littoral material plays a very important role in the bank stability. It is one of the most important things on bank stability. Based upon the grain size distribution and moisture, riverbanks have their own resistance from eroding [3]. Figure 3 summarizes the grain size varying the navigation channel which records the predominant of silk and clay material (>60%). Regarding the cohesive soils, the eroding resistance is defined by the percentage of clay in the bank material [9], certainly, the critical shear stress of the bank materials at navigation channel is given in Fig. 4 [8].

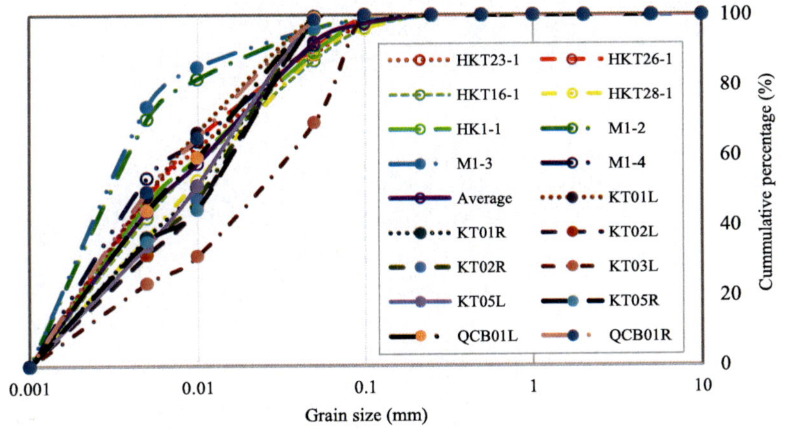

Fig. 3 Grain size distribution along the navigation channel

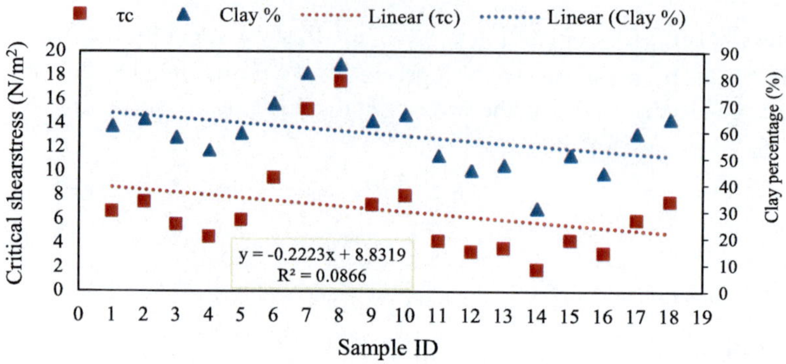

Fig. 4 Relationship of clay contribution and critical shear stress on the littoral material samples

3.2 *Ship-Induced Wave Measurement*

There were total four measurements conducted at two location separately. As intention, the monitoring measurement has focus on the wind-wave and ship-generated waves. Figures 5 and 6 illustrate the recorded data from the wave characteristics. It was found that, the wind-wave at the starting point in the Tat channel (SK01) higher than in QCB channel. The normal wave (wind-wave) in the SK01 varies from 0.2 to 0.65 m [10] while the wave height in QCB is 0.1–0.3 m. Wave height is one of the most important generating the pressure on the river bank.

Table 2 determines shear stress generated by waves recorded in navigation channel. It was seen that, the pressure of >0.6 m wave height on the riverbank is approximately 20 N/m^2.

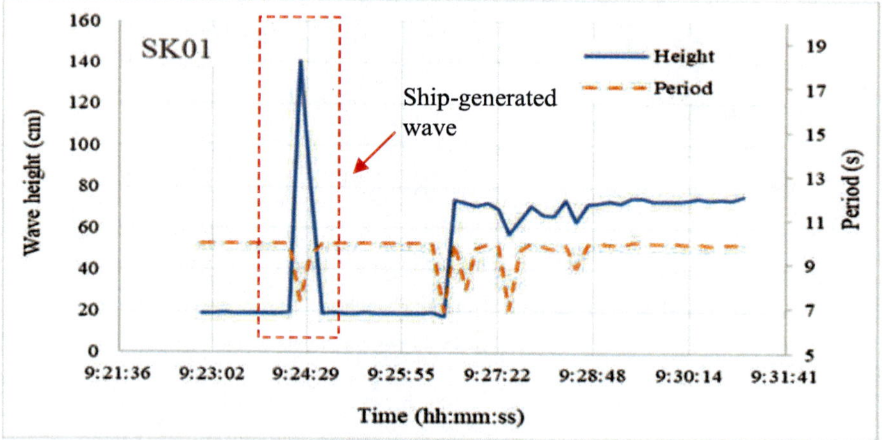

Fig. 5 Ship-generated wave measured at SK01 station

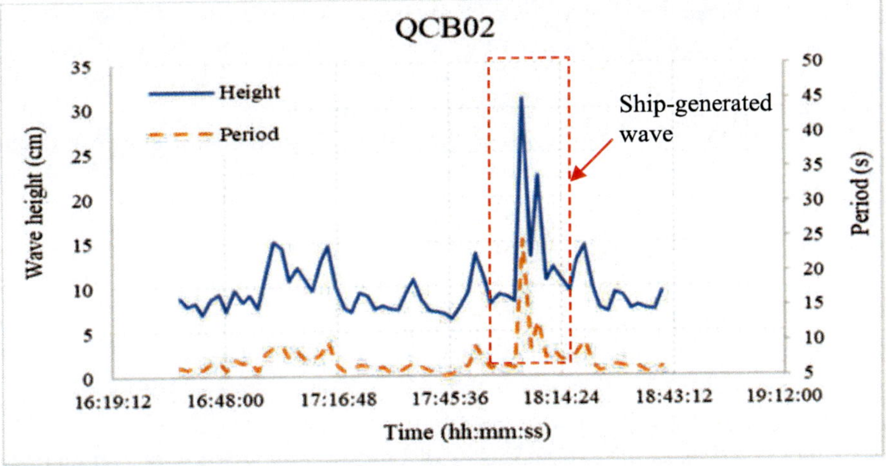

Fig. 6 Ship-generated wave measured at QCB02 station

Table 2 Calculated orbital velocity and shear stress created by ship-generated wave

Wave No.	Height (m)	Period (s)	Orbital velocity (m/s)	Shear stress (N/m^2)
1	1.406	7.308	0.691	**336.342**
2	0.813	9.322	0.935	**87.554**
3	1.108	14.895	0.563	**145.048**
4	0.600	22.627	0.730	**30.029**
5	0.400	9.762	1.629	16.444
6	0.715	9.903	1.002	**63.740**
7	0.026	4.241	222.799	0.000
8	1.226	14.768	0.628	**108.507**
9	0.149	14.902	3.590	0.745
10	0.149	14.895	3.592	0.744
11	0.149	14.908	3.588	0.745
12	0.573	6.943	1.915	**20.518**
13	0.133	8.975	6.421	0.392
14	0.224	12.830	2.795	2.047
15	1.390	14.877	0.566	**143.661**

The purpose of bold is to emphasize the highlight the value compared with resistance value in Fig. 4

4 Discussions

Soil materials and moisture play an important role in the bank stability. Contribution of the clay material has significant effect on the eroding resistance from external impacts including current, wave, and other impacts. Assuming that other impacts are beyond from this study, the bank erosion rate could be determined by the subtraction of the shear stress generated from ship movement and the critical shear stress of the bank material.

Comparing the calculated external shear stress from the ship movement, soil material along navigation channel could resist the external stress of under 0.4 m height wave generated.

5 Conclusion

High percentage of clay component plays an important role in the eroding resistance of the bank material. According the field survey boreholes data, the soft layer on the riverbanks could resist the external stress of 12–20 kN/m^2. As results, soil material along Hau's navigation channel could resist the external stress of 12–20 kN/m^2 which could be generated by the 0.4–0.6 m ship wave in respective.

References

1. Gabel, F., Lorenz, S., Stoll, S.: Effects of ship-induced waves on aquatic ecosystems. Sci. Total Environ. **601**, 926–939 (2017)
2. Dorava, J.M., Moore, G.W.: Effects of boatwakes on streambank erosion, Kenai River, Alaska. Water Resources Investigations Report, 97-4105 (1997)
3. Nanson, G.C., et al.: Experimental measurements of river-bank erosion caused by boat-generated waves on the Gordon river, Tasmania. Regulated Rivers Res. Manage. **9**(1), 1–14 (1994)
4. Sorensen, R.M., Weggel, J.R.: Development of ship wave design information. Coast. Eng. **1984**, 3227–3243 (1985)
5. Kriebel, D.L., Seelig, W.N.: An empirical model for ship-generated waves. In: Paper presented at the 7th International Conference on Mathematical and Numerical Aspects of Waves (WAVES'05) (2005)
6. Lawler, D.M.: The measurement of riverbank erosion and lateral channel change: a review. Earth Surf. Proc. Land. **18**(9), 777–821 (1993)
7. Althage, J.: Ship-Induced Waves and Sediment Transport in Göta River, Sweden (2010)
8. Clark, L.A., Wynn, T.M.: Methods for determining streambank critical shear stress and soil erodibility: implications for erosion rate predictions. Trans. ASABE **50**(1), 95–106 (2007)
9. Vinh, B.T., et al.: Measurement of critical shear stress for erosion of cohesive riverbanks. In: OCEANS 2008-MTS/IEEE Kobe Techno-Ocean, pp. 1–7. IEEE
10. The Project of Hau River's Entrance Navigation Channel, Tra Vinh Province, September 2009

Application of Information Technology Towards Natural Disaster Mitigation

Battery-Less Infrastructure Monitoring Sensor Platform

Haruichi Kanaya, Mohamed M. Mansour, Shunsuke Hatanaka, Ataru Nakashima, and Osamu Takiguchi

Abstract In Japan, the construction of extensive infrastructure such as roads, railways, airports, energy plants, and so on, became popular from the 1960s when the period of high economic growth began, and high-level investments continued for about 40 years. Recently, infrastructure serious accidents have occurred, which is caused by aging or lack of inspection are of significant social concern. Therefore, as the population declines in the future, infrastructure monitoring using wireless sensor networks (IoT, Internet of things) is an urgent issue. On the other hand, the power supply is indispensable for driving the sensor device and transferring data. In this research, focusing on microwaves (electromagnetic waves) radiated from broadcasting and/or telecommunication that have been rarely reused, a battery-less infrastructure monitoring sensor platform that uses electromagnetic waves as a power source is realized. The proposed system is composed of a circularly polarized wide-band planar antenna, rectifier circuit with impedance matching circuit, sensor module and Bluetooth low energy (BLE) module. For the feasibility study, the temperature and humidity sensors are attached on the surface of a concrete block, which is the model of the infrastructure. The 920 MHz RF signal is injected into the receiving antenna as a power source. The data of the temperature and humidity are transmitted to the smartphone or laptop computer at every 2-s interval. We can obtain the battery-less infrastructure monitoring sensor platform.

Keywords Energy harvesting · Rectenna · Bluetooth · Rectifier · Log-periodic dipole antenna

H. Kanaya (✉) · M. M. Mansour · S. Hatanaka · A. Nakashima
Kyushu University, Fukuoka, Fukuoka 819-02395, Japan
e-mail: kanaya@ed.kyushu-u.ac.jp

O. Takiguchi
ALSENS Inc., Sagamihara, Kanagawa 252-0131, Japan

H. Hazarika et al. (eds.), *Advances in Sustainable Construction and Resource Management*, Lecture Notes in Civil Engineering 144,
https://doi.org/10.1007/978-981-16-0077-7_74

1 Introduction

The construction of extensive infrastructure such as roads, railways, airports, energy plants, and so on, became popular from the 1960s when the period of high economic growth began, and high-level investments continued for about 40 years in Japan. Recently, infrastructure serious accidents have occurred, such as the fall of the ceiling of the Sasago Tunnel in Chuo Expressway in 2012, which is caused by aging or lack of inspection. And there is a sizable social concern [1]. Therefore, as the population declines in the future, infrastructure monitoring using wireless sensor networks (IoT, Internet of things) is an urgent issue [2].

On the other hand, the power supply is indispensable for driving the sensor device and transferring data. "Energy harvesting" also called "environmental power generation" means a power generation technology that harvests various types of environmental energy such as light, heat, and vibration, including solar power, and converts it into electrical energy [3–5].

In this research, focusing on microwaves (electromagnetic waves) radiated from broadcasting and/or telecommunication that have been rarely reused, a battery-less infrastructure monitoring sensor platform that uses electromagnetic waves as a power source is realized. The 920 MHz RF signal, which used as radio frequency identifier (RFID) tag system, is injected into the receiving antenna as a power source. The proposed system is composed of a circularly polarized wideband planar antenna, rectifier circuit with impedance matching circuit, sensor module, and Bluetooth low energy (BLE) module. For the feasibility study, the temperature and humidity sensors are attached on the surface of a concrete block, which is the model of the infrastructure.

2 Rectifier and Antenna Designs

In this section, the rectifier and the antenna are designed and fabricated. Rectifier and antenna designs are validated using circuit and 2D electromagnetic field (EM) simulator (ADS, Keysight), and 3D EM simulator (HFSS, Ansys), respectively [6–9].

Figure 1a shows the circuit model of the voltage doubler, which is composed of the diodes and capacitors. In this design, the SPICE model of the diode (SMS7621-079LF, Skyworks) was used for simulation. In this figure, Z_{inR} means the input impedance of the proposed voltage doubler. Figure 1b shows the circuit model of the rectifier. The equivalent circuit of the voltage doubler represents C_{L} and R_{L} parallel circuit, which is indicated with Z_{inR}, as shown in Fig. 1a. To realize the wideband impedance matching characteristic, the transmission line and two inductors (impedance: Z_{a}) are inserted before the voltage doubler. The electrodes and metal lines are modeled by using the EM simulator and co-simulated with the SPICE simulation.

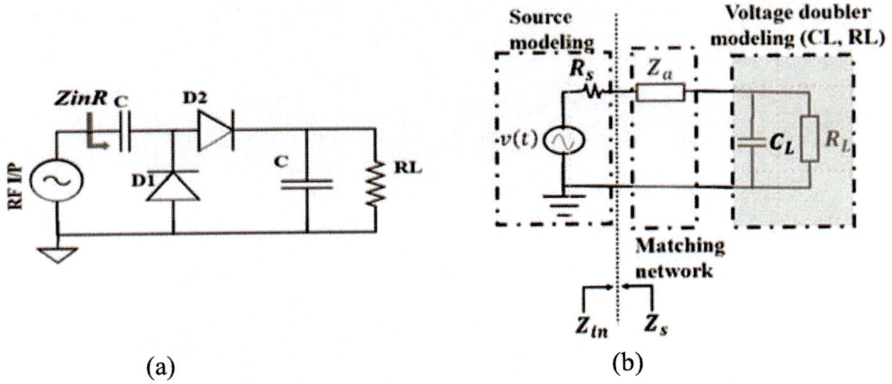

Fig. 1 Circuit model of the voltage doubler (**a**) and proposed rectifier with impedance matching circuit (**b**)

Figure 2a shows the layout of the proposed log-periodic dipole antenna (LPDA). This antenna is based on the half-wavelength dipole antenna. There are two metal lines: one is the upper metal (red), and another is the bottom metal (green). To enhance the bandwidth, both metal planes are tapered. To reduce the antenna size, we adopted the spiral structure. RF signal is injected at the center of the spiral structure with coaxial cable. The signal line of the coaxial cable is connected to the upper spiral metal and the GND line is connected to the bottom spiral metal. The dielectric constant of the substrate is 4.2. The size is 60 × 83 × 15 mm with a reflector. Figure 2b shows the simulated 3D radiation pattern of the antenna. The simulated maximum gain is 8.89 dB in the +*Z*-direction as shown in Fig. 2b.

3 Measurement Results

3.1 Rectifier Measurement

Figure 3 shows the layout and photograph of the proposed antenna. A black area shows the metal area (See Fig. 3a). A coplanar waveguide (CPW) structure, which has a GND-signal-GND (*G-S-G*) line on the top metal, is utilized. It is easy to optimize the characteristic impedance of the CPW by changing the width of the signal line or each gap size. In this configuration, the width of the center conductor is 4.7 mm and both gaps are 0.4 mm. As shown in Fig. 3b, RF signal is injected from the left side in the photograph and a coaxial connecter that has 50 Ω characteristic impedance is connected. The DC output is obtained from the right side. The size of the rectifier (Fig. 3b) is 31 × 22.5 mm.

Figure 4 shows the measurement setup and setup photograph. The signal generator and voltmeter are connected to the input and output port of the proposed rectifier.

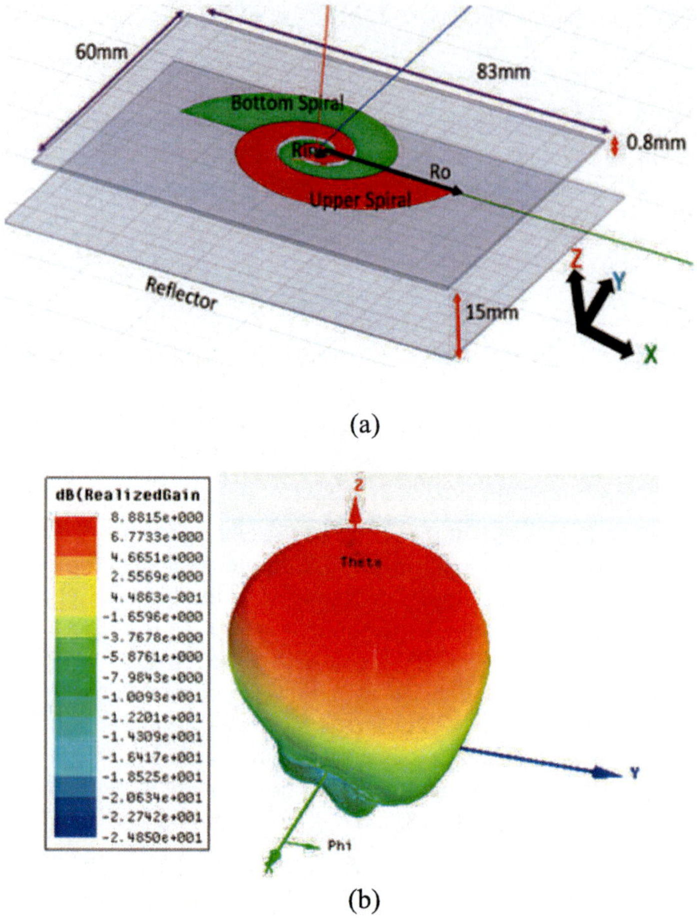

Fig. 2 Layout of the spiral LPDA (**a**) and simulated radiation pattern of the LPDA (**b**)

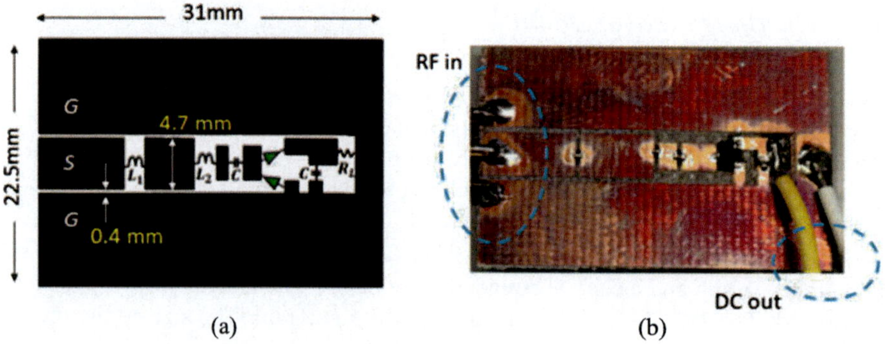

Fig. 3 Layout of the proposed rectifier (**a**) and the photograph of the rectifier (**b**)

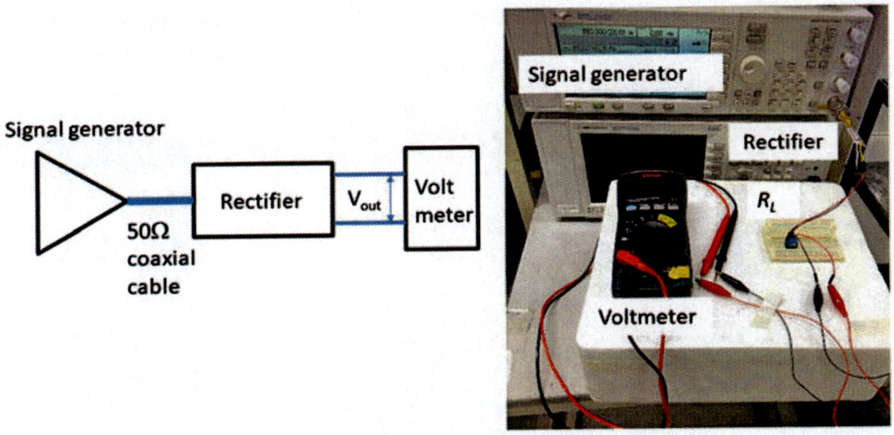

Fig. 4 Efficiency measurement setup and setup photograph

The efficiency (η) is calculated by the following equation:

$$P_{\mathrm{dc}} = \frac{V_{\mathrm{out}}^2}{R_{\mathrm{L}}} \tag{1}$$

$$\eta = \frac{P_{\mathrm{dc}}}{P_{\mathrm{rf}}} = \frac{V_{\mathrm{dc}}}{V_{\mathrm{L}}^2 \times P_{\mathrm{rf}}} \tag{2}$$

where P_{dc} is the output DC power at R_{L}, P_{rf} is the RF input power, and V_{dc} is the output DC voltage. Figure 5 shows the input power dependence of the measured η at 0.8 GHz.

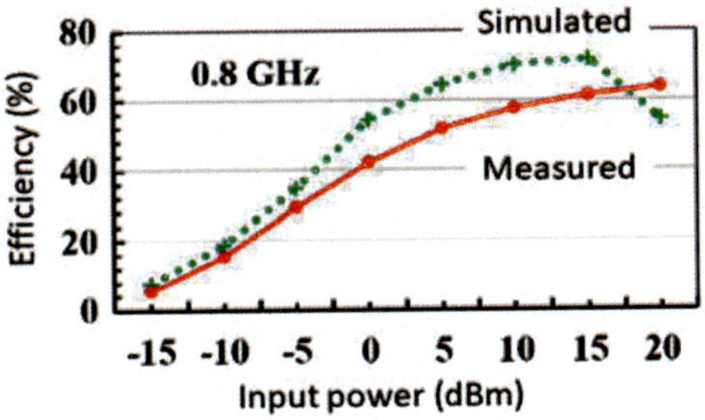

Fig. 5 Input power dependence of the measured η at 0.8 GHz

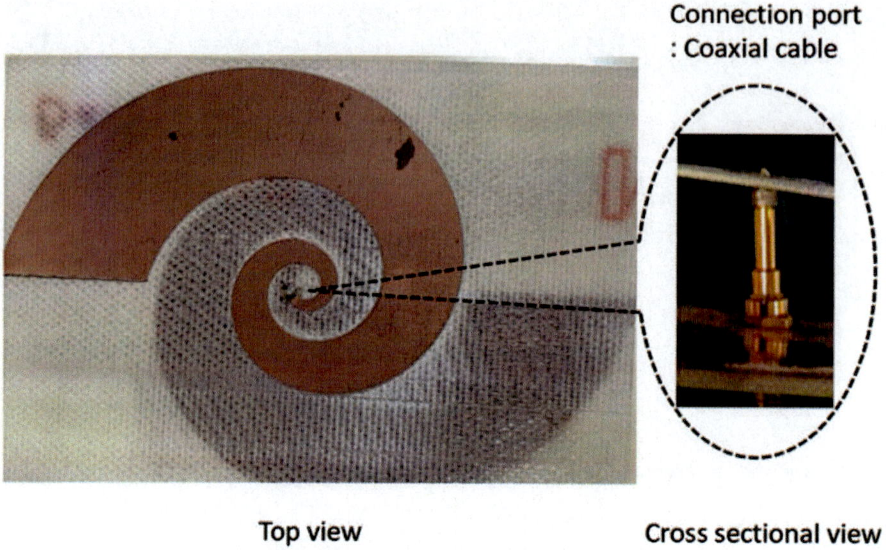

Fig. 6 Photograph of the proposed antenna

3.2 Antenna Measurement

Figure 6 shows the photograph of the proposed antenna. In the figure, the coaxial cable, which has 50 Ω input impedance, is connected from the backside. The measured antenna gain is calculated by the following Friis transmission equation [10]:

$$G_t = \left(\frac{4\pi r^2}{\lambda}\right)\frac{P_r}{P_t G_r} \tag{3}$$

where P_t and P_r are the transmitting and receiving power, G_t and G_r are the antenna gain, r is the distance between the antenna, and λ is the wavelength, respectively.

Figure 7 shows the measured antenna gain at a different frequency. The proposed antenna has 7.6 dBi gain in the measurement.

3.3 Battery-Less Sensor Measurement

Figure 8a shows the experimental setup for temperature and humidity measurements on the concrete block with a battery-less sensor system. The transmitted signal is a license-free band (920 MHz and 20 dBm). The BLE module, and temperature and humidity sensor device are in Fig. 8b. Tx (transmitter) and Rx (receiver) antenna

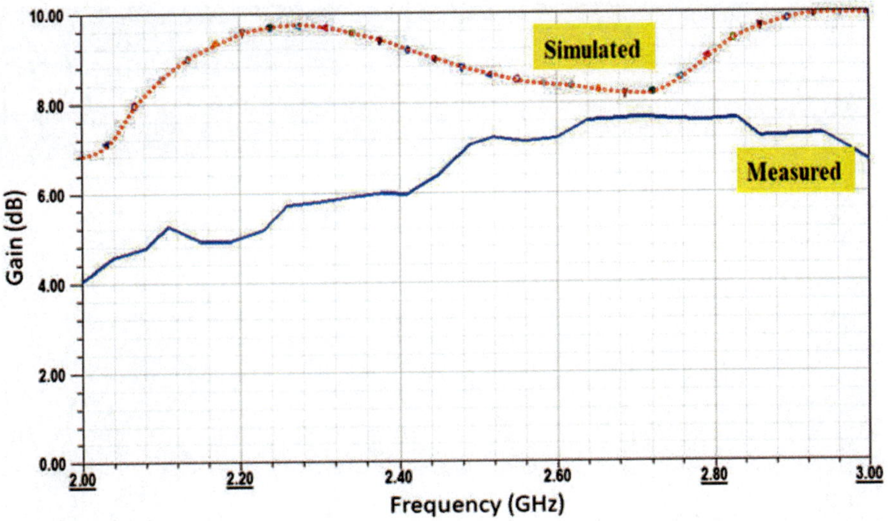

Fig. 7 Frequency dependence of the antenna gain

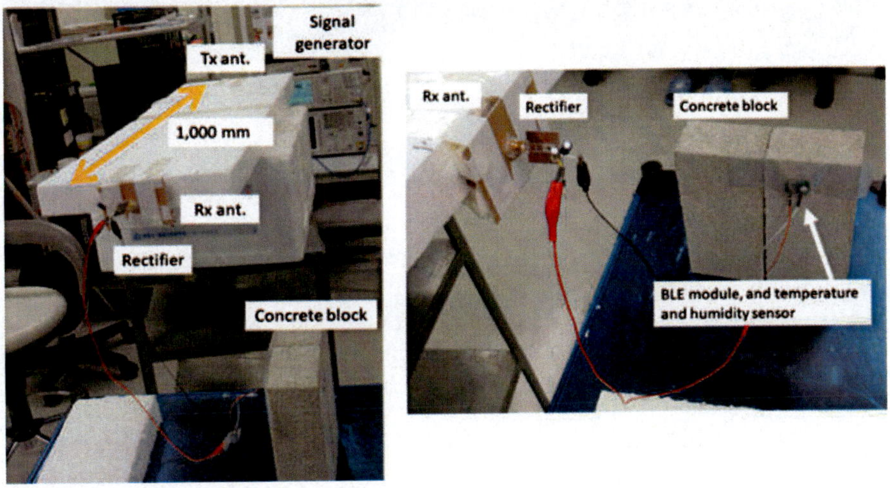

(a) (b)

Fig. 8 Photographs of the measured setup (**a**) and rectifier and sensor section (**b**)

are the same as that in Fig. 6. The data of the temperature and humidity on the concrete block are transmitted to the laptop computer every 2 s as shown in Fig. 9. If the sensor is changed, for example, a strain sensor, stress sensor, accelerometer or moisture sensor, we can realize the real-time monitoring platform for infrastructure.

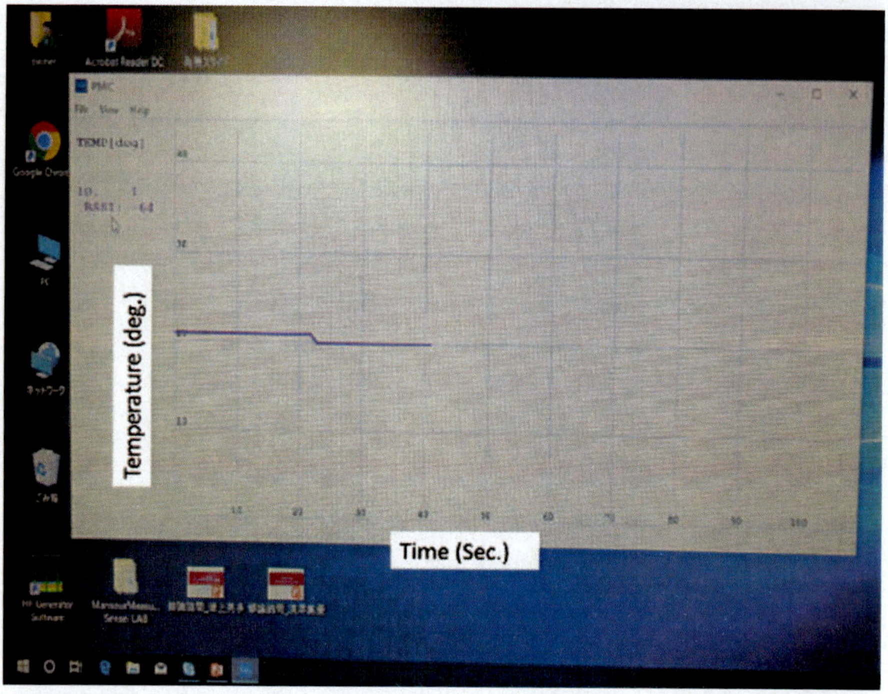

Fig. 9 Photograph of the measured result (temperature vs. time)

4 Conclusions

The battery-less infrastructure monitoring sensor platform that uses electromagnetic waves as a power source is realized. The 920 MHz RF signal is injected into the receiving antenna as a power source. The data of the temperature and humidity on the concrete block are transmitted to the laptop computer as well as the smartphone every 2 s by using BLE module. We can obtain the battery-less infrastructure monitoring sensor platform.

Acknowledgements This work was partly supported by VLSI Design and Education Center (VTEC), the University of Tokyo in collaboration with CADENCE Corporation and Keysight Technologies. This work was partly supported by KAKENHI (18K04146), CREST (JPMJCR1431), JSPS Japan, and SCOPE. This work was also partially supported by the Grant-in-Aid for the Collaborative Research Program Based on Industrial Demand, JST. This work was partially supported by the Cabinet Office (CAO), Cross-ministerial Strategic Innovation Promotion Program (SIP), "An intelligent knowledge processing infrastructure, integrating physical and virtual domains" (funding agency: NEDO).

References

1. Nikkei Electronics, 49–57 (2013)
2. Roselli, L., Mariotti, C., Mezzanotte, P., Alimenti, F., Orecchini, G., Virili, M., Carvalho, N.B.: Review of the present technologies concurrently contributing to the implementation of the Internet of Things (IoT) paradigm: RFID, green electronics, WPT and energy harvesting. In: IEEE Topical Conference on Wireless Sensors and Sensor Networks (WiSNet), pp. 1–3 (2015)
3. Kanaya, H., Tsukamoto, S., Hirabaru, T., Kanemoto, D., Pokharel, R.K., Yoshida, K.: Energy harvesting circuit on a one-sided directional flexible antenna. IEEE Microwave Wirel. Compon. Lett. **23**(3), 164–166 (2013)
4. Shinki, Y., Shibata, K., Mansour, M., Kanaya, H.: Impedance matching antenna-integrated high-efficiency energy harvesting circuit. Sensors **17**(8), 1763 (14 pages) (2017)
5. Mansour, M.M., Kanaya, H.: High-efficient broadband CPW RF rectifier for wireless energy harvesting. IEEE Microwave Wirel. Compon. Lett. **29**(4), 288–290 (2019)
6. Mansour, M.M., Kanaya, H.: Efficiency-enhancement of 2.45-GHz energy harvesting circuit using integrated CPW-MS structure at low RF input power. IEICE Trans. Electron. **102**(5), 399–407 (2019)
7. Mansour, M.M., Kanaya, H.: Novel L-slot matching circuit integrated with circularly polarized rectenna for wireless energy harvesting. Electronics **8**, 651 (10 pages) (2019)
8. Ijiguchi, T., Kanemoto, D., Yoshitomi, K., Yoshida, K., Ishikawa, A., Fukagawa, S.: Circularly polarized one-sided directional slot antenna with reflector metal for 5.8-GHz DSRC operations. IEEE Antennas Wirel. Propag. Lett. **13**, 778–781 (2014)
9. Fujita, K., Yoshitomi, K., Yoshida, K., Kanaya, H.: A circularly polarized planar antenna on flexible substrate for ultra-wideband high-band applications. AEU-Int. J. Electron. Commun. **69**(9), 1381–1386 (2014)
10. Stutzman, W.L., Thiele, G.A.: Antenna Theory and Design, 3rd edn. Wiley, USA (2013)

Developing a Sustainable System for Early Warning Against Landslides During Rainfall

Yan Liu, Hemanta Hazarika, Osamu Takiguchi, and Haruichi Kanaya

Abstract Landslides have been occurring frequently because of rainfall, which leads to serious consequences, often plugging rivers, disrupting transportation, communication, and other lifeline projects and posing a serious threat to people's life and property. Therefore, effective early warning system for rainfall-induced landslide is particularly important. This research presented the development of early warning system that includes soil moisture sensor, temperature sensor, and data transmission system. Some tests were carried out to evaluate the feasibility of the developed system under different water contents. Test with different sample depths under the same water content was also used to verify the accuracy of this system. The results show that, the sensitivity of the sensor is affected by the initial state, and it takes about 10–20 min to reach the stable measurement state. With the increase of water content in soil samples, the resistance value obtained by the system decreases accordingly, which is consistent with theoretical results. The tests of early warning system also worked well under different depths of the same soil sample. The development of the current warning system is the basis for predicting landslides in the future, and lower development costs and accurate test results make it possible to be widely adopted to both the developed and developing nations.

Keywords Landslide · Early warning system · Low cost · Water content · Sensors

1 Introduction

In recent years, landslides have been occurring frequently because of rainfall. For example, a series of landslide disasters occurred in Kumamoto Prefecture, Japan, in July 2003 due to continuous heavy rains. Among them, the landslide formed a

Y. Liu (✉) · H. Hazarika · H. Kanaya
Kyushu University, Fukuoka, Japan
e-mail: liu.yan.441@s.kyushu-u.ac.jp

O. Takiguchi
ALSENS Inc., Sagamihara, Japan

H. Hazarika et al. (eds.), *Advances in Sustainable Construction and Resource Management*, Lecture Notes in Civil Engineering 144,
https://doi.org/10.1007/978-981-16-0077-7_75

mudslide that hit villages under the action of flash floods, causing immeasurable property losses and casualties [1]. In June 2010, a landslide occurred in Gangwu Town, Guanling County in China, leading the death of 99 and burying 37 houses by the damage of heavy rain [2]. In July 2017, localized and torrential rainfall has caused a massive landslide, debris flows, and flooding of rivers resulting in devastating damage to several areas of the Fukuoka and Oita prefectures of Kyushu, Japan [3]. On October 11, 2018, a number of landslide disasters occurred in West Sumatra and North Sumatra provinces of Indonesia due to rainfall. Many houses collapsed and the death toll reached more than 27. The global landslide database recorded a total of 2,620 landslides, causing 32,322 deaths within 6 years [4]. Therefore, the early warning system for rainfall-induced landslide is particularly important in saving human lives and protecting property.

Many scholars have continued to study the stability of rainfall-induced landslide for a long time [5, 6]. In previous studies, Global Navigation Satellite System (GNSS) is one of the main technical means to continuously obtain real-time three-dimensional vector deformation of landslide surface. It has become a common system for high-precision monitoring of landslides in complex areas [7, 8]. However, professional GNSS monitoring equipment is expensive, both in terms of the equipment and in terms of its installation, meaning that their application is limited to well-funded projects. The problem of high cost makes it impossible to deploy large-scale professional GNSS monitoring equipment [9], especially in developing countries where geotechnical disasters occur frequently. And, therefore, it is crucial to build an inexpensive and sustainable landslide warning system. Recent research showed that the early warning system based on different sensors and IoT technology is practically versatile and effective; thus, it was widely used in landslide monitoring and early warning, and generated considerable related research interest. In general, a whole early warning system includes temperature sensor, soil moisture sensor, and wireless data transmit. Many scholars have conducted research on sensors to screen out more suitable and low-cost devices [10–12].

In this study, a real-time landslide early warning system consisting of temperature sensor, soil moisture sensor, and wireless communication system was developed. Tests were used to verify the feasibility of the sensor based on soil samples with different water content. The accuracy of this system was also tested with different sample depths under the same water content. The newly developed transmission system with micro-module and A-term allows us to receive sensor data in real time, whether it is on a mobile phone or a PC. Solar battery will be used in running the sensors and transferring data. The purpose of this study is to evaluate the effectiveness of this system, which is the basis for landslide model tests in the future and large-scale launch in the rainfall areas.

2 Scheme of the System

2.1 Temperature Sensor and Soil Moisture Sensor

In this research, the M5STACK-EARTH-UNIT sensor was used to test the temperature, humidity, and resistance. The sensor has the adjustable threshold (including adjustable 10 kΩ resistor), and grove interface equipped with two mounting holes. It also has the function of analog and digital output. The sensor unit is small in size (7.0 * 2.5 cm) and easy to install. The weight of the sensor is only 100 g, so that the proposed system can be used extensively in hard-reached places in a landslide-prone area. The total cost of the sensor is very cheap, making it very easy to install and use on a large scale.

2.2 Micro-Module and Mobile Router

ESP32-PICO-D4, a System-in-Package (SiP) module, which is based on ESP32, is providing complete Wi-Fi and Bluetooth functionalities. The module has a size as small as (7.00 ± 0.10) mm × (7.00 ± 0.10) mm × (0.94 ± 0.10) mm, thus requiring minimal PCB area. The module integrates a 4-MB SPI flash.

At the core of this module is the ESP32 chip, which is a single 2.4 GHz Wi-Fi and Bluetooth combo chip designed with TSMC's 40 nm ultra-low power technology. With its ultra-small size, robust performance, and low-energy consumption, ESP32-PICO-D4 is well suited for the early warning system.

A wireless LTE mobile router was used to connect to the computer. In the research, the mobile phone (or computer) can be used together with the built-in wireless LAN with Wi-Fi connection, and it can be communicated through a wired connection using a USB cable. LTE can switch between the 2.4 GHz band and the 5 GHz band. In order to use the functions of LTE, a nano-SIM card must be installed in this product. Therefore, the connection between sensors and other devices is shown in Fig. 1, real-time monitoring can be carried out through a computer or mobile phone, and the monitoring process is shown in Fig. 2. When the device is connected, we can monitor it in real time through PC and mobile phone, and the monitoring screen is shown in Fig. 3. The current device can choose to update the data every minute or every five minutes.

Detailed information of early warning system is shown in Table 1. In order to increase the sustainability of the early warning system, solar battery will be used in running the sensors and transferring data. Figure 4 shows solar cell system containing energy absorbing panel and large mobile power sources that we have installed in the outdoor measurement. For this case, the size of the solar cell can be reduced to the suitable situation and mobile battery is also needed for night measure. Solar cell will save energy in the daytime and provide energy to the mobile battery so that the system will keep working at night.

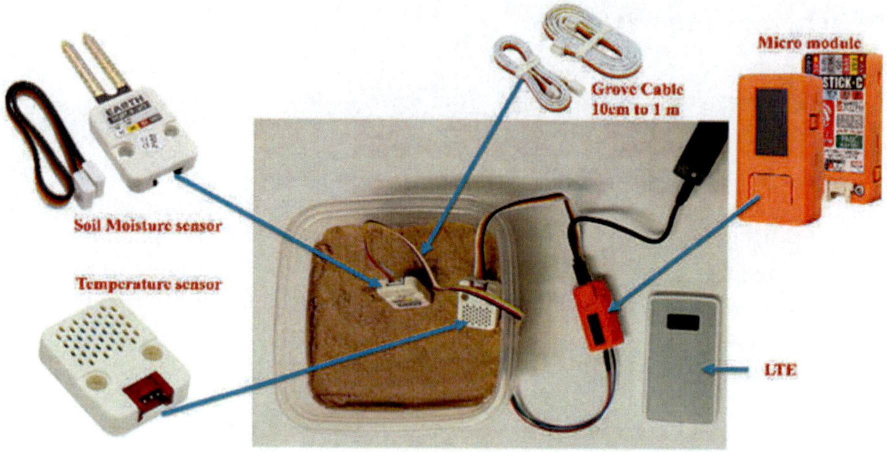

Fig. 1 Early warning system during the test

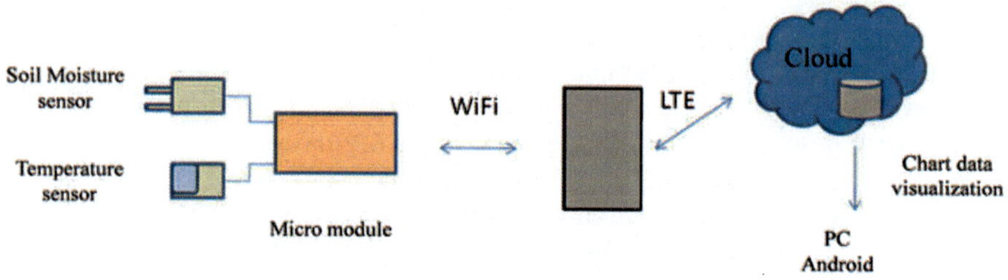

Fig. 2 Sensor test and early warning process

3 Sample Tests to Check the System

This sensor has a 12-bit digital-to-analog converter. Therefore, the test for the calibration from 100% wet to 0% dry condition showed that the output changed from 1767 to 4095. The output of 4095 means the whole system is in the air or freshly dried sandy soil, while the reading of 1767 means the sensor in the cup filled by the water. The specific change trend is shown in Fig. 5.

3.1 Analysis of the System Under Different Water Content

Rainfall-induced landslides often happen due to the rapid increase in water content in the soil. Therefore, testing the resistance at different levels of water content is the key to the early warning system. In this research, four different water contents, including 0%, 10%, 15%, and 20% were considered under the same failure condition. Toyoura

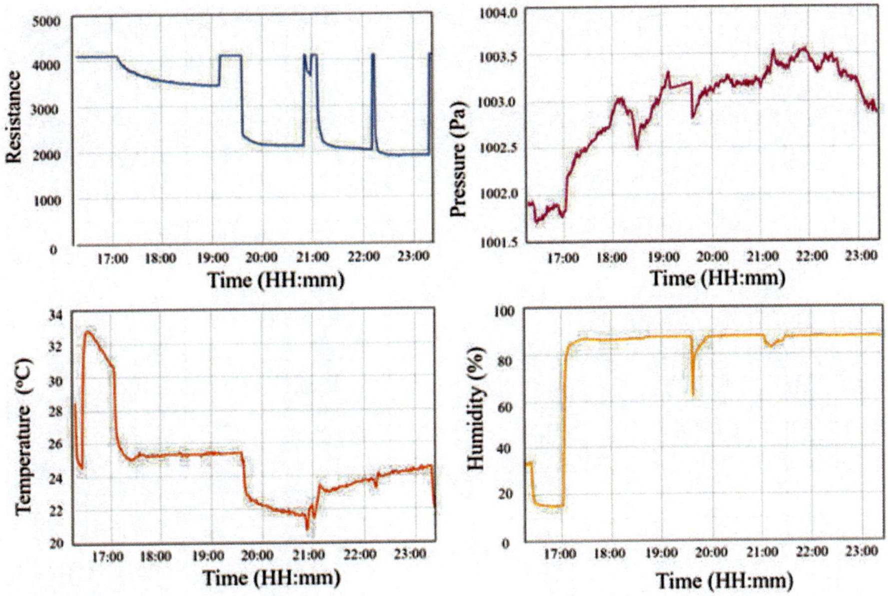

Fig. 3 Real-time monitoring

Table 1 Device information of early warning system

Device	Type	Size
Soil moisture Sensor	M5STACK-EARTH-UNIT	70 * 25 * 5 mm
SiP module	ESP32-PICO-D4	45 * 23 * 13 mm
Mobile router	PA-MP01LN	45 * 90 * 10 mm

sand was chosen in the research, and density ratio Dr = 60% in experiments. A cup, with the size of 2.5 L, was chosen in the research. The result of dry soil in case 1 is shown in Fig. 6, and the output is 4095, which is equivalent to putting the sensor in the air. While considering the influence of water content, it showed that the initial value recorded by the sensor was affected by the previous state and gave around 10–20 min to arrive a relatively stable state. With the increase of the water content (w), the resistance kept falling, 3455 for $w = 10\%$, 3175 for $w = 15\%$, while 2131 for $w = 20\%$. The trend was in line with expectations; that is, bigger water content leads to lower resistance, which also confirms the feasibility of the system.

Fig. 4 Solar cell system

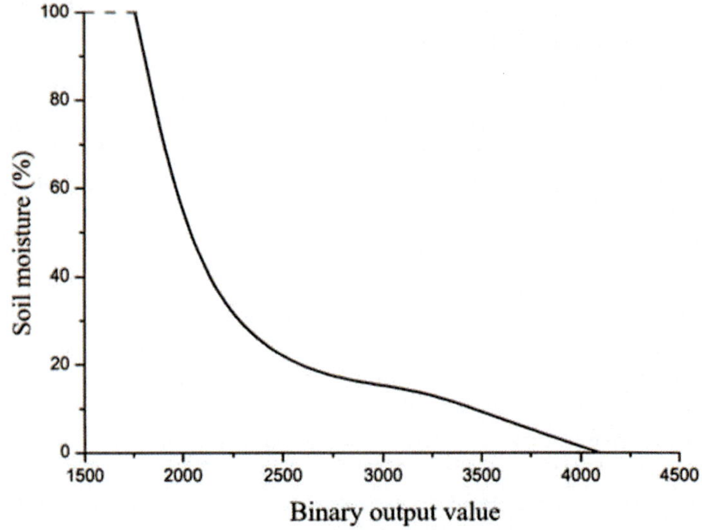

Fig. 5 Relationship between binary output value and soil moisture

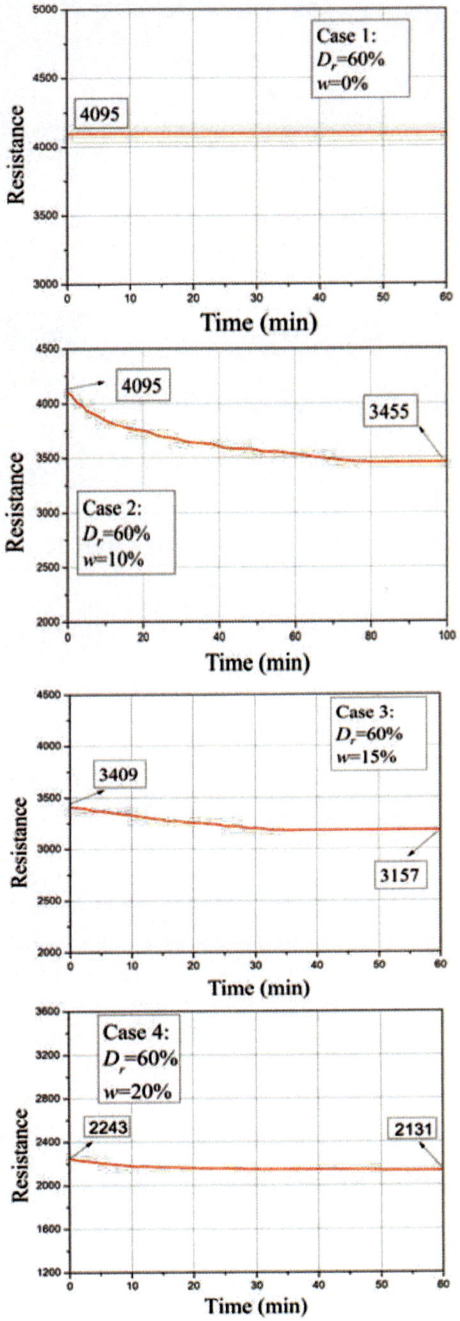

Fig. 6 Analog number under different water contents

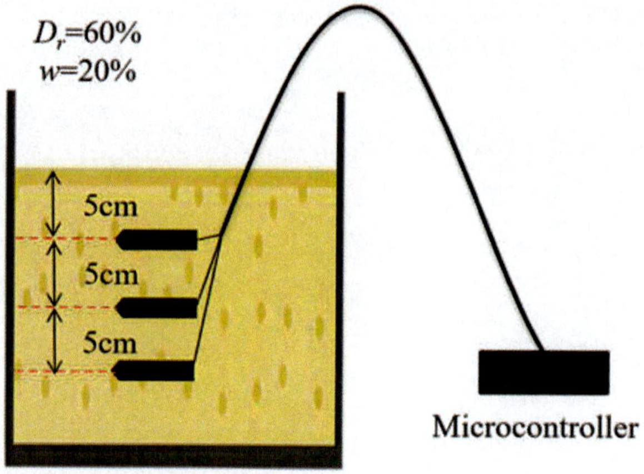

Fig. 7 Diagram of the test showing location of all sensors

3.2 Analysis of the System at Different Depths from the Ground

In order to increase the applicability of the system, it is necessary to monitor the response of the sensor in the soil samples under different depths. This research selected three depths for analysis: 5 cm, 10 cm, and 15 cm, the diagram of the test in shown in Fig. 7. The result shows that there are some differences in the resistance value at different depths, but the gap is not large, the result from lower depth (5 cm) is about 4% bigger than deeper depth (15 cm), the detailed results are shown in Fig. 8. When the test time continues to increase, the water content at the bottom gradually increases relative to the top area, which leads to a greater water content and results in a lower resistance value. In order to verify the rationality of this conjecture, the water content near the test soil was measured at the end of each experiment. The result in Table 2 found that the water content in the area near the deeper depth (15 cm) was slightly lower than the water content in the upper part (5 cm), which differed by about 3%. Therefore, this experiment verified the effectiveness of the sensor and the feasibility of the whole early warning system.

4 Conclusions and Future Scope

In this research, a low-cost and portable real-time landslide early warning sensor system has been developed. In order to validate the system's effectiveness, the sensors were tested under different water content and embedment depth in the soil sample. The results from the monitoring tests concluded that:

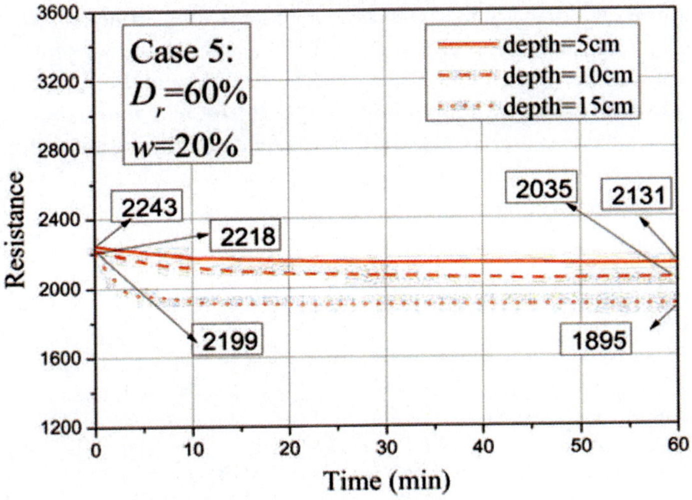

Fig. 8 Resistance under different depths

Table 2 Change of water content

Case	Depth (cm)	Initial water content (%)	Water content after experiment (%)
1	5	0	0
2	5	10	8.75
3	5	15	13.6
4	5	20	16.7
5	10	20	17.2
	15	20	17.89

(1) The sensitivity of the sensor is affected by its initial state, and it takes about 10–20 min to reach the stable measurement state. Therefore, an initial stabilizing time period is required before each test.
(2) With the increase of water content in the soil sample, the resistance value decreases, but the decrease is disproportionate with respect to water content. For example, the resistance value in dry soil is 4095, while the resistance value decreases to 2123 when the water content is 20%.
(3) There is no significant effect on the resistance value of sensors at different embedment depths. The variation observed on influence of depth is less than 4%. With elapsed time, the water percolates through the pores to lower depths, which results in a higher water content at the bottom of the soil sample, thereby influencing the results to some extent.

Furthermore, the development cost of the whole early warning system is low and affordable by countries with economies of all scale, and that is very suitable for large-scale launch. Further research needs to be done to fine tune the sensor's sensitivity and response time along with providing an alternate power source like solar power to improve its sustainability. In addition, actual field conditions of rainfall and slope stability are required to be performed for a scientific validation of the system, in order to be applicable in field.

References

1. Ministry of land and resources: 12th five-year plan for geological disaster prevention and control (2012) (In Chinese)
2. Yueping, Y., Xuanming, P.: Discussion on the instability of Qianjiangping landslide in the three gorges reservoir area. Hydrogeol. Eng. Geol. **3**, 51–54 (2017)
3. Hazarika, H., Yamamoto, S., Ishizawa, T., Danjo, T., Kochi, Y., Fujishiro, T., Okamoto, K., Matsumoto, D., Ishibashi, S.: The 2017 July Northern Kyushu torrential rainfall disaster—geotechnical and geological perspectives. In: 3rd Indo-Japan Workshop on Geotechnics for Natural Disaster Mitigation and Management, Guwahati, India, CD-ROM (2017)
4. Guochang, L.: Engineering Geology of Regional Stability. Jilin University Press (1993) (In Chinese)
5. Hazarika, H., Kazama, M., Lee, W. F.: Geotechnical Hazards from Large Earthquakes and Heavy Rainfalls. Springer (2016)
6. Tran, T.T.T., Hazarika, H., Indrawan, I.G.B., Karnawati, D.: Prediction of time to soil failure based on creep strength reduction approach. Geotech. Geol. Eng. **36**(4), 2749–2760 (2018)
7. Yang, Y., Xu, Y., Li, J., Yang, C.: Progress and performance evaluation of BeiDou global navigation satellite system: data analysis based on BDS-3 demonstration system. Sci. China Earth Sci. **61**(5), 614–624 (2018)
8. Yang, Y., Gao, W., Guo, S., Mao, Y., Yang, Y.: Introduction to BeiDou-3 navigation satellite system. Navigation **66**(1), 7–18 (2019)
9. Tu, R., Wang, R., Ge, M., Walter, T.R., Ramatschi, M., Milkereit, C., Dahm, T.: Cost-effective monitoring of ground motion related to earthquakes, landslides, or volcanic activity by joint use of a single-frequency GPS and a MEMS accelerometer. Geophys. Res. Lett. **40**(15), 3825–3829 (2013)
10. Kanaya, H., Tsukamaoto, S., Hirabaru, T., Kanemoto, D., Pokharel, R.K., Yoshida, K.: Energy harvesting circuit on a one-sided directional flexible antenna. IEEE Microwave Wirel. Compon. Lett. **23**(3), 164–166 (2013)
11. Ichihashi, M., Kanaya, H.: A simple methodology for on-chip transmission line modeling and optimization for high-speed clock distribution. Jpn. J. Appl. Phys. **58**(SB), SBBC06 (2019)
12. Mansour, M., Le, P.X., Kanaya, H.: Enhanced broadband RF differential rectifier integrated with Archimedean spiral antenna for wireless energy harvesting applications. Sensors **19**(3), 655 (2019)

Evaluation of Slope Failure and Driftwood Disaster Susceptibility Zone Using Geo-spatial Information

Lu Tian, Yasuhiro Mitani, Yuki Okajima, Hisatoshi Taniguchi, and Taiga Tabuchi

Abstract From July 5 to 6, 2017, the heavy rain in the northern Kyushu causes flooding and landslide. It also severely damages the surrounding forests, farmland, and even urban area after falling trees from hillside flowed into the rivers. Therefore, this paper selected the river area with lots of driftwood in Asakura City and Toho village in Fukuoka as the research area. Slope failure, one of the causes of driftwood, was obtained through analysis. The susceptibility zone was evaluated by using the SVM model, and the characteristics distribution of influencing factors was described using GIS. The aim is to provide useful basic information for the prevention and mitigation of driftwood disasters and forest maintenance in the future.

Keywords GIS · Driftwood · Slope failure · Support vector machine

1 Introduction

From July 5 to 6, 2017, the Northern Kyushu area suffered the strongest rainfall on record, forming a linear precipitation zone due to the stagnation of the Meiyu front near the Tsushima Strait into warm and humid air. According to the meteorological agency data, Asakura City in Fukuoka prefecture had maximum daily precipitation of 545.5 mm [1]. The forest was severely damaged. Many driftwoods flowed into plain rivers with debris flow, resulting in the expansion of the inundation area, promoting the accumulation of sediment, and blocking the part under the bridge. Even into the plains downstream, into the surrounding roads, farmland, or destroy residential areas. Therefore, the driftwood disaster has become one of the major disasters of this heavy rain in Kyushu. There are two main reasons for a large number of driftwoods. First, the reforested forests and secondary forests cover the hillsides. In excessively dense forests, the dense crowns, insufficient light, slender trunks, and shallow roots cause the trees to fall more easily [2]. Second, the impact of global warming brings increased rainfall. According to the meteorological agency data, we collected the

L. Tian (✉) · Y. Mitani · Y. Okajima · H. Taniguchi · T. Tabuchi
Kyushu University, Fukuoka 819-0395, Japan
e-mail: tian.lu.441@s.kyushu-u.ac.jp

H. Hazarika et al. (eds.), *Advances in Sustainable Construction and Resource Management*, Lecture Notes in Civil Engineering 144,
https://doi.org/10.1007/978-981-16-0077-7_76

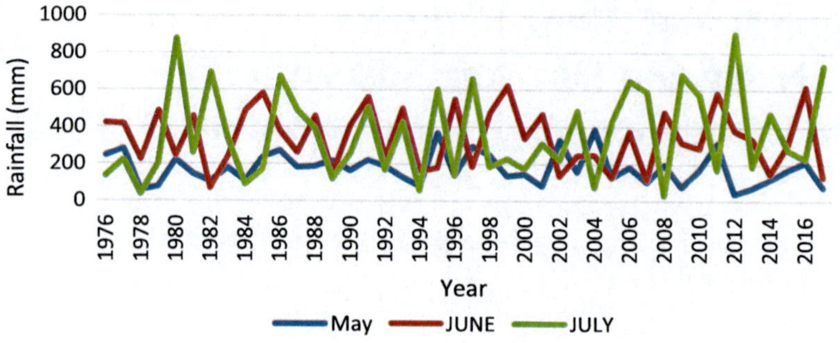

Fig. 1 Total monthly rainfall in Asakura region. *Source* Japan Meteorological Agency

monthly precipitation during the rainy season (May–July) in the Asakura area over the past 40 years, as shown in Fig. 1. As a result, the scale of the rainfall increased further, causing slope failure on the hillside.

A report on heavy rain in Northern Kyushu in July 2017 was released by MLIT [3]. In this report, the upstream of the right bank in the middle reach of the Chikugo River experienced a large number of slope failures, which resulted in driftwood outflow. By calculating the driftwoods production, they come from four forest types. The largest one derives from the hillside forest, accounting for 63% of the total, about 130,000 m^3, followed by the valley forests, riparian forest, and others, as shown in Fig. 2. Some reports have mentioned that driftwoods come from the outflow of unremoved offal timber after logging. But in Takehiko Oota's article [4], it is said that in any case, it cannot be considered that sufficient water depth will be generated on the hillside to flush these woods into the valley plain or be involved with mudslides. Even if it exists, it is small. Therefore, in this article, we do not consider that driftwoods come from logging.

In summary, we know that one of the important relationships with the driftwood disaster is slope failure in hillsides. Based on the concentrated area of slope failure and the river regions with a large amount of driftwood production on the hillside, namely Akaya River, Kita River, Ohi River, Sozu River, Sada River, and Shirakidani

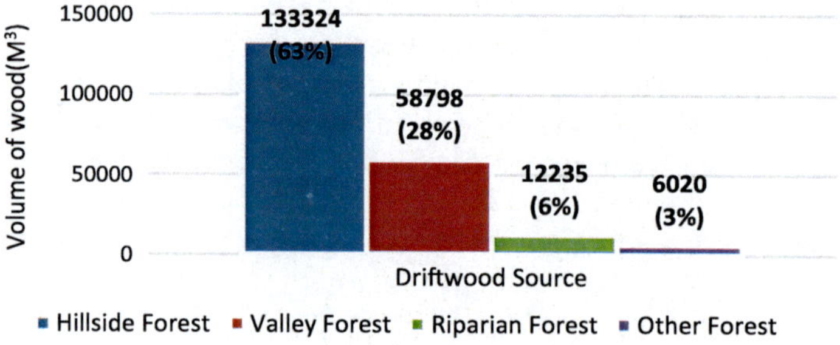

Fig. 2 Production of driftwood. *Source* MLIT

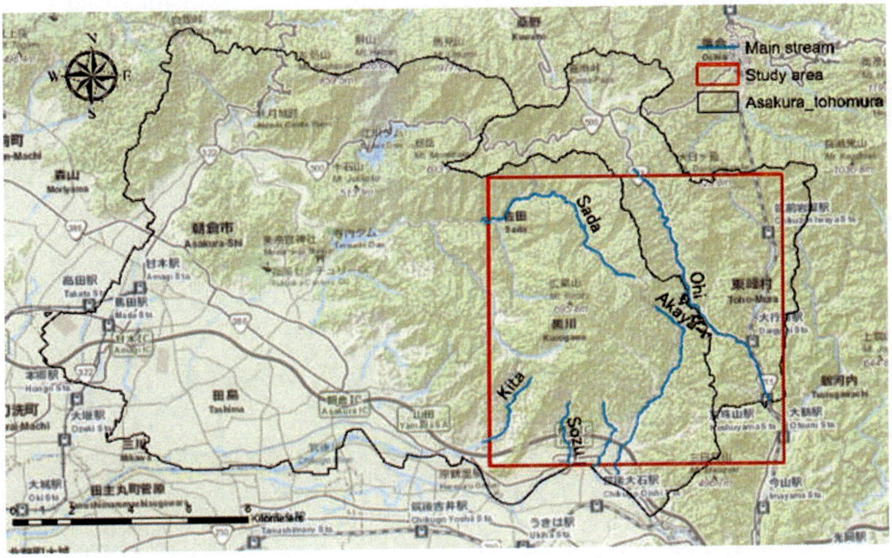

Fig. 3 Study area

River as the study area, as shown in Fig. 3. Using the support vector machine model to identify the susceptibility area of slope failure is driftwood's main source area and analyzing the influencing factors of each failure. In the future, effective basic information will be provided to enhance forest conservation and add new or improved old forest protection facilities.

2 Data

Slope failure is the result of the interaction between internal influencing factors and external factors. The selection of influencing factors and the establishment of corresponding thematic data layers are the keys to identify susceptibility areas [5]. Internal factors are internal characteristics of the material conditions, structural conditions, and spatial conditions of the hillside, while external factors are composed of natural factors and human factors. There are two principles for selecting factors: First, the factor should be closely related to the occurrence of failure; second, the factor should be able to obtain and quantify expression through an effective path. According to the statistics of Paola Reichenbach's research [6], the influencing factors can be divided into five major classes: morphological, geological, land cover, hydrological, and others. Therefore, these five classes will also be used in this research area. Each class includes a variety of subclasses, as shown in Table 1.

In morphological, elevation affects the distribution of vegetation, gradient, aspect, and plan curvature, directly affecting the slope's stability and failure probability.

Table 1 Slope failure influencing factors

Class	Factor	Scale	Source
Morphological	Elevation	10 m (DEM)	Geo-spatial Information Authority of Japan
	Gradient		
	Aspect		
	Curvature (Plan, Profile)		
	Topographic Wetness Index		
	Landform	1:200,000	Ministry of Land, Infrastructure, and Transport (MLIT)
Geological	Geo-lithological		
Land cover	Soil		
	Land use	100 m (GRID)	
	Distance to Road	1: 25,000	
	Vegetation		Biodiversity Center of Japan
Hydrological	Distance to Stream		(MLIT)
Others	Precipitation	250 m(GRID)	Japan Meteorological Agency

Profile curvature is reflected in the thickness and composition of the soil. Topographic wetness index (TWI) is used to describe basin terrain changes and its effect on soil runoff. It is commonly used to quantify topographic control on hydrological processes [10]. The factors are calculated by DEM with a resolution of 10 m. Geological factors include stratum lithology and geological structure. Different rock and soil corrosion resistance and mechanical strength are not the same. When the weight and external conditions change, it is easy to make the slope unstable. There are not many existing fault zones (geological structures) in our study area, so we do not consider this factor. Land use and roads are important factors influencing the occurrence of a failure in human activities. River and soil properties also restrict the hillside change. Finally, rainfall is the induced cause of this disaster, so we have to know the precipitation within 48 h.

Using SAR data to extract the slope failure sites on the hillside [7], the correctly extracted failure number is 2930, while the incorrect number (no failure) is 2313. Since the scale of each influencing factor obtained above is different, we need to unify them. Convert the thematic map of all factors in Table 1 to a 10 m $\times$ 10 m grid format. Therefore, the study area's grid scale is 1001 columns and 1001 rows, with 1002,001 grids.

3 Methodology

3.1 Research Process

In this study, the workflow of evaluating the slope failure susceptibility area is shown in Fig. 4. The 2930 slope failure sites and 2313 non-slope failure sites [7] were randomly divided into two parts: 50% for the training set and the remaining 50% for the testing set. The data of five major classes (14 subclasses) were used to make thematic maps. Especially in morphological factors, excluding the landform factor, the other six factors are calculated using the ArcGIS tool based on DEM data. When all the thematic maps are got, we will reclassify different levels or types according to each factor's features, such as soil, vegetation, and landform types, high or low elevation, and land-use categories. Then, combined with the slope failure sites, the distribution characteristics of damaged areas are summarized. After that, regarding the susceptibility zone, firstly, we stack 14 pieces of thematic maps for one layer. Secondly, the training set as the region of interest for image classification using the SVM model in ENVI software. Thirdly, using the testing set evaluated the result of classification and extracted the ROC curve as an accurate indicator of the model.

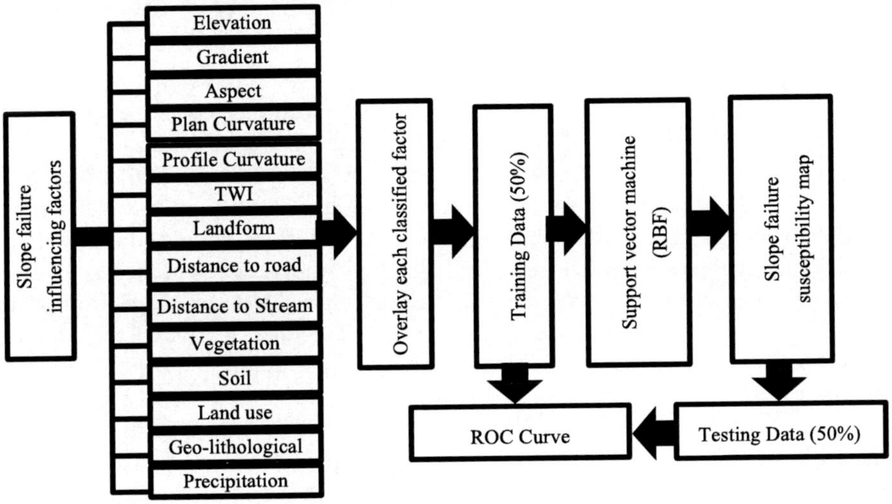

Fig. 4 Workflow of research processing

3.2 Support Vector Machine (SVM) and Accuracy Assessment Indicator

Support vector machine is a machine learning method based on traditional learning theory, where the limited sample information seeks the optimal compromise between the complexity of model and learning ability, thus obtaining the best generalization capability [8]. The core of SVM is to construct the optimal hyperplane, which separates the two kinds of data as accurately as possible and at the furthest margin. In reality, most of the data are nonlinear classification problems. The input variable is mapped to a high-dimensional space through the predetermined nonlinear mapping function, and then linear regression is carried out here. Finally, the nonlinear regression in the original space is achieved. So, this mapping function is called the kernel function, which is the function that makes a low-dimensional linearly inseparable sample map to high-dimensional space and becomes linearly separable. In this study, we used ENVI 5.3 software for image visualization environment, in which support vector machine classifier provides four basic kernel functions:

Linear kernel function $K\left(x_i, x_j\right) = x_I^T x_j.$

Polynomial kernel function $K\left(x_i, x_j\right) = \left(\gamma x_i^T x_j + r\right)^d, \gamma > 0.$

Sigmoid kernel function $K\left(x_i, x_j\right) = \tan h\left(\gamma x_i^T x_j + r\right).$

Radial basis kernel function $K\left(x_i, x_j\right) = \exp\left(-\gamma \left\| x_i - x_j^2 \right\|\right), \gamma > 0.$

Here, γ, r, and d are kernel parameters.

By summarizing the research related to SVM at domestic and overseas, researchers often choose different kernel functions for different study contents and purposes in order to get good results. According to Yu Xiaoyu's research proved that the linear kernel function is a special case of the radial basis kernel function, and the sigmoid kernel function has the work-energy approximate to RBF on some parameters [11]. Hence, the usual method is to take the radial basis kernel function as the first choice.

The prediction model's evaluation indicators are usually described from three aspects: the suitability, robustness, and accuracy. The ultimate goal is to judge the match between the predicted result and the actual condition [9]. In this research, we used the ROC curve to verify the results. ROC curve is a useful visualization tool, which used the area under the roc curve (AUC) as the standard to measure the accuracy of the model prediction. It accurately reflects the relationship between specificity and sensitivity of an analytical method and a comprehensive test accuracy representative—the larger AUC, the higher precision. On the contrary, the smaller of AUC, the lower of accuracy.

4 Results

4.1 Thematic Influencing Factor Map

The factors were reclassified to obtain the thematic maps, as shown in Fig. 5; the proportion of slope failure in each classification was calculated, respectively.

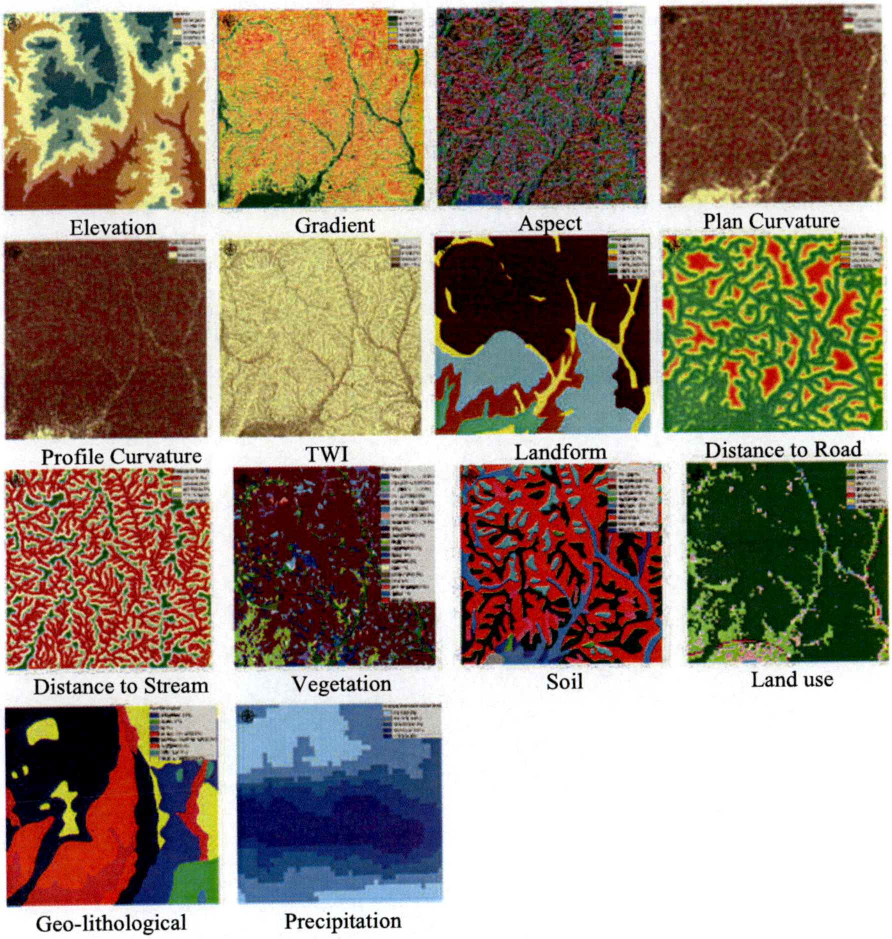

Fig. 5 Thematic influencing factor maps

4.2 *Evaluation of Slope Failure Susceptibility Area and Accuracy Assessment*

The trained SVM model was used to evaluate the slope failure susceptibility area's results, as shown in Fig. 6. This study chooses the natural breaks (Jenks) method for the susceptibility area level as the classification method in this study area: very high, high, medium, low, very low. The result of the success rate (training set) and prediction rate (testing set) of the slope failure susceptibility evaluation in the SVM model is represented by the ROC curve, as known in Fig. 7. The area under the curve (AUC) of 0.8656 and 0.8406 is reliable and accurate.

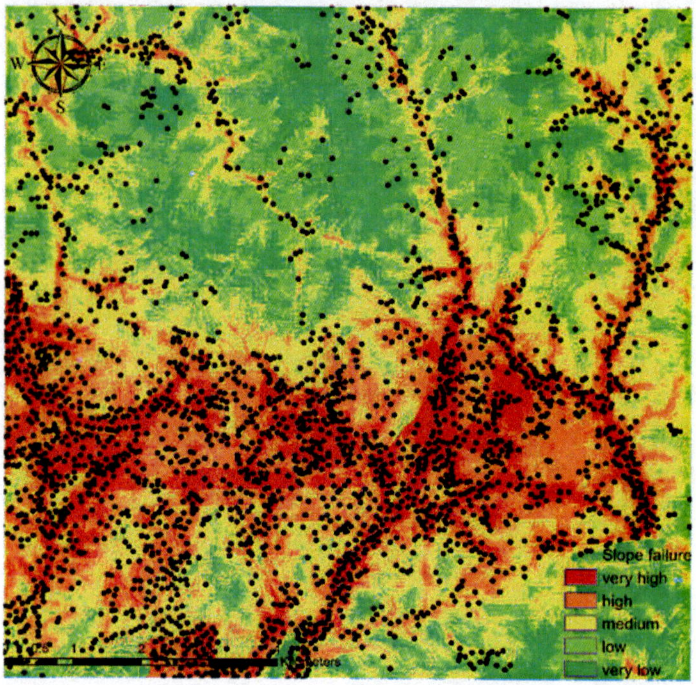

Fig. 6 Slope failure susceptibility map

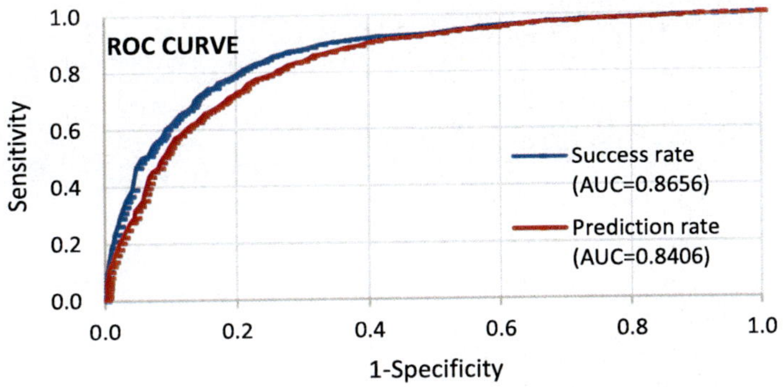

Fig. 7 Success rate and prediction rate curve (AUC: Area under the curve)

5 Discussions

5.1 Relationship Between Slope Failure and Slope Failure Influencing Factors

Based on the morphological results of the study area, we statistically analyzed the spatial distribution characteristics of slope failure on the gradient, aspect, elevation, and curvature. The slope failure was mainly concentrated on the gradient of 10°–45°, especially the steep hill of 30°–45°, with the highest slope failure density reached 27.3% (801 sites). The effect of aspect on the slope failure is not apparent, but there is a tendency for more slope failure in the southern aspect and less distribution in the northwest aspect. Due to the south direction being a sunny slope, there are usually more sunshine, strong weathering of the rocks, and the loose layer's thickness, prone to slope failure. We know from the distribution on the elevation that less than 400 m accounts for 87.5% of the total number of slope failures. The higher the altitude, the more the slope failure shows a decreasing trend. At the same time, combining with the landform, it is known that the slope failure is mainly small and medium small relief mountains, which is the same as the result of elevation. The larger the TWI value, the catchment area increases, and the gradient decreases from the top of the slope to the riverbank. In this study area, the TWI is between 1 and 21. It can be found that the larger TWI, the less the slope failure. 62.7% of them occur between 1 and 5. It is also determined that the smaller TWI, the easier it is to form a slope failure. The proportion difference between the concave slope and convex slope is not too big from the curvature, but the convex's slope failure is more than concave. When the crust is strongly uplifted, and the river's down-cut speed is greater than the slope's erosion, a convex shape will appear, and the slope stability will decrease. The spatial distribution of distances to streams and roads shows that most slope failures are very close to them. There are 1358 sites within 225 m of the stream, accounting for 46.3% of the total, and 1,722 sites within 60 m of roads, accounting for 58.8%.

Table 2 Proportion of slope failure in the susceptibility classes

Susceptibility class	No. of slope failure	Proportion (%)
Very high	779	26.59
High	1138	38.84
Medium	762	26.01
Low	215	7.34
Very low	36	1.23
Total	2930	100.01

Therefore, the impact and destruction of the slope failure on the human activity area are more serious. In the rock–soil composition, the pelitopsammite and granodiorite are the most serious. The rainfall makes the sandy and muddy rock layer easy to be washed away and moved. Granodiorite is composed of a strong weathering layer and residual soil, with a loose structure and low cohesive force. Heavy rain will seep into the slope, which will soften the soil, reduce the soil adhesion, and easily induce failure. According to the report released by the MLIT [3], Cryptomeria japonica, Chamaecyparis obtusa, Chamaecyparis pisifera are the main driftwood flowing from the hillside. The results that 60% of the slope failure occurred in this plant area. Simultaneously, the brown forest soil covered 75% of the study area, so it is also a susceptible soil type of slope failure.

5.2 *The Effect of the SVM Model on Disasters in the Study Area*

From the susceptibility map that we have produced, we can conclude that 65.4% of the slope failure sites fall into "very high" and "high" susceptible areas, and the proportion of "low" and "very low" is only 8.6% in Table 2. It shows that the SVM model well evaluates the susceptibility of slope failure in the study area. Then, the rivers with driftwoods in the hillside: Akaya River, Kita River, Ohi River (middle and downstream), Sozu River, Sada River (upstream), and Shirakidani River are included in the very high/high susceptibility area for slope failure that we have identified (Fig. 8). Therefore, it also indicates that the SVM model is better evaluated.

5.3 *Maintenance of Hillside Forest*

Firstly, we must give priority to areas with high and very high susceptibility. After removing the remaining driftwood and dumped trees, the chaotically growing secondary forests must also be removed to improve the trees' problem, not fully due to the dense canopy. Then, forests in susceptibility areas near roads and streams need

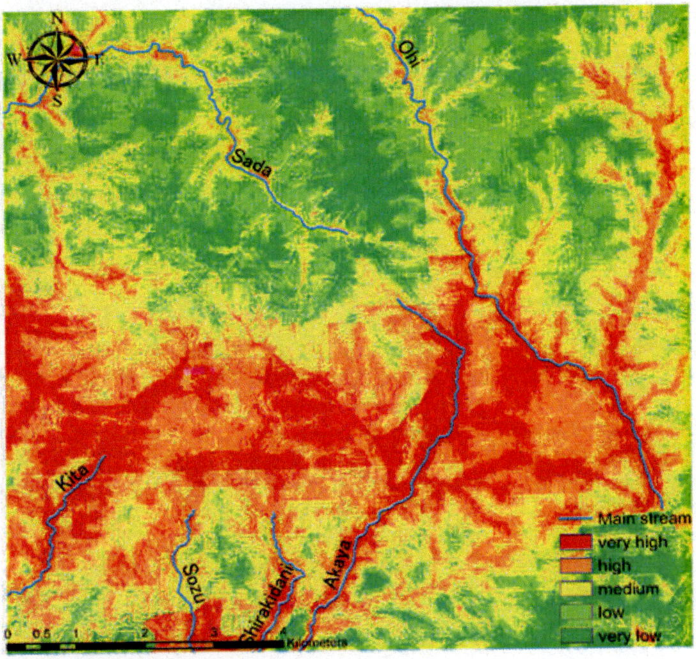

Fig. 8 Rivers in a very high/high susceptibility area

to be re-maintained to ensure people's safety and property in the plains. Based on morphology, geology, land cover, and hydrology, and other important features related to slope stability, protective measures have been formulated to prevent driftwood from flowing into the river to cause severe disasters again.

Secondly, the susceptibility is medium and low. We need to establish a sound forest management standard. In forests, human activities' impact is greater, human disturbance is relatively strong, and the damage to ecosystems is also more serious. For forests with a single stand structure dominated by commercial forests, they are more vulnerable. We should gradually adjust the stand structure in accordance with the changes in climate and ecology, and rationally plant ecological forests while maintaining commercial forests to ensure environmental stability and reduce vulnerability. In this way, there is a certain suppression effect on the driftwood disaster caused by slope failure.

6 Conclusion

In this paper, an SVM model was proposed to evaluate the susceptibility of the slope failure area of many driftwoods generated by Asakura City and Toho Village, and the accurate susceptibility zoning map was obtained. Therefore, the SVM model is a simple, efficient, and high-precision evaluation model. Its result, combined with the characteristics of various influencing factors, will become extremely useful basic information in future disaster prevention and forest maintenance work.

References

1. Cabinet Officer, Government of Japan: Disaster Report—Study on damage situation and correspondence of the heavy rain Northern Kyushu in July (2017)
2. Department of Fisheries and Forestry Bureau of Forestry Flood Control Afforestation Division in Hokkaido: Study on forests established to reduce driftwood disaster (2019)
3. Kyushu Regional Development Bureau, MLIT: River, control of soil erosion restoration technology investigation committee report- Right bank basin of Chikugo river (2017)
4. Ota, T.: Northern Kyushu Heavy Rain Disaster: Forestry and Driftwood (Disasters). Green & Life. Vol. Autumn (2017)
5. Sarkar, S., Kanungo, D.P.: An integrated approach for landslide susceptibility mapping using remote sensing and GIS. Photogram. Eng. Remote Sens. **70**(5), 617–625 (2004)
6. Reichenbach, P., Rossi, M., Malamud, B., Mihir, M., Guzzetti, F.: A review of statistically based landslide susceptibility models. Earth Sci. Rev. **180**, 60–91 (2018)
7. Kimura, S.: A Study on Utilization of SAR Data for Grasping Slope Failure Area Over Wide Area. Kyushu University (2020)
8. Li, X.Z., Kong, J.M., Wang, C.H.: Application of multi-classification support vector machine in the identifying of landslide stability. J. Jilin Univ. (Earth Sci. Ed.) **40**(3), 631–637 (2010)
9. Lin, Y.L.: Application of Kernel Learning Methods on Regional Landslide Susceptibility Assessment. Hebei University of Technology (2017)
10. Sørensen, R., Zinko, U., Seibert, J.: On the calculation of the topographic wetness index: evaluation of different methods based on field observations. Hydrol. Earth Syst. Sci. **10**, 101–112 (2006)
11. Yu, X.Y.: Study on the Landslide Susceptibility Evaluation Method Based on Multi-source Data and Multi-scale Analysis. China University of Geosciences (2016)

Innovative Case Studies in Earthquake Geotechnical Engineering

Change in Shear Wave Velocity During Consolidation and Undrained Cyclic Loading on Cemented Sand

Masataka Shiga and Takashi Kiyota

Abstract Concerns have been raised that future intense earthquakes in Japan will cause liquefaction in Pleistocene deposits beneath critical infrastructure facilities such as nuclear power plants or highways, which have not been examined by the current liquefaction assessment. In order to assess accurately in-situ liquefaction characteristic of Pleistocene deposits, shear wave velocity, V_s was introduced to quantify the effect of cementation on liquefaction characteristics. In this study, consolidation and undrained cyclic loading was applied to lightly cemented specimens while V_s is continuously measured to verify whether strengthened soil fabric degrades or not. From the results during consolidation, although the specimens with cement addiction rate, $C = 1\%$ has a higher value of stress dependency, the ones with $C =$ 3 and 5% have a value lower than that of the non-cemented specimens. After simple nonlinear regression was conducted, the result revealed that higher cemented specimens could become hard to deteriorate soil fabric during undrained cyclic loading, although non-cemented specimens maintained its soil fabric and lower cemented specimens declines it.

Keywords Pleistocene sand · Liquefaction · Shear wave velocity · Cementation

1 Introduction

In general, Pleistocene deposits are subjected to a diagenetic effect. The effect can be divided into two categories: physical stabilization and chemical stabilization. The microstructure stabilization can be described as an enhancement of the soil particle structure due to stress histories. The chemical stabilization can be described as an increase in adherence between soil particles due to cementation. When considering liquefaction of Pleistocene deposits, it is required to study the effects of cementation

M. Shiga (✉) · T. Kiyota
University of Tokyo, 4-6-1, Komaba, Meguro, Tokyo, Japan
e-mail: shiga815@iis.u-tokyo.ac.jp

H. Hazarika et al. (eds.), *Advances in Sustainable Construction and Resource Management*, Lecture Notes in Civil Engineering 144,
https://doi.org/10.1007/978-981-16-0077-7_77

and soil fabric on liquefaction properties in a manner consistent with in-situ conditions. This paper discusses the impact of cementation on liquefaction properties and shear wave velocities, V_s.

Many papers discussed the liquefaction characteristics of artificially cemented soils and Pleistocene soils collected from natural grounds. Koseki and Ohta [1] investigated the effect of different consolidation conditions on the liquefaction characteristics of a specimen mixed with Toyoura sand and bentonite clay. Amoly [2] conducted undrained cyclic loading test by using new and old deposits and concluded that older deposits would yield larger values of V_s as compared to newer ones. Porcino [3] showed the relationship between normalized number of cycle, $N_c/N_{c-\text{end}}$ and excess pore water pressure ratio, r_u and stated that the generation of r_u of cemented specimen is faster than that of uncemented sands. Another example of how liquefaction properties are affected by artificial or natural aging effects was provided by Tokimatsu and Hosaka [4] and Teachaborasinskun et al. [5].

Although laboratory testing is essential for accurate liquefaction determination, it is difficult to collect in-situ samples without disturbing soil fabric and cementation effects. Yoshimi et al. [6, 7] and Hatanaka et al. [8], for example, stated that the liquefaction strength of samples collected by tube sampling was smaller than that of samples collected by frozen sampling because freezing sampling was a method suitable for preservation of in-situ conditions. However, the freezing sampling is quite expensive and is not yet widely used. Therefore, it is desirable to develop a method that can assess in-situ liquefaction resistance of disturbed soil sample.

Among several existing methods, it was found that small-strain stiffness or V_s has been useful for evaluating soil fabric. Tokimatsu et al. [9] and Goto [10] measured changes in small-strain shear moduli during undrained cyclic loading. Koseki and Ohta [1] statically measured the small-strain characteristics of Toyoura sand during isotropic consolidation and undrained cyclic shear. They concluded that the small-strain stiffness in undrained cyclic shear became lower than that in isotropic consolidation as soil fabric deteriorates. Kiyota et al. [11] also investigated the effects of aging on liquefaction properties and small-strain stiffness using frozen samples from Pleistocene and Holocene deposits. They concluded that an increase in liquefaction resistance was observed with an increase in initial shear-strain stiffness, and degradation of aging effect was confirmed during undrained cyclic loading.

In this study, the effect of aging on V_s and liquefaction characteristics are discussed using lightly cemented sand made by adding early strength Portland cement to silica sand No. 7. The values of V_s is also continuously measured during consolidation and undrained cyclic loading, and the degree of degradation of V_s is investigated on the cemented sand.

2 Methodology

The samples used in this study is silica sand No. 7 ($D_{50} = 0.153$ mm, $G_s = 2.617$, $e_{\min} = 0.671$ and $e_{\max} = 1.145$) and early strength Portland cement, which is used as the cementing agent. Cement addition ratio, C to silica sand is 0, 1, 2, 3, and 5% by mass. The relative density, D_r of the specimens without cement are 40, 50, and 60%, and D_r of the specimens with cement are 65%.

The specimen is prepared by wet tamping method. A plastic bag with pre-measured silica sand, cement, and water are shaken for 1 min. The mixture is thrown into a plastic mold using a funnel, and it is pounded with a hammer to a desired height. The tamping is repeated four times to create a specimen 200 mm in height and 100 mm in diameter, which is cured in water for 7 days. After curing, the plastic mold is dismantled, the specimen is covered with a membrane, and is placed in strain-controlled triaxial apparatus. For the non-cemented specimen, sand and water mixed in a plastic bag is put into the mold, and the specimen is placed in the test apparatus immediately after it is shaped.

The specimen is initially consolidated at effective confinement pressure, $\sigma_c^{'}$ of 30 kPa. While maintaining $\sigma_c^{'}$ back pressure is reduced to -100 kPa and double vacuuming is applied to remove the air from the specimen, the back pressure is gradually increased to 200 kPa after finishing saturation. B-value of Skempton is confirmed to be above 0.95. Then, isotropic consolidation is performed and $\sigma_c^{'}$ is increased to 100 kPa.

A trigger and accelerometer system, originally developed by AnhDan et al. [12], is used to measure V_s transmitted through the specimens. The shear wave, in the form of a single sinusoidal wave with a frequency of 1 kHz, is generated by a pair of wave sources (actuators) attached on the top cap that are simultaneously excited in the torsional direction. A pair of accelerometers is employed to measure the arrival time of the shear wave at two different heights of specimen.

After the measurement of V_s during consolidation, undrained cyclic loading are applied with a constant amplitude of cyclic deviator stress until double amplitude of axial strain, ε_{DA} reached 5% at constant rate of 0.1%/min. The specification of all the tests is shown in Table 1.

3 Results and Discussion

3.1 Shear Wave Velocity Measured During Isotropic Consolidation

Figure 1a shows relationship between the stress parameter, $\sqrt{\sigma_1^{'}\sigma_3^{'}}$ and V_s. For the non-cemented specimens, which are represented in blue-colored lines in Fig. 1a, the values of V_s increased with increasing D_r for all the specimens. The results of

Table 1 Specification of the test

Name	Cement addiction rate, C (%)	Target D_r (%)	CSR	B-value	Achieved e	Achieved D_r	$N_{c(\mathrm{DA5})}$	$N_{c(\mathrm{DA3})}$	$N_{c(\mathrm{DA1})}$	$N_{c(u95)}$
WT_C00_2	0	40	0.17	0.990	0.956	39.9%	2.5	1.5	1.5	1.5
WT_C00_1		50	0.12	0.993	0.913	49.0%	104	102.5	99.5	100.5
WT_C00_5		60	0.23	0.990	0.860	60.1%	10.5	8.5	4.5	6.5
WT_C01_6	1	65	0.12	0.983	0.828	66.9%	(200)[a]	150	142	142.5
WT_C01_3			0.3	0.995	0.822	68.1%	6.5	2.5	0.5	2
WT_C02_2	2		0.17	0.977	0.837	64.9%	92	80.5	66	63.5
WT_C02_1			0.3	0.912	0.819	68.7%	6.5	3.5	1.5	2.5
WT_C03_2	3		0.3	0.992	0.846	63.0%	16.5	10.5	4.5	4.5
WT_C05_1	5		0.3	0.965	0.820	68.6%	N/A	29.5	16.5	12.5

[a] ε_{DA} did not reach 5% even 200 cyclic loadings applied

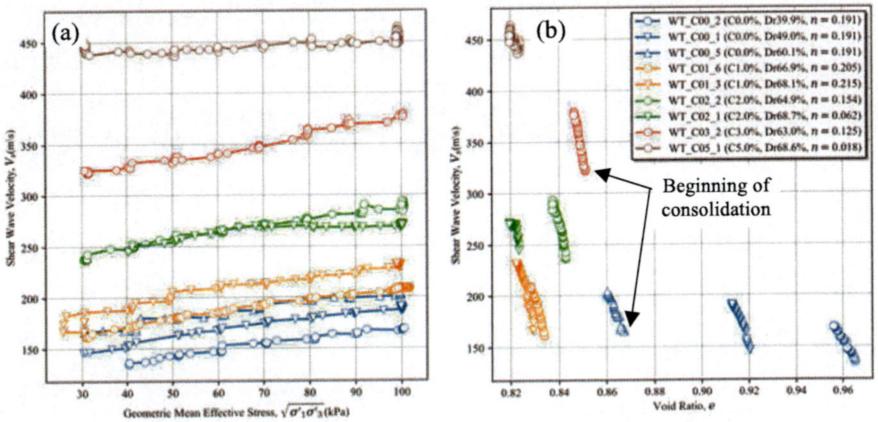

Fig. 1 **a** Relationship between $\sqrt{\sigma_1'\sigma_3'}$ and V_s, **b** Relationship between e and V_s

the non-cemented specimens are almost parallel. The values of V_s increases with increasing cement addiction rate.

Generally, V_s or dynamic shear moduli, G_d are expressed as a function of e and stress. Many previous studies used the same type of functions (Eq. (1)) for the relationship between G_d, e, and σ_c' of various soils.

$$G_d = Af(e)\sigma_c'^n \quad (1)$$

where A is constant, n is stress dependency and $f(e)$ is void ratio function. Specific forms of the void ratio function was proposed for sand [13], such as

$$f(e) = (2.17 - e)^2/(1 + e) \quad (2)$$

Using the fundamental properties of elastic mechanics, the relationship between V_s and G_d can be expressed as Eq. (3).

$$G_d = \rho V_s^2 \quad (3)$$

where ρ is a mass density. From the combination of Eqs. (1), (2), and (3), the following equation can be derived.

$$V_s = A'(B - e)/(1 + e)\sigma_c'^n \quad (4)$$

where A' and B are constants.

In this study, we use Eq. (4) to perform nonlinear regression of the results for the uncemented specimens. The resulting constants A', B, and n are found to be 214.3, 1.60, and 0.191, respectively.

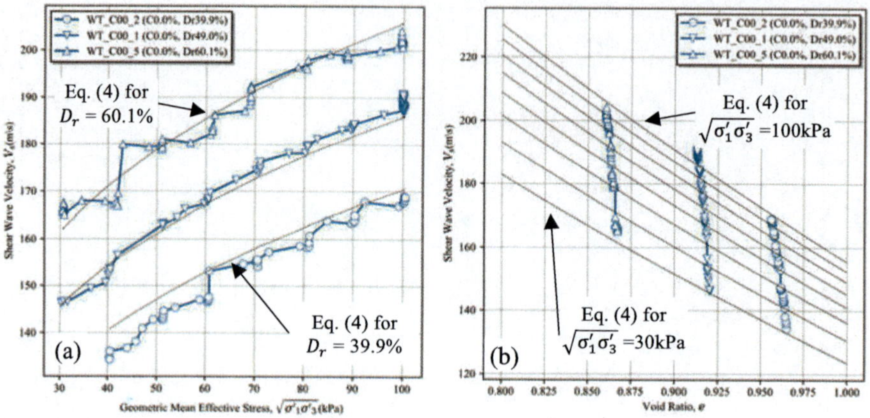

Fig. 2 **a** Relationship between $\sqrt{\sigma_1'\sigma_3'}$ and V_s with referenced curve for the result of uncemented specimen, **b** Relationship between e and V_s with referenced curve for the result of uncemented specimen

The value of stress dependency for all the specimens are shown in the legend box of Fig. 1b. Kohata et al. [14] reported that the value of stress dependency for cemented specimens became smaller than that for uncemented soils. Although the specimens with $C = 1\%$ has higher value of stress dependency, the ones with $C = 3$ and 5% has value lower than that of non-cemented specimen. This result means that the trend of increasing in the values of V_s during consolidation depends on cement addiction rate. This also infers that if the cement addition rate is less than a certain threshold, stress dependency is greater than one of the non-cemented specimens.

Figure 1b shows the relationship between void ratio, e, and V_s during isotropic consolidation. The points located at the right lower of each test dataset are the beginning of consolidation. For the specimen without cement, V_s decreased as the initial e increased.

The regression curves of the model are shown in Fig. 2a, b. Although the V_s of WT_C00_1 increases gradually even after $\sqrt{\sigma_1'\sigma_3'}$ reaches to 100 kPa, the curve fitting is generally consistent with the experimental result.

3.2 Shear Wave Velocity Measured During Undrained Cyclic Loading

Figure 3a–c show the results of typical undrained cyclic loading tests for specimens with $C = 0$, 1, and 5%, respectively. The white circles in the stress–strain relationships and stress paths indicate the points at which V_s is measured during the test. Common

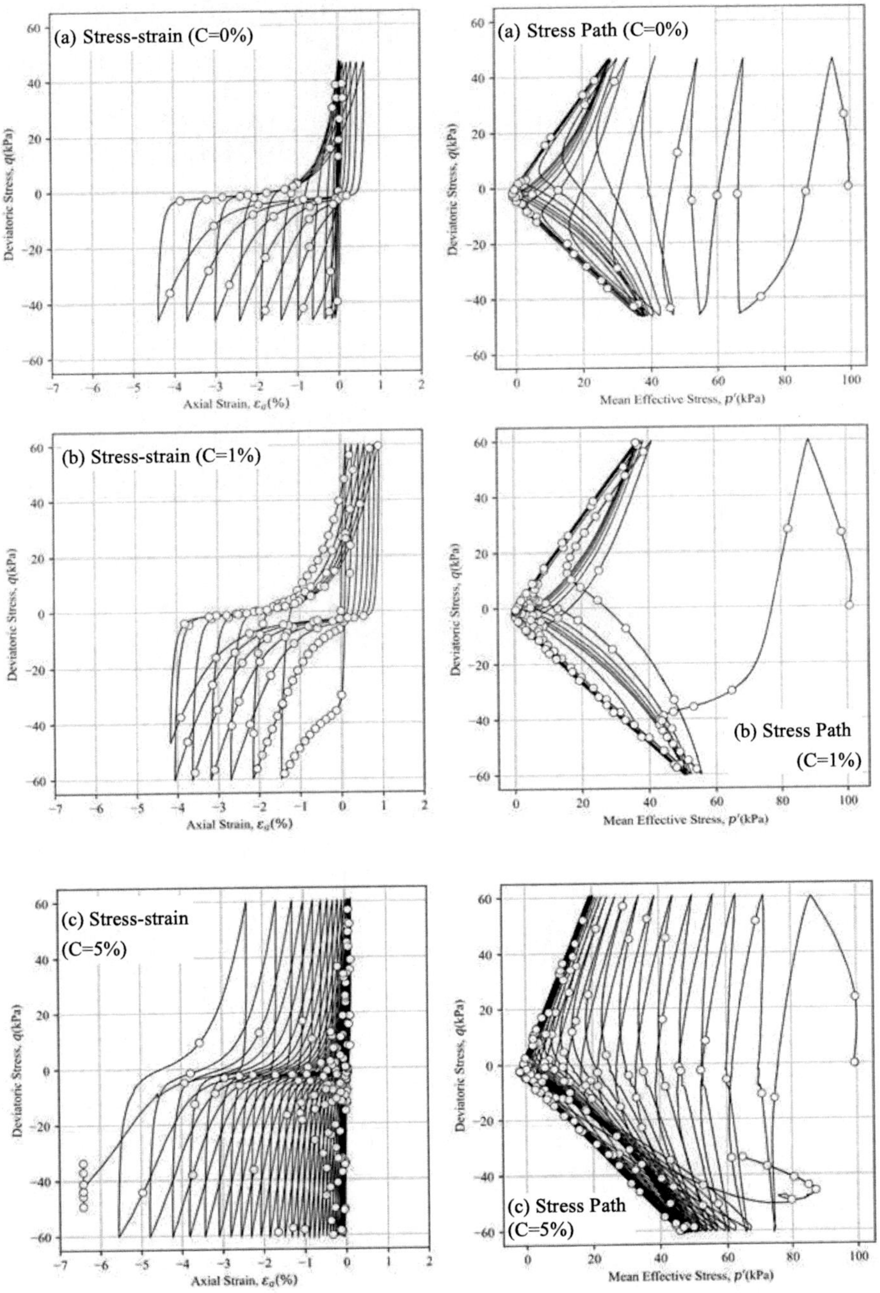

Fig. 3 Stress–strain relationship and stress path of **a** WT_C00_5, **b** WT_C01_3, and **c** WT_C05_1

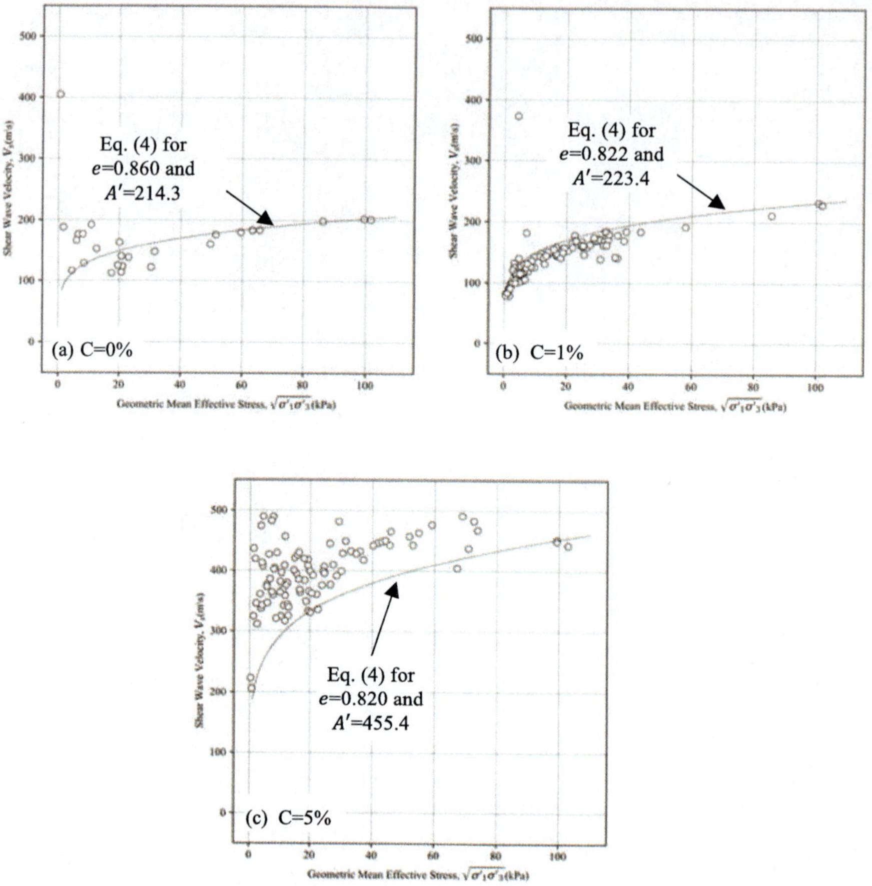

Fig. 4 Relationship between $\sqrt{\sigma_1' \sigma_3'}$ and V_s with referenced curve for the result of uncemented specimen: **a** WT_C00_5, **b** WT_C01_3 and **c** WT_C05_1

to both specimens is that the axial strain slowly increases on the extension side after the excess pore pressure becomes equal to the initial effective confining pressure.

Comparing Fig. 3b, c under the same CSR, the axial strain develops on both compression and extension side, whereas, for $C = 5\%$ (Fig. 3c), the axial strain develops only in the extension side. Considering that the liquefaction process can be roughly divided into two phases: one is the phase of increase in excess pore pressure and the other is the phase of development of strain, it can be said that the number of cycles for the former phase increases due to the increase in the cement addiction rate. In other words, it might be said that the strengthening of soil fabric associated with the addition of cement makes the rate of effective stress slower while it does not have a significant effect on the subsequent strain development.

Figures 4a–c show the change in the value of V_s during undrained cyclic loading. The white circles represent the measured data, and the lines show the relationship calculated by Eq. (4) whose parameter of A' changes to fit the value of V_s at the beginning of undrained cyclic loading while the parameters of B, n are maintained and e is referred as the void ratio of each specimen just after consolidation. Note that the calculated line represents the change in V_s with respect to $\sqrt{\sigma_1'\sigma_3'}$ for a non-cemented specimen during isotropic consolidation; hereafter, this calculated line is called the reference curve.

For the results of the non-cemented specimen shown in Fig. 4a, the values of V_s is equal to or lower than the reference curve in the range of $\sqrt{\sigma_1'\sigma_3'}$ from 20 to 100 kPa. The specimens with $C = 1\%$ shown in Fig. 4b, the values of V_s are smaller than the reference curve when the $\sqrt{\sigma_1'\sigma_3'}$ is below 80 kPa. The trend of the specimens with $C = 5\%$ is different from the above two specimens. When $\sqrt{\sigma_1'\sigma_3'}$ fell below 70 kPa, the values of V_s are plotted at a much larger position than the reference curve. These results indicate that the soil fabric liquefaction is degraded compared to that before liquefaction. In addition, the soil fabric of lightly cemented specimen during liquefaction is weaker than that of non-cemented specimen during consolidation.

Figure 5 illustrates the relationship between normalized shear wave velocity, $V_s/V_{s(\mathrm{ref})}$ and $\sqrt{\sigma_1'\sigma_3'}$. $V_s/V_{s(\mathrm{ref})}$ is the measured V_s divided by the shear wave velocity estimated by the reference curve at the same stress state. The figure means that if $V_s/V_{s(\mathrm{ref})}$ is above one, it indicates that soil fabric is strengthened comparing with that at the initiation of the loading.

As $\sqrt{\sigma_1'\sigma_3'}$ approaches zero effective stress, $V_s/V_{s(\mathrm{ref})}$ become scattered. This can be due to the weakening of the connection between the soil particles as a result of excess pore water pressure generation. This weakening reduces the transmitted energy of the propagated wave in the specimen. Further studies might be required because the decreasing of amplitude makes it difficult to detect the arrival of waves with high accuracy by means of the conventional accelerometers.

From the results, for the specimens with $C = 0\%$ and 1%, the $V_s/V_{s(\mathrm{ref})}$ is below one at lower $\sqrt{\sigma_1'\sigma_3'}$. On the other hand, the values of $V_s/V_{s(ref)}$ of the specimens with $C = 3$ and 5% exceed one in proportion to the increase in the cementing ratio, and this tendency is noticeable at low $\sqrt{\sigma_1'\sigma_3'}$. It may be suggested that the degree of liquefaction-induced degradation of the soil fabric of the specimens with strong cementation effect is exceedingly small.

4 Conclusion

In this study, a mock-up soil test with consolidation and undrained cyclic loading is prepared using silica sand and early strength Portland cement in order to clarify

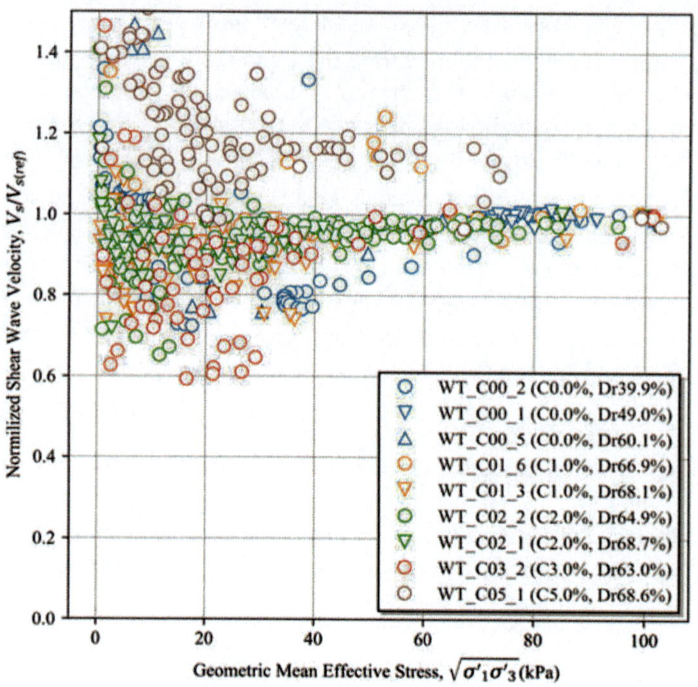

Fig. 5 Relationship between $\sqrt{\sigma'_1\sigma'_3}$ and $V_s/V_{s(\text{ref})}$

the liquefaction characteristics of lightly cemented sand. V_s and some parameters of stress and strain are measured during consolidation and undrained cyclic loading, and the effects of cement addiction rate on soil fabric and liquefaction characteristics are discussed. Conclusions are presented below.

1. Good correlations are observed between V_s, $\sqrt{\sigma'_1\sigma'_3}$ and e for the specimens with 0% addition of cement. On the other hand, although V_s increases with increasing the addition rate of cement, the rate of increase and the incremental trend vary among cementation addition ratios.
2. The empirical model of V_s obtained from the specimens with 0% addition of cement is in agreement with the experimental results.
3. The increase in the cement addition rate may influence slowing down the effective stress reduction and have little effect on strain development
4. Higher cemented specimen can become hard to deteriorate soil fabric during undrained cyclic loading.

Acknowledgements This work was supported by JSPS KAKENHI Grant Number 20H02240.

References

1. Koseki, J., Ohta, A.: Effects of different consolidation conditions on liquefaction resistance and small strain quasi-elastic deformation properties of sands containing fines. Soils Found. **41**(6), 53–62 (2001)
2. Amoly, R.S., Ishihara, K., Bilsel, J.: The relation between liquefaction resistance and shear wave velocity for new and old deposits. Soils Found. **56**(3), 506–519 (2016)
3. Porcino, D., Marcianò, V.: Bonding degradation and stress–dilatancy response of weakly cemented sands. Geomech. Geoeng. **12**(4), 221–233 (2017)
4. Tokimatsu, K., Hosaka, Y.: Effects of sample disturbance on dynamic properties of sand. Soils Found. **26**(1), 53–64 (1986)
5. Teachavorasinskun, S., Tatsuoka, F., Lo Presti, D. C. F.: Effects of the cyclic prestraining on dilatancy characteristics and liquefaction strength of sand. In: Pre-failure Deformation of Geomaterials, pp. 75–80, Balkema, Rotterdam (1994)
6. Yoshimi, Y., Hatanaka, M., Ohoka, H.: Undisturbed sampling of saturated sands by freezing. Soils Found. **18**(3), 59–73 (1978)
7. Yoshimi, Y., Tokimatsu, K., Kaneko, O., Makihara, Y.: Undrained cyclic shear strength of a dense Niigata sand. Soils Found. **24**(4), 131–145 (1984)
8. Hatanaka, M., Sugimoto, M., Suzuki, Y.: Liquefaction resistance of two alluvial volcanic soils sampled by in situ freezing. Soils Found. **25**(3), 49–63 (1985)
9. Tokimatsu, K., Yamazaki, T., Yoshimi, Y.: Soil liquefaction evaluations by elastic shear moduli. Soils Found. **26**(1), 25–35 (1986)
10. Goto, S.: Influence of freeze and thaw cycle on liquefaction resistance of sandy soils. Soils Found. **33**(4), 148–158 (1993)
11. Kiyota, T., Koseki, J., Sato, T., Tatsumi, Y.: Effects of sample disturbance on small strain characteristics and liquefaction properties of holocene and pleistocene sandy soils. Soils Found. **49**(4), 509–523 (2009)
12. AnhDan, L.Q., Koseki, J., Sato, T.: Comparison of Young's moduli of dense sand and gravel measured by dynamic and static methods. Geotech. Test. J., ASTM **25**(4), 349–368 (2002)
13. Hardin, B.O., Richart, F.E.: Elastic wave velocities in granular soils. J. Soil Mech. Found. **89**(1), 33–65 (1963)
14. Kohata, Y., Tatsuoka, F., Wang, L., Jiang, G.L., Hoque, E., Kodaka, T.: Modelling of non-linear deformation properties of stiff geomaterials. Geotechnique **47**(3), 563–580 (1997)

Contribution of Geotechnical Engineers in Restoration After Large-Scale Earthquake Disaster—Example of Aso Ohashi Bridge Replacement Project

Koji Yamashita, Shingo Tanaka, and Takefumi Yamaguchi

Abstract For the plan and design of a bridge crossing an active fault, the design including alignment change or consideration of the fault displacement can be possible by knowing the fault location with a certain accuracy by the soil investigation. This paper presents ① detection of the active fault zone, ② evaluation method for the fault activity based on ground conditions, ③ evaluation method for the excavation stability behind the bridge pier, and shows the technical knowledges obtained by a series of soil investigations for the restoration of Kumamoto Earthquake disaster.

Keywords Surface earthquake fault · Active fault · Geomorphological geological event · Direct shear test · Shear stiffness coefficient

1 Introduction

The Aso Ohashi Bridge was located in the band where the active fault was inferred below ground, and the cause of the bridge collapse may have been related to this fault displacement. Hereafter, this band is called the "inferred active fault" in this paper. At the Technical Committee set up after the earthquake, the study of bridge replacement routes, active fault surveys, and the design of new bridge were conducted in order for early restoration of the road. The authors who belong to a geotechnical consultancy firm were requested by the government for an urgent measure in cooperation with the design firm, and the following studies were carried out from the geotechnical engineer point of views.

First, a wide range of surface reconnaissance was carried out immediately after the earthquake, and the location, direction, and displacement (horizontal and vertical) were measured for cracks, steps, slope failures, etc.; and the inferred active fault locations where a large displacement would occur in the future were extracted taking account the literature survey results. Next, we examined the cause and possible future activity of the discontinuity confirmed on the cut slope shortly after starting

K. Yamashita (✉) · S. Tanaka · T. Yamaguchi
Kyushu Branch, Kiso-Jiban Consultants Co., Ltd., Fukuoka, Japan
e-mail: yamashita.koji@kiso.co.jp

H. Hazarika et al. (eds.), *Advances in Sustainable Construction and Resource Management*, Lecture Notes in Civil Engineering 144,
https://doi.org/10.1007/978-981-16-0077-7_78

the bridge replacement work based on its distribution structure, shape, properties, displacement, geological structural relationship, dating of the sandwiched material between the discontinuities, X-ray diffraction analysis and microscope observation results. Furthermore, the mechanical properties of the rock discontinuities such as columnar joints at the excavated rock behind the bridge piers were estimated by the laboratory tests. Numerical analysis stability evaluation was carried out using the ground model with the obtained the discontinuities properties.

2 Selection of Safe New Bridge Replacement Route

2.1 Basic Concept of Route Selection

In the new bridge replacement route selection, the following basic priorities were established: (1) To ensure the route is safe in consideration of Kumamoto Earthquake damages, (2) Recovery as soon as possible, (3) Function as the gateway to Aso tourism, (4) Early resolution of the separation in Minami-Aso Village.

As one of the conditions of the above priority item (1) safe route, the new bridge substructure shall avoid the inferred active fault location. Thus, it was necessary to reveal the inferred active fault distribution.

2.2 Distribution, Direction, and Displacement of Cracks on the Ground Surface

Ground movement caused by the Kumamoto Earthquake is generally distributed linearly in the ENE–WSW direction independently from the topography, on the flat to gentle slopes northeast of the former Aso Bridge. Many of the displacements showed 0.5 m or more, and damage to buildings was also observed. In particular, in the direction of the east-northeast from the former Aso Bridge, highly continuous cracks and displacements are observed in the cultivated land. They have the characteristics of right lateral displacement north side uplifting with horizontal slickensides (Fig. 1).

Generally, EW direction cracks and steps occurred linearly, in the golf course and developed residential land on the south side of the Nigori River on the southeast side from the former Aso Bridge. Graben of width of about 10–20 m is also confirmed. Some show a large vertical displacement of 1.0 m or more.

In addition, on the southwest side of the former Aso Bridge, cracks and slope failures occurred along the cliff slope along the Kurokawa River on the flat to gentle cultivated land; and cracks and steps are distributed in generally NE–SW direction independent from the terrain.

a) Lateral Displacement of 1.2m   b) Slickenside of Crack

Fig. 1 Lateral displacement and slickenside of crack at cultivated land ENE of Former Aso Bridge

2.3 In Inferred Active Fault Detection

Using surface cracks distribution, horizontal displace directions, displacement amount as indexes, the inferred active fault is detected by zoning surface earthquake fault characteristics such as (a) continuously distribution independently from the terrain, (b) a series of cracks and steps with same displacement direction, (c) concentration of structure damage and slope failures in the continuous direction of cracks and steps, and (d) the distribution of the faults reconciling to the existing literature faults [1].

The inferred active fault extends the north side of Mt. Saikaku in the northeast direction almost on the extension line of the literature fault as shown in Fig. 2. While the literature fault converges in the southeast side of the Kurokawa River, the inferred active fault found by the reconnaissance extends further northeast across the Kurokawa River. Displacement direction and displacement amount of up to 1.2 m are characteristics of the inferred active fault.

Since the new bridge will be connected to the existing roads at the longitudinal section, it was not evitable, but the new bridge was planned to cross the inferred active fault.

3 Investigation of Active Fault Found During New Bridge Replacement Construction

3.1 Investigation and Analysis

Shortly after the start of the slope cut work associated with the new replacement bridge, the discontinuity of the strata was confirmed at the slope which was different

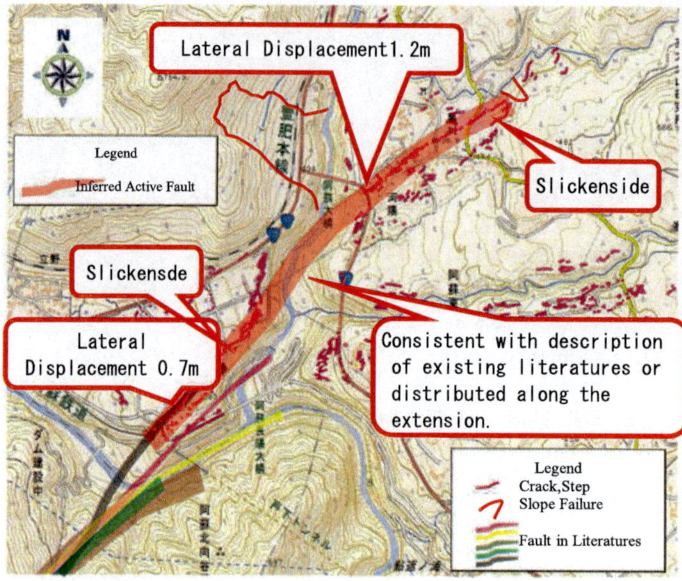

Fig. 2 Distribution of inferred active fault

from the inferred active fault location. The geological structural positional relationship, distribution form, shape, properties, displacement, etc., were investigated and examined the origin and possibility of the future activity of the discontinuity. Since it is important to conduct rational evaluation from the viewpoint of micro and macro, we conducted surveys and analyses such as literature survey, topographic geological reconnaissance, slope observation sketches during cut slope excavation, X-ray diffraction analysis of sandwiched material in the discontinuity, and microscopic observation.

3.2 *General Topography and Geology*

In the surrounding area, active fault groups extending in the southwest direction from Tateno Barranco are prominent. The graben activity at the north of the Tateno Barranco and the right lateral displacement movement of the Futagawa Fault in the southwest direction occur, and as a result, it is considered that the left echelon grabens were formed as tensile cracks [2]. In terms of the geology of the area, Tateno Lava is overlain by Akase Lava, and volcanic ash overlays Akase Lava at the cut slope on the right bank. Volcanic ash layer and rock mixed sediment are sandwiched thinly between Tateno Lava and Akase Lava.

3.3 Properties of Discontinuities

In Tateno Lava, V-shape discontinuities with large opening was confirmed at four locations (Fig. 3). Figure 4 shows the sketch of the large opening discontinuity closest to the pier.

This discontinuity is continuous in the N65E–N78E direction, accompanied by a clayey zone with a width of 2 m or less consisting of volcanic ash clayey soil and fine gravels of Tateno Lava of which diameter of φ5 mm–several cm. It is solidified throughout, and there is no significant difference in the mineral composition of the clay band and the Tateno Lava on both sides of the discontinuity based on X-ray diffraction (total rock analysis). In addition, in the X-ray diffraction (oriented method), although the peak of halloysite is observed in the clay zone, it is not observed in the Tateno Lava mass portion. Furthermore, in the microscopic observation, the

Fig. 3 Discontinuity observed at cut slope of right bank

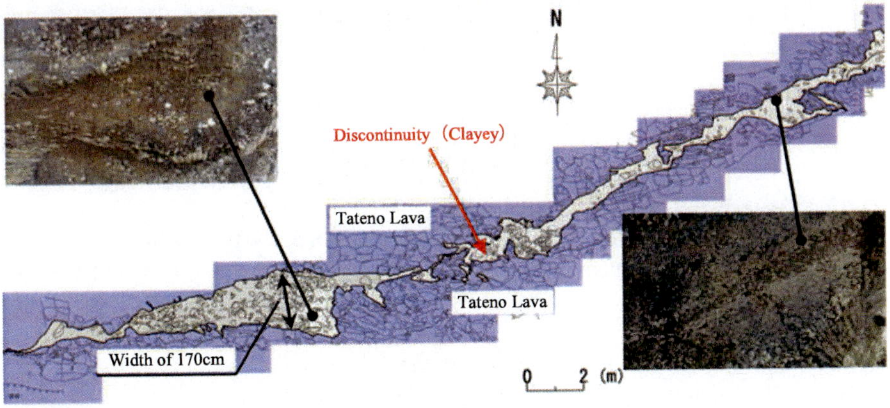

Fig. 4 Sketch of discontinuity at excavation base

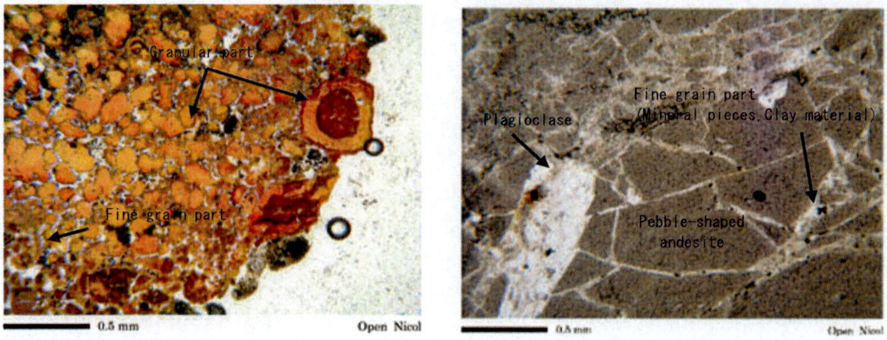

a) Volcanic Ash Clayey Soil: Fracturing Structure is not found.

b) Fine Gravel Part with Many Cracks 〔Clayey Zone〕

Fig. 5 Microscopic photographs

fractured texture is not observed except a part of fine gravels where fractured cracks are observed (Fig. 5).

3.4 *Geomorphological Geological Events*

From the active fault survey, this discontinuity is considered to be an open cleavage caused by the settlement of the Tateno Barranco. However, since some fracturing structures were observed by the microscopic observation, it is also conceivable that it was formed passively in association with the fault movement. It is also possible that it is an alteration zone caused by low-temperature alteration due to hot spring because it is mainly composed of the halloysite and there is a hot spring in the vicinity. As mentioned above, it was concluded that the fault displacement is not likely to appear in the future, from the fact that hardly any features of fault fracture zone was observed.

3.5 *Bridge Design with Safety Improvement Measures*

On the other hand, in the new bridge design, we considered the possibility still remains that displacement may occur in the future even though the displacement was not taken place at Kumamoto Earthquake based on our evaluation. Therefore, the design includes measures to improve safety of the bridge against assumed fault displacement of more than 1.4 m, the displacement of the fault by Kumamoto Earthquake, by designing the anchor bolt of the bearing so as to take over the damage to improve the resilience of the bridge, as well as widen the top end of the pier and extend the girder seating length so as not to collapse easily [3].

4 Slope Stability Analysis Considering Unusual Soil Property

4.1 Background of Analysis

Angle cut cylinder retaining wall was considered as the environmental protection measures. However, it was found that the ring beam could not be constructed due to the ground conditions. Thus, design method such as cut slope reinforcing had to be considered. Therefore, numerical analysis which had previously performed as rock slope stability evaluation was conducted, and the analysis result was adopted into the retaining wall design. In the analysis, the stability of the cut slope is governed by the deformation and strength characteristics of the columnar joints developed in the rock slope. Therefore, a discontinuous model specifically considering the discontinuity such as columnar joints in the rock was adopted for the analysis. Further, it is important to grasp the geometric distribution characteristics of the discontinuities as much as to correctly evaluate the mechanical behavior characteristics of a single discontinuity surface. Two-dimensional key block analysis of the columnar joint was carried out prior to numerical analysis, and the result was used to determine the extent of the collapse of the excavation for the pier.

4.2 Key Block Analysis

In the analysis, a columnar block which contacts three or less faces of the other blocks is defined as a key block that may slide off. A block which contacts four or more contact surfaces is assumed to be stable due to the engagement effect of the columnar blocks each other. Furthermore, secondary to fifth blocks are defined as secondary block is destabilized after the key block primary falls, tertiary block is destabilized after the secondary falls, fourth block is destabilized after the tertiary falls, the fifth block is destabilized after the fourth falls.

Figure 6 shows a two-dimensional key block analysis result diagram. From the geometric distribution form of the columnar joint, bedrock thickness that may collapse is estimated to be $t =$ about 90 cm. Considering columnar joints developed in Tateno Lava, if 90 cm thick block slide occurs at the bottom of the excavation, the upper bedrock destabilization spreading in a shape along the columnar joints can be considered.

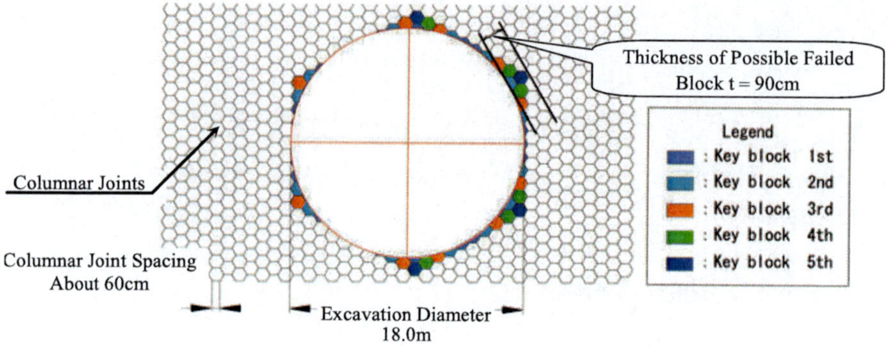

Fig. 6 Two-dimensional key block analysis diagram

4.3 Physical Properties of Columnar Joint

The relation between shear stiffness coefficient K_s and vertical stiffness coefficient K_n, is generally expressed by Formula (1) [4]. In this analysis, K_s of the columnar joint is obtained from direct shear test, and K_n is determined using Formula (1).

$$K_n = K_s \times 10 \tag{1}$$

The core with the size of 10 cm in diameter and less than 100 cm in length including the columnar joint crack was collected from the site as the direct shear test sample, followed by trimming into 10 cm length in order to fit in the tester as shown in Fig. 7. The test quantity was 3 samples 9 specimens (3 specimens per sample). The specimen joint surfaces were observed in detail, and classified in two types, i.e., smooth and rough, as shown in Fig. 8. The test is carried out at a shear rate of 0.1 mm/min under constant confined pressure. Three stage confined pressures (vertical stresses) of 500 kN/m^2, 1,500 kN/m^2, and 3,000 kN/m^2 were applied based on FEM analysis results, in consideration of the stress state acting on joints near the ground with rock bolt before and after excavation.

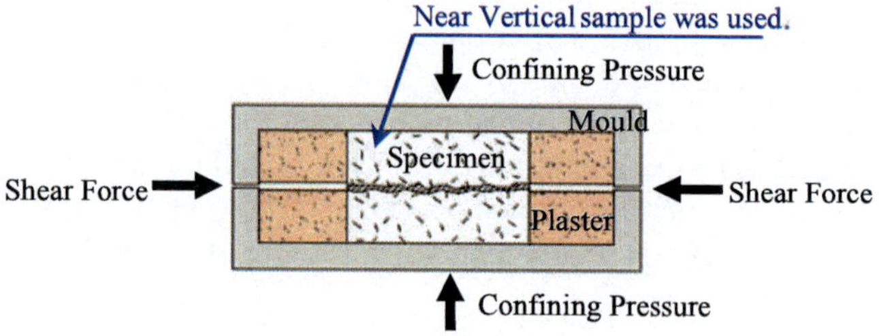

Fig. 7 Overview of direct shear test

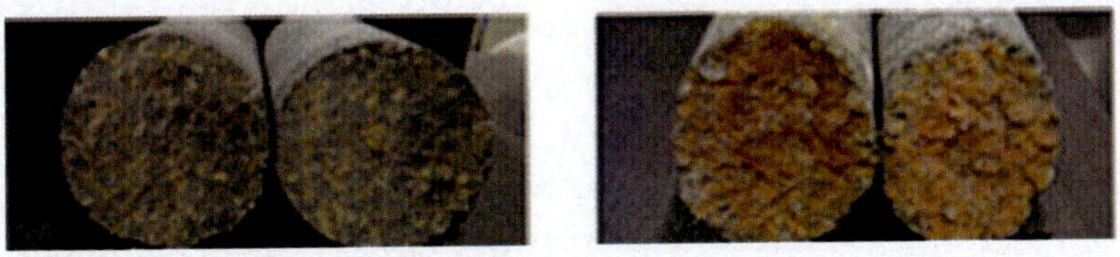

Fig. 8 Conditions of columnar joint surface (Left: Smooth, Right: Rough)

Table 1 Direct shear test results

No	Joint classification	Shear stiffness coefficient K_s (kN/m^3)				Angle of shear resistance φ (°)
		Vertical stress 200 kN/m^2	Vertical stress 500 kN/m^2	Vertical stress 1,500 kN/m^2	Vertical stress 3,000 kN/m^2	
1	Smooth	–	445,000	4,297,000	5,420,000	40
2	Smooth	–	795,000	890,000	11,428,000	51
3	Rough	–	1,125,000	2,714,300	4,833,300	70
Adopted value		512,600		2,593,500		40

Table 1 shows the direct shear test results. Shear stiffness coefficient, K_s, of rough surface joint is larger than that of smooth surface joint under the vertical stress $\sigma_n = 500$ kN/m^2. However, Ks of smooth surface joint becomes higher as σ_n becomes higher. This indicates that engagement effect due to projections of jagged irregularities is large when a vertical stress is small, and the effect disappears when the vertical stress increases. In order to perform a study (analysis) in consideration of a change in the orthogonal stress in the ground, the joint properties for numerical analysis are set as the values corresponding to the vertical stress of $\sigma_n = 200$ kN/m^2 and 1,500 kN/m^2 based on the test results.

4.4 Slope Stability Analysis

The analysis was carried out, while performing sensitivity analysis of the joint model physical properties and apparent cohesive strength, *c*, described above, with the following order: (1) existing slope analysis (normal and seismic conditions) → (2) analysis for excavation without countermeasure to understand the range of separation block → (3) reinforcing ground analysis for the validity of the reinforcement material distribution. The following shows the numerical analysis results of each process.

- The existing slope analysis with normal and seismic conditions shows stable result with 60% of joint properties or more (K_s = test value, $c \geq$ design value × 60%). It agrees with the actual site conditions.

- In the analysis for excavation without countermeasure, analysis shows unstable results when K_s is employed as the test value K_s with a vertical stress $\sigma_n =$ 1,500 kN/m^2, and 80% of the joint properties ($K_s =$ test value, $c =$ design value $\times$ 80%); and when K_s is employed as the test value with a vertical stress of $\sigma_n =$ 200 kN/m^2 and 100% of the joint properties ($K_s =$ test value, $c =$ design value $\times$ 100%). Figure 9 shows the collapse location cumulative diagram created based on the analysis results with the test value K_s at the vertical stress $\sigma_n = 1{,}500$ kN/m^2. By comparing the collapse range of Fig. 9 and the collapse mechanism assumed from the geometric distribution form of the columnar joint described above (slip surface), the 70% of the joint properties is too small, 50% of those is too much, and 60% of those agree with each other.
- In the reinforcing ground analysis, block separation is not found in reinforced ground surface with the 60% of joint physical properties (K_s test value is adopted when vertical stress $\sigma_n = 1{,}500$ kN/m^2). Although shear failure is observed in the member excavation area, the failure does not appear on the excavation surface and ground, area with $F_s < 1.0$ is not observed in the element safety factor diagram. Further, in the maximum shear strain diagram, although a slight strain increase is observed in the excavation corner portion, strain increase is not observed in the other areas, and the results show the final reinforced excavation surface is stable.

5 Summary

The technical findings obtained through these tasks are summarized below.

The inferred active fault extracted by the surface reconnaissance of the possible bridge route, when focusing on the topography, displacement direction, and the continuous direction of the crack and step, extends the north side of Mt. Saikaku in the northeast direction as the almost extension line of the literature fault. The inferred active fault is also considered to be extended further northeast across the Kurokawa River. It is right lateral fault with the maximum displacement of 1.2 m.

Discontinuities confirmed in the active fault survey are considered to be the opening gap caused by the settlement of the Tateno Barranco. However, the fracture structure was observed in microscopic observation, and there is a possibility that it was formed passively in conjunction with the fault motion. On the other hand, a lot of halloysite is detected by X-ray diffraction analysis. Since there is a hot spring in the vicinity, there is a possibility that it is a low-temperature alteration zone associated with the hot spring flowing along the fissures from the underground.

In direct shear tests with vertical stress of $\sigma_n = 500$ kN/m^2, 1,500 kN/m^2, and 3,000 kN/m^2 carried out for the stability evaluation of the excavation behind the pier, shear stiffness coefficient, K_s, of rough surface joint is larger than that of smooth surface joint under the vertical stress $\sigma_n = 500$ kN/m^2. However, K_s becomes equal or higher when vertical stress, σ_n, becomes higher. This indicates that engagement effect due to projections of jagged irregularities is large when a vertical stress is small, and the effect disappears when the vertical stress increases.

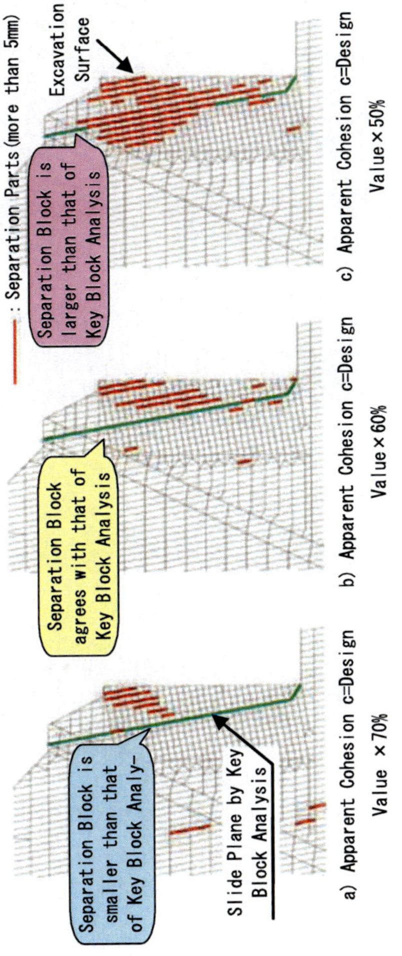

Fig. 9 Separation block cumulative diagram (Apparent Cohesive c = Design Value × 70%, 60%, 50%)

References

1. Tanaka, S., Yamashita, K., Yamaguchi, T.: Characteristics of surface earthquake faults caused by the Kumamoto earthquake (2017)
2. Chida, N.: Recent crustal movements in Central Kyushu—especially active faults in Quaternary volcanic area. Annual Study Report of Department of Education, Iwate University, vol. 39 (1979)
3. Nishida, H., Imamura, T., Hoshikuma, J.: Consideration for minimizing the effect of ground deformation in the planning and design of Aso Ohashi (tentative name). Learn from bridges damaged by the Kumamoto Earthquake, Kyushu Giho No. 65, 27–32 (2019)
4. Jiang, Y., Nakagawa, M., Esaki, T.: Quantitative evaluation of mechanical properties of the natural rock joints for analyzing behavior of structures in discontinuous rock masses. In: JSCE Proceedings No. 624/III-47, pp. 231–243 (1999)

Liquefaction Evaluation of Residential Land Using Geophysical Survey

Shotaro Higuchi, Atsushi Yashima, Yoshinobu Murata, Keizo Kariya, and Hiroshi Yokawa

Abstract Damage such as settlement and tilting of detached houses due to ground liquefaction has often been observed during recent earthquakes. In order to achieve the safe and sustainable residential environment, ground surveys are necessary to predict the damage due to ground liquefaction. However, it takes a long time and needs hard work to conduct conventional survey works. Alternative quick and accurate survey techniques are therefore required. In this study, we focused on the surface wave and passive linear array surveys, which are common geophysical survey methods. The surface wave survey can tell us the ground stiffness over the wide range. The latter one can give us the information from the ground surface down to a substantial depth. These survey techniques can give us the ground conditions more extensively and more quickly than conventional ones. From this background, we carried out geophysical explorations in Christchurch, New Zealand, where the earthquake on February 22, 2011 caused a lot of damage to detached houses due to ground liquefaction. The risk of liquefaction was then evaluated based on survey results. The evaluation methods using the F_L and P_L values were used for judging the liquefaction risk. As a result, the high liquefaction risk in severely damaged areas was reproduced in a short time by the results of surface wave and passive linear array explorations.

Keywords Liquefaction · Geophysical survey · Surface wave exploration · Passive linear array · Detached house

S. Higuchi (✉) · A. Yashima · Y. Murata · K. Kariya
Gifu University, Gifu 5011193, Japan
e-mail: z3921008@edu.gifu-u.ac.jp

H. Yokawa
Chubu University, Kasugai 4878501, Japan

H. Hazarika et al. (eds.), *Advances in Sustainable Construction and Resource Management*, Lecture Notes in Civil Engineering 144,
https://doi.org/10.1007/978-981-16-0077-7_79

1 Introduction

Damages such as settlement and tilting of detached houses due to ground liquefaction have often been observed during recent earthquakes [1]. Ground surveys are necessary to predict the damage due to ground liquefaction. Therefore, drilling surveys or weight sounding tests have been widely used to capture the ground condition in Japan. Because these survey techniques can measure the ground condition directly, the information obtained by these surveys guarantees a certain degree of accuracy. However, it takes long time and needs hard working to conduct these survey works. Therefore, quick and accurate survey techniques have been desired for many years.

In this study, we focused on the surface wave survey and the microtremor array measurement which are ones of conventional geophysical survey methods. The surface wave survey can tell us the shear wave velocity (ground stiffness) over the wide range. On the other hand, the microtremor array measurement can give us the information from the ground surface down to a substantial depth. These survey techniques can provide important information for liquefaction risk assessment more extensively and more quickly than the conventional ones. Based on the obtained ground information, the risk of ground liquefaction can be easily evaluated. The survey area targeted in this study is Christchurch, New Zealand, where the 2011 Christchurch earthquake caused severe liquefaction damage [2].

2 Surface Wave and Passive Linear Array Explorations

Active surface wave exploration using wooden hammer as source as well as passive surface wave exploration using microtremors and vehicle vibration as source were conducted.

In the active surface wave survey of this study, a new system that can calculate the shear wave velocity distribution immediately after data acquisition was developed and applied. In the microtremor array measurement, seismometers are generally arranged in a circle, equilateral triangle, or L-shape. In this study, the straight-line arrangement method of seismographs recently proposed by Kita et al. [3] was applied. Since the survey site is a regulated area after the 2011 Christchurch earthquake, there is always no vibration in the surrounding area. Therefore, along with the survey sideline, we made fast back and forth movement all times by running the survey vehicle.

The surface wave survey using velocity seismometers with a natural frequency of 4.5 Hz was carried out. At three survey areas, velocity seismometers were installed at intervals of 2 m, and surface waves were measured. The passive linear array exploration was conducted using a velocity type seismometer installed at intervals of 5 m with natural frequency of 2 Hz.

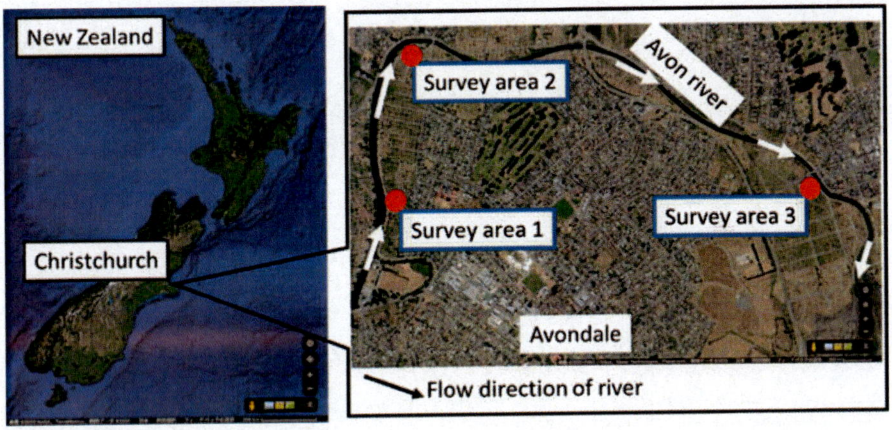

Fig. 1 Survey site

3 Target Area

We conducted active surface wave and passive linear array explorations in Christchurch, New Zealand, severely damaged by the 2011 Christchurch earthquake. The survey areas are shown in Fig. 1. In these areas, extensive house damages were observed such as tilting and settlement by ground liquefaction. Presently, these areas are designated as the red zones, and no one is not allowed to live there.

The active surface wave explorations gave us the two-dimensional shear wave velocity structure with a total extension of about 400 m within 3 h. On the other hand, the microtremor array measurement gave the shear wave velocity distribution with a total depth of 350 m within 3 h.

4 Survey Results

Through geophysical explorations in Christchurch, it is found that we can understand the two-dimensional distribution of shear wave velocity of the ground in wide area by two-dimensional surface wave explorations and the one-dimensional distribution of shear wave velocity from the ground surface down to 30 m in depth by passive linear array exploration.

The results of surface wave surveys at survey areas 1, 2, and 3 are shown in Figs. 2, 3, and 4. The results of microtremor array exploration at survey areas 1, 2, and 3 are also shown in Figs. 5, 6, and 7. From the survey results, it was found that there is a ground with a slightly lower shear wave velocity than the upper and lower layers in the vicinity of 2–4 m from the ground surface. The shear wave velocity tends to increase as the depth increases. A slight inclination of the ground formation was also understood based on the shear wave velocity distribution.

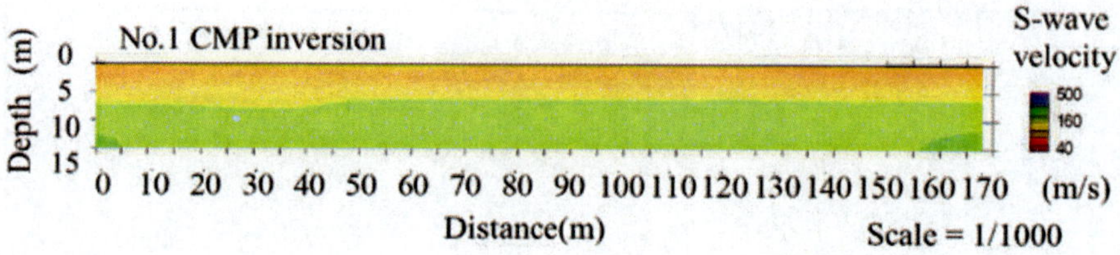

Fig. 2 Shear wave velocity at area 1

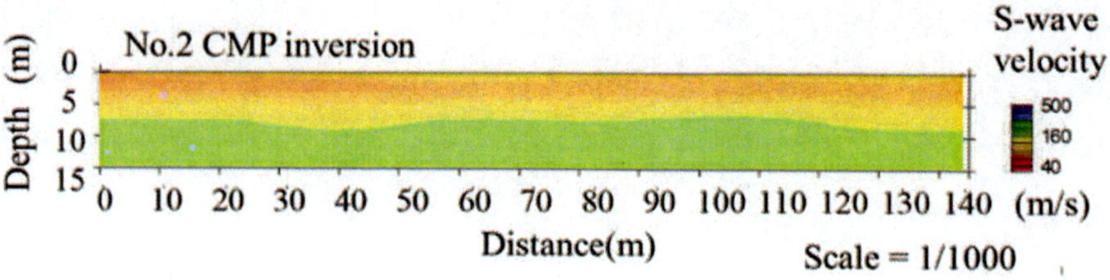

Fig. 3 Shear wave velocity at area 2

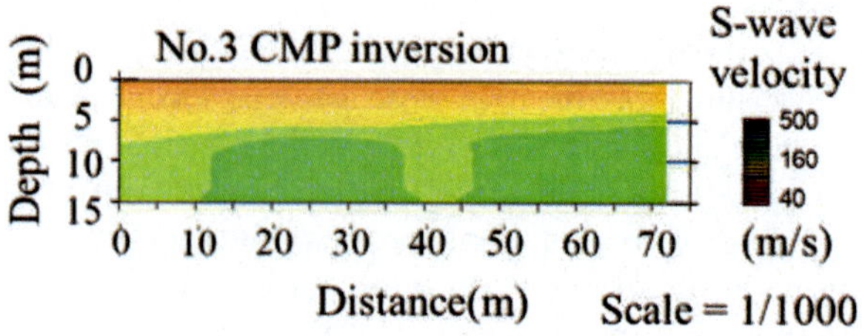

Fig. 4 Shear wave velocity at area 3

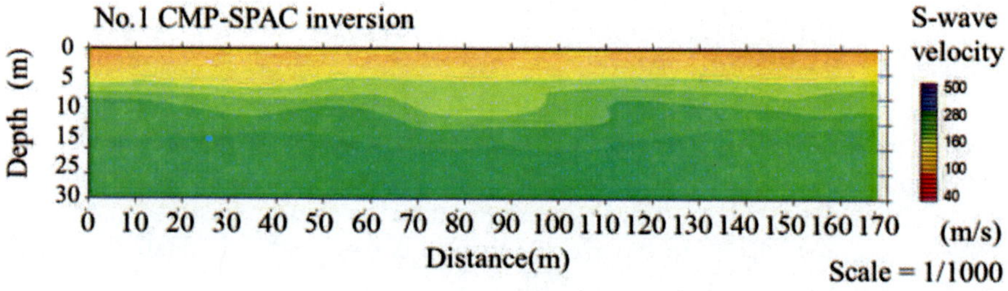

Fig. 5 Result of microtremor array exploration at area 1

5 Evaluation of Liquefaction Risk

The evaluation of liquefaction risk was performed by the F_L and P_L values based on the survey results. The evaluation equations [4] are shown in Eqs. (1) and (2),

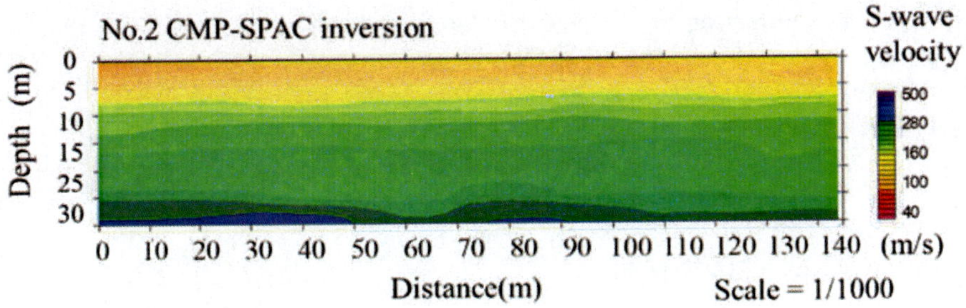

Fig. 6 Result of microtremor array exploration at area 2

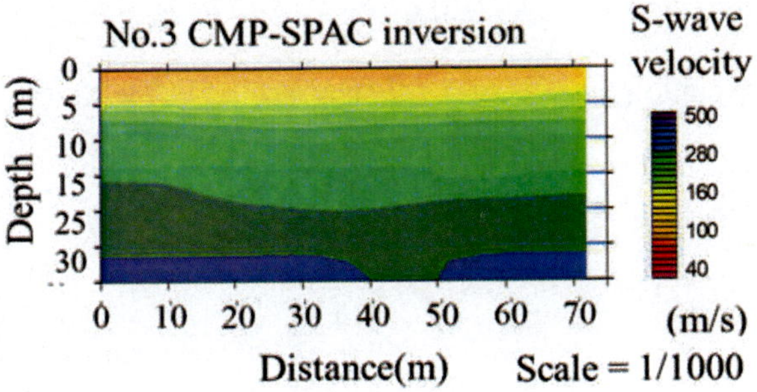

Fig. 7 Result of microtremor array exploration at area 3

$$F_L = \frac{R}{L} \tag{1}$$

$$P_L = \int_0^{20} (1.0 - F_L)(10.0 - 0.5Z)\Delta Z dz \tag{2}$$

where R is the liquefaction strength ratio, L is the cyclic shear strength ratio, Z is the depth from the ground surface, and ΔZ is the soil layer thickness. There is no risk of liquefaction if the F_L value is 1.0, or more and liquefaction takes place if the F_L value is less than 1.0.

The calculation formula of R in Eq. (1) is shown in Eqs. (3) and (4).

$$R = C_w R_L \tag{3}$$

$$\begin{aligned} R_L &= 0.0882\sqrt{\frac{N_a}{1.7}} \qquad (N_a < 14) \\ R_L &= 0.0882\sqrt{\frac{N_a}{1.7}} + 1.6 \times 10^{-6} \times (N_a - 14)^{4.5} \qquad (N_a \geq 14) \end{aligned} \tag{4}$$

where C_w is a correction coefficient based on the seismic motion characteristics and calculated by Eqs. (5) and (6). R_L is the repeated triaxial strength ratio and N_a is the N value that considers the effect of grain size.

In case of subduction-zone earthquake,

$$C_w = 1.0 \tag{5}$$

In case of inland earthquake,

$$\begin{aligned} C_w &= 1 \quad (1.0 \le R_L) \\ C_w &= 2.0 \quad (0.4 \le R_L) \\ C_w &= 3.3R_L + 0.67 \quad (1.0 < R_L \le 0.4) \end{aligned} \tag{6}$$

Next, the calculation formula of N_a is shown in Eqs. (7), (8), (9), and (10).

$$N_a = C_1 N_1 + C_2 \tag{7}$$

$$N_1 = \frac{1.7N}{\sigma_v' + 0.7} \tag{8}$$

$$\begin{aligned} C_1 &= 1 \quad (0\% \le FC < 10\%) \\ C_1 &= \frac{FC + 40}{50} \quad (10\% \le FC < 60\%) \\ C_1 &= \frac{FC}{20} - 1 \quad (60\% \le FC) \end{aligned} \tag{9}$$

$$\begin{aligned} C_2 &= 0 \quad (0\% \le FC < 10\%) \\ C_2 &= \frac{FC - 10}{18} \quad (10\% \le FC) \end{aligned} \tag{10}$$

where C_1 and C_2 are the correction coefficients based on the content of fine particles, N is the blow account obtained from the standard penetration test, N_1 is the N value converted to an effective stress of 98 kPa, σ_v' is the effective stress, and FC is the fine content ratio of soil particles with a particle size of 75 μm or less.

The calculation formula of L is shown in Eq. (11).

$$L = \frac{a}{g}\frac{\sigma_v}{\sigma_v'}\gamma_d \tag{11}$$

where a is the maximum acceleration on the ground (gal), g is the gravitational acceleration (gal), σ_v is the total stress and γ_d is the reduction factor calculated by Eq. (12).

$$\gamma_d = 1.0 - 0.015\Delta Z \tag{12}$$

In this study, we did not carry out the standard penetration tests. Therefore, the N value should be calculated based on the following empirical equation [5].

$$V_s = 80N^{\frac{1}{3}} \tag{13}$$

The maximum accelerations used for the liquefaction strength ratio shown in Eq. (11) were 220 gal and 350 gal in order to investigate the effect of the seismic intensity. The results of F_L value at survey areas 1, 2, and 3 with an acceleration of 220 gal are shown in Figs. 8, 9, and 10. The results of F_L value at survey areas 1, 2, and 3 with an acceleration of 350 gal are also shown in Figs. 11, 12, and 13. From the evaluation result based on the F_L value, it is found that the lower the shear wave velocity is, the higher the liquefaction risk is.

The P_L value was also calculated based on Eq. (2). The evaluation of P_L value was based on Iwasaki's formula [4]. The P_L value indicates the degree of liquefaction damage. The degree of liquefaction damage is judged to be high when the P_L value

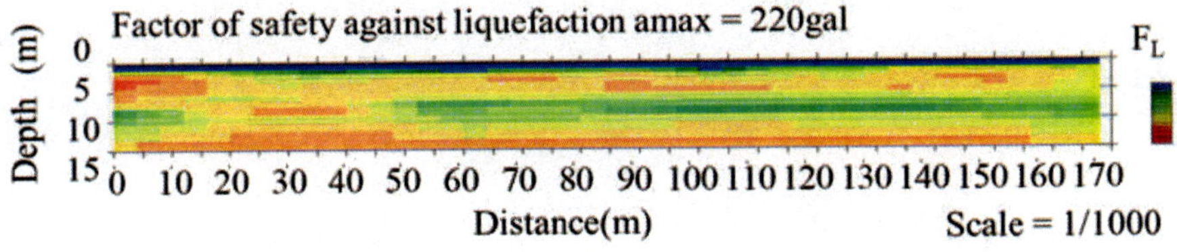

Fig. 8 Liquefaction evaluation result at survey area 1 with acceleration 220 gal

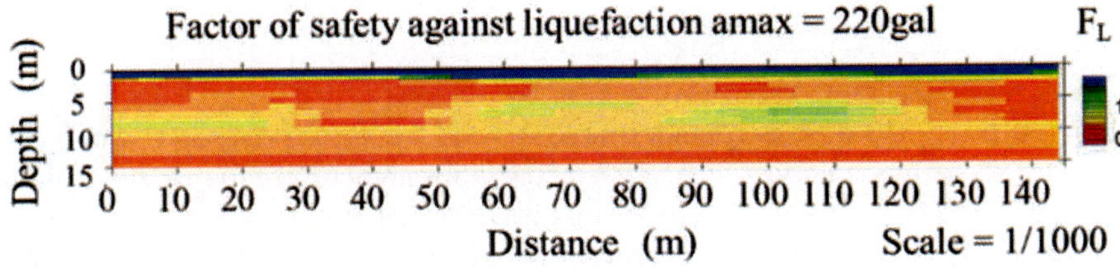

Fig. 9 Liquefaction evaluation result at survey area 2 with acceleration 220 gal

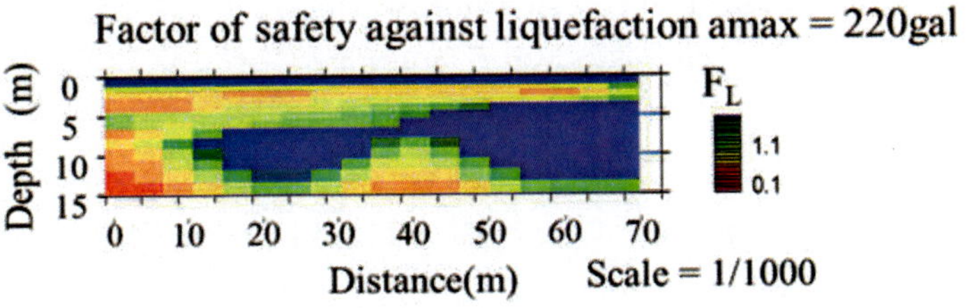

Fig. 10 Liquefaction evaluation result at survey area 3 with acceleration 220 gal

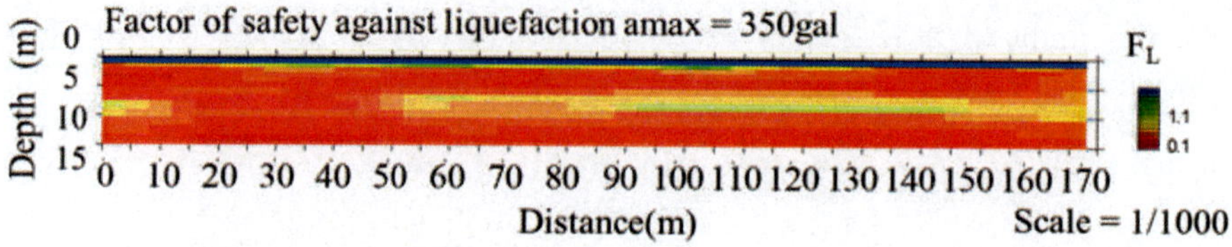

Fig. 11 Liquefaction evaluation result at survey area 1 with acceleration 350 gal

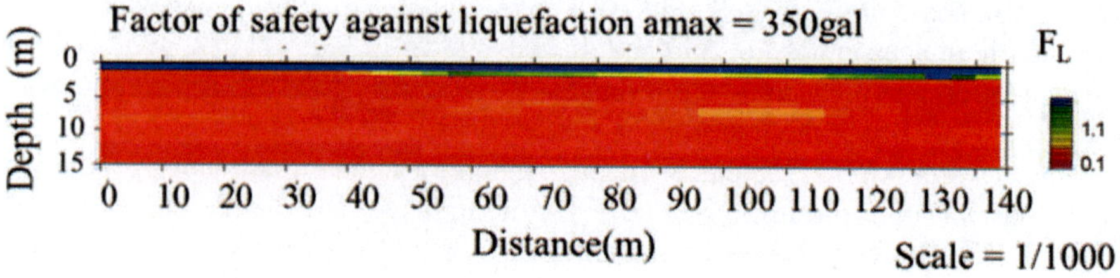

Fig. 12 Liquefaction evaluation result at survey area 2 with acceleration 350 gal

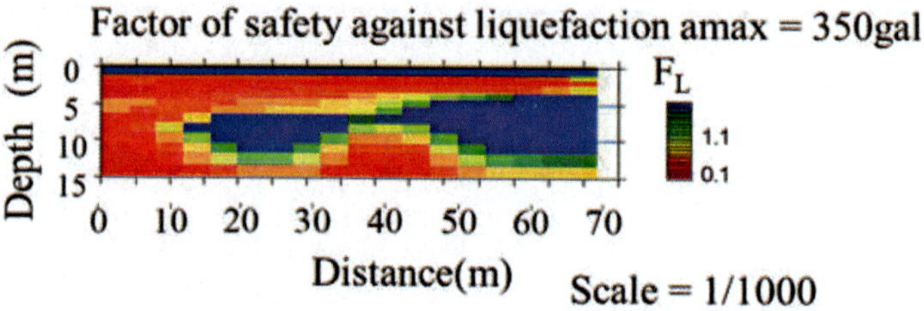

Fig. 13 Liquefaction evaluation result at survey area 3 with acceleration 350 gal

is more than 15. On the other hand, it is low when the P_L value is less than 5.The results of P_L value at survey areas 1, 2, and 3 with an acceleration of 220 gal are shown in Figs. 14, 15, and 16. The results of P_L value at survey areas 1, 2, and 3 with an acceleration of 350 gal are also shown in Figs. 17, 18, and 19. The red lines in all figures from Figs. 14, 15, 16, 17, 18 and 19 indicate the P_L value of 15, the

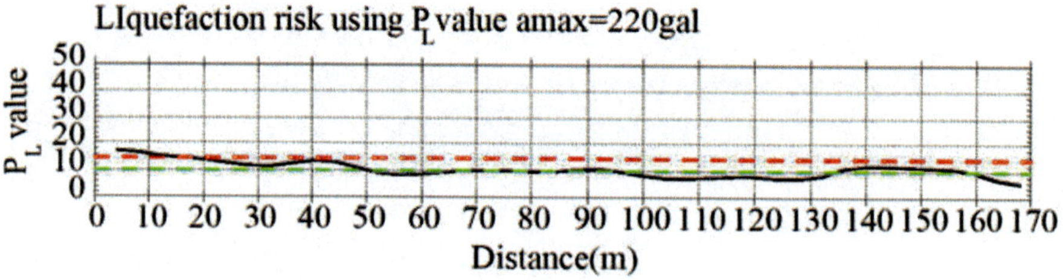

Fig. 14 P_L value at survey area 1 with acceleration 220 gal

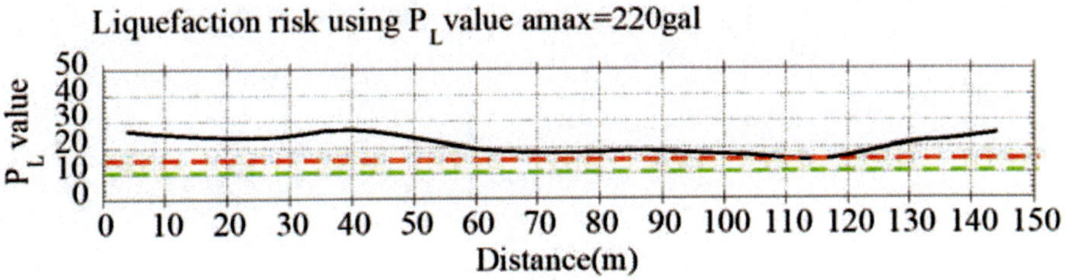

Fig. 15 P_L value at survey area 2 with acceleration 220 gal

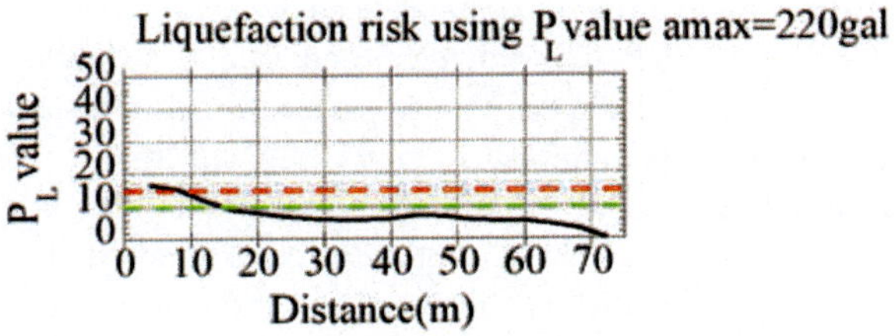

Fig. 16 P_L value at survey area 3 with acceleration 220 gal

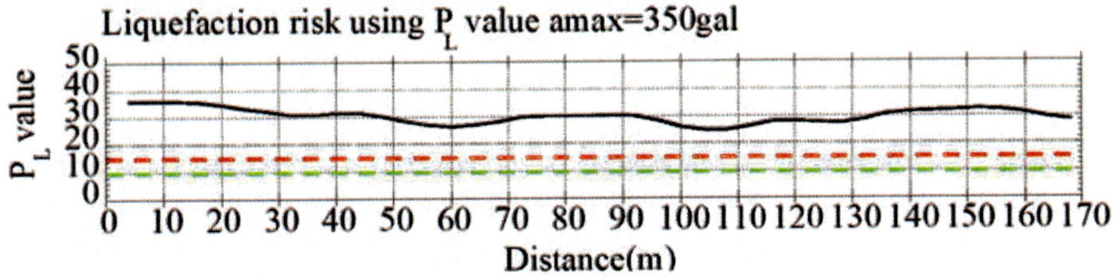

Fig. 17 P_L value at survey area 1 with acceleration 350 gal

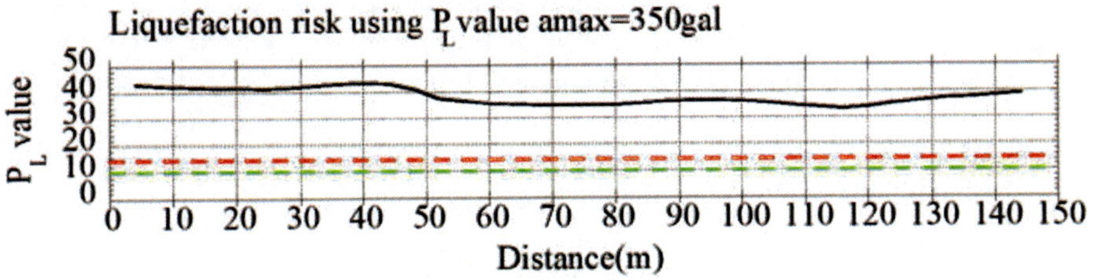

Fig. 18 P_L value at survey area 2 with acceleration 350 gal

green lines indicate the P_L value of 5, and the black line indicates the P_L value at each area. From the results of the P_L value, it is found that the degree of liquefaction risk becomes higher as the shear wave velocity becomes lower.

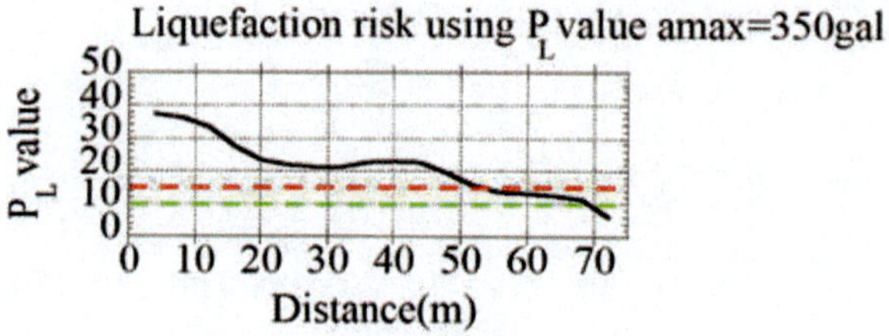

Fig. 19 P_L value at survey area 3 with acceleration 350 gal

Based on the comparison among the result of each survey area, it is found that the low degree of liquefaction risk area is widespread at area 3 shown in Figs. 10 and 13. This is due to the high shear wave velocity near the ground surface at area 3. It is understood that the liquefaction risk is greatly affected by the near-surface shear wave velocity. In order to evaluate the liquefaction risk with much higher accuracy, detailed measurements must be carried out near the ground surface. For example, the surface wave survey with the combination of simple survey technique such as Swedish weight sounding, small dynamic cone penetration test [6], or screw driver sounding [7] will give us more accurate judgment of liquefaction risk.

6 Conclusion

We evaluated the risk of ground liquefaction based on the fast and convenient geophysical survey results. From these results, the risk of ground liquefaction is found to be easily understood extensively and quickly at three survey areas in Christchurch, New Zealand.

In addition, based on the comparison among the result of each survey area, it is found that the liquefaction risk is greatly affected by the near-surface shear wave velocity. Therefore, the combination of simple survey technique such as Swedish weight sounding, small dynamic cone penetration test, or screw driver sounding with the proposed geophysical survey method will give us more accurate judgment.

Acknowledgements This study was supported by Provision of survey sites by Land Information New Zealand (LINZ) and JSPS Grant-in-Aid for Young Scientists (B) (16k18192). We would like to thank Prof. Misko Cubrinovski of University of Canterbury, New Zealand for his cooperation in on-site measurement.

References

1. Tokimatsu, K., Katsumata, K.: Liquefaction-induced damage to buildings in Urayasu City during The 2011 Tohoku Pacific Earthquake. In: Proceedings of the International Symposium on Engineering Lessons Learned from the 2011 Great East Japan Earthquake, pp. 665–674 (2012)
2. Japanese Geotechnical Society: Report of Disaster Emergency Survey Team for the 2011 New Zealand Christchurch Earthquake Damage. URL: https://www.jiban.or.jp/file/saigai-/2011christchurch.pdf
3. Kita, T., Hayashi, K., Haldun, B.: The development of a 2-dimensional microtremor survey Method BASED on SPAC method using sequential linear arrays. In: Proceedings of the Symposium on the Application of Geophysics to Engineering and Environmental Problems, pp. 115–120 (2011)
4. Iwasaki, T., Tatsuoka, F., Tokida, K., Yasuda, S.: Estimation of degree of soil liquefaction during earthquakes. Soil Mech. Found. Eng. **28**, 23–29 (1980). (in Japanese)
5. Japan Road Association: Handbook for Asphalt Pavement Part V, Seismic Design (2019) (in Japanese)
6. Murata, Y., Sato, M., Iwata, M., Sawada, K., Yashima, A.: Comparison of various dynamic sounding tests in Urayasu City (Part7: PENNY), Urayasu City Ground Survey Simultaneous Test Report Meeting, pp. 21–23 (2012) (in Japanese)
7. Sugano, Y., Yamato, S., Suemasa N., Tanaka, T., Oshima, A.: Comparative study of various static sounding tests in Urayasu City (Part7: SDS), Urayasu City Ground Survey Simultaneous Test Report Meeting, pp. 49–50 (2012) (in Japanese)